Medical Physiology

Medical Physiology

Edited by

Rodney A. Rhoades, Ph.D.
Professor and Chairman of Physiology and
Biophysics, Indiana University School of
Medicine, Indianapolis

George A. Tanner, Ph.D.
Professor of Physiology and Biophysics, Indiana
University School of Medicine, Indianapolis

Illustrations by Christopher Wikoff, C.M.I.

 LIPPINCOTT WILLIAMS & WILKINS

A **Wolters Kluwer** Company

Philadelphia · Baltimore · New York · London
Buenos Aires · Hong Kong · Sydney · Tokyo

Library of Congress Cataloging-in-Publication Data
Medical physiology / edited by Rodney A. Rhoades, George A. Tanner.
        p.   cm.
    Includes bibliographical references and index.
    ISBN 0-316-74228-7
    1. Human physiology.   I. Rhoades, Rodney A.   II. Tanner, George A.
    [DNLM: 1. Physiology.   QT 104 M4892   1994]
QP34.5.M473   1995
612—dc20
DNLM/DLC
for Library of Congress   94-24639
                                                            CIP

Printed in the United States of America
EB-M

10 9 8 7 6

Editorial: Evan R. Schnittman, Rebecca Marnhout
Production Services: Ruttle, Shaw & Wetherill, Inc.
Copyeditor: Jane Grochowski
Indexer: Indexing Research
Designer: William T. Donnelly
Cover Designer: Linda D. Willis and Patrick Newbery

Contents

Preface

Medical Physiology is intended to provide the health sciences student with a solid understanding of modern medical physiology. Our approach emphasizes broad concepts, application of knowledge, and an integrative approach that relies on thinking skills instead of memorization.

The contributing authors were selected from 13 different departments representing 9 medical schools. These authors have 10 or more years of experience teaching basic science. Several authors are course directors, chairs, former chairs, or former deans of major medical schools. Many have won teaching awards at their universities and have served or are currently serving on the National Board of Medical Examiners for Step 1 and Step 2 examinations. Each author has expertise in his or her own area and has contributed significantly to the broad knowledge of physiology through research. All these authors were willing to blend their personal writing styles to produce a unified textbook rather than a compendium of isolated chapters or a rewrite of existing monographs.

The scope of the book is midway between that of the numerous oversimplified surveys of physiology and the overwhelming, encyclopedic tomes. To achieve this, we have pruned away advanced details, of little use to the beginning health sciences student, to allow the fundamentals to be more readily grasped. Although chapters are concise, they fully ground students in these fundamentals, preparing them for more-advanced study.

We have paid special attention to making this book *usable* as well as *useful* for students:

Physiologic processes are explained in terms that beginning students can understand, without sacrificing accuracy.

Only the essentials are covered, easing the time pressure on students.

Chapters are easy to read—written for students, not for academics.

Pedagogic features are included to assist student mastery of content. These include chapter outlines, chapter objectives, and setting key terms in bold type in the text. Comprehensive review questions and annotated answers are found at the end of each chapter.

Because visual learning is critically important to today's student, numerous two-color illustrations are used to explain basic concepts. Clinical Focus Boxes, appearing throughout the book, discuss pathophysiologic conditions that illustrate key physiologic concepts, helping students appreciate the physiologic basis of clinical medicine.

We organized the book with the beginning student taking an introductory medical physiology course in mind. Chapters are grouped in parts. Part One deals with fundamental physiologic principles that apply throughout the book. Basic topics covered include the internal environment, homeostasis, feedback control, gap junctions, neuroendocrine interactions, the membrane potential, and membrane transport. Neurophysiology is briefly covered in Part Two, as well as in Chapter 3. In recognition of the fact that neurobiology is taught as a separate course in most medical schools, these chapters only summarize the essentials in critical subject areas. Major topics covered include the action potential, synaptic transmission, the special senses, control of movement, the autonomic nervous system, the homeostatic role of the central nervous system, and the higher functions of the brain.

The book progresses from cellular events to organ and body function, building on information and concepts presented in the first parts of the book. Part Three, the section on muscle, emphasizes basic mechanisms of contraction and shows that contraction is essentially the same in skeletal, smooth, and cardiac muscle. To provide a clear understanding of the cardiovascular system, in Part Four we provide a chapter on hemodynamics that simplifies and clarifies what traditionally is a difficult subject. Other chapters in Part Four discuss blood components and immunity, the electrical activity of the heart, the cardiac pump, systemic circulation, special circulations, and control mechanisms of circulatory function. Part Five emphasizes physical and chemical principles underlying the mechanics and control of breathing, ventilation, gas exchange, and pulmonary blood flow.

Part Six, the renal physiology section, describes important mechanisms for handling solutes. The processes responsible for the regulation of fluid and electrolyte balance and acid-base balance are presented. Part Seven, on gastrointestinal physiology, examines motility, secretion, and mechanisms of food absorption. A chapter on liver function is included. Part Eight covers the fundamentals of temperature regulation and exercise with an emphasis on many facets of cellular and integrative biology. The last two parts deal with endocrinology and reproduction. Part Nine examines hormonal control and integrative control of homeostasis. Important topics covered include the hypothalamus and pituitary gland and endocrine functions of the thyroid and adrenal glands. Separate chapters discuss the pancreatic hormones and regulation of carbohydrate metabolism and the endocrine regulation of calcium, phosphate, and bone metabo-

lism. Part Ten covers basic concepts in reproduction; a special chapter deals with conception, pregnancy, and fetal development.

Physiology bridges biology and medicine. It plays a unique role in medicine by providing the essential link by which cellular and molecular events are integrated into the understanding of organ and body function in health and disease. We hope this idea is conveyed to the student who uses *Medical Physiology*.

R. A. R.
G. A. T.

Contributing Authors

Nira Ben-Jonathan, Ph.D.
Professor of Anatomy and Cell Biology, University of Cincinnati College of Medicine, Cincinnati
Chapters 39, 40, and 41

Harold Glenn Bohlen, Ph.D.
Professor of Physiology and Biophysics, Indiana University School of Medicine, Indianapolis
Chapters 16 and 17

Ayus Corcia, Ph.D.
Assistant Professor of Physiology and Biophysics, Indiana University School of Medicine, Indianapolis
Chapter 2

Denis English, Ph.D.
Adjunct Associate Professor of Physiology and Biophysics, Indiana University School of Medicine; Director, Bone Marrow Transplantation Laboratory, Methodist Hospital of Indiana, Indianapolis
Chapter 11

Alon Harris, Ph.D.
Assistant Professor of Ophthalmology, Physiology, and Biophysics, Indiana University School of Medicine, Indianapolis
Chapter 32

Edith D. Hendley, Ph.D.
Professor of Molecular Physiology, Biophysics, and Psychiatry, University of Vermont College of Medicine, Burlington
Chapters 3, 5, 6, and 7

Stephen A. Kempson, Ph.D.
Professor of Physiology and Biophysics, Indiana University School of Medicine, Indianapolis
Chapter 2

Ali A. Khraibi, Ph.D.
Assistant Professor of Physiology, Mayo Medical School, Rochester, Minnesota
Chapter 24

Franklyn G. Knox, M.D., Ph.D.
Professor of Physiology and Medicine, Mayo Medical School, Rochester, Minnesota
Chapter 24

Jack L. Kostyo, Ph.D.
Professor of Physiology, University of Michigan Medical School; Associate Director, Michigan Diabetes Research and Training Center, Ann Arbor
Chapters 34, 35, and 36

Thomas C. Lloyd, Jr., M.D.
Professor of Medicine, Physiology, and Biophysics, Indiana University School of Medicine, Indianapolis
Chapter 22

Walter C. Low, Ph.D.
Professor of Physiology, Neurosurgery, and Neuroscience, University of Minnesota Medical School—Minneapolis
Chapter 3

Bruce J. Martin, Ph.D.
Associate Professor of Physiology, Medical Sciences Program, Indiana University School of Medicine, Bloomington
Chapter 32

Richard A. Meiss, Ph.D.
Professor of Physiology, Biophysics, Obstetrics, and Gynecology, Indiana University School of Medicine, Indianapolis
Chapters 4, 8, 9, and 10

Susan J. Mitchell, Ph.D.
Associate Professor of Biological Sciences, Onondaga Community College, Syracuse, New York
Chapter 5

Daniel E. Peavy, Ph.D.
Associate Professor of Physiology and Biophysics, Indiana University School of Medicine; Research Chemist, Richard L. Roudebush Veterans Affairs Medical Center, Indianapolis
Chapters 1, 33, 37, and 38

Rodney A. Rhoades, Ph.D.
Professor and Chairman of Physiology and Biophysics,
Indiana University School of Medicine, Indianapolis
Chapters 19, 20, and 21

Thom W. Rooke, M.D.
Director, Gonda Vascular Center, Mayo Medical Center,
Rochester, Minnesota
Chapters 12, 13, 14, 15, and 18

Harvey V. Sparks, Jr., M.D.
Professor of Physiology, Michigan State University
College of Human Medicine, East Lansing
Chapters 12, 13, 14, 15, and 18

George A. Tanner, Ph.D.
Professor of Physiology and Biophysics, Indiana
University School of Medicine, Indianapolis
Chapters 1, 23, and 25

Patrick Tso, Ph.D.
Professor of Physiology, Louisiana State University
School of Medicine in Shreveport
Chapters 28, 29, and 30

Christian Bruce Wenger, M.D., Ph.D.
Research Pharmacologist, Thermal Physiology and
Medicine Division, U.S. Army Research Institute of
Environmental Medicine, Natick, Massachusetts
Chapter 31

Jackie D. Wood, Ph.D.
Chairman of Physiology and Professor of Physiology
and Internal Medicine, Ohio State University College
of Medicine, Columbus
Chapters 26 and 27

PART ONE

Cellular Physiology

Notice

The indications and dosages of all drugs in this book have been recommended in the medical literature and conform to the practices of the general medical community. The medications described do not necessarily have specific approval by the Food and Drug Administration for use in the diseases and dosages for which they are recommended. The package insert for each drug should be consulted for use and dosage as approved by the FDA. Because standards for usage change, it is advisable to keep abreast of revised recommendations, particularly those concerning new drugs.

Regulation and Communication

OBJECTIVES
After studying this chapter, the student should be able to:
1. Explain the importance of a stable internal environment
2. Define *homeostasis* and give examples
3. Explain why regulation of intracellular pH, ionic strength, and calcium activity are important for cell function
4. Describe the components of a negative feedback system
5. Explain why positive feedback results in progressive change
6. Distinguish between the conditions of equilibrium and steady state
7. Describe how the nervous and endocrine systems communicate information in the body
8. Explain why and how second messengers mediate the actions of many protein and peptide hormones and why they do not mediate steroid and thyroid hormone effects

Physiology is the study of processes and functions in living organisms. It is a broad field that encompasses many disciplines and has strong roots in physics, chemistry, and mathematics. Physiologists assume that processes in the body are governed by the same chemical and physical laws that apply to the inanimate world. They attempt to describe functions in chemical, physical, or engineering terms. For example, the distribution of ions across cell membranes is described in thermodynamic terms; muscle contraction is analyzed in terms of forces and velocities; and regulation in the body is described in terms of control systems theory. Since the functions of living systems are carried out by their constituent structures, a knowledge of structure from gross anatomy to the molecular level is germane to an understanding of physiology.

The scope of physiology ranges from the activities or functions of individual molecules and cells to the interaction of our bodies with the external world. Recent years have seen many advances in our understanding of physiologic processes at the molecular and cellular levels. In higher organisms, changes in cell function always occur in the context of a whole organism, and different tissues and organs obviously affect one another. Independent activity of an organism requires coordination of function at all levels, from cellular and molecular to the organism as a whole. An important part of physiology is understanding how different parts of the body are controlled, how they interact, and how they adapt to changing conditions.

For a person to remain healthy, physiologic conditions in the body must be kept at optimal levels and closely regulated. Regulation requires effective communication. This chapter discusses a number of topics related to regulation and communication: the internal environment, homeostasis of extracellular fluid, intracellular homeostasis, negative and positive feedback, feedforward control, compartments, steady state and equilibrium, intercellular communication, nervous and endocrine control, and cell membrane transduction.

Physiologic Regulation

■ A Stable Internal Environment Is Essential for Normal Cell Function

The nineteenth-century French physiologist Claude Bernard was the first to formulate the concept of the *internal environment (milieu intérieur)*. He pointed out that multicellular organisms are surrounded by an external environment—air or water—but the cells live in a liquid internal environment—extracellular fluid. Most body cells are not directly exposed to the external world, but rather interact with it through the internal environment, which is continuously renewed by the circulating blood (Fig. 1–1).

For optimal cell, tissue, and organ function in animals, several conditions in the internal environment

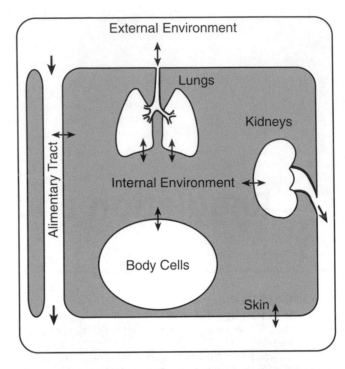

Figure 1–1 ■ The living cells of our body are surrounded by an internal environment—the extracellular fluid—and communicate with the external world through this medium. Exchanges *(arrows)* of matter and energy between the body and the external environment occur via the alimentary tract, kidneys, lungs, and skin (including the specialized sensory organs).

must be maintained within narrow limits. These include, but are not limited to: (1) oxygen and carbon dioxide tensions; (2) concentrations of glucose and other metabolites; (3) osmotic pressure; (4) concentrations of hydrogen, potassium, calcium, and magnesium ions; and (5) temperature. Departures from optimal conditions may result in disordered function.

Bernard stated that "stability of the internal environment is the primary condition for a free and independent existence." He recognized that an animal's independence from changing external conditions is related to its capacity to maintain a relatively constant internal environment. A good example is the ability of warm-blooded animals to live in different climates. Over a wide range of external temperatures, core temperature in mammals is maintained quite constant by both physiologic and behavioral mechanisms. This has a clear survival value.

■ Homeostasis Is the Maintenance of Steady States in the Body by Coordinated Physiologic Mechanisms

The key to maintaining stability of the internal environment is the presence of regulatory mechanisms in the body. In the first half of the twentieth century, the American physiologist Walter B. Cannon introduced a concept describing this capacity for physiologic self-regulation: **homeostasis,** "the maintenance of steady states in the body by coordinated physiological mechanisms."

The concept of homeostasis is helpful in understanding and analyzing conditions in the body. That steady conditions exist, even though we are composed of unstable materials and are often subject to disturbances, is evidence that regulatory mechanisms in the body maintain stability. To function optimally under a variety of conditions, the body must sense departures from normal and must engage mechanisms for restoring conditions to normal. Departures from normal may be in the direction of too little or too much, so mechanisms exist for opposing changes in either direction. For example, if blood glucose concentration is too low, the hormones glucagon, from alpha cells of the pancreas, and epinephrine, from the adrenal medulla, will increase it. If blood glucose concentration is too high, insulin from the beta cells of the pancreas will lower it by enhancing cellular uptake, storage, and metabolism of glucose. Behavioral responses also contribute to the maintenance of homeostasis; for example, a low blood glucose concentration stimulates feeding centers in the brain, driving the animal to seek food.

Homeostatic regulation of a physiologic parameter often involves several cooperating mechanisms put into action at the same time or in succession. The more important a parameter is, the more numerous and complicated are the mechanisms that keep it at the desired value (i.e., regulate it). Disease is often the result of dysfunction of homeostatic mechanisms.

The effectiveness of homeostatic mechanisms varies over a person's lifetime. Some homeostatic mechanisms are not fully developed at the time of birth. For example, a newborn infant cannot concentrate urine as well as an adult and is therefore less able to tolerate water deprivation. Homeostatic mechanisms gradually become less efficient as people age. For example, older adults are less able than younger adults to tolerate stresses such as exercise or changing weather.

■ Intracellular Homeostasis Is Essential for Normal Cell Function

The term **homeostasis** has traditionally been applied to the internal environment—the extracellular fluid—but can be quite properly applied to conditions within cells. In fact, the ultimate goal of maintaining a constant internal environment is to promote intracellular homeostasis, and toward this end conditions in the cytosol are closely regulated.

The many biochemical reactions within a cell must be regulated to provide metabolic energy and proper rates of synthesis and breakdown of cellular constituents. Metabolic reactions within cells are catalyzed by enzymes and therefore subject to several factors that regulate or influence enzyme activity. First, the catalytic activity of enzymes may be inhibited by the final product of the reactions. This is called **end-product inhibition** and is an example of negative feedback control (see below). Second, enzyme activity may be controlled by intracellular regulatory proteins, such as the calcium-binding protein calmodulin. Third, enzymes may be controlled by covalent modification, such as phosphorylation or dephosphorylation. Fourth, the structure and activity of enzymes are influenced by the ionic environment within cells, including hydrogen ion concentration (pH), ionic strength, and calcium ion concentration.

pH also affects the electrical charge of protein molecules and hence their configuration and binding properties. pH affects enzyme activities and the organization of structural proteins in cells. Cells regulate their pH via mechanisms for buffering intracellular hydrogen ions and extruding them into the extracellular fluid (Chap. 25).

The structure and activity of cell protein molecules are also affected by ionic strength. Cytosolic ionic strength depends on the total number and charge of ions per unit volume of water within cells. Cells can regulate their ionic strength by maintaining the proper mixture of ions and un-ionized molecules.

Many cells use calcium as an intracellular signal for cell activation, and therefore they must possess mechanisms for regulating cytosolic calcium concentration. Such fundamental activities as muscle contraction, secretion of neurotransmitters, hormones, and digestive enzymes, and opening and closing of ion channels are mediated via transient changes in cytosolic calcium. Cytosolic calcium in resting cells is very low, about 10^{-7} M, and far below extracellular fluid calcium concentration (about 2.5 mM). Cytosolic calcium is regulated by the binding of calcium to intracellular proteins, transport by adenosine triphosphate (ATP)-dependent calcium pumps in mitochondria and other organelles (e.g., sarcoplasmic reticulum in muscle), and extrusion of calcium via cell membrane Na^+/Ca^{2+} exchangers. Toxins or diminished ATP production can lead to an abnormally elevated cytosolic calcium. A high cytosolic calcium activates many enzyme pathways, some of which have detrimental effects and may cause cell death.

■ Negative Feedback Promotes Stability; Feedforward Control Anticipates Change

Engineers have long recognized that stable conditions can be achieved by **negative feedback control systems** (Fig. 1–2). **Feedback** is a flow of information along a **closed loop.** The components of a simple negative feedback loop include a **regulated variable, sensor** (or detector), **controller** (or comparator), and **effector.** Each component controls the next component. Various disturbances may arise within or outside the system and cause undesired changes in the regulated variable. With **negative feedback,** a regulated

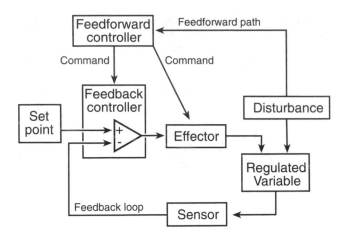

Figure 1–2 ■ Elements of negative feedback and feedforward (red) control systems. In the negative feedback control system, information flows along a closed loop. The regulated variable is sensed and information about its level is fed back to a feedback controller (comparator), which compares it to a desired value (set-point). If there is a difference, an error signal is generated, which drives the effector to bring the regulated variable closer to the desired value. A feedforward controller generates commands without directly sensing the regulated variable, although it may sense a disturbance. Feedforward controllers often operate through feedback controllers.

variable is sensed, information is fed back to the controller, and the effector acts to oppose change (hence the term *negative*).

A familiar example of a negative feedback control system is the thermostatic control of room temperature. Room temperature (regulated variable) is subjected to disturbances; on a cold day, room temperature falls. The room temperature is detected by a thermometer (sensor) in the thermostat (controller). The thermostat is set for a certain temperature **(set point).** The controller compares the actual temperature (feedback signal) to the set point temperature, and an **error signal** is generated if the former falls below the latter. The error signal activates the furnace (effector). The resulting change in temperature is monitored, and when the temperature rises sufficiently the furnace is turned off. Such a negative feedback system allows some fluctuation in room temperature. Effective communication between the sensor and effector is important in keeping these oscillations to a minimum.

Similar negative feedback systems maintain homeostasis in the body. One example is the system that regulates arterial blood pressure (Chap. 18). This system's sensors (arterial baroreceptors) are located in the carotid sinuses and aortic arch. Changes in stretch of the walls of the carotid sinus and aorta, which follow from changes in blood pressure, stimulate these sensors. Afferent nerve fibers transmit impulses to control centers in the medulla oblongata. Efferent nerve fibers send impulses from the medullary centers to

the system's effectors, the heart and blood vessels. The output of blood by the heart and the resistance to blood flow are altered in an appropriate direction to maintain blood pressure, as measured at the sensors, within a given range of values. This negative feedback control system compensates for disturbances that affect blood pressure, such as changing body position, exercise, and hemorrhage. Continuous rapid communication between the feedback elements is accomplished by nerves. Various hormones are also involved in regulating blood pressure, but their effects are generally slower and more long-lasting.

Feedforward control is another strategy used to control systems in the body, particularly when a change with time is desired. In this case, a command signal is generated, which specifies the target or goal. The moment-to-moment operation of the controller is "open loop"; that is, the regulated variable itself is not sensed. Feedforward control mechanisms often sense a disturbance and can therefore take corrective action that anticipates change. For example, heart rate and breathing increase even before a person has begun to exercise.

Feedforward control usually acts in combination with negative feedback systems. One example is picking up a pencil. The movements of the arm, hand, and fingers are directed by the cerebral cortex (feedforward controller); the movements are smooth and forces are appropriate only in part because of feedback of visual information and sensory information from receptors in the joints and muscles. Another example of this combination occurs during exercise. Respiratory and cardiovascular adjustments closely match muscular activity, so that arterial blood oxygen and carbon dioxide tensions hardly change during all but exhausting exercise. One explanation for this remarkable behavior is that exercise involves a centrally generated feedforward signal to the active muscles and the respiratory and cardiovascular systems; feedforward control, together with feedback information generated as a consequence of increased movement and muscle activity, adjusts the heart, blood vessels, and respiratory muscles.

Control system function can adapt over a period of time. Past experience and learning can change the control system's output so that it behaves more efficiently or appropriately.

Although homeostatic control mechanisms usually act for the good of the body, they are sometimes deficient, inappropriate, or excessive. Many diseases, such as cancer, diabetes, and hypertension, develop because of a defective control mechanism. Homeostatic mechanisms may have inappropriate actions, as in autoimmune diseases in which the immune system attacks the body's own tissues. Scar formation is one of the most effective homeostatic mechanisms of healing, but it is excessive in many chronic diseases, such as pulmonary fibrosis, hepatic cirrhosis, and renal interstitial diseases.

■ Positive Feedback Promotes a Change in One Direction

With **positive feedback,** a variable is sensed and action taken to reinforce change of the variable. Positive feedback does not lead to stability or regulation, but rather to the opposite, a progressive change in one direction.

One example of positive feedback in a physiologic process is the upstroke of the action potential in nerve and muscle (Fig. 1–3). Depolarization of the cell membrane to a value greater than threshold leads to an increase in sodium permeability. Positively charged sodium ions rush into the cell through membrane sodium channels and cause further membrane depolarization; this leads to a further increase in sodium permeability and more sodium entry. This snowballing event, which occurs in a fraction of a millisecond, leads to an actual reversal of membrane potential and an electrical signal (action potential) conducted along the nerve or muscle fiber membrane. The process is stopped by inactivation (closure) of the sodium channels. Another example occurs during the follicular phase of the menstrual cycle. The female sex hormone estrogen stimulates release of gonadotropin-releasing hormone, which causes release of luteinizing hormone, which in turn causes further estrogen synthesis by the ovaries. This positive feedback culminates in ovulation. A third example is calcium-induced calcium release, which occurs with each heartbeat. Depolarization of the cardiac muscle plasma membrane leads to a small influx of calcium through membrane calcium channels. This leads to an explosive release of calcium from the muscle's sarcoplasmic reticulum, which increases the cytosolic calcium level and activates the contractile machinery.

Positive feedback, if unchecked, can lead to a vicious cycle and dangerous situations. For example, a heart may be so weakened that it cannot provide adequate blood flow for itself. This leads to a further reduction in cardiac pumping ability, even less coronary blood flow, and further deterioration of cardiac function. The physician's task is sometimes to interrupt or "open" such a positive feedback loop.

■ Steady State and Equilibrium Are Separate Ideas

Physiology often involves the study of exchanges of matter or energy between different **compartments** separated by some type of membrane. For example, the whole body can be divided into two major compartments, the extracellular fluid and intracellular fluid. The **extracellular fluid** consists of all the body fluids outside of cells: interstitial fluid, lymph, blood plasma, and specialized fluids, such as cerebrospinal fluid. It constitutes the internal environment of the body. The **intracellular fluid** is comprised of the water and dissolved solutes bounded by cell plasma membranes. The blood inside capillaries and the fluid

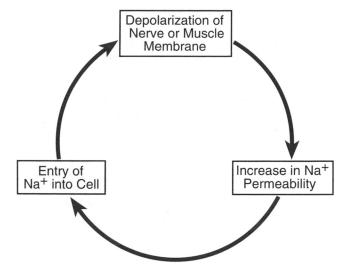

Figure 1–3 ■ Positive feedback cycle involved in the upstroke of the action potential.

in interstitial spaces are separated by the capillary endothelium. Air in the lung alveoli and blood in the pulmonary capillaries are separated by the alveolar-capillary membrane. Even within cells there is compartmentation—the interiors of organelles are separated from the cytosol by membranes. This restricts enzymes and substrates to structures such as mitochondria and lysosomes and allows for fine regulation of enzymatic reactions and a greater variety of metabolic processes.

Equilibrium is a condition in which opposing forces are balanced. When two compartments are in equilibrium, there is *no net transfer* of a particular substance or of energy from one compartment to the other. Equilibrium occurs if sufficient time for exchange has been allowed and if there is no longer a physical or chemical driving force that would favor net movement in one direction or the other. For example, in the lung, oxygen in alveolar gas diffuses into pulmonary capillary blood until the same oxygen tension is attained in both compartments. Another example is the presence of osmotic equilibrium between cells and extracellular fluid, which is a consequence of the high water permeability of most cell membranes. An equilibrium condition remains stable, if undisturbed. No energy expenditure is required to maintain an equilibrium state.

Equilibrium and steady state are sometimes confused with each other. A **steady state** is simply a condition that doesn't change with time. It indicates that the amount or concentration of a substance in a compartment is constant. If there is no net gain or net loss of a substance in a compartment, then input and output are, of course, equal, and a steady state exists. Steady state and equilibrium both suggest stable conditions, but (1) a steady state does not necessarily indicate an equilibrium condition, and (2) energy expenditure may be required to maintain a steady state. For example, in most body cells, there is a steady state for sodium ions; the amounts of sodium entering and leaving cells per unit time are equal. But intracellular and extracellular sodium ion concentrations are far from equilibrium. Sodium ions tend to move into cells down concentration and electrical gradients. The cell continuously uses metabolic energy to pump sodium ions out of the cell to maintain the cell in a steady state with respect to sodium ions.

Figure 1-4 illustrates a few simple models and serves to emphasize the distinctions between equilibrium and steady state. In Figure 1-4A, the fluid level in the sink is constant (a steady state) because the rates of inflow and outflow are equal. If we were to increase the rate of inflow (open the tap), the fluid level would rise, and with time a new steady state might be established at a higher level. In Figure 1-4B, the fluids in compartments X and Y are not in equilibrium (the fluid levels are different), but the system as a whole and each region (compartment) are in a steady state, since inputs and outputs are equal. In Figure 1-4C, the system is in a steady state and compartments X and Y are in equilibrium. Note that the term *steady state* can apply to a single or several regions (compartments); the term *equilibrium* describes the relation between at least two adjacent regions that can exchange matter or energy with each other.

Communication

Communication between cells is essential for the control and coordination of activities of cells, tissues, and organs in the body and for the maintenance of homeostasis. In the human body there are several means by which information may be transmitted, including direct communication between adjacent cells through gap junctions, autocrine and paracrine signaling, nerve

action potentials and release of neurotransmitters, and hormones produced by endocrine and nerve cells (Fig. 1-5).

■ Gap Junctions Provide a Pathway for Direct Communication between Adjacent Cells

Adjacent cells sometimes communicate directly with each other via **gap junctions,** specialized protein channels made of the protein **connexin** (Fig. 1-6). Six connexins form a half-channel, called a **connexon.** Two connexons join end to end to form an intercellular channel between adjacent cells. Gap junctions allow the flow of ions (hence electrical current) and small molecules between the cytosol of neighboring cells (Fig. 1-5). Gap junctions appear to be important in transmission of electrical signals between neighboring cardiac muscle cells, smooth muscle cells, and some nerve cells. They may also functionally couple adjacent epithelial cells. Gap junctions are thought to play a role in control of cell growth and differentiation by allowing adjacent cells to share a common intracellular environment. Often when a cell is injured, gap junctions close, thereby isolating a damaged cell from its neighbors. This isolation process may result from a rise in calcium and a fall in pH in the cytosol of the damaged cell.

■ Cells May Communicate Locally by Paracrine and Autocrine Signaling

Cells may interact via the local release of chemical substances. This is a means of communication, present in primitive living forms, that does not depend on a vascular system. With **paracrine** signaling, a chemical substance is liberated from a cell, diffuses a short distance through interstitial fluid, and acts on nearby cells. With **autocrine** signaling, a cell releases a chemical into the interstitial fluid that affects its own activity (Fig. 1-5).

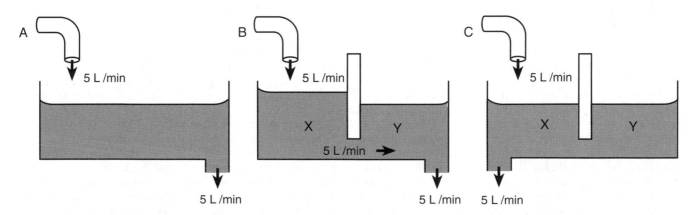

Figure 1–4 ■ Models of the concepts of steady state and equilibrium. A, B, and C depict a steady state. In C, compartments X and Y are in equilibrium. (Modified from Riggs DS: *The Mathematical Approach to Physiological Problems.* Cambridge, MIT Press, 1970, p. 169.)

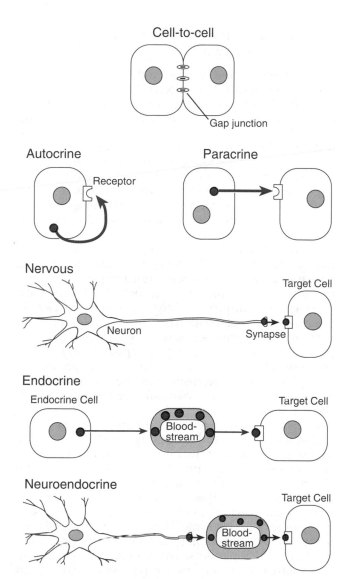

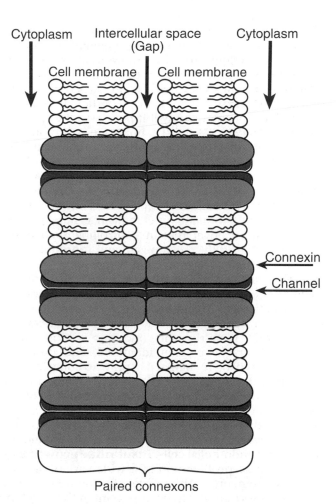

Figure 1–6 ■ Structure of gap junctions. The channel connects the cytosol of adjacent cells. Six molecules of the protein connexin form a half-channel called a connexon.

Figure 1–5 ■ Modes of intercellular signaling. Cells may communicate with each other directly via gap junctions or chemical messengers. With autocrine and paracrine signaling, a chemical messenger diffuses a short distance through the extracellular fluid and binds to a receptor on the same cell or a nearby cell. Nervous signaling involves action potentials carried rapidly, often over long distances, and the release of a neurotransmitter at a synapse. Endocrine signaling involves release of a hormone into the bloodstream and binding of the hormone to specific target cell receptors. Neuroendocrine signaling involves release of a hormone from a nerve cell and transport of the hormone by the blood to a distant target cell.

■ The Nervous System Provides for Rapid and Targeted Communication

The nervous system provides for rapid communication between body parts, with conduction times measured in milliseconds. This system is also organized for discrete activities; it provides an enormous number of private lines for sending messages from one distinct

locus to another. Conduction of information along nerves occurs via **action potentials,** and transmission between nerves or between nerves and effector structures takes place at a **synapse.** Synaptic transmission is almost always mediated by release of specific chemicals **(neurotransmitters)** from the nerve terminals (Fig. 1-5). The innervated cells have specialized protein molecules in their cell membranes **(receptors)** that selectively bind neurotransmitter. Chapter 3 discusses the actions of various neurotransmitters, and how they are synthesized and degraded. Chapters 4-6 discuss the role of the nervous system in coordinating and controlling body functions.

■ The Endocrine System Provides for Slower and More Diffuse Communication

The endocrine system produces messenger molecules called *hormones* in response to a variety of stimuli. In contrast to the nervous system, responses to hormones are often much slower (seconds to hours) in onset and effects are often more long-lasting. The endocrine system is analogous to a radio broadcasting system: Hormones are carried to all parts of the body by the bloodstream (Fig. 1–5). A particular cell can respond to a hormone only if it possesses the specific receptor ("receiver") for the hormone. Hormone effects may be discrete; for example, antidiuretic hormone increases the water permeability of kidney collecting duct cells but does not change the water permeability of other cells. They may also be diffuse, influencing practically every cell in the body; for example, thyroxine stimulates metabolism generally. Hormones play a critical role in controlling such body functions as growth, metabolism, and reproduction.

A special category of chemical messengers called *tissue growth factors* is produced by cells that are not traditional endocrine cells. These growth factors are protein molecules that influence cell division, differentiation, and cell survival. They may exert effects in an autocrine, paracrine, or endocrine fashion. Many growth factors have been identified, and probably many more will be recognized in years to come. **Nerve growth factor** enhances nerve cell development and stimulates the growth of axons. **Epidermal growth factor** stimulates the growth of epithelial cells in the skin and other organs. **Platelet-derived growth factor** stimulates the proliferation of vascular smooth muscle and endothelial cells. **Insulinlike growth factors** stimulate proliferation of a wide variety of cells and mediate many of the effects of growth hormone. Growth factors appear to be important in the development of multicellular organisms and in the regeneration and repair of damaged tissues.

■ The Nervous and Endocrine Control Systems Overlap

The separation between nervous and endocrine control systems is blurred in places. First, the nervous system exerts important controls over endocrine gland function. For example, the hypothalamus controls the secretion of hormones from the pituitary gland. Second, specialized nerve cells, called **neuroendocrine cells,** secrete hormones. Examples include the hypothalamic neurons, which liberate releasing factors that control secretion by the anterior pituitary gland, and the hypothalamic neurons, which secrete antidiuretic hormone and oxytocin into the circulation. Third, many proved or presumed neurotransmitters found in nerve terminals are also well-known hormones. These include antidiuretic hormone, cholecystokinin, enkephalins, norepinephrine, secretin, and vasoactive intestinal peptide. Thus, it is sometimes difficult to classify a particular molecule as a hormone or neurotransmitter.

■ Chemical Signals Are Processed by Cells in Different Ways

Neurotransmitters, protein and peptide hormones, and growth factors do not exert their effects by entering cells. Rather, their messages are converted (*transduced*) by cell membrane proteins into signals within the cells, a process often referred to as **signal transduction.** The first step in cell signaling is the combination of the messenger with a specific plasma membrane receptor. In some cases the membrane receptor is linked to an ion channel and signaling results from a change in membrane potential. An example is the neurotransmitter acetylcholine, which acts at the endplate of skeletal muscle fibers by binding to the nicotinic acetylcholine receptor; binding changes the configuration of the receptor complex, opening an ion channel that allows sodium and potassium ions to move through the cell membrane. The chemical message is transformed into an electrical change (membrane depolarization) in the muscle cell. The protein hormone insulin and many growth factors appear to act via cell membrane receptors with intrinsic *tyrosine kinase* activity; the enzyme catalyzes the phosphorylation of tyrosine residues of intracellular proteins, thereby transferring an external message into the cell interior.

Cell signaling by protein and peptide hormones is commonly via **second messengers** (Fig. 1–7). The first step is binding of the hormone (the "first messenger") to a specific receptor in the cell membrane. The receptor is usually coupled to a guanosine nucleotide-binding protein, called a *G protein.* The G protein in turn stimulates or inhibits an intracellular, plasma membrane-bound enzyme. If the enzyme is stimulated, many second messenger molecules are produced, thereby amplifying the signal. Second messengers are small, diffusible molecules that relay information from the cell membrane to the cell interior; they include cyclic adenosine monophosphate (cAMP), inositol 1,4,5-trisphosphate and diacylglycerol, cyclic guanosine monophosphate (cGMP), and several other chemicals. For example, when the enzyme adenylate cyclase is activated, hundreds or thousands of molecules of cAMP are formed from ATP. Subsequent steps vary with the second messenger pathway. In the case of cAMP, a cAMP-dependent protein kinase (**protein kinase A**) is activated and transfers a phosphate group from ATP to other protein molecules. This phosphorylation alters the configuration and function of these proteins. Such changes eventually produce the cell response to the first messenger, whether it be secretion, contraction or relaxation, a change in metabolism or cell membrane transport, or cell division or differentiation.

Binding of a single hormone often produces more than one kind of second messenger. For example, some receptors are coupled to the enzyme phospholipase

G PROTEINS AND DISEASE

Cell communication mechanisms are essential for survival of higher organisms. These mechanisms open important avenues for control and coordination, but they also provide opportunities for errors. Many human diseases are caused by abnormalities in various steps of the communication pathway.

G proteins function as key transducers of information across cell membranes by coupling receptors to cell effectors (Fig. 1–7). They are part of a large family of proteins that bind guanosine triphosphate (GTP) and hydrolyze this compound. G proteins are heterotrimers and consist of α, β and γ subunits, each of which is encoded by a different gene. Different G proteins stimulate or inhibit adenylate cyclase, open ion channels, increase cGMP-phosphodiesterase activity, or activate phospholipase C.

G protein abnormalities have been identified in human disease and may result from a genetic defect or abnormal function or expression.

Mutations of G proteins can result in either constitutive activation or loss of activity. Since G protein activation may change intracellular effectors involved in regulating cell growth, it is not surprising that G protein genes have the propensity to become **oncogenes,** whose products are responsible for the abnormal growth of malignant cells.

Growth hormone (GH), produced by the anterior pituitary gland, is an important regulator of growth of the skeleton during childhood. Overproduction of GH during adolescence can lead to **gigantism,** and during adulthood leads to **acromegaly** (discussed in Chap. 34). Most commonly, gigantism and acromegaly are the result of pituitary tumors involving GH-producing cells.

Regulation of GH secretion and growth of GH-producing cells involve changes in intracellular cAMP concentration. These changes in intracellular cAMP are coupled to extracellular signals by G proteins. It was recently demonstrated that basal adenylate cyclase activity and cAMP are elevated in about 40% of human pituitary GH-secreting tumors. This was shown to be due to a single amino acid substitution in the α-subunit of a G protein, resulting in constitutive activation. Thus, in this situation, abnormally high GH levels result from a simple mutation in a G protein. G protein mutations may also be involved in some thryoid, adrenal, and ovarian tumors.

Certain bacterial toxins can stimulate or inhibit G proteins by covalently modifying α-subunits. For example, the toxin of *Vibrio cholerae* binds and activates the stimulatory G protein linked to adenylate cyclase, thereby producing elevated intracellular levels of cAMP. As part of their normal control mechanism, cells lining the small intestine are stimulated to secrete fluid in response to increased cAMP. In response to cholera toxin, these cells overproduce fluid, leading to the watery diarrhea characteristic of **cholera.**

Abnormal expression or function of G proteins may play a role in the pathophysiology of a variety of disorders, including diabetes mellitus, alcoholism, cardiomyopathy, neuropsychiatric diseases, and immune system disorders.

C, which liberates two different second messengers, inositol 1,4,5-trisphosphate and diacylglycerol. Or a single hormone may activate both cAMP and other second messenger pathways. Such redundancy is thought to provide a system of checks and balances for regulating cell activities. Chapter 33 discusses second messenger systems in greater detail.

Steroid and thyroid hormones do *not* require plasma membrane receptors and second messengers to bring the signal into the cell, because they readily

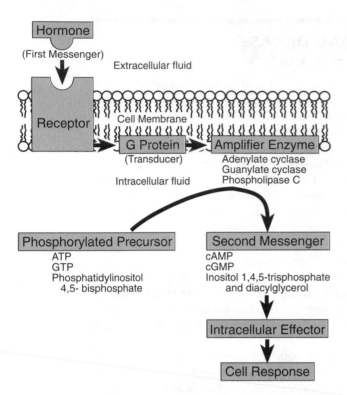

Figure 1–7 ■ Signal transduction pattern common to several second messenger systems. A protein or peptide hormone binds to a plasma membrane receptor, which stimulates or inhibits a membrane-bound amplifier enzyme via a G protein. The enzyme catalyzes the production of many second messenger molecules from a phosphorylated precursor (e.g., cAMP from ATP, cGMP from GTP, or inositol 1,4,5-trisphosphate and diacylglycerol from phosphatidylinositol 4,5-bisphosphate). The second messengers, in turn, activate protein kinases or cause other intracellular changes that ultimately lead to the cell response.

penetrate cell membranes. They bind to receptors in both the nucleus and cytoplasm. The activated receptors bind to nuclear DNA, which changes the transcription rate of specific genes. This leads to an increase or decrease in the production of messenger RNA and a change in the cell's content of specific proteins. Because protein synthesis takes a while, the expression of the effects of steroid and thyroid hormones may take many hours.

■ Coordinated Body Activity Requires Integrated Function

Body functions can be analyzed in terms of several systems, such as the nervous, muscular, cardiovascular, respiratory, renal, gastrointestinal, and endocrine systems. These divisions are rather arbitrary, however, and all systems interact and depend on each other. For example, walking involves the activity of many systems. The nervous system coordinates the movements of the limbs and body, stimulates the muscles to contract, and senses muscle tension and limb position. The cardiovascular system supplies blood flow to muscles, providing for nourishment and removal of metabolic wastes and heat. The respiratory system supplies oxygen and removes carbon dioxide. The renal system maintains an optimal blood composition. The gastrointestinal system supplies energy-yielding metabolites. The endocrine system helps adjust blood flow and the supply of various metabolic substrates to active muscle. Coordinated body activity demands integration of many systems.

Recent research demonstrates that many diseases can be explained on the basis of abnormal function at the molecular or cellular level. Diseases occur within the context of a whole organism, however, and it is important to understand how all cells, tissues, organs, and organ systems respond to a disturbance (disease process). Principles of prevention and treatment of disease must take into account mechanisms in the body at all levels. Knowledge of the body's processes and functions and homeostatic mechanisms is clearly fundamental to the intelligent practice of medicine.

■ ■ ■ ■ ■ ■ ■ ■ ■ ■ ■ **REVIEW EXERCISES** ■ ■ ■ ■ ■ ■ ■ ■ ■ ■ ■

Identify Each with a Term

1. Internal environment
2. A regulated variable is sensed, information is sent to a controller, and action is taken to *oppose* change from the desired value
3. A condition in which opposing forces exactly counteract each other
4. A protein molecule that forms gap junctions
5. A cell membrane protein that binds guanosine nucleotides and couples a hormonal signal from a receptor to an amplifier enzyme or ion channel

Define Each Term

6. Homeostasis
7. Steady state
8. Receptor
9. Paracrine signaling
10. Second messenger

Choose the Correct Answer

11. Which of the following is a homeostatic response to a low blood glucose concentration?
    a. Decreased food intake

b. Decreased secretion of glucagon
c. Increased secretion of epinephrine
d. Increased secretion of insulin

12. Steroid hormones produce their effects by:
a. Binding to cytoplasmic or nuclear receptors
b. Binding to receptors on the cell membrane
c. Stimulating hydrolysis of cell membrane lipids
d. Stimulating production of cAMP

13. A rise in cytosolic calcium:
a. causes gap junctions to close
b. is seen in dying or dead cells
c. results in contraction in muscle
d. All of the above
e. None of the above

Briefly Answer

14. Give two physiologic examples of positive feedback.

15. List five parameters of the internal environment (arterial blood plasma) that are regulated by homeostatic mechanisms.

16. List four intracellular second messengers.

17. Give two examples of neuroendocrine cells.

18. Antidiuretic hormone (ADH) increases collecting duct water permeability in the kidney using cAMP as a second messenger. List the initial events involved in the action of this hormone.

■ ■ ■ ■ ■ ■ ■ ■ ■ ■ ■ ■ ■ ANSWERS ■ ■ ■ ■ ■ ■ ■ ■ ■ ■ ■ ■ ■

1. Extracellular fluid (blood plasma, interstitial fluid, lymph)

2. Negative feedback

3. Equilibrium

4. Connexin

5. G protein

6. The maintenance of steady states in the body by coordinated physiologic mechanisms

7. A condition that does not change with time

8. A protein molecule in the cell membrane or within cells that selectively binds to a specific chemical (hormone, growth factor, neurotransmitter, drug) and produces a specific physiologic effect

9. Release into the extracellular fluid of a chemical messenger that affects nearby cells

10. A small, diffusible molecule produced when a hormone combines with a cell membrane receptor and which carries the message to the inside of the cell

11. c

12. a

13. d

14. Upstroke of action potential; release of LH prior to ovulation; calcium-induced calcium release

15. Oxygen and carbon dioxide tensions; glucose concentration; osmotic pressure; hydrogen, potassium, calcium, and magnesium ion concentrations; temperature

16. cAMP, inositol trisphosphate, diacylglycerol, cGMP, Ca^{2+}

17. Hypothalamic cells that produce releasing factors; hypothalamic cells that produce oxytocin and antidiuretic hormone

18. ADH combines with its receptor; G protein is activated; adenylate cyclase activity increases; cAMP is produced from ATP; and cAMP activates protein kinase A. The latter phosphorylates cell proteins.

Suggested Reading

Adolph EF: Physiological integrations in action. *Physiologist* 25, No. 2 (suppl):1-67, 1982.

Bernard C: *An Introduction to the Study of Experimental Medicine.* New York, Dover, 1957.

Berridge MJ: Inositol trisphosphate and calcium signalling. *Nature* 361:315-325, 1993.

Cannon WB: *The Wisdom of the Body.* New York, Norton, 1945.

Hanley RM, Steiner AL: The second-messenger system for peptide hormones. *Hosp Pract* 24, No. 8:59-70, 1989.

Houk JC: Control strategies in physiological systems. *FASEB J* 2:97-107, 1988.

Linder ME, Gilman AG: G Proteins. *Sci Am* 267:56-65, July 1992.

Rasmussen H: Disordered cell communication as the basis of human disease: Implications for 21st-century medicine. *Issues Biomed* 15:33-68, 1991.

Riggs DS: *The Mathematical Approach to Physiological Problems.* Cambridge, MA, MIT Press, 1970.

Snyder SH: The molecular basis of communication between cells. *Sci Am* 253:132-141, Oct. 1985.

C H A P T E R
■ ■ ■ ■ ■ ■ ■
2

The Cell Membrane, Membrane Transport, and Resting Membrane Potential

CHAPTER OUTLINE

I. STRUCTURE OF THE CELL MEMBRANE
   A. The cell membrane has proteins inserted in the lipid bilayer
   B. There are different types of membrane lipids
      1. Phospholipids
      2. Cholesterol
II. MECHANISMS OF SOLUTE TRANSPORT
   A. Macromolecules cross the cell membrane by vesicle fusion
      1. Endocytosis
      2. Exocytosis
   B. Passive movement of solutes tends to equilibrate concentrations
      1. Simple diffusion
      2. Diffusive membrane transport
      3. Facilitated diffusion via carrier proteins
      4. Facilitated diffusion through ion channels
   C. Solutes are moved against gradients by active transport systems
      1. Primary active transport
      2. Secondary active transport
      3. Movement of solutes across epithelial cell layers
III. MOVEMENT OF WATER ACROSS THE CELL MEMBRANE
   A. Movement of water across the cell membrane is driven by differences in osmotic pressure
   B. Many cells can regulate their volume
IV. THE RESTING MEMBRANE POTENTIAL
   A. Ion movement is driven by the electrochemical potential
   B. Net ion movement is zero at the equilibrium potential
   C. The resting membrane potential is determined by passive movement of several ions

OBJECTIVES

After studying this chapter, the student should be able to:

1. Summarize current concepts of the chemical composition and structure of the plasma membrane of a cell
2. Define *amphipathic molecule, integral membrane protein, phospholipid, ion channel,* and *ion pump*
3. Describe the characteristics of equilibrating carrier-mediated transport that distinguish it from simple diffusion
4. Explain the mechanisms that open voltage-gated and ligand-gated ion channels
5. Discuss the concept of active transport using the Na^+/K^+-ATPase pump and Na^+-coupled glucose transport as specific examples
6. Describe how epithelial cells are organized to move solutes across the epithelium
7. Define *osmosis* and *osmotic pressure* and explain how regulatory volume decrease or increase helps to maintain normal cell volume
8. Explain electrochemical potential, Nernst potential, and Goldman equation
9. Explain why the resting membrane potential is close to the Nernst potential for K^+

The intracellular fluid of living cells, the *cytosol,* has a composition very different from that of the extracellular fluid. For example, the concentrations of potassium and phosphate ions are higher inside cells than outside, whereas sodium, calcium, and chloride ion concentrations are much lower inside cells than outside. These differences are necessary for the proper functioning of many intracellular enzymes; for instance, the synthesis of proteins by the ribosomes requires a relatively high potassium concentration. The *cell membrane,* or *plasma membrane,* creates and maintains these differences by establishing a permeability barrier around the cytosol. The ions and cell proteins needed for normal cell function are prevented from leaking out and those not needed by the cell are unable to enter the cell freely. The cell membrane also keeps metabolic intermediates near where they will be needed for further synthesis or processing and retains metabolically expensive proteins inside the cell.

The cell membrane is necessarily selectively permeable. Cells need to receive nutrients to function and need to dispose of metabolic waste products. To function in coordination with the rest of the organism, cells need to receive and send information in the form of hormones and neurotransmitters. The plasma membrane possesses mechanisms to allow specific molecules to cross the barrier around the cell. A selective barrier surrounds not only the cell but also every intracellular organelle that requires an internal milieu different from that of the cytosol. The cell nucleus, mitochondria, endoplasmic reticulum, Golgi apparatus, and lysosomes are delimited by membranes similar in composition to the cell membrane.

Structure of the Cell Membrane

The first theory of membrane structure proposed that cells are surrounded by a double layer of lipid molecules, a **lipid bilayer.** This theory was based on the known tendency of lipid molecules to form lipid bilayers with very low permeability to water-soluble molecules. However, the lipid bilayer theory did not explain the selective movement of certain water-soluble compounds, such as glucose and amino acids, across the cell membrane. In 1972, Singer and Nicolson proposed the **fluid mosaic model** of the cell membrane (Fig. 2-1), a model that with minor modifications is still accepted as the correct picture of the structure of the cell membrane.

■ The Cell Membrane Has Proteins Inserted in the Lipid Bilayer

Proteins and lipids are the two major components of the cell membrane, present in about equal proportions. The different lipids are arranged in a lipid bilayer,

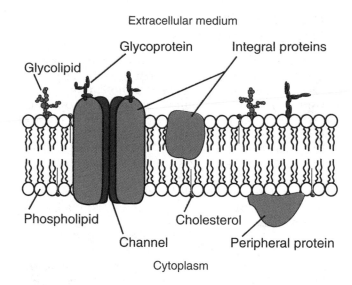

Figure 2–1 ■ Fluid mosaic model of the plasma membrane. Lipids are arranged in a bilayer. Integral proteins are embedded in the bilayer and often span it, forming a channel. Peripheral proteins do not penetrate the bilayer.

and two different types of proteins are found there. **Integral** or **intrinsic** proteins are embedded in the lipid bilayer; many span it completely, being accessible from the inside and outside of the membrane. **Peripheral** or **extrinsic** proteins do not penetrate the lipid bilayer. They are in contact with the outer side of only one of the lipid layers—either the layer facing the cytoplasm or the layer facing the extracellular medium (Fig. 2-1). Many membrane proteins have carbohydrate molecules, in the form of specific sugars, attached to the parts of the proteins that are exposed to the extracellular medium. These molecules are known as **glycoproteins.** Some of the intrinsic membrane proteins can move in the plane of the membrane, like small boats floating in the "sea" formed by the bilayer arrangement of the lipids. Other membrane proteins are anchored to the cytoskeleton inside the cell or to proteins of the extracellular matrix.

The proteins in the cell membrane serve a variety of roles. Many peripheral membrane proteins are enzymes, and many membrane-spanning integral proteins are carriers or channels for movement of water-soluble molecules and ions into and out of the cell. Another important role of membrane proteins is structural. For example, certain membrane proteins in the erythrocyte help maintain the biconcave shape of the cell. Finally, some membrane proteins serve as highly specific recognition sites or receptors on the outside of the cell membrane to which extracellular molecules, such as hormones, can bind. If the receptor is a membrane-spanning protein, it provides a mechanism for converting an extracellular signal into an intracellular response.

■ There Are Different Types of Membrane Lipids

Lipids found in cell membranes can be classified into two broad groups: those that contain fatty acids as part of the lipid molecule and those that do not. Phospholipids are an example of the first group, and cholesterol is the most important example of the second group.

Phospholipids ■ The fatty acids present in phospholipids are molecules with a long hydrocarbon chain and a carboxyl terminal group. The hydrocarbon chain can be saturated (no double bonds between the carbon atoms) or unsaturated (one or more double bonds present). The composition of fatty acids gives them some peculiar characteristics. The long hydrocarbon chain tends to avoid contact with water and is described as **hydrophobic.** The carboxyl group at the other end is compatible with water and is termed **hydrophilic.** Fatty acids are said to be **amphipathic,** because both hydrophobic and hydrophilic regions are present in the same molecule.

Phospholipids are the most abundant complex lipids found in cell membranes. They are amphipathic molecules formed by two fatty acids (normally one saturated and one unsaturated) and one phosphoric acid group substituted on the backbone of a glycerol molecule. This arrangement produces a hydrophobic area formed by the two fatty acids and a polar hydrophilic head. When phospholipids are arranged in a bilayer, the polar heads are on the outside and the hydrophobic fatty acids on the inside. It is very difficult for water-soluble molecules and ions to pass directly through the hydrophobic interior of the lipid bilayer.

Sphingolipids are phospholipids in which the backbone is sphingosine, a long amino alcohol. Sphingolipids are present in all plasma membranes in small amounts, but they are especially abundant in brain and nerve cells.

Glycolipids are lipid molecules that contain sugars and sugar derivatives (instead of phosphoric acid) in the polar head. They are located mainly in the outer half of the lipid bilayer, with the sugar molecules facing the extracellular medium.

Cholesterol ■ Cholesterol is an important component of animal cell membranes. The proportion of cholesterol in cell membranes varies from 10% to 50% of total lipids. Cholesterol has a rigid structure that stabilizes the cell membrane and reduces the natural mobility of the complex lipids in the plane of the membrane. When the amount of cholesterol increases, it is more difficult for lipids and proteins to move in the membrane. Some cell functions, such as the capability of cells of the immune system to respond to the presence of an antigen, depend on the ability of membrane proteins to move in the plane of the membrane to bind the antigen. A decrease in membrane fluidity produced by an increase in cholesterol will impair these functions.

Mechanisms of Solute Transport

All cells need to import oxygen, sugars, amino acids, and some small ions and need to export carbon dioxide, metabolic wastes, and secretions. At the same time, specialized cells require mechanisms to transport molecules such as enzymes, hormones, and neurotransmitters. The movement of large molecules is carried out by **endocytosis** and **exocytosis,** the transfer of substances into or out of the cell, respectively, by vesicle formation and vesicle fusion with the plasma membrane. Cells also have mechanisms for rapid movement of ions and solute molecules across the cell membrane. These mechanisms are of two general types: **passive movement,** which requires no direct expenditure of metabolic energy, and **active movement,** which uses metabolic energy to drive solute transport.

■ Macromolecules Cross the Cell Membrane by Vesicle Fusion

Endocytosis ■ *Endocytosis* is a general term for the process whereby a region of the plasma membrane is pinched off to form an endocytic vesicle inside the cell. During vesicle formation some fluid, dissolved solutes, and particulate material from the extracellular medium are trapped inside the vesicle and internalized by the cell. Two main types of endocytosis can be distinguished, based on the size of the endocytic vesicles and the mechanisms by which they are formed (Fig. 2–2).

The first type is **phagocytosis,** which refers to the ingestion of large particles or microorganisms and usually occurs only in specialized cells such as macrophages. An important function of macrophages in humans is to remove invading bacteria. The phagocytic vesicle (1–2 μm diameter) is almost as large as the phagocytic cell itself. Phagocytosis requires a specific stimulus. It occurs only after the extracellular particle has bound to the extracellular surface. The particle is then enveloped by expansion of the cell membrane around it (Fig. 2–2).

The second type of endocytic process is **pinocytosis,** which produces much smaller endocytic vesicles (0.1–0.2 μm diameter) than phagocytosis. Pinocytosis occurs in almost all cells. It is known as a **constitutive** process because it occurs continually and specific stimuli are not required. In further contrast to phagocytosis, pinocytosis originates with the formation of depressions in the cell membrane (Fig. 2–2). The depressions pinch off within a few minutes after they form and give rise to endocytic vesicles inside the cell. There are two types of pinocytosis. **Fluid-phase endocytosis** is the nonspecific uptake of the extracellular medium and all its dissolved solutes. The material is trapped inside the endocytic vesicle as it is pinched off inside the cell (Fig. 2–2). The amount of extracellular material internalized by this

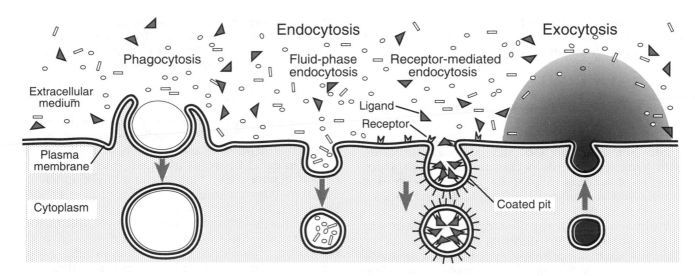

Figure 2–2 ■ Transport of macromolecules across the cell membrane via formation of vesicles. Particulate matter in extracellular fluid is engulfed and internalized by phagocytosis. During fluid-phase endocytosis, extracellular fluid and dissolved macromolecules enter the cell in endocytic vesicles that pinch off at depressions in the cell membrane. Receptor-mediated endocytosis uses membrane receptors at coated pits to bind and internalize specific solutes (ligands). Exocytosis is the release of macromolecules designed for export from the cell. These are packed inside secretory vesicles that fuse with the cell membrane and release their contents ouside the cell. (Modified from Dautry-Varsat A, Lodish HF. How receptors bring proteins and particles into cells. *Sci Am* 250:52–58, May, 1984.)

process is directly proportional to its concentration in the extracellular solution. **Receptor-mediated endocytosis** is a much more efficient process that uses receptors on the cell surface to bind specific molecules. These receptors accumulate at specific depressions known as **coated pits** because the cytosolic surface of the membrane at this site is covered with a coat of several proteins. The coated pits pinch off continually to form endocytic vesicles. This provides the cell with a mechanism for rapid internalization of a large amount of a specific molecule without the need to endocytose large volumes of the extracellular medium. The receptors also aid cell uptake of molecules present at low concentrations outside the cell. Receptor-mediated endocytosis is the mechanism by which cells take up a variety of important molecules, including hormones, growth factors, and serum transport proteins such as **transferrin** (an iron carrier). Foreign substances, such as diphtheria toxin and certain viruses, also enter cells by this pathway.

Exocytosis ■ Many cells synthesize important macromolecules that are destined for export from the cell. These molecules are synthesized in the endoplasmic reticulum, modified in the Golgi complex, and packed inside transport vesicles. The vesicles move to the cell surface, fuse with the cell membrane, and release their contents outside the cell (Fig. 2–2). There are two exocytic pathways. Some proteins are secreted continuously by the cells that make them. Secretion of mucus by **goblet cells** in the small intestine is a

specific example. In this case, exocytosis follows the **constitutive** pathway, which is present in all cells. In other cells the macromolecules are stored inside the cell in secretory vesicles. These vesicles fuse with the cell membrane and release their contents only when a specific extracellular stimulus arrives at the cell membrane. This pathway, known as the **regulated** pathway, is responsible for the rapid "on-demand" secretion of many specific hormones, neurotransmitters, and digestive enzymes.

■ Passive Movement of Solutes Tends to Equilibrate Concentrations

Simple Diffusion ■ Any solute will tend to occupy in a uniform way the entire space available to it. This movement, known as **diffusion,** is due to the spontaneous brownian movement that all molecules experience and which explains many everyday observations. Sugar diffuses in coffee, lemon in tea, and a drop of ink placed in a glass of water will diffuse and slowly color all the water. The net result of diffusion is movement of substances according to their difference in concentrations, from regions of high concentration to regions of low concentration. Diffusion is a very effective way for substances to move short distances.

The speed with which the diffusion of a solute in water occurs depends on the difference of concentration, the size of the molecules, and the possible interactions of the diffusible substrate with water. These different factors appear in Fick's law, which describes the diffusion of any solute in water. In its simplest formulation, Fick's law can be written as:

$$J = D A (C_1 - C_2)/\Delta X \tag{1}$$

where J is the flow of solute from region 1 to region 2 in the solution, D is the diffusion coefficient of the solute and takes into consideration factors such as solute molecular size and interactions of the solute with water, A is the cross-sectional area through which the flow of solute is measured, C is the concentration of the solute at regions 1 and 2, and ΔX is the distance between regions 1 and 2.

Diffusive Membrane Transport ■ Solutes can enter or leave a cell by diffusing passively across the cell membrane. The principal force driving diffusion of an uncharged solute is the difference of concentration between the inside and the outside of the cell (Fig. 2–3). In the case of an electrically charged solute, such as an ion, diffusion is also driven by the membrane potential, which is the electrical gradient across the membrane. The membrane potential of most living cells is negative inside the cell relative to the outside.

Diffusion across a membrane has no preferential direction; it can occur from the outside of the cell toward the inside or vice versa. For any substance, it is possible to measure the **permeability coefficient** (P), which gives the speed of the diffusion across a unit area of plasma membrane for a defined driving force. Fick's law for diffusion of an uncharged solute across a membrane can be written as:

$$J = PA (C_1 - C_2) \tag{2}$$

which is similar to equation 1. P includes the membrane thickness, diffusion coefficient of the solute within the membrane, and solubility of the solute in the membrane. Dissolved gases, such as oxygen and carbon dioxide, have high permeability coefficients and thus diffuse across the cell membrane rapidly. Since diffusion across the plasma membrane usually implies that the diffusing solute enters the lipid bilayer to cross it, the solute's solubility in a lipid solvent (e.g., olive oil or chloroform) compared to its solubility in water is important in determining its permeability coefficient. A substance's solubility in oil compared to its solubility in water is its **partition coefficient.** Lipophilic substances that mix well with the lipids in the plasma membrane have high partition coefficients and, as a result, high permeability coefficients and tend to cross the plasma membrane easily. Hydrophilic substances, such as ions and sugars, do not interact well with the lipid component of the membrane, have low partition coefficients and low permeability coefficients, and diffuse across the membrane more slowly.

For solutes that diffuse across the lipid part of the plasma membrane, the relation between the rate of movement and the difference in concentration between the two sides of the membrane is linear (Fig. 2–4). The higher the difference in concentration (C_1

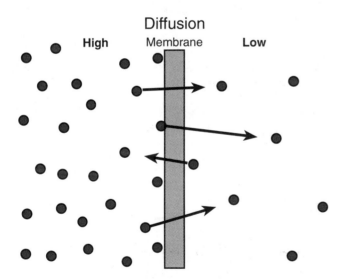

Figure 2–3 ■ Gases and lipid-soluble molecules readily diffuse through the lipid bilayer. In this example, diffusion of a solute across a cell membrane is driven by the difference in concentration on the two sides of the membrane. The solute molecules move randomly by brownian motion. Initially, random movement from left to right across the membrane is more frequent than movement in the opposite direction because there are more molecules on the left side. This results in net movement of solute from left to right across the membrane until the concentration of solute is the same on both sides of the membrane. At this point, equilibrium (no net movement) is reached, because movement of solute from left to right is balanced by equal movement from right to left.

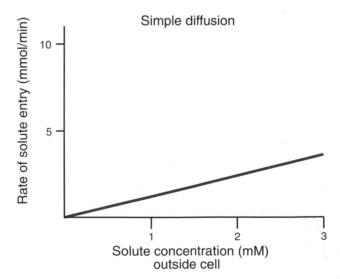

Figure 2–4 ■ Graphic representation of solute transport across a cell membrane by simple diffusion. The rate of solute entry increases linearly with extracellular concentration of the solute. Assuming no change in intracellular concentration, increasing the extracellular concentration increases the gradient that drives solute entry.

— C₂), the greater the amount of substance crossing the membrane per unit time.

Facilitated Diffusion via Carrier Proteins ■ For many solutes of physiologic importance, such as sugars and amino acids, the relation between transport rate and difference in concentration follows a curve that reaches a plateau (Fig. 2-5). Furthermore, the rate of transport of these hydrophilic substances across the cell membrane is much faster than expected for simple diffusion through a lipid bilayer. Membrane transport with these characteristics is often called **carrier-mediated transport,** because an integral membrane protein, the carrier, binds the transported solute on one side of the membrane and releases it at the other side. Although the details of this transport mechanism are unknown, it is hypothesized that the binding of the solute causes a conformational change in the carrier protein, which results in translocation of the solute (Fig. 2-6). Since there are a limited number of these carriers in any cell membrane, increasing the concentration of the solute initially uses the existing "spare" carriers to transport the solute at a higher rate than by simple diffusion. As the concentration of the solute increases further and more solute molecules bind to carriers, the transport system eventually reaches

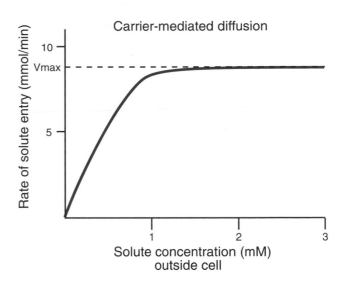

Figure 2–5 ■ Graphic representation of solute transport across a cell membrane by carrier-mediated transport. The rate of transport is much faster than that of simple diffusion (Fig. 2-4) and increases linearly as the extracellular solute concentration increases. The increase in transport is limited, however, by the availability of carriers. Once all are occupied by solute, further increases in extracellular concentration have no effect on the rate of transport. A maximum rate of transport (Vmax) is achieved, which cannot be exceeded.

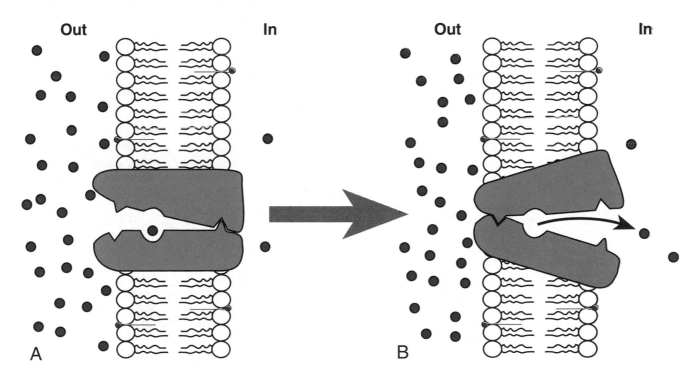

Figure 2–6 ■ Possible role of a carrier protein in facilitated diffusion of solute molecules (small circles) across a cell membrane. In this example, solute transport into the cell is driven by the high solute concentration outside compared to inside. (A) Binding of extracellular solute to the carrier, a membrane-spanning integral protein, may trigger a change in protein conformation that exposes the bound solute to the interior of the cell. (B) Bound solute readily dissociates from the carrier because of the low intracellular concentration of solute. Release of solute may allow the carrier to revert to its original conformation (A) to begin the cycle again.

saturation, when all the carriers are involved in translocating molecules of solute (Fig. 2–5). At this point, additional increases in solute concentration do not increase the rate of solute transport.

The types of carrier-mediated transport mechanisms considered here can transport a solute along its concentration gradient only, just as in simple diffusion. Net movement stops when the concentration of the solute has the same value on both sides of the membrane. At this point, with reference to equation 2, C_1 = C_2 and the value of J is 0. The transport systems function until the solute concentrations have **equilibrated.** However, equilibrium is attained much faster than with simple diffusion.

Equilibrating carrier-mediated transport systems have several characteristics: (1) They allow the transport of polar (hydrophilic) molecules at rates much higher than expected from the partition coefficient of these molecules. (2) They eventually reach saturation at high substrate concentration. (3) They have structural specificity, meaning each carrier system recognizes and binds very specific chemical structures (a carrier for D-glucose will not bind or transport L-glucose). (4) They show competitive inhibition by molecules with similar chemical structure. For example, carrier-mediated transport of D-glucose occurs at a slower rate when molecules of D-galactose also are present. This is because galactose, structurally similar to glucose, competes with glucose for the available glucose carrier proteins.

Equilibrating carrier-mediated transport, like simple diffusion, does not have directional preferences. Carrier mechanisms function equally well bringing their specific solutes into or out of the cell, depending on the concentration gradient. Net movement by equilibrating carrier-mediated transport ceases once the concentrations inside and outside the cell become equal.

Facilitated Diffusion through Ion Channels ■ Small ions, such as Na^+, K^+, Cl^-, and Ca^{2+}, also cross the plasma membrane much faster than would be expected based on their partition coefficients in the lipid bilayer. Its electrical charge makes it very difficult for the ion to move across the lipid bilayer. Fast movement of ions across the membrane is, however, an aspect of many cell functions. The nerve action potential, the contraction of muscle, the pacemaker function of the heart, and many other physiologic events are possible because of the ability of small ions to enter or leave the cell very rapidly. This movement occurs through selective **ion channels.**

Ion channels are intrinsic proteins spanning the width of the plasma membrane and are normally composed of several polypeptide subunits. Certain specific stimuli cause the protein subunits to open a **gate,** creating an aqueous channel through which the ions can move (Fig. 2–7). In this way, ions do not need to enter the lipid bilayer to cross the membrane; they are always in an aqueous medium, and when the chan-

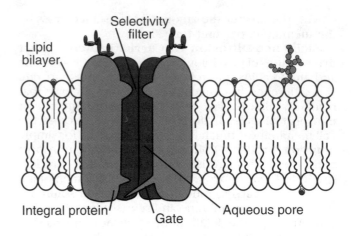

Figure 2–7 ■ Ion channels are formed between the polypeptide subunits of integral proteins that span the membrane, providing an aqueous pore through which ions can cross the membrane. Different types of gating mechanisms (see text) are used to open and close channels. Ion channels are often selective for a specific ion.

nels are open they move rapidly from one side of the membrane to the other by facilitated diffusion. Specific interactions between the ions and the sides of the channel produce an extremely rapid rate of ion movement. In fact, ion channels permit a much faster rate of solute transport, about 10^8 ions/sec, than carrier-mediated systems. Ion channels are often selective. For example, there are channels selective for Na^+, for K^+, for Ca^{2+}, for Cl^-, and for other monovalent anions and cations. It is generally assumed that some kind of ionic filter must be built into the structure of the channel (Fig. 2–7). No clear relation between amino acid composition of the channel protein and ion selectivity of the channel has been established.

Neher and Sakmann were awarded the Nobel prize in 1991 for developing the **patch clamp** technique, which revealed a great deal of information about the characteristic behavior of channels for different ions. The technique is based on the detection of the small electrical current due to ion movement when a channel is open. It is so sensitive that the opening and closing of a single ion channel can be observed (Fig. 2–8). In general, ion channels exist either fully open or completely closed, and they open and close very fast. The frequency with which a channel opens is variable, and the time the channel remains open (usually a few milliseconds) is also variable. Thus, the overall rate of ion transport across a membrane can be controlled by changing the frequency of channel opening or by changing the time each channel remains open.

Ion channels usually open in response to a specific stimulus. Ion channels can be classified into two groups according to their **gating mechanisms,** the signals that make them open or close.

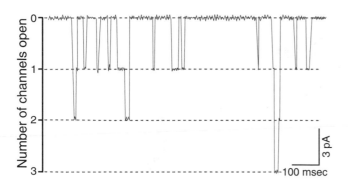

Figure 2–8 ■ Patch clamp recording from a frog muscle fiber. Ions flow through the channel when it opens, generating a current. In this experiment the current, about 3 pA, is detected as a downward deflection in the recording. When more than one channel opens, the current and the downward deflection increases in direct proportion to the number of opened channels. This record shows that up to three channels are open at any instant. (Modified from Kandel ER, Schwartz JH, Jessell TM. *Principles of Neural Science*. 3rd ed. New York, Elsevier, 1991.)

Voltage-gated ion channels open when the membrane potential changes beyond a certain threshold value. Channels of this kind are involved in the conduction of the action potential along the nerve axon and include sodium and potassium channels (Chap. 3). Voltage-dependent ion channels are found in many cell types. It is thought that some charged amino acids in the channel protein are sensitive to the transmembrane potential. Changes in the membrane potential cause these amino acids to move and induce a conformational change of the protein that opens the way for the ions.

Ligand-gated (or **chemically gated**) ion channels cannot open unless they first bind a specific agonist. The opening of the gate is produced by a conformational change in the protein induced by the ligand binding. The ligand can be a neurotransmitter arriving from the extracellular medium or an intracellular second messenger produced in response to some cell activity or hormone action and reaching the ion channel from the inside of the cell. The nicotinic acetylcholine receptor channel found in the postsynaptic neuromuscular junction (Chaps. 3 and 9) is a ligand-gated ion channel that is opened by an extracellular ligand (acetylcholine). Examples of ion channels gated by intracellular messengers also abound in nature. This type of gating mechanism allows the channel to open or close in response to events that occur at other locations in the cell. For example, a sodium channel gated by intracellular cGMP is involved in the process of vision. The sodium channel in the rod cells of the retina opens in the presence of cGMP. Light is absorbed by the protein rhodopsin and starts a complex process that culminates with the activation of phosphodiesterase enzyme, which breaks down the cGMP and causes the channel to close. In other cells there are potassium channels that open when the intracellular concentration of calcium ion increases. Other known channels respond to cAMP, ATP, inositol 1,4,5-trisphosphate, or the activated part of G proteins.

■ Solutes Are Moved Against Gradients by Active Transport Systems

The passive transport mechanisms just discussed all tend to bring the cell into equilibrium with the extracellular medium. Cells must oppose these equilibrating systems and preserve intracellular concentrations of solutes, particularly ions, that are compatible with life.

Primary Active Transport ■ Intrinsic membrane proteins that directly use metabolic energy to transport ions against a gradient of concentration or electrical potential are known as **ion pumps.** The direct use of metabolic energy to carry out transport defines a **primary active transport mechanism.** The source of metabolic energy is ATP synthesized by the mitochondria, and the different ion pumps hydrolyze ATP to ADP and use the energy stored in the third phosphate bond to carry out transport. Due to this ability to hydrolyze ATP, ion pumps also are called **ATPases.**

The most abundant ion pump in higher organisms is the sodium pump, or **Na^+/K^+-ATPase.** It is found in the plasma membrane of practically every eukaryotic cell and is responsible for maintaining the low sodium and high potassium concentrations in the cytoplasm. The sodium pump is an integral membrane protein and consists of two subunits, one large and one small. Sodium ions are transported out of the cell and potassium ions are brought in. It is known as a **P-type** ATPase because the protein is phosphorylated during the transport cycle (Fig. 2-9). The sodium pump counterbalances the tendency of sodium ions to enter the cell passively and the tendency of potassium ions to leave passively. It maintains a high intracellular potassium concentration, necessary for protein synthesis. It also plays a role in the resting membrane potential by maintaining ion gradients. The sodium pump can be inhibited either by metabolic poisons that stop the synthesis and supply of ATP or by specific pump blockers, such as the cardiac glycoside *digitalis.*

Calcium pumps, **Ca^{2+}-ATPases,** are found in the plasma membrane and in membranes of the endoplasmic reticulum and sarcoplasmic reticulum of muscle. They are also P-type ATPases. They pump calcium ions from the cytosol of the cell into the extracellular space or into these two organelles. These organelles store calcium and, as a result, help maintain a very low cytosolic concentration of this ion (Chap. 1).

Proton pumps, **H^+-ATPases,** are found in the membranes of the lysosomes and the Golgi apparatus. They pump protons from the cytosol into these organelles, maintaining the inside of the organelles more acidic

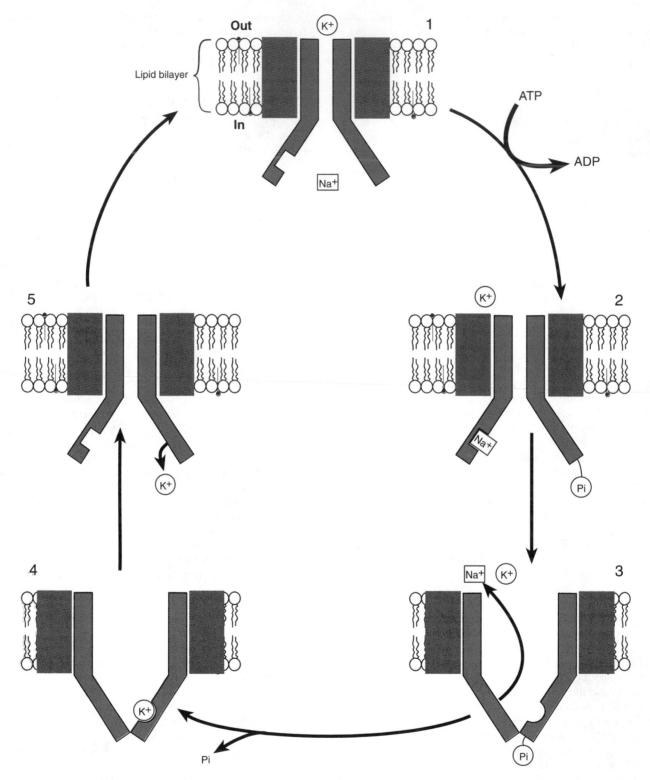

Figure 2–9 ■ The possible sequence of events during one cycle of the Na$^+$/K$^+$-ATPase. The functional form may be a tetramer of two large catalytic subunits and two smaller subunits of unknown function. Binding of intracellular Na$^+$ and phosphorylation by ATP inside the cell may induce a conformational change that transfers Na$^+$ outside the cell (steps 1–3). Subsequent binding of extracellular K$^+$ and dephosphorylation return the protein to its original form and transfer K$^+$ into the cell (steps 3–5). There are thought to be three Na$^+$ binding sites and two K$^+$ binding sites. During one cycle, three Na$^+$ are exchanged for two K$^+$ and one ATP molecule is hydrolyzed.

(at a lower pH) than the rest of the cell. Proton pumps are classified as **V-type** ATPases because many are located in intracellular vacuolar structures. A proton pump in the luminal membrane of the renal tubule is important for acidifying the tubular urine. There is an **F-type** ATPase in mitochondria. It normally functions in the opposite way to the P- and V-type ATPases. Instead of using the energy stored in ATP molecules to pump protons, it synthesizes ATP by using the energy stored in the gradient of protons produced by the respiratory chain.

The **H⁺/K⁺-ATPase** is found in the luminal membrane of the parietal cells that line the gastric mucosa. By pumping protons into the lumen of the stomach in exchange for potassium ions, this pump maintains the very low pH in the stomach that is necessary for proper digestion. It is another example of a P-type ATPase.

Secondary Active Transport ▪ The net effect of ion pumps is to maintain the different environments needed for the proper function of organelles, cells, and

organs. Metabolic energy is expended by the pumps to create and maintain the differences in ion concentrations. Besides the importance of local ion concentrations for cell function, differences in concentrations represent stored energy. An ion releases potential energy when it moves down an electrochemical gradient just as a body releases energy when falling to a lower level. This energy can be used to perform work. Cells have developed several carrier mechanisms to transport one solute against its concentration gradient by using the energy stored in the favorable gradient of another solute. Most of these mechanisms use sodium as the driver solute and use the energy of the sodium gradient to carry out the "uphill" transport of another important solute (Fig. 2–10). Since the sodium gradient is maintained by the action of the sodium pump, the function of these transport systems is also dependent on the function of the pump. Thus, although they do not directly use metabolic energy for transport, these systems ultimately depend on the proper supply of metabolic energy to the pump. They are called sec-

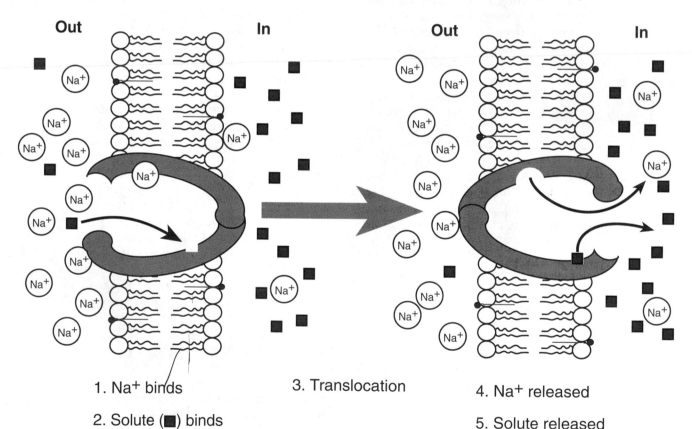

1. Na⁺ binds

2. Solute (▪) binds

3. Translocation

4. Na⁺ released

5. Solute released

Figure 2–10 ▪ Possible mechanism of secondary active transport, whereby a solute is moved against its concentration gradient by coupling it to Na⁺ moving down a favorable gradient. Binding of extracellular Na⁺ to the carrier protein may increase the affinity of binding sites for solute so that solute also can bind to the carrier even though its extracellular concentration is low. A conformational change in the carrier protein may expose the binding sites to the cytosol, where Na⁺ readily dissociates due to the low intracellular Na⁺ concentration. Release of Na⁺ decreases the affinity of the carrier for solute and forces release of the solute inside the cell, where solute concentration is already high.

CYSTIC FIBROSIS

Cystic fibrosis is the most common lethal genetic disease of Caucasians. In northern Europe and the United States, for example, about 1 child in every 2500 is born with the disease. It was first recognized clinically in the 1930s, when it appeared to be a gastrointestinal problem because patients usually died from malnutrition during the first year of life. Survival has improved as management has improved, and afflicted newborns now have a life expectancy of about 40 years. Cystic fibrosis affects several organ systems, with the severity varying enormously among individuals. Clinical features can include deficient secretion of digestive enzymes by the pancreas, infertility in males, increased concentration of chloride ions in sweat, intestinal and liver disease, and airway disease leading to progressive lung dysfunction. Involvement of the lungs determines survival: 95% of cystic fibrosis patients die from respiratory failure.

The basic defect in cystic fibrosis is a failure of chloride transport across epithelial cell membranes, particularly in the epithelial cells that line the airways. Much of the information about defective chloride transport was obtained by studying individual chloride channels with the patch clamp technique described in this chapter. One hypothesis is that all the pathophysiology of cystic fibrosis is due directly to the failure of chloride transport. In the lungs, for example, reduced secretion of chloride is usually accompanied by an abnormally high rate of sodium reabsorption. These changes retard secretion of water, so the mucus secretions that line airways become thick and sticky and the smaller airways become blocked. The thick mucus also traps bacteria, which may lead to bacterial infection. Once established, bacterial infection is very difficult to eradicate from the lungs of a cystic fibrosis patient.

It was predicted that the flawed gene in cystic fibrosis patients would normally encode either a chloride channel protein or a membrane protein that regulates chloride channels. The gene was identified in 1989 and localized to the long arm of chromosome 7. It encodes a protein of 1480 amino acids, named the **cystic fibrosis transmembrane conductance regulator** (CFTR). There is abundant evidence that CFTR contains both a chloride channel and a channel regulator. It has structural similarities to ion-transporting ATPases, which are integral membrane proteins. The CFTR protein is anchored in the plasma membrane by two hydrophobic membrane-spanning segments that also form a channel. Two regions control channel activity through interactions with nucleotides, such as ATP, present in the cell cytosol. The loss of phenylalanine from one of these binding sites is a mutation found in 70% of cystic fibrosis patients. A regulatory domain is also exposed to the cytosol and contains a number of sites that can be phosphorylated by a cAMP-dependent protein kinase. Phosphorylation appears to be one step in the activation of many proteins and may be required to open the CFTR chloride channel.

Greater understanding of the pathophysiology of airway disease in cystic fibrosis is giving rise to new therapies, and a definitive solution may be close at hand. Two approaches are undergoing clinical trials. One is gene therapy to insert a normal gene for CFTR into affected airway epithelial cells. This has the advantage of restoring both the known and unknown functions of the gene. The second approach is to bypass the defective CFTR chloride channel and stimulate other membrane chloride channels in the same cells.

ondary active transport mechanisms. Blocking the pump with metabolic inhibitors or pharmacologic blockers causes these transport systems to stop once the sodium gradient has been dissipated.

Similar to passive carrier-mediated systems, secondary active transport systems are intrinsic membrane proteins, have specificity for the solute they transport, and show saturation kinetics and competitive inhibition. They differ, however, in two respects. First, they cannot function in the absence of the driver ion, the ion that moves along its electrochemical gradient and supplies energy. Second, they transport the solute against its own concentration or electrochemical gradient. Functionally, the different secondary active transport systems can be classified into two groups: **symport** or **cotransport** systems, in which the solute being transported moves in the same direction as the sodium ion; and **antiport** or **exchange** systems, in which sodium moves in one direction and the solute moves in the opposite direction (Fig. 2–11).

Examples of symport mechanisms are the sodium-coupled sugar transport system and the several sodium-coupled amino acid transport systems found in the small intestine and the renal tubule that allow the organism to transport these nutrients even when they are present at very low concentrations in the lumen of these structures. The Na^+/glucose cotransporter in the human intestine has been cloned and sequenced.

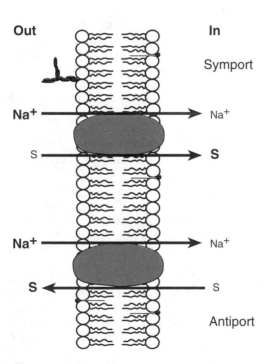

Figure 2–11 ■ Secondary active transport systems. In a symport system (top), the transported solute (S) is moved in the same direction as the Na^+ ion. In an antiport system (bottom), the solute is moved in the opposite direction to Na^+. Large and small type indicate high and low concentrations, respectively.

The protein contains 664 amino acids, and the polypeptide chain is thought to contain 12 membrane-spanning segments (Fig. 2–12). Other examples of symport systems are the sodium-coupled phosphate transporter in renal tubules and the sodium-coupled chloride transporter in the gallbladder and in the ascending limb of Henle's loop in the kidney.

The most important examples of antiporters are the Na^+/H^+ exchange and Na^+/Ca^{2+} exchange systems, found mainly in the plasma membrane of many cells. The first uses the sodium gradient to remove protons from the cell, controlling the intracellular pH and counterbalancing the production of protons in metabolic reactions. It is an **electroneutral** antiporter because there is no net movement of charge. One Na^+ enters the cell for each H^+ that leaves. The second antiporter removes calcium from the cell and, together with the different calcium pumps, helps maintain a low cytosolic calcium concentration. It is an **electrogenic** system because there is net movement of charge. Three Na^+ enter the cell and one Ca^{2+} leaves during each cycle.

Movement of Solutes across Epithelial Cell Layers ■ Epithelial cells occur in layers or sheets and are designed to allow directional movement of solutes not just across the cell membrane but from one side of the cell layer to the other. This is achieved because the cell membranes of epithelial cells have two distinct regions with different morphology and different transport systems. These regions are the **apical** (or luminal) membrane and the **basolateral** membrane facing the blood supply (Fig. 2–13). The specialized or **polarized** organization of the cells is maintained by the presence of **tight junctions** at the areas of contact between adjacent cells. The tight junctions prevent proteins on the apical membrane side from migrating to the basolateral membrane, and vice versa. Thus, the entry and exit steps for solutes can be localized to opposite sides of the cell. This is the key to **transcellular** transport across epithelial cells.

An example is the absorption of glucose in the small intestine. Glucose enters the intestinal epithelial cells via an electrogenic Na^+/glucose cotransport system in the apical membrane. The glucose molecules then move out of the cell and into the blood via a carrier-mediated passive transport mechanism in the basolateral membrane (Fig. 2–13). The sodium ions that enter the cell together with the molecules of glucose are pumped out by the Na^+/K^+-ATPase pumps that are located in the basolateral membrane only. Cells thus accomplish transcellular movement of both glucose and sodium ions.

Movement of Water across the Cell Membrane

Since the lipid part of the cell membrane is very hydrophobic, the movement of water across it is too

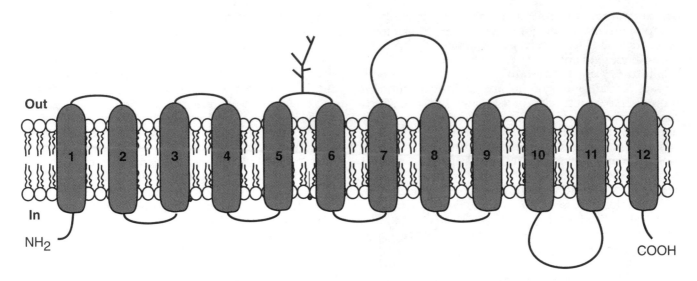

Figure 2–12 ■ Model of the secondary structure of the Na⁺/glucose cotransport protein from human intestine. The polypeptide chain of 664 amino acids passes back and forth across the membrane 12 times. Each membrane-spanning segment consists of 21 amino acids arranged in an alpha-helical conformation. Both the NH_2 and the COOH ends are located on the cytoplasmic side of the cell membrane. Carbohydrate, attached between membrane spans 5 and 6, is located on the exterior surface of the cell. In the functional protein it is likely that the membrane-spanning segments are clustered together to provide a hydrophilic pathway between them. (Modified from Wright EM. The intestinal Na⁺/glucose cotransporter. *Ann Rev Physiol* 55:575–589, 1993.)

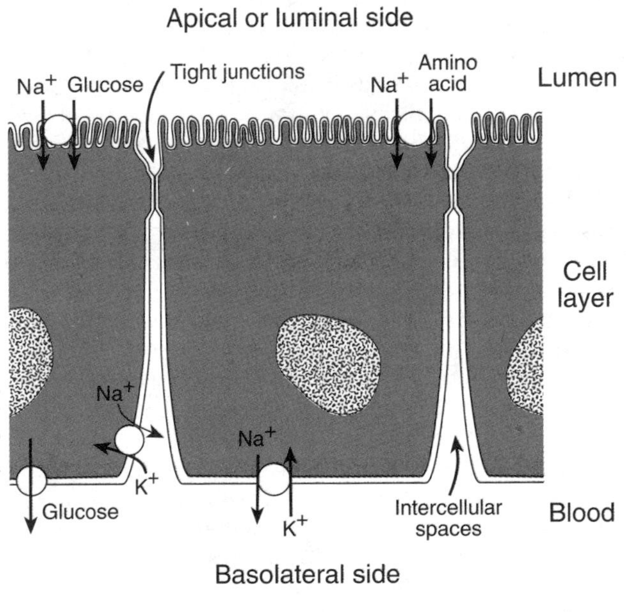

Figure 2–13 ■ Localization of transport systems to different regions of the cell membrane in epithelial cells of the small intestine. This produces a polarized cell in which entry and exit of solutes, such as glucose and Na⁺, occur at opposite sides of the cell. The result is net movement of glucose from the luminal side of the cell to the basolateral side, ensuring efficient absorption of glucose from the intestinal lumen. Like glucose, amino acid entry is restricted to the apical side of the cell.

slow to explain the speed at which water can move in and out of the cells. The partition coefficient of water into lipids is very low, thus the permeability of the lipid bilayer for water is also very low. Specific membrane proteins that function as **water channels** explain the rapid movement of water across the cell membrane.

■ Movement of Water across the Cell Membrane Is Driven by Differences in Osmotic Pressure

The spontaneous movement of water across a membrane driven by a gradient of water concentration is called **osmosis.** The water moves from an area of high concentration of water to an area of low concentration. Since concentration is defined by the number of particles per unit of volume, a solution with a high concentration of solutes has a low concentration of water, and vice versa. Osmosis can therefore be viewed as the movement of water from a solution of high water concentration (low concentration of solute) toward a solution with a lower concentration of water (high solute concentration). Osmosis is a passive transport mechanism that tends to equalize the concentrations of the solutions on both sides of every membrane.

If a cell that is normally in osmotic equilibrium is transferred to a more dilute solution, water will enter the cell, the cell volume will increase, and the solute concentration of the cytoplasm will be reduced. If the cell is transferred to a more concentrated solution water will leave the cell, the cell volume will decrease, and the solute concentration of the cytoplasm will increase. As we will see below, many cells have regulatory mechanisms that keep cell volume within a certain range. Other cells, such as mammalian erythrocytes, do not have volume regulatory mechanisms and large volume changes occur when the solute concentration of the extracellular medium is changed.

The driving force for the movement of water across the cell membrane is the difference in water concentration between the two sides of the membrane. For historical reasons, this driving force is not called the chemical gradient of water but the difference in **osmotic pressure.** The osmotic pressure of a solution is defined as the pressure necessary to stop the net movement of water across a membrane that separates the solution from pure water. When a membrane separates two solutions of different osmotic pressure, water will move from the solution with low osmotic pressure (high water concentration) to the solution of high osmotic pressure (low water concentration).

The osmotic pressure of a solution depends on the number of particles dissolved in it, the total concentration of all solutes. Many solutes, such as salts, acids, and bases, dissociate in water, so the number of particles is greater than the molar concentration. For example, NaCl dissociates in water to give Na^+ and Cl^-, so one molecule of NaCl will produce two osmotically active particles. In the case of $CaCl_2$, there are three particles per molecule. The equation giving the osmotic pressure of a solution is:

$$\pi = n\,R\,T\,C \qquad (3)$$

where π is the osmotic pressure of the solution, n is the number of particles produced by the dissociation of one molecule of solute (2 for NaCl, 3 for $CaCl_2$), R is the universal gas constant, T is the absolute temperature, and C is the concentration of the solute in moles/liter. Solutions with the same osmotic pressure are called **isosmotic.** A solution is **hyperosmotic** with respect to another solution if it has a higher osmotic pressure and **hyposmotic** if it has a lower osmotic pressure.

Equation 3, called the **van't Hoff law,** is valid only when applied to very dilute solutions, in which the particles of solutes are so far away from each other that no interactions occur between them. Generally, this is not the case at physiologic concentrations. Interactions between dissolved particles, mainly between ions, cause the solution to behave as if the concentration of particles is less than the theoretical value (n × C). A correction coefficient, called the **osmotic coefficient** of the solute, needs to be introduced in the equation. The osmotic coefficient varies with the specific solute and its concentration. It has values between 0 and 1. Thus, the osmotic pressure of a solution can be written more accurately as:

$$\pi = n\,R\,T\,\phi\,C \qquad (4)$$

where ϕ is the *osmotic coefficient.* The osmotic coefficient of NaCl is 1.00 in an infinitely dilute solution but changes to 0.93 at the physiologic concentration of 0.15 M.

The **osmolality** or **osmotic concentration** of a solution can be calculated as n × ϕ × C and is expressed in *osmoles per kilogram of H_2O.* Since R and T in equation 4 are constants, the osmotic pressure of a solution is directly proportional to the osmolality. Most physiologic solutions, such as blood plasma, contain many different solutes, and each contributes to the total osmolality of the solution. The osmolality of a solution containing a complex mixture of solutes is usually measured by freezing point depression. The freezing point of an aqueous solution of solutes is lower than that of pure water and depends on the total number of solute particles. Compared to pure water, which freezes at 0°C, a solution with an osmolality of 1 osm/kg H_2O will freeze at −1.86°C. The ease with which osmolality can be measured has led to the wide use of this parameter for comparing the osmotic pressure of different solutions.

■ Many Cells Can Regulate Their Volume

Osmolality and Tonicity ■ A solution's osmolality is determined by the total concentration of all the solutes present. In contrast, the solution's **tonicity** is determined by the concentrations of only those solutes that do not enter ("penetrate") the cell. Tonicity determines cell volume, as illustrated in the following examples. Na^+ behaves as a nonpenetrating solute because it is pumped out of cells by the Na^+/K^+-ATPase at the same rate that it enters. A solution of NaCl at 0.2 osm/kg H_2O is hyposmotic compared to cell cytosol at 0.3 osm/kg H_2O. The NaCl solution is also **hypotonic,** because cells will accumulate water and swell when placed in this solution. A solution containing a mixture of NaCl (0.3 osm/kg H_2O) and urea (0.1 osm/kg) has a total osmolality of 0.4 osm/kg H_2O and will be hyperosmotic compared to cell cytosol. The solution is *isotonic,* however, because it produces no permanent change in cell volume. The reason is that urea is a penetrating solute that rapidly diffuses into the cells until the urea concentration is the same inside and outside the cells. At this point the total osmolality both inside and outside the cells will be 0.4 osm/kg H_2O.

Volume Regulation ■ When cell volume increases due to extracellular hypotonicity (Fig. 2–14A), the response of many cells is rapid activation of transport mechanisms, which tend to decrease the cell volume. Different cells use different **regulatory volume decrease** (RVD) **mechanisms** to move solutes out of the cell and decrease the number of particles in the

A

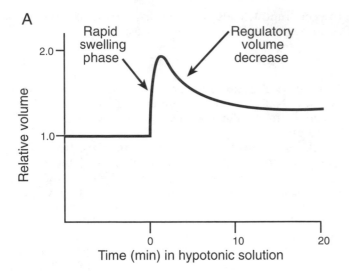

B

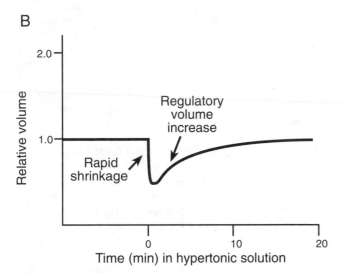

Figure 2–14 ■ Time course of changes in cell volume when a cell is placed at time 0 in either (A) hypotonic or (B) hypertonic solution. Reversal of the initial increase in cell volume in A is known as regulatory volume decrease. Transport systems for solute exit are activated and water follows movement of solute out of the cell. Reversal of the initial decrease in cell volume in B is a regulatory volume increase. Transport systems for solute entry are activated and water follows solute into the cell.

cytosol, which causes water to leave the cell. Since cells have high intracellular concentrations of potassium, many RVD mechanisms involve increased efflux of K⁺, either by stimulating the opening of potassium channels or by activating cotransport mechanisms for KCl. Other cells activate the efflux of some amino acids, such as taurine or proline. The net result is a decrease in intracellular solute content and a reduction of cell volume close to its original value (Fig. 2–14A).

When placed in a **hypertonic** solution, cells lose water and their volume decreases (Fig. 2–14B). In

many cells, a decreased volume triggers **regulatory volume increase** (RVI) **mechanisms,** which tend to increase the number of intracellular particles and bring water back into the cells. Since Na⁺ is the main extracellular ion, many RVI mechanisms involve influx of sodium into the cell. Na⁺/Cl⁻ symport, Na⁺/K⁺/2Cl⁻ symport, and Na⁺/H⁺ antiport are some of the mechanisms activated to increase the intracellular concentration of Na⁺ and increase the cell volume toward its original value (Fig. 2–14B).

Mechanisms based on an increased Na⁺ influx are effective for only a short period of time, since eventually the sodium pump will increase its activity and reduce intracellular Na⁺ to its normal value. Thus, cells that are chronically faced with hypertonic extracellular solutions have developed additional mechanisms for maintaining normal volume. These cells can synthesize specific organic solutes, enabling them to increase intracellular osmolality for a long time. The cells thus avoid altering the concentrations of ions they must maintain within a narrow range of values. The organic solutes are usually small molecules that do not interfere with normal cell function when they accumulate inside the cell. For example, cells of the medulla of mammalian kidney can increase the level of the enzyme aldose reductase when subjected to elevated extracellular osmolality. This enzyme converts glucose to an osmotically active solute, sorbitol. Brain cells can synthesize and store inositol. Synthesis of sorbitol and inositol represents different answers to the problem of increasing the total intracellular osmolality so that normal cell volume can be maintained in the presence of hypertonic extracellular fluid.

The Resting Membrane Potential

The different passive and active transport systems are coordinated in a living cell to maintain intracellular ions and other solutes at concentrations compatible with life. The consequence is that the cell does not equilibrate with the extracellular fluid but rather exists in a **steady state** with the extracellular solution. For example, intracellular Na⁺ concentration (10 mM in a muscle cell) is much lower than extracellular Na⁺ concentration (140 mM), so Na⁺ enters the cell by passive transport. The rate of Na⁺ entry is matched, however, by the rate of active transport of Na⁺ out of the cell via the Na⁺/K⁺-ATPase pump (Fig. 2–15). The net result is that intracellular Na⁺ is maintained constant and at a low level even though Na⁺ continually enters and leaves the cell. The reverse is true for K⁺, which is maintained at a high concentration inside the cell relative to the outside. The passive exit of K⁺ is matched by active entry via the pump (Fig. 2–15). Maintenance of this steady state with ion concentrations inside the cell different from extracellular concentrations is the basis for the difference in electrical

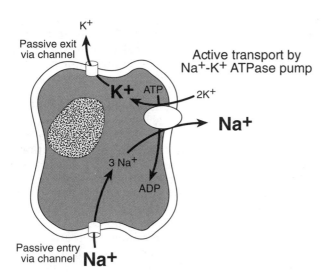

Figure 2–15 ■ The concept of a steady state. The rate at which Na+ enters the cell, moving passively down its electrochemical gradient, is matched by the rate of active transport of Na+ out of the cell via the Na+/K+-ATPase. The intracellular concentration of Na+ remains small and constant. Similarly, the rate of passive K+ exit is matched by the rate of active transport of K+ into the cell via the pump. The intracellular K+ concentration remains high and constant. During each cycle of the ATPase, two K+ are exchanged for three Na+ and one molecule of ATP is hydrolyzed to ADP. Large and small type indicate high and low ion concentrations, respectively.

potential across the cell membrane, or the **resting membrane potential.**

■ Ion Movement Is Driven by the Electrochemical Potential

If there are no differences in temperature or hydrostatic pressure between the two sides of a cell membrane, there are two forces that drive the movement of ions and other solutes across the membrane. One force is due to the difference in the concentration of a substance between the inside and the outside of the cell and arises from the tendency of every substance to move from areas of high concentration to areas of low concentration. The other force is due to the difference in electrical potential between the two sides of the membrane and applies only to ions and other electrically charged solutes. When a difference in electrical potential exists, positive ions tend to move toward the negative side while negative ions tend to move toward the positive side.

The sum of these two driving forces is called the gradient (or difference) of **electrochemical potential** across the membrane for a specific solute. It measures the tendency of that solute to cross the membrane. The expression of this force is given by:

$$\Delta\tilde{u} = RT \ln \frac{C_i}{C_o} + zF (E_i - E_o) \tag{5}$$

where $\tilde{u}$ represents the electrochemical potential ($\Delta\tilde{u}$ is the difference in electrochemical potential between two sides of the membrane), C_i and C_o are the concentrations of the solute inside and outside of the cell, respectively, E_i is the electrical potential inside the cell measured with respect to the electrical potential outside the cell (E_o), R is the universal gas constant (2 cal/mol/°K), T is the absolute temperature (°K), z is the valence of the ion, and F is the Faraday constant (23 cal/mV/mol). If the solute is not an ion and has no electrical charge, then z = 0 and the last term of the equation becomes zero. In this case, the electrochemical potential is defined only by the different concentrations of the uncharged solute, called the **chemical potential.** The driving force for solute transport becomes solely the difference in chemical potential.

■ Net Ion Movement Is Zero at the Equilibrium Potential

Net movement of an ion into or out of a cell continues as long as the driving force exists. Net movement stops and equilibrium is reached only when the driving force of electrochemical potential across the membrane becomes zero. The condition of equilibrium for any permeable ion will be $\Delta\tilde{u} = 0$. Substituting this condition into equation 5, we obtain:

$$0 = RT \ln \frac{C_i}{C_o} + zF (E_i - E_o)$$

$$E_i - E_o = \frac{-RT}{zF} \ln \frac{C_i}{C_o}$$

$$E_i - E_o = \frac{RT}{zF} \ln \frac{C_o}{C_i} \tag{6}$$

Equation 6, known as the **Nernst equation,** gives the value of the electrical potential difference ($E_i - E_o$) necessary for a specific ion to be at equilibrium. This value is known as the **Nernst equilibrium potential** for that particular ion. At the equilibrium potential, the tendency of an ion to move in one direction because of the difference in concentrations is exactly balanced by the tendency to move in the opposite direction due to the difference in electrical potential. At this point the ion will be in equilibrium and net movement zero. Assuming a physiologic temperature of 37°C and, for Na+ or K+, a value of + 1 for z, the Nernst equation can be expressed as:

$$E_i - E_o = 61 \log_{10} \frac{C_o}{C_i} \tag{7}$$

Since Na+ and K+ (and other ions) are present at different concentrations inside and outside a cell, it follows from equation 7 that the equilibrium potential will be different for each ion.

■ The Resting Membrane Potential Is Determined by Passive Movement of Several Ions

The resting membrane potential is the electrical potential difference across the cell membrane of a normal living cell in its unstimulated state. It can be measured directly with a microelectrode inserted into the cell and a reference electrode in the extracellular medium. It is determined by those ions that can cross the membrane and are prevented from attaining equilibrium by active transport systems. Potassium, sodium, and chloride ions can cross the membranes of every living cell, and each of these ions contributes to the resting membrane potential. In contrast, the permeability of the membrane of most cells to divalent ions is so low that it can be ignored in this context.

The **Goldman equation** gives the value of the membrane potential when all the permeable ions are accounted for:

$$E_i - E_o =$$
$$\frac{RT}{F} \ln \frac{P_K[K^+]_o + P_{Na}[Na^+]_o + P_{Cl}[Cl^-]_i}{P_K[K^+]_i + P_{Na}[Na^+]_i + P_{Cl}[Cl^-]_o} \qquad (8)$$

where P_K, P_{Na}, and P_{Cl} represent the permeability of the membrane to potassium, sodium, and chloride ions, respectively, and [ ] represents the concentration of the ion inside (i) and outside (o) the cell. If a certain cell is not permeable to one of these ions, then the contribution of the impermeable ion to the membrane potential will be zero. If a specific cell is permeable to an ion other than the three considered in equation 8, then that ion's contribution to the membrane potential must be included in the equation.

It can be seen from equation 8 that the contribution of any ion to the membrane potential is determined by the membrane's permeability to that particular ion. The higher the permeability of the membrane to one ion relative to the others, the more that ion will contribute to the membrane potential. The cell membranes of most living cells are much more permeable to potassium ions than to any other ion. Making the assumption that P_{Na} and P_{Cl} are zero relative to P_K, equation 8 can be simplified to:

$$E_i - E_o = \frac{RT}{F} \ln \frac{P_K[K^+]_o}{P_K[K^+]_i}$$
$$E_i - E_o = \frac{RT}{F} \ln \frac{[K^+]_o}{[K^+]_i} \qquad (9)$$

which is the Nernst equation for the equilibrium potential for K+ (see equation 6). This illustrates two important points: (1) in most cells the resting membrane potential is close to the equilibrium potential for K+; and (2) the resting membrane potential of most cells is dominated by K+ because the cell membrane is more permeable to this ion compared to the others. As a typical example, the K+ concentrations outside and inside a muscle cell are 3.5 mM and 155 mM, respectively. Substituting these values in equation 7 gives an equilibrium potential for K+ of −100 mV, negative inside the cell relative to the outside. The resting membrane potential in a muscle cell is −90 mV (negative inside). This value is close to, although not the same as, the equilibrium potential for K+.

The reason the resting membrane potential in the muscle cell is less negative than the equilibrium potential for K+ is as follows. Under physiologic conditions, there is passive entry of Na+ ions. This entry of positively charged ions has a small but significant effect on the negative potential inside the cell. Assuming intracellular Na+ to be 10 mM, the Nernst equation gives a value of +70 mV for the Na+ equilibrium potential (positive inside the cell). This is far from the resting membrane potential of −90 mV. Na+ makes only a small contribution to the resting membrane potential because membrane permeability to Na+ is very low compared to that of K+.

The contribution of Cl− ions need not be considered, because the resting membrane potential in the muscle cell is the same as the equilibrium potential for Cl−, so there is no net movement of Cl−.

In most cells, as shown above using a muscle cell as an example, the equilibrium potentials of K+ and Na+ are different from the resting membrane potential, which indicates that neither K+ nor Na+ ions are at equilibrium. As a consequence, these ions continue to cross the cell membrane via specific protein channels, and these passive ion movements are *directly* responsible for the resting membrane potential.

The Na+/K+-ATPase pump is important *indirectly* for maintaining the resting membrane potential because it sets up the gradients of K+ and Na+ that drive passive K+ exit and Na+ entry. During each cycle of the pump, two K+ ions are moved into the cell in exchange for three Na+, which are moved out (Fig. 2-15). Due to the unequal exchange mechanism, the pump's activity does not abolish the resting membrane potential. In fact, it contributes slightly to the negative potential inside the cell, because three Na+ ions are replaced by only two K+.

■ ■ ■ ■ ■ ■ ■ ■ ■ ■ ■ ■ REVIEW EXERCISES ■ ■ ■ ■ ■ ■ ■ ■ ■ ■ ■ ■

Identify Each with a Term

1. The structure that separates the cell cytoplasm from the external medium

2. A type of membrane protein that does not penetrate the lipid bilayer

3. The electrical potential difference between the two sides of a membrane for a specific ion that is at equilibrium

4. A membrane transport system that exhibits saturation kinetics

5. An ion channel that does not open unless a specific agonist binds to it

Define Each Term

6. Secondary active transport

7. Osmosis

8. Regulatory volume increase

9. Goldman equation

10. Electrochemical potential difference

Choose the Correct Answer

11. A phospholipid is an example of:
    a. A glycoprotein
    b. A steroid
    c. A water-soluble molecule
    d. An amphipathic molecule

12. Which statement about membrane phospholipids is true?
    a. A phospholipid contains cholesterol.
    b. Phospholipids move rapidly in the plane of the bilayer.
    c. Specific phospholipids are always present in equal proportions in the two halves of the bilayer.
    d. Phospholipids form ion channels through the membrane.

13. In integral proteins the segments of the polypeptide chain that span the lipid bilayer frequently:
    a. Adopt an alpha-helical configuration
    b. Contain many hydrophilic amino acids
    c. Form covalent bonds with cholesterol
    d. Contain unusually strong peptide bonds

14. The electrical potential difference necessary for a single ion to be at equilibrium across a membrane is best described by the:
    a. Goldman equation
    b. van't Hoff law
    c. Fick's law
    d. Nernst equation

15. The ion present in highest concentration inside most cells is:
    a. Sodium
    b. Potassium
    c. Calcium
    d. Chloride

16. Solute movement by active transport can be distinguished from solute transport by equilibrating carrier-mediated transport because active transport:
    a. Is saturable at high solute concentration
    b. Is inhibited by other molecules with structures similar to that of the solute
    c. Moves the solute against its electrochemical gradient
    d. Allows movement of polar molecules

17. A sodium channel that opens in response to an increase in intracellular cGMP is an example of:
    a. A ligand-gated ion channel
    b. An ion pump
    c. Sodium-coupled solute transport
    d. A peripheral membrane protein

18. During regulatory volume decrease, many cells will:
    a. Increase their volume
    b. Increase the influx of Na^+
    c. Increase the efflux of K^+
    d. Increase synthesis of sorbitol

Calculate

19. At equilibrium the concentrations of Cl^- inside and outside a cell are 10 mM and 100 mM, respectively. Calculate the equilibrium potential for Cl^- at 37°C.

20. Calculate the osmotic pressure (in atm) of an aqueous solution of 100 mM $CaCl_2$ at 27°C. Assume the osmotic coefficient is 0.86 and the gas constant (R) is 0.082 l-atm/°K/mol.

Briefly Answer

21. Explain the essential difference between electroneutral and electrogenic transport systems.

22. Why do most cells develop a resting membrane potential that is negative inside the cell?

23. Explain why the passive entry of Na^+ into a cell is useful.

24. Explain how epithelial cells of the small intestine are organized to ensure absorption of glucose from the lumen.

25. What adaptations occur in cells of the medulla of the mammalian kidney, where the extracellular medium is hypertonic?

■ ■ ■ ■ ■ ■ ■ ■ ■ ■ ■ ■ ANSWERS ■ ■ ■ ■ ■ ■ ■ ■ ■ ■ ■ ■

1. Cell (plasma) membrane
2. Peripheral or extrinsic protein
3. Nernst potential
4. Carrier-mediated

5. Ligand-gated channel

6. A carrier-mediated transport system by which a solute is moved against its electrochemical gradient. The mechanism involves coupling to another ion, usually Na^+,

which is moving down an electrochemical gradient. Solute transport is not linked directly to the energy-yielding step.

7. Spontaneous net movement of water across a membrane driven by a gradient of water concentration. Water moves from a solution with high water concentration (low solute concentration) to a solution with low water concentration (high solute concentration).

8. A mechanism triggered by a decrease in cell volume. It increases the amount of intracellular solutes so that water enters and increases cell volume toward normal.

9. The equation used to calculate the membrane potential as determined by the major ions that can cross the membrane. These ions are Na^+, K^+, and Cl^- for most living cells. The equation takes into account the ion concentrations and the membrane permeability to each ion.

10. The net driving force, or the sum of the electrical and chemical gradients, which (if greater than zero) causes net movement of a solute from one region to another or across a membrane.

11. d

12. b

13. a

14. d

15. b

16. c

17. a

18. c

19. $E_i - E_o = 61/-1 \times \log_{10} 100/10$
$$= -61 \times 1.0$$
$$= -61 \text{ mV inside the cell}$$

20. $3 \times 0.082 \times 300 \times 0.86 \times 0.1 = 6.35$ atm

21. There is no net movement of charge during electroneutral transport (e.g., Na^+/H^+ exchange), whereas electrogenic transport (e.g., Na^+/Ca^{2+} exchange, Na^+/glucose cotransport) involves charge movement.

22. The intracellular and extracellular solutions have marked differences in the concentrations of ions such as Na^+, K^+, and Cl^-. These ions, K^+ in particular, tend to cross the cell membrane to reach equilibrium. Passive exit of K^+ from the cell leaves an excess of negative ions inside the cell.

23. Entry of Na^+ down its electrochemical gradient often is coupled to the movement of another molecule or ion. Molecules important for cell functions (e.g., glucose, amino acids) can thus be moved into the cell against their electrochemical gradients. Conversely, H^+ ions can be removed from the cell if the cytoplasm becomes too acidic.

24. The cell membrane lining the lumen contains a Na^+-coupled glucose transport system that brings glucose into the cell. This system is restricted to the luminal side of the cell by the tight junctions between the cells, which prevent lateral movement of membrane proteins. Similarly, a carrier-mediated passive transport system, which allows exit of glucose, is restricted to the other side of the cell.

25. In these cells the RVI response involves increased synthesis and storage of intracellular solutes. For example, there is an increased level of aldose reductase, an enzyme required for synthesis of sorbitol from glucose, which leads to increased concentration of sorbitol inside the cells.

Suggested Reading

Armstrong WMcD: The cell membrane and biological transport. In Selkurt EE (ed): *Physiology.* 5th ed. Boston, Little, Brown, 1984, pp 1–26.

Barinaga M: Novel function discovered for the cystic fibrosis gene. *Science* 256:444–445, 1992.

Bretscher MS: The molecules of the cell membrane. *Sci Am* 253:100–108, October 1985.

Brown MS, Goldstein JL: How LDL receptors influence cholesterol and atherosclerosis. *Sci Am* 251:58–66, November 1984.

Field M: Cholera toxin, adenylate cyclase, and the process of active secretion in the small intestine: The pathogenesis of diarrhea in cholera. In Andreoli TE, Hoffman JE, Fanestil DD (eds): *Physiology of Membrane Disorders.* New York, Plenum, 1978, pp 877–899.

Finean JB, Coleman R, Michell RH: *Membranes and Their Cellular Functions.* 2nd ed. New York, Wiley, 1985.

Handler JS: Overview of epithelial polarity. *Ann Rev Physiol* 51:729–740, 1989.

Keynes RD: Ion channels in the nerve-cell membrane. *Sci Am* 240:126–135, March 1979.

Lewis SA, Donaldson P: Ion channels and cell volume regulation: Chaos in an organized system. *News Physiol Sci* 5:112–119, 1990.

Matthews R: A low-fat theory of anesthesia. *Science* 255:156–157, 1992.

Neher E: Ion channels for communication between and within cells. *Science* 256:498–502, 1992.

Rothman JE, Lenard J: Membrane asymmetry. *Science* 195:743–753, 1977.

Stein WD: *Channels, Carriers and Pumps: An Introduction to Membrane Transport.* New York, Academic, 1990.

Thorens B: Facilitated glucose transporters in epithelial cells. *Ann Rev Physiol* 55:591–608, 1993.

Unwin N, Henderson R: The structure of proteins in biological membranes. *Sci Am* 250:78–94, February, 1984.

Wright EM: The intestinal Na^+/glucose cotransporter. *Ann Rev Physiol* 55:575–589, 1993.

The Action Potential, Synaptic Transmission, and Maintenance of Nerve Function

CHAPTER OUTLINE

OBJECTIVES

After studying this chapter, the student should be able to:
1. Compare and contrast the functional properties of non-gated, voltage-gated, and ligand-gated ion channels
2. Describe the mechanisms involved in the initiation and conduction of an action potential
3. Describe the mechanisms involved in the absolute and relative refractory periods
4. Describe the mechanisms involved in the process of synaptic transmission
5. Describe the mechanisms involved in producing inhibitory and excitatory postsynaptic potentials
6. Discuss the functional significance of membrane space and time constants
7. Compare and contrast temporal and spatial summation
8. Discuss the major neurotransmitters, their synthesis, release, receptor subtypes, effects on the postsynaptic cell, and mechanism of termination of action
9. Describe the mechanisms involved in axoplasmic transport

The nervous system coordinates the activities of many other organ systems. It activates muscles for movement, controls the secretion of hormones from glands, regulates the rate of breathing, and is involved in modulating and regulating a multitude of other physiologic processes. To perform these functions the nervous system relies on neurons, which are designed for the rapid transmission of information from one cell to another by conducting electrical impulses and secreting chemical neurotransmitters. The electrical impulses propagate along the length of nerve fiber processes to their terminals, where they initiate a series of events that cause the release of chemical neurotransmitters. The release of neurotransmitters occurs at sites of synaptic contact between two nerve cells. Receptors on the postsynaptic cell membrane bind with the released transmitters. The activation of these receptors either excites or inhibits the postsynaptic neuron. Propagation of action potentials, release of neurotransmitters, and activation of receptors constitute the means whereby nerve cells communicate and transmit information to one another. In this chapter we examine the specialized membrane properties of nerve cells that endow them with the ability to produce action potentials and examine aspects of neuronal structure necessary for the maintenance of nerve cell function.

The Action Potential and Electrical Signaling by Neurons

■ Special Anatomic Features of Neurons Adapt Them for Information Communication

The shape of a nerve cell is highly specialized for the reception and transmission of information. One region of the neuron is designed to receive and process incoming information; another is designed to conduct and transmit information to other cells. The type of information that is processed and transmitted by a neuron depends on its location in the nervous system. For example, nerve cells associated with visual pathways convey information about the external environment, such as light and dark, to the brain; neurons associated with motor pathways convey information

to control the contraction and relaxation of muscles for walking. Regardless of the type of information transmitted by neurons, they transduce and transmit this information via similar mechanisms. The mechanisms depend mostly on the specialized structures of the neuron and the electrical properties of their membranes.

Emerging from the **soma** (cell body) of a neuron are fiberlike processes called **dendrites** and **axons** (Fig. 3–1). Many neurons in the central nervous system also have knoblike structures called **dendritic spines** that extend from the dendrites. The dendritic spines, dendrites, and soma receive information from other nerve cells. The axon conducts and transmits information. The end of the axon, called the **axon terminal,** contains small **vesicles** packed with **neurotransmitter** molecules. When a neuron is activated an electrical impulse, called an **action potential,** is generated in the **axon hillock** (or **initial segment**) and conducted along the axon. The action potential causes the release of neurotransmitters from the terminal. These neurotransmitters bind to receptor molecules **(receptors),** located on target cells.

The site of contact between a neuron and its target cell is called a **synapse.** Synapses are classified according to their site of contact as axospinous, axodendritic, axosomatic, or axoaxonic (Fig. 3–2). Special contacts can also exist between dendrites. These dendrodendritic contacts are typically gap junctions and are described later in this chapter.

The binding of a neurotransmitter to its receptor typically causes a flow of ions across the membrane of the postsynaptic cell. This temporary redistribution of ionic charge can lead to the generation of an action potential, which itself is mediated by the flow of specific ions across the membrane. These electrical changes, critical for the transmission of information, are the result of ions moving through cell membrane ion channels (Chap. 2).

■ Channels Allow Ions to Flow Through the Nerve Cell Membrane

Ions can flow across the nerve cell membrane through three types of ion channels: nongated, ligand-gated, and voltage-gated (Fig. 3–3). **Nongated ion channels** are always open. They are responsible for the influx of Na^+ and efflux of K^+ when the neuron is in its resting state. **Ligand-gated ion channels**

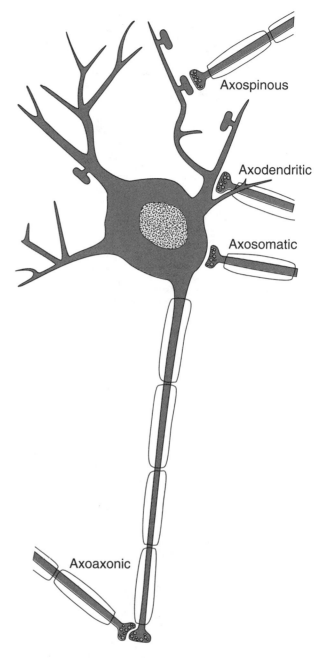

Figure 3–2 ■ Types of synapses: axospinous, axodendritic, axosomatic, and axoaxonic. The dendritic and somatic areas of the neuron integrate incoming information, while the axon conducts information in the form of electrical impulses.

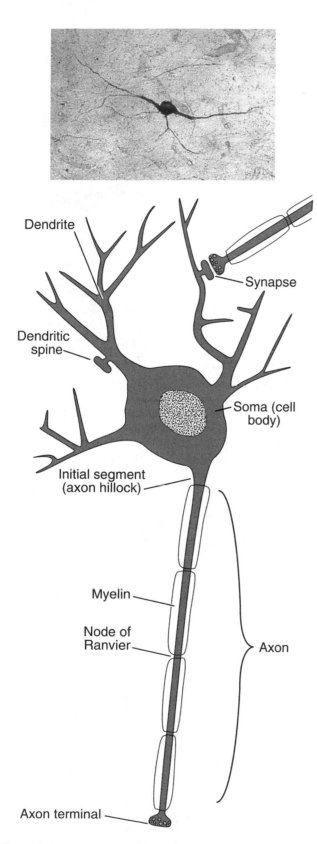

Figure 3–1 ■ Structure of the neuron. (A) Light micrograph. (B) Schematic illustration, showing soma, dendrites, dendritic spines, axon, axon terminal, and synapse.

are directly or indirectly activated by the binding of chemical neurotransmitters to membrane receptors. In this type of channel, the receptor itself forms part of the ion channel or may be coupled to the channel via a G protein and a second messenger. When chemical transmitters bind to their receptors, the associated ion channels can either open or close to permit or block the movement of specific ions across the cell mem-

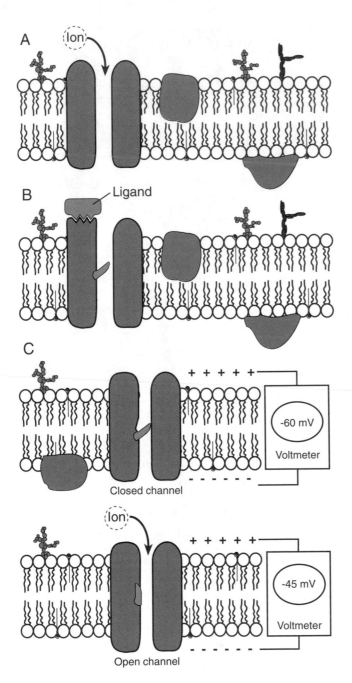

Figure 3–3 ■ (A) Nongated ion channel, which remains open, permitting free movement of ions across the membrane. (B) Ligand-gated channel, which remains closed (or open) until binding of a neurotransmitter. (C) Voltage-gated channel, which remains closed until there is a change in membrane potential.

brane. **Voltage-gated ion channels** are sensitive to the voltage difference across the membrane. In their initial resting state these channels are typically closed; they open when a critical voltage level is reached. Each type of ion channel has a somewhat unique distribution on the nerve cell membrane. Nongated ion

channels, important for the establishment of the resting membrane potential, are found throughout the neuron. Ligand-gated channels, located at sites of synaptic contact, are found predominantly on dendritic spines, dendrites, and somata. Voltage-gated channels, required for the initiation and propagation of the action potential, are found predominantly on axons and axon terminals.

■ Electrical Properties of the Nerve Membrane Affect Ion Flow

The electrical properties of the nerve membrane play important roles in the flow of ions through the membrane and in the initiation and conduction of the action potential along the axon and the integration of incoming information at dendrites and soma. These properties include membrane conductance and capacitance.

The movement of ions across the nerve membrane is driven by ionic concentration and electrical gradients (Chap. 2). The ease with which ions flow across the membrane through their channels is a measure of the membrane's **conductance;** the greater the conductance, the greater the flow of ions. Conductance is the inverse of resistance (R), which is measured in **ohms.** The conductance (g) of a membrane or single channel is measured in **siemens.** For an individual ion channel, the conductance is a constant value, determined in part by factors such as the relative size of the ion with respect to that of the channel and the charge distribution within the channel. The relationship between a single channel conductance, ionic current, and the membrane potential is described by **Ohm's law:**

$$I_{ion} = g_{ion}(E_m - E_{ion})$$
$$\text{or} \tag{1}$$
$$g_{ion} = I_{ion}/(E_m - E_{ion})$$

where I_{ion} is the ion current flow, E_m is the membrane potential, E_{ion} is the equilibrium (Nernst) potential for a specified ion, and g_{ion} is the channel conductance for an ion. Notice that if $E_m = E_{ion}$, then there is no net movement of the ion and $I_{ion} = 0$.

The conductance for a nerve membrane is the summation of all of its single channel conductances. The membrane conductance for Na^+ ion channels, g_{Na}, for example, can be represented as:

$$g_{Na} = \Sigma \; [g_{Na1} + g_{Na2} + . \; . \; .] \tag{2}$$

Likewise, the membrane conductance for K^+ ions can be expressed as the summation of the single channel conductances:

$$g_K = \Sigma \; [g_{K1} + g_{K2} + . \; . \; .] \tag{3}$$

Another electrical property of the nerve membrane that influences the movement of ions is **membrane capacitance** (C_m), the membrane's ability to store an electrical charge. Membrane capacitance is measured in units of **farads** (F). One factor that contributes to the amount of charge a membrane can store is its surface area; the greater the surface area, the greater the storage capacity. Large-diameter dendrites, therefore, can store more charge than small-diameter dendrites of the same length. As we shall see, membrane capacitance plays an important role in integration of incoming information.

■ The Nerve Membrane Can Be Modeled as an Electrical Circuit

Membrane capacitance, membrane conductances, and equilibrium potentials can be used in a model of the nerve membrane to understand its functional properties. The model shown in Figure 3–4 incorporates membrane conductances for Na^+ and K^+ ions, their associated equilibrium potentials, and the membrane capacitance for a small patch of membrane. This simplified electrical model of the resting membrane illustrates two important mechanisms by which neurons integrate incoming information: temporal and spatial summation of postsynaptic potentials. The model can be modified to illustrate the properties of active membrane that are responsible for the generation and conduction of the action potential.

■ The Action Potential Is Generated at the Initial Segment and Conducted along the Axon

Characteristics of the Action Potential ■ Depolarization of the initial segment to threshold results in the generation and propagation of the action potential. The action potential is a transient change in the membrane potential characterized by a gradual depolarization to threshold, a rapid rising phase, an overshoot, and a repolarization phase. This is followed by a brief afterhyperpolarization (undershoot) before the membrane potential again reaches resting level (Fig. 3–5A).

The action potential may be recorded by placing a microelectrode inside a nerve cell or its axon. The

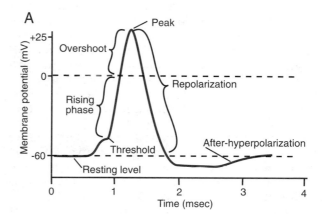

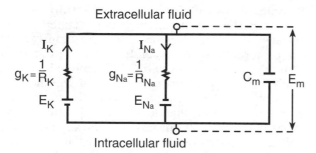

Figure 3–4 ■ Electrical model of the nerve membrane, represented by an equivalent electrical circuit with parallel conductive and capacitative branches. The conductive pathways are specific ion channels for Na^+ and K^+, with their associated conductances (g) represented by a zigzag line for a resistor (conductance = 1/resistance). The electromotive forces (Nernst equilibrium potentials) are represented by the symbol for a battery. These favor inward movement of a sodium current (I_{Na}) and outward movement of a potassium current (I_K) through the ion channels. The capacitative branch (C_m is the membrane capacitance) indicates that the membrane also acts as an insulator, because of its lipid bilayer. Note that in this simplified model the possible contribution of other ions (Cl^-, Ca^{2+}) to the membrane conductance and membrane potential is neglected.

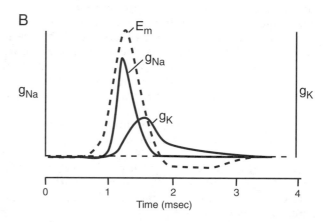

Figure 3–5 ■ (A) The action potential, with depolarization to threshold, rising phase, overshoot, peak, repolarization, afterhyperpolarization, and return to the resting membrane potential. (B) Changes in sodium (g_{Na}) and potassium (g_K) conductances associated with the action potential. The rising phase of the action potential is the result of an increase in sodium conductance, while the repolarization phase is due to a decrease in sodium conductance and a delayed increase in potassium conductance.

voltage measured is compared to that detected by a reference electrode placed outside the cell. The difference between the two measurements is a measure of the membrane potential. This technique is used to monitor the membrane potential at rest as well as during an action potential.

Action Potential Gating Mechanisms ■ Alterations in the conductance of voltage-gated Na^+ and K^+ channels are responsible for the generation of the action potential. As discussed in Chapter 2, the membrane potential when a nerve cell is at rest is near the equilibrium potential for potassium ($E_m \approx E_K$). This is because the permeability of the membrane for potassium in the resting state is much greater than that for sodium ($P_K >> P_{Na}$). The depolarizing and repolarizing phases of the action potential can also be explained by relative changes in membrane conductance (permeability) to sodium and potassium. During the rising phase of the action potential, the nerve cell membrane becomes more permeable to sodium; as a consequence, the membrane potential begins to shift more toward the equilibrium potential for sodium ($E_m \rightarrow E_{Na}$). However, before the membrane potential reaches E_{Na}, sodium permeability begins to decrease and potassium permeability increases. This change in membrane conductance again drives the membrane potential toward E_K, accounting for repolarization of the membrane.

The action potential can also be viewed in terms of the flow of charged ions through selective ion channels. These channels are closed when the neuron is at rest. When the membrane is depolarized, the voltage-gated channels begin to open. The Na^+ channel quickly opens its **activation gate** and allows Na^+ ions to flow into the cell (Fig. 3–6A). The influx of positively charged Na^+ ions causes the membrane to depolarize. In fact, the membrane potential actually reverses, with the inside becoming positive; this is called the **overshoot.** In the initial stage of the action potential, more Na^+ than K^+ channels are opened. This increase in Na^+ permeability compared to that of K^+ causes the membrane potential to move toward the equilibrium potential for Na^+.

At the peak of the action potential, the sodium conductance begins to fall as an **inactivation gate** closes. Also, more K^+ channels slowly open, allowing more positively charged K^+ ions to leave the neuron. The net effect of inactivating Na^+ channels and opening additional K^+ channels is the repolarization of the membrane (Fig. 3–6B).

As the membrane continues to repolarize, the membrane potential becomes more negative than its resting level. This **afterhyperpolarization** is a result of K^+ channels remaining open, thus allowing the continued efflux of K^+ ions. Another way to think about afterhyperpolarization is that the membrane's permeability to K^+ is higher than when the neuron is at rest. Consequently, the membrane potential is driven even more toward the K^+ equilibrium potential.

The changes in membrane potential during an action potential result from selective alterations in membrane conductance (Fig. 3–5B). These membrane conductance changes reflect the summated activity of individual voltage-gated sodium and potassium ion channels. From the temporal relationship of the action potential and the membrane conductance changes, the depolarization and rising phase of the action potential can be attributed to the increase in sodium ion conductance, the repolarization phases to both the decrease in sodium conductance and the increase in potassium conductance, and the afterhyperpolarization to the sustained increase of potassium conductance.

Initiation of the Action Potential ■ The axon hillock is the trigger zone that generates the action potential. The membrane of the initial segment contains a high density of voltage-gated sodium and potassium ion channels. When the membrane of the initial segment is depolarized, voltage-gated sodium channels are opened, permitting the influx of sodium ions. Influx of these positively charged ions further depolarizes the membrane, leading to the opening of other voltage-gated sodium channels. This cycle of membrane depolarization, sodium channel activation, sodium ion influx, and membrane depolarization is an example of positive feedback, a regenerative process (Fig. 1–3) that results in the explosive activation of many sodium ion channels when the threshold membrane potential is reached. If the depolarization of the initial segment does not reach threshold, then not enough sodium channels are activated to initiate the regenerative process. The initiation of an action potential is therefore an "all-or-none" process: it is generated completely or not at all.

Propagation and Speed of the Action Potential ■ After the action potential is generated at the initial segment, it propagates along the axon toward the axon terminals and is conducted along the axon with no decrement in amplitude. The mode in which action potentials propagate depends on whether the axon is myelinated. Schwann cells in the peripheral nervous system and oligodendrocytes in the central nervous system wrap themselves around axons to form **myelin,** layers of lipid membrane that insulate the axon and prevent the passage of ions through the axonal membrane. Between the myelinated segments of the axon are **nodes of Ranvier,** where action potentials are generated.

Na^+ and K^+ channels are distributed uniformly along the length of unmyelinated axons. When an action potential is generated at the axon hillock, the hillock acts as a "sink" where Na^+ ions enter the cell. The "source" of these Na^+ ions is the extracellular space along the length of the axon. Entry of Na^+ ions into the axon hillock causes the adjacent region to depolarize, generating an action potential. By depolarizing adjacent segments of the axon, the action potential propagates or moves along the length of the axon from point to point, like a traveling wave (Fig. 3–7A).

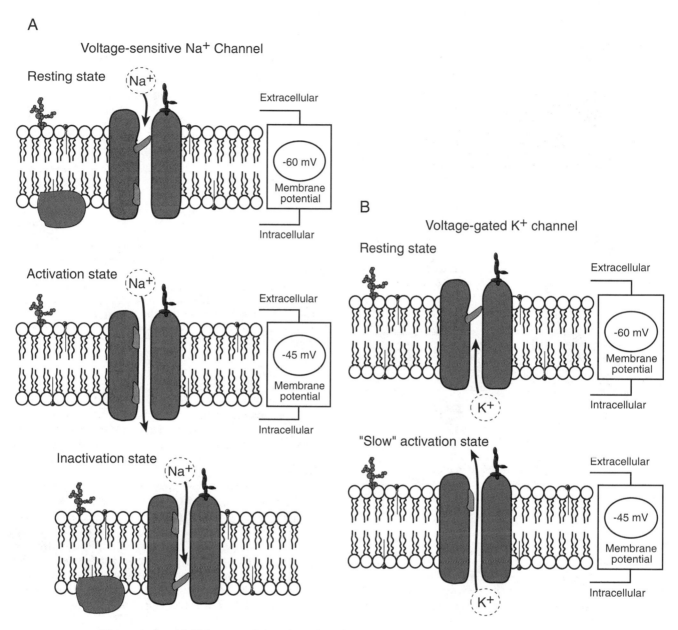

Figure 3–6 ■ (A) Voltage-gated Na⁺ channel, with resting state, activation state, and inactivation state. Na⁺ ions flow into the cell during the activation state. (B) Voltage-gated K⁺ channel, with resting state and "slow" activation state. K⁺ ions flow out of the cell when the channel is activated.

In myelinated axons, voltage-gated Na⁺ channels are highly concentrated in the nodes of Ranvier, where the myelin sheath is absent. When an action potential is initiated at the initial segment, the influx of Na⁺ ions causes the adjacent node of Ranvier to depolarize, resulting in an action potential at the node. This, in turn, causes depolarization of the next node of Ranvier and the eventual initiation of an action potential. Action potentials are successively generated at neighboring nodes of Ranvier (Fig. 3–7B). Thus the action potential in a myelinated axon appears to jump from one node to the next, a process called **saltatory con-**

duction (from the Latin *saltus,* to jump). This process results in a faster conduction velocity for myelinated than unmyelinated axons. The conduction velocity in mammals ranges from 3 to 120 m/sec for myelinated axons and 0.5 to 2.0 m/sec for unmyelinated axons.

The diameter of the axon also influences the speed of action potential conduction: larger-diameter axons have faster action potential conduction velocities than smaller-diameter axons. Just as large-diameter tubes allow a greater flow of water than small-diameter tubes, because of their decreased resistance, large-diameter axons have less cytoplasmic resistance, thus

A

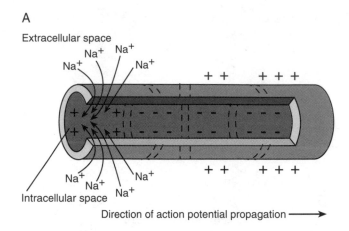

Extracellular space

Na⁺ Na⁺
Na⁺ Na⁺

+ +   + + +

+ + +

Na⁺ Na⁺
Na⁺ Na⁺

Intracellular space

Direction of action potential propagation ⟶

B

Saltatory conduction along myelinated axon

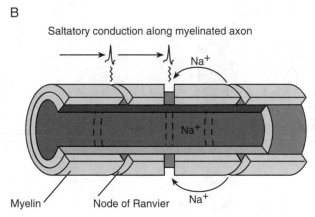

Na⁺

Na⁺

Myelin   Node of Ranvier   Na⁺

Figure 3–7 ■ (A) Propagation of the action potential in an unmyelinated axon. The initiation of an action potential in one segment of the axon depolarizes the adjacent section. (B) Propagation of the action potential in a myelinated axon. The initiation of an action potential in one node of Ranvier depolarizes the next node.

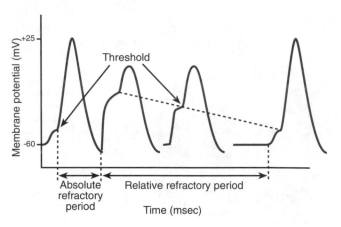

Figure 3–8

Figure 3–8 ■ Absolute and relative refractory periods. Immediately after the start of an action potential, a nerve cell is incapable of generating another impulse. This is the absolute refractory period. With time, the neuron can generate another action potential, but only at higher levels of depolarization. The period of increased threshold for impulse initiation is the relative refractory period. Note that action potentials initiated during the relative refractory period have a lower-than-normal amplitude.

permitting a greater flow of ions. This increase in ion flow in the cytoplasm causes greater lengths of the axon to be depolarized, decreasing the time needed for the action potential to travel along the axon.

Refractory Periods ■ After the start of an action potential, there are periods when the initiation of additional action potentials requires a greater degree of depolarization and when action potentials cannot be initiated at all. These are called the **relative** and **absolute refractory periods,** respectively (Fig. 3–8).

The inability to generate an action potential during the absolute refractory period is primarily due to the state of the voltage-gated Na⁺ channel. After the inactivation gate closes during the repolarization phase of the action potential, it remains closed for some time. Therefore, another action potential cannot be generated no matter how much the membrane is depolarized. The importance of the absolute refractory period is that it limits the rate of firing of action potentials.

In the relative refractory period, the inactivation gate of a portion of the voltage-gated Na⁺ channels is

open. Since these channels have returned to their initial resting state, they can now respond to depolarizations of the membrane. Consequently, when the membrane is depolarized, many of the channels open their activation gates and permit the influx of Na⁺ ions. However, because only a portion of the Na⁺ channels have returned to the resting state, depolarization of the membrane to the original threshold level activates an insufficient number of channels to initiate an action potential. With greater levels of depolarization, more channels are activated until eventually an action potential is generated. The K⁺ channels are maintained in the open state during the relative refractory period, leading to membrane hyperpolarization. By these two mechanisms the action potential threshold is increased during the relative refractory period.

Synaptic Transmission

Neurons communicate at synapses. Two types of synapses have been identified: **electrical** and **chemical synapses.** At electrical synapses, **gap junctions** connect the cytoplasm of adjacent neurons (Fig. 1–6) and permit the bidirectional passage of ions from one cell to another. Electrical synapses are uncommon in the mammalian nervous system. Typically they are found at dendrodendritic sites of contact; they are thought to synchronize the activity of neuronal populations.

■ Synaptic Transmission Usually Occurs via Chemical Transmitters

At chemical synapses, a space called the **synaptic cleft** separates the presynaptic terminal from the postsynaptic neuron (Fig. 3-9). The presynaptic terminal

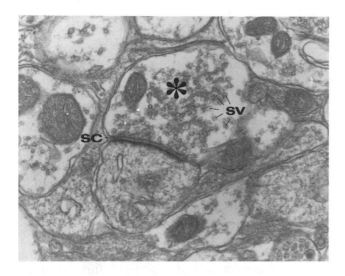

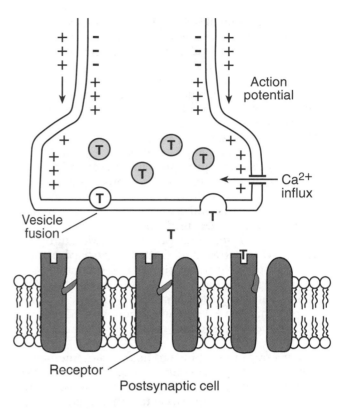

Figure 3–10 ■ Schematic drawing of transmitter release, showing fusion of synaptic vesicle with presynaptic membrane, release of transmitter into the synaptic cleft, and binding with postsynaptic receptor.

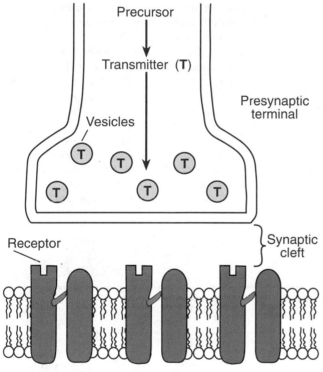

Figure 3–9 ■ (A) Electron micrograph of chemical synapse, showing presynaptic terminal (asterisk) with synaptic vesicles (SV) and synaptic cleft. (SC) separating pre- and postsynaptic membranes. Magnification 60,000×. (Courtesy of Dr. Lazaros Triarhou, Indiana University School of Medicine.) (B) Schematic drawing of chemical synapse.

is packed with numerous spheric vesicles that contain chemical neurotransmitters which are released into the synaptic cleft when an action potential enters the terminal. Once released, the chemical neurotransmit-

ter diffuses across the synaptic cleft and binds to receptors on the postsynaptic cell. The binding of the transmitter to its receptor leads to the opening (or closing) of specific ion channels, which, in turn, alter the membrane potential of the postsynaptic cell.

Release of neurotransmitters from the presynaptic terminal begins with the invasion of the action potential into the terminal axon region (Fig. 3-10). The depolarization of the terminal by the action potential causes the activation of voltage-gated Ca^{2+} channels. The electrochemical gradients for Ca^{2+} result in forces that drive Ca^{2+} into the terminal. This causes fusion of vesicles that contain neurotransmitters with the presynaptic membrane at "active zones." The neurotransmitters are then released into the cleft by **exocytosis.** Increasing the amount of Ca^{2+} that enters the terminal increases the amount of transmitter released into the synaptic cleft. The amount of transmitter molecules released by any one exocytosed vesicle is called the **quantal number.**

It is not clear how the entry of Ca^{2+} leads to the fusion of the vesicles with the presynaptic membrane. One hypothesis is that the vesicles are anchored to cytoskeletal components in the terminal by **synapsin,** a protein surrounding the vesicle. The entry of Ca^{2+}

ions into the terminal is thought to result in phosphorylation of this protein and a decrease in its binding to the cytoskeleton, thus allowing the vesicle to fuse with the presynaptic membrane of the terminal.

Receptors in the postsynaptic membrane are of two sorts. In some, the receptor forms part of the ion channel; in others, it is coupled to the ion channel via a G protein and a second messenger system.

In receptors associated with a specific G protein, a series of enzyme steps is initiated by binding of transmitter to its receptor, producing a second messenger that alters intracellular functions over a longer period of time than for direct ion channel opening. These membrane-bound enzymes and the second messengers they produce inside the target cells include **adenylate cyclase,** which produces cAMP, **guanylate cyclase,** which produces cGMP, and **phospholipase C,** which leads to the formation of two second messengers, diacylglycerol and inositol trisphosphate. Table 3–1 lists some neurotransmitters and their receptor subtypes that have been identified with direct ion channel opening, second messenger formation, or both.

When a transmitter binds to its receptor, membrane conductance changes occur, leading to depolarization or hyperpolarization. An increase in membrane conductance to Na^+ depolarizes the membrane. An increase in membrane conductance that permits the efflux of K^+ or the influx of Cl^- hyperpolarizes the membrane. In some cases, membrane hyperpolarization can occur when there is a decrease in membrane conductance that reduces the influx of Na^+.

■ Integration of Postsynaptic Potentials Occurs at Dendrites and Soma

The transduction of information between neurons in the nervous system is mediated by changes in the membrane potential of the postsynaptic cell. These membrane depolarizations and hyperpolarizations are integrated or summated and can result in activation or inhibition of the postsynaptic neuron. The alterations in the membrane potential that occur in the postsynaptic neuron initially take place in the dendrites and soma as a result of activation of afferent inputs.

Since depolarizations can lead to the excitation and activation of a neuron, they are commonly called **excitatory postsynaptic potentials (EPSPs).** In contrast, hyperpolarizations of the membrane prevent the cell from becoming activated and are thus called **inhibitory postsynaptic potentials (IPSPs).** These membrane potential changes are caused by the influx or efflux of specific ions (Fig. 3–11).

The rate at which the membrane potential of a postsynaptic neuron is altered can greatly influence the efficiency of transducing information from one neuron to the next. This is seen by considering the simplified model of the resting nerve membrane. If the activation of a synapse is assumed to lead to the influx of positively charged ions, the postsynaptic membrane will depolarize. When the influx of these ions is stopped, the membrane will repolarize back to the resting level. The rate at which it repolarizes depends on the membrane resistance per unit area, R_m ($1/g_m$), and capaci-

TABLE 3–1 ■ Types of Neurotransmitters and Receptors

Neurotransmitter	Receptor	Membrane Conductance Change	Membrane Potential Change	Second Messenger
Acetylcholine	Nicotinic	Increase g_{Na}, g_K	EPSP	
	Muscarinic M_1	Decrease g_K	EPSP	IP_3 and DAG
	Muscarinic M_2	Increase g_K	IPSP	cAMP
Norepinephrine	α_1		IPSP (CNS)	
	α_2		IPSP?	cAMP
	β_2			cAMP
Dopamine	D_1		EPSP?	cAMP
	D_2		IPSP	cAMP
Serotonin	5-HT-1A	Increase g_K	IPSP	cAMP
	5-HT-1B			
	5-HT-1C	Increase g_{Cl}	IPSP	IP_3
	5-HT-1D			
	5-HT-2	Decrease g_K	EPSP	IP_3
	5-HT-3	Increase g_{Na}, g_K	EPSP	
Glutamate	Kainate	Increase g_{Na}, g_K	EPSP	
	Quisqualate	Increase g_{Na}, g_K	EPSP	
	NMDA	Increase g_{Ca}	EPSP	IP_3 and DAG
GABA	GABA-A	Increase g_{Cl}	IPSP	
	GABA-B	Increase g_K	IPSP	cAMP?

CNS, central nervous system; IP_3, inositol 4,5-trisphosphate; DAG, diacylglycerol.

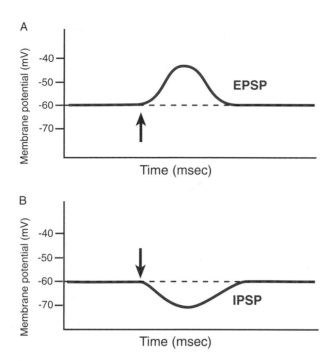

Figure 3–11 ■ (A) Depolarization of the membrane (arrow) brings a nerve cell closer to the threshold for the initiation of an action potential and produces an excitatory postsynaptic potential (EPSP). (B) Hyperpolarization of the membrane produces an inhibitory postsynaptic potential (IPSP).

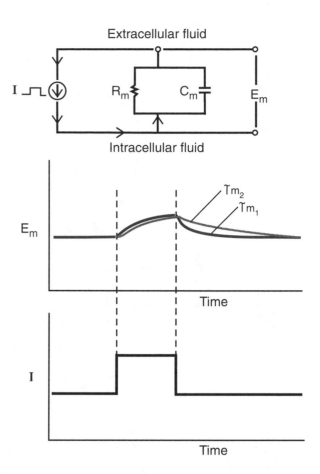

Figure 3–12 ■ The rate of decay of membrane potential varies with a given neuron's membrane time constant. The responses of two neurons to a brief application of depolarizing current are shown. Each neuron depolarizes to the same degree, but the time for return to the baseline membrane potential differs for each. Neuron 2 takes longer to return to baseline than neuron 1 because its time constant is longer ($\tau_{m2} > \tau_{m1}$).

tance per unit area, C_m (Fig. 3-12). The product of membrane resistance and capacitance is the **membrane time constant,** τ_m:

$$\tau_m = R_m \times C_m \qquad (4)$$

τ_m represents the time required for the membrane potential to decay to 37% of its initial peak value. The decay rate for repolarization is slower for longer time constants, because the increase in membrane resistance and/or capacitance results in a slower discharge of the membrane. The slow decay of the repolarization thus allows additional time for the synapse to be reactivated and depolarize the membrane. A second depolarization of the membrane can be added to that of the first depolarization. Consequently, longer periods of depolarization increase the likelihood of summating two postsynaptic potentials (Fig. 3-13). The process in which postsynaptic membrane potentials are added with time is called **temporal summation.** If the magnitude of the summated depolarizations is above a threshold value, as detected at the initial segment of a neuron, then it will generate an action potential.

The summation of postsynaptic potentials also occurs with the activation of several synapses located at different sites of contact. This is called **spatial sum-**

mation, and it can be illustrated using a modification of the simplified nerve model. In this case we shall consider several membrane patches, as shown in Figure 3-14A. These patches of membrane are connected in the cell by an internal axial resistance, R_a, which represents the resistance ions encounter when they flow through the cytoplasm. These patches are also connected by an extracellular resistance, R_e, which represents the resistance ions encounter as they flow through the extracellular space.

When a synapse is activated, causing an influx of positively charged ions, a depolarizing **electrotonic potential** develops, with maximal depolarization occurring at the site of synaptic activation. The electrotonic potential is due to the passive spread of ions in the dendritic cytoplasm and across the membrane. The amplitude of the electrotonic potential decays with distance from the site where the synapse is

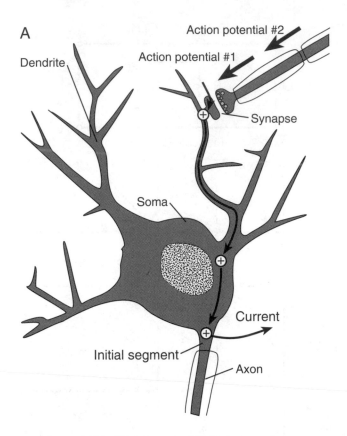

A

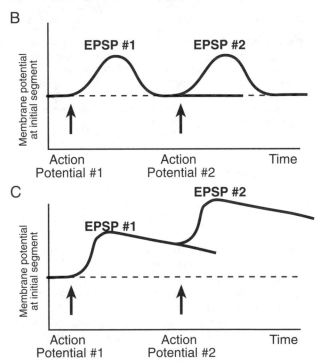

B

Figure 3–13 ■ Model of temporal summation. (A) Depolarization of a dendrite by two sequential action potentials. (B) Dendritic membrane with short time constant is unable to summate postsynaptic potentials. (C) Dendritic membrane with long time constant is able to summate membrane potential changes.

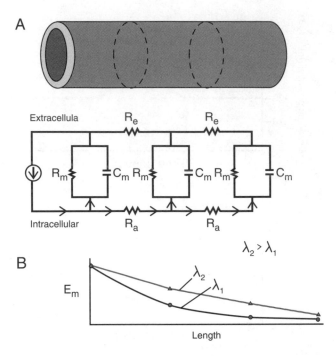

Figure 3–14 ■ (A) At the top is a dendritic process that has been divided into three sections. Beneath is an electrical model of these sections, with membrane resistance and capacitance for each section. R_e, extracellular resistance; R_a, axial cytoplasmic resistance. (B) Profile of the electrotonic membrane potential produced along the length of the dendrite. The decay of the membrane potential as it proceeds along the length of the dendrite is affected by the length constant, λ_m. Long length constants cause the electrotonic potential to decay more gradually. Profiles are shown for two dendrites with different length constants, λ_1 and λ_2. The electrotonic potential of dendrite 2 decays less steeply than that of dendrite 1 because its length constant is longer. I, current; C_m, capacitance; R_a, internal axial resistance; R_e, extracellular resistance; R_m, membrane resistance.

activated. This membrane potential profile is produced because the majority of synaptic current returns across the membrane of an adjacent patch, and the remainder of the current continues along the inside of the cell to other patches of membrane (Fig. 3–14B).

The decay of the electrotonic potential per unit length along the dendrite is determined by the **length** or **space constant**, λ_m:

$$\lambda_m = \sqrt{(r_m/r_a)} \qquad (5)$$

where r_m is the membrane resistance of a unit length (ohm·cm) and r_a is the longitudinal resistance of the cytoplasm per unit length (ohm/cm). λ_m represents the length required for the membrane potential depolarization to decay to 37% of its maximal value. Therefore, the larger the λ_m value, the smaller the decay per unit length. This implies that as the membrane resistance (R_m) increases or as the cytoplasmic axial resistance (R_a) decreases, the length constant is increased. Either condition will cause more current to flow along the cytoplasmic axial pathway, thus delivering more charge to more distant membrane patches.

By depolarizing distal patches of membrane, other electrotonic potentials that occur by activating synaptic inputs at other sites can summate to produce an even greater depolarization. Thus, the resulting postsynaptic potentials are added along the length of the dendrite. As with temporal summation, if the depolarizations resulting from spatial summation are sufficient to cause the membrane potential in the region of the initial segment to reach threshold, the postsynaptic neuron will generate an action potential (Fig. 3–15).

Because of the spatial decay of the electrotonic potential, the location of the synaptic contact has a great effect in determining whether a synapse can activate a postsynaptic neuron. For example, axodendritic synapses, located in distal segments of the dendritic tree, are far removed from the initial segment of the axon, and their activation has very little impact on the membrane potential near this trigger zone. In contrast, axosomatic synapses have a greater effect in altering the membrane potential at the initial segment, because of their proximal location.

Neurochemical Transmission

Neurons communicate with other cells by release of a chemical neurotransmitter. The transmitter is stored in synaptic vesicles and released on nerve stimulation by the process of exocytosis, following opening of voltage-gated, calcium-ion channels in the nerve terminal. Once released, the neurotransmitter binds to and stimulates its receptors briefly before being rapidly removed from the synapse, thereby allowing transmission of a new neuronal message. The most common mode of removal of the neurotransmitter following release is **high-affinity reuptake** by the presynaptic terminal. This is a carrier-mediated, sodium-dependent, secondary active transport that uses energy from the Na^+/K^+- ATPase pump. Other removal mechanisms include enzymatic hydrolysis in the synapse, or diffusion into the extracellular space (Fig. 3–16).

The details of synaptic events in chemical transmission were originally described for peripheral nervous system synapses. Central synapses appear to use similar mechanisms, with the important difference that muscle and gland cells are the targets of transmission in peripheral nerves, whereas neurons and glia make up the postsynaptic elements at central synapses.

■ There Are Three Classes of Neurotransmitters

The first neurotransmitters described were acetylcholine and norepinephrine, identified at synapses in the peripheral nervous system. Many others have since been identified, and they fall into three main classes:

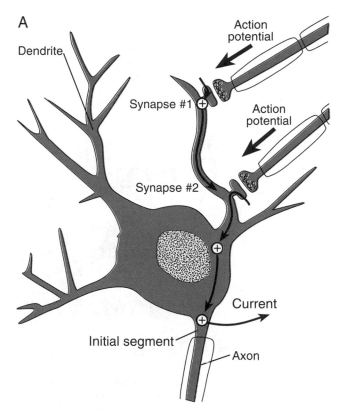

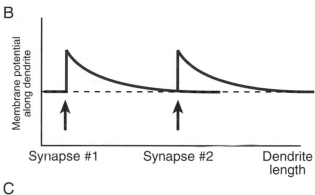

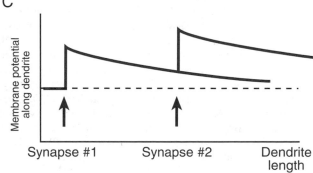

Figure 3–15 ■ Model of spatial summation. (A) Depolarization of dendrite at two spatially separated synapses. (B) Dendritic membrane with short length constant is unable to summate postsynaptic potentials. (C) Dendritic membrane with long length constant is able to summate membrane potential changes.

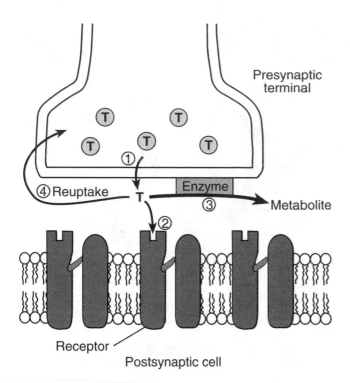

Presynaptic terminal

④ Reuptake

Enzyme

Metabolite

Receptor

Postsynaptic cell

Figure 3–16 ■ Schematic illustration of (1) neurotransmitter release, (2) binding of transmitter to receptor, (3) degradation of transmitter, and (4) reuptake of transmitter into the presynaptic terminal.

amino acids, monoamines, and polypeptides. Amino acids and monoamines are together termed **small-molecule transmitters.** The monoamines (or biogenic amines) are so named because they are synthesized from a single readily available amino acid precursor. The polypeptide transmitters (or neuropeptides) consist of a chain of amino acids varying in length from three to several dozen.

Examples of amino acid transmitters include the excitatory amino acids **glutamate acid** (GLU) and **aspartate** (ASP) and the inhibitory amino acids **glycine** (GLY) and **γ-aminobutyric** (GABA). (Note that GABA is biosynthetically a monoamine but has the features of an amino acid transmitter, not a monoaminergic one.) Examples of monoaminergic transmitters include **acetylcholine** (ACh), derived from choline; the catecholamine transmitters **dopamine** (DA), **norepinephrine** (NE), and **epinephrine** (EPI), derived from the amino acid tyrosine; and an indoleamine, **serotonin** or **5-hydroxytryptamine** (5-HT), derived from tryptophan. Examples of polypeptide transmitters include the opioids and substance P.

■ The Nerve Terminal Regulates Transmission of Amines and Amino Acids

Acetylcholine ■ Neurons that use ACh as their neurotransmitter are known as **cholinergic neurons.**

Acetylcholine is synthesized in the cholinergic neuron from choline and acetate, under the influence of the enzyme **choline acetyltransferase,** or choline acetylase. This enzyme is localized in the cytoplasm of cholinergic neurons, especially in the vicinity of storage vesicles, and is an identifying marker of the cholinergic neuron.

All of the components for synthesis, storage, and release of ACh are localized in the terminal region of the cholinergic neuron (Fig. 3–17). The storage vesicles and choline acetyltransferase are produced in the soma and are transported to the axon terminals. The rate-limiting step in ACh synthesis in the nerve terminals is the availability of choline, of which specialized mechanisms ensure a continuous supply.

Acetylcholine, together with ATP and other undefined constituents, is stored in vesicles in the axon terminals. Stored ACh is protected from enzymatic degradation and is packaged appropriately for release on nerve stimulation.

The enzyme **acetylcholinesterase** hydrolyzes ACh back to choline and acetate after release of ACh. This

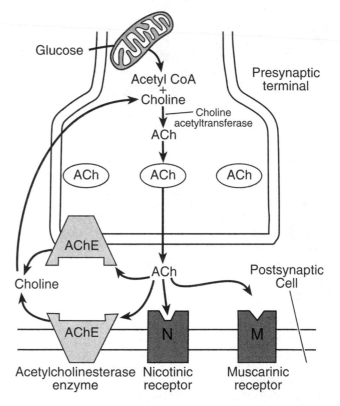

Glucose

Acetyl CoA + Choline

Choline acetyltransferase

ACh

Presynaptic terminal

ACh   ACh   ACh

AChE

Choline

ACh

Postsynaptic Cell

AChE   N   M

Acetylcholinesterase enzyme   Nicotinic receptor   Muscarinic receptor

Figure 3–17 ■ Cholinergic transmission. When an action potential invades the presynaptic terminal, ACh is released into the synaptic cleft and binds to receptors on the postsynaptic cell to activate either nicotinic or muscarinic receptors. ACh is also hydrolyzed in the cleft by the enzyme acetylcholinesterase (AChE) to produce the metabolites choline and acetate. Choline is transported back into the presynaptic terminal by a high-affinity transport process to be reused in ACh resynthesis.

enzyme is found in both presynaptic and postsynaptic cell membranes, allowing rapid and efficient hydrolysis of extracellular ACh. So efficient is this enzymatic mechanism that no ACh spills over from the synapse into the general circulation. The choline generated from ACh hydrolysis is taken back up by the cholinergic neuron by a high-affinity, sodium-dependent uptake mechanism, which ensures a steady supply of the precursor for ACh synthesis. An additional source of choline is the low-affinity transport used by all cells to take up choline from the extracellular fluid for use in the synthesis of phospholipids. Low-affinity transport is found on all parts of the neuron. The high-affinity transport of choline is unique to cholinergic neurons and is found only in the axon terminal region.

The receptors for ACh, known as **cholinergic receptors,** fall into two categories, based on the drugs that mimic or antagonize the actions of ACh on its many target cell types. In classical studies dating to the beginning of the twentieth century, the drugs **muscarine,** isolated from poisonous mushrooms, and **nicotine,** isolated from tobacco, were used to distinguish two separate receptors for ACh. Muscarine stimulates some of the receptors and nicotine stimulates all the others, so receptors were designated as either **muscarinic** or **nicotinic.** It should be noted that ACh has the actions of both muscarine and nicotine at cholinergic receptors. However, these two drugs cause fundamental differences that ACh cannot distinguish.

The **nicotinic acetylcholine receptor** is composed of five components: two α subunits and a β, γ, and δ subunit (Fig. 3–18). The two α subunits are binding sites for ACh. When ACh molecules bind to both α subunits, a conformational change occurs in the receptor, which results in an increase in channel conductance for Na^+ and K^+, leading to depolarization of the postsynaptic membrane. This is due to the strong inward electrical and chemical gradient for Na^+, which predominates over the outward gradient for K^+ ions and results in a net inward flux of positively charged ions.

The structure and function of the **muscarinic acetylcholine receptor** is quite different from that of the nicotinic ACh receptor. As many as five subtypes of muscarinic receptors have been identified. The M_1 and M_2 receptors are composed of seven membrane-spanning domains, and each exerts its actions through a G protein. Activation of M_1 receptors results in a decrease in K^+ conductance via phospholipase C, and activation of M_2 receptors causes an increase in K^+ conductance via inhibition of adenylate cyclase (Table 3-1). As a consequence, when ACh binds to an M_1 receptor it results in membrane depolarization; when it binds to an M_2 receptor it causes hyperpolarization.

Various drugs affect the release of ACh and its binding to cholinergic receptors. For example, botulinum toxin, a product of bacteria that thrive in improperly canned food products, inhibits ACh release. The re-

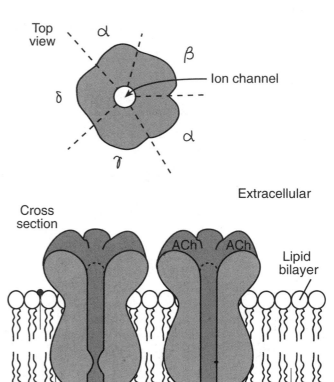

Figure 3–18 ■ Nicotinic acetylcholine receptor structure. The nicotinic receptor is composed of five subunits: two α subunits and β, γ, and δ subunits. The two α subunits serve as binding sites for ACh. Both binding sites must be occupied to open the channel, permitting sodium ion influx and potassium ion efflux.

sulting decreased ACh receptor activation can lead to respiratory distress and paralysis.

Catecholamines ■ The catecholamines are so named because they consist of a catechol moiety (a phenyl ring with two adjacent hydroxyl groups) and an ethylamine side chain. The catecholamines DA, NE, and EPI share a common pathway for enzymatic biosynthesis (Fig. 3–19). Three of the enzymes involved—tyrosine hydroxylase (TH), dopamine β-hydroxylase (DβH), and phenylethanolamine N-methyltransferase (PNMT)—are unique to catecholamine-secreting cells, and all are derived from a common ancestral gene. Dopaminergic neurons express only TH, noradrenergic neurons express both TH and DβH, and epinephrine-secreting cells express all three. Epinephrine-secreting cells include a small population of central nervous system neurons as well as the hormonal cells of the adrenal medulla, **chromaffin cells,** which secrete EPI during the fight-or-flight response (Chap. 6).

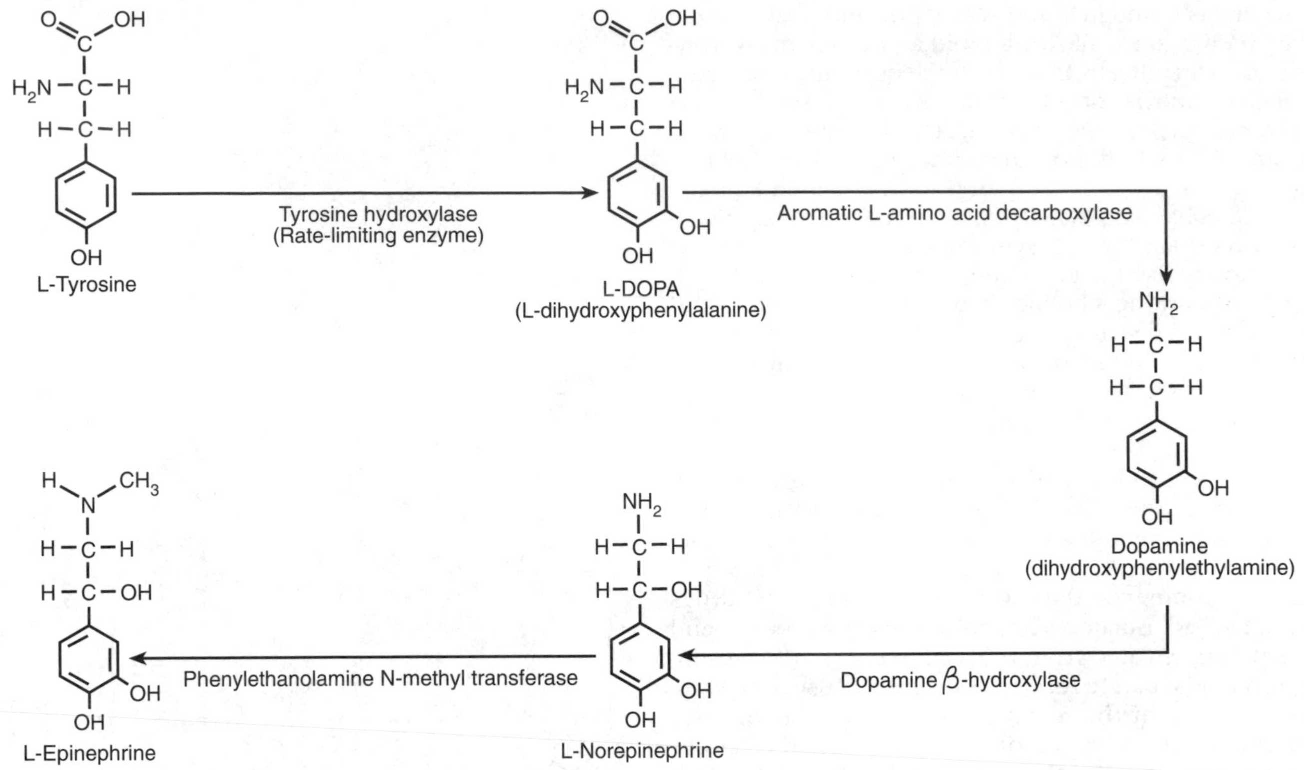

Figure 3–19 ■ The catecholamine neurotransmitters are synthesized by way of a chain of enzymatic reactions to produce L-DOPA, dopamine, L-norepinephrine, and L-epinephrine.

The rate-limiting enzyme in catecholamine biosynthesis is tyrosine hydroxylase, which converts L-tyrosine to L-3,4-dihydroxyphenylalanine (L-DOPA). Figure 3–19 illustrates the pathway for catecholamine synthesis, which requires tetrahydrobiopterine as a cofactor. Tyrosine hydroxylase is regulated by short-term activation and long-term induction. As shown in dopaminergic neurons of the basal ganglia, short-term excitation of dopaminergic neurons results in an increase in the conversion of tyrosine to DA. This phenomenon is mediated by the phosphorylation of TH via a cAMP-dependent protein kinase, which results in an increase in TH affinity for the pterin cofactor. Long-term induction is mediated by the synthesis of new TH.

A nonspecific cytoplasmic enzyme, aromatic L-amino acid decarboxylase, catalyzes the formation of dopamine from L-DOPA. Dopamine is then taken up in storage vesicles and protected from enzymatic attack. In NE- and EPI-synthesizing neurons DβH, which converts DA to NE, is found within vesicles, unlike the other synthetic enzymes, which are in the cytoplasm. In EPI-secreting cells, PNMT is localized in the cytoplasm. The PNMT adds a methyl group to the amine in NE to form EPI.

Two enzymes are involved in degrading the catechol-

amines following vesicle exocytosis. **Monoamine oxidase** (MAO) removes the amine group, and **catechol-O-methyltransferase** (COMT) methylates the 3-OH group on the catechol ring. As shown in Figures 3–20A and 3–20B, MAO is localized in mitochondria, present in both presynaptic and postsynaptic cells, whereas COMT is localized in the cytoplasm and only postsynaptically. At synapses of noradrenergic neurons in the peripheral nervous system (i.e., postganglionic sympathetic neurons of the autonomic nervous system) (Chap. 6), the postsynaptic COMT-containing cells are the muscle and gland cells and other nonneuronal tissues that receive sympathetic stimulation. In the CNS, on the other hand, most of the COMT is localized in glia rather than in postsynaptic target neurons. The catecholamine-degrading enzymes act slowly, whereas membrane-bound acetylcholinesterase rapidly inactivates ACh. The end products of catecholamine degradation are deaminated and O-methylated compounds, which are excreted in the urine in a conjugated form.

Most of the catecholamine released into the synapse (up to 80%) is rapidly removed by uptake. Once inside the neuron the transmitter enters the vesicles and is made available for recycling. In the case of EPI-secreting neurons, which are only sparsely represented in the brainstem and diencephalon, released

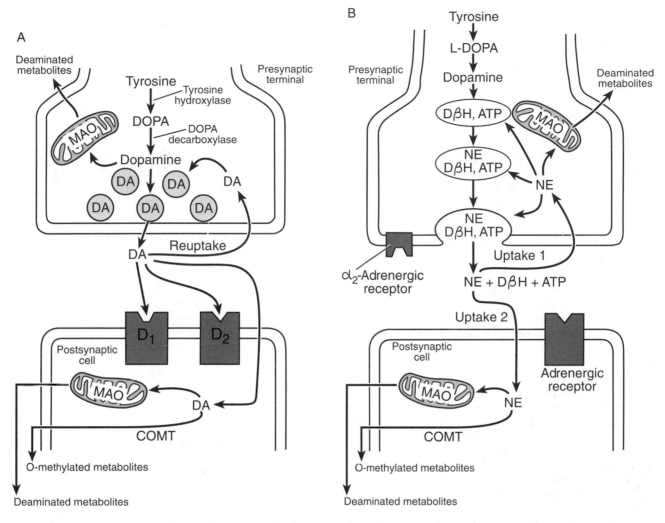

Figure 3–20 ■ (A) In dopamine-producing nerve terminals, dopamine is enzymatically synthesized from tyrosine, then taken up and stored in vesicles. The fusion of DA-containing vesicles with the terminal membrane results in the release of DA into the synaptic cleft and permits DA to bind to dopamine receptors (D₁ and D₂ receptors) on the postsynaptic cell. The termination of DA neurotransmission occurs when DA is transported back into the presynaptic terminal via a high-affinity mechanism. (B) In norepinephrine (NE)-producing nerve terminals, DA is transported into synaptic vesicles and converted into NE by the enzyme dopamine β-hydroxylase. On release into the synaptic cleft, NE can bind to postsynaptic α- or β-adrenergic receptors and presynaptic α₂-adrenergic receptors. Uptake of NE into the presynaptic terminal (uptake 1) is responsible for the termination of synaptic transmission. In the presynaptic terminal, NE is repackaged into vesicles or deaminated by mitochondrial MAO. NE can also be transported into the postsynaptic cell by a low-affinity process (uptake 2), where it is deaminated by MAO and O-methylated by catechol-O-methyl transferase (COMT).

EPI is taken up more readily into NE-secreting nerve terminals, which are found in abundance in the vicinity of the EPI-secreting nerve terminals.

In peripheral noradrenergic synapses (the sympathetic nervous system) the neuronal uptake process described above is referred to as **uptake 1,** to distinguish it from a second uptake mechanism, **uptake 2,** localized in the target cells (smooth muscle, cardiac muscle, and gland cells) (Fig. 3–20B). In contrast with

uptake 1, an active transport, uptake 2 is a facilitated diffusion mechanism, which takes up the sympathetic transmitter, NE, as well as the circulating hormone, EPI, and degrades them enzymatically by MAO and COMT localized in the target cells. In the CNS, there is little evidence of an uptake 2 of NE, but glia serve a comparable role by taking up catecholamines and degrading them enzymatically by glial MAO and COMT. Unlike uptake 2 in the peripheral nervous system, glial uptake of catecholamines has many of the characteristics of uptake 1.

The catecholamines differ substantially in their interactions with receptors; DA interacts with DA receptors and NE and EPI interact with adrenergic receptors. As many as five subtypes of DA receptors have been described in the CNS. Of these, two have been well characterized (Table 3-1). **D_1 receptors** are coupled to stimulatory G proteins (G_s), which activate adenylate cyclase, and **D_2 receptors** are coupled to inhibitory G proteins (G_i), which inhibit adenylate cyclase. Activation of D_2 receptors hyperpolarizes the postsynaptic membrane by increasing potassium conductance. A third subtype of DA receptor postulated to modulate the release of DA is localized on the cell membrane of the nerve terminal that releases DA; accordingly, it is called an **autoreceptor.**

Adrenergic receptors, stimulated by EPI and NE, are located on cells throughout the body, including the CNS and the peripheral target organs of the sympathetic nervous system (Chap. 6). Adrenergic receptors are classified as either α or β, based on the rank order of potency of catecholamines and related analogs in stimulating each type. The analogs used originally in distinguishing α- from β-adrenergic receptors are NE, EPI, and the two synthetic compounds isoproterenol (ISO) and phenylephrine (PE). Alquist, in 1948, designated as α those receptors in which EPI was highest in potency and ISO was least potent (EPI > NE > > ISO). β-receptors exhibited a different rank order: ISO was most potent and EPI either more potent or equal in potency to NE. Studies with PE further distinguished these two classes of receptors: α-receptors were stimulated by PE, whereas β-receptors were not. Since their discovery, α- and β-receptors each has been further subdivided, into α_1 and α_2 and β_1 and β_2 subtypes, on the basis of more extensive pharmacologic studies. In peripheral (sympathetic) noradrenergic neurons, an autoreceptor that modulates NE release is an α_2-receptor subtype, whereas postsynaptic receptors on the target cells are of the α_1 subtype (Fig. 3-20B). In the CNS there is no such anatomic distribution of α-receptor subtypes, since α_2-receptors have been localized postsynaptically as well as presynaptically.

Serotonin ■ Serotonin or 5-hydroxytryptamine (5-HT) is the transmitter in **serotonergic neurons.** Chemical transmission in these neurons is in several ways similar to that described for catecholaminergic neurons. Tryptophan hydroxylase, a marker of serotonergic neurons, converts tryptophan to 5-hydroxytryptophan (5-HTP), which is then converted to 5-HT by decarboxylation (Fig. 3-21).

5-Hydroxytryptamine is stored in vesicles and is released by exocytosis on nerve depolarization. The major mode of removal of released 5-HT is by a high-affinity, sodium-dependent, active uptake mechanism. There are several receptor subtypes for serotonin (Table 3-1). The 5-HT-3 receptor contains an ion channel. Activation results in an increase in sodium and

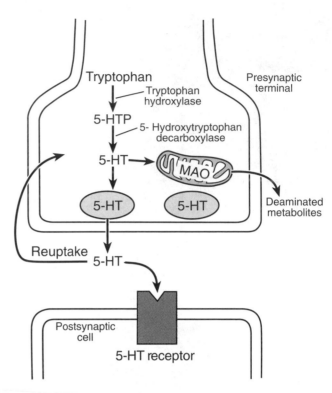

Figure 3–21 ■ 5-HT neurotransmission. 5-HT is synthesized by the hydroxylation of tryptophan to form 5-hydroxytryptophan (5-HTP) and the decarboxylation of 5-HTP to form 5-HT. On release into the synaptic cleft, 5-HT can bind to a variety of serotonergic receptors on the postsynaptic cell. Synaptic transmission is terminated when 5-HT is transported back into the presynaptic terminal for repackaging into vesicles.

potassium ion conductances, leading to EPSPs. The remaining well-characterized receptor subtypes appear to operate through second messenger systems. The 5-HT-1A receptor, for example, uses cAMP. Activation of this receptor results in an increase in K$^+$ ion conductance, producing IPSPs.

Glutamate and Aspartate ■ Both glutamate (GLU) and aspartate (ASP) serve as excitatory transmitters of the CNS. These dicarboxylic amino acids are important substrates for transaminations in all cells, but in certain neurons they also serve as neurotransmitters (i.e., they are sequestered in high concentration in synaptic vesicles, released by exocytosis, stimulate specific receptors in the synapse, and are removed by high-affinity uptake). Since GLU and ASP are readily interconvertible in transamination reactions in cells, including neurons, it has been difficult to distinguish neurons that use glutamate as a transmitter from those that use aspartate. This difficulty is further compounded by the fact that GLU and ASP stimulate common receptors. Accordingly, it is customary to refer to both as **glutamatergic neurons.**

Sources of GLU for transmission are the diet or mitochondrial conversion of α-ketoglutarate derived from the Krebs cycle (Fig. 3–22). Glutamate is stored in vesicles and released by exocytosis, where it activates specific receptors to depolarize the postsynaptic neuron. Two efficient active transport mechanisms remove GLU rapidly from the synapse. Neuronal uptake recycles the transmitter by restorage in vesicles and rerelease. Glia contain a similar, high-affinity, active transport that ensures efficient removal of excitatory transmitter from the synapse (Fig. 3–22). Glia further serve to recycle the transmitter by converting it to **glutamine,** an inactive storage form of GLU containing a second amine group. Glutamine from glia readily enters the neuron, where glutaminase removes the second amine, thereby regenerating GLU for use again as a transmitter.

At least five subtypes of GLU receptors have been described, based on the relative potency of synthetic analogs in stimulating them. Three of these, named for the synthetic analogs that best activate them—kainate, quisqualate, and N-methyl-D-aspartate (NMDA) receptors (Table 3–1)—are associated with cationic channels in the neuronal membrane. Activation of the **kainate** and **quisqualate receptors** produces excit-

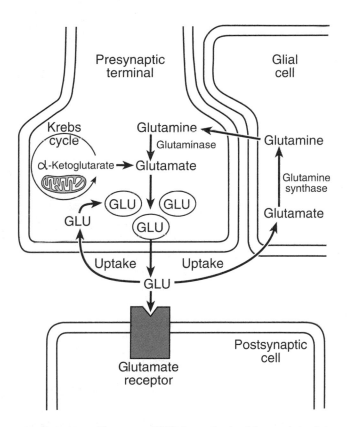

Figure 3–22 ■ Glutamate (GLU) is synthesized from α-ketoglutarate by enzymatic amination. On release into the synaptic cleft, GLU can bind to a variety of receptors. Removal of GLU is primarily by transport into glial cells, where it is converted into glutamine. Glutamine, in turn, is transported from glial cells to the nerve terminal, where it is converted to glutamate by the enzyme glutaminase.

atory postsynaptic potentials by opening ion channels that increase Na^+ and K^+ conductance. **NMDA receptor** activation increases Ca^{2+} conductance. This receptor, however, is blocked by Mg^{2+} when the membrane is in the resting state and becomes unblocked when the membrane is depolarized. Thus the NMDA receptor can be thought of as both a ligand- and a voltage-gated channel. Its gating of Ca^{2+}, particularly during ischemic disorders of the brain, is thought to be responsible for the rapid death of neurons in stroke and hemorrhagic brain disorders (see Focus Box, p. 52).

Gamma-aminobutyric Acid and Glycine ■ The inhibitory amino acid transmitters γ-aminobutyric acid (GABA) and glycine (GLY) bind to their respective receptors, causing hyperpolarization of the postsynaptic membrane. GABAergic neurons represent the major inhibitory neurons of the CNS, whereas glycine neurons are found in limited numbers, restricted only to the spinal cord and brainstem. Glycinergic transmission has not been as well characterized as transmission using GABA, thus only GABA is discussed here.

Synthesis of GABA in the neuron is by decarboxylation of GLU by the enzyme glutamic acid decarboxylase, a marker of GABAergic neurons. The GABA is stored in vesicles and released by exocytosis, leading to the stimulation of postsynaptic receptors (Fig. 3–23).

There are two types of GABA receptors: $GABA_A$ and $GABA_B$ (Table 3–1). The **$GABA_A$ receptor** is a ligand-gated Cl^- channel, and its activation produces inhibitory postsynaptic potentials by increasing the influx of Cl^- ions. It is composed of two α and two β subunits, with each subunit spanning the membrane four times. Activation requires the binding of GABA to the β subunits. The increase in Cl^- conductance is facilitated by **benzodiazepines,** drugs that bind to the α subunits. Benzodiazepines, including the widely used tranquilizer diazepam and chlordiazepoxide HCl, are used mainly to treat anxiety. Activation of the **$GABA_B$ receptor** also produces inhibitory postsynaptic potentials, but the IPSP results from an increase in K^+ conductance via activation of a G protein.

Gamma-aminobutyric acid is removed from the synaptic cleft by transport into the presynaptic terminal and glial cells (Fig. 3–23). The GABA enters the Krebs cycle in both neuronal and glial mitochondria and is converted to succinic semialdehyde by the enzyme GABA-transaminase. This enzyme is also coupled to the conversion of α-ketoglutarate to glutamate. The glutamate produced in the glial cell is converted to glutamine. As in the recycling of glutamate, glutamine is then transported into the presynaptic terminal, where it is then converted into glutamate.

■ Neuropeptides Modulate the Activity of Neurotransmitters

Neurally active peptides are stored in synaptic vesicles of neurons in the central and autonomic nervous systems, and they undergo exocytotic release in common with other neurotransmitters. In many instances

ROLE OF GLUTAMATE RECEPTORS IN NERVE CELL DEATH IN HYPOXIC/ISCHEMIC DISORDERS

Excitatory amino acids (EAA), GLU and ASP, are the neurotransmitters for more than half the total neuronal population of the CNS. Not surprisingly, most neurons in the CNS contain receptors to EAA. When transmission in glutamatergic neurons functions normally, very low concentrations of EAA appear in the synapse at any one time, due primarily to the efficient uptake mechanisms contained in both the presynaptic neuron and neighboring glial cells. In certain pathologic states, however, extraneuronal concentrations of EAA exceed the ability of the uptake mechanisms to remove them, resulting in cell death in a matter of minutes. This can be seen in severe **hypoxia,** such as during respiratory or cardiovascular failure, and in **ischemia,** where the blood supply to a region of the brain is interrupted, as in stroke. In either condition, the affected area is deprived of oxygen and glucose, which are essential for normal neuronal functions, including energy-dependent mechanisms for removal of extracellular EAA and their conversion to glutamine.

The consequences of prolonged exposure of neurons to EAA has been described as **excitotoxicity;** most of the cytotoxicity can be attributed to the destructive actions of high intracellular calcium brought about by stimulation of the various subtypes of glutamatergic receptors. One subtype, a presynaptic kainate receptor, opens voltage-gated calcium channels and promotes further release of GLU. Several postsynaptic receptor subtypes, ionotropic and metabotropic, depolarize the nerve cell and promote the rise of intracellular calcium via ligand- and voltage-gated channels and second messenger-mediated mobilization of intracellular calcium stores. The spiraling consequences of increased extracellular GLU leading to further release of GLU and of increased calcium entry leading to further mobilization of intracellular calcium bring about cell death due to inability of ischemic/hypoxic conditions to meet the high metabolic demands of excited neurons and the triggering of destructive changes in the cell by increased free calcium.

Intracellular free calcium is an activator of calcium-dependent proteases, which destroy microtubules and other structural proteins that maintain neuronal integrity. Calcium activates phospholipases, which break down membrane phospholipids and lead to lipid peroxidation and formation of oxygen-free radicals, which are toxic to cells. Another consequence of activated phospholipase is the formation of arachidonic acid and metabolites, including prostaglandins, which constrict blood vessels and further exacerbate hypoxia/ischemia. Calcium activates cellular endonucleases, leading to fragmentation of DNA and destruction of chromatin. In mitochondria, high calcium induces swelling and impaired formation of ATP via the Krebs cycle. Calcium is the primary toxic agent in EAA-induced cytotoxicity.

Proposed new treatment strategies hold promise of enhancing survival of neurons in brain ischemic/hypoxic disorders. These include drugs that block specific subtypes of glutamatergic receptors, such as the NMDA receptor, which is most responsible for promoting high calcium levels in the neuron. Other strategies include drugs that destroy oxygen-free radicals, such as 21-aminosteroids, and calcium ion channel blocking agents.

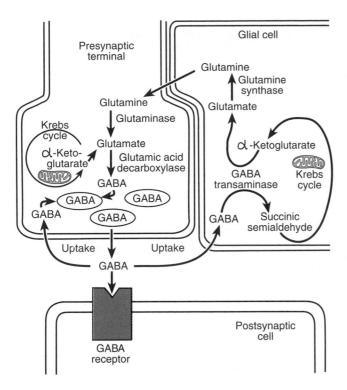

Figure 3–23 ■ Gamma-aminobutyric acid (GABA) is synthesized from glutamate by the enzyme glutamic acid decarboxylase. On release into the synaptic cleft, GABA can bind to several receptor subtypes ($GABA_A$, $GABA_B$). Removal of GABA from the synaptic cleft is primarily by uptake into the presynaptic neuron and surrounding glial cells. The conversion of GABA to succinic semialdehyde is coupled to the conversion of α-ketoglutarate to glutamate by the enzyme GABA-transaminase. In glia, glutamate is converted into glutamine, which is transported back into the presynaptic terminal for synthesis into GABA.

vesicles containing neuropeptides are colocalized with vesicles containing other transmitters in the same neuron (Table 3–2), and both can be shown to be released during nerve stimulation. Fundamental differences in the characteristics of neuropeptide-mediated transmission, however, suggest a synaptic role different from that of more classical transmitters.

TABLE 3–2 ■ Examples of Colocalization of Neurotransmitters

Low-Molecular-Weight Neurotransmitter	Neuropeptide
Acetylcholine	Vasoactive intestinal peptide
Dopamine	Enkephalin
	Cholecystokinin octapeptide
Norepinephrine	Enkephalin
	Somatostatin
Serotonin	Substance P
	Thyrotropin-releasing hormone

The list of candidate peptide transmitters continues to grow and includes well-known gastrointestinal hormones, pituitary hormones, and hypothalamic releasing factors. With the possible exceptions of substance P and the enkephalins, though, most have not met all of the criteria for neurotransmitters. As a class, the neuropeptides fall into several families of peptides, based on their origins, homologies in amino acid composition, and similarities in the response they elicit at common or related receptors. Table 3–3 lists some members of each of these families.

Peptides are synthesized as large prepropeptides in the endoplasmic reticulum and are packaged into vesicles that reach the axon terminal by **axoplasmic transport.** While in transit, the prepropeptide in the vesicle is posttranslationally modified by proteases that split it into small peptides and by other enzymes that alter the peptides by hydroxylation, amidation, sulfation, and so on. The products released by exocytosis include a neurally active peptide fragment and many unidentified peptides and enzymes from within the vesicles.

There are no mechanisms for recycling of peptide transmitters at the axon terminal. This is in contrast to more classical transmitters, where all of the mechanisms for recycling, including synthesis, storage, reuptake, and release, are contained within the terminals. Accordingly, classical transmitters are not subject to exhaustion of their supply, whereas peptide transmitters can be depleted in the axon terminal unless replenished by a steady supply of new vesicles transported from the soma.

While there is no evidence of synaptic uptake of any of the peptides, specific proteases have been localized in some cases in the synapse on external surfaces of target cells (e.g., enkephalinases that destroy leu- and met-enkephalins). The most common removal mechanism appears to be diffusion, a slow process that ensures a longer-lasting action of the peptide in the synapse and in the extracellular fluid surrounding it.

Peptides interact with specific peptide receptors located on postsynaptic target cells. They also modify the response of the coreleased transmitter interacting with its own receptor in the synapse. For this reason, they are said to be **modulators** of the actions of other neurotransmitters.

Opioids are peptides that bind to morphine receptors. They appear to be involved in the control of pain information. Opioid peptides include met-enkephalin, leu-enkephalin, dynorphin, and β-endorphin (Table 3–3). Structurally, they share homologous regions consisting of the amino acid sequence Tyr-Gly-Gly-Phe. The opioids are derived from three propeptides: **proenkephalin, proopiomelanocortin, and prodynorphin.** Proenkephalin gives rise to met- and leu-enkephalin; proopiomelanocortin gives rise to β-endorphin; and prodynorphin is the precursor of

TABLE 3–3 ■ Neuropeptides

Neuropeptide	Structure
Opioids	
Met-enkephalin	Tyr-Gly-Gly-Phe-Met-OH
Leu-enkephalin	Tyr-Gly-Gly-Phe-Leu-OH
Dynorphin	Tyr-Gly-Gly-Phe-Leu-Arg-Arg-Ile
β-endorphin	Tyr-Gly-Gly-Phe-Met-Thr-Ser-Glu-Lys-Ser-Gln-Thr-Pro-Leu-Val-Thr-Leu-Phe-Lys-Asn-Ala-Ile-Val-Lys-Asn-Ala-His-Lys-Gly-Gln-OH
Gastrointestinal peptides	
Cholecystokinin octapeptide (CCK-8)	Asp-Tyr-Met-Gly-Trp-Met-Asp-Phe-NH$_2$
Substance P	Arg-Pro-Lys-Pro-Gln-Gln-Phe-Phe-Gly-Leu-Met
Vasoactive intestinal peptide	His-Ser-Asp-Ala-Val-Phe-Thr-Asp-Asn-Tyr-Thr-Arg-Leu-Arg-Lys-Gln-Met-Ala-Val-Lys-Lys-Tyr-Leu-Asn-Ser-Ile-Leu-Asn-NH$_2$
Hypothalamic and pituitary peptides	
Thyrotropin-releasing hormone (TRH)	Pyro-Glu-His-Pro-NH$_2$
Somatostatin	Ala-Gly-Cys-Lys-Asn-Phe-Phe-Trp-Lys-Thr-Phe-Thr-Ser-Cys
Luteinizing hormone-releasing hormone (LHRH)	Pyro-Glu-His-Trp-Ser-Tyr-Gly-Leu-Arg-Pro-Gly
Vasopressin	Cys-Tyr-Phe-Gln-Asn-Cys-Pro-Arg-Gly-NH$_2$
Oxytocin	Cys-Tyr-Ile-Gln-Asn-Cys-Pro-Leu-Gly-NH$_2$

dynorphin. There are several opioid receptor subtypes. β-endorphin binds preferentially to **μ receptors;** enkephalins bind preferentially to **δ receptors;** and dynorphin binds preferentially to **κ receptors.** The enkephalins are metabolized by two enzymes: **aminopeptidase,** which hydrolyzes the Tyr-Gly bond, and **enkephalinase,** which hydrolyzes the Gly-Gly bond.

Substance P was originally isolated in the 1930s and four decades later was found to have the properties of a neurotransmitter. Substance P is an 11-amino acid polypeptide. It is found in high concentrations in the spinal cord and hypothalamus. In the spinal cord, substance P is localized in nerve fibers involved in the transmission of pain information. It slowly depolarizes neurons in the spinal cord and appears to use inositol 1,4,5-trisphosphate as a second messenger. Antagonists that block the action of substance P produce an analgesic effect. The opioid enkephalin also diminishes pain sensation, probably by presynaptically inhibiting the release of substance P.

Many of the other peptides found throughout the CNS were originally discovered in the hypothalamus as part of the neuroendocrine system. Among the hypothalamic peptides, somatostatin has been fairly well characterized in its role as a transmitter. Somatostatin is a 14-amino acid polypeptide that forms a ringlike structure by linking two cysteine residues via disulfide bonds. As part of the neuroendocrine system, this peptide inhibits the release of growth hormone by the anterior pituitary (Chap. 34). About 90% of brain somatostatin, however, is found outside the hypothalamus. Somatostatin-containing cells have been identified in the hippocampus, amygdala, other areas of the limbic system, and neocortex. Application of somatostatin to target neurons inhibits their electrical activity, but the ionic mechanisms mediating this inhibition are unknown.

Maintenance of Nerve Cell Function

■ Proteins Are Synthesized in the Soma of Neurons

The nucleus in neurons is large, and a substantial portion of the genetic information it contains is continuously transcribed. Based on hybridization studies, it is estimated that one-third of the genome in brain cells is actively transcribed, producing more mRNA than any other kind of cell in the body. Because of the high level of transcriptional activity the nuclear chromatin is dispersed. In contrast, the chromatin in non-neuronal cells in the brain, such as glia, is found in clusters on the internal face of the nuclear membrane.

Most of the proteins formed by free ribosomes and polyribosomes remain within the soma, whereas proteins formed by rough endoplasmic reticulum are exported to the dendrites and the axon. Polyribosomes and rough endoplasmic reticulum are found predominantly in the soma of neurons. Axons contain no rough endoplasmic reticulum and are unable to synthesize proteins. The smooth endoplasmic reticulum is involved in the intracellular storage of calcium. Smooth endoplasmic reticulum in neurons binds calcium and maintains the intracellular cytoplasmic concentration at a low level, about 10^{-7} M. Prolonged elevation of intracellular calcium leads to neuronal death and degeneration.

The Golgi apparatus in neurons is found only in the soma. As in other types of cells, this structure is engaged in the terminal glycosylation of proteins

THE ACTION POTENTIAL, SYNAPTIC TRANSMISSION, AND MAINTENANCE OF NERVE FUNCTION

synthesized in the rough endoplasmic reticulum. The Golgi apparatus forms export vesicles for proteins produced in the rough endoplasmic reticulum. These vesicles are released into the cytoplasm, and some are carried by axoplasmic transport to the axon terminals.

■ The Cytoskeleton Is the Infrastructure for Neuron Form

The transport of proteins from the Golgi apparatus and the highly specialized form of the neuron depend on the internal framework of cytoskeleton. The neuronal cytoskeleton is made of microfilaments, neurofilaments, and microtubules. **Microfilaments** are composed of actin, a contractile protein also found in muscle. They are 4–5 nm in diameter and are found in dendritic spines. **Neurofilaments** are found in both axons and dendrites and are thought to provide structural rigidity. They are not found in the growing tips of axons and dendritic spines, which are more dynamic structures. Neurofilaments are about the size of intermediate filaments found in other types of cells (10 nm diameter). In other cell types, however, intermediate filaments consist of one protein, whereas neurofilaments are composed of three proteins. The core of neutrofilaments consists of a 70 kd protein, similar to intermediate filaments in other cells. The two other neurofilament proteins are thought to be side arms that interact with microtubules.

Microtubules are responsible for the rapid movement of material in axons and dendrites. They are 23 nm in diameter and are composed of tubulin. In neurons, microtubules have accessory proteins, called **microtubule-associated proteins** (MAPs). Dendrites have high-molecular-weight MAPs and axons have low-molecular-weight MAPs; these MAPs are thought to be responsible for the distribution of material to dendrites or axons.

■ Mitochondria Are Important for Synaptic Transmission

Mitochondria in neurons are highly concentrated in the region of axon terminals. They produce ATP, which is required as a source of energy for many cellular processes. In the axon terminal the mitochondria not only provide a source of energy for processes associated with synaptic transmission but also provide substrates for the synthesis of certain neurotransmitter chemicals, such as the amino acid glutamate. In addition, mitochondria contain enzymes for degrading neurotransmitter molecules, such as MAO, which degrades catecholamines and 5-HT, and GABA-transaminase which degrades GABA.

■ Transport Mechanisms Distribute Material Needed by the Neuron and Its Fiber Processes

The shape of most cells in the body is relatively simple in comparison to the complexity of neurons, with their elaborate axonal and dendritic processes.

Because of the length of nerve cell processes, neurons have developed mechanisms to transport along the length of axons and dendrites the proteins, organelles, and other cellular materials needed for the maintenance of the cell. These transport mechanisms are capable of moving cellular components along fiber processes in an **anterograde** direction, away from the soma, or in a **retrograde** direction, toward the soma (Fig. 3–24). **Kinesin,** a microtubule associate protein, is involved in anterograde transport of organelles and vesicles via the hydrolysis of ATP. Retrograde transport is mediated by **dynein,** another microtubule-associated protein.

In the axon, anterograde transport occurs at both slow and fast rates. The rate of **slow axoplasmic transport** is 1–2 mm/day. Structural proteins, such as actin, neurofilaments, and microtubules, are transported at this speed. The rate of **fast axoplasmic transport** is 400 mm/day. Fast transport mechanisms are used for organelles, vesicles, and membrane glycoproteins needed at the synaptic terminal. Another feature that distinguishes fast from slow transport is that fast transport requires Ca^{2+}, glucose, and ATP and depends on oxidative metabolism. In dendrites, anterograde transport occurs at a rate of 0.4 mm/day and

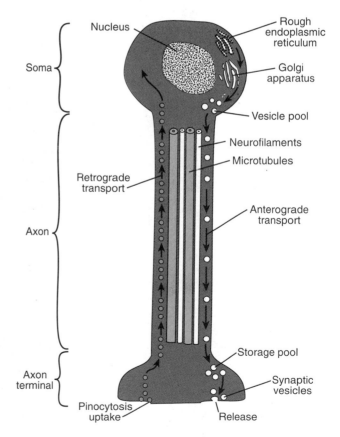

Figure 3–24 ■ Anterograde and retrograde axoplasmic transport. Anterograde transport is mediated by kinesin and retrograde transport by dynein.

requires ATP. Dendritic transport also moves ribosomes and RNA, suggesting that protein synthesis occurs within dendrites.

In retrograde axoplasmic transport, material is moved from terminal endings to the cell body. This provides a mechanism for the cell body to sample the environment around its synaptic terminals. In some neurons, maintenance of synaptic connections depends on the **transneuronal** transport of trophic (from the Greek *trophikos,* nourishing) substances, such as nerve growth factor across the synapse. After retrograde transport to the soma, nerve growth factor activates mechanisms for protein synthesis.

■ Nerve Fibers Migrate and Extend during Development and Regeneration

One of the major features that distinguishes differentiation and growth in nerve cells from these processes in other types of cells is the outgrowth of the axon from the nerve cell body in a specific direction and along a specific pathway to form synaptic connections with specific targets. Axon growth is determined largely by interactions between the growing axon and the tissue environment. At the leading edge of a growing axon is the **growth cone.** Growth cones are flat structures that give rise to protrusions called **filopodia**. Growth cones contain actin and are quite motile, with filopodia extending and retracting at a velocity of 6–10 μm/min. Newly synthesized membranes in the form of vesicles are also found in the growth cone and fuse with the growth cone as it extends. As the growth cone elongates, microtubules and neurofilaments are added to the distal end of the

fiber and partially extend into the growth cone. They are transported to the growth cone by slow axoplasmic transport.

The direction of axonal growth is dictated in part by **cell adhesion molecules** (CAMs), cell membrane glycoproteins that promote cell adhesion. Neuron-glia-CAM (N-CAM) is expressed in postmitotic neurons and is particularly prominent in growing neurites (axons and dendrites), which migrate along certain types of glial cells that provide a guiding path to target sites. The secretion of tropic (from the Greek *tropikos,* turning) factors by target cells also influences the direction of axon growth. Once the proper target site is reached and synaptic connections are formed, the processes of growth cone elongation and migration are terminated.

More axon terminals are found on target cells during development than after the nervous system has matured. Cells in the lateral geniculate nucleus of the visual system, for example, receive both ipsi- and contralateral inputs during development. In the adult, however, these cells no longer receive inputs from both eyes but from only the ipsilateral or the contralateral eye. This loss of synaptic contacts is a result of a selection process whereby the most active inputs predominate and survive and the less active synaptic contacts are lost.

Growth cones are found not only during development but also during the regeneration of nerve fibers. When a nerve fiber is cut, the distal end degenerates and the fiber segment connected to the soma develops growth cones for elongation and extension. Regeneration occurs at a rate of 1 mm/day, the rate of slow axoplasmic transport.

■ ■ ■ ■ ■ ■ ■ ■ ■ ■ ■ REVIEW EXERCISES ■ ■ ■ ■ ■ ■ ■ ■ ■ ■

Identify Each with a Term

1. The transient change in the membrane potential that is mediated by voltage-gated sodium and potassium ion channels
2. The graded change in the membrane potential that is mediated by ligand-gated ion channels and results in membrane depolarization
3. The site of cell-to-cell communication between neurons
4. The movement of cytoplasmic material from the axon terminal to the soma
5. The molecules involved in the process of synaptic transmission

Define Each Term

6. Ion channel conductance
7. Inhibitory postsynaptic potential
8. Temporal summation
9. Membrane length constant
10. Absolute refractory period
11. Ligand-gated ion channel

Choose the Correct Answer

12. What is the direction of the driving forces for the movement of Na+ ions when a nerve cell is at rest?
    a. Inward electrical gradient
    b. Outward electrical gradient
    c. Inward chemical gradient
    d. Both a and c are correct

13. What properties of the postsynaptic membrane would optimize the effectiveness of two consecutive action potentials in activating the postsynaptic neuron?
    a. Low membrane resistance
    b. High membrane resistance
    c. Low membrane capacitance
    d. Both a and c are correct

14. What properties of the postsynaptic neuron would optimize the effectiveness of two closely spaced axodendritic synapses?
    a. High membrane resistance
    b. Low membrane resistance
    c. High cytoplasmic resistance
    d. Both b and c are correct

15. Which conductance changes will produce an excitatory postsynaptic potential (EPSP)?
    a. An increase in sodium conductance
    b. A decrease in potassium conductance
    c. An increase in chloride conductance
    d. Both a and b are correct

16. Neurons that contain colocalized amine and neuropeptide transmitters show which of these characteristics?
    a. The amine, but not the peptide, will be released on nerve stimulation.
    b. The amine, but not the peptide, will bind to postsynaptic receptors.
    c. The amine, but not the peptide, will be recycled by reuptake.
    d. Vesicles containing the amine are manufactured in the nerve terminal, whereas vesicles containing the peptide are manufactured in the cell soma.

Calculate

17. a. If the extra- and intracellular concentrations of K^+ are 4 mM and 120 mM, respectively, what is the equilibrium potential for K^+? (Note that the Nernst equation at 37°C for a univalent positively charged ion is $E_{ion} = 61 \log [ion]_o/[ion]_i$.
    b. If the resting membrane potential is -60 mV, would the membrane depolarize or hyperpolarize by the opening of an ion channel that is selective for K^+ ions?

    c. If the resting membrane potential is -90 mV, what would happen to the membrane potential by the opening of K^+ channels?

18. A patient's forearm is electrically stimulated by two electrodes to activate the motor nerves that cause the thumb to move. The first electrode is placed 15 cm from the tip of the thumb and the second electrode is placed 30 cm away. The first electrode causes the thumb to move 2.5 msec after the electrical stimulus. The second electrode causes the thumb to move 4.5 msec after the stimulus. What is the conduction velocity of the motor nerves that innervate the thumb?

Briefly Answer

19. What are the conductance changes associated with (a) the rising phase of the action potential? (b) Repolarization? (c) Afterhyperpolarization?

20. What are the conductance changes associated with (a) the absolute refractory period? (b) The relative refractory period?

21. What is the role of glia in transmission at nerve terminals of (a) a catecholaminergic neuron? (b) A glutamatergic neuron? (c) A GABAergic neuron?

■ ■ ■ ■ ■ ■ ■ ■ ■ ■ ■ ■ ■ ■ **ANSWERS** ■ ■ ■ ■ ■ ■ ■ ■ ■ ■ ■ ■ ■ ■

1. Action potential
2. Excitatory postsynaptic potential
3. Synapse
4. Retrograde axoplasmic transport
5. Neurotransmitters
6. A measure of how easily ions flow through the channel. It is inversely proportional to channel resistance.
7. A hyperpolarization of the membrane mediated by ligand-gated ion channels that causes the influx of negatively charged ions or the efflux of positively charged ions
8. A postsynaptic integrative process in which membrane potential changes are added over time at a particular spatial location on a neuronal dendrite or soma.
9. A measure of the decay in amplitude of an electrotonic potential along the length of a nerve membrane. It is defined as the length at which the electrotonic potential decays to 37% of its maximum amplitude.
10. The time interval during which an axon is incapable of initiating a second action potential
11. An ion channel whose opening is regulated by binding of a ligand (e.g., hormone or transmitter) directly to the ion channel protein complex or to an associated receptor
12. d
13. b

14. a
15. d
16. c
17. a. $E_K = 61 \log ([4]/[120]) = -90$ mV
    b. Hyperpolarize
    c. No change
18. Velocity = Distance/Time = (30 cm − 15 cm)/(4.5 msec − 2.5 msec) = 75 m/sec
19. (a) Activation of voltage-gated Na^+ conductance channels (b) Inactivation of voltage-gated Na^+ conductance channels and activation of voltage-gated K^+ conductance channels (c) Maintenance of voltage-gated K^+ conductance channels in the open state
20. (a) Maintenance of the voltage-gated Na^+ channel inactivation gate in the closed state (b) Return of a critical number of voltage-gated Na^+ channels to the resting state
21. (a) Glia remove catecholamines from the synapse by high-affinity uptake and metabolize them via MAO and COMT. (b) Glia remove GLU from the synapse by high-affinity uptake and convert it to glutamine, which enters the neuron for regeneration back to GLU. (c) Glia remove GABA from the synapse by high-affinity uptake and convert it to GLU via the Krebs cycle. GLU is converted further to the storage form, glutamine, which enters the GABA neuron and is reconverted to GLU, the precursor of GABA.

Suggested Reading

Bradford HF: *Chemical Neurobiology: An Introduction to Neuro-chemistry.* New York, Freeman, 1986.

Choi DW, Rothman SM: The role of glutamate neurotoxicity in hypoxic-ischemic neuronal death. *Ann Rev Neurosci* 13: 171–182, 1990.

Eccles JC: *The Physiology of Nerve Cells.* Baltimore, Johns Hopkins Press, 1957.

Hall Z: *An Introduction to Molecular Biology.* Sunderland, MA; Sinauer, 1992.

Hodgkin AL: *The Conduction of the Nervous Impulse.* Springfield, IL, Thomas, 1964.

Horn JP: The heroic age of neurophysiology. *Hosp Prac* 27:65–74, July 15, 1992.

Kandel ER, Schwartz JH, Jessell TM: *Principles of Neural Science.* 3rd ed. New York, Elsevier, 1991.

Katz B: *The Release of Neurotransmitter Substances.* Springfield, IL, Thomas, 1969.

Meldrum B, Garthwaite J: Excitatory amino acid neurotoxicity and neurodegenerative disease. *Trends Pharm Sci* 11:379–387, 1990.

Neher E, Sakmann B: Single channel currents recorded from membrane of denervated muscle fibers. *Nature* 60:799–802, 1976.

Nicholls JG, Martin AR, Wallace BG: *From Neuron to Brain: A Cellular and Molecular Approach to the Function of the Nervous System.* 3rd ed. Sunderland, MA, Sinauer, 1992.

Shepard GM: *Neurobiology.* Oxford, Oxford University Press, 1987.

Sherrington CS: *The Integrative Action of the Nervous System.* 2nd ed. New Haven, Yale University Press, 1947.

PART TWO

Neuro-physiology

Sensory Physiology

CHAPTER OUTLINE

I. THE GENERAL PROBLEM OF SENSATION
   A. Sensory receptors translate energy from the environment into biologically useful information
      1. The nature of environmental stimuli
      2. The specificity of sensory receptors
      3. The process of sensory transduction
   B. Perception of sensory information involves encoding and decoding
      1. Encoding and transmission of sensory information
      2. Interpretation of sensory information
II. SPECIFIC SENSORY RECEPTORS
   A. Cutaneous sensation provides information via the body surface
      1. Tactile receptors
      2. Temperature sensation
      3. Pain
   B. The eye is a sensor for vision
      1. The nature of light and light optics
      2. Anatomy of the eye
      3. Optics of the eye
      4. Movements of the eye
      5. The retina and its photoreceptors
      6. Central projections of the retina
   C. The ear serves as a sensor for hearing and equilibrium
      1. The nature of sound
      2. The external ear
      3. The middle ear
      4. The inner ear
      5. Central auditory pathways
      6. Equilibrium and balance
   D. The special chemical senses detect molecules in the environment
      1. Gustatory sensation
      2. Olfactory sensation

OBJECTIVES

After studying this chapter, the student should be able to:
1. Diagram the general steps in the process of sensory transduction, starting with the external environment and ending with the central nervous system
2. Give examples of how the structure of specific sensory organs restricts their response to the preferred types of stimuli
3. Describe, in general terms, the relationship between a stimulus and the resulting generator potential and explain how changes in the generator potential affect the action potentials in the sensory nerve
4. Distinguish between rapid and slow adaptation in sensory receptors and relate the speed of adaptation to the biologic role of several specific receptors
5. List the types of tactile sensation located in the skin and name the sensory structures responsible for each type
6. Distinguish between somatic and visceral pain
7. Diagram the anatomy of the eyeball and label the principal structures involved in image formation, accommodation, and sensory transduction
8. List the cell types present in the human retina and briefly describe their function in the visual process
9. List the common and distinguishing features of rods and cones and explain how these specializations aid the visual process
10. Diagram the principal steps in the visual transduction process, from the absorption of light to the electrical response
11. Distinguish between the outer, middle, and inner ears and list the principal parts and functions of each
12. Describe the transmission of sound through the middle ear and list factors that affect its efficiency, including protective mechanisms and the effects of disease
13. Diagram the structure of the cochlea and demonstrate the path of sound through its canals
14. Describe the process by which displacements of the basilar membrane give rise to stimulation of the hair cells and explain why different sites along the basilar membrane are sensitive to different frequencies
15. Distinguish between the roles of the vestibular apparatus in sensing head position and head movement and describe the structures involved

Survival of any organism, human included, depends on having adequate information about the external environment, where food is to be found and where hazards abound. Equally important for maintaining the function of a complex organism is information about the state of numerous internal bodily processes and functions. Events in our external and internal worlds must first be translated into signals that our nervous systems can process. Despite the wide range of types of information to be sensed and acted on, a small set of common principles underlie all sensory processes.

This chapter deals with the function of the organs that permit us to gather this information, the sensory receptors. The discussion emphasizes somatic sensations, those dealing with the external aspect of the body, and does not specifically treat visceral sensations, those that come from internal organs.

The General Problem of Sensation

■ Sensory Receptors Translate Energy from the Environment into Biologically Useful Information

The essential process of sensation involves selectively sampling small amounts of energy from some restricted aspect of the environment and then using this energy to control the generation of nerve action potentials. This process is the function of **sensory receptors,** biologic structures that can be as simple as a free nerve ending or as complicated as the human eye or ear. The pattern of the sensory action potentials, together with the specific nature of the sensory receptor and its unique nerve pathways in the brain, provides an internal biologic representation of a specific component of the external world. The process of sensation is a portion of the more complex process of **perception,** in which sensory information is integrated with previously learned information and other sensory inputs to allow us to make judgments regarding the quality, intensity, and relevance of what is being sensed.

The Nature of Environmental Stimuli ■ Factors in the environment that produce an effective response in a sensory receptor are called **stimuli** and involve an exchange of energy between the environment and the receptor. Typical stimuli include **electromagnetic** quantities, such as **radiant heat** or **light; mechanical** quantities, such as **pressure, sound waves,** and other **vibrations;** and **chemical** qualities, such as **acidity** and **molecular shape** and **size.** Common to all these types of stimuli is the property of **intensity,** a measure of the energy content (or concentration, in the case of chemical stimuli) available to interact with the sensory receptor. It is not surprising, therefore, that a fundamental property of receptors is their ability to respond to different intensities of stimulation with an appro-

priate output. Also related to receptor function is the concept of **sensory modality.** This term refers to the kind of sensation, which may range from the relatively general modalities of taste, smell, touch, sight, and hearing (the traditional "five senses") to more complex sensations, such as slipperiness or wetness. Many sensory modalities are a combination of simpler sensations; the sensation of "wetness" is composed of sensations of pressure and temperature. (Try placing your hand in a plastic bag and immersing it in cold water. Although the skin will remain dry, the sensation will be one of "wetness.")

It is often difficult to communicate a precise definition of a sensory modality because of the subjective perception, or **affect,** that accompanies it. This property has to do with the psychologic feeling attached to the stimulus. Some stimuli may give rise to impressions of discomfort or pleasure that are apart from the primary sensation of, for example, "cold" or "touch." Previous experience and learning play a role in determining the affect of a sensory perception.

Some sensory receptors are classified by the nature of the signals they sense. For example, **photoreceptors** sense light and serve a visual function; **chemoreceptors** detect chemical signals and serve the senses of taste and smell, as well as detecting the presence of specific substances in the body; **mechanoreceptors** sense physical deformation, serve the senses of touch and hearing, and can detect the amount of stress in a tendon or muscle; and **thermal receptors** detect heat (or its relative lack). Other sensory receptors are classified by their "vantage point" in the body. Among these, **exteroceptors** detect stimuli from outside the body; **enteroceptors** detect internal stimuli; **proprioceptors** provide information about the positions of joints and about muscle activity and the orientation of the body in space; **nociceptors** (pain receptors) detect noxious agents.

The Specificity of Sensory Receptors ■ Most sensory receptors respond preferentially to a single kind of environmental stimulus. The usual stimulus for the eye is light; that for the ear is sound. These differences are due to a number of features that match a receptor to its preferred stimulus. In many cases there are **accessory structures,** such as the lens of the eye or the structures of the outer and middle ears, which serve to enhance the specific sensitivity of the receptor or to exclude unwanted stimuli. Often these accessory structures are under some sort of biologic **feedback** control (Fig. 4–1) so that their sensitivity may be continuously adjusted on the basis of information being received. The usual and appropriate stimulus for a receptor is called its **adequate stimulus,** and it is for this stimulus that the receptor has the lowest **threshold.** This latter term describes the lowest stimulus intensity that can be detected. It is often difficult to measure, since it can vary over time and with the presence of interfering stimuli or the action of accessory structures. Most receptors will respond to stimuli

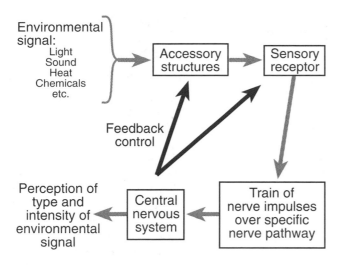

Figure 4–1 ■ The steps involved in sensory transduction. While the details vary with each type of sensory process, the overall process is similar.

other than the adequate stimulus, although the threshold for inappropriate stimuli is much higher. Gently pressing at the inner or outer corner of the eye will produce a visual sensation caused by pressure, not light; extremes of temperature may be perceived as pain. Almost all receptors can be stimulated electrically to produce sensations that mimic the one usually associated with that receptor. The central nervous pathway over which sensory information travels is also important in determining the nature of the perception; information arriving by way of the optic nerve, for example, is always perceived as light and never as sound. This is known as the concept of the **labeled line,** spoken of in older writings as the law of specific nerve energies.

The Process of Sensory Transduction ■ This section focuses on the actual function of the sensory receptor in translating environmental energy into action potentials, the fundamental unit of information in the nervous system. A device that performs such a translation is called a **transducer;** sensory receptors are biologic transducers.

A hypothetical sensory receptor is shown in Figure 4-2. This is a mechanoreceptor, in which deformation or deflection of the receptor tip gives rise to a series of action potentials in the sensory nerve fiber leading to the central nervous system (CNS). The sequence of events in the transduction process is shown in the figure. The stimulus (1) is applied at the tip of the receptor, and the deflection (2) is held constant (dotted lines). This deformation of the receptor causes a portion of its cell membrane (shaded region, 3) to become more permeable to positive ions (especially sodium). The increased permeability of the membrane leads to a localized depolarization, called the **generator potential.** At the depolarized region, sodium ions enter the cell down their electrochemical gradient; this

causes a current to flow in the extracellular medium. Because current is flowing into the cell at one place, it must flow out of the cell in another place. It does this at a region of the receptor membrane (4) called the **impulse initiation region** (or **coding region**) because here the flowing current causes the cell membrane to produce action potentials at a frequency related to the strength of the current caused by the stimulus. These currents, called **local excitatory currents,** provide the link between the formation of the generator potential and the excitation of the nerve fiber membrane. In complex sensory organs that contain a great many individual receptors, the generator potential may be called a **receptor potential,** and it may arise from several sources within the organ. Often the receptor potential is given a special name related to the function of the receptor; for example, in the ear it is called the **cochlear microphonic,** while an **electroretinogram** may be recorded from the eye. (Note that in the eye the change in receptor membrane potential associated with the stimulus of light is a **hyperpolarization,** not a depolarization.)

The production of the generator potential is of critical importance in the transduction process, because it is the step in which information related to stimulus intensity and duration is transduced. The strength (intensity) of the stimulus applied (in Fig. 4-2, the amount of deflection) determines the size of the generator potential depolarization. Varying the intensity of the stimulation will correspondingly vary the generator potential, although the changes will not usually be directly proportional to the intensity. This is called a **graded response,** in contrast to the all-or-none response of an action potential, and it causes a similar gradation of the strength of the local excitatory currents. These, in turn, determine the amount of depolarization produced in the impulse initiation region (4) of the receptor, and it is events in this region that form the next important link in the process.

Figure 4-3 shows a variety of possible events in the impulse initiation region. The threshold (dotted line) is a critical level of depolarization; membrane potential changes below this level are caused by the local excitatory currents and vary in proportion to them, while above the threshold level the membrane activity consists of locally produced **action potentials.** The lower trace shows a series of different stimuli applied to the receptor, and the upper trace shows the resulting action potentials in the impulse initiation region.

No stimulus is given at A, and the membrane voltage is at the resting potential. At B a small stimulus is applied; it produces a generator potential too small to bring the impulse initiation region membrane to threshold, and no action potential activity results. (Such a stimulus would not be sensed at all.) A brief stimulus of greater intensity is given at C; the resulting generator potential displacement is of sufficient amplitude to trigger a single action potential. As in all excitable and all-or-none nerve membranes, the action

COCHLEAR IMPLANTS

Disorders of hearing are broadly divided into the categories of **conductive hearing loss,** related to structures of the outer and middle ear; **sensorineural hearing loss,** so-called nerve deafness, dealing with the mechanisms of the cochlea and peripheral nerves; and **central hearing loss,** concerning processes that lie in higher portions of the central nervous system.

Damage to the cochlea, especially to the hair cells of the organ of Corti, produces sensorineural hearing loss by several means. Prolonged exposure to loud occupational or recreational noises can lead to hair cell damage, including mechanical disruption of the stereocilia. Such damage is localized in the outer hair cells along the basilar membrane at a position related to the pitch of the sound that produced it. Exposure to antibiotics such as streptomycin and to certain diuretics can cause rapid and irreversible damage to hair cells similar to that caused by noise, but it occurs over a broad range of frequencies. Diseases such as meningitis, especially in children, can also lead to sensorineural hearing loss.

In carefully selected patients, the use of a **cochlear implant** can restore some function to the profoundly deaf. The device consists of an external microphone, amplifier, and speech processor coupled by a plug-and-socket connection, magnetic induction, or a radio frequency link to a receiver implanted under the skin over the mastoid bone. Stimulating wires then lead to the cochlea. A single **extracochlear electrode,** applied to the round window, can restore perception of some environmental sounds and aid in lip reading but will not restore pitch or speech discrimination. A **multielectrode intracochlear implant** (with as many as 22 active elements spaced along it) can be inserted into the basal turn of the scala tympani. The linear spatial arrangement of the electrodes takes advantage of the tonotopic organization of the cochlea, and some pitch (frequency) discrimination is possible. The external processor separates the speech signal into several frequency bands that contain the most critical speech information, and the multielectrode assembly presents the separated signals to the appropriate locations along the cochlea. In some devices the signals are presented in rapid sequence, rather than simultaneously, to minimize interference between adjacent areas.

When implanted successfully, such a device can restore much of the ability to understand speech. Considerable training of the implanted patient and fine-tuning of the speech processor are necessary. The degree of restoration of function ranges from recognition of critical environmental sounds to the ability to converse over a telephone. Cochlear implants are most successful in adults who became deaf after having learned to speak and hear naturally. Success in children depends critically on their age and linguistic ability; currently, implants are being used in children as young as 2 years.

Infrequent problems with infection, device failure, and natural growth of the auditory structures may limit the usefulness of cochlear implants for some patients, and in some cases psychological and social considerations may weigh heavily in the advisability of the use of auditory prosthetic devices in general. From a technical standpoint, however, continual refinements in the design of implantable devices and the processing circuitry are extending the range of subjects who may benefit from cochlear implants, and research directed at external stimulation of higher auditory structures may eventually lead to even more effective treatments for profound hearing loss.

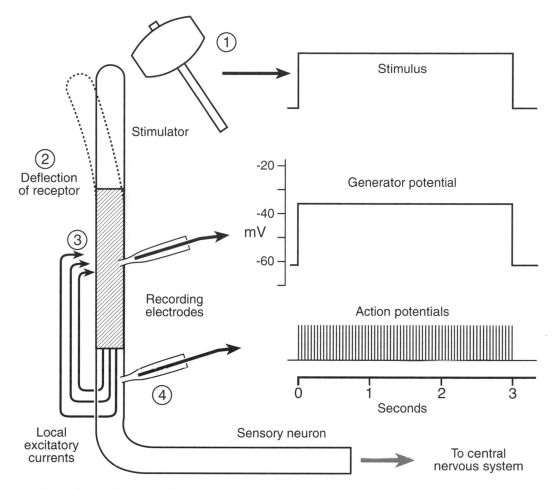

Figure 4–2 ■ The relation between an applied stimulus and the production of sensory nerve action potentials. See text for details.

potential is immediately followed by repolarization, often to a level that transiently hyperpolarizes the membrane potential because of temporarily high potassium conductance. Since the brief stimulus has been removed by this time, no further action potentials are produced. A longer stimulus of the same intensity (D) produces repetitive action potentials, because as the membrane repolarizes from the action potential, local excitatory currents are still flowing. They bring the repolarized membrane to threshold at a rate proportional to their strength. During this time interval the fast sodium channels of the membrane are being reset, and another action potential is triggered as soon as the membrane potential reaches threshold. As long as the stimulus is maintained this process will repeat itself at a rate determined by the stimulus intensity. If the intensity of the stimulus is increased (E), the local excitatory currents will be stronger and threshold will be reached more rapidly. This will result in a reduction of the time between each action potential and, as a consequence, a higher action potential frequency. This

change in action potential frequency is critical in communicating the intensity of the stimulus to the CNS.

The discussion thus far has depicted the generator potential as though it does not change while a constant stimulus is applied. Although this is approximately correct for a few receptors, most will show some degree of **adaptation.** In an adapting receptor the generator potential, and consequently the action potential frequency, will decline even though the stimulus is maintained. Example A in Figure 4–4 shows the output from a receptor in which there is no adaptation. As long as the stimulus is maintained, there is a steady rate of action potential firing. Example B shows **slow adaptation;** as the generator potential declines the interval between the action potentials increases correspondingly. C demonstrates **rapid adaptation;** the action potential frequency falls rapidly and then maintains a constant slow rate that does not show further adaptation. Responses in which there is little or no adaptation are called **tonic,** whereas those in which significant adaptation occurs are called **phasic.** In

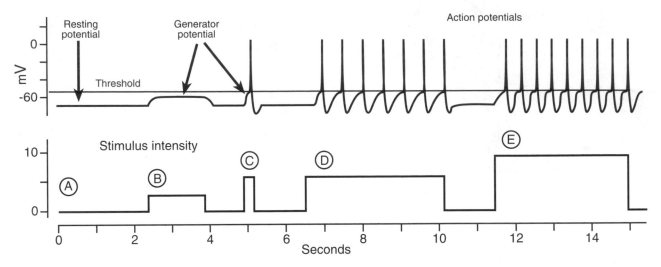

Figure 4–3 ■ Sensory nerve activity with different stimulus intensities and durations. With no stimulus (A), the membrane is at rest. A subthreshold stimulus (B) produces a generator potential too small to cause membrane excitation. A brief but intense stimulus (C) can cause a single action potential. Maintaining this stimulus (D) leads to a train of action potentials. Increasing the stimulus intensity (E) leads to an increase in the action potential firing rate.

some cases tonic receptors may be called **intensity receptors,** and phasic receptors, **velocity receptors.** Many receptors, **muscle spindles** for example, show a combination of responses; on application of a stimulus a rapidly adapting phasic response is followed by a steady tonic response. Both of these responses may be graded by the intensity of the stimulus. As a receptor adapts, the sensory input to the CNS is reduced and the sensation is perceived as less intense.

The phenomenon of adaptation is important in preventing "sensory overload," and it allows less important or unchanging environmental stimuli to be partially ignored. When a change occurs, however, the phasic response will occur again, and the sensory input will become temporarily more noticeable. Rapidly adapting receptors are also important in sensory systems that must sense the **rate of change** of a stimulus, especially when its intensity can vary over a range that would overload a tonic receptor.

Receptor adaptation can occur at several places in the transduction process. In some cases the receptor's sensitivity is changed by the action of accessory structures, as in the constriction of the pupil of the eye in the presence of bright light. This is an example of feedback-controlled adaptation; in the sensory cells of the eye, light-controlled changes in the amounts of the visual pigments also can change the basic sensitivity of the receptors and produce adaptation. As mentioned above, adaptation of the generator potential can produce adaptation of the overall sensory response. Fi-

nally, the phenomenon of **accommodation** in the impulse initiation region of the sensory nerve fiber can slow the rate of action potential production even though the generator potential may show no change. Accommodation refers to a gradual increase in threshold caused by prolonged nerve depolarization and is due to inactivation of sodium channels.

■ Perception of Sensory Information Involves Encoding and Decoding

Encoding and Transmission of Sensory Information ■ Stimuli from the environment that have been detected and partially processed by a sensory receptor, using the processes discussed above, must be conveyed to the CNS. This must be done in such a way that the complete range of the intensity of the stimulus is preserved, because this is the essence of the useful information provided by the sensory transduction process. The first step in this process is **compression.** There is an extremely wide range of intensities possible for most possible environmental stimuli; even when the receptor sensitivity is modified by accessory structures and adaptation, the range is quite large. Figure 4–5 illustrates this problem; at the left is a hundredfold range in the intensity of a stimulus. At the right is a different scale of intensities that results from the processes in the sensory receptor. In most receptors the magnitude of the generator potential is not exactly proportional to the stimulus intensity, but

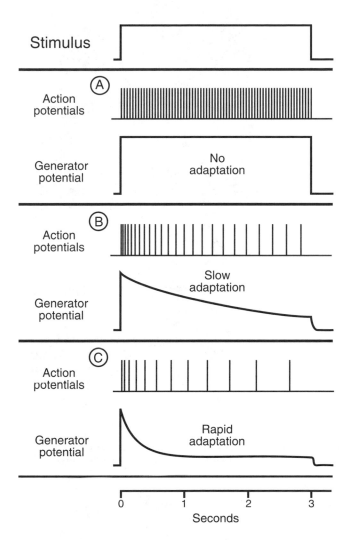

Figure 4–4 ■ Adaptation in a sensory receptor is often related to a decline in the generator potential with time. (A) The generator potential is maintained without decline, and the action potential frequency remains constant. (B) A slow decline in the generator potential is associated with slow adaptation. (C) In a rapidly adapting receptor the generator potential declines rapidly.

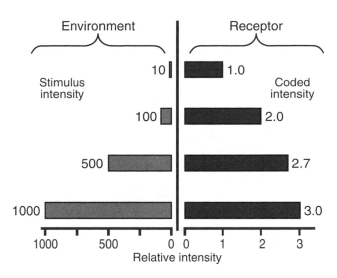

Figure 4–5 ■ Compression in a sensory process. By a variety of means, a wide range of input intensities is coded into a much narrower range of biologic responses that can be represented by variations in action potential frequency.

increases less and less as the stimulus intensity increases. The frequency of the action potentials produced in the impulse initiation region is also not proportional to the strength of the local excitatory currents; there is an upper limit to the number of action potentials per second because of the refractory period of the nerve membrane. These factors are responsible for the process of compression; changes in the intensity of a small stimulus cause a greater change in action potential frequency than the same change would cause if the stimulus intensity were high. As a result, the hundredfold variation in the stimulus is compressed into a threefold range after the receptor has processed the stimulus. Some information is neces-

sarily lost in this process, but integrative processes in the CNS can restore the information or compensate for its absence. Physiologic evidence for compression is based on the observed nonlinear (logarithmic or power function) relation between the actual intensity of a stimulus and its perceived intensity.

What remains is to transfer the sensory information from the receptor to the CNS. The encoding processes in the receptors have already provided the basis for this information transfer by producing a series of action potentials related to the stimulus intensity. A special process is necessary for the transfer because of the nature of the conduction of action potentials. As an action potential travels along a nerve fiber, it is sequentially recreated and renewed at a series of locations along the nerve. This means the duration and amplitude of the action potential depend not on that of the previous action potentials but on factors intrinsic to a given neuron. The only information that can be conveyed by a single action potential is its presence or absence. However, relationships between and among action potentials can convey large amounts of information, and this is the system found in the biologic transmission process. This system can be explained by analogy with a physical system such as that used for transmission of signals in communications systems. Figure 4-6 outlines a hypothetical **frequency-modulated** (FM) encoding, transmission, and reception system. An **input signal** provided by some physical quantity (1) is continuously measured and converted into an **electrical signal** (2), analogous to the generator potential, whose amplitude is proportional to the input signal. This signal then controls the

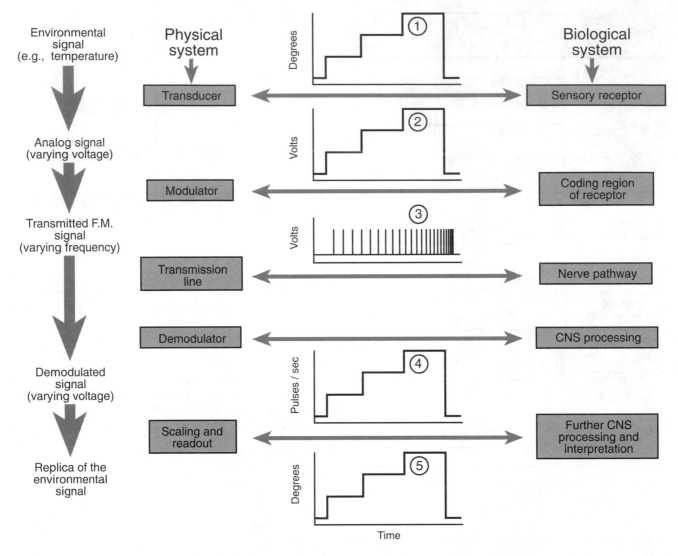

Figure 4–6 ■ Transmission of sensory information. Because signals of varying amplitude cannot be transmitted along a nerve fiber, specific intensity information is transformed into a corresponding action potential frequency, and CNS processes decode the nerve activity into biologically useful information. The steps in the process are shown at the left, with the parts of a physical system that perform them. At the right are the analogous biologic steps involved in the same process.

frequency of a pulse generator (3), as in the impulse initiation region of a sensory nerve fiber. Like action potentials, these pulses are of a constant height and duration, and the amplitude information of the original input signal is now contained in the intervals between the pulses. The resulting signals may be sent along a **transmission line** (analogous to a nerve fiber) to some distant point, where they produce an electrical voltage (4) proportional to the frequency of the arriving pulses. This voltage is a replica of the input voltage (2) and is not affected by changes in the amplitude of the pulses as they travel along the transmission line. Further scaling and processing can produce a graphic

or numeric presentation (5) of the input data. In a biologic system these latter functions are accompanied by processing and interpretation in the CNS.

Interpretation of Sensory Information ■ The interpretation of encoded and transmitted information into a useful perception requires consideration of a number of other factors. For instance, the interpretation of sensory input by the CNS depends on the neural pathway it takes to the brain. All information arriving on the **optic nerves** is interpreted as light, even though the signal may have arisen as a result of pressure applied to the eyeball. The **localization** of a cutaneous sensation to a particular part of the body also depends on the particular pathway it takes to the CNS. Often a sensation (usually pain) arising in a visceral structure (e.g., heart, gallbladder) is perceived as coming from a portion of the body surface, because developmentally related nerve fibers come from these anatomically different regions. Such a sensation is called **referred pain.**

Specific Sensory Receptors

The remainder of this chapter surveys specific sensory receptors, concentrating on the **special senses.** These traditionally include cutaneous sensation, (touch, temperature, etc.), sight, hearing, taste, and smell.

■ Cutaneous Sensation Provides Information via the Body Surface

The skin is richly supplied with sensory receptors serving the modalities of touch (light and deep pressure), temperature (warm and cold), and pain as well as the more complicated composite modalities of itch, tickle, wet, and so on. By using special probes that deliver highly localized stimuli of pressure, vibration, heat, or cold, the distribution of cutaneous receptors over the skin can be mapped. In general, areas of skin used in tasks requiring a high degree of spatial localization (e.g., fingertips, lips) have a high density of specific receptors, and these areas are correspondingly well represented in the somatosensory areas of the cerebral cortex (Chap. 7).

Tactile Receptors ■ A number of receptor types serve the sensations of touch in the skin. In regions of **hairless skin** (e.g., the palm of the hand) are found **Merkel's disks, Meissner's corpuscles,** and **pacinian corpuscles** (Fig. 4-7). The first of these are **intensity** receptors (located in the lowest layers of the epidermis) that show slow adaptation and respond to steady pressure. Meissner's corpuscles adapt more rapidly to the same stimuli and serve as **velocity** receptors. The pacinian corpuscles are very rapidly adapting **(acceleration)** receptors that are most sensitive to rapidly changing stimuli such as vibration. In regions of hairy skin, small hairs serve as accessory structures for **hair follicle receptors,** mechanoreceptors that adapt more slowly. **Ruffini's endings** (located in the dermis) are also slowly adapting receptors. Merkel's disks in areas of hairy skin are grouped into **tactile disks.** Pacinian corpuscles also serve to sense vibration in hairy skin. Nonmyelinated nerve endings, also usually found in hairy skin, appear to have a limited tactile function and may serve to sense pain.

Temperature Sensation ■ From a physical standpoint, the notions of heat and cold represent values along a temperature continuum and do not differ fundamentally except in the amount of molecular motion present. However, the familiar subjective differentia-

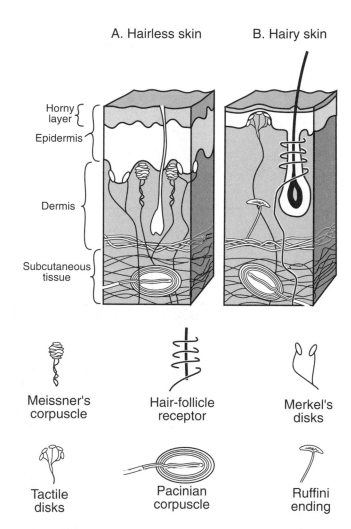

Figure 4-7 ■ Tactile receptors in the skin. See text for details. (Modified from Schmidt RF (ed): *Fundamentals of Sensory Physiology.* 2nd ed. New York, Springer-Verlag, 1981.)

tion of the temperature sense into "warm" and "cold" reflects the underlying physiology of the two populations of receptors responsible for thermal sensation.

Thermal receptors appear to be naked nerve endings supplied by either thin myelinated fibers (cold receptors) or nonmyelinated fibers (warm receptors) with low conduction velocity. Cold receptors form a population with a broad response peak at about 30°C; the warm receptor population has its peak at about 43°C (Fig. 4-8). Both sets of receptors share some common features: they are sensitive only to thermal stimulation; they have both a phasic response component that is rapidly adapting and responds only to changes in temperature (in a fashion roughly proportional to the rate of change) and a tonic (intensity) response that depends on the local temperature. The density of thermal receptors differs at different places on the body surface. They are present in much

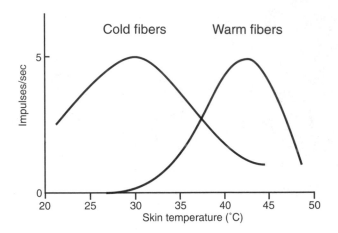

Figure 4–8 ■ Separate temperature responses of cold and warm receptors in the skin. The skin temperature was held at different values while nerve impulses were recorded from representative fibers leading from each receptor type. (Modified from Kenshalo, in Zotterman Y: *Sensory Functions of Skin in Primates.* Oxford, Pergamon, 1976.)

lower numbers than cutaneous mechanoreceptors, and there are many more cold receptors than warm receptors.

The perception of temperature stimuli is closely related to the properties of the receptors. The phasic component of the response is apparent in our adaptation to sudden immersion in, for example, a warm bath. The sensation of warmth, very apparent at first, soon fades away, and a less intense impression of the steady temperature may remain. Moving to somewhat cooler water produces an immediate sensation of cold that soon fades away. Over an intermediate temperature range (the so-called comfort zone) there is no appreciable temperature sensation. This range is approximately 30°–36°C for a small area of skin; the range is narrower when the whole body is exposed. Outside this range there is a steady temperature sensation that depends on the ambient (skin) temperature. At skin temperatures lower than 17°C **cold pain** is noted. At very high skin temperatures (above 45°C) there is a sensation of **paradoxical cold,** caused by activation of a part of the cold receptor population.

Temperature perception is subject to considerable processing by higher centers. While the perceived sensations reflect the activity of specific receptors, the phasic component of temperature perception may take many minutes to be completed, whereas the adaptation of the receptors is complete within seconds.

Pain ■ The familiar sensation of pain is not limited to cutaneous sensation; pain coming from stimulation of the body surface is called **superficial pain,** while that arising from within muscles, joints, bones, and connective tissue is called **deep pain.** These two categories comprise **somatic pain. Visceral pain** arises from internal organs and is often associated with strong contractions of visceral muscle or its forcible deformation.

Pain is sensed by a population of specific receptors called **nociceptors.** In the skin these are the free endings of thin myelinated and nonmyelinated fibers with characteristically low conduction velocities. They typically have a high threshold to mechanical, chemical, or thermal stimuli (or to these combined) of an intensity sufficient to cause tissue destruction. The skin has many more points at which pain can be elicited than it has mechanically or thermally sensitive sites. Because of the high threshold of pain receptors (compared with that of other cutaneous receptors), we are usually unaware of their existence.

There are often two components to superficial pain—an immediate, sharp, and highly localizable **initial pain** and, after a latency of about 1 second, a longer-lasting and more diffuse **delayed pain.** These two submodalities appear to be mediated by different nerve fiber endings. In addition to their normally high thresholds, both cutaneous and deep pain receptors show little adaptation, a fact that is unpleasant but biologically necessary. Deep and visceral pain appear to be sensed by similar nerve endings, which may also be stimulated by local metabolic conditions such as ischemia (lack of adequate blood flow, as may occur during the heart pain of angina pectoris).

The free nerve endings mediating pain sensation are anatomically distinct from other free nerve endings involved in the normal sensation of mechanical and thermal stimuli. The functional differences are not microscopically evident and are likely to relate to specific elements in the molecular structure of the receptor cell membrane.

■ The Eye Is a Sensor for Vision

The Nature of Light and Light Optics ■ The adequate stimulus for human visual receptors is **light,** which may be defined as electromagnetic radiation between the wavelengths of 770 nm (red) and 380 nm (violet). The familiar colors of the spectrum all lie between these limits. A wide range of intensities, from a single photon to the direct light of the sun, exists in nature.

As with all such radiation, light rays travel in a straight line in a given medium. Light rays are bent **(refracted)** as they pass between media (e.g., glass, air) that have different **refractive indices.** For a given interface, the amount of bending is determined by the angle at which the ray strikes the surface; if the angle is at 90 degrees, there is no bending, while successively more oblique rays are bent more sharply. A simple prism (Fig. 4–9A) can therefore cause a light ray to deviate from its path and travel in a new direction. An appropriately chosen pair of prisms can turn parallel rays to a common point (Fig. 4–9B). A **convex** lens may be thought of as a series of such prisms with increasingly more bending power (Fig. 4–9C, D), and

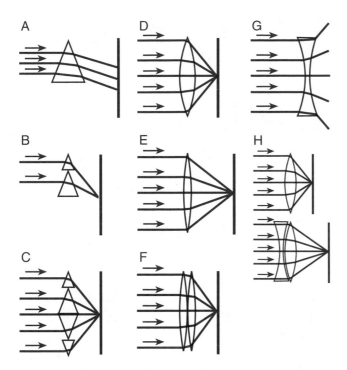

Figure 4–9 ■ How lenses control the refraction of light. A prism (A) bends the path of parallel rays of light. The amount of bending varies with the prism shape (B). A series of prisms (C) can bring parallel rays to a point. The limiting case of this arrangement is a convex (converging) lens (D). Such a lens with less curvature (E) has a longer focal length. Placing two such lenses together (F) produces a shorter focal length. A concave (negative) lens (G) causes rays to diverge. A negative lens can effectively increase the focal length of a positive lens (H).

Anatomy of the Eye ■ The human eyeball is a roughly spherical organ consisting of a number of layers and structures (Fig. 4-10). The outermost of these consists of a tough white connective tissue layer, the **sclera,** and a transparent layer, the **cornea.** Six **extraocular muscles** that control the direction of the eyeball insert on the sclera. The next layer is the **vascular coat;** its rear portion, the **choroid,** is pigmented and highly vascular, supplying blood to the outer portions of the retina (see below). The front portion contains the **iris,** a circular smooth muscle structure that forms the **pupil,** the neurally controlled aperture through which light is admitted to the interior of the eye. The iris also gives the eye its characteristic color. The transparent **lens** is located just behind the iris and is held in place by a radial arrangement of **suspensory ligaments** (or **zonule fibers**) that attach it to the **ciliary body,** a complex arrangement of smooth muscle fibers that regulates the curvature of the lens and hence its focal length. The lens is composed of many thin and interlocking layers of fibrous protein and is highly elastic. Between the cornea and the iris/lens is the **anterior chamber,** a space filled with thin clear liquid (the **aqueous humor**) similar in composition to cerebrospinal fluid. This liquid is continuously secreted by the epithelium of the **ciliary process,** located behind the iris. As the fluid accumulates, it is drained through the **canal of Schlemm** into the venous circulation. (Drainage of aqueous humor is critical. If too much pressure builds up in the anterior

such a lens (called a **converging,** or **positive,** lens) will bring an infinite number of parallel rays to a common point, called the **focal point.** A converging lens can form a **real image.** The distance from the lens to this point is its **focal length** (FL), which may be expressed in meters. A convex lens with less curvature has a longer focal length (Fig. 4-9E). Often the **diopter** (D) which is the inverse of the focal length (1/FL), is used to describe the power of a lens. For example, a lens with a focal length of 0.5 m has a power of 2 D. An advantage of this system is that dioptric powers are additive; two convex lenses of 25 D each will function as a single lens with a power of 50 D when placed next to each other (Fig. 4-9F).

A **concave** lens causes parallel rays to **diverge** (Fig. 4-9G). Its focal length (and its power in diopters) is **negative,** and it cannot form a real image. However, a convex (negative) lens placed before a positive lens can modify (i.e., lengthen) its focal length (Fig. 4-9H), which will be the algebraic sum of the two dioptric powers. This property allows the use of external lenses (eyeglasses or contact lenses) to correct optical defects in the eye.

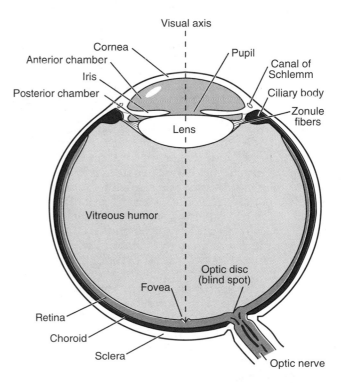

Figure 4–10 ■ The major parts of the human eye.

chamber, the internal structures are compressed and **glaucoma,** a sight-threatening condition, results.) The **vitreous humor** (or **vitreous body**), a clear gelatinous substance, fills the large cavity between the rear of the lens and the front surface of the retina. This substance is exchanged much more slowly than the aqueous humor.

The innermost layer of the eyeball is the **retina,** where the optical image is formed. This tissue contains the photoreceptor cells, called **rods** and **cones,** and a complex multilayered network of nerve fibers and cells that function in the early stages of image processing. The rear of the retina is supplied with blood from the choroid, while the front is supplied by the central artery and vein that enter the eyeball with the **optic nerve,** the fiber bundle that connects the retina with structures in the brain. The vascular supply to the front of the retina, which ramifies and spreads over the retinal surface, is visible through the lens and affords a direct view of the microcirculation; this window is useful for diagnostic purposes, even for conditions not directly related to ocular function. At the optical center of the retina, where the image falls when one is looking straight ahead, is the **macula lutea,** an area of about 1 mm² that is specialized for very sharp color vision. At the center of this is the **fovea centralis,** a depressed region about 0.4 mm in diameter, the **fixation point** of direct vision. Slightly off to the nasal side of the retina is the **optic disc,** where the optic nerve leaves the retina. There are no photoreceptor cells here, resulting in a **blind spot** in the field of vision, but because the two eyes are mirror images of each other, information from the overlapping visual field of one eye "fills in" the missing part of the image from the other eye.

Optics of the Eye ■ The image that falls on the retina is real and inverted, as in a camera. Neural processing restores the upright appearance of the field of view. The image itself can be modified by optical adjustments by the lens and the iris. Most of the refractive power (about 43 D) is provided by the curvature of the cornea, with the lens providing an additional 13–26 D, depending on the focal distance. The muscle of the ciliary body has a parasympathetic innervation. When it is fully relaxed, the lens is at its flattest and the eye is focused at infinity (actually, at anything more than 6 m away) (Fig. 4–11A). When the ciliary muscle is fully contracted, the lens is at its most curved and the eye is focused at its nearest point of distinct vision (Fig. 4–11B). This adjustment of the eye for close vision is called **accommodation.** The **near point** of vision for the eye of a young adult is about 10 cm. With age, the lens loses its elasticity and the near point of vision moves farther away, becoming approximately 80 cm at 60 years. This condition is called **presbyopia;** supplemental refractive power, in the form of external lenses (reading glasses), is required for distinct near vision.

Errors of refraction are very common and can be

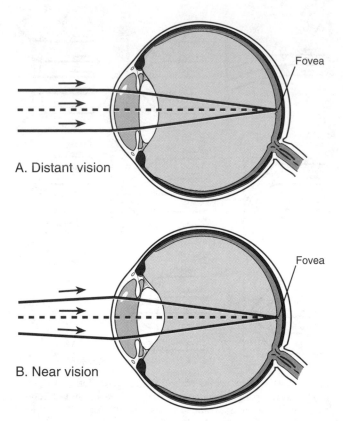

Figure 4–11 ■ The eye as an optical device. During fixation the center of the image falls on the fovea. (A) With the lens flattened, parallel rays from a distant object are brought to a sharp focus. (B) Lens curvature increases with accommodation, and rays from a nearby object are focused.

corrected with external lenses (eyeglasses or contact lenses). Farsightedness **(hyperopia)** is caused by an eyeball that is physically too short to focus on distant objects. The natural accommodation mechanism may compensate for distance vision (Fig. 4–12), but the near point will still be too far away; the use of a positive (converging) lens corrects this error. If the eyeball is too long, nearsightedness **(myopia)** results. In effect, the converging power of the eye is too great; close vision is correct, but the eye cannot focus on distant objects. A negative (diverging) lens corrects this defect. If the curvature of the cornea is not symmetric, **astigmatism** results. Objects with different orientations in the field of view will have different focal positions. Vertical lines may appear sharp and horizontal structures are blurred. This condition is corrected with the use of a **cylindric lens,** which has different radii of curvature at the proper orientations along its surfaces. Normal vision (i.e., the absence of any refractive errors) is termed **emmetropia** (literally, "eye in proper measure").

Normally the lens is completely transparent to visible light. Especially in older persons there may be a progressive increase in its opacity, to the extent that vision is obscured. This condition, called a **cataract,**

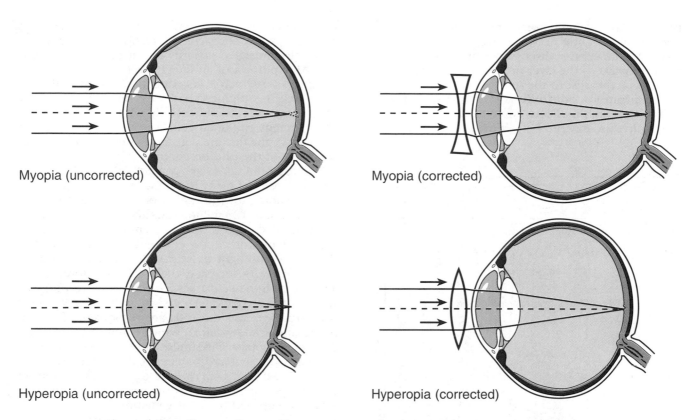

Myopia (uncorrected)

Myopia (corrected)

Hyperopia (uncorrected)

Hyperopia (corrected)

Figure 4–12 ■ The use of external lenses to correct refractive errors.

is treated by surgical removal of the defective lens. An artificial lens may be implanted in its place, or eyeglasses may be used to replace the refractive power of the lens.

The iris, which has both sympathetic and parasympathetic innervation, controls the diameter of the pupil. It is capable of a thirtyfold change in area and in the amount of light admitted to the eye. This change is under complex reflex control, and bright light entering just one eye will cause the appropriate constriction response in both eyes. As with a camera, when the pupil is constricted, less light enters, but the image is focused more sharply because the more poorly focusing peripheral rays are cut off.

Movements of the Eye ■ The extraocular muscles move the eyes. These six muscles, which originate on the bone of the **orbit** (the eye socket) and insert on the sclera, are arranged in three sets of antagonistic pairs. They are under visually compensated feedback control and produce a number of types of movement. Continuous activation of a small number of motor units produces a very small **tremor** at a rate of 30–80 cycles per second. This movement and a **slow drift** cause the image to be in constant motion on the retina, a necessary condition for proper visual function. Larger movements include rapid flicks, called **saccades,** which suddenly change the orientation of the eyeball, and large, slow movements, used in following moving objects. Organized movements of the eyes,

which occur simultaneously but not identically in both eyes, include **fixation,** the training of the eyes on a stationary object; **tracking movements,** used to follow the course of a moving target; **convergence** adjustments, in which both eyes turn inward to fix on near objects; and **nystagmus,** a series of slow and saccadic movements (part of a vestibular reflex; see below) that serve to keep the retinal image steady during rotation of the head.

Because the eyes are separated by some distance, each receives a slightly different image of the same object. This property, **binocular vision,** along with information about the different positions of the two eyes, allows **stereoscopic** vision and its associated **depth perception,** abilities that are largely lost in the case of blindness in one eye. Many abnormalities of eye movement are types of **strabismus** ("squinting"), in which the two eyes do not work together properly. Other defects include **diplopia** (double vision), when the convergence mechanisms are impaired, and **amblyopia,** when one eye assumes improper dominance over the other. Failure to correct this latter condition can lead to loss of visual function in the subordinate eye.

The Retina and Its Photoreceptors ■ The retina is a multilayered structure containing the photoreceptor cells and a complex web of several types of nerve cells (Fig. 4–13). Technically, there are 10 layers of cells in the retina, but this discussion employs a simpler four-layer scheme: pigment epithelium, photoreceptor layer, neural network layer, and ganglion cell layer. These layers are discussed in order of the deepest layer outward to the layer on top of the inner surface of the eye (nearest to the lens).

The **pigment epithelium** (Fig. 4–13) consists of cells with a high **melanin** content. This opaque material, which also extends between portions of individual

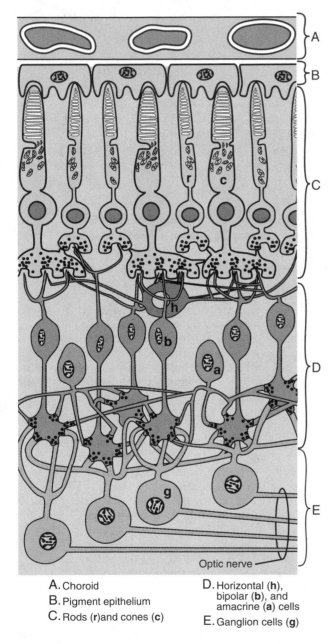

A. Choroid

B. Pigment epithelium

C. Rods (**r**)and cones (**c**)

D. Horizontal (**h**), bipolar (**b**), and amacrine (**a**) cells

E. Ganglion cells (**g**)

Figure 4–13 ■ Organization of the human retina. The various layers are described in the text. c, cone cell; r, rod cell; h, horizontal cell; b, bipolar cell; a, amacrine cell; g, ganglion cell. (Modified from Dowling JE, Boycott BB: Organization of the primate retina: Electron microscopy. *Proc Roy Soc Lond* 166:80–111, 1966.

rods and cones, prevents the scattering of stray light and thus greatly sharpens the resolving power of the retina. Its presence ensures that a tiny spot of light (or a tiny portion of an image) will excite only those receptors on which it falls directly. Persons with albinism lack this pigment and have blurred vision that cannot be corrected effectively with external lenses. The pigment epithelial cells also phagocytize bits of cell membrane that are constantly shed from the outer segments of the photoreceptors.

The rods and cones are packed tightly side-by-side, with a density of many thousands per square millimeter, depending on the region of the retina. Each eye contains about 125 million rods and 5.5 million cones. Because of the eye's mode of embryologic development, the photoreceptor cells occupy a deep layer of the retina, and light must pass through a number of overlying layers to reach them. The photoreceptors are divided into two classes. The cones are responsible for **photopic** (daytime) **vision,** which is in color **(chromatic),** and the rods are responsible for **scotopic** (nighttime) **vision,** which is not in color. Their functions are basically similar, although they have important structural and biochemical differences.

A **rod cell** (Fig. 4–14) is long, slender, and cylindric; its outer segment contains numerous disc-shaped lamellae composed of cellular membrane in which the molecules of the photopigment **rhodopsin** are embedded. The lamellae near the tip are regularly shed and replaced with new membrane synthesized at the opposite end of the outer segment. The **inner segment,** connected to the outer segment by a modified **cilium,** contains the cell nucleus, many mitochondria that provide energy for the phototransduction process, and other cell organelles. At the base of the cell is a **synaptic body** that makes contact with one or more bipolar nerve cells and liberates a transmitter substance in response to changing light levels.

Cones (Fig. 4–14) differ from rods in that they are smaller and have an outer segment that tapers to a point. There are three different photopigments associated with cone cells, of which a given cone will contain two. The pigments differ in the wavelength of light that optimally excites them. The peak spectral sensitivity for the **red-sensitive pigment** is 560 nm; for **green-sensitive pigment** it is about 530 nm; and for **blue-sensitive pigment** it is about 420 nm. At wavelengths away from the optimal the pigments still absorb light but with reduced sensitivity. Because of the interplay between light intensity and wavelength, cones with a single pigment would not be able to detect colors unambiguously. The presence of two of the three pigments in each cone removes this uncertainty. Colorblind individuals, who have a genetic lack of one or more of the pigments or of an associated transduction mechanism, cannot distinguish between the affected colors. Loss of a single color system produces **dichromatic** vision and lack of two of the sys-

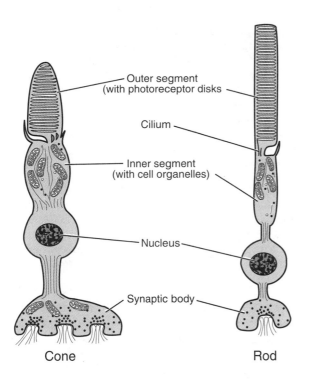

Figure 4–14 ■ Photoreceptors of the human retina. Rod and cone receptors are compared. (Modified from Davson H (ed): *The Eye: Visual Function in Man.* 2nd ed. New York, Academic, 1976.)

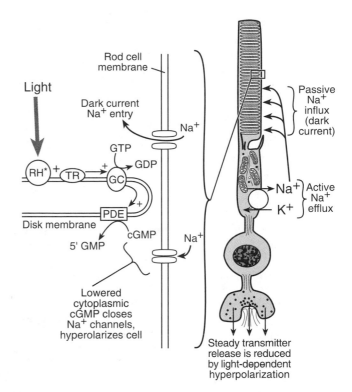

Figure 4–15 ■ The biochemical steps in visual transduction. (Right) An active Na^+/K^+ pump maintains the ionic balance of a rod cell, while Na^+ enters passively through channels in the cell membrane, causing a maintained depolarization and a dark current under conditions of no light. (Left) The amplifying cascade of reactions (which take place in the disc membrane of a photoreceptor) allows a single activated rhodopsin molecule to control the hydrolysis of 500,000 cGMP molecules. See text for details of the reaction sequence. In the presence of light, the reactions lead to the depletion of cGMP, resulting in closing of cell membrane Na^+ channels and production of a hyperpolarizing generator potential. Release of transmitter substance decreases during stimulation by light. RH*, activated rhodopsin; TR, transducin; GC; guanylate cyclase; PDE, phosphodiesterase.

tems causes **monochromatic** vision; if all three are lacking, vision is monochromatic but depends only on the rods.

The visual pigments of the photoreceptor cells convert light to a nerve signal. This process is best understood as it occurs in rod cells. In the dark, the pigment **rhodopsin** (or **visual purple**) consists of a light-trapping **chromophore** called **scotopsin** (cone cells have other **opsins**) that is chemically conjugated with **11-*cis*-retinal,** the aldehyde form of vitamin A_1. When struck by light, rhodopsin undergoes a series of rapid chemical transitions, with the final intermediate form **metarhodopsin II** providing the critical link between this reaction series and the electrical response. The end products of the light-induced transformation are the original scotopsin and an all-*trans* form of retinal, now dissociated from each other. Under conditions of both light and dark, the all-*trans* form of retinal is isomerized back to the 11-*cis* form and the rhodopsin is reconstituted. All of these reactions take place in the highly folded membranes comprising the outer segment of the rod cell.

The coupling of the light-induced reactions and the electrical response involves the activation of **transducin,** a specific G protein; the associated exchange of GTP for GDP activates a **phosphodiesterase** (Fig. 4–15). This, in turn, catalyzes the breakdown of cyclic GMP (cGMP) to 5′-GMP. When cellular cGMP levels are high (as in the dark), membrane sodium channels

are kept open and the cell is relatively depolarized. Under these conditions there is a tonic release of transmitter substance from the synaptic body of the rod cell. Reduction in the level of cGMP due to light-induced reactions causes the cell to close its sodium channels and hyperpolarize, thus reducing the release of neurotransmitter; this change is the signal that is further processed by the nerve cells of the retina to form the final response in the optic nerve. An active sodium pump maintains the cellular concentration at proper levels. A very large amplification of the light response takes place during the coupling steps; one activated rhodopsin molecule will activate approximately 500 transducins, each of which activates the hydrolysis of several thousand cGMP molecules. Under proper conditions a rod cell can respond to a single photon striking the outer segment. The processes in cone cells are similar, although there are three different opsins (with different spectral sensitivities) and the specific transducing is also different. The overall sensitivity of the transduction process is also lower.

In the light, much rhodopsin is in its unconjugated form and the sensitivity of the rod cell is relatively low. During the process of **dark adaptation,** which takes about 40 minutes to complete, the stores of rhodopsin are gradually built up, with a consequent increase in sensitivity (by as much as 25,000 times). Cone cells adapt more quickly than rods, but their final sensitivity is much less. The reverse process, **light adaptation,** takes about 5 minutes.

Bipolar cells, horizontal cells, and **amacrine cells** together comprise the neural network layer. These cells are together responsible for considerable initial processing of visual information. Because the distances between neurons here are so small, most cellular communication involves the **electrotonic spread** of cell potentials, rather than propagated action potentials. Light stimulation of the photoreceptors produces hyperpolarization that is transmitted to the bipolar cells. Some of these cells respond with a depolarization that is excitatory to the ganglion cells, whereas other cells respond with a hyperpolarization that is inhibitory. The horizontal cells also receive input from rod and cone cells but spread information laterally, causing **inhibition** of the bipolar cells on which they synapse. Another important aspect of retinal processing is **lateral inhibition.** A strongly stimulated receptor cell can, via lateral inhibitory pathways, inhibit the response of neighboring cells that are less well illuminated. This has the effect of increasing the apparent contrast at the edge of an image. Amacrine cells also send information laterally but synapse on ganglion cells.

The results of retinal processing are finally integrated by the **ganglion cells,** whose axons form the **optic nerve.** These cells are tonically active, sending action potentials into the optic nerve at an average rate of five per second even when unstimulated. Input from other cells converging on the ganglion cells modifies this rate up or down.

Many kinds of information regarding color, brightness, contrast, and so on are passed along the optic nerve. The output of individual photoreceptor cells is **convergent** on the ganglion cells. In keeping with their role in visual acuity, relatively few cone cells converge on a ganglion cell, especially in the fovea, where the ratio is nearly 1:1. Rod cells, however, are highly convergent, with as many as 300 rods converging on a single ganglion cell. While this reduces the sharpness of an image, it allows for a great increase in light sensitivity.

Central Projections of the Retina ■ The optic nerves, each carrying about 1 million fibers from each retina, enter the rear of the orbit and pass to the underside of the brain to the **optic chiasma,** where about half of the fibers from each eye "cross over" to the other side. The divided output goes through the **optic tract** to the paired **lateral geniculate bodies**

(part of the thalamus) and thence via the **geniculocalcarine tract** (or **optic radiation**) to the **visual cortex** in the **occipital lobe** of the brain (Fig. 4–16). Specific portions of each retina are mapped to specific areas of the cortex, with the foveal and macular regions having the greatest representation and the peripheral areas the least. Mechanisms in the visual cortex detect and integrate visual information such as shape, contrast, line, and intensity into a coherent visual perception.

Information from the optic nerves is also sent to the **suprachiasmatic nucleus** of the hypothalamus, where it participates in the regulation of circadian rhythms; the **pretectal nuclei,** concerned with control of visual fixation and pupillary reflexes; and the **superior colliculus,** which coordinates simultaneous bilateral eye movements, such as tracking and convergence.

■ The Ear Serves as a Sensor for Hearing and Equilibrium

The Nature of Sound ■ Sound waves are mechanical disturbances that travel through an elastic medium

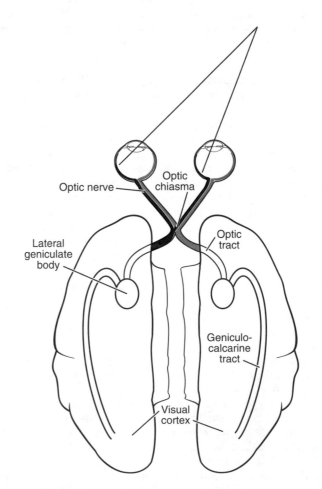

Figure 4–16 ■ The central nervous system pathway for visual information.

(usually air or water). A sound wave is produced by a mechanically vibrating structure that alternately compresses and rarefies the air (or water) in contact with it. For example, as a loudspeaker cone moves forward, air molecules in its path are forced closer together (**compression** or **condensation**), and as the cone moves back, the space between the disturbed molecules is increased (**rarefaction**). The compression (or rarefaction) of air molecules in one region causes a similar compression in adjacent regions. Continuation of this process causes the disturbance (called a **sound wave**) to spread away from the source. The speed at which the sound wave travels is determined by the elasticity of the air (the tendency of the molecules to spring back to their original positions). Assuming the sound source is moving back and forth at a constant rate of alternation (i.e., at a constant **frequency**), a propagated compression wave will pass a given point once for every cycle of the source. Because the propagation speed is constant in a given medium, at higher frequencies the compression waves are closer together; that is, more of them pass the given point every second. The distance between the compression peaks is called the **wavelength** of the sound, and it is inversely related to the frequency. A tone of 1000 cycles per second, traveling through the air, has a wavelength of approximately 34 cm, while a tone of 2000 cycles per second has a wavelength of 17 cm. Both waves, however, travel at the same speed through the air. Because the elastic forces in water are greater than those in air, the speed of sound in water is about four times as great, and the wavelength is correspondingly increased. Since the wavelength depends on the elasticity of the medium (which varies according to temperature and pressure), it is more convenient to identify sound waves by their frequency. Sound frequency is usually expressed in units of **Hertz** (Hz, or cycles per second).

Another fundamental characteristic of a sound wave is its **intensity, or amplitude.** This may be thought of as the relative amount of compression or rarefaction present as the wave is produced and propagated; it is related to the amount of energy contained in the wave. Usually the intensity is expressed in terms of **sound pressure,** the pressure the compressions and rarefactions exert on a surface of known area (expressed in dynes per square centimeter). Because the human ear is sensitive to sounds over a millionfold range of sound pressure levels, it it convenient to express the intensity of sound as the logarithm of a ratio referenced to the **absolute threshold of hearing** for a tone of 1000 Hz. This reference level has a value of 0.0002 dyne/cm², and the scale for the measurements is the **decibel** (dB) scale. In the expression

$$dB = 20 \log (P/P_{ref}), \qquad (1)$$

the sound pressure (P) is referred to the absolute reference pressure (P_{ref}). For a sound that is 10 times greater than the reference, the expression becomes

$$dB = 20 \log (0.002 / 0.0002) = 20. \qquad (2)$$

Thus any two sounds having a tenfold difference in intensity have a decibel difference of 20; a hundredfold difference would mean a 40 dB difference and a thousandfold difference would be 60 dB. Usually the reference value is assumed to be constant and standard, and it is not expressed when measurements are reported. Table 4-1 gives the sound pressure levels (SPLs) and the decibel levels for some common sounds. The total range of 140 dB shown in the table expresses a relative range of 10 million-fold. Adaptation and compression processes in the human auditory system allow encoding of most of this wide range into biologically useful information.

Sound waves that are purely **sinusoidal** in form contain all of their energy at one frequency and are perceived as **pure tones.** Complex sound waves, such as speech or music, consist of the addition of a number of simpler waveforms of different frequencies and amplitudes. The human ear is capable of hearing sounds over the range of 20-16,000 Hz, although the upper limit decreases with age. Auditory sensitivity varies with the frequency of the sound; we hear sounds most readily in the range of 1000-4000 Hz and at a sound pressure level of around 60 dB. Not surprisingly, this is the frequency and intensity range of human vocalization. The ear's sensitivity is also affected by **masking:** in the presence of background sounds or noise, the auditory threshold for a given tone rises. This may be due to refractoriness induced by the masking sound, which would reduce the number of available receptor cells.

The External Ear ■ An overall view of the human ear is shown in Figure 4-17. The **pinna,** the visible portion of the ear, is not critical to hearing in humans, although it does slightly emphasize frequencies in the range of 1500-7000 Hz and aids in localization of sources of sound. The **external auditory canal** extends inward through the **temporal bone.** Wax-secreting glands line the canal, and its inner end is sealed by the **tympanic membrane,** or **eardrum,** a

TABLE 4–1 ■ Relative Pressures of Some Common Sounds

Pressure (dy/cm²)	SPL (dB)	Sound Source	Relative Pressure
0.0002	0	Absolute threshold	1
0.002	+ 20	Faint whisper	10
0.02	+ 40	Quiet office	100
0.2	+ 60	Conversation	1,000
2.0	+ 80	City bus	10,000
20.0	+ 100	Subway train	100,000
200.0	+ 120	Loud thunder	1,000,000
2000.0	+ 140	Pain and damage	10,000,000

Modified from Gulick WA, Gescheider GA, Frisna RD: *Hearing: Physiological Acoustics, Neural Coding, and Psychoacoustics.* New York, Oxford University Press, 1989, Table 2.2, p 51.

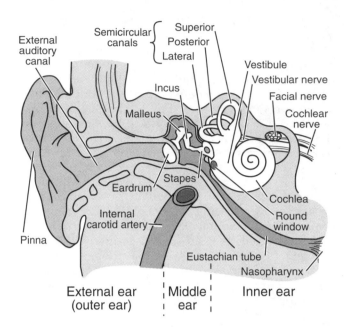

Figure 4–17 ■ Overall structure of the human ear, divided into outer, middle, and inner portions. The structures of the middle and inner ear are encased in the temporal bone of the skull; those of the inner ear (responsible for hearing and the sense of equilibrium and balance) are located in the series of passages called the bony labyrinth.

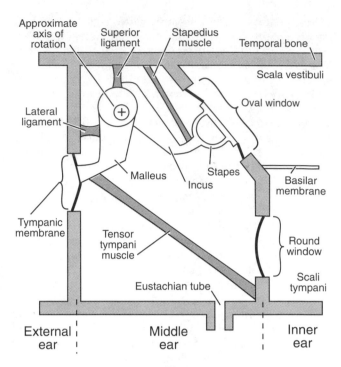

Figure 4–18 ■ Schematic diagram of the middle ear. Vibrations from the tympanic membrane are transmitted by the lever system formed by the ossicular chain to the oval window of the scala vestibuli. The anterior and posterior ligaments, part of the suspensory system for the ossicles, are not shown. The combination of the four suspensory ligaments produces a virtual pivot point (marked by a cross); its position varies with the frequency and intensity of the sound. The stapedius and tensor tympani muscles modify the lever function of the ossicular chain.

thin, oval, slightly conical, flexible membrane that is anchored around its edges to a ring of bone. An incoming pressure wave traveling down the external auditory canal causes the eardrum to vibrate back and forth in step with the compressions and rarefactions of the sound wave. This is the first mechanical step in the transduction of sound. The overall acoustic effect of the external ear structures is to produce an amplification of 10–15 dB in the frequency range broadly centered around 3000 Hz.

The Middle Ear ■ The next portion of the auditory system is an air-filled cavity (volume about 2 ml) in the mastoid region of the temporal bone (Fig. 4–18). The middle ear is connected to the pharynx by the **eustachian tube.** The tube opens briefly during swallowing and allows equalization of the pressures on either side of the eardrum. During rapid external pressure changes (as may occur in an elevator or aircraft), the unequal forces displace the eardrum; such physical deformation may cause discomfort or pain and, by restricting the motion of the tympanic membrane, may impair hearing. Blockages of the eustachian tube or fluid accumulation in the middle ear (due to an infection) can also lead to difficulties with hearing.

Bridging the gap between the tympanic membrane and the inner ear is a chain of three very small bones **(ossicles).** The first of these, the **maleus (hammer),** is attached to the tympanic membrane such that the back-and-forth movement of the eardrum causes a rocking movement of the maleus. The **incus (anvil)** connects the head of the maleus to the third bone, the **stapes (stirrup).** This last bone, through its oval **footplate,** connects to the **oval window** of the inner ear and is anchored there by an **annular ligament.**

Four separate **suspensory ligaments** hold the ossicular chain in position in the middle ear cavity. The **superior** and **lateral ligaments** lie roughly in the plane of the ossicular chain and anchor the head and shaft of the maleus. The **anterior ligament** attaches the head of the maleus to the anterior wall of the middle ear cavity, and the **posterior ligament** runs from the head of the incus to the posterior wall of the cavity. This suspensory system allows the ossicles sufficient freedom to function as a lever system to transmit the vibrations of the tympanic membrane to the oval window. This mechanism is especially important because, although the eardrum is suspended in air, the oval window seals off a fluid-filled chamber. Transmission of sound from air to liquid is inefficient; if sound waves were to strike the oval window directly, 99.9% of the energy would be reflected away and lost. Two mechanisms work to compensate for this loss. Although it varies with frequency, the ossicular chain has a lever ratio of about 1.3:1, producing a slight gain in force. In addition, the relatively large area of the tympanic membrane is coupled to the smaller area of the oval window (approximately a 17:1 ratio). These conditions result in a pressure gain of around 25 dB, largely compensating for the potential loss. Although the efficiency depends on the frequency, approximately 60% of the sound energy that strikes the eardrum is transmitted to the oval window.

Sound transmission through the middle ear is also affected by the action of two very small muscles that attach to the ossicular chain (Fig. 4–18) and aid in holding the bones in position and modifying their function. The **tensor tympani** inserts on the maleus (near the center of the tympanic membrane), passes diagonally through the middle ear cavity, and enters the **tensor canal,** in which it is anchored. Contraction of this muscle limits the vibration amplitude of the eardrum and makes sound transmission less efficient. The **stapedius muscle** attaches to the stapes near its connection to the incus and runs posteriorly to the mastoid bone. Its contraction changes the axis of oscillation of the ossicular chain and causes dissipation of excess movement before it reaches the oval window. These muscles are activated by a reflex (simultaneously in both ears) in response to moderate and loud sounds; they act to reduce the transmission of sound to the inner ear and thus to protect its delicate structures. Because this **acoustic reflex** requires up to 150 msec to operate (depending on the loudness of the stimulus), it cannot provide protection from sharp or sudden bursts of sound.

The process of sound transmission can bypass the ossicular chain entirely. If a vibrating object, such as a tuning fork, is placed against a bone of the skull (typically the mastoid), the vibrations are transmitted mechanically to the fluid of the inner ear and the normal processes there (see below) act to complete the hearing process. Bone conduction is used as a means of diagnosing hearing disorders that may arise because of lesions in the ossicular chain. Some hearing aid devices employ bone conduction to overcome such deficits.

The Inner Ear ■ The actual process of sound transduction takes place in the inner ear. The auditory structures are located in the **cochlea,** part of a cavity in the temporal bone called the **bony labyrinth** (Fig. 4–19). The cochlea (meaning "snail") is a fluid-filled spiral tube that arises from a cavity called the **vestibule,** with which the organs of balance also communicate. In the human ear the cochlea is about 35 mm long and makes about two and three quarters turns. Together with the vestibule it contains a total fluid volume of 0.1 ml. It is partitioned longitudinally into three divisions (canals) called the **scala vestibuli** (into which the oval window opens), the **scala tympani** (sealed off from the middle ear by the round window), and the **scala media** (in which the sensory cells are located). Arising from the bony center axis of the spiral (the **modiolus**) is a winding shelf called the **osseous spiral lamina;** opposite it on the outer wall of the

spiral is the **spiral ligament,** and connecting these two structures is a highly flexible connective tissue sheet, the **basilar membrane,** that runs for almost the entire length of the cochlea. The basilar membrane separates the scala tympani (below) from the scala media (above). The **hair cells,** which are the actual sensory receptors, are located on the upper surface of the basilar membrane. They are called hair cells because each has a bundle of hairlike **cilia** at the end that projects away from the basilar membrane.

Reissner's membrane, a delicate sheet only two cell layers thick, divides the scala media (below) from the scala vestibuli (above) (Fig. 4–19). The scala vestibuli communicates with the scala tympani at the apical (distal) end of the cochlea via the **helicotrema,** a small opening where a portion of the basilar membrane is missing. The scala vestibuli and scala tympani are filled with **perilymph,** a fluid that is high in sodium and low in potassium. The scala media contains **endolymph,** a fluid high in potassium and low in sodium. The endolymph is secreted by the **stria vascularis,** a layer of fibrous vascular tissue along the outer wall of the scala media. Because the cochlea is filled with incompressible fluid and is encased in hard bone, pressure changes caused by the in-and-out motion of the oval window (driven by the stapes) are relieved by an out-and-in motion of the flexible round window.

The **organ of Corti** (Fig. 4–19) is formed by structures located on the upper surface of the basilar membrane and runs the length of the scala media. It contains one row of some 3000 **inner hair cells;** the **arch of Corti** and other specialized supporting cells separate the inner hair cells from the three or four rows of **outer hair cells** (numbering around 12,000) located on the stria vascularis side. The rows of inner and outer hair cells are inclined slightly toward each other and covered by the **tectorial membrane,** which arises from the **spiral limbus,** a projection on the upper surface of the osseus spiral lamina.

Nerve fibers from cell bodies located in the **spiral ganglia** form **radial bundles** on their way to synapse with the inner hair cells. Each nerve fiber makes synaptic connection with only one hair cell, but each hair cell is served by 8–30 fibers. While the inner hair cells comprise only 20% of the hair cell population, they receive 95% of the afferent fibers. In contrast, many outer hair cells are each served by a single external spiral nerve fiber. The collected afferent fibers are bundled in the **cochlear nerve,** which exits from the inner ear via the **internal auditory meatus.** Some **efferent fibers** also innervate the cochlea. They may serve to enhance pitch discrimination and the ability to distinguish sounds in the presence of noise. Recent evidence suggests that efferent fibers to the outer hair cells may cause them to shorten (contract), thereby altering the mechanical properties of the cochlea.

The hair cells of both the inner and outer rows are similar anatomically. Both sets are supported and anchored to the basilar membrane by the **Deiters cells**

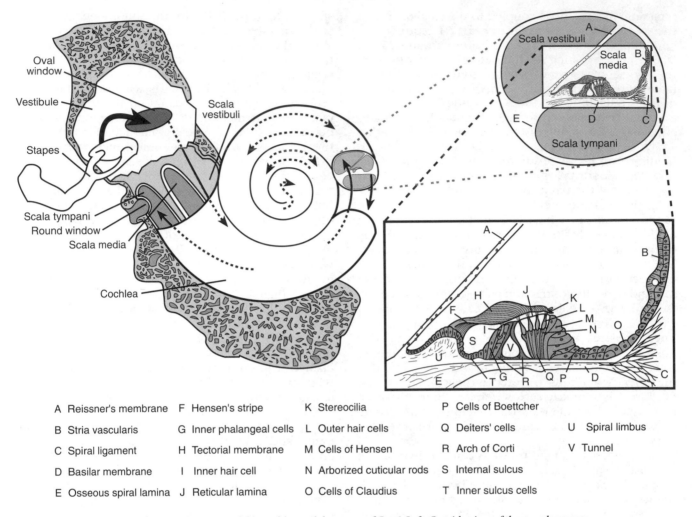

A Reissner's membrane	F Hensen's stripe	K Stereocilia	P Cells of Boettcher		
B Stria vascularis	G Inner phalangeal cells	L Outer hair cells	Q Deiters' cells	U Spiral limbus	
C Spiral ligament	H Tectorial membrane	M Cells of Hensen	R Arch of Corti	V Tunnel	
D Basilar membrane	I Inner hair cell	N Arborized cuticular rods	S Internal sulcus		
E Osseous spiral lamina	J Reticular lamina	O Cells of Claudius	T Inner sulcus cells		

Figure 4–19 ■ Structure of the cochlea and the organ of Corti. Left: Outside view of the membranous labyrinth of the cochlea. Upper right: A cross-section through one turn of the cochlea, showing the canals and membranes that make up the structures involved in the final processes of auditory sensation. Lower right: Enlargement of a cross-section of the organ of Corti, showing the relationships among the hair cells and the membranes. (Modified from Gulick WA, Gescheider GA, Frisna RD. *Hearing: Physiological Acoustics, Neural Coding, and Psychoacoustics.* New York, Oxford University Press, 1989.)

and extend upward into the scala media toward the **tectorial membrane.** Processes (the "hairs") of the outer hair cells (see below) actually touch the tectorial membrane, while those of the inner hair cells appear to come just short of contact. The hair cells make synaptic contact with afferent neurons that run through channels between Deiters cells. A chemical transmitter of unknown identity is contained in synaptic vesicles near the base of the hair cells; as in other synaptic systems, the entry of calcium ions (associated with cell membrane depolarization) is necessary for the migration and fusion of the synaptic vesicles with the cell membrane prior to transmitter release.

At the apical end of each inner hair cell is a projecting bundle of about 50 **stereocilia,** rodlike structures packed in three parallel, slightly curved rows.

Minute strands link the free ends of the stereocilia together, so the bundle tends to move as a unit. The height of the individual stereocilia increases toward the outer edge of the cell (toward the stria vascularis), giving a sloping appearance to the bundle. Along the cochlea the inner hair cells remain constant in size, while the stereocilia increase in height from about 4 μm at the basal end to 7 μm at the apical end. The outer hair cells are more elongated than the inner cells, and their size increases along the cochlea from base to apex. Their stereocilia (about 100 per hair cell) are also arranged in three rows that form an exaggerated W figure. The height of the stereocilia also increases along the length of the cochlea, and they are embedded in the tectorial membrane. The stereocilia of both types of hair cells extend from the

cuticular plate at the apex of the cell. The diameter of an individual stereocilium is uniform (about 0.2 μm) except at the base, where it decreases markedly. Individually, the stereocilia contain cross-linked and closely packed **actin filaments,** and near the tip of each is a cation-selective **transduction channel.**

When a hair bundle is deflected slightly (the threshold is less than 0.5 nm) toward the stria vascularis, minute mechanical forces open the transduction channels and cations (mostly potassium) enter the cells (Fig. 4–20). The resulting **depolarization,** roughly proportional to the deflection, causes the release of transmitter molecules, resulting in generation of afferent nerve action potentials. Approximately 15% of the transduction channels are open in the absence of any deflection, and bending in the direction of the modiolus of the cochlea results in **hyperpolarization,** increasing the range of motion that can be sensed. The hair cells are quite insensitive to movements of the stereocilia bundles at right angles to their preferred direction. The speed of response of the hair cells is remarkable: they can detect repetitive motions of up to 100,000 times per second. They can therefore provide information throughout the course of a single cycle of a sound wave. Such rapid response is also necessary for accurate localization of sound sources. When a sound comes from directly in front of a listener, the waves arrive simultaneously at both ears. If the sound

originates off to one side, it reaches one ear sooner than the other and is slightly more intense at the nearer ear. The difference in arrival time is on the order of tenths of a millisecond, and the rapid response of the hair cells allows them to provide temporal input to the auditory cortex, where the timing and intensity information is processed into a very accurate perception of the location of the sound source.

The actual transduction of sound requires an interaction among the tectorial membrane, the arches of Corti, the hair cells, and the basilar membrane. When a sound wave is transmitted to the oval window by the ossicular chain, a pressure wave travels up the scala vestibuli and down the scala tympani (Fig. 4–21). The canals of the cochlea, being encased in bone, are not deformed, and movements of the round window allow the small volume change needed for the transmission of the pressure wave. Resulting eddy currents

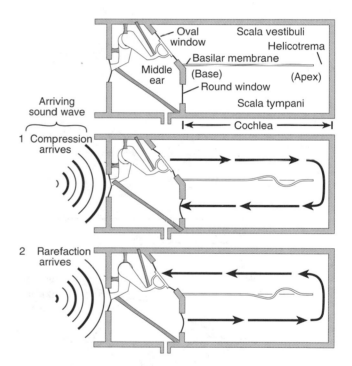

Figure 4–21 ■ Mechanics of the cochlea, showing the action of the structures responsible for pitch discrimination. (Only the basilar membrane of the organ of Corti is shown.) When the compression phase of a sound wave arrives at the tympanic membrane, the ossicles transmit it to the oval window, which is pushed inward. A pressure wave travels up the scala vestibuli and (via the helicotrema) down the scala tympani. To relieve the pressure, the round window bulges outward. Associated with the pressure waves are small eddy currents that cause a traveling wave of displacement to move along the basilar membrane from base to apex. The arrival of the next rarefaction phase reverses these processes. The frequency of the sound wave, interacting with the variation of the stiffness and width of the basilar membrane along its length, determines the characteristic position at which the membrane displacement is maximal. This effect of this localization is further detailed in Figure 4–22.

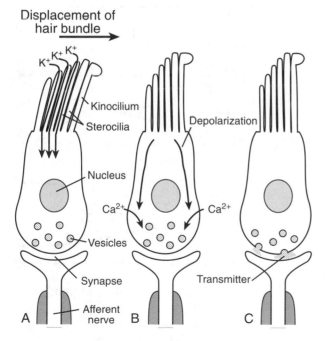

Figure 4–20 ■ Mechanical transduction in the hair cells of the ear. Deflection of the stereocilia opens apical K+ channels. The resulting depolarization allows the entry of Ca²⁺ at the basal end of the cell, causing release of the synaptic transmitter substance to excite the afferent nerve. (Adapted from Hudspeth AJ: The hair cells of the inner ear. *Sci Am* 248:54–64, 1983.)

in the cochlear fluids produce an undulating distortion in the basilar membrane. Because the stiffness and width of the membrane vary with its length (it is wider and less stiff at the apex than at the base), the membrane deformation takes the form of a so-called **traveling wave** (Fig. 4–22), which has its maximal amplitude at a position along the membrane corresponding to the particular frequency of the sound wave. Low-frequency sounds cause a maximal displacement of the membrane near its apical end (near the helicotrema), whereas high-frequency sounds produce their maximal effect at the basal end (near the oval window). As the basilar membrane moves, the arches of Corti transmit the movement to the tectorial membrane and the stereocilia of the outer hair cells (embedded in the tectorial membrane) are subjected to lateral shearing forces that stimulate the cells, and

action potentials arise in the afferent neurons. Because of the tuning effect of the basilar membrane, only hair cells located at a particular place along the membrane are maximally stimulated by a given frequency (pitch). This localization is the essence of the **place theory** of pitch discrimination, and the mapping of specific tones (pitches) to specific areas is called **tonotopic organization.** As the signals from the cochlea ascend through the complex pathways of the auditory system in the brain, the tonotopic organization of the neural elements is at least partially preserved and pitch can be spatially localized throughout the system. The sense of pitch is further sharpened by the resonant characteristics of the different-length stereocilia along the length of the cochlea and by the frequency-response selectivity of neurons in the auditory pathway. Thus the cochlea acts as both a transducer for sound waves and a

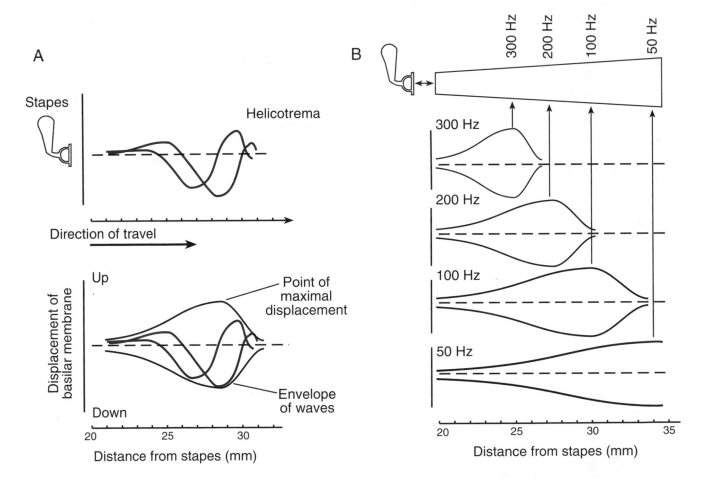

Figure 4–22 ■ Membrane localization of different frequencies. (A) The upper portion shows a traveling wave of displacement along the basilar membrane at two instants of time. Over a period of time the peak excursions of many such waves form an envelope of displacement, with a maximal value at about 28 mm from the stapes (lower portion); at this position, its stimulating effect on the hair cells will be most intense. (B) The effect of frequency. Lower frequencies produce a maximal effect at the apex of the basilar membrane, where it is the widest and least stiff. Pure tones affect a single location; complex tones affect multiple loci. (Adapted from von Békésy G: *Experiments in Hearing.* New York, McGraw-Hill, 1960.)

frequency analyzer that sorts out the different pitches so they can be separately distinguished. In the midrange of hearing (around 1000 Hz) the human auditory system can sense a difference in frequency of as little as 3 Hz.

Central Auditory Pathways ■ Nerve fibers from the cochlea enter the **spiral ganglion** of the organ of Corti; from there, fibers are sent to the **dorsal** and **ventral cochlear nuclei.** The complex pathway that finally ends at the **auditory cortex** in the superior portion of the **temporal lobe** involves several sets of synapses and considerable crossing over and intermediate processing. As with the eye, there is a spatial correlation between cells in the sensory organ and specific locations in the primary auditory cortex. In this case, the representation is called a **tonotopic map,** with different pitches being represented by different locations, even though the firing rates of the cells no longer correspond to the frequency of sound originally presented to the inner ear.

Equilibrium and Balance ■ The ear also has important nonauditory sensory functions. These senses are located in the **vestibular apparatus** (Fig. 4-23), which consists (on each side of the head) of three **semicircular canals** and two **otolithic organs,** the **utricle** (utriculus) and the **saccule** (sacculus). These structures are located in the bony labyrinth of the temporal bone. They sense rotary motion and linear acceleration. The basic sensing elements in both cases are hair cells.

Rotary motion and acceleration are sensed by the semicircular canals, hooplike tubular membranous structures (part of the membranous labyrinth). Their interior is continuous with the scala media and is filled with endolymph; on the outside, they are bathed by perilymph. The three canals on each side lie in three mutually perpendicular planes. With the head tipped

forward by about 30 degrees, the **horizontal** (or **lateral**) semicircular **canal** lies in the horizontal plane. At right angles to this are the planes of the **anterior vertical** (or **superior**) **canal** and the **posterior vertical canal,** which are perpendicular to each other. The planes of the anterior vertical canals are each at approximately 45 degrees to the midsagittal section of the head (and at 90 degrees to each other). Thus the anterior canal on one side lies in a plane parallel with the posterior canal on the other side, and the two function as a pair. The horizontal canals also lie in a common plane.

Near its junction with the utricle, each canal has a swollen portion called the **ampulla.** Each ampulla contains a **crista ampullaris,** the sensory structure for that semicircular canal; it is composed of hair cells and supporting cells encapsulated by a **cupula,** a gelatinous mass (Fig. 4-24). The cupula extends to the top of the ampulla and is moved back and forth by movements of the endolymph in the canal. This movement is sensed by displacement of the stereocilia of the hair cells. These cells are much like those of the organ of Corti, except that at the "tall" end of the stereocilia array there is one larger cilium, the **kinocilium.** All of the hair cells have the same orientation. When the stereocilia are bent toward the kinocilium, the frequency of action potentials in the afferent neu-

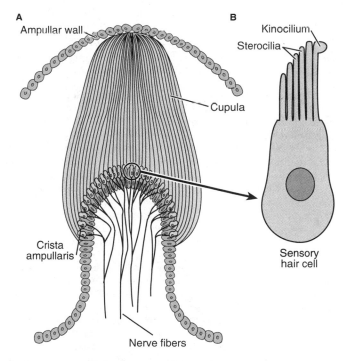

Figure 4–24 ■ The sensory structure of the semicircular canals. (A) The crista ampularis contains the hair (receptor) cells, and the whole structure is deflected by motion of the endolymph. (B) Diagram of an individual hair cell. (Modified from Selkurt EE (ed): *Basic Physiology for the Health Sciences.* 2nd ed. Boston, Little, Brown, 1982.)

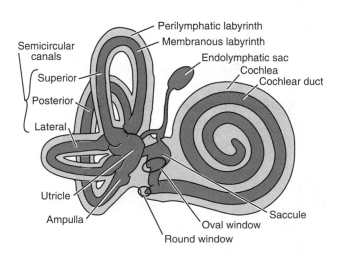

Figure 4–23 ■ The labyrinth of the inner ear, showing the semicircular canals and the utricle and saccule, organs devoted to the sensation of rotary motion and static position.

rons leaving the ampulla increases; bending in the other direction decreases the action potential frequency.

The role of the semicircular canals in sensing rotary acceleration is shown at the left in Figure 4–25. The

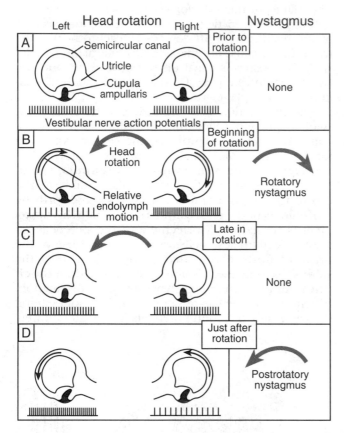

Figure 4–25 ■ The role of the semicircular canals in sensing rotary acceleration. This sensation is linked to compensatory eye movements by the vestibuloocular reflex. Only the horizontal canals are considered here. This pair of canals is shown as if one were looking down through the top of the head that was looking toward the bottom of the page. Within the ampulla of each canal is the cupula, an extension of the crista ampullaris, the structure that senses motion in the endolymph fluid in the canal. Below each canal is the action potential train recorded from the vestibular nerve. (A) The head is still, and equal nerve activity is seen on both sides. There are no associated eye movements. (B) The head has begun to rotate to the left. The inertia of the endolymph causes it to lag behind the movement, producing a fluid current that stimulates the cupulas (arrows show the direction of the relative movements). Because the two canals are mirror images, the neural effects are opposite on each side (the cupulas are bent in relatively opposite directions). The reflex action causes the eyes to move slowly to the right, opposite to the direction of rotation; they then snap back and begin the slow movement again as rotation continues. This is called rotatory nystagmus. (C) As rotation continues, the endolymph "catches up" with the canal because of fluid friction and viscosity, and there is no relative movement to deflect the cupulas. Equal neural output comes from both sides, and the eye movements cease. (D) When the rotation stops, the inertia of the endolymph causes a current in the same direction as the preceding rotation, and the cupulas are again deflected, this time in a manner opposite to that shown in B. The slow eye movements now occur in the same direction as the former rotation; this is a postrotatory nystagmus.

mechanisms linking stereocilia deflection to receptor potentials and action potential generation are quite similar to those in the auditory hair cells. Because of the inertia of the endolymph in the canals, when the position of the head is changed, fluid currents in the canals cause the deflection of the cupula and the hair cells are stimulated. The fluid currents are roughly proportional to the rate of change of velocity (i.e., to the radial acceleration), and they result in a proportional increase or decrease (depending on the direction of head rotation) in action potential frequency. As a result of the bilateral symmetry in the vestibular system, canals with opposite pairing produce opposite neural effects. The vestibular division of cranial nerve VIII passes the impulses first to the **vestibular ganglion,** where the cell bodies of the primary sensory neurons lie. The information is thus passed to the **vestibular nuclei** of the brainstem and from there to various locations involved in sensing, correcting, and compensating for changes in bodily motion.

The remaining vestibular organs, the saccule and the utricule, are also part of the membranous labyrinth. They communicate with the semicircular canals, the cochlear duct, and the endolymphatic duct. The sensory structures in these organs, called **maculae,** also employ hair cells (similar to those of the ampullar cristae) (Fig. 4–26). The macular hair cells are covered with the **otolithic membrane,** a gelatinous substance in which are embedded numerous small crystals of calcium carbonate called **otoliths.** Because the otoliths are heavier than the endolymph, tilting of the head results in gravitational movement of the otolithic membrane and a corresponding change in sensory neuron action potential frequency. As in the ampulla, the action potential frequency increases or decreases depending on the direction of displacement. The maculae are adapted to provide a steady signal in response

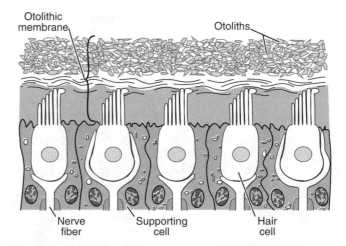

Figure 4–26 ■ The relation of the otoliths to the sensory cells in the macula of the utricle and saccule. (Adapted from Selkurt EE (ed): *Basic Physiology for the Health Sciences.* 2nd ed. Boston, Little, Brown, 1982.)

to displacement, hence they can monitor the position of the head with respect to a gravitational field. The maculae also respond proportionally to **linear acceleration.**

The vestibular apparatus is an important component in a number of reflexes that serve to orient the body in space and to preserve that orientation. Integrated responses to vestibular sensory input include balancing and steadying movements controlled by skeletal muscle, along with specific reflexes that automatically compensate for bodily motions. One such mechanism is the **vestibuloocular reflex.** If the body begins to rotate and thus to stimulate the horizontal semicircular canals, the eyes will move slowly in a direction opposite to that of the rotation and then suddenly snap back the other way (Fig. 4–25, right). This movement pattern, called **rotatory nystagmus,** aids in visual fixation and orientation and takes place even with the eyes closed. As rotation continues, the relative motion of the endolymph in the semicircular canals ceases and the nystagmus disappears. When rotation stops, the inertia of the endolymph causes it to continue in motion and again the cupulas are displaced, this time from the opposite direction. The **postrotatory nystagmus** that develops is in the same direction as the previous rotation. As long as the endolymph continues its relative movement, the nystagmus (and the sensation of rotary motion) persists. Irrigation of the ear with water above or below body temperature, as in treatment of ear infection, causes convection currents in the endolymph. The resulting unilateral **caloric stimulation** of the semicircular canal produces symptoms of vertigo, nystagmus, and nausea.

Related mechanisms involving the otolithic organs produce automatic compensations (via the postural and extraocular musculature) when the otolithic organs are stimulated by transient or maintained changes in the position of the head. If the otolithic organs are stimulated rhythmically, as by the motion of a ship or automobile, the distressing symptoms of motion sickness (vertigo, nausea, sweating, etc.) may appear. Over a period of time these symptoms lessen and disappear.

■ The Special Chemical Senses Detect Molecules in the Environment

Chemical sensation includes not only the special chemical senses described below, but also internal sensory receptor function responsible for monitoring the concentrations of gases and other chemical substances dissolved in the blood or other body fluids. Since we are seldom aware of these internal chemical sensations, they are treated throughout this book as the need and occasion arise; the account here deals only with taste and smell.

Gustatory Sensation ■ The sense of taste is mediated by multicellular receptors called **taste buds,** several thousand of which are located on folds and projections on the dorsal tongue, or **papillae.** Taste buds are located mainly on the tops of the numerous

fungiform papillae but also on the sides of the less numerous **foliate** and **vallate** papillae. The **filiform** papillae, which cover most of the tongue, usually do not bear taste buds. An individual taste bud is a spheroid collection of around 50 individual cells that is about 70 μm high and 40 μm in diameter (Fig. 4–27). Most of the cells are **sensory cells.** At their apical ends they are connected laterally by tight junctions, and they bear **microvilli** that greatly increase the surface area that they present to the environment. At their basal ends they form synapses with the **facial (VII)** and **glossopharyngeal (IX) cranial nerves.** This arrangement indicates that the sensory cells are actually **secondary receptors** (like the hair cells of the ear), since they are anatomically separate from the afferent sensory nerves. About 50 afferent fibers enter each taste bud, where they branch so that each axon may synapse with more than one sensory cell. Among the sensory cells are elongated **supporting cells** that do not have synaptic connections. The sensory cells typically have a life span of 10 days. They are continually replenished by new sensory cells formed from the **basal cells** of the lower part of the taste buds. When a sensory cell is replaced by a maturing basal cell, the old synaptic connections are broken and new ones must be formed.

The four modalities of taste, from the point of view of their receptors, are well defined, and the areas of the tongue where they are located are also rather specific, although the degree of localization depends

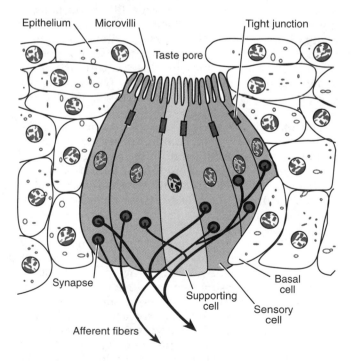

Figure 4–27 ■ The sensory and supporting cells in a taste bud. The afferent nerve fibers (of which only a few are shown) synapse with the basal areas of the sensory cells. (Modified from Schmidt RF (ed): *Fundamentals of Sensory Physiology.* 2nd ed. New York, Springer-Verlag, 1981.)

on the concentration of the stimulating substance. In general, the receptors for the **sweet** modality are located just behind the tip of the tongue; **sour** receptors are located along the sides; the **salt** sensation is localized to the tip; and the **bitter** sensation is found across the rear of the tongue. (The two so-called **accessory qualities** of taste sensation are **alkaline** [soapy] and **metallic.**) The broad surface of the tongue is not as well supplied with taste buds. Most taste experiences involve a number of different sensory modalities, such as taste, smell, mechanoreception (for texture), and temperature; artificially confining the taste sensation to only the four modalities found on the tongue (e.g., by blocking the sense of smell) greatly diminishes the range of taste perceptions.

While the receptor categories are well defined, it is much more difficult to determine what kind of stimulating chemical will produce a given taste sensation. Chemicals that produce a sour sensation are usually acids, and the intensity of the perception depends on the degree of dissociation of the acid (i.e., the number of free hydrogen ions). Most sweet substances are organic; sugars, especially, tend to produce a sweet sensation, although their thresholds vary widely. For example, sucrose is about 8 times as sweet as glucose. On the other hand, the apparent sweetness of saccharine, an artificial sweetener, is 600 times as great as that of sucrose, although it is not a sugar. Unfortunately, salts of lead are also sweet, which can lead to ingestion of toxic levels of this poisonous metal. Substances producing the bitter taste modality form a heterogeneous group. The classical bitter substance is quinine; nicotine and caffeine are also bitter, as are many of the salts of calcium, magnesium, and ammonium, the bitter taste being due to the cation portion of the salt. Sodium ions produce a salty sensation; some organic compounds, such as lysyltaurine, are even more potent in this regard than sodium chloride itself.

The intensity of a taste sensation depends on the concentration of the stimulating substance, but application of the same concentration to larger areas of the tongue produces a more intense sensation; this is probably due to facilitation involving a greater number of afferent fibers. Some taste sensations also increase with time, although taste receptors show a slow but definite adaptation. Temperature, over some ranges, tends to increase the perceived taste intensity, while dilution by saliva and serous secretions from the tongue decreases the intensity. The specificity of the taste sensation arising from a particular stimulating substance is due to the effects of specific receptor molecules on the microvilli of the sensory cells. Salty substances probably depolarize sensory cells directly, while sour substances may produce depolarization by blocking potassium channels with hydrogen ions. Bitter substances, when they bind to specific membrane receptors, appear to act through a G protein mechanism, activating phospholipase C to increase the cell

concentration of inositol trisphosphate, which promotes calcium release from the endoplasmic reticulum. Sweet substances also act through a G protein mechanism and cause increases in adenylate cyclase activity; this causes an increase in cAMP that in turn promotes phosphorylation of membrane potassium channels. The resulting decrease in potassium conductance leads to depolarization.

Olfactory Sensation ■ Compared with many other animals, the human sense of smell is not particularly acute; nevertheless, we can distinguish 2000–4000 different odors that cover a wide range of chemical species. The receptor organ for olfaction is the **olfactory mucosa,** an area of approximately 5 cm^2 located in the roof of the nasal cavity. Normally there is little air flow in this region of the nasal tract, but sniffing serves to direct air upward to this region and thus increases the likelihood of an odor being detected. The olfactory mucosa contains about 10–20 million receptor cells. In contrast to the taste sensory cells, the olfactory cells are neurons, and as such are **primary receptors.** These cells are interspersed among **supporting,** or **sustentacular, cells** (Fig. 4–28). The receptor cells terminate at their apical ends with short and thick dendrites called **olfactory rods,**

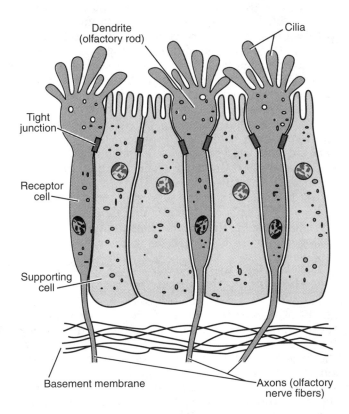

Figure 4–28 ■ The sensory cells in the olfactory mucous membrane. The axons leading from this receptor (the fila olfactoria) are part of the sensory cells, in contrast to the situation in taste receptors. (Modified from Ganong WF: *Review of Medical Physiology.* 16th ed. Norwalk, CT, Appleton & Lange, 1993.)

and each cell bears 10–20 cilia that extend into a thin covering of mucus secreted by **Bowman's glands** located throughout the olfactory mucosa. To be sensed, molecules must be dissolved in this mucous layer. The basal ends of the receptor cells form axonal processes called **fila olfactoria** that pass through the **cribriform plate** of the **ethmoid bone.** These short axons synapse with the **mitral cells** in complex spheric structures called **olfactory glomeruli** located in the **olfactory bulb,** part of the brain located just above the olfactory mucosa. Here the complex afferent and efferent neural connections for the olfactory tract are made. Approximately 1000 fila olfactoria synapse on each mitral cell, resulting in a highly convergent relationship. Lateral connections are also plentiful in the olfactory bulb, which contains **efferent fibers** thought to have an inhibitory function.

The olfactory mucosa also contains sensory fibers from the **trigeminal (V) cranial nerve.** They are sensitive to certain odorous substances, such as peppermint and chlorine, and play a role in the initiation of reflex responses (e.g., sneezing) that result from irritation of the nasal tract.

The modalities of smell are numerous and do not fall into convenient classes, though some general categories of smell, such as flowery, sweaty, or rotten, may be distinguished. Olfactory thresholds vary widely from substance to substance; the threshold concentra-tion for detection of ethyl ether is around 5.8 mg/L air, while that for methyl mercaptan (the odor of garlic) is approximately 0.5 ng/L. This represents a 10 million-fold difference in sensitivity. The basis for odor discrimination is not well understood. It is not likely that there is a receptor molecule for every possible odor substance located in the membranes of the olfactory cilia, and it appears that complex odor sensations arise from unique spatial patterns of activation throughout the olfactory mucosa. Signal transduction appears to involve binding of a molecule of an odorous substance to a specialized receptor molecule on a cilium of a sensory cell. This binding, via a membrane-based G protein mechanism, causes the eventual production of cAMP that binds to, and opens, sodium channels in the ciliary membrane. The resulting inward sodium current depolarizes the cell to produce a generator potential, which causes action potentials to arise in the initial segments of the fila olfactoria. The sense of smell shows a high degree of adaptation, some of which takes place at the level of the generator potential and some of which may be due to the action of efferent neurons in the olfactory bulb. Discrimination between odor intensities is not well defined; detectable differences may be on the order of 30% (the visual system can distinguish between light intensity levels as little as 1%).

■ ■ ■ ■ ■ ■ ■ ■ ■ ■ ■ REVIEW EXERCISES ■ ■ ■ ■ ■ ■ ■ ■ ■ ■ ■

Identify Each with a Term

1. An environmental factor that causes a response in a sense organ
2. An sensory receptor that responds to physical deformation
3. The voltage change in a sensory receptor that results from application of a stimulus
4. The decrease in sensory response while a stimulus is maintained
5. The adjustment of the lens of the eye for near vision
6. The structures in the eye responsible for image formation
7. The type of lens that causes light rays to diverge
8. The fluid that fills the anterior chamber of the eye
9. The set of six muscles that move the eyeball
10. The opaque portion of the fibrous coat of the eye
11. Photoreceptor cells responsible for color vision
12. The retinal cells whose axons are the optic nerve fibers
13. Coordinated eye movements that compensate for head rotation
14. Rapid movements that reposition the eyes quickly
15. The three small bones of the middle ear
16. The coiled structure containing the parts of the inner ear

17. The fluid that fills the scala tympani
18. The bundle of afferent nerve fibers from the organ of Corti
19. The portion of the inner ear that senses position and motion
20. A G protein found in rod cells
21. The two somatic chemical senses

Define Each Term

22. Adequate stimulus
23. Phasic receptor
24. Threshold
25. Nociceptor
26. Refraction
27. Tactile receptor
28. Iris
29. Focal point
30. Fovea centralis
31. Diopter
32. Ciliary body
33. Cornea
34. Choroid
35. Retina

36. Canal of Schlemm
37. Rhodopsin
38. Myopia
39. Diplopia
40. Scotopsin
41. Visual cortex
42. Hair cells
43. Tympanic membrane
44. Oval window
45. Endolymph
46. Tectorial membrane
47. Utricule
48. Ampulla
49. Fila olfactoria
50. Taste bud
51. Olfactory mucosa

Choose the Correct Answer

52. An increase in the action potential frequency in a sensory nerve usually signifies:
    a. Increased intensity of the stimulus
    b. Cessation of stimulation
    c. Adaptation of the receptor
    d. A constant and maintained stimulus

53. Factors that determine proper interpretation of a sensory stimulus include:
    a. A measure of its intensity
    b. Its pathway to the central nervous system
    c. The rate of change of its intensity
    d. All of the above

54. Why is the blind spot on the retina not usually perceived?
    a. It is very small.
    b. It is present only in very young children.
    c. Its location in the visual field is different in each eye.

d. Constant eye motion prevents the spot from remaining still.

55. Presbyopia is due to:
    a. Changes in the shape of the eyeball due to age
    b. Age-related loss of rod cells in the retina
    c. Changes in the elasticity of the lens due to age
    d. Loss of transparency in the lens

56. What kind of lens will compensate a myopic eye for distance vision?
    a. A positive (converging) lens
    b. A negative (diverging) lens
    c. A cylindric lens
    d. No lens is needed; the eye itself can accommodate.

57. At which location along the basilar membrane are the highest-frequency sounds detected?
    a. Nearest to the oval window
    b. Farthest from the oval window, near the helicotrema
    c. Uniformly along the membrane
    d. At the midpoint of the membrane

58. Motion of the endolymph in the semicircular canals when the head is held still will result in the perception of:
    a. Being upside down
    b. Moving in a straight line
    c. Continued rotation
    d. Being stationary and upright

Briefly Answer

59. In what ways is stimulus localization by the eyes and ears similar?
60. Why does postrotatory nystagmus occur?
61. Do the senses of taste and smell function independently? Explain.
62. Why do objects appear colorless when viewed by moonlight?
63. How do sensation and perception differ?

ANSWERS

1. Stimulus
2. Mechanoreceptor
3. Generator (receptor) potential
4. Adaptation
5. Accommodation
6. Cornea, lens, retina
7. Convex (or negative)
8. Aqueous humor
9. Extraocular muscles
10. Sclera
11. Cones
12. Ganglion cells
13. Nystagmus
14. Saccades
15. Maleus (hammer), incus (anvil), stapes (stirrup)

16. Cochlea
17. Perilymph
18. Cochlear nerve
19. Vestibular apparatus
20. Transducin
21. Olfaction (smell) and gustation (taste)
22. The appropriate stimulus for a specific sensory organ
23. A receptor that responds best to changing signals; a velocity receptor
24. The lowest intensity of a stimulus that can be sensed
25. A receptor of painful or unpleasant stimuli
26. Bending of light rays at the interface between two transparent substances
27. A receptor for touch
28. A smooth muscle structure that controls the diameter of the pupil

29. The point behind a lens at which the image is formed

30. The optical center of the retina, a region of high acuity where only cones are present

31. The reciprocal of the focal length (in meters) of a lens; its power

32. The circular structure that controls the curvature of the lens and produces aqueous humor

33. The transparent front portion of the fibrous coat of the eyeball

34. The opaque rear portion of the vascular coat of the eyeball

35. The layer of photosensors and nerve cells at the rear of the eye, on which the image is formed

36. The passageway through which the aqueous humor is drained into the venous circulation

37. The visual pigment in the rod cells, responsible for scotopic (noncolor) vision

38. Nearsightedness; the inability to focus on distant objects

39. Double vision caused by inappropriate tracking of the two eyes

40. The opsin for rod vision; in the dark, it is conjugated

41. The region of the occipital lobe of the brain where visual information is processed

42. Mechanoreceptors in the cochlea (hearing) and vesibular apparatus (equilibrium)

43. The eardrum, at the inner end of the external auditory canal

44. The opening through which sound vibrations enter the cochlea (into the scala vestibuli)

45. The fluid that fills the scala media of the cochlea and the semicircular canals

46. The membrane in the organ of Corti into which the stereocilia of the outer hair cells project

47. A portion of the vestibular apparatus that senses position with respect to gravity

48. The portion of a semicircular canal where movements of the endolymph are sensed

49. The axonlike basal ends of the olfactory receptor cells

50. A structure on the tongue or pharynx serving the sense of taste

51. The structure within a nasal cavity where the sensory cells for smell are located

52. a

53. d

54. c

55. c

56. b

57. a

58. c

59. Both require detecting small differences in signals reaching identical paired organs that are located symmetrically on the head.

60. Because of inertia, the endolymph fluid in the semicircular canals remains in motion for a period of time. The compensatory neural pathways continue to make the visual corrections appropriate for rotation.

61. They are independent at the receptor level, but the fullest interpretation of the sensations requires CNS input from both.

62. Because of the low brightness of moonlight, only rod cells are sensitive enough to respond, and they cannot sense color.

63. Sensation is the uninterpreted signal from a sensory receptor, whereas perception requires processing at higher levels of the CNS.

Suggested Reading

Ackerman D: *A Natural History of the Senses.* New York, Random House, 1990.

Gulick WA, Gescheider GA, Frisna RD: *Hearing: Physiological Acoustics, Neural Coding, and Psychoacoustics.* New York, Oxford University Press, 1989.

Hudspeth AJ: How the ear's works work. *Nature* 134:397–404, 1989.

Kinnamon SC: Taste transduction: A diversity of mechanisms. *Trends Neurosci* 11:491–496, 1988.

Koretz JF, Handelman GH: How the human eye focuses. *Sci Am* 259:92–99, 1988.

Schmidt RF (ed): *Fundamentals of Sensory Physiology,* 2nd ed. New York, Springer-Verlag, 1981.

von Békésy G: *Experiments in Hearing.* New York, McGraw-Hill, 1960.

CHAPTER

■ ■ ■ ■ ■ ■ ■

5

The Motor System

CHAPTER OUTLINE

I. THE CONSTITUENTS OF THE MOTOR SYSTEM
   A. Bones provide the frame that muscles move
   B. Skeletal muscle powers movement by contracting
      1. Multiple functions
      2. Physiologic extensors and flexors
   C. Motor portions of the nervous system control movement
      1. Peripheral afferent input
      2. Brainstem and cerebral cortex
      3. Cerebellum and basal ganglia

II. MUSCLE INNERVATION
   A. Muscle spindles and Golgi tendon organs are the major sensory organs of skeletal muscle
      1. Muscle spindles
      2. Golgi tendon organs
   B. Alpha and gamma motor neurons constitute the motor innervation of muscle
      1. Gamma motor neurons and the fusimotor system
      2. Alpha motor neurons and the motor unit
   C. Coordination between sensory and motor neurons is important in regulating muscle contraction
      1. Simultaneous activation of alpha and gamma motor neurons
      2. Sensory responses to stretch

III. THE SPINAL CORD IN THE CONTROL OF MOVEMENT
   A. The anatomic arrangement of spinal motor systems serves reflex functions
   B. The spinal cord mediates reflex activity
      1. The myotatic (stretch) reflex
      2. The inverse myotatic reflex
      3. The flexor withdrawal reflex
   C. Transection of the spinal cord causes loss of voluntary movement and sensation, but reflexes are retained

IV. THE BRAINSTEM IN THE CONTROL OF MOVEMENT
   A. All but one of the descending motor tracts originate in the brainstem
      1. The red nucleus
      2. The vestibular nuclei
      3. The reticular formation
   B. Terminations of the brainstem motor systems reflect their functions
   C. Sensory and motor systems work together to control posture

OBJECTIVES

After studying this chapter, the student should be able to:
1. Describe the mechanical and physical structures that are under the control of the motor system
2. Explain how the fusimotor system modifies sensory signaling from muscles
3. Describe the role of somatosensory systems in muscle contraction, initiating reflexes, and conscious awareness of skeletomuscular position
4. Explain how walking movements in the legs can be elicited as a programed series of coordinated movements
5. Outline the neuronal circuits used in adjusting position, equilibrium, and balance when the body is in motion
6. Explain why destruction of the spinal cord motor neurons abolishes all contractility in muscles but destruction of the corticospinal neurons does not
7. Describe the pathway by which the cerebellum uses sensory input from muscles to influence alpha motor neurons even though the cerebellum does not innervate the motor neurons
8. Account for the close coordination of sensory and motor control of a given body part by the cerebral cortex
9. Explain which motor control systems are most responsible for planning movements and initiating movements and how these control systems find their way to the final common path

Whether it be the step that takes the sprinter across the finish line first, or the elegant slow arc of a dancer's arm, all human movement results from forces generated by muscles, gravity, friction, and external loads. Muscles provide the power that moves the skeleton. The qualities of speed, accuracy, smoothness, and forcefulness that characterize our movements result from complex interactions between bones, muscles, and the nervous system that controls them. Thus, the motor system is composed of three elements: bone, muscle, and neurons. Understanding how movement is produced requires a knowledge of each of these.

The Constituents of the Motor System

■ Bones Provide the Frame that Muscles Move

Bones are the body's levers, the elements that move. The way adjacent bones articulate determines the range of movement at a joint. The arrangement of ligaments crossing the joint and the bulk of adjacent tissue also impose limits on the range of movement.

Movements are described based on the anatomic planes through which the skeleton moves and the physical structure of the joint. Joints may have one or more axes, thus permitting movement in one, two, or multiple anatomic reference planes (Fig. 5–1).

Hinge joints, such as the elbow, are uniaxial, permitting movements in the sagittal plane. The wrist is an example of a biaxial joint. The shoulder is a multiaxial joint; movement can occur in oblique planes as well as the three major planes of that joint. Flexion and extension describe movements of uniaxial joints in the sagittal plane. **Flexion** movements decrease the angle between the moving and adjacent body segments or,

alternatively, bring ventral or volar surfaces closer together. **Extension** describes movement in the opposite direction. **Abduction** moves limbs away from the

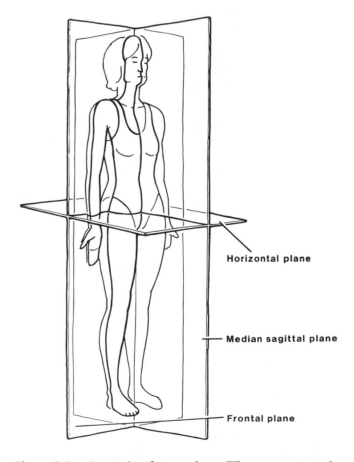

Figure 5–1 ■ Anatomic reference planes. When a person stands in the anatomic position with palms facing front, the cardinal reference planes are defined as illustrated.

Horizontal plane

Median sagittal plane

Frontal plane

midline or fingers away from the middle digit, and **adduction** does the opposite in the frontal plane.

▪ Skeletal Muscle Powers Movement by Contracting

Muscles span joints and are attached at two or more points to the bony levers of the skeleton; they provide the power that moves the body's levers. There are many different shapes: long, cylindric muscles, such as the biceps brachii, can produce rapid movements over large ranges of motion; sheetlike muscles, such as abdominal muscles, can form support walls against which other movements can take place; and pennate muscles, such as the gastrocnemius, can produce very strong contractions. Movement occurs when a muscle shortens (isotonic contraction) or when a muscle exerts tension while being lengthened by an outside force (eccentric contraction) (Chap. 9).

Since muscle contraction can produce movement in only one direction, at least two muscles opposing each other at a joint are needed to achieve motion in more than one direction. Thus, the number of muscles that span a joint and their anatomic arrangement also determine the type and range of movement at a joint.

Multiple Functions ▪ A muscle rarely acts alone; rather, groups of muscles act together to produce coordinated movement. Thus a single muscle can have different functions, depending on the movement. When a muscle produces movement by shortening it is an **agonist,** or mover; the **prime mover** is the muscle that contributes most to the movement. Muscles that oppose the action of the prime mover are **antagonists.** During both simple and skilled movements against a light load (e.g., when the load is only the body segment being moved) the antagonist is relaxed. Contraction of the agonist with concomitant relaxation of the antagonist occurs by **reciprocal innervation** (or **reciprocal inhibition**). Cocontraction of agonist and antagonist occurs during movements that require great effort. A muscle functions as a **synergist** if it contracts at the same time as the agonist and cooperates in production of movement. Synergists can aid in producing a movement (e.g., activity of both flexor carpi ulnaris and extensor carpi ulnaris are used in producing ulnar deviation of the wrist); eliminate unwanted movements (e.g., activity of wrist extensors prevents flexion of the wrist when finger flexors contract in closing the hand); or stabilize proximal joints (e.g., isometric contractions of muscles of the forearm, upper arm, shoulder, and trunk accompany a forceful grip of the hand). Stabilization differs from fixation in that it provides firmness but not immobility. Fixation, which requires strong isometric contraction of all the muscles that span a joint, is rarely used during normal movement.

Physiologic Extensors and Flexors ▪ The muscles are divided into two groups: physiologic extensors and physiologic flexors.

Physiologic extensors are also called **antigravity muscles,** which reflects their function as muscles that normally oppose the force of gravity. They play an important role in posture and locomotion. The gastrocnemius is a physiologic extensor, because it functions to raise the heel against gravity. Despite its anatomic definition, the flexor digitorum longus of the lower leg, which flexes the small toes and plantar flexes the foot, is also a physiologic extensor, because it assists in raising the body against gravity.

The common function of the physiologic extensors is vividly illustrated by extreme antigravity reactions. If the brainstem of a cat is transected between the inferior colliculus and vestibular nuclei, thereby eliminating cerebral function, the **decerebrate** animal will stand with all antigravity muscles rigidly contracted, legs rigidly extended, tail lifted slightly, and chin tipped upward. In humans in whom brain lesions occur between the cortical and brainstem motor areas, increased tone in the antigravity muscles is also observed. This is expressed as a flexed and pronated position of the arms (the flexors are the antigravity muscles of the upper limbs) and an extended, adducted posture of the legs.

Physiologic flexors are antagonistic to physiologic extensors; their common functional role is to withdraw a body part, a reaction often seen after contact with a painful stimulus. The tibialis anterior (an antagonist of the gastrocnemius), which dorsiflexes the foot, is a physiologic flexor. Despite its anatomic definition, the extensor digitorum longus of the lower leg, which extends the small toes and dorsiflexes the foot, is also a physiologic flexor. Both muscles are active in withdrawal movements of the foot.

▪ Motor Portions of the Nervous System Control Movement

In describing neural control of movement, we identify which portions of the nervous system are predominantly motor and discuss probable functions for each area. It is important to remember that even the simplest movement requires the complex interactions of many muscles as well as the attentive watchfulness of the nervous system over balance and posture. Thus, there are always multiple levels of control that are simultaneously active. The common feature of the controlling elements is that they ultimately influence the motor neurons which are connected with skeletal muscles (Fig. 5–2).

The motor neurons in the spinal cord and cranial nerve nuclei and their axons constitute the **final common path,** the path by which all central nervous activity influences the skeletal muscles. In the spinal cord, these motor neurons are located in the ventral horns of the gray matter. Their long axons exit the cord via the ventral roots and terminate in the motor endplate region of the skeletal muscle fibers. Afferent fibers that influence the final common path come from

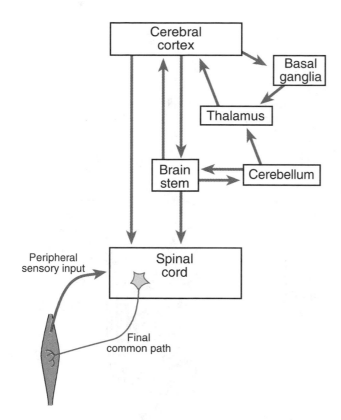

Figure 5–2 ■ Overview of the motor control system. The final common path transmits all central nervous system commands to the skeletal muscles. This path is influenced by peripheral sensory input and descending signals from the brainstem and cerebral cortex. The cerebellum and basal ganglia exert their influence on the final common path indirectly, using the brainstem and cerebral cortical pathways.

three sources: the periphery, nuclei in the brainstem, and the cerebral cortex.

Peripheral Afferent Input ■ Sensory signals influencing the final common path include those from receptors such as muscle spindles and Golgi tendon organs, the major sensory receptors of muscle, and receptors from joints, skin, and subcutaneous tissues; collectively, these pathways provide **somatosensory** input. These sensory fibers either terminate directly on the motor neurons, forming a monosynaptic pathway, or, more frequently, form polysynaptic pathways by synapsing on **interneurons,** intrinsic neurons interposed between the afferent and efferent neuron. Receptors originating in muscles and their tendons respond to changes in muscle length and tension; receptors in joints and skin respond to changes in pressure and noxious or injurious stimuli.

Brainstem and Cerebral Cortex ■ All but one of the descending afferent pathways to the motor neurons arise in the brainstem. The single exception is the **corticospinal tract,** which arises from the motor portions of the cerebral cortex. While some of the corticospinal fibers terminate directly on motor neurons, most others synapse on interneurons.

Interneurons interposed between the afferent fibers and the final common path serve an important integrative role in motor activity. Their intrinsic circuitry allows an afferent signal to influence multiple motor neurons at the same time, to coordinate agonist and antagonist muscles in a reciprocal fashion during movement and to coordinate ipsilateral and contralateral muscle groups for postural adjustments to movement.

Cerebellum and Basal Ganglia ■ Two additional structures with important roles in control of movement do not directly innervate the motor neurons. The **cerebellum** and **basal ganglia** influence the motor neurons via projections to both motor cortical areas and the brainstem.

Muscle Innervation

The muscles, joints, and ligaments are richly innervated with sensory receptors that inform the central nervous system about the position of the body in space. Skeletal muscles contain muscle spindles, Golgi tendon organs, free nerve endings, and some pacinian corpuscles. Joints contain Ruffini's endings and pacinian corpuscles, joint capsules contain nerve endings, and ligaments contain Golgi tendonlike organs. Together these are the **proprioceptors,** providing sensation from the deep somatic structures. These sensations include awareness of the position of the limbs, of movement of the limbs, and of how the limbs are moving. They provide feedback control of movements, thus permitting immediate compensation for unexpected disturbances, and participate in the learning of skilled movements by poorly understood mechanisms.

■ Muscle Spindles and Golgi Tendon Organs Are the Major Sensory Organs of Skeletal Muscle

Muscle spindles provide information about the length of muscle and velocity at which the muscle is being stretched, while the Golgi tendon organs provide information about tension. Spindles are located in the body of the muscle, in parallel with the main muscle fibers. Tendon organs are located at the border between tendon and muscle, in series with the muscle fibers (Fig. 5–3A).

Muscle Spindles ■ These sensory organs are found in almost all of the skeletal muscles and are of greatest density in small muscles serving fine movements, such as those of the hand, and in the deep muscles of the neck. The muscle spindle, named for its long **fusiform** shape, lies adjacent to and is attached at both ends to the main or **extrafusal** muscle fibers. Within the spindle's expanded middle portion is a fluid-filled capsule containing 2–12 striated muscle fi-

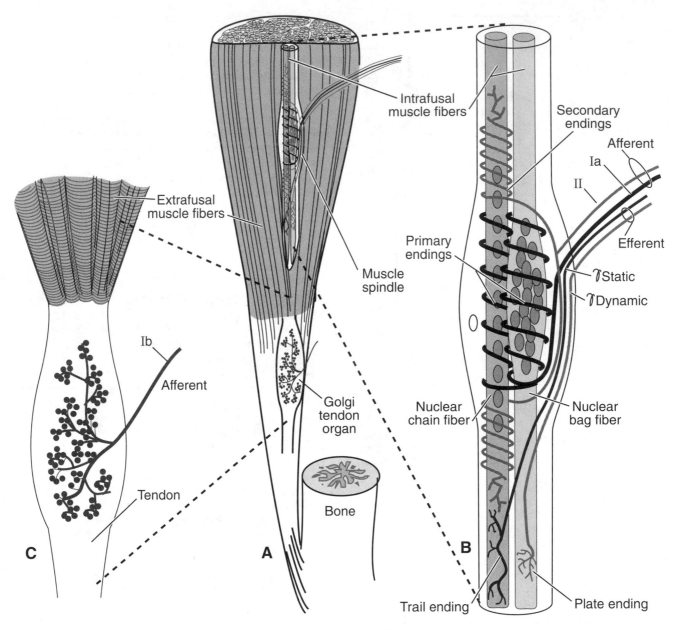

Figure 5-3 ■ The muscle spindle and Golgi tendon organ. (A) The muscle spindle lies parallel with the extrafusal muscle fibers; the Golgi tendon organ is in series with them. (B) Enlargement of the muscle spindle, showing two types of intrafusal muscle fibers, the nuclear bag and nuclear chain fibers; the sensory innervation, with a type Ia fiber ending wrapped around each intrafusal fiber and type II fiber endings on the nuclear chain fibers; and motor innervation by gamma motor neurons. The dynamic gamma motor fibers have plate nerve endings and the static fibers have trail endings. Although each spindle typically contains up to 12 intrafusal fibers, for purposes of illustration only 2 are shown. (C) Enlargement of the Golgi tendon organ, showing its attachments to a group of extrafusal muscle fibers on one end and the tendon on the other. An inner spray of fine nerve endings interspersed in collagen fiber bundles gives rise to the type Ib afferent fibers.

bers entwined by sensory nerve terminals. These **intrafusal** muscle fibers, about 300 μm long, have an array of contractile filaments at the ends of the fibers; the noncontractile midportion is filled with cell nuclei (Fig. 5-3B). Afferent nerve terminals surround the con-

tractile ends and noncontractile middle of the intrafusal fiber; motor neurons innervate the contractile ends.

There are two types of intrafusal fibers: **nuclear bag fibers,** so named for the large number of nuclei packed into their middle, and **nuclear chain fibers,** in which the nuclei are arranged in a single row. The nuclear bag fibers are subdivided further into bag$_1$ and bag$_2$ fibers, based on physiologic and histochemical criteria. There are about twice as many nuclear chain fibers as nuclear bag fibers per spindle, and the chain fibers are usually shorter than the bag fibers.

Specialized sensory fibers originating in the spindle consist of the **primary,** or **type Ia,** and the **secondary,** or **type II,** afferent fibers. Type Ia fibers are larger in diameter (12–20 μm) than type II fibers (6–12 μm) and are 2 to 3 times faster in conduction velocity. Type Ia fibers have spiral endings wrapped around the middle of the intrafusal fiber (Fig. 5-3B), and they innervate nuclear bag fibers and nuclear chain fibers. Each spindle is served by a single Ia afferent fiber and its axon branches. In contrast, as many as five type II fibers may innervate a single spindle, and they innervate only the nuclear chain fibers. Type II fibers have either spiral endings or, more rarely, spray endings, located along the contractile part of the nuclear chain fiber on either side of the type Ia spiral ending. The specialized endings of both primary and secondary afferent fibers of the muscle spindles respond to stretch by generating action potentials that convey information to the central nervous system about change in muscle length and the **velocity of stretch.** Both stop generating action potentials during a muscle twitch.

Golgi Tendon Organs ■ The Golgi tendon organs (GTOs) are long (approximately 1 mm), slender receptors encapsulated within the tendons of the skeletal muscles (Fig. 5-3A, C). The distal pole of the GTO is anchored in collagen fibers in the extracellular matrix of the tendon. The proximal pole is firmly attached to the ends of the extrafusal muscle fibers. By this arrangement, the GTO lies in series with the extrafusal muscle fibers, such that contractions of the muscle result in stretching the GTO.

Large-diameter, fast-conducting, type Ib afferent nerve fibers arise from each GTO. Contractions in the muscle stretch the GTO, thereby straightening the spray endings and generating action potentials in type Ib fibers. The neural information they convey to the central nervous system is that of **tension** in the muscle.

■ Alpha and Gamma Motor Neurons Constitute the Motor Innervation of Muscles

Alpha motor neurons innervate the extrafusal muscle fibers, and gamma motor neurons innervate the intrafusal fibers. Cells bodies of both alpha and gamma motor neurons reside in the ventral horns of the spinal cord and in nuclei of the cranial motor nerves, and both are activated by the same afferent systems. Nearly one-third of the motor nerve fibers innervating the muscles supply the intrafusal fibers, an astonishing figure that indicates a complex role for this motor control system. Intrafusal muscle fibers likewise constitute a significant portion of the total number of muscle cells, yet they contribute little or nothing to the total force generated when the muscle contracts. Rather, the contractions of intrafusal fibers serve a sensory role, as they alter the length and sensitivity of the sensory receptors.

Gamma Motor Neurons and the Fusimotor System ■ The gamma motor neurons and the intrafusal fibers they innervate are collectively referred to as the **fusimotor system.** Axons of the gamma neurons terminate in one of two types of endings, each located distal to the sensory endings on the striated poles of the spindle fibers (Fig. 5-3B). The nerve terminals are either **plate endings** or **trail endings,** and each intrafusal fiber has only one of these two types of endings. Plate endings occur predominantly on bag$_1$ fibers, whereas trail endings are seen primarily on chain fibers but also on bag$_2$ fibers. This arrangement allows for largely independent control of the nuclear bag and nuclear chain fibers in the spindle.

Fibers from gamma motor neurons containing plate endings are designated **dynamic gamma fibers,** and those containing trail endings are designated **static gamma fibers.** This functional distinction is based on experimental findings that showed that stimulation of gamma neurons with plate endings enhanced the response of type Ia afferent sensory fibers to stretch, but only during the *dynamic* phase of muscle stretching (i.e., when the muscle length was actually changing). During the steady-state or **static** phase, where the stretch was maintained at constant muscle length, stimulation of the gamma neurons with trail endings enhanced the response of type II, or secondary, afferent fibers. Static gamma neurons affect the responses of both Ia and type II afferent fibers; dynamic gamma neurons affect the response of only type Ia fibers. Furthermore, gamma fibers are distributed differently to different spindles, all of which suggests that the motor system has the ability to control and monitor muscle length more carefully in some muscles and the speed of contraction in others.

Alpha Motor Neurons and the Motor Unit ■ The alpha motor neuron, its axon, and all of the muscle fibers (extrafusal) it innervates comprise the **motor unit.** One alpha motor neuron may innervate as few as three extrafusal fibers (e.g., in the extrinsic muscles of the eye) to as many as 1700 extrafusal fibers (e.g., in the gastrocnemius muscle).

Alpha motor neurons separate into two populations according to their size. The larger cells are rapidly conducting, have a high threshold to synaptic stimulation, and are normally silent. They innervate fast twitch muscle fibers (Chap. 9). The smaller alpha motor neurons conduct action potentials at a lower velocity, have a low threshold to synaptic stimulation, and are normally tonically active. They innervate slow twitch muscle fibers.

Unlike most muscles that contain differing numbers of fast and slow twitch fibers, the motor unit is homogeneous, in that all of the innervated extrafusal fibers are of the same type—either fast or slow. This organization has important functional consequences for the production of smooth, coordinated contractions. The smallest neurons have the lowest threshold and are therefore activated first when synaptic activity is low; this produces low-force contractions in slow muscle fibers. As synaptic drive increases, the larger motor neurons are activated. This orderly process, called the **size principle,** accomplishes two functions. First, a smooth contraction is produced as increasingly faster units are recruited before the previously recruited ones reach their maximum tension. Second, fine control is possible at all tensions. At low tensions, recruitment of increasing numbers of small neurons produces small increments in tension. At higher tensions, as larger motor neurons are recruited, each contributes only a small increment to the total tension generated. A logical corollary of this arrangement is that muscles concerned with endurance, such as the antigravity muscles, contain predominantly slow muscle fibers, in accordance with their function of continuous postural support. They are tonically active even at rest, because they are so sensitive to stretch. Muscles that contain predominantly fast fibers, including many physiologic flexors, are capable of producing rapid movements and fine, skilled movements.

Muscles concerned with strength as well as endurance, such as trunk and leg muscles, contain many large motor units, and those concerned with fine, manipulative movements, such as muscles of the hands, face, and eyes, contain many small motor units. Thus the size of the motor units that supply a muscle and the composition of fast and slow fibers in the muscle are important determinants of its overall capabilities.

■ Coordination between Sensory and Motor Neurons Is Important in Regulating Muscle Contraction

The sensory nerves from muscles and tendons are primarily stretch receptors. Muscle spindles signal an increase in length by firing action potentials when the spiral endings of type Ia afferent fibers are stretched. Golgi tendon organs signal an increase in tension by firing action potentials when axons of type Ib afferent fibers are stretched. Activation of both types of receptors serves not only to provide information about muscle length and tension to the central nervous system, but also to ensure that the sensory systems remain responsive regardless of the state of muscle contraction. This is especially important in the case of muscle spindles, since muscle contraction silences their receptors.

When the muscle is at rest, the muscle spindles are slightly stretched and type Ia afferent nerves exhibit a slow rate of discharge of action potentials. Contraction of the muscle increases the firing rate in type Ib afferent fibers from Golgi tendon organs, whereas the type Ia afferent fibers cease firing as shortening of the surrounding extrafusal fibers shortens the intrafusal fibers as well, thereby slackening their tension and silencing the spirally wound type Ia neurons. If a load were placed on the spindle, the slack would be removed and the spiral endings would resume their sensitivity to stretch. This is the role assumed by the gamma motor neurons: they "reload" the spindle during muscle contraction by contracting the ends of the spindle fibers to stretch the type Ia afferent endings in the middle. They accomplish this by firing simultaneously with the alpha motor neurons when muscles contract (Fig. 5-4).

Simultaneous Activation of Alpha and Gamma Motor Neurons ■ During normal movement, alpha and gamma motor neurons are stimulated simultaneously, a phenomenon called **alpha-gamma coactivation.** Activation of the gamma system in this way allows the intrafusal fibers to shorten and thus remain sensitive over a wide range of muscle lengths. While alpha-gamma coactivation suggests a single controller for both motor systems, there is also evidence that the systems are independently controlled.

Sensory Responses to Stretch ■ Stretching the muscle, such as by attaching a weight to one end, results in loading the spindle as well as stimulating stretch receptors in muscles and tendons. Both primary (Ia) and secondary (II) spindle afferent fibers are sensitive to stretch, but because of their differing anatomic arrangement on intrafusal fibers and inherent electrical properties they respond quite differently to a stretch stimulus and its component phases (Fig. 5-5). During the onset of stretch, or **dynamic** phase, type Ia fibers have a uniform rapid discharge rate. During the steady-state, or **static** phase, the discharge rate is slower, and it shuts off entirely during the offset of the stretch stimulus. In contrast, type II fibers respond with a gradually increasing rate of discharge during stretching, which is maintained during the static stretch and decreases when the stretch is released. Type Ia fiber responses are linear—the greater the stretch, the greater the rate of firing of action potentials—and they have high sensitivity during small stretches. However, with a stretch greater than 100–200 μm their response is less sensitive and nonlinear. After a large stretch they can reset to the original high sensitivity observed with small stretches. Type II fibers are less sensitive at all degrees of stretch and do not exhibit the nonlinearities of type Ia fibers. These properties may reflect the differences in the structural properties of the regions each fiber innervates. Overall, the physiologic properties suggest that the signal carried by type Ia fibers reflects both the length of the fiber and the velocity of stretch, whereas the signal carried by the type II fibers is solely a function of length.

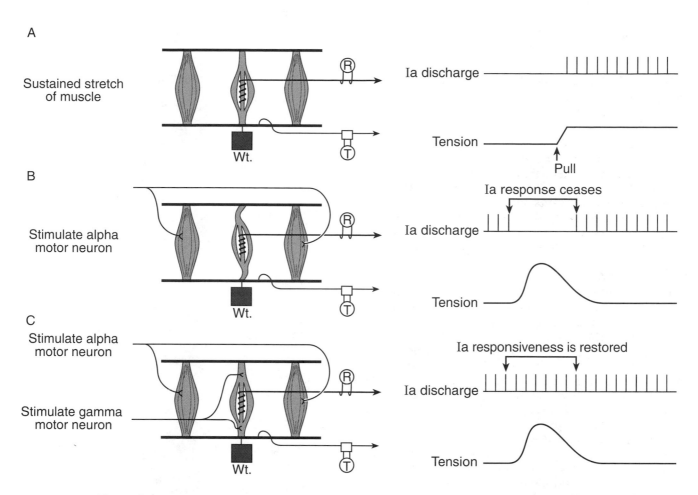

Figure 5–4 ■ Loading and unloading the muscle spindle. An intrafusal muscle fiber with its type Ia afferent nerve fiber is shown parallel with the extrafusal muscle fibers in a muscle. The firing rate (R) of the Ia fiber is continuously recorded electrically, and muscle tension (T) is recorded mechanically. (A) In the "resting" state, extrafusal and intrafusal muscle fibers are slightly stretched and there is little or no Ia discharge. When stretch is applied by pulling the attached weight, stretch of the spindle results in increased firing of the Ia fiber and a rise in muscle tension. (B) The alpha motor neurons are stimulated, resulting in active contraction of the extrafusal fibers. This shortens the spindle fibers as well and abolishes the stretch signal; the Ia afferent fibers cease firing. This inactivated state of the spindle is referred to as "unloaded." In this condition it ceases to inform the central nervous system about change in muscle length. (C) If the gamma motor neurons are stimulated at the same time as the alpha motor neurons, the polar ends of the spindle fibers shorten and allow the midportion to stretch. This stretches the surrounding Ia afferent fibers ("reloading" the spindle) and reinstitutes their responsiveness to changes in muscle length. (Modified from Kandel ER, Schwartz JH, Jessel TM. *Principles of Neural Science.* 3rd ed. New York, Elsevier, 1991.)

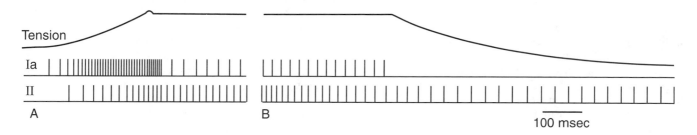

Figure 5–5 ■ Responses of a type Ia ending and a type II ending during (A) a rapidly applied stretch and (B) a slowly applied release. Both fiber types are sensitive to passive stretch. The signal from Ia fibers reflects both length of the fiber and velocity of stretch; the signal from type II fibers reflects only length.

The Spinal Cord in the Control of Movement

Muscles interact extensively in the maintenance of posture and the production of coordinated movement. The circuitry of the spinal cord automatically controls much of this interaction. Sensory feedback from muscles reaches motor neurons of related muscles and, to a lesser degree, of more distant muscles. In addition to activating local circuits, muscles and joints transmit sensory information up the spinal cord to higher centers. This information is processed and can be relayed back to spinal cord circuits through a **long loop.**

■ The Anatomic Arrangement of Spinal Motor Systems Serves Reflex Functions

Cell bodies of motor neurons of the spinal cord are grouped into pools in the ventral horns. A **pool** is defined as all the neurons that serve a particular muscle. They are organized such that those innervating the muscles of the body's axis are located mostly medially and those innervating the limbs are located mostly laterally. Neurons innervating the physiologic flexors and extensors are also segregated into pools, located in different sites at proximal and distal segments of the spinal cord (Fig. 5–6). A zone between the medial and lateral pools is also organized such that interneurons in the lateral pools project to limb motor neuron pools ipsilaterally and medial interneurons project to axial pools bilaterally. This zone also contains interneurons, which project onto motor neurons innervating the girdle muscles. The outcome of this compartmentalization of spinal cord motor systems is that whole body and limb movements for maintaining posture and balance are controlled in the medial area of the ventral horns, whereas fine manipulative movements of the extremities are controlled by the lateral areas.

Between the spinal cord's dorsal and ventral horns lies the intermediate zone, an area containing an extensive network of interneurons that interconnect motor neuron pools in well-defined circuits. Some interneurons make connections in their local cord segment; others have longer axon projections that travel in the white matter to terminate in other segments of the spinal cord. These long-axon interneurons, the **propriospinal cells,** can carry sensory information to many of the motor neurons necessary to maintain balance and to continue execution of movements. The importance of spinal cord interneurons is reflected in the fact that they comprise the majority of neurons in the spinal cord and the majority of synapses on motor neurons.

■ The Spinal Cord Mediates Reflex Activity

The spinal cord contains the circuitry to generate **reflexes,** stereotyped actions elicited in response to a stimulus applied to the periphery. Some reflexes are

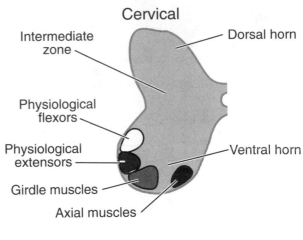

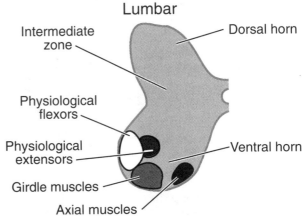

Figure 5–6 ■ The organization of motor neuron pools in the spinal cord at the cervical and lumbar levels. Pools for axial muscles are located medially; those innervating the limbs are located laterally and slightly dorsal to the axial pools. Muscles of the pelvic girdle are served by pools intermediate between the axial and limb pools. Pools controlling physiologic flexors and extensors are anatomically segregated in the spinal cord.

simple, others exceedingly complex. Even the simplest requires coordination in which prime movers contract, antagonists relax, and synergists and stabilizers become active. The circuitry in the spinal cord that accomplishes this coordination is based on formation of functional units, consisting of specific motor neurons and interneurons, ipsilateral and contralateral, whose activation as a unit results in a coordinated, stereotypical response. The circuits of these functional units under normal circumstances are heavily modulated by higher motor centers, whose descending pathways either facilitate or inhibit their activation.

A main function of reflex movement is to generate a rapid response. This is illustrated by the common occurrence of touching a hot object and withdrawing the hand well before the heat is felt. The spinal cord reflex mechanism protects the organism before the central nervous system identifies the problem.

Like all reflex arcs, the reflexes of the spinal cord require a sensor, an afferent pathway, an integrating center, an efferent pathway, and an effector. The sensory receptors for spinal reflexes are **proprioceptors** and **cutaneous receptors.** Impulses initiated in these receptors travel along **afferent nerves** to the **spinal cord,** where the motor neurons and interneurons constitute the integrating center. The final common path, or **motor neurons,** make up the efferent pathway to the effector organs, the **skeletal muscles.**

Three important types of spinal reflexes are elicited by specific sensory stimuli: muscle proprioceptors initiate the myotatic and inverse myotatic reflexes, and other peripheral receptors, particularly nociceptors, initiate the flexor withdrawal reflex.

The Myotatic (Stretch) Reflex ■ Stretching a muscle causes it to contract reflexly with very short **latency,** that period between the onset of a stimulus and the response. The short latency of this excitatory response, called the **myotatic** or **stretch reflex,** is due to its monosynaptic circuitry, where an afferent sensory neuron synapses directly on the efferent motor neuron. The receptors activated by stretch are primary muscle spindle afferent fibers. Type Ia afferent fibers carry signals to the spinal cord, where their axon branches synapse directly on motor neurons of the same **(homonymous)** muscle and on motor neurons of synergist **(heteronymous)** muscles (Fig. 5-7). These synapses are excitatory, with each type Ia fiber contacting all or almost all of the neurons of a given motor neuron pool. Monosynaptic type Ia synapses occur predominantly or solely on alpha motor neurons; gamma motor neurons seemingly lack such terminals. Collaterals of type Ia fibers also synapse on inhibitory interneurons, which inhibit motor neurons of antagonist muscles (Fig. 5-7). This synaptic pattern, called **reciprocal inhibition,** serves to coordinate muscles of opposing function around a joint. Secondary (type II) spindle afferent fibers contribute to the reflex by making monosynaptic excitatory connections to homonymous motor neurons, thus reinforcing the reflex.

Information entering the spinal cord via type Ia fibers goes to many other places as well, including many spinal interneurons. They also serve as primary afferent fibers ascending in the spinal cord as the dorsal column lemniscal pathway that projects via the thalamus to the cerebral cortical somatosensory area. This sensory mechanism imparts to the organism an awareness of the position of the limbs and trunk in space. Type Ia and Ib afferent fibers are also the primary afferent neurons in the ascending spinocerebellar and spinoolivary tracts, which convey information to the cerebellum from moment to moment on length and tension of muscles.

The myotatic reflex has two components: a phasic, intense, short-lasting component, exemplified by tendon jerks (e.g., the familiar "knee jerk"), and a tonic, less powerful, but longer-lasting component thought

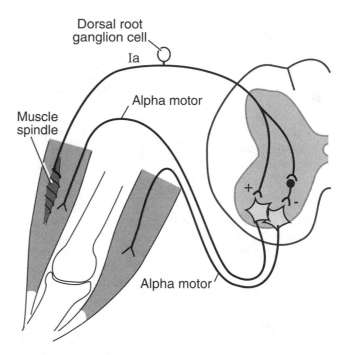

Figure 5–7 ■ Schematic representation of the circuitry of the myotatic reflex. Each Ia fiber contacts all or almost all of the neurons of a motor pool monosynaptically. A collateral branch excites an interneuron (dark cell) which inhibits motor neurons of antagonist muscles. + indicates stimulation; − indicates inhibition.

to be important for maintaining posture. These phases blend together, but one or the other may predominate depending on whether other afferent activity, such as from descending and cutaneous afferent neurons, simultaneously influences the motor response. Primary spindle afferent fibers probably mediate the tendon jerk, with secondary afferent fibers contributing mainly to the tonic phase of the reflex.

The stretch reflex performs many functions. At the most general level, it produces rapid corrections of motor output in the moment-to-moment control of movement. It also forms the basis for postural reflexes, which maintain upright posture despite a varying range of loads and/or external forces on the body. Myotatic reflexes are strongest and most readily elicited in the physiologic extensors, or antigravity muscles, and only in these muscles is the contraction sustained for as long as the stretch is applied.

The Inverse Myotatic Reflex ■ Active contraction of a muscle causes reflex inhibition of the contraction. This is called the **inverse myotatic reflex** because it produces an effect that is opposite to that of the myotatic reflex, although both are initiated by proprioceptors from the muscle. While the name accurately reflects the opposite output of the myotatic reflex, it is somewhat misleading to think of a tension-regulating system as being the inverse of a length-regulating system. It is best to think of these reflexes

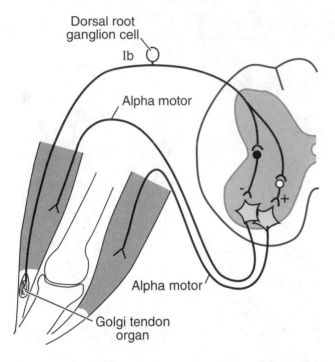

Figure 5-8 ■ Schematic representation of the circuitry of the inverse myotatic reflex. Golgi tendon organs discharge during active contraction and Ib fibers carry impulses to interneurons, which inhibit (dark cell) homonymous motor neurons and excite (light cell) motor neurons of antagonist muscles.

as being elicited by different stimuli, stretch being the stimulus for the myotatic reflex and active contraction the stimulus for the inverse myotatic reflex.

Active muscle contraction deforms Golgi tendon organs by stretching them and causing them to send discharges along type Ib fibers to the spinal cord. The fibers synapse on inhibitory interneurons that contact homonymous and heteronymous motor neurons and on excitatory interneurons that contact motor neurons of antagonists. Thus, all of the connections of type Ib fibers are through interneurons, and the longer latencies reflect the polysynaptic nature of the reflex (Fig. 5-8).

The inverse myotatic reflex, as in the case of the myotatic reflex, has a more potent influence on the physiologic extensor muscles than on the flexor muscles, suggesting that the two reflexes act together to maintain optimal conditions when using the antigravity muscles in postural adjustments. The function of the inverse myotatic reflex is best described in general terms as a tension feedback system. A current hypothesis about the conjoint function of both of these reflexes is that they contribute to the smooth generation of tension in muscle, helping to overcome some of the irregular mechanical properties of muscle by regulating muscle stiffness.

The Flexor Withdrawal Reflex ■ Cutaneous stimulation, such as touch, pressure, heat, cold, or tissue damage, can elicit a **flexor withdrawal reflex.** This consists of contraction of flexors and relaxation of extensors in the stimulated limb, which can be accompanied by contraction of the extensors on the contralateral side. Cutaneous sensory receptors send fibers into the spinal cord to synapse on interneurons in the dorsal horn, which ultimately act to excite the alpha motor neurons of flexor muscles and inhibit those of extensor muscles. Collaterals of interneurons that contact the ipsilateral alpha motor neurons cross the cord to excite extensor alpha motor neurons and inhibit flexor alpha motor neurons on the contralateral side (Fig. 5-9).

There are two types of flexor withdrawal reflexes: those that result from innocuous stimulation and those that result from potentially harmful or injurious stimulation. The first type produces a localized flexor response accompanied by slight or no limb withdrawal; the second type produces widespread flexor contraction throughout the limb and abrupt withdrawal. The function of the second type is easily seen to be protection of the individual. The endangered body part is rapidly removed and postural support of the opposite side is strengthened if that is needed (e.g., if the foot is being withdrawn). The function of the first type of reflex is less obvious, but may be considered in a general way to be a mechanism for allowing movement of a body part over or around an obstacle.

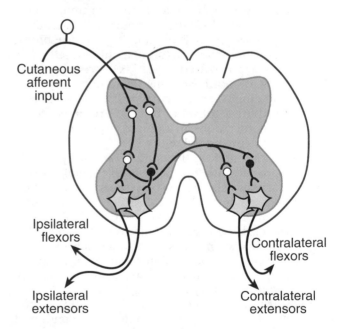

Figure 5-9 ■ Schematic representation of the circuitry of the flexor withdrawal reflex. Cutaneous stimulation elicits flexor withdrawal reflexes by activating ipsilateral flexors (via light interneurons) and inhibiting ipsilateral extensors (via dark interneurons). Contralateral muscles, which provide postural support, can also be engaged.

These reflexes, when taken together, provide for stability and postural support (the myotatic and inverse myotatic) and mobility (flexor withdrawal); that is, the reflexes provide a foundation of automatic responses on which more complicated voluntary movements may be built.

■ Transection of the Spinal Cord Causes Loss of Voluntary Movement and Sensation, but Reflexes Are Retained

When a person's spinal cord is severed, all voluntary movement and all sensation below the cut are permanently lost. After a period of **spinal shock,** characterized by absence of reflexes **(areflexia),** lasting from weeks to months, the reflexes return. The extent and time course of recovery is highly variable, depending to some extent on the nature and extent of damage to the cord. Frequently, noxious stimuli elicit reflexes early during the recovery period. Flexor withdrawal reflexes first reappear in the foot and ankle, then in the more proximal muscles. Tendon or stretch reflexes are hyperactive on their reappearance, and **clonus** may occur. Clonus is characterized by repetitive contractions and relaxations of muscle in an oscillating fashion in response to a single stimulus. There may also be strong supporting reflexes. Genital reflexes, reflex emptying of the bladder and the bowel, and outbursts of sweating may reappear following spinal cord section.

The Brainstem in the Control of Movement

For locomotion (walking) to occur, the body must be well supported and in equilibrium. The control of stance, postural adjustment, and spontaneous locomotion are functions of the brainstem, usually defined as including the medulla oblongata (medulla), pons, and mesencephalon (midbrain). The fact that all descending motor pathways but one, the corticospinal tract, originate from neurons in the brainstem underscores the importance of this part of the brain in motor control.

Studies in experimental animals (mostly cats) have demonstrated that their spinal cords contain the basic circuitry needed for coordinated walking movements. This circuitry, called the **central pattern generator** for locomotion, can bring about the properly timed alternate contraction of flexors and extensors needed for walking. The normal strategy for generating locomotion engages the central pattern generators and uses sensory feedback for control. Sensory afferent signals from muscle spindles, Golgi tendon organs, joint receptors, and sensory receptors of the skin and subcutaneous tissues play a role in modifying the temporal pattern of locomotion. Descending signals from the cortex and brainstem can alter the force of muscle contraction during the appropriate phase of stepping.

The spinal cord possesses the machinery for generating locomotion but lacks the mechanisms for control of posture and equilibrium.

The brainstem contains the mechanisms to control posture and achieve equilibrium and to initiate locomotion. Initiation of locomotion is believed to be controlled in the **mesencephalic locomotor region,** located at the level of the inferior colliculus. Motivated locomotion depends on the presence of the diencephalon and basal ganglia; the corticospinal system is necessary for fine individual movements of the distal limbs and digits. Each higher level of the nervous system acts on lower levels to produce more refined and appropriate movements.

■ All but One of the Descending Motor Tracts Originate in the Brainstem

Three brainstem nuclear groups give rise to descending motor tracts that influence motor neurons and their associated interneurons. These consist of the red nucleus, the vestibular nuclear complex, and the reticular formation (Fig. 5–10). The other major descending influence on the motor neurons is the corticospinal tract, the only volitional control in the motor system. These central pathways are complex and are best understood in terms of their anatomic connections, the motor responses they elicit when stimulated or ablated, the nature of the inputs to their nuclei in the brainstem, and the impact of their loss in neurologic diseases.

The descending brainstem pathways are organized in a manner that corresponds to the organization of spinal cord motor neuron pools. For example, some descending pathways exert effects on the antigravity muscles, and others influence the flexor muscles. Those pathways serving a role in posture, equilibrium, and balance innervate the medial pools of motor neurons associated with axial muscles and proximal limb muscles. Those serving the extremities, where finely

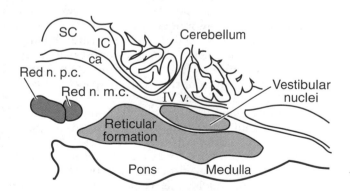

Figure 5–10 ■ Nuclei of origin of the major descending brainstem motor tracts, as seen in the monkey. SC, superior colliculus; IC, inferior colliculus; ca, cerebral aqueduct; IV v., fourth ventricle; red n. p.c., parvicellular portion of the red nucleus; red n. m.c., magnocellular portion of the red nucleus.

controlled movements are coordinated, innervate the lateral motor groups of motor neuron pools. In most cases, brainstem nuclei innervate motor neurons indirectly via interneurons, but in some cases direct connections are made monosynaptically.

The Red Nucleus ■ The red nucleus (nucleus ruber) contains a rostral parvicellular (small cells) portion and a caudal magnocellular (large cells) portion. In primates, including humans, projections from the rostral portion innervate brainstem motor neurons and projections from the magnocellular cells form the **rubrospinal tract,** which innervates spinal cord motor neurons. The major inputs to the red nucleus are from the cerebellum's **interposed nucleus** and the cerebral cortical motor and supplementary motor areas. Terminations of rubrospinal tract fibers in the spinal cord are similar to those of corticospinal terminations, a factor of interest when the functions of this system are considered. Electrophysiologic studies reveal that many rubrospinal neurons are active during locomotion, with more than half showing increases during the swing phase of stepping, when the flexors are most active. This system appears to be important for the production of movement, especially that of the distal limbs. Experimental lesions that sever primarily rubrospinal fibers produce deficits in distal limb flexion, with little change in more proximal muscles or extensors.

The Vestibular Nuclei ■ There are four major nuclei in the vestibular complex: the superior, lateral (also called Deiters), medial, and inferior vestibular nuclei. These nuclei receive a prominent input from the labyrinths in the inner ear (the vestibule), from which afferent nerves originating in the semicircular canals, utricle, and saccule convey sensory information on angular and linear acceleration of the head. They also receive a large input from higher centers and other brainstem regions. Vestibular nuclei are reciprocally connected with the superior colliculus, a protuberance on the dorsal surface of the midbrain where visual input from the retina aids in orienting the body and adjusting the gaze during movements of the head. Reciprocal connections with the vestibular nuclei are also made with the cerebellum and reticular formation. Overall, the vestibular system serves to regulate the body musculature for postural adjustments to changes in position of the head in space and with accelerations of the body.

The chief descending tract, commonly called the **vestibulospinal tract,** is the lateral vestibulospinal tract originating entirely in Deiters nucleus. Its terminal fibers are excitatory to extensor motor neurons, both alpha and gamma, and inhibitory to flexor motor neurons. Deiters nucleus is responsible for the increased tone of the antigravity muscles in decerebrate rigidity, and lesions in this area in a decerebrate animal relieve the rigidity on that side. This tract functions to drive the extensor system of the musculature, a role of some importance in the maintenance of posture.

The Reticular Formation ■ The central gray core of the brainstem contains many fiber bundles interwoven with cells of various shapes and sizes. A prominent characteristic of reticular formation neurons is that their axons project widely in ascending and descending pathways, making innumerable synaptic connections throughout the neuraxis (Chap. 7). In addition, cell bodies of the major ascending and descending monoaminergic pathways of the central nervous system originate in the reticular formation. The medial region of the reticular formation contains large neurons that project cranially to the thalamus as well as caudally to the spinal cord. Afferent input to the reticular formation comes from the spinal cord, vestibular complex, cerebellum (especially the fastigial nucleus), lateral hypothalamus, globus pallidus, tectum, and sensorimotor cortex. Its descending tracts arise in the pontine reticular formation from the nucleus reticularis pontis caudalis and nucleus reticularis pontis oralis and in the medullary reticular formation from the nucleus reticularis gigantocellularis. These tracts exert excitatory and inhibitory effects on the motor system throughout the central nervous system.

There are two subsystems of the reticular formation that are important in control of extensor motor neurons: an area in the medulla that gives rise to the **reticulospinal tract,** which is predominantly inhibitory to extensor motor neurons, and a more rostral area, which is predominantly facilitatory.

■ Terminations of the Brainstem Motor Systems Reflect Their Functions

The vestibulospinal and reticulospinal tracts descend medially in the spinal cord and terminate bilaterally in the ventromedial part of the intermediate zone, an area of long propriospinal neurons that have some bilateral distribution (Fig. 5–11). There are some direct connections with motor neurons of neck and back muscles and proximal limb muscles. These tracts, the medial descending pathway, are the main control systems of the central nervous system for maintenance of posture, coordination of movements of the body and limbs, and orientation of body and head position during movement.

The rubrospinal tract descends laterally in the spinal cord and terminates in the dorsal and lateral spinal intermediate zone, an area of short propriospinal neurons (Fig. 5–11). This tract has some monosynaptic connections directly on motor neurons to muscles of the distal extremities. These pathways supplement the medial descending pathways in postural control and provide for independent movements of the extremities.

In accordance with their medial or lateral distributions to spinal motor neurons, reticulospinal and vestibulospinal tracts are thought to be most important for control of axial and proximal limb muscles, whereas the rubrospinal (and corticospinal) tracts are

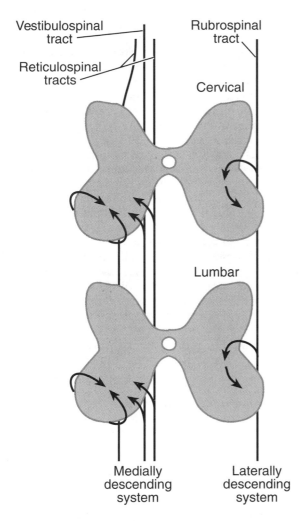

Figure 5–11 ■ Organization of the descending brainstem tracts. The vestibular and reticular tracts are thought to be most important for control of axial and proximal limb muscles. The rubrospinal tract is important for control of distal extremities.

most important for control of distal limb muscles, particularly the flexors.

■ Sensory and Motor Systems Work Together to Control Posture

The maintenance of an upright posture in humans requires active muscular resistance against gravity. For movement to occur, one must alter the posture by flexing against gravity. The balance must be maintained during movement, and this is achieved by postural reflexes initiated by several key sensory systems. Vision, the vestibular system, and the somatosensory system are important for postural reflexes. Somatosensation provides information about the position and movement of one part of the body with respect to other body parts; the vestibular system provides information about the position and movement of the head and neck with respect to the external world; and vision

provides both of these types of information as well as information about objects in the external world. Visual and vestibular reflexes interact to produce coordinated head and eye movements associated with a shift in gaze. Vestibular reflexes and somatosensory neck reflexes interact to produce reflex changes in limbs, and hence in posture. Stability can be restored after a positional disturbance by rapid postural adjustment, or a **functional stretch reflex** or **long loop reflex.** The earliest effective compensations occur at about twice the latency of the myotatic reflex, thus the functional stretch reflex is different from the monosynaptic, myotatic reflex. Whether these rapid adjustments are reflexive (automatic) or voluntary (having some characteristics of choice) is disputed. Some evidence suggests that the fastest responses to change in position are automatic, whereas the subsequent responses require a learning process over repeated trials, involving interactions of vision, somatosensation, and vestibular information with the motor systems.

The Cerebral Cortex in the Control of Movement

The highest level of motor control is exerted by the cortical areas with motor functions. It is difficult to formulate an unequivocal definition of a **motor area,** but three defining criteria may be used. An area is said to have a motor function (1) if stimulation, using minimal current strengths, elicits movements; (2) if destruction of the area results in a motor deficit; and (3) if it has output connections going directly or relatively directly (i.e., with a minimal number of intermediate connections) to the motor apparatus of the spinal cord. Some cortical areas fulfill all of these criteria and have exclusively motor functions, but many cortical areas fulfill only some of the criteria yet are involved in movement, particularly volitional movement.

■ Several Functionally Distinct Cortical Areas Are Involved in Movement

The **primary motor cortex,** Brodmann's area 4, fulfills all three criteria for a motor area. Rostral and medial to it, in Brodmann's area 6, lies the **supplementary motor cortex,** which also fulfills all three criteria. Other areas that fulfill some of the criteria include the rest of Brodmann's area 6, areas 3, 1, and 2 of the postcentral gyrus, and areas 5 and 7 of the parietal lobe, all of which contribute fibers to the corticospinal tract (Fig. 5–12).

■ The Corticospinal Tract Is the Descending Cortical Motor Pathway

Traditionally, the descending motor tract originating in the cerebral cortex was called the *pyramidal tract,* since it traverses the **medullary pyramids** on its way to the spinal cord. All other descending motor tracts emanating from the brainstem were generally grouped

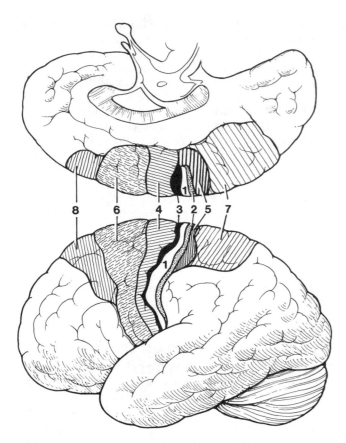

Figure 5–12 ■ Brodmann's cytoarchitectural map of the human cortex, showing areas with motor function. Area 4 is the primary motor cortex (MI); area 6 is the premotor area, containing the supplementary motor cortex (MII) on the medial wall of the hemisphere. Area 8 contains fields related to voluntary eye movements. Areas 1, 2, and 3 are primary somatosensory cortex (SI); parietal lobe areas 5 and 7 are electrically excitable and contribute fibers to the corticospinal tract.

together as the *extrapyramidal system.* These terms are no longer in use because more recent clinical, anatomic, and functional studies provided contradictory information. An example of the erroneousness of this concept is in the corticospinal tract, where some of its descending fibers do not traverse the pyramids or descend to the spinal cord. Thus, the **cortico-bulbar tract** originates in the cerebral cortex and terminates on motor neurons and interneurons in cranial nerve nuclei of the brainstem without descending in the spinal cord, in contrast with the **corticospinal tract,** which originates in the cerebral cortex and terminates on motor neurons and interneurons in the spinal cord.

Cells in Brodmann's area 4 contribute 30% of the corticospinal fibers, and the parietal lobe, especially Brodmann's areas 3, 1, and 2, contribute 40%. In primates, 10–20% of corticospinal fibers end directly on motor neurons; the others end on interneurons associated with motor neurons. Fibers from areas 3, 1, and 2 terminate in the sensory portion of the dorsal horn, and fibers from area 4 terminate in the intermediate zone and motor neuron pools of the ventral horns.

On exiting the cerebral cortex, the corticospinal tract descends through the brain along a path located between the basal ganglia and the thalamus, known as the **internal capsule,** then continues along the ventral brainstem as the **cerebral peduncles,** and on through the **pyramids** of the medulla oblongata. Most of the corticospinal fibers cross in the medullary pyramids, thus the motor cortex in each hemisphere controls the muscles on the contralateral side of the body. After crossing in the medulla, the corticospinal fibers descend in the dorsal lateral columns of the spinal cord and terminate in motor pools that control distal muscles of the limbs. Those fibers that are uncrossed travel in the ventral columns and terminate bilaterally in the motor pools and adjacent intermediate zones controlling the axial and proximal musculature. Most corticospinal fibers are small, only 1–4 μm in diameter, and half are unmyelinated, consistent with the finding that the corticospinal tract is a slowly conducting fiber tract (Fig. 5–13).

In addition to the corticospinal tract, there are other indirect influences of cortical fibers on motor function. Some cortical efferent fibers project to the reticular formation, then to the spinal cord via the reticulospinal tract; others project to the red nucleus, then to the spinal cord via the rubrospinal tract. Despite the fact that these pathways involve intermediate neurons on the way to the cord, volleys relayed through the reticular formation can reach the spinal cord motor circuitry at the same time or earlier than some volleys along the corticospinal tract.

■ All of the Cortical Motor Areas Are Topographically Organized

The muscles are represented in an orderly manner, as **somatotopic maps,** in the sensory and motor cortical areas. Those parts of the body that perform fine movements, such as the digits and the facial muscles, are controlled by neurons that occupy more cortical territory than neurons for body parts capable of only gross movements. The amount of cortical area in which a body part is represented is related to the density of peripheral innervation of that body part (Fig. 5–14).

The Primary Motor Cortex ■ This cortical area corresponds to Brodmann's area 4 in the precentral gyrus. Area 4 is structured in six well-defined layers, I–VI, with layer I closest to the pial surface. Afferent fibers terminate in layers I–V, and efferent fibers, including the giant Betz cells (up to 80 μm in diameter), arise in layers V and VI to descend as the corticospinal tract. Thalamic afferent fibers terminate in two layers. Those that carry somatosensory information end in layer IV and those from nonspecific nuclei end in layer I. Cerebellar afferents terminate in layer IV.

The primary motor cortex (MI) receives somatosensory input, both cutaneous and proprioceptive, via

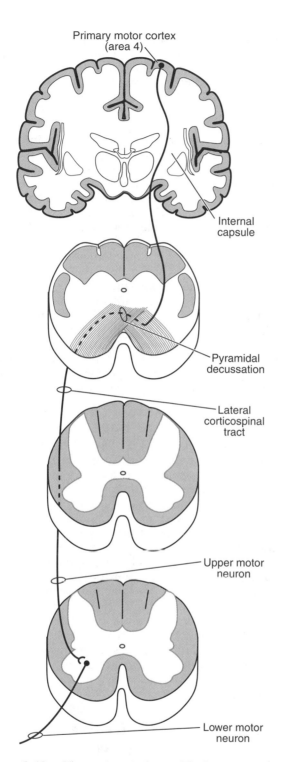

Figure 5–13 ■ The corticospinal tract. The long axons of motor neurons originating in the primary motor cortex descend through the telencephalon via the internal capsule and traverse the brainstem in a ventral path through the cerebral peduncles and the pyramids. The axons cross in the lower medulla (pyramidal decussation) to the opposite side and continue as the corticospinal tract in the spinal cord, where they synapse on motor neurons and interneurons in the ventral horns. (Modified from Kandel ER, Schwartz JH: *Principles of Neural Science.* 2nd ed. New York, Elsevier, 1985.)

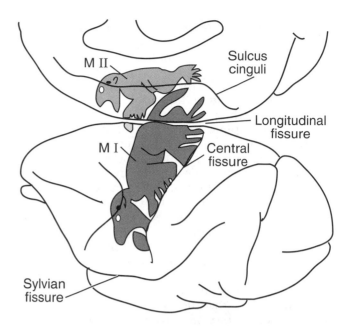

Figure 5–14 ■ The muscles of the body map onto the primary motor cortex (MI) and supplementary motor cortex (MII), as in the monkey brain. Adjacent body surfaces map to adjacent motor cortical areas of MI and MII. (Modified from Eyzaguirre C, Fidone SJ: *Physiology of the Nervous System.* 2nd ed. Chicago, Year Book, 1975.)

the ventrobasal thalamus. The cerebellum projects to MI via the red nucleus and ventrolateral thalamus. Other afferent projections come from the nonspecific nuclei of the thalamus, the contralateral motor cortex, and many other ipsilateral cortical areas. There are many fibers between the precentral (motor) and postcentral (somatosensory) gyri and many connections to the visual cortical areas. The primary motor cortex also makes reciprocal connections with all of the areas that influence neurons that give rise to the corticospinal tract.

Low-level electrical stimulation of the MI produces a twitchlike contraction of a muscle or muscles, and stronger stimuli produce responses in adjacent muscles. Movements elicited from area 4 have the lowest thresholds and are the most discrete of any movements elicited by stimulation of motor cortical areas.

Destruction of any part of the primary motor cortex leads to paralysis of the affected area i.e., the abolition of all voluntary or purposive movement. In humans there may be considerable recovery of function, but the movements are gross rather than the fine, fractionated, independently manipulated muscle movements seen in the normal state. Use of the hand recovers, but the capacity for discrete finger movements is permanently lost after a lesion that involves the arm area of MI.

Neurons in the primary motor cortex encode the capability to control muscle force, length, joint movement, and position. Because of their connections with the somatosensory cortex, they can also respond to sensory stimulation. For example, cells innervating a particular muscle may respond to cutaneous stimuli originating in the area of skin that moves when that muscle is active and/or they may respond to proprioceptive stimulation from the muscle to which they are related or from the joint that muscle moves.

This close coupling of sensory and motor functions may play a role in two cortically controlled reflexes that are important for maintaining normal body support during locomotion—the **placing and hopping reactions,** described originally in experimental animals. Placing can be demonstrated in the cat by holding it so that its limbs hang freely. Contact of any part of the animal's foot to the edge of a table elicits immediate placement of the foot on the table surface. Hopping is demonstrated by holding an animal so it stands on one leg. If the body is moved forward, backward, or to the side, the leg hops in the direction of the movement so that the foot is kept directly under the shoulder or hip, thereby stabilizing the body's position. Lesions of the contralateral precentral or postcentral gyrus abolish placing. Hopping is abolished by a contralateral lesion of the precentral gyrus.

Many efferent fibers from the primary motor cortex terminate in brain areas that contribute to ascending somatic sensory pathways. Through these connections the motor cortex controls the flow of somatosensory information to other motor control centers.

The Supplementary Motor Cortex ■ This cortical area (MII) has no clear cytoarchitectonic boundaries; that is, the shapes and sizes of cells and their processes are not obviously compartmentalized, as in the layers of MI. The MII is located on the medial surface of the hemispheres, above the cingulate gyrus, and rostral from the leg area of the primary motor cortex (Fig. 5-14). Stimulation produces movements, but they are qualitatively different from those produced in MI and greater current strength is required to elicit them. Muscle contractions elicited from MII last longer; the postures elicited may remain after the stimulation is over; and the movements are less discrete than those elicited from MI. Bilateral responses are common.

The MII is reciprocally connected with the MI, and it receives projections from other motor cortical areas. It receives a major projection from the basal ganglia via the ventrolateral thalamus, input that is essential for initiating movements and setting the postural tone in muscles which participate in movements.

Current knowledge is insufficient to describe adequately the unique role of the MII in higher motor functions. Some evidence suggests that lesions in this area produce deficits in fine finger movements and bimanual coordination. Other evidence suggests a role in complex sequences of movements, especially in preparation for a series of complex movements.

The Primary Somatosensory Cortex and Superior Parietal Lobe ■ The primary somatosensory cortex (SI; areas 3, 1, and 2 of Brodmann; Fig. 5-12) lies on the postcentral gyrus. Stimulation of this area produces movement, with the topographic representation of the body coinciding with the somatosensory representation. Thresholds of stimulation for producing movements are two to three times higher here than in MI.

The somatosensory cortex is reciprocally interconnected with MI in a somatotypically organized pattern, with arm areas of sensory cortex projecting to arm areas of motor cortex. While this area is named *sensory cortex* and the precentral area *motor cortex,* each could as easily be named *sensorimotor cortex,* since it is clear that each area has both sensory and motor functions. We use the terms *sensory* and *motor cortex* because they represent the predominant function of each area.

The superior parietal lobe (Brodmann's areas 5 and 7; Fig 5-12) also has important motor functions. In addition to contributing fibers to the corticospinal tract, it is well connected to the motor areas in the frontal lobe. Lesion studies in animals and humans suggest this area is important for the use of complex sensory information in the production of movement. It seems to process sensory information necessary for making purposive movements.

■ Cortical Motor Areas Have Complex Roles in the Control of Movement

The role played by the cortical motor areas is vast and complex. The spasticity evident after a lesion of the internal capsule indicates that the organization of the motor system, even on a cortical level, follows the broad outlines of a system of physiologic extensors and flexors and that the cortical areas have a role in posture. While a decorticate cat looks very much like a normal cat, but has no placing or hopping reactions, a decorticate monkey can right itself and move, but it is awkward and lacks fine qualities of movement. This shows that as one ascends the evolutionary scale the cortical areas subserve more and finer qualities of movement. The mechanisms that underlie the more complex aspects of movement, such as thinking about and performing skilled movements and using complex sensory information to guide movement, remain areas that are poorly understood.

The Basal Ganglia in the Control of Movement

The **basal ganglia** are a large group of nuclei that occupy space primarily in the telencephalon, but with some members in the diencephalon and brainstem. They receive input from the entire cortical mantle and direct their output selectively to the frontal lobes. From there they influence the entire motor system and play a role in posture, the execution of coordinated movements, initiation of locomotion, and preparation for movement.

■ The Striatum, Globus Pallidus, Subthalamic Nucleus, and Substantia Nigra Comprise the Basal Ganglia

The telencephalic portion of the basal ganglia consists of the **striatum,** made up of the **caudate nucleus** and the **putamen,** and the **globus pallidus.** The two parts of the striatum are histologically identical and are separated anatomically by fibers of the internal capsule. The nucleus accumbens and olfactory tubercle, both part of the limbic system (Chap. 7), are considered the ventral extensions of the striatum. The globus pallidus is made up of two quite distinct nuclei: the external segment (GPe) lies closest to the putamen, and the internal segment (GPi) lies adjacent to the GPe. These main telencephalic nuclei are interconnected with the other members of the basal ganglia, the **subthalamic nucleus** in the diencephalon and the **substantia nigra** in the mesencephalon. Two distinct populations of cells are seen in the substantia nigra: a more dorsal, tightly packed group, the pars compacta (SNc), containing the cell bodies of dopaminergic neurons that project to the striatum; and a more ventral group, the pars reticulata (SNr), whose neurons resemble neurons of GPi both histologically and functionally (Fig. 5–15). The following description, while not exhaustive, presents the picture of a system that is interconnected by many internal loops. Mastery of the anatomy presented here provides a good picture of how the basal ganglia are related to other structures; however, one should not forget that many thalamic, midbrain, and brainstem loops are not included in this brief description.

■ Input to the Basal Ganglia Comes from the Cerebral Cortex, Thalamus, and Brainstem Nuclei

A major source of input to the basal ganglia is the cerebral cortex. This projection is topographically organized so that a given cortical area projects to the striatum nearest to it and forms a partially closed loop. Other inputs to the striatum come from the centromedian nucleus of the thalamus, which projects broadly to the putamen and body of the caudate; dopaminergic neurons from the substantia nigra (SNc); serotonergic neurons from the dorsal raphe nuclei; noradrenergic neurons from the locus coeruleus; and a projection from the amygdala, a limbic structure deep in the temporal lobe.

Internally, the sole projections of the striatum are to both segments of the globus pallidus and the SNr (Fig. 5–15). Projections from the caudate terminate in the dorsomedial one-third of GPe and GPi, those of the putamen in the ventrolateral two-thirds of these nuclei. The GPe projects to the subthalamic nucleus, which in turn projects back on both segments of the globus pallidus. Dopamine-containing neurons from SNc form a functionally significant reciprocal projection, giving the striatum some of the densest dopamin-

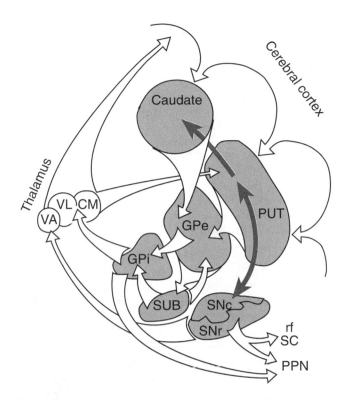

Figure 5–15 ■ Nuclei and anatomic connectivity of the basal ganglia, showing the connections of major loops. The output of the striatum (caudate and putamen) to the globus pallidus and from there to the thalamus provides a major input to the frontal cortex for initiation of movement. PUT, putamen; GPe, external pallidal segment; GPi, internal pallidal segment; SUB, subthalamic nucleus; SNc, substantia nigra pars compacta; SNr, substantia nigra pars reticulata; VA, ventral anterior thalamic nucleus; VL, ventrolateral nucleus of the thalamus; CM, centromedian nucleus of the thalamus; rf, reticular formation; SC, superior colliculus; PPN, pedunculopontine nucleus.

ergic terminal regions in the central nervous system. It is this dopaminergic system that is dysfunctional in Parkinson's disease.

■ Output of the Basal Ganglia Is Directed to the Cerebral Cortex and Brainstem

Projections out of the basal ganglia arise from the globus pallidus and the substantia nigra (SNr). Internal pallidal neurons project to the thalamus, including the centromedian (CM), ventral anterior (VA), and ventrolateral (VL) nuclei, and to the pedunculopontine nucleus (PPN) of the midbrain. Neurons in SNr project to the VA nucleus of the thalamus, to the superior colliculus, and to the reticular formation and PPN. Nuclei VA and VL of the thalamus project selectively to the frontal lobes, including the cortical areas most important for motor function (Fig. 5–15). The projection of PPN may be quite important functionally, since this nucleus is coextensive with the mesencephalic

PARKINSON'S DISEASE

We learned a lot about aging, and about a progressive neurodegenerative disease named after the physician who first described it, James Parkinson, when grandpa began to shake.

Parkinson's disease is not a natural concomitant of aging. It is caused by the loss of dopaminergic neurons in the substantia nigra in the basal ganglia. Loss of these cells produces a classic pattern of symptoms, including tremor in the resting limb, rigidity or increased tone in all muscle, akinesia, including slowness of movement, inability to initiate movement, and paucity of spontaneous movements, and loss of postural reflexes. Three major categories of neurologic disorders include these symptoms, based on the cause of neuronal loss. First there is **primary** or **idiopathic Parkinson's disease,** for which the cause of neuronal loss is not known, although there is some evidence to suggest hereditary factors. This is the only syndrome properly called Parkinson's disease (the others are referred to as parkinsonism) and the only one in which Lewy bodies (pathologic entities inside affected neurons) are found. **Secondary parkinsonism** results from infections (e.g., the great encephalitis lethargica pandemics of the post-World War I era), toxins (e.g., MPTP or manganese), drugs used in the treatment of mental disorders (e.g., neuroleptics used to treat schizophrenia), and repetitive trauma or metabolic causes. The third category of parkinsonism, sometimes called **parkinsonism-plus,** refers to clinical syndromes that include symptoms of other neurologic disorders (e.g., progressive supranuclear palsy and multiple system atrophy) as well as parkinsonism symptoms.

The onset of Parkinson's disease is insidious. Typically many symptoms go unnoticed for years. Tremor is often the presenting symptom, with a careful history likely revealing discoordination of the hands, as in the case of a roofer who begins hitting his thumb instead of the nails he's trying to drive, and decreasing size of handwriting. Treatment is primarily symptomatic, although recent studies show that Deprenyl, a monoamine oxidase B inhibitor, slows progression of the disease and may be thought of as treatment of the disease itself. Pharmacologic treatment is preferred in most cases. Drugs used for treatment are of three types: L-dopa and dopamine receptor agonists; anticholinergics (dopamine and acetylcholine are both present in the striatum and are thought to be in balance in control of striatal neurons); and amantadine, which releases dopamine stores and has anticholinergic properties. L-dopa, the immediate precursor of dopamine, which can cross the blood-brain barrier and then be converted to dopamine in the striatum, is often the primary agent for treatment. How and when to treat depends on the needs of the patient weighed against the side effects of L-dopa, which include involuntary dyskinesias and psychotic complications. The side effects are very serious, can be as disruptive as the disease itself, and occur in most patients within a few years of treatment. One must also consider that the disease is a long-term degenerative process, and pharmacologic therapy gives optimal symptomatic relief for only a limited period.

Neural transplantation—placing dopamine-containing neurons into the striatum—is another way of replacing the lost transmitter. This procedure is still in experimental stages. Stereotactic surgery, in which lesions are produced in particular nuclei of the basal ganglia circuitry, was the primary form of treatment before L-dopa came into use in the late 1960s and can still be effective in well-selected cases. Tremor and rigidity respond well to thermocoagulative lesions of small areas of the thalamus, while akinesia responds not at all.

locomotor region. This region is larger in humans than in other primates, and larger in primates than in cats, suggesting the increased importance of central descending input to locomotion in evolutionarily higher animals.

All of the cortical areas with motor functions project to the putamen, making this portion of the striatum a motor structure. These cortical projections are organized topographically in the putamen into regions related to body musculature, with leg being represented dorsally, then arm, and then, most ventrally, orofacial areas. These body representations are maintained in the projections from the putamen to GPe and GPi, whose output for this circuit is directed to the rostral portion of the ventrolateral nucleus of the thalamus. This "cortical motor loop" is partially closed by the ventrolateral nucleus projection to the supplementary motor cortex. The functional implication of the basal ganglia motor loop terminating in the supplementary motor cortex is that this area is in some way important in planning or programming of complex movements.

■ Lesions and Clinical Diseases of the Basal Ganglia Suggest Some of Its Functions

Disorders associated with basal ganglia pathology produce profound motor dysfunctions in humans. Two important disorders of the basal ganglia are Parkinson's disease and Huntington's disease. Huntington's disease is associated with severe cell loss in the striatum as well as loss of cells in the cerebral cortex. This disorder is characterized by irregular involuntary movements of the limbs **(choreiform movements),** dementia, slowness in the execution of movement **(bradykinesia),** and changes in muscle tone.

Functions of the basal ganglia are also apparent from experimental studies. Recordings from single neurons in the putamen and globus pallidus suggest a role of the basal ganglia motor circuit in control of the direction and scaling of movement amplitude and/or velocity and in the preparation to move.

The Cerebellum in the Control of Movement

The **cerebellum** or "little brain," lies caudal to the occipital lobe and is attached to the brainstem through three fiber tracts; the inferior, middle, and superior **cerebellar peduncles.** It has three main morphologic divisions—the **anterior, posterior,** and **flocculonodular lobes**—separated from each other by the primary fissure and posterolateral fissure (Fig. 5–16). The major lobes are divided into smaller lobes, each with many fissures, that permit a large cortical surface to be tucked into this small space. In addition to the cortex, the cerebellum contains three pairs of deep nuclei: the **fastigial, interpositus** (interposed), and

dentate. Input to the cerebellum comes from peripheral sensory receptors, the brainstem, and the cerebral cortex, each source sending axons to the deep nuclei and the cerebellar cortex. The cerebellar cortex projects inwardly to the deep nuclei and outside to the brainstem vestibular nuclei, which in turn project to motor areas of the cerebral cortex and the brainstem. These output projections are mainly, if not totally, to other motor control areas of the central nervous system. The connectivity and intrinsic circuitry of the cerebellum are well known, but the full scope of its functions are not well understood. In general, the functions of the cerebellum are to compare the descending motor commands with the sensory feedback signals that result from contraction of muscles and to inform the central motor control systems of a need to correct its signals accordingly during the execution of movements. It accomplishes this because of its direct sensory input from each muscle and by direct communication with cortical and brainstem motor control systems.

■ The Cerebellum Has Three Functional Divisions

The cerebellum is divided into three functional parts: the **vestibulocerebellum,** the **spinocerebellum,** and the **cerebrocerebellum.** These appear phylogenetically in this sequence during evolution, and the lateral hemispheres increase in size with increasing size of the cerebral cortex as one ascends the evolutionary scale. The three divisions have similar intrinsic circuitries, thus the function of each depends on the nature of the output nucleus to which it projects. The vestibulocerebellum is composed of the flocculonodular lobes and its connections with the vestibular nuclei. The spinocerebellum is medially placed, and it consists of the **vermis** (for "worm"), a midline structure separated from the cerebellar hemispheres by two sagittal fissures, and the medial portion of the lateral hemispheres, called the **intermediate zones.** The cerebrocerebellum contains the lateral portions of the hemispheres, or **lateral zones,** which project on the dentate nucleus (Fig. 5–16). As their names suggest, these divisions have functions related to the vestibular system, the spinal cord, and the cerebral cortex, and this is further reflected in their anatomic connectivity.

The Vestibulocerebellum ■ The flocculonodular lobe receives input from the vestibular system and many different visual areas. Output goes to the vestibular nuclei, which can in a sense be considered the fourth pair of deep cerebellar nuclei. The vestibulocerebellum functions to control equilibrium and eye movements.

The Spinocerebellum ■ Spinocerebellar pathways carrying somatosensory information terminate in the vermis and intermediate zones in two somatotopic body maps. The auditory, visual, and vestibular systems as well as sensorimotor cortex project to this portion

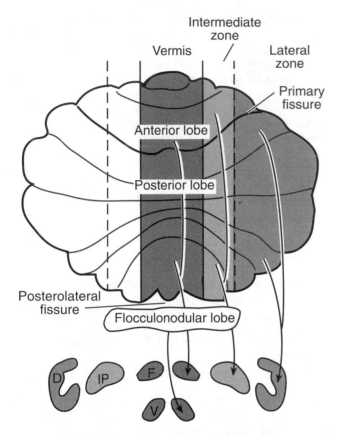

Figure 5–16 ■ A flattened view of the cerebellar cortex, showing the anterior, posterior, and flocculonodular lobes and the nuclei to which they project. The cerebellum can be divided functionally into three zones: the vestibulocerebellum, composed of the flocculonodular lobe, which projects to the vestibular nuclei (V) in the brainstem; the spinocerebellum, composed of the medial cerebellar cortex or vermis, which projects to the fastigial (F) nuclei, and the intermediate cerebellar cortex, which projects to the interposed nuclei (IP); and the cerebrocerebellum, consisting of the lateral cerebellar cortex, which projects to the dentate nuclei (D).

of the cerebellum. Output from the vermis is directed to the fastigial nuclei, which project through the inferior cerebellar peduncle to the vestibular nuclei and reticular formation of the pons and medulla. From the intermediate zones, output goes to the interposed nuclei, through the superior peduncle to the red nucleus and the ventrolateral nucleus of the thalamus, and from there to the motor cortex. It is believed that both the fastigial nuclei and the interposed nuclei contain a complete representation of the muscles of the body and that the fastigial system controls antigravity muscles in posture and locomotion, while the interposed nuclei perhaps act to control stretch reflexes and other somatosensory reflexes as well as help to damp tremor.

The Cerebrocerebellum ■ Input to the lateral zones comes exclusively from the cerebral cortex as relayed through the pons. The cortical areas that are prominent in motor control are the sources for most of this input. Output is directed to the dentate nuclei and from there via the ventrolateral thalamus to the motor and premotor cortices. Functionally, this region is important for planning and initiation of volitional movement.

■ The Intrinsic Circuitry of the Cerebellum Is Very Regular

The cerebellar cortex is composed of five types of neurons arranged regularly into neat layers. The outermost, or molecular, layer contains special interneurons: the **stellate** and **basket** cells. The next layer contains the dramatic **Purkinje cells,** with dendrites reaching upward in a dense, flat, fanlike array. These are the only projection neurons of the cerebellar cortex, and they are inhibitory. Deep to the Purkinje cells is the granular layer, containing **Golgi cells** and small local circuit neurons, the **granule cells** (Fig. 5–17). There are more granule cells in the cerebellum than there are neurons in the entire cerebral cortex!

Input fibers to the cerebellar cortex are of two sorts, the **mossy fibers** and the **climbing fibers.** Mossy fibers arise from brainstem neurons and have complex multicontact synapses on granule cells. The granule cell axons ascend to the molecular layer and then branch to travel perpendicular to the dendrites of Purkinje cells. Called **parallel fibers,** they activate a row of Purkinje cells. Mossy fibers discharge at a high tonic rate and are modulated during movement as one might expect of brainstem motor nuclei. When mossy fiber

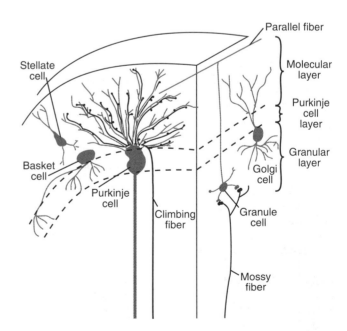

Figure 5–17 ■ The neurons of the cerebellum. The cerebellar cortex is everywhere identical in terms of the cells, the layers, and the internal connectivity.

input is of sufficient strength to drive Purkinje cells, the latter cells discharge a single action potential.

Climbing fibers arise from the **inferior olive,** a nucleus in the medulla. Each climbing fiber wraps around the body and dendrites of 1–10 Purkinje cells, making multiple contacts. Compared with the mossy fiber synapses on Purkinje cells, the climbing fibers exert a very strong synaptic influence. One action potential in a climbing fiber produces a burst of action potentials of decreasing amplitude in the Purkinje cell; this response is called a **complex spike.** Neurons that give rise to climbing fibers discharge at very low rates.

Mossy and climbing fibers are excitatory, and both give off excitatory collateral axons to the deep cerebellar nuclei before they reach the cerebellar cortex. Interneurons of the molecular layer are inhibitory and seem to focus the excitatory input from the mossy fibers. This is accomplished by producing inhibition of neighboring rows of Purkinje cells.

■ Clinical and Experimental Lesions Highlight the Functional Divisions of the Cerebellum

Lesions of the cerebellum produce impairments in muscle coordination (i.e., the coordinated contraction of agonists, antagonists, synergists, and stabilizers). This discoordination is known as limb **ataxia.** Tremor is also characteristic of cerebellar disease and is most obvious at the end of fine movements. This type of tremor is called **intention** or **action tremor.** Lesions produce changes in muscle tone, depending on the location of the lesion and the species, with decreased tone (**hypotonia**) accompanying some cerebellar lesions in humans. Patients with cerebellar lesions seem to have normal sensation, and they are not paralyzed. Indeed, standard neurologic tests of patients with lesions of the lateral cerebellar hemispheres may reveal no deficits, yet, for example, if the patient is a musician s/he may be incapacitated. An affected arm could be used, but could no longer play music. Function recovers after cerebellar lesions, with recovery somewhat less after a lesion of the deep nuclei of the cortex.

While these lesions establish a picture of the absence of cerebellar function, we are left without a firm idea of exactly what the cerebellum does and how this function is accomplished at the cellular level. Cerebellar function is sometimes described as adjusting the output of the descending motor systems. If the cerebellum is thought of as a comparator, its job might be described as comparing the intended movement with the actual movement so as to be able to correct ongoing movements. Other putative functions include modification of central programs for movement and a role in the plasticity or learning of certain motor skills.

■ ■ ■ ■ ■ ■ ■ ■ ■ ■ ■ ■ ■ **REVIEW EXERCISES** ■ ■ ■ ■ ■ ■ ■ ■ ■ ■ ■

Identify Each with a Term

1. The muscle that contributes most to a given movement
2. A reflex initiated by stretch of a muscle that causes a muscle to contract
3. Motor neurons that innervate intrafusal muscle fibers
4. The sole tract with direct projections from the cerebral cortex to the spinal cord
5. The nucleus in basal ganglia in which dopaminergic neurons reside
6. Deep cerebellar nucleus to which the cerebrocerebellum projects

Define Each Term

7. Agonist
8. Synergist
9. Reciprocal innervation
10. Intrafusal muscle fibers
11. Inverse myotatic reflex
12. Motor neuron pool
13. Brainstem
14. MII
15. Climbing fiber

Choose the Correct Answer

16. Antigravity muscles:
    a. Are physiologic flexors
    b. Are flexors of the human arm
    c. Are the stabilizers around a joint
    d. Function to withdraw a body part from a painful stimulus
    e. b and c

17. The final common path:
    a. Connects alpha motor neurons and muscle fibers
    b. Is influenced only by sensory impulses from muscles
    c. Is not essential for normal human movement
    d. Includes cells in cranial nerve nuclei and the spinal cord
    e. a and d

18. Gamma fibers:
    a. Innervate spinal motor neurons
    b. Are insensitive to passive stretch
    c. Define a class of fibers of large diameter
    d. Do not innervate extrafusal muscle fibers
    e. a and b

19. Extrafusal muscle fibers:
    a. Are innervated by spinal motor neurons
    b. Produce all the tension generated during a contraction
    c. Are all slowly contracting

d. Are likely to fatigue rapidly
e. a and b

20. Golgi tendon organs:
a. Sense length and velocity of stretch of the muscle
b. Are innervated by axons in the final common path
c. Are unloaded during active contraction
d. Are organs of sensation
e. a and c

21. Fusimotor fibers:
a. Carry sensory information to the central nervous system
b. Cause muscle fibers to contract
c. Participate in control of part of the sensory apparatus in muscles
d. Are activated only when alpha motor fibers are activated
e. b and c

22. Each alpha motor neuron:
a. Innervates both intrafusal and extrafusal muscle fibers
b. Innervates a single muscle fiber
c. Is part of the final common path
d. Is tonically active
e. c and d

23. A motor unit includes:
a. An alpha motor neuron
b. The axon from an alpha motor neuron
c. All the muscle fibers innervated by an alpha motor neuron
d. All the intrafusal muscle fibers attached to the extrafusal fibers innervated by one alpha motor neuron
e. a, b, and c

24. Which of the following is true of both intrafusal and extrafusal muscle fibers?
a. They contribute to tension generated by the muscle.
b. They are sensory organs.
c. They contract.
d. They are controlled by alpha motor neurons.
e. b and d

25. A brainstem function in control of movement is:
a. Control of posture
b. Initiating locomotion

c. Pattern generation for locomotion
d. Control of equilibrium
e. a, b, and d

26. Damage to the motor cortex (MI):
a. Causes paralysis of the affected area, which can recover
b. Produces permanent gross motor deficits
c. Causes permanent loss of hand function if the hand area is involved
d. Produces only transient loss of balance
e. a, b, and c

27. Which of the motor behaviors listed below is *not* a particular motor function of the cerebral cortex?
a. Hopping
b. Locomotion
c. Placing
d. Individual digit use
e. All of the above

Briefly Answer

28. What characteristics of skeleton and muscle help determine the range of movement at a joint?

29. What does reciprocal innervation accomplish, and what is its cellular mechanism?

30. What is feedback control of movement, and how is it accomplished?

31. Give one example of a reflex movement mediated entirely by the spinal cord circuitry in humans. Describe (a) the stimulus necessary for the movement; (b) the circuitry that mediates this movement; and (c) its function in motor control.

32. What four nuclear groups are considered part of the basal ganglia? What are the motor functions of this area?

33. What is the importance of the cerebellum for movement?

34. The basal ganglia and cerebellum do not directly innervate the spinal cord motor neurons. Trace the pathways by which these two motor systems influence the final common path.

ANSWERS

1. Prime mover
2. Myotatic or stretch reflex
3. Gamma motor neurons
4. Corticospinal tract
5. Substantia nigra
6. Dentate nucleus
7. The muscle that produces shortening during movement
8. A muscle that contracts at the same time as the agonist muscle and cooperates in production of movement
9. Synaptic activity of a type Ia fiber onto both an alpha motor neuron and an interneuron that inhibits motor neurons of antagonist muscles

10. The muscle fibers that comprise a muscle spindle
11. A reflex initiated by active contraction, which causes inhibition of the contracting muscle
12. All the neurons that serve a particular muscle
13. The portion of the brain that includes the medulla, pons, and midbrain
14. Supplementary motor cortex, a second motor cortical area located on the medial surface of the cerebral hemisphere
15. A fiber arising from cells of the inferior olive and projecting to 1-10 Purkinje cells in the cerebellum, where it makes multiple synaptic contacts
16. e

17. e

18. d

19. e

20. d

21. e

22. c

23. e

24. c

25. e

26. a

27. b

28. The range of motion at a joint depends on the type of articulation of the joint, the arrangement of muscle around the joint, and the bulk of adjacent tissue.

29. Reciprocal innervation accomplishes contraction of the agonist with simultaneous relaxation of the antagonist. This permits coordinated flexion or extension of the limb. At the neuronal level, a command issued to the spinal cord acts on neurons and interneurons to excite the agonist and inhibit the antagonist motor neurons.

30. Feedback control describes the system whereby receptors in muscles provide information about muscle con-traction. This information may be used by the central nervous system to make rapid corrections of ongoing movements.

31. Three examples are (1): stretch or myotatic reflex; (a) stretch; (b) (Fig. 5-7); (c) regulation of motor output. (2): inverse myotatic reflex; (a) active contraction; (b) (Fig. 5-8); (c) tension feedback. (3): Flexor withdrawal; (a) cutaneous stimulation; (b) (Fig. 5-9); (c) the function of this reflex is to overcome obstacles and remove the body part from harmful stimulus.

32. The striatum, globus pallidus, subthalamic nucleus, and substantia nigra are parts of the basal ganglia. The motor functions of this area are rapid, accurate, smooth movements; control of posture; initiation of movement.

33. The cerebellum is important for coordinated, smooth, accurate movements; rapid, complex sequences of movements; and control of posture.

34. Output from the basal ganglia is via the thalamus to the cerebral cortex; corticocortical projections carry the information to the corticospinal tract neurons, which innervate the final common path. Output from the cerebellum is to the brainstem motor nuclei, which, in turn, communicate the information to the final common path.

Suggested Reading

Barker D: The morphology of muscle receptors, in Hunt CC (ed): *Handbook of Sensory Physiology: Muscle Receptors.* Berlin, Springer-Verlag, 1974, pp 1–190.

Brodal A: *Neurological Anatomy in Relation to Clinical Medicine.* New York, Oxford University Press, 1981.

Brooks VB, Thach WT: Cerebellar control of posture and movement, in Brookhart JB, Mountcastle VB, and Brooks VB (eds): *Handbook of Physiology: The Nervous System.* Bethesda, American Physiological Society, 1981, pp 877–946.

DeLong MR, Georgopoulos AP: Motor functions of the basal ganglia, in Brookhart JB, Mountcastle VB, Brooks VB (eds): *Handbook of Physiology: The Nervous System.* Bethesda, American Physiological Society, 1981, pp 1017–1061.

Gowitzke BA, Milnor M (eds): *Scientific Bases of Human Movement.* Baltimore, Williams & Wilkins, 1988.

Kandel ER, Schwartz JH, Jessell TM (eds): *Principles of Neural Science.* 3rd ed. New York, Elsevier, 1991.

Kuypers HGJM: Anatomy of the descending pathways, in Brookhart JB, Mountcastle VB, Brooks VB (eds): *Handbook of Physiology: The Nervous System.* Bethesda, American Physiological Society, 1981, pp 597–666.

Matthews PCB: *Mammalian Muscle Receptors and their Central Action.* Baltimore, Williams & Wilkins, 1972.

CHAPTER

6

The Autonomic Nervous System

CHAPTER OUTLINE

I. OVERVIEW OF THE AUTONOMIC NERVOUS SYSTEM
- A. The ANS consists of two divisions, with actions that usually oppose each other
- B. Peripherally placed ganglia distinguish the ANS from the somatic nervous system
- C. Many involuntary organs have dual innervation
- D. Sensory as well as motor neurons participate in autonomic reflexes
- E. Acetylcholine and norepinephrine are neurotransmitters in the ANS
  1. Acetylcholine
  2. Norepinephrine
  3. Other neurotransmitters

II. THE SYMPATHETIC NERVOUS SYSTEM
- A. Ganglia of the sympathetic division are situated near the spinal cord
  1. Paravertebral ganglia
  2. Collateral ganglia
- B. The sympathetic division innervates all organs of the body
  1. Somatic structures
  2. Organs of the head and chest
  3. Abdominal and pelvic organs
- C. The sympathetic division is anatomically designed to permit a generalized response to stress
  1. Divergence
  2. Secretion of epinephrine

III. THE PARASYMPATHETIC NERVOUS SYSTEM
- A. Ganglia of the parasympathetic division are located in or near the organ innervated
- B. The cranial division innervates organs in the head, chest, and abdomen
  1. The third (oculomotor) nerve
  2. The seventh (facial) and ninth (glossopharyngeal) nerves
  3. The tenth (vagus) nerve
- C. The sacral division innervates the organs of the pelvic cavity

IV. CENTRAL INTEGRATION OF AUTONOMIC FUNCTION
- A. The simplest autonomic reflex has no "center" of integration
- B. ANS ganglia are sites of integration of reflexes

OBJECTIVES

After studying this chapter, the student should be able to:

1. Compare the innervation of the involuntary organs with that of skeletal muscle, in terms of anatomy and neurochemistry
2. Account for the ability of the sympathetic division to activate multiple effectors at one time and contrast this with the parasympathetic division
3. Name the neurotransmitters at preganglionic and postganglionic synapses of both divisions of the ANS; name the exceptions to the rule and describe the innervation of the adrenal medulla
4. Describe the opposing actions of the sympathetic and parasympathetic nerves on the heart, the pupil of the eye, the viscera of the gastrointestinal tract, and the urinary bladder; describe the role of the central nervous system in coordinating two opposing autonomic influences
5. Name the hormones that the adrenal cortex and adrenal medulla secrete and describe their actions in the stress response

---

Chapter 5 dealt with control of the voluntary (skeletal) muscles by the somatic motor nerves. The other peripheral targets of central nervous system (CNS) efferent nerves are the involuntary organs, and these are regulated by the autonomic nervous system (ANS). The involuntary organs include visceral (hollow) organs, exocrine glands, blood vessels, and the heart. Three types of cells in these organs are the sites of synapses with autonomic nerves: smooth muscle, cardiac muscle, and gland cells. The ANS is not the sole regulator of the involuntary organs; circulating hormones of the endocrine system and cytokines produced locally are important additional chemical influences involved in moment-to-moment regulation of their function.

Overview of the Autonomic Nervous System

The functions of the ANS fall into three major categories: maintaining homeostatic conditions in the body; coordinating the body's responses to exercise, stress, or injury; and assisting the endocrine system to regulate reproduction.

■ The ANS Consists of Two Divisions, with Actions that Usually Oppose Each Other

On the basis of anatomic, neurochemical, pharmacologic, and functional differences, the ANS is usually separated into two divisions, **sympathetic** and **parasympathetic.** A third division, the enteric nervous system, primarily concerned with gut reflexes, is considered in detail in Chapter 26.

Anatomic differences between the two divisions led to the designation of the sympathetic as the **thoracolumbar** division, because it arises from the thoracic and lumbar spinal cord, and the parasympathetic as the **craniosacral** division, because it arises from four cranial nerve nuclei of the brainstem and from nerves emerging from the sacral spinal cord.

Neurochemical and pharmacologic differences between the two divisions led to the designation of the sympathetic as **adrenergic,** for the adrenalinelike actions of sympathetic nerve stimulation, and the parasympathetic as **cholinergic,** for the acetylcholinelike actions of nerve stimulation.

The functions of the sympathetic and parasympathetic divisions are often simplified into a two-part scheme. Generally, the sympathetic division is said to preside over emergency responses of the body and the utilization of its resources and the parasympathetic division to preside over the restoration and buildup of the body's reserves and elimination of waste. It is important to understand, though, that the divisions are interdependent; neither one operates without the other's influence. Most autonomic effector organs are **dually innervated** by sympathetic and parasympathetic nerves, and the CNS coordinates both divisions in carrying out most ANS functions. In many instances the two divisions are activated in a reciprocal fashion, where the firing rate in one division is increased and in the other is decreased. An example is in control of heart rate: increased firing in the sympathetic nerves to the atria and simultaneous decrease in firing rate of the parasympathetic nerves result in increased heart rate. In some dually controlled organs the two divisions work synergistically, with each division activated simultaneously. For example, during secretion in exocrine glands of the gastrointestinal tract the parasympathetic nerves increase volume and enzyme content at the same time that sympathetic activation contributes mucus to the total secretory product. Some organs, such as the skin and vasculature, receive sympathetic but not parasympathetic innervation, in which case these effectors are regulated by the firing rate of the sympathetic nerves as well as by the presence of hormones, cytokines, and other local products of metabolism.

Anatomically, the sympathetic division is capable of firing all at once in a coordinated fashion during exercise, emotional excitement, or stress or when perceiving danger or other threats to life. The hypothalamus, on command from the limbic system, coordinates this generalized sympathetic response, known as the **fight-or-flight response,** by activating the chain ganglia of the thoracolumbar outflow. This galvanizes the body's sympathetic effectors to respond appropriately to the challenging stimulus.

■ Peripherally Placed Ganglia Distinguish the Autonomic from the Somatic Nervous System

The CNS sends out an extensive motor innervation to all organs, voluntary (somatic nervous system) and involuntary (ANS). In the somatic nervous system there is an uninterrupted path from the cell body in the ventral horn of the spinal cord or from the brainstem to the skeletal muscle effector cells; in the ANS the motor path consists of a two-neuron chain with a ganglion interposed between the CNS and the effector cells (Fig. 6–1). The autonomic motor neuron emerges from the interomediolateral horn of the spinal cord or brainstem (**preganglionic fiber**) and synapses with a ganglion cell in the periphery. The **ganglion** cell then projects its axon (**postganglionic fiber**) to the autonomic effector cells.

■ Many Involuntary Organs Have Dual Innervation

All organs, including the somatic structures, skeletal muscles, and bones, receive ANS innervation because of the blood vessels contained within them. In most cases the involuntary organs receive a dual innervation, from both the sympathetic and parasympathetic divisions. Structures innervated exclusively by the sympathetic division include the skin and the blood vessels, with the exception of the specialized blood vessels that form blood sinuses in erectile tissue of the reproductive system, including the penis, clitoris, and labia minora. The parasympathetic nerves control dilation of these blood vessels to permit erection.

■ Sensory as well as Motor Neurons Participate in Autonomic Reflexes

The ANS is traditionally regarded as a motor system, thus the sensory neurons originating in the involuntary organs are not considered part of the ANS. Nevertheless, visceral sensory neurons form the afferent limb of autonomic reflexes, much as somatic sensory neurons form the afferent limb of somatic reflexes (Fig. 6–1). Sensory neurons from involuntary organs differ from somatic sensory neurons both in the sensory information they convey and in the lack of conscious awareness of the sensations. Exceptions to unconscious sensory awareness are the feeling of fullness in the urinary bladder or stomach, when stretched, and the sensation of pain from nociceptors originating in involuntary organs.

Three types of sensory neurons serve the involuntary organs. **Chemoreceptors** detect changes in the chemical environment in an organ, such as pH, oxygen tension, glucose concentration, and fat content. **Mechanoreceptors** detect changes in wall tension in visceral organs and blood vessels, where they are called pressoreceptors or baroreceptors. **Nociceptors** transmit pain sensation from the visceral organs during injury, inflammation, ischemia, obstruction, or overdistension of the walls of the viscus. This visceral pain is poorly localized and is referred to a somatic structure rather than to the organ in which it originates, a phenomenon called **referred pain.**

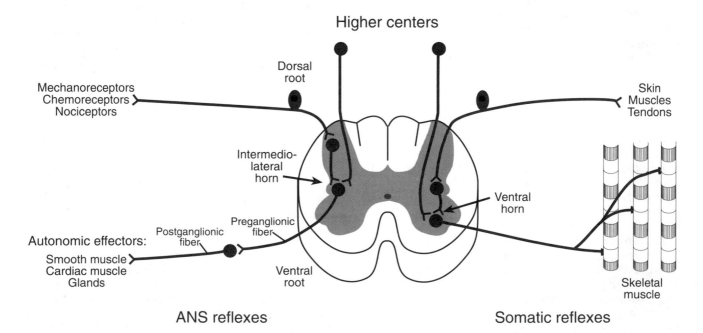

Figure 6–1 ■ Reflexes in the autonomic and somatic nervous systems. On the left are reflexes involving the involuntary organs; on the right are reflexes involving skeletal muscles.

■ Acetylcholine and Norepinephrine Are Neurotransmitters in the ANS

The neurotransmitter in the somatic nervous system is released from specialized nerve endings that make intimate contact, one terminal to one skeletal muscle fiber, at the mammalian motor endplate region. This contrasts with the ANS, where the postganglionic fibers terminate in a spray of **varicosities,** swellings enriched in synaptic vesicles, which release the transmitter into the extracellular space surrounding the effector cells (Fig. 6-2). Because of this arrangement, sympathetic and parasympathetic nerves can innervate the same organ, and the net response of the organ at any one time is determined by which receptors are stimulated and which second messengers are generated intracellularly as a result (Chap. 3).

Acetylcholine ■ Acetylcholine (ACh), the first identified chemical mediator in nerves, is the most prevalent neurotransmitter in the peripheral nervous system. It is the transmitter used at the neuromuscular junction of all skeletal muscles, all ganglionic synapses of both the sympathetic and parasympathetic divisions (and in the adrenal medulla, which receives preganglionic sympathetic innervation), and all postganglionic synapses in the parasympathetic division (Fig. 6-2). The cholinergic receptor is **nicotinic** at the first two sites and **muscarinic** at the third site.

Norepinephrine ■ Norepinephrine (NE) is the neurotransmitter for postganglionic fibers of the sympathetic division. Two notable exceptions to this general rule are nerves serving the eccrine (watery) sweat glands in the skin and certain blood vessels of the skeletal muscles, where ACh, not NE, is the transmitter released from "sympathetic" postganglionic fibers. The innervation of these two effectors is anatomically from the sympathetic division, that is, thoracolumbar in origin; however, the neurochemistry of the postganglionic fibers is cholinergic, not noradrenergic. The cholinergic receptor in the sweat glands and skeletal muscle blood vessels is *muscarinic,* as in the case of postganglionic receptors of the parasympathetic division.

Norepinephrine activates receptors at postganglionic effectors of sympathetic nerves; these have been classified as either **α-adrenergic** or **β-adrenergic** receptors, based on their pharmacologic responses to agents that mimic or block the actions of NE. In most cases, these receptors have been subclassified on the basis of further pharmacologic characterization: α_1 or α_2 and β_1 or β_2 (Chap. 3). These same adrenergic receptors and their subtypes are activated by the hormone epinephrine when it appears in the general circulation.

Other Neurotransmitters ■ As noted in Chapter 3, neurally active peptides are often colocalized with small-molecule transmitters and released simultaneously during nerve stimulation in the CNS. This is clearly the case in the ANS as well, especially in the intrinsic plexuses of the gut, where amines, amino acid transmitters, and neurally active peptides are widely distributed. In the ANS, examples of a colocalized amine and peptide are seen in the sympathetic division, where neuropeptide Y and NE are coreleased in vasoconstrictor nerves, and in the parasympathetic division, where vasoactive intestinal polypeptide (VIP) is coreleased with ACh in the secretory fibers innervating the salivary glands.

The Sympathetic Nervous System

■ Ganglia of the Sympathetic Division Are Situated near the Spinal Cord

In general, preganglionic neurons in the sympathetic division have short axons, terminating in ganglia close to the spinal column in the sympathetic trunk

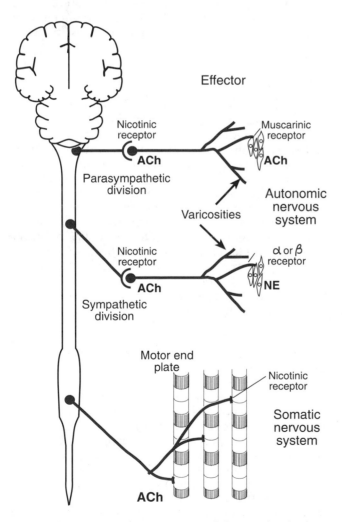

Figure 6-2 ■ Main neurotransmitters of the peripheral nervous system. ACh, acetylcholine; NE, norepinephrine.

or surrounding the descending aorta. In contrast, the preganglionic neurons of the parasympathetic division are relatively long, terminating near or within the target organ.

Preganglionic neurons of the sympathetic division originate in the thoracic (T1–T12) and upper lumbar (L1–L3) spinal cord (Fig. 6–3). These 15 pairs of preganglionic nerves give rise to postganglionic nerves that are distributed widely to all parts of the body. The latter originate in ganglia from one of two types: **paravertebral ganglia,** located next to the vertebral column, and **collateral** or **prevertebral ganglia,** overlying the descending aorta.

The collateral ganglia innervate primarily the abdominal and pelvic visceral organs, usually serving as integrating centers for reflexes involving these organs. The paravertebral ganglia innervate all of the other sympathetic effector organs and tissues, and their anatomically interconnected ganglia mediate the sympathetic nervous system's generalized response to stress.

Paravertebral Ganglia ■ The paravertebral ganglia are located in a paired chain of interconnected ganglia, the **sympathetic trunk,** located close to and on either side of the vertebral column (Fig. 6–3). These ganglia extend above and below the thoracolumbar region, where preganglionic fibers exit the spinal cord; postganglionic fibers thereby emerge from cervical and sacral as well as thoracic and lumbar levels. The trunk ganglia are comprised of 3 cervical ganglia (superior, middle, and inferior), 12 thoracic, 5 lumbar, and at least 4 sacral ganglia.

Collateral Ganglia ■ The three major collateral ganglia overlie the celiac, superior mesenteric, and inferior mesenteric arteries, where they branch away from the aorta. They are named, correspondingly, the **celiac ganglion** (or solar plexus) and the **superior mesenteric** and **inferior mesenteric ganglia.**

Preganglionic fibers in general synapse in only one ganglion: paravertebral or collateral. Those synapsing in the collateral ganglia exit the spinal cord and pass through the paravertebral ganglia at the same segmental level without synapsing there. Preganglionic fibers synapsing in the paravertebral ganglia innervate ganglion cells at the same segmental level. In addition, branches of the preganglionic axons ascend and descend to other levels of the sympathetic chain to synapse at more than one level in the chain ganglia (Fig. 6–3).

■ The Sympathetic Division Innervates All Organs of the Body

Somatic Structures ■ Postganglionic fibers emerging at each segment of the sympathetic trunk travel together with the somatic nerves at the same segmental level to provide sympathetic innervation to the skin, mucous membranes, muscles, and bones in the limbs, trunk, and head. The innervated structures include the sweat glands, pilomotor muscles, and blood vessels of the skin as well as blood vessels of the skeletal muscles and bones.

Organs of the Head and Chest ■ The **superior cervical ganglia** innervate the radial muscle of the iris (pupil dilator), the eyelids (Müller's muscle), the lacrimal (tear) glands, and the salivary glands. The **middle** and **inferior cervical ganglia** innervate organs of the chest, including the heart, lungs, trachea, and esophagus. The preganglionic fibers innervating the cervical ganglia originate in the upper thoracic spinal cord, from T1 through T5.

Abdominal and Pelvic Organs ■ The **celiac ganglion** provides sympathetic innervation to the stomach, liver, pancreas, gall bladder, small intestine, spleen, and kidney. Preganglionic fibers originate in the thoracic cord from T5 to T12. The **superior mesenteric ganglion** innervates the small and large intestine; preganglionic fibers originate primarily from the lower thoracic cord, from T10 to T12. The **inferior mesenteric ganglion** innervates the lower colon and rectum, urinary bladder, and reproductive organs. Preganglionic fibers originate from the lowest thoracolumbar segments (T12–L3).

■ The Sympathetic Division Is Anatomically Designed to Permit a Generalized Response to Stress

From moment to moment the sympathetic division exerts a continual influence on the organs it innervates. This is called **sympathetic tone** and is associated with a low rate of discharge of the sympathetic nerves serving their effectors. When required, the rate of firing to a particular organ can be increased or decreased to meet the changing needs of the organism, such as increased firing rate to the radial pupillary muscle in dim lighting or increased firing of the vasoconstrictor nerves to the skin when ambient temperature falls. On the other hand, when the organism perceives an emergency situation, the sympathetic division can activate many effectors at once, resulting in the fight-or-flight response, described at the end of this chapter. The anatomic basis for mounting a generalized response is based on the following characteristics of the sympathetic division.

Divergence ■ The number of postganglionic fibers emerging from the paravertebral ganglia is greater than the number of preganglionic neurons that emerge from the spinal cord (Fig. 6–3). This is particularly striking in the superior cervical ganglion, where an estimated 4–20 postganglionic fibers are innervated by a single preganglionic fiber. In addition, axon collaterals of preganglionic sympathetic neurons throughout the thoracolumbar cord make synaptic connections with postganglionic trunk ganglia both above and below their level of emergence from the cord. This high ratio of post- to preganglionic synapses,

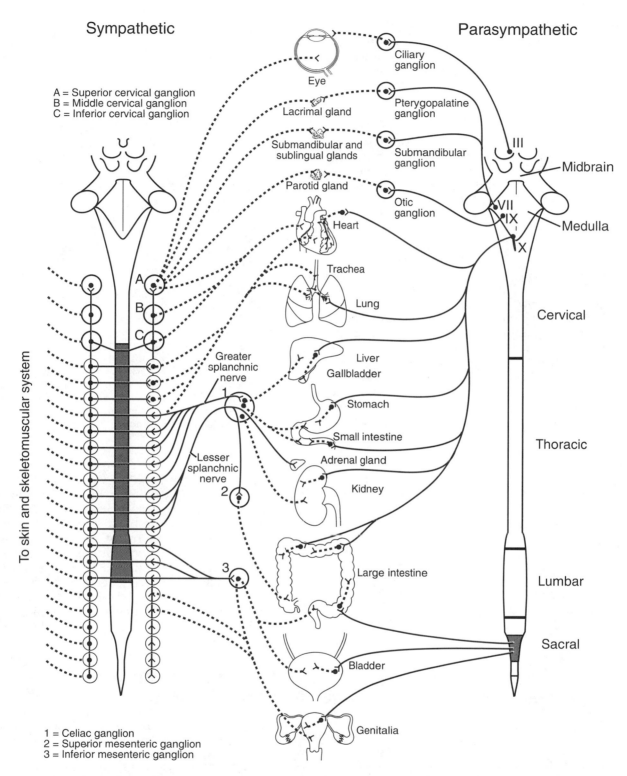

Sympathetic

A = Superior cervical ganglion
B = Middle cervical ganglion
C = Inferior cervical ganglion

To skin and skeletomuscular system

A
B
C

Greater splanchnic nerve
1
Lesser splanchnic nerve
2
3

1 = Celiac ganglion
2 = Superior mesenteric ganglion
3 = Inferior mesenteric ganglion

Parasympathetic

Ciliary ganglion
Eye

Pterygopalatine ganglion
Lacrimal gland

Submandibular and sublingual glands
Submandibular ganglion

Parotid gland
Otic ganglion

Heart

Trachea

Lung

Liver
Gallbladder

Stomach

Small intestine

Adrenal gland

Kidney

Large intestine

Bladder

Genitalia

III
VII
IX
X

Midbrain

Medulla

Cervical

Thoracic

Lumbar

Sacral

Figure 6–3 ■ The autonomic nervous system. The involuntary organs are depicted with their parasympathetic innervation (craniosacral) indicated on the right and sympathetic innervation (thoracolumbar) on the left. Preganglionic fibers are solid lines; postganglionic fibers are dashed lines. For purposes of illustration, the sympathetic outflow to the skin and skeletomuscular system is shown separately (to the far left); effectors include sweat glands, pilomotor muscles and blood vessels of the skin, and blood vessels of the skeletal muscles and bones. (Modified from Heimer L: *The Human Brain and Spinal Cord: Functional Neuroanatomy and Dissection Guide.* New York, Springer-Verlag, 1983.)

known as **divergence** (Fig. 6–4), enables the sympathetic division to activate many effectors at once. Divergence is not observed in the parasympathetic division, where more discrete activation of individual effectors usually occurs when the preganglionic fibers are stimulated.

Secretion of Epinephrine ■ In addition to divergence, the sympathetic division uses a hormonal mechanism to activate, by means of the circulatory system, all effectors endowed with adrenergic receptors, including those innervated by the sympathetic nerves. That hormone is the catecholamine **epinephrine,** which is secreted (along with a small amount of norepinephrine) by the adrenal medulla during the generalized response to stress.

The adrenal medulla, a neuroendocrine gland, forms the inner core of the adrenal gland situated on top of each kidney. Cells of the adrenal medulla are innervated by the lesser splanchnic nerve, which contains preganglionic sympathetic fibers originating in the lower thoracic spinal cord. These fibers pass through the paravertebral ganglion chain and celiac ganglion without synapsing in either site (Fig. 6–3), to terminate on the **chromaffin cells** of the adrenal medulla (Fig. 6–5). The chromaffin cells are modified ganglion cells. They synthesize catecholamines, epinephrine, and norepinephrine, in a ratio of about 8:1, using the same enzymatic mechanisms described for catecholamine-synthesizing neurons (Chap. 3) and store the catecholamines in secretory vesicles. Unlike neurons, they possess neither axons nor dendrites; rather, they are neuroendocrine cells that secrete hormone into the blood on nerve stimulation.

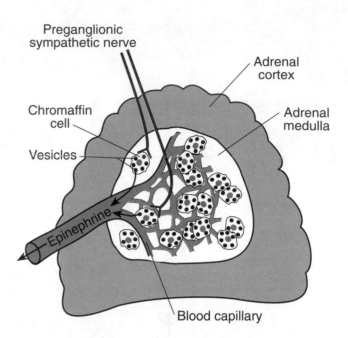

Figure 6–5 ■ Sympathetic innervation of the adrenal medulla. Preganglionic fibers originating in the lower thoracic spinal cord travel via the lesser splanchnic nerve through the celiac ganglion, without synapsing, and enter the adrenal medulla. Their axons terminate on chromaffin cells, which contain vesicles that store epinephrine. On stimulation, epinephrine is secreted directly into the blood and the general circulation.

Circulating epinephrine mimics the actions of sympathetic nerve stimulation but with greater efficacy, because epinephrine is usually more potent than norepinephrine in stimulating both α- and β-adrenergic receptors (Chap. 3). Furthermore, epinephrine stimulates adrenergic receptors on cells that receive sparse or no sympathetic innervation. These include liver and fat cells, which mobilize glucose and fatty acids, respectively, and blood cells (erythrocytes, leukocytes, and platelets), which participate in clotting and mounting immune and tissue responses to injury.

The Parasympathetic Nervous System

The parasympathetic division is comprised of a cranial portion, emanating from the brain, and a sacral portion, originating in the interomediolateral column of the sacral spinal cord (Fig. 6–3). In marked contrast to the sympathetic division, the nerves of the parasympathetic division are not normally activated as a whole, but rather are stimulated individually and independently of each other. A striking example of this discrete form of nerve stimulation is seen in the vagus nerve; one portion of its motor output, for example to the heart, can be stimulated to slow the heart rate without

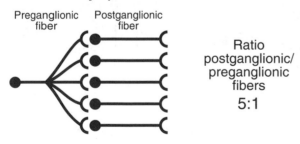

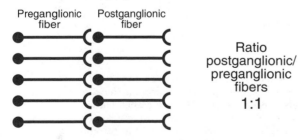

Figure 6–4 ■ Divergence in the sympathetic nervous system.

HORNER'S SYNDROME

Horner's syndrome illustrates the sympathetic and parasympathetic regulation of structures of the head. In this condition, usually unilateral, the sympathetic innervation from the superior cervical ganglion is partially or totally damaged, resulting in loss of sympathetic control of the pupil size, eyelid, blood vessels, and sweat glands on the affected side.

The symptoms of partial or complete sympathetic denervation include loss of sympathetic tone in the iris dilator muscle, resulting in **miosis,** or narrowing of the pupil. Denervation of the smooth muscle of the eyelid (Müller's muscle) leads to a relaxation or droopy eyelid on the affected side. This is known as **ptosis,** or lidlag. There is no loss of lens accommodation or secretion of aqueous humor or ability to contract or relax the pupillary constrictor muscle of the iris, as these functions are served either exclusively or primarily by parasympathetic (third cranial nerve) innervation, nor is there loss of salivary secretion, as the parasympathetic innervation (seventh and ninth cranial nerves) is the major secretory influence.

The loss of sympathetic innervation of the skin of the face is detectable by a loss of constrictor tone in blood vessels, resulting in a flushed appearance on the affected side. There is also failure of the sweat glands to secrete in response to a rise in body temperature. At such times sweating is seen on only the unaffected side of the face.

If denervation of the sympathetic nerves is complete, **supersensitivity,** or increased responsivity to adrenergic stimulants, ensues such as when epinephrine is secreted by the adrenal medulla. This develops over a period of several weeks following denervation and is the result of upregulation, that is, increase in number of adrenergic receptors in effector cells following loss of their neurotransmitter. In Horner's syndrome supersensitivity is observed most strikingly by a dilation of the pupil when subthreshold doses of catecholamine (doses that are not high enough to elicit a response in the normally innervated iris on the other side) are present in the blood.

It should be noted that sympathetically denervated receptors in the sweat glands do not exhibit supersensitivity to epinephrine, since the neurotransmitter in the eccrine sweat glands is not a catecholamine, but rather acetylcholine.

affecting the vagal output to the stomach. Furthermore, there is no evidence of divergence in the parasympathetic division; rather, there is the usual synaptic arrangement of pre- to postsynaptic connectivity (Fig. 6–4).

■ Ganglia of the Parasympathetic Division Are Located in or near the Organ Innervated

Ganglia in the parasympathetic division are either located close to the organ innervated or embedded within its walls, such as in the organs of the gastrointestinal system (Fig. 6–3). (In the gastrointestinal tract the postganglionic parasympathetic ganglion cells make up a significant portion of the neuronal cells of the intrinsic plexuses; Chap. 26). The close proximity of ganglion cells to effector organs results in a long

preganglionic fiber relative to the postganglionic fiber, leading to their designation as **terminal ganglia.**

■ The Cranial Division Innervates Organs in the Head, Chest, and Abdomen

Four of the 12 cranial nerves that emerge from the brainstem make up the cranial portion of the parasympathetic division. Their nuclei, extending from the medulla to the tectum of the midbrain, are important centers of integration of autonomic reflexes involving the organ systems they innervate. The sympathetic division and somatic nervous system are regulated along with the parasympathetic division in reflexes coordinated by these brainstem nuclei.

The Third (Oculomotor) Nerve ■ The third nerve originates in nuclei in the tectum of the midbrain, where synaptic connections with the optic nerves from the retina provide important visual input to eye reflexes. Preganglionic fibers of the third nerve synapse in the ciliary ganglion located outside the orbit, and the postganglionic fibers innervate the pupillary sphincter muscle of the iris, ciliary muscle, and ciliary body.

The Seventh (Facial) and Ninth (Glossopharyngeal) Nerves ■ The inferior and superior salivatory nuclei in the rostral medulla give rise to the motor nerves that promote salivary and lacrimal gland secretion. The secretory nerves are carried by the seventh and ninth cranial nerves, synapsing in ganglia close to the respective glands.

The Tenth (Vagus) Nerve ■ The vagus nerve has an extensive motor component emerging from the dorsal medulla. Long preganglionic fibers traveling in the vagus trunk terminate in ganglia outside the heart and lungs and in the intrinsic plexuses within the walls of the gastrointestinal tract, extending from the midesophagus all the way to the midtransverse colon. There is also a vagal innervation to the kidneys, liver, spleen, and pancreas, but the functions of some of these vagal inputs are not yet understood.

■ The Sacral Division Innervates the Organs of the Pelvic Cavity

Preganglionic fibers of the sacral division originate in the intermediolateral horns of the sacral spinal cord, emerging from segments S2, S3, and S4 (Fig. 6-3). Their preganglionic fibers synapse in ganglia in or near the pelvic organs, including the lower portion of the gastrointestinal tract (the sigmoid colon, rectum, and internal anal sphincter), the urinary bladder, and the reproductive organs.

Central Integration of Autonomic Function

Reflexes that control the involuntary organs range in complexity from a simple one involving only one neuron, to peripheral autonomic ganglia and ganglion plexuses serving as local centers of integration, to the highest level, coordinated by the CNS. In general, the higher the level of complexity, the more likely the reflex is to require coordination of sympathetic and parasympathetic divisions, together with the somatic nervous system and endocrine system.

■ The Simplest Autonomic Reflex Has No "Center" of Integration

Skin injury stimulates nociceptors that synapse in the dorsal horn of the spinal cord to mediate spinal cord reflexes and ultimately to transmit the sensation of pain to the cerebral cortex. The nociceptor fibers have axon collaterals that branch and innervate blood vessels in the adjacent region, resulting in vasodilation and reddening of the skin at the site of injury (Fig. 6-6). Conduction of impulses in this collateral axon branch is designated as **antidromic,** meaning "against the prevailing direction," because an *efferent* nerve fiber (axon collateral) is part of an *afferent* sensory neuron (nociceptor).

This reflex vasodilation associated with injury is known as the **local axon reflex,** and it differs from the usual reflex in that it does not require a center for integration. The neurotransmitter of this reflex is substance P, a neuropeptide released at the spinal cord synapse of primary afferent nociceptors as well as at their antidromic axon collaterals to the blood vessels.

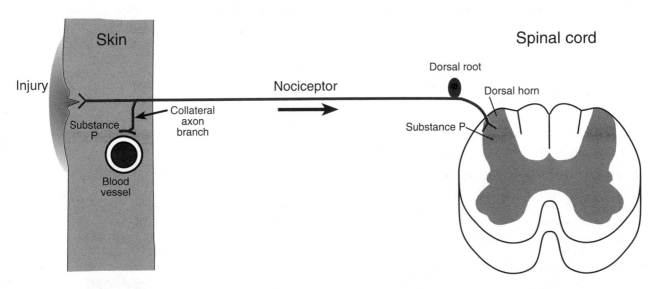

Figure 6–6 ■ The local axon reflex. Nociceptor fibers in the skin possess antidromic axon collaterals that dilate local blood vessels at the site of injury to the skin. The neurotransmitter released at the blood vessel and spinal cord synapses is substance P.

■ ANS Ganglia Are Sites of Integration of Reflexes

As noted earlier, the paravertebral chain ganglia can fire in a concerted fashion to mount a fight-or-flight response. This integrative action is based on a circuitry that includes a high degree of connectivity among ganglia within the sympathetic trunk as well as divergence of pre- to postganglionic synapses.

The collateral ganglia also serve as integrative centers for reflexes in the gastrointestinal tract (Chap. 26). Chemoreceptors and mechanoreceptors located in the gut bring sensory information to the celiac or mesenteric ganglia and initiate sympathetic reflex changes in motility (usually inhibitory) and secretion as conditions vary locally during digestion. The integrative actions of collateral ganglia are also responsible for halting motility and secretion in the gastrointestinal tract during a fight-or-flight response and, more important, for constricting the vasculature throughout the splanchnic bed. This aids in mobilizing the blood volume for preferential perfusion of the brain, muscles, heart, and lungs during strenuous physical activity.

The intrinsic plexuses of the gastrointestinal visceral wall are not only the sites of ganglionic synapse of the parasympathetic innervation, but they are a reflex integrative center where sympathetic postganglionic fibers, parasympathetic preganglionic fibers, and numerous intrinsic neurons participate in reflexes that allow motility and secretion to occur autonomously. The intrinsic plexuses also participate in centrally mediated gastrointestinal reflexes (Chap. 26).

■ The Central Nervous System Coordinates Reflexes Involving both the Somatic and Autonomic Nervous System

Reflexes Coordinated by the Spinal Cord ■ Reflexes coordinated by centers in the lumbar and sacral spinal cord include micturition (emptying the urinary bladder), defecation (emptying the rectum), and sexual reflexes (ejaculation of semen and erection of erectile tissue). Higher centers provide an important control, either enabling or inhibitory, of these reflexes. Their primary integrative center is, however, the spinal cord. The sympathetic component is from the lumbar cord, the parasympathetic is from the sacral cord, and the somatic nerves to the voluntary muscles originate in the ventral horn of the sacral spinal cord.

Reflexes Coordinated by the Brainstem ■ Physiologists have long considered the central regulation of vital functions to reside in "centers" in the brainstem, including centers for maintaining blood pressure, heart rate, and respiration. Modern neuroscience has taught us otherwise; the concept of a single center for regulating reflexes of the heart, blood vessels, and respiratory system is no longer tenable. As an example, cardiovascular regulation is known to occur at all levels of the CNS, including the medulla, hypothalamus, limbic system, and cerebral cortex, each of which has sites that respond as integrating centers when stimulated. In spite of this caveat, the centrally located cell bodies of autonomic preganglionic neurons are major regulatory components of autonomic function, since they constitute the final motor pathway of summated central influences on a given autonomic effector.

Centrally mediated autonomic reflexes are initiated by sensory signals from both peripheral and central sensory pathways. The motor responses are generated by both divisions of the ANS and in many cases by the somatic nervous system as well, thereby imparting a behavioral component to the reflex.

Autonomic reflexes coordinated in the brainstem include salivation, swallowing, vomiting, blood pressure regulation, respiration, pupillary reflexes, and lens accommodation.

Reflexes Coordinated by the Hypothalamus
■ The hypothalamus participates in homeostasis by detecting changes in extracellular fluid constituents and initiating reflexes within nuclei of the hypothalamus to bring these to their regulated level. The repertoire of hypothalamically mediated reflexes involving the ANS includes, among others, temperature regulation, regulation of homeostasis of blood constituents, regulation of blood volume, blood pressure, and heart rate, regulation of eating and drinking, reproduction, regulation of circadian rhythms and cyclic functions, and coordination of the response to stress. Because of the major role of the hypothalamus in autonomic function, it has been labeled the "head ganglion of the ANS."

The hypothalamus is also uniquely situated at the union of the brain's **limbic system,** or emotional centers of the brain (Chap. 7), with the centers that regulate the ANS and endocrine system. Information from the cerebral cortex, particularly the **association areas** (Chap. 7), reaches the hypothalamus via the limbic system, giving rise to autonomic reflexes and hormonal secretion appropriate to the perceived needs of the organism.

The Responses of Individual Organs to ANS Stimulation

As noted earlier, most involuntary organs are dually innervated by the sympathetic and parasympathetic divisions, often with opposing actions. A list of these organs and a summary of their responses to sympathetic and parasympathetic stimulation is given in Table 6–1. Note that in some cases only one division of the ANS innervates a given structure. Table 6–1 also lists the subtype(s) of adrenergic receptors currently thought to mediate the responses of effectors to NE and epinephrine (EPI). This list must be considered tentative, as the classification of receptors is continually being revised. The subtype of acetylcholine receptor in the parasympathetic division is muscarinic, since the organs receive postganglionic innervation. A de-

TABLE 6–1 ■ Responses of Effectors to Sympathetic and Parasympathetic Stimulation

Effector	Parasympathetic Stimulation	Sympathetic Stimulation	Adrenergic Receptor
Heart			
Rate	Decrease	Increase	β_1
Force	Decrease	Increase	β_1
Conduction velocity	Decrease	Increase	β_1
Blood vessels			
Skin	No innervation	Constriction	α_1
Viscera	No innervation	Constriction	α_1
Skeletal muscle	No innervation	Dilation	β_2
		Dilation	Cholinergic muscarinic
		Constriction	α_1
Gastrointestinal tract			
Walls of viscera	Contraction	Relaxation	β_2
Sphincters	Relaxation	Contraction	α_1
Glands (salivary, gastric, pancreas)	Secretion (volume, enzymes, ions)	Secretion (mucus)	α_1, α_2
Urinary bladder			
Detrusor muscle	Contraction	Relaxation	β_2
Trigone and sphincter	Relaxation	Contraction	α_1
Lung bronchioles	Constriction	Relaxation	β_2
Skin			
Pilomotor muscles	No innervation	Contraction	α_1
Eccrine sweat glands	No innervation	Secretion	Cholinergic muscarinic
Eye			
Pupil sphincter	Contraction	No innervation	
Pupil dilator	No innervation	Contraction	α_1
Ciliary muscle	Contraction (near vision)	No innervation	
	Relaxation (far vision)		
Müller's muscle	No innervation	Contraction	α_1
Lacrimal glands	Secretion	Secretion	
Genital tract	Erection	Ejaculation	α_1
Adrenal medulla	No innervation	Secretion of epinephrine	Cholinergic nicotinic
Liver	No known effect	Glycogenolysis	β_2
Adipose tissue	No innervation	Lipolysis	β_1
Kidney	No known effect	Secretion of renin	α_1, β_1

tailed discussion of the effects of autonomic nerve stimulation on individual organs is found in subsequent chapters, where these organ systems are considered.

■ The Fight-or-Flight Response Activates the Sympathetic Division

The fight-or-flight response is a summation of responses of individual effectors to sympathetic stimulation, with the exception of the sexual reflexes. This stereotypical response is coordinated by the hypothalamus, acting on input from sensory and limbic systems of the forebrain. The hypothalamus activates the sympathetic division, while generally inhibiting parasympathetic control of dually innervated organs. Added to this is the secretion of epinephrine, which stimulates all adrenergic receptors accessible to the general circulation.

■ Many Organs Participate in the Fight-or-Flight Response

Cardiovascular System. ■ Sympathetic stimulation of the heart and blood vessels results in a rise in blood pressure due to increased cardiac output and increased total peripheral resistance. There is also a redistribution of the blood flow: a decrease in flow to the splanchnic region and skin and an increase in flow to the muscles and heart. The brain, a privileged organ that does not participate in the sympathetic redistribution of blood flow, receives a continuing and undiminishing supply of oxygen and glucose at all times, including peak demand periods, such as in strenuous muscular exertion.

Respiratory System ■ The increased demand for exchange of blood gases is met by an increased rate of respiration and dilation of the bronchiolar musculature. The volume of salivary secretion is low during a fight-or-flight reaction, but the relative proportion of mucus is increased, permitting lubrication of the mouth despite increased ventilation.

Eye ■ The eyes are set for long distance vision and a wide field of view during the fight-or-flight response. This is accomplished by contraction of the iris radial muscle (pupillary dilator), relaxation of the iris sphincter muscle (pupillary constrictor), contraction of Müller's muscle of the eyelid for a wide palpebral fissure, and inhibition of the parasympathetic innervation of the ciliary muscle. Relaxation of the ciliary muscle results in flattening of the lens, as needed for far vision.

Mobilization of Metabolic Substrates ■ The increase in demand for glucose and fatty acids (a preferred muscle fuel) as metabolic substrates is met by the actions of the sympathetic nerves and circulating epinephrine on liver hepatocytes and adipose cells, respectively. Glycogenolysis mobilizes stored liver glycogen, thereby increasing plasma levels of glucose. Lipolysis in fat cells converts stored triglycerides to free fatty acids that enter the bloodstream.

Skin ■ The skin plays an important role in maintaining body temperature in the face of marked increase in heat production, by contracting muscles. The sympathetic innervation of the vasculature of the skin is well adapted to allow heat exchange in the skin (usually accomplished by vasodilation), while constricting the capacitance vessels to lower the overall blood flow to the skin. The eccrine sweat glands are another important mechanism for heat loss. Sympathetic nerve stimulation of the sweat glands results in secretion of a watery fluid; its evaporation dissipates body heat.

Frightening thoughts can trigger a fight-or-flight reaction that does not include the muscular exertion usually associated with fighting or fleeing. The skin responds by vasoconstriction, sweating, and piloerection (raising the body hairs), reactions otherwise associated with temperature regulation. Individuals in a state of intense fear therefore exhibit pallor, "cold sweat" (so designated because sweating is inappropriately associated with vasoconstricted skin), and "goosebumps" due to contraction of the pilomotor muscles. In birds and fur-bearing mammals the raising of feathers or fur during a fear reaction is of some use, since it imparts a ferocious appearance that may frighten away the predator.

■ The Adrenal Cortex Participates in the Response to Stress

Stress is perceived by sensory systems acting at the cortical level and the limbic system to initiate a response coordinated by the hypothalamus (Fig. 6–7). The same hypothalamic stimuli that lead to the sympathetic adrenal medullary response to stress also lead to the secretion of a releasing factor, **corticotropin-releasing hormone** (CRH), into the hypothalamo-hypophysial portal system (Chap. 34). The CRH

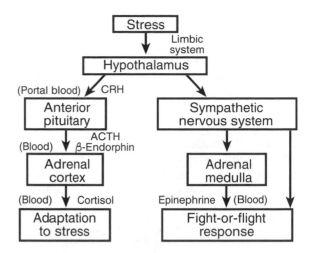

Figure 6–7 ■ The hypothalamo-pituitary-adrenal response to stress. The hypothalamus commands both the sympathetic nervous system response to stress and the endocrine response, via the hypothalamus-anterior pituitary-adrenal cortical axis. CRH, corticotropin-releasing hormone; ACTH, adrenocorticotropic hormone.

stimulates cells in the anterior pituitary gland to secrete adrenocorticotropic hormone (ACTH), and an opioid peptide, β-endorphin, into the blood. β-Endorphin raises the threshold to pain, thereby imparting an analgesic component to the stress response. Adrenocorticotropic hormone stimulates the secretion of cortisol from the adrenal cortex at the same time the adrenal medulla is stimulated to secrete epinephrine by preganglionic sympathetic fibers. The steroid hormone, cortisol, and the catecholamine hormone, epinephrine, serve to enhance the effectiveness of sympathetic nerve stimulation during the response to stress.

The actions of epinephrine have already been described; they serve, together with sympathetic nerve stimulation, to mediate the short-term response to stress. The actions of cortisol are equally widespread, including a reinforcement of sympathetic stimulation of effector organs, since cortisol potentiates the actions of catecholamines at their receptors. In addition, cortisol aids in long-term adaptation of the body to stress. Among its actions (Chap. 36), cortisol ensures replenishing of the body's metabolic reserves, by stimulating lipolysis and liver gluconeogenesis.

■ ■ ■ ■ ■ ■ ■ ■ ■ ■ ■ ■ ■ ■ **REVIEW EXERCISES** ■ ■ ■ ■ ■ ■ ■ ■ ■ ■ ■ ■ ■ ■ ■

Identify Each with a Term or Terms

1. Two additional names for the sympathetic division

2. Two additional names for the parasympathetic division

3. The sensory receptors that form the afferent limb of autonomic reflexes

4. The two major subtypes of cholinergic receptors

5. The two major subtypes of adrenergic receptors

Define Each Term

6. Divergence

7. Collateral ganglia

8. Cold sweat
9. Nerve varicosity

Choose the Correct Answer

10. The parasympathetic division has a greater influence than the sympathetic division on which of the following functions?
   a. Vasodilation
   b. Gastrointestinal motility and secretion
   c. Widening of the pupil
   d. The fight-or-flight response

11. A drug that can dilate the pupil and simultaneously block accommodation of the lens is a drug that:
   a. Stimulates adrenergic receptors
   b. Blocks adrenergic receptors
   c. Stimulates muscarinic cholinergic receptors
   d. Blocks muscarinic cholinergic receptors

12. The structure responsible for coordinating both sympathetic and parasympathetic influences on the heart rate is:
   a. The central nervous system
   b. The paravertebral ganglia
   c. The terminal ganglion in the heart
   d. The sacral spinal cord

13. Which of the following is not found in the intrinsic plexuses of the gut?
   a. Preganglionic sympathetic fibers
   b. Postganglionic sympathetic fibers
   c. Preganglionic parasympathetic fibers
   d. Postganglionic parasympathetic fibers

14. The active ingredient in hot chili peppers is a stimulant of nociceptor neurons. If applied under the skin, this substance will cause which of the following, in addition to the sensation of pain?
   a. Piloerection
   b. Sweating
   c. Vasodilation in the skin
   d. Vasoconstriction in the skin

15. During the fight-or-flight response, epinephrine reinforces all of the effects of sympathetic nerve stimulation.

Which effector of sympathetic nerve stimulation does epinephrine not influence?
   a. Constriction of blood vessels in the splanchnic bed
   b. Sweating
   c. Rise in blood pressure
   d. Dilation of bronchioles

16. Drugs that increase acetylcholine transmission, such as antiacetylcholinesterases, promote sympathetic as well as parasympathetic responses because:
   a. ACh is the transmitter at all autonomic ganglionic synapses
   b. ACh is the transmitter at all sympathetic and parasympathetic postganglionic synapses
   c. ACh receptors are located in all effector organs that receive autonomic innervation
   d. ACh is the sole neurotransmitter in the peripheral nervous system

17. To stimulate dilation of the bronchioles in an asthmatic patient without causing stimulation of the heart, it would be best to give:
   a. A selective α_1 stimulant
   b. A selective α_2 stimulant
   c. A selective β_1 stimulant
   d. A selective β_2 stimulant

Briefly Answer

18. Describe the role of the celiac ganglion as an integrating center.
19. Describe the role of the third cranial nerve in eye function.
20. Blood vessels have only sympathetic innervation, yet their caliber can vary from the resting diameter to a state of vasoconstriction or vasodilation. Explain.
21. How do emotions influence heart rate?
22. Name a sympathetically mediated function that is not a normal component of the fight-or-flight response.
23. Explain how a football player injured during a tackle may not feel pain at the time of injury.

ANSWERS

1. Thoracolumbar; adrenergic
2. Craniosacral; cholinergic
3. Chemoreceptors, mechanoreceptors, and nociceptors
4. Muscarinic and nicotinic
5. Alpha and beta
6. Seen in synapses where one presynaptic neuron stimulates more than one postsynaptic neuron
7. Sympathetic ganglia located at major branchings from the aorta; they innervate the abdominal and pelvic visceral organs
8. Sweating when the skin is vasoconstricted, such as in response to fear
9. A swelling in the axon terminal of autonomic nerves (and many CNS neurons) that contains numerous synaptic vesicles

10. b
11. d
12. a
13. a
14. c
15. b
16. a
17. d
18. The celiac ganglion receives afferent input from the gastrointestinal tract and sends efferent sympathetic postganglionic fibers to effectors in the gut. Together with interneurons in the ganglion, these neurons mediate local gut reflexes.
19. The third nerve supplies the parasympathetic innervation to the ciliary body (to promote secretion of aqueous

humor), ciliary muscle (to accommodate the lens for near vision), and pupillary sphincter muscle (to decrease pupil size).

20. The resting state in blood vessels is mediated by a low rate of firing of the sympathetic nerves. When firing ceases, the blood vessels dilate; when firing rate increases, the blood vessels constrict.

21. Emotional state is generated in the limbic system, whose connections with the hypothalamus mediate changes in heart rate, primarily by stimulating a center in the medulla that coordinates the vagus and sympathetic influences on heart rate.

22. Reflex ejaculation of semen in males is a sympathetically mediated response, but its integration center in the lumbar spinal cord is not part of the pathway from the hypothalamus to the paravertebral ganglia that is activated in a fight-or-flight response.

23. During a stress response, such as in playing football or in injury, the hypothalamic-pituitary system secretes β-endorphin as well as ACTH into the blood. ACTH mediates the adrenal cortical response in stress and endorphin is responsible, at least in part, for blocking pain transmission.

Suggested Reading

Bannister R: *Autonomic Failure: A Textbook of Clinical Disorders of the Autonomic Nervous System.* 2nd ed. Oxford, Oxford University Press, 1988.

Cannon WB: *The Wisdom of the Body.* 2nd ed. New York, Norton, 1939.

Gilman AG, Rall TW, Nies AS, Taylor P: *Goodman and Gilman's The Pharmacological Basis of Therapeutics.* 8th ed. New York, Pergamon, 1990.

Loewy AD, Spyer KM: *Central Regulation of Autonomic Functions.* New York, Oxford University Press, 1990.

Integrative Functions of the Nervous System

CHAPTER OUTLINE

I. THE HYPOTHALAMUS
   A. The hypothalamus is composed of anatomically distinct nuclei
   B. Hypothalamic nuclei are centers of physiologic regulation
   C. The hypothalamus regulates feeding behavior
   D. The hypothalamus controls the gonads and sexual activity
   E. The hypothalamus contains the "biologic clock"
II. THE ASCENDING RETICULAR ACTIVATING SYSTEM
   A. Neurons of the reticular formation exert widespread modulatory influence in the central nervous system
   B. The ascending reticular activating system mediates consciousness and arousal
   C. The electroencephalogram is a recording of the electrical activity of the surface of the brain
      1. Waves of the EEG
      2. Sleep and the EEG
III. THE FOREBRAIN
   A. The cerebral cortex is functionally compartmentalized
   B. The limbic system is the seat of the emotions
      1. Anatomy of the limbic system
      2. Monoaminergic innervation
      3. The brain's reward system
      4. Aggression and the limbic system
      5. Sexual activity
   C. Psychiatric disorders involve the limbic system
      1. Affective disorders
      2. Schizophrenia
   D. Memory and learning require the cerebral cortex and limbic system
      1. Short-term memory
      2. Long-term memory
   E. Language and speech are coordinated in specific areas of association cortex

OBJECTIVES

After studying this chapter, the student should be able to:
1. List the homeostatic functions regulated by the hypothalamus
2. Describe the mechanisms by which homeostatically regulated functions fluctuate in a diurnal pattern
3. Describe the means by which the reticular formation serves as the activating system of the forebrain
4. List the stages of sleep and describe their EEG patterns
5. Describe the anatomy of the limbic system and its monoaminergic innervation
6. Describe the role of the limbic system in aggression, sexual activity, and the brain's reward system
7. Explain the actions of antidepressant and antipsychotic drugs in ameliorating the symptoms of affective disorders and schizophrenia
8. Describe the role of the frontal cortex, hippocampus, and cerebral cortex in learning and memory

The central nervous system (CNS) receives sensory stimuli from the body or the outside world and processes that information in neural networks, or centers of integration, to mediate an appropriate response or learned experience. Centers of integration are hierarchic in nature; the higher they are placed in a caudal-to-rostral sequence, the greater the complexity of the neural network. This chapter considers functions integrated within the diencephalon and telencephalon, where emotionally motivated behaviors, appetitive drives, consciousness, sleep, memory, and cognition are coordinated.

The Hypothalamus

The **hypothalamus** coordinates autonomic reflexes of the brainstem and spinal cord. It also activates the endocrine and somatic motor systems when responding to signals generated either within the hypothalamus or brainstem or in higher centers, such as the limbic system, where the emotions and motivations are generated. The hypothalamus can accomplish this by virtue of its unique location at the interface of the limbic system with the endocrine and autonomic nervous systems.

The hypothalamus is a major regulator of homeostasis. Capillaries in the hypothalamus, unlike other brain blood vessels, possess a "leaky" blood-brain barrier, which allows the cells of hypothalamic nuclei to sample freely, from moment to moment, the composition of the extracellular fluid. It then initiates the mechanisms necessary to maintain levels of regulated constituents at a given set point, which is itself fixed within more or less narrow limits by a specific nucleus of the hypothalamus. Homeostatic functions regulated by the hypothalamus include body temperature, water and electrolyte balance, and plasma glucose level.

The hypothalamus is the major regulator of the endocrine system by virtue of its connections with the pituitary gland, the master gland of the endocrine system. These include direct neuronal innervation of the posterior pituitary lobe by specific hypothalamic nuclei and a direct hormonal connection between specific hypothalamic nuclei and the anterior pituitary lobe. **Hypothalamic hormones,** designated as **releasing factors,** reach the anterior pituitary lobe by a portal system of capillaries. Releasing factors then regulate the secretion of most hormones of the endocrine system.

■ The Hypothalamus Is Composed of Anatomically Distinct Nuclei

The hypothalamus forms the ventral half of the **diencephalon;** its rostral border is at the optic chiasm and its caudal border at the mammillary body. Above it lie the subthalamus and thalamus, which together with the hypothalamus constitute the diencephalon (Fig. 7–1).

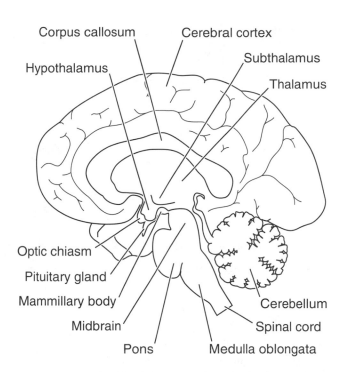

Figure 7–1 ■ A midsagittal section through the human brain, showing the most prominent structures of the brainstem, diencephalon, and forebrain. (Modified from Kandel ER, Schwartz JH, Jessel TM: *Principles of Neural Science.* 3rd ed. New York, Elsevier, 1991.)

On the basal surface of the hypothalamus, exiting from the median eminence, the pituitary stalk contains the **hypothalamo-hypophyseal portal blood vessels,** which transport releasing factors from the hypothalamus to the anterior pituitary gland (Fig. 34–3). Neurons within specific nuclei of the hypothalamus secrete releasing factors into vascular channels, which carry them to the anterior pituitary gland. They stimulate the anterior pituitary to secrete hormones that are trophic to other glands of the endocrine system (Chap. 34). The pituitary stalk also contains the axons of neuroendocrine neurons originating in the **magnocellular cells** (having large cell bodies) of the supraoptic and paraventricular hypothalamic nuclei and terminating in the posterior lobe of the pituitary gland. These neural pathways, forming the **hypothalamo-hypophyseal tract** within the pituitary stalk, represent the efferent limbs of neuroendocrine reflexes leading to secretion of the hormones vasopressin and oxytocin into the blood.

The nuclei of the hypothalamus have ill-defined boundaries, despite their customary depiction (Fig. 7–2). Many are named according to their anatomic location (e.g., anterior hypothalamic nuclei, ventromedial nucleus) or for the structures they lie next to (e.g., the periventricular nucleus surrounds the third ventricle, the suprachiasmatic nucleus lies above the optic chiasm).

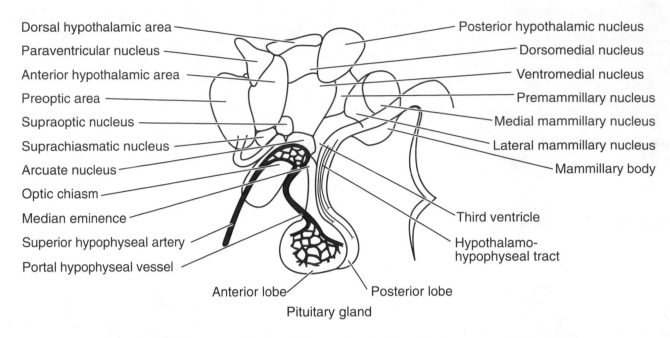

Figure 7–2 ■ Enlargement of the hypothalamus, showing the anatomic arrangement of individual nuclei and the connections of the hypothalamus with the pituitary gland. (Modified from Ganong WF: *Review of Medical Physiology.* 16th ed. Norwalk, CT, Appleton & Lange, 1993.)

The hypothalamus receives afferent inputs from all levels of the CNS. It makes reciprocal connections with the limbic system via fiber tracts in the fornix and a tract from the thalamus to the mammillary complex, the **mammillothalamic tract.** The hypothalamus also makes extensive reciprocal connections with the brainstem, including the reticular formation and the medullary centers of cardiovascular, respiratory, and gastrointestinal regulation.

Other connections of the hypothalamus are one-way rather than reciprocal. One of these carries visual information from the retina to the **suprachiasmatic nucleus** via the optic nerve. Through this retinal input, the light cues of the day/night cycle entrain, or synchronize, the "biologic clock" of the brain to the external clock. The other one-way connection is the hypothalamo-hypophyseal tract from the supraoptic and paraventricular nuclei to the posterior pituitary gland.

■ Hypothalamic Nuclei Are Centers of Physiologic Regulation

The anatomically delineated nuclei of the hypothalamus contain functional groupings of neuronal cells that regulate a number of important physiologic functions:

1. Water and electrolyte balance in magnocellular cells of the supraoptic and paraventricular nuclei (Chap. 34)
2. Secretion of hypothalamic releasing factors in the arcuate and periventricular nuclei and in **parvocellular cells** (having small cell bodies) of the paraventricular nucleus
3. Temperature regulation in the anterior and posterior hypothalamic nuclei (Chap. 31)
4. Activation of the sympathetic nervous system and adrenal medullary hormone secretion in the dorsal and posterior hypothalamus
5. Thirst and regulation of drinking in the lateral hypothalamus (Chap. 24)
6. Hunger, satiety, and regulation of feeding in the ventromedial nucleus and lateral hypothalamic area
7. Regulation of sexual behavior in the anterior and preoptic areas
8. Regulation of circadian rhythms in the suprachiasmatic nucleus

■ The Hypothalamus Regulates Feeding Behavior

The ventromedial nucleus of the hypothalamus serves as a **satiety center** and the lateral hypothalamic area serves as a **feeding center.** Together, these areas coordinate the processes that govern eating behavior and the subjective perception of satiety. These hypothalamic areas also influence secretion of hormones, particularly from the thyroid gland, adrenal gland, and pancreatic islet cells, in response to changing metabolic demands.

Lesions in the ventromedial nucleus in experimental animals result in morbid obesity due to unrestricted eating (hyperphagia). Conversely, electrical stimulation of this area results in cessation of feeding (hypophagia). Destructive lesions in the lateral hypothalamic area lead to hypophagia, even in the face of starvation; electrical stimulation of this area initiates feeding activity, even when the animal has already eaten.

Eating is a regulated function of feeding and cessation of feeding, driven by the psychologic states of hunger and satiety, the balance of hormones that regulate metabolism of foodstuffs, and the feedback signals from energy storage sites of the body (fat cells) and energy-utilizing structures (exercising muscles). In addition, there is an important genetic component to the regulation of food intake and satiety by the hypothalamus and of energy metabolism and storage by the endocrine/metabolic system.

In general, the hypothalamus regulates caloric intake, utilization, and storage in a manner that tends to maintain the body weight in adulthood. The presumptive set point around which it attempts to stabilize body weight, however, is poorly defined or maintained, as it changes readily with changes in physical activity, composition of the diet, emotional states, stress, pregnancy, and so on.

Factors that limit the amount of food ingested during a single feeding episode originate in the gastrointestinal tract and influence the hypothalamic regulatory centers. These include sensory signals carried by the vagus nerve that signify stomach filling and chemical signals giving rise to the sensation of satiety, including absorbed nutrients (glucose, certain amino acids, and fatty acids) and gastrointestinal hormones, especially cholecystokinin. Over a longer time frame, eating behavior is regulated by signals emanating from the body's food reserves and the rate of energy use, such as in muscular activity and increased metabolic rate.

The satiety center in the ventromedial hypothalamus contains receptors for monoamine and peptide neurotransmitters; manipulation of these receptors by drugs has been useful in treating obesity. Drugs that decrease appetite, including amphetamine and fenfluramine, are potent stimulators of monoaminergic transmission. Analogs of cholecystokinin, a presumptive satiety hormone, which are not peptide in nature and thus can be taken orally, are currently used as appetite suppressants in obesity.

■ The Hypothalamus Controls the Gonads and Sexual Activity

The anterior and preoptic hypothalamic areas are sites for regulating gonadotrophic hormone secretion and sexual behavior. Neurons in the preoptic area secrete gonadotropin-releasing hormone (GnRH), beginning at puberty, in response to signals that are not understood. These neurons contain receptors for gonadal steroid hormones, testosterone and/or estradiol,

which regulate GnRH secretion in either a cyclic (female) or continual (male) pattern following the onset of puberty.

At a critical period in fetal development, circulating testosterone secreted by the testes of a male fetus changes the characteristics of cells in the preoptic area that are destined later in life to secrete GnRH. These cells, which would secrete GnRH in a cyclic fashion at puberty had they not been exposed to androgens prenatally, are transformed into cells that secrete GnRH continually at a homeostatically regulated level. As a result, males exhibit a steady-state secretion rate for gonadotrophic hormones, and consequently for testosterone (Chap. 39).

In the absence of androgens in fetal blood during development, the preoptic area remains unchanged, so that at puberty the GnRH-secreting cells begin to secrete in a cyclic pattern. This pattern is reinforced and synchronized throughout the reproductive life of females by the cyclic feedback of ovarian steroids, estradiol and progesterone, on secretion of GnRH by the hypothalamus during the menstrual cycle (Chap. 40). In a sense, then, the GnRH-secreting cells of the preoptic area are by default cyclic secreters, subject to transformation into steady-state secreters if sufficient quantities of androgens appear in the fetal circulation at a critical period in development.

■ The Hypothalamus Contains the "Biologic Clock"

Many physiologic functions, including body temperature and sleep/wake cycles, vary throughout the day in a pattern that repeats itself daily. Others, such as the menstrual cycle in females, repeat themselves approximately every 28 days. Still others, such as reproductive function in seasonal breeders, repeat annually. The hypothalamus is thought to play a major role in regulating all of these biologic rhythms. Furthermore, these rhythms appear to be endogenous (within the body), because they persist even in the absence of time cues, such as day/night cycles for light and dark periods, lunar cycles for monthly rhythms, or changes in temperature and length of the day for seasonal change. Accordingly, many organisms, including humans, are said to possess an endogenous timekeeper, a so-called **biologic clock** that times the body's regulated functions.

To study biologic rhythms researchers have eliminated all external cues of time and temperature by placing their subjects in a deep cave for extended periods of time, under comfortable conditions with normal activity and routine but deprived of any information about the time of day or year. After several months of deprivation of time cues, subjects still exhibited a daily rhythm of biologic functions, such as sleep/wake periods and fluctuations in body temperature, but their free-running biologic clock seemed to operate on a day of 25, rather than 24, hours. It appears that the external signals of light and darkness perceived by the visual system serve to entrain, or synchronize, the internal clock to the rhythm of the external 24-hour day.

Most homeostatically regulated functions exhibit peaks and valleys of activity that repeat approximately daily. These are called **circadian** or **diurnal rhythms,** based on the Latin *dies* ("day") and *circa* ("approximately"). The circadian rhythms of the body are driven by the **suprachiasmatic nucleus,** a center in the hypothalamus that serves as the biologic clock of the brain. The suprachiasmatic nucleus, which influences many hypothalamic nuclei via its efferent connections, has the properties of an oscillator whose spontaneous firing patterns change dramatically during a day/night cycle. During the waking period, cells in this nucleus show high levels of metabolic activity; during sleep they shut down metabolic activity to its lowest level. In nocturnal animals (they hunt and eat by night and sleep by day), the metabolic activity of the suprachiasmatic nucleus is reversed accordingly. An important pathway influencing the suprachiasmatic nucleus is the afferent **retinohypothalamic tract** of the optic nerve, which originates in the retina and enters the brain through the optic chiasm and terminates in the suprachiasmatic nucleus of the hypothalamus. This pathway is the principal means by which light signals from the external world transmit the day/night rhythm to the brain's internal clock and thereby entrain the endogenous oscillator to the external time clock.

Figure 7–3 illustrates some of the circadian rhythms of the body. One that is experienced most vividly is alertness, which peaks in the afternoon and is lowest in the hours preceding and following sleep. Another, body temperature, ranges over approximately 2°F (about 1°C) throughout the day, with the low point occurring during sleep. Plasma levels of growth hormone increase greatly during sleep, in keeping with this hormone's metabolic role as a glucose-sparing agent during the nocturnal fast. Cortisol, on the other hand, has its highest daily plasma level prior to arising in the morning.

Other homeostatically regulated functions exhibit diurnal patterns as well; when they are all in synchrony, they function harmoniously and impart a feeling of well-being. When there is a disruption in rhythmic pattern, such as by sleep deprivation or when passing too rapidly through several time zones, the period required for reentrainment of the suprachiasmatic nucleus to the new day/night pattern is characterized by a feeling of malaise and physiologic distress. This is commonly experienced as jet lag in travelers crossing several time zones or by workers changing from day shift to night shift or vice versa. In such cases, the hypothalamus requires time to "reset its clock" before the regular rhythms are restored and a feeling of well-being ensues. The suprachiasmatic nucleus uses the new pattern of lightness/darkness,

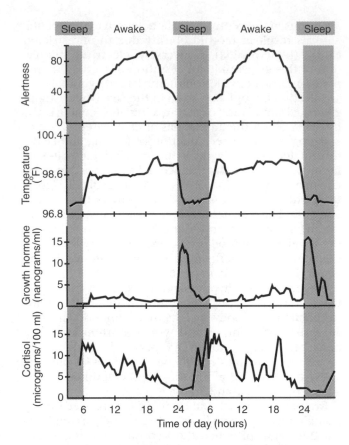

Figure 7–3 ■ Diurnal rhythms in some homeostatically regulated functions over two 24-hour periods. Alertness is measured on an arbitrary scale between sleep and most alert. (Modified from Coleman RM: *Wide Awake at 3:00 AM.* New York, WH Freeman, 1986. With permission.)

as perceived in the retina, to entrain its firing rate to a pattern consistent with the external world.

The Ascending Reticular Activating System

The brainstem contains anatomic groupings of cell bodies clearly identified as the nuclei of cranial sensory and motor nerves or as relay cells of ascending sensory or descending motor systems. The remaining cell groups of the brainstem, located in the central core, constitute a diffuse-appearing system of neurons with widely branching axons, known as the **reticular formation** (for *reticulum,* meaning "network").

■ Neurons of the Reticular Formation Exert Widespread Modulatory Influence in the Central Nervous System

As neurochemistry and cytochemical localization techniques improve, it is becoming increasingly clear that the reticular formation is not a diffuse, undefined system, but rather contains highly organized clusters of transmitter-specific cell groups that influence functions in specific areas of the CNS. The nuclei of monoaminergic neuronal systems, which innervate all parts of the CNS, are located in well-defined cell groups throughout the reticular formation.

A unique characteristic of neurons of the reticular formation is their widespread system of axon collaterals, which make extensive synaptic contacts and, in some cases, travel over long distances in the CNS. A striking example is the demonstration, using intracellular labeling of individual cells and their processes, that one axon branch descends all the way into the spinal cord, while the collateral branch projects rostrally all the way to the forebrain, making myriad synaptic contacts along both axonal pathways. Along with this widespread distribution of diverging axonal contacts there is, of necessity, a loss of specificity of neuronal signaling, since many afferent modalities contribute to the efferent output of reticular system neurons.

■ The Ascending Reticular Activating System Mediates Consciousness and Arousal

Sensory neurons bring peripheral sensory information to the CNS via specific pathways that ascend and synapse with specific nuclei of the thalamus, which in turn innervate primary sensory areas of the cerebral cortex. These pathways involve three to four synapses, starting from a receptor that responds to a highly specific sensory modality, such as touch, hearing, or vision. Each of these has, in addition, a nonspecific form of sensory transmission, in that axons of the ascending fibers send collateral branches to cells of the reticular formation (Fig. 7–4). The latter, in turn, send their axons to the nonspecific sensory projection area of the thalamus, which innervates wide areas of the cerebral cortex and limbic system.

Sensory signaling by the reticular formation is characterized by a loss of specificity for two reasons. First, numerous sensory modalities converge on one reticular formation cell, thereby losing the identity of individual modalities. Second, divergence of reticular formation axons to ascending and descending areas of the CNS leads to generalized stimulation of multiple brain regions. In the cerebral cortex and limbic system the influence of the nonspecific projections from the reticular formation is to arouse the organism. The nonspecific sensory signals reaching the cerebral cortex from the thalamus also serve to focus attention on the incoming specific sensory signals. The reticular formation also houses the neuronal systems that regulate sleep/wake cycles and consciousness. So important is the ascending reticular activating system to the state of arousal that malfunction in the reticular formation, particularly the rostral portion, can lead to loss of consciousness and coma.

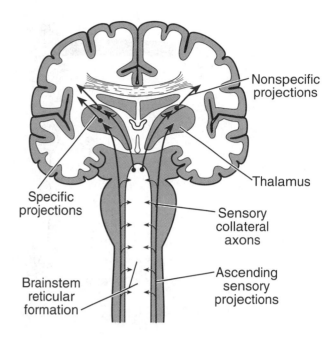

Figure 7–4 ■ The brainstem reticular formation and reticular activating system. Ascending sensory tracts send axon collateral fibers to the reticular formation. These give rise to fibers synapsing in the nonspecific nuclei of the thalamus. From there the nonspecific thalamic projections influence widespread areas of the cerebral cortex and limbic system.

■ The Electroencephalogram Is a Recording of the Electrical Activity of the Surface of the Brain

The influence of the ascending reticular formation on the brain's activity can be monitored via **electroencephalography**. The **electroencephalograph** is a sensitive recording device for picking up the electrical activity of the surface of the brain through electrodes placed on designated sites on the scalp. This noninvasive tool measures simultaneously, via multiple leads, the electrical activity of the major areas of the cerebral cortex. It is also the best diagnostic tool available for detecting abnormalities in electrical activity, such as in epilepsy, and for diagnosing sleep disorders.

The electrical activity detected by the electrodes is not generated by action potentials of cortical neurons underlying the electrode, since action potentials are carried by axons, which in the cortex project inwardly, away from the surface (Fig. 7–5). Rather, the signal they pick up consists of postsynaptic potentials generated within the dendrites of cortical neurons, as these form a dense arbor lying closest to the surface of the brain, beneath the electrode. The summated electrical potentials recorded from moment to moment in each lead are influenced greatly by the input of sensory information from the thalamus via specific and nonspecific projections to the cortical cells, as well as inputs that course laterally from other regions of the cortex.

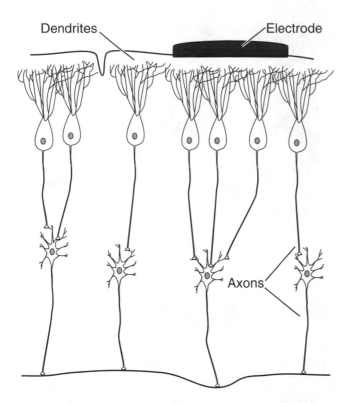

Figure 7–5 ■ Axons of cerebral cortical neurons project inwardly, perpendicular to the surface of the brain. Dendrites form a dense arbor at the brain's surface. Electrodes of the EEG measure summated postsynaptic potentials in dendrites of cortical cells, which lie closest to the electrode placed on the scalp.

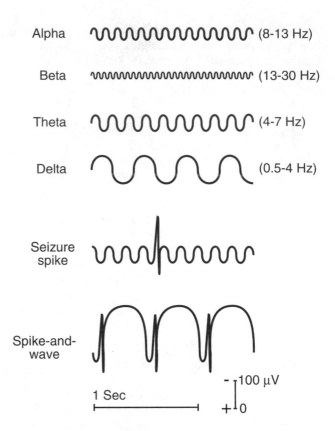

Figure 7–6 ■ The patterns of brain waves recorded by the EEG are designated alpha, beta, theta, and delta waves, based on frequency and relative amplitude. In epilepsy, abnormal spikes and large summated waves appear as many neurons are activated simultaneously.

Waves of the EEG ■ The waves recorded on an **electroencephalogram** (EEG) are described in terms of frequency, which usually ranges from less than 1 to about 30 Hz, and amplitude, or height of the wave, which usually ranges from 20 to 100 μV. Since the waves are a summation of activity in a complex network of neuronal processes, they are highly variable in a brain that never rests. However, during various states of consciousness the waves have certain characteristic patterns. At the highest state of alertness, when sensory input is greatest, the waves are of high frequency and low amplitude, as many units discharge asynchronously. At the opposite end of the alertness scale, when sensory input is at its lowest, in deep sleep, a synchronized EEG has the characteristics of low frequency and high amplitude.

Electroencephalogram wave patterns are classified according to their frequency (Fig. 7–6). **Alpha waves**, alpha rhythm ranging from 8 to 13 Hz, are observed when the person is awake but relaxed with the eyes closed. When the eyes are open, the added visual input to the cortex imparts a faster rhythm to the EEG, ranging from 13 to 30 Hz and designated as **beta waves**. The slowest waves recorded occur during sleep: **theta waves** at 4-7 Hz and **delta waves** at 0.5–4 Hz, in deepest sleep.

Abnormal wave patterns are seen in epilepsy, a neurologic disorder of the brain characterized by spontaneous discharges of electrical activity, resulting in abnormalities ranging from momentary lapses of attention, to seizures of varying severity, to loss of consciousness if both brain hemispheres participate in the electrical abnormality. The characteristic waveform signifying seizure activity is the appearance of spikes or sharp peaks, as abnormally large numbers of units fire simultaneously. Examples of spike activity occurring singly and in a spike-and-wave pattern are shown in Figure 7–6.

Sleep and the EEG ■ Sleep is regulated by the reticular formation in a manner that is not well understood, except that when manipulated by drugs or experimental lesions, the monoaminergic cell groups in the reticular formation cause profound changes in sleep. The EEG recorded during sleep reveals a continually changing pattern of wave amplitudes and frequencies, indicating that the brain remains continually active even in the deepest stages of sleep. The EEG pattern recorded during sleep varies in a cyclic fashion that repeats itself approximately every 90 minutes, starting from the time of falling asleep to awakening

7-8 hours later (Fig. 7-7). These cycles are associated with two different forms of sleep, which follow each other sequentially: **slow-wave sleep** consists of four stages of progressively deepening sleep (i.e., the subject is more and more difficult to awaken); **rapid-eye-movement (REM)** sleep is characterized by back-and-forth movements of the eyes under closed lids, accompanied by autonomic excitation. As studied by EEG recordings in sleeping subjects in the laboratory, the electrical activity of the brain varies as the subject passes through cycles of slow-wave sleep, then REM sleep, on through the night.

A sleeping episode begins with slow-wave sleep, which progresses through the four stages of increasingly deep sleep during which the EEG becomes progressively slower in frequency and higher in amplitude. Stage 4 is reached at the end of about an hour, when delta waves are observed (Fig. 7-7). The subject then passes through the same stages in reverse order, approaching stage 1 by about 90 minutes, when a REM period begins, followed by a new cycle of slow-wave sleep. The physiologic state in slow-wave sleep is characterized by decreased heart rate and blood pressure, slow and regular breathing, and relaxed muscle tone.

Stages 3 and 4, the deepest stages and presumably the most restorative, occur only in the first few sleep cycles of the night. In contrast, REM periods increase in duration with each successive cycle, so that the last few cycles consist of approximately equal periods of REM sleep and stage 2 slow-wave sleep (Fig. 7-7).

Rapid-eye-movement sleep is also known as **paradoxical sleep,** because of the seeming contradictions in its characteristics. First, the EEG exhibits unsynchronized, high-frequency, low-amplitude waves (i.e., a beta rhythm), which is more typical of the awake state than sleep (Fig. 7-7), yet the subject is as difficult to arouse as when in stage 4 slow-wave sleep. Second, the autonomic nervous system is in a state of excitation; blood pressure and heart rate are increased and breathing is irregular. In males, autonomic excitation in REM sleep includes penile erection; this has been used as a diagnostic measure in sexual impotence to determine whether failure of erection reflexes is based on a neurologic defect, in which case penile erection does not accompany REM sleep.

When subjects are awakened during a REM period, they usually report dreaming, in more or less complete detail. Accordingly, it is customary to consider REM sleep as dream sleep, particularly since the eyes move as if to follow the scenes depicted in the dream episode. Another curious characteristic of REM sleep is that most voluntary muscles are temporarily paralyzed. Two exceptions, in addition to the muscles of respiration, include the extraocular muscles, which contract rhythmically to produce the rapid eye movements, and the muscles of the inner ear, which serve to tighten the eardrum to increase acuity in hearing. Muscle paralysis is thought to be caused by a temporary disconnection of most of the motor system from its central control centers. Many of us experience this when wak-

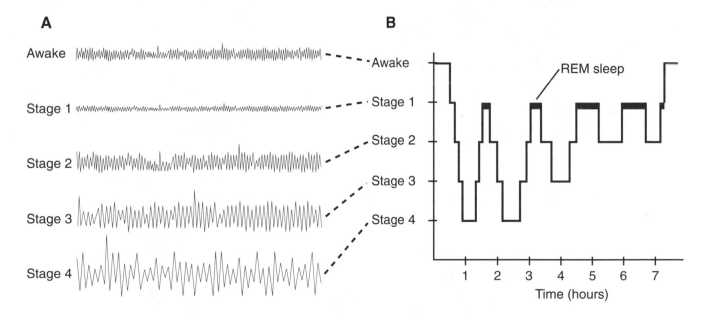

Figure 7–7 ■ The cycle of wave patterns through one sleep episode. Sleep begins with stage 1 of slow-wave sleep, then progresses through stages 2, 3, and 4 as the waves decrease in frequency and increase in amplitude. From stage 4 the EEG returns in reverse order to stage 1, at which time the first REM sleep period occurs, about 90 minutes after onset of sleep. The cycle repeats through the night, with the REM periods increasing in length and stages 3 and 4 of slow-wave sleep diminishing in length through the night. (Modified from Kandel ER, Schwartz JH, Jessel TM: *Principles of Neural Science.* 3rd ed. New York, Elsevier, 1991.)

ing from a bad dream feeling momentarily incapable of running from danger. In certain sleep disorders where skeletal muscle contraction is not temporarily paralyzed in REM sleep, subjects act out dream sequences with disturbing results, with no conscious awareness of this happening.

Sleep in humans varies with developmental stage. Newborns sleep approximately 16 hours per day, of which about 50% are spent in REM sleep. Normal adults sleep 7–8 hours per day, of which about 25% is spent in REM sleep. The percentage of REM sleep declines further with age, together with loss of the ability to achieve stages 3 and 4 of slow-wave sleep.

The Forebrain

The forebrain contains the **cerebral cortex** and the subcortical structures rostral to the diencephalon. The cortex has a rich, multilayered array of neurons and their processes (Fig. 7–5), forming columns perpendicular to the surface. The axons of cortical neurons give rise to descending fiber tracts and intra- and interhemispheric fiber tracts, which together make up the prominent white matter underlying the outer cortical gray matter. A deep sagittal fissure divides the cortex into a right and left hemisphere, each of which receives sensory input from and sends its motor output to the opposite side of the body. A set of **commissures,** or tough fibrous tracts containing axonal fibers, interconnects the two hemispheres, so that processed neural information from one side of the forebrain is transmitted to the opposite hemisphere via the commissures. The largest of these commissures is the **corpus callosum,** which interconnects the major portion of the hemispheric regions (Fig. 7–1, 7–8). The anterior and posterior ends of the hemispheres are interconnected

by several smaller commissures, including one in the anterior forebrain and several in the posterior forebrain.

Among the subcortical structures located in the forebrain are the **limbic system,** which regulates emotional response, and the **basal ganglia** (caudate, putamen, and globus pallidus), which are essential for coordinating motor activity (Chap. 5).

■ The Cerebral Cortex Is Functionally Compartmentalized

In the human brain the cerebral cortex is highly convoluted, with **gyri** (singular, *gyrus*), which project above the surface of the brain, and **sulci** (singular, *sulcus*), which form fissures that dip below the surface. Two deep fissures form prominent landmarks on the surface of the cortex; the **central sulcus** divides the frontal lobe from the parietal lobe, and the **sylvian fissure** divides the parietal lobe from the temporal lobe (Fig. 7–9). A fourth lobe, the occipital lobe, has less prominent sulci separating it from the parietal and temporal lobes. The fifth lobe, the olfactory lobe, is located on the undersurface of the frontal portion of the brain.

Topographically, the cerebral cortex is divided into areas of specialized functions, including the primary sensory areas for vision (occipital cortex), hearing (temporal cortex), somatic sensation (postcentral gyrus), and primary motor area (precentral gyrus). As shown in Figure 7–9, these well-defined areas comprise only a small fraction of the surface of the cerebral cortex. The majority of the remaining cortical area is designated as **association cortex,** where processing

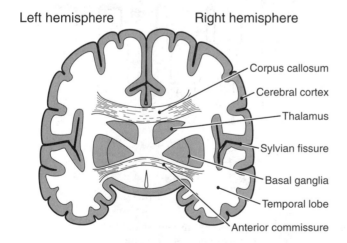

Figure 7–8 ■ The hemispheres of the forebrain and some structures deep to it, in a coronal section rostral to the brainstem. Note the corpus callosum, the major commissure that interconnects the right and left hemispheres, and the anterior commissure, in the frontal forebrain.

Left hemisphere   Right hemisphere

- Corpus callosum
- Cerebral cortex
- Thalamus
- Sylvian fissure
- Basal ganglia
- Temporal lobe
- Anterior commissure

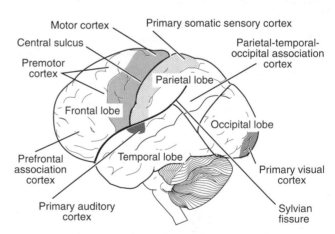

Motor cortex — Primary somatic sensory cortex
Central sulcus
Premotor cortex
Parietal lobe
Frontal lobe
Parietal-temporal-occipital association cortex
Occipital lobe
Prefrontal association cortex
Temporal lobe
Primary visual cortex
Primary auditory cortex
Sylvian fissure

Figure 7–9 ■ Four lobes of the cerebral cortex, containing primary sensory and motor areas and major association areas. (A fifth lobe, the olfactory, is located on the undersurface of the cerebral hemispheres.) The central sulcus and sylvian fissure are prominent landmarks used in defining the lobes of the cortex. Imaginary lines are drawn in to indicate the boundaries between the occipital, temporal and parietal lobes. (Modified from Kandel ER, Schwartz JH, Jessel TM: *Principles of Neural Science.* 3rd ed. New York, Elsevier, 1991.)

Patients with life-threatening, intractable epileptic seizures were treated in the past by surgical commissurotomy, or cutting of the corpus callosum and anterior commissure (Fig. 7-8). This procedure effectively cut off all neuronal communication between the left and right hemispheres and vastly improved patient status. (The procedure has since been modified to partial sectioning of the corpus callosum, with similar effectiveness.)

There was a remarkable absence of overt signs of disability following commissurotomy: patients retained their original motor and sensory functions, learning and memory, personality, talents, emotional responding, and so on. This was not unexpected, as each hemisphere has bilateral representation of most known functions and, furthermore, those ascending (sensory) and descending (motor) neuronal systems that crossed to the opposite side were known to do so at levels lower than the corpus callosum.

Notwithstanding this appearance of normalcy, following commissurotomy patients were shown to be impaired to the extent that one hemisphere literally did not know what the other was doing. It was shown further that each hemisphere processes neuronal information differently from the other and that some cerebral functions were confined exclusively to one hemisphere.

In an interesting series of studies by Nobel laureate Roger Sperry and colleagues, these patients with a so-called "split-brain" were subjected to psychophysiologic testing in which each disconnected hemisphere was examined independently. Their findings confirmed what was already known—that sensory and motor functions are controlled by cortical structures in the contralateral hemisphere; for example, visual signals from the left visual field were perceived in the right occipital lobe, and there were contralateral controls for auditory, somatosensory, and motor functions. (Note that the olfactory system is an exception, as odorant chemicals applied to one nostril are perceived in the olfactory lobe on the same side.) However, to the surprise of the scientists, they found that language ability was controlled almost exclusively by the left hemisphere. Thus, if an object was presented to the left brain via any of the sensory systems (visual, auditory, olfactory, or tactile), the subject could readily identify it by the spoken word. However, if the object was presented to the right hemisphere, the subject could not find words to identify it. This was not due to inability of the right hemisphere to perceive the object, as the subject could easily identify it among other choices by nonverbal means, such as feeling it while blindfolded. From these and other tests it became clear that the right hemisphere was mute; it could not produce language. In accordance with these findings, anatomic studies show that areas in the temporal lobe concerned with language ability, including Wernicke's area, are anatomically larger on the left hemisphere than on the right in a majority of humans, and this is seen even prenatally. Corroborative evidence of language ability in the left hemisphere is shown in stroke victims, where aphasias are most severe if the damage is on the left side of the brain.

In addition to language ability, the left hemisphere excels in mathematical ability, symbolic thinking, and sequential logic. The right hemisphere, on the other hand, excels in visuospatial ability, such as three-dimensional constructions with blocks and drawing maps, and in musical sense, artistic sense, and other higher functions that computers seem less capable of emulating. The right brain exhibits some ability in language and calculation, but at the level seen in 5–7-year-olds. It has been postulated that in early childhood both sides of the brain are capable of all of these functions, but that the larger size of the language area in the left temporal lobe favors development of that side during acquisition of language, resulting in nearly total specialization for language on the left side for the rest of one's life.

of neural information is performed at the highest levels of which the organism is capable; among vertebrates the human cortex contains the most extensive association areas. The association areas are also sites of long-term memory and control such human functions as language acquisition, speech, musical ability, mathematical ability, complex motor skills, abstract thought, symbolic thought, and other cognitive functions.

Association areas interconnect the primary sensory areas with each other via laterally coursing tracts. The **parietal-temporal-occipital association cortex** integrates neural information contributed by visual, auditory, and somatic sensory experiences. The **prefrontal association cortex** is extremely important as the coordinator of emotionally motivated behaviors, by virtue of its connections with the limbic system. Additionally, the prefrontal cortex receives neural input from the other association areas and regulates motivated behaviors by direct input to the **premotor area,** which serves as the association area of the motor cortex.

■ The Limbic System Is the Seat of the Emotions

The limbic system is comprised of large areas of the forebrain where the emotions are generated and the responses to emotional stimuli are coordinated. Understanding its functions is particularly challenging because it is a complex system of numerous and disparate elements, most of which have not been fully character-

ized. A compelling reason for studying the limbic system is that the major psychiatric disorders, including mania, depression, schizophrenia, and senile dementia, involve malfunction in the limbic system.

Anatomy of the Limbic System ■ The limbic system comprises large areas of the forebrain interconnected via circuitous pathways that link the cerebral cortex with the diencephalon. The name "limbic system," from the Latin *limbus* "border" comes from its location at the border between the cerebral cortex and the brainstem (Fig. 7–10). Originally the limbic system was considered to be restricted to a ring of structures surrounding the corpus callosum, including the olfactory system, the cingulate gyrus, parahippocampal gyrus, and hippocampus, together with the fiber tracts that interconnect them with the diencephalic components of the limbic system—the hypothalamus and anterior thalamus. Recent studies indicate that the limbic system should include additional subcortical structures that interconnect anatomically and functionally with parts of the limbic system (Fig. 7–11). These include such prominent structures as the amygdala (deep in the temporal lobe), nucleus accumbens (the limbic portion of the basal ganglia), septal nuclei (at the base of the forebrain), and habenula.

The limbic system forms circuitous loops of fiber tracts interconnecting the limbic structures, and this imparts a reverberating, long-lasting influence on brain function typical of emotional states. The main circuit (Fig. 7–12) links the hippocampus to the mammillary body of the hypothalamus by way of the fornix, the

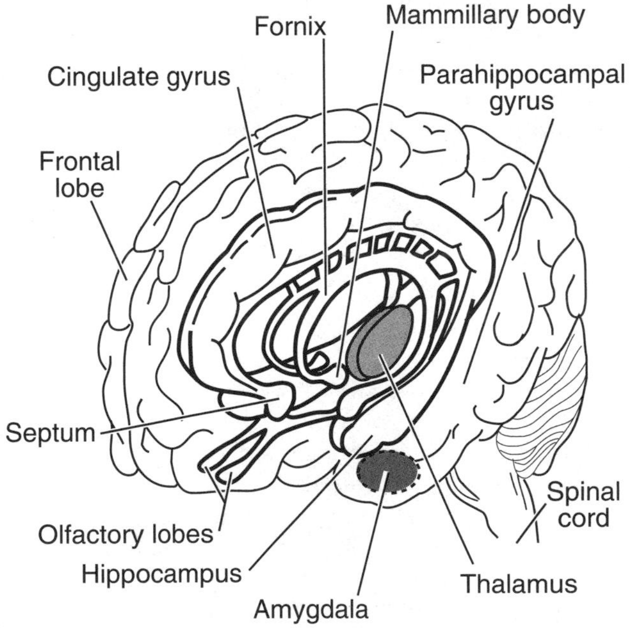

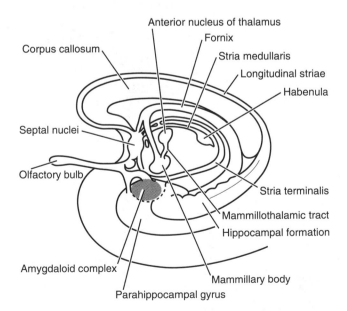

Figure 7–10 ■ The cortical and subcortical structures of the limbic system extending from the cerebral cortex to the diencephalon. (Modified from Bloom FE, Lazerson A, Hofstadter L: *Brain, Mind and Behavior.* New York, WH Freeman. Copyright © 1985 by Educational Broadcasting Group. With permission.)

Figure 7–11 ■ The fiber tracts that interconnect the structures of the limbic system and a more complete view of its subcortical components. (Modified from Truex RC, Carpenter MB: *Strong and Elwyn's Human Neuroanatomy.* 5th ed. Baltimore, Williams & Wilkins, 1964, p 446.)

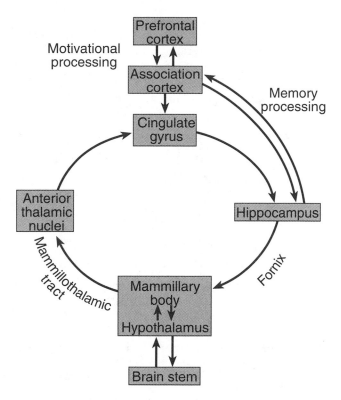

Figure 7–12 ■ The basic circuitry of key elements in the limbic system.

(Fig. 7–12). As noted above, all ascending sensory systems in the brainstem send axon collaterals to the reticular formation, which in turn innervates the limbic system, particularly via monoaminergic pathways. The reticular formation also forms the ascending reticular activating system, which serves not only to arouse the cortex but also to impart an emotional tone to the sensory information transmitted nonspecifically to the cerebral cortex.

Monoaminergic Innervation ■ Monoaminergic neurons innervate all parts of the CNS via widespread, divergent pathways starting from cell groups in the reticular formation. The limbic system and basal ganglia are particularly richly innervated by catecholaminergic (noradrenergic and dopaminergic) and serotonergic nerve terminals emanating from brainstem nuclei containing relatively few cell bodies compared with their extensive terminal projections. From neurochemical manipulation of the monoaminergic systems in the limbic system, it is apparent that they play a major role in determining emotional state.

Dopaminergic neurons are located in three major pathways originating from cell groups in either the midbrain (the substantia nigra and ventral tegmentum) or the hypothalamus (Fig. 7–13). The **nigrostriatal system** consists of neurons with cell bodies in the substantia nigra (pars compacta) and terminals in the

hypothalamus to the anterior thalamic nuclei via the mammillothalamic tract, and the anterior thalamus to the cingulate gyrus by widespread, anterior thalamic projections. To complete the circuit, the cingulate gyrus connects with the hippocampus, to enter the circuit again. Other structures of the limbic system form smaller loops within this major circuit, forming the basis for a wide range of emotional behaviors. Higher centers in the neocortex provide the limbic system with information based on previous learning and currently perceived needs. Inputs from the brainstem provide visceral and somatic sensory signals, including tactile, pressure, pain, and temperature information from the skin and sexual organs and pain information from the visceral organs.

The cortical structures making up the limbic cortex, including the olfactory lobe, cingulate gyrus, and inner aspects of the temporal lobes, are made up of **allocortex,** the phylogenetically oldest cortex sometimes called the **enterorhinal cortex** because of its direct input of olfactory stimuli from the olfactory nerves. The **neocortex,** or more highly evolved cerebral cortex, which is especially extensive in the human brain, is not considered part of the limbic system but has a direct influence via inputs from the frontal and temporal lobes to the limbic system.

At the caudal end of the limbic system, the brainstem has reciprocal connections with the hypothalamus

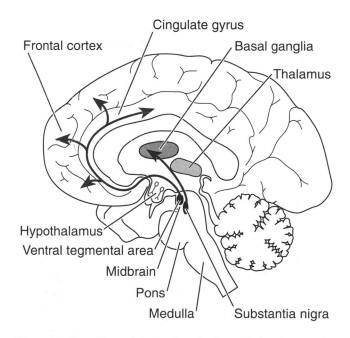

Figure 7–13 ■ The origins and projections of the three major dopaminergic systems. The nigrostriatal system originates in the substantia nigra and sends projections to the basal ganglia. The tuberoinfundibular system is completely contained in the hypothalamus. The mesolimbic/mesocortical system originates in the ventral tegmentum and sends projections to the limbic system and limbic cortex. (Modified from Heimer L: *The Human Brain and the Spinal Cord.* New York, Springer-Verlag, 1983.)

neostriatum (caudate and putamen) located in the basal ganglia. This dopaminergic pathway is essential for maintaining normal muscle tone and initiating voluntary movements (Chap. 5). The **tuberoinfundibular system** of dopaminergic neurons is located entirely within the hypothalamus, with cell bodies in the arcuate nucleus and periventricular nuclei and terminals in the median eminence on the ventral surface of the hypothalamus. The tuberoinfundibular system is responsible for the secretion of hypothalamic releasing factors into a portal system that carries them through the pituitary stalk into the anterior pituitary lobe. The **mesolimbic** and **mesocortical** system of dopaminergic neurons originates in the **ventral tegmental area** of the midbrain and innervates most structures of the limbic system (olfactory tubercles, septal nuclei, amygdala, nucleus accumbens) and limbic cortex (frontal and cingulate cortices). This dopaminergic system serves an important role in motivation and drive. For example, dopaminergic sites in the limbic system, particularly the more ventral structures such as the septal nuclei and nucleus accumbens, are associated with the brain's **reward mechanism.** Drugs that increase the transmission of brain dopamine, such as cocaine, which inhibits its reuptake, and amphetamine, which promotes release and inhibits reuptake, lead to repeated administration and abuse presumably because they stimulate the brain's reward system.

The mesolimbic/mesocortical dopaminergic system is also the site of action of **neuroleptic** drugs, agents used in the treatment of schizophrenia (see below) and other psychotic states, such as mania and Huntington's disease.

Norepinephrine-containing (noradrenergic) neurons are located in cell groups in the medulla and pons (Fig. 7-14). The medullary cell groups project to the spinal cord, where they influence cardiovascular regulation and other autonomic functions. Cell groups in the pons include the **lateral system,** which innervates the basal forebrain and hypothalamus, and the **locus coeruleus,** which sends efferent fibers to essentially all parts of the CNS (one exception is the neostriatum in the basal ganglia, whose only catecholaminergic innervation is dopaminergic).

Noradrenergic neurons innervate all parts of the limbic system and the cerebral cortex, where they play a major role in setting **mood** (sustained emotional state) and **affect** (the emotion itself; e.g., euphoria, depression, anxiety). Drugs that alter noradrenergic transmission have profound effects on mood and affect. For example, reserpine, which depletes brain norepinephrine, induces a state of depression. Drugs that enhance norepinephrine availability, such as monoamine oxidase inhibitors and inhibitors of reuptake, reverse this depression. Amphetamine and cocaine have effects on boosting noradrenergic transmission similar to those described for dopaminergic transmission; they inhibit reuptake and/or promote release of norepinephrine. Increased noradrenergic

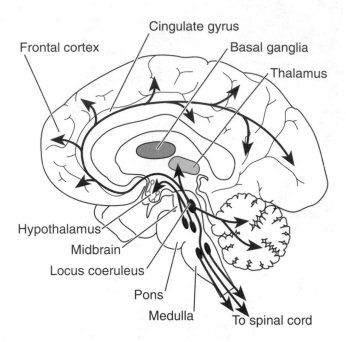

Figure 7–14 ■ The origins and projections of five of the seven cell groups of noradrenergic neurons of the brain. The depicted groups originate in the medulla and pons. Among the latter, the locus coeruleus in the dorsal pons innervates most parts of the CNS. (Modified from Heimer L: *The Human Brain and the Spinal Cord.* New York, Springer-Verlag, 1983.)

transmission results in an elevation of mood, which further contributes to their potential for abuse, despite the depression that follows when drug levels fall. Some of the unwanted consequences of administration of cocaine or amphetaminelike drugs reflect the marked increase in noradrenergic transmission, in both the periphery and the CNS. This can result in a hypertensive crisis, myocardial infarct, or stroke, in addition to marked swings in affect starting with euphoria and ending with profound depression.

Serotonergic neurons innervate all parts of the CNS. Cell bodies of these neurons are located at the midline of the brainstem (the raphe system) and more laterally placed nuclei, extending from the caudal medulla to the midbrain (Fig. 7-15). The innervated areas include those that receive noradrenergic and dopaminergic innervation, including the neostriatum.

Serotonin plays a major role in the defect underlying affective disorders (described below). Drugs that increase serotonin transmission are effective antidepressant agents.

The Brain's Reward System ■ Experimental studies beginning early in the twentieth century have demonstrated that emotional state can be altered by stimulating or creating lesions in various parts of the limbic system. Most of our knowledge comes from animal studies, but emotional feelings are reported by humans when limbic structures are stimulated during

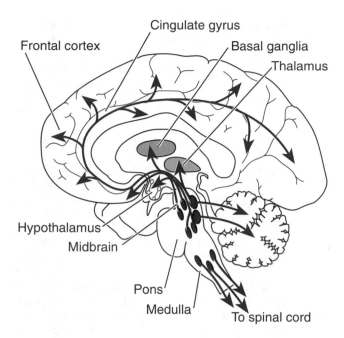

Frontal cortex

Cingulate gyrus

Basal ganglia

Thalamus

Hypothalamus

Midbrain

Pons

Medulla

To spinal cord

Figure 7–15 ■ The origins and projections of the nine cell groups of the serotonergic system of the brain. The depicted groups originate in the caudal medulla, pons, and midbrain and send projections to most regions of the brain. (Modified from Heimer L: *The Human Brain and the Spinal Cord.* New York, Springer-Verlag, 1983.)

brain surgery. Since the brain has no pain sensation when touched, subjects awakened from anesthesia after exposure of the brain have been able to communicate changes in emotional experience linked to electrical stimulation of specific areas.

Electrical stimulation of various sites in the limbic system produces either pleasurable (rewarding) or unpleasant (aversive) feelings. Electrodes have been implanted in animals' brains as a way to study this. When the electrodes are implanted in structures presumed to generate rewarding feelings and the animals are allowed to deliver current to the electrodes by pressing a bar, repeated and prolonged self-stimulation is seen. Other needs, such as food, water, and sleep, are neglected. The sites that provoke the highest rates of electrical self-stimulation are in the ventral limbic areas, including the septal nuclei and nucleus accumbens. Extensive studies of electrical self-stimulatory behavior indicate that dopaminergic neurons play a major role in mediating reward. The nucleus accumbens is of particular interest as a center of reward, as it is thought to be the site of action of addictive drugs, including opiates, alcohol, nicotine, cocaine, and amphetamine.

Aggression and the Limbic System ■ A fight-or-flight reaction, including the autonomic components (Chap. 6) and postures of rage and aggression characteristic of fighting behavior, can be elicited by electrical stimulation of sites in the hypothalamus and amygdala. If the frontal cortical connections to the limbic system are severed, the rage postures and aggressiveness become permanent, illustrating the importance of the higher centers in restraining aggression and presumably in invoking it at appropriate times. By contrast, bilateral removal of the amygdala results in a placid animal that cannot be provoked.

Sexual Activity ■ The biologic basis of human sexual activity is poorly understood because of its complexity and because findings derived from nonhuman animal studies cannot be extrapolated. The major reason for this limitation is that the cerebral cortex, uniquely developed in the human brain, plays a more important role in governing human sexual activity than the instinctive or olfactory-driven behaviors in nonhuman primates and lower mammalian species. Nevertheless, several parallels in human and nonhuman sexual activity exist, indicating that the limbic system in general coordinates sex drive and mating behavior, with higher centers exerting more or less overriding influences.

Copulation in mammals is coordinated by reflexes of the sacral spinal cord, including penile erection and ejaculation reflexes in males and engorgement of erectile tissues in females, as well as the muscular spasms of the orgasmic response. Copulatory behaviors and postures can be elicited in animals by stimulating parts of the hypothalamus, olfactory system, and other limbic areas, resulting in mounting behavior in males and lordosis (arching the back and raising the tail) in females. Ablation studies have shown that sexual behavior also requires an intact connection of the limbic system with the frontal cortex.

Olfactory cues are important in initiating mating activity in seasonal breeders. On signal from the hypothalamus, driven by its endogenous seasonal clock, its anterior and preoptic areas initiate hormonal control of gonadic function, leading to the secretion of odorants (pheromones) by the female reproductive tract that signal to the male the onset of estrus and sexual receptivity. The odorant cues are powerful stimulants, acting at extremely low concentrations to initiate mating behavior in males. The olfactory system, by virtue of its direct connections with the limbic system, facilitates the coordination of behavioral, endocrine, and autonomic responses involved in mating. Although human and nonhuman primates are not seasonal breeders (mating can occur on a continual basis) vestiges of these functions remain intact, such as the importance of the olfactory and limbic systems and the role of the hypothalamus in generating cyclic changes in ovarian function in females and continuous regulation of gonadal function in males. More important determinants of human sexual activity are the higher cortical functions of learning and memory, which serve to either reinforce or suppress the signals that initiate sexual responding, including the sexual reflexes coordinated by the sacral spinal cord.

Psychiatric Disorders Involve the Limbic System

The major psychiatric disorders, including **affective disorders** and **schizophrenia,** are disabling diseases with a genetic predisposition and no known cure. The biologic basis for these disorders remains obscure, particularly the role of environmental influences on individuals genetically predisposed to the disorder. Altered states of the brain's monoaminergic systems have been a major focus as possible underlying factors, based on extensive human studies in which neurochemical imbalances in catecholamines, acetylcholine, and serotonin have been observed. Another reason for focusing on the monoaminergic systems is that the most effective drugs in use for the treatment of psychiatric disorders are agents that alter monoaminergic transmission.

Affective Disorders ■ The affective disorders are characterized by disorders of mood and affect. They include **depression,** which can be so profound as to provoke suicide, and **manic-depressive disorder,** in which periods of profound depression are followed by periods of mania, in a cyclic pattern. Biochemical studies indicate that when depressed, these patients show decreased use of brain norepinephrine and when manic they have increased transmission of norepinephrine. Whether in depression or in mania, all patients seem to have decreased transmission of brain serotonin, suggesting that serotonin may exert an underlying permissive role in abnormal mood swings, in contrast with norepinephrine, whose transmission, in a sense, titrates the mood from highest to lowest extremes.

The most effective treatments for depression, including antidepressant drugs, monoamine oxidase inhibitors, and electroconvulsive therapy, have in common the ability to stimulate both noradrenergic and serotonergic neurons serving the limbic system, and they all require repeated application to achieve a therapeutic response. On a biochemical level, the repeated administration of antidepressant drugs and treatments to laboratory animals and in animal models of depression results in down-regulation, or decrease in number, of β-adrenergic receptors in limbic forebrain regions, together with a decrease in the second messenger, cAMP, generated by these receptors. This down-regulation has a time course similar to that needed to achieve a therapeutic response by antidepressant treatments. Furthermore, the serotonergic innervation to the limbic forebrain must be intact for these treatments to exert their biochemical actions on the β-adrenergic receptor, suggesting further that the role of serotonin in depression may be permissive.

In the long term, the most effective treatment for mania is lithium, although antipsychotic (neuroleptic) drugs, which block dopamine receptors (see below), are effective in the acute treatment of mania. The therapeutic actions of lithium remain unknown, but lithium has an important action on another receptor-mediated second messenger system that may be relevant. Lithium interferes with regeneration of phosphatidylinositol in neuronal membranes by blocking the hydrolysis of inositol-1-phosphate. Depletion of phosphatidylinositol in the membrane renders it incapable of responding to receptors that use this second-messenger system, including the α-adrenergic receptor.

Schizophrenia ■ Schizophrenia comprises a group of psychotic disorders that varies greatly in symptomatology among individuals. The features most commonly observed are thought disorder, inappropriate emotional response, and auditory hallucinations. Biochemical studies in schizophrenic patients do not readily point to a neurochemical imbalance in biogenic amines, but if the limbic system's receptors to dopamine are blocked the most troubling symptoms of schizophrenia are ameliorated. This finding suggests that the mesocortical/mesolimbic dopaminergic system functions normally in schizophrenia but may serve as a final pathway for some other (primary) cause of the disorder. When the abnormality is blocked at a point downstream from the primary event, the symptoms are no longer expressed.

Current research is focused on finding the subtype of dopamine receptor that mediates mesocortical/mesolimbic dopaminergic transmission but does not affect the nigrostriatal system, which controls motor function (Fig. 7–13). So far, neuroleptic drugs that block one almost always block the other as well, leading to unwanted neurologic side effects, including abnormal involuntary movements (tardive dyskinesia) after long-term treatment or parkinsonism in the short term.

Memory and Learning Require the Cerebral Cortex and Limbic System

Memory and learning are inextricably tied to each other, as part of the learning process involves the assimilation of new information and its commitment to memory. The most likely sites of learning in the human brain are the large association areas of the cerebral cortex, working in coordinated fashion with subcortical structures deep in the temporal lobe, including the hippocampus and amygdala. The association areas draw on sensory information received from the primary visual, auditory, somatic sensory, and olfactory cortices and on emotional feelings transmitted via the limbic system. This information is integrated with previously learned skills and stored memory, which presumably also reside in the association areas.

The learning process itself is poorly understood, but it can be studied experimentally at the synaptic level in isolated slices of mammalian brain or in more simple invertebrate nervous systems. Synapses subjected to repeated presynaptic neuronal stimulation show changes in excitability of postsynaptic neurons, including facilitation of neuronal firing, changes in patterns of release of neurotransmitters, changes in second messenger formation, and, in intact organisms, evidence that learning occurred. The phenomenon of increased excitability and altered chemical state on repeated stimulation of synapses is known as **long-term potentiation,** a condition that persists beyond cessation of electrical stimulation, as expected of learning and memory. An early event in long-term potentiation is a series of protein phosphorylations induced by receptor-activated second messengers and leading to activation of a host of intracellular proteins and altered excitability.

Much of our knowledge of human memory formation and retrieval is based on studies of patients in whom strokes, brain injury, or surgery resulted in memory disorders. Such knowledge is then examined in more rigorous experiments in nonhuman primates capable of cognitive functions. From these combined approaches it is known that the prefrontal cortex is essential for coordinating the formation of memory, starting from a learning experience in the cerebral cortex, then processing the information and communicating it to the subcortical limbic structures. The prefrontal cortex receives sensory association input from the parietal, occipital, and temporal lobes and emotional input from the limbic system, and it draws on such learned skills as language, mathematical ability, and other symbolic representations of learned experience, as stored in the other association cortices with which it communicates. The prefrontal cortex is pivotal in integrating these inputs into a form suitable for further processing elsewhere. The prefrontal cortex can thus be considered the site of "working memory," as opposed to sites that consolidate the memory and store it. After processing the new experience in light of previously acquired learning, the prefrontal cortex transmits the results via the limbic system to the hippocampus, where it is consolidated over several hours into a more permanent form that is stored in, and can be retrieved from, the association cortices.

Short-Term Memory ■ Newly acquired learning experiences can be readily recalled for only a few minutes or more. This is known as **short-term memory** and can be demonstrated by the common experience of looking up a telephone number, repeating it mentally until you finish dialing the number, then promptly forgetting it as you focus your attention on starting the conversation. Short-term memory is a product of working memory; the decision to process information further for permanent storage is based on judgment as to its importance or on whether it is associated with a significant event or emotional state. An active process involving the hippocampus must be employed to make a memory more permanent.

Long-Term Memory ■ The conversion of short-term to **long-term memory** is facilitated by repeti-

tion, by adding more than one sensory modality to learn the new experience (e.g., to write down a newly acquired fact at the same time one hears it spoken), and, even more effective, by tying the experience through the limbic system to a strong, meaningful emotional context. The role of the hippocampus in consolidating the memory is reinforced by its participation, as part of the limbic system, in generating the emotional state with which the new experience is associated. The hippocampus is not required for subsequent retrieval of the memory.

Much of our understanding of memory is garnered from clinical cases, the most notable being the case of H.M., described by Brenda Milner and colleagues, in Montreal. H.M. had sustained bilateral removal of the hippocampus as a means of treating a severe, intractable form of epilepsy. Following surgery, H.M. lost all ability to retain new memories formed after the loss of both hippocampi. This form of memory loss is known as **anterograde amnesia.** There was no evidence of loss of memories laid down prior to surgery, known as **retrograde amnesia,** nor was there loss of intellectual capacity, mathematical skills, or other cognitive functions. An extreme example of H.M.'s memory loss is that Dr. Milner had to introduce herself to H.M., as would a stranger, every time they met.

In a more recent case of interest, Larry Squire and colleagues, in San Diego, collected neuropsychologic data in a subject with anterograde amnesia, R.B., from the time of onset at age 52 following a stroke until he died 5 years later. Like H.M., R.B. exhibited anterograde amnesia without loss of cognitive functions and without signs of retrograde amnesia. Following his death, histologic examination of the brain revealed a marked, bilateral loss of the large pyramidal cells specifically in the CA1 area of the hippocampus, suggesting that these large cells of the hippocampus are essential for consolidation of short-term memory into long-term memory.

■ Language and Speech Are Coordinated in Specific Areas of Association Cortex

The ability to communicate by language verbally and in writing is one of the most difficult cognitive functions to study, since only humans are capable of these skills. The use of primates as animal models for studying language acquisition have, for the most part, yielded negative results. All of the knowledge gained has been inferred from clinical data by studying patients with **aphasias** (disturbances in producing or understanding the meaning of words) following brain injury, surgery, or other damage to the cerebral cortex.

Two areas appear to play an important role in language and speech: **Wernicke's area,** in the upper temporal lobe, and **Broca's area,** in the frontal lobe (Fig. 7–16). Both of these areas are located in association cortex, adjacent to cortical areas that are essential

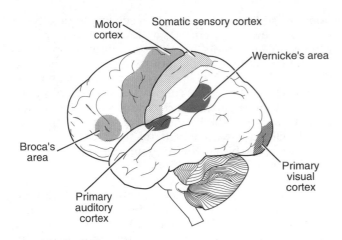

Figure 7–16 ■ Wernicke's and Broca's areas and the visual and auditory cortices.

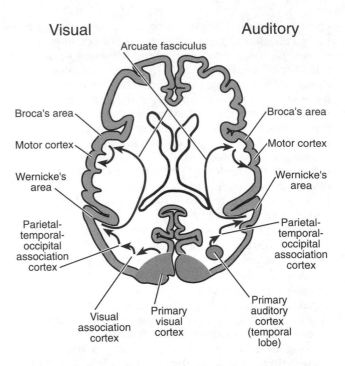

Figure 7–17 ■ The pathways by which an object is recognized and identified by the spoken word. On the left, the object is detected in the visual cortex in the occipital lobe; on the right, the object is detected by hearing in the auditory cortex of the temporal lobe. In both cases, the sensory information is processed in the secondary association areas adjacent to the primary sensory areas. The flow of neuronal processing in both instances is to the parietal-temporal-occipital association area, to Wernicke's area, to Broca's area via the arcuate fasciculus, to the motor association cortex, and from there to the motor cortex involved with the muscles of speech in the mouth and throat. (Modified from Kandel ER, Schwartz JH, Jessel TM: *Principles of Neural Science.* 3rd ed. New York, Elsevier, 1991.)

in language communication—vision, hearing, and motor production of speech. Wernicke's area is in the parietal-temporal-occipital association cortex, a major association area for processing sensory information from the somatosensory, visual, and auditory cortices (Fig. 7-16). Broca's area is in the prefrontal association cortex, adjacent to the portion of the motor cortex that regulates movement of the muscles of the mouth, tongue, and throat (i.e., the structures used in the mechanical production of speech). A fiber tract, the **arcuate fasciculus,** interconnects Wernicke's area with Broca's area (Fig. 7-17) to coordinate all aspects of understanding and executing speech and language skills.

Available clinical evidence indicates that Wernicke's area is essential for the comprehension and recognition of words and language, whereas Broca's area is essential for the mechanical production of speech. Patients with a defect in Broca's area show evidence of comprehension and accurate recognition of the word being heard or read but are not able to say the word.

In contrast, patients with damage in Wernicke's area can produce speech, but there is little evidence that the words they put together have relevant meaning.

■ ■ ■ ■ ■ ■ ■ ■ ■ ■ ■ ■ **REVIEW EXERCISES** ■ ■ ■ ■ ■ ■ ■ ■ ■ ■ ■

Identify Each with a Term

1. The hypothalamic center that signals satiety after eating
2. The brainstem system that interconnects higher and lower centers of the CNS with monoaminergic tracts and other widely diverging neuronal tracts
3. Waves of the EEG that are seen in deepest slow-wave sleep
4. Cortical areas that interconnect primary sensory and motor cortices
5. Cortical area essential for comprehension of the meanings of words
6. The monoaminergic system playing a major role in setting the affect at a particular place along the mood scale

Define Each Term

7. Circadian rhythm
8. Forebrain
9. Paradoxical sleep
10. Commissures
11. Allocortex
12. Working memory

13. Aphasia

14. Long-term potentiation

Choose the Correct Answer

15. Sensory information can reach the hypothalamus in a nonspecific manner by means of which mechanism?
    a. Direct connections from primary sensory cortical areas
    b. Specific projections from the thalamus to the hypothalamus
    c. Axon collaterals of ascending sensory neurons synapsing in the reticular formation
    d. Axon collaterals of ascending sensory neurons synapsing within the nuclei of the hypothalamus

16. The role of the hippocampus in memory function can best be described as the site where:
    a. Learning occurs prior to formation of memory
    b. Memory is stored
    c. An old memory can be recalled
    d. A new memory is consolidated and stored elsewhere

17. The cues of sunlight and darkness in a 24-hour day reach the hypothalamus by:
    a. The body's internal "clock"
    b. A direct neural pathway from the optic nerve to the suprachiasmatic nucleus
    c. The reticular formation
    d. The cerebral cortex

18. Posterior pituitary hormone secretion is mediated by:
    a. A portal capillary system from the hypothalamus to the posterior pituitary
    b. The fight-or-flight response
    c. The hypothalamo-hypophyseal tract originating from magnocellular neurons in the supraoptic and paraventricular nuclei
    d. The reticular activating system

19. The electrical activity of the brain as recorded by the EEG represents:
    a. The electrical activity in the outermost processes of cortical neurons
    b. An integrated sum of all brain neuronal activity
    c. The electrical activity of the brainstem cardiovascular and respiratory centers
    d. The emotional state (i.e., mood and affect)

20. Language and speech require the participation of both Wernicke's area and Broca's area, and the following connection between them:
    a. The thalamocortical tract
    b. The reticular activating system
    c. The prefrontal lobe
    d. The arcuate fasciculus

21. Drugs with a high potential for abuse have which of the following characteristics in common?
    a. They deplete neuronal stores of monoamine neurotransmitters.
    b. They block dopamine receptors.
    c. They stimulate specific receptors located in reward centers of the limbic system.
    d. They block electrical self-stimulation in experimental animals.

22. Name the target blocked by neuroleptic drugs.
    a. Serotonin receptors
    b. Dopamine receptors
    c. Acetylcholine receptors
    d. Norepinephrine receptors

Briefly Answer

23. The hypothalamic secretion of GnRH is cyclic in females but continual in males. Explain.

24. In what ways is REM sleep paradoxical?

25. Describe the role of the prefrontal cortex in learning and memory.

26. People commonly forget friends' birthdays, but not if the friend is the object of their affections. Explain.

■ ■ ■ ■ ■ ■ ■ ■ ■ ■ ■ ■ ■ ■ ■ ANSWERS ■ ■ ■ ■ ■ ■ ■ ■ ■ ■ ■ ■ ■ ■ ■

1. Ventromedial hypothalamic nucleus

2. Reticular formation

3. Delta waves

4. Association areas

5. Wernicke's area

6. Noradrenergic neurons

7. Fluctuations in homeostatically regulated function that repeat in a cyclic pattern every 24 hours

8. The structures of the brain located rostral to the diencephalon

9. Another term for REM sleep

10. White matter fiber tracts that interconnect the right and left hemispheres

11. The oldest part of the cerebral cortex, arising early in evolution in vertebrates

12. A process, coordinated in the prefrontal cortex, by which new experiences are brought to the prefrontal cortex from other association areas, integrated with previously acquired memory and learning, and processed into a short-term memory trace that may or may not be sent to the hippocampus for consolidation into long-term memory

13. Loss of ability to comprehend or use language and/or speech

14. A lasting increase in excitability of neurons caused by a repetitive train of presynaptic electrical stimuli

15. c

16. d

17. b

18. c

19. a

20. d

21. c

22. b

23. GnRH-secreting cells of the hypothalamus are destined to become cyclic secreters at puberty unless changed by their hormonal environment at a critical period during development. In early fetal life, testicular androgen secreted by a genetic male fetus influences these cells to become continuous secreters when they become active after puberty. In the absence of androgenic stimulation, as in a genetic female fetus, the GnRH secretory mechanism retains its cyclic characteristics.

24. Rapid-eye-movement sleep is considered paradoxical when compared with slow-wave sleep. For example, the EEG in REM sleep is similar to that in the waking state (beta rhythm) and the autonomic nervous system exhibits excitation, yet the individual is as difficult to arouse as in deepest slow-wave sleep.

25. The prefrontal cortex is considered the site of the working memory, where short-term memory is formed and either rapidly forgotten or consolidated into long-term memory in the hippocampus.

26. Sexual attraction is an intense emotional state that influences the limbic system and its functions, including learning and memory processes, which are reinforced by involving the object of affection.

Suggested Reading

Bray GA: Weight homeostasis. *Ann Rev Med* 42:205–216, 1991.

Goldman-Rakic PS: Working memory and the mind. *Sci Am* 267(3):110–117, 1992.

Kandel ER, Schwartz JH, Jessell TM: *Principles of Neural Science.* 3rd ed. New York, Elsevier, 1991.

Kryger MH, Roth T, Dement WC: *Principles and Practice of Sleep Medicine.* Philadelphia, Saunders, 1989.

Milner B: Amnesia following operation on the temporal lobes, in Witty CWM, Zangwill OL (eds): *Amnesia.* New York, Appleton-Century-Crofts, 1966, pp 109–133.

Raisman G, Brown-Grant K: The "suprachiasmatic syndrome": Endocrine and behavioral abnormalities following lesions of the suprachiasmatic nucleus in the female rat. *Proc Roy Soc London* B198:297–314, 1977.

Squire LR, Zola-Morgan S: The medial temporal lobe memory system. *Science* 253:1380–1386, 1991.

Swanson LW, Mogenson GJ: Neural mechanisms for the functional coupling of autonomic, endocrine, and somatomotor responses in adaptive behavior. *Brain Res Rev* 3:1–34, 1981.

PART THREE

Muscle Physiology

Contractile Properties of Muscle Cells

OBJECTIVES

After studying this chapter, the student should be able to:
1. List the three types of muscle and give examples of each
2. List typical functions of each of the three muscle types
3. Draw and label a skeletal muscle sarcomere at two different stages of myofilament overlap
4. Diagram the structure of the thick and thin myofilaments, labeling the proteins that make them up
5. Diagram the chemical and mechanical steps in the crossbridge cycle and explain how the crossbridge cycle results in shortening of the muscle
6. List the steps in excitation-contraction coupling in skeletal muscle and describe the roles of the sarcolemma, T tubules, sarcoplasmic reticulum, and calcium ions
7. Describe the role of ATP in the crossbridge cycle
8. List the energy sources for muscle contraction; diagram the major pathways involved; and rank the sources in terms of their relative effectiveness in supplying ATP for contraction

Muscle tissue is responsible for most of our interactions with the external world. These familiar functions include moving, speaking, and a host of other everyday actions. Less familiar, but no less important, are the internal functions of muscle: it pumps our blood and regulates its flow; it moves our food as it is being digested and causes the expulsion of wastes; and it serves as a critical regulator of numerous internal processes.

Muscle contraction is a cellular phenomenon. The shortening of a whole muscle results from the shortening of its individual cells, and the force a muscle produces is the sum of forces produced by its cells. Activation of a whole muscle involves activating its individual cells, and muscle relaxation involves a return of the cells to their resting state. The study of muscle function thus must clearly involve investigation of the cellular processes that cause and regulate muscle contraction.

As the great variety of its functions might imply, muscle as a tissue is highly diverse. But in spite of its wide range of anatomic and physiologic specializations, there is an underlying similarity in the way muscles are constructed and in their mechanism of contraction. This chapter discusses some fundamental aspects of muscle contraction expressed in all types of muscle. Chapters 9 and 10 consider the important specializations of structure and function that belong to particular kinds of muscle.

Classification and Roles of Muscle

Muscle falls naturally into categories that are related to its anatomic and physiologic properties. Within each of the major categories are subclassifications that further identify differences among specific muscle types. As with any classification scheme, some exceptions are inevitable, and some categories overlap. For this reason, three sets of criteria are in common use for distinguishing among muscle types.

■ Muscles Are Grouped in Three Major Categories

Muscles may be classified on the basis (1) of their location in relation to other body structures, (2) of their histologic (tissue) structure, or (3) the way their action is controlled. These classifications are not mutually exclusive.

Throughout these chapters on muscle, the highlighted categories in Figure 8–1 will be the preferred usage. The alternative categories are still useful, however, because in some instances they express more precisely the special attributes of a certain muscle. The inconsistencies in the classification are likewise useful in highlighting special attributes of specific muscle types.

Classifications of muscle

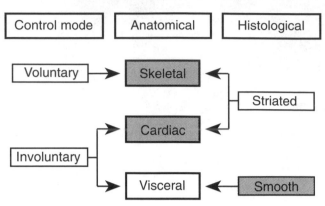

Figure 8–1 ■ Classification of muscle into its major types. The categories overlap in different ways, depending on the criteria being used.

Skeletal Muscle: Interactions with the External Environment ■ As its name implies, **skeletal muscle** is usually associated with bones of the skeleton. It is responsible for large and forceful movements, such as those involved in walking, running, and lifting heavy objects, as well as for small and delicate movements that position the eyeballs or allow the manipulation of tiny objects. Some skeletal muscle is specialized for the long-term maintenance of tension; for example, muscles of the torso involved in maintaining an upright posture can be active for many hours without undue fatigue. Other skeletal muscles, such as those in the upper arm, are better adapted for making rapid and forceful movements, but these fatigue rather rapidly when required to lift and hold heavy loads. Whatever its specialization, skeletal muscle serves as the link between the body and the external world. Much of this interaction, such as walking or speaking, is under *voluntary* control. Other actions, such as breathing or blinking the eyelids, are largely automatic, although they can be consciously suppressed for brief periods of time. All skeletal muscle is externally controlled; it cannot contract without a signal from the somatic nervous system.

Not all skeletal muscle is attached to the skeleton. The human tongue, for example, is made of skeletal muscle that does not move bones closer together. Among mammals, perhaps the most striking example of this exception is the trunk of the elephant, in which skeletal muscles are arranged in a structure capable of great dexterity even though no bones are involved in its movement.

An important secondary function of skeletal muscle is the production of body heat. This may be desirable, as when one shivers to get warm. During heavy exercise, however, muscle contraction may be a source of excess heat that must be eliminated from the body.

All skeletal muscle has a striated appearance when viewed with a light or electron microscope (Fig. 8–2). The regular and periodic pattern of the cross-striations of skeletal muscle relates closely to the way this muscle functions at a cellular level.

Smooth Muscle: Regulation of the Internal Environment ▪ Of the many processes regulating the internal state of the human body, one of the most important is the control of the movement of fluids through the visceral organs and the circulatory system. Such regulation is the task of **smooth muscle**. Smooth muscle also has many individual specializations that suit it well to particular tasks. Some smooth muscle, such as that in **sphincters**, circular bands of muscle that can stop flow in tubular organs, can remain contracted for long periods while using its metabolic energy very economically. The muscle of the uterus, on the other hand, contracts and relaxes rapidly and powerfully during birth but is normally not very active during most of the rest of a woman's life. The economical use of energy is one of the most important general features of the physiology of smooth muscle.

The contraction of smooth muscle is *involuntary*. Although contraction may occur in response to a nerve stimulus, many smooth muscles are also controlled by circulating hormones or may contract under the influence of local hormonal or metabolic influences quite independent of the nervous system. Some indirect voluntary control of smooth muscle may be possible through mental processes such as "biofeedback," but this ability is rare and is not an important aspect of smooth muscle function.

While one of the terms describing smooth muscle—*visceral*—implies its location in internal organs, much smooth muscle is located elsewhere. The muscles that control the diameter of the pupil of the eye and accommodate the eye for near vision, cause body hair to become erect (pilomotor muscles), and control the diameter of blood vessels are all examples of smooth muscles that are not visceral.

Cardiac Muscle: Motive Power for Blood Circulation ▪ Muscle provides the force that moves blood throughout the body. **Cardiac muscle** is found only in the heart. It shares with skeletal muscle a striated cell structure, but its contractions are involuntary; the heartbeat arises from within the cardiac muscle itself and is not initiated by the nervous system. The nervous system, however, does participate in regulating the rate and strength of heart muscle contractions. Chapter 10 considers the special properties of cardiac muscle.

▪ Muscles Have Specialized Adaptations of Structure and Function

All of the above should emphasize the varied and specialized nature of muscle function. Skeletal muscle, with its large and powerful contractions, smooth muscle, with its slow and economical contractions, and cardiac muscle, with its unceasing rhythm of contraction, all represent specialized adaptations of a basic cellular and biochemical system. An understanding of both the commonality and diversity of muscle is important, and it is useful to emphasize particular types of muscle when investigating a general aspect of muscle function. Often this means skeletal muscle is used as the "typical" muscle for purposes of discussion; this is done in this chapter where appropriate, with an effort to point out those features relative to muscle in general. Important adaptations of these general features found in specific muscle types are considered in Chapters 9 and 10.

The Functional Anatomy and Ultrastructure of Muscle

▪ The Ultrastructure of Muscle Provides a Key to Understanding the Mechanism of Contraction

In biology, as in architecture, it can be said that "form follows function." Nowhere is this truism more relevant than in the study of muscle. Investigations using light and electron microscopy, x-ray and light diffraction, and other modern visualization techniques have shown the complex and highly ordered internal structure of skeletal muscle, and elegant mechanical experiments have revealed how this structure determines the ways muscle functions.

Skeletal muscle is a highly organized tissue (Fig. 8–3). A whole skeletal muscle is composed of numerous muscle cells, also called **muscle fibers.** A cell can

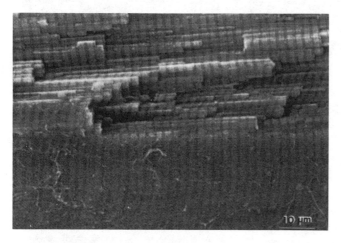

Figure 8–2 ▪ Electron micrograph of skeletal muscle. With the sarcolemma removed, the characteristic cross-striations are clearly visible. The muscle fiber has been cleaved to show individual myofibrils within the fiber. (From Sawada H, Ishikawa H, Yamada E: Surface and internal morphology of skeletal muscle: High-resolution scanning electron microscopy of frog sartorius muscle. *Tissue Cell* 10:179–190, 1978.)

Whole muscle
1x

Fasciculus
5x

Muscle fiber
500x

Myofibril
10,000x

Sarcomeres
50,000x

Myofilaments
1,000,000x

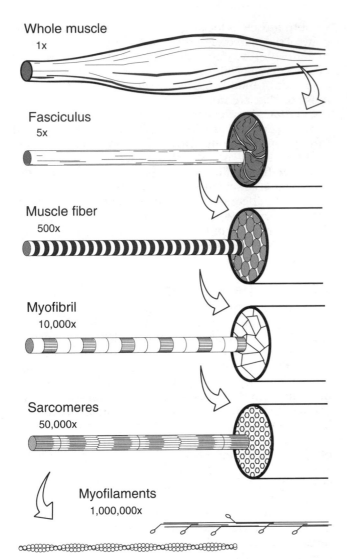

Figure 8–3 ■ Levels of complexity in the organization of skeletal muscle. The approximate amount of magnification required to visualize each level of structure is shown above each view.

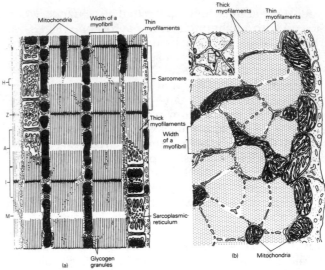

Figure 8–4 ■ The ultrastructure of skeletal muscle, based on drawings from electron micrographs. (A) Longitudinal view. (B) Cross-section. (From Rhoades R, Pflanzer R: *Human Physiology.* 2nd ed. Fort Worth, Saunders, 1992.)

be up to 100 μm in diameter and many centimeters long, especially in larger muscles. The fibers are **multinucleate,** and the nuclei occupy positions near the periphery of the fiber. Skeletal muscle has an abundant supply of **mitochondria,** which are vital for supplying chemical energy in the form of ATP to the contractile system. The mitochondria lie close to the contractile elements in the cells. Mitochondria are especially plentiful in skeletal muscle fibers specialized for the rapid and powerful contractions associated with aerobic metabolism.

Each muscle fiber is further divided lengthwise into several hundred to several thousand parallel **myofibrils.** Electron micrographs show that each myofibril has alternating light and dark bands, giving the fiber a **striated** (striped) appearance. As shown in Figures 8–4 and 8–5, the bands repeat at regular intervals.

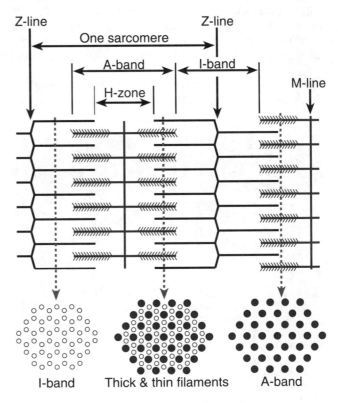

Figure 8–5 ■ Nomenclature of the skeletal muscle sarcomere. (Upper portion) The arrangement of the elements in a sarcomere. (Lower portion) Cross-sections through selected regions of the sarcomere, showing the overlap of myofilaments at different parts of the sarcomere.

Most prominent of these is a dark band called the **A band.** It is divided at its center by a narrow, lighter-colored region, called the **H zone.** In many skeletal muscles a prominent **M line** is found at the center of the H zone. Between the A bands lie the less dense **I-bands.** (The letters *A* and *I* stand for **anisotropic** and **isotropic,** respectively, and are so named because of their appearance when viewed with polarized light.) Crossing the center of the I band is a dark structure called the **Z line** (sometimes termed a **Z disk** to emphasize its three-dimensional nature). The filaments of the I band attach to the Z line and extend in both directions into the adjacent A bands. This pattern of alternating bands is repeated over the entire length of the muscle fiber. The fundamental repeating unit of these bands is called a **sarcomere** and is defined as the space between (and including) two successive Z lines (Fig. 8–5).

Closer examination of the sarcomere shows the A and I bands to be composed of two kinds of parallel structures called **myofilaments.** The I band contains **thin filaments,** made primarily of the protein **actin;** the A bands contain **thick filaments** composed of the protein **myosin.**

The Thin Myofilaments ■ Each thin (actin-containing) filament consists of two strands of macromolecular subunits entwined about each other (Fig. 8–6). The strands are composed of repeating subunits (monomers) of the globular protein **G actin,** a relatively small protein (molecular weight 41,700, diameter 5.5 nm). These slightly ellipsoid molecules are joined front to back into long chains, and these chains wind about each other to make a helical structure (called **F actin,** for *fibrous* actin) that undergoes a half-turn every seven G actin monomers. In the groove formed down the length of the helix there lies (on either side) an end-to-end series of fibrous protein molecules (molecular weight 50,000, length 38.5 nm) called **tropomyosin.** Each tropomyosin molecule extends a distance of seven G actin monomers along the F actin groove, a length that corresponds to one half-turn of the actin filament. Near one end of each tropomyosin molecule is a protein complex called **tropo-nin** composed of three attached subunits: troponin-

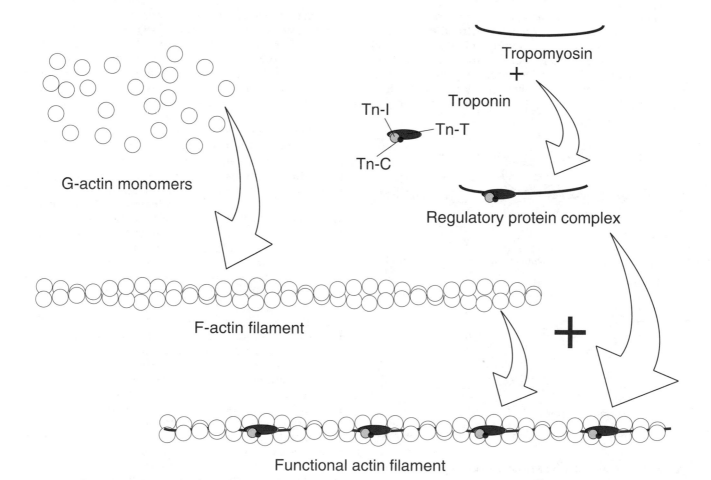

Figure 8–6 ■ Assembly of the thin (actin) filaments of skeletal muscle.

C (Tn-C), troponin-T (Tn-T), and troponin-I (Tn-I). The **Tn-C subunit** is capable of binding calcium ions; the **Tn-T subunit** attaches the complex to tropomyosin; and the **Tn-I subunit** has an inhibitory function. The troponin-tropomyosin complex regulates the contraction of skeletal muscle (see below).

The Thick Myofilaments ■ The thick (myosin-containing) filaments (Fig. 8–7) are also composed of macromolecular subunits, but they are assembled much differently from the thin filaments. The fundamental unit of the thick filaments, **myosin** (molecular weight approximately 500,000), is a complex molecule with several distinct regions. Most of the length of the molecule consists of a long, straight portion, often called the "tail" region, composed of **light meromyosin** (LMM). The remainder of the molecule, **heavy meromyosin** (HMM), consists of a protein chain that terminates in a globular head portion. The head portion, called the **S1 region** (or **subfragment**), is responsible for the enzymatic and chemical activity that results in muscle contraction. It contains an **actin binding site,** by which it can interact with the thin filament, and an **ATP binding site** that is involved in the supply of energy for the actual process of contraction. The chain portion of HMM, the **S2 region** (or **subfragment**), serves as a flexible link between the head and tail regions. Associated with the S1 region are two loosely attached peptide chains of a much lower molecular weight. One of these is called the **essential** light chain and is necessary for myosin to function. The other is called the **regulatory** light chain; it can be phosphorylated during muscle activity

and participates in the modulation of muscle function. Functional myosin molecules are paired; the tail and S2 regions are wound about each other along their length, and the two heads (each bearing its two light chains and its own ATP and actin binding sites) lie adjacent to each other. Thus the molecule, with its attached light chains, exists as a functional dimer, but the degree of functional independence of the two heads is not yet known with certainty.

Assembly of the individual myosin dimers into thick filaments involves a close packing of the myosin molecules such that their tail regions form the "backbone" of the thick filament, with the head regions extending outward in a helical fashion; a myosin head projects every 60 degrees around the circumference of the filament, with each one displaced 14.4 nm further along the filament. The effect is that of a bundle of golf clubs bound tightly by the handles, with the heads projecting from the bundle. The myosin molecules are packed such that they are tail to tail in the center of the thick filament and extend outward from the center in both directions, creating a bare zone (i.e., no heads protruding) in the middle of the filament.

Skeletal Muscle Membrane Systems ■ Muscle, in common with other types of living cells, has a system of surface and internal membranes with a number of critical functions (Fig. 8–8). A skeletal muscle fiber is surrounded on its outer surface by an electrically excitable cell membrane supported by an external meshwork of fine fibrous material. Together these layers form the cell's surface coat, the **sarcolemma.** In addition to the typical functions of any cell membrane, the sarcolemma generates and conducts action potentials much like those of nerve cells.

Contained wholly within a skeletal muscle cell is another set of membranes called the **sarcoplasmic reticulum,** a specialization of the endoplasmic reticulum. The sarcoplasmic reticulum is specially adapted for the uptake, storage, and release of **calcium ions,** which are critical in controlling the processes of contraction and relaxation. Within each sarcomere the sarcoplasmic reticulum consists of two distinct portions. The **longitudinal** portion forms a system of hollow sheets and tubes that are closely associated with the myofibrils. The ends of the longitudinal portion terminate in a system of **terminal cisternae** (or **lateral sacs**). These contain a protein, **calsequestrin,** that weakly binds calcium, and most of the stored calcium is located in this region.

Closely associated with both the terminal cisternae and the sarcolemma are the **transverse tubules (T tubules),** a set of membranes forming tubular structures that extend deep into the fiber from its surface. They represent inward extensions of the cell membrane, and their interior is continuous with the extracellular space. Although they traverse the cross-section of the fiber, they do not open into its interior. In many types of muscles they extend into the fiber at the level

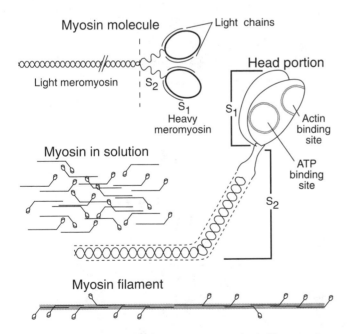

Figure 8–7 ■ Assembly of skeletal muscle thick filaments from myosin molecules.

tending into the A band from the I bands. The spacing between subsequent Z lines also depends directly on the fiber length. The length of both the thin and thick myofilaments remains constant despite changes in the fiber length. The **sliding filament theory** proposes that changes in overall fiber length are directly associated with changes in the overlap between the two sets of filaments; that is, the thin filaments telescope into the array of thick filaments. This interdigitation accounts for the change in the length of the muscle fiber and is accomplished by the interaction of the globular heads of the myosin molecules (projecting from the thick filaments) with binding sites on the actin filaments. These **crossbridges,** as the myosin heads are called, are the sites where force and shortening are produced and where the chemical energy stored in the muscle is transformed into mechanical energy. The total shortening of each sarcomere is on the order of only 1 μm or so, but a muscle contains many thousands of sarcomeres placed end to end (in series). This has the effect of multiplying all of the small sarcomere length changes into a large overall shortening of the muscle (Fig. 8-9). Similarly, the amount of force exerted by a single sarcomere is very small (a few hundred micronewtons), but, again, there are thousands of sarcomeres side by side (in parallel), resulting in the production of considerable force.

The effects of sarcomere length on force production are summarized in Figure 8-10. When the muscle is stretched beyond its normal resting length, decreased

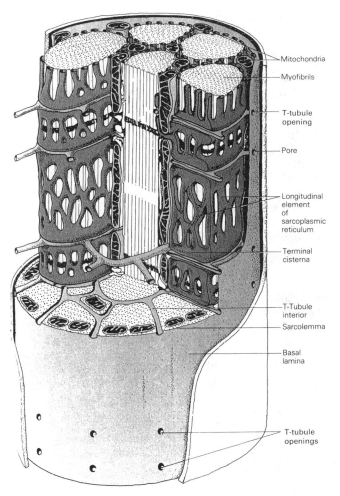

Figure 8–8 ■ The internal membrane system of skeletal muscle, responsible for communication between the surface membrane and contractile filaments. (From Rhoades R, Pflanzer R: *Human Physiology.* 2nd ed. Fort Worth, Saunders, 1992.)

of the Z line, while in others they penetrate in the region of the junction between the A and I bands (Fig. 8-8). The association of a T tubule and the two terminal cisternae at its sides is called a **triad,** a structure important in linking membrane action potentials to muscle contraction.

■ The Sliding Filament Theory Explains Muscle Contraction

The structure of skeletal muscle provides important clues to the mechanism of contraction. The width of the A bands (thick-filament areas) in striated muscle remains constant, regardless of the length of the entire muscle fiber, while the width of the I bands (thin-filament areas) varies directly with the length of the fiber. At the edges of the A band there are fainter bands whose width also varies. These represent material ex-

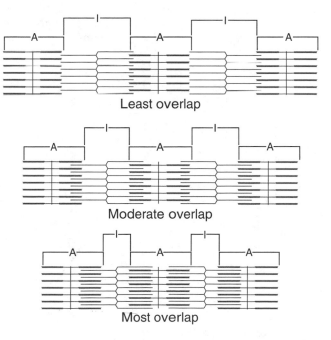

Figure 8–9 ■ The multiplying effect of sarcomeres placed in series. The overall shortening is the sum of the shortening of the individual sarcomeres.

filament overlap occurs (3.65 and 3.00 μm in Fig. 8-10). This limits the amount of force that can be produced, since a shorter length of thin I band filaments interdigitates with A band thick filaments, and fewer crossbridges can be attached. The force produced when overlap is less than maximal is directly proportional to the degree of overlap. At lengths near the normal **resting length** of the muscle (i.e., the length usually found in the body), the amount of force does not vary with the degree of overlap (2.25 and 1.95 μm in Fig. 8-10). This independence of overlap occurs because of the bare zone (the H zone) along the thick filaments at the center of the A band (where no myosin heads are present). Over this small region, increased interdigitation does not lead to an increase in the number of attached crossbridges and the force remains constant. At lengths shorter than the resting length, additional geometric and physical factors play a role in myofilament interactions. Since muscle is a "telescoping" system, there is a physical limit to the amount of shortening; as thin myofilaments penetrate the A band from opposite sides, they begin to meet in the middle and interfere with each other (1.67 μm, Fig. 8-10). At the extreme, further shortening is limited by the thick filaments of the A band being forced against the structure of the Z disks (1.27 μm, Fig. 8-10). The relationship between overlap and force at short lengths is more complex than that at longer lengths, since more factors are involved. It has also been shown that at very short lengths the effectiveness of some of the steps in the excitation-contraction coupling process is reduced. These include reduced calcium binding to troponin and some loss of action potential conduction in the T tubule system. Some of the consequences for the muscle as a whole are

apparent when the mechanical behavior of muscle is examined in more detail (Chap. 9).

■ Events of the Crossbridge Cycle Drive Muscle Contraction

The process of contraction involves a cyclic interaction between the thick and thin filaments. The steps that comprise the **crossbridge cycle** (Fig. 8-11) are attachment of thick-filament crossbridges to sites along the thin filaments, production of a mechanical movement, crossbridge detachment from the thin filaments, and subsequent reattachment of the crossbridges at different sites along the thin filaments. These mechanical changes are closely related to the biochemistry of the contractile proteins. In fact, the crossbridge association between actin and myosin actually functions as an enzyme (called **actomyosin ATPase**) that catalyzes the breakdown of ATP and releases its stored chemical energy. Most of our knowledge of this process comes from studies on skeletal muscle, but the same basic steps are followed in all types of muscle.

In resting skeletal muscle (Fig. 8-11, step 1) the interaction between actin and myosin (via the crossbridges) is very weak, and the muscle can be extended with little effort. When the muscle is activated, the

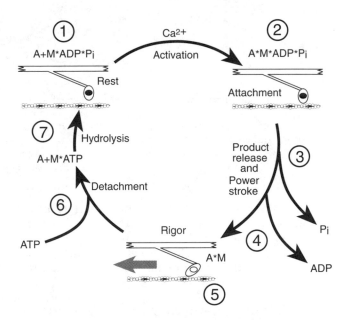

Figure 8–11 ■ The events of the crossbridge cycle in skeletal muscle. *, chemical bond; +, potential interactions. At rest (1), ATP has been bound to the myosin head and hydrolyzed, but the energy of the reaction cannot be released until the myosin head can interact with myosin (2). Release of the hydrolysis products (3) is associated with the power stroke (4); the rotated and still-attached crossbridge is now in the rigor state (5). Detachment (6) is possible when a new ATP molecule binds to the myosin head and is subsequently hydrolyzed (7). These cyclic reactions can continue as long as the ATP supply remains and activation (via Ca^{2+}) is maintained.

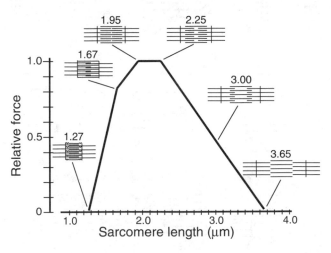

Figure 8–10 ■ The effect of filament overlap on force generation. The force a muscle can produce depends on the amount of overlap between the thick and thin filaments, because this determines how many crossbridges can interact effectively.

MUSCULAR DYSTROPHY RESEARCH

The term muscular dystrophy encompasses a variety of degenerative muscle diseases. The most common of these is Duchenne muscular dystrophy (DMD), an X-linked hereditary disease affecting mostly male children (one in 3500 live male births). It is manifested by progressive muscular weakness during the growing years, becoming apparent by age 4. There is a characteristic enlargement of the affected muscles, especially the calf muscles. This is due to a gradual degeneration and necrosis of muscle fibers and their replacement by fibrous and fatty tissue. By the age of 12 most victims are no longer ambulatory, and death usually occurs by the late teens or early twenties. The most serious defects are in skeletal muscle, but smooth and cardiac muscle are affected as well, and many victims suffer from cardiomyopathy (Chap. 10). A related (and rarer) disease, Becker muscular dystrophy (BMD), has similar symptoms but is less severe; persons with the condition often survive into adulthood.

By means of the genetic technique of chromosome mapping (using linkage analysis and positional cloning), the gene responsible for both DMD and BMD has been localized to the Xp21 region of the X chromosome, and the gene itself has been cloned. It is a very large gene of some 2.5 million base pairs; apparently because of its great size, it has an unusually high mutation rate. About one-third of cases of DMD are due to new mutations and the other two-thirds to sex-linked transmission of the defective gene. The BMD gene is a less severely damaged allele of the DMD gene.

The product of the DMD gene is dystrophin, a large protein that is absent in DMD. Aberrant forms are present in BMD. In normal muscle the function of dystrophin appears to be that of a cytoskeletal component associated with the inside surface of the sarcolemma. Muscle also contains dystrophin-related proteins that may have similar functional roles.

A disease as common and devastating as DMD has long been the focus of intensive research. Recent identification of three animals (dog, cat, mouse) in which a genetically similar condition occurs promises to offer significant new opportunities for study. The manifestation of the defect is different in each of the three animals (and also differs in some details from the human condition). The mdx mouse, although it lacks dystrophin, does not suffer the severe debilitation of the human form of the protein. Research is underway to identify dystrophin-related proteins that may help compensate for the major defect. Mice, being rapidly growing, are also suitable for studying the normal expression and function of dystrophin. Progress has been made in transplanting normal muscle cells into mdx mice, where they have expressed the dystrophin disease. Such an approach has been less successful in humans and in dogs, and the differences may hold important clues. A gene expressing a truncated form of dystrophin that has been inserted into mice using transgenic methods has corrected the myopathy. The xmd dog, which suffers a more severe and humanlike form of the disease, offers an opportunity to test new therapeutic approaches, while the cat dystrophy model shows a prominent muscle fiber hypertrophy, a poorly understood phenomenon in the human disease. Taking advantage of the differences among these models promises to shed light on many missing aspects of our understanding of a serious human disease.

---

actin-myosin interaction becomes quite strong, and crossbridges become firmly attached (step 2). Initially the crossbridges extend at right angles from each thick filament, but they rapidly undergo a change in angle of nearly 45 degrees. An ATP molecule bound to each crossbridge supplies the energy for this step. This ATP has been bound to the crossbridge in a partially broken-down form (the ADP*P_i in step 1). The myosin head to which the ATP is bound is called "charged myosin" (M*ADP*P$_i$ in step 1). When charged myosin interacts with actin, the association is represented as A*M*ADP*P$_i$ (step 2). The partial rotation of the angle of the crossbridges is associated with the final hydrolysis of the bound ATP and release of the hydrolysis products (step 3), an **inorganic phosphate ion** (P$_i$) and **ADP.** Since the myosin heads are temporarily attached to the actin filament, the partial rotation pulls the actin filaments past the myosin filaments, a movement called the **power stroke** (step 4). Following this movement (which results in a relative filament displacement of around 10 nm), the actin-myosin binding is still strong and the crossbridge cannot detach; at this point in the cycle it is termed a **rigor** crossbridge (A*M, step 5). For detachment to occur, a new molecule of ATP must bind to the myosin head (M*ATP, step 6) and undergo partial hydrolysis to M*ADP*P$_i$ (step 7). Once this happens, the newly recharged myosin head, momentarily not attached to the actin filament (step 1), can begin the cycle of attachment, rotation, and detachment again. This can go on as long as the muscle is activated, a sufficient supply of ATP is available, and the physiologic limit to shortening has not been reached. If cellular energy stores are depleted, as happens after death, the crossbridges cannot detach because of the lack of ATP, and the cycle stops in an attached state (at step 5). This produces an overall stiffness of the muscle termed "rigor," which is clinically observed as the **rigor mortis** that sets in shortly after death.

The crossbridge cycle obviously must be subject to control by the body to produce useful and coordinated muscular movements. This control involves a number of cellular processes that differ among the various types of muscle. Here, again, the case of skeletal muscle provides the basic description of the control process.

The Activation and Internal Control of Muscle Function

Control of the contraction of skeletal muscle involves many steps between the arrival of the action potential in a motor nerve and the final mechanical activity. A very important series of these steps, called **excitation-contraction coupling,** takes place deep within a muscle fiber. This is the subject of the remainder of this chapter; the very early events (communication between nerve and muscle) and the very late events (actual mechanical activity) are discussed in Chapter 9.

■ The Interaction of Calcium with Specialized Proteins Is Central to Muscle Contraction

The most important chemical link in the control of muscle protein interaction is provided by **calcium**

ions. The internal concentration of these ions is controlled by the sarcoplasmic reticulum, and changes in the internal calcium ion concentration have profound effects on the actions of the contractile proteins of muscle.

Calcium and the Troponin-Tropomyosin Complex ■ The chemical processes of the crossbridge cycle in skeletal muscle are in a state of constant readiness, even while the muscle is relaxed. Undesired contraction is prevented by a specific inhibition of the interaction between actin and myosin. This inhibition is a function of the troponin-tropomyosin complex of the thin myofilaments. When a muscle is relaxed, calcium ions are at very low concentration in the region of the myofilaments. The long tropomyosin molecules, lying in the grooves of the entwined actin filaments, interfere with the myosin binding sites on the actin molecules (Fig. 8–12). When calcium ion concentrations increase (see below), the ions bind to the troponin-C (Tn-C) subunit associated with each tropomyosin molecule. Through the action of troponin-I (Tn-I) and troponin-T (Tn-T), calcium binding causes the tropomyosin molecule to change its position slightly, uncovering the myosin binding sites on the actin filaments. The myosin (already "charged" with ATP) is allowed to interact with actin, and the events of the crossbridge cycle take place until calcium ions are no longer bound to the Tn-C subunit.

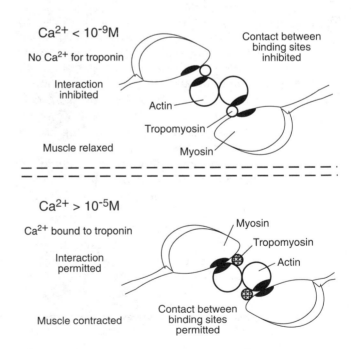

Figure 8–12 ■ The calcium switch for controlling skeletal muscle contraction. Calcium ions, via the troponin-tropomyosin system, control the unblocking of the interaction between the myosin heads (the crossbridges) and the active site on the thin filaments. The geometry of each tropomyosin molecule allows it to exert control over seven actin monomers.

The Switching Action of Calcium ■ An effective switching function requires that the transition between the "off" and "on" states be rapid and respond to relatively small changes in the controlling element. The calcium switch in skeletal muscle satisfies these requirements well. The relationship between the relative force developed and the calcium concentration in the region of the myofilaments is very steep. At a calcium concentration of 1×10^{-8} M, the interaction between actin and myosin is negligible, while an increase in the calcium concentration to 1×10^{-5} M produces essentially full-force development. This process is subject to **saturation,** after which further increases in calcium concentration lead to little increase in force. In skeletal muscle there is usually an excess of calcium ions present during activation and the contractile system is normally fully saturated. In cardiac and smooth muscle, however, only partial saturation occurs under normal conditions, and various physiologic mechanisms use control of the calcium concentration as a means of adjusting the degree of muscle activation.

The switching action of the calcium-troponin-tropomyosin system in skeletal and cardiac muscles is extended by the structure of the thin filaments, which allows one troponin molecule, via its tropomyosin connection, to control seven actin monomers. Since the calcium control in striated muscles is exercised through the thin filaments, it is termed **actin-linked regulation.** While the cellular control of smooth muscle contraction is also exercised by changes in calcium concentration, its effect is exerted on the thick (myosin) filaments. This is termed **myosin-linked regulation** and is described in Chapter 9.

■ Excitation-Contraction Coupling Links Electrical and Mechanical Events

When a nerve impulse arrives at the neuromuscular junction and its signal is transmitted to the muscle cell membrane, there ensues a rapid train of events that carries the signal to the interior of the cell, where the contractile machinery is located. The large diameter of skeletal muscle cells places interior myofilaments out of range of the immediate influence of events at the cell surface, but the T tubules, sarcoplasmic reticulum, and their associated structures act as a specialized internal communication system that allows the signal to penetrate to interior parts of the cell. The end result of electrical stimulation of the cell is the liberation of calcium ions into regions of the sarcoplasm very near the myofilaments, and this initiates the crossbridge cycle.

The process of excitation-contraction coupling, as outlined in Figure 8–13, begins in skeletal muscle with the electrical excitation of the surface (cell) membrane of the muscle fiber. An action potential sweeps rapidly down the length of the fiber. Its mode of propagation is similar to that in nonmyelinated nerve fibers, in which successive areas of membrane are stimulated

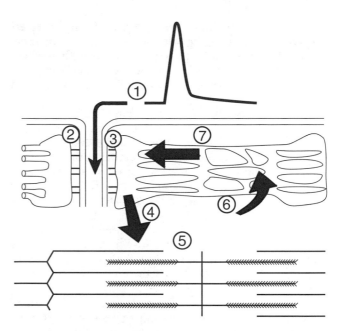

Figure 8–13 ■ The sequence of events in excitation-contraction coupling involves the cyclic movement of calcium. Action potential enters the T tubule (1). The T tubule conducts action potential inward (2). The signal crosses triad membranes (3) and calcium is released from the terminal cisternae (4). Muscle contraction occurs (5) and calcium is taken back up into the longitudinal elements of the sarcoplasmic reticulum (6). Calcium is then translocated back into the terminal cisternae (7).

by local ionic currents flowing from adjacent areas of excited membrane. Because there are no specialized conduction adaptations (e.g., myelination), this regenerative propagation is slow compared to that in the motor nerve, but its speed is sufficient to ensure the practically simultaneous activation of the entire fiber. When the action potential encounters the openings of the T tubules, it begins to propagate down the T tubule membrane. This propagation is also regenerative, resulting in numerous action potentials, one in each T tubule, traveling toward the center of the fiber. In the T tubules the velocity of the action potentials is rather low, but the total distance to be traveled is quite short.

At some point along the T tubule the action potential reaches the region of a triad. Here the presence of the action potential is communicated to the terminal cisterna of the sarcoplasmic reticulum. While the precise nature of this communication is not yet fully understood, it appears that the T tubule action potential affects a protein molecule called a **voltage sensor,** located in the region of the triad where the T tubule and sarcoplasmic reticulum membranes are closest together. Movement of the electrical charge within this voltage sensor causes specific channels in the sarcoplasmic reticulum membrane to undergo a large increase in Ca^{2+} conductance. Thus, in response to the signal arriving from the cell surface, the calcium ions stored in the terminal cisternae are rapidly released into the intracellular space surrounding the myofilaments. The calcium ions can now bind to the Tn-C molecules on the thin filaments. This allows the crossbridge cycle reactions to begin, and contraction occurs.

Even while calcium release occurs in the region of the terminal cisternae, the active transport processes in the membranes of the longitudinal elements of the sarcoplasmic reticulum pump free calcium ions from the myofilament space into the interior of the sarcoplasmic reticulum. Very soon the rapid release process stops; there is only one burst of calcium ion release for each action potential, and the continuous **calcium pump** in sarcoplasmic reticulum membrane reduces calcium in the region of the myofilaments to a very low level ($<1 \times 10^{-8}$ M). Since calcium ions are no longer available to bind to troponin, the contractile activity ceases and relaxation begins. The resequestered calcium ions are moved along the longitudinal sarcoplasmic reticulum to storage sites in the terminal cisternae, and the system is ready to be activated again.

This entire process takes place in a few tens of milliseconds and may be repeated many times each second.

Energy Sources for Muscle Contraction

Because work is done by contracting muscle, cellular processes must supply biochemical energy to the contractile mechanism. Additional energy is required to pump the calcium ions involved in the control of contraction and for other cellular functions. In muscle, as in other cells, this energy ultimately takes the form of the universal high-energy compound, ATP. A diagram of the major chemical energy relationships in muscle is shown in Figure 8–14.

■ Muscle Cells Obtain ATP from Several Sources

Although ATP is the immediate fuel for the contraction process, its concentration in the muscle cell is never high enough to sustain a long series of contrac-

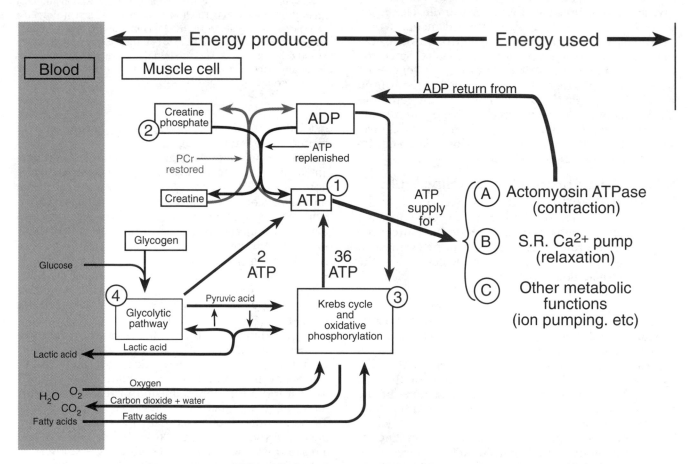

Figure 8–14 ■ The major metabolic processes of skeletal muscle. They center around the supply of ATP for the actomyosin ATPase of the crossbridges. Energy sources are numbered in order of their proximity to the actual reactions of the crossbridge cycle. Energy is used by the cell in the order A, B, C. The scheme shown here is typical for all types of muscle, although there are specific quantitative and qualitative variations.

tions. Most of the immediate energy supply is held in an "energy pool" of the compound **creatine phosphate** (or **phosphocreatine,** PCr), which is in chemical equilibrium with ATP. After a molecule of ATP has been split and yielded its energy, the resulting ADP molecule is readily rephosphorylated to ATP by the high-energy phosphate group from a creatine phosphate molecule. The creatine phosphate pool is restored by ATP from the various cellular metabolic pathways. These reactions (of which 2 and 3 are reversible and are called the Lohmann reaction) can be summarized as follows (and see also Fig. 8–14):

(1) $ATP \rightarrow ADP + P_i$ (Energy for contraction)
(2) $ADP + PCr \rightarrow ATP + Cr$ (Rephosphorylation of ATP)
(3) $ATP + Cr \rightarrow ADP + PCr$ (Restoration of PCr)

Because of the chemical equilibria involved, the concentration of PCr can fall to very low levels before the ATP concentration shows a significant decline. It has been shown experimentally that when 90% of the PCr has been used, the ATP concentration has fallen by only 10%. This situation results in a steady source of ATP for contraction that is maintained despite variations in energy supply and demand. Creatine phosphate is the most important storage form of high-energy phosphate; together with some other smaller sources, this energy reserve is sometimes called the **phosphagen pool.** A smaller contribution to the phosphagen pool, seen under conditions of extreme depletion of energy stores, is provided by the **adenylate kinase** (or **myokinase**) reaction, which can use two ADP molecules to produce one of ATP and one of AMP **(adenosine monophosphate).** Some of the AMP is removed by deamination; increased AMP concentration also facilitates the breakdown of muscle glycogen (see below).

Two major metabolic pathways supply ATP to energy-requiring reactions in the cell and to the mechanisms that replenish the creatine phosphate pool. Their relative contributions depend on the muscle type and conditions of contraction. A simplified diagram of the energy relationships of muscle is shown in Figure 8–14. The first of the supply pathways is the **glycolytic pathway,** or **glycolysis.** This is an **anaerobic** pathway; **glucose** is broken down without the use of oxygen to regenerate two molecules of ATP for every molecule of glucose consumed. Glucose for the glycolytic pathway may be derived from circulating blood glucose, or from its storage form in muscle cells, the polymer called **glycogen.** This reaction extracts only a small fraction of the energy contained in the glucose molecule, and the end product of glycolysis **(pyruvic acid,** or **pyruvate)** usually enters another cellular (mitochondrial) pathway called the **Krebs cycle.** As a result of Krebs cycle reactions, substrates are made available for **oxidative phosphorylation.** The Krebs cycle and oxidative phosphorylation are **aerobic** processes that require a continuous supply of oxygen. In this pathway an additional 36 molecules of ATP are regenerated from the energy in the original glucose molecule; the final products are carbon dioxide and water. While the oxidative phosphorylation pathway provides the greatest amount of energy, it cannot be used if the oxygen supply is insufficient; in this case glycolytic metabolism predominates.

Glucose is the preferred fuel for muscle contraction at higher levels of exercise. At maximal work levels, almost all of the energy used is derived from glucose produced by glycogen breakdown in muscle tissue and from blood-borne glucose from dietary sources. Glycogen breakdown increases very rapidly during the first tens of seconds of vigorous exercise. This breakdown, and the subsequent entry of glucose into the glycolytic pathway, is catalyzed by the enzyme **phosphorylase a.** This enzyme is itself transformed from its inactive **phosphorylase b** form by a "cascade" of protein kinase reactions whose action is in turn stimulated by the increased Ca^{2+} ion concentration and metabolite (especially AMP) levels associated with muscle contraction. Increased levels of circulating epinephrine (associated with exercise), acting through an adenylate cyclase pathway, also increase glycogen breakdown. Sustained exercise can lead to substantial depletion of glycogen stores, and this can restrict further muscle activity.

At lower exercise levels (i.e., below 50% of maximal capacity) fats may provide 50–60% of the energy for muscle contraction. Fat, the major energy storage medium in the body, is mobilized from its storage sites (in adipose tissue) to provide metabolic fuel in the form of **free fatty acids.** This process is slower than the liberation of glucose from glycogen and cannot keep pace with the very high demands of heavy exercise. Moderate activity, with brief rest periods, favors the consumption of fat as muscle fuel. Because fatty acids enter the Krebs cycle at the acetyl-CoA-citrate step, they are not involved in the reactions of glycolysis and cannot be interconverted with glucose. While complete combustion of fat yields less ATP per mole of oxygen consumed, its high energy storage capacity (the equivalent of 138 mol of ATP per mole of a typical fatty acid) makes it ideal as an energy storage medium. Depletion of body fat reserves is almost never a limiting factor in muscle activity.

In the absence of other fuels, **protein** can serve as an energy source for contraction, although this occurs mainly during dieting and starvation or under conditions of very heavy exercise. Under such conditions, proteins are broken down into amino acids that provide energy for contraction and that can be resynthesized into glucose to meet other needs.

Many of the metabolic reactions and processes supplying energy for contraction and recycling of metabolites (e.g., lactate, glucose) take place outside the muscle itself, particularly in the liver, and the products are transported to the muscle by the bloodstream. In addition to its oxygen- and carbon dioxide-carrying functions, the enhanced blood supply to exercising muscle provides for rapid exchange of the essential metabolic materials.

■ Metabolic Adaptations Allow Contraction to Continue with an Inadequate Oxygen Supply

Glycolytic (anaerobic) metabolism can provide energy for sudden, rapid, and forceful contractions of some muscles. In such cases the ready availability of glycolytic ATP compensates for the relatively low yield of this pathway, although a later adjustment must be made (see below). In most muscles, especially under conditions of rest or moderate exercise, the supply of oxygen is adequate for aerobic metabolism (fed by fatty acids and by the end products of glycolysis) to supply the energy needs of the contractile system. As the level of exercise increases, a number of physiologic mechanisms come into play to increase the blood (and thus oxygen) supply to the working muscle. At some point, however, even these mechanisms fail to supply sufficient oxygen, and the end products of glycolysis begin to accumulate. The glycolytic pathway can continue to operate because the excess pyruvic acid that is produced is converted to **lactic acid,** which serves as a temporary storage medium. The formation of lactic acid, by preventing a buildup of pyruvic acid, also allows for the restoration of the enzyme cofactor **NAD⁺,** needed for a critical step in the glycolytic pathway, so that the breakdown of glycogen can continue. Thus, ATP can continue to be produced under anaerobic conditions. The accumulation of lactic acid is the largest contributor (over 60%) to **oxygen debt,** which allows short-term anaerobic metabolism to take place despite a relative lack of oxygen (which is necessary to consume the end products, such as pyruvate, of the glycolytic pathway). Other depleted muscle oxygen stores have a smaller capacity but can still participate in oxygen debt. The largest of these other contributors is the phosphagen pool (approximately 25%). Tissue fluids (including venous blood) account for another 7%, and the protein myoglobin (see below) can hold around 2.5%.

Eventually the lactic acid must be oxidized in the Krebs cycle and oxidative phosphorylation reactions, and the other energy stores (as listed above) must be replenished. This "repayment" of the oxygen debt occurs over several minutes during recovery from heavy exercise, when the oxygen consumption (and respiration rate) remains high and depleted ATP is restored from the glucose breakdown products temporarily stored as lactic acid. As the cellular ATP levels return to normal, the energy stored in the phosphagen energy pool is also replenished.

Those muscles adapted for mostly aerobic metabolism contain significant amounts of the protein **myoglobin.** This iron-containing molecule, essentially a monomeric form of the blood protein hemoglobin (Chap. 11), gives aerobic muscles their characteristic red color. The total oxygen storage capacity of myoglobin is quite low, and it does not make a significant direct contribution to the cellular stores; all of the myoglobin-bound oxygen could support aerobic exercise for less than 1 second. However, because of its high affinity for oxygen even at low concentrations, myoglobin serves a major role in facilitating the diffusion of oxygen through exercising muscle tissue by binding and releasing oxygen molecules as they move down their concentration gradient.

Muscles of different types have varying capacities for sustaining an oxygen debt; some skeletal muscles can sustain a considerable debt, while cardiac muscle has an almost exclusively aerobic metabolism. Chapters 9 and 10 discuss specialized metabolic adaptations that are specific to skeletal, smooth, and cardiac muscles.

■ ■ ■ ■ ■ ■ ■ ■ ■ ■ ■ REVIEW EXERCISES ■ ■ ■ ■ ■ ■ ■ ■ ■ ■ ■

Identify Each with a Term

1. The fundamental repeating contractile unit of skeletal muscle
2. Muscle associated with the internal organs (except the heart)
3. The internal membrane system in muscle that holds, releases, and takes up calcium ions
4. Invaginations of the cell membrane that traverse the interior of a skeletal muscle cell
5. The series of reactions that use the energy of ATP to produce mechanical motion
6. The high-energy compound that is the immediate source of ATP for contraction
7. The temporary buildup of lactic acid and depletion of cellular energy stores due to anaerobic metabolism during heavy exercise

Define Each Term

8. Z line
9. Troponin
10. Myosin
11. Sarcolemma
12. Aerobic metabolism
13. Excitation-contraction coupling

Choose the Correct Answer

14. Which region of a skeletal muscle sarcomere changes its width during muscle contraction?
    a. The M line
    b. The Z line
    c. The A band
    d. The I band

15. Which statement best describes the movement of calcium ions during the early stages of a skeletal muscle contraction?
    a. Calcium enters the cell primarily through the sarcolemma and activates the myofilaments.
    b. Calcium ions are released from the T tubules and diffuse to the myofilaments.
    c. Calcium ions are released from the terminal cisternae of the sarcoplasmic reticulum and diffuse to the region of the myofilaments.
    d. Calcium ions in the region of the myofilaments are rapidly pumped into the terminal cisternae of the sarcoplasmic reticulum.

16. Which of the following is the basic mechanical action that results in a muscle contraction?
    a. Rapid shortening of the portion of the thin filaments that overlap the thick filaments
    b. Attachment of myosin molecule heads to active sites on the thin filaments, followed by a 45-degree rotation of the myosin heads
    c. Sliding of myosin crossbridges along the groove of the thin filaments
    d. Shortening of the myosin filaments in the center of the A band

17. The interaction of calcium ions with troponin molecules on the thin filaments:
    a. Releases inhibition of interaction of actin and myosin crossbridges by a mechanism involving the tropomyosin molecules
    b. Causes troponin to diffuse away from the thin filaments and thereby permit contraction
    c. Activates the enzymatic properties of actin and causes it to attract the thick filaments
    d. Causes a rapid shortening of the tropomyosin molecules on the thin filaments

18. When a skeletal muscle is stretched so that the overlap between thick and thin myofilaments is reduced (but not eliminated), less force can be produced. This is because
    a. Action potentials have farther to travel
    b. Fewer crossbridges can attach to the actin filaments
    c. The thin filaments become highly stretched
    d. The distance between Z lines is increased

Briefly Answer

19. What two roles does ATP play in the crossbridge cycle of skeletal muscle?

20. Why does rigor mortis set in shortly after death?

21. What is a likely result of a defect in the calcium pumping mechanism of the sarcoplasmic reticulum?

22. Give examples of the specific functions of smooth muscle.

23. What is the function of the triads in skeletal muscle?

■ ■ ■ ■ ■ ■ ■ ■ ■ ■ ■ ■ ■ ■ ■ ■ **ANSWERS** ■ ■ ■ ■ ■ ■ ■ ■ ■ ■ ■ ■ ■ ■ ■ ■

1. Sarcomere

2. Smooth (or visceral)

3. Sarcoplasmic reticulum

4. T tubule (or transverse tubule)

5. Crossbridge cycle

6. Creatine phosphate (or phosphocreatine)

7. Oxygen debt

8. The transverse line separating adjacent sarcomeres (also called Z disk)

9. The protein complex in skeletal and cardiac muscle that controls the interaction between thick and thin filaments

10. The principal protein constituent of the thick myofilaments

11. The cell membrane on the surface of muscle cells

12. An energy-yielding sequence of chemical reactions that requires the consumption of oxygen

13. The series of steps linking surface membrane activation to the chemical and mechanical events of muscle contraction

14. d

15. c

16. b

17. a

18. b

19. ATP provides the energy necessary for the power stroke of the crossbridge. Before the crossbridge can detach, another ATP molecule must bind to the myosin head.

20. When blood circulation stops, no more oxygen is supplied to the muscle, and cellular metabolism consumes all of the ATP. When the ATP is gone, crossbridges attach and cannot let go; this makes the muscle stiff and rigid.

21. Contractions would be prolonged and relaxation slower and less complete.

22. Smooth muscle, among other things, causes constriction of blood vessels, propels (and stops) the intestinal contents, causes erection of hairs, and controls the diameter of the pupil of the eye.

23. The triads are the junction between the T tubules and the terminal cisternae of the sarcoplasmic reticulum; here the signal for contraction is converted from an electrical to a chemical (calcium) message.

Suggested Reading

Aidley DJ: *The Physiology of Excitable Cells.* New York, Cambridge University Press, 1978.

Bagshaw CR: *Muscle Contraction.* 2nd ed. New York, Chapman & Hall, 1993.

Junge D: *Nerve and Muscle Excitation.* 3rd ed. Sunderland, MA, Sinauer Associates, 1992.

Keynes RD, Aidley DJ: *Nerve and Muscle.* 2nd ed. New York, Cambridge University Press, 1991.

Matthews GG: *Cellular Physiology of Nerve and Muscle.* 2nd ed. Boston, Blackwell, 1991.

Peachey LD, Adrian RH: *Handbook of Physiology: Muscle.* Bethesda, American Physiological Society, 1983.

Rüegg JC: *Calcium in Muscle Contraction: Cellular and Molecular Physiology.* 2nd ed. New York, Springer-Verlag, 1992.

Smith DS: *Muscle.* New York, Academic, 1972.

Squire JM (ed): *Molecular Mechanisms in Muscular Contraction.* Boca Raton, CRC Press, 1990.

Stein RB: *Nerve and Muscle: Membranes, Cells, and Systems.* New York, Plenum, 1980.

Woledge RC, Curtin NA, Homsher E: *Energetic Aspects of Muscle Contraction.* New York: Academic, 1985.

Skeletal and Smooth Muscle

CHAPTER OUTLINE

I. THE ACTIVATION AND CONTRACTION OF SKELETAL MUSCLE
  A. Impulse transmission from nerve to muscle occurs at the neuromuscular junction
    1. The structure of the neuromuscular junction
    2. Chemical events at the neuromuscular junction
    3. Electrical events at the neuromuscular junction
  B. Neuromuscular transmission can be blocked by toxins, drugs, and trauma

II. THE MECHANICAL PROPERTIES OF SKELETAL MUSCLE
  A. The timing of muscle stimulation is a critical determinant of contractile function
  B. Externally imposed conditions also affect contraction
    1. Isometric contraction
    2. Isotonic contraction
  C. Special mechanical arrangements allow a more precise analysis of muscle function
    1. Isometric contraction and the length-tension curve
    2. Isotonic contraction and the force-velocity curve
    3. Interactions between isometric and isotonic contractions
  D. The anatomic arrangement of muscle is a prime determinant of function
  E. Metabolic and structural adaptations fit skeletal muscle for a variety of roles
    1. Red muscle fibers and aerobic metabolism
    2. White muscle fibers and anaerobic metabolism
    3. Relation of proportion of red and white muscle fibers to muscle function
    4. Muscle fatigue and recovery

III. SMOOTH MUSCLE
  A. Anatomic adaptations of smooth muscle fit it for its special roles
    1. Structural specializations of smooth muscle
    2. Structure of smooth muscle tissues
  B. Regulation and control of smooth muscle involve many factors
    1. Innervation of smooth muscle
    2. Activation of smooth muscle contraction
    3. The role of calcium in smooth muscle contraction
    4. Biochemical control of contraction and relaxation

OBJECTIVES

After studying this chapter, the student should be able to:
1. Draw the structure of the neuromuscular junction
2. List in sequence the steps involved in neuromuscular transmission in skeletal muscle and point out the location of each step on the diagram previously drawn
3. Distinguish between an endplate potential and an action potential in skeletal muscle
4. List the possible sites for blocking neuromuscular transmission in skeletal muscle and give an example of an agent that could cause blockage at each site
5. Describe the conditions necessary to produce an isometric muscle contraction
6. Distinguish between a twitch and a tetanus in skeletal muscle
7. Draw the length-tension diagram for muscle and label the three lines representing passive (resting), active, and total force
8. Using a diagram, relate the power output of skeletal muscle to its force-velocity curve
9. Describe the antagonistic relationships between muscle shortening and lengthening for typical arrangements of skeletal and smooth muscle
10. List and distinguish between the characteristics of the regulatory systems for skeletal muscle (actin-linked) and smooth muscle (myosin-linked)
11. Diagram the cellular pathways involved in excitation-contraction coupling in smooth muscle and label the structures and substances involved
12. Diagram the biochemical pathways that control contraction and relaxation in smooth muscle
13. Distinguish between muscle relaxation from the contracted state and the phenomenon of stress relaxation and give examples of each process

Chapter 8 dealt with the mechanics and activation of the internal cellular processes that produce muscle contraction. This chapter treats muscles as organized tissues, beginning with the events leading to membrane activation by nerve stimulation and continuing with the outward mechanical expression of internal processes.

The Activation and Contraction of Skeletal Muscle

■ Impulse Transmission from Nerve to Muscle Occurs at the Neuromuscular Junction

Contraction of skeletal muscle occurs in response to action potentials that travel down somatic motor axons originating in the central nervous system. The transfer of the signal from nerve to muscle takes place at the **neuromuscular junction** (also called the **myoneural junction** or **motor endplate**). This special type of synapse has a close association between the membranes of nerve and muscle and a physiology much like that of excitatory neural synapses (Chap. 3).

The Structure of the Neuromuscular Junction ■ On reaching a muscle cell, the axon of a motor neuron typically branches into a number of terminals, which constitute the **presynaptic portion** of the neuromuscular junction. The terminals lie in grooves or "gullies" in the surface of the muscle cell, outside the muscle cell membrane, and a Schwann cell covers them all (Fig. 9-1). Within the axoplasm of the nerve terminals are located numerous membrane-bound vesicles containing **acetylcholine** (ACh). Mitochondria, associated with the extra metabolic requirements of the terminal, are also plentiful.

The **postsynaptic portion** of the junction (or **endplate membrane**) is that part of the muscle cell membrane lying immediately beneath the axon terminals. Here the membrane is formed into **postjunctional folds,** at the mouths of which are located many **nicotinic acetylcholine receptor** molecules. These are *chemically gated* ion channels that increase the cation permeability of the postsynaptic membrane in response to the binding of ACh. Between these two

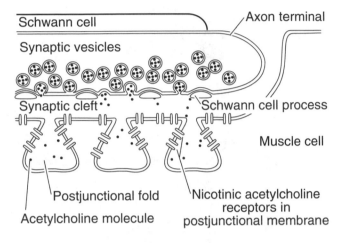

Figure 9–1 ■ Structural features of the neuromuscular junction. Processes of the Schwann cell that overlie the axon terminal wrap around under it and divide the junctional area into active zones.

regions is a narrow space called the **synaptic cleft.** Acetylcholine must diffuse across this gap to reach the receptors in the postsynaptic membrane. Also located in the synaptic cleft (and associated with the postsynaptic membrane) are **acetylcholinesterase** (AChE) molecules.

Chemical Events at the Neuromuscular Junction ■ When the wave of depolarization associated with a nerve action potential spreads into the terminal portion of a motor axon, several processes are set in motion. The lowered membrane potential causes the membrane channels to open, and external calcium ions enter the axon. The rapid rise in intracellular calcium causes the cytoplasmic vesicles of ACh to migrate to the inner surface of the axon membrane, where they fuse with the membrane and release their contents. Because all of the vesicles are of roughly the same size, they all release about the same amount— that is, a **quantum**—of transmitter. For this reason the transmitter release is called **quantal,** although normally so many vesicles are activated at once that their individual contributions are not separately identifiable.

When the ACh molecules arrive at the postsynaptic membrane after diffusing across the synaptic cleft, they bind to the ACh receptors. When two ACh molecules are bound to a receptor, it undergoes a configurational change that allows the relatively free passage of sodium and potassium ions down their respective electrochemical gradients. The binding of ACh to the receptor is reversible and rather loose; soon the ACh diffuses away and is hydrolyzed by AChE into **choline** and **acetate.** This terminates its function as a transmitter molecule, and the permeability of the ion channels returns to the resting state. The choline portion is taken up by the presynaptic axon terminal for resynthesis, and the acetate diffuses away into the extracellular fluid. These events take place over a few milliseconds and may be repeated many times per second without danger of fatigue.

Electrical Events at the Neuromuscular Junction ■ The binding of the ACh molecules to the postsynaptic receptors initiates the electrical response of the muscle cell membrane, and what was a chemical signal becomes an electrical one. The stages of the passage of the electrical signal are shown in Figure 9–2. With the opening of the postsynaptic ionic channels, sodium enters the muscle cell and potassium simultaneously leaves. Both ions share the same membrane channels; in this and several other respects the **end-plate membrane** is different from the general cell membrane of muscle and nerve. The opening of the

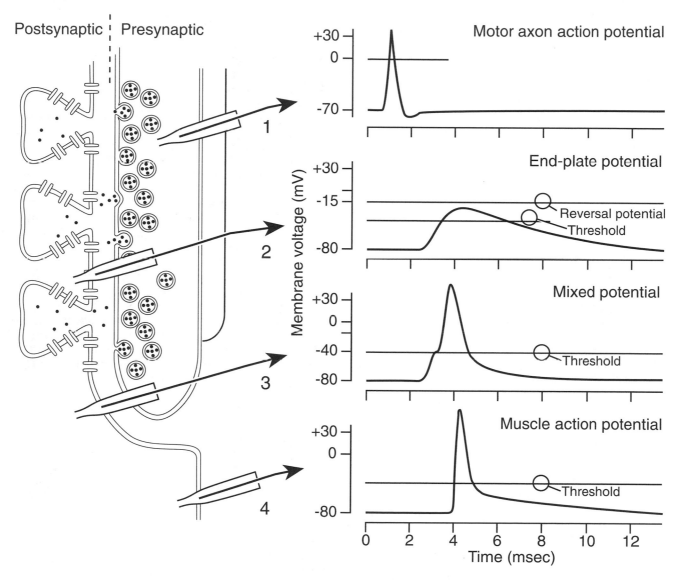

Figure 9–2 ■ Electrical activity at the neuromuscular junction. The four microelectrodes sample membrane potentials at critical regions. (These are idealized records drawn to illustrate isolated portions of the response: in an actual recording there would be considerable mixing of the responses because of the close spacing of the electrodes.) Note the time delays due to transmitter diffusion and endplate potential generation. The reversal potential is the membrane potential at which net current flow is zero (i.e., inward Na^+ and outward K^+ currents are equal).

channels depends only on the presence of the transmitter substance and not on membrane voltage, and the sodium and potassium permeability changes occur simultaneously (rather than sequentially, as they do in nerve or in the general muscle membrane). As a result of the altered permeabilities, there is a net inward current (the **endplate current**), which depolarizes the postsynaptic membrane. This voltage change is called the **endplate potential.** The voltage at which the net membrane current would become zero is called the **reversal potential** of the endplate (Fig. 9–2), although time does not permit this condition to become established, since the AChE is continuously removing transmitter molecules.

To complete the circuit, the current flowing inward at the postsynaptic membrane must be matched by a return current. This current flows through the local muscle cytoplasm (myoplasm), out across the adjacent muscle membrane, and back through the extracellular fluid (Fig. 9–3). As this endplate current flows out across the muscle membrane in regions adjacent to the endplate, it depolarizes the membrane and causes voltage-gated sodium channels to open, bringing the membrane to threshold. This leads to an action potential in the muscle membrane. The muscle action poten-

tial is propagated along the muscle cell membrane by regenerative local currents similar to those in a nonmyelinated nerve fiber. Because of the close placement of the electrodes, there is some overlap in the potentials recorded; near the region of the endplate the electrode will record a mixed potential representing contributions of both the endplate potential and the muscle action potential.

The endplate depolarization is **graded,** and its amplitude varies with the number of receptors with bound ACh. If some circumstance causes reduced ACh release, the amount of depolarization at the endplate could be correspondingly reduced. Under normal circumstances, however, the endplate potential is much more than sufficient to produce a muscle action potential; this reserve, referred to as a **safety factor,** can help to preserve function under certain abnormal conditions. The rate of rise of the endplate potential is determined largely by the rate at which ACh binds to the receptors, and indirect clinical measurements of the size and rise time of the endplate potential are of considerable diagnostic importance. The rate of decay is determined by a combination of factors, including the rate at which the ACh diffuses away from the receptors, the rate of hydrolysis, and the electrical resistance and capacitance of the endplate membrane.

■ Neuromuscular Transmission Can Be Blocked by Toxins, Drugs, and Trauma

The complex series of events making up neuromuscular transmission is subject to interference at a number of steps. **Presynaptic blockade** of the neuromuscular junction can occur if calcium does not enter the presynaptic terminal to participate in migration and emptying of the synaptic vesicles. The drug **hemicholinium** interferes with choline uptake by the presynaptic terminal and thus results in the depletion of ACh. **Botulinum toxin** interferes with ACh release.

Postsynaptic blockade can result from a variety of circumstances. Drugs that partially mimic the action of ACh can be effective blockers. Derivatives of **curare,** originally used as an arrow poison in South America, bind very tightly to ACh receptors. This binding does not result in opening of the ion channels, however, and the endplate potential is reduced in proportion to the number of receptors occupied by curare. Paralysis of the muscle results. Although the muscle can be directly stimulated electrically, nerve stimulation is ineffective. The drug **succinylcholine** blocks the neuromuscular junction in a slightly different way; this molecule binds to the receptors and causes the channels to open. Because it is hydrolyzed only very slowly by AChE, its action is long-lasting and the channels remain open. This prevents resetting of the inactivation gates of muscle membrane sodium channels near the endplate region and blocks subsequent action potentials. Drugs that produce extremely long-lasting endplate potentials are referred to as **depolarizing blockers.**

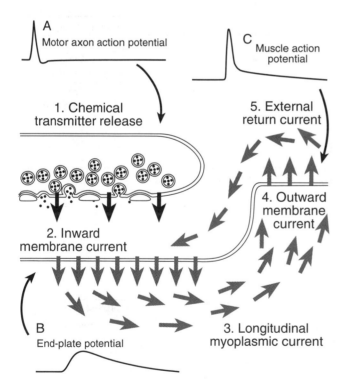

A
Motor axon action potential

C Muscle action potential

1. Chemical transmitter release

5. External return current

2. Inward membrane current

4. Outward membrane current

B
End-plate potential

3. Longitudinal myoplasmic current

Figure 9–3 ■ Ionic currents at the neuromuscular junction. The inward membrane current (2) is carried by sodium ions through the channels associated with acetylcholine receptors. The other currents are nonspecific and are carried by appropriately charged ions in the myoplasm and extracellular fluid. The muscle action potential (C) is propagated along the surface of the muscle, while the endplate potential (B) is localized to the endplate region.

BOTULINUM TOXIN AND FOCAL DYSTONIAS

Focal dystonias are neuromuscular disorders characterized by involuntary and repetitive or sustained skeletal muscle contractions that cause twisting, turning, or squeezing movements in a body part. Abnormal postures and considerable pain, as well as physical impairment, often result. Usually the abnormal contraction is limited to a small and specific region of muscles, hence the term *focal* (by itself, *dystonia* means "faulty contraction"). **Spasmodic torticollis** and **cervical dystonia** (involving neck and shoulder muscles), **blepharospasm** (eyelid muscles), **strabismus** and **nystagmus** (extraocular muscles), **spasmodic dysphonia** (vocal muscles), **hemifacial spasm** (facial muscles), and **writer's cramp** (finger muscles in the forearm) are common dystonias. Such problems are neurologic, not psychiatric, in origin, and sufferers can have severe impairment of daily social and occupational activities. The specific cause is located somewhere in the central nervous system, but usually its exact nature is unknown. There may be genetic predisposition to the disorder in some cases. Centrally acting drugs are of limited effectiveness, and surgical denervation, which carries a significant risk of permanent and irreversible paralysis, may provide only temporary relief. However, recent clinical trials using **botulinum toxin** to produce chemical denervation show significant promise in the treatment of these disorders.

Botulinum toxin is produced when the bacterium *Clostridium botulinum* grows anacrobically. It is one of the most potent natural toxins: a lethal dose for a human adult is about 2–3 μg. The active portion of the toxin is a protein with a molecular weight of around 150,000 that is conjugated with a variable number of accessory proteins. Type A toxin, the complex form most often used therapeutically, has a total molecular weight of 900,000 and is sold under the trade names Botox and Oculinum.

The toxin first binds to the cell membrane of presynaptic nerve terminals in skeletal muscles. The initial binding does not appear to produce paralysis until the toxin is actively transported into the cell, a process requiring over an hour. Once inside, it disrupts calcium-mediated acetylcholine release, producing an irreversible transmission block at the neuromuscular junction. The nerve terminals begin to degenerate and the denervated muscle fibers atrophy. Eventually new nerve terminals sprout from the axons of affected nerves and make new synaptic contact with the chemically denervated muscle fibers. During the period of denervation, which may be several months, the patient usually experiences considerable relief of symptoms. The relief is temporary, however, and the treatment must be repeated when reinnervation has occurred.

Clinically, highly diluted toxin is injected into the individual muscles involved in the dystonia. Often this is done in conjunction with electrical measurements of muscle activity (electromyography) to pinpoint the muscles involved. Patients typically begin to experience relief in a few days to a week. Depending on the specific disorder, relief may be dramatic and may last for several months or more. The abnormal contractions and associated pain are greatly reduced, speech can become clear again, eyes open and cease uncontrolled movement, and often normal activities can be resumed.

The principal side effect is a temporary weakness of the injected muscles. A few patients develop antibodies to the toxin, which renders its further use ineffective. Studies have shown that the toxin's activity is confined to the injected muscles, with no toxic effects noted elsewhere. Long-term effects of the treatment, if any, are unknown.

Compounds such as **physostigmine (eserine)** are potent inhibitors of AChE and produce a depolarizing blockade. In carefully controlled doses, they can temporarily alleviate symptoms of **myasthenia gravis,** an autoimmune condition that results in a loss of postsynaptic ACh receptors. The principal symptom is muscular weakness caused by endplate potentials of insufficient amplitude. Partial inhibition of the enzymatic degradation of ACh allows ACh to remain effective longer and thus to compensate for the loss of receptor molecules.

Under normal conditions, ACh receptors are confined to the endplate region of a muscle. If accidental denervation occurs (e.g., by the severing of a motor nerve), within several weeks the entire muscle becomes sensitive to direct application of ACh. This extrasynaptic sensitivity is due to synthesis of new ACh receptors, a process normally inhibited by the electrical activity of the motor axon. Artificial electrical stimulation has been shown experimentally to prevent the synthesis of new receptors, by regulating transcription of the genes involved. If reinnervation occurs, the extrasynaptic receptors gradually disappear. Muscle atrophy also occurs in the absence of functional innervation, which likewise can be at least partially reversed with artificial stimulation.

The Mechanical Properties of Skeletal Muscle

The variety of controlled muscular movements that humans can make is remarkable, ranging from the powerful contractions of a weightlifter's biceps to the delicate movements of the muscles that position our eyes as we follow a moving object. In spite of this diversity, the fundamental mechanical events of the contraction process can be described by a relatively small set of idealized functions that emphasize particular capabilities of muscle.

■ The Timing of Muscle Stimulation Is a Critical Determinant of Contractile Function

Skeletal muscle must be activated by the nervous system before it can begin its contraction. Through the many processes previously described, a single nerve action potential arrives at each motor nerve axon terminal. A single muscle action potential then propagates along the length of each muscle fiber innervated by that axon terminal. This leads to a single brief contraction of the muscle, a **twitch.** Though the contractile machinery may be fully activated (or nearly so) during a twitch, the amount of force produced is relatively low, because the activation is so brief that the relaxation processes begin before contraction is fully established.

The duration of the action potential in a skeletal muscle fiber is short (about 5 msec) compared to the

duration of a twitch (tens or hundreds of milliseconds, depending on muscle type, temperature, etc.). This means the absolute refractory period is also brief, and the muscle fiber membrane can be activated again long before the muscle has relaxed. Figure 9-4 shows the result of stimulating a muscle that is already active as a result of a prior stimulus. If the second stimulus is given during relaxation (Fig. 9-4B), well outside the refractory period caused by the first stimulus, significant additional force is developed. This additional force increment is associated with a second release of calcium ions from the sarcoplasmic reticulum, which reactivates thin filaments (and associated crossbridges) whose bound calcium has begun to diffuse away, back into the sarcoplasmic reticulum. When the second stimulus follows the first very closely (even before force has begun to decline), the myoplasmic calcium concentration is still high (Fig. 9-4C), and the effect of the additional calcium ions is to increase the force and to some extent the duration of the twitch, because a larger amount of calcium is present in the region of the myofilaments. If stimuli are given repeatedly and rapidly, the result is a sustained contraction, called a **tetanus.** When the contractions occur so close together that no fluctuations in force are observed, the tetanus is said to be **fused.** The repetition rate at which this occurs is the **tetanic fusion frequency,** typically 20-60 stimuli per second, with the higher rates found in muscles that contract and relax rapidly. Figure 9-5 shows these effects in a special situation, in which the interval between successive stimuli is steadily reduced and the muscle responds at first with a series of

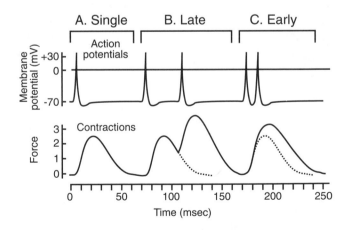

Figure 9-4 ■ Temporal summation of muscle twitches. The first contraction (A) is in response to a single action potential. The next contraction (B) shows the summed response to a second stimulus given during relaxation; the two individual responses are evident. The last contraction (C) is the result of two stimuli in quick succession. Though measured force was still rising when the second stimulus was given, the fact that there could be an added response shows that internal activation had begun to decline. In all cases the solid line in the lower graph represents the actual summed tension.

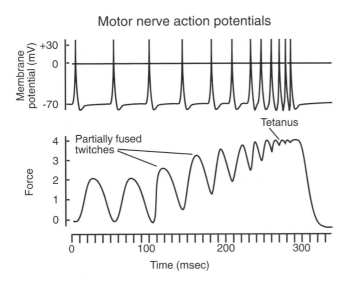

Motor nerve action potentials

Figure 9-5 ■ Fusion of twitches into a smooth tetanus. The interval between successive stimuli steadily decreases until no relaxation occurs between stimuli.

twitches that become fused into a smooth tetanus at the highest stimulus frequency. Because it involves events that occur close together in time, a tetanus is a form of **temporal summation.**

The amount of force produced in a tetanus is typically several times that of a twitch; the disparity is expressed as the **tetanus-twitch ratio.** The relaxation processes during a twitch, particularly the reuptake of calcium, begin to operate as soon as the muscle is activated, and full activation is brief (lasting less time than that required for the muscle to reach its peak force). Multiple stimuli, as in a tetanus, are needed for the full force to be expressed. Another factor is mechanical: even if the ends of a muscle are held rigidly, internal dimensional changes take place on activation. Some of this internal motion is associated with the crossbridges themselves, and the tendons at either end of the muscle make a considerable contribution. These deformable structures comprise the **series elastic component** of the muscle, and their extension takes a significant amount of time. The brief activation time of a twitch is not sufficient to extend the series elastic component fully, and not all of the potential force of the contraction is realized. Repeated activation in a tetanus allows time for the internal "slack" to be more fully taken up, and more force is produced. Muscles with a large amount of series elasticity have a large tetanus-twitch ratio. The presence of series elasticity in human muscles provides some protection against sudden overloads of a muscle and allows for a small amount of mechanical energy storage. In jumping animals, such as kangaroos, a large fraction of muscular energy is stored in the elastic tendons and contributes significantly to the economy of locomotion.

Since a skeletal muscle consists of many fibers, each supplied by its own branch of a motor axon, it is possible (and usual) that only a portion of the muscle will be activated at any one time. The pattern of activation is determined by the central nervous system and by the distribution of the motor axons among the muscle fibers. A typical motor axon branches as it courses through the muscle, and each of its terminal branches innervates a single muscle fiber. All of the fibers supplied by a single motor axon will contract together when a nerve action potential travels from the central nervous system and divides among the branches. A single motor axon and all the fibers it innervates are called a **motor unit.** Individual contractions involving only some of the fibers in a motor unit are impossible, so the motor unit is normally the smallest functional unit of a muscle. In muscles adapted for fine and precise control, only a few fibers are associated with a given motor axon; in muscles in which high force is more important, a single motor axon controls many more muscle fibers. The total force produced by a muscle is determined by the number of motor units active at any one time; as more motor units are brought into play, the force increases. This phenomenon, called **motor unit summation,** is illustrated in Figure 9-6. The force of contraction of the whole muscle is further modified by the degree of activation of each motor unit in the muscle; some may be fully tetanized, while others may be at rest or produce only a series of twitches. During a sustained contraction the pattern of activity is continually changed by the central nervous system, and the burden of contraction is shared among the fibers and motor units. This results in a smooth contraction with the force precisely controlled to produce the desired movement (or lack of it).

■ Externally Imposed Conditions Also Affect Contraction

Mechanical factors external to the muscle also influence the force and speed of contraction. For example, if a muscle is not allowed to shorten when it is stimulated, it will develop more force than it would if its length changed. If a muscle is lifting a load, its force of contraction is determined by the size of the load, not by internal factors. The speed with which a muscle shortens is likewise determined, at least in part, by external conditions.

Isometric Contraction ■ When conditions are arranged to prevent a muscle from shortening when it is activated, the muscle will express its contractile activity by pulling against its attachments and thus developing force. This type of contraction is termed **isometric** (meaning "same length"). The forces developed during an isometric contraction can be studied by attaching a dissected muscle to an apparatus similar to that shown in Figure 9-7. This arrangement provides for setting the length of the muscle and tracing a record of **force versus time.** In a twitch, isometric

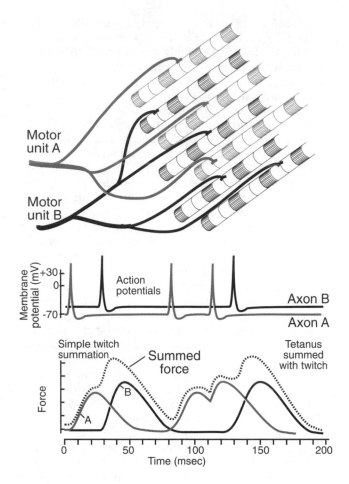

Figure 9–6 ■ Summation of motor units. Two units are shown above; their motor action potentials and twitches are shown below. In the first contraction there is a simple summation of two twitches, while in the second a brief tetanus in one motor unit sums with a twitch in the other.

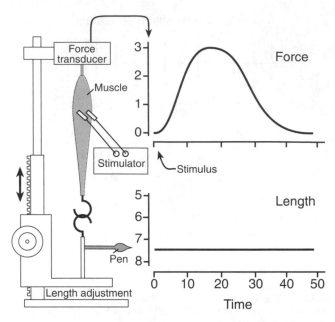

Figure 9–7 ■ Schematic diagram of simple apparatus for recording isometric contractions. The length of the muscle (marked on the graph by the pen attached near its lower end) is adjustable at rest but is held constant during contraction. The force transducer provides a record of the isometric force response to a single stimulus at a fixed length (isometric by definition). Force, length, and time units are arbitrary.

force develops relatively rapidly and subsequent isometric relaxation is somewhat slower. The duration of both contraction time and relaxation time is related to the rate at which calcium ions can be delivered to and removed from the region of the crossbridges, the actual sites of force development. During an isometric contraction no actual physical work is done on the external environment, because no movement takes place while the force is developed. The muscle, however, still consumes energy to fuel the processes that generate and maintain force.

Isotonic Contraction ■ When conditions are arranged so the muscle can shorten and exert a constant force while doing so, the contraction is called **isotonic** (meaning "same force"). In the simplest conditions, this constant force is provided by the load a muscle lifts. This load is called an **afterload,** since its presence is not apparent to the muscle until after it has begun to shorten. Recording an isotonic contraction requires modification of the apparatus used to study isometric contraction (Fig. 9–8) to allow the muscle to shorten

while lifting an afterload, provided by the attached weight. This weight is chosen to present somewhat less than the peak force capability of the muscle. When the muscle is stimulated, it will begin to develop force without shortening, since it takes some time to build up enough force to begin to lift the weight. This means that early on, the contraction is isometric (phase 1 in Fig. 9–8). After sufficient force has been generated, the muscle will begin to shorten and lift the load (phase 2). The contraction then becomes isotonic, since the force exerted by the muscle exactly matches that of the weight, and the mass of the weight does not vary. Therefore, the upper tracing in Figure 9–8 shows a flat line representing constant force, while the muscle length (lower tracing) is free to change. As relaxation begins (phase 3), the muscle lengthens at constant force, since it is still supporting the load; this phase of relaxation is isotonic, and the muscle is reextended by the weight. When the muscle has been extended sufficiently to return to its original length, conditions again become isometric (phase 4), and the remaining force in the muscle declines as it would in a purely isometric twitch. In almost all situations encountered in daily life, isotonic contraction is preceded by isometric force development; such contractions are called **mixed contractions** (isometric-isotonic-isometric).

The duration of the early isometric portion of the contraction varies, depending on the afterload. At

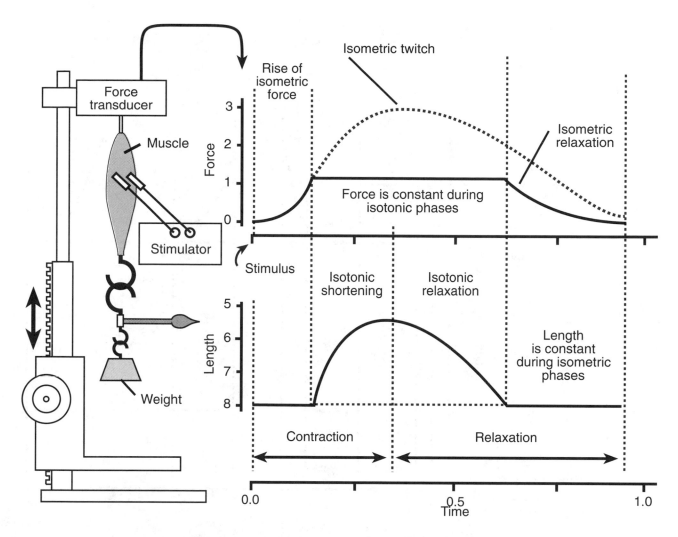

Figure 9–8 ■ Modified apparatus showing the recording of a single isotonic switch. The pen at the lower end of the muscle marks its length, and the weight attached to the muscle provides the afterload, while the platform beneath the weight prevents the muscle from being overstretched at rest. The first part of the contraction, until sufficient force has developed to lift the weight, is isometric. During shortening and isotonic relaxation the force is constant (isotonic conditions), and during the final relaxation conditions are again isometric, because the muscle can no longer lift the weight. The dotted lines in the force and length traces show the isometric twitch that would have resulted if the force had been too large (greater than 3 units) for the muscle to lift. Force, length, and time units are arbitrary.

low afterloads the muscle requires little time to develop sufficient force to begin to shorten, and conditions will be isotonic for a longer time. This is shown in Figure 9-9, which presents a series of three twitches. At the lowest afterload (weight A only), the isometric phase is briefest and the isotonic phase is longest and has the lowest force. With the addition of weight B, the afterload is doubled and the isometric phase is longer, while the isotonic phase is shorter and has twice the force. If weight C is added, the combined afterload represents more force that the muscle can exert, and the contraction is isometric for its entire duration. The speed and

extent of shortening depend on the afterload in unique ways described shortly.

In addition to the types of contraction already described, several other special sets of conditions are sometimes encountered. When the force exerted by a shortening muscle continuously increases as it shortens, the contraction is said to be **auxotonic.** Drawing back a bowstring is an example of this type of contraction. If the force of contraction decreases as the muscle shortens, the contraction is called **meiotonic.**

In the intact body, a **concentric** contraction is one in which shortening (not necessarily isotonic) takes place; in an **eccentric** contraction a muscle is ex-

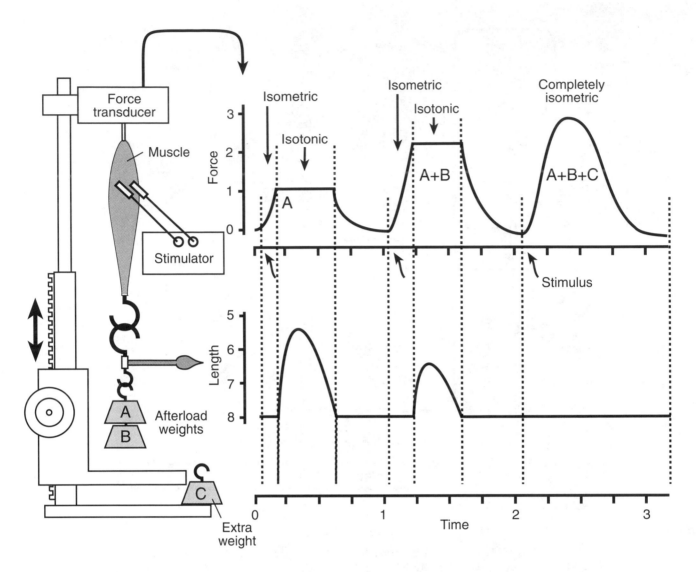

Figure 9–9 ■ A series of afterloaded isotonic contractions. The curves marked A and A + B correspond to the force and shortening records during the lifting of those weights. In each case the adjustable platform prevents the muscle from being stretched by the attached weight, and all contractions start from the same muscle length. Note the lower force and greater shortening with the lower weight (A). If weight C (total weight = A + B + C) is added to the afterload, the muscle cannot lift it, and the entire contraction remains isometric. Force, length, and time units are arbitrary.

tended (while active) by an external force. A **static contraction** results in no movement, but this may be due to partial activation (fewer motor units active) opposing a load that is not maximal. (This is different from a true isometric contraction, in which shortening is physically impossible regardless of the degree of activation.)

■ Special Mechanical Arrangements Allow a More Precise Analysis of Muscle Function

The types of contraction described above provide a basis for a better understanding of muscle function.

The isometric and isotonic mechanical behavior of muscle can be described in terms of two important relationships, the **length-tension curve** (treating isometric contraction at different muscle lengths) and the **force-velocity curve** (concerned with muscle performance during isotonic contraction).

Isometric Contraction and the Length-Tension Curve ■ An isolated muscle, since it is made of contractile proteins and connective tissue, can resist being stretched at rest. When it is very short, it is slack and will not resist passive extension. As it is made longer and longer, however, its resisting force increases more and more. Normally a muscle is protected against overextension by attachments to the skeleton or by other anatomic structures. Since the muscle has not been stimulated, this resisting force is called **passive** or **resting tension** (force). The relationship between force and length is much different in a stimulated muscle. The amount of active force (or **active tension**) a muscle can produce during an isometric contraction depends on the length at which the muscle is held. At a length that roughly corresponds to the natural length in the body, the **resting length,** the maximum force is produced. If the muscle is set to a shorter length and then stimulated, it produces less force. At an extremely short length, it produces no force at all. Also if the muscle is made longer than its optimal length, it produces less force when stimulated. This behavior is summarized in the **length-tension relationship** (or **curve**) (Fig. 9–10). The left of the upper record shows the force produced by a series of individual twitches made over the range of muscle lengths indicated at the left of the lower record. Information from these traces is plotted at the right. The total peak force from each twitch is related to each length (dotted lines). The muscle length changes only when the muscle is not stimulated and is held isometric during contraction. The resting force at each length (i.e., **passive force**) and the difference between this and the **total force** are also shown (inset). The difference between the total force and the passive force is the **active force,** which is due directly to the active contraction of the muscle. The curve shows that when

the muscle is either longer or shorter than an optimal length, it produces less force. The role of myofilament overlap is a primary factor in determining the active length-tension curve (Chap. 8), but it has also been shown that at very short lengths there is reduced effectiveness of some of the steps in the excitation-contraction coupling process. These include reduced calcium binding to troponin and some loss of action potential conduction in the T tubule system.

The functional significance of the length-tension curve varies among the different muscle types. Many skeletal muscles are confined by their skeletal attachments to a relatively short region of the curve that is near the optimal length. In these cases it is the lever action of the skeletal system, not the length-tension relationship, that is of primary importance in determining the maximal force the muscle can exert. Cardiac muscle (Chap. 10), on the other hand, normally works at lengths significantly less than optimal for force production, but its passive length-tension curve is shifted to shorter lengths. The length-tension relationship is therefore very important when considering the ability of cardiac muscle to adjust to changes in length (related to the volume of blood contained in the heart) to meet changing bodily needs. The role of the length-tension curve in smooth muscles is less clearly understood because of the great diversity among smooth muscles and their physiologic roles. For all muscle types, however, the length-tension curve has provided important information about the cellular and molecular mechanisms of contraction.

Isotonic Contraction and the Force-Velocity Curve ■ Everyday experience shows that the speed at which a muscle can shorten depends on the load that must be moved. Simply stated, light loads are lifted faster than heavy ones. Detailed analysis of this observation can provide insight into how the force and shortening of muscle are matched to the external tasks it performs, as well as how muscle functions internally to liberate mechanical energy from its metabolic stores. The analysis is performed by arranging a muscle so that it can be presented with a series of afterloads (Fig. 9–9). When it is maximally stimulated, the lighter loads are lifted quickly and the heavier loads more slowly. If the applied load is greater than the maximal force capability of the muscle (F_{max}), then no shortening will result and the contraction will be isometric. If no load at all is applied, then the muscle will shorten at its greatest possible speed (a velocity called V_{max}). The **initial velocity**—the speed with which the muscle begins to shorten—is measured at various loads. The reason initial velocity is measured is that the muscle soon begins to slow down; as it gets shorter, it moves down its length-tension curve and is capable of less force and speed of shortening. When all of the initial velocity measurements are related to each corresponding afterload lifted, an inverse relationship is obtained. This line is curved such that the relationship is steeper at low forces. This graph is

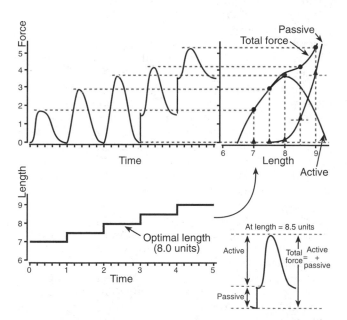

Figure 9–10 ■ Construction of the length-tension (length-force) curve for skeletal muscle. Contractions are made at a number of resting lengths, and the resting (passive) and peak (total) forces for each twitch are transferred to the graph at the right. Subtraction of the passive curve from the total curve yields the active force curve. These curves are further illustrated in the lower right corner of the figure. Force, length, and time units are arbitrary.

called the **force-velocity curve,** and when the measurements are made on a fully activated muscle it defines the upper limits of the muscle's isotonic capability. In practice, a completely unloaded contraction is very difficult to arrange, but mathematical extrapolation provides an accurate V_{max} value. Figure 9–11 shows a force-velocity curve made from such a series of isotonic contractions. The initial velocity points (A–D) correspond to the contractions shown at the top. Factors that modify muscle performance, such as incomplete stimulation (e.g., fewer motor units activated) or fatigue, result in operation *below* the limits defined by the force-velocity curve.

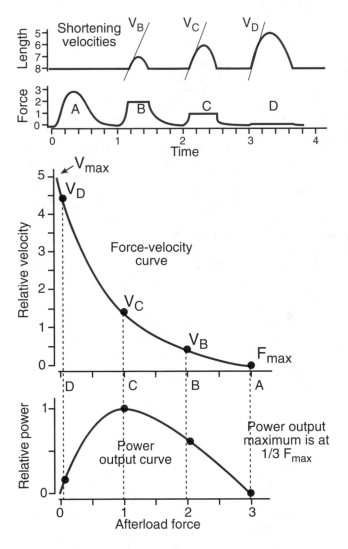

Figure 9–11 ■ Construction of the force-velocity and power output curves for skeletal muscle. Contractions at four different (decreasing left to right) afterloads are shown in the upper portion: note the differences in the amounts of shortening. The initial shortening velocity (slope) is measured (V_B, V_C, V_D) and the corresponding force and velocity points plotted on the axes below. Also shown is power output, the product of force and velocity. Note that it reaches a maximum at an afterload of about one-third of the maximal force. Force, length, and time units are arbitrary.

Consideration of the force-velocity relationship of muscle can provide insight into how it functions as a biologic motor, its primary physiologic role. For instance, V_{max} represents the maximal rate of crossbridge cycling; it is thus directly related to the biochemistry of the actin-myosin ATPase activity in a particular muscle type and can be used to compare the properties of different muscles.

Because isotonic contraction involves moving a force (the afterload) through a distance, the muscle does physical work. The rate at which it does this work is its **power output** (Fig. 9–11). The factors represented in the force-velocity curve are thus relevant to questions of muscle work and power. At the two extremes of the force-velocity curve (zero force, maximal velocity and maximal force, zero velocity) no work is done, since by definition work requires moving a force through a distance. Between these two extremes work and power output pass through a maximum at a point where the force is approximately one-third of its maximal value. The peak point on the curve represents the combination of force and velocity at which the greatest power output is produced; at any afterload force greater or smaller than this, less power can be produced. It also appears that in skeletal muscle the point of optimal power output is also the point at which muscle **efficiency** is greatest. At this point the muscle produces the largest power output for a given amount of metabolic energy input.

In terms of mechanical work, the chemical reactions of muscle are about 20% efficient; the remaining 80% of the fuel (ATP) consumed appears as heat. In some forms of locomotion, such as running, the measured efficiency is higher, approaching 40% in some cases. This apparent increase is probably due to the storage of mechanical energy (between strides) in elastic elements of the muscle and in the potential and kinetic energy of the moving body. This energy is then partly returned as work during the subsequent contraction. It has also been shown that stretching an active muscle (e.g., during running or descending stairs) can greatly reduce the breakdown of ATP, since the crossbridge cycle is disrupted when myofilaments are forced to slide in the lengthening direction. These force-velocity and efficiency relationships are important when endurance is a significant concern. Athletes who are successful in long-term physical activity have learned to optimize their power output by "pacing" themselves and adjusting the velocity of contraction of their muscles to extend the duration of exercise. Such adjustments obviously involve compromises, since not all of the many muscles involved in a particular task can be used at optimal loading and rate, and subjective factors such as experience and training enter into performance. In rapid, short-term exercise it is possible to work at an inefficient force-velocity combination to produce the most rapid or forceful movements possible. Such activity must necessarily be of more limited duration than that carried out under conditions of maximal efficiency. Examples of attempts at optimal matching of human muscles to varying loads can be found in the design of human-powered machinery, pedestrian ramps, and similar devices.

Interactions between Isometric and Isotonic Contractions ■ The length-tension curve represents the effect of length on the isometric contraction of skeletal muscle. During isotonic shortening, however, muscle length does change while the force is constant. The limit of this shortening is also described by the length-tension curve (Fig. 9–12). For example, a lightly loaded muscle will shorten farther than one starting from the same length and bearing a heavier load. If the muscle begins its shortening from a reduced length, then its subsequent shortening will be re-

duced. These relationships are diagrammed in Figure 9–12. In the case of ordinary skeletal muscle activity these limits are not usually encountered, because voluntary adjustments of the contracting muscle are usually made to accomplish a specific task. In the case of cardiac muscle (Chap. 10), however, such interrelationships between force and length are of critical importance in functional adjustment of the beating heart.

■ The Anatomic Arrangement of Muscle Is a Prime Determinant of Function

Anatomic location places important restrictions on muscle function by limiting the amount of shortening or determining the kinds of loads that will be encountered. Skeletal muscle is generally attached to bones, and bones are attached to each other. Because of the way the muscles are attached and the skeleton is articulated, the bones and muscles together constitute a lever system. This has some important consequences for the physiology of the muscles themselves and for the function of the body as a whole. In most cases (e.g., the flexion of the forearm; Fig. 9–13) the system works at a **mechanical disadvantage** with respect to

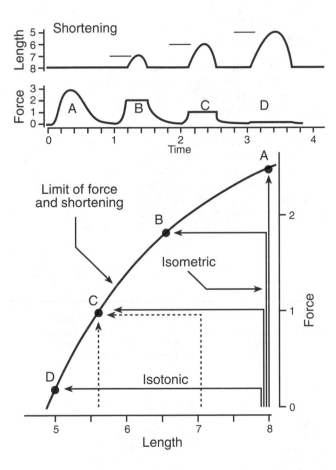

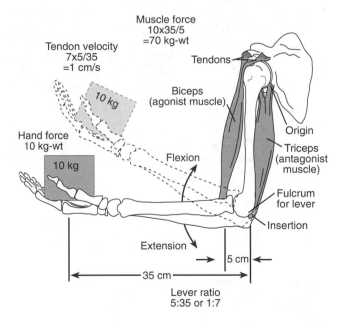

Figure 9–12 ■ The relationship between isotonic and isometric contractions. At the top are shown the contractions from Figure 9–11, with different amounts of shortening. As the lower graph shows, for contractions B, C, and D, the initial portion is isometric (the line moves upward at constant length) until the afterload force is reached. The muscle then shortens at the afterload force (the line moves to the left) until its length reaches a limit determined (at least approximately) by the isometric length-tension curve. The dotted lines show that the same final force/length point can be reached by a number of different approaches. Relaxation data, not shown on the graph, would trace out the same pathways in reverse. Force, length, and time units are arbitrary.

Figure 9–13 ■ Antagonistic arrangements and the lever system of skeletal muscle. Contraction of the biceps muscle lifts the lower arm (flexion) and elongates the triceps, while contraction of the triceps lowers the arm and hand (extension) and elongates the biceps. The bones of the lower arm are pivoted at the elbow joint (the fulcrum of the lever); the force of the biceps is applied through its tendon approximately 5 cm from the fulcrum, and the hand is 35 cm away from the elbow joint. Thus the hand will move 7 times as far (and fast) as the biceps shortens (lever ratio of 35:5, or 7:1), but the biceps will have to exert 7 times as much force as the hand is supporting. (Adapted from Rhoades RA, Pflanzer R (eds): *Human Physiology.* 2nd ed. Fort Worth, TX, Saunders, 1992.)

the force exerted. The shortening capability of skeletal muscle by itself is rather limited, and the skeletal lever system multiplies the distance over which an extremity can be moved. However, this means the muscle must exert a much greater force than the actual weight of the load being lifted (increased by the same ratio that the length change is multiplied). In the case of the human forearm, the **biceps brachii,** when moving a force applied to the hand, must exert a force at its insertion on the **radius** (and **ulna**) that is approximately seven times as great (Fig. 9-13). However, the resulting movement of the hand is approximately seven times as far and seven times as rapid as the shortening of the muscle.

Acting on its own, a muscle can only shorten. The force to relengthen it is provided externally. The arrangement of muscles into **antagonistic pairs** of **flexors** and **extensors** achieves this. For example, the shortening of the biceps is countered by the action of the triceps; the triceps, in turn, is relengthened by contraction of the biceps.

■ Metabolic and Structural Adaptations Fit Skeletal Muscle for a Variety of Roles

Specific skeletal muscles are adapted for specialized functions. These adaptations involve primarily the structures and chemical reactions that supply the contractile system with energy. The enzymatic properties (i.e., the rate of ATP hydrolysis) of the actomyosin ATPase also vary. The basic structural features of the sarcomeres and the thick/thin filament interactions are, however, essentially the same among the categories of skeletal muscle.

Chapter 8 detailed the biochemical reactions responsible for providing ATP to the contractile system. Recall that muscle fibers contain both **glycolytic** (anaerobic) and **oxidative** (aerobic) metabolic pathways, which differ in their ability to produce ATP from metabolic fuels, in particular glucose and fatty acids. Among muscle fibers, the relative importance of each pathway and the presence or absence of associated supporting organelles and structures vary. These variations form the basis for the classification of skeletal muscle fiber types (Table 9-1). A typical skeletal muscle usually contains a mixture of fiber types, but in most muscles a particular type predominates. The major classification criteria are derived from mechanical measurements of muscle function and histochemical staining techniques in which dyes specific for metabolic enzymes are used to identify individual fibers in a muscle cross-section.

Red Muscle Fibers and Aerobic Metabolism ■ The color differences among skeletal muscles arise from difference in the amount of **myoglobin** they contain. Like the related red blood cell protein **hemoglobin,** myoglobin can bind, store, and release oxygen. It is abundant in muscle fibers that depend heavily on aerobic metabolism for their ATP supply, where it facilitates oxygen diffusion (and serves as a minor auxiliary oxygen source) in times of heavy demand. Red muscle fibers are divided into **slow-twitch** and **fast-twitch** fibers on the basis of their contraction speed (Table 9-1). The differences in rates

TABLE 9–1 ■ Classification of Skeletal Muscle Fiber Types

Metabolic Type	Fast Twitch		Slow Twitch: Slow Oxidative (Red)
	Fast Glycolytic (White)	Fast Oxidative-Glycolytic (Red)	
Metabolic properties			
ATPase activity	High	High	Low
ATP source(s)	Anaerobic glycolysis	Anaerobic glycolysis/ oxidative phosphorylation	Oxidative phosphorylation
Glycolytic enzyme content	High	Moderate	Low
Number of mitochondria	Low	High	High
Myoglobin content	Low	High	High
Glycogen content	High	Moderate	Low
Fatigue resistance	Low	Moderate	High
Mechanical properties			
Contraction speed	Fast	Fast	Slow
Force capability	High	Medium	Low
Sarcoplasmic reticulum			
Ca^{2+} ATPase activity	High	High	Moderate
Motor axon velocity	100 m/sec	100 m/sec	85 m/sec
Structural properties			
Fiber diameter	Large	Moderate	Small
Number of capillaries	Few	Many	Many
Functional role in body	Rapid and powerful movements	Medium endurance	Postural/endurance
Typical example	Latissimus dorsi	In mixed-fiber muscles such as vastus lateralis	Soleus

of contraction (shortening velocity or force development) arise from differences in actomyosin ATPase activity (i.e., in the basic crossbridge cycling rate). Mitochondria are abundant in these fibers, because they contain the enzymes involved in aerobic metabolism.

White Muscle Fibers and Anaerobic Metabolism ▪ White muscle fibers, which contain little myoglobin, are fast-twitch fibers that rely primarily on glycolytic metabolism. They contain significant amounts of stored **glycogen,** which can be broken down rapidly to provide a quick source of energy. Although they contract rapidly and powerfully, their endurance is limited by their ability to sustain an oxygen debt (i.e., to tolerate the buildup of lactic acid). They require a period of recovery (and a supply of oxygen) after heavy use. White muscle fibers have fewer mitochondria than red muscle fibers, since the reactions of glycolysis take place in the myoplasm. There are indications that enzymes of the glycolytic pathway may be closely associated with the thin filament array.

Relation of Proportion of Red and White Muscle Fibers to Muscle Function ▪ The relative proportions of red and white muscle fibers fit muscles for different uses in the body. Muscles containing primarily slow-twitch oxidative red fibers are specialized for functions requiring slow movements and considerable endurance, such as maintenance of posture. Muscles containing a preponderance of fast-twitch red fibers support faster and more powerful contractions. These muscles typically contain varying mixtures of fast-twitch white fibers; their resulting ability to use both aerobic and anaerobic metabolism increases their power and speed. Muscles containing primarily fast-twitch white fibers are suited for rapid, short, powerful contraction.

Fast muscles, both white and red, not only contract rapidly but also relax rapidly. Rapid relaxation requires a high rate of calcium pumping by the sarcoplasmic reticulum, which they also have in abundance. In such muscles the energy used for calcium pumping can be as much as 30% of the total consumed. Fast muscles are supplied by large motor axons with high conduction velocities; this correlates with their ability to make very quick and rapidly repeated contractions.

Muscle Fatigue and Recovery ▪ During a period of heavy exercise, especially when working above 70% of maximal aerobic capacity, skeletal muscle is subject to **fatigue.** The speed and force of contraction are diminished, relaxation time is prolonged, and a period of rest is required to restore normal function. While there is a close correlation between the oxidative capacity of a particular muscle fiber type and its fatigue resistance, chemical measurements of fatigued skeletal muscle specimens have shown that the ATP content, while reduced, is not completely exhausted. In well-motivated subjects, central nervous system factors do not appear to play an important role in fatigue, and

transmission at the neuromuscular junction has such a large safety factor that this also does not contribute to fatigue.

Studies on isolated muscle have distinguished two different mechanisms producing fatigue. Stimulation of the muscle at a rate far above that necessary for a fused tetanus very quickly produces **high-frequency stimulation fatigue;** recovery from this condition is very rapid (a few tens of seconds). In this type of fatigue, the principal defect seems to be a failure in T tubule action potential conduction, which leads to reduced release of Ca^{2+} from the sarcoplasmic reticulum. Under most in vivo circumstances, feedback mechanisms in neural motor pathways work to reduce the stimulation to the minimum necessary for a smooth tetanus, and this type of fatigue is probably not often encountered. Prolonged or repeated tetanic stimulation produces a longer-lasting fatigue with a longer recovery time. This type of fatigue (**low-frequency fatigue**) is related to the muscle's metabolic activities. The buildup of metabolites produced by crossbridge cycling, especially *inorganic phosphate* (P_i) and *H^+ ions,* reduces calcium sensitivity of the myofilaments and the contractile force generated per crossbridge. The reduced amount of metabolic energy available to the calcium transport system in the sarcoplasmic reticulum leads to reduced Ca^{2+} pumping. As a result, relaxation time increases and there is less Ca^{2+} available to activate the contraction with each stimulus, resulting in a lowered peak force.

Smooth Muscle

The properties of skeletal muscle described thus far apply in a general way to smooth muscle. Many of the basic muscle properties are highly modified in smooth muscle, however, because of the very different functional roles it plays in the body. The adaptations of smooth muscle structure and function are best understood in the context of the special requirements of the organs and systems of which smooth muscle is an integral component. Of particular importance are the high metabolic economy of smooth muscle, which allows it to remain contracted for long periods with very little energy consumption, and the small size of its cells, which allows precise control of very small structures, such as blood vessels. Most smooth muscles are not discrete organs (like individual skeletal muscles) but are intimate components of larger organs. It is in the context of these specializations that the physiology of smooth muscle is best understood.

▪ Anatomic Adaptations of Smooth Muscle Fit It for its Special Roles

While there are major differences among the organs and systems in which smooth muscle plays a major part, the structure of smooth muscle is quite consistent at the tissue level and even more similar at the cellular level. Several typical arrangements of smooth muscle occur in a variety of locations.

Structural Specializations of Smooth Muscle ■ The variety of smooth muscle tasks—regulating and promoting movement of fluids, expelling the contents of organs, moving visceral structures—are accomplished by a few basic types of tissue structures. All of these structures are subject, like skeletal muscle, to the requirement for antagonism: if smooth muscle contracts, an external force must lengthen it again. The structures described below provide these restoring forces in a variety of ways.

The simplest organized smooth muscle arrangement is found in the muscular arteries and veins of the circulatory system. Smooth muscle cells are oriented in the circumference of the vessel so that shortening of the cells results in a reduction in vessel diameter. This reduction may be slight or may result in complete obstruction of the vessel lumen, depending on the physiologic needs of the body or organ. The orientation of the cells in the wall is helical, with a very shallow pitch. In larger muscular vessels, particularly arteries, there may be many layers of cells and the force of contraction may be quite high, but in small arterioles the muscle layer may consist of single cells wrapped around the vessel. In the case of blood vessels, the blood pressure provides the force to relengthen the cells. This type of muscle organization is extremely important, because the narrowing of a blood vessel has a powerful influence on the rate of blood flow through it (Chap. 12, 15). This circular arrangement is also prominent in the airways of the lung, where it is likewise important in regulating air flow.

A further specialization of the circular muscle arrangement is a **sphincter,** a thickening of the muscular portion of the wall of a hollow or tubular organ whose contraction has the effect of restricting flow or stopping it completely. Many sphincters, such as those in the gastrointestinal and urogenital tracts, have a special nerve supply and participate in complex reflex behavior. The muscle in sphincters is characterized by the ability to remain contracted for long periods with little metabolic cost.

Next in order of complexity is the combination of circular and longitudinal layers, as in the muscle of the small intestine. The outermost muscle layer, which is relatively thin, runs along the length of the intestine. The inner muscle layer, thicker and more powerful, has a circular arrangement. Coordinated alternating contractions and relaxations of these two layers propel the contents of the intestine, although most of the motive power is provided by the circular muscle (the longitudinal layer has important pacemaker and coordinating functions). The primary force that relengthens the shortened circular muscle in one segment is provided by circular muscle contraction "upstream" from the location in question. Figure 9–14 shows the arrangement of the intestinal musculature. When the

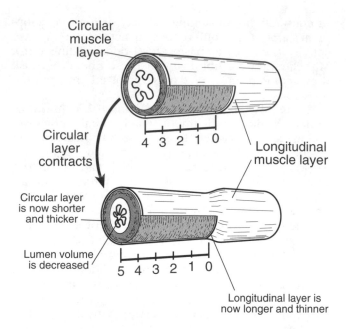

Figure 9–14 ■ Antagonistic arrangement of smooth muscle in the small intestine. Contraction of the inner circular muscle layer produces lengthening of the outer longitudinal muscle layer.

circular layer in a segment of the gut contracts, those muscle cells become shorter and increase in cross-section (and the circular layer becomes thicker). This has two effects: it produces a narrowing of the lumen of the gut and it causes elongation of the active segment. This elongation lengthens the overlying longitudinal muscle layer. Variations in this basic process cause different types of movements, such as **segmentation** and **peristalsis** (Chap. 27). Certain blood vessels and the ureters also have this type of muscle arrangement.

The most complex arrangement of smooth muscle is found in organs such as the urinary bladder and uterus. There are numerous layers and orientations of muscle fibers so that the effect of their contraction is an overall reduction of the volume of the organ. Even with such a complex arrangement of fibers, coordinated and organized contractions take place. The relengthening force in the case of such hollow organs is provided by the gradual accumulation of contents. In the urinary bladder, for example, the muscle is gradually stretched as the emptied organ fills again.

In a few instances smooth muscles are structurally similar to skeletal muscles in their arrangement. Some of the structures supporting the uterus, for example, are called ligaments; however, they contain large amounts of smooth muscle and are capable of considerable shortening. The very small cutaneous muscles that erect the hairs (the **pilomotor** muscles) are also discrete structures whose shortening is basically unidirectional. Certain areas of mesenteries also contain regions of linearly oriented smooth muscle fibers.

Structure of Smooth Muscle Tissues ■ The most notable feature of smooth muscle tissue organization, in contrast to that of skeletal muscle, is the very small size of the cells as compared to the tissue they make up. Individual smooth muscle cells (depending somewhat on the type of tissue they compose) are 100–300 μm long and 5–10 μm in diameter. When isolated from the tissue, the cells are roughly cylindric along most of their length and taper at the ends. The single nucleus is elongated and centrally located. Electron microscopy reveals that the cell margins contain many areas of small membrane invaginations, called **caveoli** (Fig. 9–15), which may play a role in increasing the surface area of the cell. Mitochondria are located at the ends of the nucleus and near the surface membrane. In some smooth muscle cells the sarcoplasmic reticulum is rather abundant, although not to the extent found in skeletal muscle. In some cases it closely approaches the cell membrane, but there is no organized transverse tubular system as in other types of muscle.

The bulk of the cell interior is occupied by three types of myofilaments: thick, thin, and intermediate. The thin filaments are similar to those of skeletal muscle but lack the troponin protein complex. The length of the individual filaments is not known with certainty because of their rather irregular organization. The thick filaments are composed of myosin molecules, as in skeletal muscle, but the details of the exact arrangement of the individual molecules into filaments are not completely understood. The thick filaments appear to be approximately 2.2 μm long, somewhat longer than skeletal muscle thick filaments (1.6 μm). The **intermediate filaments** are so named because their diameter of 10 nm lies between that of the thick and thin filaments. Intermediate filaments appear to have a cytoskeletal, rather than a contractile, function. Prominent throughout the cytoplasm are small, dark-staining areas called **dense bodies.** They are associated with the thin and intermediate filaments and are considered analogous to the Z disks of skeletal muscle. Dense bodies associated with the cell margins are often called **dense bands.** They appear to serve as anchors for thin filaments and to transmit the force of contraction to adjacent cells.

Smooth muscle lacks the regular sarcomere structure of skeletal muscle. Studies have shown some association among dense bodies down the length of a cell and a tendency of thick filaments to show a degree of lateral grouping. However, it appears that the lack of a strongly periodic arrangement of the contractile apparatus is an adaptation of smooth muscle associated with its ability to function over a wide range of lengths and to develop high forces despite a smaller cellular myosin content.

Because smooth muscle cells are so small compared to the whole tissue, some mechanical and electrical communication among them is necessary. Individual cells are coupled mechanically in a number of ways. A proposed arrangement of the smooth muscle con-

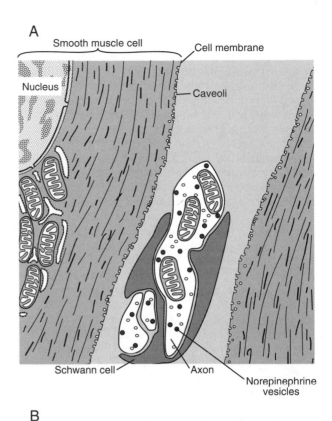

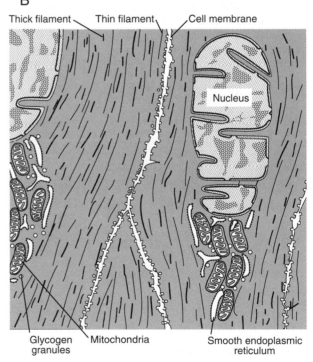

Figure 9–15 ■ Drawings from electron micrographs of smooth muscle. (A) Nerve-muscle contact between a bundle of autonomic axons and a smooth muscle cell. (B) Longitudinal section through the region of the nucleus, showing myofilaments and dense bodies. There is no specialized region of contact. (Modified from Rhoades RA, Pflanzer R (eds): *Human Physiology.* 2nd ed. Fort Worth, TX, Saunders, 1992.)

tractile and force transmission system is shown in Figure 9–16. This picture represents a consensus from many researchers and areas of investigation. Note that assemblies of myofilaments are anchored within the cell by the dense bodies and at the cell margins by the dense membrane patches. The contractile apparatus lies oblique to the long axis of the cell. When single isolated smooth muscle cells contract they undergo a "corkscrew" motion that is thought to reflect the off-axis orientation of the contractile filaments. In intact tissues the connections to adjacent cells prevent this rotation. Force appears to be transmitted from cell to cell and throughout the tissue in a number of ways. Many of the dense membrane patches are opposite one another in adjacent cells and could provide continuity of force transmission between the contractile apparatus in each cell. There are also areas of cell-to-cell contact, both lateral and end to end, where myofilament insertions are not apparent but where direct transmission of force could occur. In some places short strands of connective tissue link adjacent cells, while in other places cells are joined to the collagen and elastin fibers running throughout the tissue. These fibers, along with reticular connective tissue, comprise the **connective tissue matrix** or **stroma** found in all smooth muscle tissues. It serves to connect the cells and to give integrity to the whole tissue. In tissues that can resist considerable external force, this connective tissue matrix is very well developed and may be organized into **septa** that transmit the force of many cells.

Smooth muscle cells are also coupled electrically.

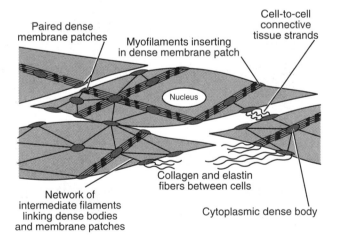

Figure 9–16 ■ Schematic view of the contractile system and cell-to-cell connections in smooth muscle. Note regions of association between thick and thin filaments that are anchored by the cytoplasmic and membrane dense bodies. A network of intermediate filaments provides some spatial organization (see, especially, left side). Several types of cell-to-cell mechanical connection are shown, including direct connections and connections to the extracellular connective tissue matrix. Structures are not necessarily drawn to scale; see text for further details.

The structure most effective in this coupling is the **gap junction** (or **nexus;** see Chap. 1). Gap junctions in smooth muscle appear to be somewhat transient structures that can form and disappear over time. In some tissues this phenomenon is under hormonal control; in the uterus, for example, gap junctions are rare during most of pregnancy, and the contractions of the muscle are weak and lack coordination. However, just prior to the onset of labor, the number and size of gap junctions increase dramatically, and the contractions become strong and well-coordinated. Shortly after the cessation of labor these gap junctions disappear and tissue function again becomes less coordinated.

The type of electrical coupling among smooth muscle cells is the basis for classifying smooth muscle into two major types. Tissues with little cell-to-cell communication that depend directly on nerve stimulation for their activation (e.g., skeletal muscle) are called **multiunit smooth muscles.** The iris of the eye is an example of a multiunit muscle. Other smooth muscle tissues have a high degree of coupling among cells, so that large regions of the tissue react as if they were a single cell. Such smooth muscle is called **unitary** or **single-unit smooth muscle,** and its cells form a **functional syncytium** (a situation in which many cells behave as one). This type of smooth muscle makes up the bulk of the muscle in the visceral organs.

■ Regulation and Control of Smooth Muscle Involve Many Factors

Smooth muscle is subject to a much more complex system of controls than is skeletal muscle. In addition to contraction in response to nerve stimulation, smooth muscle responds to hormonal and pharmacologic stimuli, the presence or lack of metabolites, cold, pressure, and stretch or touch and may be spontaneously active as well. This multiplicity of controlling factors is vital for the integration of smooth muscle into the overall body economy; while skeletal muscle achieves its primary control and accomplishes both coarse and delicate movements through interaction with the central nervous system and a relatively straightforward cellular control mechanism, the control of smooth muscle is much more closely related to the many factors involved in regulating the internal environment. It is not surprising, therefore, that many internal and external pathways have as their final effect the control of the interaction of smooth muscle contractile proteins.

Innervation of Smooth Muscle ■ Most smooth muscles have a nerve supply, usually from both branches of the autonomic nervous system. There is much diversity in this area; the muscle response to a given neurotransmitter substance depends on the type of tissue and its physiologic state. Smooth muscle does not contain the highly structured neuromuscular junctions found in skeletal muscle. **Autonomic nerve axons** run throughout the tissue; along the length of the axons are many swellings, or **varicosities,** which are the site of release of transmitter substances in response to nerve action potentials. Released molecules of excitatory or inhibitory transmitter diffuse from the nerve to the nearby smooth muscle cells, where they take effect. Since the cells are so small and numerous, relatively few are directly reached by the transmitters; those that are not stimulated by cell-to-cell communication, as described above. Neuromuscular transmission in smooth muscle is a relatively slow process, and in many tissues nerve stimulation serves mainly to modify (increase or decrease) spontaneous rhythmic mechanical activity.

Activation of Smooth Muscle Contraction ■ External factors that control the function of smooth muscle cells most often have their first influence at the cell membrane. These influences may be grouped into two major categories. The first type are agents that cause the opening or closing of cell membrane ion channels. The second type involves a second messenger (Chap. 1) that diffuses to the interior of the cell, where it causes further changes. The final common result of both mechanisms is a change in the intracellular concentration of calcium ions, which, in turn, controls the contractile process itself.

In contrast to the case in skeletal and cardiac muscle, the membrane potential of smooth muscle is subject to many external and internal influences; however, control of the cellular functions, particularly contraction, is less strongly influenced by the cell membrane.

The resting potential of most smooth muscles is approximately -50 mV. This is less negative than the resting potential of nerve and other muscle types, but it also is determined primarily by the transmembrane potassium ion gradient. The smaller potential is due primarily to a greater resting permeability to sodium ions. In many smooth muscles the resting potential varies periodically with time, producing a rhythmic potential change called a **slow wave** (Chap. 27). Action potentials in smooth muscle also have a variety of forms. In many smooth muscles the action potential is a transient depolarization event lasting approximately 50 msec. At times such action potentials will occur in rapid groups and produce repetitive membrane depolarizations that last for some time. These electrical events and related mechanical responses are illustrated in Figure 9–9. The relatively rapid **phasic** contractions (see below) are usually the result of one or more action potentials; the sustained **tonic** contraction is often only loosely related to the electrical activity of the membrane.

The ionic basis of smooth muscle action potentials is complex because of the great variety of tissues, physiologic conditions, and types of membrane channels. Since a resting membrane potential of -50 mV results in the inactivation of typical fast sodium channels, this ion is usually not the major carrier of inward current during the action potential. In most cases it has been shown that the rising (depolarizing) phase of a smooth muscle action potential is dominated by

calcium, which enters through voltage-gated membrane channels. Repolarization current is carried by potassium ions, which leave through several types of channels, some voltage-controlled and others sensitive to the internal calcium concentration. These general ionic properties are typical of most smooth muscle types, although specific tissues may have variations within this general framework. The most important common feature is the entry of calcium ions during the action potential, since this inward flux is an important source of the calcium that controls the contractile process. The mechanisms described above are called **electromechanical coupling,** since they involve changes in the membrane potential. In addition to voltage-gated calcium channels, smooth muscle also contains receptor-activated calcium channels that are opened by the binding of hormones or neurotransmitters. One such ligand-gated channel in arterial smooth muscle is controlled by ATP, which acts as a transmitter substance in some types of smooth muscle tissues.

Smooth muscle can also be activated via the generation of second messengers, such as inositol 1,4,5-trisphosphate (IP_3) (Chaps. 1, 33). This form of control is called **pharmacomechanical coupling,** since it involves chemical activators and does not depend on membrane depolarization. The IP_3 causes the release of calcium from the sarcoplasmic reticulum, which initiates contraction.

The Role of Calcium in Smooth Muscle Contraction ■ All of the processes described above are ultimately concerned with the control of muscle contraction via the pool of intracellular calcium. Figure 9–17 summarizes these mechanisms in an overall picture of calcium regulation in smooth muscle. These processes may be grouped into those concerned with **calcium entry,** intracellular **calcium liberation,** and **calcium exit** from the cell. Calcium enters the cell through several pathways, including voltage- and ligand-gated channels and a relatively small number of unregulated "**leak**" **channels** that permit the continual passive entry of small amounts of extracellular calcium. Within the cell, the major storage site of calcium is the sarcoplasmic reticulum; in some types of smooth muscle its capacity is quite small, and these tissues are strongly dependent on extracellular calcium for their function. Calcium is released from the sarcoplasmic reticulum by several mechanisms, including IP_3-induced release, a direct effect of cell depolarization or internal current flow on bound calcium, and **calcium-induced calcium release.** In this mechanism, calcium that has entered the cell via a membrane channel causes additional calcium release from the sarcoplasmic reticulum, amplifying its activating effect.

Studies in which internal calcium is continuously measured while the muscle is stimulated to contract typically reveal an initially high level of internal calcium; this activating burst most likely originates from internal (sarcoplasmic reticulum) storage. The level

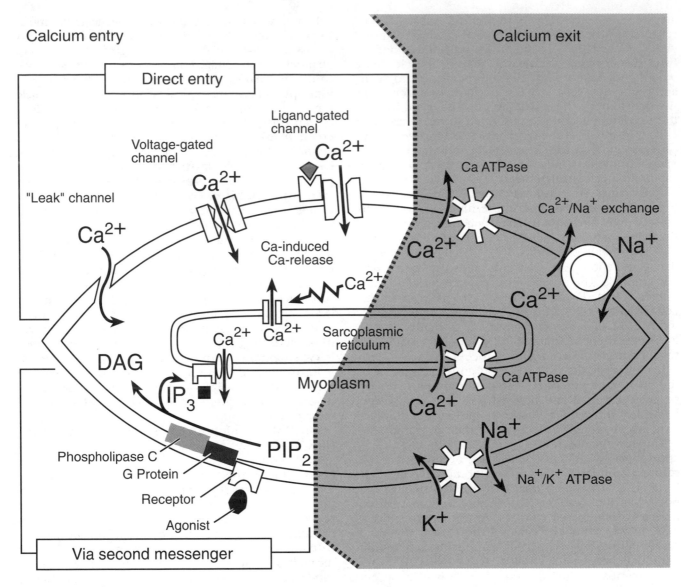

Figure 9–17 ■ Major routes of calcium entry and exit from the cytoplasm of smooth muscle. Arrows show the direction of ion movements. The ATPase reactions are energy-consuming ion pumps. The processes on the left side increase cytoplasmic calcium and promote contraction; those on the right decrease internal calcium and cause relaxation. PIP_2, phosphatidylinositol 4,5-bisphosphate; IP_3, myoinositol 1,4,5-trisphosphate; DAG, diacylglycerol. See text for details of the processes summarized here.

then decreases somewhat, although during the entire contraction it is maintained at a significantly elevated level. This sustained calcium level is the result of competition between mechanisms allowing calcium entry and those favoring its removal from the cytoplasm. Calcium leaves the myoplasm in two directions: a portion of it is returned to storage in the sarcoplasmic reticulum by an ATP-utilizing active transport system (a calcium-activated ATPase), and the rest is ejected from the cell by two principal means. The most important of these is another ATP-dependent active transport system located in the cell membrane. The second

mechanism, also located in the plasma membrane, is sodium-calcium exchange, a process in which the entry of three sodium ions is coupled to the extrusion of one calcium ion. This mechanism derives its energy from the large sodium gradient across the plasma membrane, thus it depends critically on the operation of the cell membrane Na^+/-K^+ ATPase. (The sodium-calcium exchange mechanism, relatively unimportant in smooth muscle, is of much greater consequence in cardiac muscle.)

Biochemical Control of Contraction and Relaxation ■ The contractile proteins of smooth muscle, like those of skeletal and cardiac muscle, are controlled by changes in the intracellular concentration of calcium ions. Likewise, the general features of the actin-myosin contraction system are similar in all muscle types. It is in the control of the contractile proteins themselves that important differences exist. Because in skeletal and cardiac muscle the control of contraction is associated with proteins of the thin filaments, it is called **actin-linked regulation.** The thin filaments of smooth muscle lack troponin; control of smooth muscle contraction relies instead on the thick filaments and is therefore called **myosin-linked regulation.** The fundamental mode of relaxation also differs: in actin-linked regulation the contractile system is in a constant state of **inhibited readiness,** and calcium ions remove the inhibition. In the myosin-linked regulation of smooth muscle, the role of calcium

is to cause **activation** of a resting state of the contractile system. The general outlines of this process are well understood and appear to apply to all types of smooth muscle, although a variety of secondary regulatory mechanisms are being found in different tissue types. This general scheme is shown in Figure 9–18.

When smooth muscle is at rest there is little cyclic interaction between the myosin and actin filaments, because of a special feature of the myosin molecules. As in skeletal muscle, the S2 portion of each myosin molecule (the paired "head" portion) contains four protein **light chains.** Two of these have a molecular weight of 16,000 and are called the **essential light chains;** their presence is necessary for actin-myosin interaction, but they do not appear to participate in the regulatory process. The other two light chains have a molecular weight of 20,000 and are called the

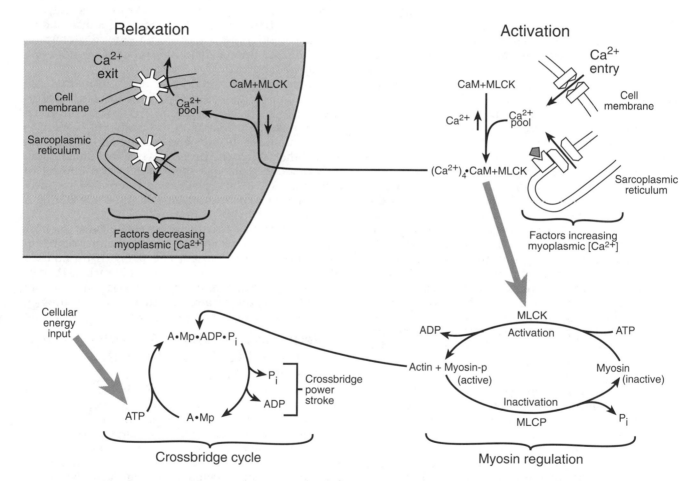

Figure 9–18 ■ Reaction pathways involved in the basic regulation of smooth muscle contraction and relaxation. Activation begins (upper right) when the cytoplasmic calcium levels are increased and calcium binds to calmodulin (CaM), activating the myosin light chain kinase (MLCK). The kinase (lower right) catalyzes the phosphorylation of myosin, changing it to an active form (Mp). The phosphorylated myosin can then participate in a mechanical crossbridge cycle (lower left) much like that in skeletal muscle, though much slower. When calcium levels are reduced (upper left), calcium leaves calmodulin, the kinase is inactivated, and the myosin light chain phosphatase (MLCP) dephosphorylates the myosin, making it inactive. The crossbridge cycle stops, and the muscle relaxes.

regulatory light chains; their role in smooth muscle is critical. These chains contain specific locations (amino acid residues) to which the terminal phosphate group of an ATP molecule can be attached in **phosphorylation;** the enzyme responsible for promoting this reaction is **myosin light chain kinase** (MLCK). When the regulatory light chains are phosphorylated, the myosin heads can interact in a cyclic fashion with actin, and the reactions of the **crossbridge cycle** (and its mechanical events) take place much as in skeletal muscle. It is important to note that the ATP molecule that donates its phosphate group to the phosphorylation of a myosin light chain is separate and distinct from the one consumed as an energy source by the mechanochemical reactions of the crossbridge cycle.

For myosin phosphorylation to occur, the MLCK itself must be activated, and this step is also subject to control. Closely associated with the MLCK is **calmodulin** (CaM), a smaller, calcium-binding protein. When four calcium ions are bound to a calmodulin molecule, it activates its associated MLCK and light chain phosphorylation can proceed. It is this MLCK-activating step that is sensitive to the cytoplasmic calcium concentration; at levels below 10^{-7} mol/L no calcium is bound to calmodulin and no contraction can take place. When cytoplasmic calcium concentration is greater than 10^{-4} mol/L, the binding sites on calmodulin are fully occupied, light chain phosphorylation proceeds at maximal rate, and contraction occurs. Between these extreme limits, variations in the internal calcium concentration can cause corresponding gradations in the contractile force. Such modulation of smooth muscle contraction is essential for its regulatory functions, especially in the vascular system.

The biochemical processes controlling relaxation in smooth muscle also differ from those in skeletal and cardiac muscle, in which a state of inhibition returns as calcium ions are withdrawn from being bound to troponin. In smooth muscle the phosphorylation of myosin is reversed by the **phosphatase** enzyme **myosin light chain phosphatase** (MLCP). The activity of this phosphatase appears to be unregulated; that is, it always functions, even while the muscle is contracting. During contraction, however, MLCK-catalyzed phosphorylation proceeds at a significantly higher rate, and phosphorylated myosin predominates. When the cytoplasmic calcium concentration falls, MLCK activity is reduced because the calcium dissociates from the calmodulin, and myosin dephosphorylation (catalyzed by the phosphatase) predominates. Because dephosphorylated myosin has a low affinity for actin, the reactions of the crossbridge cycle can no longer take place. Relaxation is thus brought about by mechanisms that lower cytoplasmic calcium concentrations or decrease MLCK activity. One such relaxing effect is mediated by phosphorylation of the MLCK enzyme itself; MLCK activity is decreased (by lowering its affinity for calmodulin) when it is phosphorylated by a cAMP-dependent protein kinase that is activated by the binding of a specific substance (e.g., norepinephrine) to a membrane receptor. Cyclic AMP has also been shown to increase the rate of calcium uptake by the sarcoplasmic reticulum pumping mechanism, a process that promotes relaxation.

In addition to myosin phosphorylation to control smooth muscle activation, secondary regulatory mechanisms are present in some smooth muscle. One of these provides long-term regulation of contraction in some tissues after the initial calcium-dependent myosin phosphorylation has activated the contractile system. For example, in vascular smooth muscle the force of contraction may be maintained for long periods. This extended maintenance of force capability, called the "**latch state,**" appears to be related to a reduction in the cycling rate of crossbridges (possibly related to reduced phosphorylation) so that each remains attached for a longer portion of its total cycle. Even during the latch state increased cytoplasmic calcium appears to be necessary for force to be maintained. Not all smooth muscle tissues can enter a latch state, however, and the details of the process are not completely understood.

Another possible secondary mechanism in some smooth muscle tissues involves the protein **caldesmon.** This molecule, also sensitive to the concentration of cytoplasmic calcium, is capable of binding to myosin at one of its ends and to actin and calmodulin at the other. While the process is not well understood, it is possible that caldesmon, under the control of calcium, could form crosslinks between actin and myosin filaments and thus aid in bearing force during a long-maintained contraction.

Other secondary regulatory mechanisms, such as direct calcium-controlled phosphorylation of myosin itself, have been proposed. It is likely that a number of such secondary mechanisms exist in various tissues, but the calcium-dependent phosphorylation of myosin light chains is the primary event in the activation of smooth muscle contraction.

■ Mechanical Activity in Smooth Muscle Is Adapted for Its Specialized Physiologic Roles

The contraction of smooth muscle is much slower than that of skeletal or cardiac muscle; it can maintain contraction far longer and relaxes much more slowly. The source of these differences lies largely in the chemistry of the interaction between actin and myosin of smooth muscle. Recall that the crossbridges of muscle form an actin-myosin enzyme system (actomyosin ATPase) that releases energy from ATP so that it may be converted into a mechanical contraction (i.e., tension or shortening). The inherent rate of this ATPase correlates strongly with the velocity of shortening of the intact muscle. Most smooth muscles require several seconds (or even minutes) to develop maximal isometric force. A smooth muscle that contracts 100 times more slowly than a skeletal muscle will have an actomyosin ATPase that is 100 times as slow. The major source of this difference in rates is due to the myosin molecules; the actin found in smooth and skeletal muscles is rather similar. There is a close association in smooth muscle between maximal shortening velocity and degree of myosin light chain phosphorylation.

A very high economy of tension maintenance, typically 300–500 times greater than that in skeletal muscle, is vital to the physiologic function of smooth muscle. *Economy,* as used here, means the amount of metabolic energy input compared to the tension produced; in smooth muscle there is a direct relationship between the isometric tension and consumption of ATP. The economy is related to the basic cycling rate of the crossbridges: early in a contraction (while tension is being developed and the crossbridges are cycling more rapidly), the energy consumption is about four times as high as in the later steady-state phase of the contraction. Compared with skeletal muscle, the crossbridge cycle in smooth muscle is hundreds of times slower, and much more time is spent with the crossbridges in the attached phase of the cycle.

The cycling crossbridges are not the only energy-utilizing system in smooth muscle. Since the cells are so small and numerous, smooth muscle tissue contains a very large cell membrane area. Maintenance of the proper ionic concentrations inside the cells requires the activity of the membrane-based ion pumps for sodium/potassium and calcium, and this ion pumping requires a significant portion of the cell's energy supply. Internal pumping of calcium ions into the sarcoplasmic reticulum during relaxation also requires energy, and the processes that result in phosphorylation of the myosin light chains consume a further portion of the cellular energy, as do the other processes of cellular maintenance and repair. Smooth muscle contains both glycolytic and oxidative metabolic pathways, with the oxidative pathway usually the most important; under some conditions a transition may temporarily be made from oxidative to glycolytic metabolism. In terms of the entire body economy the energy requirements of smooth muscle are small compared with those of skeletal muscle, but the critical regulatory functions of smooth muscle require that its energy supply not be interrupted.

Modes of Contraction ■ Smooth muscle contractile activity cannot be divided clearly into twitch and tetanus, as in skeletal muscle. In some cases smooth muscle makes rather rapid contractions, followed by complete relaxation; this is termed **phasic** activity (Fig. 9–19). In other cases smooth muscle can maintain a low level of active tension for long periods without cyclic contraction and relaxation; this maintained contraction is called **tonus** (rather than tetanus) and is typical of smooth muscle activated by hormonal, pharmacologic, or metabolic factors, whereas phasic activity is more closely associated with stimulation by neural activity.

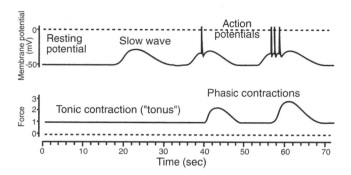

Figure 9–19 ■ Electrical and mechanical activity in smooth muscle. Tonic contractions (left) are often independent of membrane action potential activity. Note, however, the association between contraction and the rapid changes in membrane potential; these action potentials correspond to the opening and closing of voltage-gated ion channels.

The force-velocity curve for smooth muscle reflects the differences in crossbridge functions described previously. Although smooth muscle contains one-third to one-fifth as much myosin as skeletal muscle, the longer smooth muscle myofilaments and the slower crossbridge cycling rate allow it to produce as much force per unit of cross-sectional area as does skeletal muscle. Thus the maximum values for smooth muscle on the force axis would be similar, while the maximum (and intermediate) velocity values would be very different (Fig. 9–20). Furthermore, smooth muscle can have a set of force-velocity curves, each corresponding to a different level of myosin light chain phosphorylation.

Other mechanical properties of smooth muscle are also related to its physiologic roles. While its underlying cellular basis is uncertain, smooth muscle has a length-tension curve somewhat similar to that of skeletal muscle, although there are some significant differences (Fig. 9–20). At lengths at which the maximal isometric force is developed, many smooth muscles bear a substantial passive force. This is due mostly to the network of connective tissue that supports the smooth muscle cells and resists overextension; in some cases it may be partly due to residual interaction between actin and attached but noncycling myosin crossbridges. In comparison with skeletal and cardiac muscle, smooth muscle can function over a significantly greater range of lengths. It is not constrained by skeletal attachments, and it makes up a number of organs that vary greatly in volume during the course of their normal functioning. The shape of the length-tension curve can also vary with time and with the degree of distention. For example, when the urinary bladder is highly distended by its contents, the peak of the active length-tension curve can be displaced to longer muscle lengths. This means that as the muscle shortens to expel the organ's contents, it can reach

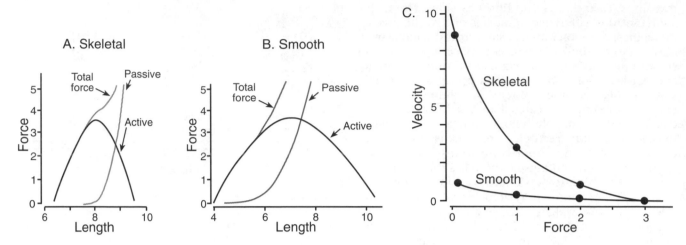

Figure 9–20 ■ Comparison between smooth and skeletal muscle mechanical characteristics. (A, B) Typical length-tension curves from skeletal and smooth muscle. Note the greater range of operating lengths for smooth muscle and the leftward shift of the passive (resting) tension curve. (C) Skeletal and smooth muscle force-velocity curves compared. While the peak forces may be similar, the maximum shortening velocity of smooth muscle is typically 100 times lower than that of skeletal muscle. Force and length units are arbitrary.

lengths at which it can no longer exert active force. After a period of recovery at this shorter length, the muscle can again exert sufficient force to expel the contents.

These reversible changes in the length-tension relationship are the result of **stress relaxation,** a property of so-called **viscoelastic** materials, of which smooth muscle has characteristics. When a viscoelastic material is stretched to a new length, it responds initially with a significant increase in force; this is an **elastic** response, and it is followed by a decline in force that initially is rapid and then continuously slows until a new steady force is reached. If a viscoelastic material is subjected to a constant force, it will elongate slowly until it reaches a new length. This phenomenon, the complement of stress relaxation, is called **creep.** In smooth muscle organs the abundant connective tissue prevents overextension. The viscoelastic properties of smooth muscle allow it to function well as a reservoir for fluids or other materials; if an organ is filled slowly, stress relaxation allows the internal pressure to adjust gradually, so that it rises much less than if the final volume had been introduced rapidly. This is illustrated in Figure 9–21 for the case of a hollow smooth muscle organ subjected to both rapid and slow infusions of liquid (since this is a hollow structure, internal **pressure** and **volume** are directly related to the **force** and **length** of the muscle fibers in the walls). The upper dashed lines in the upper panels denote the pressure that would result if the material were simply elastic rather than having the additional property of viscosity. Some of the viscoelasticity of smooth muscle is a property of the extracellular connective tissue and other substances, such as the hyaluronic acid gel, that are

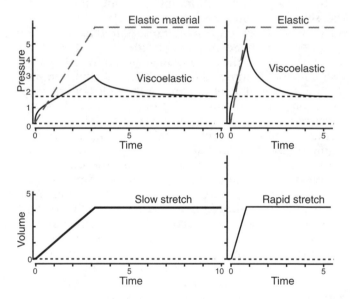

Figure 9–21 ■ Behavior of a viscoelastic material (e.g., the walls of a hollow smooth muscle-containing organ) subjected to slow (left) and rapid (right) elongation. The increase in force is proportional to the rate of extension, and at the end of the stretch the force decays exponentially to a steady level. A purely elastic material (dashed line) maintains its force without stress relaxation.

present between the cells, and some is inherent in the smooth muscle cells, probably due to the presence of noncycling crossbridges in resting tissue. One important feature of smooth muscle viscoelasticity is the tissue's ability to return to its original state following extreme extension; this is due to the tonic contractile activity present in most smooth muscles under normal physiologic conditions.

Relaxation ■ Returning from the contracted state is a complex process in smooth muscle. The central cause of relaxation is a reduction in the internal (cytoplasmic) calcium concentration, a process that is itself the result of several mechanisms. Electrical repolarization of the plasma membrane leads to a decrease in the influx of calcium ions, while the plasma membrane calcium pump and the sodium-calcium exchange mechanism (to a lesser extent) actively promote calcium efflux. Most important quantitatively is the uptake of calcium back into the sarcoplasmic reticulum. The net result of lowering the calcium concentration is to cause a reduction in the MLCK so that dephosphorylation of myosin can predominate over phosphorylation. Both the increase in calcium uptake and the MLCK activity may be subject to an additional important control mechanism, that of **β-adrenergic relaxation.** In many vascular muscles relaxation occurs in response to the presence of the hormone **norepinephrine.** Binding of this substance to cell membrane receptors causes the activation of **adenylate cyclase** and formation of cAMP (Chap. 1). Increased intracellular cAMP concentration is an effective promoter of relaxation in at least two major ways. The activity of the enzyme **cAMP-dependent protein kinase** increases as the concentration of cAMP rises. This enzyme (and perhaps also cAMP acting directly) enhances calcium uptake by the sarcoplasmic reticulum, and this results in a further lowering of the cytoplasmic calcium. At the same time, phosphorylation of MLCK itself (by the action of cAMP-dependent protein kinase) reduces its catalytic effectiveness, and myosin light chain phosphorylation is decreased just as if intracellular calcium had been lowered. Since many vascular muscles are continuously in a state of partial contraction, β-adrenergic relaxation is a physiologically important process in the adjustment of blood flow and pressure.

Relaxation is obviously also a mechanical process. Contractile force decreases as crossbridges detach, and myofilaments become free again to slide past one another. Since most smooth muscle activity involves at least some shortening, relaxation must require elongation. As with other types of muscle, an external force must be applied for lengthening to occur. In the intestine, for example, material being propelled into a recently contracted region provides the extending force. Smooth muscle relaxation (or its lack) may have important indirect consequences. Elevation of blood pressure, for example, can be caused by failure of smooth relaxation. In the uterus during labor, adequate relaxation between contractions is essential for the well-being of the fetus. During the contractions of labor the muscular walls of the uterus become quite rigid and tend to compress the blood vessels that run through them. As a result, blood flow to the fetus is restricted during uterine contractions, and failure of the muscle to relax adequately between contractions can result in fetal distress.

■ Smooth Muscle Can Adapt to Changing Conditions

A number of external influences, some not well understood, affect the growth and functional adaptation of smooth muscle. Some of these changes are vital for normal body function, while others can be part of a disease process.

Hormonally Induced Hypertrophy ■ The uterus and associated tissues are under the influence of the female sex hormones (Chap. 40, 41). During pregnancy, high levels of progesterone, later followed by high estrogen levels, promote significant changes in uterine growth and control. The mass of the muscle layers (the **myometrium**) increases as much as 70-fold, primarily through an increase in muscle cell size (**hypertrophy**) associated with a large increase in contractile protein content and associated regulatory proteins. The distention caused by the growing fetus also promotes hypertrophy. Extracellular connective tissue also increases. There is an increase in the number of cells as well, a condition called **hyperplasia.** Throughout most of pregnancy the cells are poorly coupled electrically and contractile activity is not well coordinated. As pregnancy nears term, the large increase in the number of gap junctions permits coordinated contractions that culminate in the birth process. Following delivery and the consequent hormonal and mechanical changes, the processes leading to hypertrophy are reversed and the muscle reverts to its nonpregnant state.

Other Forms of Hypertrophy ■ Chronic obstruction of hollow smooth muscle organs (e.g., the urinary bladder, intestine, portal vein) produces a chronically elevated internal pressure. This acts as a stimulus for smooth muscle hypertrophy, although the cellular mechanisms involved are not well understood. In addition to structural changes there may be alterations of the metabolic activities, contractile properties, and response to agonists. Hyperplasia is also present to some degree in these muscle adaptations, but its relative contribution is difficult to ascertain experimentally. Nonmuscular components of the organ wall (e.g., connective tissue) are also increased. These changes, especially those involving the muscle cells themselves, usually revert to near normal when the mechanical cause of the hypertrophy is removed.

Vascular smooth muscle, especially of the arteries, is also subject to hypertrophy (and hyperplasia) when it encounters a sustained pressure overload. This is an important factor in **hypertension,** or high blood pressure. An increase in blood pressure, perhaps due to chronically elevated sympathetic nervous system activity, may be present before smooth muscle hypertrophy occurs. The increase in the smooth muscle layer is a response to this stimulus, and there may be a trophic effect of the sympathetic nervous system activity as well. The resulting thickening of the vascular wall further reduces the lumen diameter and aggravates the hypertension. Lowering of blood pressure by therapeutic means can result in a return of the vessel walls to a near-normal state. Hypertension in the pulmonary vasculature is also associated with increased smooth muscle growth and with development of smooth muscle cells in areas of the arterial system that do not normally have smooth muscle in their walls.

Under some circumstances smooth muscle cells can lose most of their contractile function and become synthesizers of collagen and accumulators of low-density lipoproteins. The loss of contractile activity is accompanied by a significant loss in the number of myofilaments. Such a **phenotypic transformation** takes place, for example, in the formation of atherosclerotic lesions in artery linings. While the factors involved in initiating and sustaining this reversible transition are not well understood, they appear to involve growth-promoting substances released from platelets following endothelial injury, while circulating heparin-like substances block the transformation.

■ ■ ■ ■ ■ ■ ■ ■ ■ ■ ■ REVIEW EXERCISES ■ ■ ■ ■ ■ ■ ■ ■ ■ ■ ■

Identify Each with a Term

1. The complex structure responsible for signal transmission from nerve to skeletal muscle
2. The postsynaptic electrical response to neuromuscular transmitter action
3. Substances that block neuromuscular transmission by leaving postsynaptic membrane channels in an open state
4. The single mechanical response of a skeletal muscle to a single action potential
5. A motor axon, together with all of the skeletal muscle fibers it innervates
6. A muscle contraction in which force is kept constant
7. A graphic representation of the force capability of a muscle at various lengths
8. The reduced capability of a muscle following a long period of activity
9. A smooth muscle structure that restricts or prevents flow in tubular organs
10. The structure that allows close electrical communication between smooth muscle cells
11. The enzyme responsible for phosphorylation of regulatory light chains of smooth muscle myosin
12. The gradual decrease in force when smooth muscle is held at an extended length

Define Each Term

13. Curare
14. Tetanus
15. Isometric contraction
16. Force-velocity curve
17. Myasthenia gravis
18. Myosin light chain phosphatase
19. Tonus

Choose the Correct Answer

20. The action of acetylcholine on the postsynaptic membrane of the neuromuscular junction is to:
    a. Open chemically gated channels to chloride ions
    b. Open chemically gated channels to Na^+ and K^+ ions
    c. Cause the opening of voltage-gated Na^+ channels
    d. Cause the opening of voltage-gated K^+ channels
21. A single stimulus applied to a skeletal muscle while it is relaxing from a twitch:
    a. Will produce an action potential that will have no effect on the muscle force
    b. Will produce a second twitch whose force adds to the first
    c. Will cause a more rapid relaxation
    d. Will slow the relaxation, although no additional force will be produced
22. The smallest unit of a normally innervated skeletal muscle that can be activated by a single nerve stimulus is:
    a. A single sarcomere
    b. A small group of sarcomeres close to the neuromuscular junction
    c. A single cell, along its entire length
    d. All of the cells comprising a single motor unit
23. According to the relationships expressed by the force-velocity curve:
    a. The maximal rate of shortening can occur over a wide range of forces
    b. The highest rates of shortening occur at the lowest forces
    c. Isometric contraction is not possible
    d. There are at least two different forces at which the velocity of shortening will be the same
24. A skeletal muscle fiber that contains a large number of mitochondria and has a high myoglobin content is likely to:
    a. Fatigue rapidly
    b. Have a high velocity of shortening
    c. Be capable of sustained activity
    d. Have a predominantly anaerobic metabolism
25. An important early step in the regulation of a smooth muscle contraction is:
    a. Binding of calcium ions to calmodulin
    b. Dephosphorylation of myosin light chains by phosphatase
    c. Crossbridge interaction with myosin
    d. Calcium uptake by the sarcoplasmic reticulum

Briefly Answer

26. Why does the peak power output of a skeletal muscle not continue to increase as the force increases?

27. How does the force-velocity characteristic of a skeletal muscle set an upper limit to its performance?

28. In cellular terms, what is the significance of the peak of the isometric length-tension curve?

29. Distinguish between electromechanical and pharmacomechanical coupling in smooth muscle activation.

30. What is the role of myosin light chain phosphorylation in smooth muscle contraction?

31. What are some consequences of the high metabolic economy of smooth muscle contraction?

32. How does myosin-linked regulation of muscle contraction contrast with actin-linked regulation?

33. What is the likely role of inositol trisphosphate in the regulation of smooth muscle contraction?

■ ■ ■ ■ ■ ■ ■ ■ ■ ■ ■ ■ ■ ■ ■ ANSWERS ■ ■ ■ ■ ■ ■ ■ ■ ■ ■ ■ ■ ■ ■ ■

1. Neuromuscular junction, or motor endplate

2. Endplate potential

3. Depolarizing blockers

4. Twitch

5. Motor unit

6. Isotonic contraction

7. Length-tension curve

8. Fatigue

9. Sphincter

10. Gap junction or nexus

11. Myosin light chain kinase (MLCK)

12. Stress relaxation

13. A substance that competes with acetylcholine for postsynaptic receptors at the neuromuscular junction and thereby blocks neuromuscular transmission

14. A sustained contraction resulting from the rapid restimulation of a skeletal muscle

15. A contraction in which the length of the muscle does not change

16. A diagram expressing the inverse relationship between the shortening velocity and the force of contraction in muscle

17. An autoimmune disease of the neuromuscular junction that results in a reduction of the number of postsynaptic ACh receptors and hence impairment of neuromuscular transmission

18. The enzyme that catalyzes the dephosphorylation of the light chains of smooth muscle myosin and thus allows relaxation

19. A sustained contraction of smooth muscle that may occur without continuous nerve stimulation

20. b

21. b

22. d

23. b

24. c

25. a

26. As force increases, the speed of shortening decreases, with the net effect that the rate of doing work (the power output) decreases, until under isometric conditions no external work is done.

27. At any given level of force, there is a maximum possible velocity of shortening that cannot be exceeded. For the muscle to shorten more rapidly, the force must be decreased.

28. The overlap between the thick and thin filaments is optimal, and the number of force-producing crossbridges that can be formed is maximal.

29. Electromechanical coupling involves membrane potential changes that allow Ca^{2+} ions to enter the cell from extracellular fluid or be released internally. In pharmacomechanical coupling, agonist chemicals act through specific receptors to cause the internal production of second messengers, such as inositol trisphosphate, which in turn cause Ca^{2+} release.

30. Phosphorylation of the myosin light chains activates the interaction between myosin and actin that breaks down ATP and causes contraction.

31. High economy allows long-maintained contractions with low expenditure of energy, but the contractions are very slow, because of the very low rate of crossbridge cycling.

32. In actin-linked regulation, the control molecule (troponin) is on the thin filaments, and activation by Ca^{2+} removes the inhibition of the actin-myosin interaction. In myosin-linked regulation, Ca^{2+} binding causes activation of the resting (nonphosphorylated) myosin and thus enables actin-myosin interaction.

33. IP_3 acts as a "second messenger" between the binding of an agonist to a membrane receptor and the internal release of Ca^{2+} from the sarcoplasmic reticulum.

Suggested Reading

Cole WC, Garfield RE: Ultrastructure of the myometrium, in Wynn RM, Jollie WP (eds): *Biology of the Uterus.* 2nd ed. New York, Plenum, 1989.

Gabella G: Structure of intestinal musculature, in Schultz SG, Wood JD (eds): *Handbook of Physiology: The Gastrointestinal System I.* Bethesda, MD, American Physiological Society, 1989.

Hall ZW, Sanes JR: Synaptic structure and development: The neuromuscular junction. *Cell* 72 (suppl): 99–121, 1993.

Huszar G, Walsh MP: Biochemistry of the myometrium and cervix, in Wynn RM, Jollie WP (eds): *Biology of the Uterus*. 2nd ed. New York, Plenum, 1989.

Jankovic J, Brin MF: Therapeutic uses of botulinum toxin. *N Engl J Med* 324:1186–1194, 1991.

Kao CY: Electrophysiological properties of uterine smooth muscle, in Wynn RM, Jollie WP (eds): *Biology of the Uterus*. 2nd ed. New York, Plenum, 1989.

Meiss RA: Mechanical properties of gastrointestinal smooth muscle, in Schultz SG, Wood JD (eds): *Handbook of Physiology: The Gastrointestinal System I*. Bethesda, MD, American Physiological Society, 1989.

Rall JA: Energetic aspects of skeletal muscle contraction: Implications of fiber types, in Terjung RL (ed): *Exercise and Sport Science Reviews, 13*. New York, Macmillan, 1985.

Rüegg JC: *Calcium in Muscle Contraction*. New York, Springer-Verlag, 1992.

Shephard RJ: *Physiology and Biochemistry of Exercise*. New York, Praeger, 1985.

Stein RB: *Nerve and Muscle: Membranes, Cells, and Systems*. New York, Plenum, 1980.

Woledge RC, Curtin NA, Homsher E: *Energetic Aspects of Muscle Contraction*. New York, Academic, 1985.

Cardiac Muscle

OBJECTIVES

After studying this chapter, the student should be able to:
1. Describe the structure of cardiac muscle cells, comparing and contrasting it with that of smooth and skeletal muscle cells
2. Diagram the relationship between the action potential and a twitch in cardiac muscle and show how this prevents a tetanic contraction
3. List the possible energy sources for cardiac muscle contraction
4. Diagram the length-tension curve for cardiac muscle, showing the active and passive relationships, and indicate the range over which the muscle performs its physiologic function
5. On the length-tension diagram, indicate the pathway for an isotonic contraction of cardiac muscle and show how an increase in contractility changes the relationship between afterload and amount of shortening
6. List some possible inotropic interventions that could change cardiac contractility
7. Diagram the relationships between the afterload and the extent and speed of shortening of cardiac muscle
8. List the means and sites of calcium entry, storage, and exit in cardiac muscle cells

The muscle mass of the heart, the **myocardium,** shares characteristics of both smooth and skeletal muscle. The tissue is striated in appearance, like skeletal muscle, and the structural characteristics of the sarcomeres and myofilaments are much like those of skeletal muscle. The regulation of contraction, involving calcium control of an actin-linked troponin-tropomyosin system, is also quite like that of skeletal muscle. On the other hand, cardiac muscle is composed of many small cells, as is smooth muscle, and electrical and mechanical cell-to-cell communication is an essential feature of cardiac muscle structure and function. The mechanical properties of cardiac muscle relate more closely to those of skeletal muscle, although the mechanical performance of cardiac muscle is considerably more complex and subtle than that of skeletal muscle.

Anatomic Specializations of Cardiac Muscle

The heart is composed of several varieties of cardiac muscle tissue. The **atrial** and **ventricular** muscle masses, so named for their anatomic location, are rather similar structurally, although the electrical properties of these two areas differ significantly. The **conducting tissues** (e.g., Purkinje fibers) of the heart have a communicating function like that of nerve tissue, but actually they consist of muscle tissue that is highly adapted for the rapid and efficient conduction of action potentials, and their contractile ability is greatly reduced. Finally, there are the highly specialized tissues of the **sinoatrial** and the **atrioventricular nodes,** muscle tissue that is greatly modified into structures concerned with the initiation and conduction of the heartbeat. The discussions that follow refer primarily to the ventricular myocardium, the tissue that makes up the greatest bulk of the muscle of the heart.

■ Cardiac Muscle Cells Are Structurally Distinct from Skeletal Muscle Cells

The small size of cardiac muscle cells is one of the critical aspects in determining the function of heart muscle. The cells are approximately 10–15 μm in diameter and around 50 μm long. They do not taper at their ends, but join firmly to each other at the **intercalated disk** (Fig. 10–1). Many of the cells are **branched** and are thus attached to more than two other cells, an arrangement that aids in the lateral spread of electrical activity. Cardiac muscle cells typically have a single, centrally located **nucleus,** although some cells may contain more than one. The **cell membrane** and associated fine connective tissue structures form the **sarcolemma,** as in skeletal muscle. The sarcolemma of cardiac muscle supports the resting and action potentials and is the location of ion pumps and ion exchange mechanisms vital to cell function. Just inside the sarcolemma are regions where significant

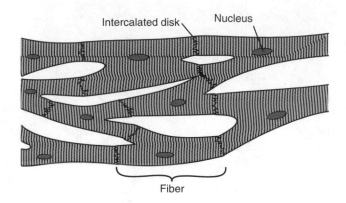

Figure 10–1 ■ The cellular structure of cardiac muscle.

amounts of calcium ions may be bound and kept from general access to the cytoplasm. This bound calcium can exchange rapidly with the extracellular space and can be rapidly freed from its binding sites by the passage of an action potential.

As in skeletal muscle, there is a **transverse tubular system** (T system), but both it and the **sarcoplasmic reticulum** are less extensive in cardiac muscle, together constituting less than 2% of the cell volume. This correlates with the small cell size and consequent reduction in diffusion distances between the cell surface membrane and contractile proteins. In cardiac muscle cells the T tubules enter the cells at the level of the Z lines, and in many cases the link between a T tubule and the sarcoplasmic reticulum is not a triad, as in skeletal muscle, but rather is a **dyad,** composed of the T tubule and the terminal cisterna of the sarcoplasmic reticulum of only one sarcomere. The small size of the sarcoplasmic reticulum also limits its calcium storage capacity, and the other source of calcium entry and exit, the sarcolemma, has an important role in the excitation-contraction coupling process in cardiac muscle.

The **sarcomeres** appear essentially like skeletal muscle, with similar A and I bands, Z disks (or lines), and M lines. The myofilaments make up almost half of the cell volume, while numerous mitochondria comprise another 30–40%; the high mitochondrial content reflects the highly aerobic nature of cardiac muscle function. The rest of the cell volume, around 15%, consists of cytosol containing numerous enzymes and metabolic products and substrates. The myofilaments are bathed in the cytosol, while the various calcium entry and exit mechanisms function to regulate the cytosolic calcium concentration and thus to regulate contractile activity.

■ Cardiac Muscle Cells Are Linked in a Functional Syncytium

Cardiac muscle tissue is a highly branched network of cells (also called **cardiac myocytes**) joined together at their intercalated disks. These structures have a dual function: careful electron-microscopic study reveals that in the region of the intercalated disk each cell sends processes deep into its neighboring cell to form an interdigitating junction with a very large surface area. **Gap junctions** in the intercalated disks function like those of smooth muscle to allow close electrical communication between cells. Also plentiful in the intercalated disk region are **desmosomes,** areas where there is a firm mechanical connection between cells. This connection, rather than an extensive extracellular connective tissue matrix as in smooth muscle, allows transmission of force from cell to cell. The intercalated disk thus allows cardiac muscle to form a **functional syncytium,** with cells acting in concert both mechanically and electrically.

The stimulus for cardiac muscle contraction arises entirely within the heart and is not dependent on its nerve supply (Chap. 13). Conduction of the action potentials is solely a function of the muscle tissue. This impulse propagation is aided by the branched nature of the cells, the intercalated disks, and specialized conductive tissue, such as the **Purkinje fibers.** These are strands of cardiac muscle cells, nervelike in external appearance, that are specialized for electrical conduction. Their contractile protein is reduced to about 20% of the cell volume and their size is increased to optimize their electrical characteristics for rapid action potential conduction. The important innervation of cardiac muscle comes from both branches of the autonomic nervous system, and these connections allow for external regulation of the heart rate and strength of contraction and provide some degree of sensory feedback.

■ Cell and Tissue Structure Allow and Require Unique Adaptations

As a result of the small size of cardiac muscle cells, the communication system described above and in Chapter 13 is necessary for organized function. The small cell size also makes each cell more critically dependent on the external environment, and cardiac function may be greatly altered by electrolyte and metabolic imbalances arising elsewhere in the body. Hormonal messengers, such as norepinephrine, also have quick access to cardiac muscle cells.

From a mechanical standpoint, the lack of skeletal attachments means cardiac muscle can function over a wide range of lengths. While the length-tension property is not of major importance in the functioning of many skeletal muscles, in cardiac muscle it forms the basis of the remarkable capacity of the heart to adjust to a wide range of physiologic conditions and requirements.

Physiologic Specializations

While cardiac muscle is essentially a striated muscle, its unique physiologic role is associated with a number of adaptations that make its function rather different from that of skeletal muscle.

■ Specialized Electrical and Metabolic Properties Control Cardiac Muscle Contraction

A more general treatment of the electrical properties of cardiac muscle is given in Chapter 13. The discussion here focuses on those electrical properties that are most closely related to controlling the mechanical function of the muscle.

The Cardiac Action Potentials ■ As in other types of muscle and nerve, the muscle cells of the heart have an excitable and selectively permeable cell membrane that is responsible for both resting potentials and action potentials. These electrical phenomena are the result of ionic concentration differences and a number of ion-selective membrane channels, some of which are voltage- and time-dependent. In cardiac muscle, however, the membrane events are more diverse and complex than in skeletal muscles and are much more closely linked to the actual details of the mechanical contraction. The closer association of electrical and mechanical events is one key to the inherent properties of cardiac muscle that suit it to its role in an organ that is largely self-regulating.

Figure 10–2 illustrates some features of the cardiac muscle action potential that pertain directly to myocar-

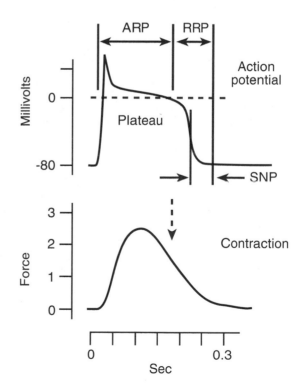

Figure 10–2 ■ A cardiac muscle action potential and isometric twitch. The duration of the action potential is such that an effective tetanic contraction cannot be produced, although a partial contraction can be elicited late in the twitch. ARP, absolute refractory period; RRP, relative refractory period; SNP, period of supranormal excitability.

dial function. Note that the duration of the action potential is quite long; in fact, it lasts nearly as long as the muscle contraction. One consequence of this is that the absolute and relative refractory periods of the membrane activity are likewise extended, and the muscle cannot be restimulated during any but the latest part of the contraction. During the repolarization phase of the action potential there is a brief period in which the muscle actually shows an increased sensitivity to stimulation. Action potentials that may be accidentally produced during this period of **supranormal excitability,** which is due to a lowered potassium conductance that persists late in the action potential (Chap. 13), are reduced in amplitude and duration and give rise to only small contractions. The physiologic significance of this period of increased excitability is that it can lead to unwanted and untimely propagation of action potentials that can seriously interfere with the normal rhythm of the heart. In general, however, the long-lasting refractoriness of the cell membrane effectively prevents the development of a tetanic contraction; any failure of cardiac muscle to relax fully after every stimulus would make it quite unsuitable to function as a pump. When cardiac muscle is stimulated to contract more frequently (equivalent to an increase in the heart rate), the action potential duration (and that of the contraction) becomes less, and consecutive twitches remain separate contraction-and-relaxation events.

It must be emphasized that contraction in cardiac muscle is not the result of stimulation by motor nerves. Cells in some critical areas of the heart generate automatic and rhythmic action potentials that are conducted throughout the bulk of the tissue. These specialized cells are called **pacemaker** cells (Chap. 13).

Excitation-Contraction Coupling ■ The rapid depolarization associated with the upstroke of the action potential is conducted down the T tubule system of the ventricular muscle, where it functions (as in skeletal muscle) to cause release of intracellular calcium ions from the sarcoplasmic reticulum. Recent work has shown that in cardiac muscle the largest part of the calcium released during rapid depolarization is from an additional intracellular location just inside the cell membrane (sarcolemma). The principal role of the sarcoplasmic reticulum is in the longer-term storage, uptake, and buffering of cytosolic calcium. The action of calcium ions on the **troponin-tropomyosin regulatory system** of the thin filaments is similar to that in skeletal muscle, but cardiac muscle differs in its cellular handling of the activator calcium. In addition to the calcium ions released from the subsarcolemmal space and the sarcoplasmic reticulum, a significant amount of calcium enters the cell from outside, during the **plateau phase** of the action potential. The principal cause of the sustained depolarization of the plateau phase is the presence of a population of voltage-gated membrane channels permeable to calcium ions (and

to sodium as well; Chap. 13). These channels open relatively slowly, and while they are open there is a net inward flux of calcium ions (called the **slow inward current**) moving down an electrochemical gradient. Although the calcium entering during an action potential does not directly affect that specific contraction, it can affect the very *next* contraction, and it does increase the cellular calcium content over time because of the repeated nature of the cardiac muscle contraction. In addition, even a small amount of Ca^{2+} entering through the sarcolemma causes the release of significant additional Ca^{2+} from the sarcoplasmic reticulum, a phenomenon called **calcium-induced calcium release** (similar to that in smooth muscle). This constant influx of calcium requires that there be a cellular system that can rid the cell of excess calcium. Regulation of cellular calcium content has important consequences for cardiac muscle function (see below).

Sources of Energy for Cardiac Muscle Function ■ In contrast to skeletal muscle, cardiac muscle does not have opportunities to rest from a period of intense activity to "pay back" an oxygen debt. As a result, the metabolism of cardiac muscle is almost entirely aerobic under basal conditions and uses free fatty acids and lactate as its primary substrates. This correlates with the high content of mitochondria in the cells and with the high cellular content of myoglobin. Under conditions of **hypoxia** (lack of oxygen) the anaerobic component of the metabolism may approach 10% of the total, but beyond that limit the supply of metabolic energy is insufficient to sustain adequate function. The substrates that provide chemical energy input to the heart during periods of increased activity consist of carbohydrates (mostly in the form of lactic acid produced as a result of skeletal muscle exercise; Chaps. 8 and 9), fats (largely as free fatty acids), and, to a small degree, ketones and amino acids. The relative amounts of the various metabolites vary according to the nutritional status of the body. Because of the highly aerobic nature of cardiac muscle metabolism, there is a strong correlation between the amount of work done and oxygen consumption. Under most conditions, the contraction of cardiac muscle in the intact heart is approximately 20% efficient, with the remainder of the energy going to other cellular processes or wasted as heat. Regardless of the dietary or metabolic source of energy, ATP (as in all other muscle types) provides the immediate energy for contraction. Like skeletal muscle, cardiac muscle contains a "rechargeable" creatine phosphate buffering system that supplies the short-term ATP demands of the contractile system.

■ Mechanical Properties of Cardiac Muscle Adapt It to Changing Physiologic Requirements

Analysis of the mechanical function of cardiac muscle differs somewhat from that of skeletal muscle contraction. Cardiac muscle, in its natural location, does not exist as separate strips of tissue with skeletal attachments at the ends. Instead, it is present as interwoven bundles of fibers in the heart walls, arranged so that its **shortening** results in a reduction of the **volume** of the heart chamber, and its **force** or **tension** results in an increase in **pressure** in the chamber. Because of geometric complexities of the intact heart and the complex mechanical nature of the blood and aorta, shortening contractions of the intact heart muscle are more nearly **auxotonic** (Chap. 9) than truly isotonic.

The experimental basis for the present understanding of cardiac muscle physiology comes largely from studies done on isolated **papillary muscles** from the right ventricles of experimental animals; this is a reasonably long, slender muscle that can serve as a representative of the whole myocardium. Isolated muscle can be arranged to function under the same sort of conditions as a skeletal muscle. Analysis of results is aided by using simple afterloads to produce isotonic contractions. Despite the limitations these simplifications impose, many of the unique properties of the intact heart can be understood on the basis of studies of isolated muscle. As the various phenomena are explained here, substitute *volume changes* for *length changes* and *pressure* for *force*. You will then be able to relate the function of the heart as a pump (Chap. 14) to the properties of the muscle responsible for its operation.

The Length-Tension Curve ■ Some aspects of the cardiac muscle **length-tension curve** (Fig. 10-3) are associated with its specialized construction and physiologic role. Over the range of lengths that represent physiologic ventricular volumes there is an appreciable resting force that increases with length; at the length at which active force production is optimal (L_o), this can amount to 10–15% of the total force. Because

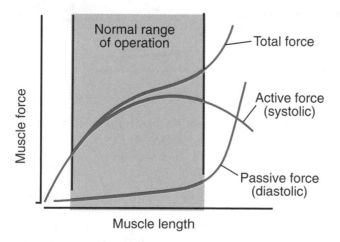

Figure 10–3 ■ The isometric length-tension curve for isolated cardiac muscle. The total force at all physiologically significant lengths includes a passive force component.

this represents muscle force that exists before contraction occurs, it is called **preload.** In the intact heart, preload sets the resting fiber length according to the intracardiac blood pressure existing prior to contraction. The passive tension rises steeply beyond the optimal length and serves to prevent overextension of the muscle (or overfilling of the heart). Note that the resting force curve is associated with the **diastolic (relaxed) phase** of the heart cycle, while the active force curve is associated with the **systolic (contraction) phase.**

The length-tension curve as presented in Figure 10-3 describes isometric behavior; since the working heart never undergoes completely isometric contractions (Chap. 14), other aspects of length-dependent behavior must be responsible for determining the effect length has on cardiac muscle function. One such aspect is the rate at which isometric force develops during a twitch. Notice the series of twitches shown in Figure 9-10; because of the constancy of the time required to reach peak force, the rate of rise of force also varies with muscle length. Other length- dependent aspects of contraction are encountered when we examine the complete contraction cycle of cardiac muscle.

The Contraction Cycle of Cardiac Muscle ■ A typical isotonic contraction of skeletal muscle (Figs. 9-8, 9-9, 9-11) can be divided into four distinct phases (Fig. 10-4). During the **isometric contraction** phase (A) the muscle force builds up to reach the afterload; **isotonic shortening** (B) then takes place, and the afterload is lifted. In the **isotonic lengthening** phase (C) of relaxation, the load stretches the muscle back to its starting length, and finally the contractile force dies away during the phase of **isometric relaxation** (D).

The **isometric contraction** and **isotonic shortening** phases (A and B, Fig. 10-5) of a typical cardiac muscle contraction are like those of skeletal muscle. However, at the peak of the shortening, the supporting platform is moved upward (by some external agency) to hold the afterload; this simulates the closing of the aortic valves at the end of the cardiac ejection phase (Chap. 14). Since the muscle is not allowed to lengthen, it undergoes **isometric relaxation** (C) at the shorter length. Some time later, the muscle is stretched back to its original length. If this is done by a constant force, it will produce an **isotonic lengthening** phase (D). Since the muscle has relaxed, only a very small force is required for the reextension. In the intact heart, this force is supplied by the returning blood.

The principal difference between these two cycles is significant: in skeletal muscle, the work done on the afterload (by lifting it) is returned to the muscle. The skeletal muscle can "retrace its steps" in a contraction, although it is not necessary that a contraction occur in this manner; there is great flexibility and variety possible in skeletal muscle activity. In cardiac muscle, the work done on the load is *not* returned to

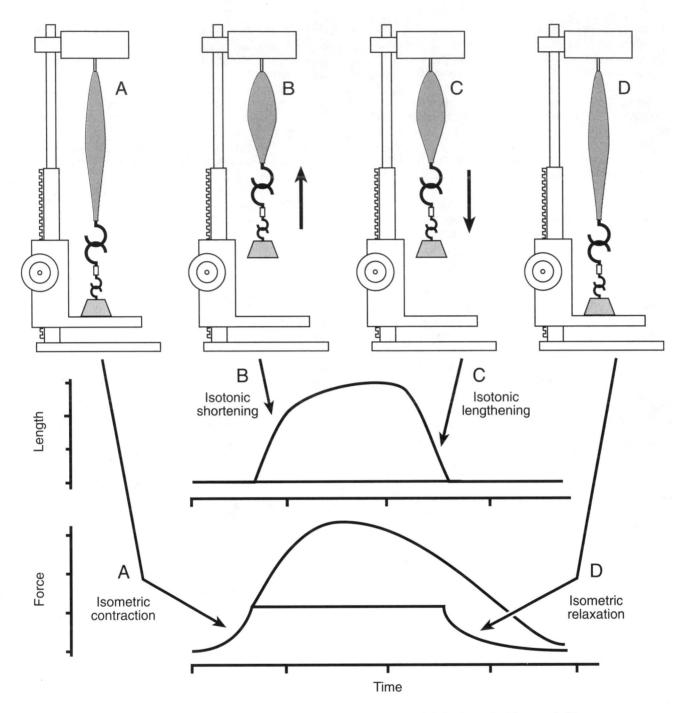

Figure 10–4 ■ An afterloaded isotonic contraction of an isolated skeletal muscle. The muscle lifts the load, and during isotonic relaxation the load stretches the muscle back to its resting length (see text for details).

the muscle but is imparted to the afterload. The heart muscle, however, is constrained by its anatomy and functional arrangements to follow different pathways during contraction and relaxation. This pattern is seen more clearly if the phases of the contraction-relaxation cycle are displayed on a length-tension diagram (Fig.

10–6). (The letters marking the phases correspond to those in Fig. 10–5.) At the beginning of the contraction (A), force increases without any change in length (isometric conditions); when the afterload is lifted (B) the muscle shortens at a constant force (isotonic conditions) to the shortest length possible for that afterload.

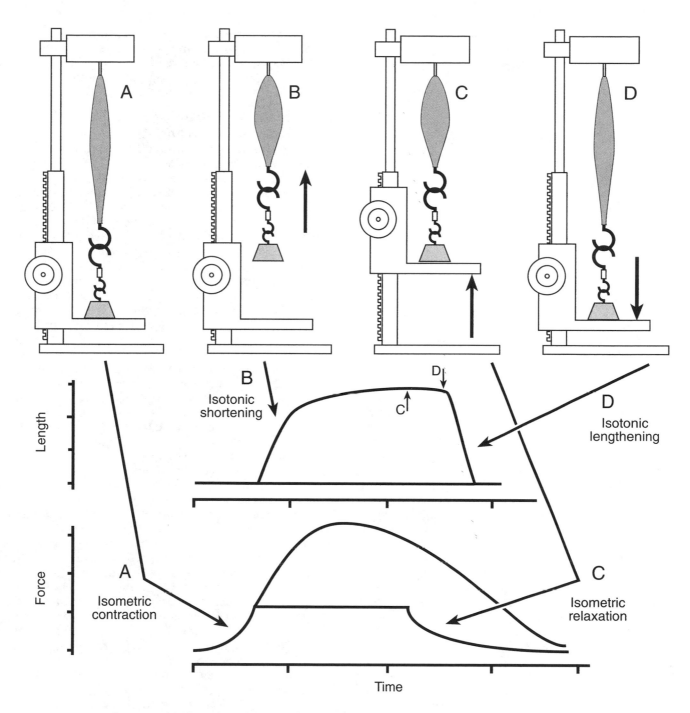

Figure 10–5 ■ Isotonic contraction of isolated cardiac muscle. The load is prevented from reextending the shortened muscle during relaxation, which thus takes place under isometric conditions (see text for details).

At the maximal extent of shortening the afterload is removed, and the muscle relaxes (C) without any change in length (isometric conditions again, but at a reduced length). With sufficient force applied to the resting muscle by some external means (D), the muscle is elongated back to its starting length. Because the muscle is unstimulated and its elastic force rises somewhat during elongation, this phase is neither strictly isotonic nor isometric. (In Fig. 10–5, the afterload weight is used for this extension, so the elongation is forced to be isotonic.) In physical terms, the area enclosed by this pathway represents work done by

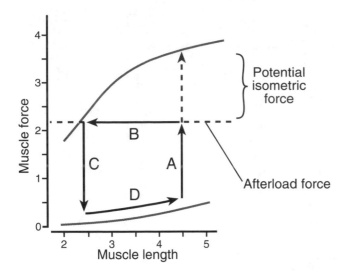

Figure 10–6 ■ An afterloaded contraction of cardiac muscle, as in Figure 10-4, plotted in terms of the length-tension curve. The limit to force is provided by the afterload and the limit to shortening by the length-tension curve. A, isometric contraction phase; B, isotonic shortening phase; C, isometric relaxation phase; D, relengthening.

the muscle on the external load. If the afterload or the starting length (or both) is changed, then a different pathway (Fig. 10-7, left) will be traced. The area enclosed will differ with changes in the conditions of contraction, reflecting differing amounts of external work delivered to the load. In skeletal muscle contractions of the sort shown in Figures 10-4 and 10-7 (right), steps A and B are reversed during relaxation. Such a contraction does no net external work, and no area is enclosed by the pathway. Only when an aortic heart valve is defective and lets the blood return to the heart does cardiac muscle undergo such a contraction.

Cardiac Muscle Self-Regulation ■ Each case in Figure 10-7 (left) demonstrates that the active portion of the length-tension curve provides the limit to shortening and thus interacts with the particular afterload chosen. That is, with lower afterloads the muscle will shorten further than it would with a higher afterload. It is important to realize that during isotonic shortening the muscle force is limited by the magnitude of the *afterload* and not by the length-tension capability of the muscle. It is the *extent* of shortening at a given afterload that is limited by the length-tension property of the muscle; this is a very important consideration when measuring cardiac performance under conditions of changing blood pressure and filling of the heart (Chap. 14). This length- and force-dependent behavior is the key to self-regulation by cardiac muscle and is the functional basis of the **Frank-Starling mechanism** (Chap. 14); when the muscle is set to a longer length at rest, the active contraction results in a greater shortening that is also more rapid and is preceded by a more rapid isometric phase. This allows

the heart to adjust its pumping to exactly the amount required to keep the circulatory system in balance. Further aspects of the self-regulating properties of cardiac muscle are treated below, after another special and vital property of the muscle has been presented.

■ Changes in Contractility Enable Further Physiologic Adjustments

Under a wide range of conditions the contractile behavior of skeletal muscle is fixed and repeatable. The peak force or shortening velocity depends primarily on muscle length or afterload, and unless the muscle is caused to fatigue, these properties will not change from contraction to contraction. For this reason skeletal muscle is said to possess fixed **contractility.** Muscle contractility, or the **contractile state** of muscle, may be defined as a certain level of functional capability (as measured by a quantity such as isometric force, shortening velocity, etc.) when it is measured at a *constant muscle length.* (Length must be constant to exclude effects due to the length-tension curve properties already discussed.) The regulation of skeletal muscle contraction to produce useful activity is primarily the task of the central nervous system, using the mechanisms of motor unit summation and partially fused tetani, as previously discussed. Cardiac muscle, while it has no motor innervation, does require a capacity for adjustment that cannot be accomplished solely by changes in afterload and starting length.

The **variable contractility** of cardiac muscle allows it to make physiologically useful adjustments to the varying demands of the circulatory system. Certain chemical and pharmacologic agents, as well as physiologic circumstances, also affect cardiac contractility. The collective term for the influence of such agents is **inotropy** ("having an effect on muscle action"). Contractility is altered by **inotropic interventions,** agents or processes that can change the functional state of cardiac muscle. Common inotropic interventions that increase contractility **(positive inotropes)** include the action of adrenergic (sympathetic nervous system) stimulation, blood-borne catecholamine hormones, drugs such as the **digitalis** derivatives, and an increase in the rate of stimulation (i.e., heart rate). Negative inotropic interventions (or **negative inotropes**) include a decrease in heart rate, disease processes such as myocarditis or coronary artery disease, and some drugs.

Effects of Inotropic Interventions ■ Figure 10-8 shows an increase in contractility plotted on the length-tension axes. It has the effect of shifting the active length-tension curve upward and to the left; relaxation and the passive curve are little affected. Careful experiments have shown that one effect of very short muscle length on muscle contraction is actually a reduction in contractility due to inefficiencies in the excitation-contraction coupling mechanism at these lengths. Such effects cannot be separated from other length-related effects on cardiac muscle func-

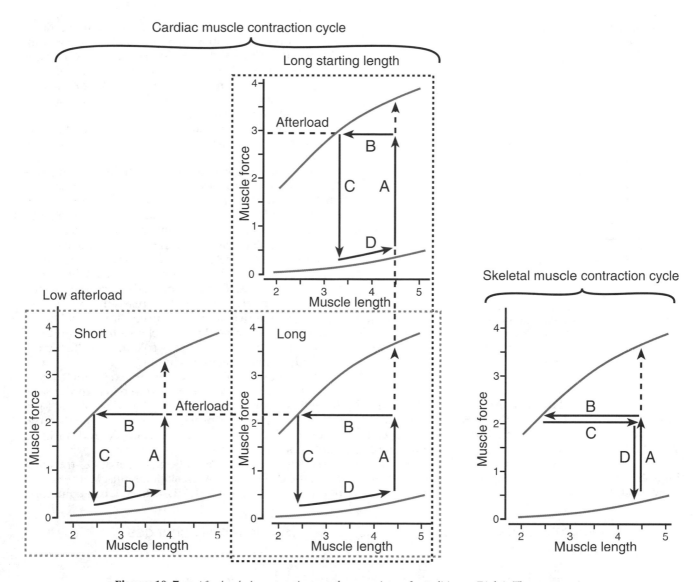

Figure 10–7 ■ Afterloaded contractions under a variety of conditions. (Right) The contraction cycle of skeletal muscle, as shown in Figure 10-4. Contraction and relaxation pathways are the same. (Left) Cardiac muscle contraction cycles, as in Figure 10-5. The horizontal box shows the effect of starting at two different initial lengths at the same afterload. The vertical box shows the effect of two different afterloads on shortening that begins at the same initial length. Increasing the afterload reduces the amount of shortening possible, as does decreasing the starting length; in both cases the limit to shortening is determined by the length-tension curve.

tions, and they are usually included without mention in the more familiar length-dependent changes in muscle performance.

An example of the similarities and difference between changes in contractility and changes in resting length is shown in the force-velocity curves in Figure 10-9. The set of curves in Figure 10-9A represents the isotonic behavior of muscle at a constant level of contractility at three different muscle lengths. The maximum force point on each curve shows the isometric length-tension effect. When a particular afterload is chosen (in this case, 0.5 units), the initial shortening

velocity varies with the starting length. This is a manifestation of the length-tension relationship that does not involve isometric contraction. The curves in Figure 10-9B represent contractions made at the same starting length but with the muscle operating at different levels of contractility. Again there is a difference in shortening velocity at a constant afterload and no tendency for the curves to converge at the low forces. These examples show only one aspect of the effects of changing contractility; those not illustrated include changes in the rate of rise of isometric force and changes in the time required to reach peak force in a

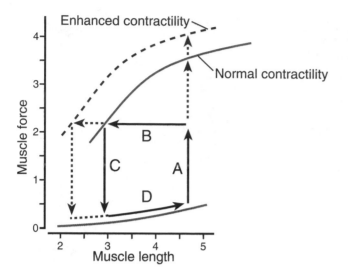

Figure 10–8 ■ The effect of enhanced contractility on the contraction cycle of cardiac muscle. When contractility is increased, the rate of rise of force is increased, the time to afterload force is decreased, and potential force is increased (A). The muscle shortens faster and further (B), while isometric relaxation (C) and relengthening (D) are little affected.

twitch. Ultimately, all of the contractile changes in the muscle result in a change in the overall performance of the heart (Chap. 14), but cardiac performance can change drastically even without changes in contractility because of length-tension effects. The need to distinguish such effects from changes in contractility (to guide treatment and therapy) has led to a search for

aspects of muscle performance that are dependent on state of contractility but independent of muscle length. The results of these studies (based on the properties of isolated muscle) are not unequivocal, because the complicated structure and function of the intact heart do not permit a reliable extrapolation of findings. Instead, a number of empiric measures (Chap. 14) have been developed from studies of the intact heart, some of which provide a reasonable and useful index of contractility.

The Cellular Basis for Contractility Changes ■ The basic determinant of the variable contractility of cardiac muscle is the calcium content in the myocardial cell. Under normal conditions the contractile filaments of cardiac muscle are only partly activated. This is because, unlike the situation in skeletal muscle, not enough calcium is released to occupy all of the troponin molecules, and not all potentially available crossbridges can attach and cycle. An increase in the availability of calcium would increase the number of crossbridges activated, thus contractility would be increased. To understand the mechanisms of contractility change, then, it is necessary to consider the factors affecting cellular calcium handling.

The processes linking membrane excitation to contraction via calcium ions are illustrated in Figure 10–10. Since this involves many possible movements and locations of calcium, the processes are considered in the order in which they would be encountered during a single contraction.

The initial event is an action potential (1) traveling along the cell surface. As in skeletal muscle, the action potential enters a T tubule (2), where it can communi-

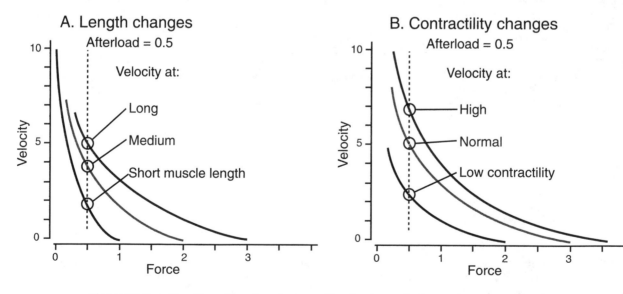

Figure 10–9 ■ The effect of length and contractility changes on the force-velocity curves of cardiac muscle. (A) Decreased starting length (with constant contractility) produces lower velocities of shortening at a given afterload. (B) Increased contractility produces increased velocity of shortening at a constant muscle length. Because of the presence of resting tension (as is characteristic of heart muscle), it is impossible to make a direct measure of a zero-force contraction at each length, although there is a tendency for the curves to converge at the lower forces.

CARDIOMYOPATHIES: ABNORMALITIES OF HEART MUSCLE

Heart disease takes many forms. While some of these are related to problems with the valves or electrical conduction system (Chaps. 13 and 14), many are due to malfunctions of the cardiac muscle itself. These conditions, called **cardiomyopathy,** result in impaired cardiac function that may range from being essentially asymptomatic to malfunctions causing sudden death.

There are several types of cardiomyopathy, and they have a number of causes. In **hypertrophic cardiomyopathy,** an enlargement of the cardiac muscle fibers occurs because of a chronic overload, such as that caused by hypertension (high blood pressure) or a defective heart valve. Such muscle may fail because its high metabolic demands cannot be met, or fatal electrical arrhythmias (Chap. 13) may develop. **Congestive** or **dilated cardiomyopathy** refers to cardiac muscle so weakened that it cannot pump strongly enough to empty the heart properly with each beat. In **restrictive cardiomyopathy** the muscle becomes so stiffened and inextensible that the heart cannot fill properly between beats. Chronic poisoning with heavy metals, such as cobalt or lead, can produce **toxic cardiomyopathy.** The skeletal muscle degeneration associated with **muscular dystrophy** (Chap. 8) is often accompanied by cardiomyopathy.

When the cause of cardiomyopathy is unknown, it is termed **idiopathic.** The cardiomyopathy arising from **viral myocarditis** is difficult to diagnose and may show no symptoms until death occurs. The action of some enteroviruses (e.g., coxsackie B virus) may cause an **autoimmune response** that does the actual damage to the muscle, and this damage may occur at the subcellular level by interfering with energy metabolism while producing little apparent structural disruption. Such conditions, which can usually be diagnosed only by direct muscle biopsy, are difficult to treat effectively, although spontaneous recovery can occur. Excessive and chronic consumption of alcohol can also cause a cardiomyopathy that is often reversible if total abstinence is maintained. In tropical regions, infection with **Chagas disease,** a parasite-borne condition, can produce chronic cardiomyopathy. The tick-borne spirochete infection called **Lyme disease** can cause heart muscle damage and lead to **heart block,** a conduction disturbance (Chap. 13).

Another important kind of cardiomyopathy arises from inadequate oxygen (blood) supply to working cardiac muscle. Following an acute episode of such **ischemia** there is a **stunned myocardium,** in which mechanical performance is reduced. **Chronic ischemia** can produce a **hibernating myocardium,** also with reduced mechanical performance. Ischemic tissue has impaired calcium handling, which can lead to destructively high levels of internal calcium. These conditions can be improved with reestablishment of an adequate oxygen supply (e.g., following clot dissolution or coronary bypass surgery), but even this is not without risk, since reperfusion of ischemic tissue can lead to production of oxygen radicals that cause significant cellular damage. Use of calcium blockers and free radical scavengers, such as vitamin E, following ischemic episodes may limit this damage.

cate with the sarcoplasmic reticulum (3) to cause calcium release. The sarcoplasmic reticulum in cardiac muscle is of much lower capacity than in skeletal muscle, and sarcoplasmic reticulum calcium release alone is insufficient to cause adequate activation of the contraction. To some extent, this is aided by a calcium-induced calcium release mechanism (4) triggered by a rise in the cytoplasmic calcium concentration (due to factors described below). More important is the presence of the action potential (5) on the cell surface (sarcolemma); the depolarization causes the opening of calcium channels, through which a strong inward

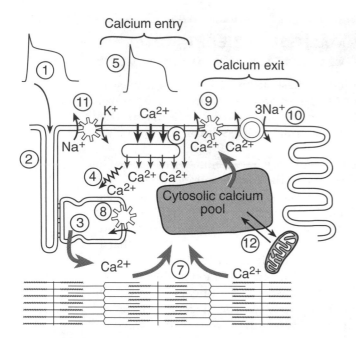

Figure 10–10 ■ The paths of calcium in and out of the cardiac muscle cell and its role in the regulation of contraction. See text for details.

calcium current flows, and the calcium ions accumulate just inside the sarcolemma (6), although some probably diffuse rapidly into the cell interior. The depolarization of the cell causes the rapid release of previously bound calcium from this subsarcolemmal site, and this calcium then diffuses the short distance to the myofilaments (7) and activates them. This amount of calcium, combined with that released from the sarcoplasmic reticulum, determines the magnitude of the myofilament activation and, hence, the level of contractility. The total cellular content of calcium, regardless of where it is stored, is referred to as the **cytoplasmic calcium pool.**

During relaxation, the cytoplasmic calcium concentration is rapidly lowered through a number of pathways. The sarcoplasmic reticulum membrane contains a vigorous Ca ATPase (8) that runs continuously and is further activated, through a protein phosphorylation mechanism, by high levels of cytoplasmic calcium. At the level of the sarcolemma two additional mechanisms work to rid the cell of the calcium that entered via previous action potentials. A membrane Ca ATPase (9) actively extrudes calcium, ejecting one calcium ion for each ATP molecule consumed. Additional calcium is removed by a Na^+/Ca^{2+} exchange mechanism (10), also located in the cell membrane. This mechanism is part of a coupled transport system in which

three sodium ions, entering the cell down their electrochemical gradient, are exchanged for the ejection of one calcium ion. Proper function of this exchange mechanism requires a steep sodium concentration gradient, maintained by the membrane Na^+/K^+ ATPase (11) located in the sarcolemma. Because the Na^+/Ca^{2+} exchange mechanism derives its energy from the sodium gradient, any reduction in the pumping action of the Na^+/K^+ ATPase leads to reduced calcium extrusion. Under normal conditions these mechanisms can maintain a 10,000-fold Ca^{2+} concentration difference between the inside and outside of the cell. Since a cardiac cell contracts repeatedly many times per minute, each beat being accompanied by an influx of calcium, the extrusion mechanisms must also work continuously to balance the incoming calcium. The mitochondria of cardiac muscle (12) are also capable of accumulating and releasing calcium, although this system does not appear to play a role in the normal functioning of the cell.

Calcium and the Function of Inotropic Agents
■ Inotropic agents usually work through changes in the internal calcium content of the cell. An increase in the heart rate, for instance, allows more separate influxes of calcium per minute, and the amount of releasable calcium in the subsarcolemmal space and sarcoplasmic reticulum increases. More crossbridges are activated and the force of isometric contraction (and other indicators of contractility) increases. This is the basis of the **force-frequency relationship,** one of the principal means of changing myocardial contractility.

An important class of therapeutic agents used to increase the contractility of failing hearts is the **cardiac glycosides.** The drug **digitalis,** used for centuries for its effects on the circulation, it typical of these agents. While some details of its action are obscure, the drug has been shown to work by inhibiting the membrane Na^+/K^+ ATPase (11). This allows the cell to gain sodium and reduce the steepness of the sodium gradient. This makes the Na^+/Ca^{2+} exchange mechanism (10) less effective, and the cell gains calcium. Since more calcium is available to activate the myofilaments, contractility increases. These effects, however, can lead to **digitalis toxicity** when the cell gains so much calcium that the capacity of the sarcoplasmic and sarcolemmal binding sites is exceeded. At this point the mitochondria (12) begin to take up the excess calcium; however, too much mitochondrial calcium interferes with ATP production. The cell, with its ATP needs already increased by enhanced contractility, is less able to pump out accumulated calcium, and the final result is a lowering of metabolic energy stores and a reduction in contractility.

■ ■ ■ ■ ■ ■ ■ ■ ■ ■ ■ ■ ■ **REVIEW EXERCISES** ■ ■ ■ ■ ■ ■ ■ ■ ■ ■ ■ ■ ■

Identify Each with a Term

1. The structure that forms the mechanical and electrical connections between the ends of cardiac muscle cells
2. The working muscle of the heart
3. The phase of the contractile cycle of heart muscle in which there is a fall in force with no length change
4. A factor or agent that changes the contractility of cardiac muscle
5. The force borne by a shortening muscle

Define Each Term

6. Period of supranormal excitability
7. Diastolic phase
8. Negative inotrope
9. Contractility
10. Force-frequency relationship
11. Purkinje fiber
12. Plateau phase
13. Dyad
14. Cardiac glycosides

Choose the Correct Answer

15. Excitation-contraction coupling in cardiac muscle:
    a. Involves Ca^{2+} as a second messenger
    b. Links the events of the action potential to muscular contraction
    c. Happens in the absence of motor nerve stimulation
    d. Involves all of the above
16. Conduction of the electrical activity throughout the muscle of the heart:
    a. Is a function of the cardiac nerves
    b. Occurs in muscle tissue only
    c. Is not always necessary for a coordinated contraction
    d. Usually occurs in the absence of muscle activation
17. The metabolic requirements of the heart:
    a. Are moderate and can be met by anaerobic metabolism

b. Result in a significant oxygen debt at high heart rates
    c. Must be met by aerobic metabolism
    d. Are usually met by consuming dietary protein
18. The force developed by shortening cardiac muscles depends on the:
    a. Afterload
    b. Speed of shortening
    c. Muscle length
    d. Frequency of contraction
19. Changes in cardiac contractility depend mainly on:
    a. Muscle length
    b. Afterload
    c. Internal ATP concentration
    d. Internal Ca^{2+} concentration
20. An increase in cardiac contractility can result in:
    a. A shift of the active length-tension curve to the right
    b. A decrease in muscle length at the end of isotonic shortening
    c. A reduction in shortening velocity at very low afterloads
    d. A failure of the muscle to relax completely

Briefly Answer

21. What characteristics of the cardiac muscle action potential act to prevent a tetanic contraction?
22. How does an inhibition of the sarcolemmal Na^+/K^+ ATPase lead to an increase in cardiac muscle contractility?
23. When cardiac muscle contracts isotonically, why is the relaxation isometric?
24. Besides isometric force, what other contractile properties of cardiac muscle are affected by changes in contractility?
25. What roles does sodium/calcium exchange play in regulating cardiac cell calcium content?

■ ■ ■ ■ ■ ■ ■ ■ ■ ■ ■ ■ ■ **ANSWERS** ■ ■ ■ ■ ■ ■ ■ ■ ■ ■ ■ ■ ■

1. Intercalated disk
2. Myocardium
3. Isometric relaxation
4. Inotropic intervention
5. Afterload
6. A brief time late in the repolarization phase when the permeability to K^+ ions is still lower than at rest, the threshold being more easily reached
7. The resting phase of the contraction cycle that occurs between beats, when cardiac muscle is relaxed
8. An agent or action that decreases the contractility of heart muscle

9. The overall state of cardiac muscle contraction capability, as measured by some mechanical output, such as isometric force or shortening velocity at a set afterload
10. The observed increase (or decrease) in cardiac contractility as a result of an increase (or decrease) in the heart rate, related to the increased calcium entry during the action potential at higher heart rates
11. Cardiac muscle strands that are specialized for conduction of the electrical activity (action potential) throughout the ventricles of the heart
12. The period of nearly constant depolarization during the action potential, largely determined by an increase in Ca^{2+} permeability, that is associated with the slow inward current

13. The association of a cardiac muscle T tubule with a single terminal cisterna of the sarcoplasmic reticulum (similar to a triad in skeletal muscle)

14. A family of drugs that increase cardiac contractility indirectly by inhibiting the sarcolemmal Na⁺/K⁺ ATPase

15. d

16. b

17. c

18. a

19. d

20. b

21. The duration of the action potential (and hence its absolute refractory period) is nearly as long at the contraction-relaxation cycle of a twitch, thus the muscle cannot be restimulated while it is still expressing significant force.

22. If the activity of the Na⁺/K⁺ ATPase in the sarcolemma is reduced, the passive entry of sodium will reduce the steepness of the sodium gradient. It is the potential energy stored in this gradient that permits the Na⁺/Ca²⁺ exchange mechanism to extrude calcium; with calcium exchange decreased, the higher intracellular calcium concentration results in increased contractility.

23. At the end of an isotonic contraction the afterload is "trapped" by the aortic valves and it cannot stretch the muscle back out. In the absence of this elongating force, muscle length remains constant (isometric) during relaxation.

24. The rate of rise of isometric force and the velocity of shortening at a given afterload both increase with an increase in contractility.

25. The actual removal of excess calcium is accompanied by this exchange mechanism (an example of secondary, or coupled, active transport).

Suggested Reading

Braunwald EG, Ross JR, Sonnenblick EH: *Mechanisms of Contraction of the Normal and Failing Heart.* Boston, Little, Brown, 1976.

Heller LJ, Mohrman DE: *Cardiovascular Physiology.* New York, McGraw-Hill, 1981.

Katz AM: *Physiology of the Heart.* 2nd ed. New York, Raven, 1992.

Noble D: *The Initiation of the Heartbeat.* Oxford, Oxford University Press, 1979.

Stern MD, Lakatta EG: Excitation-contraction coupling in the heart: The state of the question. *FASEB J* 6:3092–3100, 1992.

PART FOUR

Blood and Cardio-vascular Physiology

Blood Components, Immunity, and Hemostasis

CHAPTER OUTLINE

I. THE COMPONENTS OF BLOOD
A. Blood consists of cells and plasma
1. General properties of blood
2. Erythrocyte sedimentation rate
3. Hematocrit
B. Blood functions as a dynamic tissue
1. Transport
2. Immunity
3. Hemostasis
4. Homeostasis
C. Plasma contains many important solutes
D. There are three types of blood cell
E. Erythrocytes carry oxygen to tissues
1. Normal red cell values
2. Red cell morphology
3. Destruction of erythrocytes
4. Iron recycling
F. Platelets participate in clotting
G. Leukocytes participate in host defense
1. Neutrophils
2. Eosinophils
3. Basophils
4. Monocytes and lymphocytes
H. Blood cells are born in the bone marrow
II. THE IMMUNE SYSTEM
A. The innate immune system consists of nonspecific defenses
B. Inflammation is a multifaceted process
C. Defense mechanisms are integrated systems
D. Adaptive immunity is specific and acquired
E. The adaptive immune response involves cellular and humoral components
F. B cell immunity is mediated by immunoglobulin
1. Antibody structure
2. Antibody functions
III. HEMOSTASIS
A. Physical factors immediately act to constrain bleeding
B. Platelets form a hemostatic plug
C. Blood coagulation results in the production of fibrin
1. The coagulation cascade
2. Clot retraction and fibrinolysis

OBJECTIVES

After studying this chapter, the student should be able to:
1. Outline the cellular and molecular composition of blood
2. Describe the general functions of blood, including the main function of each of the individual cellular components
3. Differentiate the innate and adaptive immune systems
4. Define *cytokine* and *hematopoiesis*
5. Diagram the structure of a monomeric unit of IgG, labelling the Fab and Fc regions as well as the variable and constant regions of the heavy and light chains
6. Describe the role of platelets in hemostasis
7. Discuss the steps involved in the intrinsic and extrinsic pathways of blood clotting
8. Explain clot retraction and fibrinolysis

Blood is a highly differentiated, complex living tissue that pulsates through the arteries to every part of the body, interacts with individual cells via the capillary network, and returns to the heart through the venous system. Many of the functions of blood are undertaken in the capillaries, where blood flow slows dramatically, allowing efficient diffusion and transport of oxygen, glucose, and other molecules across the monolayer of endothelial cells that form the thin capillary walls.

The Components of Blood

■ Blood Consists of Cells and Plasma

Blood is an opaque, red liquid consisting of several types of cells suspended in a complex, amber fluid known as **plasma.** In the arterial circulation, blood is oxygenated and, as a result, bright red. Oxygen leaves the blood cells in the capillary network, causing the blood to develop a bluish red hue in the venous circulation. When blood is removed from the circulation of a healthy individual, it immediately clots, due to the conversion of the soluble plasma protein **fibrinogen** to **fibrin,** an insoluble protein polymer. Clotting can be prevented by adding an **anticoagulant** to the freshly removed blood or by collecting blood in a vessel containing anticoagulant. Two anticoagulants used today include heparin, which inactivates clotting factors in plasma, and citrate, which binds calcium, a cation essential for coagulation.

General Properties of Blood ■ Blood is normally confined to the circulation, including the heart and pulmonary and systemic blood vessels. Blood accounts for 6–8% of the body weight of a healthy adult. The blood volume is normally 5.0–6.0 L in men and 4.5–5.5 L in women.

The **density** (or **specific gravity**) of blood is approximately 1.050 g/ml. The density depends on the number of blood cells present and the composition of the plasma. The density of individual blood cells varies with the cell type and ranges from 1.115 g/ml for erythrocytes to 1.070 g/ml for certain leukocytes.

While blood is only slightly heavier than water, it is certainly much thicker. The **viscosity** of blood, a measure of resistance to flow, is 3.5–5.5 times that of water. Blood's viscosity increases as the total number of cells present increases and when the content of large molecules (macromolecules) in plasma increases. At pathologically high viscosity, blood flows poorly to the extremities and internal organs.

Erythrocyte Sedimentation Rate ■ Erythrocytes are the red cells of blood. Since erythrocytes have only a slightly higher density than the suspending plasma, they normally settle out of whole blood very slowly. To determine their sedimentation rate, anticoagulated blood is placed in a long, thin graduated cylinder (Fig. 11-1). As the red cells sink, they leave behind the less dense leukocytes (white blood cells) and platelets (thrombocytes) in the suspending plasma. Erythro-

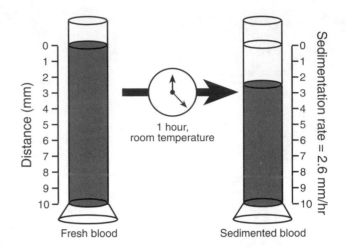

Figure 11–1 ■ Determination of the erythrocyte sedimentation rate (ESR). Fresh, anticoagulated blood is allowed to settle at room temperature in a cylindric, graduated column. After a fixed time interval (1 hour), the distance (in millimeters) from the top of the fluid column to the erythrocyte sedimentation front is measured.

cytes in blood of healthy males sediment at a rate of 2–8 mm/h; those in blood of healthy women sediment slightly faster (2–10 mm/h). The rate of red cell sedimentation is referred to as the **erythrocyte sedimentation rate** (ESR). The ESR can be an important diagnostic index, as values are often markedly elevated during infection, in patients with arthritis, and in patients with inflammatory diseases. In some diseases, such as **sickle cell anemia, polycythemia** (abnormal increase in red cell numbers), and **hyperglycemia** (elevated blood sugar), the ESR is slower than normal. The reasons for alterations in the ESR in disease states are not always clear, but the cells tend to sediment faster when the concentration of plasma proteins increases.

Hematocrit ■ Blood cells can be readily separated from the suspending fluid by simple centrifugation. When anticoagulated blood is placed in a tube that is rotated about a central point, centrifugal forces pull the blood cells from the suspending plasma. When clotted or coagulated blood is centrifuged, blood cells and fibrin are pulled from the blood **serum,** which is essentially plasma devoid of fibrinogen and certain clotting factors. The **hematocrit** is the portion of the total blood volume that is made up of cells. This value is determined by centrifugation of small capillary tubes of anticoagulated blood, to pack the cells. Hematocrit values of the blood of healthy adults are 47 ± 5% for men and 42 ± 5% for women. Decreased hematocrit values often reflect blood loss due to bleeding or deficiencies in blood cell production. Low hematocrit values indicate the presence of **anemia,** a reduction in the number of circulating erythrocytes. Increased hematocrit levels may likewise indicate a serious imbalance in the production and destruction of red cells. Increased production (or decreased rate of destruction) of erythrocytes results in polycythemia, as reflected by increased hematocrit values. Dehydration, which decreases the water content and thus the volume of plasma, also results in an increase in hematocrit. Determination of hematocrit values is a simple and important screening diagnostic procedure in the evaluation of hematologic disease.

■ Blood Functions as a Dynamic Tissue

While the cellular and plasma components of blood may act alone, they often work in concert to accomplish their functions. Working together, blood cells and plasma proteins provide a number of important functions, including **transport** of substances from one area of the body to another; defense against invading microorganisms and neoplastic disease (**immunity**); arrest of bleeding (**hemostasis**); and maintenance of a stable internal environment (**homeostasis**).

Transport ■ Blood carries a number of important substances from one area of the body to another, including oxygen, carbon dioxide, antibodies, acids and bases, ions, vitamins, cofactors, hormones, nutrients, lipids, gases, pigments, minerals, and water. Transport is one of the primary and most important functions of blood, and blood is the primary means of transport in the body. Substances can be transported free in plasma, bound to plasma proteins, or within blood cells.

Oxygen and carbon dioxide are two of the more important molecules transported by blood. Oxygen is taken up by the **red cells** in blood as the cells pass through capillaries in the lung. In tissue capillaries, red cells release oxygen, which is then used by respiring tissue cells. These cells produce carbon dioxide and other wastes, which the blood transports to organs of excretion.

Immunity ■ Blood **leukocytes** are involved in the ongoing battle against infection by microorganisms. While the skin and mucous membranes physically restrict entry of infectious agents, microbes constantly penetrate these barriers and continuously threaten internal infection. Blood leukocytes, working in conjunction with plasma proteins, are continuously on patrol for microbial invaders in the tissues and in the blood (see below). In most cases, penetrating microbes are efficiently eliminated by the sophisticated and elaborate antimicrobial systems of the blood.

Hemostasis ■ Bleeding is controlled by the process of hemostasis. Complex and efficient **hemostatic mechanisms** have evolved to stop hemorrhage after injury, and their failure can quickly lead to fatal blood loss (exsanguination). Both physical and cellular mechanisms participate in hemostasis. These mechanisms, like those of the immune system, are complex, interrelated, and essential for the survival of the organism.

Homeostasis ■ Homeostasis is a steady state that provides an optimal internal environment for cellular metabolism (Chap. 1). By maintaining pH, ion concentrations, osmolality, temperature, nutrient supply, and

vascular integrity, the blood system plays a crucial function in preserving homeostasis. Homeostasis is the desired result of normal functioning of the blood's transport, immune, and hemostatic systems.

■ Plasma Contains Many Important Solutes

Plasma is composed mostly of water (93%) with various dissolved **solutes,** including proteins, lipids (fats), carbohydrates, amino acids, vitamins, minerals, hormones, wastes, cofactors, gases, and electrolytes (Table 11-1). The solutes in plasma play crucial roles in homeostasis, such as maintenance of a normal plasma pH and osmolality.

■ There Are Three Types of Blood Cells

Blood cells include **erythrocytes** (red cells), **leukocytes** (white cells), and **platelets** (or **thrombocytes**) (Table 11-2). Each microliter (millionth of a liter) of blood contains 4-6 million erythrocytes, 4-10 thousand leukocytes, and several hundred thousand plate-

TABLE 11–1 ■ Some Components of Plasma

Class	Substance	Normal Concentration Range
Cations	Sodium (Na^+)	136-145 mEq/L
	Potassium (K^+)	3.5-5.0 mEq/L
	Calcium (Ca^{2+})	4.3-5.2 mEq/L
	Magnesium (Mg^{2+})	1.2-1.8 mEq/L
	Iron (Fe^{3+})	60-160 μg/dl
	Copper (Cu^{2+})	70-155 μg/dl
	Hydrogen (H^+)	35-45 nmol/L
Anions	Chloride (Cl^-)	98-106 mEq/L
	Bicarbonate (HCO_3^-)	23-28 mEq/L
	Lactate	0.67-1.8 mEq/L
	Sulfate (SO_4^{2-})	0.9-1.1 mEq/L
	Phosphate ($HPO_4^{2-}/H_2PO_4^-$)	3.0-4.5 mg/dl
Proteins	Total	6-8 g/dl
	Albumin	3.4-5.0 g/dl
	Globulin	2.2-4.0 g/dl
Fats	Cholesterol	150-200 mg/dl
	Phospholipids	150-220 mg/dl
	Triglycerides	145-250 mg/dl
Carbohydrates	Glucose	80-120 mg/dl
Vitamins, cofactors, and enzymes	Vitamin B_{12}	200-800 pg/ml
	Vitamin A	0.15-0.6 μg/ml
	Vitamin C	0.4-1.5 mg/dl
	2,3 diphosphoglycerate (DPG)	3-4 mmol/L
	Transaminase (SGOT)	10-40 U/ml
	Alkaline phosphatase	30-92 U/L
	Acid phosphatase	0.5-2 U/ml
Other substances	Creatinine	62-133 μmol/L
	Uric acid	0.15-0.48 mmol/L
	Blood urea nitrogen	8-25 mg/dl
	Iodine	3.5-8.0 μg/dl
	CO_2	23-30 mmol/L
	Bilirubin (total)	0.1-1.2 mg/dl
	Aldosterone	3-10 ng/dl
	Cortisol	5-18 μg/dl
	Ketones	0.2-2.0 mg/dl

TABLE 11–2 ■ Circulating Blood Cell Levels

Blood Cell Type	Approximate Normal Range
Erythrocytes (cells/μl)	$4.2–5.9 \times 10^6$
Men	$4.5–6.5 \times 10^6$
Women	$3.8–5.8 \times 10^6$
Leukocytes (cells/μl)	4,000–10,000
Neutrophils	4,000–7,000
Lymphocytes	2,500–5,000
Monocytes	100–1,000
Eosinophils	0–500
Basophils	0–100
Platelets (cells/μl)	150,000–350,000

lets. There are several subtypes of leukocytes, defined by morphologic differences (Fig. 11-2), each with vastly different functional characteristics and capabilities. Table 11-2 lists normal levels of different blood cell types.

Of the total leukocytes, 40–75% are **neutrophilic, polymorphonuclear** (multinucleated) cells, otherwise known as **neutrophils.** These phagocytic cells actively ingest and destroy invading microorganisms. **Eosinophils** and **basophils** are polymorphonuclear cells that are present in low numbers in blood (1–6% of total leukocytes) and participate in allergic hypersensitivity reactions. Mononuclear cells, including **monocytes** and **lymphocytes,** comprise 20–50% of the total leukocytes. These cells generate antibodies and mount cellular immune reactions against invading agents. The number and relative proportion of the

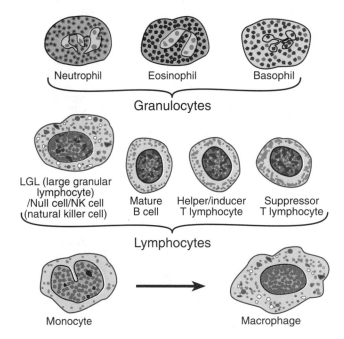

Figure 11–2 ■ Types of leukocytes in blood and tissues. All of the illustrated cells are found in the circulation except the macrophage, which differentiates from activated monocytes in tissue.

leukocyte subtypes can vary widely in different disease states. For example, the absolute neutrophil count often increases during infection, presumably in response to the infection. Eosinophil counts elevate when allergic individuals are exposed to allergens. Lymphocyte counts decrease in acquired immunodeficiency disease and during some other viral infections. For this reason, in addition to a **blood cell count,** a **differential analysis** of leukocyte subtypes, performed by microscopic examination of stained slides, can provide important clues to the diagnosis of disease.

In healthy adults, blood cells are produced in the bone marrow by **pluripotent stem cells.** The process of generating blood cells, called **hematopoiesis,** is regulated by cytokines, cellular proteins thought to act as hormones to regulate the activity of other cells. Several cytokines regulate the activity of hematopoietic stem cells, including **colony stimulating factors, growth factors,** and **interleukins.** These cytokines, collectively known as hematopoietins, also may have important effects on the progeny of the stem cells, including the mature blood cells.

■ Erythrocytes Carry Oxygen to Tissues

Erythrocytes are the most numerous cells in blood. These biconcave disks lack a nucleus and have a diameter of about 7 μm and a maximum thickness of 2.5 μm. The shape of the erythrocyte increases its surface area, increasing the efficiency of gas exchange.

The erythrocyte maintains its shape by virtue of its complex membrane skeleton, which consists of an insoluble mesh of fibrous proteins attached to the inside of the plasma membrane. This allows the erythrocyte great flexibility as the cell twists and turns through small, curved vessels. In addition to structural proteins of the membrane, several functional proteins are found in the cytoplasm of erythrocytes. These include **hemoglobin** (the major oxygen-carrying protein), antioxidant enzymes, and glycolytic systems to provide cellular energy (ATP). The cell membrane possesses ion pumps that maintain a high level of intracellular potassium and a low level of intracellular calcium and sodium.

Hemoglobin, the red, oxygen-transporting protein of erythrocytes, consists of a **globin** (or protein) portion and **heme,** the iron-carrying portion. The molecular weight of hemoglobin is 68,000. This complex protein possesses four polypeptide chains: two α-globin molecules of 141 amino acids each and two molecules of another type of globin chain (β, γ, δ, or ε), each containing 146 amino acid residues (Fig. 11-3). Four types of hemoglobin molecules can be found in human erythrocytes: embryonic, fetal, and two different types found in adults (HbA, HbA$_2$). Each hemoglobin molecule is designated by its polypeptide composition. For example, the most prevalent adult hemoglobin, HbA, consists of two α chains and two

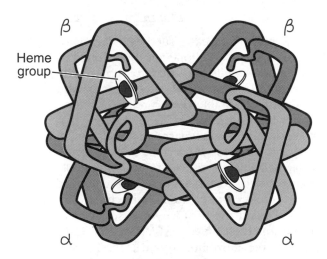

β

β

Heme group

α

α

Figure 11–3 ■ Structure of hemoglobin A. Each molecule of hemoglobin possesses four polypeptide chains, each containing iron bound to its heme group (Modified from Dickerson RE, Geis I: *The Structure and Action of Proteins.* New York, Harper & Row, 1969, p 3.)

β chains. Its formula is thus given as $\alpha_2\beta_2$. HbA$_2$, which makes up about 1.5–3% of total hemoglobin in an adult, has the subunit formula $\alpha_2\delta_2$. Fetal hemoglobin ($\alpha_2\gamma_2$) is the major hemoglobin component during intrauterine life. Its levels in circulating blood cells decrease rapidly during infancy and reach a concentration of 0.5% in adults. Embryonic hemoglobin is found earlier in development. It consists of two α chains and two ε chains ($\alpha_2\epsilon_2$). Production of ε chains ceases at about the third month of fetal development.

The production of each type of globin chain is controlled by an individual structural gene with five different loci. Mutations, which can occur anywhere in these five loci, have resulted in the production of over 550 types of abnormal hemoglobin molecules, most of which have no known clinical significance. Mutations can arise from a single substitution whithin the nucleic acid of the gene coding for the globin chain, a deletion of the codons, or gene rearrangement due to unequal crossing over between homologous chromosomes. Sickle cell anemia, for example, results from the presence of sickle cell hemoglobin (HbS), which differs from normal adult hemoglobin A because of the substitution of a single amino acid in each of the two β chains.

Oxyhemoglobin (HbO$_2$), the oxygen-saturated form of hemoglobin, transports oxygen from lungs to tissues, where the oxygen is released. When oxygen is released, HbO$_2$ becomes **reduced hemoglobin** (Hb). While oxygen-saturated hemoglobin is bright red, reduced hemoglobin is reddish blue, accounting for the difference in the color of blood in arteries and veins.

Certain chemicals readily block the oxygen-transporting function of hemoglobin. For example, carbon monoxide rapidly replaces oxygen in HbO$_2$, resulting in the formation of the stable compound **carboxyhemoglobin** (HbCO). Formation of HbCO accounts for the asphyxiating properties of carbon monoxide. Nitrates and certain other chemicals oxidize the iron in Hb from the ferrous to the ferric state, resulting in the formation of **methemoglobin** (metHb). MetHb contains oxygen bound tightly to ferric iron and, as such, is useless in respiration. **Cyanosis,** the dark blue coloration of skin associated with anoxia, becomes evident when the concentration of reduced hemoglobin exceeds 5 g/dl. Cyanosis may be rapidly reversed by oxygen if the condition is caused only by diminished oxygen supply. However, cyanosis caused by the intestinal absorption of nitrates or other toxins **(enterogenous cyanosis)** is due to the accumulation of stabilized methemoglobin and is not rapidly reversible by administration of oxygen alone.

Normal Red Cell Values ■ In evaluating patients for hematologic diseases, it is important to determine the hemoglobin concentration in the blood, the total number of circulating erythrocytes (the red cell count), and the hematocrit. From these values a number of other important blood values can be calculated, including **mean cell hemoglobin concentration** (MCHC), **mean cell hemoglobin** (MCH), **mean cell volume** (MCV), and blood **oxygen carrying capacity.**

The MCHC provides an index of the average hemoglobin content in the mass of circulating red cells. It is derived as follows:

$$MCHC = \frac{Hb\ (g/L)}{hematocrit} \qquad (1)$$

$$Ex.: \frac{150\ g/L}{0.45} = 333\ g/dl$$

Low MCHC values indicate deficient hemoglobin synthesis. High MCHC values do not occur in erythrocyte disorders, since normally the hemoglobin concentration is close to the saturation point in red cells. Note that the MCHC value is easily obtained by a simple calculation from measurements that can be made without sophisticated instrumentation.

The MCH value is an estimate of the average hemoglobin content of each red cell. It is derived as follows:

$$MCH = \frac{Blood\ hemoglobin\ (g/L)}{Red\ cell\ count\ (cells/L)} \qquad (2)$$

$$Ex.: \frac{150\ g/L}{5 \times 10^{12}\ cells/L} = 30.0 \times 10^{-12} g/cell$$

Since the red cell count is usually related to the hematocrit, the MCH is usually low when the MCHC is low. Exceptions to this rule yield important diagnostic clues.

BONE MARROW TRANSPLANTATION REGISTRIES

When a patient is afflicted with a terminal bone marrow disease, such as leukemia or aplastic anemia, often the only possibility for a cure is an allogeneic bone marrow transplant. In this procedure, healthy bone marrow cells are used to replace the patient's diseased hematopoietic system. These cells are obtained from a donor who is usually a close relative of the patient. To identify a suitable donor, relatives' blood leukocytes are screened to determine whether their antigenic pattern matches that of the patient. The antigenic composition of leukocytes in bone marrow and peripheral blood are identical, so analysis of blood leukocytes usually provides enough information to determine whether the transplanted cells will engraft successfully. If markedly different from the recipient's tissue type, transplanted leukocytes may be recognized as foreign by the patient's immune system and therefore be rejected. More commonly, sufficient differences between the engrafted cells and the host's own tissue lead to debilitating consequences as a result of **graft versus host disease** (GVHD). In GVHD, functional T lymphocytes in the proliferating graft recognize host tissue as foreign and mount an immune assault. The disease often begins with a skin rash as transplanted lymphocytes invade the derma and ends in death as the avenging lymphocytes systematically neutralize every organ system of the host.

Recent discoveries have led to useful ways to limit or prevent GVHD. These advances have decreased the morbidity of marrow transplants and have substantially increased the potential pool of bone marrow donors for a given patient. Potent immunosuppressive agents, including steroids, cyclosporin, and anti-T cell antiserum, effectively decrease the immune function of the transplanted lymphocytes. Another useful approach involves "purging"—physical removal of T cells from bone marrow prior to transplantation. T cell-depleted bone marrow is much less capable of causing acute GVHD than untreated marrow. These techniques have allowed successful transplantation of bone marrow obtained from unrelated donors. Unrelated transplants were never possible before these advances, since GVHD would almost certainly develop even when the antigenic type of the donor's leukocytes closely matched that of the recipient's. Thus many patients died for lack of a related donor. Today, transplants of unrelated marrow are common.

Many problems remain, however. One of the most serious, and the most common, is donor identification. An unrelated transplant is successful only if the donor's leukocyte antigens match closely with those of the recipient. Since there are several antigenic determinants and each can be occupied by any one of several genes, there are thousands of possible combinations of leukocyte antigens. The chance that any one individual's cells will randomly match those of another is less than one in a million. Therefore, identification of a suitable donor is a little more complicated than finding a needle in a haystack. On the other hand, these odds virtually guarantee that suitable donors are not only available but, in all probability, plentiful in the general population. Finding them is a formidable problem that often generates intense frustration when donors for terminally ill transplant candidates are not quickly identified.

To address this problem, bone marrow transplant registries have been established. In these registries, results of extensive leukocyte antigen typing are stored in a computer. Typing is performed on leukocytes isolated from a small sample of blood, so the procedure does not markedly inconvenience prospective donors. For some registries, potential donors of a specific ethnic background are targeted; in others, blood samples are obtained from as many healthy individuals as possible, regardless of their heritage. The database is searched when an individual in need of a transplant cannot identify a suitable relative. In conjunction with continued development of methods to reduce or eliminate GVHD, the expanding bone marrow transplant registries may someday allow identification of a donor for anyone who needs a bone marrow transplant.

The MCV value reflects the average volume of each red cell. It is calculated as follows:

$$MCV = \frac{Hematocrit}{Number\ of\ red\ cells\ packed} \qquad (3)$$

$$Ex.: \frac{0.450}{5 \times 10^{12}\ cells/L} = 0.090 \times 10^{-12}\ L/cell$$
$$= 90\ fl\ (1\ fl$$
$$= 10^{-15}\ L)$$

Each gram of hemoglobin can combine with and transport 1.34 ml of oxygen. Thus, the oxygen carrying capacity of 1 dl of normal blood containing 15 g of hemoglobin is $15 \times 1.34 = 20.1$ ml of oxygen.

Red Cell Morphology ■ In addition to revealing alterations in absolute values, stained blood films may provide valuable information based on the morphologic appearance of blood cells. Erythrocytes are formed from precursor blast cells in the bone marrow (Fig. 11–4). This process, termed **erythropoiesis,** is regulated by the hormone **erythropoietin,** a protein secreted by kidney cells. Changes in red cell appearance occur in a variety of pathologic conditions (Fig. 11–5). Variations in size of circular cells is referred to as **anisocytosis.** Larger-than-normal erythrocytes are termed **macrocytic;** smaller-than-normal erythrocytes are referred to as **microcytic.** Irregularities in the shape of erythrocytes is referred to as **poikilocytosis. Schistocytes** are fragments of red cells damaged during blood flow through abnormal blood vessels or cardiac prostheses. **Burr cells** are spiked erythrocytes generated by alterations in the plasma environment.

The hemoglobin content of erythrocytes is also reflected in the staining pattern of cells on dried films. Normal cells appear red-orange throughout, with a very slight central pallor due to the cell shape. **Hypochromic** cells appear pale with only a ring of deeply colored hemoglobin on the periphery. Other pathologic variations in red cell appearance include **spherocytes,** small, densely staining red cells with loss of biconcavity due to congenital or acquired cell membrane abnormalities, and **target cells,** with a densely staining central area with a pale surrounding area. Target cells are thin but bulge in the middle, in contrast to normal erythrocytes. This alteration is a consequence of **hemoglobinopathies,** mutations in the structure of hemoglobin. Target cells are observed in liver disease and after splenectomy.

Nucleated red cells are normally not seen in peripheral blood because their nucleus is lost before they move from the bone marrow into the blood. However, they appear in many blood and marrow disorders, and their presence can be of diagnostic significance. One type of nucleated red cell, the **normoblast** (Fig. 11–4), is seen in several types of anemias, especially when the marrow is actively responding to demand for new erythrocytes. In seriously ill patients, the appearance of normoblasts in peripheral blood is a grave

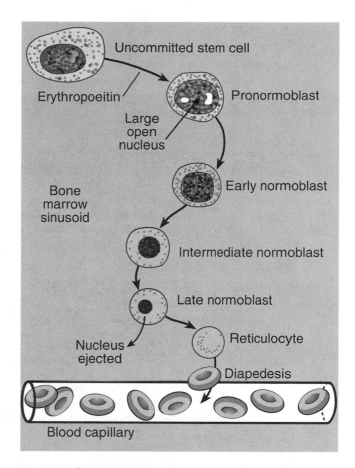

Figure 11–4 ■ Erythropoiesis. Erythrocyte production in healthy adults occurs in marrow sinusoids. Driven by the hormone erythropoietin, the uncommitted stem cell differentiates along the erythrocyte lineage, forming normoblasts (also referred to as erythroblasts or burst-forming cells), reticulocytes, and, finally, mature erythrocytes, which enter the bloodstream by the process of diapedesis.

prognostic sign preceding death, often by a number of hours. Another nucleated erythrocyte, the **megaloblast,** is seen in peripheral blood in pernicious anemia and folic acid deficiency.

Destruction of Erythrocytes ■ Red cells circulate for about 120 days after they are released from the marrow. Some of the senescent (old) red cells break up (hemolyze) in the bloodstream, but the majority are engulfed by macrophages in the monocyte-macrophage system. The hemoglobin released on destruction of red cells is metabolically catabolized and eventually reused in the synthesis of new hemoglobin. Hemoglobin released by red cells that lyse in the circulation either binds to **haptoglobin,** a protein in plasma, or is broken down to globin and heme. Heme binds a second plasma carrier protein, **hemopexin,** which, like haptoglobin, is cleared from the circulation by macrophages in the liver. In the macrophage, released hemoglobin is first broken into globin and heme. The globin portion is catabolized by proteases into constituent amino acids that are used in protein

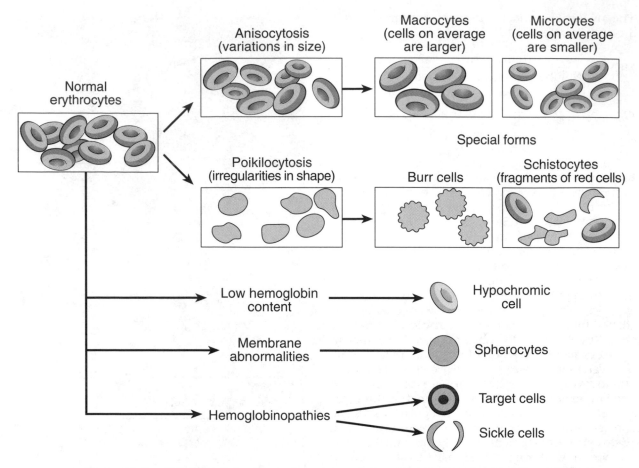

Figure 11–5 ■ Pathologic changes in erythrocyte morphology. Target cells and sickle cells result from abnormal hemoglobin content. Hypochromic cells exhibit ring staining around a pale interior as a result of their low hemoglobin content. Spherocytes are small, densely stained cells.

synthesis. Heme is broken down into free iron (Fe^{3+}) and **biliverdin,** a green substance that is further reduced to **bilirubin** (Chap. 28).

Iron Recycling ■ Most of the iron needed for new hemoglobin synthesis is obtained from the heme of senescent red cells. Iron released by macrophages is transported in the ferric state in plasma bound to the iron transporting protein, **transferrin.** Cells that need iron (e.g., for heme synthesis) possess membrane receptors to which transferrin binds. The receptor-bound transferrin is then internalized. The iron is released, reduced intracellularly to the ferrous state, and either incorporated into heme or stored as **ferritin,** a complex of protein and ferrous hydroxide. Iron is also stored as ferritin by macrophages in the liver. A portion of the ferritin is catabolized to **hemosiderin,** an insoluble compound consisting of crystalline aggregates of ferritin. The accumulation of large amounts of hemosiderin formed during periods of massive hemolysis can result in damage to vital organs, including the heart, pancreas, and liver.

The recycling of iron is quite efficient, but small amounts are continuously lost. Iron loss increases substantially in women during menstruation. Iron stores must be replenished by dietary uptake. The majority of iron in the diet is derived from heme in meat ("organic iron"), but iron can also be provided by absorption of inorganic iron by intestinal epithelial cells. In these cells, iron attached to heme is released and reduced to the ferrous form (Fe^{2+}) by intracellular flavoprotein. The reduced iron (both released from heme and absorbed as the inorganic ion) is transported through the cytoplasm bound to a transferrinlike protein. When it is released to the plasma, it is oxidized to the ferric state and bound to transferrin for use in heme synthesis.

■ Platelets Participate in Clotting

Platelets are irregularly shaped, disklike fragments of the membrane of their precursor cell, the **megakaryocyte.** Megakaryocytes shed platelets in the bone marrow **sinusoids.** From there the platelets are released to the blood, where they function in hemostasis. Several factors stimulate megakaryocytes to release platelets, including the hormone **thrombopoietin,** which is generated and released into the bloodstream when the number of circulating platelets drops. Platelets have no defined nucleus. They are one-fourth to one-third the size of erythrocytes. Platelets survive 15–45 days and possess physiologically important proteins, stored in intracellular granules, which are secreted when the platelets are activated during coagulation. The role of platelets in blood clotting is discussed below.

■ Leukocytes Participate in Host Defense

Each of the three general types of leukocytes, **myeloid, lymphoid,** and **monocytic,** follows a separate line of development from primitive cells (Fig. 11–2). Mature cells of the myeloid series are termed **granulocytes,** based on their appearance after staining with polychromatic dyes, such as **Wright stain.** While monocytes and lymphocytes may also possess cytoplasmic granules, they are not clearly visualized with stains commonly used. Thus, monocytes and lymphocytes are often referred to as **agranular leukocytes.**

The nuclei of most mature granulocytes are divided into two to five oval lobes connected by thin strands of chromatin. This nuclear separation imparts a multinuclear appearance to granulocytes, which are therefore also known as **polymorphonuclear leukocytes.** Three distinct types of granulocytes have been identified based on staining reactions of their cytoplasm with polychromatic dyes: **eosinophils, basophils,** and **neutrophils.**

Neutrophils ■ Neutrophils are usually the most prevalent leukocyte in peripheral blood. These are dynamic cells that respond instantly to microbial invasion by detecting foreign proteins or changes in host defense network proteins. Neutrophils provide an efficient defense against pathogens that have gotten past physical barriers such as the skin. Defects in neutrophil function quickly lead to massive infection and, quite often, death.

Neutrophils are amoebalike phagocytic cells. Invading bacteria induce neutrophil **chemotaxis**—migration to the site of infection. Chemotaxis is initiated by the release of **chemotactic factors** from the bacteria or by chemotactic factor generation in the blood plasma or tissues. Chemotactic factors are generated when bacteria or their products bind to circulating antibodies, by tissue cells when infected with bacteria, and by lymphocytes and platelets after interaction with bacteria.

After neutrophils chemotactically migrate to the site of infection, they engulf (**phagocytose**) the invading pathogen. Phagocytosis is facilitated when the bacteria are coated with the host defense proteins known as **opsonins.**

A burst of metabolic events occurs in the neutrophil after phagocytosis (Fig. 11–6). In the phagocytic vacuole, the bacterium is exposed to enzymes that were originally positioned on the cell surface. Thus, phagocytosis involves invagination and then vacuolization of the segment of membrane to which a pathogen is

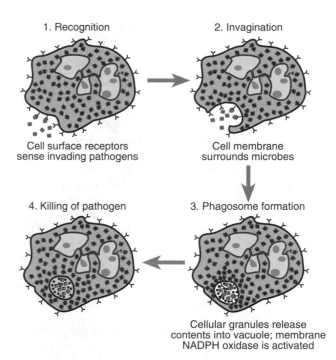

Figure 11–6 ■ Steps in phagocytosis and intracellular killing by neutrophils. (1) Cell surface receptors, including those for exposed opsonins, sense invading pathogens. (2) The neutrophil plasma membrane invaginates to surround the organisms. Bacteria are exposed to enzymes that were initially on the outer surface of the neutrophil plasma membrane. (3) A phagosome is created when the invaginated vacuole closes, trapping the bacteria inside the neutrophil. (4) Potent metabolic processes are activated to kill the ingested microbes. These include activation of the respiratory burst, resulting in generation of potent oxidants within the vacuole, and secretion of bacteria-killing enzymes into the vacuole from neutrophil granules.

bound. Membrane-bound enzymes activated when the phagocytic vacuole closes work in conjunction with enzymes secreted from intracellular **granules** into the phagocytic vacuole to destroy the invading pathogen efficiently. One important membrane-bound enzyme, **nicotinamide adenine dinucleotide phosphate (NADPH) oxidase,** produces **superoxide anion** (O_2^-). Superoxide is an unstable free radical that kills bacteria directly. Superoxide also participates in secondary free radical reactions to generate other potent antimicrobial agents, such as hydrogen peroxide. Superoxide generation in the phagocytic vacuole proceeds at the expense of reducing agents oxidized in the cytoplasm. The reducing agent, NADPH, is generated from glucose by the activity of the **hexose monophosphate shunt.** Aerobic cells generate reduced **nicotinamide adenine dinucleotide (NADH)** and ATP when glucose is oxidized to carbon dioxide. The hexose monophosphate shunt operates in neutrophils and other cells when large amounts of NADPH are needed to maintain intracellular reducing activity.

Reduction of oxygen by the NADPH oxidase that

generates superoxide in neutrophils is driven by increased availability of NADPH after phagocytosis:

$$NADPH + 2O_2 \xrightarrow[\text{NADPH Oxidase}]{} 2O_2^- + NADP^+ + H^+$$

A complex set of biochemical events unfolds after phagocytosis to activate the neutrophil NADPH oxidase, which is dormant in resting cells. The oxidase is activated by its interaction with an activated **G protein** and cytosolic molecules that are generated during phagocytosis. The NADPH oxidase is activated in a manner that allows the enzyme to secrete the toxic free radical, superoxide, into the phagocytic vacuole while oxidizing NADPH in the cell's cytoplasm. This explosion of metabolic activity, collectively termed the **respiratory burst,** leads to the generation of potent, reactive agents not otherwise generated in biologic systems. These agents are so reactive that they actually generate light (biologic chemiluminescence) when they oxidize components in the bacterial cell wall.

Other bactericidal agents and processes operate in neutrophils to ensure efficient bacterial killing. Phagocytized bacteria encounter intracellular **defensins,** cationic proteins that bind to and inhibit the replication of bacteria. Defensins and other antibacterial agents pour into the phagocytic vacuole after phagocytosis. Agents stored in neutrophil granules include **lysozyme,** a bacteriolytic enzyme, and **myeloperoxidase,** which reacts with hydrogen peroxide to generate potent, bacteria-killing oxidants. One of the oxidants generated by the myeloperoxidase reaction is hypochlorous acid (HOCl), the killing agent found in household bleach. Granules also contain **collagenase** and other **proteases.**

Eosinophils ■ Eosinophils are rare in the circulation but are easily identified on stained blood films. As the name implies, the eosinophil takes on a deep eosin color during polychromatic staining; the large, refractile cytoplasmic granules of these cells stain orange-red to bright yellow. Like neutrophils, eosinophils migrate to sites where they are needed and exhibit a metabolic burst when activated. The eosinophil participates in defense against certain parasites and is involved in allergic reactions. Exposure of allergic individuals to an allergen often results in a transient increase in eosinophil count **(eosinophilia);** infection with parasites often results in a sustained overproduction of eosinophils.

Basophils ■ Basophils are polymorphonuclear leukocytes with multiple pleomorphic, coarse, deep-staining metachromatic granules throughout their cytoplasm. These granules contain **heparin** and **histamine,** which have anticoagulant and vasodilating properties, respectively. The release of these and other mediators by basophils increases regional blood flow, facilitating the transport of other leukocytes to areas of infection and allergic reactivity or other forms of hypersensitivity.

Monocytes and Lymphocytes ■ In contrast to granulocytes, monocytes and lymphocytes are mononuclear cells. Monocytes are phagocytic cells but lymphocytes are not; both participate in multiple aspects of immunity. Monocytes were originally differentiated from lymphocytes based on morphologic characteristics. The cytoplasm of monocytes appears pale blue or blue-gray with Wright stain. The cytoplasm contains multiple fine reddish blue granules. Monocytes may be shaped like a kidney bean, indented, or shaped like a horseshoe. Frequently, however, they are rounded or ovoid. Upon activation, monocytes transform into **macrophages,** large, active mononuclear phagocytes (see below).

Morphologically, circulating lymphocytes have been assigned to two broad categories—large and small lymphocytes. In blood, small lymphocytes are more numerous than larger ones; the latter closely resemble monocytes. Small lymphocytes possess a deeply stained, coarse nucleus that is large in relation to the remainder of the cell so that often only a small rim of cytoplasm appears around parts of the nucleus. In contrast, a broad band of cytoplasm surrounds the nucleus of large lymphocytes; the nucleus of these cells is similar in size and appearance to that of the smaller lymphocytes.

The morphologic homogeneity of lymphocytes obscures their functional heterogeneity. As is discussed below, lymphocytes participate in multiple aspects of the immune response. Lymphocyte subtypes in blood are often identified based on their reaction with fluorescent monoclonal antibodies. The majority of circulating lymphocytes are **T lymphocytes** or T cells (for "thymus-dependent lymphocytes"). These cells participate in **delayed-type hypersensitivity** and other **cell-mediated immune responses** that do not depend on antibody. T cells comprise 40–60% of the total circulating pool of lymphocytes.

Subtypes of T cells have been identified using fluorescent monoclonal antibodies to specific cell surface antigens, known as CD antigens. All T cells possess the common CD3 antigen. So-called **helper T cells** possess the CD4 antigen cluster, while **suppressor T cells** lack CD4 but possess CD8. Patients with AIDS show decreased circulating levels of CD4-positive cells.

Twenty to 30% of circulating lymphocytes are **B cells,** which have immunoglobulin or antibody on their surface. B cells are bone marrow-derived lymphocytes that when immunologically activated transform into **plasma cells** that secrete immunoglobulin. Lymphocytes not characteristic of either T cells or B cells are appropriately called **null cells.** The entire scope of the function of null cells, which comprise only 1–5% of circulating lymphocytes, is unknown, but it has been established that null cells are capable of destroying tumor cells and virus-infected cells.

While B lymphocytes mediate immune responses

by releasing antibody, T lymphocytes often exert their effects by synthesizing and releasing **cytokines,** hormonelike proteins that act by binding specific receptors on their target cells. Recent research has led to the discovery of many cytokines, with activities ranging from tumor destruction (e.g., **tumor necrosis factor**) to promotion of blood cell production. Cytokines that limit viral replication in cells **(interferons),** suppress or potentiate the function of T cells, stimulate macrophages, and activate neutrophils have been identified. In some cases, these cytokines, like other hormones, can exert potent effects when supplied exogenously. For example, **colony stimulating factors** injected into cancer patients can prevent the decrease in production of leukocytes that results from administration of chemotherapeutic drugs or radiation therapy. The technology of molecular biology is employed to produce these cytokines for therapy. In this process, sections of lymphocyte DNA which contain the gene that codes for the specified cytokine are isolated and then **transfected** into a bacterial cell, fungus, or rapidly growing mammalian cell. These cells then produce the cytokine and release it into their culture supernatant, from which it can be purified, concentrated, and sterilized for injection. The biologic diversity and potency of the cytokines has opened the door to the development of a variety of new pharmacologic agents that have proved useful in the treatment of cancer, immune disorders, and other diseases.

■ Blood Cells Are Born in the Bone Marrow

Mature cells are transient residents of blood. Erythrocytes survive in the circulation for about 120 days, when they are broken down and their components recycled as discussed above. Platelets have an average life span of 15–45 days in the circulation; many, if not most, of these cells are consumed as they continuously participate in day-to-day hemostasis. The rate of platelet consumption accelerates rapidly during repair of bleeding caused by trauma. Leukocytes have a variable life span. Some lymphocytes circulate for 1 year or longer after production. Neutrophils, constantly guarding body fluids and tissues against infection, have a circulating half-life of only a few hours. Neutrophils and other blood cells must therefore be continuously replenished.

The process of blood cell generation is termed **hematopoiesis.** In healthy adults, hematopoiesis occurs only in the bone marrow. **Extramedullary hematopoiesis** (e.g., the generation of blood cells in the spleen) is observed only in some disease states, such as leukemia. Hematopoietic cells are found in high levels in the liver, spleen, and blood of the developing fetus. Shortly before birth, blood cell production gradually begins to shift to the marrow. In newborns, the hematopoietic cell content of the circulating blood is relatively high; hematopoietic cells are also found in

blood of adults, but in extremely low numbers. Large numbers of hematopoietic cells can be recovered from aspirates of the iliac crest, sternum, pelvic bones, long bones, and ribs of adults. Within the bones, hematopoietic cells germinate in extravascular sinuses, called **marrow stroma.** Circulating factors and factors released from capillary endothelial cells, stromal fibroblasts, and mature blood cells regulate the generation of immature blood cells from hematopoietic cells and the subsequent differentiation of newly formed immature cells. These factors are collectively termed **hematopoietins** (including **leukopoietins** and **erythropoietin**) or colony stimulating factors, a name derived from the initial findings of research into these factors. (Researchers succeeded in growing colonies of blood cells from cultures of isolated bone marrow cells, but only in the presence of certain proteins. These proteins were thus termed "colony stimulating factors".) Recently, researchers have attached the name **interleukin** to newly identified leukopoietins. The terms are often used interchangeably but may not have the same meaning. Thus, while all hematopoietins are cytokines, not all cytokines are hematopoietins (other cytokines regulate function of cells in other organs of the body). All interleukins are hematopoietins, but not all hematopoietins are interleukins. Many different interleukins exist, and they have been named in order of their discovery (e.g., interleukin 1, interleukin 2). In contrast, erythropoietin refers to one specific protein that promotes erythropoiesis in vitro and in vivo. Erythropoietin solutions are often injected into anemic individuals to stimulate erythrocyte production.

Blood cell production begins with the proliferation of **uncommitted** or **pluripotential stem cells.** Depending on the stimulating factors available, the progeny of pluripotential stem cells may be other uncommitted stem cells or stem cells committed to development along a certain lineage. The committed stem cells include **myeloblasts,** which form cells of the myeloid series (neutrophils, basophils, and eosinophils), **erythroblasts, lymphoblasts,** and **monoblasts** (Figs. 11–2, 11–7). Promoted by hematopoietins and other cytokines, each of these blast cells differentiates further, a process that ultimately results in the formation of mature blood cells. This is a dynamic process; the hematopoietic cells of the bone marrow are among the most actively reproducing cells of the body. Interruption of hematopoiesis (e.g., by cancer treatment) results in the eventual disappearance of granulocytes from the blood, a condition known as **granulocytopenia** (or, when specific to neutrophils, **neutropenia**), in a matter of hours. Platelets disappear next **(thrombocytopenia),** followed by erythrocytes—a sequence that reflects the circulating life span of each cell. Often, hematopoiesis can be restored after its interruption by an infusion of viable hematopoietic cells (e.g., a bone marrow transplant).

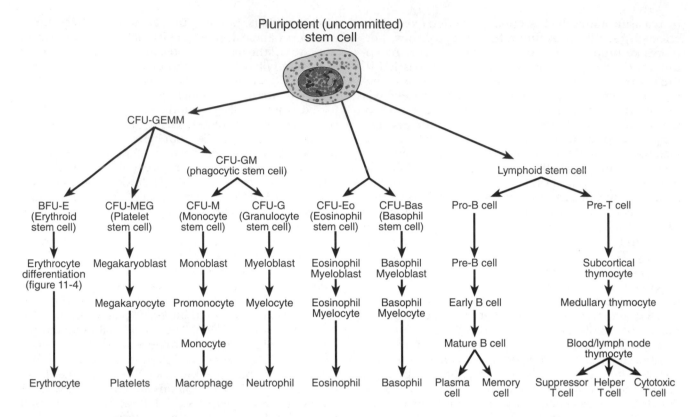

Pluripotent (uncommitted)
stem cell

CFU-GEMM

CFU-GM
(phagocytic stem cell)

Lymphoid stem cell

| BFU-E (Erythroid stem cell) | CFU-MEG (Platelet stem cell) | CFU-M (Monocyte stem cell) | CFU-G (Granulocyte stem cell) | CFU-Eo (Eosinophil stem cell) | CFU-Bas (Basophil stem cell) | Pro-B cell | Pre-T cell |

Erythrocyte differentiation (figure 11-4) — Megakaryoblast — Monoblast — Myeloblast — Eosinophil Myeloblast — Basophil Myeloblast — Pre-B cell — Subcortical thymocyte

Megakaryocyte — Promonocyte — Myelocyte — Eosinophil Myelocyte — Basophil Myelocyte — Early B cell — Medullary thymocyte

Monocyte — Mature B cell — Blood/lymph node thymocyte

| Erythrocyte | Platelets | Macrophage | Neutrophil | Eosinophil | Basophil | Plasma cell | Memory cell | Suppressor T cell | Helper T cell | Cytotoxic T cell |

Figure 11–7 ■ Hematopoiesis. All circulating blood cells are believed to be derived from a common, uncommitted bone marrow progenitor, the pluripotent stem cell. This cell differentiates along different lineages, depending on the conditions it encounters and the levels of individual hematopoietins available. CFU-GEMM, granulocyte-erythrocyte-macrophage-megakaryocyte colony forming unit; CFU-GM, granulocyte-macrophage colony forming unit; CFU-G, granulocyte colony forming unit; CFU-M, macrophage colony forming unit; BFU-E, erythroid burst forming unit; CFU-MEG, megakaryocyte colony forming unit; CFU-Eo, eosinophil colony forming unit; CFU-Bas, basophil colony forming unit.

The Immune System

Immunity, or resistance to infection, derives from the activity and intact functioning of two tightly interrelated systems, the **innate immune system** and the **adaptive immune system.** Elements of the innate, or natural, immune system include **exterior defenses,** such as skin and mucous membranes, **phagocytic leukocytes,** and serum proteins, which act nonspecifically and quickly against microbial invaders. Microbes that escape the onslaught of cells and molecules of the innate immune system face destruction by T and B cells of the adaptive immune system. Activation of the adaptive immune system results in the generation of antibodies and cells that specifically target the inducing organism or foreign molecule. Unlike the innate system, adaptive or acquired immune responses develop gradually but exhibit **memory.** Therefore, repeat exposure to the same infectious agent results in improved resistance mediated by the specific aspects of the adaptive immune system. Working together, elements of the natural and adaptive immune system provide a considerable obstacle to the establishment and long-term survival of infectious agents.

■ The Innate Immune System Consists of Nonspecific Defenses

Infectious agents cannot easily penetrate intact skin, the first line of defense against infection. Infection is a major complication when the intact skin barrier is compromised, such as by burns or trauma. Even a small needle prick can result in a fatal infection.

Natural openings to body cavities and glands are an effective entry point of infectious agents. Usually, however, these openings are protected from invasion by pathogens in at least two ways. First, they are coated with **mucus** and other **secretions** that contain **secretory immunoglobulins** as well as antibacterial enzymes, such as lysozyme. Second, organisms that invade these openings cannot easily reach the blood but instead lodge in an organ that communicates with both the exterior and interior of the body, such as the lung or the stomach. Many pathogens cannot survive the low pH of stomach acid. In the lungs, organisms face the efficient phagocytic activity of **alveolar macrophages.** These cells, derived from blood monocytes, are mobile but confined to the pulmonary capillary network. As efficient phagocytic cells, they continuously patrol the pulmonary vasculature to remove inhaled microbes.

Microbes that successfully break through these physical barriers face destruction by the **fixed macrophages** of the **monocyte-macrophage system.** These cells line the sinusoids and vasculature of many organs, including the liver, spleen, and bone marrow. The nonmobile fixed phagocytic macrophages efficiently remove foreign particles, including bacteria, from the circulation.

■ Inflammation Is a Multifaceted Process

Microbial invaders that lodge in body tissues and begin to proliferate trigger an inflammatory response (Fig. 11–8). Inflammation provides a multifaceted defense against tissue invasion by pathogens. The inflammatory response is initiated by circulating proteins and blood cells when they contact invaders in a tissue. The response results in increased blood flow to the affected tissues, which enhances delivery of immune system elements to the site. This produces redness, heat, and swelling (edema) of the affected tissue. Blood cells participating in the inflammatory response release a variety of **inflammatory mediators** that perpetuate the response. If the pathogens persist, the inflammatory response may become chronic and itself cause substantial tissue damage. Not only microbes but also proteins, chemicals, and toxins that the body recognizes as foreign can induce an inflammatory response.

Certain inflammatory mediators increase blood flow to the inflamed area. Other mediators increase capillary permeability, allowing diffusion of large molecules across the endothelium and into the infected site. These molecules may be plasma proteins or may be generated by plasma proteins or substances released by blood leukocytes. They often play important roles in eliminating the pathogenic agent or enhancing the inflammatory response. Finally, chemotactic factors produced by cells that arrived early in the inflammatory cascade cause polymorphonuclear leukocytes to migrate from the blood to the affected area. Neutrophils are an important participant in the inflammatory response. They can exert potent antimicrobial effects as well as release a variety of agents that can further amplify and perpetuate the response.

The remarkable ability of the inflammatory response to sustain itself while it generates potent cytolytic agents can result in many undesirable effects, including extensive tissue damage and pain. This has resulted in the search for and identification of a variety of **antiinflammatory agents** to control some of these undesirable effects. These agents are designed to block some of the consequences of the inflammatory response without compromising its antimicrobial efficiency. They do this by neutralizing inflammatory mediators or by preventing inflammatory cells from releasing or responding to inflammatory mediators.

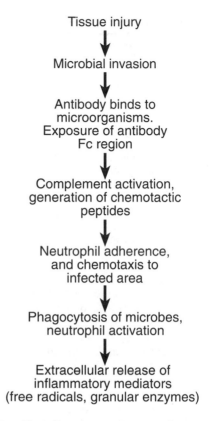

Figure 11–8 ■ The inflammatory response. Inflammation can proceed along a number of divergent pathways, each involving inflammatory cells (e.g., neutrophils) and mediators. In this illustration, a possible route of inflammation initiated by tissue injury is shown.

■ Defense Mechanisms Are Integrated Systems

As discussed above, the innate and adaptive immune systems work together in ways that obscure their differences. Indeed, consideration of these two systems as distinct, individual entities is neither justified nor correct, owing to their extensive interdependence. They are described individually only as an aid to their presentation. In this respect, it is important to define the characteristics that differentiate each system (Table 11-3). In general, responses of the innate immune system are neither **specific** nor **inducible;** that is, the response is not programmed by or directed against a specific pathogen and is not amplified as a result of previous encounters with the pathogen. The adaptive response, in contrast, is both specific and inducible; the response is set in motion by a particular pathogen and develops against that specific pathogen. The adaptive response exhibits immunologic memory, because it generates long-lived cells and molecules that react against specific inducing organisms.

While characteristics of the innate and adaptive immune system differ, each system depends on elements of the other for optimal functioning. Initiation of responses by the innate system as well as efficient phago-

TABLE 11–3 ■ Innate versus Adaptive Immunity

	Innate	Adaptive
Resistance	Not improved by repeat infection	Improved by previous infection
Specificity	Not directed toward specific pathogen	Targeted response directed by specific elements of immune system
Soluble factors	Lysozyme, complement, acute phase proteins, interferon, cytokines	Antibodies
Cells	Phagocytic leukocytes; NK cells	T cells, B cells

cytosis by neutrophils in the tissues often depends on the presence of a small amount of specific antibody in plasma. Antibody is generated by cells of the adaptive immune system in response to specific foreign molecules, called **antigens** (see below). In turn, antibodies and other mediators of the adaptive immune system depend on neutrophils and other effector agents usually associated with the innate immune system to function effectively. Thus, neither the innate nor the adaptive system stands alone, but rather depends on highly evolved interactive mechanisms to kill and remove microbial intruders of the sterile interior of the organism efficiently.

■ Adaptive Immunity Is Specific and Acquired

The adaptive immune system can be considered at three levels: the **afferent arm,** which gives the system its remarkable ability to recognize specific antigenic determinants of a wide range of infectious agents; the **efferent arm,** which supplies a cellular and molecular assault on the invading pathogens; and **immunologic memory,** which specifically accelerates and potentiates subsequent responses to the same activating agent or antigen.

The specificity of the recognition, effector, and memory aspects of the adaptive immune system derives from the specificity of antibody molecules as well as that of receptors on T and B lymphocytes. The lymphocytes of the immune system are capable of recognizing and specifically responding to hundreds of thousands of potential antigens, which may be presented, for example, as glycoproteins on the surface of bacteria, the coat protein of viruses, microbial toxins, or membranes of infected cells. Only a few circulating lymphocytes need to recognize an individual antigen initially. This initial recognition induces proliferation of the responsive cell, a process known as **clonal selection** (Fig. 11–9). Clonal selection amplifies the number of specific T or B cells (i.e., T or B lymphocytes programmed to respond to the inciting stimulus).

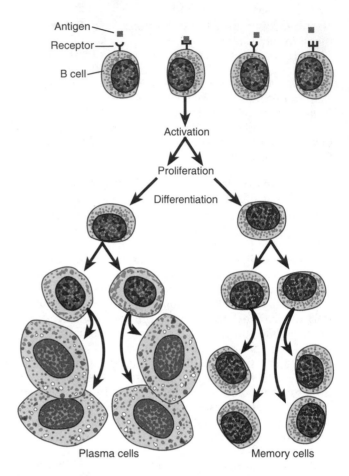

Figure 11–9 ■ Clonal selection of committed lymphocytes. In this model, only the clone of lymphocytes that has the unique ability to recognize the antigen of interest proliferates, thereby generating memory cells as well as effector cells that are specific to the inducing stimulus. This proliferation is initiated by the interaction of a specific recognition lymphocyte (afferent cell) with the antigen. Cells then proliferate and differentiate into either memory cells, which potentiate subsequent responses to the inciting antigen, or plasma cells, which secrete antibody.

While all of the cells generated after a single clone has expanded are specific for the inducing antigen, they may not all possess the same functional characteristics. Some of the daughter lymphocytes may be effector cells. For example, when B cells are activated, their progeny **plasma cells** are capable of generating antibodies. Other progeny in the expanded clone may exert an afferent recognition function and thereby function as **memory cells.** The increased number of these cells, which mimic the reactive specificity of the original lymphocytes that responded to the antigen, accelerate responsiveness when the antigen is encountered again. Memory cells thus account for one of the principal tenets of immunity: resistance is increased after initial exposure to the infectious agent. Long-term immunity to many viruses, such as influenza, measles, smallpox, and polio, can be induced by **vaccination** with a killed or mutant form of the pathogen.

■ The Adaptive Immune Response Involves Cellular and Humoral Components

Depending on the nature of the stimulus, its mode of presentation, and prior challenges to the immune system, an antigen may elicit either a **cellular** or **humoral** immune response. Both are ultimately mediated by lymphocytes, the cellular response by T cells and humoral response by B cells. As discussed above, stimulated B cells differentiate into plasma cells, which secrete antibody specific for the inciting stimulus. The antibody can be found in a variety of body fluids (or humors), including saliva, other secretions, and plasma. Cellular immunity (or cell-mediated immunity) is accomplished by activated T cells. The effector cells of this response do not secrete antibody but exert their influence by a variety of cellular mechanisms. These effector processes include direct cytotoxicity (mediated by **cytotoxic T cells**), suppression or activation of immune mechanisms in other cells (suppressor or helper T cells, respectively), and secretion of cytotoxic or immunomodulating cytokines, such as tumor necrosis factor and interleukin 2. T cells and their products may act directly or exert their effects in concert with other effector cells, such as neutrophils and macrophages.

The immune responses mediated by antibodies and T lymphocytes differ in several important respects. In a general sense, antibodies are known to induce immediate responses to antigens and thereby provoke **immediate hypersensitivity reactions.** For example, allergic or **anaphylactic** hypersensitivity results when a certain type of antibody fixed to the surface of fixed **mast cells** binds to its specific antigen. Antibody binding leads to the release of histamine and other mediators of the allergic response from intracellular granules. Immediate reactions also occur when circulating antibodies bind antigen in the tissues, thereby forming **immune complexes** that activate the **plasma complement system.** The complement system consists of a group of at least nine distinct proteins that circulate in plasma. These are activated as a cascade when the first protein recognizes preformed **immune complexes,** a large crosslinked mesh of antigen molecules bound to antibodies. In addition, complement can be activated when one of the proteins is activated by exposure to the cell wall of certain bacteria. Activation of this system results in edema, an influx of activated phagocytic cells (chemotaxis), and local inflammatory changes.

In contrast to the rapid onset of biologic responses when antigen binds antibody, the consequences of T cell activation are not noticeable until 24–48 hours after antigen challenge. During this time, the T cells that initially recognize the antigen are activated, secrete factors that recruit and activate other cells (e.g., macrophages), and release factors that damage the antigen, cells possessing the antigen, or the surrounding tissue. A common example of T cell activation is the **delayed-type hypersensitivity** reaction to purified protein derivative (PPD), a reaction used to assess prior exposure to the bacteria that cause tuberculosis. Injected under the skin of sensitive individuals, PPD elicits the familiar inflammatory reaction characterized by local erythema and edema 1–2 days later.

T cell-mediated immune responses, while slow to develop, are potent and versatile. These delayed responses provide for defense against many pathogens, including viruses, fungi, and bacteria. T cells are responsible for rejection of transplanted tissue grafts and containment of the growth of neoplastic cells. A deficiency in T cell immunity, such as that associated with AIDS, predisposes the affected patient to a wide array of serious, life-threatening infections.

■ B Cell Immunity Is Mediated by Immunoglobulin

Antibodies, or **immunoglobulins,** are glycoproteins secreted by plasma cells. They are found in high levels in plasma and other body fluids. Antibodies have the ability to bind specifically to the antigenic determinant that induced their secretion. Each antibody molecule consists of four polypeptide chains (two heavy chains and two light chains) held together as a Y-shaped molecule by one or more disulfide bridges. Each polypeptide chain consists of a highly conserved constant region and a variable region, where considerable amino acid sequence heterogeneity is found even within a single antibody class. This amino acid variability accounts for the widely diverse antigen-binding ability of antibody molecules, for it is the variable region that actually combines with the antigen. There are five major classes of immunoglobulins, based on differences in amino acid composition in the constant region of the heavy chain: immunoglobulin G (IgG), IgA, IgD, IgE, and IgM.

Antibody Structure ■ The primary structure of an immunoglobulin is illustrated in Figure 11-10. Each of the four polypeptide chain consists of a **constant** region, where the amino acid sequence is relatively the same within a given antibody class, and a **variable** region, where amino acid sequence varies widely and provides great flexibility in antigen-binding activities. The NH_2 terminal portion of the variable regions, the antigen-binding sites, are known as the **Fab** regions. Each antibody unit possesses two binding sites. The COOH terminal end of the heavy chain is termed the **Fc** region. Polypeptide fragments consisting of Fc and Fab portions of antibody molecules can be generated by protease digestion and separated by chromatography. Fc fragments are able to bind to cells such as neutrophils, monocytes, and mast cells through their **Fc receptors.** Fc receptor binding amplifies the biologic activity of antigen-bound antibody.

Antibody Functions ■ In addition to the ability to bind antigen, the antibody molecule may have a

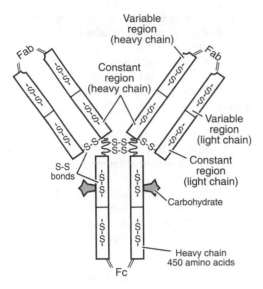

Figure 11–10 ■ Structure of a typical immunoglobulin. Each molecule consists of two heavy chains and two light chains held together in a Y configuration by disulfide bonds. Each heavy chain and light chain possesses a constant region (where the amino acid sequence of individual molecules is similar) and a variable region, where alterations in the amino acid sequence convey on the antibody its individual antigen specificity.

variety of other important biologic functions, depending on its class. Table 11–4 summarizes some characteristics and functions of the major classes of antibodies. IgG is the most prevalent antibody in serum and is responsible for induced immunity to bacteria and other microorganisms. Bound to antigen, IgG can activate serum complement, which releases several inflammatory and bactericidal mediators. At the surface of bacteria, exposed Fc portions of IgG molecules facilitate the phagocytosis of bacteria by blood phagocytes, a process called **opsonization.** IgG exists in serum as a monomer. It can cross the placenta and is secreted into colostrum, thereby protecting the fetus as well as the newborn from infection.

Unlike IgG, IgM and IgA usually exist as polymers of the fundamental Y-shaped antibody unit. In most IgA molecules, two antibody units are held together

by a **secretory piece** (J chain), a protein synthesized by epithelial cells. In this conformation, IgA is actively secreted into saliva, tears, colostrum, and mucus. IgA is thus known as **secretory immunoglobulin.** IgM is the first antibody secreted after an initial immune challenge and thus provides resistance early in the course of infection. IgM consists of five Y units. Its size and large number of antigen binding sites provide the molecule with excellent **agglutinating** ability, the ability to clump particulate antigens, such as bacteria and blood cells. Clumped antigens are efficiently and quickly removed by fixed phagocytic cells of the monocyte-macrophage system.

IgE, a monomeric antibody slightly larger than IgG, avidly binds cells that store and release mediators of allergy and anaphylaxis, including mast cells and basophils. These cells are heavily granulated. The granules contain **histamine, leukotrienes,** and other biologically active agents that increase vascular permeability, dilate blood vessels (and thereby reduce blood pressure), and contract airway smooth muscle. The granules are released when IgE, bound to mast cells at the Fc region, binds its specific antibody. The ensuing allergic response ranges from hay fever, hives, and bronchial asthma (induced by local or inhaled allergens) to systemic anaphylaxis, a potentially fatal response triggered when antigen is given systemically.

IgD, found in plasma and on the surface of some immature B cells, has no known function.

Hemostasis

The blood circulates in a high-pressure, closed system that communicates with all tissue cells and exchanges oxygen, nutrients, and wastes and provides necessary components for host defense. This communication takes place largely in the extensive and highly organized capillary bed. Disruption of the integrity of this fragile vascular bed may result from trivial tissue injury associated with normal physical activity or from massive tissue trauma due to injury or infection and quickly lead to death. Any opening in the vascular network may lead to massive bruising or blood loss if left unrepaired.

TABLE 11–4 ■ Characteristics of Different Antibody Classes

	IgG	IgA	IgM	IgD	IgE
Molecular weight ($\times 10^{-3}$)	150	150, 400	900	180	190
"Y" units/molecule	1	1–2	5	1	1
Concentration in serum (mg/dl)	600–1500	85–300	50–400	<15	0.01–0.03
Crosses placenta	+	−	−	−	−
Enters secretions	+	+ +	−	−	−
Agglutinates particles	+	+	+ + +	−	−
Allergic reactions	+	−	−	−	+ + + +
Complement fixation	+	−	+ +	−	−
Fc receptor binding to monocytes and neutrophils	+ +	−	+	−	−

To minimize bleeding and prevent blood loss after tissue injury, components of the hemostatic system are activated. The components of this dynamic, integrated system include blood platelets, endothelial cells, and plasma coagulation factors, and they may be activated on exposure to foreign surfaces during bleeding or torn tissue at the site of injury or by products released from the interior of damaged cells. Hemostasis can be viewed as four separate but interrelated events: (1) compression and vasoconstriction, which act immediately to stop the flow of blood; (2) formation of a platelet "plug"; (3) blood coagulation; and (4) clot retraction and thrombus dissolution.

■ Physical Factors Immediately Act to Constrain Bleeding

Immediately after tissue injury, blood flow through the disrupted vessel is slowed by the interplay of several important physical factors, including back pressure or compression exerted by the tissue around the injured area and vasoconstriction. The degree of compression varies in different tissue; for example, bleeding below the eye is not readily deterred because the skin in this area is readily distensible. Back pressure increases as extravasated blood accumulates. In some tissues, notably the uterus after childbirth, contraction of underlying muscles compresses blood vessels supplying the tissue and minimizes blood loss. Damaged cells at the site of tissue injury release potent substances that directly cause blood vessels to constrict, including **serotonin, thromboxane A$_2$, epinephrine,** and **fibrinopeptide B.**

■ Platelets Form a Hemostatic Plug

Platelets regulate bleeding in three stages. First, platelets form multicellular aggregates linked by protein strands at sites of openings in blood vessels. These platelet aggregates form a physical barrier that begins to limit blood loss soon after the opening occurs. Second, phospholipids on the platelet plasma membrane catalyze an enzyme, thrombin, which initiates a cascade of events ending in clot formation. Finally, platelets possess multiple storage granules, which they discharge to amplify coagulation.

Platelet activation results in the sequential responses of **adherence, aggregation,** and **secretion** (release). Adherence is initiated when one or more substances released from cells or activated in plasma at the site of hemorrhage bind to receptors in the platelet plasma membrane. Receptor binding results, via second messengers, in adherence (to both other platelets and the endothelial surface of blood vessels) and secretion.

Disruption of the endothelium at sites of tissue injury exposes a variety of proteins in the subendothelial matrix, such as **collagen** and **laminin,** which either induce or support platelet adherence. **von Willebrand factor,** a protein synthesized by endothelial cells and megakaryocytes, enhances platelet adherence by forming a bridge between cell surface receptors and collagen in the subendothelial matrix. The protein **thrombin,** which is generated by the plasma coagulation pathway, is a potent activator of platelet adherence and secretion. Ruptured cells at the site of tissue injury release adenosine diphosphate (ADP), which causes platelets to aggregate at the damaged site. The ADP, collagen, and other substances released from platelets enhance adherence, platelet aggregation, and secretion. Enhancement of blood coagulation by membranes of activated platelets and substances released from activated platelets supplies more thrombin to amplify and continue these platelet responses. Thrombin also catalyzes the formation of fibrin strands. These strands, the basic component of a blood clot, bind platelet aggregates and form a firm **hemostatic plug** that stops or minimizes bleeding from opened blood vessels (Fig. 11–11).

■ Blood Coagulation Results in the Production of Fibrin

Fibrin is a meshlike network of insoluble protein that traps red cells, leukocytes, platelets, and serum at sites of vascular damage, thereby forming a blood clot. The stable, fibrin-based blood clot eventually replaces the unstable platelet plug formed immediately after tissue injury. Fibrin is an insoluble polymer of proteolytic products of the plasma protein **fibrinogen.** Fibrin molecules are cleaved from fibrinogen by the proteolytic enzyme **thrombin,** which is generated in plasma during clotting. In the initial step of fibrin formation, thrombin cleaves four small peptides (fibrin peptides) from each molecule of fibrinogen. The resulting protein-fibrin monomer is surrounded by unaltered fibrinogen molecules. The fibrin monomers begin to polymerize as more molecules of fibrinogen are split by thrombin, resulting in an insoluble matrix of visible fibrous strands. At this stage, the clot is held together by noncovalent forces. A plasma enzyme, **fibrin stabilizing factor,** catalyzes the formation of covalent bonds between strands of polymerized fibrin, thereby stabilizing and tightening the blood clot.

The Coagulation Cascade ■ Blood clotting is mediated by sequential activation of a series of **coagulation factors,** proteins synthesized in the liver which circulate in the plasma in an inactive state. They are referred to by number (designated by a Roman numeral) in sequence based on the order of the discovery of each factor, not position in any specific activation sequence. The plasma coagulation factors and their common names are listed in Table 11–5.

Sequential activation of a series of inactive molecules resulting in a biologic response is called a **metabolic cascade.** The sequential activation of coagulation factors resulting in the conversion of fibrinogen to fibrin (and, hence, clotting) is called the **coagulation cascade.** Deficiency or deletion of any one factor of the cascade carries severe consequences. Individuals deficient in factor VIII (von Willebrand factor), for example, display prolonged bleeding time on

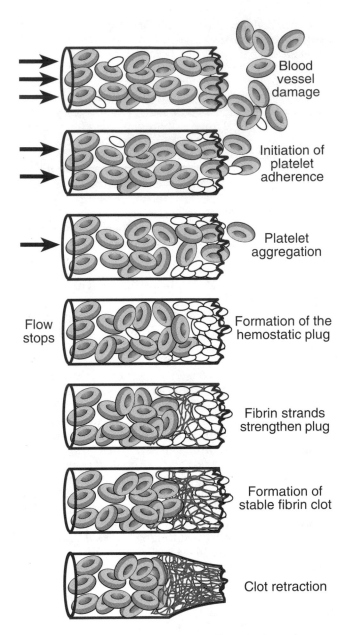

Blood vessel damage

Initiation of platelet adherence

Platelet aggregation

Flow stops

Formation of the hemostatic plug

Fibrin strands strengthen plug

Formation of stable fibrin clot

Clot retraction

Figure 11–11 ■ Formation of a platelet-fibrin clot during hemostasis. Injury to the blood vessel results immediately in a series of events, including activation of the coagulation cascade and induction of platelet adherence, aggregation, and secretion. Within minutes, a hemostatic plug, consisting of platelets and fibrin, is formed. During the next few hours, the fibrin strands stabilize, strengthening the fibrin-platelet plug and resulting in clot formation.

tissue injury due to delayed clotting. Those who lack factor VIII have hemophilia, a condition resulting in severe coagulation defects.

Two separate coagulation cascades result in coagulation in different circumstances (Fig. 11–12). The two systems are the **extrinsic coagulation pathway** and **intrinsic coagulation pathway.** The final steps to fibrin formation are common to both pathways. In the

TABLE 11–5 ■ Factors of the Clotting Cascade

Scientific Name	Common Name	Other Names
Factor I	Fibrinogen	
Factor II	Prothrombin	
Factor III	Tissue thromboplastin	Tissue factor
Factor IV	Calcium	
Factor V	Proaccelerin	Labile factor
Factor VII	Proconvertin	Serum prothrombin conversion accelerator (SPCA)
Factor VIII	Antihemophilic factor	Platelet cofactor 1
Factor IX	Christmas factor	Platelet thromboplastin component
Factor X	Stuart factor	
Factor XI	Plasma thromboplastin antecedent	
Factor XII	Hageman factor	Contact factor
Factor XIII	Fibrin stabilizing factor	

intrinsic system, all the factors required for coagulation are present in the circulation. For initiation of the extrinsic coagulation cascade, a factor extrinsic to blood but released from injured tissue, called **tissue thromboplastin** or **tissue factor** (factor III), is required. **Phospholipids** are required for activation of both coagulation pathways. Phospholipids provide a surface for the efficient interaction of several factors. A component of tissue factor provides the necessary phospholipid for the extrinsic pathway. Phospholipids required for efficient activation of the intrinsic pathway are found on platelet membranes.

Activation of both pathways results in the conversion of fibrinogen to fibrin. The final events leading to fibrin formation by both pathways result from activation of the **common pathway.** The common pathway is initiated by conversion of inactive clotting factor X to its active form, factor Xa (Fig. 11–12). Along with factor V, calcium ions, and platelet factor 3 (PF-3), factor Xa converts prothrombin to thrombin. Thrombin influences subsequent coagulation events in several ways. First, it catalyzes fibrin formation from fibrinogen. Second, thrombin activates factor XIII (to XIIIa), which physically interacts with fibrin to form a more stable clot. Thrombin also enhances the activity of clotting factors V and VIII, thus accelerating "upstream" events in the coagulation pathway. Finally, thrombin is a potent platelet and endothelial cell stimulus and enhances the participation of these cells in coagulation.

Factor X is activated during both the extrinsic and intrinsic cascades. In the extrinsic pathway, factor X is activated by a complex consisting of activated factor VII, Ca^{2+}, and factor III (tissue factor). Activation of this complex by tissue factor bypasses the requirement for coagulation factors XII, XI, IX, and VIII used in

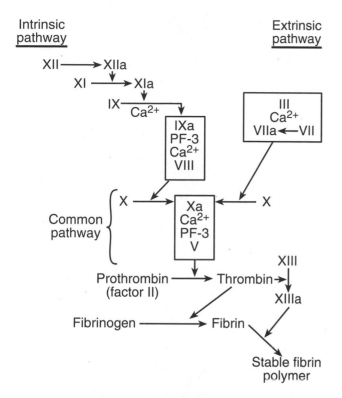

Intrinsic pathway

Extrinsic pathway

Figure 11–12 ■ Steps in the coagulation cascade. The extrinsic pathway is initiated by tissue factor (factor III) released from damaged cells. In the presence of Ca^{2+}, factor III converts factor VII to factor VIIa, which then forms a complex with factor III and Ca^{2+}. This complex converts factor X to factor Xa. In the intrinsic system, factor XII is first converted to factor XIIa following its exposure to foreign surfaces, such as subendothelial matrix. Factor XIIa initiates a cascade of events, including activation of factor X, subsequent conversion of prothrombin to thrombin, and, finally, fibrin formation. The size of the thrombus is subsequently limited by the process of clot retraction, which is mediated in part by proteases activated during the coagulation cascade.

the intrinsic pathway (Fig. 11–12). In the intrinsic pathway, clotting is initiated by activation of factor XII by contact to exposed surfaces, such as collagen in the subendothelial matrix. Efficient activation of factor XII requires a number of cofactors, including **kallikrein** and **high-molecular-weight kininogen.** In this pathway, factor X is activated by a complex consisting of factor IXa, platelet factor 3, and factor VIII.

It should be noted that distinct division of coagulation into two separate pathways is an oversimplification, and the cascade theory has been extensively modified. There are many points of interaction between the two pathways, and in no case will only one pathway account for hemostasis. For example,

thrombin generated during activation of the extrinsic pathway is an essential cofactor for factors VIII and V of the intrinsic pathway. Factor VIIa of the extrinsic pathway directly activates factor IX of the intrinsic system. Factor VII can be activated by factors XIIa, IXa, and Xa and thrombin. The many additional points of interaction are beyond the scope of this discussion, but the concept of independently acting intrinsic versus extrinsic coagulation pathways has been abandoned. However, the activity of the intrinsic system and the extrinsic system are monitored individually in clinical coagulation tests for diagnostic purposes. The test used to monitor activity of the intrinsic system is the **partial thromboplastin time** (PTT). The extrinsic system is evaluated by determination of the **prothrombin time** (PT).

To a large extent, the interaction of coagulation factors occurs on the surface of platelets and endothelial cells. While plasma can eventually clot in the absence of surface contact, localization and assembly of coagulation factors on cell surfaces amplifies reaction rates by several orders of magnitude.

Clot Retraction and Fibrinolysis ■ Given the presence of these efficient clotting amplification mechanisms, the availability of cells and factors constantly poised for activation in the clotting cascade, and the devastating consequences of uncontrolled intravascular coagulation, it is obvious that efficient mechanisms to control, limit, or reverse coagulation must be available. Several such mechanisms exist. Platelet function is strongly inhibited, for example, by the endothelial cell metabolite **prostacyclin** (PGI₂), which is generated from arachidonic acid during cellular activation. Activated endothelial cells also release **tissue plasminogen activator** (TPA), which converts plasminogen to **plasmin,** a protein that hydrolyzes fibrin and thus limits clotting. Thrombin bound to **thrombomodulin** on the surface of endothelial cells converts **protein C** to an active protease. Activated protein C and its cofactor, **protein S,** restrain further coagulation by proteolysis of factors Va and VIIIa. Furthermore, activated protein C augments fibrinolysis by blocking an inhibitor of TPA. Finally, **antithrombin III** is a potent inhibitor of proteases involved in coagulation cascade, such as thrombin. The activity of antithrombin III is accelerated by small amounts of **heparin,** a mucopolysaccharide present in the cells of many tissues. Deficiencies or abnormalities in proteins that regulate or constrain coagulation may result in **thrombotic** disorders, in which intravascular clot formation leads to severe pathologic consequences, including embolism and stroke. Such disorders have been associated with abnormalities in protein C, protein S, antithrombin III, and plasminogen.

■ ■ ■ ■ ■ ■ ■ ■ ■ ■ ■ **REVIEW EXERCISES** ■ ■ ■ ■ ■ ■ ■ ■ ■ ■ ■

Identify Each with a Term

1. The process that controls bleeding and thereby preserves the integrity of the internal environment
2. Average red cell volume
3. The percentage of volume occupied by red cells in blood
4. The formation of blood cells from primitive stem cells
5. The explosion of metabolic activity after phagocytosis
6. Hypersensitivity response mediated by T cells
7. Substance that induces platelet production by megakaryocytes
8. Coagulation factor that cleaves fibrinogen.

Define Each Term

9. Hematopoiesis
10. Pluripotential stem cell
11. Neutropenia
12. Antibody
13. Memory cell
14. Plasma cell
15. Clonal selection
16. Metabolic cascade

Choose the Correct Answer

17. Plasma devoid of fibrinogen, fibrin, and certain coagulation factors is:
    a. Plasmin
    b. Defibrin
    c. Bilirubin
    d. Precipitin
    e. Serum
18. The majority of blood leukocytes are typically:
    a. Lymphocytes
    b. Monocytes
    c. Neutrophils
    d. Basophils
    e. Eosinophils
19. All of the following are granulocytes, except:
    a. Neutrophils
    b. Eosinophils
    c. Monocytes
    d. Basophils
20. Which cytokine does not stimulate hematopoiesis?
    a. Interleukin
    b. Granulocyte colony stimulating factor
    c. Thrombopoietin
    d. Erythropoietin
    e. Tumor necrosis factor
21. The structure of hemoglobin (HbA) is:
    a. Four polypeptide chains; two α globin molecules, and two β chains
    b. Two light chains, two heavy chains, and variable and constant regions
    c. Glycerol backbone, two acylated fatty acids, and a polar head group
    d. Polymorphonuclear nucleus, cytoplasmic granules, and ruffled plasma membrane

Calculate

22. A blood sample has a hematocrit of 0.48 (48%), a hemoglobin level of 14.3 g/dl, and a red cell count of 4.7 million cells/μl. Calculate the MCHC, MCH, MCV, and oxygen carrying capacity.

Briefly Answer

23. What is the basic structure of an immunoglobulin molecule?
24. Describe the essential differences between the adaptive and innate immune responses.
25. Describe the clonal selection theory.
26. What is the complex that activates factor X during activation of the extrinsic coagulation cascade?
27. What is the complex that activates factor X during activation of the intrinsic coagulation cascade?
28. What are four functions of thrombin?

■ ■ ■ ■ ■ ■ ■ ■ ■ ■ ■ ■ ■ **ANSWERS** ■ ■ ■ ■ ■ ■ ■ ■ ■ ■ ■ ■ ■

1. Hemostasis
2. Mean cell volume
3. Hematocrit
4. Hematopoiesis
5. Respiratory burst
6. Delayed type hypersensitivity
7. Thrombopoietin
8. Thrombin
9. The process of blood cell generation
10. The uncommitted hematopoietic cell responsible for producing all hematopoietic progenitors
11. A reduction in the number of circulating neutrophils
12. An immune protein, or immunoglobulin, evoked by an antigen and characterized by reacting specifically with the inducing antigen in some recognizable way
13. The lymphocyte responsible for the adaptive immune response
14. B lymphocyte-derived cell that actively secretes immunoglobulin
15. The process by which antigen specifically induces one line of committed lymphocyte to proliferate, thereby generating specific memory cells and effector cells of the adaptive immune response

16. The sequential activation of several inactive intermediates leading to induction of some biologic function, such as coagulation

17. e

18. c

19. c

20. e

21. a

22.

$$MCHC = \frac{143 \text{ g/L}}{0.48} = 297.9 \text{ g/L}$$

$$MCH = \frac{143 \text{ g/L}}{4.7 \times 10^{12} \text{ cells/L}}$$
$$= 30.4 \times 10^{-12} \text{ g/cell}$$
$$= 30.4 \text{ pg/cell}$$

$$MCV = \frac{0.48}{4.7 \times 10^{12} \text{ cells/L}}$$
$$= 0.102 \times 10^{-12} \text{ L/cell}$$
$$= 102 \text{ fl/cell}$$

O_2 carrying capacity $= 14.3$ g Hb/dl $\times 1.34$ ml O_2/g Hb $= 19.2$ ml O_2/dl

23. Two light chains, two heavy chains arranged in a Y shape with two antigen binding sites, a variable and a constant region, and a cell-binding site (Fc site)

24. Unlike responses of the innate immune system, adaptive immune responses are improved (accelerated and/or amplified) by previous encounters of immune cells with the inciting antigen. Adaptive responses may be cell-mediated or directed by soluble antibodies. In both cases, the response is targeted specifically to the inciting antigen. Innate responses, on the other hand, are not directed toward a specific antigen. These responses are not amplified or accelerated by prior encounters with antigen. Innate responses are not mediated by antibodies or specifically targeted cells. The complement system working in conjunction with phagocytic leukocytes plays a key role in innate immune responses.

25. Receptors on the surface of afferent lymphocytes, such as B cells, specifically sense antigenic determinants on the surface of invading pathogens. Individual afferent lymphocyte recognizes only one antigenic determinant; immune diversity results from the presence of millions of recognition cells, each with a different specificity. Receptor recognition leads to cell activation and results in proliferation of the specifically responsive clone(s) of cells. Many daughter cells are generated, with a specificity identical to that of the original responding lymphocyte. Some of these progeny become quiescent memory cells that are activated during later encounters with the same antigen. Others become effector lymphocytes, which orchestrate specific responses against the challenge antigen.

26. A complex of tissue factor, including a phospholipid, Ca^{2+}, and factor VIIa

27. A complex consisting of factor IXa, platelet factor 3, Ca^{2+}, and factor VIII

28. The functions of thrombin include its ability to convert fibrinogen to fibrin, activate factor XIII, enhance the activity of factors V and VIII, and induce platelet aggregation.

Suggested Reading

Babior BM: *Hematology: A Pathophysiological Approach.* 3rd ed. New York, Churchill Livingstone, 1994.

Benjamini E: *Immunology: A Short Course.* 2nd ed. New York, Wiley-Liss, 1991.

Claman HN: The biology of the immune response. *JAMA* 268:2790–2796, 1992.

Hoffman R, Benz EJ, Shattil SJ, Furie B, Cohen HJ: *Hematology: Basic Principles and Practice.* New York, Churchill Livingstone, 1991.

Pittiglio DH, Sacher RA: *Clinical Hematology and Fundamentals of Hemostasis.* Philadelphia, FA Davis, 1987.

Rapaport SI: *Introduction to Immunology.* 2nd ed. Philadelphia, JB Lippincott, 1987.

Ratnoff OD, Forbes CS (eds): *Disorders of Hemostasis.* 2nd ed. Philadelphia, WB Saunders, 1991.

Roitt IM, Brostoff J, Male DK: *Immunology.* 2nd ed. St. Louis, CV Mosby, 1989.

CHAPTER

■ ■ ■ ■ ■ ■ ■

12

An Overview of Circulation and Hemodynamics

CHAPTER OUTLINE

I. ONCE AROUND THE CIRCULATION
II. HEMODYNAMIC PRINCIPLES OF THE CIRCULATION
   A. Poiseuille's law describes the relationship between pressure and flow
   B. Flow may be streamlined or turbulent
III. PRESSURES IN THE CARDIOVASCULAR SYSTEM
   A. The contractions of the heart produce hemodynamic pressure in the aorta
   B. A column of fluid exerts hydrostatic pressure
   C. Intravascular pressure stretches blood vessels in proportion to their compliance
   D. Mean arterial pressure depends on cardiac output and systemic vascular resistance
IV. SYSTOLIC AND DIASTOLIC PRESSURE
   A. Arterial compliance and cardiac stroke volume primarily determine pulse pressure
   B. Compliance decreases as blood vessels are stretched
V. TRANSPORT IN THE CIRCULATION
   A. Hemodynamic pressure gradients drive bulk flow; concentration gradients drive diffusion
   B. Bulk flow and diffusion are influenced by blood vessel size and number
VI. CONTROL OF THE CIRCULATION

OBJECTIVES

After studying this chapter, the student should be able to:
1. Describe the role of the circulatory system in maintaining the internal environment
2. Outline the general structure of the circulatory system
3. Describe the relationships among pressure, flow, resistance, and compliance
4. Show how Poiseuille's law can be used to help describe circulatory hemodynamics
5. Describe the factors influencing arterial pressure
6. Describe the roles of bulk flow and diffusion in circulatory transport

The physiologic and medical importance of the cardiovascular system has been apparent since William Harvey first described the circulation of blood in 1628. A properly functioning, well-regulated cardiovascular system is essential to meet the metabolic and hemodynamic demands of various physiologic and pathophysiologic situations (e.g., prolonged standing, exercise, fever, hemorrhage, labor and delivery). Unfortunately, failure of the cardiovascular system to perform normally occurs all too often; in industrialized countries the leading causes of death and morbidity include myocardial infarction, stroke, hypertension, congestive heart failure, and an assortment of other cardiovascular problems. A knowledge of the structure and function of the cardiovascular system is therefore crucial for understanding many aspects of health and disease.

In higher organisms, physiologic regulatory functions are performed by anatomically separate organs or tissues. For example, the lungs regulate gas tensions, the splanchnic organs control the uptake and metabolism of a variety of organic and inorganic substances, and the skin is especially important in temperature regulation. Because the cells of every organ and tissue require access to all of these regulating capacities, a system of transport connecting each site of regulation with the cellular environment must be present. The circulation of the blood provides this system and enables homeostasis to be achieved.

Once around the Circulation

An understanding of the circulation depends on a knowledge of the physical principles governing blood flow, but before approaching that subject a brief overview of the circulation is in order (Fig. 12–1). Contractions of the left ventricle propel blood into the aorta, large arteries and the vasculature beyond. Because of the elasticity (or compliance) of the large vessels, they are distended by each injection of blood from the heart. Between ventricular contractions the aorta and large arteries recoil, thereby continuing the flow of blood to the periphery. A number of regulatory mechanisms normally keep aortic pressure within a narrow range. This provides a pulsatile but consistent pressure, driving blood to the small arteries and arterioles. The smooth muscle in the relatively thick walls of the small arteries and arterioles can contract or relax and in doing so can cause large changes in flow to a particular organ or tissue. The pulsatile pressure that is so prominent in the aorta and large arteries is damped by the small arteries and arterioles. Pressure and flow are steady in the smallest arterioles. Blood flows from the small arteries and arterioles (resistance vessels) into the capillaries. The capillaries are numerous and small enough that red blood cells flow through them in single file. The thin capillary walls allow rapid exchange of oxygen, carbon dioxide, substrates, hormones, and other molecules. Blood leaves the capil-

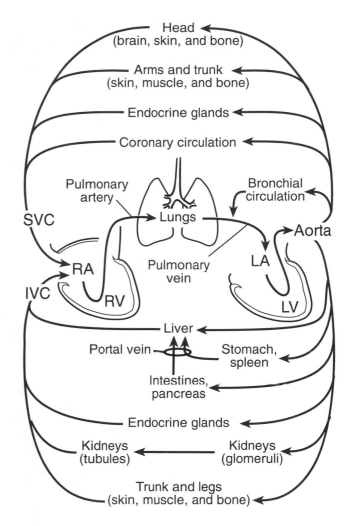

Figure 12–1 ■ Schematic diagram of the cardiovascular system. The right and left hearts are aligned in series, as are the systemic and pulmonary circulation. In contrast, the circulation of the organs other than the lungs is in parallel; that is, each organ receives blood from the aorta and returns it to the vena cava. The exceptions are the various "portal" circulations, which include the liver, kidney tubules, and hypothalamus. RA, right atrium; RV, right ventricle; LA, left atrium; LV, left ventricle; IVC, inferior vena cava; SVC, superior vena cava.

laries (exchange vessels) and enters the venules and small veins. These vessels have larger diameters and thinner walls than their companion small arteries and arterioles. Because of their larger caliber they hold a larger volume of blood. When the smooth muscle in their walls contracts, the volume of blood they contain is reduced. The pressure generated by the contractions of the left ventricle is largely dissipated by this point, and blood flows through the veins (capacitance vessels) to the right atrium at much lower pressures.

The right atrium receives blood from the largest veins, the superior and inferior vena cavae, which drain the entire body except the heart and lungs. The

EFFECT OF VASCULAR DISEASE ON ARTERIAL RESISTANCE

The pressure gradient along large and medium-sized arteries, such as the aorta and renal arteries, is usually very small, due to the minimal resistance typically provided by these vessels. However, several disease processes can produce arterial narrowing and thus increase vascular resistance. Arterial narrowing exerts a profound effect on arterial blood flow because resistance varies inversely with the fourth power of the luminal radius.

The most common such disease is **atherosclerosis,** in which plaques composed of fatty substances (including cholesterol), fibrous tissue, and calcium form in the intimal layer of the artery. Atherosclerosis is the largest cause of morbidity and mortality in the United States: myocardial infarction secondary to coronary atherosclerosis occurs more than 1 million times annually and accounts for over 700,000 deaths. Figure 12–I is an **arteriogram** from a patient with severe aortoiliac disease. The irregular contour of the aortic walls, narrowing of the iliac arteries (large arrowheads), and narrowing of the superior mesenteric artery (small arrowheads) are all caused by atherosclerosis.

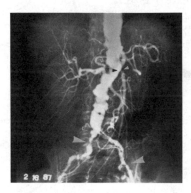

Arteriogram of the abdominal aorta and iliac arteries, demonstrating atherosclerotic changes.

Figure 12–I ■

thin wall of the right atrium allows it to stretch easily to store the steady flow of blood from the periphery. Because the right ventricle can receive blood only when it is relaxing, this storage function of the right atrium is critical. The muscle in the wall of the right atrium contracts at just the right time to help fill the right ventricle. Contractions of the right ventricle propel blood through the lungs where oxygen and carbon dioxide are exchanged in the pulmonary capillaries. Pressures are much lower in the pulmonary than in the systemic circulation. Blood then flows via the pulmonary vein to the left atrium, which functions much like the right atrium. The thick muscular wall of the left ventricle develops the high pressure necessary to drive blood around the systemic circulation.

The mechanisms that regulate all of the above anatomic elements of the circulation are the subject of the next few chapters. In this chapter we consider the physical principles on which the study of circulation is based.

Hemodynamic Principles of the Circulation

Hemodynamics is the scientific field concerned with the relationships among the physical principles governing pressure, flow, resistance, and compliance as they relate to the cardiovascular system.

■ Poiseuille's Law Describes the Relationship between Pressure and Flow

Fluid flows when a pressure gradient exists. Pressure is force applied over a surface, such as the force applied to the cross-sectional surface of a fluid at each end of a rigid tube (Fig. 12-2). The physics of fluid flow through rigid tubes is governed by the pressure gradients and the resistance to flow due to the radius and length of the tube as well as the viscosity of the fluid. All of this is summarized by Poiseuille's law. While not exactly descriptive of blood flow through elastic, tapering blood vessels, Poiseuille's law is useful in understanding blood flow. The volume of fluid flowing through a rigid tube per unit time (**flow,** or F) is proportional (proportionality constant = K) to the pressure (P) difference between the ends of the tube:

$$F = K(P_1 - P_2) = K \times \Delta P \qquad (1)$$

Assume the pressure difference, ΔP, along the rigid tube in Figure 12-2, is $100 - 10$, or 90, mm Hg, and that this produces a flow of 10 ml/min. If the pressure at each end of the tube is raised by 400 mm Hg (to 500 mm Hg and 410 mm Hg, respectively), ΔP will still be 90 mm Hg and the flow will remain 10 ml/min. This illustrates that the pressure difference, not the absolute pressure, determines the flow. If the outlet pressure (P_2) is raised to 55 mm Hg, ΔP is 45 mm Hg and the flow falls to 5 ml/min. This illustrates the proportionality between ΔP and F.

Flow is determined not only by the pressure difference, but also by the value of the proportionality constant, K. The reciprocal of K is the **resistance to flow** (R); that is, $1/K = R$. In the example in Figure 12-2, R is calculated from the observed values of ΔP and F:

$$F = K \times \Delta P = \frac{\Delta P}{R} \qquad (2)$$

and

$$R = \frac{\Delta P}{F} \qquad (3)$$

We calculate for Figure 12-2:

$$R = \frac{\Delta P}{F} = \frac{90 \text{ mm Hg}}{10 \text{ ml/min}}$$

$$= 9 \text{ mm Hg} \times \text{min/ml}$$

$$= 9 \text{ PRU}$$

Peripheral resistance unit (PRU) is frequently used in place of mm Hg × min/ml. Note that resistance is not measured directly but is calculated from pressure

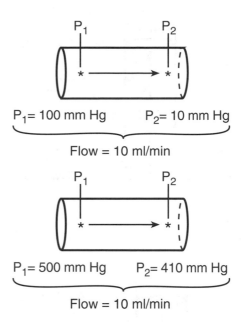

Figure 12–2 ■ Relationship between pressure and flow. Flow is proportional to the pressure difference between two points. Note that the pressure difference, not the absolute pressure, determines flow.

and flow. When fluid flows through a tube, the resistance to flow (R) is determined by the properties of both the fluid and the tube. In the case of a steady, streamlined flow of fluid through a rigid tube, Poiseuille found that these factors determine resistance:

$$R = \frac{8\eta L}{\pi r^4} \qquad (4)$$

where r is the radius of the tube, L is its length, and η is the viscosity of the fluid; 8 and π are geometrical constants. This equation shows that the resistance to blood flow increases proportionately with increases in fluid viscosity or tube length. In contrast, radius changes have a much greater influence, because resistance is inversely proportional to the fourth power of the radius (Fig. 12-3).

Poiseuille's law incorporates all of the factors influencing flow, so that:

$$F = \frac{\Delta P \pi r^4}{8\eta L} \qquad (5)$$

The most important determinants of blood flow in the cardiovascular system are ΔP and r^4. Numerous control systems exist for the sole purpose of maintaining the arterial pressure relatively constant so there is a constant force to drive blood through the cardiovascular system. In the presence of constant pressure, small changes in arteriolar radius can cause large changes in flow to a tissue or organ, because flow and radius are related by the fourth power.

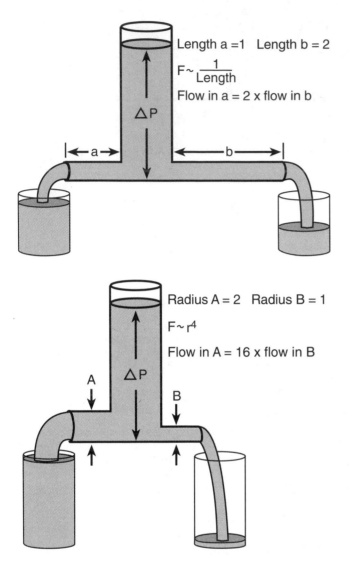

Figure 12–3 ■ Effect of length and radius of tube on flow. Because flow is determined by the fourth power of the radius, small changes in radius have a much greater effect than small changes in length. Furthermore, changes in blood vessel length do not occur over short periods of time and are not involved in physiologic control of blood flow. The pressure difference $(P_1 - P_2)$ is the result of a force applied to the top of the column of fluid (P_1) minus atmospheric pressure at the openings of the smaller tubes (P_2).

Despite the usefulness of Poiseuille's law, it is worthwhile to examine the ways the cardiovascular system does not strictly meet the criteria necessary to apply the law. First, the cardiovascular system is composed of tapering, branching, elastic tubes, rather than rigid tubes of constant diameter. Second, blood is not a strict **Newtonian fluid,** a fluid that exhibits a constant viscosity (η) regardless of flow velocity. When measured in vitro, the viscosity of blood decreases as the flow rate increases. This is explained by the observation that red cells tend to collect in the center of the lumen of a vessel as flow velocity increases (Fig. 12–4),

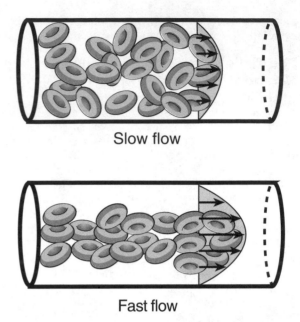

Figure 12–4 ■ Streamline blood flow. In streamline (or laminar) flow, concentric rings of fluid slip over each other, with the highest velocity occurring in the center of the blood vessel and the lowest at the vessel wall. As flow rate increases, red blood cells move toward the center of the blood vessel (axial streaming), where velocity is highest. Axial streaming of red blood cells lowers the apparent viscosity of blood.

an arrangement **(axial streaming)** that offers less resistance to flow. This is only a minor effect in the range of flow velocities encountered in most blood vessels, and we can assume that the viscosity of blood (which is three to four times that of water) is independent of velocity. However, blood viscosity does increase with hematocrit and with protein concentration.

■ Flow May Be Streamlined or Turbulent

Application of Poiseuille's law also requires that flow be streamlined. **Streamline (laminar) flow** describes the movement of fluid through a tube in concentric layers that slip past each other. The layers at the center have the fastest velocity and those at the edge of the tube have the slowest. This is the most efficient pattern of flow velocities, in that the fluid exerts the least resistance to flow in this configuration. Turbulent flow has crosscurrents and eddies, and the fastest velocities are not necessarily in the middle of the stream. A number of factors contribute to the tendency for turbulence: high flow velocity, the large diameter of the tube, high fluid density, and low viscosity. All of these factors can be combined to calculate **Reynolds number** (N_R), which quantifies the tendency for turbulence:

$$N_R = \bar{V}d\rho/\eta \qquad (6)$$

where $\bar{V}$ is the mean velocity, d the tube diameter, ρ the fluid density, and η the fluid viscosity. **Turbulent flow** occurs when N_R exceeds a critical value. This value is hardly ever exceeded in a normal cardiovascular system, but in pathologic states high-flow velocity is the most common cause of turbulence. Figure 12–5 shows that as the pressure gradient along the tube increases, flow velocity increases until streamline flow breaks into eddies and crosscurrents (i.e., turbulent flow). Once turbulence occurs, a given increase in pressure gradient causes less increase in flow because the resistance to flow is greater in a turbulent fluid. Under normal circumstances turbulent flow is found only in the aorta (just beyond the aortic valve) and in certain localized areas of the peripheral system, such as the carotid sinus. Pathologic changes in the cardiac valves or a narrowing of arteries that raise flow velocity often induce turbulent flow. Turbulent flow generates vibrations that are transmitted to the surface of the body; these vibrations **(bruits and murmurs)** can be heard with a stethoscope.

Pressures in the Cardiovascular System

■ The Contractions of the Heart Produce Hemodynamic Pressure in the Aorta

The left ventricle imparts energy to the blood it ejects into the aorta, and this energy is responsible for the blood's circuit from the aorta back to the right side of the heart. Most of this energy is in the form of potential energy, which is the pressure referred to in Poiseuille's law. This is **hemodynamic pressure,** produced by contractions of the heart and stored in the elastic walls of the blood vessels. A much smaller component of the energy imparted by contraction of the heart is kinetic energy, which is the inertial energy associated with the movement of blood. The next section describes a third form of energy, hydrostatic pressure, derived from the force of gravity on blood. When a person is standing, blood pressure is greater in the vessels of the legs than in analogous vessels in the arms because hydrostatic pressure is added to hemodynamic pressure. Often when capillary pressure is discussed, the term *hydrostatic pressure* is used to mean hemodynamic plus hydrostatic pressure. This is not strictly correct but is consistent with conventional usage.

■ A Column of Fluid Exerts Hydrostatic Pressure

A fluid standing in a container exerts pressure that is proportional to the height of the fluid above it. The pressure at a given depth depends only on the height of the fluid and its density and not on the shape of the container. This **hydrostatic pressure** is caused by

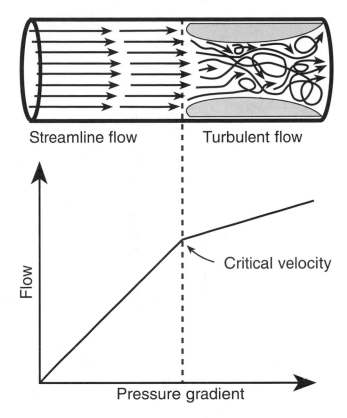

Streamline flow | Turbulent flow

Flow

Critical velocity

Pressure gradient

Figure 12–5 ■ Blood flow is proportional to the pressure gradient until a critical velocity is reached and turbulent flow results. Because energy is lost in the turbulence, flow does not increase as much for a given rise in pressure after the critical velocity is exceeded. The critical velocity is defined by the Reynolds number.

the force of gravity acting on the fluid. As mentioned above, when a person stands, the pressure in a third-order artery of the legs is higher than that in a third-order artery of the arms because hydrostatic pressure is added to hemodynamic pressure. The hydrostatic pressure difference is proportional to the height of the column of blood between the arms and legs.

The height of a column of fluid is often used as a measure of pressure. For example, the pressure at the bottom of a container containing a column of water 100 cm high is 100 cm of H_2O. The height of a column of mercury (Fig. 12–6) is frequently used for this purpose because it is dense (approximately 13 times more dense than water) and a relatively small column height can be employed to measure physiologic pressures. For example, mean arterial pressure is equal to the pressure at the bottom of a column of mercury approximately 93 mm high (abbreviated 93 mm Hg). If the same arterial pressure were measured using a column of water, the column would be approximately 4 ft high.

Two conventions are observed when measuring blood pressure. First, ambient atmospheric pressure

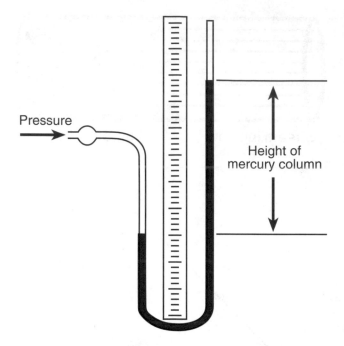

Figure 12–6 ■ Pressure can be expressed as the height of a column of mercury. It is most convenient to use mercury instead of water because its density allows use of a relatively short column. A variety of electronic and mechanical transducers are used to measure blood pressure, but the convention of expressing pressure in mm Hg persists.

is used as a zero reference, so the mean arterial pressure is actually about 93 mm Hg above atmospheric pressure. Second, all cardiovascular pressures are referred to the level of the heart. This takes into account the fact that pressures vary depending on position, because of the addition of hydrostatic to hemodynamic pressure.

■ Intravascular Pressure Stretches Blood Vessels in Proportion to their Compliance

Thus far, we have discussed pressure and flow in the cardiovascular system as if it were composed of rigid tubes. The opposite is true. Blood vessels are elastic, and they expand when the blood in them is under pressure. The degree to which a distensible vessel or container expands when it is filled with fluid is determined by the distending pressure and its compliance. **Compliance** (C) is defined by the equation:

$$C = \Delta V / \Delta P \qquad (7)$$

where ΔV is the change in volume and ΔP is the change in distending pressure. The distending pressure is equal to the pressure inside the vessel minus the pressure outside the vessel; this is called the **transmural pressure.** The more compliant a structure, the greater the change in volume for a given transmural pressure

change. The lower the compliance of a vessel, the greater the pressure that will result when a given volume is introduced. For example, each time the left ventricle contracts and ejects blood into the aorta, the aorta expands; in doing so it exerts an elastic force on the increased volume of blood that it contains. This force is measured as the pressure in the aorta. With aging, the aorta becomes less compliant, so aortic pressure rises more for a given increase in aortic volume. Veins, which have thinner walls, are much more compliant than arteries. This means that if the volume of blood in the cardiovascular system is increased, most of the increase is found in the venous system, because veins accept a large volume change with little change in pressure.

In vessels that are thin-walled and relatively permeable (e.g., capillaries and small venules) the transmural pressure difference may force fluid out of the vessels and into the interstitial space. This fluid eventually returns from the interstitial space to the systemic circulation via another set of vessels, the **lymphatics.** This movement of fluid from the systemic and pulmonary circulation into the interstitial space and then back to the systemic circulation via the lymphatic vessels is referred to as the **lymphatic circulation** (Chap. 16).

■ Mean Arterial Pressure Depends on Cardiac Output and Systemic Vascular Resistance

The aorta and large arteries fill with blood until the pressure created by their distention is sufficient to drive the blood out of them (and into the smaller blood vessels) at a flow rate equal to that of inflow from the heart (Fig. 12-7). Flow (F) through the aorta and large arteries, which is equal to the cardiac output (CO) in the steady state, is proportional to the pressure difference between the aorta or large arteries (**mean**

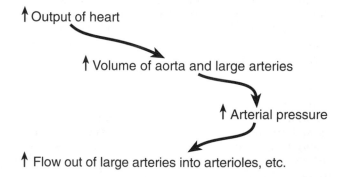

Figure 12–7 ■ Relationship between flow, pressure, and blood vessel volume. The flow of blood from the heart into the aorta is equal to that out of the arteries and into the rest of the systemic circulation only when the aorta and large arteries are expanded enough to raise arterial pressure to a level adequate to push blood on into the small arteries and vessels beyond. Whenever a mismatch between flow out of the heart and flow out of the aorta occurs, large vessel volume (and thus arterial pressure) changes very rapidly and brings the two flows back into equality.

LYMPHEDEMA

In normal individuals, many liters of fluid pass through the capillary walls and into the interstitial space each day, in part because of the transmural pressure gradients. The fluid is eventually returned to the circulation via **lymphatic vessels,** which collect the interstitial fluid, filter it through the **lymph nodes,** and return it to the vena cava via the **thoracic duct.** Unfortunately, a number of congenital and acquired disease conditions can destroy the delicate lymphatic vessels. When this occurs, interstitial fluid accumulates in the affected limbs, causing **lymphedema** (Fig. 12–IIA, patient's right leg). In some cases the lymphedema can become truly massive and debilitating (Fig. 12–IIB, patient's left leg). Lymphedema is typically treated by elevating or "pumping" (with special pneumatic devices) the leg to remove excess fluid, then fitting the leg with a tight-fitting elastic stocking to prevent reaccumulation of fluid. The compression from the stocking opposes the normal transcapillary pressure gradient and helps force fluid back through any lymphatic vessels that may remain open.

A

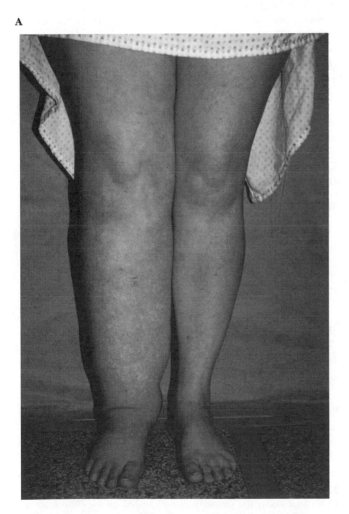

B

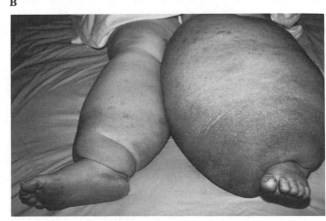

Figure 12–II ■ (A) Lymphedema. (B) Massive lymphedema.

arterial pressure, P_a) and the right atrium (right atrial pressure, $\bar{P}_{Ra}$):

$$\bar{P}_a - \bar{P}_{Ra} \sim F \text{ or } CO \qquad (8)$$

Since right atrial pressure is normally close to zero and mean arterial pressure is much higher (e.g., 93 mm Hg), right atrial pressure is often ignored:

$$\bar{P}_a \sim CO \qquad (9)$$

Mean arterial pressure is also proportional to the **systemic vascular resistance** (SVR), and the inclusion of this term gives the standard relationship between pressure and flow:

$$\bar{P}_a = CO \times SVR \qquad (10)$$

Systematic vascular resistance is the total resistance to flow offered by the blood vessels of the systemic circulation. Physiologic changes in SVR are primarily caused by changes in radius of small arteries and arterioles, the resistance vessels of the systemic circulation.

The relationship in equation 10 states that whenever the cardiac output or systemic vascular resistance increases, the mean arterial pressure must increase proportionally.

Systemic vascular resistance may be calculated from cardiac output and mean arterial pressure, because both of these variables can be measured. Cardiac output and systemic vascular resistance are the variables that are actually regulated physiologically, however, and mean arterial pressure is controlled by adjusting them.

Systolic and Diastolic Pressure

Thus far, we have discussed only mean arterial pressure, despite the fact that the pumping of blood by the heart is a cyclic event. In a resting individual the heart ejects blood into the aorta about once every second (i.e., the heart rate is about 60 beats/min). The phase during which cardiac muscle contracts is called **systole** (from the Greek for "a drawing together"). During atrial systole, the pressure in the atria increases and pushes blood into the ventricles. During ventricular systole, pressure in the ventricle rises and the blood is pushed into the pulmonary artery or aorta. During **diastole** ("a drawing apart") the cardiac muscle relaxes and the chambers fill from the venous side. Because of the pulsatile nature of the cardiac pump, pressure in the arterial system rises and falls with each heart beat. The large arteries are distended when the pressure within them is increased (during systole), and they recoil when the ejection of blood falls during the latter phase of systole and ceases entirely during diastole. This recoil of the arteries sustains the flow of blood into the

distal vasculature when there is no ventricular input of blood into the arterial system.

The difference between **systolic pressure** (P_s) and **diastolic pressure** (P_d) is the **pulse pressure.** Mean arterial pressure ($\bar{P}_a$), determined mathematically as indicated in Figure 12–8, is approximately one-third the pulse pressure added to the diastolic pressure:

$$\bar{P}_a = P_d + \frac{(P_s - P_d)}{3} \qquad (11)$$

Mean arterial pressure is closer to diastolic pressure, instead of halfway between diastolic and systolic pressures, because the duration of diastole is longer than that of systole.

■ Arterial Compliance and Cardiac Stroke Volume Primarily Determine Pulse Pressure

Each ventricular systole causes an increase in arterial pressure that is proportional to the volume of blood ejected and inversely proportional to the aortic compliance. The greater the stroke volume the greater the pulse pressure, and the lower the aortic compliance the greater the pulse pressure. Other factors may exert a lesser effect on pulse pressure; for example, pressure waves reflected from small vessels add algebraically to the pressure determined by arterial compliance and stroke volume. Because of reflected waves, femoral artery systolic pressure in a supine person is actually a few mm Hg greater than aortic systolic pressure. Diastolic and mean pressure are 1–2 mm Hg lower in the femoral artery.

■ Compliance Decreases as Blood Vessels Are Stretched

Compliance was previously defined as:

$$C = \Delta V/\Delta P$$

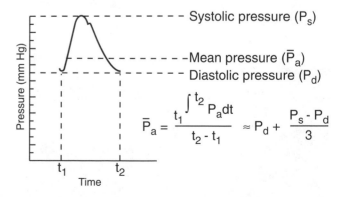

Figure 12–8 ■ Definition of mean arterial pressure. Mean pressure is the area under the pressure curve divided by the time interval. This can be approximated as one-third the pulse pressure plus the diastolic pressure.

It is therefore the slope of the line relating changes in vessel volume to changes in vessel pressure:

$$\Delta V = C \times \Delta P$$

Compliance is constant only if the relationship between volume and pressure is linear, as depicted in Figure 12-9A. In reality, as volume increases the pressure increases disproportionately because the vessel becomes progressively stiffer; that is, compliance decreases with increases in volume. This occurs because the connective tissue elements located primarily in the outer layer (**adventitia**) of the aorta and large arteries is brought into play as these vessels are stretched. The adventitia stiffens and resists further stretching, reducing compliance. As a result, when the mean volume of blood in the aorta increases, a given stroke volume produces a larger pulse pressure.

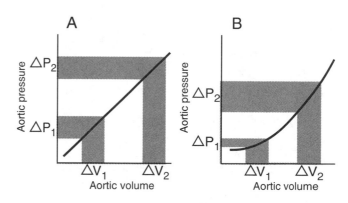

Figure 12–9 ■ Aortic compliance. (A) If the aortic wall were to exhibit linear compliance, a given increase in its volume ($\Delta V_1 = \Delta V_2$) would always result in the same increase in pressure ($\Delta P_1 = \Delta P_2$) regardless of the starting volume. (B) In reality, the compliance of the wall decreases as the aorta is inflated. For this reason, the same increase in volume (ΔV_1 or ΔV_2) starting at a higher initial volume results in a greater increase in pressure ($\Delta P_2 > \Delta P_1$).

Transport in the Circulation

The cardiovascular system employs two types of transport: **diffusion** and **bulk flow.**

■ Hemodynamic Pressure Gradients Drive Bulk Flow; Concentration Gradients Drive Diffusion

Blood circulation is an example of transport by **bulk flow.** Bulk flow requires a pressure difference, with the fluid (in this case blood) flowing from an area of high to low pressure. This is an efficient means of transport over long distances, such as those between the legs and the lungs. **Diffusion** is accomplished by the random movement of individual molecules and is an effective transport mechanism over short distances. Diffusion occurs at the level of the capillaries, where the distances between blood and the surrounding tissue are short. The net transport of molecules by diffusion can occur within hundredths of a second or less when the distances involved are no more than a few microns. On the other hand, minutes or hours would be needed for diffusion to occur over millimeters or centimeters. The cardiovascular system thus depends on the energy provided by hemodynamic pressure gradients to move materials over long distances (bulk flow) and the energy provided by concentration gradients to move material over short distances (diffusion).

■ Bulk Flow and Diffusion Are Influenced by Blood Vessel Size and Number

The aorta has the largest diameter of any artery, and the subsequent branches become progressively smaller down to the capillaries. Although the capillaries are the smallest blood vessels, the total cross-sectional area of the lumens of all systemic capillaries (there are several billion) greatly exceeds that of the lumen of the aorta (approximately 2000 cm^2 versus 7 cm^2). In a steady state, the total blood flow is equal at any point along the circulation; that is, the total flow through the aorta and the capillaries is the same. Because the combined cross-sectional area of the capillaries is much greater and the total flow is the same, the velocity of flow in the capillaries is much lower. The slower movement of blood through the capillaries provides maximum opportunity for diffusional exchange of substances between blood and extracellular fluid. In contrast, blood moves quickly in the aorta, where bulk flow, not diffusion, is important.

Control of the Circulation

The normal cardiovascular system is capable of providing appropriate blood flow to each of the organs and tissues of the body under a wide range of conditions. This is done by maintaining arterial blood pressure within normal limits, adjusting the output of the heart to the appropriate level, and adjusting the resistance to blood flow and capillary exchange in specific organs and tissues to meet special functional needs. Regulation of arterial pressure, cardiac output, and regional blood flow and capillary exchange is achieved using a variety of neural, hormonal, and local mechanisms. In complex situations (e.g., standing or exercise) multiple mechanisms may interact to regulate the cardiovascular response. In abnormal situations (e.g., heart failure) these regulatory mechanisms, which have evolved to handle normal events, may be inadequate to restore proper function. The next few chapters describe these regulatory mechanisms in detail.

■ ■ ■ ■ ■ ■ ■ ■ ■ ■ ■ REVIEW EXERCISES ■ ■ ■ ■ ■ ■ ■ ■ ■ ■ ■

Identify Each with a Term

1. The force exerted by a fluid on a unit of area
2. The law that defines a relationship between pressure and flow
3. Vibrations caused by turbulent blood flow
4. The component of the vascular system that returns interstitial fluid to the circulation
5. The difference between systolic and diastolic blood pressure

Define Each Term

6. Newtonian fluid
7. Compliance
8. Bulk flow

Choose the Correct Answer

9. Which of the following must be regulated with precision at the cellular level for an organism to survive?
    a. Partial pressure of gases
    b. Concentration of organic substances
    c. Concentration of inorganic ions
    d. Temperature
    e. All of the above
10. Which of the following is false regarding homeostasis and higher organisms?
    a. The lungs regulate gas tensions.
    b. The skin is relatively unimportant in regulating body temperature.
    c. The splanchnic organs control the uptake of organic/inorganic substances.
    d. The cardiovascular system produces a link between the various organs and tissues.
11. Flow through a tube is proportional to:
    a. The square of the radius
    b. The square root of the length
    c. The fourth power of the radius
    d. The square of the length
    e. The square root of the radius

12. Which of the following is false?
    a. Poiseuille's law defines kinetic energy.
    b. Hemodynamic pressure pushes blood through the circulation.
    c. Hydrostatic pressure is produced by gravitational energy.
    d. The pressure in the legs of a standing person is higher than the pressure in the arms because hydrostatic and hemodynamic pressure are added.
13. Which of the following is true?
    a. Pulse pressure equals diastolic pressure minus systolic pressure.
    b. Mean arterial pressure equals one-half the pulse pressure plus diastolic pressure.
    c. Compliance equals the change in pressure over the change in volume.
    d. Mean arterial pressure equals one-third the pulse pressure plus diastolic pressure.
    e. Systolic pressure minus one-third the pulse pressure equals the diastolic pressure.

Calculate

14. The volume of an aorta is increased by 30 ml and pressure is increased from 80 to 120 mm Hg. What is its compliance?
15. Flow through a tube is 100 ml/min, the pressure at the proximal end is 100 mm Hg, and the pressure at the distal end is 80 mm Hg. What is the resistance, in peripheral resistance units?
16. If systolic pressure is 120 and diastolic pressure is 90, what is the mean arterial pressure?

Briefly Answer

17. How does the circulatory system help maintain homeostasis in higher organisms?
18. How do the effects of changes in viscosity and tube length (on vascular resistance) differ from the effects of changes in tube radius?

■ ■ ■ ■ ■ ■ ■ ■ ■ ■ ■ ■ ■ ■ ANSWERS ■ ■ ■ ■ ■ ■ ■ ■ ■ ■ ■ ■ ■ ■

1. Pressure
2. Poiseuille's law or Ohm's law
3. Bruits and murmurs
4. Lymphatic system
5. Pulse pressure
6. A fluid that exhibits a constant viscosity regardless of flow velocity
7. The change in volume produced by a change in pressure; this provides an assessment of the "stretchability" of an object
8. Fluid movement from an area of high to low pressure
9. e
10. b
11. c

12. a (Poiseuille's law defines pressure energy)
13. d
14.

$$\text{Compliance} = \frac{\Delta V}{\Delta P}$$

$$= \frac{30 \text{ ml}}{40 \text{ mm Hg}}$$

$$= 0.750 \text{ ml/mm Hg}$$

15.

$$\text{Resistance} = \frac{20 \text{ mm Hg}}{100 \text{ ml/min}} = 0.2 \text{ PRU}$$

16.

$$\text{Pulse pressure} = 120 - 90 = 30 \text{ mm Hg}$$

$$\frac{30}{3} = 10 \text{ mm Hg}$$

$$\text{Pa} = 10 \text{ mm Hg} + 90 \text{ mm Hg}$$

$$= 100 \text{ mm Hg}$$

17. The circulation helps maintain homeostasis by providing a link between the various organs. For example, waste products made in one part of the body can be excreted by the kidneys and oxygen obtained through the lungs can be distributed through the rest of the body.

18. Vascular resistance is directly proportional to viscosity or tube length but inversely proportional to the fourth power of the tube radius.

Suggested Reading

Burton AC: Why have a circulation? in *Physiology and Biophysics of the Circulation.* 2nd ed. Chicago, Year-Book, 1982, pp 3-14.

Cournand A: Air and blood, in Fishman AP, Richards DW. (eds): *Circulation of the Blood: Men and Ideas.* Bethesda, American Physiological Society, 1982, pp 3-70.

Hamilton WF, Richards DW: The output of the heart, in Fishman AP, Richards DW (eds): *Circulation of the Blood: Men and Ideas.* Bethesda, American Physiological Society, 1982, pp 71-126.

Milnor WR: The circulatory system, in *Cardiovascular Physiology.* New York, Oxford University Press, 1990, pp 3-18.

Milnor, WR: Principle of hemodynamics, in *Cardiovascular Physiology.* New York. Oxford University Press, 1990, pp 171-218.

Rowell LB: General principles of vascular control, in *Human Circulation: Regulation During Physical Stress.* New York, Oxford University Press, 1986, pp 8-43.

CHAPTER

■ ■ ■ ■ ■ ■ ■

13

The Electrical Activity of the Heart

CHAPTER OUTLINE

I. THE IONIC BASIS OF CARDIAC ELECTRICAL ACTIVITY: THE CARDIAC MEMBRANE POTENTIAL
   A. The cardiac membrane potential depends on transmembrane sodium, potassium, and calcium gradients
   B. Changes in membrane ionic conductances cause the ventricular action potential
      1. Early depolarization: selective influx of sodium
      2. Late depolarization (plateau): selective influx of calcium
      3. Repolarization: selective efflux of potassium
      4. Resting membrane potential: high permeability to potassium
   C. Changes in membrane ionic conductances cause the pacemaker potential of the sinoatrial and atrioventricular nodes.
   D. Changes in membrane ionic conductances cause the refractory period
   E. Neurotransmitters and other ligands can influence membrane ion conductance

II. THE INITIATION AND PROPAGATION OF CARDIAC ELECTRICAL ACTIVITY
   A. Excitation starts in the sinoatrial node because sinoatrial cells reach threshold first
   B. The action potential is propagated by local currents created during depolarization
   C. Current usually spreads from the sinoatrial node to the atrial muscle to the atrioventricular node to the Purkinje system to the ventricular muscle
      1. Slow conduction through the atrioventricular node
      2. Rapid conduction through the ventricles

III. THE ELECTROCARDIOGRAM
   A. The heart is a dipole because of its electrical activity
      1. Measurement of the voltage associated with a dipole
      2. Changes in dipole magnitude and direction
   B. Portions of the electrocardiogram are associated with electrical activity in specific cardiac regions
      1. The P wave and atrial depolarization
      2. The PR segment and Atrioventricular conduction
      3. The QRS complex and ventricular depolarization
      4. The T wave and ventricular repolarization
      5. Determining the cardiac dipole from the electrocardiogram

OBJECTIVES

After studying this chapter, the student should be able to:
1. Describe how cardiac membrane potentials can be generated by selective ionic movement
2. Describe the mechanisms behind depolarization and repolarization of myocardial cells
3. Describe how cardiac electrical activity is initiated within and propagated throughout the myocardium
4. Describe how various neurotransmitters, drugs, and other agents can affect the ventricular action potential and initiation and/or propagation of cardiac electrical activity
5. Define *dipole* and explain how changes in the cardiac dipole give rise to the electrocardiogram
6. Explain how the electrocardiogram provides clinically useful information about the heart

The heart will beat in the absence of any nervous connections because the electrical (pacemaker) activity that generates the heartbeat resides within the heart itself. After initiation, the electrical activity is conducted through a network of specialized cells and tissues and spreads throughout the heart. The electrical activity reaches every cardiac cell rapidly with the correct timing, thus enabling coordinated contraction of individual cells.

The electrical pacemaker and conduction properties of cardiac cells depend on the ionic gradients across their selectively permeable membranes. This chapter describes how these ionic gradients and changes in membrane permeability result in the electrical activity of individual cells and how this electrical activity is propagated throughout the heart.

The Ionic Basis of Cardiac Electrical Activity: The Cardiac Membrane Potential

In ventricular muscle cells, the resting membrane potential (phase 4, Fig. 13–1a) is stable at approximately -90 mV relative to the outside of the cell. When the cell is appropriately stimulated, a transient change in the membrane potential occurs. First, there is a rapid increase (phase 0) from -90 mV to $+20$ mV (**depolarization**). This is followed by a slight decline (phase 1) to a plateau (phase 2), when the membrane potential is close to 0, and then a rapid return (phase 3, **repolarization**) of potential to the resting value (phase 4). Phases 0–3 describe the ventricular muscle action potential.

In contrast to ventricular cells, cells of the sinoatrial (SA) and atrioventricular (AV) nodes exhibit a progressive depolarization during phase 4 (or diastole) called the **pacemaker potential** (Fig. 13–1B). When the membrane potential reaches a certain characteristic voltage (the threshold potential), there is a sudden and rapid depolarization (phase 0) to approximately $+20$ mV. The membrane subsequently repolarizes (phase 3) without going through a plateau phase, and the pacemaker potential resumes. Other myocardial cells combine various characteristics of the electrical activity of these two cell types. Atrial cells, for exam-

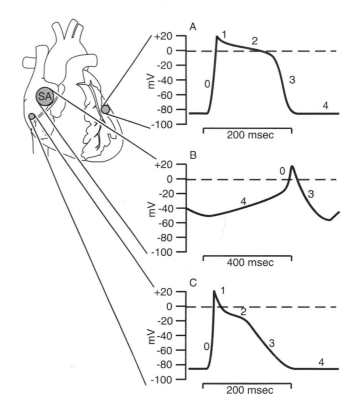

Figure 13–1 ■ Cardiac action potentials. Action potentials (mV) are recorded from (A) ventricular, (B) sinoatrial, and (C) atrial cells. Note difference in time scale of B. Zero through four refer to the various action potential phases (see text).

ple (Fig. 13–1C), have a steady diastolic resting membrane potential (phase 4) but lack a definite plateau (phase 2).

■ The Cardiac Membrane Potential Depends on Transmembrane Sodium, Potassium, and Calcium Gradients

The membrane potential of a cardiac cell depends on concentration differences in Na^+, K^+, and Ca^{2+} across the cell membrane and ion channel permeability. Some Na^+, K^+, and Ca^{2+} channels depend on membrane voltage to determine whether they are opened or closed (voltage-gated channels), and others depend on the concentration of a neurotransmitter, hormone, or metabolite (or drug) to determine their state (ligand-gated channels). Table 13-1 lists the channel proteins that are important in generating cardiac cell action potential.

The ion concentration gradients that determine transmembrane potentials are created and maintained by active transport. The transport of Na^+ and K^+ is accomplished by the cell membrane Na^+-K^+ ATPase (Chap. 2). Calcium is transported partially by means of a Ca^{2+} ATPase and partially by a carrier that uses energy derived from the Na^+ electrochemical gradient (Na^+-Ca^{2+} exchange). If the energy supply of myocardial cells is restricted by inadequate coronary blood flow, ATP synthesis (and, in turn, active transport) may be impaired. This can produce a reduction in ionic concentration gradients and eventually disrupt the electrical activity of the heart.

The magnitude of the intracellular potential depends on the relative permeability of the membrane to Na^+, Ca^{2+}, and K^+. When the membrane is much more permeable to K^+ than to Na^+ or Ca^{2+} (as occurs in the resting state), the measured potential is close to that which would exist if the membrane were permeable only to K^+ (**potassium equilibrium potential**). In contrast, when the membrane is more permeable to Na^+ than to other ions (as occurs at the peak of phase 0 of the action potential), the measured potential is closer to the potential that would exist if the membrane were permeable only to Na^+ (**sodium equilibrium potential**) (Fig. 13-2). As with sodium, an increase in the membrane permeability to Ca^{2+} causes Ca^{2+} to enter the cell and changes the intracellular charge in a positive direction. Specific changes in the number of open channels conducting the various ions are therefore responsible for changes in membrane permeability and the different phases of the action potential.

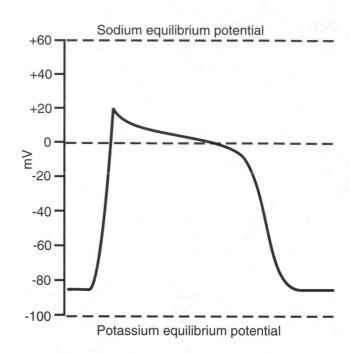

Figure 13-2 ■ Effect of ionic permeability on membrane potential, primarily determined by the relative permeability of the membrane to Na^+ and K^+ and Ca^{2+}. Relatively high permeability to K^+ places the membrane potential close to the K^+ equilibrium potential, and relatively high permeability to Na^+ places it close to the Na^+ equilibrium potential. The same is true for Ca^{2+}. An equilibrium potential is not specified for Ca^{2+} because, unlike Na^+ and K^+, it changes during the action potential. This is because cytosolic Ca^{2+} concentration changes approximately fivefold during excitation. During the plateau of the action potential, the equilibrium potential for Ca^{2+} is approximately +90mV. Membrane permeability to Na^+, K^+, and Ca^{2+} depends on channel proteins (Table 13-1).

TABLE 13-1 ■ Selected Cardiac Membrane Channels

Name	Voltage (V) or Ligand (L)-Gated	Functional Role
Voltage-gated Na^+ channel (fast)	V	Phase 0 of action potential (permits influx of Na^+)
Voltage-gated Ca^{2+} channel (slow)	V	Phase 2 of action potential (permits influx of Ca^{2+} when membrane is depolarized); early pacemaker potential of nodal cells. β-adrenergic agents increase the probability of channel opening and raise Ca^{2+} influx. Acetylcholine lowers the probability of channel opening.
K^+ channel (inward rectifier; i_{K1})	V	Maintains resting membrane potential (phase 4) by permitting outflux of K^+ at highly negative membrane potentials
K^+ channel (transient outward rectifier; i_{to})	V	Contributes briefly to phase 1 by permitting outflux of K^+ at positive membrane potentials
K^+ channel (delayed outward rectifiers; i_K)	V	Causes phase 3 of action potential by permitting outflux of K^+ after a delay when membrane depolarizes
ATP-sensitive K^+ channel	L	Inhibits activation by permitting outflux of K^+ when ATP is low
Acetylcholine-activated K^+ channel	L	Vagal stimulation opens channels that hyperpolarize resting heart and shorten phase 2

i, current.

■ Changes in Membrane Ionic Conductances Cause the Ventricular Action Potential

Figures 13–3 and 13–4 depict the membrane changes that occur during an action potential in ventricular cells.

Early Depolarization: Selective Influx of Sodium ■ The initial upswing of the action potential (phase 0) occurs when electrical current from an external source or the adjacent cell membrane brings the membrane potential to **threshold,** the membrane potential at which the cell will fire an action potential. At threshold, Na^+ permeability suddenly increases (due to opening of Na^+ channels). Since permeability to Na^+ exceeds that to K^+, the membrane potential approaches the Na^+ equilibrium potential (i.e., the inside of the cell becomes positively charged relative to the outside). The change in the membrane potential from rest away from the negative K^+ equilibrium potential and toward the Na^+ equilibrium potential is called **depolarization.**

Phase 1 of the ventricular action potential is related to a decrease in the number of open Na^+ channels

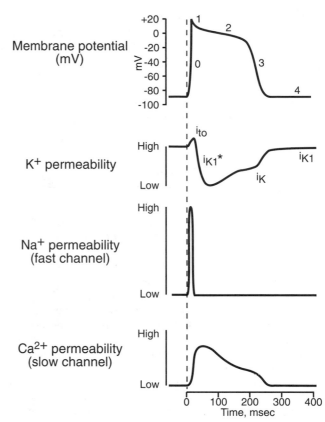

Figure 13–4 ■ Changes in cation permeabilities during a Purkinje fiber action potential (compare with Fig. 13–3). The rise in action potential (phase 0) is caused by rapidly increasing Na^+ current. Na^+ current falls rapidly because voltage-gated Na^+ channels are inactivated. K^+ current rises briefly because of opening of i_{to} channels and then falls precipitously because i_{K1} channels are closed by depolarization (* closing of i_{K1} channels). Ca^{2+} channels are opened by depolarization and are responsible along with closed i_{K1} channels for phase 2. K^+ current begins to increase because i_K channels are opened by depolarization, after a delay. Once repolarization occurs, Ca^{2+} channels close. Reopened i_{K1} channels maintain phase 4.

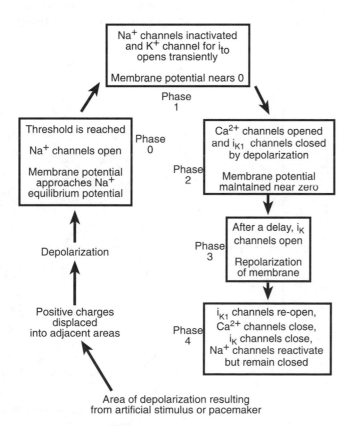

Figure 13–3 ■ Events associated with the ventricular action potential. i_{K1}, inward rectifier K^+ channel; i_K, delayed outward rectifier K^+ channel; i_{to}, transient outward rectifier K^+ channel.

and an opening of transient outward rectifier K^+ channels. These changes tend to repolarize the membrane slightly.

Late Depolarization (Plateau): Selective Influx of Calcium ■ The plateau of phase 2 results from a combination of low membrane permeability to K^+ due to closed K^+ channels (inward rectifier potassium channels) and an increased permeability to Ca^{2+} due to opening of Ca^{2+} channels.

Repolarization: Selective Efflux of Potassium ■ The return of the membrane potential to the resting state (phase 3, or repolarization) reflects the closing of Ca^{2+} channels and the opening of delayed outward rectifier K^+ channels. This relative increase in permeability to K^+ drives the membrane potential toward the K^+ equilibrium potential.

Resting Membrane Potential: High Permeability to Potassium ■ The resting (diastolic) membrane potential (phase 4) of ventricular cells is maintained primarily by K^+ channels that are open at highly negative membrane potentials. They are called inward rectifier K^+ channels because when the membrane is depolarized (e.g., by the opening of voltage gated Na^+ channels) they no longer permit outward movement of K^+.

■ Changes in Membrane Ionic Conductances Cause the Pacemaker Potential of the Sinoatrial and Atrioventricular Nodes

When the electrical activity of a cell from the SA or AV node is compared to that of a ventricular muscle cell, three important differences are observed: (1) the presence of a pacemaker potential; (2) the slow rise of the action potential; and (3) the lack of a well-defined plateau. The pacemaker potential results primarily from a steady increase in the membrane permeability to Ca^{2+} and, more important, Na^+ during diastole (phase 4). Calcium moves in through the voltage-gated Ca^{2+} channel early in diastole (Table 13–1); Na^+ enters late in diastole via an as yet unidentified channel. The increased permeability to Ca^{2+} and Na^+ moves the membrane potential in a positive direction toward the Na^+ and Ca^{2+} equilibrium potentials; that is, it becomes less negative. An action potential is triggered when threshold is reached. This action potential rises more slowly than the ventricular action potential because fast Na^+ channels play an insignificant role in it. Instead, further opening of slow Ca^{2+} channels is primarily responsible for the upstroke of the action potential in nodal cells. The absence of a well-defined plateau probably occurs because K^+ permeability increases late in the action potential in nodal cells. This pulls the membrane potential toward the K^+ equilibrium potential. The decrease in K^+ permeability during early diastole permits the Ca^{2+} and Na^+ currents to cause the pacemaker potential.

■ Changes in Membrane Ionic Conductances Cause the Refractory Period

As discussed in Chapter 10, cardiac muscle cells display long refractory periods, so they cannot be tetanized by fast, repeated stimulation. A prolonged refractory period eliminates the possibility that a sustained contraction might occur and prevent the cyclic contractions required to pump blood. The refractory period begins with the downswing of phase 1 and continues until nearly the end of phase 3 (Fig. 10–2). This occurs because the Na^+ channels that opened to cause phase 0 then close and become inactive for more than 100 msec.

■ Neurotransmitters and Other Ligands Can Influence Membrane Ion Conductance

The normal pacemaker cells are under the influence of **parasympathetic** (vagus) and **sympathetic** (cardioaccelerator) nerves. The vagus nerves release acetylcholine and the cardioaccelerator nerves release norepinephrine at their terminals in the heart. Acetylcholine slows the heart rate (a condition known as **bradycardia**) by reducing the rate of spontaneous depolarizations of the pacemaker cells (Fig. 13–5), thereby increasing the time required to reach threshold and initiate an action potential. Acetylcholine exerts this effect by increasing the number of open K^+ channels and decreasing the number of open Ca^{2+} channels, which holds the membrane potential closer to the K^+ equilibrium potential. In contrast, norepinephrine causes an increase in the slope of the pacemaker potential, so that threshold is reached more rapidly and the heart rate increases (a condition known as **tachycardia**). Norepinephrine increases the slope of the pacemaker potential by opening Na^+ and Ca^{2+} channels and closing K^+ channels, which results in faster movement of the membrane potential toward the Na^+ equilibrium potential. Norepinephrine and acetylcholine exert their effects on K^+ and Ca^{2+} channels in the heart via G protein-mediated events.

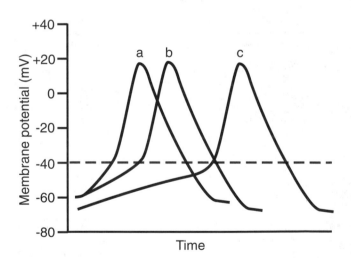

Figure 13–5 ■ Sinoatrial cell membrane potential as a function of time. Normal pacemaker potential (b) is affected by norepinephrine (a) and acetylcholine (c). The dashed line indicates threshold potential. The more rapidly rising pacemaker potential in the presence of norepinephrine (a) results from increasing Na^+ permeability. The hyperpolarization and slower rising pacemaker potential in the presence of acetylcholine results from decreasing Na^+ permeability and increased K^+ permeability due to opening of acetylcholine-activated K^+ channels.

The Initiation and Propagation of Cardiac Electrical Activity

Cardiac electrical activity is normally initiated and spread in an orderly fashion. The heart is said to be a **functional syncytium,** because the excitation of one cardiac cell eventually results in the excitation of all cells.

■ Excitation Starts in the Sinoatrial Node because Sinoatrial Cells Reach Threshold First

Excitation of the heart normally begins in the SA node (Fig. 13–1) because the pacemaker potential of this tissue reaches threshold before the pacemaker potential of the AV node (the pacemaker rate of the SA node is about 60–100 per minute versus about 15–35 per minute for the AV node). Many cells of the SA node reach threshold and depolarize almost simultaneously, creating a voltage difference between these depolarized cells and nearby polarized resting cells.

■ The Action Potential Is Propagated by Local Currents Created during Depolarization

As Na^+ enters a cell during phase 0, the positive charges repel intracellular K^+ ions into nearby areas where depolarization has not yet occurred. Potassium is even driven into adjacent resting cells through low-resistance areas of the intercalated disks (the end-to-end junctions of myocardial cells) called **gap junctions.** This movement of K^+ depolarizes these adjacent areas, causing threshold to be reached. The cycle of depolarization to threshold, Na^+ entry, and subsequent displacement of positive charges into nearby areas explains the spread of electrical activity. Excitation proceeds as succeeding cycles of local ion current and action potential move out of the SA node and across the atria. This process is called the **propagation of the action potential.** It requires 60–90 msec to excite all regions of the atria (Fig. 13–6), with excitation proceeding at a speed of approximately 1 m/sec. Atrial conduction may occur preferentially through specialized pathways in the atria, but the extent to which this occurs remains unknown.

■ Current Usually Spreads from the Sinoatrial Node to the Atrial Muscle to the Atrioventricular Node to the Purkinje System to the Ventricular Muscle

A fibrous, nonconducting connective tissue ring separates the atria from the ventricles everywhere except at the AV node. Transmission of electrical activity from the atria to the ventricles therefore occurs only through the AV node. Action potentials in the atrial

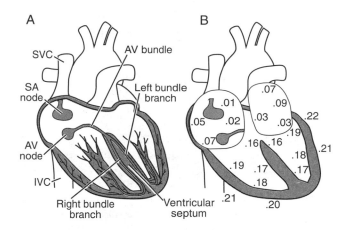

Figure 13–6 ■ Timing of excitation of various areas of the heart (in fractions of a second).

muscle adjacent to the AV node produce local ion currents that invade the node and trigger intranodal action potentials.

Slow Conduction through the Atrioventricular Node ■ The propagation of the action potential by local current flow continues within the AV node, but at a much slower velocity (0.05–0.1 m/sec). The slow conduction is partially explained by the small size of the nodal cells. Less current is produced by depolarization of a small nodal cell (as opposed to a large atrial or ventricular cell), and the relatively smaller current causes neighboring cells to depolarize more slowly, thus decreasing the rate at which electrical activation spreads. Other significant factors are the slow upstroke of the action potential (because it depends on slow Ca^{2+} channels) and possibly weak electrical coupling because of relatively few gap junctions. Propagation of the action potential through the AV node takes approximately 120 msec. Excitation then proceeds through the AV bundle (bundle of His), the left and right bundle branches, and the Purkinje system.

The AV node is the weak link in the excitation of the heart. Inflammation, hypoxia, parasympathetic neural activity, and certain drugs (e.g., digitalis, beta-blockers, and calcium entry blockers) can cause failure of the AV node to conduct some or all of the atrial depolarizations to the ventricles. On the other hand, its tendency to conduct slowly is sometimes of benefit in pathologic situations in which atrial depolarizations are too frequent and/or uncoordinated, as in atrial flutter or fibrillation. In these conditions, not all of the electrical impulses that reach the AV node are conducted to the ventricles, and the ventricular rate tends to stay below the level at which diastolic filling is impaired (Chap. 14).

Rapid Conduction through the Ventricles ■ The Purkinje system is composed of specialized cardiac muscle cells of large diameter and high conduction velocities (up to 2 m/sec) that rapidly propagate electrical activity throughout the subendocardium of both ventricles. Depolarization then proceeds from endocardium to epicardium (Fig. 13-6). The conduction velocity through ventricular muscle is 0.3 m/sec, and complete excitation of both ventricles takes approximately 75 msec.

The Electrocardiogram

The electrocardiogram (ECG) is a continuous record of cardiac electrical activity obtained by placing sensory electrodes on the surface of the body and recording the voltage differences generated by the heart. The equipment amplifies these voltages and causes a pen to deflect in proportion to them on a paper moving under it at 25 mm/sec. When the electrocardiograph is properly calibrated, a 1 mV voltage difference between two points on the body produces a 1 cm deflection of the pen. There are several combinations of points on the body from which the ECG is routinely recorded. The **standard (bipolar) limb leads** record the potential (voltage) differences between the right and left arm (lead I), the right arm and left leg (lead II), and the left arm and left leg (lead III) (Fig. 13-7). For each of these leads, the pen makes an upward deflection when the second-named point is positive relative to the first-named point (e.g., when the left arm is positive to the right arm). Since the arms and legs act as conductors, the electrodes can be attached to them at any location.

■ The Heart Is a Dipole because of Its Electrical Activity

To interpret the ECG it is necessary to understand the behavior of electrical potentials in a three-dimensional conductor of electricity. Consider what happens when wires are run from the positive and negative terminals of a battery into a dish containing salt solution (Fig. 13-8). Positively charged ions flow toward the negative wire (negative pole) and negatively charged ions simultaneously flow in the opposite direction toward the positive wire (positive pole). The combination of two poles that are equal in magnitude and opposite in charge, and located in close physical proximity to one another, is called a **dipole.** The flow of ions (current) is greatest in the region between the two poles, but some current flows at every point; this reflects that fact that voltage differences exist everywhere in the solution.

Measurement of the Voltage Associated with a Dipole ■ What points encircling the dipole in Figure 13-8 have the greatest voltage difference between them? Points A and B do, because A is closest to the positive pole and B is closest to the negative pole. Positive charges are drawn from the area around point B by the negative end of the dipole, which is relatively near. The positive end of the dipole is relatively distant and therefore has little ability to attract negative charges from point B (although it can draw negative

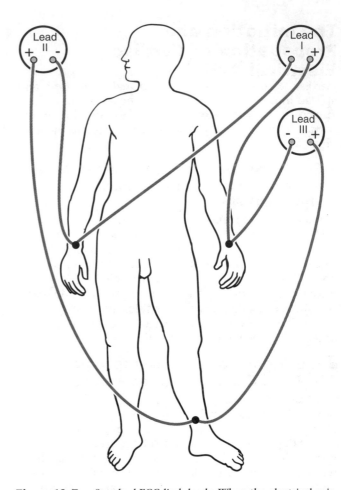

Figure 13–7 ■ Standard ECG limb leads. When the electrical axis is directed to the left or inferiorly (i.e., toward the positive end of each lead), an upward deflection occurs.

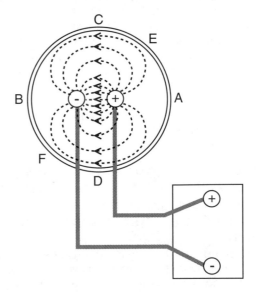

Figure 13–8 ■ Creation of a dipole in a tub of salt solution. The dashed lines indicate current flow; the current flows from the positive to the negative poles. See text for description.

charges from point A). As positive charges are drawn away, point B is left with a negative charge (or negative voltage). The opposite happens between the positive end of the dipole and point A, leaving A with a net positive charge (or voltage). Points C and D have no voltage difference between them because they are equally distant from both poles and are therefore equally influenced by positive and negative charges. Any other two points on the circle, E and F for example, have a voltage difference between them that is less than that between A and B and greater than that between C and D. This is also true of other combinations of points, such as A and C, B and D, and D and F. Voltage differences exist in all cases and are determined by the relative influences of the positive and negative ends of the dipole.

Changes in Dipole Magnitude and Direction ■ What would happen if the dipole were to change its orientation relative to points C and D? Figure 13-9 diagrams an apparatus in which electrodes from a voltmeter are placed at the edges of a dish of salt solution in which the dipole can be rotated. This solution is analogous to that depicted in Figure 13-8, except the dipole position is changed relative to the electrodes, instead of the electrode being changed relative to the dipole. Figure 13-9 shows the changes in measured voltage that occur if the dipole is rotated 90 degrees to point the positive end of the dipole directly at C and the negative end at D. The measured voltage increases slowly as the dipole is turned and is maximal when the dipole reaches the new position. In each position the dipole sets up current fields like

those shown in Figure 13-8. The voltage measured depends on how the electrodes are positioned relative to those currents. This figure also shows that the voltage between C and D will decrease to a new steady-state level as the voltage applied to the wires by the battery is decreased. These imaginary experiments illustrate two characteristics of a dipole that determine the voltage measured at distant points in a volume conductor: **direction** of the dipole relative to the measuring points and **magnitude** (voltage) of the dipole.

■ Portions of the Electrocardiogram Are Associated with Electrical Activity in Specific Cardiac Regions

Consider the origin of the cardiac voltage changes that are recorded as the ECG. At rest, myocardial cells have a negative charge inside and a positive charge outside the cell membrane. As the cell depolarizes, the depolarized portion becomes negative on the outside, whereas the region ahead of the depolarized portion remains positive on the outside (Fig. 13-10). When the

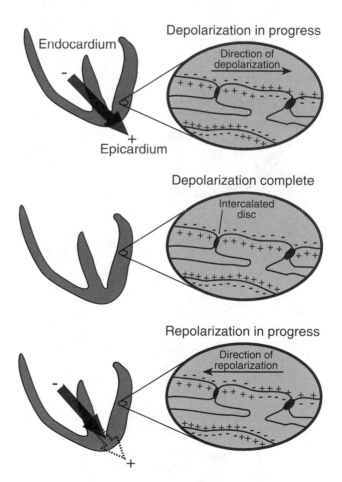

Figure 13-10 ■ Cardiac dipoles. Partially depolarized or repolarized myocardium creates a dipole. Arrows show the direction of depolarization (or repolarization). Dipoles are present only when myocardium is undergoing depolarization or repolarization.

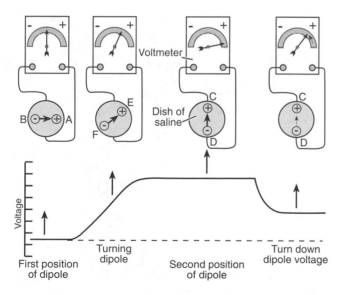

Figure 13-9 ■ In a salt solution, the dipole can be represented as a vector having a length and direction determined by the dipole magnitude and position, respectively. In this example, electrodes for the voltmeter are at points C and D. When a vector is directed parallel to a line between C and D, the voltage is maximum. If the magnitude of the vector is decreased, the voltage decreases.

entire myocardium depolarizes, no voltage differences exist between any areas outside the cell. When the cells in a given region depolarize synchronously, as occurs during normal excitation, that portion of the heart becomes a dipole. The depolarized portion constitutes the negative side and the yet-to-be depolarized portion the positive side of the dipole. The tub of salt solution is analogous to the rest of the body, so the heart is a dipole

in a volume conductor. With electrodes located at various points around the volume conductor (i.e., the body), the voltage resulting from the dipole generated by the electrical activity of the heart can be measured.

Consider the voltage changes produced by a two-dimensional model in which the body serves as a volume conductor and the heart generates a continuously changing dipole (Fig. 13–11). An electrocardiographic

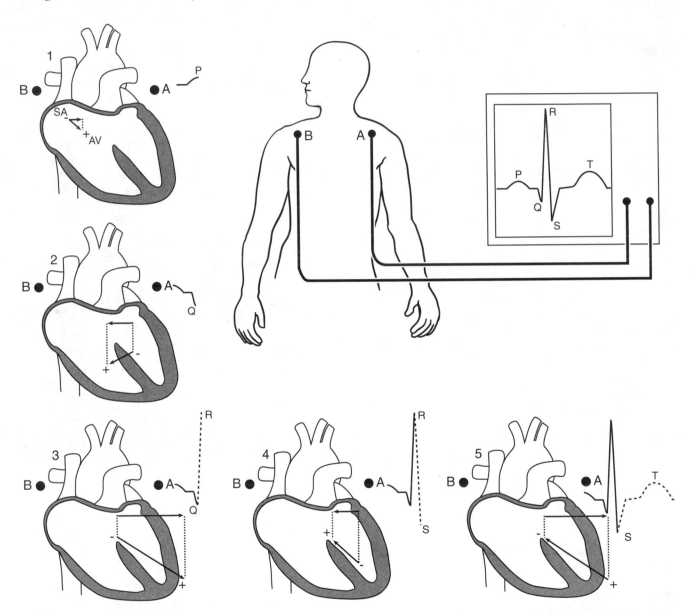

Figure 13–11 ■ Sequence of major dipoles giving rise to electrocardiographic waveforms. The solid arrows are vectors that represent the magnitude and direction of a major dipole. The magnitude is proportional to the mass of myocardium involved. The direction is determined by the orientation of polarized and depolarized regions of the myocardium. The vertical dashed lines project the vector onto the A-B coordinate; it is this component of the vector that is sensed and recorded (open arrow). In panel 5 the orientation of the solid arrow reflects the direction of repolarization (as opposed to depolarization). The head of this arrow represents the yet-to-be repolarized region of the myocardium (negative), and the tail represents the region that has already repolarized (positive). By convention, the projected vector (open arrow) points so that its head is directed toward the more positive electrode (A) as opposed to the less positive electrode (B).

recorder (a voltmeter) is connected between points A and B (lead I). By convention, when point A is positive relative to point B the ECG is deflected upward, and when B is positive relative to A a downward deflection results. The solid arrows show (in two dimensions) the magnitude (voltage) and direction of the actual dipoles. The lengths of the arrows are proportional to the magnitude, which is related to the mass of myocardium generating the dipole. The open arrows show the magnitude of the dipole component that is parallel to the line between points A and B (the recorder electrodes); this component determines the voltage that will be recorded.

The P Wave and Atrial Depolarization ■ Atrial excitation results from a wave of depolarization that originates in the SA node and spreads over the atria, as indicated in panel 1 of Figure 13-11. The dipole generated by this excitation has a magnitude proportional to the mass of the atrial muscle involved and a direction indicated by the solid arrow. The head of the arrow points toward the positive end of the dipole, where the atrial muscle is not yet depolarized. The negative end of the dipole is located at the tail of the arrow, where depolarization has already occurred. Point A is therefore positive relative to point B, and there will be an upward deflection of the ECG as determined by the magnitude and direction of the dipole. Once the atria are completely depolarized, no voltage difference exists between A and B, and the voltage recording returns to 0. The voltage change associated with atrial excitation appears on the ECG as the **P wave.**

The PR Segment and Atrioventricular Conduction ■ As depolarization moves slowly through the AV node, the mass of tissue involved is too small to create a dipole with sufficient magnitude to produce a voltage difference on the ECG. Atrioventricular node conduction occurs during the interval between the P wave and the onset of ventricular depolarization; this period is referred to as the PR segment. (Note: When a Q wave is present at the start of ventricular depolarization, it would be correct to call this period the PQ segment; however, by convention the term *PR segment* is used to describe the period of AV node conduction.)

The QRS Complex and Ventricular Depolarization ■ The depolarization wave emerges from the AV node and travels along the AV bundle (bundle of His), bundle branches, and Purkinje system; these tracts extend down the interventricular septum. The initial direction of the resulting dipole is shown in panel 2 of Figure 13-11. Point B is positive relative to point A because the left side of the septum depolarizes before the right side. The small downward deflection produced on the ECG is the **Q wave.** The normal Q wave is often so small that it is not apparent.

The wave of depolarization spreads via the Purkinje system across the inside surface of the ventricle. Depolarization of the ventricular muscle proceeds from the inner muscle layer (endocardium) to the outer layer (epicardium). The muscle mass of the left ventricle is much greater than that of the right ventricle, and the net dipole during this phase has the direction indicated in panel 3. The deflection of the ECG is upward because point A is positive relative to point B, and it is large because of the great mass of tissue involved. This upward deflection is the **R wave.**

The voltage returns to zero as all of the ventricular muscle becomes depolarized and the dipole disappears. The last portion of the ventricle to depolarize is near the atria; the direction of the dipole associated with this phase is shown in panel 4. Point B is positive compared with point A, and the deflection on the ECG is downward. This final deflection is the **S wave.**

The Q, R, and S waves together are known as the **QRS complex** and show the progression of ventricular muscle depolarization. When all of the ventricular muscle is depolarized no potential differences (dipoles) exist and the ECG is said to be **isoelectric;** that is, the voltage is zero. At this point all of the ventricular muscle cells are in the plateau phase (phase 2) of the action potential (Fig. 13-12).

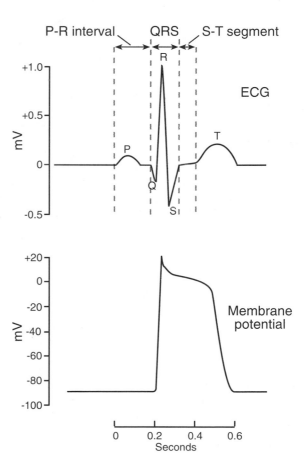

Figure 13–12 ■ Timing of ventricular membrane potential and ECG. Note that the ST segment occurs during the plateau of the action potential.

PREEXCITATION SYNDROME

Occasionally the AV node fails to provide an appropriate delay in the action potential as it passes from the atria to the ventricles. When this occurs, the ventricles are activated sooner than expected, causing **preexcitation.** The most common form of preexcitation is the **Wolff-Parkinson-White syndrome,** which results when an **accessory pathway** bypasses the AV node. Accessory bypass fibers not only conduct electrical impulses from the atria to the ventricle (antegrade conduction) but in some circumstances also conduct impulses from the ventricle to the atria (retrograde conduction); this arrangement can predispose the heart to "reciprocating tachycardias," in which electrical impulses are conducted down one pathway (either the AV node or the bypass tract) and back up the other. The resulting "circus movement" of electrical activity back and forth between the atria and the ventricles may produce a rapid, ineffective heart beat. Dangerous tachycardias can also occur if **atrial fibrillation** or **flutter** develops; as the rapid atrial impulses bypass the AV node they may trigger ventricular contractions at a rate that is too fast to allow adequate diastolic filling.

Although certain drugs can be useful in the treatment of Wolff-Parkinson-White syndrome, generally the most effective form of treatment is ablation of the accessory pathway using either direct surgical disruption of the bypass tract or electrical destruction of the tract using special catheters positioned near the bypass tract, through which electrical current can be discharged. The heat associated with electrical discharge can destroy tissues adjacent to the catheter, including the bypass tract.

The T Wave and Ventricular Repolarization ■ Repolarization, like depolarization, generates a dipole because the voltage of the depolarized area is different from that of the repolarized areas. The dipole associated with atrial repolarization does not appear as a separate deflection on the ECG because it occurs during ventricular depolarization and is masked by the greater deflection (QRS complex) that is present at this time. Ventricular repolarization occurs well after the QRS complex and is not as orderly as ventricular depolarization. Unlike depolarization, repolarization proceeds from the epicardial to the endocardial surface of the ventricles. The reasons for this difference are not certain. One possible explanation is that cold blood returning from the lungs and extremities cools the endocardial surface, thus slowing repolarization and prolonging the plateau of the ventricular muscle action potential. As a consequence, repolarization begins at the warmer epicardial surface and proceeds inward. Although the direction of repolarization (epicardium to endocardium) is opposite that associated with depolarization (Fig. 13-10), the polarity of the dipole is the same. This happens because the epicardium of the ventricle is positive on the outside (having repolarized first) at a time when the endocardium is still negative. As in depolarization, point A is positive with respect to point B, and an upward deflection occurs on the ECG. This deflection is the **T wave** (panel 5, Fig. 13-11).

Determining the Cardiac Dipole from the Electrocardiogram ■ As explained above, changes in the magnitude and direction of the cardiac dipole will cause changes in a given ECG lead. By looking at several leads simultaneously the ECG can be used to determine the direction and magnitude of the cardiac dipole at any given moment. For purposes of illustration, consider **Einthoven's triangle** (Fig. 13-13), which can be constructed from standard leads I, II, and III (Fig. 13-7) and may be rearranged graphically into a set of coordinates as shown in Figure 13-14. If a QRS complex is simultaneously observed in each lead, the direction and magnitude of the dipole at any instant can be calculated from the QRS deflections. For example, Figure 13-14 shows a method of determining dipole direction and magnitude for the QRS complexes shown. For simplicity, the sample calculation will be made during the moment the QRS deflection is maximal, although similar calculations could be made for any point on the complex. First, the magnitude of the deflection is measured in appropriate units; since the deflection is upward in all the leads, its magnitude is plotted on the line for each lead toward its positive pole. Next, a perpendicular line (dotted) is drawn from each plotted point. The point where all three of these perpendiculars intersect defines a vector that represents the magnitude and direction of the cardiac dipole at that point in time.

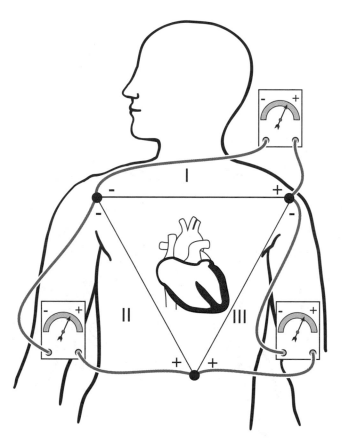

Figure 13–13 ■ Einthoven's triangle. Einthoven codified analysis of electrical activity of the heart by proposing that certain conventions be followed. The heart is considered to be at the center of a triangle, each corner of which serves as the location for an electrode for two leads to the electrocardiographic recorder. The three resulting leads are I, II, and III. By convention, one electrode causes an upward deflection of the recorder when it is under the influence of a positive dipole relative to the other electrode.

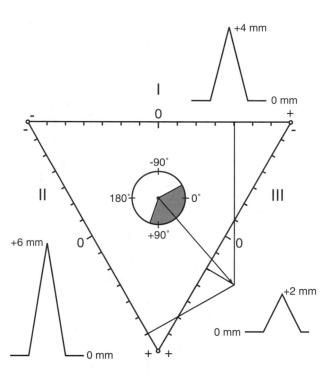

Figure 13–14 ■ Projection of QRS measurements onto standard lead coordinates and subsequent determination of instantaneous cardiac vector. The direction of the QRS vector is normally between − 30 and + 110 degrees (the shaded area).

■ Unipolar Leads Are Used to Record the Electrocardiogram from Standard Sites

Several unipolar leads are routinely recorded along with the standard bipolar ones. The designation "unipolar" means that most of the electrical activity is due to voltage differences on only one side of the lead. For example, the unipolar chest leads are obtained by connecting the three limb leads and recording the difference between their combined voltage and the voltage at various points on the chest (Fig. 13-15). The voltage changes due to the sum of the three limb electrodes are considered to be zero, and the voltage differences recorded by the unipolar leads are therefore the result of voltage changes on the surface of the chest. This permits more detailed examination of the electrical activity generated by various parts of the heart by measuring voltages from nearby areas of the chest wall. By convention, an upward deflection of the pen indicates that the unipolar (exploring) elec-

trode is positive with respect to the three combined electrodes. Unipolar chest leads are designated V_1-V_6, and they are traditionally placed over the areas of the chest shown in Figure 13-16.

Three other routinely recorded leads are aVR, aVL, and aVF ("a" stands for augmented). These give the voltages between electrodes on two of the limbs (averaged to yield a reference potential) and the third limb. For example, lead aVF gives the potential difference between the left and the right arm (interconnected and averaged) and the left foot. The exploring electrode in this case is the one attached to the left foot. The use of multiple leads ensures that the magnitude and direction of the cardiac dipole can be accurately determined no matter how it changes with time.

■ The Electrocardiogram Permits Detection and Diagnosis of Irregularities in Heart Rate and Rhythm

The ECG provides information about the rate and rhythm of excitation as well as the pattern of conduction of excitation throughout the heart. The following illustrations of irregularities of cardiac rate and rhythm are by no means comprehensive; they were chosen

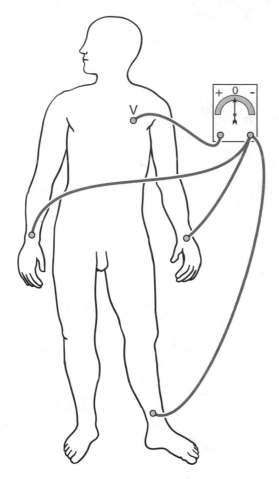

Figure 13–15 ■ Unipolar chest lead. Three limb leads are combined to give the reference voltage (zero) for the unipolar chest lead (V). The unipolar lead is placed on each of the positions shown in Figure 13-16.

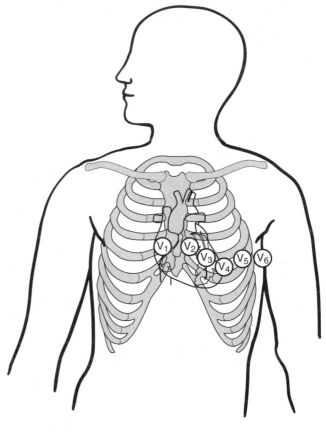

Figure 13–16 ■ Positions of the unipolar chest leads. V_1 is just to the right of the sternum in the fourth intercostal space. V_2 is just to the left of the sternum in the fourth interspace. V_4 is in the fifth interspace in the midclavicular line. V_3 is midway between V_2 and V_4. V_5 is in the fifth interspace in the anterior axillary line. V_6 is in the fifth interspace in the midaxillary line.

to describe basic physiologic principles. Disorders of cardiac rate and rhythm are referred to as **arrhythmias.**

Figure 13-17 compares the ECG from a normal individual with those from persons with variations in the cardiac rhythm. Panel A is a 12-lead ECG from an individual with normal sinus rhythm. The P wave (atrial excitation) is always followed by a QRS complex of uniform shape and size. The PR interval (beginning of the P wave to beginning of the QRS complex) is 0.18 seconds (normal, 0.10–0.20 seconds). This indicates that the conduction velocity of the action potential from the SA node to the ventricular muscle is normal. The average time between R waves (successive heart beats) is about 0.8 seconds, making the heart rate approximately 75 beats/min.

Panel B shows a normal finding in the ECG of a child with **respiratory sinus arrhythmia,** which is an increase in the heart rate with inspiration and a decrease with expiration. The presence of a P wave

before each QRS complex indicates that these beats originate in the SA node. Intervals between successive beats of 0.85 , 0.92, 1.02, 1.36, and 1.20 seconds correspond to heart rates of 71, 65, 59, 44, and 50 beats/min. The interval between the beginning of the P wave and the end of the T wave is uniform, and the change in the interval between beats is primarily accounted for by the variation in time between the end of the T wave and the beginning of the P wave. Although the heart rate changes, the interval during which electrical activation of the atria and ventricles occurs does not change nearly as much as the interval between beats. Respiratory sinus arrhythmia is caused by cyclic changes in sympathetic and parasympathetic neural activity to the SA node that accompany respiration. It is observed in individuals with healthy hearts.

Panel C shows an ECG during excessive stimulation of the parasympathetic nerves. The stimulation releases acetylcholine from nerve endings in the SA node; this suppresses the pacemaker activity, slows

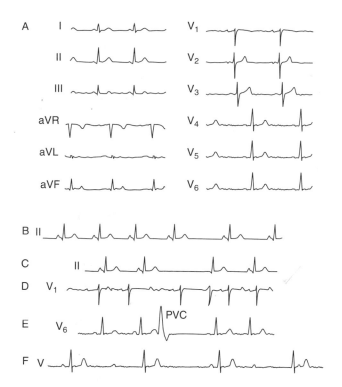

Figure 13–17 ■ Normal and abnormal ECGs. (A) Normal 12-lead ECG. (B) Respiratory sinus arrhythmia (lead II). (C) Sinus arrest with vagal escape (lead II). (D) Atrial fibrillation (lead V₁). (E) Premature ventricular complex (lead V₆). (F) Complete atrioventricular block (precordial lead).

carditis, hypertension, and hyperthyroidism, but sometimes occurs in otherwise normal individuals.

The ECG in panel E shows a **premature ventricular complex** (PVC). Following a normal QRS complex, a complex of increased voltage and longer duration occurs. This premature complex is not preceded by a P wave and may be followed by a pause before the next normal P wave and QRS complex. The premature ventricular excitation is initiated by an **ectopic focus,** an area of pacemaker activity in other than the SA node. In the example of excessive parasympathetic stimulation in panel C, the ectopic focus was in the AV node. In panel E it is probably in the Purkinje system or ventricular muscle, where an aberrant pacemaker reaches threshold before being depolarized by the normal wave of excitation (ectopic foci can also be in the atria). Once the ectopic focus triggers an action potential, the excitation is propagated over the ventricles and may reach the AV node in a retrograde fashion. The abnormal pattern of excitation accounts for the greater voltage, change of dipole direction, and longer duration (inefficient conduction) of the QRS complex. Retrograde conduction often dies out in the AV node, so the atria are not excited. The normal atrial excitation (P wave) occurs but is hidden by the abnormal QRS complex. This P wave does not result in ventricular excitation, because the AV node is still refractory when the impulse arrives. As a consequence, the next "scheduled" ventricular beat is missed. The prolonged interval following a premature ventricular beat is the **compensatory pause.**

Premature beats can also arise in the atria. In this case, the P wave is abnormal but the QRS complex is normal. Premature beats are often called extrasystoles, frequently a misnomer because there is no "extra" beat. In some cases, the premature beat is interpolated between two normal beats and the premature beat is indeed "extra."

In panel F, both P waves and QRS complexes are present, but they appear to occur independently. This is **atrioventricular block.** The AV node fails to conduct impulses from the atria to the ventricles, and since this is the only electrical connection between these areas the pacemaker activities of the two become entirely independent. In this particular example the atrial rate is near 50 while the ventricular rate is only 33 beats/min. The atrial pacemaker is probably in the SA node and the ventricular pacemaker is probably in some portion of the atrioventricular conducting system. In certain conditions the PR interval is lengthened, but all atrial excitations are eventually conducted to the ventricles. This is **first-degree atrioventricular block.** When some, but not all, of the atrial excitations are conducted by the AV node, it is **second-degree atrioventricular block.** If atrial excitation never reaches the ventricles, as in this example, it is **third-degree (complete) atrioventricular block.**

the heart rate, and increases the distance between P waves. The third QRS complex is not preceded by a P wave because the ventricles were excited by an impulse that originated in the AV node. This occurs because acetylcholine suppressed the activity of the SA node below that of the AV node (sinus arrest), making the AV node the primary pacemaker. When a QRS complex is recorded without a preceding P wave, it reflects the fact that ventricular excitation has occurred without a preceding atrial contraction. This phenomenon is called **nodal escape.** The interval between the last atrial beat and the escape beat is 1.82 seconds, equivalent to a heart rate of 33 beats/min and consistent with the observation that the AV node pacemaker has a firing rate between 15 and 35 beats/min.

The ECG in panel D is from a patient with **atrial fibrillation.** In this condition atrial systole does not occur because the individual atrial cells are electrically excited at random. As a result, there are always some excited cells among those located near the AV node, and each nodal cell will be excited as soon as its refractory period ends. The result is a rapid and irregular ventricular rate. Atrial fibrillation is associated with numerous disease states, such as cardiomyopathy, peri-

■ The Electrocardiogram Provides Three Types of Information about the Ventricular Myocardium

The ECG provides information about the pattern of excitation of the ventricles, changes in the mass of electrically active ventricular myocardium, and abnormal dipoles resulting from injury to the ventricular myocardium. It provides no direct information about the mechanical effectiveness of the heart; other tests are used to study the efficiency of the heart as a pump (Chap. 14).

The Pattern of Ventricular Excitation ■ Disease or injury can affect the pattern of ventricular depolarization and produce an abnormality in the QRS complex. Figure 13–18 shows a normal QRS complex (Fig. 13–18A) and two examples of complexes that have been altered by impaired conduction. In 13–18B, the AV bundle branch to the right side of the heart is not conducting (i.e., there is right bundle branch block) and depolarization of right-sided myocardium is therefore dependent on delayed electrical activity coming from the normally depolarized left side of the heart. The resulting QRS complex has an abnormal shape (because of aberrant electrical conduction) and is prolonged (because of the increased time necessary to depolarize the heart fully). In Figure 13–18C the AV bundle branch to the left side of the heart is not conducting (i.e., there is left bundle branch block), also resulting in a wide, deformed QRS complex.

Changes in the Mass of Electrically Active Ventricular Myocardium ■ The recording in Figure 13–19 show the effect of right ventricular enlargement on the ECG. The increased mass of right ventricular muscle changes the direction of the major dipole during ventricular depolarization, resulting in large R waves in lead V_1. The large S waves in lead I and the large R waves in lead aVF are also characteristic of shift in the dipole of ventricular depolarization to the right. This illustrates how a change in the mass of

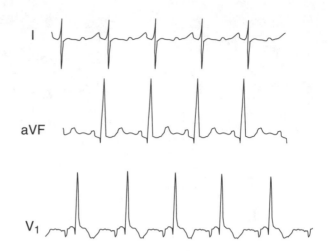

Figure 13–19 ■ The ECG (lead I, aVF, and V_1) of a patient with right ventricular hypertrophy.

excited tissue can affect the amplitude and direction of the QRS complex.

Figure 13–20 shows the effects of atrial hypertrophy on the P waves of lead III (panel A) and the altered QRS complexes in leads V_1 and V_5 associated with left ventricular hypertrophy (panel B). Left ventricular hypertrophy rotates the direction of the major dipole associated with ventricular depolarization to the left, causing large S waves in V_1 and large R waves in V_5.

Abnormal Dipoles Resulting from Ventricular Myocardial Injury ■ If a portion of the ventricular myocardium fails to receive sufficient blood flow (a condition known as **myocardial ischemia**) the supply of ATP may decrease below the level required to maintain the active transport of ions across the cell membrane, and this can affect the ECG. There are normally two periods when no dipoles exist: the interval between the completion of the T wave and the onset of the P wave, during which repolarization is complete (diastole); and the interval between the end of the QRS complex and the onset of the T wave, during which depolarization is complete (systole).

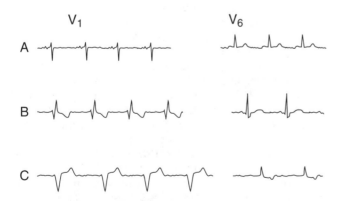

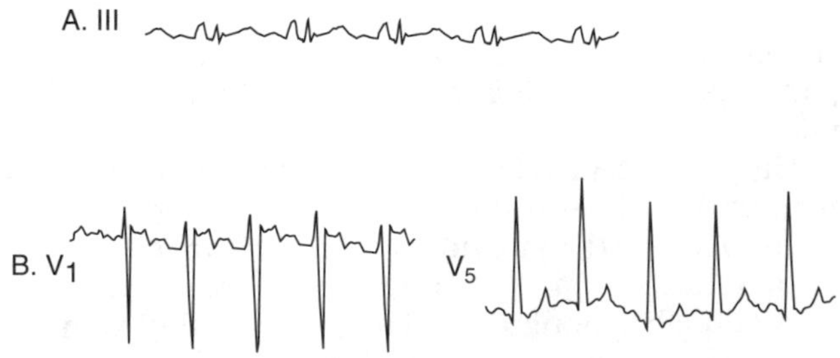

Figure 13–18 ■ The ECG (lead V_1 and V_6) of patients with (A) normal QRS complex; (B) right bundle branch block; and (C) left bundle branch block.

Figure 13–20 ■ (A) Large P waves (lead III) caused by atrial hypertrophy. (B) Altered QRS complex (leads V_1 and V_5) produced by left ventricular hypertrophy.

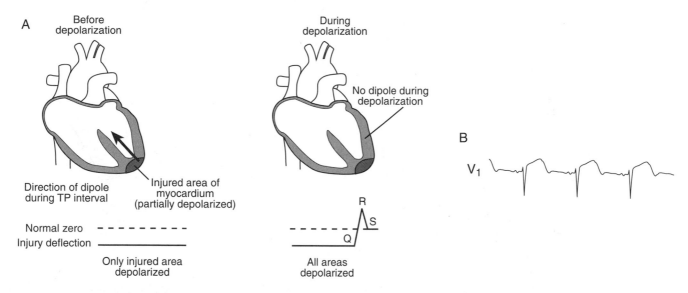

Figure 13–21 ■ (A) Electrocardiogram changes in myocardial injury. Dark shading depicts depolarized ventricular tissue. ST segment elevation can occur with myocardial injury. The apparent zero baseline of the ECG before depolarization is below zero because of partial depolarization of the injured area (shading). After depolarization (during the action potential plateau) all areas are depolarized and true zero is recorded. Because zero baseline is set arbitrarily (on the ECG recorder), a depressed diastolic baseline (TP segment) and an elevated ST segment cannot be distinguished. Regardless of the mechanism, this is referred to as an elevated ST segment. (B) The ECG (lead V₁) of a patient with acute myocardial infarction.

These periods are normally recorded as zero voltages on the ECG. When blood flow is reduced enough to lower ATP, the cells in the region partially depolarize to a lower resting membrane potential, although they are still capable of action potentials. As a consequence, a dipole develops during diastole (described as the TP interval) in injured hearts because of the voltage differences between normal (polarized) and abnormal (partially polarized) tissue. However, this dipole is arbitrarily recorded on the ECG as zero. No dipole occurs during the ST interval because depolarization is uniform and complete in both injured and normal tissue (this is the plateau period of ventricular action potentials). Because the ECG is designed so that the TP interval is recorded as zero voltage, the true zero during the ST interval is recorded as positive or negative deflection (Fig. 13–21). These deflections during the ST interval are of major clinical utility in the diagnosis of cardiac injury.

■ ■ ■ ■ ■ ■ ■ ■ ■ ■ ■ REVIEW QUESTIONS ■ ■ ■ ■ ■ ■ ■ ■ ■ ■ ■

Identify Each with a Term

1. The period when the ventricular cells are depolarized and voltage changes very slowly
2. The membrane protein responsible for the ventricular resting membrane potential
3. The change in membrane potential away from the resting potential and toward the sodium equilibrium potential
4. Two poles that are equal in magnitude but opposite in direction and which are located in close physical proximity to one another
5. The electrocardiographic wave caused by atrial depolarization

Define Each Term

6. Pacemaker potential
7. Depolarization threshold
8. Absolute refractory period
9. QRS complex
10. Atrioventricular block

Choose the Correct Answer

11. Depolarization occurs during which phase of the action potential?
    a. Phase 0
    b. Phase 1

  c. Phase 2
  d. Phase 3
  e. Phase 4

12. What is most responsible for phase 0 of a cardiac nodal cell?
  a. Voltage-gated Na^+ channels
  b. K^+ leak channels
  c. Cl^- channels
  d. Voltage-gated Ca^{2+} channels
  e. Pacemaker channels

13. Which of the following is true?
  a. Pacemaker cells in the SA node are affected by sympathetic but not parasympathetic innervation.
  b. Norepinephrine increases the slope of the pacemaker potential by opening K^+ channels and closing Ca^{2+} channels.
  c. At the end of diastole, the cells of the SA node normally reach threshold before other tissues in the heart.
  d. Acetylcholine causes tachycardia by stimulating the SA node cells to reach threshold faster.
  e. The long refractory period of cardiac muscle enables tetanus to occur when necessary for optimal myocardial function.

14. The ECG can be used to determine or detect:
  a. The direction of depolarization
  b. Arrhythmias
  c. Ventricular hypertrophy
  d. Myocardial ischemia
  e. All of the above

15. Atrial repolarization normally occurs during the:
  a. P wave
  b. QRS complex
  c. ST segment
  d. T wave
  e. Isoelectric period

Calculate

16. Draw the mean cardiac vector in the frontal plane, based on the following ECG tracings for leads, I, II, and III.

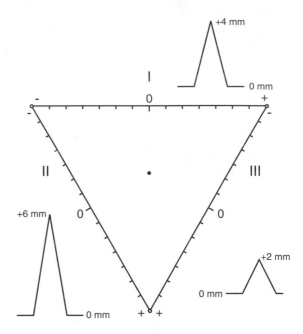

Briefly Answer

17. How does the delay in ventricular activation caused by slow electrical conduction through the AV node contribute to normal cardiac function?

18. A 72-year-old man with an atrial rate of 80 beats/min develops third-degree (complete) AV block. A pacemaker site located just below the AV node (in the bundle of His) triggers ventricular activity, but at a rate of only 20 beats/min. (a) How will the shape of the QRS complex change when the beats originate in the His bundle, as opposed to the atria? (b) To ensure an adequate heart rate, a temporary electronic pacemaker lead is attached to the apex of the right ventricle and the heart is paced by electrically stimulating this site at a rate of 70 beats/min. How will the shape of the QRS complex change when the beats originate in the right ventricular apex, as opposed to the atria?

19. (a) How might atrial hypertrophy affect the ECG? (b) Would left ventricular hypertrophy be expected to affect the ECG?

■ ■ ■ ■ ■ ■ ■ ■ ■ ■ ■ ■ ■ ■ **ANSWERS** ■ ■ ■ ■ ■ ■ ■ ■ ■ ■ ■ ■ ■ ■

1. Plateau of action potential
2. Inward rectifier K^+ channel
3. Depolarization
4. Dipole
5. P wave
6. The spontaneous, progressive depolarization of nodal cells occurring during phase 4 of the action potential
7. The membrane potential at which Na^+ permeability suddenly increases, initiating an action potential
8. The time during which no stimulus of any magnitude can trigger an action potential

9. The deflection on the ECG caused by depolarization of the ventricles

10. Failure of the P waves to conduct in a normal fashion through the AV node and trigger a QRS complex. In first-degree AV block, all the impulses are conducted, but only after a delay. In second-degree AV block, some, but not all, of the impulses are conducted. In third-degree AV block, no impulses are conducted.

11. a

12. d

13. c

14. e

15. b

16.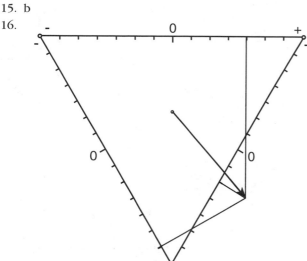

17. For the heart to function properly, blood must be pumped first from the atria into the ventricles and then from the ventricles into the aorta and pulmonary artery. If the atria and ventricles both depolarized (and contracted) at the same time, the blood could not flow from the atria into the ventricles. Delaying the wave of depolarization in the AV node enables the atria to pump blood into the ventricles before ventricular depolarization and contraction occurs. The atria and ventricles therefore pump sequentially rather than simultaneously.

18. (a) The form of the QRS will be similar, since electrical excitation of the ventricles occurs over essentially the same pathways in both situations (i.e., His bundle to bundle branches to Purkinje system to myocardium). (b) The morphology will be markedly different from normal, since depolarization now originates in the right ventricle and propagates in a retrograde fashion. Because the right side of the heart depolarizes before the left, the configuration of the QRS may resemble that seen with left bundle branch block, another situation in which the right side of the heart depolarizes before the left.

19. (a) Atrial hypertrophy will increase the magnitude (and possibly the duration) of the P wave. It will have no effect on the QRS or T waves. (b) The QRS in left ventricular hypertrophy is similar in shape to that seen in normals, but the voltage is significantly greater.

Suggested Reading

Fozzard HA, Jennings RB, Haber E, Katz AM, Morgan HE: *The Heart and Cardiovascular System.* New York, Raven, 1986, chap. 1,27,30,32.

Friedman HH: *Diagnostic Electrocardiography and Vectorcardiography.* 3rd ed. New York, McGraw-Hill, 1985.

Katz AM: *Physiology of the Heart.* 2nd ed. New York, Raven, 1992, chap. 18–20.

Katz LN, Hellerstein HD: Electrocardiography, in Fishman AP, Richards DW (eds): *Circulation of the Blood: Men and Ideas.* Bethesda, American American Physiological Society, 1982, pp. 265–354.

Sperelakis N: *Physiology and Pathophysiology of the Heart.* Boston, Martinus Nijhoff, 1989.

CHAPTER

■ ■ ■ ■ ■ ■ ■

14

The Cardiac Pump

OBJECTIVES

After studying this chapter, the student should be able to:
1. Define the cardiac cycle and describe its events
2. Correlate clinical findings with the electrical and mechanical
   events occurring during the cardiac cycle
3. Discuss the determinants of cardiac stroke volume and their
   effects on cardiac output
4. Discuss the determinants of heart rate and their effects on
   cardiac output
5. Describe the techniques used to measure cardiac output
6. Define *cardiac work* and discuss the factors that determine it

---

The heart consists of a series of four separate chambers (two atria and two ventricles) that use one-way valves to direct blood flow. Its ability to pump blood depends on the integrity of the valves and proper cyclic contraction and relaxation of the muscular walls of the four chambers. An understanding of the cardiac cycle is a prerequisite for understanding the performance of the heart as a pump.

The Cardiac Cycle

The **cardiac cycle** refers to the sequence of electrical and mechanical events occurring in the heart during a single beat and the resulting changes in pressure, flow, and volume in the various cardiac chambers. The functional interrelationships of the cardiac cycle described below are represented in Figure 14-1.

■ Sequential Contractions of the Atria and Ventricles Pump Blood through the Heart

The cycle of events described here occurs almost simultaneously in the right and left heart; the main difference is that the pressures are higher on the left side. We focus on the left side of the heart, beginning with electrical activation of the atria.

Atrial Systole and Diastole ■ The P wave of the electrocardiogram (ECG) reflects atrial depolarization, which initiates **atrial systole.** Contraction of the atria "tops off" ventricular filling with a final small volume of blood from the atria (a wave). Under resting conditions, atrial systole is not essential for ventricular filling, and in its absence ventricular filling is only slightly reduced. However, when increased cardiac output is required, as during exercise, the absence of atrial systole can limit ventricular filling and stroke volume. This happens in patients with atrial fibrillation, whose atria do not contract synchronously. The P wave is followed by an electrically quiet period, during which atrioventricular (AV) node transmission occurs (PR segment). During this electrical pause the mechanical events of atrial systole and ventricular filling are concluded before the excitation and contraction of the ventricles begins. Atrial diastole follows atrial systole and occurs during ventricular systole. As the left

atrium relaxes, blood enters the atrium from the pulmonary veins. Simultaneously, blood enters the right atrium from the superior and inferior vena cavae. The gradual rise in left atrial pressure during atrial diastole reflects its filling (v wave). The small pressure oscillation early in atrial diastole (c wave) is caused by movements of the heart associated with ventricular contraction.

Ventricular Systole ■ The QRS complex reflects excitation of ventricular muscle and the beginning of **ventricular systole** (Fig. 14-1). As ventricular pressure rises above atrial pressure, the left **atrioventricular (mitral) valve** closes. Contraction of the papillary muscles prevents the mitral valve from everting into the left atrium and enables the valve to prevent the regurgitation of blood into the atrium as ventricular pressure rises. The aortic valve does not open until left ventricular pressure exceeds aortic pressure. During the interval when both mitral and aortic valves are closed, the ventricle contracts **isovolumetrically** (i.e., the ventricular volume does not change). The contraction causes ventricular pressure to rise, and when ventricular pressure exceeds aortic pressure (at approximately 80 mm Hg) the aortic valve opens and allows blood to flow from the ventricle into the aorta. At this point ventricular muscle begins to shorten, reducing the volume of the ventricle. When the rate of ejection begins to fall (see the aortic blood flow record in Fig. 14-1) the aortic and ventricular pressures decline. Ventricular pressure actually decreases slightly below aortic pressure prior to closure of the aortic valve, but flow continues through the aortic valve because of the inertia imparted to the blood by ventricular contraction. (Think of a rubber ball connected to a paddle by a rubber band. The ball continues to travel away from the paddle after you pull back because the inertial force on the ball exceeds the force generated by the rubber band.)

Ventricular Diastole ■ Ventricular repolarization (producing the T wave) initiates ventricular relaxation or **diastole.** When the ventricular pressure drops below the atrial pressure, the mitral valve opens, allowing the blood accumulated in the atrium during systole to flow rapidly into the ventricle; this is the so-called first (rapid) phase of ventricular filling. Both pressures continue to decrease—the atrial pressure

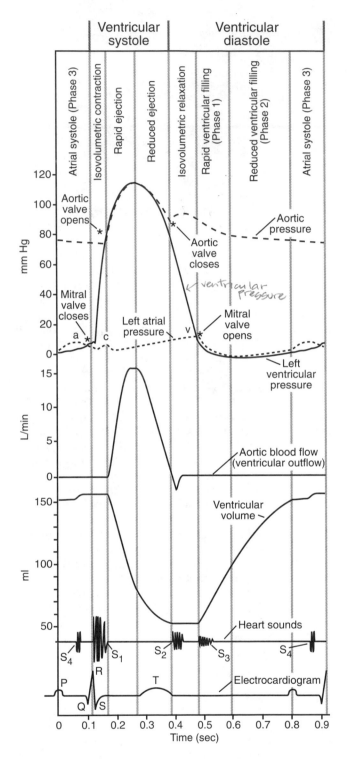

Figure 14–1 ■ Timing of various events in the cardiac cycle.

filling slows as ventricular and atrial pressures converge (second phase). Atrial systole tops off ventricular volume during the third phase of ventricular filling.

■ Pressures, Flows, and Volumes in the Cardiac Chambers, Aorta, and Great Veins Can Be Matched with the Electrocardiogram and Heart Sounds

The pressures, flows, and volumes in the cardiac chambers, aorta, and great veins can be studied in conjunction with the ECG and heart sounds to yield an understanding of the coordinated activity of the heart. Ventricular diastole and systole can be defined in terms of both electrical and mechanical events. In electrical terms, ventricular systole is defined as the period between the QRS complex and the end of the T wave. In mechanical terms, it is the period between the closure of the mitral valve and the subsequent closure of the aortic valve. In either case ventricular diastole comprises the remainder of the cycle. The **first (S_1)** and **second (S_2) heart sounds** signal the beginning and end of mechanical systole. The first heart sound (usually described as a "lub") occurs as the ventricle contracts and ventricular pressure rises above atrial pressure (Fig. 14–2), causing the atrioventricular valves to close. The relatively low-pitched sound associated with their closure is caused by vibrations of the valves and walls of the heart that occur as a result of their elastic properties when the flow of blood through the valves is suddenly stopped. In contrast, the aortic and pulmonary valves close at the end of the ventricular systole, when the ventricles relax and pressures in the ventricles fall below those in the arteries. The elastic properties of the aortic and pulmonary valves produce the second heart sound, which is relatively high-pitched (typically described as a "dup"). Mechanical events other than vibrations of the valves and nearby structures contribute to these two sounds, especially S_1; these factors include movement of the great vessels and turbulence of the rapidly moving blood. The second heart sound is often heard as having two components, the first corresponding to aortic valve closure and the second to pulmonary valve closure. This so-called **splitting** usually widens with inspiration and sometimes disappears with exhalation.

A **third heart sound (S_3)** may result from vibrations during the first phase of ventricular early diastole. It is usually due to rapid passive ventricular filling and can signify abnormal ventricular function. Although it may be heard in normal children and adolescents, its appearance in a patient older than 35 years usually signals the presence of a cardiac abnormality. A **fourth heart sound (S_4)** may be heard during atrial systole. It is caused by blood movement due to atrial contraction and, like S_3, is more common in patients with abnormal hearts.

because of emptying into the ventricle and the ventricular pressure because of continued ventricular relaxation (which, in turn, draws more blood from the atrium). About midway through ventricular diastole,

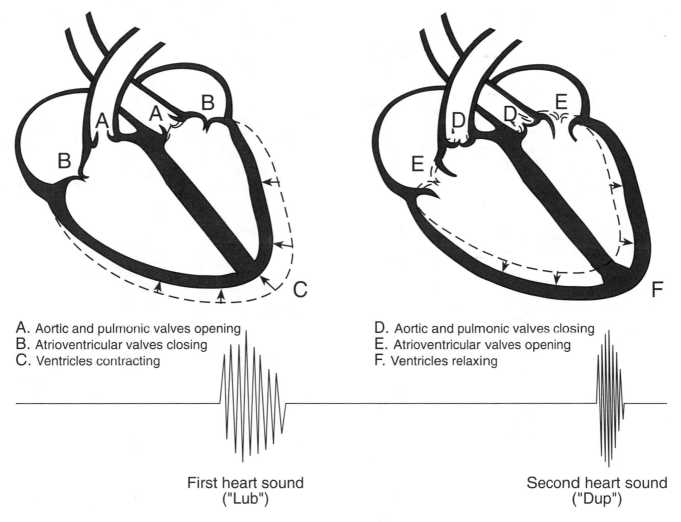

A. Aortic and pulmonic valves opening
B. Atrioventricular valves closing
C. Ventricles contracting

D. Aortic and pulmonic valves closing
E. Atrioventricular valves opening
F. Ventricles relaxing

First heart sound
("Lub")

Second heart sound
("Dup")

Figure 14–2 ■ Production of the heart sounds. At the bottom is a phonocardiogram, which shows the sounds heard with closure of the atrioventricular valves (first heart sound) and the aortic and pulmonary valves (second heart sound).

Cardiac Output

Cardiac output is defined as the volume of blood ejected from the heart per unit time. The usual resting values quoted for adult humans are 5–6 L/min, or approximately 8% of body weight per minute. (Cardiac output divided by body surface area is called the **cardiac index.** When it is necessary to normalize the value to compare the cardiac output among individuals of different sizes, either cardiac index or cardiac output divided by body weight can be used.) Cardiac output (CO) is the product of heart rate (HR) and **stroke volume** (SV) (the volume of blood ejected with each beat):

$$CO = SV \times HR \qquad (1)$$

Stroke volume also may be defined as the volume of blood in the ventricle at the end of diastole (**end-**

diastolic volume) minus the volume of blood in the ventricle at the end of systole (**end-systolic volume**).

If heart rate remains constant, cardiac output increases in proportion to stroke volume, and vice versa. Table 14–1 outlines the factors that influence cardiac output.

■ Stroke Volume Is One Variable that Determines Cardiac Output

Stroke volume is the result of the balance between force of contraction and afterload. Force of contraction is discussed in terms of end-diastolic fiber length, contractility, and hypertrophy. **Afterload** (the force against which the ventricle must contract to eject blood) is considered in terms of ventricular radius and ventricular systolic pressure. Because the pressure drop across the aortic valve is normally small, aortic

TABLE 14–1 ▪ Factors Influencing Cardiac Output

I. Stroke volume
  A. Force of contraction
    1. End-diastolic fiber length (heterometric autoregulation, Starling's law, ventricular function curves, preload)
    2. Contractility
      a. Sympathetic stimulation with norepinephrine (and epinephrine); mediated via β_1-receptor
      b. Drugs (digitalis, anesthetics, toxins, etc.)
      c. Disease (coronary artery disease, myocarditis, etc.)
      d. Homeometric autoregulation
    3. Hypertrophy
  B. Afterload
    1. Ventricular radius
    2. Aortic pressure
II. Heart rate (and pattern of electrical excitation)

From Sparks HV Jr, Rooke TW. *Essentials of Cardiovascular Physiology.* Minneapolis, University of Minnesota Press, 1987.

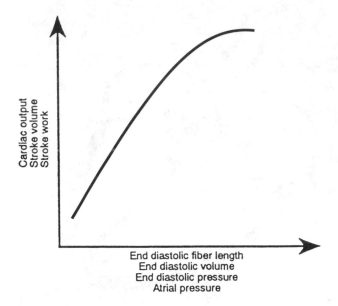

Figure 14–3 ▪ Ventricular function curve. Several combinations of variables can be used to plot the ventricular function curve, depending on the assumptions made. For example, cardiac output can be substituted for stroke volume if heart rate is constant, and stroke volume can be substituted for stroke work if arterial pressure is constant. End-diastolic fiber length and volume are related by laws of geometry, and volume is related to pressure by ventricular compliance.

pressure is often used as a substitute for ventricular pressure in such considerations.

Effects of End-Diastolic Fiber Length ▪ The relationship between ventricular end-diastolic fiber length (preload) and stroke volume is known as **Starling's law of the heart,** or **heterometric autoregulation.** Within limits, increases in the left ventricular end-diastolic fiber length augment the ventricular force of contraction, which increases the stroke volume. This reflects the relationship between length of a muscle and force of contraction (Chap. 10). After reaching an optimal diastolic fiber length, stroke volume no longer increases with further stretching of the ventricle (Fig. 14–3).

End-diastolic fiber length is determined by end-diastolic volume, which is dependent on end-diastolic pressure. End-diastolic pressure is the force that expands the ventricle to a particular volume. For a given ventricular **compliance** (change in volume caused by a given change in pressure), higher end-diastolic pressures increase both diastolic volume and fiber length. Furthermore, the end-diastolic pressure depends on the degree of ventricular filling during ventricular diastole, which is influenced largely by atrial pressure. Thus the curve in Figure 14–3 can be plotted with end-diastolic volume, end-diastolic pressure, or atrial pressure as the abscissa, instead of end-diastolic fiber length.

The ordinate on the plot of Starling's law (Fig. 14–3) can also be a variable other than stroke volume. For example, if heart rate remains constant, cardiac output can be substituted for stroke volume. The effect of arterial pressure on stroke volume can also be taken into account by plotting **stroke work** on the ordinate. An increase in arterial pressure (afterload) decreases stroke volume by increasing the force that opposes the ejection of blood during systole. To correct for this, stroke volume times mean arterial pressure, instead of stroke volume alone, can be plotted on the ordinate. Stroke work is a good approximation of the external work of the heart. If stroke work is on the ordinate, any increase in the force of contraction that

results in increased pressure or stroke volume shifts the stroke work curve upward and to the left. If stroke volume were the dependent variable, a change in the performance of the heart causing increased pressure would not be expressed by a change in the curve.

A plot using any combination of these variables is called a **ventricular function curve,** and it expresses Starling's law. This relationship is also called heterometric autoregulation because it is intrinsic to the heart (autoregulation) and is elicited by changes in ventricular fiber length (heterometric).

Starling's law explains the remarkable balancing of the output between the two ventricles. If the right heart were to pump 1% more blood than the left heart each minute without a compensatory mechanism, the entire blood volume of the body would be displaced into the pulmonary circulation in less than 2 hours. A similar error in the opposite direction would likewise displace all of the blood volume into the systemic circuit. Fortunately, heterometric autoregulation prevents this. If the right ventricle pumps slightly more blood than the left ventricle, left atrial filling (and pressure) will increase. As left atrial pressure increases, left ventricular pressure and left ventricular end-diastolic fiber length increase both the force of contraction and the stroke volume of the left ventricle. If the stroke volume rises too much, the left heart begins to pump more blood than the right heart and left atrial pressure drops; this decreases left ventricular filling and reduces stroke volume. The process continues until left heart output is exactly equal to right heart output (Fig. 14–4).

The descending limb of the ventricular function curve, analogous to the descending limb of the length-tension curve (Chap. 10), is probably never reached in a living heart, because the resistance to stretch increases as the end-diastolic volume reaches the limit for optimum stroke volume. Further enlargement of the ventricle would require end-diastolic pressures that do not occur. Due to increased resistance to stretch **(decreased compliance),** the atrial pressures necessary to produce further filling of the ventricles are probably never reached. The limited compliance therefore prevents optimal sarcomere length from being exceeded. In heart failure, the ventricles can dilate beyond the normal limit because they become less resistant to stretch **(increased compliance).** However, even under these conditions optimal sarcomere length is not exceeded. Instead, the sarcomeres appear to realign so that there are more of them in series, allowing the ventricle to dilate without stretching sarcomeres beyond their optimal length.

Effects of Contractility ■ Factors other than end-diastolic fiber length can influence the force of ventricular contraction; there is a family of curves relating stroke volume (or work) to end-diastolic fiber length under different conditions. For example, increases in sympathetic nerve activity increase the force of contraction for a given end-diastolic fiber length (Fig. 14-5). The increase in force of contraction causes more blood to be ejected against a given aortic pressure and thus raises stroke volume. A change in the force of contraction at a constant end-diastolic fiber length reflects a change in the **contractility** of the heart. (The cellular mechanisms governing contractility are discussed in Chap. 10). A shift in the ventricular function curve to the left indicates increased contractility (i.e., more force and/or shortening occurring at the same initial fiber length), and shifts to the right

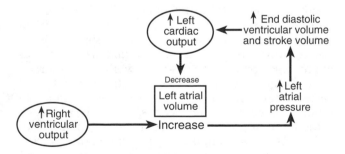

Figure 14–4 ■ Starling's law keeps the output of the left and right hearts balanced. Heart rate is assumed to be constant. The compensatory decline in the initial increase in left atrial volume caused by increased right ventricular output is indicated by "decrease." In the new steady state, left atrial volume is less than the initial high volume present immediately after the increase in right ventricular output but greater than the steady-state atrial volume present before the increase in right ventricular output.

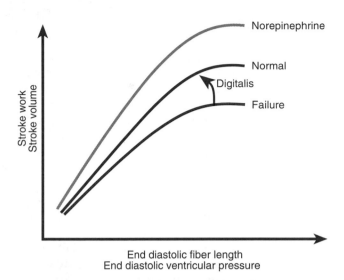

Figure 14–5 ■ Effect of norepinephrine and heart failure on the ventricular function curve. Norepinephrine raises ventricular contractility (i.e., stroke volume and/or stroke work are elevated at a given end-diastolic fiber length). In heart failure, contractility is decreased, so that stroke volume and/or stroke work are decreased at a given end-diastolic fiber length. Digitalis raises the intracellular calcium ion concentration and restores the contractility of the failing ventricle.

indicate decreased contractility. When an increase in contractility is accompanied by an increase in arterial pressure, the stroke volume may remain constant and the increased contractility will not be evident by plotting the stroke volume against the end-diastolic fiber length. However, if stroke work is plotted, a leftward shift of the ventricular curve is observed (Fig. 14-5). A ventricular function curve with stroke volume on the ordinate can be used to indicate changes in contractility only when arterial pressure does not change.

During heart failure, the ventricular function curve is shifted to the right, causing a particular end-diastolic fiber length to be associated with less force of contraction and/or shortening and a smaller stroke volume. As described in Chapter 10, cardiac glycosides, such as digitalis, tend to normalize contractility; that is, they shift the ventricular curve of the failing heart back to the left (Fig. 14-5).

Another influence on contractility is homeometric autoregulation, discussed below.

Effects of Hypertrophy ■ The force of contraction is also increased by **myocardial hypertrophy.** Repeated bouts of increased cardiac output (as occur with physical exercise) or a sustained elevation of arterial pressure result in increased synthesis of contractile proteins and enlargement of cardiac myocytes. The latter is the result of increased numbers of parallel myofilaments, increasing the number of actomyosin crossbridges that can be formed in parallel. As each cell

CONGESTIVE HEART FAILURE

Heart failure occurs when the heart is unable to pump blood at a rate sufficient to meet the body's metabolic needs. One possible consequence of heart failure is that blood may "back up" on the atrial/venous side of the failing ventricle, leading to engorgement and distention of veins (and the organs they drain) as the venous pressure rises. The signs and symptoms typically associated with this occurrence constitute **congestive heart failure** (CHF). This syndrome can be limited to the left ventricle (producing pulmonary venous distention, pulmonary edema, and symptoms such as dyspnea or cough) or right ventricle (producing symptoms such as pedal edema, abdominal edema or **ascites,** and hepatic venous congestion) or may affect both ventricles. The causes of heart failure are numerous and include acquired and congenital conditions, such as valvular disease, myocardial infarction, assorted infiltrative processes (e.g., amyloid or hemochromatosis), inflammatory conditions (i.e., myocarditis), and various types of **cardiomyopathies** (a diverse assortment of conditions in which the heart becomes pathologically dilated, hypertrophied, or stiff). The treatment of heart failure hinges on treating the underlying problem, when possible, and judicious use of medical therapy. Medical treatment may include **diuretics** to reduce the venous fluid overload, **cardiac glycosides** (e.g., digoxin) to improve myocardial contractility, and **afterload reducing agents** (i.e., arterial vasodilators) to reduce the load against which the ventricle must contract.

Heart transplantation is becoming an increasingly viable option for severe, intractable, unresponsive heart failure. More than 13,000 patients worldwide have received new hearts for end-stage heart failure.

---

enlarges, the ventricular wall thickens and is capable of greater force development. There are several isoforms of cardiac actin and myosin with differing properties that translate at the cellular level into differing contractile performances. The expression of these different isoforms may be a key determinant of the long-term adaptation of the heart to stress.

Effects of Afterload ■ The second determinant of stroke volume is afterload (Table 14–1), the force against which the ventricular muscle fibers must shorten. In normal circumstances, afterload can be equated to the aortic pressure during systole. If arterial pressure is suddenly increased, a ventricular contraction (at a given level of contractility and end-diastolic fiber length) produces a lower stroke volume. This decrease can be predicted from the force-velocity relationship of cardiac muscle (Chap. 10). The shortening velocity of ventricular muscle decreases with increasing load, which means that for a given duration of contraction (reflecting the duration of the action potential) the lower velocity results in less shortening and a decrease in stroke volume (Fig. 14–6).

Fortunately, the heart can compensate for the decrease in left ventricular stroke volume produced by increased afterload. Although a sudden rise in systemic arterial pressure causes the left ventricle to eject less

blood per beat, the output from the right heart remains constant. Left ventricular filling subsequently exceeds its output. As the end-diastolic volume and fiber length of the left ventricle increase, the ventricular force of contraction is enhanced. A new steady state is quickly reached in which the end-diastolic fiber length is increased and the previous stroke volume is maintained. Within limits, additional compensation also occurs. Over the next 30 seconds, the end-diastolic fiber length returns toward the control level, while the stroke volume is maintained despite the increase in aortic pressure. If arterial pressure times stroke volume (stroke work) is plotted against end-diastolic fiber length, it is apparent that stroke work has increased for a given end-diastolic fiber length. This leftward shift of the ventricular function curve indicates an increase in contractility. Because there is a change in the force of contraction occurring independent of end-diastolic fiber length, this phenomenon is called **homeometric autoregulation** (same-length self-regulation). Homeometric autoregulation is a relatively minor influence, however, and causes only a small increase in contractility compared with stimuli that are discussed below.

Effects of Ventricular Radius ■ The ventricular radius influences stroke volume because of the rela-

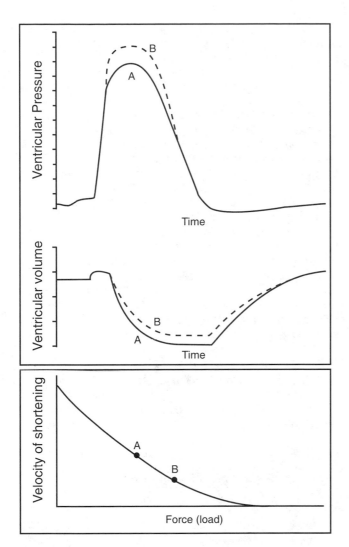

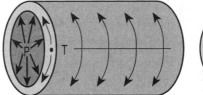

Figure 14–7 ■ Pressure and tension in a cylindric blood vessel. The tension tends to open an imaginary slit along the length of the blood vessel. Laplace's law relates pressure, radius, and tension as described in the text.

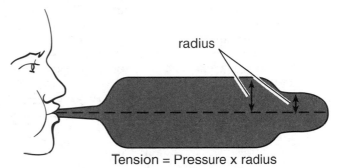

Tension = Pressure x radius

Figure 14–8 ■ Effect of radius of a cylinder on tension. In a balloon the tension in the wall is proportional to the radius since the pressure is the same everywhere inside the balloon. The tension is lower in the portion of the balloon with the smaller radius.

Figure 14–6 ■ Effect of aortic pressure on ventricular function. Ventricular pressure, ventricular volume, and force-velocity relationship are shown for (A) normal and (B) elevated aortic pressure. The slower velocity of shortening means less ventricular emptying and a lower stroke volume.

tionship between ventricular pressures (P_v) and ventricular wall tension (T). For a hollow structure such as a ventricle, by the Laplace equation:

$$P_v = T \left(\frac{1}{r_1} + \frac{1}{r_2} \right) \qquad (2)$$

where r_1 and r_2 are the radii of curvature for the ventricular wall. Figure 14-7 shows this relationship for a simpler structure, in which curvature occurs in only one dimension (i.e., a cylinder). In this case, r_2 approaches infinity. Therefore:

$$P_v = T\frac{1}{r_1} \text{ or } T = P_v \times r_1 \qquad (3)$$

The internal pressure expands the cylinder until it is exactly balanced by the wall tension. The larger the radius, the larger the tension needed to balance a particular pressure. For example, in a long balloon that has an inflated part with a large radius and an uninflated parted with a much smaller radius, the pressure inside the balloon is the same everywhere, yet the tension in the wall is much higher in the inflated part because the radius is much greater (Fig. 14-8). This general principle also applies to noncylindric objects, such as the heart and tapering blood vessels.

When the ventricular chamber enlarges, the wall tension required to balance a given intraventricular pressure increases. As a result, the force resisting ventricular wall shortening (afterload) likewise increases with ventricular size. Despite the effect of increased radius on afterload, an increase in ventricular size (within physiologic limits) raises both wall tension and stroke volume. This occurs because the positive effects of adjustment in sarcomere length overcompensate for the negative effects of increasing ventricular radius. However, if a ventricle becomes pathologically dilated, the myocardial fibers may be unable to generate enough tension to raise pressure to the normal systolic level, and the stroke volume may fall.

■ Heart Rate Is the Other Variable that Influences Cardiac Output

Heart rate can vary from less than 50 beats/min in a resting, physically fit individual to greater than 200 beats/min during maximal exercise. If stroke volume is held constant, increases in heart rate cause proportional increases in cardiac output. Normally, more than half the rise in cardiac output produced by strenuous exercise is due to increased heart rate.

In considering the influence of heart rate on cardiac output it is important that changing heart rate primarily changes the duration of ventricular diastole; that is, as heart rate increases, the duration of ventricular diastole decreases much more than the duration of systole. As the duration of diastole decreases, filling of the ventricles also decreases.

If heart rate decreases below the normal value (due to a change in sympathetic or parasympathetic nervous activity to the heart, athletic conditioning, diseases affecting the Sinoatrial [SA] or AV node, drug effects, or other causes) cardiac output is reduced very little, because as the heart rate falls the duration of ventricular diastole increases, and the longer duration of diastole results in greater filling. The resulting elevated end-diastolic fiber length increases stroke volume, which compensates for the decreased heart rate. This balance works until the heart rate is below 20 beats/min, at which point additional increases in end-diastolic fiber length cannot augment stroke volume further because the peak of the ventricular function curve has been reached.

Effects of Increased Rate Due to Electronic Pacing ■ If an electronic pacemaker is attached to the right atrium and the heart rate is increased by electrical stimulation, surprisingly little increase in cardiac output results; this is because as the heart rate increases, the interval between beats shortens and the duration of diastole decreases. The decrease in diastole leaves less time for ventricular filling, producing a shortened end-diastolic fiber length, which subsequently reduces both the force of contraction and the stroke volume. The increased heart rate is therefore offset by the decrease in stroke volume. When the rate increases above 180 beats/min secondary to an abnormal pacemaker, stroke volume begins to fall because of poor diastolic filling. A person with abnormal tachycardia (e.g., caused by an ectopic ventricular pacemaker) may therefore have a reduction in cardiac output despite an increased rate.

Events in the myocardium compensate to some degree for the decreased time available for filling. First, increases in heart rate reduce the duration of the action potential and thus the duration of systole, so the time available for diastolic filling decreases less than would otherwise be true. In addition, faster heart rates are accompanied by an increase in the force of contraction, which tends to maintain stroke volume. The increased contractility is sometimes called **treppe** or

the **staircase phenomenon.** It is classified as homeometric autoregulation because the increase in contractile force occurs at a constant end-diastolic fiber length. These internal adjustments are not very effective, and by themselves would be insufficient to permit increases in heart rate to raise cardiac output.

Effects of Increased Rate Due to Changes in Autonomic Nerve Activity ■ Increased heart rate usually occurs because of decreased parasympathetic and increased sympathetic neural activity. The release of norepinephrine by sympathetic nerves not only increases the heart rate (Chap. 13) but also dramatically increases the force of contraction (Fig. 14–5). Furthermore, norepinephrine increases conduction velocity in the heart, resulting in a more efficient and rapid ejection of blood from the ventricles. These effects, summarized in Figure 14–9, maintain the stroke volume as heart rate increases. Thus when heart rate increases physiologically due to an increase in sympathetic nervous system activity (as during exercise), cardiac output increases proportionately over a broad range.

■ Influences on Stroke Volume and Heart Rate Regulate Cardiac Output

Stroke volume is affected by the contractile force of the ventricular myocardium and by the force opposing ejection (the aortic pressure or afterload). Contractile

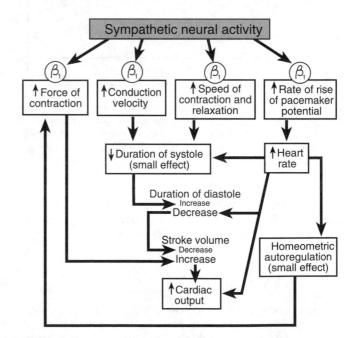

Figure 14–9 ■ Effects of increased sympathetic neural activity on heart rate, stroke volume, and cardiac output. Various effects of norepinephrine on the heart compensate for the decreased duration of diastole and hold stroke volume relatively constant, so that cardiac output increases with increasing heart rate. "Increase" and "Decrease" in small type denote quantitatively less important effects than the same words in large type.

force depends on ventricular end-diastolic fiber length (Starling's law) and myocardial contractility. Contractility is influenced by four major factors: (1) norepinephrine release from sympathetic nerves and, to a much lesser extent, norepinephrine and epinephrine release from the adrenal medulla; (2) certain hormones and drugs, including glucagon, isoproterenol, and digitalis (which increase contractility) and anesthetics (which decrease contractility); (3) disease states, such as coronary artery disease, myocarditis (Chap. 10), bacterial toxemia, and alterations in plasma electrolytes and acid-base balance; and (4) homeometric autoregulation, including increased contractility with increased heart rate and increased contractility with increased mean arterial pressure. When accompanied by changes in myocardial contractility, changes in heart rate also play a major role in determining cardiac output.

The Measurement of Cardiac Output

The ability to measure output accurately is essential for performing physiologic studies involving the heart and managing clinical problems in patients with heart disease or failure.

■ Cardiac Output Can Be Measured Using Variations on the Fick Principle

The Fick principle is based on the conservation of mass and is best understood by considering the measurement of a volume of liquid in a beaker (Fig. 14–10). If the beaker contains an unknown volume of liquid, the volume can be determined by dispersing a known quantity of dye throughout it and then measuring the concentration of dye in a sample of liquid. The quantity of dye (known) in the liquid is equal to the concentration of dye in the liquid (measured) times the volume of liquid (unknown):

Amount of dye (A)
= Concentration of dye (C)   (4)
   × Volume of liquid (V).

This is a rearranged version of the definition of concentration:

$$C = A/V \text{ and } V = A/C \qquad (5)$$

In the case of cardiac output, the goal is to measure the volume of blood flowing through the heart per unit of time.

The Dye Dilution Method ■ When the dye dilution or **indicator dilution method** is used, a known amount of dye solution is injected into the circulation and the blood downstream is serially sampled after the dye has had a chance to mix (Fig. 14–11). The dye is usually injected on the venous side of the circulation

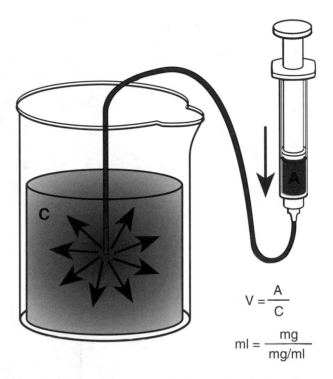

$$V = \frac{A}{C}$$

$$ml = \frac{mg}{mg/ml}$$

Figure 14–10 ■ Measurement of volume using the dye dilution method. The volume (V) of liquid in the beaker equals the amount (A) of dye divided by the concentration (C) of the dye after it has dispersed uniformly in the liquid.

(often into the right ventricle or pulmonary artery, but occasionally directly into the left ventricle) and sampling is performed from a distal artery. The resulting concentration of dye in the distal arterial blood (C) changes with time. First the concentration rises as the dye particles carried by the fastest-moving blood reach the arterial sampling point. Concentration rises to a peak as the majority of dye particles arrive and then falls off as the slower ones arrive. Before the slowest ones arrive, the fastest come around again via the shortest pathways (recirculation). The method for accounting for recirculation is described later. Assuming no recirculation (as in Figure 14–11), the average concentration of dye can be determined by measuring the dye concentration continuously from its first appearance (t_1) until its disappearance (t_2). The average concentration $(\bar{C})$ over that period is determined and cardiac output is calculated as:

$$CO = \frac{A}{\int_{t_1}^{t_2} C dt} = \frac{A}{\bar{C}(t_2 - t_1)} \qquad (6)$$

Note the similarity between this equation and the one for calculating volume in a beaker. On the left is volume per minute (rather than volume in equation 5) and on

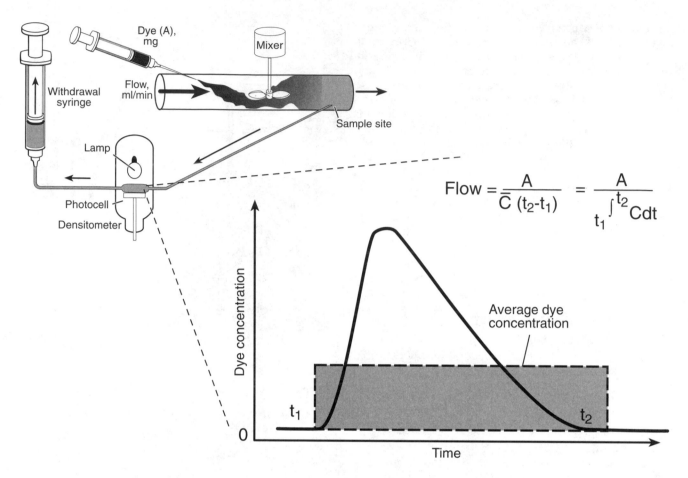

$$Flow = \frac{A}{\bar{C}\,(t_2 - t_1)} = \frac{A}{\int_{t_1}^{t_2} C dt}$$

Figure 14–11 ■ Dye dilution method for determining flow through a tube. The volume per minute flowing in the tube equals the quantity of dye injected divided by the average dye concentration (C) at the sample site, multiplied by the time between the appearance (t_1) and disappearance (t_2) of the dye. Note the analogy between this time-dependent measurement (volume/time) and the simple volume measurement in Figure 14-10.

the right is amount of dye in the numerator and the product of concentration (C) and time (dt) (rather than concentration, as in the earlier equation) in the denominator. Concentration, volume, and amount appear in both equations, but in this equation time is present in the denominator on both sides.

Before the entire arterial concentration curve needed to calculate the average concentration can be obtained, the dye begins to recirculate, giving a falsely elevated value for arterial concentration (Fig. 14–12). To correct for this, the downslope of the curve is assumed to be semilogarithmic and the arterial value is extrapolated.

The Oxygen Uptake/Consumption Method
■ Another way the conservation of mass is used to calculate cardiac output involves the continuous entry of oxygen into the blood via the lungs (Fig. 14–13). In a steady state the oxygen leaving the lungs (per unit time) via the pulmonary veins must equal the oxygen entering the lungs via the (mixed) venous blood and respiration (in a steady state, the amount of

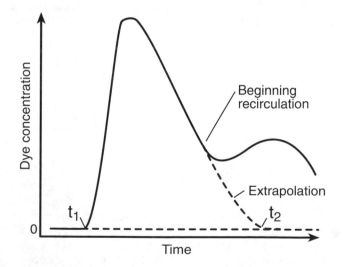

Figure 14–12 ■ Effect of recirculation on the dye concentration curve. Recirculation of dye prevents direct use of the concentration curves to calculate cardiac output. The initial part of the curve is extrapolated to get an estimate of the true dye concentration curve.

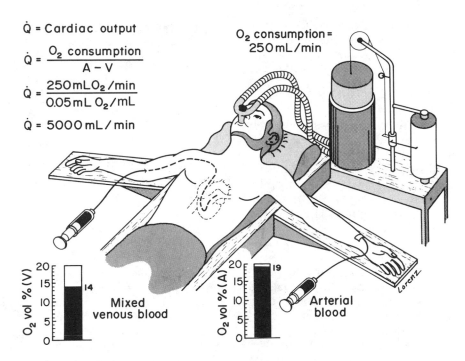

$\dot{Q}$ = Cardiac output

$\dot{Q}$ = $\dfrac{O_2 \text{ consumption}}{A - V}$

$\dot{Q}$ = $\dfrac{250 \text{ mL } O_2/\text{min}}{0.05 \text{ mL } O_2/\text{mL}}$

$\dot{Q}$ = 5000 mL/min

O_2 consumption = 250 mL/min

O_2 vol % (V) — Mixed venous blood — 14

O_2 vol % (A) — Arterial blood — 19

Figure 14–13 ■ Calculation of cardiac output using the oxygen uptake/consumption method. Oxygen is the "dye" that is "added" to the mixed venous blood. For oxygen, 1 vol % = 1 ml/100 ml blood.

oxygen replaced by respiration is equal to the amount consumed by body metabolism):

| O_2 in blood coming out of lungs | = | O_2 in blood and air coming into lungs | (7) |

or:

| Pulmonary vein O_2 output | = | Pulmonary artery (mixed venous) O_2 input + O_2 added by respiration (equal to O_2 consumption) | (8) |

The pulmonary vein oxygen output is also equal to the pulmonary vein oxygen content (same as arterial oxygen content, or aO_2) multiplied by the cardiac output (CO). Mixed venous oxygen input is likewise defined as mixed venous content ($\bar{v}O_2$) multiplied by cardiac output. Using $\dot{V}O_2$ for oxygen consumption, it follows by substitution that:

$$(CO)\,(aO_2) = [(CO)\,(\bar{v}O_2)] + \dot{V}O_2 \qquad (9)$$

which rearranges to:

$$CO = \frac{\dot{V}O_2}{aO_2 - \bar{v}O_2} \qquad (10)$$

Systemic arterial blood oxygen content, pulmonary arterial (mixed venous) blood oxygen content, and oxygen consumption can all be measured, and therefore cardiac output can be calculated. The theory behind this method is more sound than that behind the dye dilution method because it avoids the need for extrapolation. However, because the cardiac catheterization required to measure pulmonary artery oxygen content is avoided, the dye dilution method is more popular. The two methods agree very well in a wide variety of circumstances.

The Thermodilution Method ■ In most clinical situations, cardiac output is measured using a variation of the dye dilution method called **thermodilution.** A **Swan-Ganz catheter** (a soft, flow-directed catheter with a balloon at the tip) is placed into a large vein and threaded through the right atrium and ventricle so that its tip lies in the pulmonary artery. The catheter is designed so a known amount of ice-cold saline solution can be injected into the right side of the heart via a side port in the catheter. This solution decreases the temperature of the surrounding blood. The magnitude of the decrease in temperature depends on the volume of blood that mixes with the solution, which depends on cardiac output. A thermistor on the catheter tip (located downstream in the pulmonary artery) measures the fall in blood temperature. Using calculations similar to those described for the dye dilution method, the cardiac output can be determined.

■ Other Techniques Are Currently Used to Measure Cardiac Output

A variety of other clinical techniques, many of which use imaging modalities, can be used to measure or estimate cardiac output.

Radionuclide Techniques ■ A radioactive substance (usually technetium − 99) can be made to circulate throughout the vascular system by attaching (tagging) it to red blood cells or albumin. The radiation (gamma rays) emitted by the large pool(s) of blood in the cardiac chambers is measured using a specially designed **gamma camera.** The emitted radiation is proportional to the amount of technetium bound to the blood (easily determined by sampling the tagged blood) and the volume of blood in the heart. Using computerized analysis, the amount of radiation emitted by the left (or right) ventricle during various portions of the cardiac cycle can be determined (Fig. 14-14A, B). The amount of blood ejected with each heart beat (stroke volume) is determined by comparing the amount of radiation measured at end systole with that at end diastole; multiplying this by heart rate yields cardiac output.

Echocardiography ■ Echocardiography (ultrasound cardiography) provides two-dimensional, real-time images of the heart; in addition, the velocity of blood flow can be determined by measuring the Doppler shift (change in sound frequency) that occurs when the ultrasound wave is reflected off moving blood. Echocardiography can therefore be used to measure changes in ventricular chamber size (Fig. 14-14C, D), aortic diameter, and aortic blood flow velocity occurring throughout the cardiac cycle. With this information cardiac output may be estimated in one of two ways. First, the change in ventricular volume occurring with each beat (stroke volume) can be determined and multiplied by the heart rate. Second, the average aortic blood flow velocity can be measured (just above or below the aortic valve) and multiplied by the measured aortic cross-sectional area to give aortic blood flow (which is nearly identical to cardiac output).

Computed Tomography ■ *Cine (or ultrafast)* computed tomography provides cross-sectional views of the heart during different phases of the cardiac cycle (Fig. 14-14E, F). Stroke volume (and cardiac output) can be calculated using the same principles described for radionuclide or echocardiographic modalities. When ventricular volume changes are estimated from cross-sectional data, assumptions are made about ventricular geometry. Although these assumptions can lead to errors in calculation of cardiac output, these methods have proven to be highly useful.

The Energetics of Cardiac Function

The heart converts chemical energy in the form of ATP into mechanical work and heat. The relationship between (1) the supply of oxygen and nutrients needed to synthesize ATP and (2) the output of mechanical work by the heart is at the center of many clinical problems.

■ Cardiac Energy Production Depends Primarily on Oxidative Phosphorylation

The sources of energy for cardiac muscle function were described in Chapter 10. Although the major source of energy for the formation of ATP is oxidative phosphorylation, glycolysis can also briefly compensate for a transient lack of aerobic production of ATP when a portion of the heart receives too little oxygen, as during brief coronary artery occlusion.

Oxidative phosphorylation in the heart can use either carbohydrates or fatty acids as metabolic substrates. The formation of ATP depends on a steady supply of oxygen via coronary blood flow. Oxygen delivery by coronary blood flow is therefore the single most important determinant of an adequate supply of ATP for the mechanical, electrical, and metabolic energy needs of cardiac cells. Furthermore, cardiac oxygen consumption is an accurate measure of the use of energy by the heart. Coronary blood flow is discussed in Chapter 17.

As in skeletal muscle, ATP is in near equilibrium with phosphocreatine, which adds to the storage capacity of high-energy phosphate and speeds its transport from mitochondria to actomyosin crossbridges.

■ Cardiac Energy Consumption Depends on External and Internal Cardiac Work

Cardiac energy consumption (which is equivalent to cardiac oxygen consumption) results in the release of energy in the form of external work and heat, also called internal work.

■ External Cardiac Work Can Be Described in Terms of "Stroke Work," the Product of Pressure and Stroke Volume

As previously described, the external mechanical work per beat (stroke work) can be calculated as the mean ventricular systolic pressure multiplied by the volume moved or stroke volume. This is analogous to lifting a weight a certain distance and calculating the mechanical work as the force times the distance. Although mean arterial pressure and mean ventricular systolic pressure differ, mean arterial pressure is usually used as an approximation to calculate external work.

A small component of external work (usually ∼ 10%)

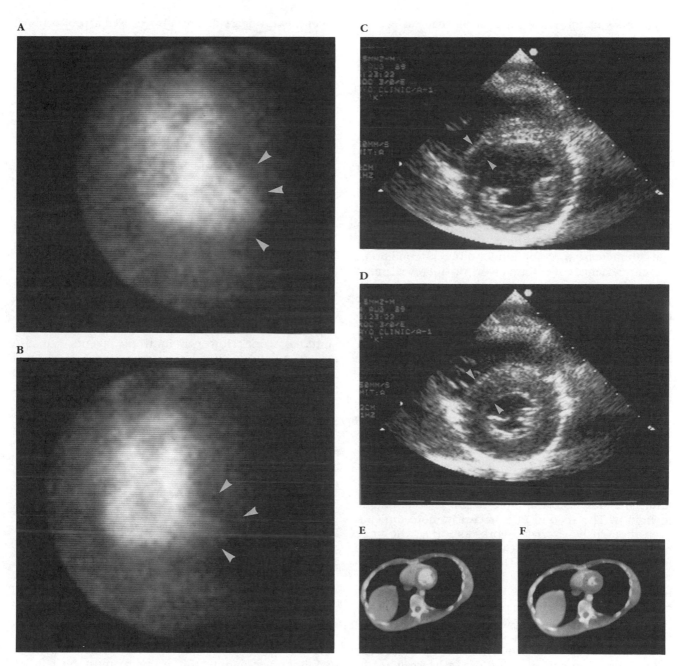

Figure 14–14 ■ (A, B) Radionuclide angiogram. The white arrows in A show the boot-shaped left ventricle as it appears during cardiac diastole when it is maximally filled with radionuclide-labeled blood. In B, much of the apex appears to be missing (white arrows) because cardiac systole has caused the blood to be ejected as the intraventricular volume decreases. (C, D) Two-dimensional echocardiogram. In this cross-sectional view, the left ventricle appears as a ring. Wall thickness is indicated by white arrows. In diastole (C), the ventricle is large and the wall is thinned; during systole (D), the wall thickens and the ventricular size decreases. (E, F) Ultrafast (cine) computed tomography. The ventricular size and wall thickness can be assessed during diastole and systole, and the change in ventricular size can be used to calculate cardiac output.

is kinetic work. This is the energy imparted to blood in the form of flow velocity as it is ejected with each beat. We do not elaborate on this component of external work because it is of little importance in most situations.

■ Internal Work Is Done to Enable Cardiac Contractions

Cardiac contractions involve many events that do not result in external work. These include transporting ions across membranes, internal mechanical events, such as developing force by stretching series elasticity (Chap. 10), overcoming internal viscosity, and rearranging the muscular architecture of the heart as it contracts. These activities, known as internal work, use far more energy (perhaps five times as much) than external work. The energy used to perform internal work leaves the heart as heat. Thus, the consumption of oxygen results in two forms of energy production by the heart: external work and heat.

Cardiac Efficiency ■ The efficiency of the heart at performing useful work can be estimated by dividing the external work of the heart by the energy equivalent of the oxygen consumed. Only 5–10% of the energy liberated by cardiac oxygen consumption is used for external work under resting conditions. This percentage can increase threefold under different conditions, therefore changes in external work do not reveal much about changes in energy consumption in the heart. This is because internal work, the other major determinant of oxygen consumption and thereby cardiac efficiency, varies independently of external work. As we shall see, large increases in internal work can occur in the absence of changes in external work. When this happens, oxygen consumption increases and efficiency decreases. The difference between pressure and volume work illustrates this point.

"Pressure Work" versus "Volume Work"■ Most of the cardiac energy devoted to internal work is used to maintain the force of contraction (and thus ventricular pressure) rather than to eject the blood. The importance of this is seen by comparing two tasks: lifting a 20-pound weight from the floor to a table and lifting the weight to the table height and continuing to hold it. The second task is clearly more difficult, even though the external work done (i.e., the force multiplied by the distance that the object was moved) in each case is the same. The ventricles not only develop the pressure required to move the blood but must maintain the pressure during systole. This takes far more energy than the external work alone as calculated from ventricular pressure and stroke volume. In fact, if the external work of the heart is raised by increasing stroke volume but not mean arterial pressure, the oxygen consumption of the heart increases very little. Alternatively, if the mean arterial pressure (and thus the ventricular pressure) is increased, the oxygen consumption per beat goes up much more. In other words, "pressure work" by the heart is far more expensive in terms of oxygen consumption than is "volume work." This makes sense, since internal work uses far more energy than does external work.

Afterload ■ The discussion of pressure work versus volume work emphasizes the importance of afterload as a determinant of energy use and oxygen consumption by the heart. An increase in cardiac radius, as

occurs with heart failure, also causes a proportional increase in internal work and energy use, independent of any change in external work.

Heart Rate ■ Thus far we have considered only the energetic events associated with a single cardiac contraction. It follows that the case of a single contraction is multiplied by the heart rate to give the energy use per unit time. There is another important consideration related to heart rate. Much of the internal work of the heart occurs during isovolumetric contraction, when force is being developed but no external work is being done. If heart rate is increased, the energy expended in isovolumetric contraction increases proportionately. This means that increasing cardiac output by increasing heart rate is more energetically costly than the same increase by means of stroke volume, which increases external work.

Contractility ■ The energetic consequences of altered myocardial contractility can be explained in terms of external and internal work. Inotropic agents (e.g., norepinephrine) may increase pressure work by raising arterial pressure and thereby increase internal work. However, inotropic agents can also cause the heart to do the same stroke work at a smaller end-diastolic volume. This reduces afterload and reduces internal work. During exercise, increased contractility causes end-diastolic volume to decrease despite the increase in venous return. This lowers the contribution of ventricular radius to afterload and thereby avoids the inefficiency of an increase in end-diastolic volume.

■ The "Double Product" (Heart Rate Times Blood Pressure) Is Used Clinically to Estimate the Energy Requirements of Cardiac Work

A useful index of the cardiac oxygen consumption is the product of mean arterial pressure (more easily measured than the mean ventricular pressure) and heart rate. Since the duration of systole (and therefore the length of time the left ventricle must maintain the mean arterial pressure) does not generally change, this index is usually valid. The heart rate gives the number of times per minute this pressure must be held. A slightly more accurate prediction of oxygen consumption can be obtained by multiplying heart rate by systolic arterial pressure **(double product).**

These calculations do not take into account the dependence of ventricular pressure development on the radius of the ventricle. The Laplace relationship states that the heart must develop more tension (force) to generate the same pressure as the radius increases. The extra energy required by pathologically dilated hearts is not reflected in either of the estimates of oxygen consumption outlined above.

■ ■ ■ ■ ■ ■ ■ ■ ■ ■ ■ ■ **REVIEW EXERCISES** ■ ■ ■ ■ ■ ■ ■ ■ ■ ■ ■ ■

Identify Each with a Term

1. The product of stroke volume and heart rate

2. The law defining the relationship between ventricular end-diastolic fiber length (preload) and stroke volume

3. Mean arterial pressure multiplied by stroke volume

4. The relationship between force of contraction and a *given* end-diastolic fiber length

5. A variation in the dye dilution method of measuring cardiac output, in which temperature changes are used

Define Each Term

6. Cardiac cycle

7. Cardiac index

8. Ventricular function curve

9. Afterload

10. Double product

Choose the Correct Answer

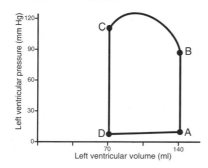

11. The figure above shows another way to plot the changes in pressure and volume in the left ventricle during one complete cardiac cycle. Which of the following is true?

    a. Point A marks the beginning of isovolumetric contraction.
    b. Point B marks the closure of the aortic valve.
    c. Point C marks the opening of the mitral valve.
    d. Point D marks the beginning of ventricular ejection.

12. Which statement is incorrect?

    a. Ventricular systole can be defined in terms of mechanical or electrical events.
    b. The first heart sound occurs at the beginning of ventricular diastole.
    c. The aortic and pulmonary valves typically open and close at approximately the same time.
    d. In older patients the presence of a third (S_3) or fourth (S_4) heart sound usually indicates a cardiac abnormality.

13. Which statement is true?

    a. The output of the left ventricle usually exceeds that of the right ventricle.
    b. The relationship between end-diastolic fiber length and stroke volume is caused by changes in sympathetic nerve activity.

    c. During heart failure, the ventricular function curve shifts to the right.
    d. Cardiac glycosides do *not* shift the ventricular function curve.

Data for the Following Two Questions

Drug A causes a 33% increase in stroke volume and no change in systolic aortic blood pressure. Drug B causes a 33% increase in systolic aortic blood pressure and no change in stroke volume. Neither drug changes heart rate.

14. Drug A increases the external work (stroke work) of the left ventricle _____ drug B.
    a. More than
    b. Less than
    c. By the same amount as

15. Drug A increases the internal work of the left ventricle _____ drug B.
    a. More than
    b. Less than
    c. By the same amount as

16. In a normal cardiac cycle:
    a. The first heart sound coincides with the beginning of ventricular ejection
    b. The second heart sound coincides with the end of ventricular ejection
    c. The highest left ventricular pressure is reached just as the aortic valve closes
    d. Atrial systole occurs during ventricular isovolumetric relaxation

Calculate

17. Given the following data, calculate the cardiac output and stroke work.

Volume in ventricle at end of diastole:	130 ml
Volume in ventricle at end of systole:	60 ml
Heart rate:	70 beats/min
Mean arterial blood pressure:	90 mm Hg

18. The following data were collected in an athletic 70-kg man during heavy exercise. What is the cardiac output and stroke volume?

Oxygen consumption:	4 L/min
Arterial oxygen content:	19 ml/100 ml blood
Mixed venous oxygen content:	3 ml/100 ml blood
Heart rate:	180 beats/min

Briefly Answer

19. Consider a patient with mitral regurgitation of moderate severity. What influence might you expect this valvular dysfunction to have on cardiac output (forward flow) when measured by: (a) dye dilution (catheter in the right ventricle and sampling site in a peripheral artery)?

(b) thermodilution (injection in the right ventricle, sampling in the pulmonary artery)? (c) the Fick principle, calculated using steady-state oxygen consumption and arterial venous oxygen difference? (d) stroke volume times heart rate, where stroke volume is obtained using an imaging modality such as radionuclide angiography, echocardiography, or ultrafast CT?

20. Discuss how each of the following influences stroke volume: (a) Reduction in afterload (b) Ventricular hyper-

trophy (c) An increase in arterial blood pressure (d) Stenosis (narrowing) of the aortic valve (e) An increase in end-diastolic volume (f) Heart failure (g) Digitalis (h) Intensive stimulation of the vagus nerves

21. What effect (increase, decrease, or no change) will each of the following have on left ventricular afterload (assuming moderate severity for each)? (a) Aortic stenosis? (b) Mitral regurgitation? (c) Arterial hypertension? (d) Aortic regurgitation? (e) Mitral stenosis?

■ ■ ■ ■ ■ ■ ■ ■ ■ ■ ■ ■ ■ ■ ANSWERS ■ ■ ■ ■ ■ ■ ■ ■ ■ ■ ■ ■ ■ ■ ■

1. Cardiac output

2. Starling's law of the heart

3. Stroke work

4. Contractility

5. Thermodilution

6. The sequence of electrical and mechanical events occurring in the heart during a single beat and the resulting changes in pressure, flow, and volume in the various cardiac chambers

7. Cardiac output divided by body surface area

8. A plot of cardiac function, such as stroke work, stroke volume, or cardiac output, against a preload variable, such as end-diastolic fiber length, end-diastolic volume, or end-diastolic pressure

9. The force against which the ventricle must contract

10. The product of heart rate and systolic blood pressure

11. a—Point A marks the beginning of isovolumetric contraction. This follows ventricular filling (D to A) and is followed by ventricular ejection (B to C). C to D is isovolumetric relaxation.

12. b

13. c

14. c—Because stroke work is the multiple of stroke volume and systolic aortic pressure (or mean arterial pressure), it is the same in both cases.

15. b—Because drug A causes an increase in volume work, whereas drug B causes an increase in pressure work, drug A increases internal work (and oxygen requirements) less than drug B.

16. b—Closure of the aortic and pulmonic valves ends ventricular ejection. The second heart sound signals this event.

17. Cardiac output:

$$CO = SV \times HR$$
$$SV = EDV - ESV$$
$$SV = (130 - 60) \text{ ml}$$
$$CO = 70 \text{ ml} \times 70 \text{ min}^{-1} = 4.9 \text{ L/min}$$

Stroke work:

$$SW = SV \times MAP$$
$$SW = 70 \text{ ml} \times 90 \text{ mm Hg} = 6300 \text{ ml} \times \text{mm Hg}$$

18. Cardiac output:

$$CO = \frac{\dot{V}O_2}{aO_2 - \bar{v}O_2}$$
$$= \frac{4000 \text{ ml } O_2/\text{min}}{190 \text{ ml } O_2/L - 30 \text{ ml } O_2/L}$$
$$= 25 \text{ L/min}$$

Stroke volume:

$$SV = \frac{CO}{HR} = \frac{25 \text{ L/min}}{180/\text{min}} = 139 \text{ ml}$$

19. (a) After the injection of dye into a normal heart, the concentration curve obtained by sampling a peripheral artery shows a rapid rise, distinct peak, and steady fall until recirculation occurs (Fig. 14–12). In a heart with significant mitral regurgitation, the back-and-forth movement of dye between the left ventricle and left atrium delays its delivery into the peripheral circulation, resulting in a lower, more drawn out curve. Extrapolation to separate the true curve from the recirculation artifact is more difficult and dye dilution is therefore less accurate in this setting. (b) Thermodilution performed in the pulmonary artery is relatively unaffected by mitral regurgitation. (c) The Fick principle/oxygen consumption method is unaffected by mitral regurgitation. (d) Stroke volume times heart rate using imaging modalities to study changes in left ventricular size occurring over the cardiac cycle can be seriously affected by mitral regurgitation because of the difficulty in separating **forward flow** from **regurgitant flow.** For example, if the amount of blood regurgitated during each beat is 30% of the stroke volume (regurgitant fraction), then a left ventricle with a stroke volume of 100 ml would eject only 70 ml into the aorta; the other 30 ml would be ejected into the left atrium. Assuming a heart rate of 70 beats/min, the correct (forward) cardiac output would be 70 beats/min times 70 ml, or 4.9 L/min, not 100 ml times 70 beats/min, or 7 L/min.

20. (a) Reduction of afterload results in increased stroke volume. The decrease in required force development allows ventricular muscle to shorten more rapidly (the force velocity relationship) and thereby increase ejection of blood i.e., stroke volume). (b) Ventricular hypertrophy provides the wall with more actin and myosin filaments in parallel and so increases total force develop-

ment by the ventricular wall. The result is increased stroke volume. (c) Increased arterial blood pressure elevates afterload. This reduces velocity of shortening and the ejection of blood. Stroke volume is decreased. (d) Stenosis of the aortic valve increases outflow resistance and the pressure gradient across the aortic valve. This means ventricular pressure is increased for any given ejection rate of blood. This increase in afterload results in a decrease in stroke volume by means of the same mechanism described in c. (e) Increased end-diastolic volume increases sarcomere length and force of contraction. This results in an increase in stroke volume. Because the heart works only on the ascending limb of the length-tension curve, the answer does not have to include the possible effects of being on the descending limb of the length tension curve. (f) Heart failure results from a reduction in the heart's capacity to eject a normal stroke volume under normal conditions. (g) Digitalis raises contractility of a failing heart, thereby increasing stroke volume. (h) Stimulation of the vagus nerves exerts the parasympathetic fibers to the heart, markedly slowing heart rate. Although there may be some decrease in contractility, this is more than compensated by the increased time for ventricular filling, which by means of Starling's law raises stroke volume.

21. (a) Aortic stenosis would increase afterload (see 20). (b) Sudden mitral regurgitation would result in some decrease in afterload. Regurgitation through the mitral valve is a second pathway for the exit of blood during systole, so ventricular pressure would be somewhat lower. Compensatory mechanisms, including increased blood volume, would tend to elevate ventricular filling and thereby counteract the effect. (c) Arterial hypertension increases afterload. (d) Aortic regurgitation results in significant dilitation of the ventricle, which, because of the Laplace relationship, increases afterload. (e) Mitral stenosis results in decreased filling of the left ventricle. After compensation, including increased blood volume and peripheral vasoconstriction, there would be little change in afterload.

Suggested Reading

Fozzard HA, Jennings RB, Harber E, Katz AM, Morgan HE: *The Heart and Cardiovascular System.* Raven Press New York, 1986, chap. 2, 3, 19–22, 40.
Hamilton WF, Richards DW: The output of the heart, in Fishman AP, Richards DW (eds); *Circulation of the Blood: Men and Ideas.* Bethesda, American Physiological Society, 1982, pp 71–126.
Katz AM: *Physiology of the Heart.* 2nd ed. New York, Raven Press 1992, chap. 4–6, 13, 15–17.
Milnor WR: The circulatory system and principles of hemodynamics, in *Cardiovascular Physiology.* New York, Oxford University Press, 1990, chap. 4.
Sperelakis N: *Physiology and Pathophysiology of the Heart.* 2nd ed. Boston Kluwer, 1989.

CHAPTER

■ ■ ■ ■ ■ ■ ■

15

The Systemic Circulation

CHAPTER OUTLINE

I. THE NORMAL RANGE OF ARTERIAL PRESSURE
   A. Age, race, gender, diet, weight, and other factors affect blood pressure
II. PHYSIOLOGIC CHANGES IN BLOOD IN BLOOD PRESSURE
   A. Stroke volume, heart rate, and total peripheral resistance interact to affect mean arterial and pulse pressures
   B. Pulse pressure is altered by changes in mean arterial pressure and compliance
III. MEASUREMENT OF ARTERIAL PRESSURE
   A. The routine method for measuring human blood pressure is by an indirect procedure using a sphygmomanometer
   B. Indirect methods of measuring arterial pressure may be subject to artifacts
IV. SYSTEMIC VASCULAR RESISTANCE
   A. The small arteries, arterioles, and capillaries account for 90% of vascular resistance
   B. Blood viscosity, vessel length, and vessel radius affect resistance
   C. Sources of resistance in the systemic circulation are arranged in series and in parallel circuits
V. BLOOD VOLUME
   A. Three-quarters of the blood in the systemic circulation is on the venous side
   B. Small changes in venous pressure can cause large changes in venous volume
   C. Central venous pressure can be measured using invasive means and provides information on central blood volume
   D. Approximately half of the total blood volume is central blood volume
   E. Cardiac output is extremely sensitive to changes in central blood volume
   F. Certain hemodynamic stresses activate compensatory mechanisms that maintain central blood volume and cardiac output
      1. Increase in volume
      2. Redistribution of volume
      3. Changes in venous compliance

OBJECTIVES

After studying this chapter, the student should be able to:
1. Define *normal blood pressure* and the factors that affect it
2. Explain how blood pressure is measured
3. Discuss the relationships between stroke volume, heart rate, systemic vascular resistance, mean arterial pressure, pulse pressure, and vascular compliance
4. Discuss the factors affecting systemic vascular resistance, including blood viscosity, vessel length, and vessel radius
5. Review the function of veins as a storage site for circulating blood and explain how small shifts in venous volume can affect cardiac preload and output
6. Discuss the relationship(s) between hemodynamic "stresses" (e.g., upright posture) and changes in venous compliance and/or blood volume

An understanding of the major systemic hemodynamic variables, such as pressure, systemic vascular resistance, and blood volume, is a prerequisite to understanding the regulation of arterial pressure and blood flow to individual tissues, discussed in subsequent chapters.

The Normal Range of Arterial Pressure

The noninvasive measurement of arterial pressure gives values for systolic and diastolic arterial pressure. As with all physiologic variables, values for individuals are distributed around a mean value. The range of blood pressures in the population as a whole is rather broad; changes in a given patient are of diagnostic importance. Normal arterial blood pressure is approximately 120 mm Hg systolic and 80 mm Hg diastolic (usually written 120/80).

■ Age, Race, Gender, Diet, Weight, and Other Factors Affect Blood Pressure

In Western societies, arterial pressure is dependent on age. For example, Figure 15-1 shows the mean values for systolic and diastolic pressure for females at different ages. The standard deviations illustrate that measured blood pressure is extremely variable at any age. Actuarial data suggest that when blood pressure is elevated above certain value, it is associated with

significant excess cardiovascular morbidity and mortality. A reasonable upper limit for normal blood pressure according to age is:

Age	Systolic	Diastolic
17-40	140	90
41-60	150	90
>60	160	90

The higher systolic pressure with age reflects, in part, diminishing arterial compliance.

Race, gender, diet, weight, pregnancy, and individual behavioral factors (e.g., smoking, alcohol intake, and certain medications) can also influence blood pressure. Because of this variability, the World Health Organization (WHO) sets a higher arbitrary limit (160/95 mm Hg) on the upper limits of normal for blood pressure than do insurance companies. Many of these (and other) variables may affect blood pressure transiently, therefore sometimes several determinations (over time and under various conditions) are required to make an accurate assessment.

Physiologic Changes in Blood Pressure

Physiologic changes in systolic and diastolic pressure can be explained in terms of mean arterial pressure and pulse pressure. **Mean arterial pressure** is determined by cardiac output and systemic vascular resistance. **Pulse pressure** is determined largely by stroke volume and aortic compliance; its sensitivity to aortic compliance means it is also dependent on aortic volume, because aortic compliance decreases as volume increases. Chapter 12 discusses compliance and systemic vascular resistance as well as systolic, diastolic, mean, and pulse pressure. The methods for calculating mean arterial pressure and pulse pressure from systolic and diastolic pressure are given in Chapter 12.

■ Stroke Volume, Heart Rate, and Total Peripheral Resistance Interact to Affect Mean Arterial and Pulse Pressures

When cardiac output (CO) increases in the face of a constant systemic vascular resistance (SVR), mean arterial pressure (MAP) increases according to the formula CO = MAP ÷ SVR. The effect of the increase in cardiac output on pulse pressure depends on two factors. First, if the increase in cardiac output is achieved by raising stroke volume, this raises pulse pressure. Second, if increased cardiac output elevates mean arterial pressure, aortic compliance is reduced. Reduced aortic compliance causes increased pulse pressure even if stroke volume is constant. It should be apparent that these interactions are quite complex. For this reason we begin by considering situations in which heart rate or stroke volume changes with *no* change in cardiac output or mean arterial pressure,

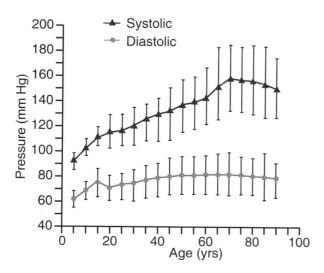

Figure 15–1 ■ Mean value for arterial systolic and diastolic pressure of women according to age. The vertical bars show the standard deviation around each mean value. Systolic pressure increases with age, rapidly until age 20 and then more slowly until age 50. The leveling off of systolic pressure after age 70 could be the result of the excess mortality associated with high pressures. (Modified from Diem D, Lentner C: *Documenta Geigy Scientific Tables.* Basel, JR Geigy, 1970, p 553.)

then we return to the more complex interactions that occur when mean arterial pressure changes.

An increase in heart rate does not change the *mean* arterial pressure if cardiac output remains constant. The decrease in stroke volume that occurs in this situation results in a diminished pulse pressure; the diastolic pressure increases while the systolic pressure decreases around an unchanged mean arterial pressure (Fig. 15-2). An increase in stroke volume with no change in cardiac output likewise causes no change in mean arterial pressure. The increased stroke volume produces a rise in pulse pressure; systolic pressure increases and diastolic pressure decreases.

Another way to think about these events is depicted in Figure 15-3A. The first two pressure waves have a diastolic pressure of 80 mm Hg, systolic pressure of 120 mm Hg, and mean arterial pressure of 93 mm Hg. Heart rate is 72 beats/min. After the second beat, heart rate is slowed to 60 beat/min but stroke volume is increased sufficiently to result in the same cardiac output. The longer time interval between beats allows the diastolic pressure to fall to a new (lower) value of 70 mm Hg. The next systole, however, produces an increase in systolic pressure because of the ejection of a greater stroke volume, so systolic pressure rises to 130 mm Hg. The pressure then falls to the new (lower) diastolic pressure and the cycle is repeated. Mean arterial pressure does not change because the increased pulse pressure is distributed evenly around the same mean. Figure 15-3B shows what would happen if stroke volume increased with no change in heart rate: the increased stroke volume would occur at the time of the next expected beat, and the diastolic pressure would be, as for previous beats, 80 mm Hg. After a transition beat, the increased stroke volume would result in an elevation in systolic pressure to 140 mm Hg, after which the pressure would fall to a new diastolic pressure of 90 mm Hg. In this new steady state, systolic, diastolic, and mean arterial pressure would all be higher.

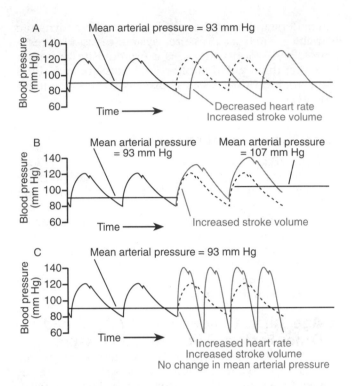

Figure 15–3 ■ (A) Effect of increased stroke volume on arterial pressure with constant (solid lines) cardiac output. Systemic vascular resistance remains constant. When cardiac output is held constant by lowering heart rate, there is no change in mean arterial pressure (93 mm Hg) and systolic pressure increases while diastolic pressure decreases. (B) If the heart rate remains constant and cardiac output increases, mean arterial pressure increases (to 107 mm Hg), as do systolic, diastolic, and pulse pressures. (C) Effect of increased heart rate *and* stroke volume with no change in mean arterial pressure, because of decreased systemic vascular resistance. After the first two beats, stroke volume and heart rate are increased. Pulse pressure increases around an unchanged mean arterial pressure, so systolic pressure is higher and diastolic pressure lower than control.

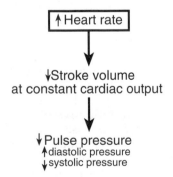

Figure 15–2 ■ Effect of increased heart rate on pulse pressure at constant cardiac output and systemic vascular resistance. Mean arterial pressure does not change, since cardiac output and systemic vascular resistance do not change. Pulse pressure decreases because stroke volume is less.

An important example of a physiologic event that can affect blood pressure is **dynamic exercise** (e.g., running or swimming). Dynamic exercise produces little change in mean arterial pressure because the increase in cardiac output is balanced by a decrease in systemic vascular resistance. Pulse pressure increases due to the increased stroke volume; the lower systemic vascular resistance results in a relatively lower diastolic pressure, and the increase in stroke volume results in a higher systolic pressure (Figs. 15-3B, 15-4). These examples demonstrate that mean, systolic, and diastolic pressure can change independently of each other and that these changes can be predicted from heart rate, stroke volume, and systemic vascular resistance.

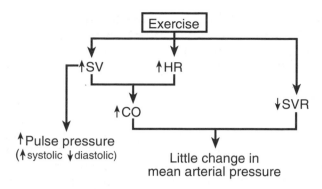

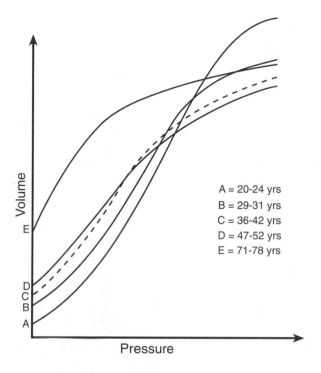

Figure 15–4 ■ Effect of dynamic exercise on mean arterial pressure and pulse pressure. Heart rate (HR) and stroke volume (SV) increase, resulting in an increase in cardiac output (CO). However, dilation of resistance vessels in skeletal muscle lowers systemic vascular resistance (SVR), balancing the increase in cardiac output so that little change in mean arterial pressure occurs.

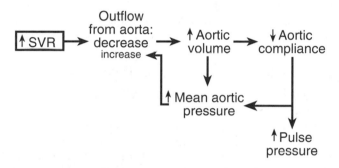

Figure 15–6 ■ Effect of aging on vascular compliance. The curves illustrate the relationship between pressure and volume for aortas of humans in different age groups. In older aortas, because of decreased compliance, a given increase in volume causes a larger increase in pressure. (Modified from Hallock, P. and Benson, I.C., Studies on the elastic properties of human isolated aorta. *J Clin Invest* 16:595, 1937.)

Figure 15–5 ■ Effect of increased systemic vascular resistance (SVR) on mean arterial and pulse pressures. Increased SVR impedes outflow from the aorta and large arteries, increasing their volume and pressure. The increase in aortic pressure brings the outflow from the aorta back to its original value, but at a higher aortic volume. The larger volume lowers aortic compliance and thereby raises pulse pressure at a constant stroke volume. "Increase" in smaller type indicates a secondary change.

■ Pulse Pressure Is Altered by Changes in Mean Arterial Pressure and Compliance

When systemic vascular resistance increases, flow out of the larger arteries transiently decreases. If cardiac output is unchanged, the volume in the aorta and large arteries increases (Fig. 15-5). Mean arterial pressure also increases, until it is sufficient to drive the blood out of the larger vessels and into the smaller vessels at the same rate as it enters from the heart (i.e., cardiac output). At a higher volume (and mean arterial pressure) aortic compliance is lower, therefore pulse pressure is greater for a given stroke volume (Fig. 12-9). The net result is an increase in mean arterial, systolic, and diastolic pressures; the extent to which

pulse pressure increases depends on how much arterial compliance decreases with the rise in mean arterial pressure and aortic volume. The fall in compliance for a given increase in mean arterial pressure is greater in older than in younger individuals (Fig. 15-6). This explains the higher pulse and systolic pressures often observed in older individuals with modest elevations in systemic vascular resistance.

We can now return to a discussion of the effect of increased cardiac output on mean arterial and pulse pressure. We will consider increases in cardiac output caused by either increased heart rate or stroke volume, with no change in systemic vascular resistance. Whenever cardiac output increases in the face of a constant systemic vascular resistance, mean arterial pressure increases. We have already seen that increased arterial pressure leads to decreased aortic compliance. If the increased cardiac output results from elevated heart rate, with no change in stroke volume, pulse pressure will increase because of the reduction in aortic compliance. If cardiac output increases because of heightened stroke volume, pulse pressure will increase even more, because both increased volume of ejected blood and decreased aortic compliance raise pulse pressure.

Measurement of Arterial Pressure

Arterial blood pressure can be measured by direct or indirect (noninvasive) methods. In the laboratory or hospital setting a cannula can be placed in an artery and the pressure measured directly using electronic transducers. In clinical practice, however, blood pressure is usually measured indirectly.

■ The Routine Method for Measuring Human Blood Pressure Is by an Indirect Procedure Using a Sphygmomanometer

The **sphygmomanometer** uses an inflatable cuff that is wrapped around the patient's arm (Fig. 15–7) and inflated so that the pressure in it exceeds systolic blood pressure. The external pressure compresses the artery and cuts off blood flow into the limb. The external pressure is measured by the height of the mercury in the manometer, which is connected to the cuff. The air in the cuff is then slowly released until blood can spurt past the occlusion at the peak of systole. Blood spurts past the point of partial occlusion at high velocity, resulting in turbulence. The vibrations associated with the turbulence are in the audible range, so a stethoscope (placed over the brachial artery) is used to detect a noise due to the turbulent flow of the blood pushing under the cuff (**Korotkoff sound**). The pressure corresponding to this first appearance of blood pushing under the cuff is the systolic pressure. As pressure in the cuff continues to fall, the brachial artery resumes its normal shape and both the abnormal turbulence and noise cease. The pressure at which the noise ceases is taken to be the diastolic pressure.

■ Indirect Methods of Measuring Arterial Pressure May Be Subject to Artifacts

The width of the inflatable cuff is an important factor that can affect pressure measurements. A cuff that is too narrow will give a falsely high pressure, because the pressure in the cuff is not fully transmitted to the underlying artery. Ideally, cuff width should be approximately 1.5 times the diameter of the limb at the measurement site. In the elderly (or those who have "stiff" or hard-to-compress blood vessels from other causes, such as arteriosclerosis) additional external pressure may be required to compress the blood vessels and stop flow. This extra pressure gives a falsely high estimate of blood pressure. Obesity may likewise contribute to an inaccurate assessment if the cuff used is too small.

Systemic Vascular Resistance

Systemic vascular resistance (SVR) is the frictional resistance to blood flow provided by all of the vessels between the large arteries and right atrium, including the small arteries, arterioles, capillaries, venules, small veins, and veins.

■ The Small Arteries, Arterioles, and Capillaries Account for 90% of Vascular Resistance

The relative importance of the various segments contributing to the systemic vascular resistance is appreciated by observing the profile of the pressure drop along the vascular tree (Fig. 15–8). Very little change in pressure occurs in the aorta and large arteries. Ap-

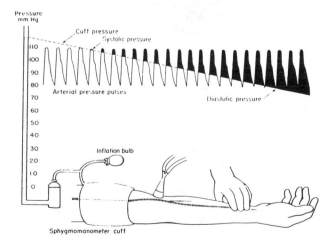

Figure 15–7 ■ Relationship between true arterial pressure and blood pressure as measured with a sphygmomanometer. (From Rushmer RF: *Cardiovascular Dynamics.* 3rd ed. Philadelphia, Saunders, 1970, p 155.)

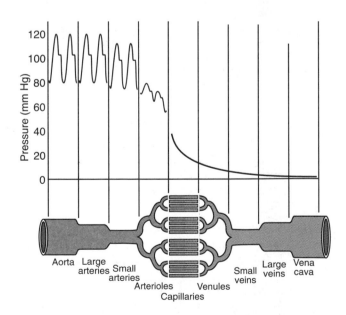

Figure 15–8 ■ Pressure in different vessels of the systemic circulation. Pulse pressure is greatest in the aorta and large arteries. The greatest drop in pressure occurs in the arterioles.

proximately 70% of the pressure drop occurs in the small arteries and arterioles, and another 20% occurs in the capillaries. Contraction and relaxation of the smooth muscle in the walls of small arteries and arterioles cause changes in vessel diameter, which, in turn, influence blood flow.

■ Blood Viscosity, Vessel Length, and Vessel Radius Affect Resistance

To understand the importance of smooth muscle in the control of systemic vascular resistance, consider the role of each factor in Poiseuille's law (Chap. 12):

$$R = 8\eta L/\pi r^4 \tag{1}$$

Viscosity (η) increases with hematocrit (Fig. 15-9), especially when the hematocrit is above the normal range of 38–54%. An increase in viscosity raises vascular resistance and thereby limits flow. Since oxygen delivery depends on oxygen content of the blood and the flow, the limitation in flow can negate the increase in oxygen content resulting from the increased number of red blood cells. In individuals with **polycythemia** (increased mass or number of red blood cells) less oxygen may actually be delivered to tissues during exercise because of increased viscosity; this occurs despite the enhanced oxygen-carrying capacity provided by the extra blood cells. A normal hematocrit reflects a good balance between sufficient red blood cells for oxygen transport and the drag exerted by red blood cells on flow.

Despite the potential effect of blood viscosity on resistance, hematocrit does not change very much under most physiologic conditions and is usually not an important determinant of vascular resistance. The length (L) of blood vessels likewise does not change significantly and is therefore not important as a physiologic determinant of vascular resistance. *The remaining influence, vessel radius (r), is thus the major determinant of systemic vascular resistance.* Since resistance is proportional to r^4, small changes in the radius cause relatively large changes in vascular resistance. For example, during exercise the vascular resistance to skeletal muscle may decrease by 25-fold. This fall in resistance results from a 2.2-fold increase in resistance vessel radius (i.e., $2.2^4 = 25$). Vessel radius is determined primarily by the contractile activity of smooth muscle in the vessel wall (Chap. 16).

■ Sources of Resistance in the Systemic Circulation Are Arranged in Series and in Parallel Circuits

Systemic vascular resistance is the net result of the resistance offered by many vessels arranged both in series and in parallel, and it is worth considering the effects of vessel arrangement on total resistance. Resistances in series are simply summed (Fig. 15-10); for example:

$$SVR = R_{small\ arteries} + R_{arterioles} \\ + R_{capillaries} + R_{venules} + R_{small\ veins} \tag{2}$$

For resistances in parallel, the reciprocals of the parallel resistances are summed (Fig. 15-11). Those who remember their physics may have already recognized that the pressure difference is analogous to potential (as is blood flow to current); the ratio of potential to current is electrical resistance. Resistances in series and in parallel are treated the same way in the analysis of electrical and hydraulic circuits.

Blood Volume

The blood volume is distributed among the various portions of the circulatory system according to the pattern shown in Figure 15-12. Total blood volume in an adult is about 5 L.

■ Three-Quarters of the Blood in the Systemic Circulation Is on the Venous Side

Approximately 80% of the total blood volume is located in the **systemic circulation** (i.e., the total volume minus the volume in the heart and lungs). Sixty percent of the total blood volume (or 75% of the systemic blood volume) is located on the venous side of the circulation. The blood present in the arteries and capillaries is only 20% of total blood volume. Since most of the systemic blood volume is in veins, it is not surprising that changes in systemic blood volume primarily reflect changes in venous volume.

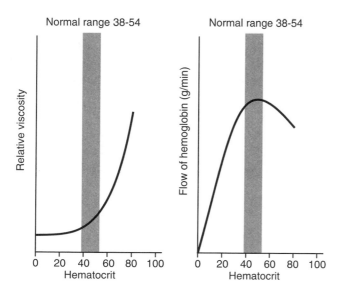

Figure 15–9 ■ Effect of hematocrit on blood viscosity. Above-normal hematocrits produce a sharp increase in viscosity. This raises resistance, so there is little increase in the delivery of hemoglobin and oxygen when the hematocrit rises above the normal range.

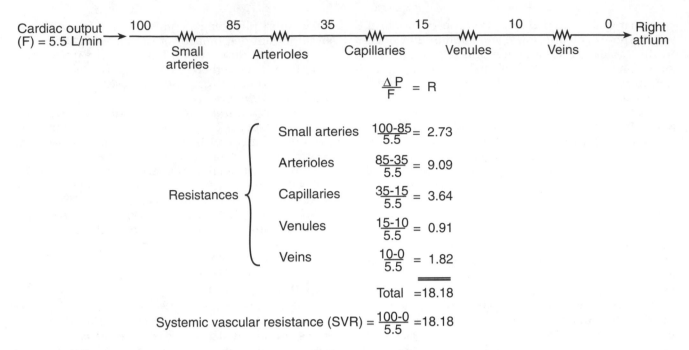

Figure 15–10 ■ Resistances (R) in series. Resistances in series are added to obtain the total resistance. In this case the resistances of consecutive vascular segments are added to obtain the systemic vascular resistance. ΔP is the difference in pressure across the resistance; F, cardiac output.

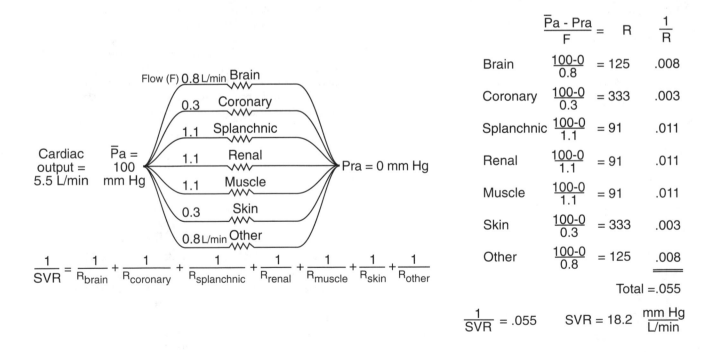

Figure 15–11 ■ Resistances in parallel. The resistances of the parallel organ circuits cannot be added directly to obtain systemic vascular resistance (SVR). Instead, the reciprocals of each organ resistance must be added to obtain the reciprocal of SVR. $\bar{P}a$, mean arterial pressure; Pra, right atrial pressure.

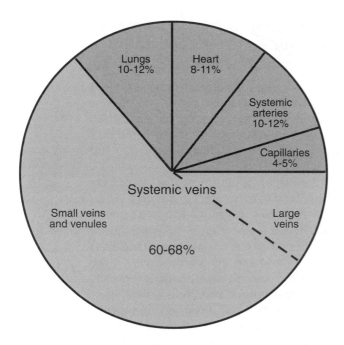

Figure 15–12 ■ Blood volumes of various elements of the circulation in the resting individual.

Small Changes in Venous Pressure Can Cause Large Changes in Venous Volume

Veins are approximately 20 times more compliant than arteries; small changes in venous pressure are therefore associated with large changes in venous volume. If 1000 ml of blood is infused into the circulation, 95% (or 950 ml) ends up in veins and only 5% (50 ml) in arteries.

Central Venous Pressure Can Be Measured Using Invasive Means and Provides Information on Central Blood Volume

Unfortunately, measurements of the peripheral venous pressure, such as the pressure in an arm or leg vein, are subject to too many influences (e.g., partial occlusion caused by positioning or venous valves) to be very helpful in most clinical situations. **Central venous pressure** can be measured by placing a catheter in the superior vena cava, inferior vena cava, or right atrium via a large vein. Changes in central venous pressures are a good indicator of central blood volume because the compliance of the intrathoracic vessels tends to be constant. In certain situations, however, such as in tricuspid valve insufficiency, the physiologic meaning of central venous pressure is changed because right ventricular pressure is transmitted to the right atrium during ventricular systole. The use of central venous pressure as an estimate of central blood volume therefore depends on the assumption that the

right heart is capable of pumping normally. Also, central venous pressure does not necessarily reflect left atrial or left ventricular filling pressure. Abnormalities in right or left heart function or in pulmonary vascular resistance can make it difficult to predict left atrial pressure from central venous pressure.

Approximately Half of the Total Blood Volume Is Central Blood Volume

In considering the venous side of the circulation, it is useful to divide the blood volume into an intrathoracic *(central)* blood volume and an extrathoracic (peripheral) blood volume. The central blood volume includes the blood in the superior vena cava and intrathoracic portions of the inferior vena cava, right atrium and ventricle, pulmonary circulation, and left atrium; this constitutes approximately 50% of the total blood volume. From a functional standpoint, the most important components of the extrathoracic blood volume are those in the extremities and abdominal cavity. The blood in the neck and head is less important because there is far less blood in these regions and the blood volume inside the cranium cannot change much because the skull is rigid. In contrast, the veins of the extremities and abdominal cavity are of considerable functional importance because large changes in blood volume can occur.

Cardiac Output Is Extremely Sensitive to Changes in Central Blood Volume

Consider what happens if blood is infused into the inferior vena cava of a normal subject. As this occurs, the volume of blood returning to the chest (venous return) is transiently greater than that leaving it (cardiac output). This difference between the input and output of blood produces an increase in the right atrial pressure, which in turn increases right ventricular end-diastolic fiber length and therefore right heart stroke volume. Flow through the lungs to the left heart will subsequently rise and cause a proportional increase in left heart stroke volume. The process will raise cardiac output until it equals the sum of the venous return to the heart (via the inferior and superior vena cava) plus the infused blood. This example illustrates how central blood volume is a critical determinant of the cardiac output.

Certain Hemodynamic Stresses Activate Compensatory Mechanisms that Maintain Central Blood Volume and Cardiac Output

Two important factors influence the central blood volume: changes in total blood volume and changes in the distribution of total blood volume between intrathoracic and extrathoracic regions.

Increase in Volume ■ An increase in total blood volume can occur because of an infusion of fluid, retention of salt and water by the kidneys, or a shift

ARTERIAL DISEASE

Disease processes such as atherosclerosis can reduce the diameter of most large and medium-sized arteries, causing an increase in arterial resistance and a subsequent decrease in blood flow. The signs and symptoms resulting from atherosclerotic disease depend on which arteries are narrowed (**stenotic**) and the severity of the reduction in blood flow. Areas and organs commonly affected by atherosclerosis include the heart, brain, and legs.

Coronary artery disease is the most common serious manifestation of atherosclerosis. When the stenotic lesions are relatively mild, flow may be inadequate only when the myocardial demand for blood is high, such as during exercise. If blood flow is inadequate to meet the metabolic needs of a particular tissue, the tissue is said to be **ischemic.** In the heart, short periods of **ischemia** may produce chest pain (**angina**) or heart failure. As the disease progresses and the coronary stenosis becomes more severe, ischemia tends to occur at increasingly lower cardiac workloads, eventually resulting in **angina at rest.** In cases of severe stenosis and/or complete occlusion of the coronary arteries, blood flow may become inadequate to maintain myocardial viability, resulting in cell/tissue death (**infarction**). Millions of people in the United States are affected by coronary disease; of these, more than 1 million will experience myocardial infarction each year, and 700,000 will ultimately die from infarction, making this the leading cause of death in the nation.

Stenoses in the carotid or vertebral arteries can lead to ischemia and infarction (**stroke** or **cerebrovascular accident**) involving the brain. Strokes are the third leading cause of death in the United States and a leading cause of significant disability.

As with the heart, mild arterial disease involving the legs usually becomes symptomatic only when the demand for blood is high such as during exercise involving the lower extremities. Muscle ischemia produces pain called **claudication,** which typically resolves rapidly when the patient rests. As the disease becomes more severe symptoms may progress to include **rest pain** and, ultimately, limb infarction with gangrene.

In all of these cases, blood flow to the affected organ may be preserved by the development of collateral arteries, which can carry blood around the stenotic or occluded segments of arteries. When collateral flow is inadequate to meet needs, blood flow may be improved with **angioplasty** (using a balloon catheter, laser, etc.) or **bypass surgery** (using autologous vein or synthetic material to route blood around a blockage). More than 1 million revascularization procedures using these techniques are performed annually.

---

in fluid from the interstitial space to plasma. A decrease in blood volume can occur because of hemorrhage, losses through sweat or other body fluids, or the transfer of fluid from plasma into the interstitial space. In the absence of compensatory events, changes in blood volume result in proportional changes in both central blood volume and extrathoracic blood volume. For example, a moderate hemorrhage (10% of blood volume) with no distribution shift would cause a 10% decrease in central blood volume. The reduced intrathoracic blood volume would, in the absence of compensatory events, lead to decreased

filling of the ventricles with diminished stroke volume and cardiac output.

Redistribution of Volume ■ Shifts in the distribution of blood volume occur because of changes in the transmural pressures of the vessels and/or changes in the compliance of the extrathoracic or intrathoracic vessels. The best physiologic example of a change in transmural pressure occurs whenever an individual stands up. Standing increases the pressure in the blood vessels of the legs because it creates a vertical column of blood between the heart and the blood vessels of the leg. The arterial and venous pressure at the ankles during standing can easily be increased by 130 cm (4.3 ft) of water (blood), which is almost 100 mm Hg higher than in the recumbent position. The increased transmural pressure (outside pressure is still atmospheric) results in little distention of arteries because of their low compliance but results in considerable distention of veins because of their high compliance. In fact, approximately 550 ml of blood is needed to fill the stretched veins of the legs and feet when an average person stands up.

The extra blood is obtained from the central blood volume by the following sequence of events (Fig. 15–13). When a person stands, blood continues to be pumped by the heart at the same rate and stroke volume for one or two beats. However, much of the blood reaching the legs remains in the veins as they become passively stretched to their new size by the increased venous (transmural) pressure. This decreases the return of blood to the chest. Since cardiac output exceeds venous return for a few beats, the central blood volume falls (as do the end-diastolic fiber length, stroke volume, and cardiac output). Once the veins of the legs reach their new steady-state volume, the venous return again equals cardiac output. The equality between venous return and cardiac output occurs even though the central blood volume is reduced by 550 ml. The new cardiac output and venous return are decreased (relative to what they were before standing) because of the reduction in central blood volume.

Changes in Venous Compliance ■ Volume shifts associated with standing would not occur without the

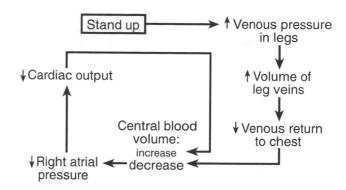

Figure 15–13 ■ Effect of standing on central blood volume and cardiac output. Standing causes increased hydrostatic pressure in the legs. On the venous side, this pressure markedly distends the veins as they fill with blood, lowering venous return. Decreased venous return lowers central blood volume, decreasing right atrial pressure and cardiac output. The decrease in cardiac output balances the diminished venous return and limits further decreases in central blood volume. "Increase" in smaller type indicates a secondary change.

high compliance of veins. These volume shifts involve changes in central blood volume, which, in turn, affect stoke volume and cardiac output. Changes in central blood volume are the result of transient imbalances between cardiac output and venous return. These principles are used in later chapters to explain the overall performance of the cardiovascular system.

■ ■ ■ ■ ■ ■ ■ ■ ■ ■ ■ REVIEW EXERCISES ■ ■ ■ ■ ■ ■ ■ ■ ■ ■ ■

Identify Each with a Term

1. The venous blood contained in the thorax
2. The venous pressure as measured at the right atrium
3. If total resistance (R) = R_A + R_B + R_C, then R_A, R_B, and R_C are arranged in _____
4. If 1/total resistance = $1/R_A$ + $1/R_B$ + $1/R_C$, then R_A, R_B, and R_C are arranged in _____
5. The fractional volume of red blood cells in the blood

Define Each Term

6. Systemic vascular resistance
7. Korotkoff sound
8. Dynamic exercise
9. Sphygmomanometer
10. Polycythemia

Choose the Correct Answer

11. Which of the following statements about the determinants of blood pressure is false?
    a. Pulse pressure is primarily determined by stroke volume and aortic compliance.
    b. An increase in heart rate will change mean arterial pressure if cardiac output remains constant.
    c. Mean arterial pressure is determined solely by cardiac output and systemic vascular resistance.
    d. If stroke volume increases and cardiac output remains constant, diastolic pressure decreases.

12. Which statement about blood pressure measurement using a sphygmomanometer is false?
    a. A cuff that is too narrow will give a falsely low blood pressure.
    b. Patients with arteriosclerosis (or other factors that make the blood vessels "stiff") may have falsely elevated blood pressures.
    c. Obesity may cause a falsely high estimate of blood pressure.
    d. The blood pressure cuff's width should ideally be 1.5 times the diameter of the limb at the site of measurement.

13. Which of the following statements is true?
    a. Blood flow is primarily regulated by changes in the resistance of large arteries.
    b. An increase in blood viscosity causes a decrease in vascular resistance.

c. Changes in blood vessel radius influence vascular resistance less than changes in blood viscosity.
d. Seventy percent of the pressure drop along the vasculature occurs in the small arteries and arterioles.

14. Which statement is false?
   a. Cardiac output is extremely sensitive to changes in central blood volume.
   b. The blood volume inside the cranium cannot change very much because the skull is rigid.
   c. The diameter of leg veins normally decreases when a person stands.
   d. Approximately 550 ml of additional blood is required to fill the leg veins when a person stands.

Calculate

15. Assume a subject has a normal blood pressure of 125/75 (measured in a large artery). (a) What is the pulse pressure? (b) Estimate the mean arterial pressure.

16. The normal heart has a mean right atrial pressure of 1-5 mm Hg (assume 3 mm Hg). Assume a patient has an arterial blood pressure of 150/90 mm Hg. (a) What is the pulse pressure? (b) What is the estimated mean arterial pressure? (c) What is the average gradient forcing blood through the vasculature? (d) Suppose the patient develops tricuspid valve regurgitation and as a result the mean right atrial pressure subsequently increases to 13 mm Hg (with no change in arterial blood pressure, systemic vascular resistance, or stroke volume). How will this affect the pressure gradient forcing blood through the vasculature?

Briefly Answer

17. If mean arterial pressure increases (due to an increase in systemic vascular resistance) and stroke volume, heart rate, and cardiac output remain constant, what happens to pulse pressure?

18. What effect on flow would standing have if arteries and veins were rigid (rather than compliant)?

ANSWERS

1. Central blood volume
2. Central venous pressure
3. Series
4. Parallel
5. Hematocrit
6. The frictional resistance to blood flow provided by all of the vessels between the aorta and the right atrium
7. The sound made when turbulent blood flow passes under a pneumatic pressure cuff placed around a limb
8. Exercise (e.g., running, swimming) that causes an increase in heart rate and stroke volume and a decrease in systemic vascular resistance
9. A device used to measure blood pressure in the limbs noninvasively
10. An increased mass or number of red blood cells
11. b
12. a
13. d
14. c

15.(a) 125 mm Hg − 75 mm Hg = 50 mm Hg
   (b) 50 mm Hg/3 = 16.7 mm Hg + 75 mm Hg = 92 mm Hg
16.(a) 60 mm Hg
   (b) 110 mm Hg
   (c) 107 mm Hg
   (d) If right atrial pressure increases to 13 mm Hg, then the pressure gradient falls to 110 − 13 = 97 mm Hg. If all other hemodynamic factors remain unchanged (which would be unlikely in this situation) systemic blood flow will fall in proportion to the decrease in pressure gradient.
17. If arterial compliance remains constant, pulse pressure will not change. However, an increase in mean arterial pressure will tend to stretch the large arteries and decrease their compliance (Figs. 15-5, 15-6); ejecting the same stroke volume into the less compliant arteries will therefore increase the pulse pressure.
18. No effect. As with a siphon, whatever pressure was added to the venous side of the circulation would also be added to the arterial side; thus the increased pressure would exactly balance. This is discussed in more detail in Chapter 18.

Suggested Reading

Genest J, Kuchel O, Hamet P, Cantin M: *Hypertension.* 2nd ed. New York, McGraw-Hill, 1983, chaps. 44, 45.
Milnor WR: The circulatory system and principles of hemodynamics, in *Cardiovascular Physiology.* New York, Oxford University Press, 1990, chaps. 2, 6.
Pickering G: Systemic arterial hypertension, in Fishman AP, Richards DW (eds): *Circulation of the Blood: Men and Ideas.* Bethesda, American Physiological Society, 1982, pp 3–70.
Rowell LB: General principles of vascular control, in *Human Circulation Regulation during Physical Stress.* New York, Oxford University Press, 1986, chaps. 2, 3.

The Microcirculation and the Lymphatic System

CHAPTER OUTLINE

I. THE ARTERIAL MICROVASCULATURE
  A. Arterioles regulate resistance by contraction of vascular smooth muscle
  B. Vessel wall tension and intravascular pressure interact to determine vessel diameter

II. THE CAPILLARIES
  A. Exchange between blood and tissue occurs in capillaries
  B. Passage of molecules through the capillary wall occurs both between capillary endothelial cells and through them

III. THE VENOUS MICROVASCULATURE
  A. Venules collect blood from capillaries
  B. The venular microvasculature acts as a blood reservoir

IV. THE LYMPHATIC VASCULATURE
  A. Lymphatic vessels collect excess tissue water and plasma proteins
  B. Lymph fluid is mechanically collected into lymphatic vessels from tissue fluid between cells

V. VASCULAR AND TISSUE EXCHANGE OF SOLUTES
  A. The large number of microvessels provides a large vascular surface area for exchange
  B. The microvessels must be permeable to lipid- and water-soluble molecules for exchange to occur
  C. The large number of microvessels minimizes the diffusion distance between cells and blood
  D. The interstitial space between cells is a complex environment of water- and gel-filled areas
  E. The rate of diffusion depends on permeability and concentration differences
  F. The extraction of molecules from blood is influenced by membrane permeability, surface area, and blood flow

VI. TRANSCAPILLARY EXCHANGE OF FLUID
  A. The osmotic forces developed by plasma proteins oppose filtration of fluid from capillaries
  B. Leakage of plasma proteins into tissue increases filtration of fluid from blood to tissue

OBJECTIVES

After studying this chapter, the student should be able to:

1. Define the primary functions of the arteriolar, capillary, venular, and lymphatic sections of the microcirculation
2. List the anatomic structures through which lipid- and water-soluble molecules diffuse from blood to tissue and explain why they are different
3. Describe the major mechanical factors and energy source that influence diffusion exchange across capillary walls
4. List the four physical forces that provide the energy source for filtration and absorption of fluid across capillary walls; describe how the fluid exchange process is influenced by capillary wall permeability to plasma proteins and water
5. Define how the pressure at any point in the microcirculation is influenced by events in vessels preceding and following the area; describe what conditions will cause the pressure to increase or decrease
6. Define myogenic and metabolic regulation of blood flow in organs and provide several examples of situations in which each will alter blood flow
7. Describe how the sympathetic nervous system communicates with individual microvessels; describe conditions in which local regulation of the microcirculation interferes with sympathetic control of the vessels.

The part of the circulation where nutrients, water, gases, hormones, and waste products are exchanged between the blood and cells is known as the **microcirculation,** and this exchange process is its most important function. Virtually every cell in the body is in close contact with a microvessel, such that there are tens of thousands of microvessels per gram of tissue. The lens and cornea are exceptions, since their nutritional needs are supplied by the fluids in the eye.

A second major function of the microcirculation is to regulate vascular resistance to maintain an adequate arterial pressure. Regulation of vascular resistance to preserve the arterial pressure and allowing each tissue to receive sufficient blood flow to sustain metabolism are sometimes in conflict. Often the temporary compromise is to preserve the mean arterial pressure by increasing arterial resistance at the expense of reduced blood flow to most organs, other than the heart and brain; ultimately, however, the tissue exchange function must be restored. The mean arterial pressure is determined by the product of vascular resistance and cardiac output (Chap. 12). If all of the resistance vessels of the body were to dilate maximally, even maximum cardiac output would not maintain an arterial pressure more than 60–70% of normal. Therefore, regulation of vascular resistance in the microcirculation is an important aspect of total health.

The microvasculature is considered to begin where the smallest arteries enter organs and to end where the smallest veins, the **venules,** exit them. In between are the microscopic arteries, the **arterioles,** and the capillaries. Depending on the size of the animal, the largest arterioles have an inner diameter of 100–400 μm and the largest venules a diameter of 200–800 μm. The arterioles divide into progressively smaller vessels, such that each section of the tissue has its own specific microvessels. The branching pattern typical of the microvasculature of different major organs and how it relates to organ function are discussed in Chapter 17.

The Arterial Microvasculature

The larger arteries have a low resistance to blood flow and function primarily as conduits (Chap. 15). However, as arteries approach the organ they perfuse, they divide into many small arteries both just outside and within the organ. These smallest of arteries and the arterioles of the microcirculation constitute the resistance blood vessels; together they regulate about 70–80% of the total vascular resistance. Constriction of the resistance vessels maintains the relatively high vascular resistance in organs and is caused by the release of norepinephrine from the sympathetic nervous system, except in the heart and brain. The resistance blood vessels are also influenced by their physical and chemical environment and hormones in blood (see below).

■ Arterioles Regulate Resistance by Contraction of Vascular Smooth Muscle

The vast majority of arterioles, whether large or small, are a tube of **endothelial cells** surrounded by a connective tissue **basement membrane,** a single or double layer of vascular smooth muscle cells, and a thin outer adventitial layer (Fig. 16-1). The vascular smooth muscle cells around the arterioles are 70–90 μm long. A single muscle cell will not completely encircle a larger vessel but may encircle a smaller vessel almost two times (Fig. 16-1). Vascular smooth muscle cells wrap spirally around the arterioles at approximately a 90-degree angle to the long axis of the vessel. This arrangement is efficient because the tension developed by the vascular smooth muscle cell can be almost totally directed to maintaining or changing vessel diameter.

■ Vessel Wall Tension and Intravascular Pressure Interact to Determine Vessel Diameter

The arterioles are primarily responsible for regulating vascular resistance and blood flow. Arteriolar ra-

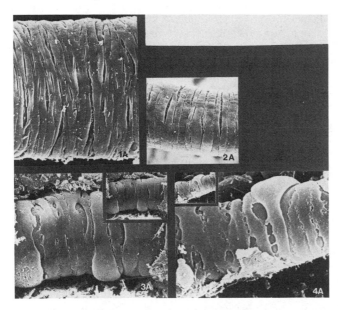

Figure 16–1 ■ Scanning electron micrographs, showing how vascular smooth muscle cells wrap around various-sized arterioles. A cell only partially passes around large- (1A) and intermediate-diameter (2A) arterioles but completely encircles the smaller arterioles (3A, 4A). 1A, 2A, and the small insets of 3A and 4A are at the same magnification. The enlarged views of 3A and 4A are at 10 times greater magnification. (Modified from Miller BR, Overhage JM, Bohlen HG, Evan AP: Hypertrophy of arteriolar smooth muscle cells in the rat small intestine during maturation. *Microvasc Res* 29:56–69, 1985.)

dius is determined by the transmural pressure gradient and wall tension, as expressed in Laplace's law (Chap. 14). Changes in wall tension developed by arteriolar smooth muscle cells directly alter vessel radius. This alters vascular resistance according to Poiseuille's law, which states that a vessel's resistance is related to the inverse of the fourth power of the vessel radius (Chap. 12). Studies indicate that most arterioles can dilate 60–100% from their resting diameter and can maintain a 40–50% constriction for long periods. Therefore, large decreases and increases in vascular resistance and blood flow are well within the capability of the microscopic blood vessels.

The Capillaries

■ Exchange between Blood and Tissue Occurs in Capillaries

While the arterioles and small arteries regulate the vascular resistance and blood flow, **capillaries** provide for most of the exchange between blood and tissue cells. The capillaries are perfused by the smallest of arterioles, the **terminal arterioles,** and their outflow is collected by the smallest venules. The capillary is an endothelial tube surrounded by a basement membrane composed of dense connective tissue (Fig. 16–2). Capillaries in mammals do not have vascular smooth muscle cells and

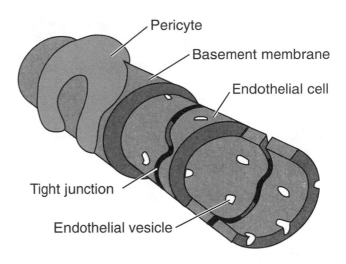

Figure 16–2 ■ The various layers of a mammalian capillary. The junctions between endothelial cells are partially fused by tight junctions. Water-soluble molecules pass through pores formed where tight junctions stop. Vesicle formation and diffusion of lipid-soluble molecules through the endothelial cell are other pathways for exchange.

are unable to change their inner diameter appreciably. However, the **pericytes** (Rouget cells), wrapped around the outer basement membrane, may be a primitive form of vascular smooth muscle cell and may add structural integrity to the capillary.

Capillaries, with inner diameters of about 4–8 μm, are the smallest vessels of the vascular system. Although they are small in diameter and individually have a high vascular resistance, the parallel arrangement of many thousands of capillaries per cubic millimeter of tissue minimizes their collective resistance. For example, in skeletal muscle, small intestine, and brain, capillaries account for only about 15% of the total vascular resistance of each organ, even though a single capillary has a resistance higher than that of the entire organ's vasculature.

The capillary lumen is so small that red blood cells fold as they pass through and virtually fill the entire lumen. The small diameter of the capillary and the thin endothelial wall minimize the diffusion path for molecules from the capillary core to the tissue just outside the vessel. In fact, the diffusion path is so short that most inorganic ions can pass through the capillary wall is less than 2 msec.

■ Passage of Molecules through the Capillary Wall occurs Both between Capillary Endothelial Cells and through Them

The exchange function of the capillary is intimately linked to the structure of its endothelial cells and basement membrane. Lipid-soluble molecules, such as oxygen and carbon dioxide, readily pass through the lipid components of endothelial cell membranes. Water-soluble molecules, however, must diffuse through water-filled pathways formed in the capillary wall at the junctions between endothelial cells. These pathways are called **pores:** they are not cylindric holes but complex passageways formed by the irregular junctions between adjacent endothelial cells. The complexity of these passages is increased because the cell membranes of adjacent endothelial cells are partially fused together by **tight junctions** (Fig. 16-2). The fused cell membranes exclude a water-filled pathway, preventing diffusion of water-soluble molecules, except where pores form. The capillaries of the brain and spinal cord have virtually continuous tight junctions between adjacent endothelial cells, consequently only the smallest water-soluble molecules pass through the capillary walls. The basement membrane is also part of the diffusion barrier but primarily restricts movement of very large molecules. For example, a very large organic molecule used to study capillary permeability, horseradish peroxidase, occasionally passes through the pores between endothelial cells, only to be stopped by the basement membrane.

An alternative pathway for water-soluble molecules through the capillary wall is the cell membrane vesicles (Fig. 16-2). Vesicles form on both the luminal and tissue side of the capillary wall by pinocytosis, and exocytosis occurs on the other side. The vesicles appear to migrate randomly between the outer and inner surfaces of the endothelial cell. Even the largest molecules may cross the capillary wall in this way. The importance of transport by vesicles to the overall process of transcapillary exchange remains to be defined.

The Venous Microvasculature

■ Venules Collect Blood from Capillaries

After the blood has passed through the capillaries, it enters the venules, endothelial tubes usually surrounded by a monolayer of vascular smooth muscle cells. In general, the vascular muscle cells of the venules are much smaller in diameter but longer than those of the arterioles. This may reflect the fact that venules operate at intravascular pressures of 10-16 mm Hg, compared to 30-70 mm Hg in arterioles, and do not need a powerful muscular wall. The smallest venules are unique in that they are more permeable to large and small molecules than are the capillaries. In fact, it is probable that much of the exchange of large water-soluble molecules occurs as the blood passes through small venules.

■ The Venular Microvasculature Acts as a Blood Reservoir

In addition to their blood collection and exchange functions, the venules are an important component of the blood reservoir system in the venous circulation. At rest, approximately two-thirds of the total blood volume is within the venous system, and perhaps more than half of this volume is within venules. Although the blood moves within the venous reservoir, it moves slowly, much like water in a reservoir behind a river dam. Although the venules and veins account for only 15-20% of systemic vascular resistance, a very low vascular resistance compared to that of small arteries and arterioles, their reservoir function is extremely important. If venule radius is increased or decreased, the volume of blood in tissue can change up to 20 ml/kg tissue; thus, the volume of blood readily available for circulation would increase by more than 1 liter in a 70-kg person. Such a large change in available blood volume can substantially improve venous return of blood to the heart or compensate for blood loss caused by hemorrhage or dehydration. For example, the volume of blood typically removed from blood donors is about 500 ml, or about 10% of the total blood volume; usually no ill effects are experienced, in part because the venules and veins decrease their reservoir volume to restore the circulating blood volume.

The Lymphatic Vasculature

■ Lymphatic Vessels Collect Excess Tissue Water and Plasma Proteins

Lymphatic vessels are unique microvessels that form an interconnected system of simple endothelial tubes within tissues. The gastrointestinal tract organs, liver, and skin have the most extensive lymphatic systems, and the central nervous system may not contain any lymph vessels. The lymphatic system typically begins as blind-ended tubes, or **lymphatic end-bulbs,** which drain into the meshwork of interconnected lymphatic vessels (Fig. 16-3). Although lymph collection begins in the end-bulbs, lymph collection from tissue also occurs in the interconnected lymphatic vessels, by the same mechanical process. A schematic drawing of the lymphatic system in the small intestine (Fig. 16-4) demonstrates the complexity of lymphatic branching. The villus lacteals are special adaptations of lymphatic end-bulbs. Note that lymph collection from the submucosal and muscle layers of this tissue must occur primarily in tubular lymphatic vessels, because few, if any, end-bulbs are present in these layers.

In all tissues with a lymphatic system, the lymphatic vessels coalesce into increasingly more developed and larger collection vessels. These larger vessels in the tissue and the macroscopic lymphatic vessels outside the organs have contractile cells similar to vascular smooth muscle cells. In connective tissues of the mesentery and skin, even the simplest of lymphatic vessels and end-bulbs spontaneously contract, perhaps due to contractile endothelial cells. Even if the lymphatic bulb or vessel cannot itself contract, compression of these lymphatic structures by movements of the host organ

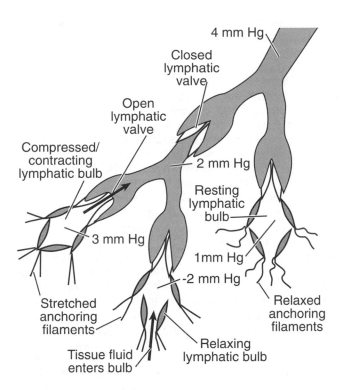

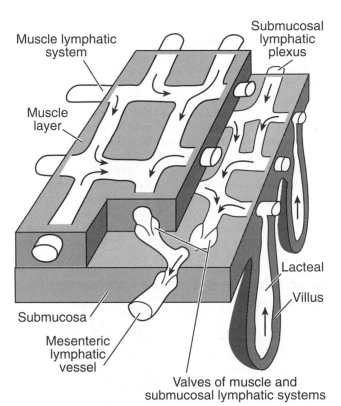

Figure 16–3 ■ Schematic representation of lymphatic vessel anatomy and functions. The contraction-relaxation cycle of lymph bulbs (bottom) is the fundamental process that removes excess water and plasma proteins from the interstitial spaces. Lymphatic pressures along the lymphatic vasculature are generated by lymphatic vessel contractions and organ movements.

Figure 16–4 ■ The arrangement of lymphatic vessels in gastrointestinal organs. The intestinal lymphatic vessels are unusual in that lymphatic valves are normally restricted to vessels about to exit the organ, whereas valvular structures exist throughout the lymphatic system of skin and skeletal muscle. (Modified from Unthank JL, Bohlen HG: Lymphatic pathways and role of valves in lymph propulsion from small intestine. *Am J Physiol* 254:G389–G398, 1988.)

(e.g., intestinal movements or skeletal muscle contractions) changes lymphatic vessel size.

■ Lymph Fluid Is Mechanically Collected into Lymphatic Vessels from Tissue Fluid between Cells

In all organ systems except the brain, more fluid is filtered than is absorbed by the capillaries, and plasma proteins diffuse into the interstitial spaces. By removing the fluid, the lymphatic vessels also remove proteins. This is an essential function because the protein concentration in plasma is higher than that in tissue fluid and only some form of mechanical transport can return the protein to the plasma. The ability of lymphatic vessels to change diameter, whether initiated by the lymphatic vessel or by forces generated within a contractile organ, is important for lymph formation and protein removal. In the smallest lymphatic vessels, and to some extent in the larger lymphatic vessels in a tissue, the endothelial cells are overlapped rather than fused together as in blood capillaries (Fig. 16–3). The overlapped portions of the cells are attached to **anchoring filaments,** which extend into the host organ tissues. Anchoring filaments probably help spread apart the endothelial cells when the lymphatic

vessels relax after a compression or contraction, so that tissue water and molecules carried in the fluid can easily enter the lymphatic vessels. This movement of fluid from tissue to the lymphatic vessel lumen is passive. When compressed or actively contracted lymphatic vessels passively relax, the pressure in the lumen becomes slightly lower than in the interstitial space, so that tissue fluid enters the lymphatic vessel. Once the interstitial fluid is in a lymphatic vessel, it is called **lymph.** When the lymphatic end-bulb or vessel again actively contracts or is compressed, the overlapped cells are mechanically sealed to hold the lymph. The pressure developed forces the lymph into the downstream segment of the lymphatic system. The anchoring filaments are stretched during this process, so that the overlapped cells can again be parted during relaxation of the lymphatic vessel and allow fluid to enter the lymphatic vessel lumen.

Active and passive compression of lymphatic bulbs and vessels also provide the forces needed to propel the lymph back to the venous side of the blood circulation. To maintain directional lymph flow, microscopic end-bulbs and lymphatic vessels as well as large lymphatic vessels have one-way valves (Fig. 16–3) that allow lymph to flow only from the tissue toward the progressively larger lymphatic vessels and finally into the venous system in the chest cavity. Lymphatic pressures are only a few mm Hg in the end-bulbs and smallest lymphatic vessels and as high as 10–20 mm Hg during contraction of larger lymphatic vessels. This progression from lower to higher lymph pressures is possible only because as each lymphatic segment contracts, it develops a slightly higher pressure than in the next lymphatic vessel and the lymphatic valve momentarily opens to allow lymph flow. When the activated lymphatic vessel relaxes, its pressure is again lower than that in the next vessel and the lymphatic valve closes. This sequence of events is illustrated in Figure 16–3.

Vascular and Tissue Exchange of Solutes

■ The Large Number of Microvessels Provides a Large Vascular Surface Area for Exchange

The overall branching structure of the microvasculature is a treelike branching system, with major trunks dividing into progressively smaller branches. This occurs for both the arteriolar and venular microvasculature, such that actually two "trees" exist, one to perfuse the tissue through arterioles and one to drain the tissue through venules. The increase in numbers of vessels through successive branches of the microvasculature dramatically increases the surface area of the microvasculature. The surface area is determined by the length, diameter, and number of vessels. In the small intestine, for example, the total surface area of the capillaries and smallest venules would be more than 10 cm² for a 1-cm³ cube of tissue. The large surface area of the capillaries and smallest venules is important, because the vast majority of exchange of nutrients, wastes, and fluid occurs across these tiny vessels.

■ The Microvessels Must Be Permeable to Lipid and Water-Soluble Molecules for Exchange to Occur

Lipid-soluble molecules, such as gases and lipids, can pass through the cell membranes of capillary and venular endothelial cells. The rate at which they diffuse through the cell barrier is directly proportional to their lipid solubility and concentration difference (partial pressure difference, in the case of gases) across the vessel wall and inversely proportional to their molecular weight. The major lipid-soluble compounds in moment-to-moment exchange are oxygen and carbon dioxide gases. Their lipid solubility and low molecular weight ensure that they readily permeate the walls of arterioles, capillaries, and venules. Another major class

of lipid-soluble compounds is the various lipids used for metabolic fuel, such as fatty acids. Their larger molecular weights, however, do not allow them to pass through the vessel walls as rapidly as the low-molecular-weight gases.

Water-soluble molecules cannot readily pass through the cell membranes of vascular endothelial cells and must traverse the vessel wall through the water-filled pores between adjacent endothelial cells (Fig. 16–2). These spaces or pores are not entirely filled with water; they are partially filled with a matrix of small fibers of submicron dimensions. The potential importance of this **fiber matrix** is that the size of the spaces in the matrix, rather than the size of the physical hole between cells, determines the maximum size of the molecules that readily diffuse through the pore. Due to the combination of the small spaces between endothelial cells and in the basement membrane plus the fiber matrix in these spaces, the vessel wall behaves as if only about 1% of the total surface area were available for exchange of water-soluble molecules. The majority of pores permit only molecules with a radius less than 3–6 nm to pass through the vessel wall.

This **small-pore system** is complemented by large pores, which allow molecules with a radius up to about 30 nm to pass through the vessel wall. For every large pore, there are several hundred small pores, although the ratio of large to small pores is variable among organs. The small-pore system allows passage of water, all inorganic ions, and the majority of small organic molecules, such as glucose and amino acids. Large water-soluble molecules, such as plasma proteins, can only pass through the large-pore system. Because of the small number of large pores, the vessel wall is very impermeable to large molecules. A second type of large-pore system, in addition to the occasional large gaps between endothelial cells, may be the previously mentioned membrane vesicles in endothelial cells (Fig. 16–2). These vesicles may shuttle tiny amounts of plasma and tissue fluid across the endothelial cell and, in doing so, passively carry any large molecules dissolved in the fluid across the vessel wall.

■ The Large Number of Microvessels Minimizes the Diffusion Distance between Cells and Blood

The spacing of microvessels in the tissue determines the distance molecules must diffuse from the blood to the interior of tissue cells. In the example shown in Figure 16–5A, a single capillary provides all the nutrients to a cell and the concentration of blood-borne molecules across the cell interior is represented by the density of dots at various locations. Diffusion distances are important because as the molecules travel further from the capillary, their concentration decreases, because the volume into which diffusion occurs increases as the square of the distance. In addition, some of the molecules may be consumed by different components of the cell, which further re-

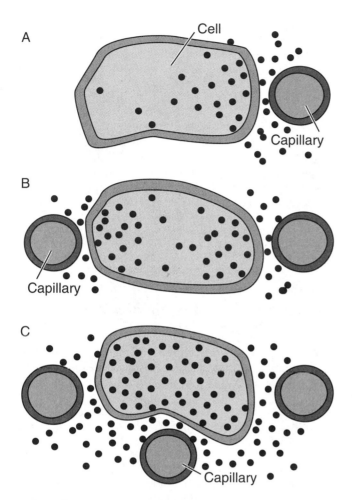

Figure 16–5 ■ The effect of the number of perfused capillaries on cell concentration of blood-borne chemicals (dots). (A) With one capillary, the left side of the cell has a low concentration. (B) The concentration can be substantially increased if a second capillary is perfused. (C) Perfusion of three capillaries about the cell increases concentrations throughout the cell.

duces the concentration. If there is a capillary on either side of a cell (Fig. 16–5B) the left side of the cell has a much higher concentration of molecules because of the reduction in diffusion distances caused by the new source of blood-borne molecules. Therefore, increasing the number of blood-perfused microvessels can dramatically reduce diffusion distances from a given point inside a cell to the nearest capillary, and doing so minimizes the dilution of molecules within the cells caused by large diffusion distances. As an example of physiologic events that change capillary spacing, exercise allows the arterioles to perfuse nearly all of the available capillaries in muscle, as in Figure 16–5B. Regular exercise induces the growth of new capillaries in skeletal muscle. This condition is shown in Figure 16–5C, in which three capillaries contribute to the nutrition of the cell and substantially elevate cell concentrations of molecules derived from blood. Likewise,

decreasing the number of capillaries perfused with blood by constricting arterioles or obliterating capillaries, as in diabetes mellitus, can lengthen diffusion distances and decrease exchange.

■ The Interstitial Space between Cells Is a Complex Environment of Water- and Gel-Filled Areas

As molecules diffuse from the microvessels to the cells, or vice versa, they must pass through the interstitial space that forms the extracellular environment between cells. This space is not a simple water-filled system but contains strands of collagen and elastin. Hyaluronic acid, a high-molecular-weight unbranched polysaccharide, and proteoglycans, complex polysaccharides bound to polypeptides, are embedded in the tissue fiber strands. These large molecules are arranged in complex coils with large amounts of water within the coils. To some extent, these large molecules and associated water may allow the interstitial space to behave as alternating areas with a gellike consistency, interspersed with water-filled regions. The gel phases are thought to represent areas where diffusion of solutes is restricted and where molecules may be excluded from some of the water held in the gel phase. The implication of such a gel and water system is that the effective concentration of molecules in the free interstitial water is higher than expected, because the molecules are restricted to readily accessible water-filled areas. Diffusion of water-soluble molecules is slowed by the circuitous pathway a molecule must move in the maze of the interstitial gel and water-filled spaces. There is also a distinct possibility that the relative amount of gel and water phases can be altered such that the diffusion barrier of the extracellular space can be changed.

■ The Rate of Diffusion Depends on Permeability and Concentration Differences

Diffusion is by far the most important means for moving solutes across capillary walls. The rate of diffusion of a solute between blood and tissue is given by Fick's permeability equation (Chap. 2):

$$Js = P (Cb - Ct) \qquad (1)$$

where Js is expressed in mol/min $\times$ 100 g tissue, P is the permeability coefficient, Cb is the blood concentration, and Ct is the tissue concentration. P is usually measured under conditions where neither the surface area of the vasculature nor the diffusion distance is known but the tissue mass can be determined. The permeability coefficient is directly related to the **diffusion coefficient** of the solute in the capillary wall and the vascular surface area available for exchange and is inversely related to the diffusion distance (ΔX). The surface area and diffusion distance are both determined in part by the number of microvessels perfused by blood. In everyday life, the diffusion coefficient is relatively constant unless the capillaries are damaged, but the number of perfused capillaries and blood and tissue concentrations of solutes are constantly changing. Therefore, the diffusion distance and surface area for exchange can be influenced by physiologic events, as can be the concentrations in the tissue and blood. In this context, microvascular exchange is dynamically altered by many physiologic events and is a complex process.

The magnitude of the difference in concentrations between the tissue and blood is influenced by many processes that occur simultaneously and influence each other. It is important to remember that the diffusion rate depends on the *difference* between the high and low concentrations, not their specific concentrations. For example, if the cell consumes a particular solute the concentration in the cell will decrease, and for a constant concentration in blood plasma the diffusion gradient will enlarge to increase the rate of diffusion. If the cell ceases to use as much of a given solute, the concentration in the cell will increase but the rate of diffusion will decrease. Both of these examples assume that more than sufficient blood flow exists to maintain a relatively constant concentration in the microvessel. However, in many cases this may not be true. For example, as blood passes through the tissues, approximately one-fourth to one-third of the oxygen contained in arterial blood is extracted by the tissue before it reaches the capillaries. There is usually ample oxygen in the capillary blood to maintain aerobic metabolism, but if tissue metabolism is increased and blood flow is not appropriately elevated, the tissue will exhaust the available oxygen from the blood while it is in the microvessels. The consequences are a decreased aerobic metabolic rate, which forces the cell to resort to anaerobic glycolysis to provide cell energy. This scenario routinely occurs when skeletal muscles begin to contract and blood flow has not yet been appropriately increased to meet the increased oxygen demand.

■ The Extraction of Molecules from Blood Is Influenced by Membrane Permeability, Surface Area, and Blood Flow

As a result of diffusional losses and gains of molecules as blood passes through the tissues, the concentrations of various molecules in venous blood can be very different from those in arterial blood. The extraction (E) of material from blood perfusing a tissue can be calculated from the arterial (Ca) and venous (Cv) blood concentration as:

$$E = (Ca - Cv)/Ca \qquad (2)$$

If the blood loses material to tissue, the value of E is positive and has a maximum value of 1 if all material

is removed from arterial blood (Cv = 0). An E value of 0 (Ca = Cv) indicates that no loss or gain occurred. A negative E value indicates that tissue added material to blood (Cv > Ca). The total mass of material lost or gained by the blood can be calculated as:

$$\text{Amount lost or gained} = E \times F \times Ca \qquad (3)$$

where F is blood flow. While this equation is useful to estimate the total amount of material exchanged between tissue and blood, it does not allow a direct determination of how changes in vascular permeability and exchange surface area would influence the extraction process. The extraction can be related to the permeability (P) and surface area (A) available for exchange as well as the blood flow:

$$E = 1 - e^{-PA/F} \qquad (4)$$

This equation implies that extraction increases when either permeability or exchange surface area increases or blood flow decreases. Extraction decreases when permeability and surface area decrease or blood flow increases. As a consequence, physiologically induced changes in the number of perfused capillaries, which alters surface area, and changes in blood flow are important determinants of overall extraction and, therefore, exchange processes. In everyday life, the blood flow and total perfused surface area usually change in the same direction, although by different relative amounts. For example, surface area is usually able, at most, to double or be reduced by about half, but blood flow can increase three- to fivefold, and even greater amounts in skeletal muscle, or decrease by about half in most organs, yet maintain viable tissue. The net effect is that extraction is rarely more than doubled or decreased by one-half relative to the resting value in the majority of organs. This is still an important range, because changes in extraction can compensate for reduced blood flow or accentuate exchange when blood flow is increased.

Transcapillary Exchange of Fluid

To force blood through the microvessels, the heart pumps blood into the elastic arterial system and provides the pressure needed to move the blood. This hemodynamic (or hydrostatic) pressure, while absolutely necessary, allows filtration of water through pores if the hydrostatic pressure on the blood side of the pore is greater than on the tissue side. The capillary pressure is different in each organ, ranging from about 15 mm Hg in intestinal villus capillaries to 55 mm Hg in the kidney glomerulus. The interstitial hydrostatic pressure ranges from slightly negative to 8–10 mm Hg and in most organs is substantially less than capillary pressure.

■ The Osmotic Forces Developed by Plasma Proteins Oppose Filtration of Fluid from Capillaries

The primary defense against excessive fluid filtration is the **colloid osmotic pressure** generated by plasma proteins. The plasma proteins are too large to pass readily through the vast majority of water-filled pores of the capillary wall. In fact, more than 90–95% of these large molecules are retained in the blood during passage through the microvessels of most organs. The colloid osmotic pressure is conceptually similar to osmotic pressures for small molecules generated across the selectively permeable cell membranes and primarily depends on the number of molecules in solution. The major plasma protein that impedes filtration is albumin, because it has the highest molar concentration of all plasma proteins. The colloid osmotic pressure of plasma proteins is typically 16–20 mm Hg in mammals when measured using a membrane that prevents diffusion of all large molecules. The colloid osmotic pressure offsets the capillary hydrostatic blood pressure such that the net filtration force is only slightly positive or negative. If the capillary pressure is sufficiently low, the balance of colloid and hydrostatic pressures is negative and tissue water is absorbed into the capillary blood.

■ Leakage of Plasma Proteins into Tissue Increases Filtration of Fluid from Blood to Tissue

A small amount of plasma protein enters the interstitial space; these proteins and perhaps native proteins of the space generate the **tissue colloid osmotic pressure.** This pressure of 2–5 mm Hg offsets part of the colloid osmotic pressure in the plasma. This is, in a sense, a filtration pressure that opposes the blood colloid osmotic pressure. As discussed earlier, plasma proteins in the interstitial fluid are returned to the plasma by the lymphatic vessels.

■ Hydrostatic Pressure in Tissue Can Either Favor or Oppose Water Filtration from Blood to Tissue

The hydrostatic pressure on the tissue side of the endothelial pores is the **tissue hydrostatic pressure.** This pressure is determined by the volume of water in the interstitial space and the tissue's distensibility. Tissue hydrostatic pressure can be increased by external compression, such as with support stockings or muscle contraction. There is substantial debate about exactly what tissue hydrostatic pressure is in various tissues during resting conditions. It is probable that tissue pressure is slightly below atmospheric pressure ("negative") during normal hydration of the interstitial space and becomes positive when excess water is in the interstitial space. In a very real sense, tissue hydrostatic pressure is a filtration force when negative and absorption force when positive.

If water is removed from the interstitial space, the hydrostatic pressure becomes very negative and opposes further fluid loss (Fig. 16–6). However, if water is added to the interstitial space, the tissue hydrostatic pressure is increased very little. Over a range of tissue fluid volumes, there is a margin of safety (Fig. 16–6) that prevents excessive tissue hydration or dehydration. If the tissue volume exceeds a certain range, **edema** exists. In extreme situations, the tissue swells with fluid to the point that pressure dramatically increases and strongly opposes capillary filtration. The ability of tissues to allow substantial changes in interstitial volume to occur with only small changes in pressure indicates that the interstitial space is rather distensible.

■ The Balance of Filtration and Absorption Forces Regulates Exchange of Fluid Between Blood and Tissue

The role of hydrostatic and colloid osmotic pressures in determining fluid movement across capillaries was first postulated by the English physiologist Ernest Starling at the end of the nineteenth century. In the 1920s, the American physiologist Eugene Landis obtained experimental proof for Starling's hypothesis. The relationship is defined for a single capillary by the **Starling-Landis equation:**

$$J_v = K_h\, A\, \{(P_c - P_t) - \sigma\,(COP_p - COP_t)\} \qquad (5)$$

where J_v is the net volume of fluid moving across the capillary wall per unit of time (cubic microns per minute) and K_h is the **hydraulic conductivity for**

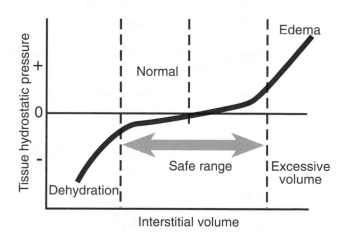

Figure 16–6 ■ Variations in tissue hydrostatic pressure as interstitial water volume is altered. Under normal conditions, tissue pressure is slightly negative (subatmospheric), but an increase in volume can cause the pressure to be positive. If the interstitial tissue volume exceeds the "safe range," high hydrostatic pressures will exist and edema will be present. Dehydration of the tissue can cause very negative tissue hydrostatic pressures.

HYPERTENSION AND THE MICROCIRCULATION

Hypertension, or high blood pressure, occurs in about 25% of the population and appears be a disease whose genetic predisposition is more likely to be expressed in the dietary and social conditions of an industrial society. Although once associated with aging, hypertension can begin at any age. For example, some young children of hypertensive patients have higher-than-normal arterial pressure. The major form of hypertension is known as **idiopathic** or **essential hypertension** and represents about 95% of all cases. The primary physical problem in the cardiovascular system once hypertension is sustained is an elevated vascular resistance in all organs. Cardiac output is typically normal once hypertension is established but tends to be elevated during the early episodic or labile phase, when arterial pressure alternates between normal and elevated. Once hypertension is established, it is a disease of the microvasculature, particularly the arteriolar microvasculature. For the typical increase in resistance that increases arterial blood pressure by 30–50 mm Hg, the arteriolar vasculature need constrict only 4–7%. In addition, there is evidence of an increased incidence of temporary and, perhaps, permanent closure of small arterioles, which would also elevate resistance. The vascular smooth muscle cells of arteries are hypertrophied, but hypertrophy of these cells around arterioles is minor. However, all vascular smooth muscle cells are exceptionally responsive to norepinephrine, and increased sympathetic nervous system activity occurs in hypertensive humans. In addition, the vessel walls are less distensible than normal, which may help them tolerate very high pressures throughout the arterial and most of the arteriolar circulation. Despite the fact that increased microcirculatory resistance is the physical cause of hypertension, the microcirculation functions remarkably well to meet tissue nutritional demands. Edema formation, a common problem with acute elevation of microvascular pressure, is a rare complication of sustained hypertension, primarily because the increased precapillary vascular resistance helps maintain normal capillary blood pressures. Most complications are caused by accelerated atherosclerosis and overload on the left ventricular muscle due to high arterial pressure. Although a cure for hypertension has not been found, interventions to decrease sympathetic nervous system activity, decrease vascular responses to norepinephrine, and suppress the functions of the renin-angiotensin hormonal system (Chap. 24) can restore a normal arterial pressure. Discontinuing medication in hypertensive patients is a particular problem because hypertension usually returns in days to several months, often, at a higher pressure than before treatment began.

water, which is the fluid permeability of the capillary wall, expressed as cubic microns per minute per square micron of capillary surface area per mm Hg pressure difference. The value of K_h increases up to fourfold from the arterial to the venous end of a typical capillary. P_c is the capillary hydrostatic pressure and P_t is the tissue hydrostatic pressure. COP_p and COP_t represent the plasma and tissue colloid osmotic pressures, and σ is the **reflection coefficient for plasma proteins.** This coefficient is necessary because the plasma proteins are slightly permeable to the microvascular wall, preventing full expression of the two colloid osmotic pressures. The value of σ is 1 when molecules cannot cross the membrane and is 0 when molecules freely cross the membrane. Typical σ values for plasma proteins in the microvasculature exceed 0.9 in most organs except the liver or spleen, which have capillaries that are very permeable to plasma proteins. The reflection coefficient is normally relatively constant but can be decreased dramatically by hypoxia, inflammatory processes, and tissue injury. This leads to increased hydrostatic fluid filtration because the effective colloid osmotic pressure is reduced when the vessel wall becomes more permeable to plasma proteins.

The capillary exchange of fluid is bidirectional, since

capillaries and venules may filter or absorb fluid depending on the balance of hydrostatic and colloid osmotic pressures. It is certainly possible that filtration occurs primarily at the arteriolar end of capillaries, where filtration forces exceed absorptive forces. It is equally likely that absorption of water occurs in the venular end of the capillary and small venules, since hydrostatic blood pressure is dissipated by the friction of blood flow in the capillary. Based on directly measured capillary hydrostatic and plasma colloid osmotic pressures, the entire length of the capillaries in skeletal muscle filters slightly all of the time, while the lower capillary pressures in the intestinal mucosa and brain primarily favor absorption along the entire capillary length.

The extrapolation of fluid filtration or absorption for a single capillary to fluid exchange in a whole tissue is difficult because of regional variations in microvascular pressures, possible filtration and absorption of fluid in vessels other than capillaries, and physiologically and pathologically induced variations in the available surface area for capillary exchange. Therefore, for whole organs, a measurement of total fluid movement relative to the mass of the tissue is used. To take into account the various hydraulic conductivities and total surface area of all vessels involved, the volume (ml) of fluid moved per minute for a change of 1 mm Hg in capillary pressure for each 100 g of tissue is determined. This value is called the **capillary filtration coefficient** (CFC), though it is very likely that fluid exchange occurs in many vessels, particularly venules, in addition to capillaries. Values for CFC in tissues such as skeletal muscle and small intestine are typically in the range of 0.025–0.16 ml/min/mm Hg/100 g. Though these values may seem small, in a 24-hour period an amount of water equal to the plasma blood volume is filtered from the blood to the lymphatic system and then returned to the venous system.

The CFC replaces the hydraulic conductivity (K_h) and capillary surface area (A) terms in the Starling-Landis equation for filtration across a single capillary. The CFC can change if permeability to water, the surface area (determined by the number of perfused microvessels), or both are altered. For example during intestinal absorption of lipids both capillary permeability to water and perfused surface area increase, dramatically increasing CFC. In contrast, the skeletal muscle vasculature seems to increase CFC primarily because of increased perfused capillary surface area during exercise, and only small increases in water permeability occur.

The hydrostatic and colloid osmotic pressure differences across capillary walls (called Starling forces) clearly cause movement of water and dissolved solutes into the interstitial spaces. These movements are, however, normally quite small and contribute only a very minor extent to tissue nutrition. The vast majority of solutes transferred to the tissues move across capillary walls by simple diffusion, not by bulk flow of fluid.

Regulation of Microvascular Pressures

The pressures, both hydrostatic and colloid osmotic, involved in transcapillary exchange depend on the concentration of plasma proteins and how the microvasculature dissipates the prevailing arterial and venous pressures. Plasma protein concentration is determined largely by the rate of protein synthesis in the liver, which makes the vast majority of plasma proteins. Reduced plasma protein concentration occurs in disorders that decrease protein synthesis, such as liver diseases and malnutrition, and kidney diseases in which plasma proteins are filtered into the urine and lost. A lowered plasma colloid osmotic pressure favors filtration of plasma water and gradually causes serious edema. Edema formation in the abdominal cavity (**ascites**) can allow large quantities of fluid to collect in and grossly distend the abdominal cavity.

■ Capillary Pressure Is Determined by both the Resistance of and Blood Pressure in Arterioles and Venules

Capillary pressure (P_c) is not constant and is influenced by four major variables: precapillary (R_{pre}) and postcapillary (R_{post}) resistances and arterial (P_a) and venous (P_v) pressures. Pre- and postcapillary resistances can be calculated by the pressure dissipated across the respective vascular regions divided by the total tissue blood flow (F), which is equal for both regions:

$$R_{pre} = [P_a - P_c] / F \qquad (6)$$
$$R_{post} = [P_c - P_v] / F \qquad (7)$$

In the majority of organ vasculatures, the precapillary resistance is three to four times higher than the postcapillary resistance. This has a substantial effect on capillary pressure.

To demonstrate the effect of pre- and postcapillary resistances on capillary pressure, we use the equations for the pre- and postcapillary resistances to solve for blood flow:

$$F = [P_a - P_c] / R_{pre} = [P_c - P_v] / R_{post} \qquad (8)$$

If the two equations to the right of the flow term are solved for capillary pressure, P_c, an equation to determine capillary pressure is obtained:

$$P_c = \frac{[(R_{post} / R_{pre}) \times P_a] + P_v}{[1 + (R_{post}/R_{pre})]} \qquad (9)$$

This indicates that the effect of arterial pressure (P_a) on capillary pressure is determined by the ratio of post- to precapillary resistance, rather than the magnitude of either resistance. In addition, venous pressure substantially influences capillary pressure. Both pressure effects are also influenced by the denominator. At a typical post- to precapillary resistance ratio of 0.16, the denominator will be 1.16, which allows about 80% of a change in venous pressure to be reflected back to the capillaries. The post- to precapillary resistance ratio increases during the arteriolar vasodilation that accompanies increased tissue metabolism. The decreased precapillary resistance and minimal change in postcapillary resistance increase capillary pressure. Since the balance of hydrostatic and colloid osmotic pressures is usually -2 to $+2$ mm Hg, a 10–15 mm Hg increase in capillary pressure during maximum vasodilation can cause a profound increase in filtration. The increased filtration associated with microvascular dilation is usually associated with a large increase in lymph production, which removes excess tissue fluid.

■ Capillary Pressure Is Reduced when the Sympathetic Nervous System Increases Arteriolar Resistance

When sympathetic nervous system stimulation causes a substantial increase in precapillary resistance and a proportionately smaller increase in postcapillary resistance, the capillary pressure can decrease up to 15 mm Hg and greatly increase absorption of tissue fluid. This is a very useful process, because it can add approximately 1 L of fluid to the blood in a normal adult human, which can compensate for vascular volume loss during sweating, vomiting, or diarrhea. Since water is lost by any of these processes, the plasma proteins are concentrated because they are not lost. This increases plasma colloid osmotic pressure, which further accentuates absorption and acts to restore plasma volume.

Regulation of Microvascular Resistance

The vascular smooth cells around arterioles and venules respond to a wide variety of physical and chemical stimuli, altering the diameter and resistance of the microvessels. Here the various physical and chemical conditions in tissues that influence the muscle cells of the microvasculature are considered.

■ Myogenic Vascular Regulation Allows Arterioles to Respond to Changes in Intravascular Pressure

Vascular smooth muscle is able to contract rapidly when stretched and conversely, to reduce actively developed tension when passively shortened. In fact, vascular smooth muscle may be able to contract or relax when the force load on the muscle is increased or decreased, respectively, even though the initial muscle length is not substantially changed. These responses are known to persist as long as the initial stimulus is present, unless vasoconstriction so greatly reduces blood flow that the tissue becomes seriously hypoxic.

In everyday life a vessel's ability to constrict when intravascular pressure is increased or dilate as the pressure is decreased, called **myogenic regulation,** has several benefits. First, and perhaps most important, blood flow can be indirectly regulated when the arterial pressure is too high or low for appropriate tissue perfusion. Second, the myogenic response is rapidly activated (within 2 seconds) by changes in arterial and venous pressure. This allows rapid modifications in microvascular resistance before appreciable changes in the tissue chemical and metabolic environment occur. Third, the arteriolar vasculature demonstrates very strong constrictor responses when venous pressure is elevated by more than about 5–10 mm Hg above the typical resting venous pressure. The elevation of venous pressure results in an increase in capillary and arteriolar pressures, the former leading to the rapid development of edema. The arteriolar constriction lowers the transmission of arterial pressure to the capillaries and small arterioles to minimize the risk of edema, but at the expense of a decreased blood flow. The myogenic response to elevated venous pressure may be due to venous pressures transmitted backward through the capillary bed to the arterioles and, perhaps, to some type of response initiated by venules and transmitted to arterioles, perhaps through endothelial cells or local neurons.

■ Tissue Metabolism Influences Arteriolar Diameter

A reduction in arterial pressure by compression of the artery to an organ is associated with microvascular vasodilation, which helps maintain blood flow. If the arterial pressure is increased, the vasculature constricts spontaneously, even if the sympathetic nervous system is not activated. The general concept behind both responses is that the vasculature responds to depletion or accumulation of certain key chemicals in a tissue. It attempts to maintain a constant supply of nutrients and washout of waste products when the change in arterial pressure initially alters the blood flow. Part of the response is also related to myogenic regulation.

The role of oxygen in local vascular regulation has been studied extensively. Oxygen is not stored in appreciable amounts in tissues and must be constantly replenished by blood flow, since oxygen is consumed by aerobic metabolism. An increase in metabolic rate would decrease the tissue oxygen concentration and possibly directly signal vascular muscle to relax. Examples of the changes in oxygen concentration (or partial pressure) around arterioles (periarteriolar space), in the capillary bed, and around large venules during skeletal muscle contractions are shown in Figure 16–7. At rest, venular blood oxygen tensions are usually higher than in the capillary bed, because venules acquire oxygen that diffuses out of the nearby arterioles. Although both periarteriolar and capillary bed tissue

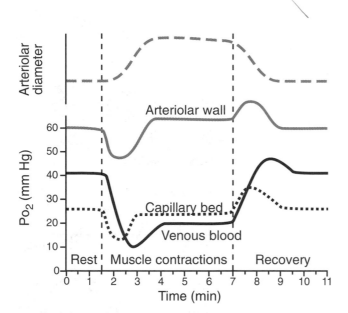

Figure 16–7 ■ Arteriolar dilation and changes in arteriolar wall, venous blood, and capillary bed partial pressure of oxygen during and after skeletal muscle contractions. The initial decrease in all oxygen tensions at the start of contractions reflects increased oxygen usage, which is not replenished by increased blood flow until the arterioles dilate. As arteriolar dilation occurs, arteriolar wall and capillary bed oxygen tensions are substantially restored but venous blood has a low oxygen tension. During recovery, oxygen tensions transiently increase above resting values, because blood flow remains temporarily elevated as oxygen usage is rapidly lowered to normal. (Modified from Lash JM, Bohlen HG: Perivascular and tissue PO₂ in contracting rat spinotrapezius muscle. *Am J Physiol* 252:H1192–H1202, 1987.)

oxygen tensions decrease at the onset of contractions, both are restored as arteriolar dilation occurs. The oxygen tension in venular blood rapidly and dramatically decreases at the onset of skeletal muscle contractions and demonstrates very little recovery despite increased blood flow. The sustained decline in venular blood oxygen tension probably reflects increased extraction of oxygen from the blood. It is important that neither at rest nor during muscle contractions is the oxygen content of venular blood a trustworthy indicator of the oxygen status of the capillary bed. In contrast to venules, many arterioles have a normal to slightly increased periarteriolar oxygen concentration during skeletal muscle contractions, because the increased delivery of oxygen through elevated blood flow offsets the increased use of oxygen by tissues immediately around the arteriole. Therefore, as long as blood flow is allowed to increase substantially, it is unlikely that oxygen availability at the arteriolar wall is a major factor in the sustained vasodilation during increased metabolism.

Recent studies indicate that the vascular smooth muscle cell is not particularly responsive to oxygen concentrations (tensions) likely to be found routinely in tissue. Only unusually low or high oxygen concentrations (tensions) seem to be associated with changes in vascular smooth muscle tension. However, depletion of oxygen from an organ's cells as well as an increased metabolic rate does cause the release of adenine nucleotides, free adenosine, Krebs cycle intermediates, and, in hypoxic conditions, lactic acid. This provides a large potential source of various molecules, most of which cause vasodilation at physiologic concentrations, to influence the regulation of blood flow. If adequate oxygen is available, increased aerobic metabolism liberates a larger-than-normal amount of carbon dioxide, which produces hydrogen ions from carbonic acid. In addition, if the oxygen supply to the tissue cannot meet the aerobic demands, lactic acid is released from anaerobic metabolism. An increase in hydrogen ion concentration, resulting from either carbon dioxide or acidic metabolic intermediates, causes vasodilation, but usually only transient increases in venous blood and interstitial tissue acidity occur if blood flow through an organ with increased metabolism is allowed to increase appropriately.

■ Endothelial Cells Are Capable of Releasing Chemicals that Cause Relaxation and Constriction of Arterioles

A potentially important contributor to local vascular regulation is released by the endothelial cells. This substance, **endothelial-derived relaxing factor** (EDRF), is released from all arteries, microvessels, veins, and even lymphatic endothelial cells. It appears to be nitric oxide (NO) or a related compound formed by the action of the nitric oxide synthase on arginine. Endothelial-derived relaxing factor causes relaxation of vascular smooth muscle by inducing an increase in cyclic guanosine monophosphate (cGMP). It also suppresses platelet activation and reduces adhesion of leukocytes to endothelial cells.

Compounds such as acetylcholine, histamine, and adenine nucleotides (ATP, ADP) released into the interstitial space, as well as hypertonic conditions, shear forces on the vessel lumen caused by blood flow, and hypoxia cause the release of EDRF. Adenosine, one of the most potent naturally occurring vasodilators, does not appear to act through the release of EDRF. One of the best naturally occurring inhibitors of EDRF is hemoglobin, which reacts with and destroys EDRF. Virtually any type of endothelial damage decreases formation of EDRF.

Endothelial cells also release one of the most potent vasoconstrictor agents, the peptide **endothelin.** Extremely small amounts are released under natural conditions. The role of endothelin during normal events is unknown, but increased release of this vasoconstrictor during tissue and vascular injury may further decrease blood flow.

■ The Sympathetic Nervous System Regulates Blood Pressure and Flow by Constricting the Microvessels

Though the microvasculature has an impressive ability to use local control mechanisms to adjust vascular resistance based on the physical and chemical environment of the tissue and vasculature, the dominant regulatory system is related to the regulation of the arterial blood pressure. As is explained in Chapter 18, the arterial blood pressure is regulated moment to moment by the baroreceptor system, acting through the effects of the sympathetic nervous system on the cardiac output and systemic vascular resistance. The sympathetic nervous system communicates with the resistance vessels and venous system through the release of norepinephrine onto the smooth muscle cells in vessel walls. Norepinephrine is released in close proximity to the vasculature and then diffuses to the vasculature. The sympathetic nerves form an extensive meshwork of axons over the exterior of the microvessels such that all vascular smooth muscle cells are likely to receive norepinephrine. Since the diffusion path is short (<1 μm), norepinephrine rapidly reaches the vascular muscle and activates α receptors, and constriction begins within 2-5 seconds. The actions of the sympathetic nervous system must occur quickly because rapid changes in body position or sudden exertion require immediate responses to maintain or increase arterial pressure. If an organ's metabolic rate is substantially increased or myogenic mechanisms are activated, the local control mechanisms can partially or completely override the constrictor effects of the sympathetic nervous system.

■ Certain Organs Control Their Blood Flow via Autoregulation and Functional Hyperemia

If the arterial blood pressure to an organ is decreased such that blood flow is compromised, the vascular resistance decreases and blood flow returns to approximately normal. If arterial pressure is elevated, flow initially is increased but the vascular resistance increases and restores the blood flow toward normal. This is known as **autoregulation of blood flow.** The regulation of autoregulation appears to be primarily related to metabolic and myogenic control, although some input from EDRF mechanisms and local neural reflexes is possible. The cerebral and cardiac vasculature, followed closely by the renal vasculature, are most able to regulate blood flow. Skeletal muscle and intestinal vasculatures exhibit less well-developed autoregulation. An example of autoregulation, based on data from the cerebral vasculature, is shown in Figure 16-8. Note that the arterioles continue to dilate at arterial pressures below 60 mm Hg, when blood flow begins to decrease markedly as arterial pressure is further lowered. The vessels clearly cannot dilate sufficiently to maintain blood flow at very low arterial

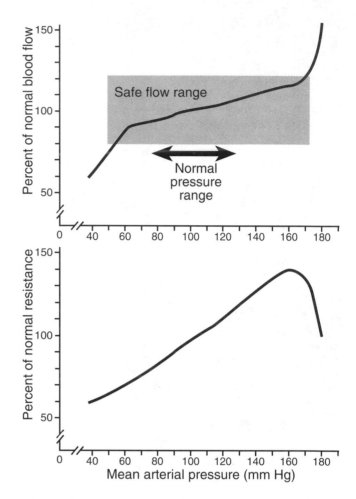

Figure 16-8 ■ Autoregulation of blood flow and vascular resistance as mean arterial pressure is altered. The safe range for blood flow is about 80-125% of normal and usually occurs at arterial pressures of 60-160 mm Hg due to active adjustments of vascular resistance. At pressures above about 160 mm Hg, vascular resistance decreases, since the pressure forces dilation to occur; at pressures below 60 mm Hg, the vessels are fully dilated and resistance cannot be appreciably further decreased.

pressures. At greater-than-normal arterial pressures, the arterioles constrict. If the mean arterial pressure is elevated appreciably above 150-160 mm Hg, the vessel walls cannot maintain sufficient tension to oppose passive distention by the high arterial pressure. The end result is excessive blood flow and very high microvascular pressures, eventually leading to edema. In everyday life, only the cerebral and cardiac vasculature exhibit very impressive autoregulatory abilities. The other vasculatures are so dominated by the vascular control imposed by the sympathetic nervous system that their autoregulatory abilities are suppressed.

In processes thought to use mechanisms similar to those of autoregulation, blood flow is increased to support increased metabolism. The greatest increase in blood flow occurs in skeletal muscle during exercise, and increments of 4- to 20-fold above resting conditions are possible. Similarly, cardiac blood flow can increase 4- to 5-fold during strenuous exercise. For both cardiac and skeletal muscle, the increased sympathetic nervous system activity associated with exercise does not appreciably interfere with the metabolically induced vasodilation in active tissues. The absorption of food nutrients, particularly fatty acids, in the small intestine is a very potent stimulus to increase blood flow by 50–100%. Even while you read this line, the increased metabolism in the areas of the brain used for vision and language processing is accompanied by an increased blood flow. Each of these events ensures that the organ receives adequate blood flow to suit its increased metabolic requirements during increased metabolic activity of the organ.

■ ■ ■ ■ ■ ■ ■ ■ ■ ■ ■ ■ REVIEW EXERCISES ■ ■ ■ ■ ■ ■ ■ ■ ■ ■ ■ ■

Identify Each with a Term

1. The vessels that primarily regulate vascular resistance and blood flow

2. The fibrous material between adjacent endothelial cells that influences capillary permeability

3. An excess of water in the interstitial space between cells

4. The maintenance of a relatively constant blood flow in the brain and heart as the arterial pressure changes

Define Each Term

5. Microvascular basement membrane

6. Colloid osmotic pressure

7. Autoregulation of blood flow

Choose the Correct Answer

8. The most important function of the microcirculation is:
   a. Exchange of nutrients and wastes between blood and tissue
   b. Filtration of water through capillaries
   c. Regulation of vascular resistance
   d. Autoregulation of blood flow

9. When a lipid-soluble and water-soluble molecule pass through a capillary wall by simple diffusion:
   a. The molecules pass through the same anatomic pathways, which include both intra- and extracellular passages
   b. Both molecules pass through the cell cytoplasm of the endothelial cells or water spaces between cells
   c. Lipid-soluble molecules can pass through the endothelial cell membranes, but the water-soluble molecules cannot
   d. Lipid-soluble molecules pass through a water-filled channel, just like water-soluble molecules

10. Venules function to collect blood from the tissue and:
    a. Act as a substantial source of resistance to regulate blood flow
    b. Serve as a living reservoir for blood in the cardiovascular system
    c. Are virtually impermeable to both large and small molecules
    d. Are about the same diameter as the arterioles

11. The interstitial space can best be described as:
    a. A water-filled space with a low plasma protein concentration
    b. A viscous space with a large plasma protein concentration
    c. A space with alternating gel and liquid areas with a low plasma protein concentration
    d. A space primarily filled with gel-like material and a small amount of liquid

12. An increase in pressure in arterioles, which are capable of myogenic vascular regulation, typically results in:
    a. A sustained dilation of the vessel as the increased pressure expands the vessel
    b. A transient dilation followed by sustained constriction of the vessel
    c. An increase in blood flow as the endothelial cells release endothelial-derived relaxing factor
    d. An increase in capillary pressure, whether or not the arterioles exhibit a myogenic response

13. An arteriole with a damaged endothelial layer will not:
    a. Constrict when intravascular pressure is increased
    b. Dilate when adenosine is applied to the vessel wall
    c. Constrict in response to norepinephrine
    d. Dilate in response to adenosine diphosphate or acetylcholine

Calculate

14. What would the tension, in dyne/cm, be at an arteriolar pressure of 70 mm Hg, tissue hydrostatic pressure of 1 mm Hg, and vessel diameter of 0.01 cm? Note that 1 mm Hg pressure is equal to 1330 dyne/cm^2.

15. For an arterial blood content of 20 ml oxygen per 100 ml blood and venous blood content of 15 ml oxygen per 100 ml of blood, what is the extraction of oxygen, and how much oxygen is transferred from blood to tissue if the blood flow is 200 ml/min?

16. Assume plasma proteins have a reflection coefficient of 0.9, plasma colloid osmotic pressure is 24 mm Hg and tissue colloid osmotic pressure is 4 mm Hg. Will fluid be filtered or absorbed by the capillary if capillary hydrostatic pressure is 23 mm Hg and tissue hydrostatic pressure is 1 mm Hg?

17. What would the capillary pressure be if arterial pressure is 100 mm Hg, venous pressure is 5 mm Hg, and precapillary resistance is five times higher than postcapillary resistance?

Briefly Answer

18. What are the two major steps used by lymphatic vessels to remove excess fluid and plasma proteins from tissue?

19. Under what circumstances might the oxygen concentration (tension) in the tissue and around the microvessels cause dilation or constriction of arterioles?

20. Describe the sequence of events leading to arteriolar constriction when the sympathetic nervous system is activated.

21. How does an increase or decrease in the number of perfused capillaries influence exchange between blood and tissue in the microvasculature?

■ ■ ■ ■ ■ ■ ■ ■ ■ ■ ■ ■ ■ ■ **ANSWERS** ■ ■ ■ ■ ■ ■ ■ ■ ■ ■ ■ ■ ■ ■

1. Arterioles

2. Fiber matrix

3. Edema

4. Autoregulation of blood flow

5. The fibrous connective tissue layer that supports the endothelial cells of capillaries and both the endothelial and vascular smooth muscle cells of arterioles and venules

6. The pressure generated by large molecules in an aqueous solution when force is used to attempt to filter water from the solution through a membrane that is not permeable to the large molecules

7. Used to describe the attempt by the vasculature of most organs to maintain a constant blood flow if the arterial pressure is locally decreased or raised

8. a

9. c

10. b

11. c

12. b

13. d

14. Tension = (70 mm Hg − 1 mm Hg) × (0.01 cm /2) × 1333 = 459 dyne/cm

15. Extraction = (20 ml/100 ml − 15 ml/100 ml) / (20 ml/100 ml) = 0.25
    Amount of Exchange = (20 ml/100 ml - 15 ml/100 ml) × 200 ml/min = 10 ml of oxygen/min

16. Fluid will be filtered, because the balance of hydrostatic pressures is 22 mm Hg (capillary hydrostatic pressure − tissue hydrostatic pressure) and is greater than the balance of colloid osmotic pressures, which is 18 mm Hg [reflection coefficient × (plasma colloid osmotic pressure − tissue colloid osmotic pressure)].

17.
$$P_{cap} = \frac{[(R_{post} / R_{pre}) \times P_a] + P_v}{1 + (R_{post} / R_{pre})}$$
$$= \frac{[(1/5) \times 100] + 5}{1 + (1/5)} = 20.8 \text{ mmHg}$$

18. First, contraction of the lymphatic vessels or compression by an external force moves lymph in the lumen into the next lymph vessel. Second, when the lymphatic vessel recoils, hydrostatic pressure in the lymphatic vessel falls below that in the tissue and tissue fluid is forced into the lumen of the lymphatic vessel.

19. A very low oxygen tension in the tissue and around microvessels causes vasodilation. This can happen during a major increase in tissue metabolism, when blood flow is inadequate for the tissue's metabolic needs, or when there is inadequate oxygen in the blood. Very high oxygen tensions in and around vessels causes constriction of the arterioles and can occur if metabolism suddenly decreases, blood flow is too high for the metabolic needs of the tissue, or gas with a high oxygen content is used for breathing.

20. First, the nerve releases norepinephrine when an action potential occurs. Second, the norepinephrine diffuses to the vascular smooth muscle cell. Third, norepinephrine binds to α receptors on vascular smooth muscle, initiating the processes of contraction. Fourth, the activated muscle cells develop sufficient tension to constrict the vessel against the intravascular pressure.

21. An increased number of perfused capillaries decreases the average distance from capillaries to the cell interior and increases capillary surface area. Together, these factors result in a higher intracellular concentration of the diffusing molecules. A loss of perfused capillaries increases the average distance from capillaries to the cell interior and decreases the area available for exchange. The result is a decrease in the cell concentration of diffusing molecules.

Suggested Reading

Bevan JA, Halpern W, Mulvany MJ (eds): *The Resistance Vasculature*. Totowa, NJ, Humana, 1990, pp 373-402.

Bundgaard M: Functional implication of structural differences between consecutive segments of microvascular endothelium. *Microcirc Endothel Lymphatics* 4:113-142, 1988.

Harper SL, HG Bohlen, MJ Rubin: Arterial and microvascular contributions to cerebral cortical autoregulation in rats. *Am J Physiol* 246:H17-H24, 1984.

Johnson PC: The myogenic response in the microcirculation and its interaction with other control systems. *J Hypertension* 7:S33-S39, 1989.

Lash JM, Bohlen HG: Perivascular and tissue PO₂ in contracting rat spinotrapezius muscle. *Am J Physiol* 252:H1192-H1202, 1987.

Mellander S, Bjornberg J: Regulation of vascular smooth muscle tone and capillary pressure. *News Physiol Sci* 7:113-119, 1992.

Milnor WR: *Cardiovascular Physiology*. New York, Oxford University Press, 1972, pp 290-356.

Moncada S: From endothelium-dependent relaxation to the L-arginine: NO pathway. *Blood Vessels* 27:208-217, 1990.

Unthank JL, Bohlen HG: Lymphatic pathways and role of valves in lymph propulsion from small intestine. *Am J Physiol* 254:G389-G398, 1988.

Special Circulations

OBJECTIVES

After studying this chapter, the student should be able to:

1. Discuss the relationship between cardiac metabolism and coronary blood flow
2. Explain how changes in arterial blood pressure and cerebral metabolic rate influence blood flow to the brain
3. Describe how changes in oxygen availability, release of vasoactive metabolites, and osmolality of the interstitial environment interact to regulate intestinal blood flow during food absorption
4. List the two sources of blood flow into the liver and describe why their arrangement is unusual compared to that of other organ vasculatures
5. Discuss the range of blood flows observed in skeletal muscle and how this range affects the function of muscle and the cardiovascular system
6. Explain why blood flow in the skin primarily serves temperature regulation rather than cell metabolism
7. Describe blood flow through the fetus and how it differs from that in the adult

This chapter is devoted to the unique anatomic and physiologic properties of the vasculatures in the heart, brain, small intestine, liver, skeletal muscle, and skin. The features of the vasculature are related to the specific functions and specialized needs of the host organ. The specialized vascular anatomy and physiology of the fetus and placenta during fetal life and the circulatory changes that occur at birth are also presented. The pulmonary and renal circulations are discussed in Chapters 20 and 23, respectively.

Coronary Circulation

■ The Work Done by the Heart Determines Its Oxygen Use and Blood Flow Requirement

The vasculature of the heart, the **coronary circulation,** is unique because even when the body is at rest, it must supply more oxygen per gram of tissue than is received by an equal mass of skeletal muscle during vigorous exercise. This is readily appreciated from the data in Table 17-1, which provides a comparison of flow per 100 g of tissue and total flow for all the organ circulations discussed. Coronary blood flow can

normally increase about fourfold to supply additional oxygen, such as is needed by the heart muscle during exercise (Table 17-1). This constitutes the **coronary blood flow reserve.** Heart tissue extracts almost the maximum amount of oxygen from blood during resting conditions. Because the heart's ability to use anaerobic glycolysis to provide energy is very limited, the only practical way to increase energy production is to increase blood flow and oxygen delivery. Coronary blood flow increases with the workload demands on the heart.

■ Cardiac Blood Flow Decreases during Systole and Increases during Diastole

As in skeletal muscle, blood flow through the left ventricle decreases to a minimum when the muscle contracts, because the small blood vessels are compressed. Blood flow in the left coronary artery during cardiac systole is only 10-30% of that during an equivalent period of diastole, when the heart musculature is relaxed and most of the blood flow occurs. The compression effect of systole is minimal in the right ventricle, probably due to the lower pressures developed by a smaller muscle mass (Fig. 17-1). Blood flow in the wall of the left ventricle is not uniformly de-

TABLE 17–1 ■ Blood Flow and Oxygen Consumption of the Major Systemic Organs Estimated for a 70-kg Human

Organ	Mass (kg)	Flow (ml/100 g/min)	Total Flow (ml/min)	Oxygen Use (ml/100 g/min)	Total Oxygen Use (ml/min)
Heart					
Rest	0.4–0.5	60–80	250–300	7.0–9.0	25–40
Exercise		200–300	1,000–1,200	25.0–40.0	65–85
Brain	1.4	50–60	750	4.0–5.0	50–60
Small Intestine					
Rest	3	30–40	1,500	1.5–2.0	50–60
Absorption		45–70	2,200–2,600	2.5–3.5	80–110
Liver					
Total	1.8–2.0	100–130	1,400–1,500	13.0–14.0	180–200
Portal		70–90	1,100	5.0–7.0	
Hepatic Artery		30–40	350	5.0–7.0	
Muscle					
Rest	28.0	2–6	750–1,000	0.2–0.4	60
Exercise		40–100	15,000–20,000	8.0–15.0	2,400–?
Skin					
Rest	2.0–2.5	1–3	200–500	0.1–0.2	2–4
Exercise		5–15	1,000–2,500		

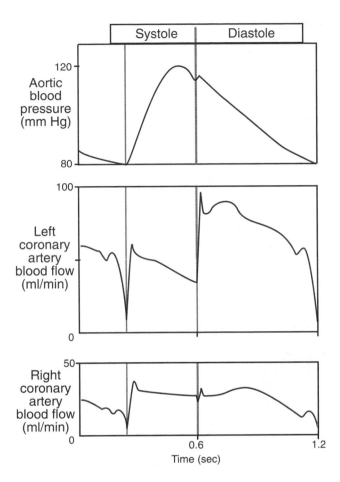

Figure 17–1 ■ Aortic blood pressure and left and right coronary blood flows during the cardiac cycle. Note that left coronary artery blood flow decreases dramatically during the isovolumetric phase of systole, just prior to opening of the aortic valve. Left coronary artery blood flow remains lower during systole than during diastole because of compression of the coronary blood vessels in the contracting myocardium. The left ventricle receives most of its arterial blood inflow during diastole. Right coronary artery blood flow tends to be sustained during both systole and diastole because lower intraventricular pressures are developed by the contracting right ventricle and consequently there is less compression of coronary blood vessels. (Adapted from Gregg DE, Khouri EM, Rayford CR: Systemic and coronary energetics in the resting unanesthetized dog. *Circ Res* 16:102–113, 1965, and Lowensohn HS, Khouri EM, Gregg DE, Pyle RL, Patterson RE: Phasic right coronary artery blood flow in conscious dogs with normal and elevated right ventricular pressures. *Circ Res* 39:760–766, 1976.)

creased during systole. Changes in blood flow during the cardiac cycle are of academic interest in normal people and have no obvious deleterious effects. However, in people whose coronary arteries are compromised and blood flow during diastole is inadequate, an increased heart rate decreases the time spent in diastole, such that blood flow is compromised.

The heart musculature is perfused from the outside (**epicardial**) surface to the inside (**endocardial**) surface. Microvascular pressures are dissipated by blood flow friction as the vessels pass through the heart tissue. Therefore, the mechanical compression of systole has more effect on the blood flow through inner layers of the heart where compressive forces are higher and microvascular pressures are lower. This is a problem particularily during heart diseases of virtually all types, and the vast majority of tissue impairment occurs in the endocardial layers.

■ Coronary Vascular Resistance Is Primarily Regulated by Responses to Heart Metabolism

Animal studies indicate that about 75% of total coronary vascular resistance occurs in vessels with inner diameters less than about 200 μm. This is supported by clinical measurements in humans showing that little pressure dissipation occurs in normal coronary arteries. The majority of the coronary resistance vessels, the small arteries and arterioles, are surrounded by cardiac muscle cells and are exposed to chemicals released by cardiac cells into the interstitial environment. Many of these cause dilation of the coronary arterioles. For example, adenosine, derived from breakdown of ATP in cardiac cells, is a potent vasodilator, and its release increases whenever cardiac metabolism is increased or blood flow to the heart is experimentally or pathologically decreased. Blockade of the vasodilator actions of adenosine with theophylline, however, does not prevent coronary vascular dilation when cardiac work is increased, blood flow is suppressed, or the arterial blood is depleted of oxygen. Therefore, while adenosine is likely an important contributor to cardiac vascular regulation, there are obviously other potent regulatory agents. Vasodilatory prostaglandins, H^+, CO_2, endothelial-derived relaxing factor, and decreased availability of oxygen, as well as myogenic mechanisms, are all capable of contributing to coronary vascular regulation. No single mechanism adequately explains dilation of coronary arterioles and small arteries when the metabolic rate of the heart is increased or pathologic or experimental means are used to restrict blood flow.

The coronary arteries and arterioles are innervated by the sympathetic nervous system and can be constricted by norepinephrine, whether released from nerves or carried in the arterial blood. However, the importance of the sympathetic nervous system and circulating norepinephrine in causing coronary vascular constriction, via α receptors, is controversial. Only the larger coronary arteries may have α receptors, which induce vascular constriction when activated by norepinephrine. Smaller vessels may have β receptors, which cause vasodilation in response to epinephrine released by the adrenal medulla during sympathetic activity. In addition, epinephrine increases the metabolic rate of the heart via β receptors. This, in turn, leads to dilatory stimuli that potentially could overcome vasoconstriction. The overall concept evolving

TREATMENT OF ATHEROSCLEROTIC CORONARY ARTERY DISEASE

The primary problem in atherosclerotic coronary artery disease is mechanical occlusion such that blood flow is inadequate for even moderate activity. The vast majority of coronary artery occlusions occur in vessels of sufficient diameter that a vascular bypass can be inserted around the obstruction with a segment of vein or systemic artery. While there may be immediate improvement, open-chest heart surgery is a major procedure that requires substantial recovery time. An alternative approach is to thread a catheter with an inflatable balloon at the tip through a peripheral artery, usually the femoral artery, into the impaired coronary artery. The balloon is inflated to enlarge the occluded vessel. This procedure, known as **percutaneous transluminal coronary angioplasty,** can be done in conscious but sedated patients and avoids the risks of general anesthesia and extensive surgery. In some cases, a metallic or fiber mesh tube is placed in the expanded vessel to maintain patency. The primary vascular risks with both bypass surgery and angioplasty are clot formation in the affected vessels leading to heart muscle ischemia and restenosis over a 1–5-year period. The benefits are usually improved blood flow at rest and increased (but seldom normal) ability of the coronary vasculature to elevate blood flow above resting conditions. Nonsurgical treatment of coronary artery disease usually involves dietary and pharmacologic means to lower blood lipid and cholesterol concentrations to reverse atherosclerotic plaques partially, exercise to increase cardiac and body efficiency and, perhaps, stimulate development of or enlarge collateral vessels, and acute and chronic pharmacologic dilation of the coronary vasculature. As with surgical methods, ability to increase blood flow is improved but usually remains below normal. Since the vast majority of work activities in everyday life do not require maximum coronary blood flow, even a modest increase in coronary blood flow reserve can dramatically increase the patient's tolerance for moderate work and exercise.

from animal studies is that the sympathetic nervous system suppresses the decrease in coronary vascular resistance during exercise despite the metabolic effects of epinephrine mentioned. In exercising animals, α receptor blockade leads to an increase in coronary blood flow. Whether a similar effect occurs in exercising people remains to be determined.

■ Coronary Vascular Disease Limits Cardiac Blood Flow and Cardiac Work

Pathology of the coronary vasculature is the cause of death in about one-third of the population in industrialized societies. Progressive occlusion of coronary arteries by atherosclerotic plaques and the acute formation of blood clots in damaged coronary arteries are life-threatening because the metabolic needs of the cardiac muscle cannot be supplied by the blood flow. As the plaque or clot partially occludes the vessel lumen, vascular resistance is increased and blood flow would decrease if smaller coronary vessels did not dilate to restore a relatively normal blood flow at rest. In doing so, the reserve for dilation of these vessels is compromised. While this usually has no effect at rest,

when cardiac metabolism is increased the decreased ability to increase blood flow can limit cardiac performance. In many cases, inadequate blood flow is first noticed as chest pain **(angina pectoris)** originating from the heart and a feeling of shortness of breath during exercise or work. The vascular occlusion can cause conditions ranging from impaired contractile ability of the cardiac muscle, which limits cardiac output and therefore tolerance to everyday work and exercise, to death of the muscle tissue, a **cardiac infarct.**

If the problem is not severe, drugs can be used to cause coronary vasodilation or decreased cardiac work to reduce blood flow requirements, or both. If the arterial pressure is higher than normal, various approaches are used to lower the blood pressure and thus decrease the heart's work load and oxygen needs. In addition to pharmacologic treatment, mild to moderate exercise, depending on the status of the coronary disease, is often advised. Exercise stimulates the development of collateral vessels in the heart, improves the overall performance of the cardiovascular system, and increases the efficiency of the body during exercise. This latter effect lowers the cardiac output needed for a given task and thus decreases the heart's metabolic energy requirement. These lifestyle changes and pharmacologic interventions are intended to improve the coronary blood flow and lower the heart's metabolic needs to the remaining capabilities of the coronary vasculature.

■ Collateral Vessels Interconnect Sections of the Cardiac Microvasculature

One of the probable biologic adaptations to slowly developing coronary vascular disease is the enlargement of **collateral blood vessels** between either the left and right coronary arterial systems or parts of each system. In healthy human hearts, collateral arterial vessels are rare, but arteriolar collaterals (internal diameter < 100 μm) do occur in small numbers. In some patients with coronary vascular disease, expansion of existing collateral vessels and limited formation of new collaterals seem to be the primary mechanisms for improvement of collateral vessel function. These changes provide a partial bypass for blood flow to areas of muscle whose primary supply vessels are impaired. Endocardial arteriolar collaterals usually enlarge more than epicardial collaterals. In part, the greater collateral enlargement in the endocardium compared to the epicardium may be due to the lower pressure reaching the endocardial vessels. Therefore, less force is available to perfuse the deep layers of muscle, and flow is likely to be more compromised. The exact mechanism responsible for the development of collateral vessels is unknown. However, periods of inadequate blood flow to the heart muscle caused by experimental flow reduction do stimulate collateral enlargement in animals. It is assumed that in humans with coronary vascular disease who develop functional collateral vessels, the mechanism is related to occasional or even sustained periods of inadequate blood flow. Whether or not routine exercise programs aid in the development of collaterals in normal humans is debatable; the benefits of exercise may be by other mechanisms, such as enlargement of the primary perfusion vessels and reduction of atherosclerosis.

A coronary vascular problem that can exist even though the coronary arteries appear relatively normal is a compromised ability to increase blood flow in a pathologically hypertrophied heart. Although the heart may have a relatively normal coronary blood flow per mass of tissue, the ability to increase blood flow is often substantially decreased. The typical mild form of cardiac muscular hypertrophy that occurs with exercise training tends to increase the number of coronary capillaries and enlarge existing vessels and does not appear to compromise vasodilation of resistance vessels. The much greater cardiac muscle hypertrophy associated with semilunar valve stenosis or systemic hypertension results in an impaired ability to increase blood flow. The mechanisms currently thought to be responsible for the impairment are compression of the smaller blood vessels and decreased distensibility of these vessels, particularly if the systemic arterial blood pressure is chronically elevated.

Cerebral Circulation

■ Brain Blood Flow is Virtually Constant Despite Changes in Arterial Blood Pressure

The cerebral circulation shares many of the physiologic characteristics of the coronary circulation. Both the heart and brain tissues have a high metabolic rate (Table 17–1), extract a large amount of oxygen from blood, and have a very limited ability to use anaerobic glycolysis for metabolism, and the vessels have a limited ability to constrict in response to sympathetic nerve stimulation. As described in Chapter 16, the brain and coronary vasculatures have an excellent ability to *autoregulate* blood flow at arterial pressures from about 50–60 mm Hg to about 150–160 mm Hg. The vasculature of the brainstem exhibits the most precise autoregulation, with good but less precise regulation of blood flow in the cerebral cortex.

The exact mechanisms responsible for cerebral vascular autoregulation are unknown. Identification of a specific chemical to cause cerebral autoregulation has not been possible. For example, when blood flow is normal, regardless of the arterial blood pressure, little extra adenosine, K^+, H^+, or other vasodilator metabolites are released and brain tissue P_{O_2} remains relatively constant. The brain vasculature does exhibit myogenic vascular responses, but whether this mechanism is a major contributor to autoregulation is unknown. Animal studies indicate that both the cerebral arteries and arterioles are involved in cerebral vascular autoregulation and other types of naturally occurring vascular responses (see below). In fact, the arteries can change their resistance almost proportionately to the arterioles during autoregulation. This may occur in part because cerebral arteries exhibit myogenic vascular responses and because they are partially to fully embedded in the brain tissues and would likely be influenced by the same vasoactive chemicals in the interstitial environment as affect the arterioles.

■ Brain Microvessels Are Sensitive to CO_2 and H^+

The cerebral vasculature dilates in response to increased CO_2 and H^+ and constricts if either is decreased. Both of these substances are formed when cerebral metabolism is increased by nerve action potentials, such as during normal brain activation. Also, interstitial K^+ is elevated when a large number of action potentials are fired. Therefore, the 10–30% increase in blood flow in areas of brain excited by peripheral nerve stimulation, mental activity, or visual activity may be related to these three substances released from active nerve cells. The cerebral vasculature also dilates when the oxygen content of arterial blood is reduced, but the vasodilatory effect of elevated CO_2 is much more powerful.

■ Cerebral Blood Flow Is Remarkably Insensitive to Hormones and Sympathetic Nerve Activity

Neither circulating vasoconstrictor and vasodilator hormones nor the release of norepinephrine at sympathetic nervous system terminals on cerebral blood vessels plays much of a role in regulation of cerebral blood flow. The blood-brain barrier effectively prevents constrictor and dilator agents in blood plasma from reaching the vascular smooth muscle. Though the cerebral arteries and arterioles are fully innervated by the sympathetic nervous system, stimulation of these nerves produces only mild vasoconstriction in the majority of cerebral vessels. If, however, sympathetic activity to the cerebral vasculature is permanently interrupted, the cerebral vasculature has a decreased ability to autoregulate blood flow at high arterial pressures and the integrity of the blood-brain barrier is more easily disrupted. Therefore, some aspect of the sympathetic nerve activity other than routine regulation of vascular resistance is important for maintenance of normal cerebral vascular function. This may occur because of a trophic factor that promotes the health of endothelial and vascular smooth muscle cells in the cerebral microvessels.

■ The Cerebral Vasculature Adapts to Chronic High Blood Pressure

Cerebral vascular resistance increases in chronic hypertension and allows cerebral blood flow, and presumably capillary pressures, to be normal. The adaptation of cerebral vessels to sustained hypertension allows them to maintain vasoconstriction at arterial pressures that would overcome the contractile ability of a normal vasculature (Fig. 17-2). The mechanisms that allow the cerebral vasculature to adjust the autoregulatory range upward appear to be hypertrophy of the vascular smooth muscle and a mechanical constraint to vasodilation, due to more muscle tissue or connective tissue or both. The drawback to such adaptation is partial loss of the ability to dilate and regulate blood flow at low arterial pressures. This occurs because the passive structural properties of the resistance vessels decrease the vessel diameter and increase resistance to flow at subnormal pressures. In fact, the lower pressure limit of constant blood flow (autoregulation) can be almost as high as the normal mean arterial pressure (Fig. 17-2). This can be problematic, because if the arterial blood pressure is rapidly lowered to normal the person may faint due to inadequate brain blood flow secondary to passive vessel constriction.

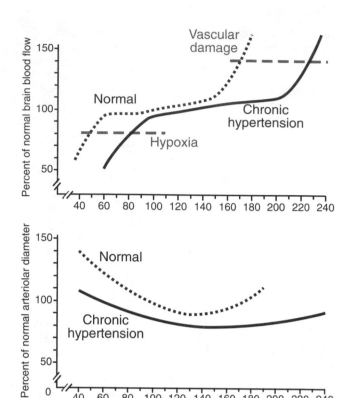

Figure 17–2 ■ Chronic hypertension is associated with a rightward shift in the arterial pressure range over which autoregulation of blood flow occurs (upper panel) because for any given arterial pressure, resistance vessels of the brain have smaller than normal diameters. As a consequence, hypertensive subjects can tolerate high arterial pressures that would cause vascular damage in normal subjects, but they risk reduced blood flow and brain hypoxia at low arterial pressures that are easily tolerated by normal subjects.

Fortunately, a gradual reduction in arterial pressure over weeks or months returns autoregulation to a more normal pressure range.

Small Intestine Circulation

The small intestine has two major functions: to propel partially digested food using contractions of visceral smooth muscle layers and to absorb the nutrients and water in food into the mucosal villi. Small arteries and veins penetrate the muscular wall of the bowel and form a microvascular distribution system in the submucosa (Fig. 17-3). The muscle layers receive small arterioles from the submucosal vascular plexus, and other small arterioles then continue into individual vessels of the deep submucosa around glands and the villi. About 70% of the intestinal vascular resistance is controlled by small arteries and larger arterioles preceding the separate muscle and submucosal-mucosal vasculatures. The small arterioles of the muscle and submucosal-mucosal layers can partially adjust blood flow to

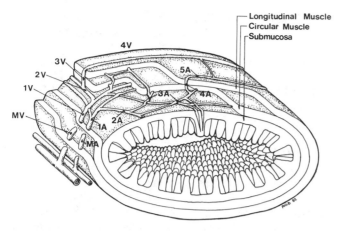

Figure 17–3 ■ The intestinal vasculature is unusual because three very different tissues—muscle, submucosal layer, and mucosal layer—are served by branches from a common vasculature located in the submucosa. Most of the intestinal vascular resistance is regulated by small arteries and arterioles preceding the separate muscle and submucosal and mucosal vasculatures. (Modified from Bret Connors: Quantification of the architectural changes observed in intestinal arterioles from diabetic rats. Ph.D. diss., Indiana University Medical School, Indianapolis, 1993.)

high lipid solubility absorbed by the villus epithelium enter the capillary blood and may exchange from this blood as it passes through venules to the incoming blood in arterioles. As a consequence, materials absorbed by the villus may be partially retained in the tissue. This may contribute to the interstitial osmolality of about 600 mOsm/kg H_2O near the villus tip, compared to 400 mOsm/kg H_2O near the villus base during food absorption. A second consequence is that oxygen in arteriolar blood diffuses into blood at a lower oxygen concentration in venules or capillaries. The oxygen is short-circuited from the villus tip as the venous blood removes incoming oxygen. Measurements of oxygen tension in the villus tip indicate a tissue oxygen tension of about half that at the villus base. These differences in tissue osmolality and Po_2 from villus tip to base also may be partially caused by higher rates of transport and metabolism by epithelial cells near the tip of the villus. The gradient of villus Po_2 and osmolality substantially increases during food absorption. Though blood flow is increased at least 50–100%, absorption is such a highly metabolic event that near hypoxic and hyperosmotic interstitial conditions persist.

suit the needs of small areas of tissue. Compared to other major organ vasculatures, the intestinal circulation has a poorly developed response to locally decreased arterial pressure, and blood flow usually declines because resistance does not adequately decrease. However, elevation of venous pressure outside the intestine causes sustained myogenic constriction; in this regard, the intestinal circulation equals or exceeds similar regulation in other organ systems. Intestinal motility has little effect on the overall intestinal blood flow, probably because the increases in metabolic rate are so small. In contrast, the 30–100% increase in metabolic rate during absorption of food can increase blood flow by 20–80% (Table 17-1). The complex events that take place during food absorption and their possible effect on blood flow regulation are discussed below.

■ The Microvasculature of Intestinal Villi Has a High Blood Flow and Unusual Exchange Properties

The intestinal mucosa receives about 60–70% of the total intestinal blood flow. Blood flows of 70–100 ml/min/100 g in this specialized tissue are probable and much higher than the average blood flow for the total bowel wall (Table 17-1). This blood flow can exceed those through resting blood flow in the heart and brain. The mucosa is composed of individual projections of tissue called **villi,** which are approximately cylindric in humans. The villus provides an anatomic structure that allows **countercurrent flow,** a flow of blood in opposite directions in inflow arterioles and outflow venules. Materials of low molecular weight or

■ Food Absorption Requires a High Blood Flow to Support Metabolism of the Mucosal Epithelium

Lipid absorption causes a greater increase in intestinal blood flow (**absorptive hyperemia**) and oxygen consumption than either carbohydrate or amino acid absorption. During absorption of all three classes of nutrients, the mucosa releases adenosine and CO_2 and oxygen is depleted. These conditions, as well as release of other vasodilatory substances, could contribute to the absorptive hyperemia. The hyperosmotic lymph and venous blood that leaves the villus to enter the submucosal tissues around the major resistance vessels is also a major contributor to the absorptive hyperemia. The intestinal vasculature is one of the most responsive vasculatures to hyperosmolality.

The mucosal capillaries are arranged in a dense network just beneath the mucosal epithelium, and virtually every cell is in contact with a portion of a villus capillary. The capillaries of the villus are unusual in that portions of the cytoplasm are missing, so that the two opposing surfaces of the endothelial cell membranes appear to be fused. These areas of fusion, or **closed fenestrae,** generally face toward the epithelial cells and facilitate uptake of absorbed materials by the capillaries. In addition, intestinal capillaries have a higher capillary filtration coefficient than other major organ systems, and this probably enhances the uptake of water absorbed by the villi (Chap. 16). However, large molecules, such as plasma proteins, do not easily cross the fenestrated areas because the reflection coefficient for the bowel vasculature is greater than 0.9, about the same as in skeletal muscle (Chap. 16).

■ Low Capillary Pressures in Intestinal Villi Aid in Absorption of Water

Though the mucosal layer of the small intestine has a high blood flow both at rest and during absorption of food, the capillary blood pressure is usually 13–18 mm Hg and seldom higher than 20 mm Hg during food absorption. Therefore, plasma colloid osmotic pressure is higher than capillary blood pressure favoring absorption of water brought into the villi. During lipid absorption, the plasma protein reflection coefficient for the overall intestinal vasculature is decreased from a normal value of over 0.9 to about 0.7. It is assumed that most of the decrease in reflection coefficient occurs in the mucosal capillaries. This lowers the ability of plasma proteins to counteract capillary filtration; the highest rates of intestinal lymph formation normally occur during fat absorption.

■ Sympathetic Nerve Activity Can Greatly Decrease Intestinal Blood Flow and Venous Volume

The intestinal vasculature is richly innervated by the sympathetic nervous system. Major reductions in gastrointestinal blood flow occur whenever sympathetic nerve activity is increased, such as during strenuous exercise or periods of pathologically low arterial blood pressure. This is an important vascular response, because at rest about one-fourth of the cardiac output perfuses the gastrointestinal organs. Suppression of this blood flow by the sympathetic nervous system allows the needs of the gastrointestinal system to be sacrificed so that more vital functions can be supported with the available cardiac output. Gastrointestinal blood flow can be so dramatically decreased by a combination of low arterial blood pressure and sympathetically mediated vasoconstriction that mucosal tissue damage can result. In less drastic situations and during severe stress, constriction of venules and veins in the gastrointestinal system is a means to increase the effective circulating blood volume by moving blood from the abdominal venous vasculature to the general circulation.

Hepatic Circulation

The functions of the liver are discussed in Chapter 30. The human liver has a large blood flow—1.2–1.5 L/min, about 25% of resting cardiac output. It is perfused by both arterial blood through the hepatic artery and venous blood that has passed through the stomach, small intestine, pancreas, spleen, and portions of the large intestine. The venous blood arrives via the hepatic portal vein and accounts for about 67–80% of the total liver blood flow (Table 17–1). The remaining 20–33% of the total flow is through the hepatic artery (Table 17–1). About half of the oxygen used by the liver is derived from venous blood, even though one-third to one-half of the available oxygen has been removed by the splanchnic organs (Table 17–1). This does not cause a problem because the liver tissue has an unusually good ability to extract oxygen from blood. The remainder of the oxygen used is extracted from the much smaller blood flow provided by the hepatic artery. The liver has a high metabolic rate and is a large organ (Table 17–1), consequently it has the largest total oxygen consumption at rest of all organ systems.

■ The Liver Acinus Is a Complex Microvascular Unit with Mixed Arteriolar and Venular Blood Flow

The liver vasculature is arranged into subunits that allow the arterial and portal blood to mix and provide nutrition for the liver cells. Each subunit is called an **acinus** and is about 300–350 μm long and wide. In humans, usually three acini occur together. The core of each acinus (Fig. 17–4) is perfused by a single **terminal portal venule,** and capillaries **(sinusoids)** originate from the portal venule. The endothelial cells of the capillaries have fenestrated regions with discrete openings that allow easy exchange between the plasma and interstitial environment. The capillaries do not have a basement membrane, which partially contributes to the high permeability of the liver vasculature. The **terminal hepatic arteriole** to each acinus is paired with the terminal portal venule at the acinus core, and blood flow in the arteriole and venule jointly perfuse the capillaries. The intermixing of the arterial and portal blood tends to be intermittent, as the vascular smooth muscle of the small arteriole alternately constricts to decrease flow and then relaxes to increase flow. This prevents arteriolar pressure from causing a sustained reversed flow in the venular sinusoids, where pressures are 7–10 mm Hg. The sinusoidal capillaries are drained by the **terminal hepatic venules** at the outer margins of each acinus; usually at least two hepatic venules drain each acinus.

■ Regulation of Hepatic Arterial and Portal Venous Blood Flows Requires an Interactive Control System

Regulation of portal venous and hepatic arterial blood flow is interactive because hepatic arterial flow increases and decreases reciprocally with the portal venous blood flow. This mechanism, known as the **hepatic arterial buffer response,** can compensate or buffer about one-fourth of the portal blood flow change. Exactly how this is accomplished is still under investigation, but vasodilatory metabolite accumulation, possibly adenosine, during decreased portal flow and increased metabolite removal during elevated portal flow are suspected to influence the resistance of the hepatic arterioles.

One might suspect that during digestion, when gastrointestinal blood flow and therefore portal venous blood flow are increased, the gastrointestinal hormones released into portal venous blood would influence hepatic vascular resistance. However, at con-

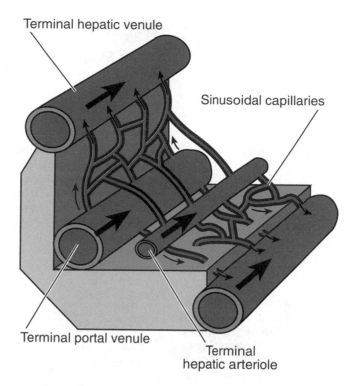

Figure 17–4 ■ A single liver acinus, the basic subunit of liver structure, is perfused from the inside to the outside by a terminal portal venule and a terminal hepatic arteriole. Most of the mixture of portal venous and arterial blood occurs in the sinusoidal capillaries formed from the terminal portal venule. Usually two terminal hepatic venules drain the sinusoidal capillaries at the external margins of each acinus.

centrations in portal venous blood equivalent to those during digestion, none of the major hormones appears to influence hepatic blood flow. Therefore, the increased hepatic blood flow during digestion would appear to be determined primarily by vascular responses of the gastrointestinal vasculatures.

The vascular resistances of the hepatic arterial and portal venous vasculatures are increased during sympathetic nervous system activation and the buffer mechanism is suppressed. When the sympathetic nervous system is activated, about half of the blood volume of the liver can be expelled into the general circulation. Since up to 15% of the total blood volume is in the liver, constriction of the hepatic vasculature can increase significantly the circulating blood volume during times of cardiovascular stress.

Skeletal Muscle Circulation

■ Skeletal Muscle Metabolism and Blood Flow Are Greatly Increased During Exercise

The skeletal muscle circulation is important because of the large mass of tissue involved—30–40% of body weight. Its resistance can be changed over a large range (Table 17–1). At rest, the skeletal muscle vasculature accounts for about 25% of systemic vascular resistance; the dominant mechanism controlling this resistance is the sympathetic nervous system. Resting skeletal muscle has a remarkably low oxygen consumption per 100 g of tissue, but its large mass makes its metabolic rate a major contributor to the resting oxygen consumption (Table 17–1). During maximal dilation associated with high-performance aerobic exercise, skeletal muscle blood flow can increase 10–20-fold or more (Table 17–1), and comparable increases in metabolic rate occur. Under such circumstances, total muscle blood flow can amount to about three times the normal cardiac output (Table 17–1); obviously, cardiac output must increase during exercise to maintain the normal to increased arterial pressure. With severe hemorrhage or other baroreceptor-induced reflexes, skeletal muscle vascular resistance can easily double due to increased sympathetic nervous activity and dramatically reduce blood flow with minimal risk to the muscle cells. The increased vascular resistance helps preserve arterial blood pressure when cardiac output is compromised. In addition, contraction of the skeletal muscle venules and veins forces blood in these vessels to enter the general circulation and help restore a depleted blood volume. In effect, the skeletal muscle vasculature can either place major demands on the cardiopulmonary system during exercise or perform as if expendable during a cardiovascular crisis, so that absolutely essential tissues can be perfused with the available cardiac output.

■ Regulation of Muscle Blood Flow Depends on Many Types of Mechanisms to Improve Oxygen Delivery

As mentioned in Chapter 16, there are many potential local regulatory mechanisms to adjust blood flow to the metabolic needs of the tissue. For fast-twitch muscle types, which primarily depend on anaerobic metabolism, the accumulation of hydrogen ions from lactic acid is a potentially major contributor to the small amount of vasodilation that occurs. In slow-twitch skeletal muscles, which can easily increase oxidative metabolic requirements by over 10–20 times during heavy exercise, it is not hard to imagine that whatever causes metabolically linked vasodilation is in ample supply at high metabolic rates. During rhythmic muscle contractions the blood flow during the relaxation phase can be very high, and it is unlikely that the muscle becomes significantly hypoxic during exercise. As evidence, studies in humans and animals indicate that lactic acid formation, an indication of some degree of hypoxia and anaerobic metabolism, is present only during the first several minutes of submaximal exercise. Once the vasodilation and increased blood flow associated with exercise are established, after 1–2 minutes, the microvasculature is probably capable of maintaining ample oxygen for most workloads, perhaps up to 75–80% of maximum performance, because remarkably little additional lactic acid is accumulated in blood. This may mean the tissue oxygen content does decrease but the reduction does not compromise the high aerobic metabolic rate except with the most demanding forms of exercise. A model based on actual measurements of P_{O_2} in animals during muscle contractions (Fig. 16–7) demonstrates the various changes in oxygen tension during and after a period of muscle contractions. To ensure the best possible supply of nutrients, particularly oxygen, even mild exercise causes sufficient dilation to perfuse virtually all of the capillaries, rather than just 60–70% of capillaries, as occurs at rest. However, there is no doubt that near-maximum to maximum exercise exhausts the microvasculature's ability to supply tissue oxygen requirements and hypoxic conditions rapidly develop and limit the performance of the muscles. The burning sensation in and fatigue of muscle tissue during maximum exercise or any time muscle blood flow is inadequate to supply adequate oxygen is partially a consequence of hypoxia.

Dermal Circulation

■ The Skin Has a Microvascular Anatomy to Support both Tissue Metabolism and Dissipation of Heat

The anatomic structure of the skin vasculature differs with body location. In all areas, an arcade of arterioles exists at the boundary of the dermis, the deepest skin layer, and the subcutaneous tissue over fatty tissues and skeletal muscles (Fig. 17–5). From this arteriolar arcade, arterioles ascend through the dermis into the superficial layers of the dermis, adjacent to the epidermal layers. These arterioles form a second network in the superficial dermal tissue and perfuse the extensive capillary loops that extend upward into the dermal papillae just beneath the epidermis. The der-

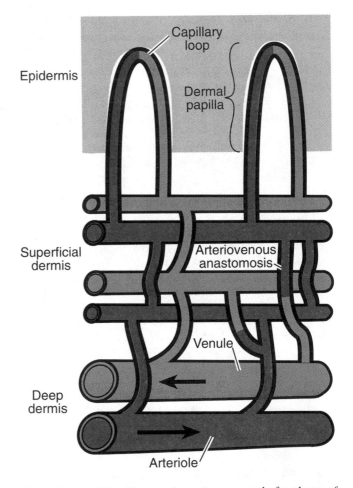

Figure 17–5 ■ The skin vasculature is composed of a plexus of large arterioles and venules in the deep dermis, which send branches to the superficial plexus of smaller arterioles and venules. Arteriovenous anastomoses allow direct flow from arterioles to venules and greatly increase blood flow when dilated. The capillary loops into the dermal papillae beneath the epidermis are perfused and drained by microvessels of the superficial dermal vasculature.

mal vasculature also provides the vessels that surround the hair follicles, sebaceous glands, and sweat glands. Sweat glands derive virtually all sweat water from blood plasma and are surrounded by a dense capillary network in the deeper layers of the dermis. As is explained in Chapter 31, neural regulation of the sweating mechanism not only causes the formation of sweat but also substantially increases skin blood flow. All of the capillaries from the superficial skin layers are drained by venules, which form a venous plexus in the superficial dermis and eventually drain into many small veins and large venules beneath the dermis. The vascular pattern just described is modified in the tissues of the hand, feet, ears, nose, and some areas of the face in that direct vascular connections between arterioles and venules, **arteriovenous anastomoses,** occur primarily in the superficial dermal tissues (Fig. 17-5). In comparison, relatively few arteriovenous anastomoses exist in the major portion of human skin over the limbs and torso. If a great amount of heat must be dissipated, dilation of the arteriovenous anastomoses allows substantially increased skin blood flow to warm the skin and thus, increase heat loss to the environment. This allows vasculatures of the hands and feet and, to a lesser extent, the facial vasculature to lose heat efficiently in a warm environment.

■ Skin Blood Flow Is Important in Body Temperature Regulation

The skin is a large organ, representing 10-15% of total body mass. The primary functions of the skin are protection of the body from the external world and dissipation or conservation of heat during body temperature regulation. The skin has one of the lowest metabolic rates in the body and requires relatively little blood flow for purely nutritive functions. Consequently, despite its large mass it does not place a major flow demand on the cardiovascular system. However, regulation of body temperature requires that warm blood from the body core be carried to the external surface, where heat transfer to the environment can occur. Therefore, at typical indoor temperatures and during warm weather, skin blood flow is usually far in excess of the needs for tissue nutrition. The reddish color of the skin during exercise in a warm environment reflects the large blood flow and dilation of skin arterioles and venules (Table 17-1). The increase in skin blood flow probably occurs through two main mechanisms. First, an increase in body core temperature causes a reflex increase in activity of sympathetic cholinergic nerve fibers, which release acetylcholine. Acetylcholine released near sweat glands leads to the breakdown of a protein to form bradykinin, a potent dilator of skin blood vessels. Simply increasing skin temperature will cause the blood vessels to dilate. This can result from heat applied to the skin from the external environment, heat from underlying active skeletal muscle, or increased blood temperature as it enters the skin. Total blood flows of 5-8 L/min have been estimated in humans during vigorous exercise in a hot environment. During mild to moderate exercise in a warm environment, skin blood flow can equal or exceed blood flow to the skeletal muscles. Exercise tolerance can therefore be lower in a warm environment, because the sum of the skin and muscle blood flows can approach or exceed maximum cardiac output.

The vast majority of humans live in cool to cold regions, where body heat conservation is imperative. The sensation of cool or cold skin or a lowered body core temperature elicits a reflex increase in sympathetic nerve activity, which causes vasoconstriction of skin blood vessels. The increased skin vascular resistance decreases skin blood flow to a minimum. Heat loss is minimized because the skin becomes a poorly perfused insulator, rather than a heat dissipator. So long as the skin temperature is above about 50-55°F, the neurally induced vasoconstriction is sustained. However, at lower tissue temperatures, the vascular smooth muscle cells lose their contractile ability and the vessels passively dilate to various extents. The reddish color of the hands, face, and ears on a cold day demonstrates increased blood flow and vasodilation due to low temperatures. To some extent, this cold-mediated vasodilation is useful in that it lessens the chance of cold injury to exposed skin. However, were this process to include most of the body surface, such as occurs when the body is submerged in cold water or inadequate clothing is worn, heat loss would be rapid and hypothermia would result. Chapter 31 discusses skin blood flow and temperature regulation.

Fetal and Placental Circulation

■ The Placenta Has Maternal and Fetal Circulations and Allows Exchange between the Mother and Fetus

The development of a human fetus depends on nutrient, gas, water, and waste exchange in the maternal and fetal portions of the placenta. The human **fetal placenta** is supplied by the umbilical arteries, which branch from the internal iliac arteries, and is drained by the **umbilical vein** (Fig. 17-6). The microvasculature of the fetal portion of the human placenta is composed of structures that resemble a fir tree: an array of large villuslike branches extend from a central core, and the number and length of branches decreases as the core approaches the deep structures of the maternal placenta. In some cases, the terminal core portions of the fetal structures attach to the maternal tissues. During development of the fetal placenta, the fetal tissues invade and cause partial degeneration of the maternal endometrial lining of the uterus. The

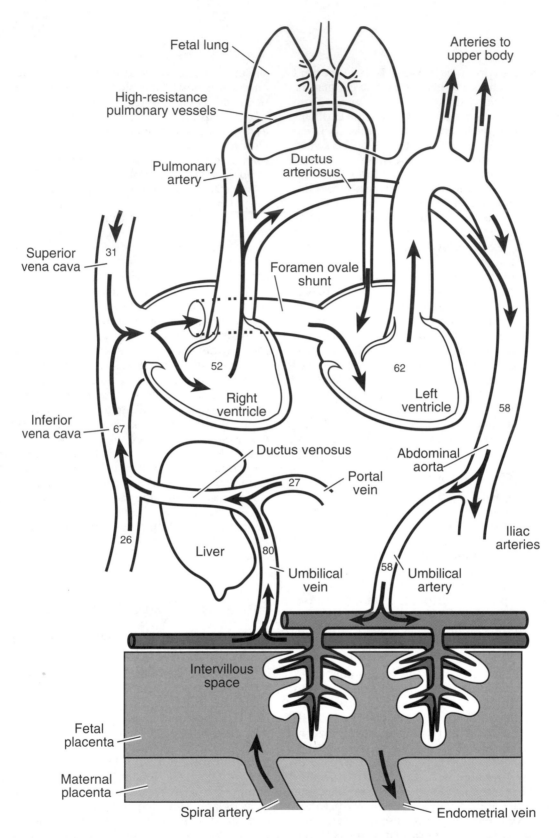

Figure 17–6 ■ Schematic drawing of the left and right sides of the fetal heart, which are separated to emphasize that a right to left shunt of blood occurs through the open foramen ovale in the heart and a second right to left shunt occurs through the ductus arteriosus. The numbers in vessels represent the percent saturation of blood with oxygen in the fetal circulation. Mixing of oxygenated umbilical vein blood with fetal venous blood lowers the oxygen content of blood entering the heart. Mixing is not complete, however, and the higher oxygenated blood from the placenta preferentially passes through the foramen ovale, allowing improved oxygenation of the coronary and cerebral vascultures (see text). Arrows indicate the direction of blood flow. Closure of the ductus venosus, foramen ovale, ductus arteriosus, and placental vessels at birth and dilation of the pulmonary vasculature establish the adult circulation pattern.

result, after about 10–16 weeks of gestation, is an **intervillous space** between fetal placental villi that is filled with maternal blood. There are no maternal structures in the intervillous space that resemble the usual microvessels, but there is rather a cavernous blood-filled space. The intervillous space is supplied by 100–200 **spiral arteries** of the maternal endometrium and is drained by the **endometrial veins.** During gestation, the spiral arteries simultaneously enlarge in diameter and lose their vascular smooth muscle coat, so it is the arteries preceding them that actually regulate blood flow through the placenta. At the end of gestation, the total maternal blood flow to the intervillous space is approximately 600–1000 ml/min, which represents about 15–25% of resting cardiac output. In comparison, the fetal placenta has a blood flow of about 600 ml/min, which represents about half of the fetal cardiac output.

The Placental Vasculature Permits Very Efficient Exchange of O_2 and CO_2

In a normal placenta, there is virtually no exchange of fetal and maternal blood, although a fraction of 1 ml/day may inadvertently exchange between the fetus and mother. The absence of blood exchange requires that all movements of molecules between fetal and maternal blood occur through the fetal placental tissues. The fetal placenta allows simple diffusion for small molecules, passive carrier-mediated exchange for glucose, and active transport of amino acids and iron. Virtually any compound in maternal blood can cross into the fetal circulation, and this allows antibodies from the mother to establish passive immunity against most infectious organisms in the newborn infant. Noxious substances and the vast majority of legal and illegal drugs can gain access to the fetus, all too often resulting in fetal abnormalities.

Placental exchange between mother and fetus is not perfect. This is not a problem for nutrients used to build the fetal body and support metabolism, because the rates at which these compounds are needed are low. However, special fetal adaptations are required for gas exchange, particularly oxygen, because of the limitations of passive exchange across the placenta. The P_{O_2} of maternal arterial blood is about 80–100 mm Hg and of incoming blood in the umbilical artery is about 20–25 mm Hg. This difference in oxygen tension provides a large driving force for exchange; the result is an increase in the fetal blood P_{O_2} to 30–35 mm Hg in the umbilical vein. Fortunately, **fetal hemoglobin** carries as much or more oxygen at its normally low P_{O_2} than adult hemoglobin carries at a P_{O_2} two to three times higher. In addition, fetal blood has about 20% more hemoglobin than adult blood, per volume. The net result is that the fetus has sufficient oxygen to support its metabolism and growth but does so at very low oxygen tensions, using the unique properties of fetal hemoglobin. After birth, when much more efficient oxygen exchange can occur in the lung, the newborn gradually replaces the red cells containing fetal hemoglobin with red cells that contain adult hemoglobin.

The Absence of Lung Ventilation Requires a Unique Circulation through the Heart and Body of the Fetus

After the umbilical vein leaves the fetal placenta, it passes through the abdominal wall at a site that will in neonatal life be the umbilicus (navel). The umbilical vein enters the liver's portal venous circulation, although the bulk of the oxygenated venous blood passes directly through the liver in the **ductus venosus** (Fig. 17-6). As the low-oxygen-content venous blood from the lower body and the high-oxygen-content placental venous blood mix in the inferior vena cava, the oxygen content of the blood returning from the lower body is increased to a level about twice that of venous blood returning from the upper body in the superior vena cava. The two streams of blood from the superior and inferior vena cava do not completely mix as they enter the right atrium. The net result is that oxygen-rich blood from the inferior vena cava passes through the open **foramen ovale** in the atrial septum to the left atrium, while the upper body blood generally enters the right ventricle just as in the adult. The preferential passage of oxygenated venous blood into the left atrium and minimal amount of venous blood returning from the lung to the left atrium allows blood in the left ventricle to have an oxygen content about 20% higher than that in the right ventricle. This is advantageous because this relatively high-oxygen-content blood supplies the coronary vasculature, head, and brain.

Blood pumped by the right ventricle into the pulmonary artery is substantially shunted into the aorta through the **ductus arteriosus.** However, this blood, with a lower oxygen content, empties into the aorta after the ascending aorta has sent oxygen-rich blood to the head, upper body, and coronary circulations. This ensures that the heart and brain receive blood with a higher oxygen content than does the remainder of the body. For ductus arteriosus blood to enter the initial part of the descending aorta, the right ventricle must develop a higher pressure than the left ventricle, the exact opposite of circumstances in the adult. Perfusion of the collapsed lungs of the fetus is minimal because the pulmonary vasculature has a high resistance. The elevated pulmonary resistance occurs because the lungs are not inflated and probably because the pulmonary vasculature has the unusual characteristic of vasoconstriction at low oxygen tensions. The result for the heart is that the right ventricular muscle is hypertrophied at birth, but this hypertrophy gradually subsides over a period of months due to the lowering of the pulmonary resistance. However, the higher systemic arterial pressure due to increased vascular resistance after birth leads to substantial hyper-

trophy of the left ventricular muscle over the same time period.

The Transition from Fetal to Neonatal Life Involves a Complex Sequence of Cardiovascular Events

After the fetus is delivered and the initial ventilatory movements cause the lungs to expand with air, pulmonary vascular resistance is dramatically decreased. At this point, the right ventricle can perfuse the lungs and the circulation pattern in the fetus switches to that of adults. The vascular resistance of the lungs dramatically decreases on inflation and allows the right ventricle to perfuse the pulmonary circulation. The net result is that now oxygen-rich blood returns to the left atrium. The increased oxygen tension in the aortic blood may provide the signal for closure of the ductus arteriosus, although suppression of vasodilator prostaglandins cannot be discounted. In any event, the ductus arteriosus constricts to virtual closure and over time becomes anatomically fused. The closure of the ductus arteriosus allows passage of the entire right ventricle cardiac output through the lungs to the left atrium. This increases the left atrial pressure, as does increased resting distention of the left ventricle due to the higher blood flow and increased arterial resistance of the systemic vasculature. The net effect is that left atrial pressure exceeds right atrial pressure, and this passively pushes the tissue flap on the left side of the foramen ovale against the open atrial septum. In time, the tissues of the atrial septum fuse, but an anatomic passage (probe patent) that is probably only passively sealed can be documented in some adults. The ductus venosus in the liver is patent for several days after birth but gradually closes and within 2–3 months is obliterated. After the fetus begins breathing, the fetal placental vessels and umbilical vessels undergo progressive vasoconstriction to force placental blood into the fetal body and minimize the possibility of fetal hemorrhage through the placental vessels.

The final act of gestation is separation of the fetal and maternal placenta as a unit from the lining of the uterus. The separation process begins almost immediately after the fetus is expelled, but external delivery of the placenta can require up to 30 minutes. The separation occurs along the **decidua spongiosa,** a maternal structure, and requires that blood flow in the spiral arteries be stopped. The cause of the placental separation may be mechanical, as the uterus surface area is greatly reduced by removal of the fetus and shear forces are placed on the placental structures. Normally about 500–600 ml of maternal blood is lost in the process of placental separation. However, as maternal blood volume increases 1000–1500 ml during gestation, normally this blood loss is not of significant concern.

■ ■ ■ ■ ■ ■ ■ ■ ■ ■ ■ REVIEW EXERCISES ■ ■ ■ ■ ■ ■ ■ ■ ■ ■ ■

Identify Each with a Term

1. The increase in blood flow that can occur when heart workload is increased
2. The increase in intestinal blood flow during digestion and absorption of food molecules
3. The basic tissue and vascular unit of the liver that receives both arteriolar and portal venular blood flows
4. The tissue area between the fetal and maternal placenta that is filled with maternal blood

Define Each Term

5. Countercurrent flow
6. Hepatic arterial buffer response
7. Arteriovenous anastomoses

Choose the Correct Answer

8. Coronary blood flow would be expected to follow which of following response patterns if the work of the heart increased?
   a. Minor increase in blood flow but substantial increase in oxygen extraction from the coronary blood
   b. Increase in blood flow in approximately the same proportion as the relative increase in heart work
   c. No appreciable change in coronary blood flow unless heart work at least doubles
   d. Increase in blood flow and oxygen usage not in proportion to the increase in heart work

9. As the arterial pressure is raised and lowered during the course of everyday events, blood flow through the brain would be expected to:
   a. Change in the same direction as the arterial blood pressure, due to the limited autoregulatory ability of the cerebral vessels
   b. Change in a direction opposite to the change in mean arterial pressure
   c. Remain about constant, since cerebral vascular resistance changes in the same direction as arterial pressure
   d. Widely fluctuate, as both arterial pressure and brain neural activity status change

10. Which of the following is *not* a unique property of the intestinal microvasculature compared to other organ system vasculatures?
    a. The arterioles that primarily regulate vascular resistance are located far from the cells with a high metabolic rate.
    b. The high interstitial osmolality created during absorption may be a major stimulus for vasodilation.

c. The three distinct tissue layers have separate and independent microvasculatures.

d. The tissue surrounds all the arterioles but the arteries lie outside the organ.

11. Which of the following special circulations has the widest range of blood flows as part of its contributions to both regulation of systemic vascular resistance and modification of resistance to suit the organ's various metabolic needs?
    a. Heart
    b. Brain
    c. Small intestine
    d. Skeletal muscle

12. Which of the following sequences is a possible anatomic path for a red blood cell passing through a fetus from and back to the placenta? (some intervening structures are left out)
    a. Umbilical vein, right ventricle, ductus arteriosus, pulmonary artery
    b. Ductus venosus, foramen ovale, right ventricle, ascending aorta
    c. Spiral artery, umbilical artery, left ventricle, umbilical artery

d. Right ventricle, ductus arteriosus, descending aorta, umbilical artery

Briefly Answer

13. Progressive occlusion of a coronary artery often causes less severe clinical problems than an acute occlusion of the vessel that would decrease blood flow by the same amount. How would collateral circulation vessels influence both of these situations?

14. How does chronic hypertension affect the range of arterial pressure over which the brain circulation can maintain relatively constant blood flow?

15. What types of vessels are primarily responsible for regulation of coronary blood flow, and how does contraction of the heart muscle interact with these vessels to influence blood flow during systole and diastole?

16. Describe the arterial and venous blood flow into and out of the liver acinus. Why is this arrangement unique compared to other microvascular beds?

17. Why during fetal life is the oxygen content of blood sent to the upper body higher than that sent to the lower body?

ANSWERS

1. Coronary blood flow reserve

2. Absorptive hyperemia

3. Liver acinus

4. Intervillous space

5. Occurs when arterial and venous or arteriolar and venular blood flows in opposite directions in adjacent vessels, a common flow situation throughout the body that may be particularly important in the intestinal villi and kidney medulla

6. The reciprocal change in hepatic arterial blood flow when portal venous blood flow changes

7. Direct connections between arteries (or arterioles) and veins (or venules), which act as a shunt to bypass the capillaries; such vessels are found almost exclusively in the skin of the face, hands, and feet

8. b

9. c

10. c

11. d

12. d

13. For equivalent reductions in blood flow, progressive occlusion of a coronary artery often causes less severe clinical problems than acute occlusion because downstream vessels have time to enlarge and develop collateral vessels. Rapid onset of occlusion may stimulate the use and growth of collateral vessels, but their ability to improve flow would initially be limited.

14. The autoregulatory range is shifted to higher pressures, because the arteries and arterioles increase their resistance. These functional and structural changes increase the arterial pressure at which autoregulation of blood flow occurs but increase the lowest pressure at which blood flow can be maintained.

15. The small arteries and all of the arterioles are the primary vessels that regulate coronary blood flow. Most of these vessels are embedded in heart muscle tissue and are compressed by contraction of the heart muscle. This lowers blood flow during systole, but the heart muscle relaxation during diastole allows blood flow to increase.

16. The liver acinus is perfused by an arteriole from the hepatic artery but actually receives most of its blood flow from venules perfused by the portal vein. All other vascular beds, except that in the pituitary gland, receive only arterial blood. The venous drainage of the capillaries in the liver acinus is very similar to venular anatomy throughout the body.

17. Oxygenated blood from the placenta does not become fully mixed with blood returning for the superior vena cava and is mechanically diverted through the foramen ovale in the left atrium. Consequently, the oxygen content of blood in the ascending aorta is significantly greater than that in the ductus arteriosus. The upper body, brain, and coronary arteries are perfused by vessels that branch from the aorta before the ductus arteriosus empties less well oxygenated blood into the aorta to perfuse the lower body and fetal placenta.

Suggested Reading

Battaglia FC, Meschia G: *An Introduction to Fetal Physiology.* New York, Academic, 1986.

Bohlen HG, Maass-Moreno R, Rothe CF: Hepatic venular pressures of rats, dogs, and rabbits. *Am J Physiol* 261:G539–G547, 1991.

Chilian WM, Eastham CL, Layne SM, Marcus ML: Small vessel phenomena in the coronary microcirculation: Phasic intramyocardial perfusion and coronary microvascular dynamics. *Prog Cardiovasc Dis* 31:17–38, 1988.

Faraci FM, Baumbach GF, Heistad DD: Cerebral circulation: Humoral regulation and effects of chronic hypertension. *J Am Soc Nephrol* 1:53–57, 1990.

Feigl EO: Coronary autoregulation. *J Hypertension* 7:S55–S58, 1989.

Johnson JM: Nonthermoregulatory control of human skin blood flow. *J Appl Physiol* 61:1613–1622, 1986.

Laughlin MH, Armstrong RB: Muscle blood flow during locomotory exercise. *Exercise Sport Sci Rev* 13:95–136, 1985.

Lautt WW, Greenway CV: Conceptual review of the hepatic vascular bed. *Hepatology* 7:952–963, 1987.

Milnor WR: *Cardiovascular Physiology.* New York, Oxford University Press, pp. 387–428. 1992.

Paulson OB, Waldemar G, Schmidt JF, Strandgaard S: Cerebral circulation under normal and pathological conditions. *Am J Cardiol* 63:2C–5C, 1989.

Rowell LB: *Human Cardiovascular Control.* New York, Oxford University Press, 1993.

Schaper W, Gorge G, Winkler B, Schaper J: The collateral circulation of the heart. *Prog Cardiovasc Dis* 31:57–77, 1988.

Shepherd AP, Granger DN (eds): *Physiology of the Intestinal Circulation.* New York, Raven, 1984.

Control Mechanisms in Circulatory Function

OBJECTIVES

After studying this chapter, the student should be able to:
1. Describe the role of autonomic (sympathetic and parasympathetic) nerves in the control of the circulatory system
2. Describe how neural control of the cardiovascular system is regulated and integrated
3. Define the roles of various cardiovascular reflex "receptors," including baroreceptors, volume receptors, chemoreceptors, and pain receptors
4. Describe several illustrative corticohypothalamic responses (e.g., the fight-or-flight response, fainting, blushing, the diving response) and use them to achieve a better understanding of the integration involved in neurocirculatory control
5. Describe the regulatory effects of certain hormones on the cardiovascular system
6. Describe how mechanical, neural, and hormonal mechanisms are integrated during a "complex" cardiovascular maneuver such as standing

The mechanisms controlling the circulation can be divided into neural control mechanisms, hormonal control mechanisms, and local control mechanisms. Cardiac performance and vascular tone at any time are the result of integration of all three control mechanisms. To some extent this categorization is artificial, because each of the three categories affects the other two. This chapter deals with neural and hormonal mechanisms; local mechanisms are covered in Chapter 16.

Central blood volume and arterial pressure are normally maintained within narrow limits by neural and hormonal mechanisms. Adequate central blood volume is necessary to assure a proper cardiac output, and constant arterial blood pressure allows maintenance of tissue perfusion in the face of changes in regional blood flow. Neural control involves sympathetic and parasympathetic branches of the autonomic nervous system. Blood volume and arterial pressure are monitored by stretch receptors in the heart and arteries. Afferent nerve traffic from these receptors is integrated with other afferent information in the medulla, which leads to activity in sympathetic and parasympathetic nerves that adjusts cardiac output and systemic vascular resistance to maintain arterial pressure. Sympathetic nerve activity and, even more important, hormones such as antidiuretic hormone (ADH, or vasopressin), angiotensin, aldosterone, and atrial natriuretic peptide serve as effectors for the regulation of salt and water balance and blood volume. Neural control of cardiac output and systemic vascular resistance is most important in the minute-to-minute regulation of arterial pressure, whereas hormones are most important in the long-term regulation of arterial pressure.

In some situations, factors other than blood volume and arterial pressure regulation strongly influence cardiovascular control mechanisms. These situations include the defense response, diving, thermoregulation, standing, and exercise.

Autonomic Neural Control of the Circulatory System

Afferent information from the cardiovascular system (as well as other organ systems) and the external environment is integrated at multiple levels in the brain. It then generates efferent signals to the effector tissue of the cardiovascular system: atrial and ventricular muscle, pacemaking and conducting tissues of the heart, and vascular smooth muscle. All neural output to these tissues is via the parasympathetic and sympathetic components of the autonomic nervous system.

The autonomic control of the heart and blood vessels was described in Chapter 6. Briefly, the heart is innervated by parasympathetic (vagus) and sympathetic (cardioaccelerator) nerve fibers (Fig. 18-1). The parasympathetic fibers release acetylcholine, which binds to muscarinic receptors of the sinoatrial node, atrioventricular node, and specialized conducting tissues. Stimulation of parasympathetic fibers causes slowing of heart rate and conduction velocity. The ventricles are only sparsely innervated by parasympathetic nerve fibers, and stimulation of these fibers has little direct effect on cardiac contractility. Some cardiac parasympathetic fibers end on sympathetic nerves and inhibit the release of norepinephrine from sympathetic nerve fibers. Sympathetic fibers to the heart release noradrenaline, which combines with β_1-adrenergic receptors in the sinoatrial node, atrioventricular node and specialized conducting tissues, and atrial and ventricular muscle. Stimulation of these fibers causes increased heart rate, increased conduction velocity, and increased contractility. The two divisions of the autonomic nervous system tend to oppose each other in their effects on the heart, and activities along these two pathways usually change in a reciprocal manner. Tables 18-1 and 18-2 summarize the cardiovascular effects of stimulating each division.

Most arteries, arterioles, venules, and veins, with the exception of those of the external genitalia, receive sympathetic innervation only (Fig. 18-1). The usual transmitter is noradrenaline, which binds to α_1-adrenergic receptors and causes contraction of vascular smooth muscle and vasoconstriction. Circulating epinephrine, released from the adrenal medulla, combines with β_2-adrenergic receptors in vascular smooth muscle cells, especially coronary and skeletal muscle arterioles, producing vascular smooth muscle relaxation and vasodilation. Blood vessels of skeletal muscle receive sympathetic cholinergic innervation in addi-

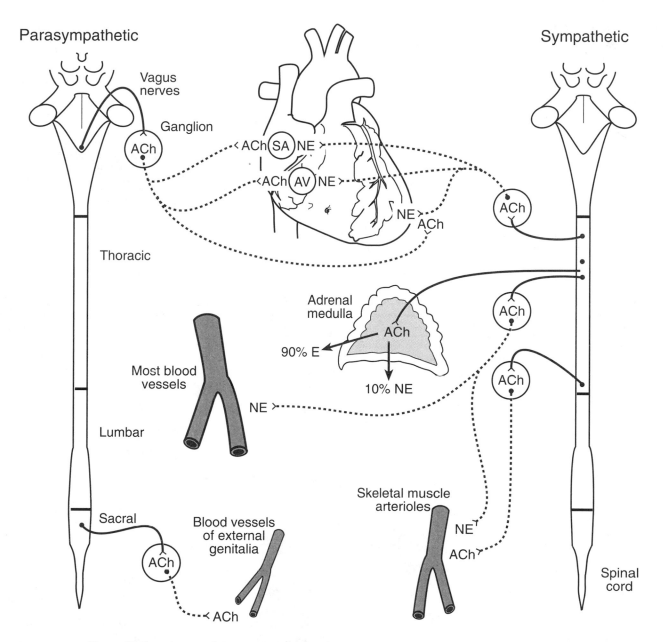

Figure 18–1 ■ Autonomic innervation of the cardiovascular system. ACh, acetylcholine; NE, norepinephrine; E, epinephrine; SA, sinoatrial node; AV, atrioventricular node.

tion to their sympathetic adrenergic innervation; stimulation of these cholinergic fibers via muscarinic receptors leads to smooth muscle relaxation and vasodilation.

The Central Nervous System in the Control of the Cardiovascular System

The central nervous system influences and/or regulates the activity of the autonomic nerves to the cardio-

vascular system. Interactions between the central and autonomic nervous systems take place at several anatomic sites.

■ Reflex Loops and Neural Integration Occur in the Spinal Cord

Humans have some spinal reflexes of cardiovascular significance. For example, the stimulation of pain fibers entering the spinal cord below the level of a cord transection can cause reflex vasoconstriction and increased blood pressure. In addition, neural pathways from higher centers converge at the level of the spinal

TABLE 18–1 ■ Cardiovascular Effects of Sympathetic Stimulation, by Receptor Type

Receptor Type	Location	Action	Physiologic Agonist
α_1	Smooth muscle of arteries and veins	Contraction	Norepinephrine from postganglionic sympathetics
α_2	Sympathetic nerve varicosities	Inhibition of norepinephrine release	Norepinephrine from postganglionic sympathetics
	Central nervous system	Inhibition of sympathetic neural output	Norepinephrine from central neurons
β_1	Sinoatrial node	Increased heart rate	Norepinephrine from postganglionic sympathetics
	Atrial and ventricular muscle	Increased conduction velocity and contractility	Norepinephrine from postganglionic sympathetics
	Atrioventricular node and Purkinje system	Increased conduction velocity	Norepinephrine from postganglionic sympathetics
β_2	Smooth muscle of arteries and veins	Relaxation	Epinephrine from adrenal medulla

These are the predominant locations, actions, and agonists but there are others. For example, atrial and ventricular muscle have α_1 receptors that bind norepinephrine, but their physiologic importance is minor.

TABLE 18–2 ■ Cardiovascular Effects of Parasympathetic Stimulation, by Receptor Type

Receptor Type	Location	Action	Physiologic Agonist
Nicotinic	Postganglionic neurons of autonomic nervous system	Excitation of postganglionic neurons	Acetylcholine from preganglionic neurons
Muscarinic	Sinoatrial node	Slowing of heart rate	Acetylcholine from postganglionic parasympathetics
	Atrioventricular node and specialized conducting tissues	Slowing of conduction velocity	Acetylcholine from postganglionic parasympathetics
	Arterioles of skeletal muscle[2]	Vasodilation[1]	Acetylcholine from postganglionic sympathetics
	Arterioles of heart, brain, and external genitalia*	Vasodilation[1]	Acetylcholine from postganglionic parasympathetics
	Most other blood vessels*	Vasodilation[1]	None known; acetylcholine does not ordinarily circulate, and there are no cholinergic nerves to these vessels
	Sympathetic varicosities	Inhibition of norepinephrine release	Acetylcholine from postganglionic parasympathetics

[1]In many cases, the muscarinic receptors that mediate vasodilation may be located in the plasma membrane of endothelial cells, not vascular smooth muscle cells.
[2]Of less importance in humans than some other species.

cord, allowing the final integration of various neural activities to occur there.

■ The Medulla Is a Major Area for Cardiovascular Reflex Integration

The major area for the integration of the simple reflex behavior for the cardiovascular system is the medulla oblongata. The medullary cardiovascular neurons are grouped into three distinct pools that lead to activation of sympathetic neurons to the blood vessels, sympathetic neurons to the heart, and parasympathetic neurons to the heart. The first two are often called the **vasomotor center** and the third the **cardioinhibitory center.** All three neuron pools interconnect extensively; it is on this level that the "dual innervation" firing of parasympathetic and sympathetic fibers to the heart is actually integrated. Because of this integration, the entire area may be called the

medullary cardiovascular center, which does not distinguish between the various subdivisions.

■ Corticohypothalamic Pathways Are the Highest Centers for Cardiovascular Control

The highest levels of organization in the autonomic nervous system are the **corticohypothalamic** pathways, which orchestrate cardiovascular correlates to different patterns of behavior. For example, in cats stimulation of certain areas in the hypothalamus produces a characteristic rage response (fight-or-flight response; Chap. 6), with spitting, clawing, tail lashing, arched back, and so on. Accompanying characteristic cardiovascular responses, including tachycardia and elevated blood pressure, are also invoked. If certain connections between the hypothalamus and the cerebral cortex are severed, cats will exhibit this rage reaction even when they are not being threatened. Their reaction apparently is due to the loss of pathways that normally exert inhibitory influences on the hypothalamic center. The defense response and other examples of combined corticohypothalamic behavior and cardiovascular events are detailed below. The important points here are that multiple connections between the cerebral cortical, hypothalamic, medullary, and spinal levels of the autonomic nervous system exist and that corticohypothalamic pathways integrate many behavioral and cardiovascular patterns involved in such activities as fighting, eating, diving, and thermoregulation.

Reflex Behavior in the Cardiovascular System

The most important reflex behavior of the cardiovascular system originates in the stretch or mechanoreceptors located in various blood vessels and the atria of the heart. The stretch receptors are positioned in the walls of these tissues such that increases in volume in the lumen will stretch the receptors along with the rest of the wall. Because of the high compliance of the atria, large volume changes are accompanied by relatively small pressure changes; the stretch receptors are therefore called **volume receptors.** Stretch receptors in the arteries (where compliance is lower) are called **pressure receptors, pressoreceptors,** or **baroreceptors** because small changes in volume accompany relatively large changes in pressure.

■ Baroreceptors Are Stretch Receptors that Sense Arterial Pressure

The **carotid sinus** and **aortic baroreceptors** are exceedingly important in the rapid, short-term regulation of arterial blood pressure. The carotid sinus is a slight dilation of the internal carotid artery located near its origin above the bifurcation of the common carotid artery. Baroreceptors are located in the wall of the carotid sinus. The aortic baroreceptors are located in the wall of the arch of the aorta and function much like the carotid sinus baroreceptors. The latter have been studied in greater detail because they are more accessible; their effects on the medullary cardiovascular center are emphasized here. The afferent nerve fibers from the carotid sinus baroreceptors run to the medullary cardiovascular center and higher areas of the brain. These receptors are innervated by the sinus nerves, which are branches of cranial nerve IX (glossopharyngeal nerves); the aortic baroreceptors send impulses centrally via vagus nerve fibers.

The Effects of Pressure Changes on Baroreceptors ■ Elevated mean arterial pressure increases pressure in the carotid sinuses and stretches carotid sinus baroreceptors, which, in turn, increase their firing rate. The increased action potential traffic to the medullary cardiovascular center subsequently increases parasympathetic neural activity to the heart and de-

creases sympathetic neural activity to the heart and resistance vessels (primarily arterioles) (Fig. 18–2). These changes in neural activity result in a decrease in cardiac output and resistance. Since mean arterial pressure is the product of systemic vascular resistance and cardiac output (Chap. 12), mean arterial pressure is returned toward the control value. This completes a negative feedback loop by which increases in mean arterial pressure can be attenuated. Conversely, decreases in carotid sinus pressure (and decreased stretch of the baroreceptors) increase sympathetic and decrease parasympathetic neural activity, resulting in increased heart rate, stroke volume, and systemic vascular resistance; this returns blood pressure toward control. If the fall in mean arterial pressure is very large, there is an increase in sympathetic neural activity to veins. This causes contraction of the venous smooth muscle and reduces venous compliance. The decreased venous compliance shifts blood toward the central blood volume, increasing right atrial pressure and, in turn, stroke volume.

The Effects of Increased Baroreceptor Firing on Various Organs ■ Not all resistance vessels are equally affected by the sympathetic neural discharge initiated by the baroreceptor reflex. For example, brain and coronary arteries are less affected than skeletal muscle or splanchnic or cutaneous arteries. Brain blood flow decreases only a small amount and coronary blood flow actually increases (despite the increased sympathetic vasoconstrictor activity), due to the increased metabolic activity of the heart caused

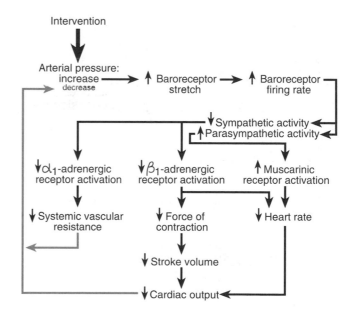

Figure 18–2 ■ Effect of the baroreceptor reflex on increased arterial pressure. An intervention elevates arterial pressure (either mean arterial pressure or pulse pressure), changing baroreceptor stretch and initiating the reflex. The final effect of reduced systemic vascular resistance and cardiac output is to return arterial pressure toward the level existing before intervention.

by the increased heart rate and stroke volume. Because of this differential effect, the maintenance of blood pressure by the baroreceptor reflex does not interfere appreciably with the blood supply to the two most vital organs: the heart and the brain.

Pressure Range for Baroreceptors ■ The effective range of the arterial baroreceptor mechanism (Fig. 18-3) is approximately 40 mm Hg (where the receptor stops firing) to 180 mm Hg (where the firing rate reaches a maximum). The firing rate of the baroreceptors increases with pulse pressure for a given mean arterial pressure. Aortic baroreceptors do not fire until a pressure of approximately 100 mm Hg is reached.

Adaptation of Baroreceptors ■ An important property of the baroreceptor reflex is that it adapts over a period of 1–2 days to the prevailing mean arterial pressure. Consequently, if the mean arterial pressure is artificially held at an elevated level, the tendency for the baroreceptors to initiate a decrease in cardiac output and systemic vascular resistance will quickly disappear. This occurs in part because sustained hypertension causes a reduction in the rate of baroreceptor firing for a given mean arterial pressure (Fig. 18-3). This is an example of **receptor adaptation.** There is also a "resetting" of the reflex in the central nervous system. Consequently, the baroreceptor mechanism is the "first line of defense" in the maintenance of normal blood pressure (and makes possible the rapid control of blood pressure needed with changes in posture and exercise), but it cannot provide for long-term control of blood pressure.

■ Volume Receptors Are Stretch Receptors that Sense Atrial Blood Volume

The atrial stretch (volume) receptors are located at the junction of the atria and the great veins emptying into them—the vena cava in the case of the right atrium and the pulmonary veins in the case of the left atrium. These receptors increase their firing rate when the volume of the atria increases and when the atria contract. The afferent fibers arising from these receptors travel in the vagus nerves to the medulla. The atrial volume receptors project to the medullary cardiovascular center, as do the arterial baroreceptors. Stretching of the atrial receptors results in increased sympathetic neural activity to the sinoatrial node (increased heart rate) and decreased sympathetic activity to the kidneys (Fig. 18-4). When venous return is suddenly augmented, activation of these stretch receptors causes elevated cardiac output and renal blood flow, because heart rate and cardiac filling are increased and renal vascular resistance is reduced. The lower renal vascular resistance causes increased glomerular filtration and, along with other mechanisms, raises salt and water excretion.

The atrial volume receptors also project to the area

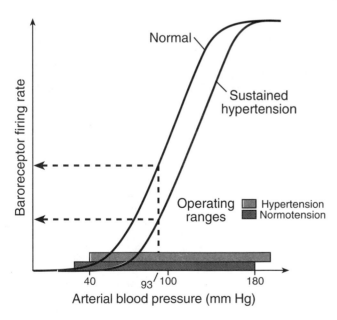

Figure 18–3 ■ Effect of mean arterial pressure on baroreceptor nerve firing rate. Under normal conditions, a mean arterial pressure of 93 mm Hg is near the midrange of the firing rates for the nerves. Sustained hypertension causes the operating range to shift to the right, putting 93 mm Hg at the lower end of the firing range for the nerves.

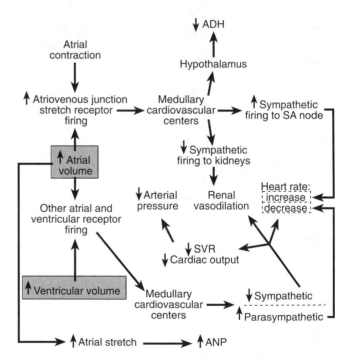

Figure 18–4 ■ Effect of increased atrial and ventricular volume on the cardiovascular system. Stretching of volume receptors causes renal vasodilation, increased atrial natriuretic peptide (ANP) secretion, and decreased antidiuretic hormone (ADH) secretion. It also causes peripheral vasodilation (decreased systemic vascular resistance; SVR) and decreased ventricular contractility. Effects on heart rate are variable because of conflicting effects between atriovenous junction receptors and other receptors distributed throughout the atria and ventricles.

HYPOTENSION

Baroreceptors, volume receptors, chemoreceptors, and pain receptors all help maintain adequate blood pressure during various forms of hemodynamic stress, such as standing and exercise. However, in the presence of certain cardiovascular abnormalities, these mechanisms may fail to regulate blood pressure appropriately; when this occurs, a person may experience transient or sustained **hypotension.** As a practical definition, hypotension exists when symptoms are caused by low blood pressure, and in extreme cases hypotension may cause weakness, lightheadedness, or even syncope.

Hypotension may be due to **neurogenic** or **non-neurogenic** factors. Neurogenic causes include **autonomic dysfunction** or **failure** (which can occur in association with other central nervous system abnormalities, such as Parkinson's disease, or may be secondary to systemic diseases that can damage the autonomic nerves, such as diabetes or amyloidosis), vasovagal hyperactivity, hypersensitivity of the carotid sinus, drugs with sympathetic stimulating or blocking properties, and tumors that produce adrenergically active compounds (e.g., pheochromocytoma). Non-neurogenic causes of hypotension include vasodilation (due to drugs such as alcohol, vasodilating drugs, or fever), cardiac disease (e.g., cardiomyopathy, valvular disease), or reduced blood volume (secondary to hemorrhage, dehydration, or other causes of fluid loss). In many patients multiple factors are involved. The treatment of symptomatic hypotension is to eliminate the underlying cause whenever possible; in some cases this leads to satisfactory results. When this is impossible, other adjunctive measures may be necessary, especially when the symptoms are disabling. Common treatment modalities include avoidance of factors that precipitate hypotension (e.g., sudden changes in posture, hot environments, alcohol, certain drugs, large meals); volume expansion (using salt supplements and/or medications with salt-retaining/volume-expanding properties); and mechanical measures (including tight-fitting thigh-elastic compression stockings to prevent the blood from pooling in the veins of the legs on standing). Unfortunately, even when these measures are employed, some patients continue to have severe, debilitating symptoms due to hypotension.

---

of the hypothalamus, where release of ADH is controlled (Chap. 34). Antidiuretic hormone causes the kidney to retain water, which contributes to increasing blood volume. Stretch of atrial receptors results in decreased ADH release, which, along with the increased glomerular filtration rate mentioned above, contributes to water excretion and reduced blood volume. A reduction in blood volume along with the increase in cardiac output resulting from increased heart rate and ventricular filling returns atrial volume toward normal. This completes a negative feedback loop that began with increased atrial volume.

Other stretch receptors in the atria and ventricles have the same effect as arterial baroreceptors: an increase in stretch causes an increase in parasympathetic outflow and a decrease in sympathetic outflow. These changes lower heart rate and cause reduced systemic vascular resistance by dilating peripheral arterioles.

Because of the opposing effects of different atrial stretch receptors, the overall effect of increased atrial volume on heart rate is variable. A sudden increase in venous return and atrial volume tends to elevate heart rate transiently. A more sustained increase in atrial volume is usually accompanied by increased cardiac output and systemic arterial pressure. The combination of increased atrial volume and arterial pressure results in reduced heart rate and systemic vascular resistance.

In summary, atrial stretch always leads to reduced systemic vascular resistance but to variable effects on heart rate, although usually the steady-state effect on heart rate parallels the effect of baroreceptor stretch receptors on heart rate, so the two types of stretch receptors reinforce each other.

■ Chemoreceptors Detect Changes in Pco_2, pH, and Po_2

Another cardiovascular reflex pathway begins with excitation of the **chemoreceptors** located peripherally in the **carotid** and **aortic bodies** or centrally in the medulla. The carotid and aortic bodies are specialized structures located in approximately the same areas as the carotid and aortic baroreceptors. They are primarily sensitive to elevated P_{CO_2} and decreased pH and P_{O_2}. The peripheral chemoreceptors exhibit an increased firing rate when (1) the P_{O_2} of the arterial blood is low; (2) P_{CO_2} or H^+ of arterial blood is increased, (3) the flow through the bodies is very low or stopped; or (4) a chemical is given that blocks oxidative metabolism in the chemoreceptor cells. The central medullary chemoreceptors increase their firing rate primarily in response to elevated P_{CO_2} and H^+. The increased firing of both peripheral and central chemoreceptors leads (via the medullary cardiovascular center) to profound peripheral vasoconstriction as well as increased heart rate and cardiac output. The blood pressure is markedly elevated. If respiratory movements are voluntarily stopped the vasoconstriction is even more intense, but a striking bradycardia and decreased cardiac output occur. This is part of the **diving response,** discussed below. As in the case of the baroreceptor reflex, the coronary and cerebral circulation are not subject to the same degree of vasoconstrictor effects and exhibit vasodilation, due to local metabolic effects.

In addition to its importance when arterial blood gases are abnormal, chemoreceptor drive is important in the cardiovascular response to hemorrhagic hypotension. As blood pressure falls, blood flow through the carotid and aortic bodies decreases and chemoreceptor neural firing increases. This probably occurs because the decrease in blood flow lowers the P_{O_2} of these chemoreceptors; another possibility is that a substance which stimulates the chemoreceptors is not washed out when flow is decreased.

■ Pain Receptors Produce Significant Reflex Responses that Affect the Cardiovascular System

Reflex cardiovascular responses to pain also occur. Two different responses to pain may be seen. In the first and most common reaction, pain causes increased sympathetic activity to the heart and blood vessels coupled with decreased parasympathetic activity to the heart. These events lead to increases in cardiac output, systemic vascular resistance, and mean arterial pressure. An example of this reaction is the elevated blood pressure that normally occurs when an extremity is placed in ice water. This is the **cold pressor response,** and the increase in blood pressure produced by this challenge is exaggerated in some forms of hypertension. A second type of response is produced by deep pain. The stimulation of deep pain fibers associated with crushing injuries, disruption of joints, testicular trauma, or distension of the abdominal organs results in diminished sympathetic neural activity and enhanced parasympathetic activity with decreased cardiac output and systemic vascular resis-

tance and a drop in blood pressure. This hypotensive response contributes to certain forms of shock.

Injection of bradykinin, 5-hydroxytryptamine (serotonin), certain prostaglandins or alkaloids, and various other compounds into the coronary arteries supplying the posterior regions of the ventricles also causes reflex bradycardia and hypotension. This reflex could play a role in producing the bradycardia and hypotension that can occur in response to myocardial infarction.

Integrated Corticohypothalamic Responses in Cardiovascular Control

Corticohypothalamic response patterns are integrated in the hypothalamus, and the stimulation of certain areas of the hypothalamus leads to distinct behavioral and cardiovascular response patterns.

■ The Fight-or-Flight Response Leads to Specific Cardiovascular Changes

An example of corticohypothalamic response patterns is the **defense reaction.** The rage response of cats described earlier is an exaggerated defense reaction. In general, the behavioral pattern exhibited includes increased skeletal muscle tone and general alertness. There is increased sympathetic neural activity to blood vessels and the heart. The sympathetic cholinergic fibers innervating skeletal muscle arterioles are activated; acetylcholine, rather than norepinephrine, is released, causing vasodilation of skeletal muscle. The result of this cardiovascular response is to increase cardiac output (by increasing both heart rate and stroke volume), blood pressure, and skeletal muscle blood flow and to reduce flow to the splanchnic and renal vascular beds. This anticipatory behavioral redistribution of blood flow to muscle in preparation for either fight or flight may provide an all-important competitive edge in lower animals, where split seconds separate survivors from those less fortunate. Emotional situations often provoke the defense reaction in humans, but it is usually not accompanied by the muscle exercise that would follow a true life-or-death situation. Much current speculation centers on the question of whether the dissociation of the cardiovascular component of the defense reaction from the behavioral (fight-or-flight) component is deleterious.

■ Fainting Is the Cardiovascular Response to Certain Emotional Stimuli

Vasovagal syncope (fainting) is the somatic and cardiovascular response to certain emotional experiences. Stimulation of specific areas of the cerebral cortex can lead to a sudden relaxation of skeletal muscles, depression of respiration, and loss of consciousness. The cardiovascular events accompanying these somatic changes include profound parasympathetic bradycardia and removal of sympathetic vasoconstrictor tone (with a dramatic drop in heart rate, cardiac output, and systemic vascular resistance). The resultant decrease in mean arterial pressure results in unconsciousness because of lowered cerebral blood flow. Vasovagal syncope appears in lower animals as the "playing dead" response typical of the opossum.

■ Blushing Results from Increased Skin Blood Flow

The cardiovascular response associated with embarrassment **(blushing)** is an increase in skin blood flow, which follows the sudden removal of the normal level of sympathetic vasoconstrictor activity.

■ The Diving Response Is Present in Humans

The cardiovascular response to diving is best observed in seals and ducks but is also present to some degree in humans. An experienced diver can exhibit intense slowing of the heart (parasympathetic) and peripheral vasoconstriction (sympathetic) when his or her face is submerged in cold water. During the dive this is reinforced by the chemoreceptor reflex, which, in the absence of respiratory movements, causes the same cardiovascular response. The arterioles of the brain and heart are not constricted, therefore cardiac output is distributed to those organs. This heart-brain circuit makes use of the oxygen stored in the blood that would normally be used by the other tissues, especially skeletal muscle. Once the diver surfaces, the heart rate and cardiac output increase dramatically and peripheral vasoconstriction is replaced by vasodilation, restoring nutrient flow and washing out accumulated waste products.

■ Behavioral Conditioning May Affect Cardiovascular Responses

Cardiovascular responses can be conditioned (as can other autonomic responses, such as those observed in Pavlov's famous experiments). These responses must originate in the cerebral cortex. Both classical and operant conditioning techniques have been used to raise and lower the blood pressure and heart rate of animals. Humans can also be taught to alter their heart rate and blood pressure, using a variety of behavioral techniques (e.g., biofeedback).

Behavioral conditioning of cardiovascular responses has significant clinical implications. Animal and human studies indicate that psychological stress can raise blood pressure, increase atherogenesis, and predispose to fatal cardiac arrhythmias. These effects are thought to result from an inappropriate defense response (see above). Other studies have shown beneficial effects of behavior patterns designed to introduce

a sense of relaxation and well-being. Some clinical regimens for the treatment of cardiovascular disease take these facts into account.

■ Interactions Occur between Different Cardiovascular Responses

Several of the corticohypothalamic responses override the baroreceptor reflex. For example, the defense reaction causes the heart rate to rise above normal despite a simultaneous rise in arterial pressure. In such circumstances the neurons connecting the hypothalamus to medullary areas inhibit the baroreceptor reflex and allow the corticohypothalamic response to predominate. Furthermore, the various cardiovascular response patterns do not necessarily occur in isolation, as previously described. Many response patterns interact, reflecting the extensive neural interconnections between all levels of the central nervous system and interaction with various elements of the local control systems. For example, the baroreceptor reflex interacts with thermoregulatory responses; cutaneous sympathetic nerves serve temperature regulation (Chap. 31) but also serve the baroreceptor reflex. At moderate levels of heat stress, the baroreceptor reflex causes cutaneous arteriolar constriction despite elevated core temperature. With severe heat stress, however, the baroreceptor reflex is no longer effective in causing arteriolar vasoconstriction and as a result pressure regulation may fail.

Hormonal Control of the Cardiovascular System

Various hormones play a role in the control of the cardiovascular system. Important sites of hormone secretion include the adrenal medulla, posterior pituitary, kidney, and cardiac atria.

■ Circulating Catecholamines Come Primarily from the Adrenal Medulla

When the sympathetic nervous system is activated, the adrenal medulla releases epinephrine (>90%) and norepinephrine (<10%) into the blood. The small changes in the circulating levels of norepinephrine have relatively little effect when compared with the direct release of norepinephrine from nerve endings nestled close to the target cells. Increased circulating epinephrine, however, contributes to skeletal muscle vasodilation during the defense reaction and exercise. In this situation, epinephrine binds to the β_2-adrenergic receptors of skeletal muscle arteriolar smooth muscle cells and causes relaxation.

A comparison of the responses to infusions of epinephrine and norepinephrine illustrates not only the difference in the effects of the two hormones but also the differences in the reflex response each elicits (Fig. 18-5). Epinephrine and norepinephrine have similar direct effects on the heart, but norepinephrine elicits

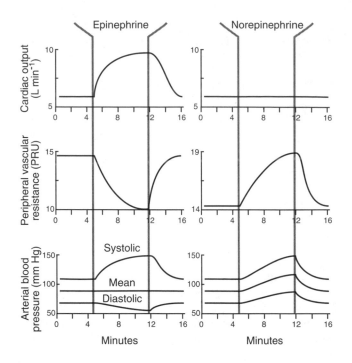

Figure 18–5 ■ Comparison of the effects of intravenous infusions of epinephrine and norepinephrine (see text). (Modified from Rowell LB. *Human Circulation: Regulation during Physical Stress.* New York, Oxford University Press, 1986.)

a powerful baroreceptor reflex because it causes peripheral vasoconstriction and increased mean arterial pressure. The reflex masks some of the direct cardiac effect(s) of norepinephrine by markedly increasing parasympathetic tone. In contrast, epinephrine causes peripheral vasodilation in skeletal muscle and splanchnic beds; the baroreceptor reflex is not elicited and the direct cardiac effects of epinephrine are evident.

At very high concentrations, epinephrine binds to α-adrenergic receptors and causes peripheral vasodilation, but this level of epinephrine is probably never reached except when it is administered as a drug.

Denervated organs, such as transplanted hearts, are dramatically influenced by circulating levels of epinephrine and norepinephrine. This increased sensitiv-

ity to neurotransmitters is referred to as **denervation hypersensitivity.** Several mechanisms contribute to denervation hypersensitivity, including the lack of sympathetic nerve endings to take up circulating norepinephrine and epinephrine actively, leaving more transmitter available for binding to receptors. In addition, denervation results in increased numbers of neurotransmitter receptors in target cells (**up-regulation** of receptors). Circulating levels of norepinephrine and epinephrine increase with exercise, and because of denervation hypersensitivity transplanted hearts can perform almost as well as normal hearts.

■ The Renin-Angiotensin-Aldosterone System Helps Regulate Blood Pressure and Volume

Control of total blood volume is extremely important in the regulation of arterial pressure. Because changes in total blood volume lead to changes in central blood volume, the long-term influence of blood volume on ventricular end-diastolic volume and cardiac output is paramount. Cardiac output, in turn, strongly influences arterial pressure. Hormonal control of blood volume depends on a set of hormones that regulates salt and water intake and output and red blood cell formation.

A variety of stimuli related to reduced arterial pressure and blood volume causes release of **renin** from the kidney. Renin is a proteolytic enzyme that catalyzes the conversion of angiotensinogen, a plasma protein, to angiotensin I (Fig. 18–6). Angiotensin I is then converted to angiotensin II by **angiotensin-converting enzyme** (ACE), primarily in the lungs. Angiotensin II is a powerful arteriolar vasoconstrictor, and in some circumstances it is present in plasma in concentrations sufficient to cause vasoconstriction and influence peripheral resistance. It reduces renal sodium excretion by increasing its reabsorption by proximal tubules. Angiotensin II also causes the release of **aldosterone** from the adrenal cortex. One of the effects of aldosterone is to reduce renal excretion of sodium, the major cation of the extracellular fluid. Retention of sodium paves the way for increasing blood volume. Renin, angiotensin, aldosterone, and the factors that control

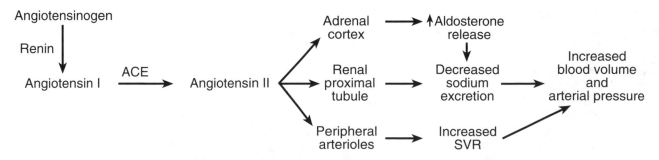

Figure 18–6 ■ The renin-angiotensin-aldosterone system plays an important role in the regulation of arterial blood pressure and blood volume. ACE, angiotensin-converting enzyme.

their release and formation are discussed in Chapter 24. The renin-angiotension-aldosterone system is important in the normal maintenance of blood volume and blood pressure. It is critical when salt and water intake are reduced.

Angiotensin II is a significant vasoconstrictor in a number of circumstances. It plays an important role in increasing systemic vascular resistance as well as blood volume in individuals on a low-salt diet. If an ACE inhibitor is given to such individuals, blood pressure falls. Renin is released during blood loss, even before blood pressure falls, and angiotensin II increases systemic vascular resistance. Rarely, renal artery stenosis causes hypertension that can be attributed solely to elevated renin and angiotensin II levels. In addition, the renin-angiotensin system plays an important (but not unique) role in maintaining elevated pressure in more than 60% of those with essential hypertension. Angiotensin II directly contracts vascular smooth muscle, but also augments norepinephrine release from sympathetic nerves and sensitizes vascular smooth muscle to the constrictor effects of norepinephrine.

In patients with congestive heart failure, renin and angiotensin II are increased and contribute to increased systemic vascular resistance as well as increased sodium retention, which lead to increased blood volume.

■ The Posterior Pituitary Releases Antidiuretic Hormone in Response to Specific Cardiovascular Stimuli

Antidiuretic hormone (ADH) is released by the posterior pituitary gland under the control of the hypothalamus. The three important classes of stimuli leading to ADH release are increased plasma osmolality, decreased atrial stretch receptor firing, and various types of stress, such as physical injury or surgery. Antidiuretic hormone is a vasoconstrictor but is not ordinarily present in plasma in high enough concentrations to exert much direct effect on blood vessels; however, in special circumstances (e.g., hemorrhagic shock) it probably contributes to increased systemic vascular resistance. Antidiuretic hormone normally exerts its major effect on the cardiovascular system by causing the retention of water by the kidneys (Chap. 23); this is an important part of the neurohumoral mechanisms that regulate blood volume.

■ Atrial Natriuretic Peptide Helps Regulate Blood Volume

Atrial natriuretic peptide is a 28-amino acid polypeptide synthesized and stored in the atrial muscle cells and released into the bloodstream when the atria are stretched. By increasing sodium excretion (Chap. 24), it decreases blood volume. Increased atrial natriuretic peptide (along with decreased aldosterone and ADH) may be partially responsible for the reduction in blood volume that occurs with chronic weightlessness or prolonged bedrest. These situations cause an increase in central blood volume and atrial stretch. Atrial natriuretic peptide secretion increases, leading to increased sodium excretion and a reduction in blood volume.

■ Erythropoietin Controls the Production of Erythrocytes

The final step in blood volume regulation is production of erythrocytes. **Erythropoietin** is a hormone produced by the kidneys that causes the bone marrow to increase production of red blood cells, thus increasing the total mass of circulating red cells. The stimuli for increased erythropoietin release include hypoxia and reduced hematocrit. If ADH and aldosterone secretion are increased, salt and water retention is enhanced, which results in an increase in plasma volume. The increased plasma volume (in the face of a constant volume of red blood cells) results in a lower hematocrit. The decrease in hematocrit stimulates erythropoietin release, which increases red blood cell synthesis and red cell mass and therefore balances the increase in plasma volume caused by aldosterone and ADH.

Cardiovascular Control during Standing

An integrated view of the cardiovascular system requires an understanding of the relationships among cardiac output, venous return, and central blood volume and how these relationships are influenced by the interaction among various neural, hormonal, and other control mechanisms. Consideration of the responses to standing erect provides an opportunity to explore these elements in detail.

■ Standing Requires a Complex Cardiovascular Response

In general, standing causes intravascular pressures above the diaphragm to decrease and pressures below the diaphragm to increase (Fig. 18-7). With quiet standing, normal right atrial pressure typically decreases from approximately 3 to 0 mm Hg and venous pressure in the feet increases by approximately 80 mm Hg. (Because the right atrium tends to undergo little variation in pressure during postural changes, it is often used as the "zero" pressure). The increased venous pressure below the diaphragm causes expansion of the leg venous volume; the blood for this increase comes from the cardiac output that is trapped in the legs by the expanding veins. Thus, for a few seconds after standing, venous return to the heart is lower than cardiac output, and during this time blood is withdrawn from the central blood volume. For example, when a 70-kg individual stands, the central blood volume is reduced by approximately 550 ml. If no compensatory mechanisms existed, this would significantly reduce cardiac filling and cause a more than

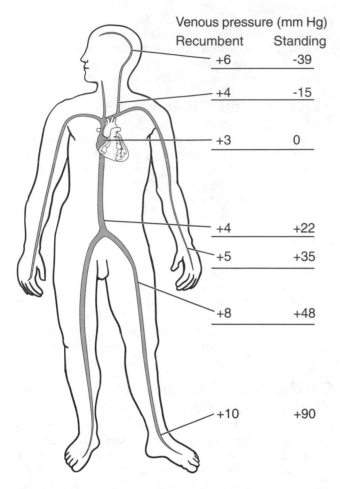

Venous pressure (mm Hg)	
Recumbent	Standing
+6	-39
+4	-15
+3	0
+4	+22
+5	+35
+8	+48
+10	+90

Figure 18–7 ■ Venous pressures in the recumbent and standing positions. In this individual, standing places a hydrostatic pressure of approximately 80 mm Hg on the feet. Right atrial pressure is lowered because of the reduction in central blood volume. The negative pressures above the heart with standing do not actually occur because once intravascular pressure drops below atmospheric pressure, the veins collapse. These are the pressures that would exist if the veins remained open.

60% decrease in stroke volume, cardiac output, and blood pressure; the resulting fall in cerebral blood flow would probably cause loss of consciousness. An adequate cardiovascular response to the changes caused by upright posture **(orthostasis)** is therefore absolutely essential to our lives as bipeds.

The major cardiovascular adjustments to upright posture are reflex increases in rate and force of contraction of the heart, the muscle and respiratory pumps, constriction of arterioles, and long-term adjustments in blood volume.

■ Standing Elicits Autonomic Nervous System Responses

The decreased central blood volume caused by standing includes decreased atrial volume and left and right end-diastolic volumes. Reduced atrial volume diminishes firing rate of atrial stretch (volume) receptors. Reduced left ventricular end-diastolic volume decreases stroke volume and cardiac output, leading to decreased firing of aortic arch and carotid baroreceptors. The combined reduction in firing of volume and pressure receptors results in a reflex increase in sympathetic nerve activity from the medullary cardiovascular center to the heart. Heart rate generally increases approximately 10–20 beats/min when an individual stands. The increased sympathetic nerve activity to the ventricular myocardium shifts the ventricle to a new function curve, so despite the lowered ventricular filling, stroke volume is decreased to only 50–60% of the recumbent value. In the absence of the compensatory increase in sympathetic nerve activity, stroke volume would fall much more. These cardiac adjustments mean that cardiac output is reduced to 60–80% of the recumbent value. The increase in sympathetic activity also causes arteriolar constriction and increased systemic vascular resistance. The effect of these compensatory changes in heart rate, ventricular contractility, and systemic vascular resistance is to maintain mean arterial pressure. In fact, mean arterial pressure may be increased slightly above the recumbent value.

A good question at this point would be: How is increased sympathetic nerve activity maintained if the mean arterial pressure reaches a value near that of the recumbent value? In other words, why doesn't the baroreceptor nerve firing (and thus sympathetic nerve activity) return to recumbent levels if the mean arterial pressure returns to the recumbent value? There are two reasons. First, although the mean arterial pressure returns to the same level (or even higher), the pulse pressure remains reduced because the stroke volume is decreased to 50–60%. As indicated earlier, the firing rate of the baroreceptors depends on both mean arterial and pulse pressure. The reduced pulse pressure means the baroreceptor firing frequency remains lower even if the mean arterial pressure is slightly higher. Another reason is the continued low central blood volume, which means the volume receptors of the atria fire less frequently and therefore cause increased sympathetic activity via the medullary cardiovascular center. Some investigators believe it is the decreased stretch of the atrial receptors that provides the primary steady-state afferent information for the reflex cardiovascular response to standing.

The heart and brain do not participate in the arteriolar constriction caused by increased sympathetic nerve activity during standing, therefore the blood flow supply of oxygen and nutrients to these two vital organs is maintained.

■ Muscle and Respiratory Pumps Help Maintain Central Blood Volume

Although standing would appear to be a perfect situation for increased venoconstriction (which could return some of the blood volume from the legs to the central blood volume), reflex venoconstriction is a relatively minor part of the response to standing. For example, substantially more venoconstriction can occur in response to the dramatic baroreceptor reflex activation that results from severe hemorrhage. However, two other mechanisms act to return blood from the legs to the central blood volume. The more important of these is the **muscle pump** (Fig. 18–8). If the leg muscles periodically contract while an individual is standing, stroke volume increases to the recumbent value very quickly, because of an increase in central blood volume that results from increased venous return. The muscles swell as they shorten, which compresses nearby veins. Because of the venous valves in the limbs, the blood in the compressed veins can flow only toward the heart. The combination of contracting muscle and venous valves provides a very effective muscle pump that transiently increases venous return relative to cardiac output. Blood volume is shifted from the legs to the central blood volume, and end-diastolic volume is increased. Even very mild exercise, such as walking, returns central blood volume and stroke volume to recumbent values (Fig. 18–9).

The **respiratory pump** is another mechanism that acts to enhance venous return transiently and restore central blood volume (Fig. 18–10). Quiet standing for 5–10 minutes invariably leads to sighing. This exaggerated respiratory movement lowers intrathoracic pressure more than usually occurs with inspiration. The fall in intrathoracic pressure raises the transmural pressure of the intrathoracic vessels, causing these vessels to expand. Contraction of the diaphragm simultaneously raises intraabdominal pressure, which compresses the abdominal veins. Since the venous valves prevent the backflow of blood into the legs, the raised intraabdominal pressure forces blood toward the intrathoracic vessels (which are expanding due to the lowered intrathoracic pressure). The seesaw action of the respiratory pump tends to displace extrathoracic blood volume toward the chest and raise right atrial pressure and stroke volume. Figure 18–11 provides an overview of the main cardiovascular events associated with a short period of standing.

■ Capillary Filtration during Standing Works to Reduce Central Blood Volume

During quiet (minimum muscular movement) standing for 10–15 minutes, the baroreceptor reflex effects on the heart and arterioles are insufficient to prevent a lowered arterial pressure. Prolonged quiet standing causes an individual to faint as the mean arterial pressure and cerebral blood flow fall. This vascular decompensation occurs because of effects caused by elevated capillary hydrostatic pressure. The hydrostatic column of blood above the capillaries of the legs and feet raises capillary filtration. Over a period of 30 minutes, a loss of 10% of the blood volume into the interstitial space

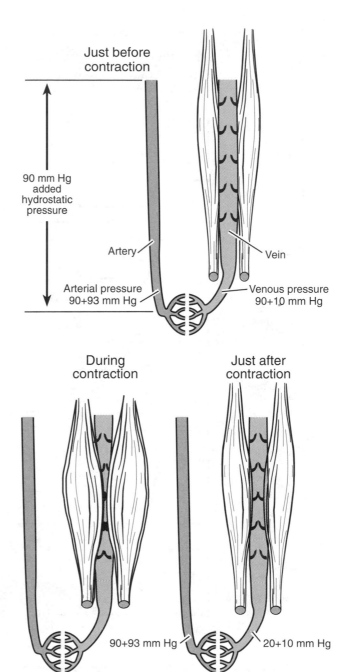

Figure 18–8 ■ The muscle pump, which increases venous return and decreases venous volume. The valves (which are closed after contraction) break up the hydrostatic column of blood, thus lowering venous (and capillary) hydrostatic pressure.

can occur. This loss, coupled with the 550 ml displaced by redistribution from the central blood volume into the legs, causes central blood volume to fall to a level so low that reflex sympathetic nerve activity cannot maintain cardiac output and mean arterial pressure; diminished cerebral blood flow and a loss of consciousness result.

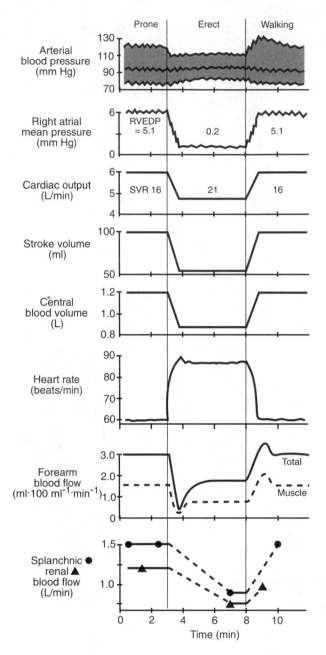

Figure 18–9 ■ Quiet standing causes the changes shown in the middle panel. The muscle pump is activated by contracting leg muscles in the right panel. Note that the muscle pump restores right atrial pressure and cardiac output by increasing venous return to the recumbent level. The fall in heart rate and rise in peripheral blood flow (forearm and splanchnic) associated with activation of the muscle pump reflect the reduction in baroreceptor reflex activity. RVEDP, right ventricular end-diastolic pressure; SVR, systemic vascular resistance. (Modified from Rowell LB. *Human Circulation: Regulation during Physical Stress.* New York, Oxford University Press, 1986.)

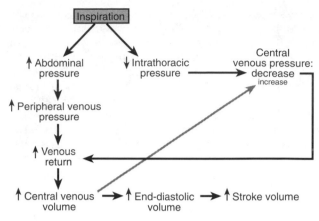

Figure 18–10 ■ The respiratory pump. Inspiration leads to an increase in venous return and stroke volume.

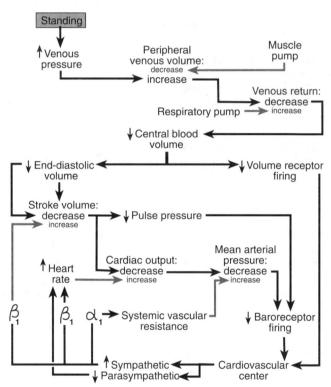

Figure 18–11 ■ Cardiovascular events associated with standing. Small type represents compensatory events that return variables toward the recumbent values. α_1 and β_1 refer to adrenergic receptor types.

Arteriolar constriction due to the increased reflex sympathetic nerve activity tends to reduce capillary hydrostatic pressure. This alone does not bring capil-

lary hydrostatic pressure back to normal because it does not affect the hydrostatic pressure exerted on the capillaries from the venous side. The most important factor counteracting increased capillary hydrostatic pressure is the muscle pump. The alternate compression and filling of the veins as the muscle pump works means the venous valves are closed a good deal of the time. When the valves are closed, the hydrostatic column of blood in the leg veins at any point is only as high as the distance between the proximate valves. The combination of the muscle pump and arteriolar constriction reduces but does not prevent net filtration.

The myogenic response of arterioles to increased transmural pressure also acts to oppose filtration. As discussed above, raising the transmural pressure stretches vascular smooth muscle and stimulates it to contract. This is especially true for the myocytes of precapillary arterioles. The elevated transmural pressure associated with standing causes a myogenic response and decreases the number of open capillaries. The fewer the open capillaries, the lower the filtration rate for a given capillary hydrostatic pressure imbalance.

In addition to the factors cited above, other safety factors against edema play an important role in preventing excessive translocation of plasma volume into the interstitial space (Chap. 16). These factors, along with the neural and myogenic responses and the muscle and respiratory pumps, play a significant role during the seconds and minutes following standing (Fig. 18–12).

■ Long-term Responses Defend Venous Return during Prolonged Upright Posture

In addition to these relatively short-term cardiovascular responses, there are equally important long-term

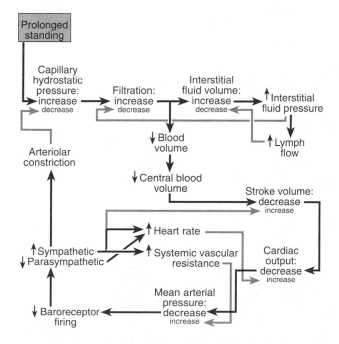

Figure 18–12 ■ With prolonged standing, capillary filtration reduces venous return. Without the compensatory events that result in the changes shown in small type, prolonged standing would inevitably lead to syncope.

adjustments. For example, patients confined to bed or astronauts not subject to the force of gravity do not intermittently trap venous blood or filter capillary fluid into their lower extremities. Over time their average central blood volume (and pressure) will therefore be greater than it would be if they were ambulating under normal conditions. The body responds to weightlessness or bedrest by reducing blood volume; this response begins during the first day and is quite dramatic after a week. Looking at it another way, maintaining an erect posture in the earth's gravitational field results in an increase in blood volume. This increase, proportioned between the extrathoracic and intrathoracic vessels, augments stroke volume during standing. In the absence the increment in blood volume caused by intermittent erect posture, standing is extremely difficult or impossible because of **orthostatic hypotension** (diminished blood pressure associated with standing). In addition to weightlessness and bedrest there are many other causes of orthostatic hypotension, including hemorrhage and dysfunction of the autonomic nervous system.

Long term regulation of blood volume is initiated by changes in plasma volume. Figure 18–13 depicts several components of plasma volume regulation by showing their response to a moderate (approximately 10%) blood loss, which is easily compensated in normal individuals. Plasma is a part of the extracellular compartment and is subject to the factors that regulate the size of that space. The osmotically important electrolytes of the extracellular fluid are the sodium ion and its main partner, the chloride ion. Control of the extracellular fluid volume is determined by the balance between the intake and excretion of sodium and water. Sodium excretion is much more closely regulated than sodium intake. Excretion of sodium can be controlled by altering the glomerular filtration rate, the plasma aldosterone concentration, and a variety of other factors, including angiotensin II and atrial natriuretic peptide (Chap. 24). The important determinant of glomerular filtration is glomerular capillary pressure, which is dependent on precapillary (afferent arteriolar) and postcapillary (efferent arteriolar) resistance and arterial pressure. Decreased mean arterial pressure and/or afferent arteriolar constriction tends to result in lowered glomerular capillary pressure, less filtration of fluid into the nephron, and lower sodium excretion. Changes in glomerular capillary pressure are primarily the result of changes in sympathetic nerve activity and plasma angiotensin II and atrial natriuretic peptide concentrations.

Aldosterone acts on the distal nephron to cause increased reabsorption of sodium and thereby decrease its excretion. Aldosterone released from the adrenal cortex is increased by (among other things) angiotensin II. Water intake is determined by thirst and the availability of water. Excretion of water is strongly influenced by ADH. Both thirst and ADH release are increased by increases in plasma osmolality,

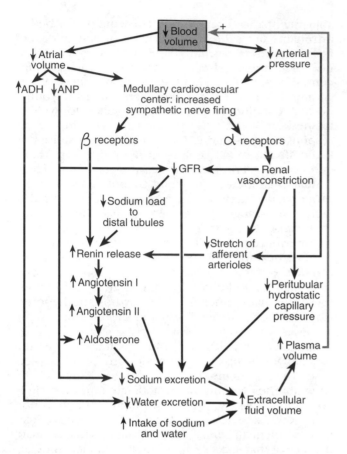

Figure 18–13 ■ Blood loss influences sodium and water excretion by the kidney via a number of pathways. All of these pathways, combined with increased intake of sodium and water, restore the extracellular fluid volume and eventually blood volume. These responses occur later than those shown in Figures 18–10 to 18–12. The pathways responsible for stimulating increased intake of sodium and water are not shown.

which are sensed by certain areas of the hypothalamus. Like increased osmolality, decreased stretch receptor activity results in enhanced thirst and ADH release.

Consider how these physiologic variables are altered by an upright posture to produce an increase in the extracellular fluid volume. Renal arteriolar vasoconstriction associated with the increased sympathetic activity produced by standing reduces the glomerular filtration rate. This results in a decrease in filtered sodium into the nephron and tends to decrease sodium excretion. The increased sympathetic nerve activity to the kidney also triggers the release of renin, which increases circulating angiotensin II and, in turn, aldosterone release. The decrease in central blood volume associated with standing reduces atrial stretch receptor activity, causing increased release of ADH from the posterior pituitary. Thus both sodium and water are retained, and thirst is increased. The precise quantities of water and sodium can be adjusted to maintain the correct osmolality of the plasma.

The distribution of extracellular fluid between plasma and interstitial compartments is determined by the balance of hydrostatic and colloid osmotic forces across the capillary wall. Retention of sodium and water tends to dilute the plasma proteins, which decreases plasma colloid osmotic pressure and favors filtration of fluid from the plasma into the interstitial fluid. However, increased synthesis of plasma proteins by the liver occurs, so a portion of the retained sodium and water contributes to an increase in plasma volume.

Finally, the increase in plasma volume (in the absence of any change in total red cell volume) decreases hematocrit, which can stimulate erythropoietin release and erythropoiesis. This helps red blood cell volume to keep pace with plasma volume.

■ ■ ■ ■ ■ ■ ■ ■ ■ ■ ■ ■ **REVIEW EXERCISES** ■ ■ ■ ■ ■ ■ ■ ■ ■ ■ ■

Identify Each with a Term

1. Type of adrenergic receptor found in the plasma membrane of vascular smooth muscle; it causes contraction when stimulated

2. The highest levels of organization in the autonomic nervous system

3. "Stretch" receptors located in low-compliance arteries

4. "Stretch" receptors located in the high-compliance cardiac atria

5. The increase in number of neurotransmitter receptors following denervation

Define Each Term

6. Orthostatic hypotension

7. Erythropoietin

8. Renin

9. Angiotensin converting enzyme

10. Fight-or-flight (defense) response

Choose the Correct Answer

11. Which reflex is mediated by pain receptors?
    a. Diving response
    b. Cold pressor response
    c. Adaptation to chronic hypertension
    d. Adaptation to chronic hypotension

12. Which statement about the cardiovascular system is false?
    a. The blood vessels receive innervation from the sympathetic and parasympathetic systems, while the heart receives sympathetic innervation only.
    b. Arteries of skeletal muscle receive sympathetic fibers that release norepinephrine and others that release acetylcholine.

c. All neural output to the heart and vascular smooth muscle is via the parasympathetic and sympathetic components of the autonomic nervous system.

d. The designation of a cholinergic receptor as either muscarinic or nicotinic depends on its specific binding properties.

13. The adrenergic receptor type found primarily in the plasma membrane of cardiac cells, which increases the rate of cardiac pacemaker cell depolarization when stimulated, is called:
a. α_1
b. α_2
c. β_1
d. β_2

14. Which of the following does *not* occur when acetylcholine binds to muscarinic receptors?
a. Heart rate slows
b. Cardiac conduction velocity slows
c. Norepinephrine release from sympathetic nerve terminals is enhanced
d. Atrial nodal potassium channels open, causing hyperpolarization

15. Which statement about baroreceptors is false?
a. Carotid sinus and aortic baroreceptors are exceedingly important in the rapid, short-term regulation of arterial blood pressure.
b. Aortic baroreceptors do not fire until a pressure of approximately 100 mm Hg is reached.
c. Baroreceptors adapt over 1-2 days to the prevailing mean arterial pressure.
d. Baroreceptor activity usually causes a significant drop in cerebral and coronary blood flow.

16. Which statement about chemoreceptors is false?
a. Chemoreceptors are located in the carotid and aortic bodies and in the medulla.

b. Increased P_{O_2} causes an increase in the peripheral chemoreceptor firing rate.
c. The diving response is partially mediated by chemoreceptors.
d. Chemoreceptor drive is important in the cardiovascular response to hemorrhagic hypotension.

17. Which response is best explained by intense parasympathetic stimulation accompanied by the withdrawal of sympathetic stimulation?
a. Fight-or-flight response
b. Vasovagal syncope
c. Blushing
d. Diving response

18. Antidiuretic hormone is released from the posterior pituitary in response to all of the following *except:*
a. Increased plasma osmolality
b. Increased temperature
c. Decreased atrial stretch receptor firing
d. Various types of cardiovascular stress

Briefly Answer

19. Describe the effect of each of the following (increase or decrease) on the potential *fall* in blood pressure that occurs with standing. (a) Diuretics; (b) α blockers; (c) β blockers; (d) Elastic compression stockings on legs; (e) Lumbar sympathectomy or blockade (i.e., interruption of the sympathetic nerves to the legs); (f) Salt tablets; (g) Standing waist-high in water.

20. Some individuals lose consciousness when the region of the carotid sinus is rubbed or pressed (**carotid massage**). Describe the mechanism by which this occurs.

21. Patients with renal artery stenosis (and therefore decreased renal artery pressure and stretch) may develop severe hypertension. Describe the mechanism by which this occurs.

■ ANSWERS ■

1. α_1
2. Corticohypothalamic pathways
3. Pressure or baroreceptors
4. Volume receptors
5. Up-regulation
6. Diminished blood pressure associated with standing
7. A renal hormone that induces bone marrow to produce red blood cells
8. An enzyme of renal origin that catalyzes the conversion of angiotensinogen to angiotensin I
9. An enzyme of (primarily) pulmonary origin that catalyzes the conversion of angiotensin I to angiotensin II
10. A behavioral response to anticipated aggression in which cardiac output, blood pressure, and skeletal muscle blood flow are increased while the blood flow to organs such as the kidneys and gastrointestinal tract is decreased
11. b

12. a
13. c
14. c
15. d
16. b
17. b
18. b
19. (a) Increase; (b) Increase; (c) Increase; (d) Decrease; (e) Increase; (f) Decrease; (g) Decrease.
20. Carotid massage stretches the carotid baroreceptors, which leads to a decrease in sympathetic and increase in parasympathetic activity. In certain individuals, the resultant decrease in heart rate, cardiac output, and systemic vascular resistance may cause hypotension sufficient to impair cerebral circulation and cause fainting. (See Fig. 18–2.)
21. Renal artery stenosis mimics renal vasoconstriction by lowering renal artery perfusion pressure and renal blood

flow. The kidney responds by releasing renin, which increases fluid retention (by decreasing sodium excretion) and causes generalized vasoconstriction (via the production of angiotensin). These two effects raise blood pressure, often to dangerously high levels. Drugs that inhibit the effects of renin by blocking the formation of angiotensin II (converting enzyme inhibitors) are clinically useful in this situation.

Suggested Reading

Heymans CJF, Folkow B: Vasomotor control and the regulation of blood pressure, in Fishman AP, Richards DW (eds). *Circulation of the Blood: Men and Ideas.* Bethesda, American Physiological Society, 1982, pp 407–486.

Katz AM: *Physiology of the Heart.* 2nd ed. New York, Raven, 1992.

Milnor WR: The circulatory system and principles of hemodynamics, in *Cardiovascular Physiology.* New York, Oxford University Press, 1986, chaps. 7, 8.

Rowell LB: *Human Circulation: Regulation During Physical Stress.* New York, Oxford University Press, 1986.

PART FIVE

Respiratory Physiology

C H A P T E R

■ ■ ■ ■ ■ ■ ■

19

Ventilation and the Mechanics of Breathing

CHAPTER OUTLINE

I. THE RELATIONSHIPS OF STRUCTURE AND FUNCTION IN THE LUNG
   A. The airway tree divides repeatedly to increase total cross-sectional area
   B. Lung internal surface area is increased by its architecture
   C. The vascular and airway trees merge, forming a blood-gas interface

II. LUNG VOLUMES AND VENTILATION
   A. The diaphragm is the main muscle of breathing
   B. Air flow is due to a change in alveolar pressure
   C. Spirometry is used to measure lung volumes and air flow
   D. Not all of the inspired air reaches the alveoli
   E. Alveolar ventilation is the amount of fresh air that reaches alveoli
   F. V_A can be measured indirectly from expired carbon dioxide

III. THE MECHANICS OF BREATHING
   A. Compliance is a measure of lung distensibility
   B. Surfactant lowers surface forces and stabilizes alveoli
   C. Airway resistance decreases air flow in the lung
   D. Airways become compressed during forced expiration
   E. Lung compliance affects the equal pressure point
   F. Work is required to expand the lung and chest wall

OBJECTIVES

After studying this chapter, the student should be able to:
1. Describe how lung architecture increases the internal surface area of the lung
2. Describe the blood-gas interface
3. Describe how the interaction of the lung and chest wall leads to a negative pleural pressure
4. Describe how spirometry is used to assess lung function
5. Describe how alveolar ventilation influences carbon dioxide levels in the blood
6. Describe anatomic, alveolar, and physiologic dead space
7. Define *compliance* and state how it affects functional residual capacity
8. State the roles of pulmonary surfactant and alveolar interdependence in lung function
9. Define airway resistance and list factors that affect it
10. Define dynamic airway compression and describe how airways are compressed during forced expiration

The process of transferring oxygen to cells and removing carbon dioxide is known as **respiration.** Respiration takes place in two stages. The first stage, known as **gas exchange,** involves the transfer of oxygen and carbon dioxide between the atmosphere and blood in the pulmonary capillaries and then between the systemic blood and the metabolic active tissue. The second stage, known as **cellular respiration,** consists of a series of complex metabolic reactions that break down molecules of food, releasing carbon dioxide and energy. Recall that oxygen is required in the final step of cellular respiration to serve as an electron acceptor in the process by which cells obtain energy.

The lungs are the organs of gas exchange starting at birth, when the newborn takes its first breath—the breath that marks the beginning of an independent existence. With the first inrush of air, a series of events are set in motion that allows the newborn to switch from a dependent placental life-support system to an independent air breathing system. A breath in, a breath out, 12–15 times every minute, seems a simple and unimpressive process on which to build the entire human gas exchange system that supplies all of the oxygen to the trillions of cells in the human body. This simplicity is deceptive, however, because breathing is amazingly responsive to changes in blood chemistry, mood, level of alertness, and body activity. Moreover, the human lung is so efficiently designed to remove carbon dioxide and to supply oxygen to tissues that gas exchange rarely limits our activity. For example, a marathon runner who staggers across the 26-mile finish line in less than 3 hours or someone who swims the English Channel in record time rarely is limited by pulmonary gas exchange. The reason is that gas exchange can increase more than 20-fold to meet the body's energy demands. These examples of human activity not only underscore the functional capacity of the lungs, but also illustrate the important role respiration has played in the success and extraordinary adaptability of animal species.

The functions of the respiratory system can be divided into ventilation and the mechanics of breathing; blood flow to the lungs; gas transfer and transport to the tissues; and control of breathing. This chapter discusses the mechanical aspects of the respiratory system—the physical design of the lung, ventilation, and the actual mechanics of breathing. (Chapter 20 discusses the pulmonary circulation and the matching of airflow and bloodflow in the lung. Chapter 21 discusses gas uptake and transport. Chapter 22 deals with basic breathing rhythms, breathing reflexes, and the integrated control of breathing.)

The Relationships of Structure and Function in the Lung

Movement of oxygen and carbon dioxide in and out of cells occurs by simple diffusion. In unicellular organisms, the process of gas exchange is very simple and requires no special respiratory structure. In more complex organisms, however, cells deep in the body cannot exchange carbon dioxide and oxygen by simple diffusion between cells and the environment because the diffusion distance is too long. As a result, special gas exchange organs have developed. These include tracheal tubes in insects, gills in fish, and lungs in air-breathing animals.

■ The Airway Tree Divides Repeatedly to Increase Total Cross-Sectional Area

The human gas exchange organ consists of two lungs, each divided into several lobes. The lungs are comprised of two treelike structures, the **vascular tree** and the **airway tree,** which are embedded in highly elastic connective tissue. The vascular tree consists of arteries and veins connected by capillaries (Chap. 20). The airway tree consists of a series of hollow branching tubes that decrease in diameter at each branching (Fig. 19-1A). The main airway (**trachea**) branches into two **bronchi,** each entering a lung. Within each lung, these bronchi branch many times into progressively smaller bronchi, which in turn form **bronchioles.**

An idealized model of the airway tree is presented in Figure 19-1B. The first 17 generations (i.e., trachea and 16 airway branches) make up the **conducting zone.** The trachea, bronchi, and bronchioles of the conducting zone have three important functions: (1) to warm and humidify inspired air; (2) to distribute air to the depths of the lungs; and (3) to serve as part of the body's defense system (removal of dust, bacteria, and noxious gases from the lung). The first four generations of the conducting zone are subjected to the effect of thoracic pressure and contain a considerable amount of cartilage to prevent airway collapse when positive intrathoracic pressures occur during forced expiration. In the trachea and main bronchi, the cartilage consists of U-shaped rings. Further down, in the lobar and segmental bronchi, the cartilaginous rings give way to small plates of cartilage. In the bronchioles, the cartilage disappears altogether. The smallest airways in the conducting zone are the **terminal bronchioles.** Both the bronchioles and terminal bronchioles are suspended by elastic tissue in the lung parenchyma, and the elasticity of the lung tissue helps keep these airways open. The conducting zone has its own separate circulation, the **bronchial circulation,** which originates from the descending aorta and drains into the pulmonary veins. No gas exchange occurs in the conducting zone.

The last seven generations make up the **respiratory zone,** the site of gas exchange. The respiratory zone is composed almost entirely of alveoli. Like the conducting zone, the respiratory zone has its own separate and distinct circulation, the **pulmonary circulation.** It receives all of the cardiac output, hence blood flow is very high. Red blood cells can pass through the pulmonary capillaries in less than 1 second.

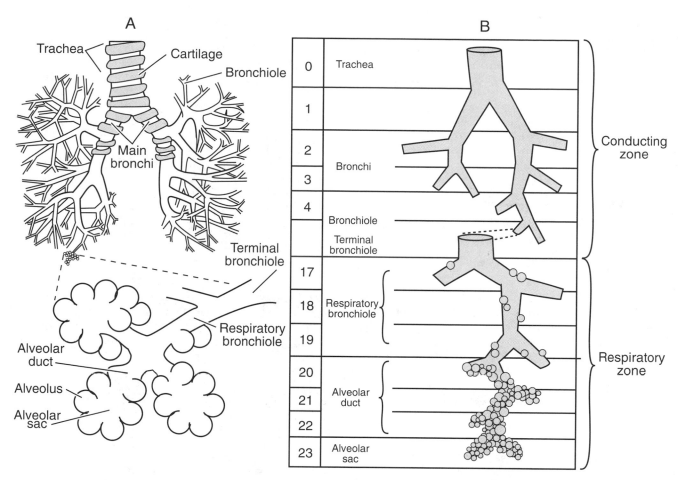

Figure 19–1 ■ (A) The airway tree consists of a series of highly branched hollow tubes that become narrower, shorter, and more numerous as they penetrate the deeper parts of the lung. (B) Functional aspects shown in an idealized model of the airway tree. The first 17 generations of branches conduct air to the deeper parts of the lungs and comprise the conductive zone. The last 7 generations participate in gas exchange and comprise the respiratory zone. (B is modified from Weibel ER: *Morphometry of the Human Lung.* Berlin, Springer-Verlag, 1963.)

■ Lung Internal Surface Area Is Increased by Its Architecture

The increase in internal surface area is accomplished by formation of outpockets from the small airways to form **alveoli** (Fig. 19–1). Each alveolus is surrounded by a network of capillaries that brings blood into close proximity with air inside the alveolus. Oxygen diffuses across the thin wall of the alveolus into the blood and carbon dioxide diffuses from the blood into the alveolus. The adult lung contains 300–500 million alveoli, with a combined internal surface area of approximately 75 m², roughly the size of a tennis court. This represents one of the largest biologic membranes in the body. The area increases as the number of alveoli increases from birth to adolescence (Table 19–1). After adolescence, alveoli increase only in size and if damaged have limited ability to repair themselves.

TABLE 19–1 ■ Alveolar Number and Surface Area in the Human Lung

Age	Number of Alveoli (10^6)	Alveolar Surface Area (m²)	Skin Surface Area (m²)
Birth	24	2.8	0.2
8 years	300	32.0	0.9
Adult	300	75.0	1.8

■ The Vascular and Airway Trees Merge, Forming a Blood-Gas Interface

In the respiratory zone, the vascular and airway trees merge to form a thin interface where air and blood are brought into close contact. The **alveolar-capillary membrane** separates the blood in the pulmonary capillaries from the gas in the alveoli and is referred to as the **blood-gas interface** (Fig. 19–2). This interface is exceedingly thin, in some places less than 0.5 μm. The blood-gas interface is composed of alveolar epithelium, an interstitial fluid layer, and capillary endothelium. Air is brought to one side of the interface by **ventilation** (i.e., movement of air to and from the alveoli). Blood is brought to the other side of the interface by the pulmonary circulation. Oxygen and carbon dioxide cross the blood-gas interface during gas exchange by diffusion.

Lung Volumes and Ventilation

Before examining how air gets into the lung, a brief discussion of the muscles and pressures involved in breathing is helpful. The lungs are housed in an airtight chest cavity, the **thoracic cavity,** and are separated from the abdomen by a large dome-shaped muscle, the **diaphragm.** The thoracic cage is made up of 12 pairs of **ribs,** a **sternum,** and a set of internal and external intercostal muscles that lie between the ribs. The rib cage is hinged to the vertebral column, allowing it to be raised and lowered during breathing (Fig. 19–3). The space between the lungs and chest wall is the **pleural space,** which contains a thin layer of fluid (about 10 μm thick) that functions, in part, as a lubricant so the lungs can slide against the chest wall.

■ The Diaphragm Is the Main Muscle of Breathing

Inflation of the lungs is due to contraction of the diaphragm, which is composed of skeletal muscle. Consider the breathing cycle. The main muscle used

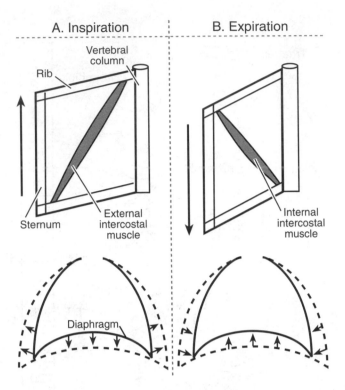

Figure 19–3 ■ Movement of the diaphragm and rib cage changes thoracic volume, which allows the lungs to inflate (A) and deflate (B). (Modified from West JB: *Respiratory Physiology.* 2nd ed. Baltimore, Williams & Wilkins, 1979.)

during the inspiratory phase is the diaphragm. When the diaphragm contracts, the thoracic cavity is expanded, and the lungs inflate automatically. This enlargement is accomplished in two ways (Fig. 19–3). First, when the diaphragm (which is attached to the lower ribs and sternum) contracts, the abdominal contents are pushed down, thus enlarging the thoracic cavity in the vertical plane. Second, when the diaphragm descends and pushes down on the abdominal contents, it also pushes the rib cage outward, further enlarging the cavity. Since the chest cavity is airtight, an increase in thoracic volume causes the **pleural pressure** (pressure in the pleural space) to decrease, which causes the lung to expand and fill with air.

The effectiveness of the diaphragm in changing thoracic volume is related to the strength of its contraction and its dome-shaped configuration when relaxed. With a normal breath, the diaphragm moves only about 1–2 cm, but with forced inspiration a total excursion of 10–12 cm can occur. The effectiveness of the diaphragm in enlarging the thoracic cavity can be impeded by obesity, pregnancy, and tight clothing around the abdominal wall. Damage to the phrenic nerves (the diaphragm is innervated by two phrenic nerves, one to each lateral half) can also lead to paralysis of the diaphragm. When one of the phrenic nerves is damaged, that portion of the diaphragm moves up rather than down during inspiration.

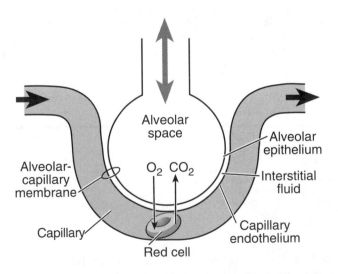

Figure 19–2 ■ Schematic view of an alveolus and its blood supply. Thick arrows indicate directions of blood flow and thin arrows indicate direction of flow of O_2 and CO_2 across the alveolar-capillary membrane (blood-gas interface), a diffusion distance equal to approximately 0.5 μm.

During forced inspiration, in which a large volume of air is taken in, the **external intercostal** muscles are used in addition to the diaphragm (Fig. 19-3). They are innervated by the intercostal nerves, and their contraction raises the anterior end of the rib cage, causing the rib cage to be pulled upward and outward. To see how this works, stand sideways in front of a mirror and take a deep breath. The chest wall moves upward while the sternum moves outward, thereby enlarging the thoracic cavity.

The last group of muscles involved in inspiration is the **accessory muscles,** which also become active with forced breathing. These include the scalene muscle in the neck and the sternocleidomastoids, which are inserted into the top of the sternum. These accessory muscles are brought into play during deep and heavy breathing, such as exhaustive exercise, and are used to elevate the upper rib cage.

Breathing out is a much simpler process than breathing in. At the end of a normal inspiration, the diaphragm relaxes, the rib cage drops, the thoracic volume decreases, and the lungs deflate. Although expiration is purely passive during normal breathing, with exercise or forced expiration the expiratory muscles become active. These muscles include those of the **abdominal wall** and the **internal intercostal muscles** (Fig. 19-3). Contraction of the abdominal wall pushes the diaphragm upward into the chest and the internal intercostal muscles pull the rib cage down, thereby reducing thoracic volume. These accessory respiratory muscles are important and necessary for such functions as coughing, straining, vomiting, and defecating. They are also important in endurance running and are one of the reasons competitive long-distance runners, as part of their training program, often do exercises to strengthen their abdominal muscles.

■ Air Flow Is Due to a Change in Alveolar Pressure

When discussing respiration, pressures are expressed in centimeters of water (cm H_2O), because small pressures are involved. A pressure of 1 cm H_2O is equal to 0.74 mm Hg (or 1 mm Hg = 1.36 cm H_2O). At sea level, barometric pressure (P_B) is equal to 760 mm Hg. In the middle of inspiration, the pressure inside the alveoli can be -2 cm H_2O. The negative sign indicates not that the pressure is negative but that it is 2 cm *below* barometric pressure. Conversely, in the middle of expiration the pressure inside the alveoli can be 3 cm H_2O. This means the pressure is 3 cm H_2O *above* barometric pressure. When relative pressures are used in respiration it is important to remember that P_B is set at zero. If airway pressure is zero, pressure inside the airway equals atmospheric pressure. Unless otherwise specified, the pressures of breathing are relative pressures and the units are cm H_2O. A list of symbols and abbreviations used in respiratory physiology is shown in Table 19-2.

TABLE 19–2 ■ Symbols and Terminology Used in Respiratory Physiology

Primary symbols	
C	Concentration or compliance
D	Diffusion
F	Fractional concentration of a gas
f	Respiratory frequency
P	Pressure or partial pressure
Q	Volume of blood
Q̇	Volume of blood per unit time (blood flow or perfusion)
R	Gas exchange ratio or resistance
S	Saturation
V	Gas volume
V̇	Volume of gas per unit time (gas flow or ventilation)
Secondary symbols	
A	Alveolar
a	Arterial
aw	Airway
B	Barometric
c	Capillary
DS	Dead space
E	Expired
I	Inspired
el	Elastic
L	Lung
p	Physiologic
pa	Pulmonary artery
pc	Pulmonary capillary
pc′	Pulmonary end capillary
pl	Pleural
pv	Pulmonary venous
pw	Pulmonary wedge
s	Shunt
st	Static
t	Time
tm	Transmural
v	Venous
v̄	Mixed venous
Examples of combinations	
C_L	Lung compliance
D_{LCO}	Lung diffusing capacity for carbon monoxide
$C\bar{v}_{O_2}$	Concentration of oxygen in mixed venous blood
P_B	Barometric pressure
P_{CO_2}	Partial pressure of carbon dioxide
P_A	Alveolar pressure
P_{O_2}	Partial pressure of oxygen
Pa_{O_2}	Partial pressure of oxygen in arterial blood
Pa_{CO_2}	Partial pressure of carbon dioxide in alveoli
PA_{O_2}	Partial pressure of oxygen in alveoli
PI_{O_2}	Partial pressure of oxygen in inspired gas
$P\bar{v}_{O_2}$	Partial pressure of oxygen in mixed venous blood
PE_{CO_2}	Partial pressure of carbon dioxide in expired gas
$P(A-a)_{O_2}$	Alveolar-arterial difference in partial pressure of oxygen
P_{pl}	Pleural pressure
Sa_{O_2}	Saturation of hemoglobin with oxygen in arterial blood
R_{aw}	Airway resistance
FI_{O_2}	Fractional concentration of inspired oxygen
V_{DS}	Volume of dead space air
$\dot{V}_A/\dot{Q}$	Alveolar ventilation-perfusion ratio
$\dot{V}_A$	Alveolar ventilation
$\dot{V}_E$	Expired minute ventilation
$\dot{V}_{O_2}$	Oxygen consumption per minute

Note: A dot above a primary symbol denotes flow per unit time.

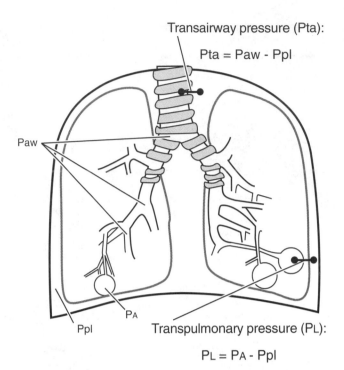

Transairway pressure (Pta):

Pta = Paw - Ppl

Paw

P_A

Ppl

Transpulmonary pressure (PL):

PL = PA - Ppl

Figure 19–4 ■ Three important pressures involved in breathing: alveolar pressure (PA), pleural pressure (Ppl), and transmural pressure (Ptm). There are two transmural pressures that change during inflation and deflation of the lungs: transpulmonary pressure (PL) and transairway pressure (Pta). Both PL and Pta can be defined as the pressure *inside* minus the pressure *outside*. In both cases the pressure outside is pleural pressure (Ppl). (Modified from Selkurt EE: *Physiology.* 5th ed. Boston, Little, Brown, 1984.)

Three important pressures are associated with breathing and air flow (Fig. 19-4): **pleural pressure** (Ppl), the pressure in the pleural fluid between the lung and chest wall; **alveolar pressure** (PA), the pressure inside the alveoli; and **transmural pressure** (Ptm), the pressure difference across an airway or across the lung wall, defined as the pressure inside the wall minus the pressure outside the wall. There are two major transmural pressures to consider in breathing. First is **transpulmonary pressure** (PL), the pressure gradient across the lung wall (Fig. 19-4). Transpulmonary pressure is measured by subtracting pleural pressure from alveolar pressure: PL = PA − Ppl. Transpulmonary pressure keeps the lungs from collapsing, and an increase in PL is responsible for inflating the lungs. The second transmural pressure important in the mechanics of breathing is **transairway pressure** (Pta), the pressure difference across the airways: Pta = Paw − Ppl. Transairway pressure is very important in keeping the airways open during forced expiration. One way to remember transairway or transpulmonary pressure is "*In* minus *Out*," where Ppl is the pressure outside the lung or airway.

Why is pleural pressure negative or subatmospheric? The lung and chest wall are both **elastic** (i.e., capable of being stretched and then recoiling to unstretched configuration, like a spring). At the end of expiration, the lung and chest wall are stretched in equal but opposite directions (Fig. 19-5A). The elastic lung has the potential to recoil inwardly, and the elastic chest wall has the potential to recoil outwardly. These opposing forces cause the pleural pressure to drop below

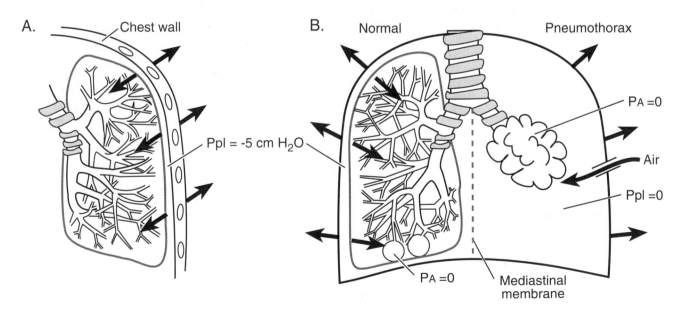

A. Chest wall

Ppl = -5 cm H$_2$O

B. Normal / Pneumothorax

PA = 0

Air

Ppl = 0

PA = 0

Mediastinal membrane

Figure 19–5 ■ (A) The stretched lung and chest wall pull in equal but *opposite* directions at the end of a normal expiration (the force is equal but opposite in direction). Consequently, pleural pressure (Ppl) becomes less than atmospheric pressure. (B) Rupture of the lung or chest wall results in pneumothorax. The lungs collapse because there is no transpulmonary pressure to keep them inflated. Note that the mediastinal membrane prevents the other lung from collapsing. (Modified from Selkurt EE: *Physiology.* 5th ed. Boston, Little, Brown, 1984.)

atmospheric pressure. Pleural pressure is negative during quiet breathing and becomes more negative with deep inspiration. Only during forced expiration does pleural pressure become positive.

An example of the elastic recoil is seen when the chest wall is cut open during thoracic surgery. The stretched lungs collapse immediately (recoil inward), and the stretched rib cage simultaneously expands outward (recoil outward). Since the normal pleural pressure is subatmospheric, air rushes into the pleural cavity any time the chest wall or lung is punctured. As a result, the pressure difference across the lung is

eliminated ($P_L = 0$), and the stretched lung collapses, a condition known as a **pneumothorax** (Fig. 19-5B). A pneumothorax can occur when the chest wall is punctured (e.g., knife or gunshot wound) or when the lung is ruptured from an abscess or severe coughing. A pneumothorax can be performed clinically by inserting a sterile needle between the rib cage and injecting nitrogen.

Pressure changes during a normal breathing cycle are illustrated in Figure 19-6. At the end of expiration, the respiratory muscles are relaxed and there is no air flow. At this point, alveolar pressure is zero (equal to atmospheric pressure or barometric pressure), and the average pleural pressure is −5 cm H_2O. Transpulmonary pressure is therefore 5 cm H_2O [0 − (−5) = 5]. Remember that the transpulmonary pressure is always positive in normal breathing, and positive transpulmonary pressure is sometimes referred to as the distending pressure that keeps the lungs inflated. The more positive transpulmonary pressure becomes, the more the lung inflates.

Inflation of the lungs is an active, muscular process initiated by contraction of the diaphragm. If we start inhalation from the end of a maximal expiration, the chest can be felt to expand as we breathe in. At no time, as our lungs fill, do we feel the need to close our epiglottis to keep the air in. This is because air is held in our lungs by only a slight pressure difference. In the example shown in Figure 19-6A, pleural pressure goes from −5 to −8 cm H_2O. One of the basic characteristics of fluids, such as air, is that the pressures between two regions tend to equilibrate. So as pleural pressure decreases, transpulmonary pressure increases, and the lungs and alveoli inflate. Alveolar pressure (P_A), therefore, drops below atmospheric pressure (Fig. 19-6B). This produces a pressure difference between the mouth and alveoli ($P_B − P_A$), which causes air to rush into the alveoli. Air flow stops at the end of inspiration because alveolar pressure again equals atmospheric pressure (Fig. 19-6C). This sequence of events is summarized in Figure 19-7.

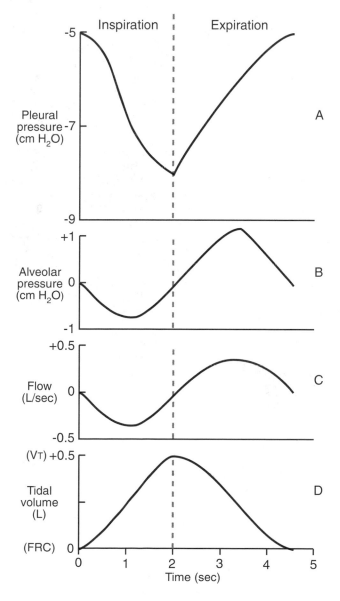

Figure 19-6 ■ Dynamics of a normal breathing cycle (12 breaths/min). The inspiratory time (2 seconds) is less than expiratory time (3 seconds). This is due in part to a higher air flow resistance during expiration, as indicated by a higher alveolar pressure change during expiration (1.2 cm H_2O) than during inspiration (0.8 cm H_2O).

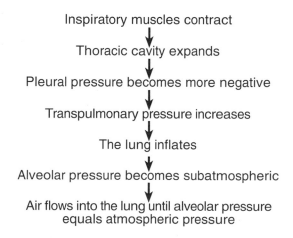

Inspiratory muscles contract

Thoracic cavity expands

Pleural pressure becomes more negative

Transpulmonary pressure increases

The lung inflates

Alveolar pressure becomes subatmospheric

Air flows into the lung until alveolar pressure equals atmospheric pressure

Figure 19-7 ■ Sequence of events during the inspiratory phase of the breathing cycle.

During expiration, the inspiratory muscles relax, the rib cage drops, pleural pressure becomes less negative, transpulmonary pressure decreases, and the stretched lungs and their alveoli deflate. Alveolar pressure becomes greater than atmospheric pressure and pushes air out of the alveoli through the airways until alveolar pressure equals atmospheric pressure. Remember, expiration is completely passive (muscles relax and the stretched lung recoils with normal breathing at rest). Only during forced expiration (e.g., during exercise) do the expiratory intercostal and abdominal muscles contract and actively compress the thoracic cavity.

■ Spirometry Is Used to Measure Lung Volumes and Air Flow

The volume of air that is breathed in or out of the lungs can be measured using a **spirometer** (Fig. 19–8). This device is a simple volume recorder consisting of a double-walled cylinder in which an inverted bell is immersed in water to form a seal. The bell is attached by a pulley to a pen that writes on a rotating drum. When air enters the spirometer from the lungs, the bell rises. Because of the pulley arrangement, the pen is lowered. Thus, a downward pen deflection represents expiration and an upward pen deflection represents inspiration. The recording is known as a **spirogram.** The slope of the spirogram on the moving drum measures rate of air flow and the amplitude of the pen deflection measures the volume of air. Volume is plotted on the vertical axis (ordinate or y-axis) and time on the horizontal axis (abscissa or x-axis).

The volume of air entering or leaving the lungs during a single breath is called **tidal volume** (V_T). Under resting conditions, V_T is approximately 500 ml

in the average adult and represents only a fraction of the air in the lungs. The maximum amount of air in both lungs at the end of a maximal inhalation is **total lung capacity** (TLC) and is approximately 6 L in a young adult. Another important spirometry measurement is the volume of air remaining in the lungs at the end of a normal tidal volume (end of expiration), the **functional residual capacity** (FRC). Note the use of "volume" in the first term and "capacity" in the next two. Volume is used when only one volume is involved and capacity is used when a volume can be broken down into two or more smaller volumes; for example, **expiratory reserve volume** (ERV) plus **residual volume** (RV) equals FRC. The various lung volumes and capacities are summarized in Table 19–3. The maximum volume of air that can be exhaled after a maximum inspiration is the **vital capacity** (VC). When exhalation is performed as rapidly and as forcibly as possible, this volume is called **forced vital capacity** (FVC) and is about 5 L in an adult (Fig. 19–8). Forced vital capacity is one of the most useful lung measurements to assess ventilatory function of the lung. To measure FVC, the subject inspires maximally and then exhales into the spirometer as forcefully, rapidly, and completely as possible.

From FVC, two additional determinations can be obtained. One is FEV_1, the forced expiratory volume of air exhaled in 1 second (Fig. 19–9). This volume has the least variability of the measurements obtained from a forced expiratory maneuver and is considered one of the most reliable spirometry measurements. Another useful way of expressing FEV_1 is as a percentage of FVC (i.e., $FEV_1/FVC \times 100$), which corrects for differences in lung size. Normally, FEV_1 is 80% of FVC

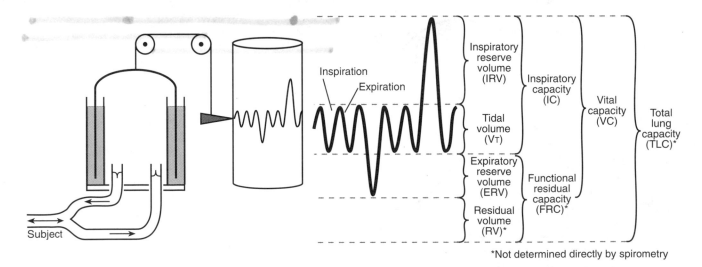

*Not determined directly by spirometry

Figure 19–8 ■ Lung volumes are measured with a spirometer. (Left) Cross-section of spirometer. With expiration, the pen shows a downward deflection. (Right) Lung volumes and capacities. Note that total lung capacity, residual volume, and functional residual capacity cannot be measured directly with the spirometer. (Modified from Selkurt EE: *Physiology.* 5th ed. Boston, Little, Brown, 1984.)

TABLE 19–3 ■ Abbreviations and Definitions Used in Pulmonary Function Testing*

ERV	Expiratory reserve volume: maximum volume of air exhaled at the end of a tidal volume (1.2 liters)
FRC	Functional residual capacity: volume of air remaining in lungs at the end of a tidal volume (2.4 liters)
IC	Inspiratory capacity: maximum volume of air inhaled after a normal expiration (3.6 liters)
IRV	Inspiratory reserve volume: maximal volume of air inhaled at end of normal inspiration (3.1 liters)
FVC	Forced vital capacity: maximum volume of air that can be forcefully exhaled after inhalation to total lung capacity (4.8 liters)
FEV_1	Timed forced expiratory volume: maximum volume of air forcibly exhaled in 1 second (3.8 liters)
$FEV_1/FVC\%$	Forced expired volume-forced vital capacity ratio: percent of forced vital capacity forcibly exhaled in 1 second (80%)
FEF_{25-75}	Forced midexpiratory flow: maximal midflow rate, measured by drawing line between points representing 25% and 75% of forced vital capacity (4.7 liters per second)
RV	Residual volume: volume of air remaining in lungs after maximal expiration (1.2 liters)
TLC	Total lung capacity: volume of air within both lungs at the end of a maximal inhalation (6.0 liters)
RV/TLC	Residual volume-total lung capacity ratio: percent of total lung capacity composed of residual volume (20%)
EPP	Equal pressure point: the point at which pressure inside the airway equals the outside pressure
V_T	Tidal volume: volume of air inhaled or exhaled with each breath (0.5 liters)
VC	Vital capacity: maximum volume of air that can be inhaled or exhaled (4.8 liters); forced vital capacity is commonly used to test lung mechanics
V_D/V_T	Dead space-tidal volume ratio: fraction of tidal volume made up of dead space (30%)

*Values in parentheses represent a typical value for a young adult male

(i.e., 80% of an individual's FVC can be exhaled in the first second). Forced vital capacity and FEV_1 are important measurements in the diagnosis of certain types of lung diseases.

A second measurement obtained from FVC, **forced expiratory flow** (FEF_{25-75}) (Fig. 19-9), has the greatest sensitivity in terms of detecting early air flow obstruction. This measurement represents the expiratory flow rate over the middle half of the forced vital capacity (between 25% and 75%). It is obtained by identifying the 25% and 75% volume points of FVC (Fig. 19-9), measuring the time between these two points, and calculating the flow rate (L/sec).

Because the lungs cannot be emptied completely following forced expiration, neither RV nor FRC can be measured directly by simple spirometry. Instead they are measured indirectly using a dilution technique involving helium, an inert and insoluble gas that is not readily taken up by the lungs. The subject breathes through a special mouthpiece connected to a spirometer filled with 10% helium in air (Fig. 19-10). The initial 10% concentration of helium in the spirometer is designated as C_1, and the volume of air in the spirom-

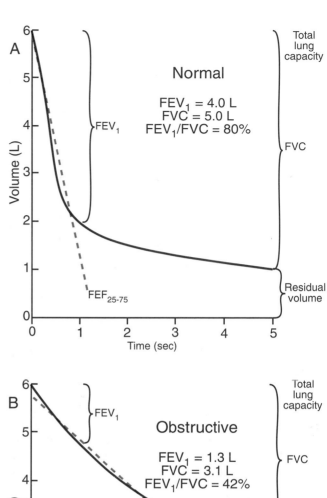

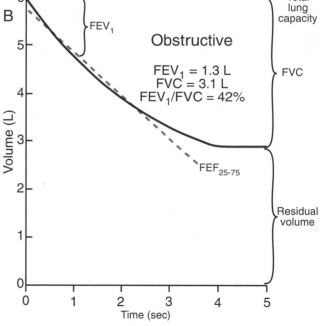

Figure 19–9 ■ Spirogram of forced vital capacity (FVC). The subject inspired maximally to total lung capacity and then exhaled forcefully and as completely as possible into the spirometer. From forced vital capacity two other measurements can be obtained: the forced expiratory volume in 1 second (FEV_1) and the flow rate over the middle half of the forced vital capacity (FEF_{25-75}). Measurements of FVC, FEV_1, and FEF_{25-75} are used to detect obstructive and restrictive disorders. Expiratory flow is obstructed in an obstructive disorder, whereas in a restrictive disorder lung inflation is restricted. In an obstructive disorder (e.g., emphysema) FVC, FEV_1, FEV_1/FVC, and FEF_{25-75} are all reduced. In a restrictive disorder (e.g., fibrosis) lung volumes (TLC, FRC, RV) are reduced.

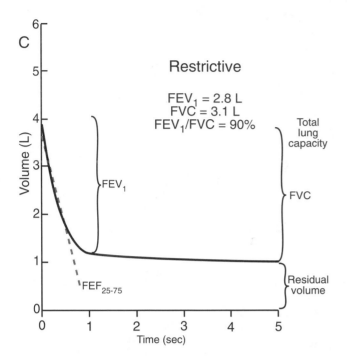

Restrictive

$$FEV_1 = 2.8\ L$$
$$FVC = 3.1\ L$$
$$FEV_1/FVC = 90\%$$

Figure 19–9 ■ Continued

eter is V_1. Since the lung initially contains no helium, the helium concentration in the lungs becomes the same as in the spirometer after equilibration after the subject breathes the helium-air mixture. The concentration of helium in the lung after equilibration is C_2, and the unknown volume in the lung is V_2. By conservation of mass, $C_1V_1 = C_2V_1 + C_2V_2$. By rearranging:

$$V_2 = \frac{V_1(C_1 - C_2)}{C_2} \qquad (1)$$

It is important to start the test at precisely the right time. If the test begins at the end of a normal tidal volume (end of expiration), the volume of air remaining in the lung represents FRC. However, if the test begins at the end of a FVC and the subject starts breathing the helium-air mixture, then the test measures RV.

The helium dilution technique is an excellent test for the measurement of FRC and RV in normal individuals, but has a major limitation in patients whose lungs are poorly ventilated because of plugged airways or high airway resistance. In these diseased lungs, helium gives a falsely low FRC value. An entirely different approach to measure FRC that overcomes this problem of trapped gas in the lung is the use of the **body plethysmograph,** or body box (Fig. 19–11). The subject is seated comfortably in an airtight box in which changes in pressure and volume can be measured very accurately. The subject breathes against a closed mouthpiece at a particular lung volume. During expiration against a closed mouthpiece, the pressure in the lung increases, the chest volume decreases, and the surrounding volume decreases. Because the body box is airtight, the decreased chest volume is reflected by a decrease in surrounding volume. An inspiratory effort against a closed mouthpiece produces just the opposite effect: the pressure inside the lung decreases as the chest expands. In Figure 19–11, the lung volume is determined by applying Boyle's law, which states that in a gas at a constant temperature the product of pressure and volume is a constant.

■ Not All of the Inspired Air Reaches the Alveoli

Ventilation is a dynamic process. If 500 ml of air are inspired with each breath (V_T) and the breathing rate

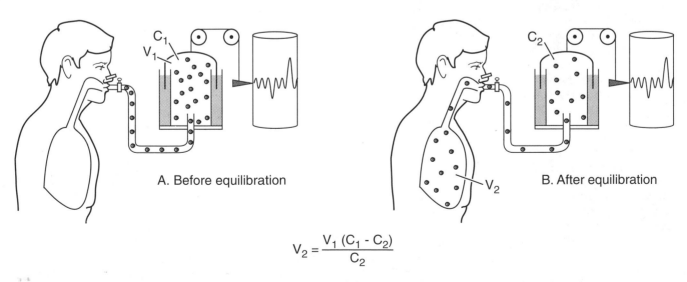

A. Before equilibration

B. After equilibration

$$V_2 = \frac{V_1(C_1 - C_2)}{C_2}$$

Figure 19–10 ■ The helium dilution technique used to measure FRC and RV. Dots represent helium before and after equilibration. C, concentration; V, volume. (Modified from Rhoades R, Pflanzer R: *Human Physiology.* 2nd ed. Fort Worth: Saunders, 1992.)

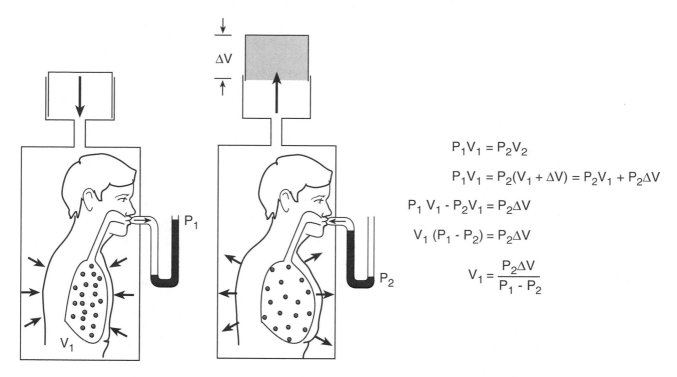

Figure 19–11 ■ Measurement of functional residual capacity (FRC) with body plethysmograph. The subject sits in the airtight box and breathes air through a mouthpiece. The mouthpiece is closed at the end of a normal expiration and the subject continues to breathe against a closed mouthpiece. P_1 is the lung pressure at the end of expiration and P_2 at the end of inspiration. ΔV is the volume of change accompanying inspiration and expiration and is due to changes in thoracic volume against a closed mouthpiece.

$$P_1 V_1 = P_2 V_2$$

$$P_1 V_1 = P_2(V_1 + \Delta V) = P_2 V_1 + P_2 \Delta V$$

$$P_1 V_1 - P_2 V_1 = P_2 \Delta V$$

$$V_1 (P_1 - P_2) = P_2 \Delta V$$

$$V_1 = \frac{P_2 \Delta V}{P_1 - P_2}$$

(f) is 14 times a minute, then the total volume of inspired air that enters the lungs each minute is 500 × 14 = 7000 ml/min, or 7 L/min. This volume of air is known as **expired minute ventilation** ($\dot{V}E$) and can be represented by:

$$\dot{V}E = V_T \times f \qquad (2)$$

Note that $\dot{V}E$ (expired minute ventilation) is used to indicate minute ventilation ($\dot{V}$) and is based on the assumption that the volume of air exhaled equals the volume inhaled. This is not quite true, because there is more oxygen consumed than carbon dioxide produced. This difference, for all practical purposes, is ignored.

Not all of the inspired air enters the alveoli for gas exchange. The tidal volume is distributed between the conducting airways and alveoli. Since gas exchange occurs only in the alveoli and not in the conducting airways, a fraction of the minute ventilation is wasted air. For each 500 ml of air inhaled, approximately 150 ml remain in the conducting airways and do not participate in gas exchange (Fig. 19–12). This volume of wasted air is known as **anatomic dead space volume** (VD). Picture what occurs during a normal breathing cycle. A normal tidal volume of 500 ml is expired.

During the next inspiration, another 500 ml are taken in, but the first 150 ml of air entering the alveoli is wasted air (old alveolar gas left behind). Thus, only 350 ml of fresh air reach the alveoli and 150 ml are left in the conducting airways. The normal ratio of dead space volume to tidal volume (VD/VT) is in the range of 0.25–0.35. In the example shown, the ratio (150/500) is 0.30, which means 30% of the tidal volume or 30% of the minute ventilation does not participate in gas exchange and constitutes wasted air.

Dead space is not confined to the conducting airways alone. Any time alveolar air does not participate in gas exchange, it also becomes wasted air. For example, if inspired air is distributed to alveoli that have no blood flow, this constitutes dead space, referred to as **alveolar dead space** (Fig. 19–13). Alveolar dead space is not confined to alveoli in which no blood flow occurs. Gas exchange units that have reduced blood flow exchange less inspired air than usual for normal gas exchange (Figure 19–13). Thus, any portion of alveolar air that is in excess of that needed to maintain normal gas exchange constitutes alveolar dead space. Thus, dead space may be either anatomic or alveolar in nature. The sum of the two types of dead space is the **physiologic dead space** (physiologic dead space = anatomic + alveolar).

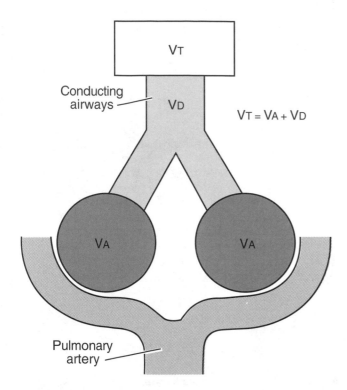

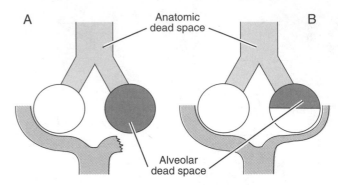

Figure 19–13 ■ Dead space volume can occur in alveoli as alveolar dead space. (A) There is no blood flow to an alveolar region. (B) There is reduced blood flow. In both cases there is a portion of alveolar air that does not participate in gas exchange; this constitutes alveolar dead space. Note that physiologic dead space equals alveolar dead space plus anatomic dead space. (Modified from Selkurt EE: *Physiology.* 5th ed. Boston, Little, Brown, 1984.)

Figure 19–12 ■ Distribution of tidal volume (VT) between conducting airways and alveolar space. VD is dead space volume in the conducting airways and constitutes anatomic dead space. VA is the volume of alveolar air from a tidal volume, not the total volume of air in the alveolar space. (Modified from Selkurt EE: *Physiology.* 5th ed. Boston, Little, Brown, 1984.)

Dead space, anatomic dead space, alveolar dead space, and physiologic dead space are terms that often lead to confusion. Basically, the terms denote the volume of inspired gas that does not participate in gas exchange. In one case, there is a fraction of VT that does not reach the alveoli (anatomic dead space). In another, a fraction of the VT reaches the alveoli, but there is reduced or no blood flow (alveolar dead space). Physiologic dead space represents the sum of anatomic and alveolar dead space. In normal individuals physiologic dead space is approximately the same as anatomic dead space. Only in lung disease does physiologic dead space increase, because of the lung's inability to match blood flow and ventilation in various lung regions. This relationship between alveolar venti-

lation and pulmonary blood flow in the lung is discussed in more detail later in Chapter 20.

■ Alveolar Ventilation Is the Amount of Fresh Air that Reaches Alveoli

The volume of fresh air actually reaching the alveoli is known as **alveolar ventilation** ($\dot{V}A$). The value of $\dot{V}A$ is determined by subtracting dead space volume from tidal volume and multiplying the result by breathing frequency: $\dot{V}A = (VT - VD) \times f$. For example, the volume of air entering the alveoli from a normal tidal volume of 500 ml with a breathing rate of 14 breaths/min is (500 ml − 150 ml) × 14 = 4900 ml/min. Alveolar ventilation is one of the most important determinants in gas exchange. This is because minute ventilation does not represent the amount of oxygen reaching the alveoli, since the amount of dead space varies. For instance, a swimmer using a snorkel breathes through a tube that increases dead space volume. Similarly, a patient connected to a mechanical ventilator has increased dead space volume. Indeed, if minute ventilation is held constant, then alveolar ventilation is decreased with snorkel breathing or mechanical ventilation.

The significance of dead space volume, minute ventilation, and alveolar ventilation is readily seen in Table 19–4. Subject A's breathing is rapid and shallow, B's

TABLE 19–4 ■ Influence of Breathing Patterns on Alveolar Ventilation

Subject	Tidal Volume (ml)	×	Frequency (breaths/min)	=	Minute Ventilation (ml/min)	−	Dead Space Ventilation (ml/min)	=	Alveolar Ventilation (ml/min)
A	150	×	40	=	6000	−	6000 (150 × 40)	=	0
B	500	×	12	=	6000	−	1800 (150 × 12)	=	4200
C	1000	×	6	=	6000	−	900 (150 × 6)	=	5100

breathing is normal, and C's breathing is slow and deep. Each subject has the same total minute ventilation (i.e., each is moving the same volume of total air in and out per minute and has the same dead space volume), but each has marked differences in alveolar ventilation. Subject A has no alveolar ventilation and would die in a matter of minutes, while subject C has an alveolar ventilation greater than normal. The important message from these examples is that increasing the *depth* of breathing is far more effective in elevating alveolar ventilation than increasing the frequency of breathing. This is important in exercise. In most exercise situations, increased ventilation is accomplished by increases in the depth of breathing more than in the rate. For example, a well-trained athlete can increase alveolar ventilation during light exercise with little or no increase in breathing frequency.

■ V̇A Can Be Measured Indirectly from Expired Carbon Dioxide

Alveolar ventilation is difficult to determine because anatomic dead space is not easily measured. Often, dead space is approximated for a seated subject by assuming that dead space (ml) is equal to the subject's weight in pounds (e.g., a subject who weighs 170 lb would have a dead space volume of 170 ml). This assumption is fairly reliable for healthy individuals. In the pulmonary laboratory, alveolar ventilation can be calculated from the volume of expired carbon dioxide and fractional concentration of carbon dioxide in the alveolar gas (Fig. 19–14). Since no gas exchange occurs

in the conducting airways and the inspired air contains essentially no carbon dioxide, all of the expired carbon dioxide originates from alveoli: Therefore:

$$\dot{V}_{CO_2} = \dot{V}_A \times F_{ACO_2} \tag{3}$$

where $\dot{V}_{CO_2}$ is the volume of carbon dioxide expired per minute and F_{ACO_2} is the fractional concentration of carbon dioxide in alveolar gas. Rearranging

$$\dot{V}_A = \frac{\dot{V}_{CO_2}}{F_{ACO_2}} \tag{4}$$

This equation is known as the **alveolar ventilation equation.** The carbon dioxide concentration can be obtained by sampling the last portion of the tidal volume during expiration, **end-tidal volume,** which contains alveolar gas.

Alveolar ventilation can also be determined from the partial pressure of carbon dioxide in the blood. However, before proceeding, we review some basic physical gas laws. First is **Dalton's law**, which states that each gas in a mixture is independent of other gases present and exerts a pressure according to its own concentration. The pressure of each gas is its **partial pressure,** or **gas tension.** The total pressure, or barometric pressure, is the sum of the individual partial pressures of the gases present and can be written as:

$$P_B = P_X + P_Y + P_Z \tag{5}$$

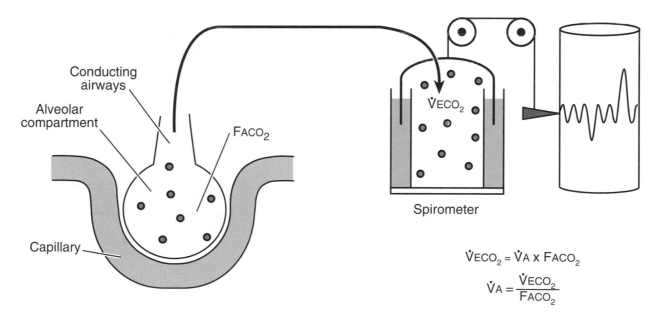

Figure 19–14 ■ Carbon dioxide in the expired air can be used to calculate alveolar ventilation, since all of the CO₂ in the expired air originates from the alveoli. The concentration of CO₂ in the alveoli is represented by dots and is designated as fractional concentration of CO₂ in alveolar gas (F$_{ACO_2}$).

where P_B is barometric (total) pressure and $P_X + P_Y + P_Z$ are partial pressures of individual gases X, Y, and Z. The partial pressure of any gas can be determined by measuring barometric pressure and the fractional concentration of the gas. For example, the partial pressure of oxygen (P_{O_2}) is determined as follows

$$P_{O_2} = P_B \times F_{O_2} \qquad (6)$$

where P_{O_2} is the partial pressure of oxygen in mm Hg, P_B is barometric pressure in mm Hg, and F_{O_2} is the fractional concentration of oxygen. Table 19-5 lists normal partial pressures of respiratory gases in different locations in the body.

At sea level, the partial pressure of oxygen is 160 mm Hg (760 mm Hg $\times$ 0.21). However, when air is inspired, it is warmed to 37°C by the nasal passage and conducting airways. As the air is warmed, water evaporates from the respiratory epithelium lining and the air becomes saturated with water vapor (water in the gas phase). Water vapor pressure does not change with barometric pressure but stays constant with a given temperature. Thus, a partial pressure (P_{H_2O}) of 47 mm Hg is exerted at 37°C. Pulmonary physiologists ignore the water vapor pressure of inspired air by considering the gas only after it has been saturated with water vapor, referred to as **tracheal air.** Thus about 6% of the tracheal air (47/760 mm Hg) is made up of water vapor.

Since alveolar partial pressure of CO_2 (P_{ACO_2}) is equal to F_{ACO_2} times total alveolar gas pressure:

$$\dot{V}_{ECO_2} = \dot{V}_A \times P_{ACO_2} \times K \qquad (7)$$

where K is a constant that takes into account the different units and the conversion of F_{ACO_2} to P_{ACO_2} (1/K = 0.863). Rearranging:

$$P_{ACO_2} = \frac{\dot{V}_{ECO_2} \times 0.863}{\dot{V}_A} \qquad (8)$$

Therefore:

$$\dot{V}_A = \frac{\dot{V}_{ECO_2} \times 0.863}{P_{ACO_2}} \qquad (9)$$

Since P_{ACO_2} is in equilibrium with the partial pressure of carbon dioxide in the arterial blood (P_{aCO_2}), the latter value can be used to calculate $\dot{V}_A$, as follows:

$$\dot{V}_A = \frac{\dot{V}_{ECO_2} \times 0.863}{P_{aCO_2}} \qquad (10)$$

It is important to recognize the inverse relationship between $\dot{V}_A$ and P_{aCO_2} (Fig. 19-15). If alveolar ventilation is halved, P_{aCO_2} will double (assuming a steady state and constant carbon dioxide production). This decrease in alveolar ventilation is called **hypoventilation.** Conversely, increased ventilation, referred to as **hyperventilation,** leads to a fall in P_{aCO_2}. The two examples shown in Figure 19-15 illustrate that $\dot{V}_A$ is an important determinant of P_{aCO_2}.

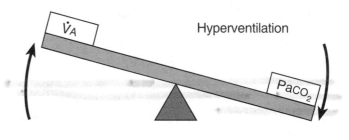

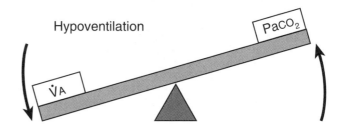

Figure 19-15 ■ Alveolar ventilation is an important determinant of arterial P_{CO_2}. If alveolar ventilation is halved arterial P_{CO_2} will double, and vice versa. (Modified from Selkurt EE: *Physiology.* 5th ed. Boston, Little, Brown, 1984.)

TABLE 19–5 ■ Partial Pressure and Percentages of Respiratory Gases at Sea Level (Barometric Pressure = 760 mm Hg)

Gas	Ambient Dry Air		Moist Tracheal Air		Alveolar Gas		Systemic Arterial Blood		Mixed Venous Blood	
	mm Hg	%	mm Hg	%	mm Hg	%	mm Hg	%	mm Hg	%
O_2	160	21	150	20	102	14	102	14	40	6
CO_2	0	0	0	0	40	5	40	5	46	6
Water Vapor	0	0	47	6	47	6	47	6	47	7
N_2	600	79	563	74	571	75[a]	571	75	571	81
Total	760	100	760	100	760	100	760	100	704[b]	100

[a]Alveolar P_{N_2} increased by 1% because R < 1
[b]Total pressure in venous blood is reduced because P_{O_2} decreased more than P_{CO_2} increased

CHRONIC OBSTRUCTIVE PULMONARY DISEASE (COPD)

Chronic bronchitis, emphysema, and asthma are some of the major pathophysiologic disorders of the lung. **Bronchitis** is an inflammatory condition that affects one or more bronchi. **Emphysema** stems from overdistention of alveoli and loss of lung elastic recoil. **Asthma** is marked by spasmodic contraction of smooth muscle in the bronchi. Although the pathophysiology and etiology of chronic bronchitis, emphysema, and asthma are quite different, they all are chronic obstructive pulmonary diseases (COPD). The hallmark of COPD is the slowing down of air movement during forced expiration, which leads to a decrease in FVC and FEV_1. Chronic obstructive pulmonary disease is characterized by chronic obstruction of the small airways. There are three ways air flow can become obstructed: excessive mucus production (as occurs in bronchitis), airway narrowing due to bronchial spasms (as in asthma), airway collapse during expiration (as occurs in emphysema). The latter stems from the loss of lung elastic recoil. In severe COPD, air is trapped in the lungs during forced expiration, leading to abnormally high residual volume.

Chronic bronchitis and emphysema often coexist. Bronchitis leads to excessive mucus production, which plugs the small airways. A cough is produced to clear the excess mucus from the airways; on repeated exposure to a bronchial irritant, such as tobacco smoke, a persistent cough develops. As a result, alveoli become overdistended and can rupture from excessive pressure, especially in plugged airways.

Late in COPD, patients often experience hypoxemia (low blood oxygen) and hypercapnia (high blood carbon dioxide). These two conditions, especially hypoxemia, lead to pulmonary hypertension (high blood pressure in the pulmonary circulation), which causes right heart failure. Based on changes in blood oxygen levels, COPD patients can be categorized into two types. Those exhibiting predominantly emphysema are referred to as "pink puffers" because their oxygen levels are usually satisfactory and their skin remains pink. They develop a puffing style of breathing. Those manifesting predominantly chronic bronchitis are called "blue bloaters" because low oxygen levels give their skin a blue cast and fluid retention from heart failure gives them a bloated appearance.

Chronic obstructive pulmonary disease is by far the most common chronic lung disease in the United States. More than 10 million Americans suffer from chronic bronchitis and emphysema, and COPD is the fifth leading cause of death in the United States. The single most common cause of chronic bronchitis and emphysema is tobacco smoke.

If alveolar ventilation increases, P_{AO_2} will also increase. Doubling alveolar ventilation, however, will not lead to a doubling of P_{AO_2}. The quantitative relationship between alveolar ventilation and P_{AO_2} is more complex than that for P_{ACO_2}, for two reasons. First, the inspired P_{O_2} is not zero. Second, the **respiratory exchange ratio** (R), defined as the ratio of the volume of carbon dioxide exhaled to the volume of oxygen taken up (V_{CO_2}/V_{O_2}), is usually less than 1, which means more oxygen is removed from the alveolar gas per unit time than carbon dioxide is added.

The alveolar partial pressure of oxygen (P_{AO_2}) can be calculated by using the **alveolar air equation:**

$$P_{AO_2} = P_{IO_2} - P_{ACO_2} [F_{IO_2} + (1 - F_{IO_2})/R] \tag{11}$$

where P_{IO_2} is the partial pressure of inspired oxygen (moist tracheal air), P_{ACO_2} is the partial pressure of carbon dioxide in the alveoli, F_{IO_2} is the fractional concentration of oxygen in the inspired air, and R is the respiratory exchange ratio. When R = 1, the complex term in the bracket equals 1. If a subject breathes 100% oxygen for a brief period of time, F_{IO_2} = 1.0, and the correction factor in the brackets also reduces to 1. The reason is that there is no nitrogen in the alveoli. In a normal resting individual breathing air at sea level, with an R of 0.82 and Pa_{CO_2} of 40 mm Hg, and $PI_{O_2} = 150$ mm Hg, Pa_{O_2} is calculated as follows:

$$Pa_{O_2} = 150 \text{ mm Hg} - 40 \text{ mm Hg}$$
$$[0.21 + (1 - 0.21)/0.82]$$
$$Pa_{O_2} = 150 \text{ mm Hg} - 40 \text{ mm Hg} \times [1.2]$$
$$Pa_{O_2} = 102 \text{ mm Hg}$$

Since PI_{O_2} and FI_{O_2} stay fairly constant and Pa_{CO_2} equals Pa_{CO_2} the alveolar air equation can be simplified to $Pa_{O_2} = 150 - 1.2 (Pa_{CO_2})$.

The Mechanics of Breathing

The vascular and airway trees are embedded in elastic tissue. When the lungs are inflated, this elastic component must be stretched. The degree of lung expansion at any given time during inflation is proportional to transpulmonary pressure. How well a lung inflates and deflates with a change in transpulmonary pressure is a function of its elastic property. Obviously, a stiff lung that is not easily stretched will not inflate to the same volume as a highly distensible one for the same pressure change.

Lung tissue elasticity is due to elastin and collagen fibers enmeshed around the alveolar walls, adjacent bronchioles, and small blood vessels. Elastin fibers are highly distensible and can be stretched to almost double their resting length. Collagen fibers resist stretch and limit lung expansion at high lung volumes. As the lungs expand during inflation there is an unfolding and rearrangement of the fiber network around alveoli, small blood vessels, and small airways. A good analogy is stretching a nylon stocking. When the stocking is stretched, there is not much change in individual fiber length, but the unfolding and rearrangement of the nylon mesh allows the stocking to be stretched.

There are three terms used to assess the elastic properties of the lung: **distensibility,** the ease with which the lung can be inflated; **stiffness,** the ability to resist stretch; and **elastic recoil,** the ability of a stretched lung to recoil back to its original position. Stiffness and elastic recoil are directly related to each other: the stiffer the material, the greater the ability of the stretched material to return to its unstretched position and hence the greater the elastic recoil. For example, when a stiff spring is stretched, it literally snaps back once it is released. Overstretching causes materials to lose their elastic recoil. In the case of nylon stockings, loss of elastic recoil causes the stocking to sag or become baggy. In the same way, a lung that loses its elastic recoil also becomes "baggy": it is easy to inflate but is very difficult to deflate, because of its inability to recoil back.

■ Compliance Is a Measure of Lung Distensibility

The elastic properties of a lung can be determined by measuring the changes in lung volume that occur with changes in pressure and plotting them as a **pressure-volume curve.** A simple analogy is the inflation of a balloon with a syringe (Fig. 19–16). For each change in pressure, the balloon inflates to a new volume. The slope of the line is known as **compliance**

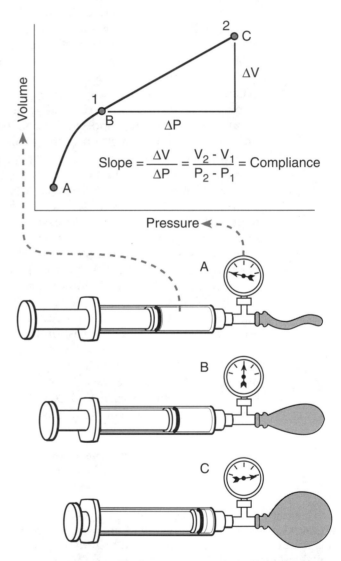

Figure 19–16 ■ Elastic properties of the lungs (distensibility, stiffness, and elastic recoil) are measured from a pressure-volume curve. A simple analogy is inflating a balloon and measuring the change in pressure and volume. For each change in pressure (shown by movement of the arrow in the manometer dial), the balloon inflates to a new volume. Compliance is defined as the slope of the line between any two points on a pressure-volume curve. The unit of measure for lung compliance is L/cm H_2O. (Modified from Rhoades R, Pflanzer R: *Human Physiology.* 2nd ed. Fort Worth, Saunders College Publishing, 1992.)

($C_L = \Delta$ volume/Δ pressure). Compliance is a measure of distensibility. A similar pressure-volume curve can be generated for the human lung by measuring a simultaneous change in lung volume with a spirometer and a change in pleural pressure with a pressure gauge (Fig. 19-17). Since the esophagus passes through the pleural space, changes in pleural pressure can be obtained indirectly from the pressure in the esophagus by using a balloon catheter. Under conditions of no air flow, alveolar pressure is zero. In practice, a pressure-volume curve is obtained by having the subject first inspire to total lung capacity and then expire slowly, periodically interrupting air flow, and simultaneously recording lung volume and pleural pressure (Fig. 19-17). Under these conditions, the volume change per unit pressure change ($\Delta V/\Delta P$) is called **static compliance,** because no air flow is occurring. Thus, the slope of the pressure-volume curve at any given point is lung compliance (C_L), a measure of lung distensibility. The reciprocal of compliance ($1/C_L$) provides a measure of lung stiffness.

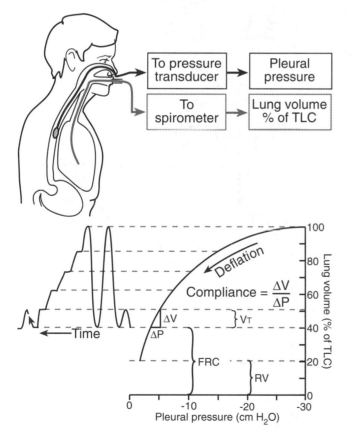

Figure 19–17 ■ Lung compliance is determined from a pressure-volume curve. The subject first inspires maximally to total lung capacity (TLC) and then expires slowly, while air flow is periodically stopped so pleural pressure and lung volume can be measured simultaneously. Note that pleural pressure is determined by measuring esophageal pressure. (Modified from Selkurt EE: *Physiology.* 5th ed. Boston, Little, Brown, 1984.)

Compliance depends on a number of factors that include not only elastic properties but also lung volume, lung size, and surface forces inside the alveoli. Since the pressure-volume curve of the lung is nonlinear, compliance is not the same at all lung volumes but is decreased at high lung volumes and increased at low lung volumes. In the midrange of the pressure-volume curve, lung compliance is about 0.2 L/cm H_2O. A mouse lung will have a smaller volume change per unit of pressure change than an elephant lung. To correct for different sizes, lung compliance is normalized by dividing compliance by FRC to give a **specific compliance.** The specific compliance of an adult lung is about 0.08 L/cm H_2O (0.2 L/cm H_2O/2.5 L), while that of a newborn is about 0.06 L/cm H_2O (0.005 L/cm H_2O/0.08 L). Normalizing lung compliance allows comparisons between lung sizes; when specific compliance is used, a similar value of about 0.08 L/cm H_2O is found for a wide spectrum of adult mammalian lungs, ranging from mouse to elephant.

What is the functional significance of an abnormally high or abnormally low compliance? Low compliance (Fig. 19-18) indicates a stiff lung and means more work is required to bring in a normal volume of air. Such stiffness can result from lung injuries (by infection, toxins, or environmental insults). The lungs become fibrotic, lose their distensibility, and become stiffer. A highly compliant lung is also detrimental. In **emphysema,** the elastic tissue has been damaged, usually by being overstretched from chronic coughing. Patients with emphysema have a very high lung compliance; they have no problem inflating the lungs but have extreme difficulty exhaling air. The poor elastic recoil is responsible, and extra work is required to get air out of the lungs.

Just as the lung has elastic properties, so does the chest wall. The chest wall is pulled in by the inward

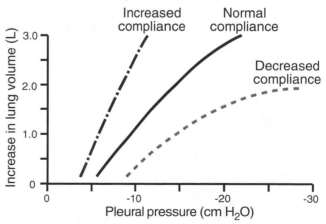

Figure 19–18 ■ Pressure-volume relationships in a patient with increased compliance and a patient with decreased compliance. Patients with obstructive disorders have high lung compliance and those with restrictive disorders have low compliance.

elastic recoil of the lung, while lung expansion is aided by the outward elastic recoil of the chest wall. The elastic recoil of the chest wall is such that if the chest were unopposed by the recoil of the lung, it would expand to about 70% of total lung capacity. This volume represents the resting position of the chest wall unopposed by the lung. If the chest wall is mechanically expanded beyond its resting position, it will recoil inward. At volumes less than 70% of total lung capacity, the recoil of the chest wall is directed outward and is opposite the elastic recoil of the lung. The outward elastic recoil of the chest wall is greatest at residual volume, whereas the inward elastic recoil of the lung is greatest at total lung capacity. The increased stiffness of the chest wall at low lung volumes is a major determinant of residual volume in young people. The lung

volume at which lung and chest are at equilibrium (i.e., equal elastic recoil but in opposite directions) is the FRC. A change in the elastic properties of either lung or chest wall has a marked effect on FRC. For example, if the elastic recoil of the lung is increased (i.e., lower C_L), then a new equilibrium is established between the lung and chest wall, resulting in a decreased FRC. Similarly, if the elastic recoil of the chest wall is increased, FRC is higher than normal.

Lung compliance has an effect on ventilation and causes air to be unevenly distributed. In a normal upright individual, compliance at the top part of the lung is less than at the base. This difference in compliance between the apex and base, known as **regional compliance,** is due to gravitational effects (Fig. 19-19). These effects occur because the lung is ap-

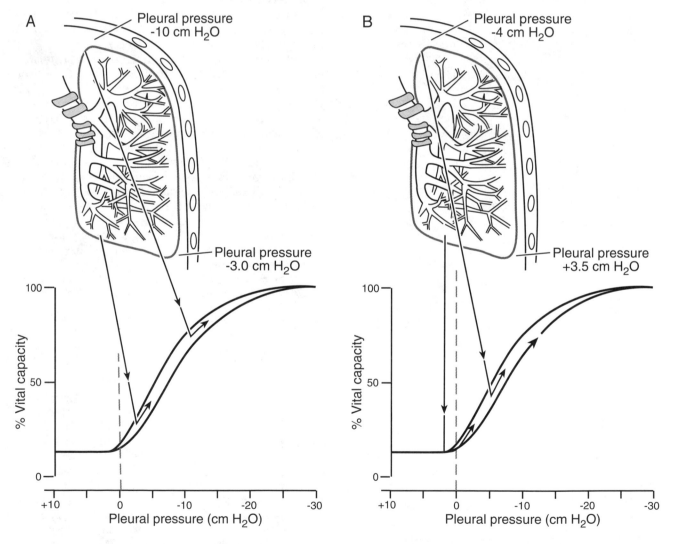

Figure 19–19 ■ Regional compliance affects the distribution of air flow. (A) Because of gravity, pleural pressure in the upright lung is more negative at the apex than at the base. At functional residual capacity the lower part of the lung is more compliant. (B) At low lung volumes (residual volume) pleural pressure at the apex is less negative and is actually positive at the base. As a result, the apical region is on the steeper part of the pressure-volume curve and is more compliant than the base. (Adapted from West JB: *Respiratory Physiology.* 4th ed. Baltimore, Williams & Wilkins, 1990, p 98.)

proximately 80% water and gravity exerts a "downward pull" on them, which results in a lower pleural pressure (i.e., more negative) at the apex than at the base.

Because of the gravitational effects on pleural pressure, the transpulmonary pressure in an upright lung is greater at the apex of the lung than at the base (Fig. 19–19A). The higher transpulmonary pressure at the apex causes the alveoli to be more expanded and leads to regional differences in compliance. At any given volume from the FRC and above, the apex of the lung is stiffer than at the base. This means the base of the lung has both a larger change in volume for the same pressure change and a smaller resting volume than at the apex. In other words, as one takes a breath in from FRC, a greater portion of the tidal volume will go to the base of the lung, consequently greater ventilation occurs at the base.

At low lung volumes, a change in regional ventilation occurs (Fig. 19–19B). At lung volumes approaching residual volume, the pleural pressure at the base of the lung actually exceeds the pressure inside the airways, leading to airway closure. At residual volume, the base is compressed, and ventilation in the base is impossible until pleural pressure falls below atmospheric pressure. By contrast, the apex of the lung is in a more favorable portion of the compliance curve and ventilates well. This means the first portion of the breath taken in from residual volume enters alveoli in the apex. Thus, the distribution of ventilation is inverted at low lung volumes (i.e., the apex is better ventilated than the base).

■ Surfactant Lowers Surface Forces and Stabilizes Alveoli

Another property that markedly affects lung compliance is surface tension at the air-liquid interface of the alveoli. The surface of the alveolar membrane is moist and is in contact with air, producing a large air-liquid interface. Surface tension (measured in dyne/cm) arises because water molecules are more strongly attracted to one another than to air molecules. In alveoli surface tension produces an inwardly directed force that tends to reduce surface area.

Surface forces of the lung can be studied by examining a pressure-volume curve. As seen in Figure 19–20, volume is plotted as a function of transpulmonary pressure in an excised lung that has been inflated and deflated in a stepwise fashion, first with air and then with saline. With air-filled lungs, the gas-liquid interface creates surface tension. With saline-filled lungs, however, the air-liquid interface and consequently surface tension are eliminated.

Two important observations can be seen by comparing these two pressure-volumes curves. First, the slope of the deflation limb of the saline curve is much steeper than that of the air curve. This means that when surface tension is eliminated, the lung is far more compliant (more distensible). Second, the different areas to the

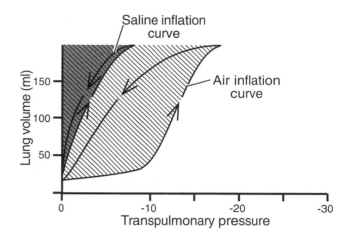

Figure 19–20 ■ Inflating and deflating an excised lung with saline minimizes surface tension, resulting in increased compliance. Two pressure-volume curves for a lung inflated to the same volume, first with air and then with saline, are shown. The differences in the two curves result because surface tension contributes significantly to lung compliance. The shaded areas are equal to the work done in inflating the lung.

left of saline and air inflation curves show that surface tension significantly contributes to the work required to inflate the lungs. Since the area to the left of each curve is equal to work, which can be defined as force (change in pressure) times distance (change in volume), the elastic force and surface forces can be separated. The area to the left of the saline inflation curve is the work required to overcome the elastic properties of the lung alone. The area to the left of the air inflation curve is the work required to overcome both elastic and surface forces. Subtracting the area to the left of the saline curve from the area to the left of the air curve shows that approximately two-thirds of the work required to inflate the lungs is needed to overcome surface forces. Thus, lung distensibility as well as elastic recoil are affected by surface forces.

Surface tension has important ramifications in maintaining alveolar stability. In a sphere such as an alveolus, surface tension produces a force that pulls inward and creates a pressure. The relationship between surface tension and pressure inside a sphere is shown in Figure 19–21. The pressure developed in an alveolus is given by Laplace's law (Chap. 12) for a sphere, which states that transmural pressure is equal to twice the surface tension (T) divided by the radius (r): $P = 2T/r$.

Since alveoli are interconnected and vary in diameter, Laplace's law assumes functional importance in the lung (Fig. 19–21A). In the example shown in the figure, surface tension is constant at 50 dyne/cm. An unstable condition results because pressure in the smaller alveolus is greater than in the larger one. This causes smaller alveoli to collapse, especially at low lung volumes, a phenomenon known as **atelectasis.** When atelectasis occurs larger alveoli become overdistended. Thus, two questions arise: How does the nor-

A.   No surfactant

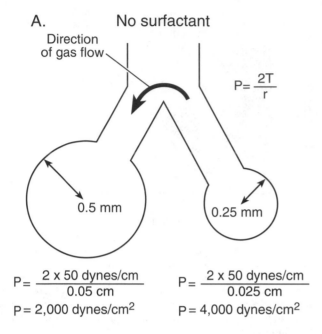

Direction of gas flow

$P = \dfrac{2T}{r}$

0.5 mm

0.25 mm

$P = \dfrac{2 \times 50 \text{ dynes/cm}}{0.05 \text{ cm}}$

$P = 2{,}000 \text{ dynes/cm}^2$

$P = \dfrac{2 \times 50 \text{ dynes/cm}}{0.025 \text{ cm}}$

$P = 4{,}000 \text{ dynes/cm}^2$

B.   Surfactant

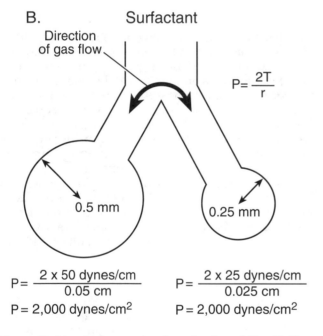

Direction of gas flow

$P = \dfrac{2T}{r}$

0.5 mm

0.25 mm

$P = \dfrac{2 \times 50 \text{ dynes/cm}}{0.05 \text{ cm}}$

$P = 2{,}000 \text{ dynes/cm}^2$

$P = \dfrac{2 \times 25 \text{ dynes/cm}}{0.025 \text{ cm}}$

$P = 2{,}000 \text{ dynes/cm}^2$

Figure 19–21 ■ Surface tension alters alveolar stability. (A) If surface tension remains constant (50 dyne/cm) then alveoli that are interconnected but vary in diameter cannot coexist. Pressure in the smaller alveolus is greater than in the larger alveolus, which causes air from the smaller alveolus to empty into the larger alveolus. At low lung volumes, the smaller alveoli tend to collapse. (B) Surfactant promotes alveolar stability by lowering surface tension proportionately more in the smaller alveolus. Pressures in the two alveoli are equal, therefore alveoli of different diameters can coexist.

mal lung keep alveoli open? How do alveoli of different sizes coexist when interconnected?

The answer lies in the fact that the alveolar surface tension is not constant at 50 dyne/cm, as seen in other biologic fluids. The alveolar surface lining is coated with a special surface-active agent, **surfactant,** which not only lowers surface tension but also changes surface tension with changes in alveolar diameter (Fig. 19–21B). Pulmonary surfactant is a lipoprotein rich in phospholipid. The principal agent responsible for its surface tension-reducing properties is **dipalmitoyl-phosphatidylcholine** (DPPC). The functional importance of surfactant can be demonstrated by using a surface tension balance (Fig. 19–22). Surface tension is measured by placing a platinum strip connected to a force transducer into a liquid trough. Surface force is created by forming a meniscus on each side of the platinum flag. The greater the contact angle of the meniscus, the greater the surface force. The surface film is repeatedly compressed and expanded (simulating lung expansion-compression) by a movable barrier that just skims the surface of the liquid. Surface tension of pure water is 72 dyne/cm, a value that is *independent* of the surface area of water in the balance. Therefore, when the surface film is expanded and compressed, surface tension does not change. When a detergent is added, surface tension is drastically reduced but again is independent of surface area. When lung surfactant is added, however, surface tension not only is reduced but also changes in a nonlinear fashion with surface area (Fig. 19–22). Thus, surfactant has a detergentlike property and also lowers surface tension with surface area. Therefore, surfactant makes it possible for alveoli of different diameters that are connected in parallel to coexist and be stable at low lung volumes, by lowering surface tension proportionately more in the smaller alveoli.

Surfactant has considerable stability. It works by generating a film that reduces surface tension at the gas-liquid interface. When the molecules are compressed during lung deflation, surfactant causes a decrease in surface tension. At low lung volumes, when the molecules are tightly compressed, some surfactant is squeezed out of the surface and forms micelles (Fig. 19–23). On expansion (reinflation), new surfactant is required to form a new film that is spread on the alveolar surface lining. The spreading of new film is one of the reasons the expansion limb of the surface tension area curve is not the same as that of the compression limb. This difference in the inflation and deflation curve, called **hysteresis,** is due to the fact that molecules are not arranged in the same fashion during inflation as in deflation.

Often during quiet and shallow breathing surface area remains fairly constant, and the spreading of surfactant is impaired. A deep sigh or yawn causes the lungs to inflate to a larger volume and new surfactant molecules to spread onto the gas-liquid interface. Patients recovering from anesthesia are often encouraged

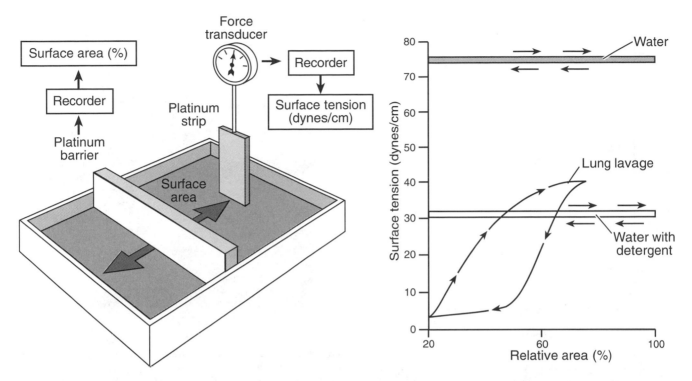

Figure 19–22 ■ Surface tension balance. When distilled water is placed in the surface tension balance, surface tension is *independent* of area. The addition of a detergent reduces surface tension, but it is still independent of area. When a lung lavage mixture is placed in the surface tension balance, surface tension not only *changes* with area but is proportionately lower with decreased area. (Modified from Selkurt EE: *Physiology.* 5th ed. Boston, Little, Brown, 1984.)

to breathe deeply to enhance the spreading of surfactant. Patients who have undergone abdominal or thoracic surgery often find it too painful to breathe deeply; poor surfactant spreading results, causing part of their lungs to become atelectatic.

Surfactant is synthesized in the **alveolar type II cell.** The alveolar epithelium consists basically of two cell types: alveolar type I and alveolar type II (or alveolar type II pneumocyte) (Fig. 19–24). The ratio of these cells in the epithelial lining is about 1:1, but type I cells occupy approximately two-thirds of the surface area. Type II cells seem to form aggregates around the alveolar septa. They are rich in mitochondria, compared with type I cells, and are metabolically quite active. Electron-dense **lamellar inclusion bodies** are a distinguishing feature of the type II cell. These lamellar inclusion bodies, rich in surfactant, are thought to be the storage sites for surfactant.

The process of surfactant metabolism is shown in Figure 19–25. Substrates (glucose, palmitate, and choline) are taken up from the circulating blood and synthesized by alveolar type II cells. Stored surfactant from the lamellar inclusion bodies is discharged onto the alveolar surface. The turnover of surfactant is high because of the continual renewal of surfactant to the alveolar surface during each expansion of the lung. The high rate of replacement of surfactant probably accounts for the active lipid synthesis that occurs in the lung.

Since the lung is one of the last organs to develop, the synthesis of surfactant appears rather late in gestation. In humans, surfactant appears at about week 34 (term is 40 weeks). Regardless of the total duration of gestation in any mammalian species, the process of lung maturation seems to be "triggered" about the time gestation is 85–90% complete. Clearly, the fetal lung is endowed with a special regulatory mechanism to control the timing and appearance of surfactant. Failure of proper lung maturation during the perinatal period is still a major cause of death in neonates. The lung may be structurally intact but is functionally immature, because adequate amounts of surfactant are not available to reduce surface tension and stabilize surface forces during breathing. Premature birth and certain hormonal disturbances (e.g., those seen in diabetic pregnancies) interfere with the normal control and timing of lung maturation. These infants have immature lungs at birth, which often leads to **infant respiratory distress syndrome** (IRDS). Breathing is extremely labored because surface tension is high, making it difficult to inflate the lungs. Because of the high surface tension, these infants develop pulmonary edema and atelectasis. They are at high risk until the lung becomes mature enough to secrete surfactant.

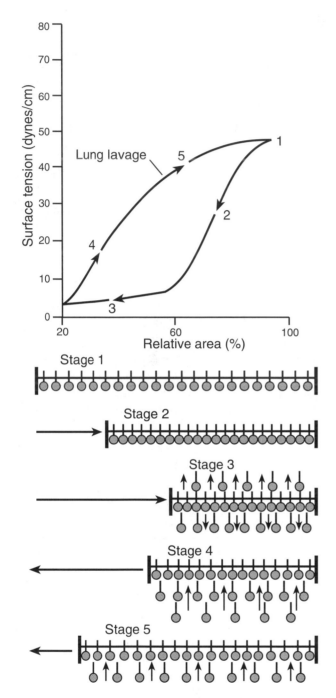

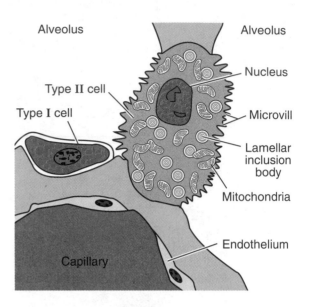

Figure 19–24 ■ Schematic drawing of the two types of alveolar cells and their relationship to alveoli and capillaries. Alveolar type I cells comprise most of the alveolar surface. Alveolar type II cells are located in the corner between two adjacent alveoli. Also shown are endothelial cells that line the pulmonary capillaries.

Figure 19–23 ■ Surfactant function at the alveolar level. Surfactant molecules are compressed during lung deflation. At stage 3, surfactant molecules form micelles and are removed from the surface. On lung inflation, new surfactant is spread onto the surface film (stage 4). Turnover is high because of continual replacement of surfactant during lung expansion.

Infant respiratory distress syndrome is still the major cause of death in newborns in North America.

In addition to lowering alveolar surface tension and promoting alveolar stability, surfactant helps prevent edema in the lung. The inward-contracting force that tends to collapse alveoli also tends to lower interstitial pressure, which "pulls" fluid from the capillaries. Pulmonary surfactant reduces this tendency by lowering surface forces. Some pulmonary physiologists think that keeping the lungs dry may be the major role of surfactant, especially in adults.

Another mechanism that plays a role in promoting alveolar stability is **interdependence,** or mutual support of adjacent lung units. Alveoli (except those next to the pleural surface) are all interconnected with surrounding alveoli and thus support each other. Studies have shown that this type of structural arrangement, with many connecting links, prevents the collapse of adjacent alveoli. For example, if a lung unit tended to collapse, large expanding forces would be developed by surrounding units. Thus, interdependence can play a role in preventing atelectasis as well as in opening up lungs that have collapsed. It seems to be more important in adults than in neonates, because neonates have fewer interconnecting links.

■ Airway Resistance Decreases Air Flow in the Lung

Two basic types of air flow occur in the lung. **Turbulent air flow** occurs in the large airways (the trachea and large bronchi) and at high flow rates. It consists of completely disorganized patterns of air flow, resulting in a sound that can be heard with inhalation

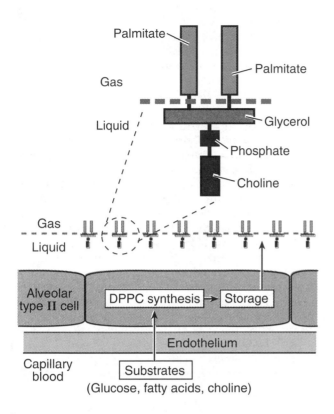

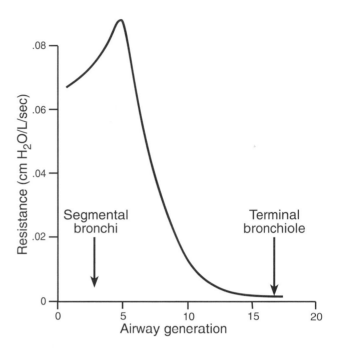

Figure 19–26 ■ Resistance to air flow is greatest in large and medium-sized airways. The major site of resistance is in lobar and segmental bronchi, down to about the seventh generation of airway branches. (Modified from Selkurt EE: *Physiology.* 5th ed. Boston, Little, Brown, 1984.)

Figure 19–25 ■ Schematic illustration of biosynthesis and secretion of surfactant. Substrates are taken up by alveolar type II cells from the pulmonary capillaries and are synthesized into dipalmitoylphosphatidycholine (DPPC). Surfactant is stored in lamellar inclusion bodies and subsequently discharged onto the alveolar surface. Surfactant is oriented perpendicular to the gas-liquid interface of the alveolar surface; the polar end is immersed in the liquid phase, while the nonpolar portion is upright in the gas phase. (Modified from Selkurt EE: *Physiology.* 5th ed. Boston, Little, Brown, 1984.)

and exhalation. The faster and deeper the breathing, the more noise produced from turbulence. In contrast, **laminar flow** is characterized by streamlined flow that runs parallel to the sides of the airways. Laminar flow is silent because layers of air molecules slide over each other. This type of flow occurs mainly in the small peripheral airways, where air flow is exceedingly slow.

Airway resistance is defined as the ratio of driving pressure (ΔP) to air flow ($\dot{V}$). For total airway resistance (Raw), the driving pressure is the pressure difference between airway opening (Pao) and the alveoli (PA)

$$Raw = \frac{Pao - PA}{\dot{V}} \qquad (12)$$

where resistance is expressed as cm H_2O/L/sec.

The major site of airway resistance is in the medium-sized bronchi (lobar and segmental) and bronchi down to about the seventh generation (Fig. 19-26). Based

on Poiseuille's equation (Chap. 12), in which the radius is raised to the fourth power, one would expect the major site of resistance to be located in the narrow airways (the bronchioles), which have the smallest radius. Pulmonary physiologists long thought this to be true. However, measurements show that only 10–20% of total airway resistance can be attributed to the small airways (those less than 2 mm in diameter). This apparent paradox results because so many small airways are arranged in parallel and their resistances are added as reciprocals. This means the resistance of each individual bronchiole is relatively high, but their greater number results in a large total cross-sectional area, causing their total *combined resistance* to be low. The fact that small airways account for such a low percentage of Raw causes a serious problem in diagnosing airway disorders, because airway diseases most often begin in the small airways. Early detection is difficult because changes in airway resistance cannot be detected until the disease becomes severe.

The smaller airways, like lung tissue, are capable of compression or distention. Since bronchi and smaller airways are embedded in lung parenchyma, they are connected by "guy wires" to surrounding tissue. Thus, a major factor affecting airway diameter, especially bronchioles, is lung volume. Figure 19-27 shows the effect of lung volume on airway resistance. As the lung enlarges, airway diameter increases, and airway resistance (Raw) falls. Conversely, at low lung volumes,

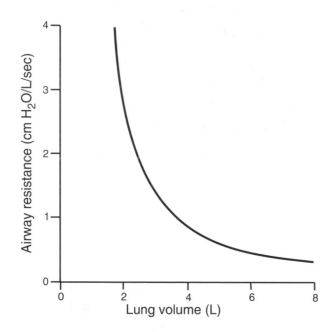

Figure 19–27 ■ Airway resistance as a function of lung volume. Airways are compressed at low lung volumes, resulting in increased airway resistance. (Modified from Selkurt EE: *Physiology.* 5th ed. Boston, Little, Brown, 1984.)

airways are compressed, diameter decreases, and airway resistance rises. Note that the inverse relationship between lung volume and airway resistance is nonlinear. At lower lung volumes, Raw rises sharply.

Bronchial smooth muscle tone also affects airway resistance. A change in smooth muscle tone will change airway diameter. The smooth muscles in the airway, from the trachea down to the terminal bronchioles, are under autonomic control. **Parasympathetic** stimulation of the cholinergic postganglionic fibers causes bronchial constriction as well as increased mucus secretion. **Sympathetic** stimulation of adrenergic fibers causes dilation of bronchial and bronchiolar airways and inhibition of glandular secretion. Drugs such as isoproterenol and epinephrine, which stimulate β_2-adrenergic receptors in the airways, cause dilation. These drugs alleviate bronchial constriction and are often used to treat asthmatic attacks. Environmental insults, such as breathing chemical irritants, dust, or smoke particles, can cause a reflex constriction of the airways. Increased P_{CO_2} in the conducting airways can cause a local dilation. More important, a decrease in P_{CO_2} causes airway smooth muscle to constrict.

Gas density also affects airway resistance. This is seen most dramatically in deep-sea diving, in which the diver breathes air whose density may be greatly increased because of increased pressure. As a result of increased resistance, a large pressure gradient is required just to move a volume of air equivalent to tidal volume. One reason a helium-oxygen mixture is often breathed during diving as a substitute for air is that helium is less dense and consequently makes

breathing easier. The fact that density has such a marked effect on airway resistance again indicates that the medium-sized airways are the main site of resistance and that air flow is primarily turbulent flow.

■ Airways Become Compressed during Forced Expiration

Airway resistance does not change much during normal quiet breathing. During maximal breathing, however, it is markedly increased, especially during forced expiration. The marked change seen during forced expiration is due to compression of the airways. This can be demonstrated with a **flow-volume curve,** which shows the relationship between air flow and lung volume during a forced vital capacity measurement (Fig. 19–28). The curve is generated by having a subject inspire maximally to total lung capacity and then exhaling to residual volume as forcibly, rapidly, and completely as possible. During forced vital capacity, flow rises very rapidly to a maximal value and then decreases linearly over most of expiration as lung volume decreases. A family of flow-volume curves can be generated by varying expiratory efforts over the entire vital capacity. For example, if a person exhales from total lung capacity but exhalation is initially slow and then forced, curve B in Figure 19–28 might be obtained. The maximal flow rate would be somewhat reduced, but over much of expiration the flow-volume curve would be almost superimposed on the curve obtained from a sustained forced expiration (curve A in Fig. 19–28). If the subject exhaled slowly and halfway

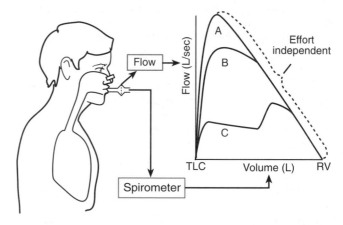

Figure 19–28 ■ Flow rate may be independent of effort during forced expiration. The flow-volume curve shows the relationship between lung volume (x-axis) and air flow (y-axis) during forced expiration. (A) During maximal forced expiration, air flow rises rapidly to a maximum before lung volume has changed much. Then air flow decreases linearly as lung volume decreases to residual volume (RV). (B) Expiration is initially slow and then forced. (C) Expiratory effort is submaximal. Both B and C show that regardless of the starting volumes forced expiration, expiratory flow cannot exceed the maximum value for that volume. The expiratory portions of the flow-volume curves are effort-independent because of dynamic airway compression.

through expiration exhaled as fast as possible, curve C in Figure 19–28 might be obtained. The latter part of the flow-volume curve is superimposed on the first curve when the subject exhales forcibly, rapidly, and as completely as possible. These curves show that air flow increases progressively with effort at lung volumes close to total lung capacity. However, flow is independent of effort over 75% of the flow-volume curve. This means that over most of a forced vital capacity, flow is virtually independent of effort, whether effort is great or small. This suggests that resistance to flow increases with the increase in driving pressure. The increase in resistance occurs because the airways are compressed, which effectively limits flow rate.

How do airways become compressed during forced expiration? The mechanism is related to changes in transairway pressure and the compressibility of the airways (Fig. 19–29). Before inspiration, pleural pressure is −5 cm H_2O and airway pressure is zero (no air flow). Transairway pressure (Paw − Ppl), the pressure that holds the airway open, is 5 cm H_2O. At the start of a maximal inspiration, pleural pressure decreases to −7 cm H_2O, and alveolar pressure falls to −2 cm H_2O. The difference between alveolar pressure and

pleural pressure is still 5 cm H_2O [(−2−(−7)=5]. However, there is a pressure drop from mouth to alveoli because of resistance to air flow, so the transairway pressure will vary up the airway. At the end of maximal inspiration, pleural pressure decreases further, to −12 cm H_2O, and airway pressure is again zero because of no air flow. Thus, during maximal inspiration, airway resistance actually decreases because transairway pressure increases, which enlarges the diameter of the small airways.

On forced expiration, two striking changes occur (Fig. 19–29). First, pleural pressure is no longer negative but rises above atmospheric pressure and can increase up to +30 cm H_2O. Second, alveolar pressure becomes greater than pleural pressure. The added pressure in the alveoli is due to the elastic recoil of the lungs. The pressure difference between alveolar pressure and pleural pressure is 8 cm H_2O because at the beginning of expiration lung volume has not appreciably decreased. Note that at the beginning of forced vital capacity, there is a pressure drop again along the airway because of airway resistance. Airway pressure falls progressively (due to resistance to air flow) from the alveolar region toward the airway opening (the mouth). The transmural pressure gradient along the airways reverses and tends to compress the airways. For example, at a point inside the airway where the pressure is 19 cm H_2O, the transmural pressure would be −11 cm H_2O, which would tend to close the airway. At some point along the airway, the airway pressure would equal pleural pressure and transairway pressure would be zero. This is the **equal pressure point** (Fig. 19–30). Theoretically, the equal pressure point (EPP) divides the airways into an **upstream segment** (from alveoli to EPP) and a **downstream segment** (EPP to the mouth). In the downstream segment, the airway pressure falls below pleural pressure and the transairway pressure becomes negative. As a result, airways in the downstream segment are compressed or collapsed. The large airways (the trachea and bronchi) are protected from compression because they are supported by cartilage. However, smaller airways without this structural support are easily compressed and collapsed.

■ Lung Compliance Affects the Equal Pressure Point

The EPP is established when peak flow is achieved; the position of the EPP then becomes fixed. This is important because it means a further increase in pleural pressure with forced expiration will produce more compression on the downstream segment but have little effect on air flow in the upstream segment (from alveoli to EPP). The driving pressure for air flow, once the EPP is established, is no longer the difference between alveolar pressure and mouth pressure but is alveolar pressure minus pleural pressure (Fig. 19–30). Under such conditions, flow is independent of downstream airway resistance.

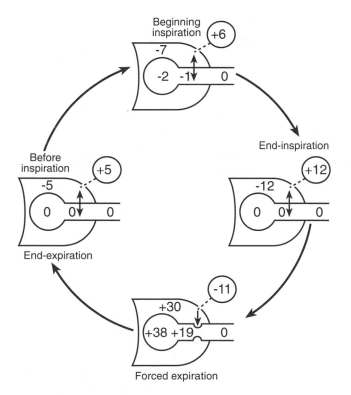

Figure 19–29 ■ Airways are compressed during forced expiration. Transairway pressure is +5 cm H_2O before inspiration and reaches +12 cm H_2O at the end of inspiration. It becomes negative in the downstream airway, and the small airways are compressed during forced expiration. (Modified from Selkurt EE: *Physiology.* 5th ed. Boston, Little, Brown, 1984.)

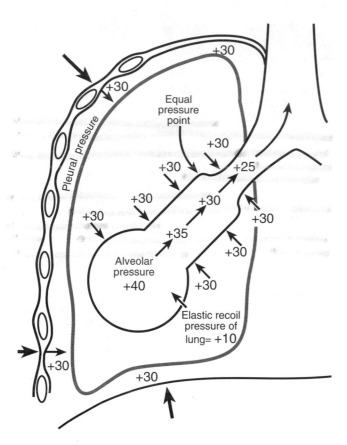

Figure 19–30 ■ With forced expiration, airway pressure falls progressively from the alveoli to the mouth. When the pressure inside the airway equals the pressures outside the airway (pleural pressure) an equal pressure point (EPP) is established. From the EPP to the trachea airways are subjected to compression during forced expiration. (Modified from Selkurt EE: *Physiology.* 5th ed. Boston, Little, Brown, 1984.)

Two basic conclusions follow from this. First, regardless of how forceful the expiratory effort is flow cannot be increased, because as pleural pressure rises alveolar pressure also increases. As a result, the driving pressure (alveolar minus pleural pressure) remains constant. This explains why 75% of the forced vital capacity is not dependent on effort. Second, maximal flow rates are determined mainly by the elastic recoil of the lung, because it is this force that generates the alveolar-pleural pressure difference. As lung volume decreases, so does elastic recoil force. The decrease in elastic recoil is the main reason maximal flow falls so rapidly at low lung volumes.

The effect of elastic recoil on expiratory air flow is best seen when a normal lung is compared with an emphysematous lung, with abnormally high compliance (Fig. 19–31). In both instances, pleural pressure rises to +30 cm H_2O with forced expiration. In the normal lung, the elastic recoil pressure of 10 cm H_2O is added to produce an alveolar pressure of +40 cm H_2O. On forced vital capacity, there is a progressive

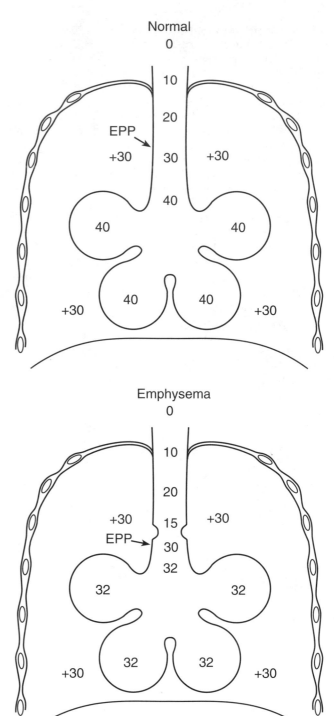

Figure 19–31 ■ Loss of elastic recoil causes a downward shift of the equal pressure point toward the alveoli. In the normal lung, elastic recoil adds 10 cm H_2O pressure, to produce an alveolar pressure of 40 cm H_2O. In the emphysematous lung the elastic recoil pressure is low and adds very little pressure to the alveolus. As a result, the equal pressure point is shifted closer to the alveoli and more airways are compressed.

fall in airway pressure. Because of the elastic recoil, the normal lung has "added" pressure to the smaller airways that keeps their pressures *above* pleural pressure, and less collapse occurs. However, with emphysema the lung has a diminished elastic recoil and adds only 2 cm H_2O to alveolar pressure, resulting in an alveolar pressure of $+32$ cm H_2O. With expiratory effort, flow proceeds along the progressive pressure gradient. However, in this case the pressure inside the smaller airways falls below intrapleural pressure well before the airway leaves the chest cavity, resulting in collapsed airways. As a result, flow stops temporarily in these collapsed airways. Airway pressure upstream to the collapsed segment then rises to equal alveolar pressure, causing the airways to open again. This process repeats itself continually, leading to a "wheeze." In lungs with abnormally high compliance, the position of the EPP with forced expiration is established further down in the smaller airways, where there is no cartilage to keep them distended. Therefore, these patients are much more vulnerable to compression and collapse. The greatest problem for a patient with emphysema is not getting air into the lung but getting it out. Consequently, they tend to breathe at higher lung volumes. This makes the lung stiffer (increases elastic recoil), which reduces airway resistance and enables the patient to exhale more easily.

■ Work Is Required to Expand the Lung and Chest Wall

During inspiration, muscular work is involved in expanding the thoracic cavity, inflating the lungs, and overcoming airway resistance. Since work can be measured as force times distance, the amount of work involved in breathing can be expressed as a change in lung volume (distance) multiplied by the change in transpulmonary pressure (force). Thus, work (W) is equal to the product of pressure (P) and volume (V). If pressure changes during work, then the product is replaced by an integral and is defined by the equation:

$$W = \int PdV \qquad (13)$$

During work, energy is expended with muscular contraction to create a force (transpulmonary pressure) to inflate the lungs. When a greater transpulmonary pressure is required to bring more air into the lungs, more muscular work and hence greater energy is required.

A pressure-volume curve can be used to determine the work required for breathing (Fig. 19-32). In the normal lung point A represents pleural pressure at the end of expiration, when air flow is zero and the lung volume is at functional residual capacity (FRC). Point C represents pleural pressure at the end of inspiration, when flow is again zero, and the lung volume is 1 liter *above* functional residual capacity. Line AEC would, therefore, be the compliance of the lung. The total

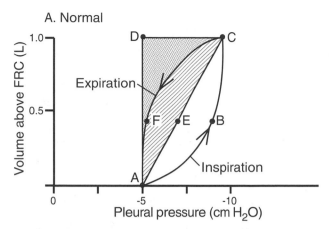

A. Normal

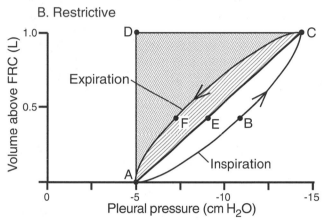

B. Restrictive

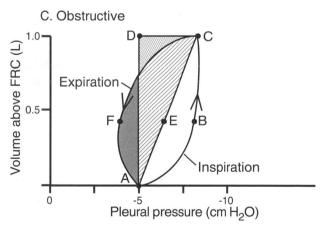

C. Obstructive

Figure 19–32 ■ Work of breathing is determined from a pressure-volume curve. Pleural pressure is plotted on the x-axis and volume at any given pressure on the y-axis. Area AECDA represents work done to overcome just the elastic forces. Area ABCEA represents work done to overcome airway resistance.

work of breathing is given by the area ABCDA. Of this, the work done to overcome just the elastic and surface forces (i.e., to distend the lungs) would be given by the area AECDA. Area ABCEA (the clear area) represents the work needed to overcome airway resistance. The greater the airway resistance or flow rate, the more negative the pleural pressure excursion between A and C becomes and the larger the area. Area AECFA (the hatched area) is work required to overcome airway resistance during expiration. This area usually falls within area AECDA and represents the portion of energy stored in the lungs at the end of inspiration due to elastic recoil. The differences between areas AECFA and AECDA represent the elastic energy not recovered during expiration and dissipated as heat.

In a normal person at rest, the energy needed for breathing represents approximately 5% of the body's total energy expenditure. During heavy exercise, only about 20% of the total energy expenditure is involved in breathing. Breathing is very efficient and is most economical when elastic and resistive forces yield the lowest work. It is important to remember that with higher flow rates, a larger work area (ABCEA) is needed to overcome airway resistance. In contrast, with larger tidal volumes, a larger work area (AECDA) is needed to overcome elastic and surface forces. Thus, patients with stiff lungs (low compliance) economize by taking rapid and shallow breaths.

Patients with severe airway obstruction tend to do the opposite: they take deeper breaths and breathe more slowly, to reduce the work. Despite this tendency, patients with obstructive disease still expend a considerable portion of their basal energy for respiratory function. The reason is that during expiration, work must be done by the expiratory muscles to overcome the increased airway resistance due to airway compression (Fig. 19–32). These different breathing patterns help minimize the amount of work required for breathing under these conditions.

■ ■ ■ ■ ■ ■ ■ ■ ■ ■ ■ REVIEW EXERCISES ■ ■ ■ ■ ■ ■ ■ ■ ■ ■ ■

Identify Each with a Term

1. The pressure that keeps the lungs inflated
2. The volume of air that remains in the lungs after maximal expiration
3. The maximal volume of air a person can breathe in after breathing in a normal tidal volume
4. Collapse of a lung resulting from leakage of air into the pleural space
5. Change in lung volume per unit pressure change
6. The volume of fresh air that reaches the alveoli per unit time
7. The volume of air flowing per second during the middle half of a forced vital capacity

Define Each Term

8. Forced vital capacity (FVC)
9. Atelectasis
10. Regional compliance
11. Equal pressure point (EPP)
12. Obstructive disorder
13. Alveolar interdependence

Choose the Correct Answer

14. Pleural pressure is the most negative at:
    a. Residual volume
    b. Functional residual capacity
    c. End of tidal volume
    d. Total lung capacity
15. Transpulmonary pressure is equal to:
    a. Alveolar pressure minus pleural pressure
    b. Airway pressure minus pleural pressure
    c. Atmospheric pressure minus alveolar pressure
    d. Pleural pressure minus atmospheric pressure
16. The lung and chest wall have equal and opposite recoil tendencies at:
    a. Total lung capacity
    b. Functional residual capacity
    c. Residual volume
    d. Inspiratory reserve volume
17. An airway would most likely collapse under which of the following conditions?
    a. Paw = 35 cm H_2O
       Ppl = 30 cm H_2O
    b. Paw = 15 cm H_2O
       Ppl = 20 cm H_2O
    c. Paw = 0 cm H_2O
       Ppl = −5 cm H_2O
    d. Paw = 20 cm H_2O
       Ppl = 0 cm H_2O
18. Which of the following is a normal value for FEV_1/FVC?
    a. 20%
    b. 50%
    c. 60%
    d. 80%
19. In which of the following pulmonary disorders is the lung most likely to show a steep compliance curve?
    a. Edema
    b. Fibrosis
    c. Emphysema
    d. Congestion
20. A threefold increase in the radius of a small peripheral airway would most likely increase air flow by a factor of:
    a. 9
    b. 16
    c. 27
    d. 81
21. In an obstructive lung disorder, which of the following is likely to increase?
    a. TLC
    b. FEF_{25-75}
    c. FVC
    d. FEV_1/FVC

22. Minute ventilation represents:
    a. Total air moved in and out of the lung
    b. Total air moved minus dead space volume
    c. Alveolar ventilation minus dead space volume
    d. Tidal volume divided by frequency of breathing

23. A subject with frequency of 10 breaths/min, vital capacity of 6 liters, minute ventilation of 8 L/min, and functional residual capacity of 3 liters has a tidal volume of:
    a. 500 ml
    b. 600 ml
    c. 800 ml
    d. 650 ml

24. An abnormally compliant lung would:
    a. Be easier to inflate
    b. Be easier to deflate
    c. Require more work to inflate
    d. Have greater elastic recoil
    e. All of the above

25. During breath holding at residual volume with the glottis open, alveolar pressure will equal:
    a. Pleural pressure
    b. Transpulmonary pressure
    c. Airway pressure minus pleural pressure
    d. Atmospheric pressure

26. Which patient has the highest minute ventilation?

	Tidal Volume	Frequency	Dead Space Volume
a.	150 ml	40 breaths/min	150 ml
b.	500 ml	12 breaths/min	150 ml
c.	1000 ml	6 breaths/min	150 ml
d.	Same in all cases		

Calculate

27. A patient inspires 500 ml from a spirometer. Pleural pressure before inspiration is -5 cm H_2O and -10 cm H_2O at the end of inspiration. What is the compliance of this patient's lung?

28. At one point during inspiration, alveolar pressure is -1 cm H_2O, pleural pressure is -7 cm H_2O, airway pressure is -4 cm H_2O, and atmospheric pressure is 750 mm Hg. What is the transpulmonary pressure?

29. An alveolus has a radius of 0.25 mm and a surface tension of 25 dyne/cm. What is the pressure in the alveolus in dyne/cm²?

30. A person has a forced vital capacity of 5 liters, functional residual capacity of 3 liters, and residual volume of 1.2 liters. What is the person's total lung capacity?

31. A subject has an alveolar ventilation of 5 L/min, a frequency of 10 breath/min, and a tidal volume of 700 ml. What is the subject's dead space ventilation?

Briefly Answer

32. Why is FVC so useful as a test in lung spirometry?
33. How does the lung fill and empty during breathing?
34. What role does elastic recoil play in breathing?
35. How do lung size, elastic recoil, and gravity affect lung compliance?
36. What is the functional importance of surfactant?

■ ■ ■ ■ ■ ■ ■ ■ ■ ■ ■ ■ ■ ■ ■ ■ **ANSWERS** ■ ■ ■ ■ ■ ■ ■ ■ ■ ■ ■ ■

1. Transpulmonary pressure (P_L)
2. Residual volume (RV)
3. Inspiratory reserve volume (IRV)
4. Pneumothorax
5. Compliance (C_L)
6. Alveolar ventilation
7. Forced midexpiratory flow (FEF_{25-75})
8. The volume of air a person can exhale as forcibly and rapidly as possible after a maximal inspiration
9. A condition in which alveoli are unstable and collapse
10. Compliance of a lung region, such as the base or apex of the lung
11. The point in the airway where airway pressure equals pleural pressure
12. A pathophysiologic condition of the lung in which air flow out of the lung is obstructed; three disorders (asthma, bronchitis, and emphysema) constitute chronic obstructive pulmonary disease (COPD)
13. A structural interdependence that prevents adjacent alveoli from collapsing
14. d

15. a
16. b
17. b
18. d
19. c
20. d
21. a
22. a
23. c
24. a
25. d
26. d
27. $C_L = \dfrac{\Delta V}{\Delta P} = \dfrac{0.5\ L}{5\ cm} = 0.10\ L/cm\ H_2O$
28. P_L = Alveolar pressure minus pleural pressure
    $= -1 - (-7)$ cm H_2O
    $= 6$ cm H_2O
29. $P = \dfrac{2 \times T}{Radius} = \dfrac{2 \times 25\ dyne/cm}{0.025\ cm} = 2000\ dyne/cm^2$

30. $TLC = VC + RV = 5\ L + 1.2\ L = 6.2\ L$
31. $\dot{V}_D = \dot{V} - \dot{V}_A = 7\ L/min - 5\ L/min = 2\ L/min$
32. Forced vital capacity is one of the most useful measurements in lung spirometry because it varies least of all the measurements obtained from a forced expiration and serves as a basis for three other measurements (FEV_1, FEV_1/FVC, and FEF_{25-75}) that provide important information about air flow and elastic properties of the lung.
33. The lung inflates when the chest cavity expands, causing pressure inside the alveoli to drop below atmospheric pressure, which creates a pressure gradient that causes air to flow into the lung. The lung deflates when the chest cavity decreases in size, causing pressure inside the alveoli to rise above atmospheric pressure and creating a pressure gradient that causes air to flow out of the lung.
34. Elastic recoil of the lung is essential for the exhalation of air from the lung during expiration. Elastic recoil also provides pressure that, in addition to alveolar pressure, keeps airways open during forced expiration.
35. A very small lung will be less compliant than a very large lung. A stiff lung (one with good elastic recoil) will be less compliant than a lung with poor elastic recoil. The base of the lung is more compliant than the apex of the lung.
36. Surfactant lowers surface tension in the alveoli, making the lung more compliant and thereby reducing the work required to inflate it. Also, because it reduces surface tension surfactant prevents edema by reducing the tendency for the alveoli to absorb fluid from the pulmonary capillaries. At low lung volumes, surfactant promotes alveolar stability.

Suggested Reading

Baum GL, Wolinsky E (eds): *Textbook of Pulmonary Disease.* Boston, Little, Brown, 1983.

Forster RE II, Dubois AB, Brisco WA, Fisher AB: *The Lung: Physiological Basis for Pulmonary Function Tests.* 3rd ed. Chicago, Year Book, 1986.

Guenter CA, Welch MH: *Pulmonary Medicine.* Philadelphia, Lippincott, 1982.

Haagsman HP, Van Golde IMG: Synthesis and assembly of lung surfactant. *Annu Rev Physiol* 53:441–464, 1991.

Nunn SF: *Applied Respiratory Physiology.* 3rd ed. Boston, Butterworth, 1987.

Pulmonary Circulation and Ventilation-Perfusion Ratio

OBJECTIVES

After studying this chapter, the student should be able to:
1. Compare and contrast the pulmonary and systemic circulations
2. Explain why pulmonary vascular resistance falls with increased cardiac output
3. Explain the effects of lung volume on pulmonary vascular resistance
4. Describe hypoxic pulmonary vasoconstriction
5. Describe how alveolar surface tension and alveolar pressure affect fluid balance in the pulmonary capillaries
6. Describe pulmonary edema and the consequences of edema in the lungs
7. Describe how gravity affects blood flow in an upright person
8. Describe the relationships between alveolar pressure, pulmonary arterial pressure, and pulmonary venous pressure in three zones of the lung
9. Describe $\dot{V}A/\dot{Q}$ ratio
10. Describe shunts and venous admixture
11. Describe the three functions of bronchial circulation

The heart drives two separate and distinct circulatory systems in the body: pulmonary circulation and systemic circulation. The pulmonary circulation is analogous to the entire systemic circulation. Like the systemic circulation, the pulmonary circulation receives all of the cardiac output and therefore cannot be considered part of regional circulation in the sense that renal, hepatic, or coronary circulation can. A change in pulmonary vascular resistance has the same implication for the right ventricle as a change in systemic vascular resistance has for the left ventricle.

Functional Organization of the Pulmonary Circulation

■ The Pulmonary Circulation Has Several Functions

The pulmonary arterial system branches in the same treelike manner as do the airways. Each time the airway branches, the arterial tree branches such that the two parallel each other (Fig. 20–1). The primary function of the pulmonary circulation is to bring systemic venous blood (i.e., mixed venous blood) into intimate contact

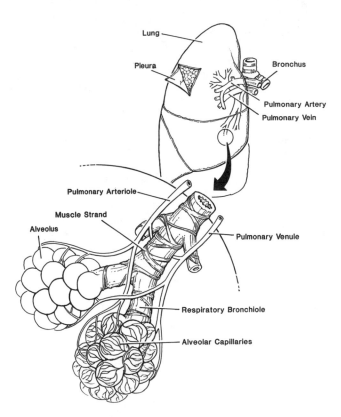

Figure 20–1 ■ The vascular tree follows the airway tree. (A) Systemic venous blood flows through the pulmonary arteries into the lung capillaries for oxygenation and then back to the heart, to be pumped into the systemic circulation. (B) Each alveolus is surrounded by a mesh of capillaries. As the blood passes through the capillaries it takes up oxygen and gives up carbon dioxide.

with alveoli for gas exchange. In addition, the pulmonary circulation has three secondary nonrespiratory functions: it serves as a blood reservoir, a filter, and a metabolic organ. More than 40% of the lung weight is due to blood in the pulmonary blood vessels. The total blood volume of the pulmonary circulation (main pulmonary artery to left atrium) is approximately 500 ml, or 10% of the total circulating blood volume (5000 ml). The pulmonary vascular blood volume, except for the amount in the lung capillaries, is divided equally between arterial and venous vessels. The blood volume in the lung capillaries is approximately equal to the stroke volume of the right heart under most physiologic conditions.

Pulmonary vessels protect the body against obstruction of important vessels in other organs. Obstruction can occur when emboli—fat globules, air, or blood clots—enter the systemic venous blood after surgery or injury. Small pulmonary arterial vessels and capillaries trap these emboli and prevent them from obstructing the vital coronary, cerebral, and renal vessels. Endothelial cells lining the pulmonary vessels release fibrinolytic substances that help dissolve thrombi. Pulmonary capillaries absorb emboli, especially air emboli. However, if a clot occludes a large pulmonary vessel or emboli are extremely numerous and lodge all over the pulmonary arterial tree, gas exchange is impaired, which can cause death.

Another important activity of the pulmonary circulation is the metabolism of vasoactive hormones. One such substance is angiotensin I, which is activated and converted to angiotensin II by angiotensin-converting enzyme (ACE) located on the surface of the pulmonary capillary endothelial cells. Activation is extremely rapid, and 80% of angiotensin I can be converted to angiotensin II during a single passage through the pulmonary vasculature. In addition to being a potent vasoconstrictor, angiotensin II has other important actions in the body (Chap. 24). Metabolism of vasoactive hormones by the pulmonary circulation appears to be rather selective. Some vasoactive hormones, including bradykinin, serotonin, and the prostaglandins E_1, E_2 and $F_{2\alpha}$, are inactivated by pulmonary endothelial cells. Other prostaglandins, such as A_1 and A_2, pass through the lungs unaltered. Norepinephrine is inactivated, but epinephrine, histamine, and vasopressin (ADH) pass through the pulmonary circulation unchanged. With acute lung injury (e.g., oxygen toxicity, fat emboli), lungs can release histamine, prostaglandins, and leukotrienes, which can cause vasoconstriction of pulmonary arteries and cause pulmonary endothelial damage.

■ The Pulmonary Circulation Differs from the Systemic Circulation

The pulmonary circulation, in contrast to the systemic circulation, is a low-pressure, low-resistance system with highly compliant vessels. The pulmonary artery and its branches have much thinner walls than the aorta. The pulmonary artery is much shorter and contains less elastin and smooth muscle in its walls. The pulmonary arterioles are also thin and contain little smooth muscle and consequently have less ability to constrict than the thick-walled, highly muscular systemic vessels. The pulmonary veins are thin-walled and highly compliant and also contain very little smooth muscle compared to their counterparts in the systemic circulation.

Pulmonary and systemic capillaries also differ. Unlike the systemic capillaries, which are often arranged as a network of tubular vessels with some interconnections, the pulmonary capillaries mesh together in the alveolar wall so that blood flows as a thin sheet. It is therefore misleading to refer to pulmonary capillaries as a capillary network; instead, they comprise a dense **capillary bed.** The walls of the capillary bed are exceedingly thin, consequently a whole capillary bed

can collapse if local alveolar pressure exceeds capillary pressure.

Another striking contrast between the systemic and pulmonary circulations is the pressure profile. The pulmonary circulation is a low-pressure but high-flow system. Figure 20–2, a schematic diagram of the pulmonary and systemic circulation, shows normal values for intravascular pressures. Mean pulmonary arterial pressure is 15 mm Hg, compared to 93 mm Hg in the aorta. The driving pressure for pulmonary flow is the small difference (10 mm Hg) between the mean pressure in the pulmonary artery (15 mm Hg) and the pressure in the left atrium (5 mm Hg). These pulmonary pressures are measured using a Swan-Ganz catheter, basically a thin, flexible tube with an inflatable rubber balloon surrounding the distal end (Fig. 20–3). The balloon is inflated by injecting a small amount of air through the proximal end. Although the Swan-Ganz

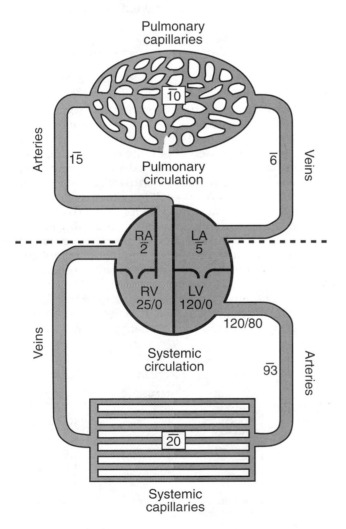

Figure 20–2 ■ Blood pressures differ in pulmonary and systemic circulations. Unlike systemic circulation, pulmonary circulation is a low-pressure, low-resistance system. Pressures are given in mm Hg; a bar over the number indicates mean pressure.

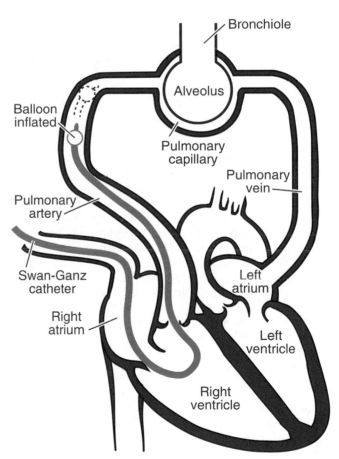

Figure 20–3 ■ Wedge pressure is measured by wedging a catheter into a small pulmonary artery. The catheter is inserted through a peripheral vein in the systemic circulation, through the right heart, and into the pulmonary artery. The wedged catheter temporarily occludes blood flow in that part of the vascular bed. The wedge pressure is a measure of downstream pressure, which is pulmonary venous pressure. Pulmonary venous pressure reflects left atrial pressure.

catheter is used for several pressure measurements, most useful is the pulmonary arterial **wedge pressure** (Fig. 20-3). To measure wedge pressure, the catheter tip with balloon inflated is "wedged" into a small branch of the pulmonary artery. When blood flow is interrupted by the inflated balloon, the tip of the catheter measures **downstream pressure.** The downstream pressure in the occluded arterial branch represents pulmonary **venous pressure,** which reflects left atrial pressure. Changes in pulmonary-venous and left atrial pressure have a profound effect on gas exchange, and pulmonary wedge pressure provides an indirect measure of these important pressures.

Pulmonary Vascular Resistance

■ Pulmonary Vascular Resistance Falls with Increased Cardiac Output

The right ventricle pumps mixed venous blood through the pulmonary arterial tree and then through the alveolar capillaries where oxygen is taken up and carbon dioxide is removed. From the alveolar capillaries blood is pumped through the pulmonary veins and then on to the left atrium. A difficult concept to understand is how all of the cardiac output is pumped through pulmonary circulation at a much lower pressure than through the systemic circulation. In Figure 20-2, the pressure gradient (10 mm Hg) across the pulmonary circulation is extremely small, which drives the same volume of blood (5 L/min) as that driven through the systemic circulation with a pressure gradient of almost 100 mm Hg. Clearly, pulmonary vascular resistance is extremely low; in fact, it is about one-tenth that of systemic vascular resistance. The difference is due, in part, to the enormous number of small muscular arteries (resistance vessels) and the dilated condition of these resistance vessels. In comparison, systemic arterioles and precapillary sphincters are partially constricted at rest. A good way to remember the difference is that the pulmonary circulation is normally dilated and the systemic circulation is normally constricted. Remember (Chap. 12) that vascular resistance (R) is equal to the pressure gradient (ΔP) divided by blood flow ($\dot{Q}$) (R = ΔP/flow). If blood flow is 5 L/min, then pulmonary vascular resistance equals 2 mm Hg/L/min [(15 − 5)/5].

An equally unique feature of the pulmonary vasculature is its ability to decrease resistance when cardiac output increases. When pulmonary arterial pressure rises with increased cardiac output, there is a substantial decrease in pulmonary vascular resistance (Fig. 20-4). Similarly, increasing pulmonary venous pressure causes pulmonary vascular resistance to fall. These responses are very different from those of the systemic circulation. Two local mechanisms are re-

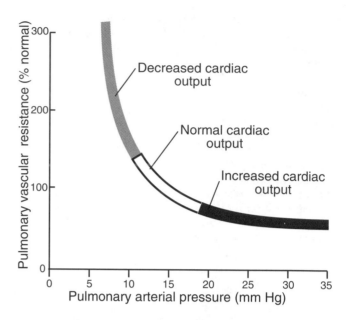

Figure 20-4 ■ Pulmonary vascular resistance falls as pulmonary arterial pressure rises. Note that if cardiac output increases, pulmonary vascular resistance decreases.

sponsible. First, under normal conditions some capillaries are closed because of the low perfusion pressure (Fig. 20-5). When blood flow increases the pressure rises and these collapsed vessels are opened, thus lowering overall resistance. This process of opening capillaries is called **capillary recruitment** and is the primary mechanism for the fall in pulmonary vascular resistance when cardiac output increases. The second mechanism is **capillary distension,** widening of capillary segments, which occurs because the pulmonary capillaries are exceedingly thin and highly compliant. Capillary recruitment and distension have other protective functions. High capillary pressure is a major threat to the lungs. It can lead to **pulmonary edema,** an abnormal accumulation of fluid at the blood-gas interface that impairs gas exchange. When cardiac

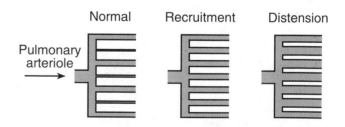

Figure 20-5 ■ Two mechanisms are responsible for decreasing pulmonary vascular resistance when pulmonary arterial pressure increases. In the normal condition, not all available capillaries are perfused. Capillary recruitment (opening up of previously closed vessels) results in perfusion of an increased number of vessels. Distension (increase in caliber of vessels) also results in a lower resistance and higher blood flow.

output increases from a resting level of 5 L/min to 25 L/min with vigorous exercise, the decrease in vascular resistance not only minimizes the load on the right heart but also keeps the capillary pressure low and prevents excess fluid from leaking out of the pulmonary capillaries and thereby flooding the alveoli. The fall in pulmonary vascular resistance with increased cardiac output has two other beneficial effects. It opposes the tendency of blood velocity to speed up with increased flow rate, thereby maintaining adequate time for pulmonary capillary blood to become saturated with oxygen. It also results in an increase in capillary surface area, which enhances the diffusion of oxygen into pulmonary capillary blood.

■ Lung Volumes Affect Pulmonary Vascular Resistance

Pulmonary vascular resistance is also markedly affected by lung volume. Since pulmonary capillaries have little structural support, they can be easily distended or collapsed, depending on the pressure surrounding them. It is the change in transmural pressure (pressure inside the capillary minus pressure outside the capillary) that influences vessel diameter. From a functional point of view, pulmonary vessels can be classified into two types: **extra-alveolar vessels** (pulmonary arteries and veins) and **alveolar vessels** (arterioles, capillaries, and venules). The extra-alveolar vessels are subjected to pleural pressure, and any changes in pleural pressure affect pulmonary vascular resistance in these vessels by changing transmural pressure. Alveolar vessels, on the other hand, are subjected primarily to alveolar pressure.

At high lung volumes the pleural pressure is more negative. Transmural pressure in the extra-alveolar vessels increases and they become distended (Fig. 20–6A). However, at high lung volumes alveolar diameter increases, causing transmural pressure in alveolar vessels to decrease. As the alveolar vessels become compressed, pulmonary vascular resistance increases. At low lung volumes, pulmonary vascular resistance also increases, due to a more positive pleural pressure, which compresses the extra-alveolar vessels. Since alveolar and extra-alveolar vessels can be viewed as two groups of resistance vessels connected in series, their resistance is additive at any lung volume. Pulmonary vascular resistance (Fig. 20–6B) gives a U-shaped curve, with pulmonary vascular resistance lowest at functional residual capacity (FRC) and increasing at both higher and lower lung volumes.

Since smooth muscle plays a key role in determining the caliber of extra-alveolar vessels, drugs can also cause a change in resistance. Serotonin, norepinephrine, histamine, thromboxane A$_2$, and leukotrienes are potent vasoconstrictors, particularly at low lung volumes when the vessel walls are already compressed. Drugs that can relax smooth muscle in the pulmonary

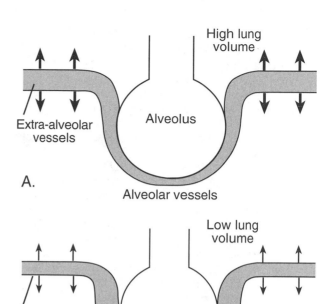

A.

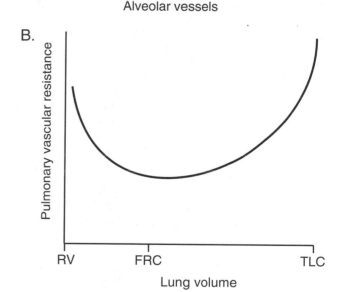

B.

Figure 20–6 ■ Pulmonary vessel diameter is affected at low and high lung volumes. (A) At high lung volumes alveolar vessels are compressed. Note that extra-alveolar vessels are actually *distended* at high lung volumes because of the lower pleural pressure. At low lung volumes, the extra-alveolar vessels are compressed due to pleural pressure. (B) Total pulmonary vascular resistance forms a U-shaped curve with resistance lowest at functional residual capacity (FRC).

circulation include adenosine, acetylcholine, prostacyclin (prostaglandin I$_2$), and isoproterenol. Pulmonary resistance is virtually unaffected by autonomic nerves.

■ Low Oxygen Increases Pulmonary Vascular Resistance

Although changes in pulmonary vascular resistance are accomplished mainly by passive mechanisms, resistance can be increased by low oxygen in the alveoli (**alveolar hypoxia**) and low oxygen in the mixed venous blood (hypoxemia). In systemic vessels hypoxemia causes vasodilation, but in pulmonary vessels hypoxemia and/or alveolar hypoxia causes vasoconstriction of small pulmonary arteries (Fig. 20-7A). This unique phenomenon of hypoxia-induced pulmonary vasoconstriction is accentuated by hypercapnia (high carbon dioxide) and a low blood pH. The precise mechanism is not known, but hypoxia can act directly on pulmonary vascular smooth muscle cells.

There are two types of alveolar hypoxia encountered in the lung, with different implications for pulmonary vascular resistance. In **regional hypoxia** (Fig. 20-7A), vasoconstriction is localized to a specific region of the lung and serves to divert blood away from a poorly ventilated region (e.g., bronchial obstruction). The resulting hypoxia-induced vasoconstriction diverts blood flow away from this region to minimize effects on gas exchange. When alveolar hypoxia no longer exists, the vessels dilate and blood flow is restored. **Generalized hypoxia** causes vasoconstriction throughout the lung, leading to a marked rise in resistance and pulmonary arterial pressure. Generalized hypoxia occurs when the partial pressure of alveolar oxygen (P_{AO_2}) is decreased with high altitude or the chronic hypoxia seen in certain types of respiratory diseases (e.g., emphysema, cystic fibrosis). Prolonged generalized hypoxia can become pathologic and cause pulmonary hypertension (high pulmonary arterial pressure), which leads to hyperplasia/hypertrophy of smooth muscle cells and a narrowing of arterial walls. Pulmonary hypertension causes a substantial increase in workload on the right heart and often leads to right heart hypertrophy.

Generalized hypoxia plays an important nonpathophysiologic role before birth. In the fetus, pulmonary vascular resistance is extremely high due to generalized hypoxia. As a result, less than 15% of cardiac output goes to the lung; the remainder is diverted to the left side of the heart via the foramen ovale and to the aorta via the ductus arteriosus. When alveoli are oxygenated on the newborn's first breath, vascular smooth muscle relaxes, vessels dilate, vascular resistance falls dramatically, the foramen ovale and ductus arteriosus close, and pulmonary blood flow increases enormously.

Fluid Exchange in the Pulmonary Capillaries

■ Removal of Alveolar Fluid Is Enhanced by Low Capillary Pressure

Startling forces, which govern the exchange of fluid across capillaries in the systemic circulation (Chap.

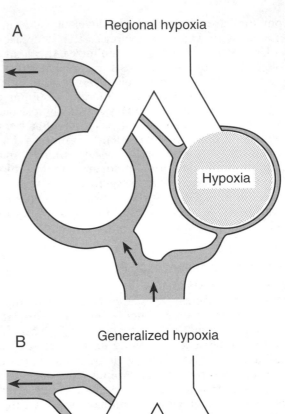

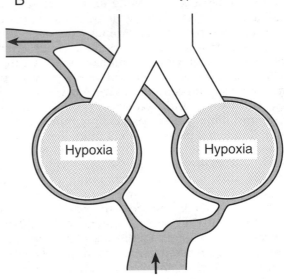

Figure 20-7 ■ Alveolar hypoxia causes precapillary vasoconstriction. (A) With regional hypoxia, blood flow is diverted away from poorly ventilated regions. There is very little change in pulmonary arterial pressure. (B) In generalized hypoxia, which can occur with high altitude or disease, precapillary constriction occurs throughout the lung. There is a marked increase in pulmonary arterial pressure.

16), also operate in the pulmonary capillaries. Net fluid transfer across the pulmonary capillaries depends on the *difference* between hydrostatic pressure and colloid osmotic pressures inside and outside the capillaries. In the pulmonary circulation, two additional forces (surface tension and alveolar pressure) play a role in fluid transfer. The force of alveolar surface tension (Chap. 19) pulls inward, which tends to lower

FREE RADICAL-INDUCED LUNG INJURY

An "oxygen paradox" has long been recognized in biology but has only recently been well understood: Oxygen is essential for life, but too much oxygen or inappropriate metabolism of oxygen can be toxic to cells and the organism. Cellular ATP is derived by reducing molecular oxygen to form water. This is accomplished by adding four electrons in a controlled reduction by the mitochondrial electron transport system. Approximately 98% of the oxygen taken up by the cell enters the mitochondria, where it is reduced to water. However, "leaks" in the mitochondrial electron transport system allow oxygen to accept less than four electrons, forming a **free radical.** A free radical is any atom, molecule, or group of molecules with an unpaired electron in its outermost orbit. The reactions leading to form reactive oxygen metabolites include **superoxide ion** ($O_2 + \bar{e} \rightarrow O_2^-$), **hydrogen peroxide** ($O_2 + 2\bar{e} + 2H^+ \rightarrow H_2O_2$), and **hydroxyl radical** ($O_2^- + H_2O_2 + H^+ \rightarrow O_2 + H_2O + \cdot OH$). In the above reactions the single unpaired electron in the free radical is denoted by a dot. The $\cdot OH$ radical is the most reactive and most damaging to cells. Hydrogen peroxide (H_2O_2), while not a free radical, is also reactive to tissues and has the potential to generate the hydroxyl radical ($\cdot OH$). Superoxide ion, hydrogen peroxide, and the hydroxyl radical are collectively called **reactive oxygen species** (ROS). In addition to free radicals produced by "leaks" in the mitochondrial transport system, ROS can be formed from the metabolism of cytochrome P450, the production of NADPH, and the metabolism of arachidonic acid. Under normal conditions, ROS are neutralized by the protective enzymes **superoxide dismutase, catalase,** and **peroxidases** and no damage occurs. However, when ROS are greatly increased, they overwhelm the protective enzyme systems and damage cellular function by oxidizing membrane lipids, oxidizing cellular proteins and DNA, and oxidizing enzymes.

The lung is a primary target organ for free radical-induced injury, with the pulmonary circulation the major affected site. Free radicals damage the pulmonary capillaries, causing them to become leaky and leading to pulmonary edema. In addition to intracellular production, ROS are produced during inflammation and episodes of oxidant exposure (i.e., oxygen therapy or breathing ozone and nitrogen dioxide from polluted air). During the inflammatory response, neutrophils become sequestered and activated; they undergo a respiratory burst (which produces free radicals) and release catalytic enzymes. This release of free radicals and catalytic enzymes is designed to kill bacteria, but endothelial cells can become damaged in the process, causing edema.

Paraquat, an agricultural herbicide, is another source of free radical-induced injury to the lung. Crop dusters and migrant workers are particularly at risk because of exposure to paraquat through the lungs and skin. Tobacco or marijuana that has been sprayed with paraquat and is smoked can also produce lung injury from ROS.

Ischemia-reperfusion injury, another cause of free radical-induced injury in the lung, usually occurs as a result of a clot being lodged in the pulmonary circulation. Tissues beyond the clot become ischemic (no flow, hypoxic, acidotic, and hypoglycemic). During the ischemic phase cellular ATP decreases and products such as hypoxanthine accumulate. When the clot dissolves, blood flow is established and ischemic vessels are reperfused. During the reperfusion phase, hypoxanthine, in the presence of oxygen, is converted to xanthine. Xanthine oxidation, catalyzed by the enzyme xanthine oxidase, located on the pulmonary endothelium, results in the production of ROS (Xanthine $+ O_2 \xrightarrow[\text{Oxidase}]{\text{Xanthine}} H_2O_2 +$ Urate). During the reperfusion phase neutrophils also become sequestered and activated in these vessels. Thus the pulmonary vasculature and surrounding lung parenchyma become damaged by a double hit of free radicals—those produced by xanthine oxidation and those from activated neutrophils.

In summary, reactive oxygen species, namely superoxide ion, hydrogen peroxide, and the hydroxyl radical, are a major cause of acute lung injury. Although the source of free radicals is diverse, the primary target tissue (pulmonary vessels) is rather specific, causing pulmonary edema and hypertension.

interstitial pressure and draw fluid into the interstitial space. In contrast, alveolar pressure tends to compress the interstitial space so that interstitial pressure is increased (Fig. 20–8). Mean capillary hydrostatic pressure is normally approximately 8–10 mm Hg, which is lower than the plasma colloid osmotic pressure (25 mm Hg). This is functionally important because it favors the net absorption of fluid.

Although mean capillary hydrostatic pressure is lower than plasma colloid osmotic pressure, alveolar surface tension tends to offset this advantage and results in a net force that still favors a small continuous flux of fluid out of the capillaries and into the interstitium. This excess fluid travels through the interstitium to the perivascular and peribronchial spaces in the lung, where it then passes into the lymphatic channels (Fig. 20–8). Compared to most organs, the lungs have an extensive lymphatic system. The lymphatics are not found in the alveolar-capillary area but are strategically located near the terminal bronchioles to drain off excess fluid. Lymphatic channels, like small pulmonary

vessels, are held open by tethers from surrounding connective tissue. Total lung lymph flow is about 0.5 ml/min and is propelled toward the hilum by smooth muscle lining the lymphatic walls and by ventilatory movements.

■ Fluid Imbalance Leads to Pulmonary Edema

Pulmonary edema, which markedly impairs gas exchange, occurs when excess fluid enters the lung interstitium and alveoli. This occurs when the rate of filtration exceeds the rate of fluid removal, which can result from an increase in capillary hydrostatic pressure, an increase in capillary permeability, an increase in surface tension, or a decrease in plasma colloid osmotic pressure. Increased capillary hydrostatic pressure most often occurs when there is an abnormally high venous pressure (mitral stenosis or left heart failure). The other major cause of pulmonary edema is increased capillary permeability, which causes both excess fluid and plasma proteins to flood the interstitium and alveoli, making edema even more severe. This occurs with pulmonary vascular injury, usually from oxidant damage (i.e., oxygen therapy, ozone toxicity), an inflammatory reaction (endotoxins), or neurogenic shock (i.e., head injury). A decrease in colloid osmotic pressure occurs when plasma protein concentration is reduced (e.g., starvation). Loss of surfactant causes high surface tension and is another serious cause of pulmonary edema seen in acute respiratory distress syndrome (ARDS). Loss of surfactant causes a rise in alveolar surface tension, which causes lowering of interstitial hydrostatic pressure, resulting in an increase of capillary fluid entering the interstitium.

Pulmonary edema is a serious problem because it hinders gas exchange and eventually causes arterial P_{O_2} to fall below normal ($Pa_{O_2} < 85$ mm Hg) and arterial P_{CO_2} to rise above normal ($Pa_{CO_2} > 45$ mm Hg). The abnormally low arterial P_{O_2} is termed **hypoxemia** and the abnormally high arterial P_{CO_2} is termed **hypercapnia.** Pulmonary edema also obstructs small airways, thereby increasing airway resistance. With pulmonary edema, the lungs become stiffer (i.e., decreased compliance) because of interstitial swelling and increased surface tension. The increased lung stiffness, together with airway obstruction, greatly

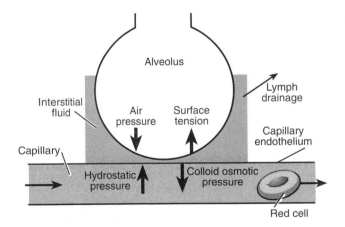

Figure 20–8 ■ Fluid exchange in the pulmonary capillaries. Fluid movement in and out of capillaries depends on the net difference between hydrostatic and colloid osmotic pressure. Two additional factors are involved in the pulmonary capillaries: a) alveolar surface tension, which enhances filtration; and b) alveolar air pressure, which opposes filtration. The relatively low pulmonary capillary hydrostatic pressure helps keep the alveoli "dry" and prevents pulmonary edema.

increases the work of breathing. Treatment of pulmonary edema is directed first toward reducing pulmonary capillary hydrostatic pressure, accomplished by decreasing blood volume with a diuretic drug, increasing left ventricular function with digitalis, or administering a vasodilator drug, which causes systemic vasodilation.

Fresh water drowning is often associated with aspiration of water into the lungs, but the cause of death is not pulmonary edema but rather ventricular fibrillation. The low capillary pressure that normally keeps the alveolar-capillary membrane free of excess fluid becomes a severe disadvantage when fresh water accidentally enters the lung. None of the water that enters the lungs can be aspirated, because it rapidly enters the pulmonary capillary circulation via the alveoli due to the much higher colloid osmotic pressure than capillary hydrostatic pressure. The hypotonic environment causes the plasma to be diluted and red cells to burst (hemolysis). The resulting elevation of plasma K^+ levels and depression of Na^+ levels alter the electrical activity of the heart. Ventricular fibrillation often occurs, due to the combined effects of these electrolyte changes and to hypoxemia. In salt water drowning, the aspirated sea water is hypertonic, which leads to increased plasma Na^+ and pulmonary edema. The cause of death in this case is asphyxia.

Blood Flow Distribution in the Lung

■ Gravity Affects Blood Flow Distribution in the Lungs

As mentioned above, blood accounts for approximately half of lung weight. The effects of gravity on the distribution of blood and blood flow are quite dramatic. In an upright individual the gravitational pull on the fluid in the pulmonary blood vessels is downward. Since these vessels are highly compliant, blood volume and flow are greater at the bottom of the lung (base) than at the top (apex). The pulmonary vessels can be compared to a continuous column of fluid. The difference in arterial pressure between the top and bottom of the lung is about 30 cm in height. Since the heart is situated midway between the top and bottom of the lungs, the arterial pressure at the top of the lung (15 cm above the heart) is about 11 mm Hg less, while at the bottom of the lungs (15 cm below the heart) the pressure is 11 mm Hg greater (15 cm = 150 mm divided by 13.6 mm H_2O) than the mean pressure in the middle of the lung. The low arterial pressure at the apex causes capillaries to collapse, thereby reducing flow, while at the base capillaries are distended, thereby augmenting blood flow.

In the upright human, pulmonary blood flow increases almost linearly from apex to base (Fig. 20-9). Since blood flow distribution is affected by gravity,

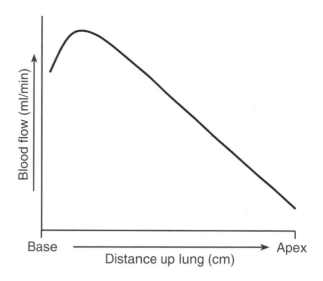

Figure 20–9 ■ The distribution of pulmonary blood flow in the upright human lung is affected by gravity. Note that pulmonary blood flow increases from the apex to the base of the lung.

it can be altered by changes in body positions. For example, when a subject is lying down, blood flow is distributed relatively evenly from apex to base. Measurement of blood flow in a subject suspended upside-down would reveal an apical blood flow exceeding basal flow in the lung. Exercise tends to offset the gravitational effects in an upright individual. The cardiac output increases, pulmonary arterial pressure increases, and blood flow is increased in the upper region via capillary recruitment, thereby minimizing regional differences.

Since gravity causes capillary beds to be underperfused in some regions (apex) and overperfused in others (base), the lungs are often divided into zones to illustrate the uneven distribution of capillary perfusion (Fig. 20–10). **Zone 1** occurs when alveolar pressure is greater than pulmonary arterial pressure; pulmonary capillaries collapse and little or no flow occurs. Zone 1 is ventilated but not perfused (no blood flow through the pulmonary capillaries), leading to an increase in alveolar dead space (Chap. 19). Zone 1 is usually very small or nonexistent in a normal individual because the pulsatile pulmonary arterial pressure is sufficient to keep the capillaries partially open at the apex of the lungs. Zone 1 may easily be created by conditions that elevate alveolar pressure or decrease pulmonary arterial pressure. For example, a zone 1 condition can be created when an individual is placed on a mechanical ventilator, since alveolar pressure increases with high ventilation pressure. Hemorrhage or low blood pressure can create a zone 1 condition by lowering pulmonary arterial pressure. A zone 1 condition can also be created in astronauts during blast-off. The rocket acceleration makes the gravitational pull great enough to cause arterial pressure in the top part of the lung to fall. In zone 1, alveolar pressure (P_A) is

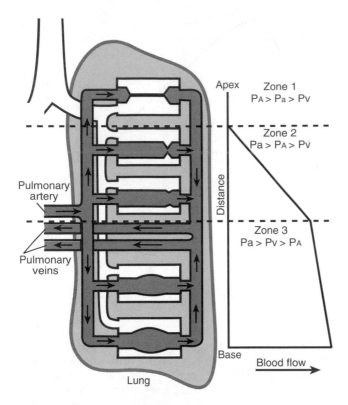

Figure 20–10 ■ The lungs can be divided into three zones, depending on the relationship among pulmonary arterial pressure (Pa), pulmonary venous pressure (Pv), and alveolar pressure (PA).

greater than pulmonary arterial pressure (Pa), but Pa is still greater than pulmonary venous pressure (Pv). This is represented as PA > Pa > Pv.

In Figure 20-10, a **zone 2** condition occurs in the middle of the lung, where pulmonary arterial pressure, because of the increased hydrostatic effect, is greater than alveolar pressure. Venous pressure is less than alveolar pressure. In a zone 2 condition, blood flow is determined not by the arterial minus venous pressure, but by the difference between arterial pressure and alveolar pressure. The pressure gradient in zone 2 is represented as Pa > PA > Pv. This means venous pressure in zone 2 has no effect on flow unless it exceeds alveolar pressure. For example, lowering venous pressure will not increase blood flow in zone 2. In **zone 3,** venous pressure exceeds alveolar pressure and flow is determined by the usual arterial-venous pressure difference. The increase in blood flow down this region is due primarily to capillary distension.

■ Ventilation and Blood Flow Are Not Uniformly Matched in the Lungs

Thus far we have assumed that if ventilation and blood flow are normal, gas exchange will also be normal. Unfortunately, this is not the case. Even though total ventilation and blood flow (i.e., cardiac output) may be normal, there are regions in the lung where

ventilation and blood flow are not matched, so that a certain fraction of the pulmonary arterial blood is not oxygenated.

The matching of air flow and blood flow is best examined by considering **regional ventilation-perfusion ratios** in the lungs. This ratio compares the flow of alveolar air to the flow of blood in lung regions. Since in a resting normal individual alveolar ventilation ($\dot{V}A$) is approximately 4 L/min and total pulmonary blood flow ($\dot{Q}$) is about 5 L/min, the normal $\dot{V}A/\dot{Q}$ ratio is 0.8 (there are no units, since this is a ratio). We have already seen that gravity can cause regional differences in alveolar ventilation (Chap. 19), causing air flow to be greater at the base of the lungs. In an upright person the base of the lungs is better ventilated and perfused than the apex. The increased ventilation is due to the downward gravitational pull, which makes the apex stiffer than the base. The increased blood flow at the base is also due to gravity, which increases the perfusion pressure. Air flow and blood flow are illustrated in Figure 20–11. Three points are apparent from this figure. First, ventilation and blood flow are both gravity-dependent, consequently air flow and blood flow increase down the lung. Second, blood flow shows a fivefold variation between the top and bottom of the lung, while ventilation shows a twofold variation. This causes gravity-dependent regional variations in the $\dot{V}A/\dot{Q}$ ratio that range from 0.7 at the bottom of the lung to 3 or higher at the top in the normal upright human. The $\dot{V}A/\dot{Q}$ ratio does not vary much over the lower two-thirds of the lung. Only near the top of the lung, where blood flow decreases to low levels, does $\dot{V}A/\dot{Q}$ rise steeply. Third, blood flow is proportionately greater than ventilation at the base, while ventilation is proportionately greater than blood flow at the apex.

The crucial factor in gas exchange is the matching of regional ventilation and blood flow, as opposed to total ventilation and total pulmonary blood flow. The distribution of $\dot{V}A/\dot{Q}$ in the normal adult is shown in Figure 20–12. Even in the normal lung, most of the ventilation and perfusion goes to lung units with a $\dot{V}A/\dot{Q}$ ratio around 1 instead of 0.8. At the apical region, where the $\dot{V}A/\dot{Q}$ ratio is high, there is overventilation with respect to blood flow. At the base, where the ratio is low, the opposite occurs (i.e., overperfused with respect to ventilation). In the latter case, a fraction of the blood passes through the capillaries at the base of the lungs without becoming fully oxygenated. The effect of $\dot{V}A/\dot{Q}$ imbalance on blood gases is shown in Figure 20–13. P_{ACO_2} is low at the apex because more carbon dioxide is "blown off," due to overventilation with respect to blood flow. Regional differences in $\dot{V}A/\dot{Q}$ ratios tend to localize some types of diseases to the top and bottom parts of the lung. For example, tuberculosis tends to be localized in the apical region of the lung because it provides a more favorable environment (i.e., higher oxygen levels for *Mycobacterium tuberculosis*).

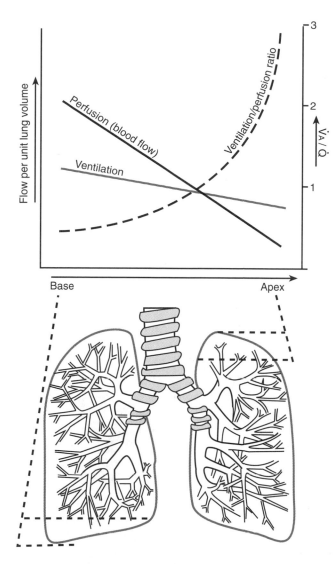

Figure 20–11 ■ Air flow and blood flow in the upright lung. Both ventilation and perfusion are gravity-dependent. Note that the ventilation-perfusion ratio increases up the lung.

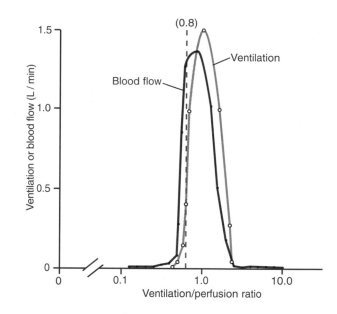

Figure 20–12 ■ Normal distribution of ventilation to perfusion in a healthy young adult. The y-axis represents blood flow, or ventilation (L/min). The ventilation-perfusion ratio is shown on the x-axis, plotted on a logarithmic scale. The overall normal $\dot{V}_A/\dot{Q}$ ratio is 0.8, shown by the slashed vertical line. (Adapted from Nunn J. F.: *Applied Respiratory Physiology.* 3rd ed. London, Butterworth, 1987.)

	$\dot{V}_A$ $\dot{Q}$ (L / min)		$\dot{V}_A / \dot{Q}$	Pa_{O_2} Pa_{CO_2} (mm Hg)		
	0.25	0.07	3.6	130	28	Apex
Base	0.8	1.3	0.62	88	42	

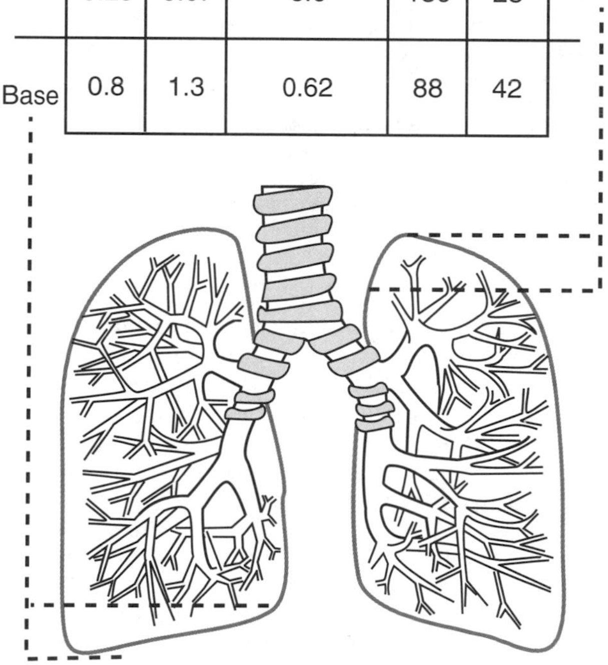

Figure 20–13 ■ Regional differences of ventilation-perfusion ratios affect the blood gases at both the apical and basal regions of the lung.

Shunts and Venous Admixture

The lung's plumbing (air flow and blood flow) is not perfect. On one side of the alveolar-capillary membrane there is wasted air (i.e., physiologic dead space) and on the other side there is wasted blood. Wasted blood refers to any fraction of the mixed venous blood that does not get oxygenated. The mixing of unoxygenated blood with oxygenated blood is known as **venous admixture** (Fig. 20–14). There are two causes: shunting of blood past alveoli (often simply called "shunt") and a low $\dot{V}_A/\dot{Q}$ ratio.

■ Blood Is Shunted Past Alveoli

There are two types of shunts: **right-to-left anatomic shunt** and **alveolar shunt.** A right-to-left anatomic shunt occurs when blood bypasses alveoli

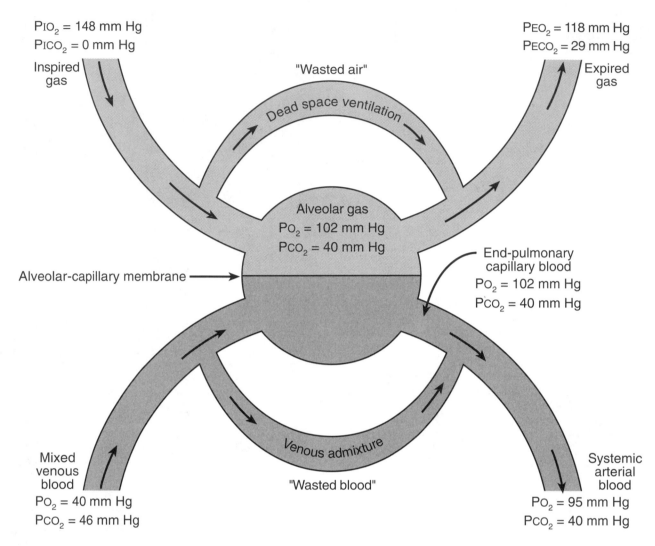

P_{IO_2} = 148 mm Hg
P_{ICO_2} = 0 mm Hg

Inspired gas

"Wasted air"

P_{EO_2} = 118 mm Hg
P_{ECO_2} = 29 mm Hg

Expired gas

Dead space ventilation

Alveolar gas
P_{O_2} = 102 mm Hg
P_{CO_2} = 40 mm Hg

End-pulmonary capillary blood
P_{O_2} = 102 mm Hg
P_{CO_2} = 40 mm Hg

Alveolar-capillary membrane

Venous admixture

"Wasted blood"

Mixed venous blood
P_{O_2} = 40 mm Hg
P_{CO_2} = 46 mm Hg

Systemic arterial blood
P_{O_2} = 95 mm Hg
P_{CO_2} = 40 mm Hg

Figure 20–14 ■ Functional representation of gas exchange in the lung. The plumbing on both sides of the alveolar-capillary membrane is imperfect: on one side there is "wasted air" and on the other side there is "wasted blood." The wasted air constitutes physiologic dead space and the wasted blood (venous admixture) constitutes physiologic shunt. (Modified from Selkurt EE: *Physiology.* 5th ed. Boston, Little, Brown, 1984.)

through an anatomic channel, such as from the right to left heart through a septal defect in the ventricle or from a branch of the pulmonary artery connecting directly to the pulmonary vein. Bronchial circulation also constitutes shunted blood because bronchial venous blood (deoxygenated blood) drains into the pulmonary veins (oxygenated blood). All individuals have some degree of anatomic shunting. Venous admixture due to an anatomic shunt represents about 1–2% of cardiac output in the normal individual and is due almost exclusively to the bronchial circulation. This amount can increase up to 15% of cardiac output with some bronchial diseases, and in certain congenital disorders the right-to-left anatomic shunt can account for up to 50% of cardiac output.

An alveolar shunt occurs when a portion of the cardiac output goes through the regular pulmonary capillaries but does not come into contact with alveolar air. With an alveolar shunt there is no abnormal anatomic connection and the blood does not bypass the alveoli. Rather, blood that passes through is not oxygenated because of poor diffusion. Alveolar shunting occurs with lobar atelectasis, pneumonia, and pulmonary edema.

The sum of a right-to-left anatomic shunt and alveolar shunt equals physiologic shunt and represents the total venous admixture due to shunting. Physiologic shunt is analogous to physiologic dead space; the two are compared in Table 20–1. Normal individuals have some degree of physiologic dead space as well as physiologic shunt.

TABLE 20–1 ■ Comparison of Shunts and Dead Space

Shunts		Dead Space	
Anatomic	$\dot{Q}_s$	Anatomic	$\dot{V}_D$
	+		+
Alveolar	$\dot{Q}_s$	Alveolar	$\dot{V}_D$
Physiologic shunt (Calculated total "wasted blood": venous admixture)	$\dot{Q}_{sp}$	Physiologic dead space (Calculated total "wasted air")	$\dot{V}_{Dp}$

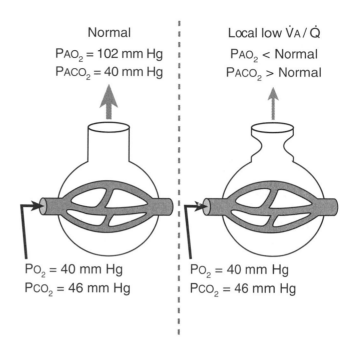

Figure 20–15 ■ Airway obstruction causes a localized low regional $\dot{V}_A/\dot{Q}$ ratio. A partially blocked airway causes this region to be underventilated with respect to blood flow. Note the alveolar gas composition. A localized low $\dot{V}_A/\dot{Q}$ ratio leads to venous admixture and will increase the physiologic shunt.

■ Some Blood Is Not Oxygenated with a Low V̇A/Q̇

The second cause of venous admixture is a low $\dot{V}_A/\dot{Q}$ ratio. This can be regional (i.e., gravity-dependent) or due to other types of local variations. Recall that in the normal upright human the $\dot{V}_A/\dot{Q}$ ratio varies regionally from about 0.7 at the bottom to 3 or higher at the top (Fig. 20–11). This regional variation in $\dot{V}_A/\dot{Q}$ ratio causes some venous admixture, especially at the base of an upright lung.

Venous admixture can also occur with a local low $\dot{V}_A/\dot{Q}$. This occurs with a partially obstructed airway (Fig. 20–15). Blood flow is normal through the region supplied by the blocked airway but ventilation is decreased. This results in a lung region that is underventilated with respect to blood flow, a condition called **localized hypoventilation.** A fraction of the blood passing through this local hypoventilated region is not fully oxygenated, resulting in an increase in venous admixture.

In summary, venous admixture results from shunting (right-to-left anatomic shunt or alveolar shunt) and/or a low $\dot{V}_A/\dot{Q}$ (regional or localized). Approximately 20% of the venous admixture in the normal lung comes from a right-to-left shunt (i.e., bronchial circulation) and 80% from regional variations in $\dot{V}_A/\dot{Q}$ (i.e., the low $\dot{V}_A/\dot{Q}$ at the base of the lung).

Bronchial Circulation

The conducting airways have a separate circulation (i.e., bronchial circulation) distinct from the pulmonary circulation (Chap. 19). Bronchial circulation distributes blood to the conducting airways and adjacent supporting structures of the lungs. The primary function of the bronchial circulation is to nourish the walls of the conducting airways and surrounding tissue. Bronchial circulation has a secondary function: it warms and humidifies incoming air, minimizing evaporative loss from the alveolar surface. Under normal conditions the bronchial circulation does not supply blood to the terminal respiratory units (respiratory bronchioles, alveolar ducts, and alveoli), which receive their blood supply from the pulmonary circulation. Venous return from bronchial circulation is by two routes: bronchial veins and pulmonary veins. About half of the bronchial blood flow returns to the right atrium by way of the bronchial veins, which empty into the azygous vein. The remainder returns through small bronchopulmonary anastomoses into the pulmonary veins. (Remember, bronchial circulation is a source of anatomic shunt in normal individuals.)

Bronchial arterial pressure is approximately the same as aortic pressure, and bronchial vascular resistance is much higher than resistance in the pulmonary circulation. Bronchial blood flow is approximately 1% of cardiac output, but in certain inflammatory disorders of the airways (e.g., chronic bronchitis) it can be approximately 1–10% of cardiac output.

The bronchial circulation is the only portion of the circulation in the adult lung that is capable of undergoing **angiogenesis,** the formation of new vessels. This is extremely important in providing collateral circulation to the lung parenchyma, especially when pulmonary circulation is compromised or lost. When the pulmonary circulation is obstructed by a clot or embolus, the development of new blood vessels by the bronchial circulation keeps the parenchyma alive when the pulmonary blood flow is stopped.

■ ■ ■ ■ ■ ■ ■ ■ ■ ■ ■ ■ REVIEW EXERCISES ■ ■ ■ ■ ■ ■ ■ ■ ■ ■ ■ ■

Identify each with a term

1. Abnormal collection of interstitial fluid in the lungs
2. Low oxygen level in the blood
3. High carbon dioxide level in the blood
4. The ratio of alveolar ventilation to pulmonary capillary blood flow
5. A lung region where pulmonary arterial pressure is greater than pulmonary venous pressure and pulmonary venous pressure is always greater than pulmonary alveolar pressure

Define Each Term

6. Hypoxic vasoconstriction
7. Capillary recruitment ↓ vaso regist
8. Zone 1 condition P_A P_a P_v
9. Extra-alveolar vessels pulm art & veins
10. Wedge pressure in pulm art measure downstream pressor

Choose the Correct Answer

11. The pulmonary circulation can be characterized as a:
    a. High-pressure, high-flow, high-resistance system
    b. Low-pressure, high-flow, low-resistance system
    c. High-pressure, high-flow, low-resistance system
    d. Low-pressure, low-flow, high-resistance system
12. Pulmonary vascular resistance is:
    a. Decreased at low lung volumes
    b. Decreased by breathing low oxygen
    c. Unaffected by high lung volumes
    d. Decreased with increased pulmonary arterial pressure
    e. Increased by parasympathetic stimulation
13. The pulmonary and systemic circulations have the same:
    a. Mean pressure
    b. Vascular resistance
    c. Compliance
    d. Flow per minute
    e. Hypoxic vasoconstriction response
14. The effect of gravity on the pulmonary circulation in an upright individual will cause:
    a. Blood flow to be evenly distributed throughout the lung
    b. Arterial pressure to be less than alveolar pressure in zone 3
    c. Higher capillary pressure in the base of the lung
    d. Lower vascular resistance in the apex of the lung
15. A patient lying on his back and breathing normally has a mean left atrial pressure of 7 cm H_2O, mean pulmonary

arterial pressure of 15 cm H_2O, cardiac output of 4 L/min, and anteroposterior chest depth of 15 cm, measured at the xiphoid. Most of his lung is perfused under which of the following conditions?
    a. Zone 1
    b. Zone 2
    c. Zone 3
    d. Zone 4
16. Lowering pulmonary venous pressure will have the greatest effect on regional blood flow in:
    a. Zone 1
    b. Zone 2
    c. Zone 3
    d. Zones 1 and 2
17. In a normal standing subject:
    a. Ventilation is greater at the top of the lungs than at the bottom
    b. Perfusion is greater at the top of the lungs than at the bottom
    c. V̇/Q̇ ratio is greater at the top of the lungs than at the bottom
    d. a and c are correct
18. The differences in regional ventilation and perfusion in a standing subject are largely brought about by:
    a. The different regional gas compositions
    b. Higher airway resistance at the top of the lung
    c. The pyramidal shape of the lung
    d. The effects of gravity

Calculate

19. If cardiac output is 5 L/min and mean pulmonary arterial and left atrial pressures are 20 and 5 mm Hg, respectively, what is the pulmonary vascular resistance?
20. If the apical portion of the lung is 20 cm above the heart, what pressure (in mm Hg) must the right ventricle produce to pump blood to the top of the lung?
21. If pulmonary vascular resistance is 4 mm Hg/L/min and cardiac output is 5 L/min, what is the driving pressure of the pulmonary circulation?

Briefly Answer

22. Why does pulmonary vascular resistance increase at low and high lung volumes?
23. List three secondary (nonrespiratory) functions of the pulmonary circulation.
24. What is the difference between regional and generalized hypoxia?
25. Describe what happens to pulmonary vascular resistance with exercise.

■ ■ ■ ■ ■ ■ ■ ■ ■ ■ ■ ■ ANSWERS ■ ■ ■ ■ ■ ■ ■ ■ ■ ■ ■ ■

1. Pulmonary edema
2. Hypoxemia
3. Hypercapnia
4. Ventilation/perfusion (V̇A/Q̇) ratio

5. Zone 3
6. Contraction of smooth muscle cells in pulmonary vessels due to hypoxia
7. A mechanism in which capillary segments are opened (recruited) with an increased cardiac output
8. A condition in which there is no blood flow in a lung region that occurs when alveolar pressure exceeds pulmonary arterial pressure
9. Vessels outside the alveolar region that are exposed to pleural pressure
10. The pressure measurement obtained by "wedging" a Swan-Ganz catheter into a branch of the pulmonary artery
11. b
12. d
13. d
14. c
15. c
16. c
17. c
18. d
19. $R = \dfrac{\Delta P}{\text{Flow}} = \dfrac{20 \text{ mm Hg} - 5 \text{ mm Hg}}{5 \text{ L/min}}$
    $= 3 \text{ mm Hg/L/min}$

20. $20 \text{ cm} = 200 \text{ mm} \times \dfrac{1 \text{ mm Hg}}{13.6 \text{ mm}}$
    $= 14.7 \text{ mm Hg}$

21. $\Delta P = R \times \text{Flow} = 4 \text{ mm Hg/L/min} \times 5 \text{ L/min} = 20 \text{ mm Hg}$

22. Pulmonary vascular resistance is high at low lung volumes because extra-alveolar vessels are compressed by pleural pressure. It is high at high lung volumes because the alveolar vessels become compressed.

23. The pulmonary circulation acts as a blood reservoir, a filter for the blood, and a metabolic organ (synthesis of lung surfactant lipids, activation and deactivation of vasoactive hormones).

24. Both involve a decrease in alveolar P_{O_2}, which causes precapillary vasoconstriction of the pulmonary vessels. Regional hypoxia is localized and is functionally important in redirecting blood flow away from poorly ventilated regions. Generalized hypoxia is a response to an abnormal lowering of the alveolar P_{O_2} in the entire lung (e.g., high altitude or disease). With generalized hypoxia, pulmonary arterial vasoconstriction occurs throughout the lung and causes pulmonary hypertension, which can eventually lead to right heart hypertrophy.

25. Pulmonary vascular resistance falls with exercise, primarily because of capillary recruitment.

Suggested Reading

Fishman AP: *The Pulmonary Circulation.* Philadelphia, University of Pennsylvania Press, 1989.
Fishman AP: Pulmonary circulation, in Fishman AP, Fisher AB (eds): *Handbook of Physiology: The Respiratory System.* vol. 1, section 3. Baltimore, Waverly, 1985.
Forster RE II, Dubois AB, Briscow WA, Fisher AB: *The Lung: Physiological Basis of Pulmonary Function Tests.* 3rd ed. Chicago, Year Book, 1986.
Nunn JF: *Applied Respiratory Physiology.* 3rd ed. London, Butterworth, 1987.
Said SI: *The Pulmonary Circulation and Acute Lung Injury.* New York, Futura, 1985.

CHAPTER

■ ■ ■ ■ ■ ■ ■

21

Gas Transfer and Transport

OBJECTIVES

After studying this chapter, the student should be able to:
1. Describe Fick's law of diffusion for gases
2. Describe how pulmonary capillary blood flow affects gas transfer in the lung
3. Define diffusing capacity and discuss factors that affect its interpretation
4. Describe the relationship between the partial pressure of oxygen in the blood and the amount physically dissolved
5. State the two forms in which oxygen is transported by the blood
6. Describe the three forms of carbon dioxide in the blood
7. Describe the Bohr and Haldane effects
8. Define hypoxemia
9. Describe four respiratory causes of hypoxemia

Gas Diffusion and Uptake

There are two types of gas movement in the lung, **bulk flow** and **diffusion.** Gas movement in the airway, from the trachea down to the alveoli, is by bulk flow, analogous to water coming out of a faucet in that all of the molecules move as a unit. The driving pressure (ΔP) for bulk flow in the lung is barometric pressure (P_B) at the mouth minus alveolar pressure (P_A). Movement of gases in the alveoli and across the alveolar-capillary membrane is by diffusion (Chap. 2).

Recall that partial pressure or gas tension can be determined by measuring barometric pressure and the fractional concentration of the gas (Dalton's law; Chap. 19). For example, the partial pressure of oxygen (P_{O_2}) in a dry gas is determined as $P_{O_2} = P_B \times F_{O_2}$, where P_{O_2} is the partial pressure of oxygen in mm Hg, P_B is barometric pressure in mm Hg, and F_{O_2} is the fractional concentration of oxygen. At sea level, P_{O_2} is 160 mm Hg (760 mm Hg $\times$ 0.21). It is important that F_{O_2} does not change with altitude and is essentially the same as at sea level. Thus, the decreased P_{O_2} at altitude is due to a decrease in P_B, not to a change in F_{O_2} (Fig. 21-1).

An exception to Dalton's law is the partial pressure of water vapor, known as the **water vapor pressure.** Water vapor pressure at saturation is influenced exclusively by temperature rather than by P_B. At normal body temperature (37°C), water vapor pressure is 47 mm Hg. As long as body temperature remains constant, the vapor pressure also remains constant, even when the P_B or partial pressure of other gases is changed. Thus allowances for water vapor pressure must be made when the partial pressures of gases are to be determined inside the airways and lungs. Dalton's law may be modified as follows for gases inside the body: $P_X = (P_B - P_{H_2O}) \times F_X$, where P_X is the partial pressure of gas X, P_{H_2O} is the water vapor pressure, and F_X is the fractional concentration of gas X.

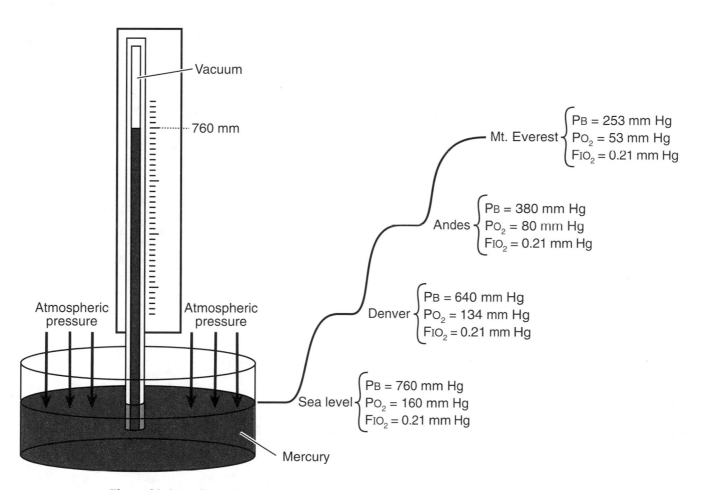

Figure 21–1 ■ Effects of altitude on P_{O_2}. The height of the column of mercury that is supported by air pressure decreases with increasing altitude, because of a fall in barometric pressure (P_B). Since the fractional concentration of inspired O_2 (F_{IO_2}) does not change with altitude the decrease in P_{O_2} is due entirely to a decrease in P_B.

TABLE 21–1 ■ Composition of Respiratory Gases at Sea Level

Gas	Dry Air Partial Pressure (mm Hg)	Moist Tracheal Air Partial Pressure (mm Hg)	Moist Alveolar Air Partial Pressure (mm Hg)
Nitrogen	600	563	571*
Oxygen	160	150	102
Carbon dioxide	0	0	40
Water Vapor	0	47	47
Total	760	760	760

*Alveolar nitrogen increases because the normal respiratory exchange ratio is 0.8.

The partial pressures of the respiratory gases are shown in Table 21-1. Note the change in P_{O_2} from the mouth to alveoli. At sea level, P_{O_2} for dry ambient air is 160 mm Hg. Because of water vapor, P_{O_2} and P_{N_2} in the tracheal air are decreased. The partial pressures for oxygen and nitrogen in the tracheal air are:

$$P_{O_2} = (760 - 47 \text{ mm Hg}) \times 0.21 \quad (1)$$
$$= 150 \text{ mm Hg}$$

$$P_{N_2} = (760 - 47 \text{ mm Hg}) \times 0.79 \quad (2)$$
$$= 563$$

At the alveolar level, the P_{O_2} is 102 mm Hg. The drop in oxygen tension from trachea to alveoli is due mainly to the removal of oxygen by the pulmonary capillary blood, but also to the increase in carbon dioxide concentration. The fractional concentration of oxygen in the alveolar gas ($F_{A_{O_2}}$) is 0.143. Thus, normal alveolar oxygen tension ($P_{A_{O_2}}$) is 102 mm Hg [(760 − 47 mm Hg) × 0.143]. It is important to remember to subtract P_{H_2O} from P_B before converting gas fractions to partial pressures in the lung. Gas diffusion is a function partial pressure difference of the individual gases. For example, oxygen diffuses across the alveolar-capillary membrane due to the difference in P_{O_2} between the alveoli and pulmonary capillaries (Fig. 21-2). The partial pressure difference for oxygen is referred to as the **oxygen diffusion gradient,** and in the normal lung the initial oxygen diffusion gradient is $P_{A_{O_2}}$ minus mixed venous P_{O_2} ($P_{\bar{v}_{O_2}}$) = 102 − 40 = 62 mm Hg. The initial diffusion gradient across the alveolar-capillary membrane for carbon dioxide is 6 mm Hg ($P_{\bar{v}_{CO_2}} - P_{A_{CO_2}}$).

Dalton's law deals with dry gases or gases saturated with water vapor. When gases are exposed to a liquid, such as plasma, gas molecules move into the liquid and exist in a dissolved state. The dissolved gases also exert a partial pressure. A gas will continue to dissolve in the liquid until the partial pressure of the dissolved gas equals the partial pressure above the liquid. **Henry's law** states that at equilibrium the amount of gas dissolved in a liquid at a given temperature is directly proportional to the partial pressure of the gas phase. Henry's law only accounts for the gas that is physically dissolved and not for chemical combination of gases

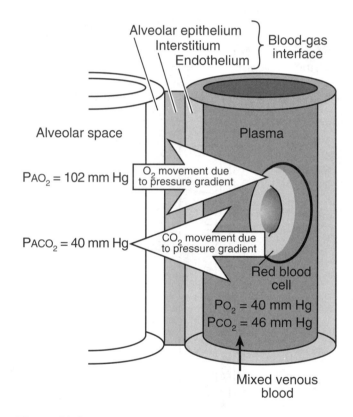

Figure 21–2 ■ Diffusion path of gases crossing the alveolar-capillary membrane (blood-gas interface). $P_{\bar{v}_{CO_2}}$, mean partial pressure of CO_2 in mixed venous blood coming into the lungs; $P_{A_{O_2}}$ alveolar partial pressure for O_2; $P_{A_{CO_2}}$, alveolar partial pressure for CO_2.

(e.g., oxygen with hemoglobin). Oxygen that binds to hemoglobin no longer exerts a partial pressure.

■ Respiratory Gases Cross the Alveolar-Capillary Membrane by Diffusion

Gas diffusion in the lung can be described by **Fick's law** (Chap. 2), which states that the volume of gas per minute diffusing across a membrane ($\dot{V}_{gas}$) is directly proportional to the membrane surface area (A_s), diffusion coefficient of the gas (D), and partial pressure difference (ΔP) of the gas and is inversely proportional to membrane thickness (T). The principles of Fick's law are illustrated in Figure 21–3 and can be summarized as:

$$\dot{V}_{gas} = \frac{A_s \times D \times \Delta P}{T} \qquad (3)$$

The diffusion coefficient of a gas is directly proportional to its solubility and inversely related to the square root of the molecular weight of the gas.

$$D \sim \frac{solubility}{\sqrt{MW}} \qquad (4)$$

Thus, a smaller molecule or one that is more soluble will diffuse at a faster rate. For example, the diffusion coefficient for carbon dioxide is about 20 times greater than that of oxygen, which means carbon dioxide will diffuse at a faster rate even though it is a larger molecule. The reason is that carbon dioxide is much more soluble than oxygen. Fick's law also states that rate of gas diffusion is inversely related to membrane thickness. This means that for the same diffusion gradient, the diffusion of a gas will be halved if membrane thickness is doubled. Finally, Fick's law states that rate of diffusion is directly proportional to surface area (A_s). If two lungs have the same oxygen diffusion gradient and membrane thickness but one has twice the alveolar-capillary surface area, the rate of diffusion will be twice as great. Under steady-state conditions, approximately 250 ml of oxygen per minute is transferred to the pulmonary circulation ($\dot{V}O_2$) while 200 ml of carbon dioxide per minute is removed ($\dot{V}CO_2$). The ratio of $\dot{V}CO_2/\dot{V}O_2$ is the respiratory exchange ratio (R) and is approximately 0.8.

■ Capillary Blood Flow Limits Oxygen Uptake from Alveoli

Pulmonary capillary blood flow has a marked affect on gas transfer across the alveolar-capillary membrane. The time required for the plasma and red cells to move through the capillary, referred to as **transit time,** is approximately 0.75 seconds, during which time the gas tension in the blood equilibrates with the alveolar gas tension. Transit time can change dramatically with cardiac output. When cardiac output increases, blood flow through the pulmonary capillaries increases and transit time therefore decreases.

The effect of blood flow on the rate of gas uptake is illustrated in Figure 21–4, which plots partial pressure of three individual gases in the blood versus time in the pulmonary capillaries. In the first case, a trace amount of nitrous oxide (laughing gas), a common dental anesthetic, is breathed. Nitrous oxide (N_2O) is chosen because it diffuses across the alveoli and dissolves in the blood but does not combine with hemoglobin. The partial pressure in the blood rises very rapidly and virtually reaches equilibrium with the

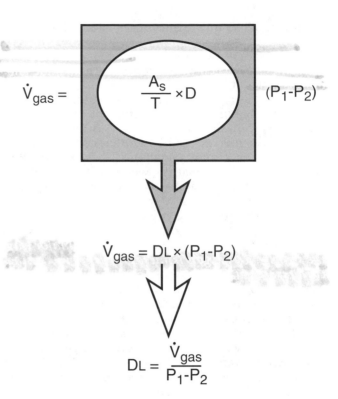

Figure 21–3 ■ The diffusing capacity of the lung (DL) consists of rearrangement of Fick's law to include membrane surface area, diffusion coefficient, and membrane thickness. $\dot{V}_{gas}$, ml of gas transferred/min; P_1, alveolar partial pressure of the test gas; P_2, pulmonary capillary partial pressure of the test gas. (Modified from Selkurt EE: *Physiology.* 5th ed. Boston, Little, Brown, 1984.)

partial pressure of nitrous oxide in the alveoli by the time the blood is one-tenth of the way along the capillary. At this point, the pressure gradient for nitrous oxide is zero ($PA_{N_2O} = PC_{N_2O}$); further net flux for nitrous oxide is zero, and no additional nitrous oxide is transferred. The only way the transfer of nitrous oxide can be increased is by increasing blood flow. The amount of nitrous oxide that can be taken up by blood is entirely limited by flow, not by the diffusion properties of the gas. Therefore, the net transfer of nitrous oxide is **perfusion-limited.**

Figure 21–4 shows the rate of equilibration for carbon monoxide (CO), which also readily diffuses across the alveolar-capillary membrane. Unlike nitrous oxide, carbon monoxide has a very strong affinity for hemoglobin. As the red cell moves through the pulmonary capillary, carbon monoxide rapidly diffuses across the alveolar-capillary membrane and into the blood, where it readily binds to hemoglobin. Because most of the diffused carbon monoxide is chemically bound, its partial pressure in the plasma (PCO) is very low. Consequently, equilibrium for carbon monoxide across the alveolar-capillary membrane is never reached, and the amount of carbon monoxide transferred to the blood is therefore limited by the diffusion properties of the alveolar-capillary membrane, not by blood flow. Thus, the transfer of carbon monoxide is **diffusion-limited.**

Figure 21–4 shows that the equilibration rate for oxygen lies between those of carbon monoxide and nitrous oxide. Oxygen also combines with hemoglobin, but not as readily as carbon monoxide because it has a lower binding affinity. As blood moves along the pulmonary capillary, the rise in P_{O_2} is much greater than the rise in P_{CO} because of differences in binding affinity. Under resting conditions, the capillary P_{O_2} equilibrates with alveolar P_{O_2} when the blood is about one-third of the way along the capillary. Beyond this point, there is no additional transfer of oxygen. Under normal conditions, oxygen transfer is like that of nitrous oxide and is limited primarily by blood flow in the capillary (perfusion-limited). Thus, *the major way the transfer of oxygen is increased is by increasing cardiac output.* Not only does cardiac output increase capillary blood flow, but the increase in pressure increases the diffusion surface area by opening up more capillary beds by recruitment.

The transit time at rest is normally about 0.75 seconds, during which capillary oxygen tension equilibrates with alveolar oxygen tension. Ordinarily, this process takes only about a third of the available time, leaving a wide margin of safety to ensure that alveolar and end-capillary P_{O_2} equilibrate. With heavy exercise, the transient time may be reduced to one-third of a second (Fig. 21–4). The important point is that when a normal subject exercises, there is still time to oxygenate the blood fully. Consequently, pulmonary end-capillary P_{O_2} rarely falls with exercise. In abnormal situations in which there is a thickening of the alveolar-capillary membrane so that oxygen diffusion is impaired, end-capillary P_{O_2} may not reach equilibrium with alveolar P_{O_2}. In this case, there is measurable difference between alveolar and end-capillary P_{O_2}.

Diffusing Capacity

■ Diffusing Capacity Measures the Amount of Gas Transfer

Oxygen enters the blood via alveoli by diffusion. The amount of oxygen the lungs can transfer to the blood is termed **diffusing capacity** (D_{LO_2}), expressed in ml O_2/min/mm Hg. For example, if 250 ml of oxygen per minute is transferred from the lungs to the blood and the average alveolar-capillary P_{O_2} difference is 14 mm Hg, then D_{LO_2} is 18 ml/min/mm Hg. Diffusing capacity provides a quantitative assessment of the diffusion properties of the lung. These include lung surface area (A_s), blood-gas interface thickness (T), and diffusion coefficient (D) for the gas in the lungs. In practice, direct measurement of A_s, T, and D in an intact lung is impossible. To circumvent this problem, Fick's law can be rewritten as shown in Figure 21–3, where the three terms are combined and called diffusing capacity.

Measurement of D_{LO_2} is technically very difficult, since the oxygen diffusion gradient continually changes along the pulmonary capillaries. The initial gradient is 62 mm Hg ($P_{A O_2} - P\bar{v}_{O_2} = 102$ mm Hg $- 40$ mm Hg) and the final gradient is zero. The problem is further complicated by the fact that the decrease in the oxygen diffusion gradient is nonlinear, and the average oxygen diffusion gradient can be determined only by using calculus. To circumvent these difficulties, carbon monoxide is used in trace amounts in the pulmonary function laboratory to measure diffusing capacity. Carbon monoxide offers several advantages for measuring D_L: it is limited by diffusion and

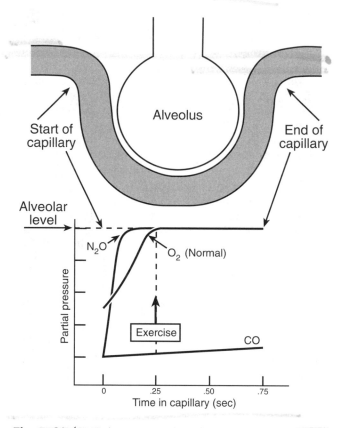

Figure 21–4 ■ Pulmonary capillary blood flow has a dramatic effect on the transfer of O_2 across the alveolar-capillary membrane. Two other test gases—nitrous oxide (N_2O) and carbon monoxide (CO) are illustrated for comparison. The top of the vertical axis indicates the maximal gas tension of the blood when it is equilibrated with the gas in the alveoli. When trace amounts of N_2O are breathed, the partial pressure of N_2O in the pulmonary capillary quickly reaches equilibrium with that of the alveoli. The transfer of this gas is limited by blood flow (perfusion limited). In contrast, when trace amounts of CO are breathed, the partial pressure of CO in the pulmonary capillary is virtually unchanged and never reaches equilibrium with alveolar P_{CO}, because all of the CO that diffuses across the alveolar-capillary membrane is taken up by the hemoglobin. Thus, the transfer in this case is limited by diffusion. The profile for oxygen is more like that of N_2O, which means its transfer across the blood-gas interface is normally limited by blood flow. (Modified from West JB: *Respiratory Physiology.* Baltimore, Williams & Wilkins, 1979.)

not by blood flow; there is essentially no carbon monoxide in the venous blood; and the affinity of carbon monoxide for hemoglobin is 210 times greater than that of oxygen, which causes the partial pressure of carbon monoxide to remain essentially zero in the pulmonary capillaries. Thus, one need only measure carbon monoxide uptake in ml/min ($\dot{V}CO$) and partial pressure in alveoli (P_{ACO}). The equation then becomes:

$$D_L = \frac{\dot{V}CO}{P_{ACO}} \quad (5)$$

The most common technique for making this measurement is called the **single-breath technique.** The patient inhales a single dilute mixture of carbon monoxide and holds his or her breath for approximately 10 seconds. By determining the percentage of carbon monoxide in the alveolar gas at the beginning and the end of 10 seconds and by measuring lung volume, one can calculate $\dot{V}CO$. The single-breath test is very reliable. The normal resting value for D_{LCO} is age-, sex-, and height-dependent and ranges from 20 to 30 ml/min/mm Hg. Diffusing capacity decreases with pulmonary edema or a loss of alveolar membrane (e.g., emphysema).

■ Hemoglobin and Pulmonary Capillary Blood Volume Alter Diffusing Capacity

Two factors that affect interpretation of the diffusing capacity are blood hematocrit and pulmonary capillary blood volume, both of which affect D_L in the same direction (i.e., a decrease in either the hematocrit or capillary blood volume will lower the diffusing capacity in an otherwise normal lung). For example, if two individuals have the same diffusion properties of the lung but one is anemic (reduced hematocrit), the anemic individual would have the appearance of decreased diffusion capacity. A change in cardiac output during the single breath test also affects diffusing capacity. An abnormally low cardiac output lowers the capillary blood volume, which causes a decrease in diffusion capacity in an otherwise normal lung.

Gas Transport by the Blood

■ Oxygen Is Transported in Two Forms

The primary function of gas exchange is the **oxygenation** of venous blood. Oxygen transport is the means by which blood carries oxygen from alveoli to metabolically active tissues, where oxygen is used for energy production. Oxygen is transported to the tissues in two forms: combined with hemoglobin (Hb) in the red cell or physically dissolved in the blood. More than 95% of the oxygen is carried by hemoglobin and only 5% physically dissolved.

Hemoglobin consists of four oxygen-binding heme sites (iron-containing porphyrin molecules) and a globular protein chain. Oxygen binds rapidly and reversibly

to hemoglobin. When oxygen binds with hemoglobin it is called **oxyhemoglobin** (HbO_2); the unoxygenated form is called **deoxyhemoglobin** (Hb). Each gram of hemoglobin can bind 1.34 ml of oxygen. The reaction $Hb + O_2 \rightleftharpoons HbO_2$ is reversible and is a function of the partial pressure of oxygen in the blood. In the lungs, where P_{O_2} is high, the reaction is shifted to the right to form oxyhemoglobin. In the tissue, where P_{O_2} is low, the reaction is shifted to the left; oxygen is unloaded from hemoglobin and becomes available to the cells. The maximum amount of oxygen that can be carried by hemoglobin is called the **oxygen carrying capacity** and is normally about 20 ml O_2/100 ml of blood. This value is calculated assuming a normal hematocrit of 15 g Hb/100 ml of blood (1.34 ml/g Hb × 15 g Hb/100 ml blood = 20.1 ml O_2/100 ml of blood). **Oxygen content** is the amount of oxygen actually bound to hemoglobin (whereas capacity is the amount that can potentially be bound). The **percent saturation** of hemoglobin is calculated as:

$$\text{percent } O_2 \text{ saturation} = \frac{O_2 \text{ content}}{O_2 \text{ capacity}} \times 100 \quad (6)$$

Thus, oxyhemoglobin saturation (S_{O_2}) is the ratio of the quantity of oxygen *actually bound* to the quantity that can be potentially bound. For example, if oxygen content is 16 ml O_2/100 ml blood and oxygen capacity is 20 ml O_2/100 ml blood, then the blood is 80% saturated. Normally, arterial blood is about 97% saturated with oxygen.

Blood P_{O_2}, O_2 saturation, and oxygen content are three closely related indices of oxygen transport. The relationship between P_{O_2}, oxygen saturation, and oxygen content is illustrated by the **oxyhemoglobin equilibrium curve** (Fig. 21–5), a sigmoidal curve (S-shaped) over a range of arterial oxygen tensions from 0 to 100 mm Hg. The shape of the curve results because the hemoglobin **affinity** for oxygen increases progressively as blood P_{O_2} rises.

The shape of the oxyhemoglobin equilibrium curve indicates several physiologic advantages. The plateau region of the curve is the **loading phase,** where oxygen is loaded onto hemoglobin to form oxyhemoglobin in the lung. The plateau describes how oxygen saturation and content remain fairly constant despite wide fluctuations in alveolar P_{O_2}. For example, if P_{AO_2} were to rise from 100 to 120 mm Hg, hemoglobin would become only slightly more saturated (97% → 98%). For this reason, oxygen content cannot be raised appreciably by hyperventilation. The steep region of the curve, the **unloading phase,** allows large quantities of oxygen to be released (dissociated) at the lower capillary P_{O_2} that prevails at the tissue level without large changes in P_{O_2}. In brief, oxygen can saturate hemoglobin under high partial pressures in the lung, and oxyhemoglobin can give up (dissociate) large

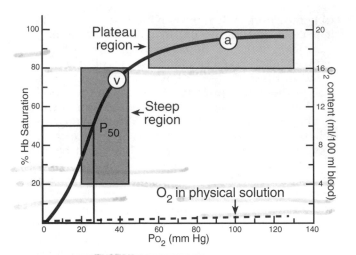

Figure 21–5 ■ Oxygen-hemoglobin equilibrium curve. a, arterial; v, venous; P_{50}, partial pressure of O_2 required to saturate 50% of the hemoglobin with oxygen.

amounts of oxygen with small changes in P_{O_2} at the tissue level.

Changing the binding affinity shifts the oxyhemoglobin equilibrium curve to the right or left of normal (Fig. 21–6). The P_{O_2} at which 50% of the hemoglobin is saturated (P_{50}) is an indicator of the binding affinity of hemoglobin for oxygen and will determine the relative position of the curve. The normal P_{50} for arterial blood ranges from 26 to 28 mm Hg. A high P_{50} signifies a decrease in hemoglobin's affinity for oxygen and results in a rightward shift in the oxyhemoglobin dissociation curve; a low P_{50} signifies the opposite and shifts the curve to the left. A shift in the P_{50} in either direction has only a small effect on the "loading" of oxygen in the normal lung, because the loading occurs at the plateau phase. However, that is not the case for the steep phase.

Body temperature, arterial carbon dioxide tension, and arterial pH affect the binding affinity of hemoglobin. A rise in P_{CO_2}, a fall in pH, and a rise in temperature all shift the curve to the right (Fig. 21–6). The effect of carbon dioxide and hydrogen on the affinity of hemoglobin for oxygen is known as the **Bohr effect.** A shift of the oxyhemoglobin equilibrium curve to the right is physiologically advantageous at the tissue level (steep phase) because the affinity is lowered (increased P_{50}). A rightward shift enhances the unloading of oxygen for a given P_{O_2} in the tissue, and a leftward shift increases the affinity of hemoglobin for oxygen, thereby lowering the ability to release oxygen to the tissues. A simple way to remember the functional importance of these shifts is that an exercising muscle is hot and acidic and has a high P_{CO_2}, all of which favor unloading more oxygen to the metabolically active muscle cells.

Red cells contain 2,3-diphosphoglycerate (2,3-DPG), an organic phosphate, that also can affect oxyhemoglo-

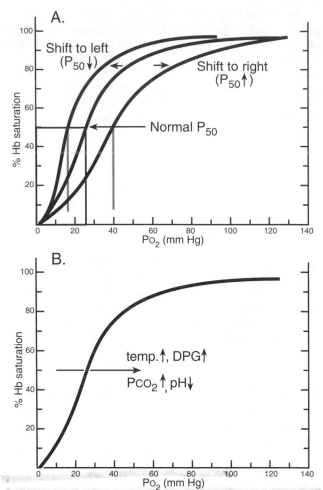

Figure 21–6 ■ Blood temperature, arterial pH, and arterial P_{CO_2} affect the binding affinity of O_2 for hemoglobin (Hb) and cause a shift in the O_2 equilibrium curve. (A) A shift in the O_2 equilibrium curve affects P_{50}. (B) A rightward shift of the curve favors the unloading of O_2 from Hb at the tissue level. This occurs with rise in body temperature, increase in arterial P_{CO_2}, or decrease in pH. Note that an increase in red cell levels of 2,3-diphosphoglycerate (DPG) will also shift the curve to the right.

bin affinity. In the red cell, 2,3-DPG levels are much higher than in other cells because the erythrocyte lacks mitochondria and uses a pathway that converts much of the formed 1,3-DPG to 2,3-DPG. An increase in 2,3-DPG facilitates unloading of oxygen from the red cell at the tissue level (shifts the curve to the right). An increase in red cell 2,3-DPG occurs with exercise and hypoxia (e.g., high altitudes, chronic lung disease).

There are two important facts to remember regarding the relationship between P_{O_2}, oxygen saturation, and oxygen content. First, Pa_{O_2} is related to the amount of oxygen dissolved in blood and not to the amount bound to hemoglobin. Second, oxygen content, rather than P_{O_2} or saturation, is what keeps us alive and serves as a better gauge for oxygenation. For example,

HYPOXIA-INDUCED PULMONARY HYPERTENSION

Hypoxia has opposite effects on the pulmonary and systemic circulations, relaxing systemic vascular smooth muscle while eliciting pulmonary vasoconstriction. Hypoxic pulmonary vasoconstriction (HPV) is the major mechanism for matching regional blood flow to regional ventilation in the lung. It automatically adjusts regional pulmonary capillary blood flow in response to alveolar hypoxia and prevents blood from perfusing poorly ventilated regions in the lung. However, when hypoxia affects all parts of the lung for several days, it causes pulmonary hypertension but has no effect on systemic circulation. Hypoxia induced pulmonary hypertension occurs in individuals who live at a high altitude (8,000–12,000 ft) and in those with chronic obstructive pulmonary disease (COPD), especially emphysemic patients.

With chronic hypoxia-induced pulmonary hypertension, the pulmonary artery undergoes a major remodeling process. There is increased wall thickening due to hypertrophy and hyperplasia of vascular smooth muscle of the normal muscular artery, along with an increase in connective tissue. These structural changes occur in large and small arteries. Also, there is abnormal extension of smooth muscle into peripheral vessels where muscularization is not present, which is especially pronounced in the precapillary segment. Finally, there is eventual obliteration of small pulmonary arteries and arterioles. Because of the hypoxic vasoconstriction and vascular remodeling, the elevated pulmonary pressure and increased pulmonary vascular resistance associated with hypoxia-induced hypertension causes right heart hypertrophy. A striking feature of vascular remodeling is that both the pulmonary artery and pulmonary vein contract with hypoxia but only the arterial side undergoes a major remodeling process. The postcapillary segments and veins are spared of the structural changes seen with hypoxia.

an individual can have a normal arterial P_{O_2} and a normal saturation but reduced oxygen content. This is seen in patients who have a decreased number of circulating red cells (anemia). An anemic patient who has a 50% reduction in hemoglobin concentration (7.5 g/100 ml instead of 15 g/100 ml) will have a normal arterial P_{O_2} and S_{aO_2}, but oxygen content will be reduced to half normal. The usual oxyhemoglobin equilibrium curve does not show changes in blood oxygen content, since the vertical axis is saturation. If the vertical axis is changed to oxygen content (ml O_2/100 ml blood), then changes in content are seen (Fig. 21–7). The shape of the oxygen equilibrium curve does not change, but the curve moves down to reflect the reduction in oxygen content.

Blood volume also affects oxygen content of the blood. In two individuals with a circulating blood volume of 3 liters and 5 liters, respectively, the P_{O_2} of blood will be the same but oxygen content will differ. A good analogy is the comparison between a bicycle and a truck tire. Both can have the same air pressure (e.g., 30 Psi), but the amount of air in each tire is different.

Carbon monoxide also interferes with oxygen trans-

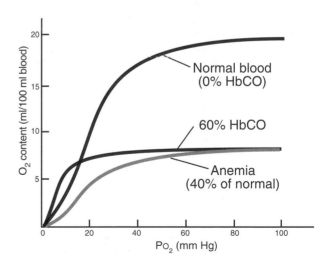

Figure 21–7 ■ Anemia affects the O_2 equilibrium curve. Severe anemia can lower O_2 content to 40% of normal. Blood O_2 content of an individual exposed to carbon monoxide is shown for comparison. When the blood is 60% saturated with carbon monoxide (HbCO), O_2 content of the blood is reduced below 10 ml/100 ml of blood. Note the leftward shift of the O_2 hemoglobin (Hb) dissociation curve when CO binds with Hb.

port by competing for the same binding sites. Carbon monoxide binds to hemoglobin to form **carboxyhemoglobin** (HbCO). The reaction (Hb + CO $\rightleftharpoons$ HbCO) is reversible and is a function of carbon monoxide tension. This means that breathing higher concentrations of carbon monoxide will favor the reaction to the right. Breathing fresh air will favor the reaction to the left, which will cause carbon monoxide to be released from the hemoglobin. A striking feature of carbon monoxide is a binding affinity about 210 times that of oxygen: carbon monoxide will bind with the same amount of hemoglobin as oxygen at a partial pressure 210 times lower than that of oxygen. For example, breathing normal air (21% O_2) contaminated with 0.1% CO would cause half the hemoglobin to be saturated with carbon monoxide and half with oxygen. With the high affinity of hemoglobin for carbon monoxide, relatively small exposure to carbon monoxide can result in the formation of large amounts of carboxyhemoglobin. Arterial P_{O_2} will still be normal, but oxygen content will be greatly reduced. This is seen in Figure 21-7, which shows the effect of carbon monoxide on the oxyhemoglobin dissociation curve. When the blood is 60% saturated with carbon monoxide (carboxyhemoglobin) the oxygen content is reduced to less than 10 ml/100 ml blood. Also, the presence of carbon monoxide shifts the curve to the left; it is more difficult to unload or release oxygen to the tissues.

Carbon monoxide is dangerous for several reasons. First is its greater binding affinity for hemoglobin. Second, it is an odorless, colorless, and nonirritating gas, which makes it virtually impossible to detect. Third, the arterial P_{O_2} is normal, which prevents any feedback mechanism to indicate that oxygen content is low. Fourth, there are no physical signs of hypoxemia (i.e., cyanosis or bluish color around lips and fingers) because the blood stays bright cherry red when carbon monoxide binds with hemoglobin. Thus, a person can be exposed to a lethal concentration of carbon monoxide and have oxygen content reduced to a level that causes anoxia to the tissues without being aware of the danger. The brain is one of the first organs affected by lack of oxygen. Carbon monoxide can alter reaction time, cause blurred vision, and, if severe enough, cause collapse into unconsciousness. The best treatment for carbon monoxide poisoning in an emergency is breathing 100% oxygen or a mixture of 95% oxygen and 5% carbon dioxide. Since oxygen and carbon monoxide compete for the same binding site on the hemoglobin molecule, breathing a high oxygen concentration will drive off the carbon monoxide and favor the formation of oxyhemoglobin. The addition of 5% carbon dioxide to the inspired gas stimulates ventilation and causes hyperventilation, which enhances the release of carbon monoxide from hemoglobin. (Carbon monoxide is not stored in the body and can be loaded and unloaded from hemoglobin as a function of P_{CO}).

■ Carbon Dioxide Is Transported in Three Forms

Figure 21-8 illustrates the processes involved in carbon dioxide transport. Carbon dioxide is carried in the blood in three forms: physically dissolved in the plasma (10%), as bicarbonate ions in the plasma (60%), and as **carbamino proteins** (30%). The difference between the tissue P_{CO_2} and blood P_{CO_2} drives carbon dioxide into the blood. Only a small amount stays as dissolved carbon dioxide in the plasma, but this fraction is important because the plasma P_{CO_2} serves as the driving pressure for carbon dioxide movement into the red cell. The bulk of the carbon dioxide diffuses into the red cell, where it forms carbonic acid, or **carbamino hemoglobin.** In the red cell, carbonic acid is formed by the following reaction:

$$CO_2 + H_2O \underset{CA}{\rightleftharpoons} H_2CO_3 \rightleftharpoons H^+ + HCO_3^- \qquad (7)$$

This reaction would take place very slowly were it not accelerated about 1000 times in the red cell by the enzyme **carbonic anhydrase** (CA). The enzyme is also found in the renal tubular cells, gastrointestinal mucosa, and muscle, but its activity is highest in the red cell.

Carbonic acid (H_2CO_3) readily ionizes in the red cell to form bicarbonate (HCO_3^-), which diffuses out of the red cell, and H^+, which cannot readily move out because of the relative impermeability of the membrane to H^+. To maintain electrical neutrality, therefore,

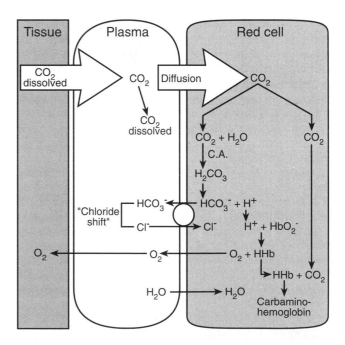

Figure 21–8 ■ Carbon dioxide is transported in three forms: physically dissolved, as HCO_3^-, and as carbamino hemoglobin in the red cell. The uptake of CO_2 favors the release of O_2. (Modified from Selkurt EE: *Physiology.* 5th ed. Boston, Little, Brown, 1984.)

Cl^- diffuses into the red cell from the plasma (Fig. 21-8). The chloride movement is known as the **chloride shift** and is facilitated by a chloride-bicarbonate exchanger in the red cell membrane. Most of the H^+ formed from the ionization of H_2CO_3 is buffered by hemoglobin: $H^+ + HbO_2^- \rightleftharpoons HHb + O_2$. As H^+ binds to hemoglobin, it reduces oxygen binding and shifts the oxyhemoglobin equilibrium curve to the right. Thus, the unloading of oxygen from the hemoglobin in the tissues favors the carriage of carbon dioxide, while in the pulmonary capillaries the oxygenation of hemoglobin favors the unloading of carbon dioxide.

Carbamino hemoglobin occurs in the red cell from the reaction of carbon dioxide with free amine groups (NH_2) on the hemoglobin molecule:

$$CO_2 + Hb\ NH_2 \rightleftharpoons Hb\ NHCOOH \qquad (7)$$

Deoxygenated hemoglobin can bind much more carbon dioxide in this way than can oxygenated hemoglobin. Although major reactions regarding carbon dioxide transport occur in the red blood cell, the bulk of the carbon dioxide is actually carried in the plasma in the form of bicarbonate.

A carbon dioxide equilibrium curve (Fig. 21-9) can be constructed in a fashion similar to that for oxygen content or saturation. Two features regarding the carbon dioxide content curve stand out. One is that it is nearly a straight line, with no plateau or steep regions, as seen in the oxygen content curve. Second, the presence of oxygen will shift the curve to the right. This is known as the **Haldane effect,** and its advantage is that it allows the blood to load more carbon dioxide

at the tissues and unload more carbon dioxide in the lungs. Major differences are observed between the carbon dioxide equilibrium curve and oxygen equilibrium curves (Fig. 21-10). First, 1 liter of blood can hold much more carbon dioxide than oxygen. Second, the carbon dioxide dissociation curve is steeper and more linear. Because of this solubility, and carbon dioxide's solubility, the blood can load and unload large amounts of carbon dioxide with a small change in gas tension. This is important not only in gas exchange but also in the regulation of acid-base balance.

Respiratory Causes of Hypoxemia

Under normal conditions the hemoglobin is fully saturated with oxygen in blood leaving the pulmonary capillaries, and the capillary P_{O_2} equals alveolar P_{O_2}. However, the blood that leaves the lungs (via the pulmonary veins) and returns to the left side of the heart has a lower P_{O_2} than pulmonary capillary blood. As a result, the systemic arterial blood has an average oxygen tension (Pa_{O_2}) of 95 mm Hg and is only 98% saturated (Fig. 21-11). This difference between

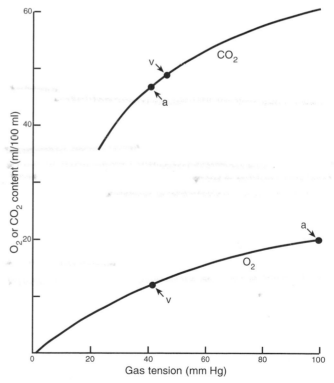

Figure 21–10 ■ Major differences can be observed between O_2 and CO_2 content curves. The carrying capacity of blood for CO_2 (in all its forms) is much greater than for O_2, and the increased steepness and linearity of the CO_2 curve allow the lung to remove large quantities of CO_2 from the blood with a small CO_2 difference. a, arterial blood; v, venous blood.

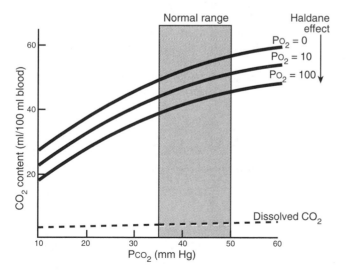

Figure 21–9 ■ Carbon dioxide equilibrium curve. Unlike that of the oxyhemoglobin dissociation curve, which is S-shape, the CO_2 content curve for blood is relatively linear. Note that the presence of O_2 causes a rightward and downward shift of the curve (Haldane effect). a, arterial blood; v, venous blood.

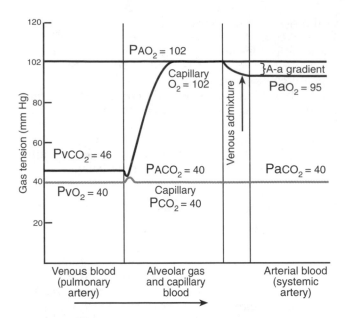

Figure 21–11 ■ Oxygen and CO_2 tensions in the pulmonary artery, pulmonary capillary, and systemic arterial blood. Note that the Po_2 leaving the pulmonary capillary is equilibrated with alveolar Po_2. However, the systemic arterial Po_2 is not equal to alveolar Po_2. This results in an alveolar-arterial (A-a) O_2 gradient, and is caused by venous admixture.

alveolar tension (Pao_2) and arterial oxygen tension (Pao_2) is the **alveolar-arterial oxygen gradient,** or A-a O_2 gradient. Alveolar Po_2 is normally 100–102 mm Hg and arterial Po_2 85–95 mm Hg. Thus, a normal A-a O_2 gradient can range from 10 to 15 mm Hg. These values are obtained by using blood gas measurements and the alveolar gas equation [$Pao_2 = Fio_2 (PB − 47) − 1.2 (Paco_2)$; Chap. 19]. The A-a O_2 gradient arises in the normal individual because of wasted ventilation, a right-to-left shunt (i.e., bronchial circulation), and regional variations of the $\dot{V}a/\dot{Q}$ ratio. Approximately one-third of the normal A-a O_2 gradient is due to wasted ventilation and to a right-to-left anatomic shunt and two-thirds is due to regional variations of the $\dot{V}a/\dot{Q}$ ratio. In some pathophysiologic disorders, the A-a O_2 gradient can be greatly increased; a value above 15 mm Hg is considered abnormal and usually leads to hypoxemia; that is, low arterial blood oxygen due to a low Po_2 or a low oxygen content in the arterial blood. The normal ranges of blood gases are shown in Table 21–2. Values for Pao_2 below 85 mm Hg would be considered hypoxemic, and a $Paco_2$ value less than 35 or greater than 48 indicates hypocapnia and hypercapnia, respectively. A pH value for arterial blood that is less than 7.35 or greater than 7.45 is called acidemia or alkalemia, respectively.

The causes of hypoxemia (Table 21–3) are classified as respiratory and nonrespiratory. Nonrespiratory causes are not common in adults. Right-to-left intracardiac shunts are seen in the perinatal period but are

TABLE 21–2 ■ Arterial Blood Gases

Measurement	Normal Range*
Pao_2	85–95 mm Hg
$Paco_2$	35–48 mm Hg
Sao_2	94–98%
pH	7.35–7.45
HCO_3^-	23–28 mEq/L

*Normal range at sea level. Normal values are age-dependent.

rare in adults. Decreased inspired oxygen pressure (Pio_2) in the hospitalized patient rarely causes hypoxemia in adults. A decrease in Fio_2 may occur during the administration of anesthesia as a result of improper connections of the oxygen supply lines or a sealed mask that is cut off from adequate ventilation. Ascent to a high altitude can cause hypoxemia from a low Pio_2. Of the nonrespiratory causes, reduced oxygen content (i.e., anemia or carbon monoxide poisoning) is the most important cause in adults.

Compared to nonrespiratory causes, respiratory dysfunction is by far the most important cause of hypoxemia. Of the four respiratory causes listed in Table 21–3, the least important is a **diffusion impairment.** This occurs when the diffusion distance across the alveolar-capillary membrane is increased or the permeability of the alveolar-capillary membrane is decreased (e.g., pulmonary edema or fibrosis). Next least important is a physiologic shunt, which can be due to a right-to-left anatomic shunt or an alveolar shunt.

By far the two most important causes of hypoxemia are **generalized hypoventilation** and a **low $\dot{V}a/\dot{Q}$ ratio.** Generalized hypoventilation occurs when total ventilation is depressed. This can arise from a chronic obstructive pulmonary disorder (e.g., emphysema) or depressed respiration. In the latter case it occurs in comatose patients from a head injury or drug overdose. Since total ventilation is depressed there is also a significant increase in arterial Pco_2 with a concomitant decrease in arterial pH. In generalized hypoventilation, total ventilation is insufficient to maintain normal sys-

TABLE 21–3 ■ Pathophysiologic Causes of Hypoxemia

Causes	Effect on $P(A\text{-}a)O_2$ Gradient
Respiratory	
Diffusion impairment	Increased
Physiologic shunt	Increased
Generalized hypoventilation	Normal
Local low $\dot{V}a/\dot{Q}$	Increased
Nonrespiratory	
Intracardiac right-to-left shunt	Increased
Decreased Pio_2, low PB, low Fio_2	Normal
Reduced oxygen content (anemia and carbon monoxide poisoning)	Normal

temic arterial P_{O_2} and P_{CO_2}. Two features that distinguish generalized hypoventilation are a high Pa_{CO_2} and a normal A-a O_2 gradient. The latter is due to the fact that the alveolar and arterial P_{O_2} are equally reduced. If a patient has a low Pa_{O_2}, high Pa_{CO_2}, and normal A-a O_2 gradient, the cause of hypoxemia is entirely due to generalized hypoventilation. Thus, the best corrective measure is to place the patient on a mechanical ventilator (to correct generalized hypoventilation and return arterial P_{O_2} and P_{CO_2} to normal). Administering supplemental oxygen to a patient with generalized hypoventilation will correct hypoxemia but not hypercapnia, because total ventilation is still depressed.

A local $\dot{V}_A/\dot{Q}$ imbalance is by far the most common cause of hypoxemia (about 90% of cases). A variation in $\dot{V}_A/\dot{Q}$ ratio can be regional (gravity-dependent) or local (not gravity-dependent) (Chap. 20). Hypoxemia

due to a low $\dot{V}_A/\dot{Q}$ ratio is almost entirely the result of a local cause rather than regional variation. The most frequent cause is **localized hypoventilation,** which stems from a partially obstructed airway. A fraction of the blood that passes through these lung units does not get fully oxygenated, resulting in an increase in venous admixture. Only a small amount of venous admixture is required to lower systemic arterial P_{O_2}, due to the nature of the oxyhemoglobin equilibrium curve. This can be seen from Figure 21–12, which depicts oxygen content from three groups of alveoli

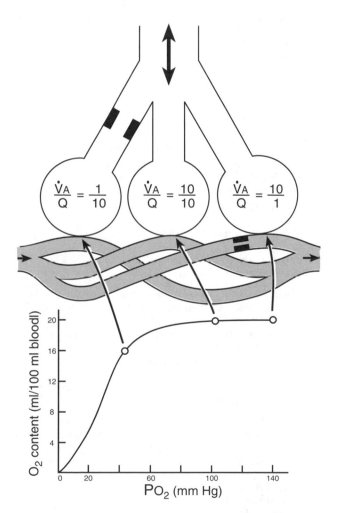

Figure 21–12 ■ Depression of Pa_{O_2} and O_2 content caused with venous admixture. Because of the S-shaped O_2-Hb equilibrium curve, a lung unit with a high $\dot{V}_A/\dot{Q}$ ratio will have little effect on arterial P_{O_2} and O_2 content. However, mixing blood from a lung unit with low $\dot{V}_A/\dot{Q}$ will have a dramatic effect on oxygenated blood leaving the lung. (Modified from Selkurt EE: *Physiology.* 5th ed. Boston, Little, Brown, 1984.)

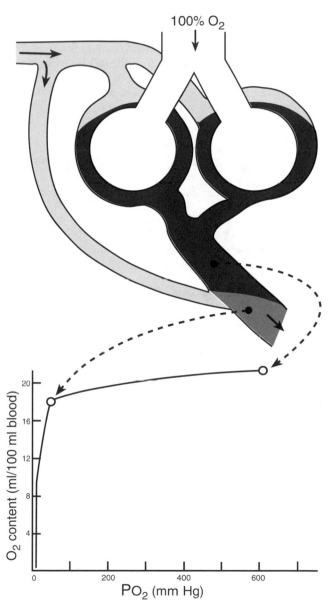

Figure 21–13 ■ A shunt can be diagnosed by having the subject breathe 100% O_2 for 15 minutes. P_{O_2} in systemic arterial blood in a patient with a shunt does not increase above 150 mm Hg. The shunted blood is not exposed to 100% O_2 and the venous admixture, with its low O_2 content, reduces arterial P_{O_2}.

with a low, normal, and high $\dot{V}A/\dot{Q}$. The oxygen content of the blood leaving these lung units is 16.0, 19.5, and 20.0 ml/100 ml of blood, respectively. Although PAO_2 is very high in a lung unit with a localized high $\dot{V}A/\dot{Q}$, relatively little oxygen is added to the blood, whereas blood leaving a lung region with a low $\dot{V}A/\dot{Q}$ will have a significant effect on oxygen content. Blood from these units has a lower oxygen content than that from units with a normal $\dot{V}A/\dot{Q}$ ratio, due to the nonlinear shape of the oxygen equilibrium curve.

A local $\dot{V}A/\dot{Q}$ imbalance is easily diagnosed. The subject breathes 100% oxygen for 15 minutes. If the increase in PaO_2 after this period is *greater than* 150 mm Hg, then hypoxemia is due to a local low $\dot{V}A/\dot{Q}$ ratio. This is because breathing 100% oxygen compensates for the low $\dot{V}A/\dot{Q}$, and blood leaving these units is oxygenated, thereby increasing arterial PO_2. Breathing 100% oxygen also helps distinguish between a shunt and a localized low $\dot{V}A/\dot{Q}$ as a cause of hypoxemia: the arterial PO_2 will not change dramatically if hyp-

oxemia is due to a shunt, because shunted blood is not exposed to the 100% oxygen (Fig. 21-13). A general rule of thumb in distinguishing between these two causes of hypoxemia is as follows. If the arterial PO_2 is less than 150 mm Hg after breathing 100% oxygen for 15 minutes, then the cause of hypoxemia is a shunt. If it is greater than 150 mm Hg, then the cause is a localized low $\dot{V}A/\dot{Q}$ ratio.

In summary, there are four basic respiratory disturbances that cause hypoxemia. Examining the A-a O_2 gradient or $PaCO_2$ and/or breathing 100% oxygen distinguishes the four types. For example, if a patient has a low PaO_2, high $PaCO_2$, and normal A-a O_2 gradient, the cause of hypoxemia is generalized hypoventilation. If the PaO_2 is low and the A-a O_2 gradient is high, then the cause can be due to a shunt, localized low $\dot{V}A/\dot{Q}$, or a diffusion block. A diffusion impairment is the least likely cause and is characterized by a low PaO_2, high A-a O_2 gradient, and high $PaCO_2$ and can be deduced if the other three are eliminated.

■ ■ ■ ■ ■ ■ ■ ■ ■ ■ ■ **REVIEW EXERCISES** ■ ■ ■ ■ ■ ■ ■ ■ ■ ■ ■

Identify Each with a Term

1. Movement due to random movement of gas molecules
2. PO_2 difference across the alveolar-capillary membrane
3. The volume of gas transferred per minute per mm Hg pressure difference across the alveolar-capillary membrane
4. The maximum amount of oxygen that can be carried by hemoglobin
5. The form in which carbon dioxide is carried by hemoglobin
6. The effect of oxygen on the carbon dioxide content curve
7. The mixing of venous blood with oxygenated blood

Define Each Term

8. Hypoxemia
9. A-a O_2 gradient
10. Oxygen saturation
11. P_{50}
12. Carboxyhemoglobin
13. Carbonic anhydrase
14. Bohr effect

Choose the Correct Answer

15. A patient whose hemoglobin is 98% saturated with oxygen but whose arterial oxygen content is 10 ml O_2/100 ml blood is most likely to have:
    a. A low lung diffusing capacity
    b. Anemia
    c. An arteriovenous shunt
    d. Alveolar hypoxia
16. In normal individuals the major cause of A-a O_2 gradient is that:

a. The diffusing capacity for oxygen is less than that for carbon dioxide
b. There is a regional spread of the $\dot{V}A/\dot{Q}$ ratio in the lung
c. No gas exchange occurs in the conducting airways
d. There is a right-to-left shunt

17. A low lung diffusing capacity (DL) could be caused by:
    a. Increased diffusion distance
    b. Decreased capillary blood volume
    c. Decreased surface area
    d. All of the above

18. With respect to oxygen and carbon dioxide transport:
    a. The slopes of the oxygen and carbon dioxide content curves are similar
    b. 100 ml of blood can carry more oxygen than carbon dioxide for the same partial pressure
    c. The presence of carbon dioxide favors oxygen binding to hemoglobin
    d. The presence of oxygen lowers carbon dioxide content in the blood

19. In an otherwise normal individual who has lost enough blood to decrease his hemoglobin concentration from the normal 15 g/100 ml of blood to 10 g/100 ml of blood, which of the following is expected to decrease?
    a. Arterial PO_2
    b. Arterial hemoglobin saturation
    c. Arterial oxygen content
    d. All of the above

20. Which of the following would *not* favor the unloading of oxygen from hemoglobin in tissues?
    a. Decrease in P_{50}
    b. Decrease in tissue pH
    c. Increase in 2,3-DPG levels
    d. Increase in tissue PCO_2

21. Arterial P_{O_2} will decrease with:
    a. Anemia
    b. Carbon monoxide poisoning
    c. Hemorrhage
    d. Alveolar hypoxia
    e. All of the above

22. Which of the following sets of arterial blood gas data is consistent with the presence of localized hypoventilation?
    a. Pa_{O_2} = 130 mm Hg; Pa_{CO_2} = 40 mm Hg
    b. Pa_{O_2} = 98 mm Hg; Pa_{CO_2} = 30 mm Hg
    c. Pa_{O_2} = 95 mm Hg; Pa_{CO_2} = 40 mm Hg
    d. Pa_{O_2} = 60 mm Hg; Pa_{CO_2} = 40 mm Hg

23. Which of the following ranges of hemoglobin saturation from systemic venous to systemic arterial blood represents normal resting condition?
    a. 25-75%
    b. 40-75%
    c. 40-95%
    d. 60-97%
    e. 75-97%

24. A 54-year-old individual sustains third-degree burns in a house fire. Respiratory rate is 30/min, Hb = 17 g/dL, arterial P_{O_2} is 95 mm Hg and arterial O_2 saturation is 50%. The most likely cause of his low oxygen saturation is:
    a. Airway obstruction
    b. Carbon monoxide poisoning
    c. Pulmonary edema
    d. Fever

25. A patient's Pa_{CO_2} is 68 mm Hg and Pa_{O_2} is 50 mm Hg. This is accompanied by a normal A-a O_2 gradient. These findings are consistent with a:

a. Shunt
b. Low $\dot{V}_A/\dot{Q}$ ratio
c. Diffusion impairment
d. Generalized hypoventilation

Calculate

26. A patient's Pa_{CO_2} is 45 mm Hg, Pa_{O_2} is 70 mm Hg, pH is 7.30, and PA_{O_2} is 100 mm Hg. What is the A-a O_2 gradient?

27. Blood volume is approximately 7% of body weight. Calculate total oxygen content of blood in a normal 60-kg woman whose blood is 95% saturated and whose oxygen carrying capacity is 20 ml O_2/100 ml blood.

28. A patient inspired a gas mixture containing a trace amount of carbon monoxide and then held his breath for 10 seconds. During breath holding, the alveolar P_{CO_2} averaged 0.5 mm Hg and carbon monoxide uptake was 10 ml/min. Calculate his lungs' diffusing capacity for carbon monoxide ($D_{L_{CO}}$).

29. A patient has a diffusing capacity for oxygen of 25 ml/min/mm Hg, an arterial P_{O_2} of 100 mm Hg, and an oxygen consumption of 200 ml/min. Calculate the oxygen diffusion gradient.

Briefly Answer

30. What limits oxygen transfer in the lung?

31. How does carbon dioxide affect the oxygen dissociation curve?

32. What is the most common cause of hypoxemia?

■ ■ ■ ■ ■ ■ ■ ■ ■ ■ ■ ■ ■ ■ ■ ■ **ANSWERS** ■ ■ ■ ■ ■ ■ ■ ■ ■ ■ ■ ■ ■ ■ ■ ■

1. Diffusion
2. Oxygen diffusing gradient
3. Diffusing capacity
4. Oxygen capacity
5. Carbamino hemoglobin
6. Haldane effect
7. Venous admixture
8. A low arterial P_{O_2} (< 85 mm Hg) or low oxygen content
9. The P_{O_2} difference between alveoli and systemic arterial blood
10. A term to express O_2 content/O_2 capacity
11. The P_{O_2} required to make the blood hemoglobin 50% saturated with oxygen
12. A complex of carbon monoxide and hemoglobin
13. An enzyme that catalyzes the hydration of carbon dioxide to form carbonic acid
14. Displacement of the oxyhemoglobin dissociation curve by a change in P_{CO_2} or pH
15. b
16. b

17. d
18. d
19. c
20. a
21. d
22. d
23. e
24. b
25. d
26. A-a O_2 gradient = PA_{O_2} − Pa_{O_2}

    = 100 mm Hg − 70 mm Hg

    = 30 mm Hg

27. Total oxygen content = 798 ml O_2
    (O_2 content = O_2 capacity × Blood volume × % Sat)

28. D_L = 20 ml/min/mm Hg

    = $\dfrac{\dot{V}_{CO}}{Pa_{CO}}$ = $\dfrac{10\ \text{ml/min}}{0.5\ \text{mm Hg}}$

29. A-a O_2 diffusion gradient = 8 mm Hg

    O_2 consumption = Diffusing capacity × O_2 diffusion gradient

    O_2 diffusion gradient = $\dfrac{\text{Oxygen consumption}}{\text{Diffusing capacity}}$

    $= \dfrac{200 \text{ ml/min}}{25 \text{ ml/min/mm Hg}}$

30. Oxygen transfer is limited primarily by blood flow, capillary blood volume, and hematocrit.

31. The presence of carbon dioxide will cause a rightward shift in the oxygen dissociation curve and favor the unloading of oxygen from hemoglobin.

32. A low regional $\dot{V}_A/\dot{Q}$ ratio

Suggested Reading

Forster RE II, Dubois AB, Brisco WA, Fisher AF: *The Lung: Physiological Basis for Pulmonary Function Tests.* 3rd ed. Chicago, Year Book, 1986.

Murray JF: *The Normal Lung.* 2nd ed. Philadelphia, WB Saunders, 1986.

Nunn JF: *Applied Respiratory Physiology.* 3rd ed. London, Butterworths, 1987.

Robin ED: *Of Men and Mitochondria: Coping with Hypoxic Dysoxia. Am Rev Respir Dis* 122:517, 1980.

West JB: *Ventilation/Blood Flow and Gas Exchange.* 3rd ed. Oxford, Blackwell, 1977.

CHAPTER

■ ■ ■ ■ ■ ■ ■

22

The Control of Breathing

CHAPTER OUTLINE

I. GENERATION OF THE BREATHING PATTERN
   A. Two major groups of cells in the medulla are associated with breathing
   B. Abrupt loss of suppression of inspiratory drive synchronizes the onset of inspiration
   C. After rising progressively, inspiratory activity is switched off to initiate expiration
   D. Expiration is divided into two phases
   E. Various strategies adjust breathing to meet demand
   F. Muscles of the upper airways are also under phasic control

II. REFLEXES FROM LUNGS AND CHEST WALL
   A. Reflexes from the lung are associated with three classes of receptors
      1. Slowly adapting receptors
      2. Rapidly adapting receptors
      3. C fiber endings
   B. Chest wall proprioceptors provide information about movement and muscle tension

III. CONTROL OF BREATHING BY H$^+$, CO$_2$ AND O$_2$
   A. Cells of the medulla respond to local H$^+$
   B. Cerebrospinal fluid pH depends on its bicarbonate concentration and P$_{CO_2}$
   C. Peripheral chemoreceptors respond to P$_{O_2}$, P$_{CO_2}$, and pH
   D. Significant interactions occur among the chemoresponses

IV. BREATHING DURING SLEEP
   A. Sleep changes the breathing pattern
   B. Sleep changes the responses to respiratory stimuli
   C. Arousal mechanisms protect the sleeper
   D. Upper airway patency may be compromised during sleep

OBJECTIVES

After studying this chapter, the student should be able to:
1. Explain how the pattern of breathing is generated and draw a block diagram of the major neural components with their connections
2. Explain the difference between feedforward and feedback control and give examples of each as related to breathing
3. Describe phasic control of upper airways
4. List the three major afferent nerve receptor types in the lung, give their locations, outline and contrast their major properties, and list their most important reflex effects
5. Describe the formation of cerebrospinal fluid and explain its buffering properties
6. Contrast the central and peripheral chemoreceptors
7. Graph the effects of hypercapnia and hypoxia on minute ventilation, including the effects of interactions of one stimulus with the other
8. Cite important changes in breathing that occur during sleep

Unlike the pumping of blood by the heart, there is no single pacemaker to set the basic rhythm of breathing and no single muscle devoted solely to the task of tidal air movement. Instead, breathing depends on cyclic excitation of many muscles that can influence the volume of the thorax. Control of that excitation is the result of multiple neuronal interactions that involve all levels of the nervous system. Furthermore, the muscles used for breathing must often be used for other purposes as well. For example, talking while walking requires that some muscles simultaneously attend to the tasks of posturing, walking, phonation, and breathing. The controller that handles this so tidily is sophisticated indeed. Because it is impossible to study extensively the subtleties of such a complex system in humans, much of what is known about control of breathing has been obtained from studies in other species. Much, however, remains unexplained.

Control of upper and lower airway muscles that affect airway patency is integrated with control of the "pumping" muscles that bring about tidal air movements. Airway control has received emphasis recently as its clinical importance has become clearer, particularly with regard to disturbances of breathing during sleep.

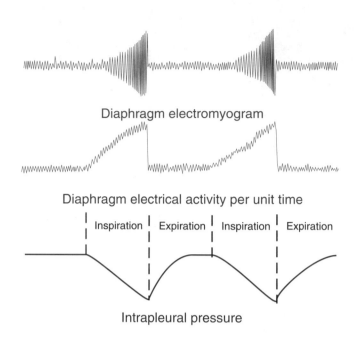

Figure 22–1 ■ Waveforms of diaphragm electrical activity and intrapleural pressure during quiet breathing. During inspiration the number of active muscle fibers and the frequency at which each fires increase progressively, leading to a mirror-image fall in intrapleural pressure as the diaphragm descends.

Generation of the Breathing Pattern

During quiet breathing, inspiration is brought about by a progressive increase in activation of inspiratory muscles, most importantly the diaphragm (Fig. 22–1). This nearly linear increase in activity with time causes the lungs to fill at a nearly constant rate until tidal volume has been reached. The end of inspiration is associated with a rapid decrease in excitation of inspiratory muscles, after which expiration occurs passively by elastic recoil of the lungs and chest wall. Some excitation of inspiratory muscles resumes during the first part of expiration, and this slows the initial rate of exhalation. As more ventilation is demanded, for example during exercise, other inspiratory muscles (external intercostals, cervical muscles) are recruited. In addition, expiration becomes an active process through use, most notably, of muscles of the abdominal wall. The neural basis of these breathing patterns depends on the generation and subsequent tailoring of cyclic changes in activity of cells primarily located in the medulla oblongata.

■ Two Major Groups of Cells in the Medulla Are Associated with Breathing

Although the pattern of breathing changes, breathing activity is not ended by transection of the brain between the pons and medulla oblongata or by section of the vagus nerves, which return afferent information from peripheral receptors. Breathing movements do stop following transection at the junction of the me-

dulla and the spinal cord; however, cyclic activity of respiration-related cells of the medulla continues even when the medulla is isolated. Thus, the medulla is both necessary and sufficient for generation of a basic breathing rhythm.

Breathing-related cells in the medulla have been identified by noting the relationship of their activity to mechanical events of the breathing cycle. Two different aggregates of cells have been found. Their anatomic locations are indicated in Figure 22–2. One, called the **dorsal respiratory group** (DRG) because of its dorsal location in the region of the nucleus tractus solitarius, predominantly contains cells that are active during inspiration. The other, the **ventral respiratory group** (VRG), is a column of cells in the general region of the nucleus ambiguous that extends caudally nearly to the bulbospinal border and cranially nearly to the bulbopontine junction. The VRG contains both inspiration- and expiration-related neurons. Both groups contain cells projecting ultimately to the bulbospinal motor neuron pools. The DRG and VRG are bilaterally paired, but there is cross-communication such that they behave in synchrony; as a consequence, respiratory movements are symmetric.

It is likely that neural networks forming the **central pattern generator for breathing** are contained within the DRG-VRG aggregate, but the exact anatomic and functional description remains uncertain. Central pattern generation probably does not arise from a single pacemaker or by reciprocal inhibition of two pools

Pons

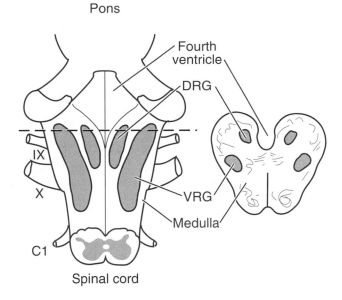

Figure 22–2 ■ Diagrams of the dorsal aspect of the medulla and a cross-section taken in the region of the fourth ventricle, showing the general locations of the dorsal (DRG) and ventral (VRG) respiratory groups. C1, first cervical nerve: X, vagus nerve, IX, glossopharyngeal nerve.

of cells, one having inspiratory- and the other expiratory-related activity. Instead, the progressive rise and abrupt fall of inspiratory motor activity associated with each breath can be modeled most parsimoniously by the starting, stopping, and resetting of an integrator of background ventilatory drive. Although sometimes inadequate in the details, an integrator-based theoretical model as described below is suitable for a first understanding of respiratory pattern generation.

■ Abrupt Loss of Suppression of Inspiratory Drive Synchronizes the Onset of Inspiration

Many different signals (e.g., volition, anxiety, musculoskeletal movements, pain, chemosensor activity, and hypothalamic temperature) provide a background ventilatory drive to the medulla. Inspiration begins by the abrupt disinhibition of a group of cells, probably within the medullary reticular formation, that integrate this background drive. Integration results in a progressive rise in the output of the integrator neurons, which, in turn, excites a similar rise in activity of inspiratory premotor neurons of the DRG-VRG complex. The rate of rise of activity of inspiratory neurons, and thus the rate of inspiration itself, can be influenced by changing the characteristics of the integrator or by changing the excitatory activity it summates. Inspiration is ended by abruptly switching off the rising excitation of inspiratory neurons. The integrator is reset to zero before the beginning of each inspiration so that activity of the inspiratory neurons begins each breath from a very

low level. Figure 22–3 is a schematic representation of this network.

■ After Rising Progressively, Inspiratory Activity Is Switched Off to Initiate Expiration

Two groups of neurons, probably located within the VRG, seem to serve as an inspiratory off switch (Fig. 22–3). Switching occurs abruptly when the sum of excitatory inputs to the off-switch reaches a threshold. Adjustment of the threshold level is one of the ways in which depth of breathing can be varied. Two important excitatory inputs to the off switch are a progressively increasing activity from the integrator's rising output and an input from lung stretch receptors, whose afferent activity increases progressively with rising lung volume. (The first of these is what allows the medulla to generate a breathing pattern on its own; the second is one of many reflexes that influence breathing.) Once the critical threshold is reached, off switch neurons apply a powerful inhibition to the ventilatory drive integrator. The integrator is thus reset by its own rising activity. Other inputs, both excitatory and inhibitory, act on the off switch and change its threshold. For example, chemical stimuli, such as hypoxemia and hypercapnia, are inhibitory, thereby raising the threshold and causing larger tidal volumes.

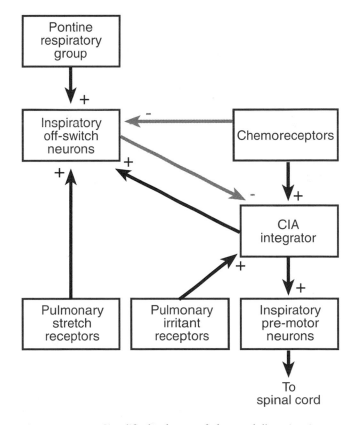

Figure 22–3 ■ Simplified schema of the medullary inspiratory pattern generator. CIA, central inspiratory activity.

An important excitatory input to the off switch comes from a group of spatially dispersed neurons in the rostral pons called the **pontine respiratory group.** Electrical stimulation in this region causes variable effects on breathing, dependent not only on the site of stimulation but also on the phase of the respiratory cycle in which the stimulus is applied. It is believed that the pontine respiratory group may serve to integrate many different autonomic functions in addition to breathing.

■ Expiration Is Divided into Two Phases

Shortly after the abrupt termination of inspiration, some activity of inspiratory muscles resumes. This serves to control expiratory air flow, reducing it from that which would occur by recoil of a passive system. This effect is greatest early in expiration and recedes as lung volume falls. Inspiratory muscle activity is essentially absent in the second phase of expiration, which includes continued passive recoil during quiet breathing or activation of expiratory muscles if more than quiet breathing is required. The duration of expiration is determined by the intensity of inhibition of activity of inspiratory-related cells of the DRG-VRG complex. Inhibition is greatest at the start of expiration and falls progressively until it is insufficient to prevent the onset of inspiration. The progressive fall of inhibition amounts to a decline of threshold for initiation of switching from expiration to inspiration. As might be expected, the rate of decline of inhibition and the occurrence of events that trigger the onset of inspiration are subject to a number of influences. The duration of expiration can be controlled not only by neural information arriving during expiration but also in response to the pattern of the preceding inspiration. How the details of the preceding inspiration are stored and later recovered is unresolved.

■ Various Strategies Adjust Breathing to Meet Demand

The basic pattern of breathing generated in the medulla is extensively modified by a number of control mechanisms. In general, physiologic control often employs several ways to achieve a desired end; the use of multiple control strategies is perhaps nowhere more apparent than in the control of breathing. Multiplicity provides a greater capability for control under a larger number of conditions. Furthermore, it provides interactions that may temper each other and provides for backup in case of failure. The set of strategies for control of a given variable such as minute ventilation typically includes individual schemes that differ in several respects, including choices of sensors and effectors, magnitudes of effects, speeds of action, and optimum operating points. Use of multiple strategies in the control of breathing can be illustrated by considering some of the ways breathing changes in response to exercise. Exercise imposes a demand for oxygen

acquisition and carbon dioxide removal that is met by regulation of breathing such that there is increased minute ventilation accompanied by very little, if any, change in arterial blood gas composition. How is this done? In the presence of unchanged blood gas values during exercise, more is required than mere regulation of ventilation in response to differences between prevailing arterial blood gas composition and fixed set points.

Perhaps the simplest strategies are feedforward strategies, in which breathing responds to some component of exercise but without recognition of how well the response meets the demand. One such mechanism would be for the central nervous system simply to vary the activity of the medullary pattern generator in parallel with, and in proportion to, the excitation of the muscles employed in exercise. Another prospective feedforward scheme involves sensing the magnitude of the carbon dioxide load delivered to the lungs by systemic venous return and then driving ventilation in response to the magnitude of that load. (There is much experimental evidence in support of this mechanism but the identity of the required intrapulmonary sensor remains uncertain.) Still another recognized feedforward mechanism is enhancement of breathing in response to increased activity of receptors in skeletal joints as joint motion increases with exercise.

Although feedforward methods bring about changes in the appropriate direction, they do not provide control in response to the difference between desired and prevailing conditions, as can be done with feedback control. For example, if Pa_{CO_2} deviates from a reference point, say 40 mm Hg, ventilation could be adjusted by feedback control to reduce the discrepancy. This well-known control system, diagrammed according to the principles given in Chapter 1, is shown in Figure 22–4. Unlike feedforward control, feedback control requires a sensor, a reference (set point), and a comparator that together generate an error signal which drives the effector. Negative feedback systems provide good control in the presence of considerable variations of other properties of the system, such as lung stiffness or respiratory muscle strength. They can, if sufficiently sensitive, act quickly to reduce discrepancies from reference points to very low levels. Too much sensitivity, however, may lead to instability and undesirable excursions of the regulated variable.

Other schemes involve minimization or optimization. For example, there is some evidence that rate and depth of breathing are adjusted so as to require minimum work expenditure for ventilation of a given magnitude. That is, the controller decides whether to use a large breath with its attendant large elastic load or more frequent smaller breaths with their associated higher resistive load. This requires afferent neural information about lung volume, rate of change of volume, and transpulmonary pressures, which can be provided by lung and chest wall mechanoreceptors. During exercise, such a controller would act in concert

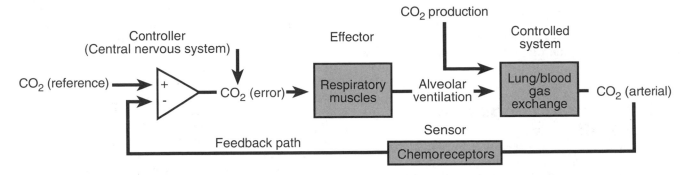

Figure 22–4 ■ Conceptual model of negative feedback control of arterial CO_2. Variations in CO_2 production lead to changes in arterial CO_2 that are sensed by chemoreceptors. The chemoreceptor signal is subtracted from a reference value and the absolute value of the difference taken as an input by the central nervous system. This is passed on to respiratory muscles as a new minute ventilation. The loop is completed as the new ventilation alters blood gas composition through the mechanism of blood-lung gas exchange.

with, among other things, the feedback control of carbon dioxide described earlier. As a final example, an optimization model using two pieces of information is illustrated in Figure 22-5. In this example, breathing is adjusted to minimize the sum of the muscle effort of breathing and the sensory "cost" of tolerating a raised Pa_{CO_2}.

■ Muscles of the Upper Airways Are also under Phasic Control

Muscles of the nose, pharynx, and larynx are controlled by the same rhythm generator that controls the chest wall muscles. But unlike the inspiratory ramplike rise of stimulation of chest wall muscles, excitation of upper airway muscles quickly reaches a plateau and is sustained until inspiration is ended. Flattening of the expected ramp excitation waveform probably is the result of progressive inhibition by the rising afferent activity of airway stretch reflexes as lung volume increases. Excitation during inspiration causes contraction of upper airway muscles, airway widening, and reduced resistance from nostrils to larynx. On the other hand, during the first phase of quiet expiration, when expiration is slowed by renewed inspiratory

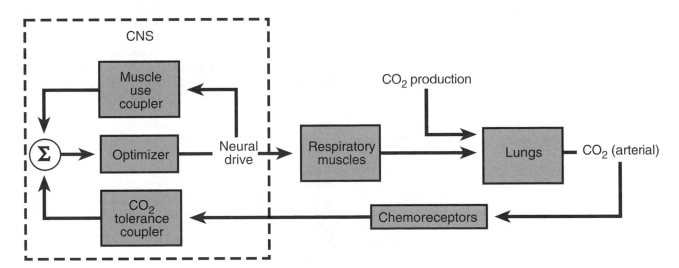

Figure 22–5 ■ Conceptual schema of an optimization controller. The components inside the dashed box constitute the controller. This is a strategy for breathing in which the conflicting desires to maintain chemical homeostasis and to minimize respiratory effort are resolved by selecting an optimum ventilation. The muscle use and CO_2 tolerance couplers convert neural drive and the output of the chemoreceptors to a form interpreted by the neural optimizer as a cost to be minimized. This scheme is described in detail in Poon CS: Ventilatory control in hypercapnia and exercise: Optimization hypothesis. *J Appl Physiol* 62:2447-2459, 1987, which may be consulted for this and other insights into use of control system methods in studying the regulation of breathing.

muscle activity, there is also expiratory braking caused by active adduction of the vocal cords. During the hyperpnea of exercise, however, the cords are abducted during expiration and expiratory resistance is reduced.

Reflexes from Lungs and Chest Wall

Reflexes arising from the periphery provide feedback telling, as it were, how breathing is going, or a signal causing a new course to be taken by the muscles of breathing. The following section considers reflexes that arise from the lung and chest wall. Among reflexes influencing breathing, the lung and chest wall mechanoreceptors and the chemoreflexes responding to blood pH and gas tension changes are the most widely recognized. But many other reflexes, less well explored, also influence breathing. Most are not covered in this chapter. Some examples of reflexes not described further include those induced by changes in arterial blood pressure, cardiac stretch, epicardial irritation, sensations in the airway above the trachea, skin injury, and visceral pain.

■ Reflexes from the Lung Are Associated with Three Classes of Receptors

Pulmonary receptors can be divided into three groups: **slowly adapting receptors, rapidly adapting receptors,** and **C fiber endings.** Afferent fibers of all three types lie predominantly in the vagus nerves, although some pass with the sympathetic nerves to the spinal cord. The role of the sympathetic afferents is uncertain and is not considered further here.

Slowly Adapting Receptors ■ These receptors, which are served by myelinated afferent fibers, lie within the smooth muscle layer of conducting airways. Because they respond to airway stretch, they are also called **pulmonary stretch receptors.** Slowly adapting receptors fire in proportion to applied airway transmural pressure, and their usual role is to sense lung volume. When stimulated, an increased firing rate is sustained as long as stretch is imposed. That is, they adapt slowly. Stimulation of these receptors causes an excitation of the inspiratory off switch and a prolongation of expiration. Because of these two effects, inflating the lungs with a sustained pressure at the mouth terminates an inspiration in progress and prolongs the time before a subsequent inspiration occurs. This sequence is known as the **Hering-Breuer reflex** (or **lung inflation reflex**). The Hering-Breuer reflex probably plays a more important role in infants than in adult humans. In adult humans, particularly in the awake state, this reflex may be overwhelmed by more prominent central control. Because increasing lung volume stimulates slowly adapting receptors, which then excite the inspiratory off switch, it is easy to see how they could be responsible for a feedback signal

that results in cyclic breathing. But as already mentioned, feedback from vagal afferents is not necessary for cyclic breathing to occur. Instead, feedback modifies a basic pattern established in the medulla. In humans, the effect may be to shorten inspiration when tidal volume is larger than normal. The most important role of slowly adapting receptors in humans is probably to participate in regulation of expiratory time, expiratory muscle activation, and functional residual capacity (FRC). Stimulation of slowly adapting receptors also relaxes airway smooth muscle, reduces systemic vasomotor tone, increases heart rate, and, as previously noted, influences laryngeal muscle activity.

Rapidly Adapting Receptors ■ These receptors, which also are served by myelinated afferent fibers, lie beneath the surface of the larger conducting airways. They frequently are called **irritant receptors** because they respond to irritation of the airway by touch or chemicals. They are also stimulated by histamine, serotonin, and prostaglandins liberated locally in response to allergy and inflammation. They are stimulated by lung inflation and deflation, but their firing rate rapidly declines when a volume change is sustained. Because of this rapid adaptation, bursts of activity occur that are in proportion to the change of volume and the rate at which that change occurs. Acute congestion of the pulmonary vascular bed also stimulates these receptors, but, unlike the effect of inflation, their activity may be sustained when congestion is maintained. Background activity of these receptors is inversely related to lung compliance, and on that basis they are thought to serve as sensors of compliance change. Rapidly adapting receptors are probably nearly inactive in normal quiet breathing. Based on what stimulates them, their role would seem to be to sense the onset of pathologic events. In spite of considerable information about what stimulates them, the effect of their stimulation remains controversial. As a general rule, stimulation causes excitatory responses such as cough, gasping, and prolonged inspiration time.

C Fiber Endings ■ C fiber endings belong to unmyelinated nerves. There are two populations of C fiber endings in the lung. One group, **pulmonary C fibers,** is located in the region of the alveoli and is accessible from the pulmonary circulation. These are sometimes called **juxtapulmonary capillary receptors** or **J receptors.** A second group, **bronchial C fibers,** is accessible from the bronchial circulation and, consequently, is located in airways. Like the rapidly adapting receptors, both groups serve a nociceptive role. They are stimulated by lung injury, large inflations, acute pulmonary vascular congestion, and certain chemical agents that have been used to study them. Pulmonary C fibers are sensitive to mechanical events but not so sensitive to products of inflammation, whereas the opposite is true of bronchial C fibers. Their activity excites breathing, and they probably provide a background excitation to the medulla. When stimulated, they cause rapid shallow breathing, bronchoconstriction, increased airway secretion, and cardiovascular depression (bradycardia, hypotension). **Apnea** (cessation of breathing) and a marked fall in systemic vascular resistance occur when they are stimulated acutely and severely. Abrupt reduction of skeletal muscle tone is an intriguing effect that follows intense stimulation of pulmonary C fibers, the homeostatic significance of which remains unexplained.

■ Chest Wall Proprioceptors Provide Information about Movement and Muscle Tension

Joint, tendon, and muscle spindle receptors (collectively called **proprioceptors**) may play roles in breathing, particularly when more than quiet breathing is called for or when breathing efforts are opposed by increased airway resistance or reduced lung compliance. Muscle spindles are present in considerable numbers in the intercostal muscles but are rare in the diaphragm. It has been proposed, but not fully verified, that muscle spindles may adjust breathing effort by sensing the discrepancy between tensions of the intrafusal and extrafusal fibers of the intercostal muscles; if a discrepancy exists, information from the spindle receptor is used to alter contraction of the extrafusal fiber and reduce the discrepancy. This mechanism would provide increased motor excitation when movement is opposed. There is also evidence that chest wall proprioceptors play a major role in the perception of breathing effort, but other sensory mechanisms may also be involved.

Control of Breathing by H+, CO₂, and O₂

Breathing is profoundly influenced by the hydrogen ion concentration and respiratory gas composition of blood. The general rule is that breathing activity is inversely related to arterial blood P_{O_2} but directly related to P_{CO_2} and H^+. Figures 22–6 and 22–7 show the ventilatory responses of a typical person when alveolar P_{CO_2} and P_{O_2} are individually varied by controlling the composition of inspired gas. Responses to carbon dioxide and, to a lesser extent, blood pH depend on sensors in the brainstem as well as sensors in the carotid arteries and aorta. In contrast, responses to hypoxia are brought about only by stimulation of arterial receptors.

■ Cells of the Medulla Respond to Local H+

Ventilatory drive is exquisitely sensitive to P_{CO_2} of blood perfusing the brain. The source of this chemosensitivity has been localized to bilaterally paired groups of cells just below the surface of the ventrolateral medulla immediately caudal to the pontomedullary junction. Each side contains a rostral and a caudal chemosensitive zone, separated by an intermediate

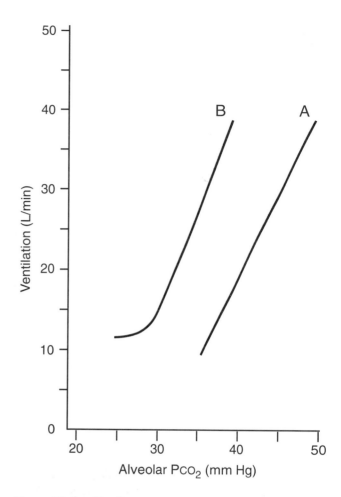

Figure 22–6 ■ Ventilatory responses to increasing alveolar CO_2 tension imposed by varying the composition of inspired gas. Curve A represents the response when alveolar P_{O_2} was held at 100 mm Hg or greater so as to essentially eliminate O_2-dependent activity of the chemoreceptors. Curve B represents the response when alveolar P_{O_2} was held at about 50 mm Hg to provide an overlying hypoxic stimulus. Note that hypoxia increases the slope of the line in addition to changing its location. (Modified from Nielsen M, Smith H: Studies on the regulation of respiration in acute hypoxia. *Acta Physiol Scand* 24:293-313, 1952.)

zone in which the activity of the caudal and rostral groups converge and in which they may be integrated with regulation of other autonomic functions. Exactly which cells exhibit chemosensitivity is unknown, but they are not the same as those of the DRG-VRG complex. The chemosensitive neurons respond to the H^+ of the surrounding interstitial fluid. This critical H^+ concentration is a function of blood P_{CO_2} and the bicarbonate content of cerebrospinal fluid (CSF).

■ Cerebrospinal Fluid pH Depends on Its Bicarbonate Concentration and P$_{CO_2}$

Cerebrospinal fluid is formed mainly by the choroid plexuses of the ventricular cavities of the brain. The epithelium of the choroid plexus provides a barrier

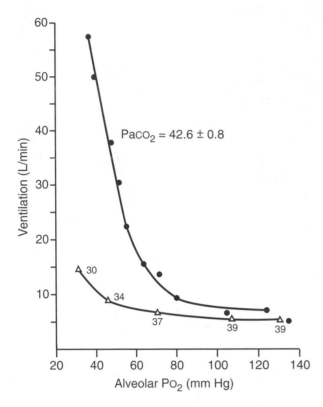

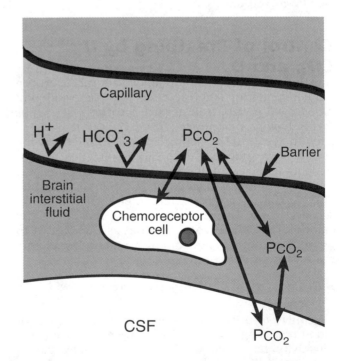

Figure 22–7 ■ Ventilatory response to hypoxia caused by progressively lowering inspired oxygen content. Pa_{CO_2} was held at 42.6 mm Hg by adding CO_2 to the inspired air. If this had not been done (lower curve), hypocapnia secondary to the hypoxic hyperventilation would have reduced the ventilatory response. The numbers next to the lower curve report measured Pa_{CO_2} at each point on the curve. (Adapted from Loeschcke HH, Gertz KH: Einfluß des O_2-Druckes in der Einatmungsluft auf die Atemtätigkeit des Menschen, geprüft unter Konstanthaltung des alveolaren CO_2-Druckes. *Arch Ges Physiol [Pflügers Arch]* 267:460–477, 1958.)

marked changes in plasma concentrations of those cations. Bicarbonate serves as the only significant buffer in CSF, but the mechanism by which bicarbonate concentration is controlled is controversial. Most proposed regulatory mechanisms invoke active transport of one or more ionic species by the epithelial and endothelial membranes. Because of the relative impermeabilities of the choroidal epithelium and capillary endothelium, changes in H^+ concentration of blood are poorly reflected in CSF. In contrast, molecular carbon dioxide diffuses readily, therefore blood P_{CO_2} can influence the pH of CSF. Hence, the pH of CSF is primarily determined by its bicarbonate concentration and P_{CO_2}. The relative ease of movement of molecular carbon dioxide in contrast to hydrogen ions and bicarbonate is depicted in Figure 22–8.

In normal individuals, the P_{CO_2} of CSF is about 6 mm Hg higher than that of arterial blood, approximating that of brain tissue. The pH of CSF, normally slightly below that of blood, is held within narrow limits. Cerebrospinal fluid pH changes very little in states of metabolic acid-base disturbances (Chap. 25)—only about 10% of that in plasma. In respiratory acid-base disturbances, however, the change in pH of the CSF may exceed that of blood. During chronic acid-base disturbances, the bicarbonate concentration of CSF changes in the same direction as in blood, but again the changes may be unequal. In metabolic disturbances, the CSF bicarbonate changes are about

between blood and CSF that severely limits passive movement of large molecules, charged molecules, and inorganic ions. However, choroidal epithelium actively transports a number of substances, including ions, and this active transport participates in determining the composition of CSF. Cerebrospinal fluid formed by the choroid plexuses is exposed to brain interstitial fluid across the surface of the brain and spinal cord, with the result that the composition of CSF away from the choroid plexuses is closer to that of interstitial fluid than it is to CSF as first formed. Brain interstitial fluid is also separated from blood by the blood-brain barrier (capillary endothelium), which has its own transport capability. Because of the properties of the limiting membranes, CSF is essentially protein-free, but it is not just a simple ultrafiltrate of plasma. Cerebrospinal fluid differs most notably from an ultrafiltrate by its lower bicarbonate and higher sodium and chloride ion concentrations. Potassium, magnesium, and calcium ion concentrations also differ somewhat from plasma and, furthermore, change very little in the face of

Figure 22–8 ■ The relative ease of movement of H^+, HCO^-_3, and molecular CO_2 between capillary blood, brain interstitial fluid, and CSF. The acid-base status of the chemoreceptive cells can be changed quickly only by changing Pa_{CO_2}.

40% of those in blood, but with respiratory disturbances CSF and blood bicarbonate changes are essentially the same. When acute acid-base disturbances are imposed, CSF bicarbonate changes more slowly than does blood bicarbonate, and it may not reach a new steady state for hours or days. As already noted, the mechanism of bicarbonate regulation is unsettled. Irrespective of how it occurs, the bicarbonate regulation that occurs with acid-base disturbances is important, because by changing buffering it influences the response to a given P_{CO_2}.

■ Peripheral Chemoreceptors Respond to PO_2, PCO_2, and pH

The **carotid bodies,** located bilaterally near the bifurcations of the common carotid arteries and served by the glossopharyngeal nerve, and the **aortic bodies,** located along the ascending aorta and served by the vagus, are specialized tissues that detect changes in arterial blood PO_2, PCO_2, and pH. Like the medullary chemosensors, they are stimulated by increasing PCO_2. About 40% of the effect of Pa_{CO_2} on ventilation is brought about by the peripheral chemoreceptors, while the rest is brought about by the central sensors. Unlike the central sensor, the peripheral chemoreceptors are sensitive to rising arterial blood H^+ and falling PO_2. They alone cause the stimulation of breathing by hypoxia; hypoxia in the brain has little effect on breathing unless severe, at which point breathing is depressed. Carotid chemoreceptors play a more prominent role than aortic chemoreceptors; because of this and their greater accessibility they have been studied in greater detail. The discharge rate of the carotid chemoreceptors (and the resulting minute ventilation) is approximately linearly related to Pa_{CO_2}. The linear behavior of the receptor is reflected in the linear ventilatory response to carbon dioxide illustrated in Figure 22-6. When expressed using pH, the response curve is no longer linear but shows a progressively increasing effect as pH falls below normal. This occurs because pH is a logarithmic function of H^+, so the absolute change in H^+ per unit change in pH is greater when brought about at a lower pH.

The response to oxygen depends on arterial Pa_{O_2} rather than oxygen content. Therefore, anemia or carbon monoxide poisoning, both of which reduce oxygen content for a given PO_2, have little effect on the response curve. The shape of the response curve is not linear; instead, hypoxia is of increasing effectiveness as PO_2 falls below about 90 mm Hg. The behavior of the receptor is reflected in the ventilatory response to hypoxia illustrated in Figure 22-7. The shape of the curve relating ventilatory response to PO_2 resembles that of the oxyhemoglobin dissociation curve when plotted upside down. As a result, the ventilatory response to hypoxemia is approximately linearly related to percent hemoglobin saturation of arterial blood.

The nonlinearities of the ventilatory responses to PO_2 and pH, and the relatively low sensitivity across the normal ranges of these variables, cause ventilatory changes to be apparent only when PO_2 and pH deviate significantly from the normal range, most notably when in the directions of hypoxia or acidemia. In contrast (compare Figs. 22-6 and 22-7), ventilation is sensitive to PCO_2 within the normal range and carbon dioxide is normally the dominant chemical regulator of breathing through use of both the central and peripheral chemosensors.

There is a strong interaction among stimuli, which causes the slope of the carbon dioxide response curve to increase if determined under hypoxic conditions (Fig. 22-6) and causes the response to hypoxia to be directly related to the prevailing PCO_2 and pH (Fig. 22-7). As will be discussed in the next section, these interactions, and interaction with the effects of the central carbon dioxide sensor, profoundly influence the integrated chemoresponses to a primary change in arterial blood composition.

The carotid and aortic bodies also can be strongly stimulated by certain chemicals, most notably cyanide ion and other poisons of the metabolic respiratory chain. Changes in blood pressure have only a small effect on activity of the chemoreceptors, but activity can be stimulated if arterial pressure falls below about 60 mm Hg. This effect is more prominent in the aortic bodies than in the carotid bodies. Afferent activity of the peripheral chemoreceptors is under some degree of efferent control capable of influencing responses by as yet unclear mechanisms. Afferent activity from the chemoreflexes is also centrally modified in its effects by interactions with other reflexes, such as the lung stretch reflex and the systemic arterial baroreflex. The magnitudes and roles of these interactions in human breathing are not well detailed, but they should be remembered as examples of the complex interactions of cardiorespiratory control. Interactions among chemoreflexes, however, are easily demonstrated.

■ Significant Interactions Occur among the Chemoresponses

The effect of PO_2 on the response to carbon dioxide, and vice versa, has already been noted. By virtue of this interdependence, a response to hypoxia must inexorably be reduced by the subsequent increased ventilation, unless Pa_{CO_2} is somehow maintained, because Pa_{CO_2} ordinarily falls as ventilation is stimulated. An even greater restraint of the stimulating effect of hypoxia occurs by way of the central chemosensors, which respond more potently than the peripheral receptors to low Pa_{CO_2}. The sequence of events in the response to hypoxia (e.g., ascent to high altitude) exemplifies interactions among chemoresponses.

Let us pursue this example a bit further. If 100% oxygen is given to the subject newly arrived at high altitude, ventilation is quickly restored to its sea level value. Over the next few days ventilation in the absence of supplemental oxygen progressively rises further, but it is no longer restored to sea level value by

SLEEP APNEA SYNDROMES

The analysis of multiple physiologic variables recorded during sleep (**polysomnography**) is an important method for research into the control of breathing that has had increasing use in clinical evaluation of sleep disturbances. In normal sleep there is reduced dilatory upper airway muscle tone and there may be brief intervals during which there are no breathing movements. Some people, typically overweight and predominantly males, exhibit more severe disruption of breathing, referred to as **sleep apnea syndrome.** Sleep apneas are divided into two broad classes, obstructive and central. In **central sleep apnea,** breathing movements cease for a longer than normal interval. In **obstructive apnea,** the fault seems to lie in failure of the pharyngeal muscles to open the airway during inspiration. This may be the result of decreased muscle activity, but the obstruction is often made worse by an excessive amount of neck fat with which the muscles must contend. With obstructive apnea, progressively larger inspiratory efforts eventually overcome the obstruction and air flow is temporarily resumed, usually accompanied by loud snoring. Many patients exhibit both central and obstructive apneas. In both types, hypoxemia and hypercapnia develop progressively during the apneic intervals. Frequent episodes of repeated hypoxia may lead to pulmonary and systemic hypertension and to myocardial distress; the accompanying hypercapnia is thought to be a cause of the morning headache these patients often experience. There may be partial arousal at the end of apneic periods, which leads to a disrupted sleep, so that the poorly rested subject is unusually sleepy during the day. Indeed, daytime sleepiness, often leading to dangerous situations, is probably the most common and most debilitating symptom. The cause of this disorder is multivariate and often obscure, but mechanically assisted ventilation during sleep often results in marked symptomatic improvement.

---

breathing oxygen. Rising ventilation while acclimatizing to altitude could be explained by a reduction of blood and CSF bicarbonate concentrations, which would reduce the initial increase in pH created by the increased ventilation. This would allow the hypoxic stimulation to be less strongly opposed. While it may be one operative mechanism, the foregoing is not the full explanation of altitude acclimatization. Cerebrospinal fluid pH is not fully restored to normal, and the increasing ventilation raises Pa_{O_2} while further lowering Pa_{CO_2}, changes that should reduce the stimulus to breathe. In spite of much inquiry, the reason for the persistent hyperventilation of the altitude-acclimatized subject, the full explanation for altitude acclimatization, and the explanation for the failure of increased ventilation of the acclimatized subject to be relieved promptly by restoring a normal Pa_{O_2} remain mysteries.

Metabolic acidosis is caused by accumulation of nonvolatile acids and, because of the increase in blood H^+, initiates and sustains hyperventilation by stimulating the peripheral chemoreceptors. Because of the restricted movement of H^+ into CSF, the fall in blood pH cannot directly stimulate the central chemoreceptors. The central effect of the hyperventilation brought about by decreased pH at the peripheral chemorecep-

tor is a paradoxical rise of CSF pH (i.e., an alkalosis due to the reduced Pa_{CO_2}) that actually restrains the hyperventilation. With time, CSF bicarbonate concentration is adjusted downward, although it changes less than does that of blood, and the pH of CSF remains somewhat higher than blood. Ultimately, ventilation increases more than it did initially as the paradoxical CSF alkalosis is removed.

Respiratory acidosis (accumulation of carbon dioxide) is rarely due to elevated environmental carbon dioxide, although this occurs in submarine mishaps, while exploring wet limestone caves, and in physiology laboratories where response to carbon dioxide is measured. Under these conditions the response is a vigorous increase in minute ventilation proportional to the Pa_{CO_2}; Pa_{O_2} actually rises slightly and arterial pH falls slightly, but these have relatively little effect. If mild hypercapnia can be sustained over a few days the intense hyperventilation subsides, probably as CSF bicarbonate is raised. More commonly, respiratory acidosis is the result of failure of the controller to respond to carbon dioxide (e.g., during anesthesia, following brain injury, and in some patients with chronic obstructive lung disease) or is a result of failure of the breathing apparatus to provide adequate ventilation at an acceptable effort, as may be the case in other patients with obstructive lung disease. In subjects breathing room air, hypercapnia caused by reduced alveolar ventilation is accompanied by significant hypoxia and acidosis. If the hypoxic component alone is corrected, for example by breathing oxygen enriched air, there may be a significant reduction in the ventilatory stimulus that results in greater underventilation, further hypercapnia, and more severe acidosis. A more appropriate treatment is to provide mechanical assistance to restore adequate ventilation.

Breathing during Sleep

People spend about one-third of their lives in sleep, and disorders of sleep and of breathing during sleep are common and often of physiologic consequence. Chapter 7 described the two different neurophysiologic sleep states: rapid eye movement (REM) sleep and non-rapid eye movement (NREM) sleep, or slow-wave sleep. Sleep is not a condition imposed on wakefulness by the appearance of an inhibitory stimulus, but the result of withdrawal of a wakefulness stimulus that arises from the brainstem reticular formation. This wakefulness stimulus is one component of the tonic excitation of brainstem respiratory neurons, and one would predict correctly that sleep results in a general depression of breathing. There are, however, other changes, and the effects of REM and NREM sleep on breathing differ.

■ Sleep Changes the Breathing Pattern

During NREM sleep, breathing frequency and inspiratory flow rate are reduced and minute ventilation falls. In part this reflects the reduced physical activity during sleep, but because there is a small (about 3 mm Hg) rise in Pa_{CO_2}, there must also be a change in either the sensitivity or the set point of the carbon dioxide controller. In the deepest stage of NREM sleep (stage IV) breathing is slow, deep, and very regular. However, in stages I and II the depth of breathing sometimes varies periodically. The explanation is that in light sleep, removal of the wakefulness stimulus varies over time in a periodic fashion: when removed, sleep is deepened and breathing is depressed; when returned, breathing is excited not only by the wakefulness stimulus but also by the carbon dioxide retained during the interval of sleep. This periodic pattern of breathing, illustrated in Fig. 22-9, is known as **Cheyne-Stokes breathing.**

In REM sleep, breathing frequency varies erratically while tidal volume varies little. The net effect on alveolar ventilation is probably a slight reduction, but this is achieved by averaging intervals of frank tachypnea (excessively rapid breathing) with intervals of apnea. Unlike NREM sleep, the variations during REM sleep do not reflect a changing wakefulness stimulus but instead represent responses to increased central ner-

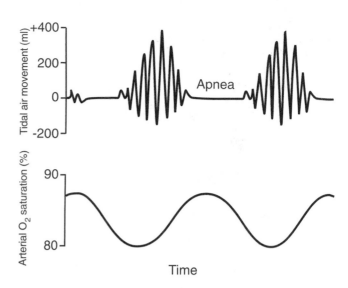

Figure 22–9 ■ Cheyne-Stokes breathing and its effect on arterial O_2 saturation. Cheyne-Stokes breathing occurs frequently during sleep, especially in subjects at high altitude, as in this example. In the presence of preexisting hypoxemia secondary to high altitude or other causes, the apneic periods may result in further falls of O_2 saturation to dangerous levels. Falling Po_2 and rising Pco_2 during the apneic intervals ultimately induce a response and breathing returns, reducing the stimuli and leading to a new apneic period.

vous system activity of behavioral, rather than autonomic or metabolic, control systems.

■ Sleep Changes the Responses to Respiratory Stimuli

Responsiveness to carbon dioxide is reduced during sleep. In NREM sleep the reduction in sensitivity seems to be secondary to a reduction in the wakefulness stimulus and its tonic excitation of the brainstem, rather than to a suppression of the chemosensory mechanisms themselves. It is important to note that breathing remains responsive to carbon dioxide during NREM sleep, although at a less sensitive level, and that the carbon dioxide stimulus may provide the major background brainstem excitation in the absence of the wakefulness stimulus or behavioral excitation. Hence, pathologic alterations in the carbon dioxide chemosensory system may profoundly depress breathing during NREM sleep.

During intervals of REM sleep in which there is little sign of increased activity in other domains, the breathing response to carbon dioxide is slightly reduced, resembling the response during NREM sleep. However, during intervals of increased activity, response to carbon dioxide during REM sleep is markedly reduced; at those times breathing seems to be under the control of the behavioral control system. It is interesting that subservience of breathing to the behavioral control system during REM sleep, rather than to carbon dioxide, is similar to the way breathing is controlled during speech.

Ventilatory responses to hypoxia are probably reduced during both NREM and REM sleep, especially in subjects who have high sensitivity to hypoxia during wakefulness. There does not seem to be a difference between the effects of NREM and REM sleep on hypoxic responsiveness, and the irregular breathing of REM sleep is unaffected by hypoxia.

Both NREM and REM sleep cause an important change in responses to airway irritation. Specifically, a stimulus that causes cough, tachypnea, and airway constriction during wakefulness will cause apnea and airway dilation during sleep unless the stimulus is sufficiently intense to cause arousal.

The limited information available suggests that the lung stretch reflex is unchanged or somewhat enhanced during arousal from sleep, but the effect of stretch receptors on upper airways during sleep may be important (see below).

■ Arousal Mechanisms Protect the Sleeper

Several stimuli cause arousal from sleep or, in less intensity, shift to a lighter sleep stage without frank arousal. In general, arousal from REM sleep is more difficult than from NREM sleep. In humans, hypercapnia is a more potent arousal stimulus than is hypoxia, the former requiring a PaCO$_2$ of about 55 mm Hg and the latter requiring a PaO$_2$ less than 40 mm Hg. Airway irritation and airway occlusion induce arousal readily in NREM sleep but much less readily during REM sleep. All of these arousal mechanisms probably are effective through activation of a reticular arousal mechanism similar to the wakefulness stimulus. They serve a very important role in protecting the sleeper from airway obstruction, alveolar hypoventilation of any cause, and

entrance into the airways of irritating substances. Recall that cough depends on the aroused state and without arousal airway irritation leads to apnea. It should be obvious that wakefulness altered by other than natural sleep, such as during drug-induced sleep, brain injury, or anesthesia, leaves the individual exposed to risk because arousal from those states is impaired or blocked. Indeed, it has been said that from a teleological point of view the most important role of sensors of the respiratory system may be to cause arousal from sleep.

■ Upper Airway Patency May Be Compromised During Sleep

There is a general reduction in skeletal muscle tone during sleep that is particularly prominent during REM sleep. Muscles of the larynx, pharynx, and tongue share in this relaxation, with the result that the upper airway is variably obstructed.

Furthermore, airway muscle relaxation may be enhanced somewhat by the increased effectiveness of the lung inflation reflex. A common consequence of this airway narrowing is snoring, but in many individuals, most often men, the degree of obstruction may at times be sufficient to cause essentially complete occlusion. In such individuals an intact arousal mechanism prevents disaster, and this sequence is not in itself unusual or abnormal. In some subjects, however, obstruction is more often complete and more frequent and the arousal threshold may be raised. Repeated obstruction leads to significant hypercapnia and hypoxemia, and repeated arousals cause sleep deprivation that leads to excessive daytime sleepiness, often interfering with daily activity.

■ ■ ■ ■ ■ ■ ■ ■ ■ ■ ■ ■ REVIEW EXERCISES ■ ■ ■ ■ ■ ■ ■ ■ ■ ■ ■

Identify Each with a Term

1. A group of medullary neurons that show activity during both the inspiratory and expiratory phases of breathing
2. The reflex in which lung inflation causes prolongation of the interval prior to the onset of the next inspiration
3. Airway receptors whose activity is more importantly determined by rate of change of lung volume than by volume itself
4. The most important chemical buffer of pH change in cerebrospinal fluid
5. Site of the formation of cerebrospinal fluid

Define Each Term

6. Inspiratory off switch neurons
7. Pontine respiratory group
8. Feedforward control of breathing
9. Pulmonary stretch receptor
10. Central chemoreceptors

Choose the Correct Answer

11. Generation of the basic cyclic pattern of breathing in the CNS requires participation of:
    a. The pontine respiratory group
    b. Vagal afferent input to the pons
    c. Vagal afferent input to the medulla
    d. An inhibitory loop in the medulla
    e. An intact spinal cord

12. Quiet expiration is associated with:
    a. A brief early burst by inspiratory neurons
    b. Active abduction of the vocal cords
    c. An early burst of activity by expiratory muscles
    d. Reciprocal inhibition of inspiratory and expiratory centers
    e. Increased activity of slowly adapting receptors

13. The ventilatory response to hypoxia:
    a. Is independent of PaCO$_2$
    b. Is more dependent on aortic than carotid chemoreceptors

c. Is exaggerated by hypoxia of the medullary chemoreceptors
d. Bears an inverse linear relationship to arterial oxygen content
e. Is a sensitive mechanism for control of breathing in the normal range of blood gases

14. Which of the following is not a consequence of stimulation of lung C fiber endings?
a. Bronchoconstriction
b. Apnea
c. Rapid shallow breathing
d. Systemic vasoconstriction
e. Skeletal muscle relaxation

15. Which of the following is true about cerebrospinal fluid?
a. Its protein content is equal to that of plasma.
b. Its P_{CO_2} equals that of systemic arterial blood.
c. It is freely accessible to blood hydrogen ions.
d. Its composition is essentially that of a plasma ultrafiltrate.
e. Its pH is a function of Pa_{CO_2}.

16. Non-rapid eye movement sleep is characterized by:
a. A fall in Pa_{CO_2}
b. A tendency for breathing to vary in a periodic fashion
c. Facilitation of the cough reflex
d. Heightened ventilatory responsiveness to hypoxemia
e. Greater skeletal muscle relaxation than REM sleep

17. Which of the following is not true during sleep?
a. Airway irritation evokes apnea.
b. Airway irritation evokes cough.
c. Airway irritation evokes arousal.
d. Airway occlusion evokes arousal.
e. Hypercapnia evokes arousal.

18. Negative feedback control systems:
a. Would not apply to the regulation of Pa_{CO_2}
b. Anticipate future events
c. Give the best control when most sensitive
d. Are ineffective if the properties of the controlled system change
e. Are not necessarily stable

19. With regard to the control of minute ventilation by carbon dioxide:
a. About 80% of the effect of Pa_{CO_2} is mediated by the peripheral chemoreceptors
b. Central effects are mediated by direct effects on cells of the DRG-VRG complex
c. Sensitivity of the control system is inversely related to the prevailing Pa_{O_2}
d. This mechanism is less sensitive than is control in response to oxygen
e. Transection of cranial nerves IX and X at the skull would have no effect

20. Which of the following relationships is best approximated by a straight line sloping downward from left to right?
a. Minute ventilation as a function of arterial pH
b. Minute ventilation as a function of arterial oxygen percent saturation
c. Carotid chemoreceptor firing frequency as a function of Pa_{CO_2}
d. Minute ventilation as a function of Pa_{O_2} while Pa_{CO_2} is held constant
e. Arterial pH as a function of arterial H^+

Briefly Answer

21. What is the potential role of chest wall muscle spindles in the control of breathing?
22. Contrast the acute ventilatory response to inhaled carbon dioxide with that of intravenous infusion of an acid solution.
23. Describe interactions between the carbon dioxide and oxygen chemoresponses.
24. Give some examples of the control of upper airway muscles that bear on the overall control of breathing.
25. How does breathing in REM and NREM sleep differ?

ANSWERS

1. Ventral respiratory group
2. Hering-Breuer inflation reflex
3. Rapidly adapting receptors
4. Bicarbonate ion
5. Choroid plexus
6. A group of medullary neurons that receive input from several sources and which when sufficiently stimulated cause an abrupt termination of inspiration and a resetting of a proposed central integrator
7. A group of neurons in the rostral pons active in integrating various autonomic activities, including breathing, that provides an important input to the respiratory off switch neurons
8. A control strategy in which breathing is varied in parallel with commands for another activity (e.g., running) or in response to some component of that activity, but without recourse to comparisons of the breathing response to desired goals (set points) and use of a closed feedback loop
9. Slowly adapting stretch receptors located in the walls of extra- and intrapulmonary airways whose afferent fibers are myelinated; responsible for the Hering-Breuer inflation reflex
10. Cells lying just below the surface of the ventrolateral medulla immediately caudal to the pons whose activity is excited by increase in local interstitial fluid H^+, leading to stimulation of breathing
11. d
12. a
13. d

14. d

15. e

16. b

17. b

18. e

19. c

20. b

21. Muscle spindles may adjust breathing effort by sensing the discrepancy between tension of the intrafusal and extrafusal fibers of the intercostals; this could provide for greater effort when movement is opposed.

22. Inhaled carbon dioxide lowers pH at the central and peripheral chemosensitive sites and breathing is stimulated at both locations. Because of the blood-brain barrier, acid infusion primarily stimulates the peripheral chemoreceptors, with little direct effect on the central chemoreceptors. The resulting hyperventilation causes hypocapnia and a higher pH at the central chemoreceptors, which reduces the central ventilatory stimulus. Thus, with inhaled carbon dioxide both sets of chemosensors stimulate breathing, but with acid infusion peripheral stimulation is restrained by central inhibition.

23. Hypoxia enhances the ventilatory response to hypercapnia and vice versa. Thus, a hypoxic stimulus may cause increased ventilation, which causes hypocapnia, reducing the net hypoxic ventilatory response to less than would have occurred had Pa_{CO_2} been held constant.

24. Activation of pharyngeal muscles during inspiration pulls tongue and pharyngeal tissue away from airway; the vocal cords are abducted during inspiration and adducted during expiration (except in active expiration, where they are abducted); upper airway motor tone is decreased during both REM and NREM sleep; and lung stretch receptors modulate the position of the vocal cords.

25. There is relative hypoventilation, especially in NREM sleep. In light NREM sleep, breathing may vary periodically as a wakefulness stimulus is intermittently restored and withdrawn, but in deep NREM sleep breathing is deep and regular. In REM sleep breathing frequency varies erratically and there may be periods of apnea. Tidal volume during REM sleep is less than during NREM sleep. Responses to hypercapnia and hypoxia are reduced in both modes, but especially during tachypneic intervals in REM sleep.

Suggested Reading

Cherniak NS, Widdicombe JG (eds): *Handbook of Physiology: The Respiratory System—Control of Breathing.* Bethesda, American Physiological Society, 1986, parts 1, 2.

Hornbein TF (ed): *Regulation of Breathing.* (*Lung Biology in Health and Disease.* vol. 17, parts 1, 2.) New York, Marcel Dekker, 1981.

Kryger MH: Sleep disorders. Clin Chest Med 6:553–731, 1985.

Martin RJ (ed): *Cardiorespiratory Disorders During Sleep.* 2nd ed. Mt. Kisco, NY, Futura, 1990.

Pallot DJ (ed): *Control of Respiration.* New York, Oxford University Press, 1983.

Poon CS: Ventilatory control in hypercapnia and exercise: Optimization hypothesis. *J Appl Physiol* 62:2447–2459, 1987.

von Euler C: On the central pattern generator for the basic breathing rhythmicity. *J Appl Physiol* 55:1647–1659, 1983.

PART SIX

Renal Physiology and Body Fluids

C H A P T E R

■ ■ ■ ■ ■ ■ ■

23

Kidney Function

CHAPTER OUTLINE

I. **FUNCTIONAL RENAL ANATOMY**
  A. The kidney is divided into cortex and medulla
  B. The nephron is the basic unit of renal structure and function
  C. Not all nephrons are alike
  D. The kidneys have a rich blood supply and innervation
  E. The juxtaglomerular apparatus is the site where renin is produced

II. **OVERALL ASSESSMENT OF KIDNEY FUNCTION**
  A. Renal clearance equals urinary excretion rate divided by plasma concentration
  B. Inulin clearance equals the glomerular filtration rate
  C. The endogenous creatinine clearance is used clinically to estimate glomerular filtration rate
  D. Plasma creatinine concentration can be used as an index of glomerular filtration rate
  E. Para-aminohippurate clearance nearly equals renal plasma flow
  F. Net tubular reabsorption or secretion of a substance can be calculated from filtered and excreted amounts
  G. The glucose titration study assesses renal glucose reabsorption
  H. The tubular transport maximum for PAH provides a measure of functional proximal secretory tissue

III. **RENAL BLOOD FLOW**
  A. The kidneys have an enormous blood flow
  B. Blood flow is higher in the renal cortex and lower in the renal medulla
  C. The kidneys have a great ability to regulate their blood flow
  D. Renal sympathetic nerves and various hormones change renal blood flow

IV. **GLOMERULAR FILTRATION**
  A. The glomerular filtration membrane has three layers
  B. Size, shape, and electrical charge affect the filterability of macromolecules
  C. Glomerular filtration rate is determined by Starling forces
  D. The pressure profile along a glomerular capillary is distinctly different from the usual capillary

OBJECTIVES

After studying this chapter, the student should be able to:
1. List the functions of the kidneys
2. Identify the parts of the nephron and collecting duct system and the blood vessels of the kidneys
3. Explain how glomerular filtration rate is measured, the nature of the glomerular filtrate, and the factors that affect filtration rate
4. Describe how renal blood flow can be determined from the clearance of *p*-aminohippurate and discuss the factors that affect renal blood flow
5. Write the equations used to calculate rates of net tubular reabsorption or secretion of a substance; explain the changing patterns of excretion of glucose and *p*-aminohippurate when plasma concentrations of these substances are increased
6. Discuss the magnitude and mechanisms of solute and water reabsorption in the proximal convoluted tubule, loop of Henle, and distal nephron; draw cell models for transport in the proximal tubule, thick ascending limb, distal convoluted tubule, and collecting duct principal cell
7. Describe the active tubular secretion of organic anions and organic cations in the proximal tubule and passive transport via non-ionic diffusion
8. Describe the cellular action of antidiuretic hormone on the collecting duct epithelium
9. Discuss the countercurrent mechanisms responsible for production of an osmotically concentrated urine; explain how an osmotically dilute urine is formed

The kidneys play a dominant role in regulating the composition and volume of the extracellular fluid. They normally maintain a stable internal environment by excreting in the urine appropriate amounts of many substances. These substances include waste products and foreign compounds, but also many useful substances that are present in excess because of eating, drinking, or metabolism. This chapter considers the basic renal processes that determine excretion of various substances.

The kidneys perform a variety of important functions:

1. They regulate the osmotic pressure (osmolality) of the extracellular fluids by excreting an osmotically dilute or concentrated urine.
2. They regulate the concentrations of numerous ions in the plasma, including sodium, potassium, calcium, magnesium, chloride, bicarbonate, phosphate, and sulfate.
3. They play an essential role in acid-base balance by excreting hydrogen ions when there is excess acid, or bicarbonate when there is excess base.
4. They regulate the volume of the extracellular fluid by controlling sodium and water excretion.
5. They help regulate arterial blood pressure by adjusting sodium excretion and producing various substances (e.g., renin) that can affect blood pressure.
6. They eliminate waste products of metabolism, including urea (the main nitrogen-containing end product of protein metabolism in humans),

uric acid (an end product of purine metabolism), and creatinine (an end product of muscle metabolism).

7. They remove many drugs (e.g., penicillin) and foreign or toxic compounds.
8. They are the major sites of production of certain hormones, including erythropoietin (Chap. 11) and 1,25-dihydroxy vitamin D_3 (Chap. 38).
9. They degrade several polypeptide hormones, including insulin, glucagon, and parathyroid hormone.
10. They synthesize ammonia, which plays a role in acid-base balance (Chap. 25). During prolonged fasting, renal synthesis of glucose contributes significantly to the maintenance of blood glucose level. They synthesize substances that affect renal blood flow and sodium excretion, including arachidonic acid derivatives (prostaglandins, thromboxane A_2) and kallikrein (a proteolytic enzyme that results in production of kinins).

Functional Renal Anatomy

To understand renal function, a good grasp of renal anatomy is essential.

■ The Kidney Is Divided into Cortex and Medulla

Each kidney in an adult weighs about 150 g and is roughly the size of one's fist. If the kidney is sectioned (Fig. 23–1), two regions are seen: an outer part, called the **cortex,** and an inner part, called the **medulla.**

The cortex typically is reddish brown and has a granulated appearance. All of the convoluted tubules, cortical collecting ducts, and glomeruli (see below) are located in the cortex. The medulla is lighter in color and has a striated appearance that results from the parallel arrangement of loops of Henle, medullary collecting ducts, and blood vessels of the medulla. The medulla can be further subdivided into an **outer medulla,** which is closer to the cortex, and an **inner medulla,** which is farther from the cortex.

The human kidney is organized into a series of **lobes,** usually 8-10 in number. Each lobe consists of a pyramid of medullary tissue plus the cortical tissue overlying its base and covering its sides. The tip of the **medullary pyramid** forms a **renal papilla.** Each renal papilla drains its urine into a **minor calyx.** The minor calices unite to form a **major calyx,** and the urine then flows into the **renal pelvis.** From there, the urine is propelled by peristaltic movements down the **ureters** to the **urinary bladder,** which stores the urine until the bladder is emptied. The medial aspect of each kidney is indented in a region called the **hilum,** where the ureter, blood vessels, nerves, and lymphatics enter or leave the kidney.

■ The Nephron Is the Basic Unit of Renal Structure and Function

Each human kidney contains about one million **nephrons** (Fig. 23-2), which consist of a **renal cor-**

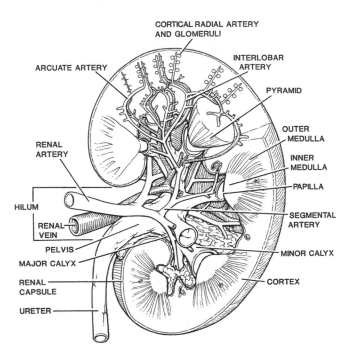

Figure 23–1 ■ Diagram of the human kidney, sectioned vertically. (From Smith HW: *Principles of Renal Physiology.* Copyright © 1956 by Oxford University Press, Inc.; renewed in 1984 by Homer Wilson Smith. Reprinted by permission of Oxford University Press, Inc.)

puscle and a **renal tubule.** The renal corpuscle consists of a tuft of capillaries, the **glomerulus,** surrounded by **Bowman's capsule.** The part of the tubule nearest the glomerulus is the **proximal tubule.** This is subdivided into a **proximal convoluted tubule** and **proximal straight tubule.** The straight portion heads toward the medulla, away from the surface of the kidney. The **loop of Henle** consists of the proximal straight tubule, **thin limb,** and **thick ascending limb.** The next segment, the short **distal convoluted tubule,** is connected to the collecting duct system by **connecting tubules.** Several nephrons drain into a **cortical collecting duct,** which passes into an **outer medullary collecting duct.** In the inner medulla, **inner medullary collecting ducts** unite to form large **papillary ducts.**

The collecting ducts perform the same types of operations as the renal tubules, so they are often considered to be part of the nephron. However, the collecting ducts and nephrons differ in embryologic origin, and since the collecting ducts form a branching system there are many more nephrons than collecting ducts. The entire renal tubule and collecting duct system consists of a single layer of epithelial cells surrounding fluid (urine) in the tubule or duct lumen. Cells in each segment have a characteristic histologic appearance. Each segment has unique transport properties (see below).

■ Not All Nephrons Are Alike

Three groups of nephrons are distinguished, based on the location of their glomeruli in the cortex: **superficial, midcortical,** and **juxtamedullary nephrons.** The juxtamedullary nephrons, whose glomeruli lie in the cortex next to the medulla, comprise about one-eighth of the total nephron population. They differ in several ways from the other nephron types, including a longer loop of Henle, which may dip deep into the medulla, a long thin limb (both descending and ascending portions), a larger glomerulus, a lower renin content, different tubular permeability and transport properties, and a different type of postglomerular blood supply. Figure 23-2 illustrates superficial and juxtamedullary nephrons; note the long loop of the juxtamedullary nephron.

■ The Kidneys Have a Rich Blood Supply and Innervation

Each kidney is typically supplied by a single **renal artery** that branches into anterior and posterior divisions, which give rise to a total of five **segmental arteries.** The segmental arteries branch into **interlobar arteries** (Fig. 23-1), which pass toward the cortex between the kidney lobes. At the junction of cortex and medulla, the interlobar arteries branch to form **arcuate arteries.** These, in turn, give rise to smaller **cortical radial arteries (interlobular arteries),** which pass through the cortex toward the surface of the kidney. Several short, wide, muscular **afferent**

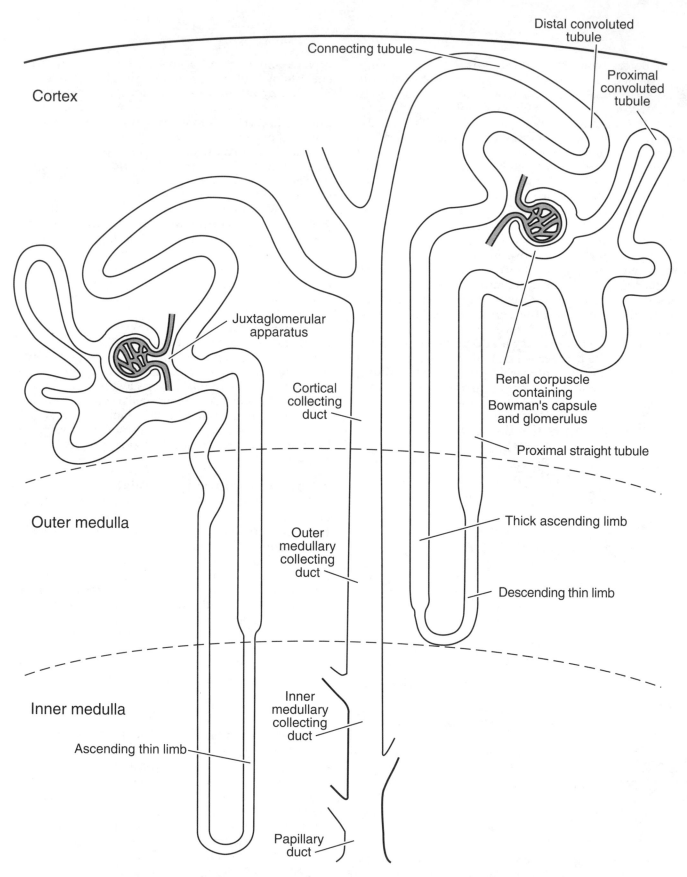

Figure 23–2 ■ Diagram of parts of the nephron and collecting duct system. On the left is a long-looped juxtamedullary nephron; on the right is a superficial cortical nephron. (Modified from Kriz W, Bankir L: A standard nomenclature for structures of the kidney. *Am J Physiol* 254:F1–F8, 1988.)

arterioles arise from the cortical radial arteries. Each afferent arteriole gives rise to a **glomerulus.** The glomerular capillaries are followed by an **efferent arteriole.** The efferent arteriole then divides into a second capillary network, the **peritubular capillaries,** that surrounds the kidney tubules. Venous vessels, in general, lie parallel to the arterial vessels and have similar names.

The blood supply to the medulla is derived from the efferent arterioles of juxtamedullary glomeruli. These vessels give rise to two patterns of capillaries: peritubular capillaries, which are similar to those in the cortex, and **vasa recta** (Fig. 23–3), straight, long capillaries. Some vasa recta reach deep into the inner medulla. In the outer medulla, descending and ascending vasa recta are grouped in vascular bundles and are in close contact with each other. This greatly facilitates exchange of substances between blood flowing into and out of the medulla.

The kidneys are richly innervated by sympathetic nerve fibers, which travel to the kidneys mainly in thoracic spinal nerves X, XI, and XII and lumbar spinal nerve I. Stimulation of sympathetic fibers causes constriction of renal blood vessels and a fall in renal blood flow. Sympathetic nerve fibers also innervate tubular cells and may cause an increase in sodium reabsorption by a direct action on these cells. In addition, stimulation of sympathetic nerves increases the release of renin by the kidneys. Afferent (sensory) renal nerves are stimulated by mechanical stretch or various chemicals in the renal parenchyma.

Renal lymphatics drain the kidney, but little is known about their functions.

■ The Juxtaglomerular Apparatus Is the Site Where Renin Is Produced

Each nephron forms a loop, and the thick ascending limb touches the vascular pole of the glomerulus (Fig. 23–2). At this site is the **juxtaglomerular apparatus,** a region that in cross-section (Fig. 23–4) often appears as a roughly triangular region bounded by the macula densa and the afferent and efferent arterioles as they enter and leave the glomerulus. It contains three major cell types: macula densa cells, extraglomerular mesangial cells, and granular cells. The **macula densa** (dense spot) consists of densely crowded tubular epithelial cells on the side of the thick ascending limb that faces the glomerular tuft; these cells monitor the composition of the fluid in the tubule lumen at this point. The **extraglomerular mesangial cells** are continuous with mesangial cells of the glomerulus; they may transmit information from macula densa cells to the granular cells. The **granular cells** are modified vascular

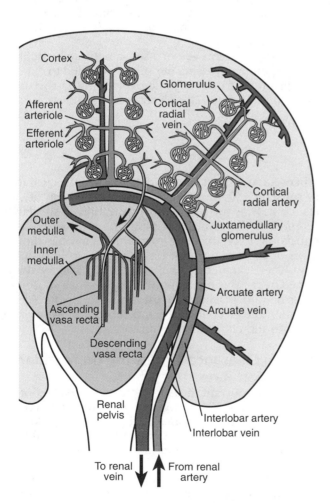

Figure 23–3 ■ Diagram of the blood vessels in the kidney; peritubular capillaries are not shown. (Modified from Kriz W, Bankir L: A standard nomenclature for structures of the kidney. *Am J Physiol* 254:F1–F8, 1988.)

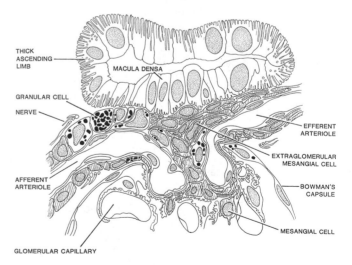

Figure 23–4 ■ Histologic appearance of the juxtaglomerular apparatus. A cross-section through a thick ascending limb is on top, and part of a glomerulus is below. The juxtaglomerular apparatus consists of macula densa, extraglomerular mesangial cells, and granular cells. (From Taugner R, Hackenthal E: *The Juxtaglomerular Apparatus: Structure and Function.* Berlin, Springer, 1989.)

smooth muscle cells with an epithelioid appearance, located mainly in the afferent arterioles close to the glomerulus. These cells synthesize and release **renin,** a proteolytic enzyme that results in angiotensin formation (Chap. 24).

Overall Assessment of Kidney Function

Three processes are involved in forming urine: glomerular filtration, tubular reabsorption, and tubular secretion (Fig. 23–5). **Glomerular filtration** involves filtration of plasma in the glomerulus. The filtrate collects in the urinary space of Bowman's capsule and then flows downstream through the tubule lumen, where its composition and volume are altered by tubular activity. **Tubular reabsorption** involves transport of substances out of tubular urine; these substances are then returned to the capillary blood, which surrounds the kidney tubules. **Tubular secretion** involves transport of substances into the tubular urine. For example, many organic anions and cations are taken up by the tubular epithelium from the blood surrounding the tubules and added to the tubular urine. Some substances (e.g., hydrogen ions, ammonia) are produced in the tubular cells and secreted into the tubular urine. The terms *reabsorption* and *secretion* indicate movement out of or into tubular urine, respectively. Tubular transport (reabsorption, secretion) may be either active or passive, depending on the particular substance and other conditions.

Excretion refers to what is eliminated in the urine.

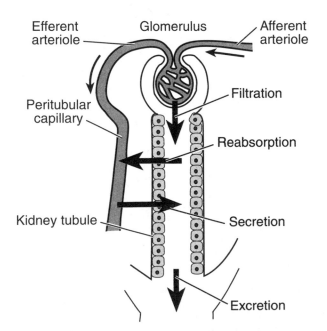

Figure 23–5 ■ Processes involved in urine formation; a highly schematic nephron and its associated blood vessels are shown.

In general, the amount excreted is given by the following expression:

$$\text{Excreted} = \text{Filtered} - \text{Reabsorbed} + \text{Secreted} \qquad (1)$$

The functional state of the kidneys can be evaluated using a number of tests based on the renal clearance concept (see below). These tests allow measurement of rates of glomerular filtration, renal blood flow, and tubular reabsorption or secretion of various substances. Some of these tests, such as glomerular filtration rate, are routinely used to evaluate kidney function.

■ Renal Clearance Equals Urinary Excretion Rate Divided by Plasma Concentration

A useful way of looking at kidney function is to think of the kidneys as clearing substances from the plasma. When a substance is excreted in the urine, a certain volume of plasma is, in effect, freed (or cleared) of that substance.

The **renal clearance** of a substance can be defined as the volume of plasma per minute needed to excrete the quantity of the substance appearing in the urine in a minute's time. The clearance formula is:

$$C_X = \frac{U_X \times \dot{V}}{P_X} \qquad (2)$$

where X is the substance of interest; C_X is the clearance of substance X; U_X is the urine concentration of substance X; P_X is the plasma concentration of substance X; and $\dot{V}$ is the urine flow rate. The product $U_X \times \dot{V}$ equals the excretion rate per minute and has dimensions of amount per unit time (e.g., mg/min or mEq/day). The clearance of a substance can easily be determined by measuring the concentrations of a substance in urine and plasma and the urine flow rate (urine volume/time of collection) and substituting these values into the clearance formula.

■ Inulin Clearance Equals the Glomerular Filtration Rate

A most important measurement in the evaluation of kidney function is the glomerular filtration rate (GFR), the rate at which plasma is filtered by the kidney glomeruli. If we had a substance that was cleared from the plasma only by glomerular filtration, it could be used to measure GFR.

The ideal substance to measure GFR is **inulin,** a fructose polymer with a molecular weight of about 5000. Inulin is suitable for measuring GFR because it is freely filterable by the glomeruli; it is not reabsorbed or secreted by the kidney tubules; it is not synthesized, destroyed, or stored in the kidney; it is nontoxic; and its concentration in plasma and urine can be determined by simple analytic methods.

The principle behind the use of inulin is illustrated in Figure 23-6. The amount of inulin (IN) filtered per unit time (the **filtered load**) is equal to the product of the plasma inulin concentration (P_{IN}) × GFR. The rate of inulin excretion is equal to $U_{IN} \times \dot{V}$. Since inulin is not reabsorbed, secreted, synthesized, destroyed, or stored by the kidney tubules, the filtered inulin load equals the rate of inulin excretion. The equation (Fig. 23-6) can be rearranged by dividing by the plasma inulin concentration. The expression $U_{IN}\dot{V}/P_{IN}$ is defined as **inulin clearance.** Therefore, inulin clearance equals GFR.

Normal values for inulin clearance or GFR (corrected to a body surface area of 1.73 m²) are 110 ± 15 (SD) ml/min for young adult women and 125 ± 15 ml/min for young adult men. In newborns, even when corrected for body surface area, GFR is low, about 20 ml/min/1.73 m² body surface area. Adult values (when corrected for body surface area) are attained by the end of the first year of life. Glomerular filtration rate declines after the age of 45–50 years an average of 13 ml/min per decade in men.

If GFR is 125 ml plasma/min, then the volume of plasma filtered in a day is 180 L (125 ml/min × 1440 min/day). Plasma volume in a 70-kg young adult man is only about 3 liters, so the kidneys filter the plasma some 60 times in a day. The glomerular filtrate contains many valuable constituents (salts, water, metabolites), most of which are reabsorbed by the kidney tubules.

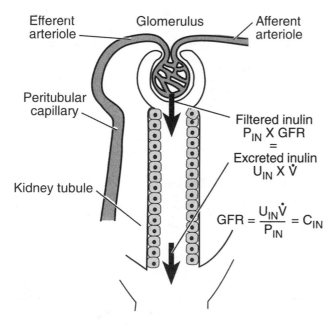

Figure 23–6 ■ Principle behind the measurement of glomerular filtration rate (GFR). P_{IN} = plasma inulin concentration, U_{IN} = urine inulin concentration, $\dot{V}$ = urine flow rate, C_{IN} = inulin clearance.

■ The Endogenous Creatinine Clearance Is Used Clinically to Estimate Glomerular Filtration Rate

Inulin clearance is the gold standard for measuring GFR and is used whenever highly accurate measurements of GFR are desired. It is not common, however, to use inulin in the clinic. Inulin must be infused intravenously and the bladder is usually catheterized, since short urine collection periods are used; these procedures are inconvenient. It would be simpler to use an endogenous substance (i.e., one native to the body) that is only filtered, is excreted in the urine, and normally has a stable plasma value that can be accurately measured. There is no such known substance, but creatinine comes close. Creatinine is an end product of muscle metabolism, a derivative of muscle creatine phosphate. It is produced continuously in the body and is excreted in the urine. Long urine collection periods (e.g., a few hours) can be used, since creatinine concentrations in the plasma are normally stable and creatinine doesn't have to be infused; this obviates the need to catheterize the bladder. Plasma and urine concentrations can be measured using a simple colorimetric method. The **endogenous creatinine clearance** is calculated from the formula:

$$C_{CREATININE} = \frac{U_{CREATININE} \times \dot{V}}{P_{CREATININE}} \quad (3)$$

There are two potential drawbacks to using creatinine to measure GFR. First is that creatinine is not only filtered but secreted by the human kidney. This elevates urinary excretion of creatinine, normally causing a 20% increase in the numerator of the clearance formula. The second drawback is due to errors in measuring plasma creatinine concentration. The colorimetric method usually used also measures other plasma substances, such as glucose, leading to a 20% increase in the denominator of the clearance formula. Since both numerator and denominator are 20% too high, the two errors cancel, so the endogenous creatinine clearance affords a good approximation of GFR when it is about normal. When GFR in an adult has been reduced to about 20 ml/min because of renal disease, the endogenous creatinine clearance may overestimate the GFR by as much as 50%. This results from higher plasma creatinine levels and increased tubular secretion of creatinine. Drugs that inhibit tubular secretion of creatinine or elevated plasma concentrations of chromogenic (color-producing) substances other than creatinine may cause the endogenous creatinine clearance to underestimate GFR.

■ Plasma Creatinine Concentration Can Be Used as an Index of Glomerular Filtration Rate

Because the kidneys continuously clear creatinine from the plasma by excreting it in the urine, the GFR and plasma creatinine concentration are inversely related. Figure 23–7 shows the steady-state relation (i.e., when creatinine production and excretion are equal) between these variables. Halving the GFR from a normal value of 180 L/day to 90 L/day results in a doubling of plasma creatinine concentration from 1 to 2 mg/dl after a few days. Reducing GFR from 90 L/day to 45 L/day results in a greater increase in plasma creatinine, from 2 to 4 mg/dl. Figure 23–7 shows that with low GFR values, small changes in GFR lead to much greater changes in plasma creatinine concentration than occur at high values of GFR.

The inverse relation between GFR and plasma creatinine concentration allows the use of plasma creatinine concentration as an index of GFR, provided certain caveats are kept in mind. First, it takes a certain amount of time for changes in GFR to produce detectable changes in plasma creatinine concentration. Second, plasma creatinine concentration is also influenced by muscle mass. A young, muscular man will have a higher plasma creatinine concentration than an elderly woman with reduced muscle mass. Third, some drugs inhibit tubular secretion of creatinine, and this leads to a raised plasma creatinine concentration even though GFR may be unchanged.

The relation between plasma creatinine concentration and GFR is one example of how plasma concentration of a substance can depend on GFR. The same relation is observed for a number of other substances whose excretion depends on the GFR. For example,

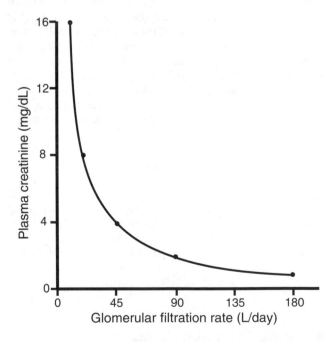

Figure 23–7 ■ Inverse relation between plasma creatinine concentration and glomerular filtration rate (GFR). If GFR is decreased by one-half, plasma creatinine concentration is doubled when production and excretion of creatinine are in balance in a new steady state.

when GFR falls, the plasma concentration of urea (or blood urea nitrogen; BUN) rises in a similar fashion.

■ Para-aminohippurate Clearance Nearly Equals Renal Plasma Flow

Renal blood flow (RBF) can be determined from measurements of renal plasma flow (RPF) and blood hematocrit, using the following formula:

$$RBF = RPF/(1 - Hematocrit) \qquad (4)$$

The hematocrit is easily determined by centrifuging a blood sample. Renal plasma flow is estimated by measuring the clearance of the organic anion p-amino-hippurate (PAH), infused intravenously. It is filtered and also vigorously secreted, so it is nearly completely cleared from all of the plasma flowing through the kidneys. The renal clearance of PAH, at low plasma PAH levels, approximates the renal plasma flow.

The formula used to calculate the true value of the renal plasma flow is:

$$RPF = C_{PAH}/E_{PAH} \qquad (5)$$

where C_{PAH} is the PAH clearance and E_{PAH} is the extraction ratio (Chap. 16) for PAH—the arterial plasma PAH concentration (P^a_{PAH}) minus renal venous plasma PAH concentration (P^{rv}_{PAH}) divided by the arterial plasma PAH concentration. This formula is derived as follows. In the steady state, the amounts of PAH per unit time entering and leaving the kidneys are equal. The PAH is supplied to the kidneys in the arterial plasma and leaves the kidneys in urine and renal venous plasma, or:

$$PAH \text{ entering kidneys} = PAH \text{ leaving kidneys}$$
$$RPF \times P^a_{PAH} = U_{PAH} \times \dot{V} \qquad (6)$$
$$+ RPF \times P^{rv}_{PAH}$$

Rearranging, we get:

$$RPF = U_{PAH} \times \dot{V}/(P^a_{PAH} - P^{rv}_{PAH}) \qquad (7)$$

If we divide the numerator and denominator of the right side of the equation by P^a_{PAH}, the numerator becomes C_{PAH} and the denominator becomes E_{PAH}.

If we assume extraction of PAH is 100% ($E_{PAH} = 1.00$), then the RPF equals the PAH clearance. When this assumption is made, the renal plasma flow is usually called the **effective renal plasma flow** and the blood flow calculated is called the **effective renal blood flow.** However, extraction of PAH by normal human kidneys at suitably low plasma PAH concentrations is not 100% but averages about 91%. Assuming 100% extraction underestimates the true renal plasma flow by about 10%. To calculate the true renal plasma flow or blood flow, it is necessary to cannulate the renal vein to measure its plasma PAH concentration, a procedure not often done.

Net Tubular Reabsorption or Secretion of a Substance Can Be Calculated from Filtered and Excreted Amounts

The rate at which the kidney tubules reabsorb a substance can be calculated if we know how much is filtered and how much is excreted per unit time. If the filtered load of a substance exceeds the rate of excretion, the kidney tubules must have reabsorbed the substance. The equation is:

$$T_{reabs} = P_X \times GFR - U_X \times \dot{V} \qquad (8)$$

where T is the tubular transport rate.

The rate at which the kidney tubules secrete a substance is calculated from the equation

$$T_{secr} = U_X\dot{V} - P_X \times GFR \qquad (9)$$

Note that the quantity excreted exceeds the filtered load, because the tubules secrete X.

In the above two equations, it is assumed the substance X is freely filterable. If substance X is bound to plasma proteins, it is necessary to correct the filtered load for the filterable fraction. For example, about 40% of plasma calcium is bound to plasma proteins, and this fraction is therefore not filterable. Many drugs are bound to plasma proteins, with the extent of binding varying with such factors as the nature of the molecule, plasma drug concentration, plasma protein concentration, and plasma pH.

The above two equations for quantitating tubular transport rates yield the *net* rate of reabsorption or secretion of a substance. It is possible for a single substance to be both reabsorbed and secreted; the equations do not give unidirectional reabsorptive and secretory movements, but only the net transport.

The Glucose Titration Study Assesses Renal Glucose Reabsorption

Insights into the nature of glucose handling by the kidneys can be derived from a **glucose titration** study (Fig. 23-8). The plasma glucose concentration is elevated to higher and higher levels by infusion of glucose-containing solutions. Inulin is administered to permit the measurement of GFR and calculation of the filtered glucose load (plasma glucose concentration × GFR). The rate of glucose reabsorption is determined from the difference between the filtered load and rate of excretion. At normal plasma glucose levels (about 100 mg/dl), all of the filtered glucose is reabsorbed and none is excreted. When the plasma glucose concentration exceeds a certain value (about 200 mg/dl in Fig. 23-8), significant quantities of glucose appear in the urine; this plasma concentration is called the **glucose threshold.** Further elevations in plasma glu-

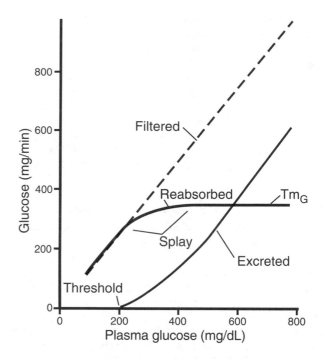

Figure 23-8 ■ Glucose titration experiment in a normal man. The plasma glucose concentration was elevated by infusing glucose-containing solutions. The amount of glucose filtered per unit time (top line) is determined from the product of the plasma glucose concentration and glomerular filtration rate (measured with inulin). Excreted glucose (bottom line) is determined by measuring the urine glucose concentration and flow rate. Reabsorbed glucose was calculated from the difference between filtered and excreted glucose. Tm_G = tubular transport maximum for glucose.

cose lead to progressively more excreted glucose. Glucose appears in the urine because the filtered amounts of glucose exceed the capacity of the tubules to reabsorb it. At very high filtered glucose loads, the rate of reabsorption of glucose reaches a constant maximal value, called the **tubular transport maximum** (Tm) for glucose (G). At Tm_G, the limited number of tubule glucose carriers are all saturated and transport glucose at the maximal rate.

The glucose threshold is not a fixed plasma concentration but depends on three factors: GFR, Tm_G, and amount of **splay.** A low GFR leads to an elevated threshold, because the filtered glucose load is reduced and the kidney tubules can reabsorb all of the filtered glucose despite an elevated plasma glucose concentration. A reduced Tm_G lowers the threshold, because the tubules have a diminished capacity to reabsorb glucose. Splay is the rounding of the glucose reabsorption curve; Figure 23-8 shows that tubular glucose reabsorption does not abruptly attain Tm_G when plasma glucose is progressively elevated. One reason for splay is that not all nephrons have the same filtering and reabsorbing capacities. Thus, nephrons with relatively high filtration rates and low glucose reabsorptive rates excrete glucose at a lower plasma concentration than nephrons with relatively low filtration rates and high reabsorptive rates. An increase in splay results in a decrease in glucose threshold.

In uncontrolled **diabetes mellitus,** plasma glucose levels are abnormally elevated, so more glucose is filtered than can be reabsorbed. Urinary excretion of glucose **(glucosuria)** produces an **osmotic diuresis.** A diuresis is an increase in urine output; in osmotic diuresis, the increased urine flow results from excretion of osmotically active solute. Diabetes (from the Greek for "a syphon") gets its name from this increased urine output.

■ The Tubular Transport Maximum for PAH Provides a Measure of Functional Proximal Secretory Tissue

Para-aminohippurate is secreted only by proximal tubules in the kidneys. At low plasma PAH concentrations, the rate of secretion increases linearly with the plasma PAH concentration. At high plasma PAH concentrations, the secretory carriers are saturated and the rate of PAH secretion stabilizes at a constant maximal value, called the **tubular transport maximum for PAH** (Tm_{PAH}). The Tm_{PAH} is directly related to the number of functioning proximal tubules and therefore provides a measure of the mass of proximal secretory tissue. Figure 23-9 illustrates the pattern of filtration, secretion, and excretion of PAH observed when the plasma PAH concentration is progressively elevated by intravenous infusion.

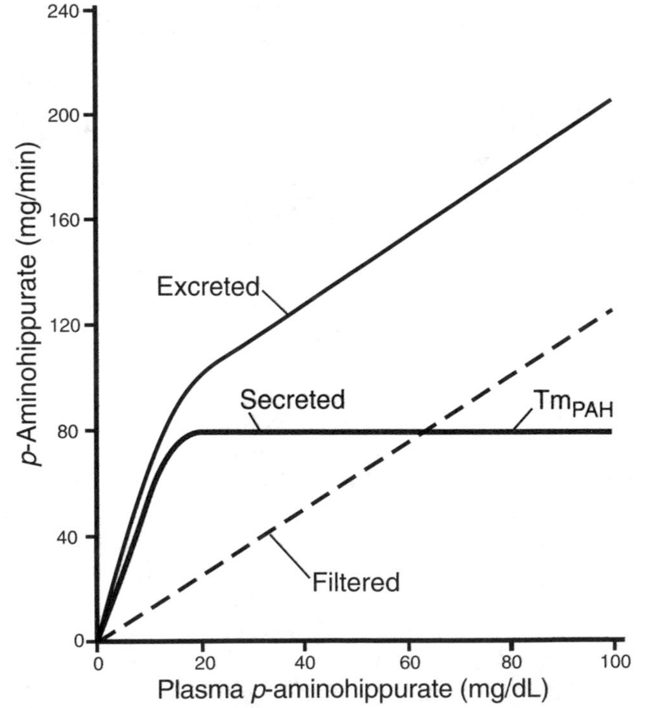

Renal Blood Flow

■ The Kidneys Have an Enormous Blood Flow

In resting, healthy, young adult men, renal blood flow averages about 1.2 L/min. This amounts to about 20% of the cardiac output (5–6 L/min). Both kidneys together weigh about 300 g, so blood flow per gram of tissue averages about 4 ml/min. This rate of perfusion exceeds that of all other organs in the body, except the neurohypophysis and carotid bodies. The high blood flow to the kidneys is not due to excessive metabolic demands. Although the kidneys use about 8% of total resting oxygen consumption, they receive much more oxygen than they need. Renal extraction of oxygen is low, and renal venous blood is characterized by its bright red color (due to a high oxyhemoglobin content). The high blood flow is necessary for a high GFR and therefore reflects the critical role of the kidneys in regulating the internal environment.

■ Blood Flow Is Higher in the Renal Cortex and Lower in the Renal Medulla

Blood flow rates in different parts of the kidney differ (Fig. 23–10). Blood flow is highest in the cortex, averaging about 4–5 ml/min/g tissue. The high cortical

Figure 23–9 ■ Rates of excretion, filtration, and secretion of *p*-aminohippurate (PAH) as a function of plasma PAH concentration. More PAH is excreted than is filtered; the difference represents secreted PAH.

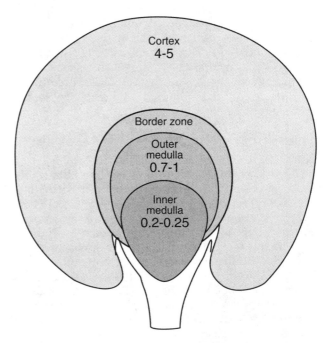

Figure 23–10 ■ Blood flow rates (in ml/min/g tissue) in different parts of the kidney. (Modified from Thurau K, Renal circulation. In Best CH, Taylor NB (eds): *The Physiological Basis of Medical Practice,* 8th ed. Baltimore, Williams & Wilkins, 1966.)

blood flow permits a high rate of filtration in the glomeruli. Blood flow (per gram of tissue) is about 0.7-1 ml/min in the outer medulla and 0.20-0.25 ml/min in the inner medulla. The relatively low blood flow in the medulla helps maintain a hyperosmolar environment in this region of the kidney (see below).

■ The Kidneys Have a Great Ability to Regulate Their Blood Flow

Despite changes in mean arterial blood pressure, renal blood flow is kept at a relatively constant level. This is called **autoregulation** (Chap. 16). It is an intrinsic property of the kidneys and is observed even in an isolated, perfused kidney. Not only renal blood flow but also glomerular filtration rate is autoregulated.

Figure 23-11 describes autoregulation. In the face of changes in arterial blood pressure over the range of 80-180 mm Hg, renal blood flow and filtration rate stay relatively constant. Vessels upstream to the glomerulus (cortical radial arteries and afferent arterioles) constrict when blood pressure is raised and dilate when blood pressure is lowered, thereby maintaining relatively constant glomerular blood flow and capillary pressure. Below or above the autoregulatory range

of pressures, blood flow and filtration rate change appreciably with arterial blood pressure.

Two explanations for renal autoregulation are commonly accepted. According to the **myogenic hypothesis,** an increase in pressure stretches blood vessel walls. The increase in wall tension causes smooth muscle in the vessel walls to contract, decreasing the lumen diameter and increasing resistance. Decreased blood pressure causes the opposite changes. According to the **tubuloglomerular feedback hypothesis** (Fig. 23-12), the transient increase in GFR resulting from an increase in blood pressure leads to increased solute delivery to the macula densa. This produces an increase in the tubular fluid sodium chloride concentration at this site and increased sodium chloride reabsorption by macula densa cells. By mechanisms that are still uncertain, this leads to constriction of the nearby afferent arteriole (perhaps adenosine is the vasoconstrictor agent). Blood flow and filtration rate are lowered to a more normal value. The tubuloglomerular feedback mechanism is a negative feedback system that stabilizes renal blood flow and glomerular filtration rate.

If delivery of sodium chloride to the macula densa is increased experimentally by perfusing the lumen of the loop of Henle, filtration rate in the perfused nephron decreases. This suggests that the purpose of tubuloglomerular feedback may be to control the amount of sodium presented to distal nephron segments. Regulation of sodium delivery to distal parts of the nephron is important because these segments have a limited capacity to reabsorb this ion.

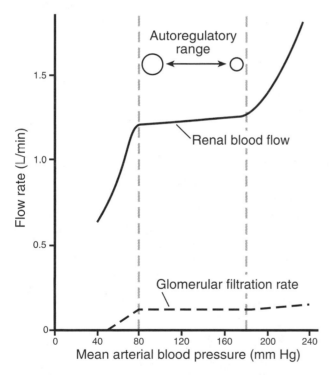

Figure 23-11 ■ Renal autoregulation. In the autoregulatory range, renal blood flow and glomerular filtration rate stay relatively constant despite changes in arterial blood pressure. This is accomplished by changes in the resistance (caliber) of preglomerular blood vessels. The circles indicate that vessel caliber is smaller when blood pressure is high and larger when blood pressure is low. Since resistance to blood flow varies as the 4th power of vessel radius, changes in caliber are greatly exaggerated in this figure.

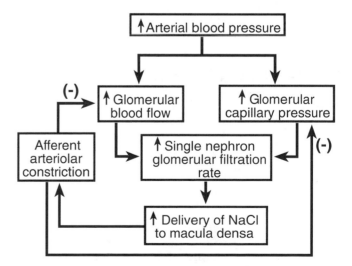

Figure 23-12 ■ Tubuloglomerular feedback hypothesis. When single nephron glomerular filtration is increased, for example because of an increase in arterial blood pressure, then more NaCl is delivered to and reabsorbed by the macula densa. This leads to constriction of the nearby afferent arteriole. This negative feedback mechanism plays a role in autoregulation of renal blood flow and glomerular filtration rate.

Renal autoregulation minimizes the impact of changes in arterial blood pressure on sodium excretion. Without renal autoregulation, increases in arterial blood pressure would lead to dramatic increases in GFR and potentially serious losses of sodium chloride and water from the extracellular fluid.

■ Renal Sympathetic Nerves and Various Hormones Change Renal Blood Flow

Renal blood flow may be changed by stimulation of renal sympathetic nerve fibers or a variety of hormonal substances. Sympathetic nerve stimulation causes renal vasoconstriction and a consequent fall in renal blood flow. Renal sympathetic nerves are activated under stressful conditions, including fright, hemorrhage, cold, pain, deep anesthesia, and heavy exercise. In these conditions, the fall in renal blood flow may be viewed as an emergency mechanism that makes more of the cardiac output available to perfuse other organs, such as the brain and heart, which are more important for immediate (short-term) survival.

A number of substances cause vasoconstriction in the kidneys, including adenosine, angiotensin II, antidiuretic hormone, endothelin, epinephrine, norepinephrine, and thromboxane A_2. Other substances cause vasodilation in the kidneys, including acetylcholine, atrial natriuretic peptide, dopamine, histamine, kinins, nitric oxide, and prostaglandins E_2 and I_2. Some of these substances (e.g., prostaglandins E_2 and I_2) are produced locally in the kidneys. An increase in sympathetic nerve activity or plasma angiotensin II concentration stimulates the production of renal vasodilator prostaglandins. These prostaglandins then oppose the pure constrictor effect of sympathetic nerve stimulation or angiotensin II and thereby reduce the fall in renal blood flow, preventing renal damage.

Glomerular Filtration

Glomerular filtration involves **ultrafiltration** of plasma. This term reflects the fact that the glomerular filtration membrane is an extremely fine molecular sieve that allows filtration of small molecules but restricts the passage of macromolecules (e.g., the plasma proteins).

■ The Glomerular Filtration Membrane Has Three Layers

The barrier through which the filtrate passes from glomerular capillary blood into the space of Bowman's capsule is the **glomerular filtration membrane** (Fig. 23–13). It consists of three layers. The first, the capillary endothelium, is called the **lamina fenestra,** because it contains pores or windows (fenestrae). At about 50–100 nm in diameter, these pores are too large to restrict the passage of plasma proteins. The second layer, the **glomerular basement membrane,** consists of a meshwork of fine fibrils embedded in a

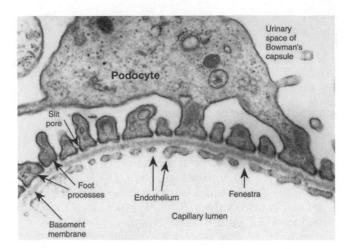

Figure 23–13 ■ Electron micrograph showing the three layers of the glomerular filtration membrane, capillary endothelium, basement membrane, and podocytes. (Courtesy of Dr. Andrew P. Evan, Indiana University.)

gel-like matrix. The third layer is the **visceral layer of Bowman's capsule.** This layer is composed of **podocytes** ("foot cells"), epithelial cells with numerous, long processes that terminate in foot processes that rest on the outer layer of the basement membrane. The space between adjacent foot processes, called a **slit pore,** is about 20 nm wide and bridged by a thin membrane, the **filtration slit diaphragm.** Pores have been observed in the filtration slit diaphragm with the electron microscope. The glomerular filtrate most likely passes between the cells, through the fenestrae and slit pores.

■ Size, Shape, and Electrical Charge Affect the Filterability of Macromolecules

The permeability properties of the glomerular filtration membrane have been studied by determining how well molecules of different sizes pass through it. Table 23–1 lists a number of molecules that have been tested. Molecular radii were calculated from diffusion coefficients. The concentration of the molecule in the glomerular filtrate (fluid collected from Bowman's capsule) is compared to its concentration in plasma water; a ratio of unity indicates complete filterability and a ratio of zero indicates complete exclusion by the glomerular filtration membrane.

Molecular size is an important factor affecting filterability. All molecules with weights less than 10,000 are freely filterable, provided they are not bound to plasma proteins. Molecules with weights greater than 10,000 experience more and more restriction to passage through the glomerular filtration membrane. Very large molecules (e.g., molecular weight 100,000) cannot get through at all. Most plasma proteins are quite large molecules, so they are not appreciably filtered.

TABLE 23–1 ■ Restrictions to Glomerular Filtration of Molecules

Substance	Molecular Weight	Molecular Radius (nm)	[Filtrate]/ [Plasma Water]
Water	18	0.10	1.00
Glucose	180	0.36	1.00
Inulin	5,000	1.4	1.00
Myoglobin	17,000	2.0	0.75
Hemoglobin	68,000	3.3	0.03
Cationic dextran*		3.6	0.42
Neutral dextran		3.6	0.15
Anionic dextran		3.6	0.01
Serum albumin	69,000	3.6	0.001

*Dextrans are high-molecular-weight glucose polymers available in cationic (amine), neutral (uncharged), or anionic (sulfated) forms. Adapted from Pitts RF: *Physiology of the Kidney and Body Fluids.* 3rd ed. Chicago, Year Book, 1974; and Brenner BM, Bohrer MP, Baylis C, Deen WM: Determinants of glomerular permselectivity: Insights derived from observations *in vivo. Kidney Int* 12:229–237, 1977.

From studies with molecules of different sizes, it has been calculated that the glomerular filtration membrane *behaves* as though it were penetrated by cylindric pores of about 7.5–10 nm diameter. No one has ever seen pores of these dimensions in electron micrographs of the glomerular filtration membrane, though.

Molecular shape influences the filterability of macromolecules. For a given molecular weight, a slender and flexible molecule will pass through the glomerular filtration membrane more easily than a spheric, nondeformable molecule.

Electrical charge influences the passage of macromolecules through the glomerular filtration membrane, because this membrane bears fixed negative charges. Glomerular endothelial cells and podocytes have a negatively charged surface coat (glycocalyx), and the glomerular basement membrane contains negatively charged sialic acid, sialoproteins, and heparan sulfate. These negative charges impede the passage of negatively charged macromolecules by electrostatic repulsion and favor the passage of positively charged macromolecules by electrostatic attraction. This is supported by the finding that the filterability of dextran is lowest for anionic dextran, intermediate for neutral dextran, and highest for cationic dextran (Table 23–1).

The net negative charge on serum albumin at physiologic pH is an important factor (in addition to its large molecular size) that reduces its filterability. In some glomerular diseases, there is a loss of fixed negative charges from the glomerular filtration membrane. This results in increased filtration of serum albumin and, therefore, abnormal amounts of protein in the urine (**proteinuria**).

It has long been debated which layer of the glomerular filtration membrane is the main barrier to filtration of macromolecules. It seems likely that the glomerular basement membrane is the principal size-selective barrier. The major electrostatic barriers are probably posed by the layers closest to the capillary lumen, the lamina fenestra, and the innermost part of the basement membrane.

■ Glomerular Filtration Rate Is Determined by Starling Forces

Glomerular filtration rate depends on the balance of hydrostatic and colloid osmotic pressures acting across the glomerular filtration membrane, the Starling forces (Chap. 16) and therefore is determined by the same factors that affect fluid movement across capillaries in general. In the glomerulus, the driving force for filtration of fluid out of the capillary is the glomerular capillary hydrostatic pressure (P_{GC}). This pressure ultimately depends on the pumping of blood by the heart, an action that raises the blood pressure on the arterial side of the circulation. Filtration is opposed by the hydrostatic pressure in the space of Bowman's capsule (P_{BS}) and by the colloid osmotic pressure (COP) exerted by plasma proteins in glomerular capillary blood. Since the glomerular filtrate is virtually protein-free, we neglect the colloid osmotic pressure of fluid in Bowman's capsule. The **net ultrafiltration pressure gradient** (UP) is equal to the difference between the pressures favoring and opposing filtration:

$$\begin{aligned} GFR &= K_f \times UP \\ &= K_f \times (P_{GC} - P_{BS} - COP) \end{aligned} \tag{10}$$

where K_f is the **glomerular ultrafiltration coefficient** (discussed below). Estimates of average, normal values for pressures in the human kidney are: P_{GC}, 55 mm Hg; P_{BS}, 15 mm Hg; COP, 30 mm Hg. From these values, we calculate a net ultrafiltration pressure gradient of +10 mm Hg.

■ The Pressure Profile along a Glomerular Capillary Is Distinctly Different from the Usual Capillary

Figure 23–14 shows how pressures change along the length of a glomerular capillary, in contrast to those seen in a capillary in other vascular beds (here, skeletal muscle). Note that average capillary hydrostatic pressure in the glomerulus is much higher (55 vs. 25 mm Hg) than in a skeletal muscle capillary. Also, capillary hydrostatic pressure declines very little (perhaps 1–2 mm Hg) along the length of the glomerular capillary. This is because the glomerulus contains many (30–50) capillary loops in parallel, which makes the resistance to blood flow in the glomerulus very low. In the skeletal muscle capillary, there is a much higher resistance to blood flow, which results in an appreciable fall in capillary hydrostatic pressure with distance. Finally, note that in the glomerulus the colloid osmotic pressure increases substantially along the length of the capillary, because a large volume of filtrate (about 20% of the entering plasma flow) is

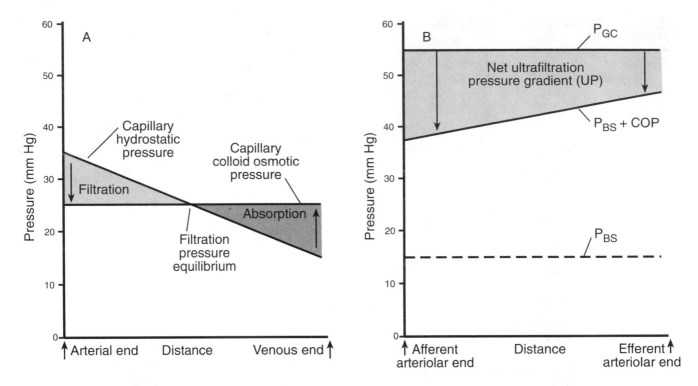

Figure 23–14 ■ Pressure profiles along a skeletal muscle capillary and a glomerular capillary. (A) In the "typical" skeletal muscle capillary, filtration occurs at the arterial end and absorption at the venous end of the capillary. Interstitial fluid hydrostatic and colloid osmotic pressures are neglected here because they are roughly equal and counterbalance each other. (B) In the glomerular capillary, glomerular hydrostatic pressure (P_{GC}) (top line) is high and declines only slightly with distance. The bottom line gives the hydrostatic pressure in Bowman's capsule (P_{BS}). The middle line is the sum of P_{BS} and the glomerular capillary colloid osmotic pressure (COP). The difference between P_{GC} and P_{BS} + COP is equal to the net ultrafiltration pressure gradient (UP). In the normal human glomerulus, filtration probably occurs along the entire capillary. Assuming that K_f is uniform along the length of the capillary, filtration rate would be highest at the afferent arteriolar end and lowest at the efferent arteriolar end of the glomerulus.

pushed out of the capillary and the proteins remain in the circulation. The increase in colloid osmotic pressure opposes outward movement of fluid. In the skeletal muscle capillary, the colloid osmotic pressure hardly changes with distance, since little fluid moves across the capillary wall. In the "average" skeletal muscle capillary, outward filtration occurs at the arterial end and absorption occurs at the venous end. At some point along the skeletal muscle capillary there is no net fluid movement; this is the point of so-called **filtration pressure equilibrium.** Filtration pressure equilibrium probably is not attained in the normal human glomerulus; in other words, outward filtration of fluid probably occurs all along the glomerular capillaries.

Several Factors Can Affect Glomerular Filtration Rate

Glomerular filtration rate may be altered by changing K_f and the pressures acting across the glomerular filtration membrane.

Glomerular Ultrafiltration Coefficient ■ The glomerular ultrafiltration coefficient (K_f) is the glomer-

ular equivalent of the capillary filtration coefficient encountered in Chapter 16. It depends on both the hydraulic conductivity (fluid permeability) and surface area of the glomerular filtration membrane. In chronic renal disease, functioning glomeruli are lost; this leads to a reduction in surface area available for filtration and a fall in GFR. Acutely, a variety of drugs and hormones appear to change glomerular K_f and thus alter GFR, but the mechanisms are not completely understood.

Glomerular Capillary Hydrostatic Pressure ■ Glomerular capillary hydrostatic pressure (P_{GC}) is the driving force for filtration; it depends on the arterial blood pressure and the resistances of upstream and downstream blood vessels. Because of autoregulation, P_{GC} and GFR are maintained at relatively constant values when arterial blood pressure is varied from 80 to 180 mm Hg. Below a pressure of 80 mm Hg, however, P_{GC} and GFR decrease, and GFR ceases at a blood pressure of about 40–50 mm Hg. One of the classic signs of hemorrhagic or cardiogenic shock is an absence of urine output, which is due to an inadequate GFR.

DIALYSIS AND TRANSPLANTATION

Chronic renal failure can result from a large variety of diseases but is most often due to inflammation of the glomeruli (glomerulonephritis) or urinary reflux and infections (pyelonephritis). Renal damage may occur over many years and may be undetected until a considerable loss of nephrons has occurred. When glomerular filtration rate has declined to 5% of normal or less, the internal environment becomes so disturbed that patients usually die within weeks or months if they are not dialyzed or provided with a functioning kidney transplant.

Most of the signs and symptoms of renal failure can be relieved by **dialysis,** the separation of smaller molecules from larger molecules in solution by diffusion of the small molecules through a selectively permeable membrane. Two methods of dialysis are commonly used to treat patients with severe, irreversible ("end-stage") renal failure.

In **continuous ambulatory peritoneal dialysis** (CAPD), the peritoneal membrane, which lines the abdominal cavity, acts as a dialyzing membrane. About 1-2 liters of a sterile glucose/salt solution is introduced into the abdominal cavity and small molecules (e.g., potassium ions and urea) diffuse into the introduced solution, which is then drained and discarded. The procedure is usually done several times every day.

Hemodialysis is more efficient in terms of rapidly removing wastes. The patient's blood is pumped through an artificial kidney machine. The blood is separated from a balanced salt solution by a cellophanelike membrane, and small molecules can diffuse across this membrane. Excess fluid can be removed by applying pressure to the blood and filtering it. Hemodialysis is usually done three times a week (4-6 hours per session), in a medical facility or at home.

Dialysis can allow patients with otherwise fatal renal disease to live useful and productive lives. A number of physiologic and psychological problems persist, however, and bone disease, disorders of nerve function, hypertension, atherosclerotic vascular disease, and disturbances of sexual function are lingering problems. There is a constant risk of infection and, with hemodialysis, clotting and hemorrhage. Dialysis does not maintain normal growth and development in children. Anemia (due primarily to deficient erythropoietin production by damaged kidneys) was once a problem but can now be treated with recombinant human erythropoietin.

Renal transplantation is the only real cure for patients with end-stage renal failure. It may restore complete health and function. In 1993, about 8000 kidney transplant operations were performed in the United States. At present, about 90% of kidneys grafted from a living donor related to the recipient function for 1 year; about 80% of kidneys from unrelated (cadaver) donors function for 1 year.

Several problems complicate kidney transplantation. The immunologic rejection of the kidney graft is a major problem; powerful drugs to inhibit graft rejection are used, but these compromise immune defense mechanisms so that unusual and difficult-to-treat infections often develop. The limited supply of donor organs is also a major unsolved problem; there are many more patients who would benefit from a kidney transplant than there are donors. Finally, the cost of transplantation (or dialysis) is quite high. Fortunately for people in the United States, Medicare currently covers the cost of dialysis and transplantation, but these lifesaving therapies are beyond the reach of most people in poorer countries.

The caliber of afferent and efferent arterioles can be altered by a variety of hormones and by sympathetic nerve stimulation, leading to changes in P_{GC}, glomerular blood flow, and GFR. Some hormones act preferentially on afferent or efferent arterioles. Afferent arteriolar dilation increases glomerular blood flow and P_{GC} and therefore produces an increase in GFR. Afferent arteriolar constriction produces the exact opposite effects. Efferent arteriolar dilation increases glomerular blood flow but leads to a fall in GFR because P_{GC} is decreased. Constriction of efferent arterioles increases P_{GC} and decreases glomerular blood flow. With modest efferent arteriolar constriction, GFR increases because of the increased P_{GC}. With extreme efferent arteriolar constriction, however, GFR decreases because of the marked decrease in glomerular blood flow.

Hydrostatic Pressure in Bowman's Capsule ■ Hydrostatic pressure in Bowman's capsule (P_{BS}) depends on the input of glomerular filtrate (GFR) and the rate of removal of this fluid by the tubule. This pressure opposes filtration. It also provides the driving force for fluid movement down the tubule lumen. If there is obstruction anywhere along the urinary tract, for example because of stones, ureteral obstruction, or prostate enlargement, then pressure upstream to the block is increased and GFR consequently falls. If tubular reabsorption of water is inhibited, pressure in the tubular system is increased because an increased pressure head is needed to force a large volume flow through the loops of Henle and collecting ducts. Consequently, a large increase in urine output caused by a diuretic drug may be associated with a tendency for GFR to fall.

Glomerular Capillary Colloid Osmotic Pressure ■ The COP opposes glomerular filtration. Dilution of the plasma proteins (e.g., by intravenous infusion of a large volume of isotonic saline) lowers the plasma COP and leads to an increase in GFR. Part of the reason glomerular blood flow has important effects on GFR is that the COP profile is changed along the length of a glomerular capillary. Consider, for example, what would happen if glomerular blood flow were low. Filtering a small volume out of the glomerular capillary would lead to a sharp rise in COP early along the length of the glomerulus. As a consequence, filtration would soon cease and GFR would be low. On the other hand, a high blood flow would allow a high rate of filtrate formation with a minimal rise in COP. In general, then, renal blood flow and GFR change hand in hand, but the exact dependence of GFR on renal blood flow depends on the magnitude of the other factors that affect GFR.

■ Several Factors Contribute to the High Glomerular Filtration Rate in the Human Kidney

The rate of plasma ultrafiltration in the kidney glomeruli (180 L/day) far exceeds that in all other capil-lary beds, for several reasons. First, the filtration coefficient is unusually high in the glomeruli. Compared to most other capillaries, the glomerular capillaries behave as though they had more pores per unit surface area; consequently, they have an unusually high hydraulic conductivity. The total glomerular filtration membrane area, about 2 m², is quite large. Second, capillary hydrostatic pressure is higher in the glomeruli than in any other capillaries. Third, the high rate of renal blood flow helps sustain a high GFR by limiting the rise in colloid osmotic pressure. This favors filtration along the entire length of the glomerular capillaries. In summary, glomerular filtration is high because the glomerular capillary blood is exposed to a large porous surface and there is a high transmural pressure gradient.

Tubular Transport in the Proximal Convoluted Tubule

Glomerular filtration is a rather nonselective process, since both useful and waste substances are filtered. By contrast, tubular transport is selective; different substances are transported by different mechanisms. Some substances are reabsorbed, others are secreted, and still others are both reabsorbed and secreted. For most, the amount excreted in the urine depends in large measure on the magnitude of tubular transport. Transport of various solutes and water differs in the various nephron segments. Here we describe transport along the nephron and collecting duct system, starting with the proximal convoluted tubule.

The proximal convoluted tubule comprises the first 60% of the length of the proximal tubule. Because the proximal straight tubule is inaccessible to study in vivo, most quantitative information about function in the living animal is confined to the convoluted portion. Studies on isolated tubules in vitro indicate that both segments of the proximal tubule are functionally similar. The proximal tubule is responsible for reabsorbing all of the filtered glucose and amino acids, reabsorbing the largest fraction of the filtered sodium, potassium, calcium, chloride, bicarbonate, and water, and secreting various organic anions and cations.

■ The Proximal Convoluted Tubule Reabsorbs about 70% of the Filtered Water

The fraction of filtered water reabsorbed along the nephron can be determined by measuring the degree to which inulin is concentrated in tubular fluid, using the **kidney micropuncture** technique. Samples of tubular fluid from surface nephrons are collected and analyzed and the site of collection is identified by nephron microdissection. Because inulin is filtered but not reabsorbed by the kidney tubules, as water is reabsorbed the inulin becomes more and more concentrated. For example, if half of the filtered water is reabsorbed by a certain point along the tubule, the concentration of inulin in tubular fluid (TF_{IN}) is twice the plasma inulin concentration (P_{IN}). The fraction of filtered water reabsorbed by the tubules is equal to $(SNGFR - \dot{V}_{TF})/SNGFR$, where SN (single nephron) GFR gives the rate of filtration of water and $\dot{V}_{TF}$ is the rate of tubular fluid flow at a particular point. The SNGFR can be measured from the single nephron inulin clearance and is equal to $TF_{IN} \times \dot{V}_{TF}/P_{IN}$. From these relations:

$$\begin{array}{l} \% \text{ of filtered water} \\ \text{reabsorbed} \end{array} = \frac{[1 - (1/(TF_{IN}/P_{IN})]}{} \times 100 \quad (11)$$

Figure 23-15 shows how the TF_{IN}/P_{IN} ratio changes along the nephron in normal rats. In fluid collected from Bowman's capsule, the inulin concentration is identical to that in plasma (inulin is freely filterable), so the concentration ratio starts at unity. By the end of the proximal convoluted tubule, the ratio is a little higher than 3, indicating that about 70% of the filtered water is reabsorbed in the proximal convoluted tubule. The ratio is about 5 at the beginning of the distal tubule, indicating that 80% of the filtered water is reabsorbed up to this point. This suggests that the loop of Henle reabsorbs 10% (80% − 70%) of the filtered water. The urine/plasma inulin concentration ratio in the ureter is greater than 100, indicating that more than 99% of the filtered water is reabsorbed. This is not a fixed percentage, but can vary widely, depending on conditions.

■ Proximal Tubular Fluid Is Essentially Isosmotic to Plasma

Samples of fluid collected from the proximal convoluted tubule are always essentially isosmotic to plasma (Fig. 23-16). This is a consequence of the high water permeability of this segment. Overall, 70% of filtered solutes and water are reabsorbed along the proximal convoluted tubule.

Sodium salts are the major osmotically active solutes in the plasma and glomerular filtrate. Since osmolality does not change appreciably with proximal tubule length, it is not surprising that sodium concentration

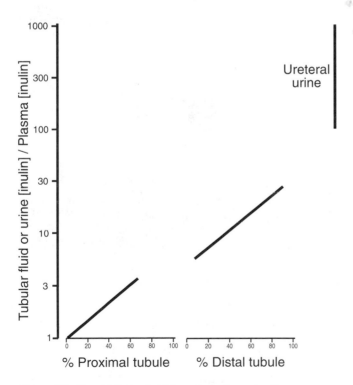

Figure 23–15 ■ Tubular fluid (or urine)-plasma inulin concentration ratios as a function of collection site (data from micropuncture experiments in rats). The increase in this ratio depends on the extent of tubular water reabsorption. The distal tubule is defined in these studies as beginning at the macula densa and ending with the junction of the tubule with a collecting duct (it includes distal convoluted tubule, connecting tubule, and initial part of the collecting duct). (Modified from Giebisch G, Windhager E: Renal tubular transfer of sodium, chloride, and potassium. *Am J Med* 36:643–669, 1964.)

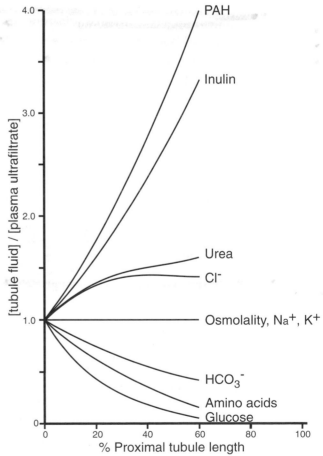

Figure 23–16 ■ Tubular fluid-plasma ultrafiltrate concentration ratios for various solutes as a function of proximal tubule length. All values start at a ratio of 1, since the fluid in Bowman's capsule (0% proximal tubule length) is a plasma ultrafiltrate.

also does not change under ordinary conditions (Fig. 23-16).

If an appreciable quantity of nonreabsorbed solute is present (e.g., the sugar alcohol mannitol), proximal tubular fluid sodium concentration falls to values below the plasma concentration. This is evidence that sodium can be reabsorbed against a concentration gradient and is an active process. The fall in proximal tubular fluid sodium concentration increases diffusion of sodium into the tubule lumen and results in reduced net sodium and water reabsorption. This leads to increased excretion of sodium and water, an osmotic diuresis.

Two major anions, chloride and bicarbonate, accompany sodium in plasma and glomerular filtrate. Bicarbonate ions are preferentially reabsorbed along the proximal convoluted tubule, leading to a fall in tubular fluid bicarbonate concentration. This occurs mainly because of hydrogen ion secretion (Chap. 25). The chloride ion lags behind; as water is reabsorbed, chloride ion concentration rises (Fig. 23-16). This creates a tubular fluid-to-plasma concentration gradient that favors diffusion of chloride out of the tubule lumen. Outward movement of chloride in the late proximal convoluted tubule creates a lumen-positive, small (1-2 mV) transepithelial potential difference that favors passive reabsorption of sodium ions.

Figure 23-16 shows that the potassium concentration hardly changes along the proximal convoluted tubule. If potassium were not reabsorbed, its concentration would increase as much as does that of inulin. The fact that the concentration ratio for potassium remains about unity in this nephron segment indicates that 70% of filtered potassium is reabsorbed along with 70% of the filtered water.

The concentrations of glucose and amino acids fall steeply in the proximal convoluted tubule. This nephron segment and the proximal straight tubule are responsible for complete reabsorption of these substances. Separate, specific reabsorptive mechanisms reabsorb glucose and various amino acids.

The concentration of urea rises along the proximal tubule, but not as much as does the inulin concentration, because about 50% of the filtered urea is reabsorbed. The concentration of PAH in proximal tubular fluid rises above that of inulin because of PAH secretion.

In summary, though the osmolality (total solute concentration) doesn't change detectably along the proximal convoluted tubule, it is clear that the concentrations of individual solutes vary widely. The concentrations of some substances (glucose, amino acids, bicarbonate) fall, others (inulin, urea, chloride, PAH) rise, and still others (sodium, potassium) do not change. By the end of the proximal convoluted tubule only about one-third of the filtered sodium, water, and potassium is left; almost all of the filtered glucose, amino acids, and bicarbonate have been reabsorbed; and a number of solutes destined for excretion (PAH,

creatinine, inulin, urea) have been concentrated in the tubular fluid.

■ Sodium Reabsorption Is the Major Driving Force for Reabsorption of Solutes and Water in the Proximal Convoluted Tubule

Figure 23-17 is a model of a proximal convoluted tubule cell. Sodium enters the cell from the lumen across the apical cell membrane, and is pumped out across the basolateral cell membrane by a Na^+/K^+ ATPase. The sodium and accompanying anions and water are then taken up by the blood surrounding the tubules. Filtered sodium salts and water are thus returned to the circulation.

At the luminal cell membrane (brush border) of the proximal tubule cell, sodium enters the cell down a combined electrical and chemical potential gradient. The inside of the cell is about −70 mV compared to tubular fluid and intracellular sodium concentration is about 30-40 mEq/L, compared to a tubular fluid concentration of about 140 mEq/L. Sodium entry into the cell occurs via a number of symport and antiport mechanisms. Sodium is reabsorbed together with glucose, amino acids, phosphate, and other solutes by

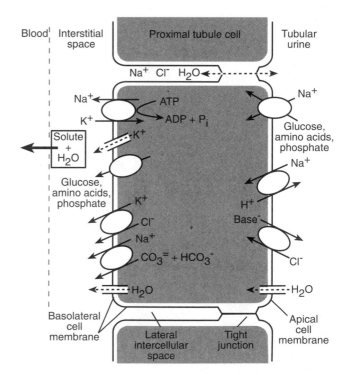

Figure 23-17 ■ Cell model for transport in the proximal tubule. The apical (luminal) cell membrane in this nephron segment has a large surface area for transport because of the numerous microvilli that form the brush border (not shown). Glucose, amino acids, phosphate, and numerous other substances are transported by separate carriers.

way of separate, specific cotransporters. The downhill (energetically speaking) movement of sodium into the cell drives the uphill transport of these solutes. In other words, glucose, amino acids, phosphate, and so on are reabsorbed by secondary active transport. Sodium is also reabsorbed across the luminal cell membrane in exchange for hydrogen ions. The Na^+-H^+ exchanger is also a secondary active transport mechanism; the downhill movement of sodium into the cell energizes the uphill secretion of hydrogen ions into the lumen. This mechanism is important in acidification of the urine (Chap. 25). Chloride may enter the cells by way of a luminal cell membrane chloride-base (formate or oxalate) exchanger.

Once inside the cell, sodium is pumped out the basolateral side by a vigorous Na^+/K^+ ATPase that keeps intracellular sodium concentration low. This membrane ATPase pumps three sodium ions out of the cell and two potassium ions into the cell and splits one ATP molecule for each cycle of the pump. Potassium ions pumped into the cell diffuse out the basolateral cell membrane mostly through a potassium channel. Glucose and amino acids, accumulated in the cell because of active transport across the luminal cell membrane, exit across the basolateral cell membrane by way of separate, sodium-independent facilitated diffusion mechanisms. Bicarbonate exits together with sodium by an electrogenic mechanism; the carrier transports one bicarbonate and one carbonate ion for each sodium ion. Chloride may leave the cell by way of an electrically neutral K-Cl cotransporter.

Reabsorption of sodium and accompanying solutes establishes an osmotic gradient across the proximal tubule epithelium that is the driving force for water reabsorption. Since the water permeability of the proximal tubule epithelium is extremely high, only a very small gradient (a few mOsm/kg H_2O) is needed to account for the observed rate of water reabsorption. Some experimental evidence indicates that proximal tubular fluid is very slightly hyposmotic to plasma; since the osmolality difference is so small it is still proper to consider the fluid as essentially isosmotic to plasma. Water crosses the proximal tubule epithelium through the cells and between the cells (tight junction, lateral intercellular space).

The final step in overall reabsorption of solutes and water is uptake by the peritubular capillaries. This involves the usual Starling forces that operate across capillary walls. Recall that blood in the peritubular capillaries was previously subjected to filtration in the kidney glomeruli. A protein-free filtrate was filtered out of the glomeruli, and consequently the protein concentration (colloid osmotic pressure) of the blood in the peritubular capillaries is quite high. This provides an important driving force for uptake of reabsorbed fluid. The hydrostatic pressure in the peritubular capillaries (a pressure that opposes capillary uptake of fluid) is low, because the blood has passed through upstream resistance vessels. The balance of pressures

acting across peritubular capillaries favors the uptake of reabsorbed fluid.

The Proximal Tubule Secretes Organic Ions

The proximal tubule, both convoluted and straight portions, secretes a large variety of organic anions and cations. Many of these compounds are endogenous compounds, drugs, or toxins. Table 23–2 lists a few secreted organic ions.

The organic anions are mainly carboxylates and sulfonates (carboxylic and sulfonic acids in their protonated forms). A negative charge on the molecule appears to be important for secretion of these compounds. An example of an organic anion that is actively secreted in the proximal tubule is PAH. Its transport shows a maximal rate at high plasma PAH concentrations (Fig. 23–9) and other organic anions compete with it for secretory transport. Figure 23–18 shows a cell model for secretion of PAH. α-Ketoglutarate accumulates in the cell by a secondary active transport mechanism in the basolateral cell membrane. The downhill movement of sodium into the cell provides the energy for this uptake. Intracellular α-ketoglutarate then exchanges with extracellular PAH, thereby driving the active accumulation of PAH into the cell; this

TABLE 23–2 ■ Some Organic Compounds Secreted by the Kidney Tubules[a]

Compound	Use
Organic anions	
Phenol red (phenolsulfonphthalein)	pH indicator dye
p-Aminohippurate (PAH)	Measurement of renal plasma flow and proximal tubule secretory mass
Penicillin	Antibiotic
Probenecid (Benemid)	Inhibitor of penicillin secretion and uric acid reabsorption
Furosemide (Lasix)	"Loop" diuretic drug
Acetazolamide (Diamox)	Carbonic anhydrase inhibitor
Creatinine[b]	Normal end product of muscle metabolism
Organic cations	
Histamine	Vasodilator, stimulator of gastric acid secretion
Cimetidine	Drug for treatment of duodenal and gastric ulcers
Cisplatin	Cancer chemotherapeutic agent
Norepinephrine	Neurotransmitter
Quinine	Antimalarial drug
Tetraethylammonium (TEA)	Ganglion blocking agent
Creatinine[b]	Normal end product of muscle metabolism

[a]The above list includes only a few of the large variety of organic anions and cations secreted by kidney proximal tubules.
[b]Creatinine is an unusual compound, since it is secreted by both organic anion and cation mechanisms. The creatinine molecule bears both negatively and positively charged groups at physiologic pH, and this property may enable it to interact with both secretory mechanisms.

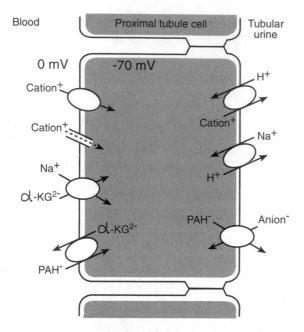

Figure 23–18 ■ Cell model for secretion of organic anions (PAH) and organic cations in the proximal tubule. α-KG^{2-}, α-ketoglutarate.

is an example of **tertiary active transport.** At the luminal membrane, PAH moves downhill into the urine in an electrically neutral fashion by exchanging for an inorganic anion (e.g., chloride) or another organic anion. Secretion of PAH therefore takes place in two membrane steps, both of them sites for competition among different organic anions.

Organic cations are mainly amine and ammonium compounds. Entry into the cell across the basolateral membrane appears to be downhill, by a facilitated diffusion step or diffusion through a channel. Active secretion into the lumen may be driven by an organic cation-hydrogen ion antiporter (Fig. 23–18).

In addition to being actively secreted, some organic compounds may move passively across the tubular epithelium. Organic anions can accept hydrogen ions and organic cations can release hydrogen ions, so their charge is influenced by pH. The nonionized (uncharged) form, if it is lipid-soluble, can diffuse through the lipid bilayer of cell membranes down concentration gradients. The ionized (charged) form passively penetrates cell membranes with difficulty.

Consider, for example, the carboxylic acid probenecid (pK$_a$ = 3.4). This compound is filtered by the glomeruli and secreted by the proximal tubule. As hydrogen ions are secreted into the tubular urine (Chapt. 25), the anionic form (A$^-$) is converted to the nonionized acid (HA). The concentration of nonionized acid is also increased because of water reabsorption. A concentration gradient for passive reabsorption across the tubule wall is created, and appreciable quantities of probenecid are passively reabsorbed. This occurs in most parts of the nephron, but particularly

those where pH gradients are largest and where water reabsorption has resulted in the greatest concentration (i.e., the collecting ducts). Excretion of probenecid is enhanced by making the urine more alkaline (by administering NaHCO$_3$) and by increasing urine output (by drinking water).

Finally, a few organic anions and cations are also actively reabsorbed. For example, uric acid is both secreted and reabsorbed in the proximal tubule. Normally, the amount of uric acid excreted is equal to about 10% of the filtered uric acid, so reabsorption predominates. In gout, plasma levels of uric acid are increased. One treatment for gout is to promote urinary excretion of uric acid by giving drugs that inhibit its tubular reabsorption.

Tubular Transport in the Loop of Henle

The loop of Henle includes several distinct segments with different structural and functional properties. As noted earlier, the proximal straight tubule has transport properties similar to those of the proximal convoluted tubule. The thin descending, thin ascending, and thick ascending limbs of Henle's loop all display different permeabilities and transport properties.

■ Descending and Ascending Limbs Differ Strikingly in Water Permeability

Tubular fluid entering the loop of Henle is isosmotic to plasma, but fluid leaving the loop is distinctly hypoosmotic. Fluid collected from the earliest part of the distal convoluted tubule has an osmolality of about 100 mOsm/kg H$_2$O, compared to 285 mOsm/kg H$_2$O in plasma. This is because more solute than water is reabsorbed by the loop of Henle. The loop of Henle reabsorbs about 20% of filtered sodium, 25% of filtered potassium, 30% of filtered calcium, 65% of filtered magnesium, and 10% of filtered water. The descending limb of Henle's loop is water-permeable. The ascending limb, however, is relatively water-impermeable. Since solutes are reabsorbed along the ascending limb and water cannot follow, fluid along the ascending limb becomes more and more dilute. Deposition of these solutes (mainly sodium salts) in the interstitium of the kidney medulla is critical in the operation of the urinary concentrating mechanism (see below).

■ The Luminal Cell Membrane of the Thick Ascending Limb Contains a Na-K-2Cl Cotransporter

Figure 23–19 is a model of a thick ascending limb cell. Sodium enters the cell across the luminal cell membrane by electrically neutral Na-K-2Cl cotransport. The downhill movement of sodium into the cell results in secondary active transport of chloride and potassium ions. Sodium is pumped out the basolateral

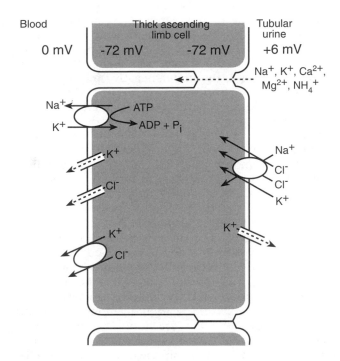

Figure 23–19 ■ Cell model for ion transport in the thick ascending limb.

cell membrane by a vigorous Na⁺/K⁺ ATPase. Potassium ions recycle back into the lumen at the luminal cell membrane. Chloride leaves through the basolateral side by a K-Cl cotransporter or chloride-conducting channels. The luminal cell membrane is predominantly permeable to potassium ions, and the basolateral cell membrane is predominantly permeable to chloride ions. Diffusion of these ions out of the cell produces a transepithelial potential difference, with the lumen about +6 mV compared to interstitial fluid around the tubules. This potential difference drives small cations (Na^+, K^+, Ca^{2+}, Mg^{2+}, and NH_4^+) out of the lumen, between the cells. The tubular epithelium is extremely impermeable to water; there is no measurable water reabsorption along the ascending limb despite a large transepithelial gradient of osmotic pressure.

Tubular Transport in the Distal Nephron

The so-called **distal nephron** includes several distinct segments: distal convoluted tubule, connecting tubule, and cortical, outer medullary, and inner medullary collecting ducts. Transport in the distal nephron differs from that in the proximal tubule in several ways. First, the proximal tubule reabsorbs large quantities of salt and water; it has a high capacity for transport. By contrast, the distal nephron reabsorbs less and has a smaller capacity. Typically, for example, the distal nephron reabsorbs 9% of the filtered sodium and 19% of the filtered water. Second, transport of salt and

water in the proximal tubule occurs along small gradients. By contrast, the distal nephron can establish steep gradients. The final urine can have an osmolality almost one-tenth that of plasma, and the urine sodium concentration may be as low as 1 mEq/L. Third, the distal nephron has a "tight" epithelium, whereas the proximal tubule has a "leaky" epithelium (Chap. 2). The leakiness of the proximal tubule explains why it cannot establish large gradients for small ions. Fourth, sodium and water reabsorption in the proximal tubule are normally closely coupled, because epithelial water permeability is always high. By contrast, sodium and water reabsorption can be uncoupled in the distal nephron, since water permeability may be low and variable. Proximal reabsorption overall can be characterized as a coarse operation that reabsorbs large quantities of salt and water along small gradients. By contrast, distal reabsorption is a finer process.

The collecting ducts are at the end of the nephron system, and what happens there largely determines the excretion of sodium, potassium, and hydrogen ions and water. Transport in the collecting ducts is finely tuned by hormones. Specifically, aldosterone increases sodium reabsorption and potassium and hydrogen ion secretion, and antidiuretic hormone increases water reabsorption at this site.

■ The Luminal Cell Membrane of the Distal Convoluted Tubule Contains a Na-Cl Cotransporter

Figure 23–20 is a model of the distal convoluted tubule cell. In this nephron segment, sodium and chloride are transported from the lumen into the cell by

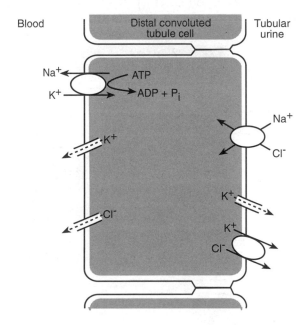

Figure 23–20 ■ Cell model for ion transport in the distal convoluted tubule.

a Na-Cl cotransporter. Sodium is pumped out the basolateral side by the Na^+/K^+ ATPase. Potassium and chloride may leave the cell by an electrically neutral K-Cl cotransporter. The water permeability of the distal convoluted tubule is low and is not changed by antidiuretic hormone.

■ The Cortical Collecting Duct Is an Important Site Regulating Potassium Excretion

Under normal circumstances, most of the excreted potassium is secreted by the cortical collecting ducts. With great potassium excess (e.g., high-potassium diet), the cortical collecting ducts may secrete so much potassium that the amount of potassium excreted exceeds the filtered amount. With severe potassium depletion, the cortical collecting ducts reabsorb potassium.

Potassium secretion appears to be a function primarily of the collecting duct principal cell (Fig. 23–21). Potassium secretion involves active uptake by a Na^+/K^+ ATPase in the basolateral cell membrane, followed by diffusion of potassium through luminal membrane potassium channels. Outward diffusion of potassium from the cell is favored by concentration gradients and opposed by electrical gradients. Note that the electrical gradient opposing potassium exit from the cell is smaller across the luminal cell membrane than across the basolateral cell membrane; this favors movement of potassium into the lumen rather than back into the blood. The luminal cell membrane potential difference is low (e.g., 20 mV, cell inside negative) because this membrane has a high sodium permeability and is depolarized by sodium diffusing into the cell. Recall (Chap. 3) that entry of sodium ions into a (nerve) cell causes membrane depolarization.

The magnitude of potassium secretion is affected by several factors (Fig. 23–21). First, the activity of the Na^+/K^+ ATPase is a key factor affecting secretion; the greater the pump activity, the higher the rate of secretion. Second, the lumen-negative transepithelial electrical potential promotes secretion of this cation. Third, an increase in permeability of the luminal cell membrane to potassium favors secretion. Fourth, a high fluid flow rate through the collecting duct lumen maintains the gradient for potassium secretion and therefore promotes potassium secretion. The hormone aldosterone promotes potassium secretion by several actions, discussed in Chapter 24.

Sodium entry into the collecting duct cell is by diffusion through a sodium ion channel (Fig. 23–21). Entry of sodium through this channel is rate-limiting for overall sodium reabsorption and is increased by aldosterone. Medullary collecting duct sodium reabsorption is so avid in cases of sodium deprivation that the final urine sodium concentration can be lowered to 1 mEq/L.

Intercalated cells are scattered among collecting duct principal cells; they are important in acid-base transport (Chap. 25). A potassium pump may be present in the luminal cell membrane of intercalated cells; this may be responsible for potassium reabsorption when animals are on a potassium-deficient diet.

Urinary Concentration and Dilution

The human kidney can form urine with a total solute concentration greater or lower than that of plasma. Maximum and minimum urine osmolalities in humans are about 1200–1400 mOsm/kg H_2O and 30–40 mOsm/kg H_2O. We consider in the remainder of this chapter the mechanisms involved in producing osmotically concentrated or dilute urine.

■ The Ability to Concentrate Urine Osmotically Is an Important Adaptation to Life on Land

When the kidneys form an osmotically concentrated urine, they save water for the body. The kidneys have the task of getting rid of excess solutes (e.g., urea, various salts), which requires the excretion of solvent (water). Suppose, for example, we excrete 600 mOsm of solutes per day. If we were only capable of excreting urine that is isosmotic to plasma (approximately 300 mOsm/kg H_2O), we would need to excrete 2.0 L H_2O/day. If we can excrete the solutes in a urine that is four times more concentrated than plasma (1200 mOsm/kg H_2O), then only 0.5 L H_2O/day would be required. By excreting solutes in an osmotically concentrated urine, the kidneys in effect saved 2.0−0.5 = 1.5 L H_2O for the body. The ability to concentrate the urine decreases the amount of water we are obliged to find and drink each day.

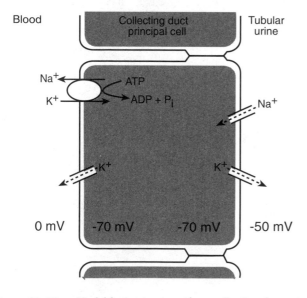

Figure 23–21 ■ Model for ion transport by a collecting duct principal cell.

■ The Antidiuretic Hormone Promotes Excretion of an Osmotically Concentrated Urine

Changes in urine osmolality are normally brought about largely by changes in plasma levels of antidiuretic hormone (ADH). In the absence of ADH, the kidney collecting ducts are relatively water-impermeable. Reabsorption of solute across a water-impermeable epithelium leads to an osmotically dilute urine. In the presence of ADH, collecting duct water permeability is increased. Since the medullary interstitium is hyperosmotic (see below), water reabsorption in the medullary collecting ducts can lead to production of an osmotically concentrated urine.

A model for the action of ADH on cells of the collecting duct is shown in Figure 23–22. When plasma osmolality is increased, plasma ADH levels increase. The hormone binds to a specific receptor in the basolateral cell membrane. By way of a guanine nucleotide stimulatory protein, the membrane-bound enzyme adenylate cyclase is activated. This enzyme catalyzes formation of cyclic AMP (cAMP) from ATP. Cyclic AMP then activates a cAMP-dependent protein kinase (protein kinase A) that phosphorylates other proteins. This leads to insertion, by exocytosis, of water channel-containing intracellular membrane vesicles into the luminal cell membrane. The resulting increase in number of luminal membrane water channels leads to an increase in water permeability. Water can then move out of the duct lumen through the cells, and the urinary solutes become concentrated.

■ A Gradient of Increasing Osmolality Is Seen in the Kidney Medulla

In 1951, Wirz, Hargitay, and Kuhn demonstrated that there is a gradient of osmolality in the kidney medulla (Fig. 23–23) and suggested a hypothesis to account for the origin of this gradient. In their experiments, they deprived rats of water for 1–2 days, anesthetized them, and rapidly removed and froze their kidneys. They cut the frozen kidneys into slices of cortex, outer medulla, and inner medulla. Under microscopic observation, they recorded the temperature at which ice crystals in the tissue melted. The melting point for ice (freezing point for liquid water) is linearly but inversely related to osmolality. They found that the cortical tissue was isosmotic to plasma. In the medulla, however, there was a gradient of osmolality that increased toward the tip of the papilla. In these water-deprived animals, which had high plasma levels of ADH, the final urine had the same osmolality as kidney tissue from the tip of the papilla. These experiments provided strong support for the **countercurrent hypothesis.**

■ The Loops of Henle Are Countercurrent Multipliers and the Vasa Recta Are Countercurrent Exchangers

The kidney medulla contains two countercurrent mechanisms (Fig. 23–24). The term *countercurrent*

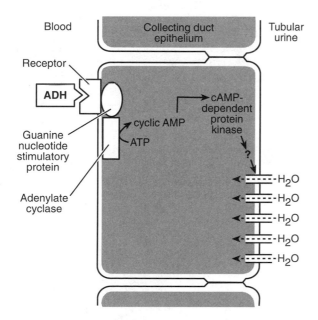

Figure 23–22 ■ Model for the action of ADH on the epithelium of the collecting duct. Note that the ADH receptor is on the basolateral side but the water permeability increase occurs on the luminal side.

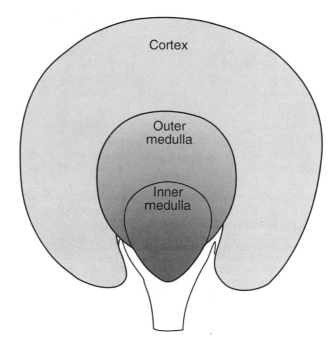

Figure 23–23 ■ Osmotic gradient in the medulla of water-deprived rats. The cortex appears to be isosmotic with plasma. Starting at the junction of cortex and medulla, there is a gradient of increasing osmolality, with the highest osmolalities observed at the tip of the papilla. (Modified from Wirz H, Hargitay B, Kuhn W: Lokalisation des Konzentrierungsprozesses in der Niere durch direkte Kryoskopie. *Helv Physiol Acta* 9:196–207, 1951.)

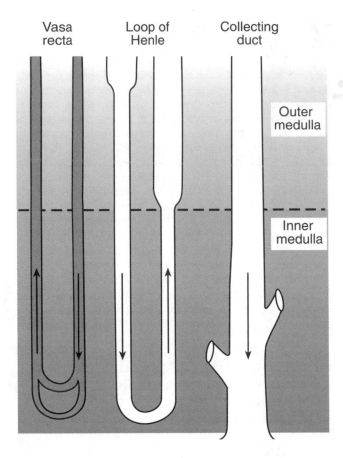

Figure 23–24 ▪ Elements of the urinary concentrating mechanism. The vasa recta are countercurrent exchangers, the loops of Henle are countercurrent multipliers, and the collecting ducts are osmotic equilibrating devices. Note that most loops of Henle and vasa recta do not reach the tip of the papilla, but turn at higher levels in the outer and inner medulla. Also, there are no thick ascending limbs in the inner medulla.

Countercurrent multiplication is the process whereby a small gradient established at any level of the loop of Henle is increased (multiplied) into a much larger gradient along the axis of the loop. The osmotic gradient established at any level is called the **single effect.** The single effect involves movement of solute out of the water-impermeable ascending limb, solute deposition in the medullary interstitium, and withdrawal of water from the descending limb. Since the fluid entering the next, deeper level of the loop is now more concentrated, repetition of the same process leads to an **axial gradient** of osmolality along the loop. The extent to which countercurrent multiplication can establish a large gradient along the axis of the loop depends on several factors, including the magnitude of the single effect, the rate of fluid flow, and the length of Henle's loop. The larger the single effect, the larger the axial gradient. Impaired solute removal, for example due to inhibition of active transport by thick ascending limb cells, leads to a reduced axial gradient. If flow rate through the loop is too high, not enough time is allowed for establishment of a significant single effect, and consequently the axial gradient is reduced. Finally, if the loops are long, there is more opportunity for multiplication, and a larger axial gradient can be established.

Countercurrent exchange is a common process in the vascular system. In many vascular beds, arterial and venous vessels lie close to each other, and exchange of heat or materials can occur between these vessels. For example, because of countercurrent exchange of heat between blood flowing toward and away from its feet, a penguin can stand on ice and yet maintain a warm body (core) temperature. Countercurrent exchange between descending and ascending vasa recta in the kidney reduces dissipation of the solute gradient in the medulla. The descending vasa recta tend to give up water to the more concentrated interstitium; this water is taken up by the ascending vasa recta, which come from more concentrated regions of the medulla. In effect, then, much of the water in the blood short-circuits across the tops of the vasa recta and does not flow deep into the medulla, where it would tend to dilute the accumulated solute. The ascending vasa recta tend to give up solute as the blood moves toward the cortex. Solute enters the descending vasa recta and therefore tends to be trapped in the medulla. Countercurrent exchange is a purely passive process; it helps maintain a gradient established by some other means.

indicates a flow of fluid in opposite directions in adjacent structures. First, the loops of Henle are **countercurrent multipliers.** Fluid flows toward the tip of the papilla along the descending limb of the loop and toward the cortex along the ascending limb of the loop. The loops of Henle set up the osmotic gradient in the medulla. Establishing a gradient requires work; the energy source is metabolism, which powers active transport of sodium out of the thick ascending limb. The vasa recta are **countercurrent exchangers.** Blood flows in opposite directions along juxtaposed descending (arterial) and ascending (venous) vasa recta, and solutes and water are exchanged between these capillary blood vessels. The vasa recta help maintain the gradient in the medulla by purely passive processes. The collecting ducts act as **osmotic equilibrating devices;** depending on the plasma level of ADH, the collecting duct urine is allowed to equilibrate more or less with the hyperosmotic medullary interstitium.

▪ Operation of the Urinary Concentrating Mechanism Requires Integrated Functioning of Loops of Henle, Vasa Recta, and Collecting Ducts

Figure 23–25 summarizes the mechanisms involved in producing an osmotically concentrated urine. A maximally concentrated urine, with an osmolality of

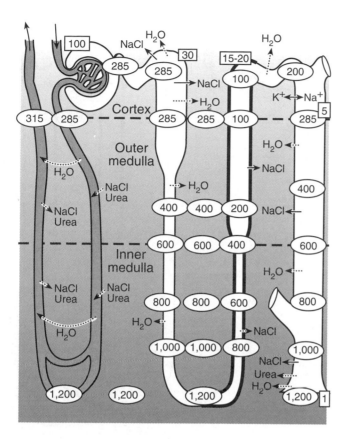

Figure 23–25 ■ Summary of movements of ions, urea, and water in the kidney during production of a maximally concentrated urine (1200 mOsm/kg H_2O). Numbers in ovals give osmolality in mOsm/kg H_2O. Numbers in boxes give relative amount of water present at each level of the nephron. Solid arrows indicate active transport; dashed arrows indicate passive transport. The heavy outlining along the ascending limb of Henle's loop indicates relative water-impermeability.

1200 mOsm/kg H_2O and a low urine volume (1% of the original filtered water), is being excreted.

About 70% of filtered water is reabsorbed along the proximal convoluted tubule, so 30% of the original filtered volume enters the loop of Henle. As discussed earlier, proximal reabsorption of water is essentially an isosmotic process, so fluid entering the loop is isosmotic. As the fluid moves along the descending limb of Henle's loop it becomes more and more concentrated. This rise in osmolality could in principle be due to two processes: movement of water out of the descending limb because of the hypertonicity of the medullary interstitium; or entry of solute from the medullary interstitium. The relative importance of these processes may depend on the animal species. For most efficient operation of the concentrating mechanism, water removal should be predominant, so only this process is depicted in Figure 23–25. The removal of water along the descending limb leads to a rise in sodium chloride concentration in the loop fluid to a value higher than in the interstitium.

When the fluid enters the ascending limb, it enters water-impermeable segments. Sodium chloride is transported out of the ascending limb and deposited in the medullary interstitium. In the thick ascending limb, sodium transport is active and is powered by a vigorous Na^+/K^+ ATPase. In the thin ascending limb, sodium chloride reabsorption appears to be mainly passive. It occurs because the sodium chloride concentration in the tubular fluid is higher than in the interstitium and because the passive permeability of the thin ascending limb to sodium is high. There is also some evidence for a weak active sodium pump in the thin ascending limb. The net addition of solute to the medulla by the loops is essential for the osmotic concentration of urine in the collecting ducts.

Fluid entering the distal convoluted tubule is hyposmotic compared to plasma (Fig. 23–25), because of the removal of solute along the ascending limb. In the presence of ADH, the cortical collecting ducts become water-permeable, and water is passively reabsorbed into the cortical interstitium. The high blood flow to the cortex rapidly carries away this water, so there is no detectable dilution of cortical tissue osmolality. Before the tubular fluid reenters the medulla it is isosmotic and reduced to about 5% of the original filtered volume. The reabsorption of water in the cortical collecting ducts is important for the overall operation of the urinary concentrating mechanism. If this water were not reabsorbed in the cortex, an excessive amount of water would enter the medulla. It would tend to wash out the gradient in the medulla and lead to an impaired ability to concentrate the urine maximally.

All nephrons drain into collecting ducts that pass through the medulla. In the presence of ADH, the medullary collecting ducts are permeable to water. Water moves out of the collecting ducts into the more concentrated interstitium. In the presence of high levels of ADH, the fluid equilibrates with the interstitium, and the final urine becomes as concentrated as the tissue fluid at the tip of the papilla.

Many different models for the countercurrent mechanism have been proposed; each must take into account the principle of conservation of matter (mass balance). In the steady state, the inputs of water and every solute must equal their respective outputs. This principle must be obeyed at every level of the medulla. Figure 23–26 presents a simplified scheme that applies the mass balance principle to the medulla as a whole. It provides some additional insights into the countercurrent mechanism. Notice that fluids entering the medulla (from the proximal tubule, descending vasa recta, and cortical collecting ducts) are isosmotic; they all have an osmolality of about 285 mOsm/kg H_2O. Fluid leaving the medulla in the urine is hyperosmotic. It follows from mass balance considerations that somewhere a hyposmotic fluid has to leave the medulla; this occurs in the ascending limb of Henle's loop. The input of water into the medulla must equal its out-

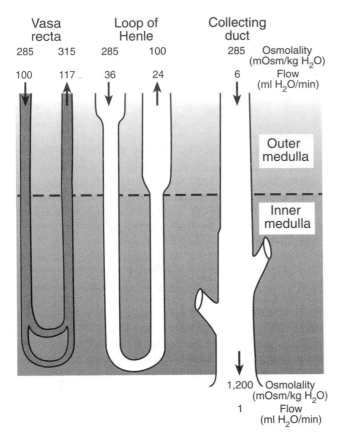

Figure 23–26 ■ Mass balance considerations for the medulla as a whole. Note that in the steady state, the input of water and solutes must equal their respective output. Water input into the medulla from the cortex (100 + 36 + 6 = 142 ml/min) equals water output from the medulla (117 + 24 + 1 = 142 ml/min). Solute input (28.5 + 10.3 + 1.7 = 40.5 mOsm/min) is likewise equal to solute output (36.9 + 2.4 + 1.2 = 40.5 mOsm/min).

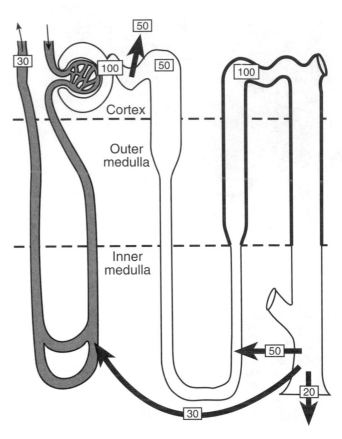

Figure 23–27 ■ Movements of urea along the nephron. The numbers indicate relative amounts (100 = filtered urea), not concentrations. The heavy outline from the thick ascending limb to the outer medullary collecting duct indicates relatively urea-impermeable segments. Urea is added to the inner medulla by its collecting ducts; most of this urea reenters the loop of Henle, and some is removed by the vasa recta.

put. Since water is added to the medulla along the descending limbs of Henle's loops and the collecting ducts, this water must be removed at an equal rate. The ascending limbs of Henle's loops cannot remove the added water, since they are water-impermeable. The water is removed by the vasa recta; this is why ascending exceeds descending vasa recta blood flow (Fig. 23–26). The blood leaving the medulla is hyperosmotic because it drains a region of high osmolality and does not instantaneously equilibrate with the medullary interstitium. Any description of the concentrating mechanism must include all three elements in the medulla (loops of Henle, vasa recta, and collecting ducts), because all work in concert.

■ Urea Plays a Special Role in the Concentrating Mechanism

It has been known for many years that animals or humans on a low-protein diet have an impaired ability to concentrate the urine maximally. A low-protein diet is associated with a decreased concentration of urea in the kidney medulla.

Figure 23–27 shows how urea is handled along the nephron. The proximal convoluted tubule is fairly permeable to urea and reabsorbs about 50% of the filtered urea. Fluid collected from the distal convoluted tubule, however, has as much urea as the amount filtered. Therefore, urea is secreted in the loop of Henle.

The thick ascending limb, distal convoluted tubule, connecting tubule, cortical collecting duct, and outer medullary collecting duct are relatively urea-impermeable. As water is reabsorbed along cortical and outer medullary collecting ducts, the urea concentration rises. The result is the delivery to the inner medulla of a concentrated urea solution. A concentrated solution has chemical potential energy and can do work.

The inner medullary collecting duct is relatively urea-permeable. Furthermore, the urea permeability of its terminal portion is increased by ADH. Urea diffuses into the inner medulla. Some of the urea reenters the loop and, in effect, is recycled, so that a high urea concentration builds up in the inner medulla. Urea may also be added to the inner medulla from calyceal urine. Urea accounts for about half of the osmolality in the inner medulla. It does osmotic work by abstracting water from the descending limb of Henle's loop, thereby resulting in an increase in sodium chloride concentration in descending limb fluid. This produces a gradient for sodium chloride to move passively out of the thin ascending limb. The outward movement of sodium chloride drives the countercurrent multiplier in the inner medulla and establishes the gradient that permits maximum concentration of the urine in the collecting ducts. Thus, a concentrated urea solution entering the inner medulla indirectly provides the energy that drives the countercurrent multiplier in this region.

■ A Dilute Urine Is Excreted When Plasma ADH Levels Are Low

Figure 23-28 depicts kidney osmolalities during excretion of a dilute urine, as occurs when plasma ADH levels are very low. Tubular fluid is diluted along the ascending limb and becomes even more dilute as solute is reabsorbed across the relatively water-impermeable distal portions of the nephron and collecting ducts. Since as much as 15% of filtered water is not reabsorbed, a high urine flow rate results. In these circumstances, the osmotic gradient in the medulla is reduced but not abolished. The decreased gradient probably results from several factors, including an increased medullary blood flow and reduced addition of urea to the medulla.

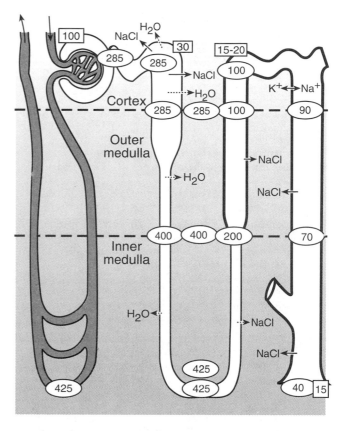

Figure 23–28 ■ Osmotic gradients during excretion of an osmotically very dilute urine. The collecting ducts are relatively water-impermeable (heavy outlining) because ADH is absent. Note that the medulla is still hyperosmotic, but less so than in a kidney producing an osmotically concentrated urine.

■ ■ ■ ■ ■ ■ ■ ■ ■ ■ ■ ■ ■ **REVIEW EXERCISES** ■ ■ ■ ■ ■ ■ ■ ■ ■ ■ ■ ■ ■

Identify Each with a Term

1. The second messenger that mediates the increase in collecting duct water permeability produced by ADH
2. Abnormal excretion of protein in the urine
3. Relative constancy of renal blood flow and glomerular filtration rate despite changes in arterial blood pressure
4. Cell that synthesizes renin
5. The amount of a substance per unit time that is filtered by the glomerulus

Define Each Term

6. Osmotic diuresis
7. Glomerular filtration membrane
8. Ultrafiltration
9. Renal clearance
10. Juxtamedullary nephron
11. Countercurrent multiplication

Choose the Correct Answer

12. Which of the following causes an *increase* in renal glucose threshold?
    a. A decrease in glomerular filtration rate
    b. A decrease in the number of proximal tubule glucose carriers
    c. Administration of a drug that inhibits glucose reabsorption
    d. An increase in splay in the glucose reabsorption curve
    e. None of the above

13. Probenecid is a lipid-soluble, weak acid that undergoes nonionic diffusion in the kidneys. Which of the following would promote urinary excretion of this substance?
    a. Abstaining from all fluids
    b. Acidifying urine by ingesting NH_4Cl tablets
    c. Administering a drug that inhibits tubular secretion of organic anions
    d. Alkalinizing the urine by ingesting sodium bicarbonate

14. Which of the following provides the most accurate measure of glomerular filtration rate?
    a. Blood urea nitrogen (BUN)
    b. Endogenous creatinine clearance
    c. Inulin clearance
    d. PAH clearance
    e. Plasma creatinine concentration

15. The main driving force for water reabsorption by the proximal tubule epithelium is:
    a. Active reabsorption of amino acids and glucose
    b. Active reabsorption of sodium
    c. Active reabsorption of water
    d. Pinocytosis
    e. The high colloid osmotic pressure in the peritubular capillaries

16. In a person with severe central diabetes insipidus (deficient production or release of antidiuretic hormone), urine osmolality and flow rate is typically about:
    a. 50 mOsm/kg H_2O, 18 L/day
    b. 50 mOsm/kg H_2O, 1.5 L/day
    c. 300 mOsm/kg H_2O, 1.5 L/day
    d. 300 mOsm/kg H_2O, 18 L/day
    e. 1200 mOsm/kg H_2O, 0.5 L/day

Calculate

17. The following determinations were made on a single glomerulus: nephron glomerular filtration rate, 42 nl/min; glomerular capillary hydrostatic pressure, 50 mm Hg; hydrostatic pressure in Bowman's capsule, 12 mm Hg; average glomerular capillary colloid osmotic pressure, 24 mm Hg. What is the glomerular ultrafiltration coefficient?

18. The following data were collected in a renal clearance study on a young adult, 60-kg woman (body surface area = 1.65 m²): plasma inulin concentration, 0.40 mg/ml; urine inulin concentration, 8.00 mg/ml; plasma glucose concentration, 5.00 mg/ml; urine glucose concentration, 40.0 mg/ml; plasma sodium concentration, 135 mEq/L; urine sodium concentration, 67.5 mEq/L; urine flow rate, 5.00 ml/min. (a) What is the glomerular filtration rate? (b) Is the glomerular filtration rate normal? (c) Has the glucose threshold been reached? (d) What is the glucose clearance, and why is it less than the GFR? (e) What is the rate of tubular reabsorption of glucose? (f) What fraction (percentage) of the filtered

water was reabsorbed by the kidney tubules? (g) What fraction (percentage) of the filtered sodium was excreted?

19. The following data were collected in a renal clearance study on a young adult, 70-kg man (body surface area = 1.73 m²): plasma inulin concentration, 0.30 mg/ml; urine inulin concentration, 7.50 mg/ml; arterial plasma PAH concentration, 0.020 mg/ml; renal venous plasma PAH concentration, 0.002 mg/ml; urine PAH concentration, 2.64 mg/ml; fraction of plasma PAH bound to plasma proteins, 0.20; urine flow rate, 5.00 ml/min; hematocrit ratio, 0.45. (a) What is the glomerular filtration rate? (b) What is the effective renal plasma flow? (c) What is the PAH extraction ratio? (d) What is the true renal plasma flow? (e) What is the true renal blood flow? (f) What fraction of the plasma flowing through the kidneys is filtered? (This is called the filtration fraction [FF]) (g) What is the rate of PAH secretion by the kidney tubules?

Briefly Answer

20. Indicate how the following affect (increase, decrease, or no change) glomerular capillary pressure (P_{GC}), glomerular blood flow (GBF), and glomerular filtration rate (GFR): (a) afferent arteriolar constriction; (b) afferent arteriolar dilation; (c) modest efferent arteriolar constriction; (d) extreme efferent arteriolar constriction; (e) efferent arteriolar dilation; (f) an equal increase in resistance in blood vessels upstream and downstream to the glomerulus.

21. Indicate the nephron segment(s) associated with each statement: (a) reabsorbs filtered glucose; (b) important site of potassium secretion; (c) luminal cell membrane contains a Na-K-2Cl cotransporter; (d) reabsorbs 70% of the filtered sodium and water; (e) secretes organic anions and organic cations; (f) luminal cell membrane that contains a Na-Cl cotransporter; (g) water permeability that is increased by antidiuretic hormone; (h) tubular fluid that is always essentially isosmotic to plasma.

22. What is the importance of renal autoregulation?

23. What is the significance of fixed negative charges in the glomerular filtration membrane?

24. What special role does urea play in osmotic concentration of the urine in the inner medulla?

■ ■ ■ ■ ■ ■ ■ ■ ■ ■ ■ ANSWERS ■ ■ ■ ■ ■ ■ ■ ■ ■ ■ ■

1. Cyclic AMP

2. Proteinuria

3. Renal autoregulation

4. Granular cell

5. Filtered load

6. An increase in urine flow due to excretion of an osmotically active solute

7. The barrier separating glomerular capillary blood from the urinary space of Bowman's capsule

8. Filtration through a membrane with exceedingly fine pores

9. The volume of plasma per unit time needed to supply the quantity of a substance excreted in the urine per unit time; $U_x \times \dot{V}/P_x$

10. A nephron with a long loop of Henle whose glomerulus lies deep in the cortex next to the medulla

11. An energy-demanding process that sets up a solute (osmotic) gradient along the length of two oppositely flowing streams, as in Henle's loop in the kidney medulla

12. a

13. d

14. c

15. b

16. a

17. $K_f = GFR/[P_{GC} - P_{BS} - COP] = 42$ nl/min $\div$ [50 − 12 − 24 mm Hg] = 3 nl/min/mm Hg

18. (a) GFR = $C_{IN} = U_{IN}\dot{V}/P_{IN} = 8$ mg/ml × 5 ml/min $\div$ 0.4 mg/ml = 100 ml plasma/min; (b) Yes; (c) Yes; (d) $C_G = U_G\dot{V}/P_G = 40$ mg/ml × 5 ml/min $\div$ 5 mg/ml = 40 ml plasma/min; because glucose is reabsorbed by the kidney tubules; (e) $T_G = P_G GFR - U_G\dot{V} = 5$ mg/ml × 100 ml/min − 40 mg/ml × 5 ml/min = 300 mg/min; (f) % water reabsorbed = $[1 - 1/(U_{IN}/P_{IN})]$ × 100 = [1 − 1/(8.00/0.40)] × 100 = 0.95 (95%); (g) $U_{Na}\dot{V}/P_{Na}GFR = 67.5$ mEq/L × 5 ml/min $\div$ 135 mEq/L × 100 ml/min = 0.025 (2.5%)

19. (a) 125 ml plasma/min; (b) $C_{PAH} = U_{PAH}\dot{V}/P_{PAH} = 2.64$ mg/ml × 5 ml/min $\div$ 0.020 mg/ml = 660 ml plasma/min; (c) $E_{PAH} = (P^a_{PAH} - P^{rv}_{PAH})/P^a_{PAH} = (0.020 - 0.002)/0.020 = 0.90$ (or 90%); (d) $C_{PAH}/E_{PAH} = 660$ ml/min/0.90 = 733 ml plasma/min; (e) RBF = RPF/(1 − Hematocrit) = 733 ml/min $\div$ 0.55 = 1.33 L/min; (f) FF = GFR/RPF = 125/733 = 0.17 (17%); (g) $T_{PAH} = U_{PAH}\dot{V} - P_{PAH}GFR = 2.64$ mg/ml × 5 ml/min − 0.8 × 0.020 mg/ml × 125 ml/min = 11.2 mg/min

20. (a) Decrease all three; (b) increase all three; (c) increase, decrease, increase; (d) increase, decrease, decrease; (e) decrease, increase, decrease; (f) no change, decrease, decrease

21. (a) Proximal tubule; (b) cortical collecting duct; (c) thick ascending limb; (d) proximal convoluted tubule; (e) proximal tubule; (f) distal convoluted tubule; (g) collecting duct; (h) proximal tubule

22. Renal autoregulation reduces the effect of changes in arterial blood pressure on renal blood flow and GFR and hence on the filtered sodium load. This minimizes the impact of blood pressure changes on sodium excretion and thereby reduces changes in extracellular fluid volume.

23. Serum albumin and many other plasma proteins bear a net negative charge at physiologic pH. Fixed negative charges in the glomerular filtration membrane electrostatically repel these plasma proteins and contribute to the low filterability of these macromolecules. A loss of fixed negative charges may contribute to increased filtration of plasma proteins and proteinuria in some glomerular diseases.

24. Urea accumulates in the inner medulla and effects the osmotic withdrawal of water from the thin descending limbs; the result is an increase in tubular fluid sodium chloride concentration. The sodium chloride can then diffuse passively out of the thin ascending limb; this produces a single effect, which can be multiplied into an axial gradient of osmolality along the depth of the inner medulla. Urea, therefore, provides the chemical potential energy for establishing an osmotic gradient in the medulla and enhances the ability to concentrate the urine osmotically.

Suggested Reading

Brenner B, Coe FL, Rector FC Jr: *Renal Physiology in Health and Disease.* Philadelphia, Saunders, 1987.

Hladky SB, Rink TJ: *Body Fluid and Kidney Physiology.* London, Edward Arnold, 1986.

Jamison RL, Maffly RH: The urinary concentrating mechanism. *N Engl J Med* 295:1059–1067, 1976.

Koeppen BM, Stanton BA: *Renal Physiology.* St. Louis, Mosby, 1992.

Kriz W, Bankir L: A standard nomenclature for structures of the kidney. *Am J Physiol* 254:F1–F8, 1988.

Marsh DJ: *Renal Physiology.* New York, Raven, 1983.

Rose BD: *Clinical Physiology of Acid-Base and Electrolyte Disorders.* 4th ed. New York, McGraw-Hill, 1994.

Schafer JA: Salt and water absorption in the proximal tubule. *Physiologist* 25:95–103, 1982.

Seldin DW, Giebisch G (eds): *The Kidney: Physiology and Pathophysiology.* 2nd ed. New York, Raven, 1992.

Valtin H, Schafer JA: *Renal Function.* 3rd ed. Boston, Little, Brown, 1994.

Vander AJ: *Renal Physiology.* 4th ed. New York, McGraw-Hill, 1991.

Windhager EE (ed): *Handbook of Physiology: Renal Physiology.* New York, Oxford University Press, 1992.

CHAPTER

■ ■ ■ ■ ■ ■ ■

24

Regulation of Fluid and Electrolyte Balance

CHAPTER OUTLINE

  I. FLUID COMPARTMENTS OF THE BODY
   A. The intracellular compartment contains about two-thirds and the extracellular compartment about one-third of total body fluid
   B. The two fluid compartments are normally in osmotic equilibrium
 II. MEASUREMENT OF VOLUME OF BODY FLUID COMPARTMENTS BY THE INDICATOR DILUTION PRINCIPLE
III. WATER BALANCE
   A. Antidiurectic hormone is critical in the control of renal water reabsorption and plasma osmolality
   B. Habit and the thirst mechanism govern water intake
 IV. SODIUM BALANCE
   A. Sodium balance is maintained by the regulation of sodium intake and excretion
   B. Many factors influence renal sodium excretion
    1. Renal hemodynamic and physical factors
    2. The renin-angiotensin-aldosterone system
    3. Atrial natriuretic peptide
    4. Renal prostaglandin and kinin systems
    5. Renal sympathetic nerves
  V. POTASSIUM BALANCE
   A. Plasma potassium directly affects potassium excretion
   B. A potent aldosterone negative feedback system regulates potassium in the extracellular fluid
 VI. CALCIUM, MAGNESIUM, AND PHOSPHATE BALANCE
   A. Parathyroid hormone decreases renal excretion of calcium
   B. Magnesium is reabsorbed in the proximal convoluted tubule and Henle's loop
   C. Phosphate excretion is regulated primarily by the proximal tubules
    1. Transport maximum and phosphate excretion
    2. Parathyroid hormone and phosphate transport maximum

OBJECTIVES

After studying this chapter, the student should be able to:
1. Define *intracellular fluid compartment, extracellular fluid compartment,* and *osmotic equilibrium*
2. List the major ions in the intracellular and extracellular fluid compartments
3. Describe the measurement of compartment volume
4. Describe how thirst and antidiuretic hormone together regulate water balance
5. List the regulators of renal sodium excretion and describe their mechanisms of action
6. Describe the regulation of extracellular fluid potassium concentration
7. List the hormones that regulate calcium ion concentration in the extracellular fluid; indicate where in the nephron they act to influence renal calcium reabsorption
8. Describe the factors that produce changes in phosphate transport
9. Describe the genesis and events of the micturition reflex

A principal function of the kidneys is to regulate the volume, composition, and osmolality of body fluids. The kidney regulates the composition of the circulating plasma, which then determines the composition of the extracellular fluid that surrounds the cells and ultimately influences the intracellular fluid compartment. It is important to define the various fluid compartments to understand how a specific renal action may influence one or more of these compartments.

Fluid Compartments of the Body

Water is the major constituent of all fluid compartments of the body. **Total body water** is about 40 liters in an average, young adult 70-kg man, or roughly 60% of total body weight. The percentage of water varies from one tissue to another. For example, adipose tissue contains only about 10% water, whereas muscle contains about 75% water.

Total body water can be divided into intracellular and extracellular fluid compartments.

■ The Intracellular Compartment Contains about Two-Thirds and the Extracellular Compartment about One-Third of Total Body Fluid

The **intracellular fluid** compartment contains approximately two-thirds of the total body water: 25

liters of 40 liters in a 70-kg man (Fig. 24-1). Two liters of red blood cell water are included in this fluid compartment. The principal metabolic reactions of the body occur inside the cells and thus within the intracellular fluid compartment.

As shown in Table 24-1, the intracellular fluid contains high concentrations of potassium and phosphate and moderate concentrations of magnesium and sulfate ions as compared with the extracellular fluid. Low concentrations of sodium, bicarbonate, and chloride also exist in the intracellular fluid. In addition, the cells contain approximately three times the concentration of protein that exists in the plasma.

The **extracellular fluid** compartment contains all fluids outside the cells and includes about 3 liters of plasma. The amount of extracellular fluid is about 15 liters (about one-third of total body fluid) in a 70-kg man (Fig. 24-1). In addition to blood plasma, the extracellular fluid includes interstitial fluid, lymph, cerebrospinal fluid, gastrointestinal secretions, aqueous humor, sweat, urine, and peritoneal and pleural fluids.

The extracellular fluid compartment contains high concentrations of sodium, chloride, and bicarbonate and low concentrations of potassium, calcium, phosphate, sulfate, magnesium, and organic acid anions (Table 24-1). Plasma and, to a lesser extent, interstitial fluid are rich in proteins. Plasma proteins are an important determinant of plasma colloid osmotic pressure, which is mainly responsible for pre-

TABLE 24–1 ■ Electrolyte Composition of the Body Fluids

Electrolytes	Plasma (mEq/L)	Plasma Water (mEq/kg H_2O)	Interstitial Fluid (mEq/kg H_2O)	Intracellular Fluid (Skeletal Muscle) (mEq/kg H_2O)
Anions				
Protein	17	18	—	45
Cl^-	103	111	117	3
HCO_3^-	25	27	28	7
Others	8	9	9	155
Total	153	165	154	210
Cations				
Ca^{2+}	5	5.4	3	1
Na^+	142	153	145	10
Mg^{2+}	2	2.2	2	40
K^+	4	4.3	4	159
Total	153	165	154	210

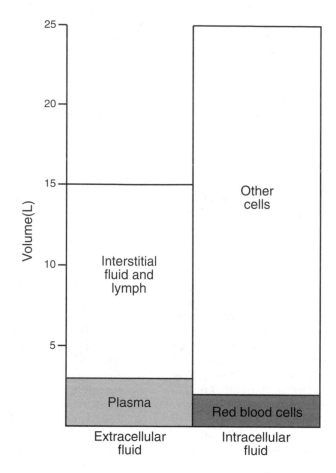

Figure 24–1 ■ The extracellular and intracellular fluid compartments. The fluid in both compartments represents the total body fluid and is equal to 40 liters.

venting fluid loss from the blood into the interstitial spaces.

■ The Two Fluid Compartments Are Normally in Osmotic Equilibrium

Two basic relations affect the movement of water from one compartment to another. First, the osmolalities (osmotic activities) in the various body compartments are almost always at equilibrium, therefore intracellular and interstitial fluid compartments and plasma osmolalities are about equal. When osmolality changes in one of these compartments, water shifts so as to restore and maintain the intercompartmental osmotic equilibrium. This occurs rapidly because of the high water permeability of cell membranes. Second, when water is lost or gained, each compartment of the body loses or gains an amount of water proportional to its volume such that the osmolality becomes the same in all compartments.

Measurement of Volume of Body Fluid Compartments by the Indicator Dilution Principle

The indicator dilution principle (see also Chap. 14) is expressed as:

$$
\frac{\text{Volume (ml)}}{} = \frac{\text{Amount of test substance injected}}{\text{Concentration/ml of sample fluid}} \quad (1)
$$

For the successful application of the indicator dilution principle, the test substance must be distributed evenly and must stay confined within the compartment being measured; must not change the volume of the compartment; must not be metabolized or synthesized in the body; and must not be toxic.

Substances used to measure total body water must distribute in extracellular and intracellular fluid compartments. Deuterium oxide (heavy water) or tritiated water is usually injected intravenously to estimate total body water. After an equilibration period of 30–60 minutes, a blood sample is withdrawn and the concentration of the indicator in plasma determined. For example, if 40 ml of deuterium oxide is injected and the concentration is found to be 0.001 ml/ml of plasma after equilibration, then:

$$
\begin{aligned}
\text{Total body water} &= \frac{40}{0.001} \\
&= 40,000 \text{ ml} \\
&= 40 \text{ liters}
\end{aligned}
$$

The ideal substance for measuring the volume of the extracellular fluid compartment would distribute quickly and uniformly outside the cells and in the plasma. Unfortunately, there is no such substance, hence the exact and true volume of the extracellular compartment has never been measured. A reasonable estimate can, however, be determined from the volume of distribution of some substances by applying the indicator dilution principle. These substances can be divided into polysaccharides, such as inulin, mannitol, and sucrose, and ions, such as radioactive sodium, radioactive chloride, radioactive sulfate, thiocyanate, and thiosulfate. A disadvantage of polysaccharides is that they diffuse slowly from plasma into interstitial fluid due to their large molecular size. A disadvantage of ions is that they can enter the cells and thus give an overestimate of the volume of the extracellular compartment.

Two substances often used to measure plasma volume are Evans blue, a dye that binds to serum albumin, and radioiodinated serum albumin. Plasma volume is measured from the distribution volume of albumin. If plasma volume and hematocrit are known, total blood volume can be calculated (Chap. 11).

There are no substances that can be distributed evenly and exclusively inside all cells when injected intravenously. Therefore, intracellular fluid volume cannot be measured directly, but can be calculated. The volume of the intracellular compartment is the difference between total body water and the volume of the extracellular fluid compartment.

Like intracellular fluid volume, interstitial fluid volume cannot be measured directly, but can be calculated. Interstitial fluid volume is the difference between extracellular fluid and plasma volumes.

Water Balance

Water balance represents the difference between water intake and water loss. The kidneys are the major sites of regulation of water output. They adjust water excretion to maintain normal body fluid volumes. At normal temperature and activity, approximately 60% of the daily loss of water is excreted in the urine (Table 24–2). When ambient temperature is significantly higher than normal room temperature, the amount of water lost by sweating during normal activity increases to more than 40% of total daily loss of water. During prolonged heavy exercise, more than 75% of the total daily loss of water may occur through sweating. In Table 24–2 **insensible loss** refers to the amount of water lost daily by evaporation from the skin and respiratory tract.

During heavy exercise, water loss increases in two significant ways. First, body temperature rises and the rate of sweating increases the water loss by this route as much as 50-fold. Second, exercise increases the rate of breathing and consequently the insensible water loss through the respiratory tract increases in proportion to the rate of ventilation. Under both conditions of elevated extrarenal water loss, renal water loss decreases to compensate for the increased sweating and insensible water losses (Table 24–2).

■ Antidiuretic Hormone Is Critical in the Control of Renal Water Reabsorption and Plasma Osmolality

Antidiuretic hormone (ADH) is a polypeptide formed in the supraoptic and paraventricular nuclei of the hypothalamus. It is then transferred to and stored in large secretory granules in nerve endings in the posterior pituitary gland. Figure 24–2 illustrates this anatomic arrangement.

Plasma osmolality is the major stimulus that initiates the release of stored ADH. When osmolality rises, **osmoreceptor cells** in the hypothalamus shrink and excite the supraoptic and paraventricular nuclei to send impulses through the hypothalamohypophysial tract to the posterior pituitary, causing ADH release. Antidiuretic hormone acts on the collecting ducts of the kidney and causes increased water reabsorption (Chap. 23). The result is an increase in urine osmolality and decrease in plasma and extracellular fluid osmolality. Conversely, when a person drinks a large amount of water, plasma osmolality decreases and ADH release is suppressed. In the absence of ADH, the collecting

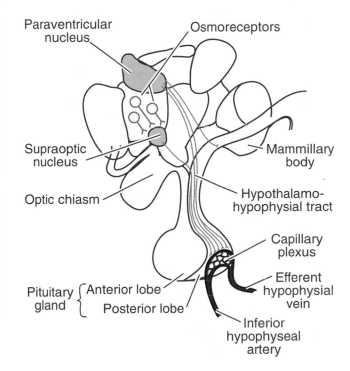

Figure 24–2 ■ Sagittal section through the pituitary and anterior hypothalamus. Antidiuretic hormone (ADH) is formed primarily in the supraoptic nucleus and to a lesser extent in the paraventricular nucleus of the hypothalamus. It is then transported down the hypothalamohypophysial tract and stored in secretory granules in the posterior pituitary, where it can be released into the blood.

TABLE 24–2 ■ Daily Loss of Water (ml)

	Normal Temperature and Normal Activity	High Temperature and Normal Activity	Prolonged and Heavy Exercise
Urine	1400	1200	500
Sweat	100	1400	5000
Feces	100	100	100
Insensible losses	700	600	1000
Total	2300	3300	6600

ducts are impermeable to water. Thus, the kidneys excrete the excess water in a dilute urine. The increased water excretion increases plasma osmolality back to normal levels. Figure 24-3 shows the relationship between plasma ADH concentration and plasma osmolality.

Plasma ADH is also influenced by changes in blood volume (Fig. 24-4). Loss of more than 10% of normal blood volume produces a striking increase in plasma ADH to levels much higher than those needed to produce a maximally concentrated urine. Antidiuretic hormone (vasopressin) also has blood pressure-raising properties when present at very high concentrations, since it constricts blood vessels; with massive blood loss these pressor effects help compensate for a low blood pressure.

The receptors that sense changes in blood volume are diffuse. A decrease in pressure is sensed by arterial

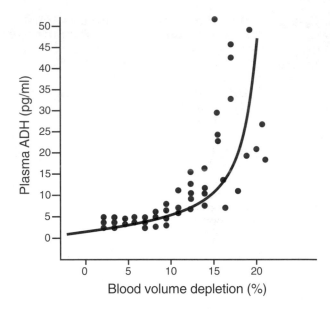

Figure 24–4 ■ Relationship between plasma antidiuretic hormone (ADH) and blood volume depletion in the rat. Note that a large decrease in blood volume causes a striking increase in plasma ADH. High plasma levels of ADH have a vasoconstrictor effect and help counteract a fall in blood pressure. (Modified from Dunn FL, Brennan TJ, Nelson AE, Robertson GL: *J Clin Invest* 52:3212-3219, 1973.)

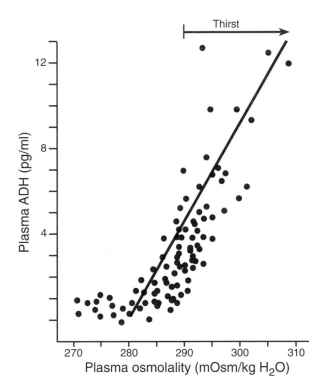

Figure 24–3 ■ Relationship between plasma antidiuretic hormone (ADH) level and plasma osmolality in normal humans. Decreases in plasma osmolality were produced by drinking water and increases by dehydration. Plasma ADH levels were measured by radioimmunoassay. At plasma osmolalities below 280 mOsm/kg H_2O, plasma ADH is decreased to low or undetectable levels. Above this threshold, plasma ADH increases linearly with plasma osmolality. Normal plasma osmolality is about 285-287 mOsm/kg H_2O; consequently, changes in plasma osmolality result in increases or decreases of plasma ADH and concomitant changes in urine osmolality. Note that the thirst threshold is attained at plasma osmolality of 290 mOsm/kg H_2O; the thirst mechanism "kicks in" only when there is an appreciable water deficit. (Modified from Robertson GL, Aycinena P, Zerbe RL: Neurogenic disorders of osmoregulation. *Am J Med* 72:339-353, 1982.)

baroreceptors (carotid sinus, aortic arch) and stimulates ADH release. Decreased stretch of the left atrium and pulmonary veins within the pericardium, by way of the vagal afferents, also reflexly stimulates ADH release. A decrease in renal blood flow stimulates renin release and increases angiotensin II production; angiotensin II then stimulates ADH release by acting on the brain. Conversely, an increase in blood volume inhibits ADH release. If blood volume is shifted to the thorax (e.g., lying down, peripheral vasoconstriction in cold weather, whole-body water immersion), stretch receptors in the atria signal the hypothalamus to release less ADH. The result is excretion of a large volume of dilute urine, a **water diuresis.**

In some situations, blood volume may be normal or greater than normal (e.g., congestive heart failure) and yet plasma ADH levels are elevated due to a decrease in effective arterial (circulating) blood volume. Effective arterial blood volume is decreased whenever cardiac output is inadequate to meet the metabolic requirements of the tissues. In this situation, high plasma ADH levels cause the kidneys to save water by forming an osmotically concentrated urine. Thirst is also experienced because of the inadequate effective arterial blood volume. The result is that water is taken in but is conserved by the kidneys, so plasma osmolality (and sodium concentration) falls below normal.

In severe **central diabetes insipidus,** characterized by the complete absence of ADH, no decrease in blood volume can be detected despite an increase in

HYPONATREMIA

Hyponatremia, defined as a plasma sodium concentration less than 130 mEq/L, is the most common disorder of body fluid and electrolyte balance in hospitalized patients. It is usually associated with hypoosmolality, which is not surprising since sodium salts are the major extracellular osmoles. Hyponatremia can also occur with a normal or elevated plasma osmolality.

Drinking even large quantities of water (20 L/day) rarely produces frank hyponatremia because of the large capacity of the kidneys to excrete dilute urine. If, however, plasma ADH level for some reason is not reduced when plasma osmolality is decreased or if the capacity of the kidneys to dilute the urine is impaired, then hyponatremia may ensue even with normal water intake.

Hyponatremia with hypoosmolality can occur in the presence of a decreased, normal, or even increased total body sodium. Hyponatremia and decreased body sodium content may be seen with increased sodium loss, such as with vomiting, diarrhea, burns, and diuretic therapy. The decreased extracellular fluid volume that occurs stimulates the thirst mechanism (increased water intake) and release of ADH (decreased water excretion), resulting in relatively more water than sodium in the plasma. Hyponatremia and a normal body sodium content are seen in hypothyroidism, glucocorticoid deficiency, and the syndrome of inappropriate antidiuretic hormone secretion (SIADH). SIADH may be caused by a tumor that secretes ADH independent of plasma osmolality; the result is renal conservation of water and dilution of the body fluids. Hyponatremia and increased total body sodium are seen in edematous states, such as congestive heart failure, hepatic cirrhosis, and nephrotic syndrome. In these disorders, a decrease in effective circulating blood volume reflexly stimulates ADH release and thirst. Excretion of solute-free water may also be impaired because of decreased delivery of tubule fluid to diluting sites along the nephron. More water than sodium is added to the body, producing hyponatremia.

Hyponatremia and hypoosmolality can cause a variety of symptoms, including muscle cramps, lethargy, fatigue, disorientation, headache, anorexia, nausea, agitation, hypothermia, seizures, and coma. These symptoms, mainly neurologic, are a consequence of swelling of brain cells as plasma osmolality falls. Excessive swelling of the brain may be lethal or may cause permanent brain damage. Treatment usually involves treating the underlying cause: giving sodium chloride to volume-depleted patients and restricting water intake in patients who have a normal blood volume or those who are edematous.

Hyponatremia should be corrected slowly and with constant monitoring, since too aggressive treatment can be harmful.

Hyponatremia in the presence of increased plasma osmolality is seen in hyperglycemic patients with uncontrolled diabetes mellitus. Glucose movement into cells is impaired, and the elevated plasma glucose concentration causes abstraction of water out of the intracellular compartment by osmosis. The plasma sodium is thereby diluted.

Hyponatremia and a normal plasma osmolality are seen with so-called **pseudohyponatremia.** This occurs when the plasma lipid or serum protein concentration is greatly elevated. These large molecules do not significantly increase plasma osmolality but may occupy a significant volume of plasma, resulting in a lower plasma sodium concentration. The concentration of sodium dissolved in plasma water is, however, normal.

urine volume as much as 10-fold. In these patients, water intake increases and compensates for the excretion of large volumes of dilute urine. If the patient is prevented from drinking, serious decreases in blood volume can result and circulatory shock might develop. In **nephrogenic diabetes insipidus,** plasma ADH levels are usually above normal but the collecting ducts are partially or completely unresponsive to ADH. An increase in water intake compensates for the excretion of large volumes of dilute urine. Acquired nephrogenic diabetes insipidus is relatively common whereas hereditary nephrogenic diabetes insipidus is rare.

The **syndrome of inappropriate ADH secretion** is characterized by the release of large amounts of ADH into the blood and consequent water retention. This leads to hypoosmolality of the body fluids and, since sodium is the main extracellular solute, a decrease in extracellular fluid sodium concentration. A reduced plasma sodium level (**hyponatremia**) usually represents the retention of excess water rather than loss of sodium.

■ Habit and the Thirst Mechanism Govern Water Intake

Drinking is largely by habit, by which we provide for future water needs. Most people drink enough that under ordinary circumstances they don't become truly thirsty. Thirst, a conscious desire to drink water, is mainly an emergency mechanism that comes into play when there is a perceived water deficit. The thirst mechanism can maintain water balance even in the absence of ADH, as in diabetes insipidus. The **thirst center** is appropriately located close to the supraoptic nuclei in the lateral preoptic area of the hypothalamus. When the osmolality of the extracellular fluid around neurons of the thirst center is high, the cells shrink. This sends a signal to the cerebral cortex, resulting in a desire to drink water. The mechanism of stimulation of the cells of the thirst center is similar to that of the osmoreceptor cells that trigger the secretion of ADH from the posterior pituitary gland.

The major stimulus initiating the sensation of thirst is intracellular dehydration. An increase in the osmolality of the extracellular fluid is the most common cause of intracellular dehydration; frequently this is due to a rise in extracellular sodium levels. High plasma levels of angiotensin II lead to increased thirst and drinking, as do hemorrhage and low cardiac output. Patients with low cardiac output have high circulating levels of angiotensin II and often experience thirst. The sensation of thirst is enhanced by dryness of the mouth, which can be induced by inhibition of saliva formation.

If either ADH or thirst fails to function, the other can often maintain a reasonably regulated extracellular fluid sodium concentration. If, however, both ADH and thirst are blocked, changes in sodium intake can produce significant variations in plasma sodium con-

centration. When both systems are intact, a 20-fold increase in sodium intake (from 10 to 200 mEq/day) produces an increase of only 2% in plasma sodium concentration (from 140 to 143 mEq/L). When both fail, the same increase of sodium intake produces an 11% increase in the plasma concentration of sodium (from 137 to 154 mEq/L).

Sodium Balance

■ Sodium Balance Is Maintained by the Regulation of Sodium Intake and Excretion

Sodium is the most abundant positive ion of the extracellular fluid compartment (including plasma) (Table 24-1). The osmotic pressure in the extracellular fluid compartment is mainly determined by sodium. Therefore, the concentration of the sodium ion is an important determinant of the osmolality of both the intracellular and extracellular compartments, since they are in osmotic equilibrium.

The amount of sodium in the extracellular fluid is determined by a balance between sodium intake and sodium excretion. Extrarenal losses of sodium can be significant under certain circumstances, such as during severe and prolonged exercise with profuse sweating and during massive vomiting and/or diarrhea. Under normal circumstances, however, extrarenal sodium losses are not significant. Therefore, for practical purposes, urinary sodium excretion represents total sodium excretion. In a healthy human, the daily intake of sodium is approximately equal to urinary sodium excretion. Properly functioning kidneys, when challenged by volume expansion or increased sodium load, respond by **natriuresis** (an increase in sodium excretion) and **diuresis.** When faced by volume depletion or a reduced sodium intake, the kidneys respond with antinatriuresis and antidiuresis. In certain pathophysiologic states, the kidneys fail to regulate sodium excretion properly, resulting in a state of volume depletion, at one extreme, or edema, at the other. Table 24-3 lists some of the physiologic and pathophysiologic conditions often associated with high or low urinary sodium excretion.

Most people consume more salt than they need. A **sodium appetite,** a craving for salt, is seen in salt-deprived humans and animals. This mechanism is analogous to thirst. Animals that have been depleted of salt and then given access to different salt solutions choose the one that is appropriate for repletion of their salt stores and maintenance of their sodium balance. Humans with Addison's disease have a depleted sodium store because their adrenal cortices do not secrete sufficient aldosterone. These patients crave salt and usually consume salty foods, thus maintaining a fairly normal extracellular sodium concentration and extracellular fluid volume.

TABLE 24–3 ■ Conditions or Physiologic States Associated with Natriuresis and Antinatriuresis

Natriuresis
  Volume-expanded states
    High sodium intake
    Syndrome of inappropriate antidiuretic hormone secretion
  Volume-depleted states
    Addison's disease (adrenocortical insufficiency)
    Renal salt wasting
    Diuretic abuse
Antinatriuresis
  Edematous states
    Heart failure
    Constrictive pericarditis
    Chronic liver disease
    Nephrotic syndrome
    Acute glomerulonephritis
    Idiopathic edema
  Nonedematous states
    Hemorrhage
    Low sodium intake
    Diuretic withdrawal
    Acute mineralocorticoid administration
    Nonrenal sodium losses (diarrhea, vomiting, profuse sweating)

■ Many Factors Influence Renal Sodium Excretion

Under steady-state conditions, the kidneys excrete the same amount of sodium as is ingested and thus maintain a constant level of total sodium in the body. Normally functioning kidneys can vary sodium excretion in direct proportion to sodium intake over a wide range. When sodium intake is low, the kidneys markedly reduce sodium excretion. When sodium intake is high, glomerular filtration rate increases and tubular sodium reabsorption decreases. The result is increased sodium excretion and the maintenance of normal sodium balance. Although the kidneys can detect a change in body sodium or water status, several days are required to reestablish sodium balance, whereas water balance can be regulated in minutes to hours. In normal subjects, the kidneys maintain sodium balance and extracellular fluid compartment volume within a narrow range despite significant and wide varying ranges of salt and water intake or extrarenal losses. Many mechanisms work together homeostatically to control sodium and water reabsorption. These include hemodynamic forces in the kidney, intrarenal physical forces, and hormonal and neural factors. These mechanisms are summarized in Table 24–4.

Renal Hemodynamic and Physical Factors ■ Variations in renal perfusion pressure result in changes in sodium and water excretion. Acute increases in renal perfusion pressure produce increases in sodium and water excretion (**pressure natriuresis** and **pressure diuresis,** respectively), even with no detectable changes in glomerular filtration rate or renal blood flow, due to reductions in tubular sodium and water reabsorption. The pressure natriuresis and diuresis responses may play an important role in regulation of

TABLE 24–4 ■ Factors Affecting Tubular Sodium Reabsorption

Hemodynamic and physical factors
  Medullary blood flow
  Renal perfusion pressure
  Peritubular capillary Starling forces
Hormonal factors
  Renin-angiotensin-aldosterone system
  Atrial natriuretic peptide
  Renal prostaglandin system
  Renal kinin system
Renal sympathetic nerve activity

sodium and water excretion, extracellular fluid volume, and arterial pressure. Elevations in renal perfusion pressure cause a decrease in renin release, an increase in kinin and prostaglandin release, and an increase in renal interstitial hydrostatic pressure; these changes contribute to the natriuresis.

The balance of Starling forces (intrarenal physical forces) is responsible for the transport of sodium and water across the peritubular capillary wall. Increases in peritubular capillary hydrostatic pressure and/or decreases in colloid osmotic pressure decrease the absorption of sodium and water, whereas decreases in peritubular capillary hydrostatic pressure and/or increases in colloid osmotic pressure favor the absorption of sodium and water. The net pressure in the peritubular capillaries is about 10 mm Hg in favor of uptake of reabsorbed fluid.

Under certain conditions, changes in sodium excretion are associated with alterations in Starling forces. During volume expansion with isotonic saline, there is a decrease in peritubular capillary colloid osmotic pressure (due to reduced plasma protein concentration) and an increase in peritubular capillary hydrostatic pressure, renal interstitial hydrostatic pressure, and renal sodium excretion. Administration of vasodilators, such as acetylcholine and bradykinin, causes an increase in sodium excretion associated with an elevated renal interstitial hydrostatic pressure.

The Renin-Angiotensin-Aldosterone System ■ The renin-angiotensin-aldosterone system plays an important part in the renal regulation of sodium homeostasis. Renin is stored in and released from granules in the juxtaglomerular cells (Fig. 23–4). When released, renin acts on a plasma protein (angiotensinogen or renin substrate) and splits off a decapeptide, angiotensin I. When angiotensin I circulates through the lungs, two amino acids are split off in a reaction catalyzed by the angiotensin converting enzyme, to form angiotensin II (Fig. 24–5).

The components and possible pathways by which the renin-angiotensin-aldosterone system increases tubular sodium reabsorption are illustrated in Figure 24–5. Angiotensin II increases sodium reabsorption through a direct stimulatory effect on tubular sodium transport and its renal hemodynamic action as a vasoconstrictor. It also stimulates aldosterone secretion

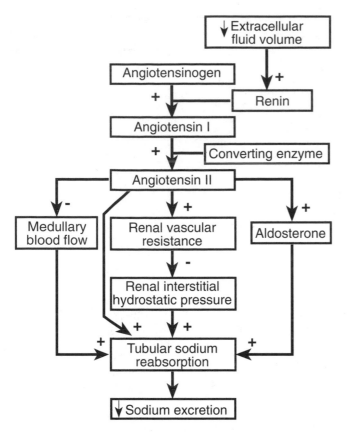

Figure 24–5 ■ Mechanisms whereby the renin-angiotensin-aldosterone system affects sodium excretion. (Modified from Knox FG, Granger JP: Control of sodium excretion: The kidney produces under pressure. *News Physiol Sci* 2:26–29, 1987.)

from the zona glomerulosa of the adrenal cortex. Aldosterone acts on the collecting ducts to increase sodium reabsorption. In addition, angiotensin II promotes thirst and drinking and thus has an effect on fluid intake. It causes an increase in renal vascular resistance and consequently a decrease in renal blood flow. This is accompanied by a fall in glomerular filtration rate and a reduction in renal interstitial hydrostatic pressure; these changes result in decreased sodium excretion.

The release of renin and the formation of angiotensin are often associated with states of volume or sodium depletion, such as hemorrhage and low sodium intake. Reductions in renal perfusion pressure can trigger the release of renin and the formation of angiotensin. For example, renal artery stenosis leads to increased levels of renin and angiotensin in the plasma.

Atrial Natriuretic Peptide ■ Atrial natriuretic peptide (ANP) plays an important role in sodium balance and regulation of extracellular fluid volume. It is synthesized and stored in atrial cells. Expansion of extracellular and intravascular fluid volumes distends the atria, stimulating ANP release. Atrial natriuretic peptide causes relaxation of vascular smooth muscles,

suppression of the renin-angiotensin-aldosterone system, and a brisk natriuresis and diuresis. The end result is a decrease in plasma volume, systemic vascular resistance, mean arterial pressure, and cardiac output. Possible pathways through which ANP accomplishes these effects are illustrated in Figure 24–6.

Pathophysiologic states associated with increased levels of plasma ANP are listed in Table 24–5. Plasma levels of ANP are elevated in patients with impaired ventricular function and mitral valve disease. In patients with coronary heart disease or cardiomyopathy, a linear relationship is found between the ANP level in the plasma and left ventricular filling pressure or mean pulmonary artery pressure. Patients with elevated cardiac filling pressure, with or without congestive heart failure, have elevated plasma levels of ANP. Plasma levels of ANP are also elevated in patients with spontaneous tachyarrhythmias, hypertension and cardiac hypertrophy, and chronic renal failure.

Renal Prostaglandin and Kinin Systems ■ Prostaglandins are arachidonic acid derivatives that are present in almost all tissues. Prostaglandin E_2 (PGE_2) plays an important role in increasing sodium and water excretion in states of elevated renal perfusion pressure. Renal production of PGE_2 is elevated during acute volume expansion, and it has a direct inhibitory effect on renal tubule sodium transport.

Kinins, small peptides that can cause vasodilation,

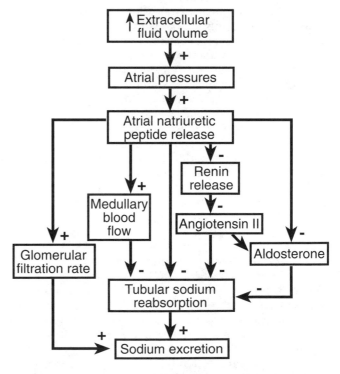

Figure 24–6 ■ Mechanisms whereby atrial natriuretic peptide affects sodium excretion. (Modified from Knox FG, Granger JP: Control of sodium excretion: The kidney produces under pressure. *News Physiol Sci* 2:26–29, 1987.)

TABLE 24–5 ■ Conditions or Pathophysiologic States Associated with Elevations in Plasma Levels of Atrial Natriuretic Peptide

Impairment of ventricular function with elevated atrial pressure
Coronary heart disease with ventricular dysfunction
Cardiomyopathy
Valvular heart disease
Hypertension with cardiac hypertrophy
Spontaneous tachyarrhythmias with increased atrial pressure
Chronic renal failure with volume overload
Conn's syndrome (primary hyperaldosteronism)

are found in many tissues, including the kidneys. Bradykinin is a potent vasodilator that causes a significant increase in capillary permeability. The role of intrarenal kinins in the regulation of sodium excretion is controversial. Bradykinin, like most vasodilators, produces an increase in urinary sodium and water excretion.

Renal Sympathetic Nerves ■ Activation of the renal sympathetic nerves increases sodium reabsorption and decreases sodium excretion. Stimulation of the renal sympathetic nerves causes constriction of afferent arterioles and a consequent decrease in glomerular filtration rate. The result is a decrease in sodium excretion, because the tubules can reabsorb sodium more completely when the filtered load is reduced. Also, renal nerves act directly on the tubules to increase sodium reabsorption. In addition, activation of the sympathetic nervous system increases the release of renin and hence the formation of angiotensin II, which leads to sodium and water conservation.

Acute renal denervation produces an increase in sodium and water excretion. Chronically, however, no differences in sodium excretion are found between renal innervated and denervated animals fed a diet containing normal amounts of sodium. It appears that intact renal nerves are required for sodium conservation and balance during sodium restriction. Therefore, the renal nerves might not be important for the regulation of sodium when sodium intake is normal but become critical for the maintenance of sodium balance during sodium restriction.

Potassium Balance

Potassium is the most abundant positive ion in the intracellular fluid compartment (Table 24–1). Intracellular and extracellular potassium concentrations usually (but not always) change in the same direction. Therefore, the discussion of potassium balance mainly focuses on extracellular fluid and plasma potassium concentration. Insulin, β-adrenergic agonists, pH, and bicarbonate concentration may influence the acute regulation of potassium in response to sudden changes in plasma potassium concentration. However, these mechanisms, which can affect potassium distribution in minutes, may not be important in long-term regula-

tion of potassium balance. Two well-recognized mechanisms appear to function in long-term regulation of renal potassium excretion and balance: plasma potassium concentration and aldosterone.

■ Plasma Potassium Directly Affects Potassium Excretion

About 90% of the filtered load of potassium is reabsorbed by the proximal convoluted tubule and loop of Henle. Therefore, by the time the tubular fluid reaches the distal convoluted tubule, only about 10% of the original filtered load of potassium remains. The fine-tuning of potassium excretion occurs in the cortical collecting ducts, where potassium is secreted into the tubular fluid.

Figure 24–7 shows the relationship between plasma potassium concentration and potassium excretion observed without alterations in aldosterone, sodium intake, or other variables known to be involved in the regulation of potassium excretion. When plasma potassium is increased above the average normal level (about 4.2 mEq/L), potassium excretion is dramatically elevated; when the concentration is reduced below this level, the effect on potassium excretion is weak. Increasing plasma potassium concentration is estimated to be about three times as effective as aldosterone in increasing potassium excretion in response to elevations in potassium intake. Increases in extracellular potassium concentration may have a direct effect

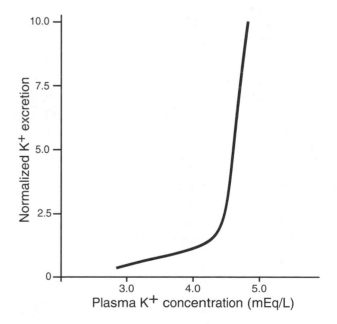

Figure 24–7 ■ Renal potassium excretion plotted as a function of plasma potassium concentration under steady-state conditions. The animals used in this study were adrenalectomized and maintained on normal levels of intravenously infused adrenal cortical hormones. (Modified from Young DB: Quantitative analysis of aldosterone's role in potassium regulation. *Am J Physiol* 255:F811–F822, 1988.)

on cortical collecting duct cells to increase basolateral uptake of potassium. Elevated levels of Na$^+$/K$^+$ ATPase may be responsible for the increased basolateral uptake of potassium. The augmented uptake can lead to increased intracellular potassium and elevation of potassium secretion into the tubular lumen, resulting in increased potassium excretion.

■ A Potent Aldosterone Negative Feedback System Regulates Potassium in the Extracellular Fluid

Aldosterone is a major component in the chronic regulation of potassium balance. Although the plasma potassium concentration response is more effective than aldosterone in altering potassium excretion, aldosterone is the only factor that can maintain a simultaneous balance of potassium and sodium over a wide range of intake of both ions.

Aldosterone is secreted from the zona glomerulosa of the adrenal gland in response to increases in extracellular potassium levels. Angiotensin II, which itself increases aldosterone secretion, also enhances the relationship between plasma potassium concentration and aldosterone secretion. Aldosterone acts on the cortical collecting ducts to increase potassium secretion into the tubular fluid and thus increase potassium excretion. Aldosterone may also cause a shift of potassium ions from the extracellular to the intracellular fluid compartment.

When the adrenal glands are destroyed, aldosterone and other adrenal hormones disappear from the body and **Addison's disease** develops. Unless aldosterone or another mineralocorticoid is administered, potassium levels increase markedly, ultimately precipitating cardiac arrest. By contrast, a tumor in the zona glomerulosa can lead to the secretion of large quantities of aldosterone, a condition known as **primary aldosteronism.** Plasma potassium concentrations decrease in some patients to the point of hyperpolarization of nerve membranes, leading to paralysis.

Calcium, Magnesium, and Phosphate Balance

The kidneys are the major regulatory organs of calcium balance. Calcium intake is approximately 900 mg/day and mainly comes from milk and milk products in the diet. Calcium is poorly absorbed by the intestine; about 750 mg/day is excreted in the feces. The remaining calcium (~150 mg/day) is excreted in the urine. Renal regulation of calcium is critical in maintaining plasma calcium levels within a narrow range (4.5–5.0 mEq/L).

About 40% of plasma calcium is bound to plasma proteins (mainly serum albumin) and therefore is not filtered through the glomerular capillaries. The remaining 60% consists of ionized calcium, which is physiologically active and represents about 50% of total plasma calcium, and un-ionized calcium, which

represents about 10% of total plasma calcium and is complexed to anions such as citrate, phosphate, bicarbonate, and sulfate.

Ionized calcium and the diffusible calcium complexes are filtered at the glomerular membrane. Of the 100% of the filtered load of calcium in Bowman's space, only 0.5–2.0% remains in the urine. The percentages of filtered calcium left in the tubule fluid in each segment of the nephron are illustrated in Figure 24–8. Like sodium, the majority of the filtered load of calcium (approximately 60%) is reabsorbed in the proximal tubule. About 20% of the filtered calcium is reabsorbed by the thick ascending limb of the loop of Henle and about 5–10% is reabsorbed in the distal tubule. This reabsorption is active and can occur against an electrochemical gradient. Under normal conditions, about 5% of the filtered load of calcium is reabsorbed in the collecting duct by passive diffusion. Abstraction of water out of the lumen causes an increase in luminal calcium concentration and the consequent diffusion of calcium ions out of the collecting ducts.

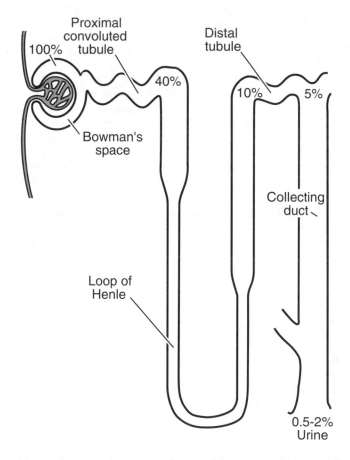

Figure 24–8 ■ Profile of the percentages of the filtered load of calcium that remains as the tubular fluid moves through the different segments of the nephron. Approximately 0.5–2.0% of the filtered load is excreted in the urine.

■ Parathyroid Hormone Decreases Renal Excretion of Calcium

The parathyroid glands respond within minutes to slight changes in extracellular fluid calcium concentration. A decrease in the ionized calcium ion concentration of the extracellular fluid increases the rate of secretion of parathyroid hormone. During pregnancy or lactation there is a slight, chronic decrease in extracellular calcium, which leads to hypertrophy of the parathyroid glands and an increase in parathyroid hormone synthesis and secretion. When the ionized calcium ion concentration of the extracellular fluid increases, parathyroid hormone secretion is reduced. Such conditions may be encountered during an increase in calcium or vitamin D intake and during bone resorption caused by disuse of the bones; the latter is seen in prolonged bedrest or exposure to weightlessness in space.

The numerous actions of parathyroid hormone are discussed in Chapter 38; here we note that parathyroid hormone stimulates calcium reabsorption in the thick ascending limb and distal tubule, thus decreasing urinary calcium excretion. Parathyroid hormone exerts its intracellular effects via the second messenger system, cyclic AMP. Figure 24-9 illustrates parathyroid hormone's renal effects with elevated plasma calcium levels. Note that parathyroid hormone is not the only regulator of calcium balance. Calcitonin is a hormone produced by the thyroid gland that, among other actions, affects renal reabsorption of calcium. Most re-

cent reports suggest that the physiologic role of calcitonin on the nephron is to conserve calcium. However, other reports indicate that calcitonin decreases renal reabsorption of calcium. It is believed that calcitonin acts as an acute regulator of extracellular calcium concentrations; its chronic effects appear to be negligible. The calcitonin mechanism can be overridden by the very powerful parathyroid hormone mechanism, which ultimately regulates extracellular calcium concentration. Surgical removal of the thyroid glands halts calcitonin secretion, yet no significant changes in extracellular calcium ion concentration are observed. The production and actions of calcitonin and other regulators of calcium balance are more fully discussed in Chapter 38.

■ Magnesium Is Reabsorbed in the Proximal Convoluted Tubule and Henle's Loop

Similar to calcium, magnesium is found in the plasma as a free divalent ion (Mg^{2+}), in complexes with substances such as phosphate, oxalate, and sulfate, or bound to plasma proteins. The normal magnesium concentration in humans is 1.7–2.3 mEq/L. Approximately 80% of plasma magnesium is filterable through the glomerular membrane (about 60% as the ion and 20% in complexed form). The remaining 20% is bound to plasma proteins and is not filterable. The kidneys filter approximately 2 g of magnesium per day. Approximately 95% of the filtered amount is reabsorbed (1.9 g) and approximately 5% is excreted in the urine.

In the intracellular fluid compartment, magnesium is the second most abundant cation, after potassium (Table 24-1). An average human body contains about 24 g of magnesium (about 2000 mEq), the majority of which is sequestered in bones and the intracellular fluid compartment (about 1980 mEq). Only about 15–20 mEq is found in the extracellular fluid. This small amount is constantly filtered and mainly regulated by the kidneys. Some similarities between magnesium and calcium in terms of their distribution in the body, the form in which they exist, and their regulation are apparent.

About 25% of the filtered load of magnesium is reabsorbed in the proximal convoluted tubule. Overall, there is far less magnesium reabsorbed here than sodium, potassium, calcium, or water. The proximal tubule epithelium is rather impermeable to magnesium under normal conditions, so there is little passive movement of that ion. The major site of magnesium reabsorption is the loop of Henle (mainly the thick ascending limb) (Fig. 24-10), which reabsorbs about 65% of the filtered magnesium. Reabsorption in the loop is both active and passive; passive reabsorption occurs between the cells of the thick ascending limb and is driven by the lumen-positive transepithelial potential difference. Changes in magnesium excretion result mainly from changes in loop transport. The distal

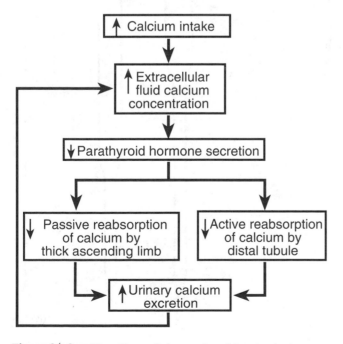

Figure 24–9 ■ The effects of changes in calcium intake on extracellular fluid calcium concentration and the negative feedback that regulates this concentration through alterations in urinary calcium excretion.

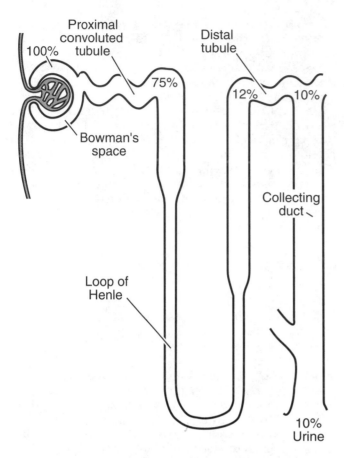

Figure 24–10 ■ Profile of the percentages of the filtered load of magnesium remaining in the tubular fluid as it moves through the different segments of the nephron. The loop of Henle, specifically the thick ascending limb, is the major site of reabsorption of filtered magnesium.

convoluted tubule and collecting duct system reabsorb a small fraction of filtered magnesium and under normal circumstances appear to play a minor role in controlling magnesium excretion.

■ Phosphate Excretion Is Regulated Primarily by the Proximal Tubules

Of the average phosphorus consumption of 1 g/day, approximately 70% (about 700 mg/day) is absorbed from the gastrointestinal tract and the rest (about 300 mg/day) is excreted in the feces. Under normal steady-state conditions, the kidneys filter 6 g/day of phosphate and reabsorb 5.3 g/day. The amount that appears daily in the urine is equal to the amount absorbed by the intestine (700 mg/day). Therefore, in normal adults the net daily phosphate balance is usually zero.

In body fluids, phosphate occurs in either organic or inorganic form. Most of the phosphate in cells is organic. Plasma contains lipid phosphates, organic ester phosphates, and inorganic phosphates (Pi). In plasma, Pi exists in two ionic forms, divalent (HPO_4^{2-}) and monovalent ($H_2PO_4^-$) phosphate. At a normal

plasma pH of 7.4, about 80% of Pi is in the divalent form and 20% is in the monovalent form. The normal concentration of plasma Pi is maintained within narrow limits of 3.0–4.5 mg/100 ml (~1.7–2.6 mEq/L) in adults and 4.0–5.0 mg/100 ml (~2.3–2.9 mEq/L) in children.

Plasma Pi is largely filtered, and 5–20% of the filtered load remains in the final urine (Fig. 24–11). With normal dietary phosphate intake and intact parathyroid glands, 80% of Pi is reabsorbed as the tubular fluid moves from Bowman's capsule to the terminal part of the collecting ducts. Figure 24–11 shows the percentages of the filtered load of Pi that remain in each segment of a superficial nephron. The proximal convoluted tubule is the major site of Pi reabsorption; it reabsorbs 60–70% of the filtered load. The loop of Henle does not appear to play an important part in Pi reabsorption. Between the beginning and end of the loop of Henle, the amount of Pi in the tubular fluid falls by about 5%. Under normal conditions, this reabsorption of 5% of the filtered load occurs in the proximal straight tubule. Between the thick ascending limb of the loop of Henle and the terminal end of the

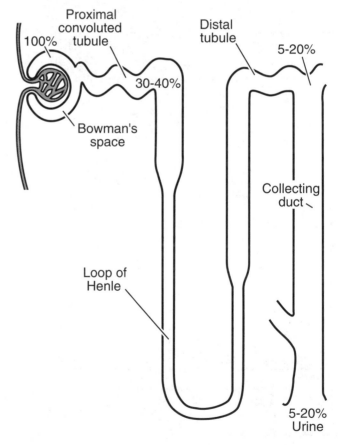

Figure 24–11 ■ Profile of the percentages of the filtered load of phosphate remaining in the tubular fluid as it moves through different segments of the nephron. Approximately 5–20% of the filtered load is excreted in the urine.

collecting duct, a minute amount of Pi is reabsorbed (~3% of the filtered load). The process of Pi transport is mainly unidirectional reabsorption with negligible backflux and essentially no secretion.

Transport Maximum and Phosphate Excretion ■ The renal tubule Pi transport system plays a major role in determining plasma Pi level. Inorganic phosphate reabsorption is precisely regulated according to the body's needs. Transport is characterized by a maximum rate (Tm) at which the ion is reabsorbed.

Figure 24–12 illustrates the relationship between the filtered load of Pi and the amount reabsorbed and excreted. At low plasma Pi concentrations and hence low filtered Pi loads, essentially all of the filtered Pi is reabsorbed by the kidney tubules. Above threshold, some of the filtered phosphate escapes in the urine. With high filtered Pi loads, the rate of reabsorption reaches a Tm of about 0.1 mmol/min. This means that a maximum of 0.1 mmol/min of filtered Pi is reabsorbed; any additional filtered Pi is excreted. With a normal dietary intake of phosphate, the filtered load of Pi is above the threshold level and Tm remains constant. Under these conditions, plasma Pi level is maintained at about 1 mmol/L.

Parathyroid Hormone and Phosphate Transport Maximum ■ Parathyroid hormone inhibits tubular Pi reabsorption in the proximal tubule and increases Pi excretion. In thyroparathyroidectomized (TPTX) animals, parathyroid hormone is absent and reabsorption of Pi in the proximal tubule increases

significantly, leading to a decrease in urinary Pi excretion and an increase in plasma Pi levels. In patients with primary hyperparathyroidism, parathyroid hormone secretion is elevated and plasma levels of Pi are low; however, steady-state urinary Pi excretion is not markedly increased, being dependent largely on alimentary Pi absorption.

Dietary Pi restriction leads to complete disappearance of Pi from the urine. This is associated with an increased phosphate Tm and reabsorption of almost 100% of the filtered load of Pi. Parathyroid hormone does not appear to play a major role, since TPTX animals and hypoparathyroid patients also adapt to dietary Pi restriction. Conversely, the reabsorption of Pi is lower in animals maintained on a high-phosphate diet. The increased urinary Pi excretion in response to a high-phosphate diet is related to a decrease in Pi Tm.

The Ureter and Urinary Bladder and the Micturition Reflex

Urine transport to the bladder via the ureters begins with electrical activity at pacemaker sites in the pelvis of each kidney. This initiates ureteral peristaltic contractions that propel the urine from the renal pelvis to the bladder. The ureteral peristaltic wave has a velocity of about 3 cm/sec and occurs from once every 10 seconds to once every 2–3 minutes. The result is movement of urine into the bladder, where it enters in spurts in synchrony with each peristaltic contraction.

Parasympathetic stimulation increases and sympathetic stimulation decreases the frequency of ureteral peristaltic contractions. The ureters are supplied with abundant pain fibers. Blockade of a ureter by stones leads to an intense reflex constriction of the ureter associated with severe pain. The pain may initiate a sympathetic reflex back to the kidney (**ureterorenal reflex**) that ultimately causes a reduction in urine output. This is an important mechanism to reduce urine flow significantly when a ureter is blocked.

The ureters enter the bladder obliquely through the **trigone,** a small triangular area near the bladder neck; the urethra leaves the bladder at the bottom corner of the trigone. There is no sphincter at the junction of the ureters and bladder; however, the oblique penetration of the ureters through the trigone and the compression on the ureters caused by the pressure in the bladder around the area of penetration keep the ureters collapsed except when peristaltic waves occur. This anatomic arrangement also prevents backflow of urine into the ureters when pressure builds in the bladder during **micturition** (urination). The proper and efficient transport of urine from the ureter into the bladder requires a relatively low intravesical pressure during the phase of bladder filling.

During storage, the bladder adapts to the increasing

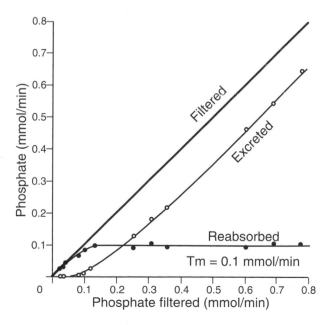

Figure 24–12 ■ Reabsorption and excretion of phosphate as functions of filtered load. A transport maximum (Tm) for phosphate is observed at high filtered phosphate loads. (Modified from Pitts RF: *Physiology of the Kidney and Body Fluids.* 3rd ed. Chicago, Year Book, 1976, p 78.)

volume of urine by maintaining a low intravesical pressure, while the urethra maintains a pressure higher than that in the bladder. The bladder does this by using its plasticity (or "stress-relaxation"): when it is stretched, the tension that is initially produced is not maintained. When the bladder is well filled, its pressure rises; with voiding, a drop in intraurethral pressure precedes the elevation in intravesical pressure and urine flows out until the bladder is emptied.

Bladder Distension Initiates a Stretch Reflex

Contraction of the bladder muscle, called the **detrusor** muscle, is primarily responsible for emptying the bladder during micturition. When the urine volume exceeds 400 ml, the radius of the bladder hardly increases, while the wall tension keeps rising. The result is a sharp increase in intravesical pressure. This activates stretch receptors in the bladder wall, resulting in a strong sense of fullness and urge to urinate.

The bladder is well supplied with sensory, parasympathetic, and sympathetic nerves (Fig. 24–13). The dominant neural inputs are the parasympathetic fibers to the bladder and bladder neck (posterior urethra). These nerve fibers originate at the level of the second through fourth sacral segments of the spinal cord and reach the bladder wall through the **pelvic nerves.** They synapse in the bladder wall, where postganglionic fibers innervate the bladder muscle cells. Sympathetic innervation of the bladder is via the **hypogastric nerves.** The external sphincter is innervated by somatic efferent fibers that arise from motor neurons in the anterior horn cells of the second through the fourth sacral segments of the spinal cord and travel via the **pudendal nerves.** Sensory fibers also originate in the bladder wall and posterior urethra and transmit information to the spinal cord about distension of the bladder and the presence of urine in the posterior urethra. These sensory fibers travel to the cord via pelvic and hypogastric nerves.

Complex Neural Mechanisms Influence the Micturition Reflex

Voluntary control of urination develops after birth. In children younger than 2–3 years of age, urination is a reflex that depends entirely on bladder stretch. In adults, micturition is mainly a spinal cord reflex that can be facilitated or inhibited by higher brain centers. When the intravesical pressure increases as a result of bladder filling, a stretch reflex is initiated by receptors in the bladder wall. Sensory signals travel to the sacral region of the spinal cord through the pelvic nerves and back to the bladder through the parasympathetic fibers of the same nerves. As the bladder fills and the intravesical pressure increases, bladder wall contractions occur more often and become more powerful. The first urge to void is felt when the urine volume in the bladder is about 150 ml. The normal urine

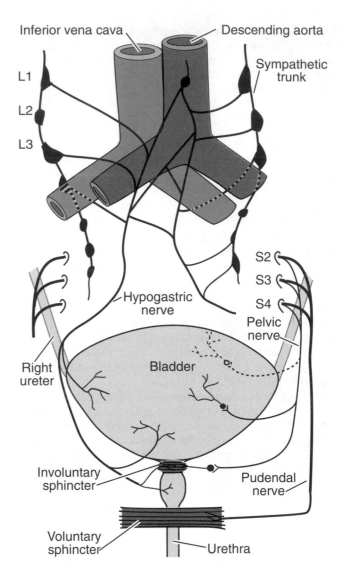

Figure 24–13 ■ The innervation of the urinary bladder. The parasympathetic pelvic nerves arise from the S2 to S4 segments of the spinal cord and supply motor fibers to the bladder musculature and internal (involuntary) sphincter. Sympathetic motor fibers supply the bladder via the hypogastric nerves. The pudendal nerves supply somatic motor innervation to the external (voluntary) sphincter. Sensory afferents (broken lines) from the bladder travel mainly in the pelvic nerves but also to some extent in the hypogastric nerves. (Modified from Anderson JE. Grant's Atlas of Anatomy. 8th ed. Baltimore: Williams & Wilkins, 1983.)

volume in the bladder that initiates a reflex contraction in an adult is about 300–400 ml.

When the intravesical pressure rises enough, a **micturition reflex** is initiated and the high pressure in the bladder forces the bladder neck (posterior urethra) to open. After opening, the stretch of the bladder neck itself initiates afferent signals that travel to the sacral segments of the spinal cord and inhibit neural signals over the pudendal nerves that normally maintain constriction of the external sphincter. If the inhibition reflex is stronger than the voluntary constrictor signals acting on the external sphincter from the brain, then urination occurs.

■ There Are Three Major Pathologic Causes of Abnormalities of Micturition

Due to the nature of the sacral micturition reflex and the inhibitory and facilitatory influences that can be exerted on the reflex by supraspinal centers, three major potential dysfunctions that can affect the bladder, and consequently the micturition reflex, can be predicted. These abnormalities can result from destruction of the afferent sensory nerves from the bladder to the spinal cord; destruction of both afferent and efferent nerves of the bladder; and interruption of inhibitory and facilitatory influences of brain stem and cerebral cortex centers on the micturition reflex.

Destruction of the afferent sensory nerves of the bladder interrupts the transmission of stretch signals from the bladder to the spinal cord, therefore impairing the micturition reflex. The bladder does not empty as frequently, it gets overfilled, and urine drips through the urethra a few drops at a time. The bladder is characterized by distension due to overfilling. It becomes thin-walled and hypotonic; however, some contractions related to the intrinsic ability of smooth muscle cells to respond to stretch remain. Crushing injuries to the sacral region of the spinal cord can cause this condition. Syphilis can cause constrictive fibrosis around the sacral dorsal roots, rendering the bladder afferent nerve fibers nonfunctional.

When both afferent and efferent nerves are de-stroyed, the bladder becomes acutely flaccid and distended. Over time, the muscle of the denervated bladder becomes active and contractions are restored. Bladder contractions can then expel urine through the urethra a few drops at a time. The denervated bladder shrinks and its wall becomes hypertrophied.

When the spinal cord is severed above the sacral region, the micturition reflex takes a few days to several weeks to become functional. It appears that the sudden loss of facilitatory signals from higher centers is responsible for the complete suppression of micturition reflexes when the spinal cord is transected. Between the time of spinal damage and the return of micturition reflex function, the bladder must be emptied periodically by urethral catheterization to prevent physical injury. It is interesting that some paraplegic patients can train themselves to initiate urination by scratching or pinching their thighs or stimulating the skin in the genital area.

Another micturition abnormality is the so-called **uninhibited neurogenic bladder** or **spastic neurogenic bladder,** characterized by hyperactivity of the **micturition reflex.** The bladder wall becomes hypertrophied and its holding capacity is reduced. This condition may be caused by damage in the spinal cord or brainstem of the tracts that carry the inhibitory signals to the sacral centers. The result is an imbalance between the supraspinal facilitatory and inhibitory signals to the sacral centers in favor of the facilitatory signals. Therefore, the micturition reflex remains excited and a very small amount of urine initiates micturition. Hyperactivity of the micturition reflex is enhanced by bladder wall infections.

■ ■ ■ ■ ■ ■ ■ ■ ■ ■ REVIEW EXERCISES ■ ■ ■ ■ ■ ■ ■ ■ ■ ■ ■

Identify Each with a Term
1. The hormone that is synthesized and stored in atrial cells
2. The hormone that plays important roles in both sodium and potassium regulation
3. The nephron segment where phosphate reabsorption is regulated
4. The stimulus that initiates the stretch reflex and the urge to urinate

Define Each Term
5. Interstitial fluid volume
6. Hyponatremia
7. Uninhibited neurogenic bladder

Choose the Correct Answer
8. When measuring extracellular fluid compartment, if a substance that is very slow to diffuse into the interstitial fluid compartment is infused intravenously and a blood sample is taken for measurement of the concentration

of this substance before complete equilibrium is reached, the measurement would reflect:
   a. An underestimation of the volume of the extracellular fluid compartment
   b. An overestimation of the volume of the extracellular fluid compartment
   c. The exact volume of the extracellular fluid compartment
   d. None of the above
9. Which of the following volumes must be measured indirectly?
   a. Total body water
   b. Plasma volume
   c. Extracellular fluid volume
   d. Interstitial fluid volume
10. Which of the following statements is correct?
   a. Antidiuretic hormone (ADH) is secreted by the adrenal glands.
   b. The main site of action of ADH is the proximal tubule.
   c. Antidiuretic hormone causes an increase in sweating.
   d. When released, ADH acts on the collecting ducts to cause increased water reabsorption.

11. Which of the following increases thirst?
    a. Hypotonic volume expansion
    b. Increased plasma levels of atrial natriuretic peptide
    c. Cardiac failure
    d. Decreased levels of angiotensin II in plasma

12. Which of the following decreases sodium reabsorption?
    a. Increased renal vascular resistance
    b. Increased plasma aldosterone level
    c. Increased renal sympathetic nerve stimulation
    d. Exposure to zero gravity and weightlessness

13. Which of the following statements is correct?
    a. Aldosterone causes decreased sodium reabsorption in the cortical collecting duct.
    b. Angiotensin II causes a decrease in renal blood flow.
    c. The extracellular and intracellular fluid compartments are not in osmotic equilibrium.
    d. Renal artery stenosis usually leads to reduced activity of the renin-angiotensin system.

14. Which of the following statements is correct?
    a. During reductions in sodium intake, the renin-angiotensin system plays a critical role in maintaining a normal glomerular filtration rate.
    b. During cardiac failure, there is a significant decrease in sodium and fluid retention.
    c. Edema is defined as an excess in fluid accumulation in the intravascular compartment.
    d. Significant edema develops when the interstitial fluid pressure is negative.

15. Which of the following statements is correct?
    a. The fine-tuning of potassium excretion occurs in the proximal tubule.
    b. Potassium can be filtered or reabsorbed only in the kidney.
    c. The main site of action of aldosterone regarding potassium transport is the proximal tubule.
    d. Potassium is the major intracellular cation.

16. Which of the following statements is correct?
    a. Approximately 90% of plasma calcium is ultrafilterable.
    b. Calcium transport in the thick ascending limb is voltage-dependent.
    c. Under normal conditions, about 20% of the filtered load of calcium remains in the urine.
    d. Parathyroid hormone causes a decrease in the tubular reabsorption of calcium.

17. Which of the following statements is correct?
    a. The filtered load of phosphate is equal to the glomerular filtration rate times the concentration of phosphate (Pi) in whole blood.
    b. The proximal tubule is the major site of Pi reabsorption.
    c. The amount of Pi that remains in the urine is determined by tubular secretion of Pi.
    d. The loop of Henle is the major site of Pi reabsorption.

18. Which of the following statements is correct?
    a. The ureters are usually open even when peristaltic waves are not occurring.
    b. The ureters have a well-defined sphincter where they enter the bladder.
    c. Stretch receptors in the bladder wall initiate the micturition reflex.
    d. The sympathetic nerve fibers are largely responsible for micturition.

19. Which of the following statements is incorrect?
    a. Generally, the supraspinal centers keep the micturition reflex inhibited except when there is a desire to urinate.
    b. The denervated bladder loses all contractions forever.
    c. When the spinal cord is transected above the sacral region, the micturition reflex remains functional.
    d. Paraplegic patients can train themselves to initiate urination by scratching their thighs.

Calculate

The following data were collected from a normal adult human: extracellular fluid volume, 15 liters; intracellular fluid volume, 25 liters; total body water, 40 liters; osmolality of each compartment, 300 mmol/kg H_2O. Two liters of 3% (1020 mOsm/kg H_2O) sodium chloride solution was infused intravenously in this person. After osmotic equilibrium has been reached:

20. What is the extracellular fluid osmolality?

21. What is the intracellular fluid osmolality?

22. What is the extracellular fluid volume?

23. What is the total body water?

24. What is the intracellular fluid volume?

Briefly Answer

25. How does angiotensin II help restore sodium and fluid balance after significant hemorrhage?

26. Why are the atria a strategic location for storage of atrial natriuretic peptide?

27. What happens to plasma sodium concentration in hyperglycemic patients with uncontrolled diabetes mellitus?

28. What are the compliance characteristics of the urinary bladder and their benefits?

■ ■ ■ ■ ■ ■ ■ ■ ■ ■ ■ ANSWERS ■ ■ ■ ■ ■ ■ ■ ■ ■ ■ ■

1. Atrial natriuretic peptide (ANP)

2. Aldosterone

3. The proximal tubule

4. Bladder distension

5. The difference between extracellular and plasma volumes

6. Plasma sodium concentration < 130 mEq/L

7. A condition characterized by hyperactivity of the micturition reflex, hypertrophy of bladder wall, and reduction in bladder holding capacity

8. a

9. d

10. d

11. c

12. d

13. b

14. a

15. d

16. b

17. b

18. c

19. b

20. ECF osmolality = ICF osmolality at equilibrium = Total solute/Total body water = (300 mOsm/kg H_2O × 40 L + 1020 mOsm/kg H_2O × 2 L)/(40 L + 2 L) = 334 mOsm/kg H_2O

21. Same answer as for question 20, i.e., 334 mOsm/kg H_2O

22. ECF volume = Amount of ECF solute / ECF osmolality = (15 L × 300 mOsm/kg H_2O + 2 L × 1020 mOsm/kg H_2O / 334 mOsm/kg H_2O = 19.6 L

23. Total body water = 40 L + 2 L = 42 L

24. ICF volume = Total body water − ECF volume = 42 L − 19.6 L = 22.4 L

25. With significant hemorrhage, angiotensin II is formed and increases sodium reabsorption through its direct effect on renal tubular transport, renal hemodynamic action as a vasoconstrictor, and effect on aldosterone secretion, which causes increased sodium reabsorption. In addition, angiotensin II promotes thirst and drinking, and therefore increases fluid intake.

26. Atrial natriuretic peptide (ANP) is released into the circulation when the cardiac atria are distended and help the body get rid of a volume load. During volume expansion, atrial pressure increases, resulting in atrial distension and ANP release, natriuresis and diuresis, and the reduction of blood volume toward normal levels.

27. Glucose levels are high in the extracellular fluid in patients with uncontrolled diabetes, since glucose does not enter most cells when there is a lack of insulin. The high levels of glucose cause the abstraction of water out of the intracellular compartment by osmosis; the result is hyponatremia.

28. Compliance of the bladder wall is defined as the change in intravesical volume divided by the change in pressure. At urine volumes less than 300 ml, bladder compliance is very high and increases in volume lead to relatively small increases in intravesical pressure, minimal distension, weak initiation of the stretch reflex, and storage of urine. At urine volumes above 300 ml, bladder compliance is very low and increases in volume lead to relatively great increases in intravesical pressure, considerable distension, strong initiation of the stretch reflex, and the urge to urinate.

Suggested Reading

Brenner B, Coe FC, Rector FC Jr: *Renal Physiology in Health and Disease.* Philadelphia, Saunders, 1987.

Brenner B, Rector FC Jr: *The Kidney.* Philadelphia, Saunders, 1986.

Hladky SB, Rink TJ: *Body Fluid and Kidney Physiology.* London, Edward Arnold, 1986.

Klahr S, Massry SG: *Contemporary Nephrology.* 5th ed. New York, Plenum, 1989.

Massry SG, Glassock RJ: *Textbook of Nephrology.* 2nd ed. Baltimore, Williams & Wilkins, 1989.

Pitts RF: *Physiology of the Kidney and Body Fluids.* 3rd ed. Chicago, Year Book, 1976.

Rose BD: *Clinical Physiology of Acid-Base and Electrolyte Disorders.* 4th ed. New York, McGraw-Hill, 1994.

Schrier RW: *Renal and Electrolyte Disorders.* 3rd ed. Boston, Little, Brown, 1986.

Seldin DW, Giebisch G: *The Kidney: Physiology and Pathophysiology.* 2nd ed. New York, Raven, 1992.

CHAPTER

■ ■ ■ ■ ■ ■ ■

25

Acid-Base Balance

CHAPTER OUTLINE

I. REVIEW OF ACID-BASE CHEMISTRY
   A. Acids dissociate to release hydrogen ions in solution
   B. The acid dissociation constant K_a shows the strength of an acid
   C. pK_a is a logarithmic expression of K_a
   D. pH is inversely related to [H+]
   E. The Henderson-Hasselbalch equation relates pH to the ratio of the concentrations of conjugate base and acid
   F. Buffers promote stability of pH

II. PRODUCTION AND REGULATION OF HYDROGEN IONS IN THE BODY
   A. Metabolism is a constant source of hydrogen ions
   B. Incomplete metabolism of carbohydrates and fats produces nonvolatile acids
   C. Metabolism of proteins generates strong acids
   D. On a mixed diet, net acid gain threatens pH
   E. Many buffering mechanisms protect and stabilize blood pH

III. CHEMICAL REGULATION OF pH
   A. Chemical buffers are the first to defend pH
   B. A pK_a of 6.8 makes phosphate a good buffer
   C. Proteins are excellent buffers
   D. The bicarbonate/carbon dioxide buffer pair is crucial in pH regulation
      1. Forms of carbon dioxide
      2. The $CO_2/H_2CO_3/HCO_3^-$ equilibria
      3. The Henderson-Hasselbalch equation for HCO_3^-/CO_2
      4. An "open" buffer system
   E. All buffers are in equilibrium with the same [H+]

IV. RESPIRATORY REGULATION OF pH

V. RENAL REGULATION OF pH
   A. Renal acid excretion equals the sum of urinary titratable acid and ammonia minus urinary bicarbonate
   B. Hydrogen ions are added to urine as it flows along the nephron
      1. Acidification in the proximal convoluted tubule
      2. Acidification in the loop of Henle
      3. Acidification in the distal nephron
   C. Reabsorption of filtered HCO_3^- restores lost HCO_3^- to the blood

OBJECTIVES

After studying this chapter, the student should be able to:

1. Define *acid, base, acid dissociation constant, pH,* and *buffer* and write the Henderson-Hasselbalch equation for the HCO_3^-/CO_2 buffer pair

2. List the metabolic processes that produce and consume hydrogen ions and explain why the body is threatened by net acid gain on a mixed diet

3. List the chemical buffers present in cells, extracellular fluid, bone, and urine

4. Explain why the HCO_3^-/CO_2 buffer pair is so important in acid-base physiology even though its pK is somewhat remote from normal plasma pH

5. Describe the respiratory responses to an increase or decrease of arterial blood pH or P_{CO_2}.

6. Describe the three processes involved in renal acidification: reabsorption of filtered bicarbonate, excretion of titratable acid, and excretion of ammonia; draw cell models and indicate where along the nephron these processes take place and which processes generate new bicarbonate

7. Describe the effects on renal H+ excretion of changes in intracellular pH, arterial blood P_{CO_2}, carbonic anhydrase activity, sodium reabsorption, plasma potassium concentration, and plasma aldosterone levels

8. Describe the mechanisms that maintain stability of intracellular pH

9. List the four simple acid-base disturbances, and for each describe the primary defect, changes in arterial blood chemistry, some common causes, chemical buffering processes, and respiratory and renal compensations

10. From blood acid-base data and a patient history, identify the type of acid-base disturbance present

CHAPTER OUTLINE (Continued)

- D. Excretion of titratable acid and ammonia generates new bicarbonate
  1. Excretion of titratable acid
  2. Excretion of ammonia
- E. Several factors influence renal excretion of hydrogen ions
  1. Intracellular pH
  2. Arterial blood P_{CO_2}
  3. Carbonic anhydrase activity
  4. Sodium reabsorption
  5. Plasma potassium concentration
  6. Aldosterone

VI. REGULATION OF pH INSIDE THE CELL

VII. DISTURBANCES OF ACID-BASE BALANCE

- A. Respiratory acidosis results from accumulation of carbon dioxide
  1. Chemical buffering
  2. Respiratory response
  3. Renal compensation
- B. Respiratory alkalosis results from excessive loss of carbon dioxide
  1. Chemical buffering
  2. Respiratory response
  3. Renal compensation
- C. Metabolic acidosis results from gain of noncarbonic acid or loss of bicarbonate
  1. Chemical buffering
  2. Respiratory compensation
  3. Renal compensation
- D. Metabolic alkalosis results from gain of strong base or bicarbonate or loss of noncarbonic acid
  1. Chemical buffering
  2. Respiratory compensation
  3. Renal compensation
- E. Clinical evaluation of acid-base disturbances requires a comprehensive study

---

Every day, metabolic reactions in the body produce and consume many moles of hydrogen ions. Yet, the hydrogen ion concentration ($[H^+]$) of most body fluids is very low and is kept within narrow limits. For example, the $[H^+]$ of arterial blood is normally 35–45 nmol/L (pH 7.45–7.35). Normally we stay in acid-base balance; inputs and outputs of acids and bases are matched so that $[H^+]$ stays relatively constant both outside and inside cells.

Most of this chapter concerns the regulation of $[H^+]$ in extracellular fluid, since it is easier to analyze than intracellular fluid and is the fluid used in the clinical evaluation of acid-base balance. In practice, systemic arterial blood is used as the reference for this purpose. Measurements on whole blood with a pH meter give values for the $[H^+]$ (strictly speaking, the H^+ activity) of plasma, and so provide an extracellular fluid pH measurement.

Review of Acid-Base Chemistry

An **acid** is a substance that can release, or donate, H^+, and a **base** is a substance that can combine with, or accept, H^+.

■ Acids Dissociate to Release Hydrogen Ions in Solution

When an acid (generically written as *HA*) is added to water it dissociates reversibly according to the following reaction: $HA \rightleftharpoons H^+ + A^-$. The species A^- is a base, since it can combine with a H^+ to form HA. In other words, when an acid dissociates, it yields a free H^+ and its conjugate base. (*Conjugate* means "joined in a pair.")

■ The Acid Dissociation Constant K_a Shows the Strength of an Acid

At equilibrium, the rate of dissociation of an acid to form $H^+ + A^-$ and the rate of association of H^+ and base A^- to form HA are equal. The equilibrium constant (K_a), also called the ionization constant or acid **dissociation constant,** is given by the expression:

$$K_a = \frac{[H^+] \times [A^-]}{[HA]} \qquad (1)$$

The higher the acid dissociation constant, the more an acid is ionized and the greater is its strength. Hydrochloric acid (HCl) is an example of a **strong acid.** It has a very high K_a and is practically completely ionized in aqueous solutions. Other strong acids include sulfuric acid (H_2SO_4), phosphoric acid (H_3PO_4), and nitric acid (HNO_3).

An acid with a low K_a is a **weak acid.** For example, in a 0.1 mol/L solution of acetic acid ($K_a = 1.8 \times 10^{-5}$) in water, most (99%) of the acid is nonionized, so very little (1%) is present as acetate and H^+. The acidity (concentration of free H^+) of this solution is low. Other weak acids are lactic acid, carbonic acid (H_2CO_3), ammonium ion (NH_4^+), and dihydrogen phosphate ($H_2PO_4^-$).

■ pK_a Is a Logarithmic Expression of K_a

Acid dissociation constants vary widely and often are very small numbers. It is convenient to convert K_a to a logarithmic form, defining pK_a as: $pK_a = \log_{10}(1/K_a) = -\log_{10} K_a$. In aqueous solution, each acid has a characteristic pK_a, which varies slightly with temperature and the ionic strength of the solution.

Note that pK_a is *inversely* proportional to acid strength. A strong acid has a high K_a and a low pK_a. A weak acid has a low K_a and a high pK_a.

■ pH Is Inversely Related to [H⁺]

[H⁺] is often expressed in pH units. The following equation defines pH: $pH = \log_{10}(1/[H^+]) = -\log_{10}[H^+]$, where [H⁺] is in mol/L. Note that *pH is inversely related to [H⁺]*. Each whole number on the pH scale represents a 10-fold (logarithmic) change in acidity. A solution with a pH of 5 has 10 times the [H⁺] of a solution with a pH of 6.

■ The Henderson-Hasselbalch Equation Relates pH to the Ratio of the Concentrations of Conjugate Base and Acid

For a solution containing an acid and its conjugate base, we can rearrange the equilibrium expression as:

$$[H^+] = K_a \times \frac{[HA]}{[A^-]} \tag{2}$$

If we take the negative logarithms of both sides:

$$-\log[H^+] = -\log K_a + \log \frac{[A^-]}{[HA]} \tag{3}$$

Substituting pH for $-\log[H^+]$ and pK_a for $-\log K_a$, we get:

$$pH = pK_a + \log \frac{[A^-]}{[HA]} \tag{4}$$

This is known as the *Henderson-Hasselbalch equation.* It shows that the pH of a solution is determined by the pK_a of the acid and the ratio of the concentrations of conjugate base A⁻ and acid HA.

■ Buffers Promote Stability of pH

The stability of pH is protected by the action of buffers. A **pH buffer** is something that *minimizes* the change in pH produced when an acid or base is added. Note that a buffer *does not prevent* a pH change.

A **chemical pH buffer** is a mixture of a weak acid and its conjugate base (or a weak base and its conjugate acid). Following are examples of buffers:

Weak Acid		Conjugate Base
H_2CO_3 (carbonic acid)	$\rightleftharpoons$	$HCO_3^- + H^+$ (bicarbonate)
$H_2PO_4^-$ (dihydrogen phosphate)	$\rightleftharpoons$	$HPO_4^{2-} + H^+$ (monohydrogen phosphate)
NH_4^+ (ammonium ion)	$\rightleftharpoons$	$NH_3 + H^+$ (ammonia)

Generally speaking, the equilibrium expression for a buffer pair can be written in terms of the Henderson-Hasselbalch equation:

$$pH = pK_a + \log \frac{[\text{conjugate base}]}{[\text{acid}]} \tag{5}$$

For example, for $H_2PO_4^-/HPO_4^{2-}$

$$pH = 6.8 + \log \frac{[HPO_4^{2-}]}{[H_2PO_4^-]} \tag{6}$$

The effectiveness of a buffer—how well it reduces pH changes when an acid or base is added—depends on its concentration and pK_a. A good buffer is present in high concentrations and has a pK_a close to the desired pH.

Figure 25–1 shows a titration curve for the phosphate buffer system. As a strong acid or strong base is progressively added to the solution (shown on the

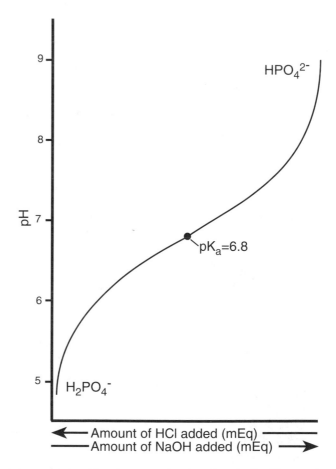

Figure 25–1 ■ Titration curve for phosphate buffer. The pK_a for $H_2PO_4^-$ is 6.8. A strong acid (HCl) (right to left) or strong base (NaOH) (left to right) was added and the resulting solution pH recorded (y-axis). Note that buffering is best (i.e. the change in pH on addition of a given amount of acid or base is least) when the solution pH is equal to the pK_a of the buffer.

x-axis), the resulting pH is recorded (shown on the y-axis). Going from right to left, as strong acid is added, H^+ combines with the basic form of phosphate: $H^+ + HPO_4^{2-} \rightarrow H_2PO_4^-$. Going from left to right, as strong base is added, OH^- ions combine with H^+ released from the acid form of the phosphate buffer: $OH^- + H_2PO_4^- \rightarrow HPO_4^{2-} + H_2O$. These reactions lessen the fall or rise in pH.

At the pK_a of the phosphate buffer, the ratio $[HPO_4^{2-}]/[H_2PO_4^-]$ equals 1 and the titration curve is flattest (the change in pH for a given amount of an added acid or base is at a minimum). In most cases, pH buffering is effective when the solution pH is within 1 unit (plus or minus) of the buffer pK_a. Beyond that range, the pH shift that a given amount of acid or base produces may be quite large, so the buffer becomes relatively ineffective.

Production and Regulation of Hydrogen Ions in the Body

Acids are continuously produced in the body and threaten the normal pH of the body fluids. Physiologically speaking, acids fall into two groups: H_2CO_3 (carbonic acid) and all other acids (noncarbonic acids, or so-called nonvolatile or fixed acids).

The distinction between these groups arises because H_2CO_3 is in equilibrium with the volatile gas, CO_2, which can leave the body via the lungs. The concentration of H_2CO_3 in arterial blood is therefore set by respiratory activity. By contrast, noncarbonic acids in the body are nonvolatile. Breathing does not directly affect their concentrations in the blood. Noncarbonic acids are buffered in the body and must be excreted by the kidneys.

■ Metabolism Is a Constant Source of Hydrogen Ions

A normal adult produces about 300 liters of CO_2 daily from the metabolism of foodstuffs. Carbon dioxide from tissues enters the capillary blood, where it reacts with water to form H_2CO_3, which dissociates instantly to yield H^+ and HCO_3^-: $CO_2 + H_2O \rightleftharpoons H_2CO_3 \rightleftharpoons H^+ + HCO_3^-$. A huge amount of H_2CO_3 can form from metabolically produced CO_2. Indeed, if H_2CO_3 were allowed to accumulate, much H^+ would also accumulate, and the pH would rapidly fall to lethal levels.

Fortunately, however, H_2CO_3 produced from metabolic CO_2 does not normally accumulate. Instead it is converted to CO_2 and water when it reaches the lung capillaries, and the CO_2 is expired. In the lung, the reactions reverse:

$$H^+ + HCO_3^- \rightleftharpoons H_2CO_3 \rightleftharpoons H_2O + CO_2$$

As long as CO_2 is expired as fast as it is produced, the concentration of H_2CO_3 in the blood does not change, and neither does the overall pH.

■ Incomplete Metabolism of Carbohydrates and Fats Produces Nonvolatile Acids

Normally, carbohydrates and fats are completely oxidized to CO_2 and water. Acid-base balance is maintained as long as CO_2 is expired as fast as it is produced. If carbohydrates and fats are incompletely oxidized, however, nonvolatile acids accumulate in the body (see below). Incomplete oxidation of carbohydrates occurs when the tissues do not receive enough oxygen, such as during heavy exercise or hemorrhagic or cardiogenic shock. In such states, glucose metabolism yields much lactic acid ($pK_a = 3.9$), which dissociates into lactate and H^+, lowering the blood pH. Incomplete fatty acid oxidation occurs in uncontrolled diabetes mellitus, starvation, and alcoholism and produces ketone body acids (acetoacetic and β-hydroxybutyric acids). These acids have pK_a values around 4–5. At blood pH they mostly dissociate into their anions and H^+, making the blood more acidic.

■ Metabolism of Proteins Generates Strong Acids

Metabolism of dietary proteins is a major source of H^+. Oxidation of proteins and amino acids produces strong acids, such as H_2SO_4, HCl, and H_3PO_4. Oxidation of sulfur-containing amino acids (methionine, cysteine, cystine) produces H_2SO_4, and oxidation of cationic amino acids (arginine, lysine, and some histidine residues) produces HCl. Oxidation of phosphorus-containing proteins and phosphoesters in nucleic acids produces H_3PO_4.

■ On a Mixed Diet, Net Acid Gain Threatens pH

A diet containing both meat and vegetables results in net production of acids, largely from protein oxidation. To some extent, acid-consuming metabolic reactions balance H^+ production. Foodstuffs also contain basic anions, such as citrate, lactate, and acetate. When these are oxidized to CO_2 and water, H^+ ions are consumed (or what amounts to the same—bicarbonate ions are produced). The balance of acid-forming and acid-consuming metabolic reactions results in net production of about 1 mEq H^+/kg body weight/day in an adult who eats a mixed diet. In vegetarians, there is a net production of base, because vegetables contain large amounts of organic anions. Since most people in the United States eat a mixed diet, metabolic acid gain is stressed in this chapter.

■ Many Buffering Mechanisms Protect and Stabilize Blood pH

Despite constant threats to acid-base homeostasis, a healthy person maintains a normal blood pH. Figure 25–2 shows some of the ways blood pH is kept normal

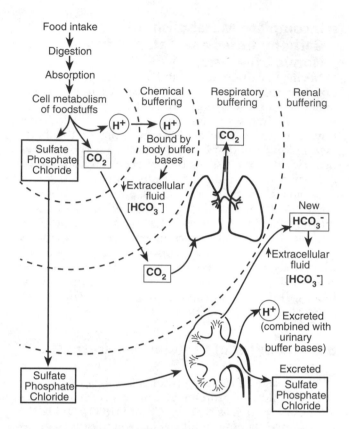

Figure 25–2 ■ A normal blood pH is maintained by the concerted action of chemical buffers, the respiratory system, and the kidneys. On a mixed diet, pH is threatened by production of strong acids (sulfuric, hydrochloric, and phosphoric) mainly as a result of protein metabolism. These strong acids are buffered in the body by chemical buffer bases, such as extracellular fluid bicarbonate. The kidneys eliminate hydrogen ions (combined with urinary buffers) and anions in the urine. At the same time, they add new bicarbonate to the extracellular fluid to replace the bicarbonate consumed in buffering strong acids. The respiratory system disposes of CO_2.

despite daily net acid gain. The key buffering agents are chemical buffers plus the lungs and kidneys.

Chemical buffers in extra- and intracellular fluids and bone are the first line of defense of blood pH. Chemical buffering minimizes a change in pH but cannot remove acid or base from the body.

The **respiratory system** is the second line of defense of blood pH. Normally, breathing removes CO_2 as fast as it forms. Large loads of acid stimulate breathing, which removes CO_2 from the body and thus lowers the $[H_2CO_3]$ in arterial blood, reducing the acidic shift in blood pH.

The **kidneys** are the third line of defense of blood pH. Although chemical buffers can bind H^+ and the lungs can change the $[H_2CO_3]$ of blood, the burden of removing excess H^+ falls directly on the kidneys. Hydrogen ions are excreted in combination with urinary buffers. At the same time, the kidneys add new HCO_3^- to the extracellular fluid to replace HCO_3^- used to buffer strong acids. The kidneys also excrete

TABLE 25–1 ■ Major Chemical pH Buffers in the Body

Buffer	Reaction
Extracellular fluid	
Bicarbonate/CO_2	$CO_2 + H_2O \rightleftharpoons H_2CO_3 \rightleftharpoons H^+ + HCO_3^-$
Inorganic phosphate	$H_2PO_4^- \rightleftharpoons H^+ + HPO_4^{2-}$
Plasma proteins (Pr)	$HPr \rightleftharpoons H^+ + Pr^-$
Intracellular fluid	
Cell proteins (e.g., hemoglobin [Hb])	$HHb \leftarrow H^+ + Hb^-$
Organic phosphates	Organic $HPO_4^- \rightleftharpoons H^+ +$ Organic PO_4^{2-}
Bicarbonate/CO_2	$CO_2 + H_2O \rightleftharpoons H_2CO_3 \rightleftharpoons H^+ + HCO_3^-$
Bone	
Mineral phosphates	$H_2PO_4^- \rightleftharpoons H^+ + HPO_4^{2-}$
Mineral carbonates	$HCO_3^- \rightleftharpoons H^+ + CO_3^{2-}$

the anions (phosphate, chloride, sulfate) that are liberated from strong acids. The kidneys affect blood pH more slowly than other buffering mechanisms.

Chemical Regulation of pH

The body contains many conjugate acid-base pairs that act as chemical buffers (Table 25–1). In the extracellular fluid, the main chemical buffer pair is HCO_3^-/CO_2. Plasma proteins and inorganic phosphate are also extracellular fluid buffers. Cells also have large buffer stores, particularly proteins and organic phosphate compounds, and contain HCO_3^-, although at a lower concentration than in extracellular fluid. Bone contains large buffer stores, specifically salts of phosphate and carbonate.

■ Chemical Buffers Are the First to Defend pH

When an acid or base is added to the body, the buffers just mentioned bind or release H^+ and so minimize the change in pH. Buffering in extracellular fluid occurs rapidly—in minutes. Acid or base also enters cells and bone, but generally more slowly—over hours. This allows cell buffers and bone to share in buffering.

■ A pK_a of 6.8 Makes Phosphate a Good Buffer

The pK_a for phosphate ($H_2PO_4^- \rightleftharpoons H^+ + HPO_4^{2-}$) is 6.8, quite close to the desired blood pH of 7.4, making it a good buffer. In the extracellular fluid, phosphate is present as inorganic phosphate. Because its concentration is low (about 1 mmol/L), it plays a minor role in extracellular buffering.

Phosphate is an important intracellular buffer, however, for two reasons. First, cells contain large amounts of phosphate in such organic compounds as adenosine triphosphate (ATP), adenosine diphosphate (ADP), and creatine phosphate. Although these compounds primarily serve in energy metabolism, they also act as pH buffers. Second, intracellular pH is generally lower than extracellular pH and is closer to the pK_a of phosphate. (The cytosol of skeletal muscle, for example, has a pH of 6.9.) Phosphate is thus more effective in this environment than in one with pH 7.4. Bone has large phosphate salt stores, which also help in buffering.

■ Proteins Are Excellent Buffers

Proteins are the largest buffer pool in the body and are excellent buffers. Proteins are **amphoteric**—can function as both acids and bases. They contain many ionizable groups, which can release or bind H^+. Serum albumin and plasma globulins are the major extracellular protein buffers, present mainly in the blood plasma. Cells also have large protein stores. Recall that the buffering properties of hemoglobin play an important role in the transport of CO_2 and O_2 by the blood (Chap. 21).

■ The Bicarbonate/Carbon Dioxide Buffer Pair Is Crucial in pH Regulation

For several reasons, the HCO_3^-/CO_2 buffer pair is especially important in acid-base physiology. First, its components are abundant: the concentration of HCO_3^- in plasma or extracellular fluid normally averages 24 mmol/L. Although the concentration of dissolved CO_2 is lower (1.2 mmol/L), metabolism provides a nearly limitless supply. Second, despite a pK of 6.10, somewhat far from the desired plasma pH of 7.40, this buffer pair is quite effective because the system is "open." Third, this buffer pair is controlled by the lungs and kidneys.

Forms of Carbon Dioxide ■ Carbon dioxide exists in the body in several different "forms": as gaseous CO_2 in the lung alveoli and as dissolved CO_2, H_2CO_3, HCO_3^-, carbonate (CO_3^{2-}), and carbamino compounds in the body fluids. Carbonate is present at appreciable concentrations only in rather alkaline solutions, so we ignore it here. We also ignore any CO_2 that is bound to proteins in the carbamino form. The most important forms are gaseous CO_2, dissolved CO_2, H_2CO_3, and HCO_3^-.

The $CO_2/H_2CO_3/HCO_3^-$ Equilibria ■ Dissolved CO_2 in pulmonary capillary blood equilibrates with gaseous CO_2 in the lung alveoli. Consequently, the partial pressures of CO_2 (P_{CO_2}) in alveolar air and systemic arterial blood are normally identical. The concentration of dissolved CO_2 ($[CO_2]_d$) is related to P_{CO_2} by the solubility coefficient for CO_2 (0.03 mmol CO_2/L/mm Hg P_{CO_2}, for plasma at 37°C): $[CO_2]_d = 0.03 \times P_{CO_2}$. If P_{CO_2} is 40 mm Hg, then $[CO_2]_d$ is 1.2 mmol/L.

In aqueous solutions, $CO_{2(d)}$ reacts with water to form H_2CO_3: $CO_{2(d)} + H_2O \rightleftharpoons H_2CO_3$. The reaction to the right is called the **hydration reaction** and the reaction to the left is called the **dehydration reaction.** These reactions are slow if uncatalyzed. In many tissues and cells, such as the kidney, pancreas, stomach, and red blood cells, the reactions are catalyzed by **carbonic anhydrase,** a zinc-containing enzyme. At equilibrium, $CO_{2(d)}$ is greatly favored; at body temperature, the ratio of $[CO_2]_d$ to $[H_2CO_3]$ is about 400:1. If $[CO_2]_d$ is 1.2 mmol/L, then $[H_2CO_3]$ equals 3 µmol/L. H_2CO_3 dissociates instantaneously into H^+ and HCO_3^-: $H_2CO_3 \rightleftharpoons H^+ + HCO_3^-$. The Henderson-Hasselbalch expression for this reaction is:

$$pH = 3.5 + \log \frac{[HCO_3^-]}{[H_2CO_3]} \tag{7}$$

Note that H_2CO_3 is a fairly strong acid ($pK_a = 3.5$). Its low concentration in body fluids (at the µmol/L level) lessens its impact on acidity.

The Henderson-Hasselbalch Equation for HCO_3^-/CO_2 ■ Because $[H_2CO_3]$ is so low and hard to measure, and since $[H_2CO_3] = [CO_2]_d/400$, we can use $[CO_2]_d$ to represent the acid in the Henderson-Hasselbalch equation:

$$\begin{aligned} pH &= 3.5 + \log \frac{[HCO_3^-]}{[CO_2]_d/400} \\ &= 3.5 + \log 400 + \log \frac{[HCO_3^-]}{[CO_2]_d} \\ &= 6.1 + \log \frac{[HCO_3^-]}{[CO_2]_d} \end{aligned} \tag{8}$$

We can also use $0.03 \times P_{CO_2}$ in place of $[CO_2]_d$:

$$pH = 6.1 + \log \frac{[HCO_3^-]}{0.03 \, P_{CO_2}} \tag{9}$$

This form of the Henderson-Hasselbalch equation is useful for understanding acid-base problems. Note that the "acid" in this equation appears to be $CO_{2(d)}$ but is really H_2CO_3 "represented" by CO_2. Hence, this equation is valid only if $CO_{2(d)}$ and H_2CO_3 are in equilibrium with each other, which they usually (but not always) are.

Many clinicians prefer to use $[H^+]$ rather than pH. The following expression results if we take antilogarithms of the Henderson-Hasselbalch equation:

$$[H^+] = 24 \, P_{CO_2}/[HCO_3^-] \tag{10}$$

In this expression, $[H^+]$ is expressed in nmol/L, $[HCO_3^-]$ in mmol or mEq/L, and P_{CO_2} in mm Hg. If the P_{CO_2} is 40 mm Hg and plasma $[HCO_3^-]$ is 24 mmol/L, $[H^+]$ is 40 nmol/L.

■ An "Open" Buffer System

As previously noted, the pK of the HCO_3^-/CO_2 system (6.10) is quite far from 7.40, the pH we want to maintain in arterial blood. From this, one might view this buffer pair as rather poor. On the contrary, it is remarkably effective because it operates in an "open" system. That is, the two buffer components can be added to or removed from the body.

The HCO_3^-/CO_2 system is open in several ways. First, metabolism provides an endless source of CO_2, which can replace any H_2CO_3 consumed by a base added to the body. Second, the respiratory system can change the amount of CO_2 in body fluids by hyper- or hypoventilation. Finally, the kidneys can change the amount of HCO_3^- in the extracellular fluid by forming new HCO_3^- when excess acid has been added to the body or excreting HCO_3^- when excess base has been added.

How the kidneys and respiratory system influence blood pH by operating on the HCO_3^-/CO_2 system is described below. For now, the advantages of an open buffer system are demonstrated by example (Fig. 25–3). Suppose we have 1 L of blood containing 24 mmol of HCO_3^- and 1.2 mmol of dissolved CO_2 (P_{CO_2} = 40 mm Hg). Using the special form of the Henderson-Hasselbalch equation just described, we find that the pH of the blood is 7.40:

$$pH = 6.10 + \log \frac{[HCO_3^-]}{0.03 \, P_{CO_2}}$$
$$= 6.10 + \log \frac{[24]}{[1.2]} \qquad (11)$$
$$= 7.40$$

Suppose we now add 10 mmol of HCl, a strong acid. HCO_3^- is the major buffer base in the blood plasma (we will neglect the contributions of other buffers). From the reaction $H^+ + HCO_3^- \rightleftharpoons H_2CO_3 \rightleftharpoons H_2O + CO_2$, we predict that the $[HCO_3^-]$ will fall by 10 mmol and that 10 mmol of $CO_{2(d)}$ will form. If the system were closed and no CO_2 could escape, the new pH would be:

$$pH = 6.10 + \log \frac{[24 - 10]}{[1.2 + 10]} = 6.20 \qquad (12)$$

This is an intolerably low, indeed fatal, pH.

Fortunately, however, the system is open, and CO_2 can escape via the lungs. If all of the extra CO_2 is expired and the $[CO_2]_d$ is kept at 1.2 mmol/L, then the pH would be:

$$pH = 6.10 + \log \frac{[24 - 10]}{[1.2]} = 7.17 \qquad (13)$$

This pH is low but is compatible with life.

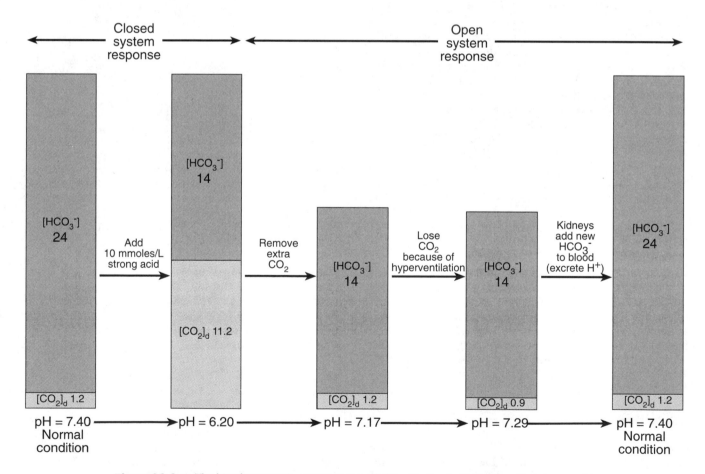

Figure 25–3 ■ The bicarbonate/CO_2 system is remarkably effective in buffering added strong acid in the body, because it operates in an open system. [HCO_3^-] and [CO_2]$_d$ are in mmol/L. (Adapted from Pitts RF: *Physiology of the Kidney and Body Fluids.* 3rd ed. Chicago, Year Book, 1974.)

Still another mechanism promotes the escape of CO_2. An acidic blood pH stimulates breathing, which can make the P_{CO_2} lower than 40 mm Hg. If P_{CO_2} falls to 30 mm Hg ($[CO_2]_d = 0.9$ mmol/L), the pH would be:

$$pH = 6.10 + \log \frac{[24 - 10]}{[0.9]} = 7.29 \qquad (14)$$

The system is also open at the kidneys and new HCO_3^- can be added to the plasma to correct the plasma $[HCO_3^-]$. Once the blood pH is normal, the stimulus for hyperventilation disappears.

■ All Buffers Are in Equilibrium with the Same [H+]

We have discussed the various buffers separately, but in the body they all work together. In any body fluid compartment, all buffers are in equilibrium with the same [H+]. This is known as the **isohydric principle** (*isohydric* means "same H+"). For plasma, for example, we can write:

$$
\begin{aligned}
pH &= 6.80 + \log \frac{[HPO_4^{2-}]}{[H_2PO_4^-]} \\
&= 6.10 + \log \frac{[HCO_3^-]}{0.03 \, P_{CO_2}} \qquad (15) \\
&= pK_{protein} + \log \frac{[proteinate^-]}{[H\text{-}protein]}
\end{aligned}
$$

If an acid or base is added to such a complex mixture of buffers, all buffers take part in buffering and shift from one form (acid or base) to the other. The relative importance of each buffer depends on its pK and its amount.

The isohydric principle underscores the fact that it is the concentration ratio for any buffer pair, along with its pK, that sets the pH. We can focus on the concentration ratio for one buffer pair, and all other buffers will automatically adjust their ratios according to the pH and their pK values.

The rest of this chapter emphasizes the role of the HCO_3^-/CO_2 buffer pair in setting blood pH. However, other buffers are present and active. The HCO_3^-/CO_2 system is emphasized because physiologic mechanisms (lungs and kidneys) regulate pH by acting on components of this buffer system.

Respiratory Regulation of pH

Reflex changes in ventilation help defend blood pH. By changing the P_{CO_2}, and hence $[H_2CO_3]$, of the blood, the respiratory system can rapidly and profoundly affect blood pH. As discussed in Chapter 22, a fall in blood pH stimulates ventilation, primarily by acting on peripheral chemoreceptors. An elevated arterial blood P_{CO_2} is a powerful stimulus to increase ventilation; it acts on both peripheral and central chemoreceptors, but primarily on the latter. Carbon dioxide diffuses into brain interstitial and cerebrospinal fluids, where it causes a fall in pH, which stimulates chemoreceptors in the medulla oblongata. When ventilation is stimulated, more CO_2 is "blown off" by the lungs, thereby making the blood less acidic. Conversely, a rise in blood pH inhibits ventilation; the consequent rise in blood $[H_2CO_3]$ reduces the alkaline shift in blood pH. Respiratory responses to disturbed blood pH begin within minutes and are maximal in about 12–24 hours.

Renal Regulation of pH

The kidneys play a critical role in maintaining acid-base balance. They remove H+ if there is excess acid in the body or HCO_3^- if there is excess base. As discussed earlier, the challenge with a mixed diet is to remove excess acid. The strong acids produced by metabolism are first buffered by body buffer bases, particularly HCO_3^-. The kidneys then must eliminate H+ in the urine and restore the depleted HCO_3^-.

Little of the H+ excreted in the urine is present as free H+. For example, if the urine has its lowest possible pH value (4.5), [H+] is only 0.03 mEq/L. With a typical daily urine output of 1–2 liters, the amount of acid the body must dispose of daily (roughly 1 mEq/kg, or 70 mEq for a 70-kg person) obviously is not excreted in the free form. Most of the H+ combines with urinary buffers to be excreted as titratable acid and as NH_4^+.

Titratable acid is measured from the amount of strong base (NaOH) needed to bring the urine pH back to the pH of the blood (usually 7.40). It represents the amount of H+ ions excreted combined with urinary buffers such as phosphate, creatinine, and other bases. The largest component of titratable acid is normally phosphate ($H_2PO_4^-$).

Hydrogen ions secreted by the renal tubules also combine with the free base NH_3 and are excreted as NH_4^+. Ammonia (which includes both NH_3 and NH_4^+) is produced by the kidney tubule cells and secreted into the urine. Because the pK_a for NH_4^+ is high (9.0), most of the ammonia in the urine is present as NH_4^+. For this reason too, NH_4^+ is not appreciably titrated when titratable acid is measured, so urinary ammonia is measured by a separate, often chemical, method.

■ Renal Acid Excretion Equals the Sum of Urinary Titratable Acid and Ammonia Minus Urinary Bicarbonate

In stable acid-base balance, net acid excretion by the kidneys equals the net rate of H+ addition to the body by metabolism or other processes, assuming other routes of loss of acid or base (e.g., gastrointestinal losses) are small and can be neglected, which normally is the case. The net loss of H+ in the urine can be calculated from the following equation, which shows typical values:

$$
\begin{aligned}
\text{Renal acid excretion} = {} & \text{Urinary titratable acid} \\
& + \text{ urinary ammonia} \\
& - \text{ Urinary HCO}_3{}^+ \\
70 \text{ mEq/day} = {} & 24 \text{ mEq/day} \\
& + 48 \text{ mEq/day} \\
& - 2 \text{ mEq/day}
\end{aligned}
\tag{16}
$$

Excretion of $HCO_3{}^-$ in the urine represents a loss of base from the body; therefore, it must be subtracted from the measurement of acid loss. Urinary ammonia (as $NH_4{}^+$) ordinarily accounts for about two-thirds of excreted H^+ and titratable acid for about one-third. Since the amount of free H^+ excreted is negligible, this is omitted from the equation.

■ Hydrogen Ions Are Added to Urine as It Flows Along the Nephron

As the urine flows along the tubule, from Bowman's capsule on through the collecting ducts, three processes occur: filtered $HCO_3{}^-$ is reabsorbed; titratable acid is formed; and ammonia is added to the tubular urine. All three processes involve H^+ secretion (urinary acidification) by the tubular epithelium. The nature and magnitude of these processes vary in different nephron segments. Figure 25–4 summarizes measurements of tubular fluid pH along the nephron and shows ammonia movements in various nephron segments.

Acidification in the Proximal Convoluted Tubule ■ The pH of the glomerular ultrafiltrate, at the beginning of the proximal tubule, is identical to that of the plasma from which it is derived (7.4). Hydrogen ions are secreted by the proximal tubule epithelium into the tubule lumen mainly via a Na^+-H^+ exchanger in the brush border membrane. This results in a fall of tubular fluid pH to about 6.7 at the end of the proximal convoluted tubule (Fig. 25–4). There are two reasons the drop in pH is modest: buffering of secreted H^+ and the high permeability of the proximal tubule epithelium to H^+. The glomerular filtrate and tubule fluid contain abundant buffer bases, especially $HCO_3{}^-$, which soak up secreted H^+, thereby minimizing a fall in pH. The proximal tubule epithelium is rather "leaky" to H^+, so any gradient from urine to blood, established by H^+ secretion, is soon limited by diffusion of H^+ out of the tubule lumen into the blood surrounding the tubules.

Most of the H^+ secreted by the nephron is secreted in the proximal convoluted tubule and used to bring about the reabsorption of filtered $HCO_3{}^-$ (see below). Secreted H^+ ions are also buffered by filtered phosphate to form titratable acid. Finally, ammonia is produced by proximal tubule cells, mainly from glutamine. It is secreted into the tubular urine by diffusion of NH_3, which then combines with a secreted H^+ to form $NH_4{}^+$, or via the brush border membrane Na^+-H^+

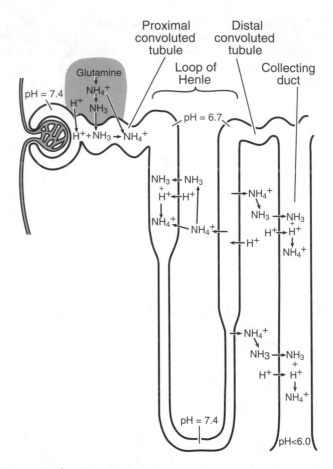

Figure 25–4 ■ The pH of tubular urine decreases along the proximal convoluted tubule, rises along the descending limb of Henle's loop, falls along the ascending limb, and reaches its lowest values in the collecting ducts. Ammonia (= NH_3 + $NH_4{}^+$) is chiefly produced in proximal tubule cells and is secreted into the tubular urine. $NH_4{}^+$ is reabsorbed in the thick ascending limb and accumulates in the kidney medulla. NH_3 diffuses into acidic collecting duct urine, where it is trapped as $NH_4{}^+$.

exchanger, which can operate in a $Na^+ - H^+$ exchange mode.

Acidification in the Loop of Henle ■ Along the descending limb of the loop of Henle, the pH of tubular fluid rises (from 6.7 to 7.4). This is explained by an increase in intraluminal $[HCO_3{}^-]$ caused by water reabsorption. Ammonia is secreted along the descending limb.

The tubular fluid is acidified by secretion of H^+ along the ascending limb via a Na^+-H^+ exchanger. Along the thin ascending limb, ammonia is passively reabsorbed. Along the thick ascending limb, $NH_4{}^+$ is mostly actively reabsorbed by the Na-K-2Cl cotransporter in the luminal cell membrane ($NH_4{}^+$ substitutes for K^+). Some $NH_4{}^+$ may be passively reabsorbed between cells in this segment; the driving force is the lumen positive transepithelial electrical potential difference. Recent evidence suggests that ammonia may undergo countercurrent multiplication in the loop of Henle; this leads to an increasing ammonia concentration gradient in the kidney medulla.

Acidification in the Distal Nephron ■ The distal nephron (distal convoluted tubule and collecting duct) secretes far fewer H+ ions than proximal portions of the nephron. Hydrogen ions are secreted primarily via an electrogenic H+ ATPase or an electroneutral H+-K+ ATPase. Also by contrast with proximal portions, the distal nephron is lined by "tight" epithelia, so little secreted H+ diffuses out of the tubule lumen, making steep urine-to-blood pH gradients possible (Fig. 25–4). Final urine pH is typically around 6 but may be as low as 4.5.

The distal nephron usually reabsorbs almost completely the small quantities of HCO_3^- that were not reabsorbed by more proximal nephron segments. Considerable titratable acid forms as the urine is acidified. Ammonia, which was reabsorbed by the loop of Henle and accumulated in the medullary interstitium, diffuses as lipid-soluble NH_3 into collecting duct urine and combines with secreted H+. The collecting duct epithelium is impermeable to the lipid-insoluble NH_4^+, so ammonia is "trapped" in an acidic urine and excreted as NH_4^+ (Fig. 25–4).

The intercalated cells of the collecting duct are involved in acid-base transport and are of two major types: an acid-secreting **α-intercalated cell** and a bicarbonate-secreting **β-intercalated cell.** These two cell types have opposite polarities (Fig. 25–5). A more acidic blood pH results in insertion of cytoplasmic H+ pumps into the luminal cell membrane of α-intercalated cells and enhanced H+ secretion. If the blood is made alkaline, bicarbonate secretion by β-intercalated cells is increased. The amount of bicarbonate secreted is ordinarily very small compared to the amounts filtered and reabsorbed, so bicarbonate secretion is neglected in the remaining discussion.

■ Reabsorption of Filtered HCO_3^- Restores Lost HCO_3^- to the Blood

HCO_3^- is freely filtered at the glomerulus, about 4500 mEq/day. Urinary loss of even a small portion of this HCO_3^- would lead to acidic blood and impair the body's ability to buffer its daily load of metabolically produced H+. The kidney tubules have the important task of recovering the filtered HCO_3^- and returning it to the blood.

Figure 25–6 shows how HCO_3^- filtration, reabsorption, and excretion normally vary with plasma $[HCO_3^-]$. The y-axis of this graph is unusual in that amounts of HCO_3^- per minute are factored by the glomerular filtration rate (GFR). The data are expressed thus because the maximal rate of tubular reabsorption of HCO_3^- varies with filtration rate. The amount of HCO_3^- reabsorbed per unit time is calculated as the difference between filtered and excreted amounts. The low plasma $[HCO_3^-]$ levels were achieved by having the subjects ingest NH_4Cl tablets. NH_4Cl is converted in the liver into urea, H_2O, and HCl, and the

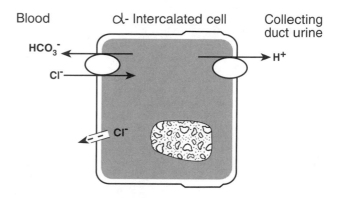

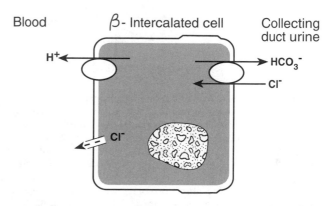

Figure 25–5 ■ The α-intercalated cell secretes H+ via an electrogenic H+ ATPase (or electroneutral H+-K+ ATPase) and adds HCO_3^- to the blood via a basolateral cell membrane Cl^--HCO_3^- exchanger. The β-intercalated cell has the opposite polarity and secretes HCO_3^-.

HCl consumes plasma HCO_3^-. As the graph shows, at low plasma concentrations of HCO_3^- (below about 26 mEq/L), all of the filtered HCO_3^- is reabsorbed. This makes good sense: if the blood is too acidic, the kidneys conserve filtered HCO_3^-.

If the plasma $[HCO_3^-]$ rises to high levels, for example because of intravenous infusion of solutions containing $NaHCO_3$, then filtered HCO_3^- exceeds the reabsorptive capacity of the tubules, and some HCO_3^- will be excreted in the urine (Fig. 25–6). This also makes good sense: if the blood is too alkaline, the kidneys excrete HCO_3^-. This loss of base would return the pH of the blood to its normal value.

At the cellular level (Fig. 25–7), filtered HCO_3^- is not reabsorbed directly across the tubule's luminal cell membrane as is, for example, glucose. Instead, filtered HCO_3^- is reabsorbed indirectly via H+ secretion, in the following way. (About 90% of the filtered HCO_3^- is reabsorbed in the proximal convoluted tubule, so events at this site are emphasized.) H+ is secreted into the tubule lumen via the Na+-H+ exchanger in the luminal membrane. It combines with filtered HCO_3^- to form H_2CO_3. Carbonic anhydrase in the luminal membrane (brush border) of the proximal tubule catalyzes

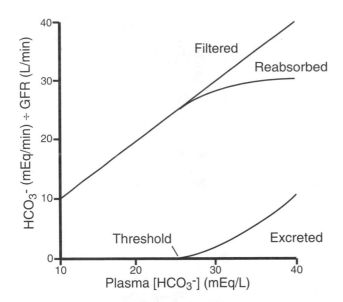

Figure 25–6 ■ Decreases in plasma bicarbonate concentration were produced by ingestion of NH_4Cl and increases by intravenous infusion of bicarbonate-containing solutions. All of the filtered bicarbonate was reabsorbed below a plasma concentration of about 26 mEq/L. Above this value (threshold), appreciable quantities of filtered bicarbonate were excreted in the urine. (Adapted from Pitts RF, Ayer JL, Schiess WA: The renal regulation of acid-base balance in man. III. The reabsorption and excretion of bicarbonate. *J Clin Invest* 28:35–44, 1949.)

the dehydration of H_2CO_3 to CO_2 and water in the lumen. The CO_2 diffuses back into the cell.

Inside the cell, the hydration of CO_2 (catalyzed by intracellular carbonic anhydrase) yields H_2CO_3, which instantaneously forms H^+ and HCO_3^-. The H^+ is secreted into the lumen, and the HCO_3^- ion moves into the blood surrounding the tubules. The HCO_3^- transporter in the basolateral cell membrane of proximal tubule cells carries one CO_3^{2-} and one HCO_3^- for each Na^+.

The reabsorption of filtered HCO_3^- does not result in H^+ excretion or the formation of any "new" HCO_3^-. The secreted H^+ is not excreted, because it combines with filtered HCO_3^- (i.e., is indirectly reabsorbed). There is no net addition of HCO_3^- to the body in this operation: it is simply a recovery or reclamation process.

■ Excretion of Titratable Acid and Ammonia Generates New Bicarbonate

When H^+ is excreted as titratable acid and ammonia, new HCO_3^- is formed and added to the blood. New bicarbonate replaces the HCO_3^- used to buffer the strong acids produced by metabolism.

The formation of new HCO_3^- and the excretion of H^+ are like two sides of the same coin. If we assume that H_2CO_3 is the source of H^+:

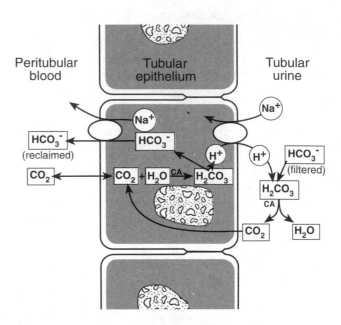

Figure 25–7 ■ Filtered bicarbonate combines with secreted H^+ and is reabsorbed indirectly. Carbonic anhydrase (CA) is present in the cells and, in the proximal tubule, on the brush border.

$$CO_2 + H_2O \rightleftharpoons H_2CO_3 \nearrow^{H^+ \text{ (urine)}}_{\searrow HCO_3^- \text{ (blood)}}$$

Loss of H^+ in the urine is equivalent to adding new bicarbonate to the blood. The same is true if H^+ is lost from the body via another route, such as vomiting acidic gastric juice. This process leads to a rise in plasma $[HCO_3^-]$. *Conversely, loss of bicarbonate from the body is equivalent to adding H^+ to the blood.*

Excretion of Titratable Acid ■ Figure 25–8 is a cell model for the formation of titratable acid. In this figure, $H_2PO_4^-$ is the titratable acid formed. H^+ and HCO_3^- form in the cell from H_2CO_3. The secreted H^+ combines with the basic form of the phosphate—HPO_4^{2-}—to form the acid phosphate, $H_2PO_4^-$. The secreted H^+ replaces one of the Na^+ ions accompanying the basic phosphate. The new HCO_3^- generated in the cell moves into the blood together with Na^+. For each mEq of H^+ excreted in the urine as titratable acid, 1 mEq of new HCO_3^- is added to the blood. This process gets rid of H^+ in the urine, replaces HCO_3^-, and restores a normal blood pH.

The amount of titratable acid excreted depends on two factors: urine pH and the availability of buffer. If the urine pH is lowered, more titratable acid can form. The supply of phosphate and other buffers is usually limited. To excrete large amounts of acid, the kidneys must rely on increased ammonia excretion.

Excretion of Ammonia ■ Figure 25–9 is a cell model for the excretion of ammonia. Ammonia is syn-

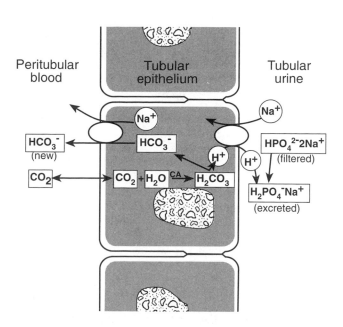

Figure 25–8 ■ Titratable acid (e.g., $H_2PO_4^-$) is formed when H^+ is secreted into the tubular urine. For each mEq of titratable acid excreted, 1 mEq of new bicarbonate is added to the peritubular capillary blood.

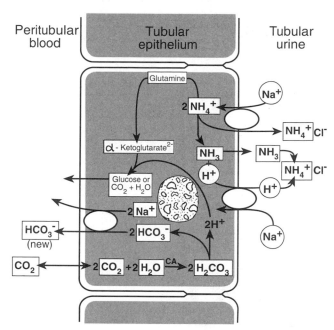

Figure 25–9 ■ Ammonium ions are formed from glutamine in the cell and are secreted into the tubular urine (top). H^+ from H_2CO_3 (bottom) is consumed when α-ketoglutarate is converted into glucose or CO_2 and H_2O. New bicarbonate ions are added to the peritubular capillary blood—1 mEq for each 1 mEq of NH_4^+ excreted in the urine.

thesized in the kidney tubule cells mainly through the deamidation and deamination of the amino acid glutamine:

$$\text{Glutamine} \xrightarrow[\text{Glutaminase}]{NH_4^+} \text{Glutamate}^- \xrightarrow[\text{Glutamate dehydrogenase}]{NH_4^+} \alpha\text{-ketoglutarate}^{2-}$$

As discussed earlier, ammonia is secreted into the urine by two mechanisms. As NH_3, it diffuses into the tubular urine; as NH_4^+, it substitutes for H^+ on the Na^+-H^+ exchanger. In the lumen, NH_3 combines with secreted H^+ to form NH_4^+, which is excreted.

For each mEq of H^+ excreted as ammonia, 1 mEq of new HCO_3^- is added to the blood. The hydration of CO_2 in the tubule cell produces H^+ and HCO_3^-, as was described earlier. H^+ is consumed when the anion α-ketoglutarate is converted into CO_2 and water or into glucose in the cell. The new HCO_3^- returns to the blood along with Na^+.

If excess acid is added to the body, urinary ammonia excretion is increased, for two reasons. First, a more acidic urine traps more ammonia (as NH_4^+) in the urine. Second, renal ammonia synthesis from glutamine increases over a period of several days. Enhanced renal ammonia synthesis and excretion is a lifesaving adaptation, because it allows the kidneys to remove large H^+ excesses and add more new bicarbonate to the blood. Also, the excreted NH_4^+ can substitute in the urine for Na^+ and K^+, thereby diminishing loss of these

cations. With severe metabolic acidosis (see below), ammonia excretion may increase almost 10-fold.

■ Several Factors Influence Renal Excretion of Hydrogen Ions

A number of factors influence renal excretion of H^+, including intracellular pH, arterial blood P_{CO_2}, carbonic anhydrase activity, sodium reabsorption, plasma potassium concentration, and aldosterone (Fig. 25–10). Some of these factors may help produce the proper renal response to a disturbed acid-base balance. On the other hand, abnormal renal excretion of an acid or base may be a side effect of another disturbance (e.g., a disorder in K^+ balance or aldosterone secretion).

Intracellular pH ■ The pH in kidney tubule cells is a key factor influencing the secretion, and, therefore, excretion of H^+. A fall in pH (increased [H^+]) enhances H^+ secretion. A rise in pH (decreased [H^+]) lowers H^+ secretion.

Arterial Blood P_{CO_2} ■ An increase in P_{CO_2} increases the formation of H^+ from H_2CO_3. This leads to enhanced renal H^+ secretion and excretion, a useful compensation for any condition in which the blood contains too much H_2CO_3 (see the discussion of respiratory acidosis). A decrease in P_{CO_2} results in lowered H^+ secretion and consequently less complete reabsorption of filtered HCO_3^- and loss of base in the urine

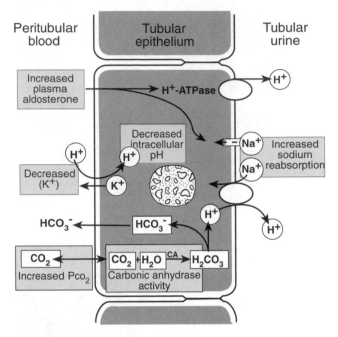

Figure 25–10 ■ Factors that lead to increased hydrogen ion secretion by the kidney tubule epithelium (see text).

(a useful compensation for respiratory alkalosis, discussed below).

Carbonic Anhydrase Activity ■ The enzyme carbonic anhydrase catalyzes two key reactions in urinary acidification: the hydration of CO_2 in the cells, forming H_2CO_3 and yielding H^+ for secretion and the dehydration of H_2CO_3 to CO_2 in the proximal tubule lumen, an important step in the reabsorption of filtered HCO_3^-. If carbonic anhydrase is inhibited (usually by a drug), large amounts of filtered HCO_3^- may escape reabsorption. This leads to a fall in blood pH.

Sodium Reabsorption ■ Sodium reabsorption is closely linked to H^+ secretion. In the proximal tubule, the two ions are directly linked, both being transported by the Na^+-H^+ exchanger in the luminal cell membrane. The relation is less direct in the collecting ducts. Enhanced Na^+ reabsorption in the ducts leads to a more negative intraluminal electrical potential, which favors H^+ secretion by its H^+ ATPase. The avid renal reabsorption of Na^+ seen in states of volume depletion is accompanied by a parallel rise in urinary H^+ excretion.

Plasma Potassium Concentration ■ Changes in plasma $[K^+]$ influence renal excretion of H^+. A fall in plasma $[K^+]$ favors the movement of K^+ from body cells into interstitial fluid (or blood plasma) and a reciprocal movement of H^+ into cells. In the kidney tubule cells, these movements lower intracellular pH and increase H^+ secretion. Potassium depletion also stimulates ammonia synthesis by the kidneys. The result is the complete reabsorption of filtered HCO_3^- and enhanced generation of new HCO_3^- as more titratable acid and ammonia are excreted.

The overall result of these events is that hypokalemia (or reduced body K^+ stores) leads to an increased plasma $[HCO_3^-]$ (metabolic alkalosis). Hyperkalemia (or excess K^+ in the body) results in the opposite changes: an increase in intracellular pH, decreased H^+ secretion, incomplete reabsorption of filtered HCO_3^-, and a fall in plasma $[HCO_3^-]$ (metabolic acidosis).

Aldosterone ■ Aldosterone stimulates the collecting ducts to secrete H^+ by three actions. First, it directly stimulates the H^+ pump. Second, it enhances Na^+ reabsorption, which leads to a more negative intraluminal potential, and consequently promotes H^+ secretion by the electrogenic H^+ ATPase. Third, it promotes potassium secretion. This leads to hypokalemia, and, as just discussed, enhanced renal H^+ secretion. Hyperaldosteronism results in enhanced renal H^+ excretion and an alkaline blood pH; the opposite occurs with hypoaldosteronism.

Regulation of pH Inside the Cell

The intra- and extracellular fluids are linked by exchanges across cell membranes of H^+, HCO_3^-, various acids and bases, and CO_2. By stabilizing extracellular fluid pH, the body helps protect intracellular pH.

If H^+ ions were passively distributed across cell membranes, intracellular pH would be lower than that seen in most body cells. In skeletal muscle cells, for example, we can calculate from the Nernst equation (Chap. 2) and a membrane potential of -90 mV that cytosolic pH should be 5.9 if extracellular fluid pH is 7.4; actual measurements, however, indicate a pH of 6.9. From this discrepancy, it is clear that H^+ ions are not at equilibrium across the cell membrane and the cell must use active mechanisms to extrude H^+.

Cells are typically threatened by acidic metabolic end products and by the tendency for H^+ to diffuse into the cell down the electrical gradient (Fig. 25–11). H^+ is extruded by a Na^+-H^+ exchanger, which is present in nearly all body cells. This exchanger is electrically neutral; it exchanges one H^+ for one Na^+. Active extrusion of H^+ keeps the internal pH constant.

The activity of the Na^+-H^+ exchanger is regulated by intracellular pH and a variety of hormones and growth factors (Fig. 25–12). Not surprisingly, an increase in intracellular $[H^+]$ stimulates the exchanger. However, this is not only because of more substrate (H^+) for the exchanger: H^+ also stimulates the exchanger by protonating an activator site on the cytoplasmic side of the exchanger. This makes the exchanger more effective in dealing with the threat of intracellular acidosis. Many hormones and growth factors, via intracellular second messengers, activate various protein kinases that stimulate or inhibit the Na^+-H^+ exchanger. In this way, they produce changes in intracellular pH, which may lead to changes in cell activity.

Besides extruding H^+, the cell can deal with acids

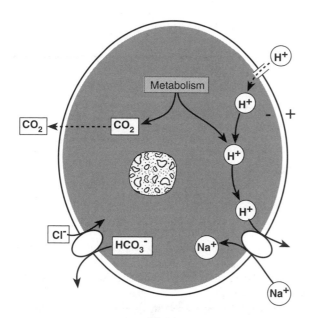

Figure 25–11 ■ Cells usually maintain a constant intracellular pH. The cell is acidified because of production of hydrogen ions from metabolism and influx of hydrogen ions from the extracellular fluid (favored by the inside negative membrane potential). To maintain a stable intracellular pH, the cell must extrude hydrogen ions at a rate matching their input. Many cells also possess various HCO_3^- transporters, which defend against excess acid or base.

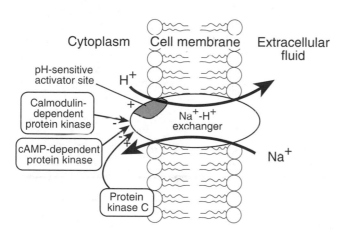

Figure 25–12 ■ The plasma membrane Na^+-H^+ exchanger plays a key role in regulating intracellular pH. Many hormones and growth factors, acting via intracellular second messengers and protein kinases, can increase ($+$) or decrease ($-$) activity of the exchanger.

and bases in other ways. In some cells, various HCO_3^- transporting systems (e.g., sodium-dependent and sodium-independent chloride-bicarbonate exchangers) may be present in plasma membranes. These exchangers may be activated by changes in intracellular pH. Cells have large stores of protein and organic phosphate buffers, which can bind or release H^+. Various chemical reactions in cells can also use up or release H^+. For example, conversion of lactic acid to

CO_2 and water or to glucose effectively disposes of acid. In addition, various cell organelles may sequester H^+. For example, a H^+ ATPase in endosomes and lysosomes pumps H^+ out of the cytosol into these organelles. In summary, ion transport, buffering mechanisms, and metabolic reactions all ensure a relatively stable intracellular pH.

Disturbances of Acid-Base Balance

Table 25–2 summarizes normal values for the pH (or [H^+]), P_{CO_2}, and [HCO_3^-] of arterial blood plasma. A blood pH below 7.35 ([H^+] greater than 45 nmol/L) indicates **acidemia.** A blood pH above 7.45 ([H^+] less than 35 nmol/L) indicates **alkalemia.** The range of pH values compatible with life is approximately 6.8-7.8 ([H^+] = 160-16 nmol/L).

Four **simple acid-base disturbances** may lead to an abnormal blood pH: respiratory acidosis, respiratory alkalosis, metabolic acidosis, and metabolic alkalosis. The term *simple* indicates that there is a single primary cause for the acid-base disturbance. **Acidosis** is an abnormal process that tends to produce acidemia. **Alkalosis** is an abnormal process that tends to produce alkalemia. If there is too much or too little CO_2, then a **respiratory disturbance** is present. If there is too much or too little HCO_3^-, then a **metabolic (or nonrespiratory) disturbance** of acid-base balance is present. Table 25–3 summarizes the changes in blood pH (or [H^+]), plasma [HCO_3^-], and P_{CO_2} in each of the four simple acid-base disturbances.

In considering acid-base disturbances, it is helpful to recall the Henderson-Hasselbalch equation for HCO_3^-/CO_2:

$$pH = 6.10 + \log \frac{[HCO_3^-]}{0.03\ P_{CO_2}} \qquad (17)$$

If the primary problem is a change in [HCO_3^-] or P_{CO_2}, the pH can be brought closer to normal by changing the other member of the buffer pair *in the same direction.* For example, if P_{CO_2} is primarily decreased, a decrease in plasma [HCO_3^-] will minimize the change in pH. In various acid-base disturbances, the lungs adjust the blood P_{CO_2} and the kidneys adjust the plasma [HCO_3^-] to reduce departures of pH from nor-

TABLE 25–2 ■ Normal Arterial Blood Plasma Acid-Base Values

	Mean	Range*
pH	7.40	7.35-7.45
[H^+], nmol/L	40	45-35
P_{CO_2}, mm Hg	40	35-45
[HCO_3^-], mEq/L	24	22-26

*The range extends from 2 SD below to 2 SD above the mean and encompasses 95% of the population of healthy people.

TABLE 25–3 ■ Directional Changes in Arterial Blood Plasma Values in the Four Simple Acid-Base Disturbances

	Arterial Plasma				
Disturbance	pH	[H⁺]	[HCO₃⁻]	Pco₂	Compensatory Response
Respiratory acidosis	↓	↑	↑	⬆	Kidneys increase H⁺ excretion
Respiratory alkalosis	↑	↓	↓	⬇	Kidneys increase HCO_3^- excretion
Metabolic acidosis	↓	↑	⬇	↓	Alveolar hyperventilation
Metabolic alkalosis	↑	↓	⬆	↑	Alveolar hypoventilation

Heavy arrows indicate the main effect.

mal. These adjustments, called **compensations,** generally do not bring about a normal blood pH.

■ Respiratory Acidosis Results from Accumulation of Carbon Dioxide

Respiratory acidosis is an abnormal process characterized by CO_2 accumulation. The CO_2 buildup pushes the following reactions to the right:

$$CO_2 + H_2O \rightleftharpoons H_2CO_3 \rightleftharpoons H^+ + HCO_3^-$$

Blood [H_2CO_3] increases, leading to an increase in [H⁺] or a fall in pH. Respiratory acidosis is usually caused by failure to expire metabolically produced CO_2 at an adequate rate. This leads to accumulation of CO_2 in the blood and a fall in blood pH. This disturbance may be due to a decrease in overall alveolar ventilation (hypoventilation) or, as occurs commonly in lung disease, a mismatch between ventilation and perfusion. Respiratory acidosis also occurs if a person breathes CO_2-enriched air.

Chemical Buffering ■ In respiratory acidosis, more than 95% of the chemical buffering occurs in cells. The cells contain many proteins and organic phosphates that can bind H⁺. For example, hemoglobin (Hb) in red blood cells combines with H⁺ from H_2CO_3, thereby minimizing the increase in free H⁺. Recall (Chap. 21) the buffering reaction: $H_2CO_3 + HbO_2^- \rightleftharpoons HHb + O_2 + HCO_3^-$. This reaction raises the plasma [HCO_3^-]. In *acute* respiratory acidosis, chemical buffering in the body leads to an increase in plasma [HCO_3^-] of about 1 mEq/L for each 10 mm Hg increase in Pco₂ (Table 25–4).

An example illustrates how chemical buffering reduces a fall in pH during respiratory acidosis. Suppose Pco₂ increased from a normal value of 40 mm Hg to 70 mm Hg ($[CO_2]_d = 2.1$ mmol/L). If there were no body buffer bases that could accept H⁺ from H_2CO_3 (i.e., if there were no measurable increase in [HCO_3^-]) the resulting pH would be 7.16:

$$pH = 6.10 + \log \frac{[24]}{[2.1]} = 7.16 \qquad (18)$$

In acute respiratory acidosis, a 3 mEq/L increase in plasma [HCO_3^-] occurs with a 30 mm Hg rise in Pco₂ (Table 25–4). Therefore the pH is 7.21:

$$pH = 6.10 + \log \frac{[24 + 3]}{[2.1]} = 7.21 \qquad (19)$$

The pH of 7.21 is closer to a normal pH, because body buffer bases (mainly intracellular buffers) such as proteins and phosphates combine with H⁺ ions liberated from H_2CO_3.

Respiratory Response ■ Respiratory acidosis produces a rise in Pco₂ and a fall in pH and is often associated with hypoxia. These changes tend to stimulate breathing (Chap. 22) and thereby diminish the severity of the acidosis. In other words, a person would be worse off if the respiratory system did not reflexly respond to the abnormalities in blood Pco₂, pH, and Po₂.

Renal Compensation ■ The kidneys compensate for respiratory acidosis by adding more H⁺ to the urine and adding new HCO_3^- to the blood. The increased Pco₂ stimulates renal H⁺ secretion, which allows the reabsorption of all filtered HCO_3^-. Excess H⁺ is excreted as titratable acid and NH_4^+; these processes add new HCO_3^- to the blood, causing plasma [HCO_3^-] to rise. This compensation takes several days to develop fully.

TABLE 25–4 ■ Compensatory Responses in Acid-Base Disturbances[a]

Respiratory acidosis	
Acute	1 mEq/L increase in plasma [HCO_3^-] for each 10 mm Hg increase in Pco₂ [b]
Chronic	4 mEq/L increase in plasma [HCO_3^-] for each 10 mm Hg increase in Pco₂ [c]
Respiratory alkalosis	
Acute	2 mEq/L decrease in plasma [HCO_3^-] for each 10 mm Hg decrease in Pco₂ [b]
Chronic	4 mEq/L decrease in plasma [HCO_3^-] for each 10 mm Hg decrease in Pco₂ [c]
Metabolic acidosis	1.3 mm Hg decrease in Pco₂ for each 1 mEq/L decrease in plasma [HCO_3^-] [d]
Metabolic alkalosis	0.7 mm Hg increase in Pco₂ for each 1 mEq/L increase in plasma [HCO_3^-] [d]

[a]These are empirically determined average changes measured in people with simple acid-base disorders.
[b]This change is mainly because of chemical buffering.
[c]This change is mainly because of renal compensation.
[d]This change is because of respiratory compensation.
[From Valtin, H, Gennari, FJ. *Acid-base disorders. Basic concepts and clinical management*. Boston: Little, Brown and Company, 1987.]

VOMITING AND METABOLIC ALKALOSIS

Vomiting of gastric acid juice results in metabolic alkalosis and fluid and electrolyte disturbances. Gastric acid juice contains about 0.1 mol/L HCl. When HCl is secreted by the stomach parietal cells and lost to the outside, there is a net gain of bicarbonate in the blood plasma. The bicarbonate, in effect, replaces lost plasma Cl^-. Ventilation is inhibited by the alkaline pH, resulting in a rise in P_{CO_2}. This respiratory compensation, however, is limited, because hypoventilation leads to a rise in P_{CO_2} and a fall in P_{O_2}, both of which stimulate breathing.

The logical renal compensation for metabolic alkalosis is enhanced excretion of bicarbonate. In people with persistent vomiting, however, the urine is sometimes acidic and renal bicarbonate reabsorption is enhanced, thereby maintaining an elevated plasma bicarbonate level. This situation arises because vomiting is accompanied by losses of extracellular fluid and potassium. Fluid loss leads to a decrease in effective arterial blood volume and engagement of mechanisms that reduce sodium excretion, such as decreased glomerular filtration rate and increased plasma aldosterone levels (Chap. 24). Renal tubular Na^+-H^+ exchange is stimulated, since there is a deficiency of reabsorbable anion (i.e., Cl^-) to accompany reabsorbed Na^+. Potassium depletion develops because of loss of K^+ in the vomitus, decreased food intake, and, most important quantitatively, enhanced renal K^+ excretion. Renal K^+ loss is promoted by an increased plasma aldosterone level. Hypokalemia develops, which may enhance renal H^+ secretion. The kidneys reabsorb filtered bicarbonate completely and maintain the metabolic alkalosis.

Treatment for the acid-base disturbance is, first, to eliminate the cause of vomiting. Correction of the alkalosis by administering an organic acid, such as lactic acid, does not make sense, since this acid would simply be converted to CO_2 and H_2O, which does not address the Cl^- deficit. The extracellular fluid volume depletion and Cl^- and K^+ deficits can be corrected by administering isotonic saline and appropriate amounts of KCl. After Na^+, Cl^-, water, and K^+ deficits have been replaced, excess bicarbonate (accompanied by surplus Na^+) will be excreted in the urine, and the kidneys will return blood pH to normal.

With *chronic* respiratory acidosis, plasma $[HCO_3^-]$ increases, on average, by 4 mEq/L for each 10 mm Hg rise in P_{CO_2} (Table 25-4). This rise exceeds that seen with acute respiratory acidosis because of the renal addition of HCO_3^- to the blood. A person with chronic respiratory acidosis and a P_{CO_2} of 70 mm Hg would be expected to have an increase in plasma HCO_3^- of 12 mEq/L. The blood pH would be 7.33:

$$pH = 6.10 + \log \frac{[24 + 12]}{[2.1]} = 7.33 \qquad (20)$$

With chronic respiratory acidosis, time for renal compensation is allowed, so blood pH (in this example 7.33) is much closer to normal than is observed during acute respiratory acidosis (pH 7.21; see above).

■ Respiratory Alkalosis Results from Excessive Loss of Carbon Dioxide

Respiratory alkalosis is most easily understood as the opposite of respiratory acidosis; it is an abnormal process causing the loss of too much CO_2. This loss causes blood $[H_2CO_3]$ and thus $[H^+]$ to fall (pH rises). Alveolar hyperventilation causes respiratory alkalosis. Metabolically produced CO_2 is flushed out of the alveolar spaces more rapidly than it is added by the pulmonary capillary blood. This causes alveolar and arterial P_{CO_2} to fall. Hyperventilation and respiratory alkalosis can be caused by voluntary effort, anxiety, direct stimulation of the medullary respiratory center by an abnormality (e.g., meningitis, fever, aspirin intoxication), or hypoxia due to severe anemia or high altitude.

Chemical Buffering ■ As with respiratory acidosis, during respiratory alkalosis more than 95% of chemical buffering occurs within cells. Cell proteins and organic phosphates liberate H^+ ions, which are added to extracellular fluid and lower the plasma $[HCO_3^-]$; this reduces the alkaline shift in pH.

With *acute* respiratory alkalosis, plasma $[HCO_3^-]$ falls by about 2 mEq/L for each 10 mm Hg drop in P_{CO_2} (Table 25-4). For example, if P_{CO_2} drops from 40 to 20 mm Hg ($[CO_2]_d = 0.6$ mmol/L) plasma $[HCO_3^-]$ falls by 4 mEq/L, and the pH will be 7.62:

$$pH = 6.10 + \log \frac{[24 - 4]}{[0.6]} = 7.62 \qquad (21)$$

If plasma $[HCO_3^-]$ had not changed, the pH would have been 7.70:

$$pH = 6.10 + \log \frac{[24]}{[0.6]} = 7.70 \qquad (22)$$

Respiratory Response ■ Although hyperventilation causes respiratory alkalosis, it also causes changes (a fall in P_{CO_2} and a rise in blood pH) that inhibit ventilation and therefore limit the extent of hyperventilation.

Renal Compensation ■ The kidneys compensate for respiratory alkalosis by excreting HCO_3^- in the urine, thereby getting rid of base. A reduced P_{CO_2} reduces H^+ secretion by the kidney tubule epithelium. As a result, some of the filtered HCO_3^- is not reabsorbed. When the urine becomes more alkaline, titratable acid excretion vanishes, and little ammonia is excreted. The enhanced output of HCO_3^- causes plasma $[HCO_3^-]$ to fall.

Chronic respiratory alkalosis is accompanied by a 4 mEq/L fall in plasma $[HCO_3^-]$ for each 10 mm Hg drop in P_{CO_2} (Table 25-4). For example, in a person with chronic hyperventilation and a P_{CO_2} of 20 mm Hg, the blood pH is

$$pH = 6.10 + \log \frac{[24 - 8]}{[0.6]} = 7.53 \qquad (23)$$

This pH is closer to normal than the pH of 7.62 seen with acute respiratory alkalosis. The difference between the two situations is largely due to renal compensation.

■ Metabolic Acidosis Results from Gain of Noncarbonic Acid or Loss of Bicarbonate

Metabolic acidosis is an abnormal process characterized by a gain of acid (other than H_2CO_3) or loss of HCO_3^-. Either causes plasma $[HCO_3^-]$ and pH to fall. If a strong acid is added to the body, the reactions

$$H^+ + HCO_3^- \rightleftharpoons H_2CO_3 \rightleftharpoons H_2O + CO_2$$

are pushed to the right. The added H^+ consumes HCO_3^-. If a great deal of acid is infused rapidly, P_{CO_2} rises, as the equation predicts. This increase occurs only transiently, however, because the body is an open system, and the lungs expire CO_2 as it is generated. P_{CO_2} actually falls below normal, since an acidic blood pH stimulates ventilation (Fig. 25-3).

Many conditions can produce metabolic acidosis, including renal failure; uncontrolled diabetes mellitus; lactic acidosis; ingestion of acidifying agents such as NH_4Cl; abnormal renal excretion of HCO_3^-; and diarrhea. In renal failure, the kidneys cannot excrete H^+ fast enough to keep up with metabolic acid production, and in uncontrolled diabetes mellitus, production of ketone body acids increases. Lactic acidosis results from tissue hypoxia. Ingested NH_4Cl is converted into a strong acid, HCl, in the liver. Diarrhea causes loss of alkaline intestinal fluids.

Chemical Buffering ■ Excess acid is chemically buffered in extra- and intracellular fluids and bone. In metabolic acidosis, roughly half of the buffering occurs in cells and bone. HCO_3^- is the principal buffer in the extracellular fluid.

Respiratory Compensation ■ The acidic blood pH stimulates the respiratory system to lower blood P_{CO_2}. This action lowers blood $[H_2CO_3]$ and so tends to alkalinize the blood, opposing the acidic shift in pH. Metabolic acidosis is accompanied on average by a 1.3 mm Hg fall in P_{CO_2} for each 1 mEq/L drop in plasma $[HCO_3^-]$ (Table 25-4). Suppose, for example, that the infusion of a strong acid causes the plasma $[HCO_3^-]$ to drop from 24 to 12 mEq/L. If there were no respiratory compensation and the P_{CO_2} did not change from its normal value of 40 mm Hg, the pH would be 7.10:

$$pH = 6.10 + \log \frac{[12]}{[1.2]} = 7.10 \qquad (24)$$

With respiratory compensation, the P_{CO_2} falls by 16 mm Hg (12 × 1.3) to 24 mm Hg ($[CO_2]_d = 0.72$ mmol/L) and pH is 7.32:

$$pH = 6.10 + \log \frac{[12]}{[0.72]} = 7.32 \qquad (25)$$

which is closer to normal than a pH of 7.10. The respiratory response develops promptly (within minutes) and is maximal after 12-24 hours.

Renal Compensation ■ The kidneys respond to metabolic acidosis by adding more H^+ to the urine. Since the plasma $[HCO_3^-]$ is primarily lowered, the filtered load of HCO_3^- drops and the kidneys can accomplish the complete reabsorption of filtered HCO_3^- (Fig. 25-6). More H^+ is excreted as titratable acid and NH_4^+. With chronic metabolic acidosis, the kidneys make more ammonia and can therefore add more new HCO_3^- to the blood, to replace lost HCO_3^-.

If the underlying cause of metabolic acidosis is corrected, then normal kidneys can correct the blood pH in a few days.

■ Metabolic Alkalosis Results from Gain of Strong Base or Bicarbonate or Loss of Noncarbonic Acid

Metabolic alkalosis is an abnormal process characterized by the gain of a strong base or HCO_3^- or the loss of an acid (other than H_2CO_3). Plasma $[HCO_3^-]$ and pH rise; P_{CO_2} rises because of respiratory compensation. These changes are opposite those seen in metabolic acidosis (Table 25–3). A variety of situations can produce metabolic alkalosis, such as ingestion of antacids, vomiting of gastric acid juice, or enhanced renal H^+ loss (e.g., due to hyperaldosteronism or hypokalemia).

Chemical Buffering ■ Chemical buffers in the body limit the alkaline shift in blood pH by releasing H^+ as they are titrated in the alkaline direction. About one-third of the buffering occurs in cells.

Respiratory Compensation ■ The respiratory compensation for metabolic alkalosis is hypoventilation. An alkaline blood pH inhibits ventilation. Hypoventilation raises the blood P_{CO_2} and $[H_2CO_3]$, thereby reducing the alkaline shift in pH. A 1 mEq/L rise in plasma $[HCO_3^-]$ caused by metabolic alkalosis is accompanied by a 0.7 mm Hg rise in P_{CO_2} (Table 25–4). If, for example, the plasma $[HCO_3^-]$ rose to 40 mEq/L, what would the plasma pH be with and without respiratory compensation? With respiratory compensation, the P_{CO_2} should rise by 11.2 mm Hg (0.7 × 16) to 51.2 mm Hg ($[CO_2]_d = 1.54$ mmol/L). The pH is 7.51:

$$pH = 6.10 + \log \frac{[40]}{[1.54]} = 7.51 \qquad (26)$$

Without respiratory compensation, the pH would be 7.62:

$$pH = 6.10 + \log \frac{[40]}{[1.2]} = 7.62 \qquad (27)$$

Respiratory compensation for metabolic alkalosis is limited because hypoventilation leads to hypoxia and CO_2 retention, both of which increase breathing.

Renal Compensation ■ The kidneys respond to metabolic alkalosis by lowering the plasma $[HCO_3^-]$. The plasma $[HCO_3^-]$ is primarily raised, so more HCO_3^- is filtered than can be reabsorbed (Fig. 25–6); in addition, bicarbonate is secreted in the collecting ducts. Both of these changes lead to increased urinary HCO_3^- excretion. If the cause of the metabolic alkalosis is corrected, the kidneys can often restore the plasma $[HCO_3^-]$ and pH to normal in 1–2 days.

■ Clinical Evaluation of Acid-Base Disturbances Requires a Comprehensive Study

Acid-base data should always be interpreted in the context of other information about a patient. A complete history and physical examination provide important clues to possible reasons for an acid-base disorder.

To identify an acid-base disturbance from laboratory values, it is best to look first at the pH. A low blood pH indicates acidosis and a high blood pH alkalosis. If acidosis is present, for example, it could be either respiratory or metabolic. A low blood pH and elevated P_{CO_2} point to respiratory acidosis and a low pH and low plasma $[HCO_3^-]$ to metabolic acidosis. If alkalosis is present, it could be either respiratory or metabolic. A high blood pH and low plasma P_{CO_2} indicate respiratory alkalosis and a high blood pH and high plasma $[HCO_3^-]$ metabolic alkalosis.

Whether the body is responding appropriately to a simple acid-base disorder can be judged from the values in Table 25–4. Inappropriate values suggest that more than one acid-base disturbance may be present. Patients sometimes have two or more of the four simple acid-base disturbances at the same time, in which case they have a **mixed acid-base disturbance.**

■ ■ ■ ■ ■ ■ ■ ■ ■ ■ ■ ■ **REVIEW EXERCISES** ■ ■ ■ ■ ■ ■ ■ ■ ■ ■ ■ ■

Identify Each with a Term
1. An arterial blood pH greater than 7.45
2. An abnormal process that tends to produce an acidic blood pH
3. The principle which states that all buffers in a solution are in equilibrium with the same $[H^+]$
4. Simultaneous presence of two or more acid-base disturbances
5. The major chemical buffer pair in extracellular fluid
6. Nephron segment that secretes the most H^+
7. Nephron segment that can establish the steepest pH gradient

Define Each Term
8. Henderson-Hasselbalch equation
9. Respiratory acidosis
10. Respiratory alkalosis
11. Metabolic acidosis
12. Metabolic alkalosis

Choose the Correct Answer

13. Most of the H+ secreted by the kidney tubules is:
    a. Consumed in the reabsorption of filtered bicarbonate
    b. Excreted as ammonium ions
    c. Excreted as free hydrogen ions
    d. Excreted as titratable acid

14. Which of the following processes adds "new" bicarbonate to the blood (i.e., results in net addition of bicarbonate to plasma)?
    a. Excretion of ammonia
    b. Excretion of titratable acid
    c. Reabsorption of filtered bicarbonate
    d. All of the above
    e. a and b

15. Which of the following results in increased secretion and excretion of H+ by the kidneys?
    a. Administration of a carbonic anhydrase inhibitor
    b. A decrease in plasma aldosterone
    c. A decrease in renal sodium reabsorption
    d. An increase in arterial blood P_{CO_2}
    e. Hyperkalemia

16. Which of the following contribute(s) to the regulation of intracellular (cytosolic) pH?
    a. Buffering by intracellular proteins and organic phosphates
    b. Cell membrane bicarbonate transporters
    c. Cell membrane Na+-H+ exchanger
    d. Cell metabolic reactions
    e. All of the above

Calculate and Briefly Answer

17. The following acid-base measurements were obtained from a person with uncontrolled diabetes mellitus: arterial blood plasma pH, 7.28; arterial blood P_{CO_2}, 22 mm Hg. (a) What is the plasma [H+]? (b) What is the plasma [HCO_3^-]? (c) What type of acid-base disturbance is present? (d) What explains the fall in plasma [HCO_3^-]? (e) What explains the fall in arterial blood P_{CO_2}? (f) If there had been no compensatory change in P_{CO_2}, what would the pH be?

18. In the person described in question 17, the following measurements on the urine were recorded: titratable acid, 60 mEq/day; ammonia, 200 mEq/day; bicarbonate, 0 mEq/day; pH, 5. (a) What accounts for the high rate of ammonia excretion? (b) What is the net acid excretion by the kidneys? (c) How much new bicarbonate did the kidneys generate?

19. A 3-year-old child was brought to the hospital with a cough, respiratory distress, and cyanosis. Physical examination suggested a lower respiratory tract infection. A throat swab was positive for *Staphylococcus*. On admission, arterial P_{O_2} was 29 mm Hg, P_{CO_2} 75 mm Hg, and pH 7.24. (a) What is the plasma bicarbonate concentration? (b) What type of acid-base disturbance is present? After 48 hours of treatment in the hospital (humidified oxygen to breathe, antibiotics), arterial P_{CO_2} was 53 mm Hg and pH 7.39. (c) What is the new plasma bicarbonate concentration? (d) Explain why the plasma bicarbonate hardly changed, despite the 22 mm Hg drop in P_{CO_2}.

20. The following acid-base measurements were obtained from a person with chronic pulmonary disease: arterial plasma pH, 7.35; arterial blood P_{CO_2} 60 mm Hg. (a) What is the [H+]? (b) What is the plasma [HCO_3^-]? (c) What type of acid-base disturbance is present? (d) What accounts for the increase in plasma [HCO_3^-]? Is this increase appropriate (see Table 25–4)? The subject had severe diarrhea for several days and the following values were then recorded: arterial blood pH, 7.30; arterial blood P_{CO_2}, 50 mm Hg; plasma [HCO_3^-], 24 mEq/L. (e) What combination of simple acid-base disturbances is present?

21. During an American expedition to Mt. Everest in 1981*, acid-base values for a climber at the summit (8848 m above sea level) were estimated to be: arterial blood pH, 7.76; arterial blood P_{O_2}, 28 mm Hg; arterial blood P_{CO_2}, 7.5 mm Hg; plasma [HCO_3^-], 10 mEq/L. Prior to the ascent to the summit, the climber had spent more than 2 weeks camped on the mountain slope to acclimatize to the high altitude. (a) What type of acid-base disturbance is present? (b) How do you explain the fall in plasma [HCO_3^-]? (c) Did the subject hyperventilate at the summit, and why?

*West JB, Lahiri S: *High Altitude and Man.* Bethesda, American Physiological Society, 1984.

■ ■ ■ ■ ■ ■ ■ ■ ■ ■ ■ ■ ANSWERS ■ ■ ■ ■ ■ ■ ■ ■ ■ ■ ■ ■

1. Alkalemia

2. Acidosis

3. Isohydric

4. Mixed acid-base disturbance

5. Bicarbonate/CO_2

6. Proximal convoluted tubule

7. Collecting duct

8. $$pH = pK_a + \log \frac{[A^-]}{[HA]}$$

9. An abnormal process characterized by CO_2 accumulation

10. An abnormal process causing the loss of too much CO_2

11. An abnormal process characterized by gain of acid (other than H_2CO_3) or loss of HCO_3^-

12. An abnormal process characterized by gain of a strong base or HCO_3^- or loss of acid (other than H_2CO_3)

13. a

14. e

15. d

16. e

17. (a) [H+] = $10^{-7.28}$ = 52 nmol/L; (b) 7.28 = 6.10 + log ([HCO_3^-]/0.66), so [HCO_3^-] = 10 mEq/L; (c) meta-

bolic acidosis; (d) buffering of ketone body acids; (e) hyperventilation stimulated by decreased pH; (f) pH = 6.10 + log ([10]/1.20) = 7.02

18. (a) Increased ammonia synthesis and increased trapping of NH_4^+ by acidic urine; (b) 260 mEq/day; (c) 260 mEq/day

19. (a) 31 mEq/L; (b) respiratory acidosis; (c) 31 mEq/L; (d) the kidneys increased the plasma bicarbonate to compensate for chronic respiratory acidosis

20. (a) 45 nmol/L; (b) 32 mEq/L; (c) chronic respiratory acidosis; (d) renal compensation; yes; (e) metabolic acidosis and chronic respiratory acidosis

21. (a) Respiratory alkalosis; (b) renal compensation for chronic respiratory alkalosis, chemical buffering, and lactic acid production due to exertion and hypoxia; (c) yes; hypoxia stimulated ventilation via peripheral chemoreceptors

Suggested Reading

Alpern RJ, Stone DK, Rector FC Jr: Renal acidification mechanisms, in Brenner BM, Rector FC Jr (eds): *The Kidney.* 4th ed. Philadelphia, WB Saunders, 1991, pp 318–379.

Boron WF: Control of intracellular pH, in Seldin DW, Giebisch G (eds): *The Kidney: Physiology and Pathophysiology.* 2nd ed. New York, Raven, 1992, pp 219–263.

Cogan MG, Rector FC Jr: Acid-base disorders, in Brenner BM, Rector FC Jr (eds). *The Kidney.* 4th ed. Philadelphia, WB Saunders, 1991, pp 737–804.

Gamble JL Jr: *Acid-Base Physiology: A Direct Approach.* Baltimore, Johns Hopkins University Press, 1982.

Gennari FJ, Maddox DA: Renal regulation of acid-base homeostasis: Integrated response, in Seldin DW, Giebisch G (eds): *The Kidney: Physiology and Pathophysiology.* 2nd ed. New York, Raven, 1992, pp 2695–2732.

Halperin ML, Kamel KS, Ethier JH, et al: Biochemistry and physiology of ammonium excretion, in Seldin DW, Giebisch G (eds). *The Kidney: Physiology and Pathophysiology.* 2nd ed. New York, Raven, 1992, pp 2645–2679.

Hamm LL, Alpern RJ: Cellular mechanisms of renal tubular acidification, in Seldin DW, Giebisch G (eds): *The Kidney: Physiology and Pathophysiology.* 2nd ed. New York, Raven, 1992, pp 2581–2626.

Knepper MA, Packer R, Good DW: Ammonium transport in the kidney. *Physiol Rev* 69:179–249, 1989.

Lowenstein J: *Acid and Basics: A Guide to Understanding Acid-Base Disorders.* New York, Oxford University Press, 1993.

Madshus IH: Regulation of intracellular pH in eukaryotic cells. *Biochem J* 250:1–8, 1988.

Rose BD: *Clinical Physiology of Acid-Base and Electrolyte Disorders.* 4th ed. New York, McGraw-Hill, 1994.

Schuster VL: Function and regulation of collecting duct intercalated cells. *Annu Rev Physiol* 55:267–288, 1993.

Valtin H, Gennari FJ: *Acid-Base Disorders: Basic Concepts and Clinical Management.* Boston, Little, Brown, 1987.

Winters RW, Dell RB: *Acid-Base Physiology in Medicine: A Self-Instruction Program.* 3rd ed. Boston, Little, Brown, 1982.

PART SEVEN

Gastro-intestinal Physiology

Gastrointestinal Neurophysiology

Nerves of the gastrointestinal tract control contraction of the muscle coats, secretion, and absorption across the mucosal lining and blood flow inside the walls of the esophagus, stomach, intestines, and gallbladder. Depending on the kind of neurotransmitter released, the neurons can make the muscles contract or may inhibit muscle contraction. Secretion of water, electrolytes, and mucus into the lumen and absorption from the lumen are determined by the innervation. The amount of blood flow within the wall and the distribution of flow between the muscle layers and mucosa are also controlled by nervous activity.

The musculature, mucosal epithelium, and blood vessels are called effector systems. Overall behavior in any region of the digestive tract at a given time reflects the integrated activity of the effector systems. In addition to stimulating or inhibiting the activity of the effector systems, the nervous system coordinates their activity. For example, mucosal secretion and muscle contractions are coordinated in a timed sequence during propulsion of the contents in the intestinal lumen. Secretion of water, electrolytes, and mucus occurs first, followed by propulsive contractions of the muscles. While this is happening, the nervous system acts to increase blood flow that supplies water and electrolytes in support of the secretory behavior. The neurally evoked secretory activity flushes materials from the mucosa and lubricates the luminal lining in preparation for muscular propulsion of the material out of the segment of bowel.

The nervous system in the digestive tract is like a minibrain that contains a library of programs for different forms of behavior. A specific neural program determines gastrointestinal behavior in the fed (**postprandial**) state and another program determines behavior in the **fasting** state. Upon intake of a meal, the program for the fasting state is switched off and the postprandial program comes into action. During vomiting, propulsion in the small intestine is reversed to move the contents rapidly toward the stomach. This **emetic** program can be called up from the nerve circuits in the intestine either by commands from the brain or by local sensory detection of noxious substances in the lumen.

Innervation of the Gastrointestinal Tract

■ Sympathetic, Parasympathetic, and Enteric Divisions of the Autonomic Nervous System Innervate the Gastrointestinal Tract

The gastrointestinal tract is innervated by the autonomic nervous system (Chap. 6; Fig. 26-1) and by sensory nerves from the spinal cord and brainstem. Although sensory nerves may occupy the same paths as autonomic nerves, they are not components of the autonomic nervous system. The autonomic innervation of the gastrointestinal tract is derived from the sympathetic, parasympathetic, and enteric divisions of the autonomic nervous system. Sympathetic and parasympathetic pathways transmit signals from the central nervous system to the digestive tract. This is the extrinsic innervation of the gastrointestinal tract. Neurons of the enteric division make up the local control circuits (minibrain) in the gastrointestinal tract itself. This is the intrinsic innervation. The overall innervation of the digestive tract involves interconnections between the brain, spinal cord, and enteric nervous system (Fig. 26-1).

The Enteric Nervous Sytem ■ Cell bodies of neurons of the enteric nervous system are located in ganglia, positioned inside the walls of the gastrointestinal tract (Fig. 26-1). The structure, function, and neurochemistry of enteric ganglia differ significantly from that of the sympathetic ganglia. Whereas sympathetic ganglia function mainly as relay distribution centers for signals transmitted from the central nervous system, the ganglia of the enteric nervous system form a system that integrates and processes information like

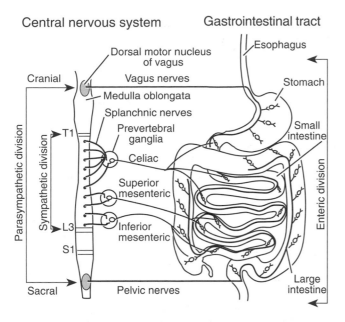

Figure 26–1 ■ Organization of the gastrointestinal tract innervation, derived from the sympathetic, parasympathetic, and enteric divisions of the autonomic nervous system. (Modified from Wood JD, Wingate DL: Gastrointestinal neurophysiology, in *The Undergraduate Teaching Project in Gastrointestinal and Liver Disease.* American Gastroenterological Society.)

the brain and spinal cord. The enteric nervous system is sometimes called the "little brain in the gut."

■ The Enteric Nervous System Is Organized Like an Independent Integrative System

A conceptual model for the enteric nervous system is shown in Figure 26–2. Like the brain and spinal cord, the enteric nervous system is an independent integrative system that contains three functional cate-

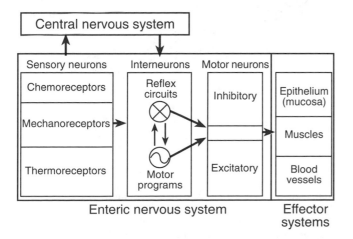

Figure 26–2 ■ Conceptual model of the enteric nervous system.

gories of neurons: sensory neurons, interneurons, and motor neurons.

Sensory Neurons ■ The cell bodies of sensory neurons are found in nodose ganglia of the vagal nerves, dorsal root ganglia of the spinal cord, and ganglia of the enteric nervous system. Sensory afferent fibers of the nodose and dorsal root ganglia carry information from the gastrointestinal tract to the central nervous system. Most of the fibers in the vagal nerves are sensory fibers. Sensory neurons belonging to the enteric nervous system provide information to the local processing circuits and send fibers to the prevertebral ganglia and central nervous system. The gastrointestinal tract has mechano-, chemo-, and thermoreceptors. Mechanoreceptors supply the enteric nervous system and central nervous system with information on the amount of stretch in the wall of the gastrointestinal tract and on the movement of the luminal contents as they brush the mucosal surface. Chemoreceptors generate information on the concentration of glucose, the osmolarity, and the pH of the luminal contents. Thermoreceptors in the intestine supply the brain with information used to regulate body temperature.

Interneurons ■ Millions of synaptic connections between interneurons form the information-processing circuitry of the enteric nervous system. These synapses occur between axons and cell bodies, between axons and dendrites, and from axon to axon. As in the brain, the interneuronal microcircuits account for the higher functions associated with an integrative nervous system.

Motor Neurons ■ Motor neurons, both excitatory and inhibitory, innervate the effector systems. Excitatory motor neurons release neurotransmitters that stimulate the effector cells, whereas inhibitory motor neurons release neurotransmitters that suppress the activity of the effectors.

Reflexes ■ Reflex responses are a behavior of the effector systems evoked by activation of sensory neurons. In a reflex circuit, sensory neurons are connected with interneurons and interneurons are connected to motor neurons that innervate cells of the effector system (Fig. 26–2). The pattern of behavior of the effector systems in a particular reflex response is always the same, due to the interconnection of the neurons that make up the reflex circuit. For example, distension of the intestine is detected by sensory neurons that initiate a reflex consisting of contraction of the circular muscle coat above the site of distension and reflex inhibition of the circular muscle below the distension. This pattern of behavior is the same each time the mechanosensors are activated by stretch of the intestinal wall.

Motor Programs ■ In addition to processing signals from sensory neurons in reflex responses, connections between interneurons form circuits that are responsible for programmed behavior of the effector systems. They form the **pattern-generating circuitry** (motor programs) that drive the motor neurons for control of repetitive cyclic behaviors, such as the cyclic patterns of intestinal motility observed after a meal. Program circuits determine the sequence of events in stereotyped repetitions of motor outflow to the effector system. Programmed motor behavior, unlike reflex behavior, does not require sensory input to start the program, and feedback from sensory neurons is unnecessary for the sequencing of the steps in the program. For behaviors generated by programmed motor circuits (e.g., walking, chewing, swimming), the complete motor program may be set in motion by input signals from a single neuron called a command neuron. Interneurons are also involved in linking together the subsets of circuitry that control each of the individual effector systems. These interconnections work to achieve coordination of the behavior of the individual effectors. This is necessary for optimal performance of the total organ in meeting the moment-to-moment needs of the body as a whole.

■ Vagal Nerves Send Command Signals from the Brain to Enteric Integrative Circuits

There are as many neurons in the enteric nervous system as in the spinal cord. This large number of neurons required to control digestive processes would greatly expand the central nervous system if located there. Rather than locating the neural control circuits in the brain and transmitting all control information over long transmission lines, vertebrate animals have evolved with most circuits for automatic control situated close to the effector systems.

Local integrative circuits of the enteric nervous system perform many operations independent of central nervous system input. Subsets of local circuits are preprogrammed to control distinct patterns of behavior of each effector system and to coordinate the activity of multiple systems. The motor neurons are final common transmission pathways for a variety of different programs and reflex circuits.

The vagal nerves are the major transmission lines from the brain to the enteric nervous system. Instead of controlling individual motor neurons, efferent vagal signals are commands for activation of large blocks of integrated circuits in the enteric nervous system. This explains the strong influence of a small number of efferent vagal fibers on motility and other effector systems over a large region of the stomach or intestine. The neurophysiology of the interactions between the brain and gastrointestinal tract is like other, better understood, control systems in which activation of single command neurons releases coordinated motor behavior. In this respect, the enteric nervous system is analogous to a microcomputer with its own independent software, whereas the brain is like a larger, more sophisticated computer that receives information from and issues commands to the enteric computer.

Structure of the Enteric Nervous System

The enteric nervous system is found in the walls of the esophagus, stomach, small and large intestine, and gallbladder. It consists of ganglia, primary interganglionic fiber tracts, and secondary and tertiary fiber projections to the effector systems (Fig. 26-3).

■ Enteric Neurons Are Organized into Two Major Plexuses

Cell bodies of the neurons are clustered in ganglia (Fig. 26-3). Fibers projecting from the cell bodies to other ganglia make up the interganglionic fiber tracts. These are all unmyelinated C fibers about 0.5 μm in diameter. The enteric ganglia together with the interganglionic connectives form two ganglionated plexuses. One of these is the **myenteric plexus (Auerbach's plexus)** and the other is the **submucous plexus (Meissner's plexus)**. The ganglionated plexuses are continuous around the circumference of the gastrointestinal tract and along its length and are prominent components of the enteric nervous system.

The myenteric plexus is located between the longitudinal and circular muscle layers of most regions of the gastrointestinal tract. It is a two-dimensional array of neurons, ganglia, and interganglionic fiber tracts situated in close apposition to the longitudinal muscle. The neuronal cell bodies are in a single layer in the ganglia. This histologic organization is probably an adaptation to the mechanical forces and deformation that occur during contraction of the muscle or expan-

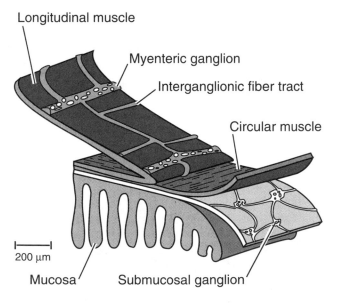

Figure 26–3 ■ Structural relations of the enteric nervous system. (Modified from Wood JD: The enteric nervous system, in Johnson LR, Christensen J, Jackson M et al (eds). *Physiology of the Gastrointestinal Tract.* 2nd ed. New York, Raven, pp 67–109.)

sion and stretching of the wall as the lumen fills. Most of the motor neurons to the circular and longitudinal muscle coats are found in the myenteric plexus. The submucous plexus is situated in the submucosal space between the mucosa and circular muscle coat. Histologically, this plexus is similar to the myenteric plexus, but the ganglia tend to be smaller and contain fewer neuronal cell bodies. The submucous plexus is most prominent in the small and large intestine. It contains motor neurons that innervate the intestinal crypts and villi. Some neurons in the submucous ganglia send fibers to the myenteric plexus and also receive synaptic input from axons projecting out of the myenteric plexus. Although the two plexuses are separated, interconnections bind the two networks into a functionally unified nervous system.

■ Enteric Ganglia Include Dogiel Type I, II, and III Neurons

The German neuroanatomist A. S. Dogiel described three types of enteric ganglion cells—Dogiel types I, II, and III—found in both plexuses (Fig. 26-4).

Dogiel type I neuron cell bodies have many short, stublike processes and a single long process. These flat neurons have processes extending from the cell body mainly in the circumferential and longitudinal planes of the wall. The short processes are presumed to be dendrites that receive synaptic input; the long process is an axon. The axon of some Dogiel type I neurons projects for relatively long distances through interganglionic fiber tracts and many rows of ganglia. Still, the projections are not long: the longest known projections of any enteric ganglion cell are only about 2-3 cm. These are interneurons. The axon of other Dogiel type I neurons in the intestines projects to the circular muscle coat or mucosa. These are motor neurons.

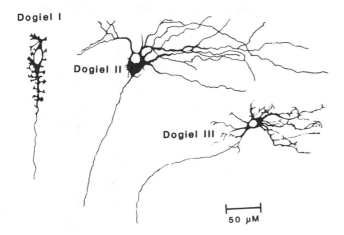

Figure 26–4 ■ Three types of neurons in the enteric nervous system.

The cell bodies of **Dogiel type II** neurons have smooth surfaces with long and short processes arising in a variety of configurations (Fig. 26-4). The long processes may extend through interganglionic fiber tracts across several rows of ganglia in the circumferential direction, oral direction, or aborally. Shorter processes may project only within the home ganglion.

Dogiel type III neurons are similar to Dogiel type II neurons, but there are more processes and more of the processes are shorter. The intermediate-length processes terminate in the same or adjacent ganglia, where, presumably, they synapse with other neurons.

■ Different Types of Enteric Neurons Have Specific Neurotransmitters and Connections

The enteric nervous system uses 20 or more different neurotransmitters for chemical signaling and transfer of information as it carries on its normal activities. These include acetylcholine and a variety of biogenic amines and peptides. Often, these substances are colocalized in the neurons. Acetylcholine, for example, may be stored together with a peptidergic neurotransmitter or amines, and peptides may be costored in various combinations. Specific neurotransmitters are localized to neurons of particular Dogiel type morphology.

Dogiel Type I Neurons ■ Subpopulations of Dogiel type I ganglion cells contain 5-hydroxytryptamine, vasoactive intestinal peptide, or enkephalin. The long process of Dogiel type I neurons contains 5-hydroxytryptamine and projects in the aboral direction for long distances through many ganglia and interganglionic fiber tracts (Fig. 26-5). These axons have smooth surfaces when they are in the fiber tracts and a varicose or beadlike appearance when passing through a ganglion. The neurotransmitter is stored in and released from synaptic vesicles that occupy the varicosities. Dogiel type I neurons containing 5-hydroxytryptamine in the intestinal myenteric plexus also send their processes to the submucous plexus. These are thought to be interneurons.

The projections of axons of Dogiel type I neurons with vasoactive intestinal peptide and nitric oxide in the myenteric plexus are in the aboral direction of the intestinal myenteric plexus, but they travel for only a few millimeters before bending and entering the circular muscle coat. The axons of some of the Dogiel type I neurons containing vasoactive intestinal peptide in the submucous plexus innervate the intestinal crypts. Dogiel type I myenteric neurons containing vasoactive intestinal peptide are inhibitory motor neurons to the circular muscle. Those containing this peptide colocalized with acetylcholine in the submucous plexus are excitatory motor neurons to the intestinal crypts. These neurons are called **secretomotor neurons,** because they stimulate the crypts to secrete water, electrolytes, and mucus.

A subpopulation of Dogiel type I neurons in the myenteric plexus that contain the peptide substance P and acetylcholine are excitatory motor neurons. Release of these neurotransmitters from these neurons stimulates contraction of gastrointestinal muscles.

Dogiel Type II and III Neurons ■ Substance P is also found in multipolar neurons with Dogiel type II and III morphology in both the myenteric and submucous plexuses of the intestine. Most of the projections of substance P-containing neurons are within the ganglion containing the cell soma or no further away than one or two ganglia (Fig. 26–5). These are believed to be interneurons.

Functional Relations of Enteric Neurons

■ Neuroeffector Junctions Are Not Structurally Specialized in the Intestine

Neuroeffector junctions are the sites where neurotransmitters released from axons of motor neurons act on the effector cell. Chapter 9 describes the highly specialized structure of the neuromuscular junction in skeletal muscles. Neuroeffector junctions in the digestive tract are simpler structures than the motor endplates of skeletal muscle. Most motor axons in the gastrointestinal tract do not release neurotransmitters from terminals as such; instead, they are released from varicosities that occur along the axons (Fig. 26–6). The neurotransmitter is released from the varicosities during propagation of the action potential along the axon. Once released, it diffuses over relatively long distances before reaching the effector cells. This structural organization is an adaptation for simultaneous application of a chemical neurotransmitter to a large number of effector cells from a small number of motor axons.

■ Excitatory Motor Neurons Initiate Muscle Contraction and Secretion from Glands

Excitatory motor neurons release neurotransmitters that evoke contractions of the muscles and secretion of water and electrolytes from the crypt glands of the intestine. Mechanisms of excitation-secretion coupling in intestinal crypts are presented in Chapter 28. Two mechanisms of excitation-contraction coupling are involved in the neural initiation of muscle contraction in the gastrointestinal tract. Transmitters from excitatory motor axons may trigger muscle contraction by depolarizing the muscle membrane to the threshold for discharge of action potentials or by direct release of calcium from intracellular stores. The neurally evoked depolarizations of the muscle membrane potential are called **excitatory junction potentials.** Direct release of calcium by the neurotransmitter is called pharmacomechanical coupling (Chap. 9). In this case, occupation of receptors on the muscle cell membrane by the neurotransmitter leads to release of intracellular calcium and calcium-triggered contraction independent of any changes in membrane potential (Chap. 27).

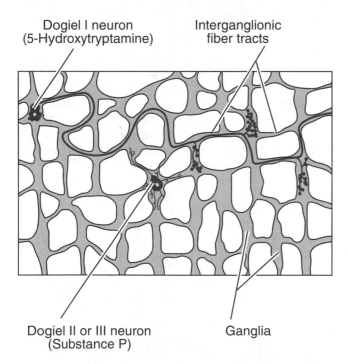

Figure 26–5 ■ Different morphologic types of enteric neurons have specific neurotransmitters and projections. Dogiel type I interneurons containing 5-hydroxytryptamine send a single axon to synapse with other neurons in several ganglia along its path. Dogiel type II neurons containing substance P have many short axons that synapse with other neurons in the same ganglion.

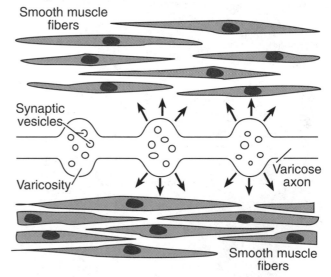

Figure 26–6 ■ Neuromuscular junctions in the gastrointestinal tract have varicose axons, which release neurotransmitters in many directions to reach multiple muscle fibers.

■ Inhibitory Motor Neurons Suppress Muscle Contraction and Secretion from Glands

Neurotransmitters released from inhibitory motor neurons produce **inhibitory junction potentials (IJPs)** in the muscle cell membranes and suppress contractile activity of the muscles. Inhibitory junction potentials are hyperpolarizing potentials that move the membrane potential away from the threshold for discharge of action potentials, thereby reducing the excitability of the muscle fiber. In gastrointestinal muscles, IJPs are generated by the opening of potassium channels in the muscle cell membrane and the resulting efflux of potassium from the cell. Vasoactive intestinal peptide and nitric oxide are inhibitory neurotransmitters at neuromuscular junctions in the digestive tract.

Electrical and Synaptic Behavior of Enteric Neurons

Enteric neurons, like neurons elsewhere, use propagated action potentials as a mechanism for transmitting coded signals from one part of the neuron to the other. Transmission of signals from neuron to neuron in the neural circuits is achieved by mechanisms of synaptic transmission similar to those in circuits of the brain and spinal cord (Chap. 3).

■ The Number of Open Potassium Channels Determines Membrane Conductance and Electrical Potential

The resting membrane potential of enteric neurons is largely determined by potassium channels. The electrical equivalent of a membrane with potassium channels is a variable resistor with the amount of resistance determined by the number of open channels (Fig. 26-7). The more open channels, the lower the electrical resistance of the membrane. The electrochemical gradient for potassium ($E_m - E_K$; Chap. 2) is the driving force for outward electrical current across the resistor shown in the electrical equivalent model of Figure 26-7 or through the open channels in the adjacent illustration of potassium channels.

The resting membrane potential of enteric neurons is normally less than the potassium equilibrium potential, which is estimated to be −90 mV. The potassium equilibrium potential is the membrane potential that would result if all of the potassium channels in the membrane were opened while all other ionic channels were closed. A resting membrane potential less than the potassium equilibrium potential is significant because it provides for modulation in either the hyperpolarizing (inhibitory) or depolarizing (excitatory) direction. The hyperpolarizing direction is determined by opening of potassium channels in response to a neurotransmitter substance and the depolarizing direc-

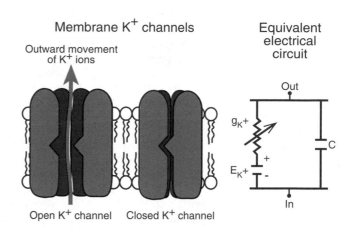

Figure 26–7 ■ Potassium channels are important determinants of the membrane potential and resistance in AH/type 2 neurons. A structural model of open and closed potassium channels is shown on the left. On the right is the electrical circuit model for a membrane with potassium channels. The electronic symbol for a resistor of adjustable value (gK+) represents ionic conductance in the population of potassium channels at any moment in time. The resistance and capacitance (C) of the membrane are connected in parallel. The number of open potassium channels determines the resistance of the membrane and size of the membrane potential. Opening of increased numbers of potassium channels reduces the resistance and hyperpolarizes the membrane potential. The driving force for the current across the resistor in the electrical circuit or the potassium current through the open channels is the electrochemical gradient for potassium ions (E_{K+} = K+ equilibrium potential ∼ −90 mV). (Modified from Wood JD, Wingate DL: Gastrointestinal neurophysiology, in *The Undergraduate Teaching Project in Gastrointestinal and Liver Disease.* American Gastroenterological Society.)

tion by the closure of potassium channels or opening of sodium or calcium channels.

Any increase or decrease in transmembrane potassium movement due to opening or closing of the channels results in changes in both membrane potential and electrical resistance of the neuronal membrane. Closure of potassium channels and decreased potassium movement results in depolarization of the membrane and an increase in resistance. Opening leads to hyperpolarization of the membrane and a decrease in membrane resistance.

■ Enteric Neurons Are Classified as S/Type 1 or AH/Type 2 Based on Their Electrical Behavior

S/type 1 and AH/type 2 are the two most common kinds of electrical behavior found in enteric neurons. The names are derived from terms applied by the researchers who first described the behavior of the neurons. One group, led by G. D. S. Hirst, called neurons with multiple fast excitatory postsynaptic potentials "S type," whereas R. A. North called them "type 1". The first group used "AH neuron" to describe neurons with long-lasting afterhyperpolarization following an action potential; North called them "type 2." The terms are combined in recognition of the pioneering work and because the electrophysiologic behavior of each was later found to be associated with a specific morphologic type. Neurons with S/type 1 behavior have the morphology of Dogiel type I neurons, whereas AH/type 2 neurons are most often Dogiel type II.

S/Type 1 Neurons ■ S/type 1 neurons are distinguished by the characteristics of their resting membrane potential, their membrane resistance, and the way they discharge action potentials. They have lower resting membrane potentials than AH/type 2 neurons. The low resting potentials are associated with higher membrane resistance, which reflects decreased potassium conductance.

S/type 1 neurons have increased excitability, seen as repetitive discharge of action potentials in response to the injection of depolarizing current through the microelectrode (Fig. 26-8). As the depolarization is increased by injecting stronger current, the frequency of the repetitive discharge increases. Application of tetrodotoxin, which prevents opening of sodium channels in nerve cell membranes, abolishes the action potentials. This indicates that the rising phase of the action potentials in S/type 1 neurons depends exclusively on opening of sodium channels.

AH/Type 2 Neurons ■ AH/type 2 neurons have higher resting potentials associated with lower membrane resistance than S/type 1 neurons. This reflects a greater number of open potassium channels and higher potassium conductance than in S/type 1 cells. Excitability of AH/type 2 neurons is lower that of S/type 1 neurons. Injection of depolarizing current in these cells evokes only one or two action potentials, and only at the beginning of the current pulse (Fig.

26-8). Repetitive discharge of spikes does not occur in response to current injection when the neuron is in its resting state. However, excitatory neurotransmitters can convert AH/type 2 neurons from low- to high-excitability state with repetitive discharge like that in S/type 1 neurons (see below). The action potentials of AH/type 2 neurons are resistant to blockade by tetrodotoxin. This indicates that ions other than sodium carry the inward current for the rising phase of the action potential. Opening of calcium channels and an increase in the inward movement of calcium ions accounts for part of the current that produces the rising phase of the action potential.

The action potentials in AH/type 2 neurons are followed by hyperpolarization of the membrane potential, lasting for several seconds (Fig. 26-9). This is called **afterhyperpolarization** (AH) and is a distinguishing characteristic of AH/type 2 cells.

AH/type 2 neurons are sometimes in a state of inexcitability with high resting membrane potentials that are close to the potassium equilibrium potential and low membrane resistance, indicative of very high resting potassium conductance. Injection of depolarizing current does not evoke action potential discharge. Although the neurons do not fire action potentials when in this state, they may respond with synaptic potentials to inputs from other neurons in the circuit.

Excitability can be restored by exposure to particular chemical messenger substances. The cells may discharge action potentials with afterhyperpolarizing potentials, if the concentration of the messenger is low, or may discharge repetitively, like an S/type 1 neuron, if the concentration of the messenger is higher. These cells are probably components of circuits that are not used continuously by the system. When the intestine does not require operation of the circuit, it is inactive due to the low excitability state

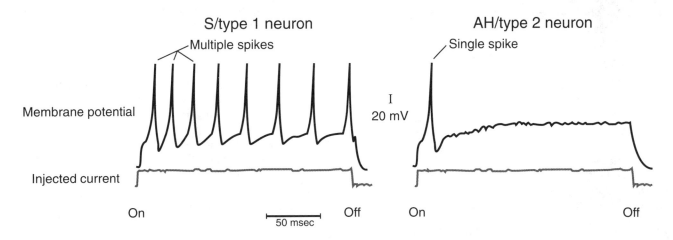

Figure 26–8 ■ Repetitive discharge of action potentials occurs during membrane depolarization in S/type 1 enteric neurons but not AH/type 2 neurons. (Modified from Wood JD, Wingate DL: Gastrointestinal neurophysiology, in *The Undergraduate Teaching Project in Gastrointestinal and Liver Disease.* American Gastroenterological Society.)

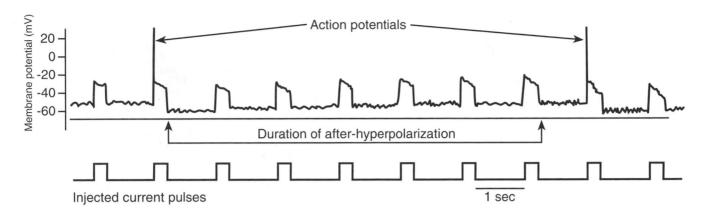

Figure 26–9 ■ Action potentials in AH/type 2 enteric neurons are followed by long-lasting after-hyperpolarization of the membrane potential. The top trace is an intracellular recording of changes in electrical potential across the membrane during repeated injection of square pulses of depolarizing current through the microelectrode into the neuron. The bottom trace is a record of injected current. Membrane depolarization by the injected current does not trigger an action potential during the afterhyperpolarization. (Modified from Wood JD, Wingate DL: Gastrointestinal neurophysiology, in *The Undergraduate Teaching Project in Gastrointestinal and Liver Disease.* American Gastroenterological Society.)

of its component neurons. The circuit is called to activity by a chemical messenger that boosts the neuronal excitability within the circuit.

AH/type 2 electrical behavior is found mainly in multipolar neurons with Dogiel type II and III morphology. It is found in the myenteric and submucous plexuses of the small and large intestine but is not seen in myenteric neurons of the body of the stomach in experimental animals. Neurons with AH/type 2 behavior are found in the antrum of the stomach.

■ Long-Lasting Afterhyperpolarization Is Associated with Action Potentials in AH/Type 2 Neurons

The afterhyperpolarization associated with action potentials in AH/type 2 neurons can last as long as 12 seconds (Fig. 26–9). Afterhyperpolarizing potentials summate when two action potentials are fired in close sequence, producing a larger potential than when only one spike occurs. Excitability of the neuron is reduced during afterhyperpolarization, and depolarizing current pulses (Fig. 26–9) are unable to evoke action potentials during hyperpolarization. The afterhyperpolarization effectively prolongs the relative refractory period of an action potential. This results in an automatic mechanism that prevents repetitive spike discharge, because the intervals between spikes cannot be much less than the duration of the afterhyperpolarization.

A decrease in membrane resistance occurs during afterhyperpolarization. This reflects opening of ion channels and increased membrane conductance. The ionic mechanism for afterhyperpolarization involves both calcium and potassium channels. **Voltage-gated**

calcium channels and **calcium-dependent potassium channels** are present in the membrane of the cell bodies of AH/type 2 neurons. Voltage-gated calcium channels are opened when the membrane potential is depolarized. Calcium-dependent potassium channels are opened by free intracellular calcium. The higher the levels of free calcium in the cytoplasm, the greater the number of open potassium channels.

Voltage-gated calcium channels open during depolarization of the action potential, and a fraction of the inward ionic current during the rising phase of the action potential is carried by calcium. Calcium entry during the spike results in elevation of intraneuronal calcium, which leads to opening of the calcium-dependent potassium channels. When these channels are opened, potassium moves down its electrochemical gradient through the channels and out of the neuron. This results in hyperpolarization of the membrane in association with a decrease in membrane resistance. The afterhyperpolarization decays as the internal calcium becomes sequestered into intracellular storage sites or is pumped out of the neuron, and the calcium-dependent potassium channels are closed again.

■ Excitatory Postsynaptic Potentials, Inhibitory Postsynaptic Potentials, and Presynaptic Inhibition Are Synaptic Events in Enteric Neurons

The general mechanism of chemically mediated synaptic transmission is the same in the enteric nervous system as elsewhere in the body (Chap. 3). Excitatory postsynaptic potentials (EPSPs), inhibitory postsynaptic potentials (IPSPs), and presynaptic inhibition are the principal synaptic events in the enteric nervous system.

Fast EPSPs ■ Fast EPSPs are depolarizing potentials of less than 50 msec duration (Fig. 26–10). They occur in S/type 1 and AH/type 2 neurons of the myenteric and submucous plexuses. All fast EPSPs are mediated by acetylcholine acting at nicotinic cholinergic postsynaptic receptors. Fast EPSPs are involved in the rapid transfer and transformation of neurally coded information between elements of the integrative circuitry of the system.

The first fast EPSP in Figure 26–10 did not reach threshold for discharge of an action potential, whereas the second one did reach threshold and trigger an action potential. Fast EPSPs do not reach threshold when the neuronal membranes are hyperpolarized during slow IPSPs. Fast EPSPs are more likely to reach spike threshold when the membranes are depolarized during slow EPSPs. This is an example of **neuromodulation,** whereby the input-output relations of a neuron to one input is modified by a second synaptic input.

Slow EPSPs ■ Slow EPSPs occur in myenteric and submucous ganglion cells and in AH/type 2 and S/type 1 neurons, but they are most prominent in AH/type 2 neurons. Slow EPSPs are characterized by a slowly rising membrane depolarization that is sustained for several seconds to minutes after termination of release of the neurotransmitter from the presynaptic terminal (Fig. 26–11). The membrane depolarization is associated with increased membrane resistance and marked enhancement of excitability, which is apparent as a long-lasting train of action potentials during the slow depolarization.

The ionic mechanism for the slow EPSPs involves closure of potassium channels, revealed experimentally as depolarization of the membrane potential and

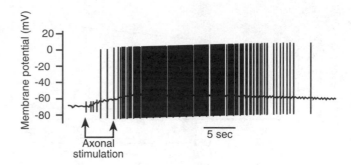

Figure 26–11 ■ Record of a slow EPSP, showing prolonged membrane depolarization accompanied by a long-lasting train of action potentials. (Modified from Wood JD, Wingate DL: Gastrointestinal neurophysiology, in *The Undergraduate Teaching Project in Gastrointestinal and Liver Disease.* American Gastroenterological Society.)

increase in the membrane resistance. AH/type 2 neurons, as described earlier, have high resting membrane potentials and low membrane resistances that result from a large proportion of open potassium channels. The number of open potassium channels is a direct function of the intraneuronal concentration of calcium. The neurotransmitters for slow EPSPs decrease the number of open potassium channels by decreasing the level of calcium in the neuron. In resting AH/type 2 neurons, open calcium channels permit continuous influx of this ion. Binding of the neurotransmitters to their postsynaptic receptors leads to closure of the calcium channels, which, in turn, results in a fall in cytoplasmic calcium and secondary reduction in the number of open potassium channels. Binding of the neurotransmitter also prevents opening of voltage-activated calcium channels during the rising phase of the action potential. Consequently, the afterhyperpolarization following the spikes in AH/type 2 neurons is suppressed during the slow EPSP.

Suppression of the afterhyperpolarization is part of the mechanism that permits repetitive spike discharge at increased frequencies during the enhanced excitability. Overall, the cell bodies of AH/type 2 neurons function between extremes of low and high excitability, due to the mechanisms of slow synaptic excitation. The low excitability state is associated with high resting potassium conductance, which is dependent on availability of cytoplasmic calcium. The probability of action potential discharge in this state is low. If a spike does occur, additional calcium enters the cell and opens more potassium channels. This produces the afterhyperpolarization that prevents repetitive spike discharge. Conversion to high excitability occurs when the transmitters for the slow EPSP reduce cytoplasmic calcium, along with a parallel decrease in potassium conductance. The conversion also suppresses postspike afterhyperpolarization by reducing calcium entry during the action potential. Additional molecular transformations in the membrane account for

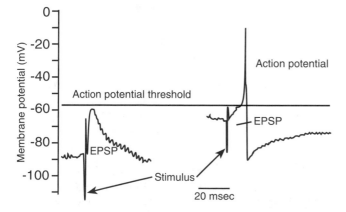

Figure 26–10 ■ Fast EPSPs are depolarizing synaptic potentials with durations measured in milliseconds. Fast EPSPs trigger action potentials when the depolarization exceeds threshold. (Modified from Wood JD, Wingate DL: Gastrointestinal neurophysiology, in *The Undergraduate Teaching Project in Gastrointestinal and Liver Disease.* American Gastroenterological Society.)

the conversion from low excitability to enhanced excitability.

Several chemical messenger substances produce responses like slow EPSPs when applied to enteric neurons experimentally. These substances include peptides and amines, and acetylcholine, acting at muscarinic receptors, is also involved. Other possible neurotransmitters are 5-hydroxytryptamine, substance P, vasoactive intestinal peptide, gastrin-releasing peptide, cholecystokinin, calcitonin, gene-related peptide and histamine. These substances are believed to function as neurotransmitters in the brain as well as in the enteric nervous system. In the enteric nervous system, receptors for these substances are localized to subpopulations of neurons that are activated only if the substances that fit its specific receptors are present. The receptors for all of the excitatory substances are coupled to the same ionic channels and evoke the same response when activated. This is a mechanism for selectivity in the activation of specific subsets of neurons.

Slow EPSPs are a mechanism for long-lasting activation or inhibition of gastrointestinal effector systems (Fig. 26–12). The prolonged discharge of spikes during a slow EPSP drives the release of neurotransmitter

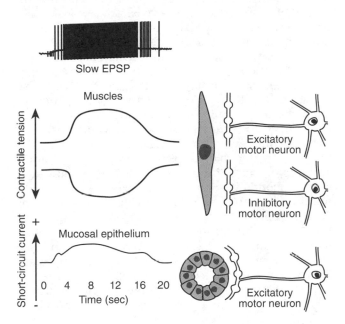

Figure 26–12 ■ The functional significance of slow EPSPs is prolonged activation or inhibition of gastrointestinal effector cells. Slow EPSPs in excitatory motor neurons to the muscles or mucosal epithelium result in prolonged muscle contraction or mucosal crypt secretion. Stimulation of secretion in experiments is seen as an increase in ion movement (short-circuit current). Slow IPSPs in inhibitory motor neurons to the muscles result in prolonged inhibition of contractile activity, which is observed as decreased contractile tension. (Modified from Wood JD, Wingate DL: Gastrointestinal neurophysiology, in *The Undergraduate Teaching Project in Gastrointestinal and Liver Disease.* American Gastroenterological Society.)

from the neuron's axon for the duration of the spike discharge. This results in prolonged inhibition or excitation at neuronal synapses and neuroeffector junctions. Contractile responses of the muscles and secretory responses of the mucosal epithelium are sluggish events that last several seconds from start to completion. The trainlike discharge of spikes during slow EPSPs is the neural correlate of long-lasting responses of the effector systems in the functioning gastrointestinal tract. Slow EPSPs in excitatory motor neurons to the intestinal muscles or the mucosa result in prolonged muscle contraction or prolonged secretion from the glands. Slow EPSPs in inhibitory motor neurons to the muscles result in prolonged suppression of contraction. This is observed experimentally as a decrease in contractile tension (Fig. 26–12).

Slow IPSPs ■ Slow IPSPs are slowly developing hyperpolarizations of the membrane potential that last several seconds after termination of the release of the neurotransmitter. The hyperpolarization is associated with decreased membrane resistance and decreased excitability (Fig. 26–13). Binding of the inhibitory neurotransmitter to its postsynaptic receptor results in opening of potassium channels and diffusion of potassium out of the neuron. This, in turn, hyperpolarizes the membrane potential toward the potassium equilibrium potential of -90 mV.

Slow IPSPs are the inverse of slow EPSPs: they act to hyperpolarize the membrane potential below the action potential threshold and decrease the probability of spike discharge. Part of their function may be to reverse the hyperexcitable condition associated with slow EPSPs and return the membrane to the low excitability state. They are found in cell bodies of the myenteric and submucous plexuses of the intestine.

Several chemical messenger substances produce responses like slow IPSPs when applied to enteric neurons experimentally. Some possible neurotransmitters

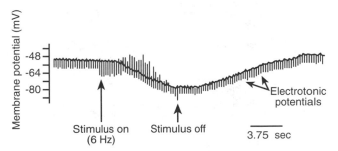

Figure 26–13 ■ Record of a slow IPSP, showing membrane hyperpolarization associated with decreased input resistance. Decreased amplitude of electrotonic potentials (downward deflections) produced by intraneuronal injection of pulses of hyperpolarizing current reflects decreased input resistance during the slow IPSP. (Modified from Wood JD, Wingate DL: Gastrointestinal neurophysiology, in *The Undergraduate Teaching Project in Gastrointestinal and Liver Disease.* American Gastroenterological Society.)

are opioid peptides (enkephalin), cholecystokinin, galanin, somatostatin, adenosine, and norepinephrine. These substances, which are present in the brain and gastrointestinal tract, produce hyperpolarization, decreased membrane resistance, and opening of potassium channels.

Presynaptic Inhibition ■ Presynaptic inhibition is suppression of neurotransmitter release from axons. It is mediated by chemical messengers acting at receptors on the axon. In the enteric nervous system, this occurs at fast and slow excitatory synapses and neuroeffector junctions. Presynaptic inhibition may involve axoaxonal transmission, in which release of a neurotransmitter from one axon suppresses release of transmitter from another axon (Fig. 26–14). Presynaptic inhibition can also be in the form of autoinhibition and occur at neural synapses and neuroeffector junctions. In autoinhibition, transmitter released from an axon accumulates around its release site and acts at local presynaptic receptors to reduce further release. This is a negative feedback mechanism that automatically regulates the concentration of neurotransmitter

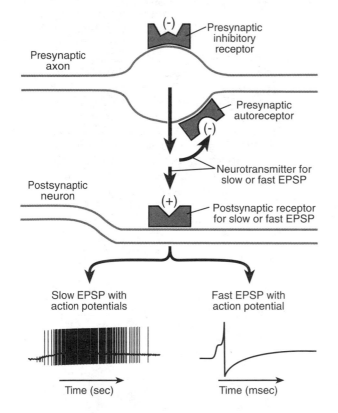

Slow EPSP with action potentials

Fast EPSP with action potential

Time (sec)

Time (msec)

Figure 26–14 ■ Presynaptic inhibitory receptors are found on axons at neurotransmitter release sites for both slow and fast EPSPs. Different neurotransmitters act through the presynaptic inhibitory receptors to suppress axonal release of the transmitters for slow and fast EPSPs. Presynaptic autoreceptors are involved in a special form of presynaptic inhibition whereby the transmitter for slow or fast EPSPs accumulates at the synapse and acts on the autoreceptor to suppress further release of the neurotransmitter. (+), excitatory receptor; (−), inhibitory receptor.

at the junction. Presynaptic inhibition can also be brought about by substances released from endocrine or other non-neuronal cells into the milieu surrounding the synaptic circuits.

Like other forms of synaptic transmission in the enteric nervous system, several chemical messengers act at presynaptic receptors to inhibit release of neurotransmitters. Possible neurotransmitters are norepinephrine, 5-hydroxytryptamine, histamine, acetylcholine, and opioid peptides (enkephalins).

■ Gating Mechanisms Control the Spread of Information within the Enteric Plexuses

The gastrointestinal tract is basically a hollow tube, with the effector systems distributed around the circumference and along the length. Much of the behavior of the gastrointestinal tract involves the spread of activity in a specific direction along the length of the tube. For instance, peristaltic propulsion may travel over long lengths of intestine in the aboral direction (Chap. 27). Neural control of such propagating behavioral events requires gating mechanisms for determining both direction and distance of travel of the event.

Slow Synaptic Gating Mechanisms ■ Slow EPSPs underlie a gating mechanism that controls the spread of action potentials between the axons or dendrites arising from opposite poles of the cell body of a multipolar neuron. Figure 26–15 is a model of the slow synaptic gating mechanism found in Dogiel type II or III neurons.

Intracellular recording in AH/type 2 neurons has demonstrated that action potentials propagating toward the cell body in one of its processes usually do not trigger an action potential in the membrane of the cell soma. An action potential in the cell body is followed by the characteristic afterhyperpolarization, and this prevents firing of the cell body by any additional incoming spikes. In the resting state of the multipolar neuron, it is improbable that an inbound spike in one of its processes will produce an action potential in the cell body. This probability is greatly increased during the slow EPSP, when excitability is enhanced, membrane resistance is increased, and afterhyperpolarization is suppressed.

The membrane of the cell body of the multipolar neuron behaves like a "closed gate" to the transfer of spike signals between its dendrites and axons when it is in the low excitability state. In this case, spike information is confined to a single dendrite (Fig. 26–15, top). The gate is opened and signals are transferred across the somal membrane to other axons during the slow EPSP. A closed somal gate isolates the initial segments of each axon or dendrite, such that spike discharge in one initial segment does not influence others attached elsewhere around the cell body.

The increase in membrane resistance during the slow EPSP increases the space constant (Chap. 3) of the somal membrane and facilitates electrotonic

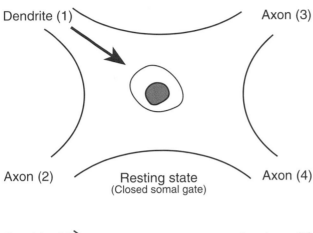

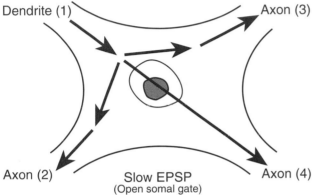

Figure 26–15 ■ Slow EPSPs are part of a gating mechanism for the spread of action potentials in neural networks of the enteric nervous system. They provide a gating mechanism that controls propagation of spike information between the axons and dendrites at opposite sides of the cell body. The probability that the cell body will be fired by an action potential arriving in a dendrite is low in the absence of the slow EPSP but increases during the slow EPSP. The somal membrane behaves like a closed gate to the transfer of spike information (top) when it is in the low excitability state. In this case, spike information is confined to dendrite 1. The gate is opened (bottom) and spike information is transferred to axons 2, 3, and 4 during the slow EPSP. (Modified from Wood JD, Wingate DL: Gastrointestinal neurophysiology, in *The Undergraduate Teaching Project in Gastrointestinal and Liver Disease.* American Gastroenterological Society.)

spread of the action potential from the dendrites into the cell body. This, together with the enhanced excitability of the somal membrane, "opens the somal gate," resulting in transfer of spike signals across the cell body to axons at other poles of the neuron. When this happens, these axons fire synchronously with the cell body and the action potentials are conducted away from the cell body and distributed to regions of the enteric networks lying further along or around the intestine.

Presynaptic Mechanism of Gating ■ Presynaptic mechanisms of gating of the spread of neural signals use presynaptic inhibition to prevent transmission

across synapses. Enteric nerve fibers do not project very far along the intestinal wall before ending as a synapse with another neuron. This means neuronal signals for propagated behavioral events must cross many synapses for the event to be propagated and that each consecutive synapse must transmit information unfailingly for propagation to continue. Since presynaptic inhibition can effectively stop transmission at the synapses, activation of presynaptic receptors in the circuits within a particular region of intestine would be expected to bring the propagated event to a halt. Virtually all of the synapses in the neural circuits of the intestine have presynaptic inhibitory receptors at the release sites for neurotransmitters. This results in a continuum of points along the bowel where presynaptic mechanisms can gate the distance, as well as the direction, of travel of neural signals within the neural plexuses.

Extrinsic Innervation of the Gastrointestinal Tract

Unified nervous control of digestive processes involves both the brain and the enteric nervous system. The enteric nervous system is analogous to an intelligent computer terminal linked to a master computer. Automatic control is handled by resident programs of the intelligent terminal with selection of the appropriate program determined by instructional commands from the master computer. Unlike the simpler circuits of the enteric nervous system, the central nervous system has vast memory stores of information as well as updated information on the homeostatic state of the body as a whole. Commands are transmitted to the digestive tract based on processing of this information in the brain. These commands are mostly adaptive for the well-being of the animal, but they may also be abnormal and result in digestive disease. The final common pathways for output from the brain's higher processing centers to the gastrointestinal tract leave from the medulla oblongata and spinal cord.

The relation between emotional state and gastrointestinal malaise is well known. Emotional stress can result in conditions such as diarrhea, gastrointestinal pain known as irritable bowel syndrome, and gastric ulcers. These are but a few examples of a variety of other conditions initiated by the information-processing functions of the brain and manifest in the digestive tract through brain-gut connections.

■ The Central Components of the Vagal Nerves Are in the Medulla Oblongata

The efferent fibers of the cranial division of the parasympathetic nervous system travel in the vagal nerves to the gastrointestinal tract. The projections to the stomach and intestines originate from neurons located in the **dorsal motor nucleus** of the vagus (Fig. 26-1).

IRRITABLE BOWEL SYNDROME

Irritable bowel syndrome (IBS) is a diagnosis made by excluding all other disorders that could possibly explain the symptoms. Patient complaints include some or all of the following: abdominal pain relieved by having a bowel movement; looser and more frequent bowel movements that may be watery and explosive, associated with cramping abdominal pain; bloated and distended abdomen or the sensation that the abdomen is swollen with gas; mucus, but no blood, in the stool; a sensation of incomplete evacuation as cramping pain persists. Typically the patient recalls the first incident happening in the teenage years, often at the time of an examination in high school or going out on the first date.

The cluster of symptoms that defined IBS cannot be explained by an underlying organic or structural abnormality, as determined by a battery of tests to exclude an obstruction, parasites, inflammatory bowel disease, and so on. Generally IBS is construed as a disorder of small and large intestinal motility resulting from abnormality of nervous mechanisms. The root causes of IBS are poorly understood, in large part due to incomplete knowledge of the intricacies of nervous control.

Twenty-four-hour motility monitoring in IBS patients often reveals normal small intestinal motility (normal migrating motor complex) during sleep with abnormal conversion to a pattern of high-amplitude random contractions on waking. Unlike disorders such as gastric ulcer, heartburn, and inflammatory bowel disease, patients do not report being wakened by their symptoms at night. Emotional stress, such as driving in London traffic, may exacerbate symptoms, and ambulatory monitoring shows this to be associated with altered small bowel motility.

A hallmark of IBS is supersensitivity to distension of the large intestine. Inflation of a balloon in the colon or small intestine evokes painful sensations in IBS patients at significantly lower distending pressures than in normal subjects. This extends to other regions of the digestive tract in some IBS patients, who may complain of an "irritable esophagus." In these individuals, the esophagus also appears hypersensitive to balloon distension. Either of two hypothetic abnormalities of sensory neurophysiology is invoked to account for the heightened sensitivity to distension. One hypothesis is that increased sensitivity of intestinal mechanoreceptors to stretch of the intestinal wall results in transmission of inaccurate information to the brain, where it is processed as noxious and interpreted as painful. The second hypothesis is that mechanosensors accurately signal distension of the wall, but abnormal processing in the brain shunts the information to consciousness as the sensation of pain.

The equivalent of IBS in the stomach and upper intestine is known as functional dyspepsia or nonulcer dyspepsia. The symptoms are related to unexplained derangements of function in the stomach and duodenum that, as in IBS, cannot be attributed to organic influences or systemic diseases, such as peptic ulcer. Epigastric pain and/or discomfort—sensation of postprandial fullness, bloating, nausea, and vomiting—are the symptoms of functional dyspepsia. As with IBS, disordered neural control of motility and perhaps sensory neurophysiology are suspected but unproved as the underlying cause.

In Western societies, IBS and functional dyspepsia are found mainly in young to middle-aged women with an average age of 33 years; the ratio of the diagnosis in women to men is 15:1. The symptoms may worsen in the postluteal phase of the menstrual cycle and lessen after the onset of menses. Tests for defined gastrointestinal diseases are normal in these patients. More than 15% of the general population experiences IBS or functional dyspepsia. Together these disorders comprise the greatest number of referrals to gastroenterologists.

A brain-gut connection that expresses emotional stress as altered gastrointestinal function symptomatic of IBS is often stated anecdotally and assumed by practitioners and laypersons alike. Nevertheless, this connection has not been confirmed unequivocally by well-controlled studies. Patients with psychological disorders such as hysteria, phobias, anxiety disorder, or depression have no higher incidence of IBS than control groups presenting with other disease entities. On the other hand, psychological stress has been found to alter gastrointestinal motility in non-IBS subjects and to exacerbate symptoms of motility disorders in IBS patients. Some studies suggest that IBS patients show greater reactivity to psychological or physical stress than normal subjects; however, there is considerable overlap between the IBS group and normal subjects.

The idiopathic nature of IBS precludes application of a rational strategy to the development of effective therapy. Many treatments (e.g., acetylcholine receptor blockers, antidepressants, high-fiber diets) have been tried in the management of IBS, but none has convincingly and repeatedly shown improvement over placebo in IBS clinical trials. A very high response to placebo, varying from 30% to 60% of patients, is a common outcome in many drug trials. This favorable response to placebo underscores the importance of the brain-gut connection and the benefits of a successful patient-physician encounter. Explanation of the illness, reassurance that no serious disease exists, and empowering the patient improve the likelihood of a placebo response to innocuous treatments such as a high-fiber diet.

The medulla oblongata receives large amounts of sensory information directly from the upper gastrointestinal tract. These signals are transmitted in afferent fibers of the vagal nerves. The first synapse in the sensory pathway is in the **nucleus tractus solitarius,** situated in the medulla close to the dorsal motor nucleus of the vagus. Neurons in the nucleus tractus solitarius synapse with efferent neurons in the dorsal vagal nucleus.

The vagally connected sensory receptors in the digestive tract inform the brain about a variety of important parameters, such as glucose concentration in the lumen, pH of the lumen, osmolality of the lumen, movement of material past mucosal mechanoreceptors, and the level of contractile tension or stretch within the wall. Much of this information is processed by synaptic circuits in the medullary nuclei. Stimulation of vagal afferent fibers often evokes contractile and secretory responses in the upper gastrointestinal tract. These are called **vagovagal reflexes** because the sensory signals are carried in vagal afferent fibers, processing occurs in the vagally related circuits in the medulla, and the outflow back to the digestive tract is along efferent vagal fibers.

■ Efferent Signals in the Vagal Nerves May Stimulate or Inhibit Gastrointestinal Effectors

The efferent signals carried by the vagal nerves to the gastrointestinal tract are involved in initiation of digestive processes in anticipation of food intake and following a meal. This involves stimulation of some effector systems and inhibition of others. The sight and smell of desirable food and the presence of ingested food evoke salivary secretion, gastric acid secretion, and small intestinal secretion. These are excitatory responses initiated by descending signals in the vagal nerves. These secretory events of the digestive process occur in parallel with relaxation of contractile tone in the lower esophageal sphincter, cardia, and fundus of the stomach. Relaxation of the muscles results from activation of inhibitory motor neurons.

■ The Intestine Receives Sympathetic Innervation through Prevertebral Ganglia

Sympathetic nervous signals to the gastrointestinal tract generally inhibit digestive processes. Activation of sympathetic pathways suppresses motility and secretion, decreases blood flow to the intestine, and contracts the muscle in the various sphincters. These effects are produced by the release of the neurotransmitter norepinephrine from sympathetic postganglionic fibers. The cell bodies for the sympathetic postganglionic fibers reside in the prevertebral ganglia (Fig. 26-1). These fibers innervate enteric neurons, intramural blood vessels, the intestinal crypts, and the smooth muscles of sphincters.

■ Sympathetic Nerves Form Inhibitory Synapses at the Interface with the Enteric Nervous System

The synaptic interface between the postganglionic sympathetic fibers and the enteric nervous system is at presynaptic α_2-adrenergic receptors. Norepinephrine released from sympathetic fibers suppresses the release of excitatory neurotransmitters at enteric synapses and neuroeffector junctions.

Suppression of synaptic transmission by the sympathetic nerves occurs at fast and slow excitatory synapses and is responsible for inactivation of the neural circuits that generate intestinal motor behavior. Activation of the sympathetic inputs allows only continuous discharge of inhibitory motor neurons to the nonsphincteric muscles. The overall effect is to produce a state of paralysis of intestinal motility in conjunction with reduced intestinal blood flow. This state is called **paralytic ileus** when it persists to the point of being pathologic (Chap. 27).

■ Enteroenteric Inhibitory Reflexes Are Mediated through Prevertebral Ganglia

Prevertebral sympathetic ganglia receive synaptic inputs from sources other than preganglionic sympathetic fibers. Mechanosensory neurons, with cell bodies in the enteric nervous system, send fibers from the intestine to the prevertebral ganglia. The mechanoreceptor inputs from the intestine use acetylcholine as an excitatory neurotransmitter at the synapses on the neurons in the prevertebral ganglia. The cholinergic fibers evoke fast EPSPs, which, like fast EPSPs in the enteric ganglia, are mediated by nicotinic receptors.

In addition to relaying signals from the spinal cord to the intestine, the prevertebral ganglia are part of a reflex loop designed for rapid transfer of signals between separated regions of bowel (Fig. 26–16). Propagating action potentials in the enteric nervous system seldom travel more than a few centimeters before encountering a synapse. Consequently, signaling over long distances up or down the tube is slow and susceptible to interruption. Connections that relay information through prevertebral ganglia are adaptations that compensate for this. Exchange of information between widely separated regions of bowel occurs over **extraintestinal pathways** that bypass the synaptic circuits in the intramural nervous system.

The reflex pathways between the intestine and prevertebral ganglia are responsible for **enteroenteric reflexes.** These are responsible when overdistension in one region of the intestine results in inhibition of

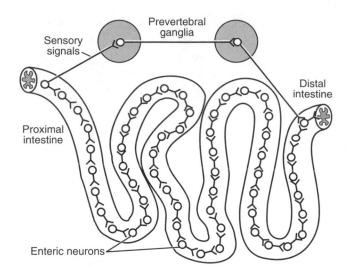

Figure 26–16 ■ Prevertebral sympathetic ganglia have neural connections for rapid transfer of signals between separated regions of the gastrointestinal tract. Sensory signals to the ganglia from one region of intestine (e.g., proximal intestine) are relayed as inhibitory signals in postganglionic sympathetic fibers to another region (e.g., distal intestine). The prevertebral ganglia and their connections provide extraintestinal pathways that bypass the synaptic circuits of the enteric nervous system, thereby increasing the speed of communication between widely separated regions of bowel. (Modified from Wood JD, Wingate DL: Gastrointestinal neurophysiology, in *The Undergraduate Teaching Project in Gastrointestinal and Liver Disease.* American Gastroenterological Society.)

motility in a distant region. The pathway is illustrated in Figure 26–16. Distension of the intestinal wall activates stretch receptors, which send action potentials in sensory fibers that synapse with neurons in the prevertebral ganglia. The sensory input activates the sympathetic neurons, which project back to the intestine, where their fibers release norepinephrine. Norepinephrine acts to shut down the synaptic circuits in the enteric nervous system that initiate motility. This results in a state of inhibition or paralysis in the affected region of intestine.

■ ■ ■ ■ ■ ■ ■ ■ ■ ■ REVIEW EXERCISES ■ ■ ■ ■ ■ ■ ■ ■ ■ ■

Identify Each with a Term
1. Enteric neurons with long-lasting hyperpolarizing afterpotentials, large resting potentials, and low input resistance
2. The enteric plexus found between longitudinal and circular muscle coats of the intestine
3. Enteric neurons with a single long axon and short, stublike dendrites
4. The division of the autonomic nervous system with preganglionic neurons in the medulla of the brain
5. Suppression of release of neurotransmitters from axons mediated by action of chemical messengers at receptors on the axon

Define Each Term
6. Enteric ganglion
7. Sympathetic preganglionic neurons
8. Galanin
9. Dogiel type II neurons
10. Vagovagal reflex

Choose the Correct Answer

11. Which of the following is a gastrointestinal effector?
    a. Mucosal epithelium
    b. Vagus nerve
    c. Slow synaptic excitation
    d. Dogiel type II neurons

12. The extrinsic innervation of the intestine consists of:
    a. Dogiel type I and II neurons
    b. Neurons in the myenteric plexus
    c. The enteric nervous system
    d. The sympathetic and parasympathetic divisions of the autonomic nervous system

13. Secretomotor neurons to the intestinal crypts and villi are found in the:
    a. Myenteric plexus
    b. Medulla of the brain
    c. Sympathetic ganglia
    d. Submucous plexus

14. Cell bodies of sensory neurons of the digestive tract are found in:
    a. The nodose ganglia of the vagal nerves
    b. The dorsal root ganglia of the spinal cord
    c. Both a and b

15. Opening of potassium channels in enteric neurons:
    a. Results in slow synaptic excitation
    b. Depolarizes the membrane
    c. Increases the input resistance of the membrane
    d. Is involved in inhibitory synaptic potentials

16. Enteric slow EPSPs are characterized by:
    a. Membrane hyperpolarization
    b. Enhanced excitability
    c. Enhanced hyperpolarizing afterpotentials
    d. Decreased membrane resistance

17. The synaptic interface between postganglionic sympathetic fibers and the enteric nervous system:
    a. Is at α_2-adrenergic receptors on the musculature
    b. Involves release of acetylcholine
    c. Is at presynaptic α_2-adrenergic receptors
    d. Is a site of excitatory input to myenteric and submucosal plexus neurons

18. Prevertebral sympathetic ganglia:
    a. Are relay sites for enteroenteric inhibitory reflexes
    b. Receive inputs from the vagus nerves
    c. Are relay sites for vagovagal excitatory reflexes
    d. Are found within the gastrointestinal wall

19. Which of the following neurotransmitter substances produces responses like slow EPSPs when applied to enteric neurons?
    a. Galanin
    b. Opioid peptides
    c. Substance P
    d. Adenosine

Briefly Answer

20. Why is the enteric nervous system called the "little brain in the gut"?

21. What are the properties of S/type 1 enteric neurons?

22. Why does the input resistance of an AH/type 2 neuron increase during a slow EPSP?

23. What is the functional significance of hyperpolarizing afterpotentials in AH/type 2 enteric neurons?

24. What kind of ionic channels open to produce hyperpolarizing afterpotentials in AH/type 2 enteric neurons?

25. How do neuroeffector junctions in the digestive tract differ from neuromuscular junctions (motor endplates) in skeletal muscles?

■ ■ ■ ■ ■ ■ ■ ■ ■ ■ **ANSWERS** ■ ■ ■ ■ ■ ■ ■ ■ ■ ■ ■ ■

1. AH/type II neurons
2. Myenteric plexus
3. Dogiel type I neurons
4. Parasympathetic division
5. Presynaptic inhibition
6. A cluster of closely grouped nerve cell bodies located in the wall of the gastrointestinal tract
7. Neurons in the intermediolateral column of the spinal cord that project their axons to synapse with postganglionic sympathetic neurons in sympathetic ganglia
8. A peptide that behaves like an inhibitory neurotransmitter at postsynaptic receptors on enteric neurons
9. Neurons in the enteric nervous system that have smooth multipolar cell bodies with several long processes extending from the cell body
10. Reflexes for which sensory signals are transmitted in vagal afferent fibers to be processed in vagally related circuits in the medulla oblongata, with the return outflow carried back to the digestive tract in efferent vagal fibers
11. a

12. d
13. d
14. c
15. d
16. b
17. c
18. a
19. c
20. The structure, function, and neurochemistry of enteric ganglia are like those of the central nervous system but unlike autonomic ganglia elsewhere in the body. The enteric nervous system behaves like a brain that processes sensory information and controls the outflow of signals in motor neurons to the gastrointestinal effector systems. Its neurons have the same kind of electrical and synaptic behavior as the central nervous system, and it has the same large variety of neurotransmitter substances as the brain and spinal cord.
21. S/type 1 enteric neurons discharge action potentials repetitively when the membrane potential is depolarized. The frequency of action potential discharge in-

creases in direct relation to the amount of membrane depolarization. The rising phase of the action potential is due exclusively to an increase in inward sodium current. S/type 1 neurons do not show long-lasting hyperpolarizing afterpotentials following discharge of an action potential.

22. Activation of postsynaptic receptors results in decreased influx of calcium ions and closure of calcium-activated potassium channels in the membrane.

23. Hyperpolarizing afterpotentials lengthen the relative refractory period for discharge of action potentials, thereby, limiting the frequency at which the neuron can discharge action potentials.

24. Hyperpolarizing afterpotentials in AH/type 2 enteric neurons are produced by opening of calcium-activated potassium channels.

25. Neuroeffector junctions in the gastrointestinal tract do not have a complex structure like that found at motor endplates in skeletal muscles. The motor axons do not release their neurotransmitters at a single end terminal. Instead, the motor axons are varicose, with release of neurotransmitter occurring from each of a string of varicosities along the length of the axon. A relatively large distance separates motor axons from the muscle cells. Neurotransmitters are released in all directions from the axonal varicosities and diffuse over long distances to excite or inhibit a large number of individual smooth muscle fibers simultaneously.

Suggested Reading

Furness JB, Costa M: *The Enteric Nervous System.* New York, Churchill Livingstone, 1987.

Wood JD: Electrical and synaptic behavior of enteric neurons, in Wood JD, (ed): *Handbook of Physiology: The Gastrointestinal System—Motility and Circulation.* Bethesda, American Physiological Society, 1989, pp 465–517.

Wood JD: Physiology of the enteric nervous system, in Johnson LR, Christensen J, Jackson M (eds): *Physiology of the Gastrointestinal Tract.* New York, Raven, 1987, pp 67–109.

Wood JD: Physiology of the enteric nervous system, in Johnson LR, Christensen J, Jacobson G, et al (eds): *Physiology of the Gastrointestinal Tract.* 3rd ed., New York, Raven, 1994.

Gastrointestinal Motility

OBJECTIVES

After studying this chapter, the student should be able to:
1. Explain the importance of inhibitory motor neurons in the control of contractile behavior of intestinal smooth muscle
2. Relate the mechanisms involved in the production of physiologic and pathophysiologic ileus in the intestine
3. Describe the behavior of the muscles of the intestine during peristaltic propulsion
4. Distinguish between driver networks, excitatory interneurons, inhibitory interneurons, motor neurons, and synaptic gates in the neural circuit for peristalsis
5. Describe the motor behavior of the lower esophageal sphincter during and after a swallow
6. Explain how function of the proximal stomach differs from that of the distal stomach in the determination of rate of gastric emptying
7. Describe the interdigestive migrating motor complex
8. Contrast the movement of material in the small and large intestine
9. Describe the mechanisms for maintenance of fecal continence
10. Describe the neural deficit in Hirschsprung's disease

The motility in the different regions of the gastrointestinal tract reflects coordinated contractions and relaxations of the smooth muscle. Contractions are organized to produce the propulsive forces that move digesta along the tract and to mix ingested foodstuffs with digestive enzymes and bring nutrients into contact with the mucosa for efficient absorption. Relaxation of spontaneous tone in the smooth muscle allows sphincters to open and ingested material to be accommodated in reservoirs such as the stomach. The enteric nervous system, together with its input from the central nervous system, organizes motility into routine patterns of efficient behavior suited to differing digestive states, (e.g., fasting and processing of a meal) as well as abnormal patterns (e.g., during emesis).

The Physiology of Gastrointestinal Motility

■ The Gastrointestinal Muscle Behaves Like an Electrical Syncytium

The circular muscle coat accounts for the bulk of smooth muscle in the stomach and intestines. It is the main generator of forces required for propulsive contractions. The longitudinal muscle accounts for less mass than the circular muscle and in the colon of some species, including humans, does not occupy the entire circumference of the bowel. Contractions of the circular muscle coat in all regions of the digestive tract are triggered by action potentials generated by the muscle fiber membrane. The longitudinal muscle does not always generate action potentials; in the absence of action potentials, contractions are produced by neurotransmitters acting through pharmacomechanical coupling (Chap. 26). The muscle fibers of the circular muscle coat behave as if they are electrically connected, so that action potentials cross the fiber boundaries and travel from muscle fiber to muscle fiber. Low-resistance electrical junctions, called **gap junctions** (Fig. 1-6), account for the fiber-to-fiber coupling.

Each action potential, as it sweeps through a muscle fiber, evokes a **phasic contraction,** a transient, "twitchlike" contraction. These differ from **tonic contractions,** during which contractile tension is maintained for prolonged periods. If action potentials pass through the fibers at a repetitive rate of sufficient frequency, the twitchlike contractions can summate to give the appearance of a tonic contraction.

■ Electrical Slow Waves Trigger Action Potentials in Intestinal Muscle

Electrical **slow waves** trigger action potentials in the circular muscle of the intestine. Circular muscle fibers possess the voltage-gated ionic channels necessary for production of action potentials. Electrical slow waves depolarize the membranes of the muscle fibers to the threshold for opening of voltage-gated calcium channels, which carry the inward current for generation of action potentials. Electrical slow waves are rhythmic cycles of depolarization and repolarization that resemble sinusoidal waveforms when recorded from the intestine (Fig. 27-1). They occur spontaneously and are always present. Action potentials occur at the crest of the depolarization phase of the slow wave and are followed by the associated muscle contraction. Consequently, the highest frequency of phasic contractions is the same as the frequency of the slow waves. The frequency of slow waves is highest in the duodenum (about 12/min in humans) and decreases in descending steps, with the distal ileum having the lowest frequency.

Slow waves are also present in the large intestine,

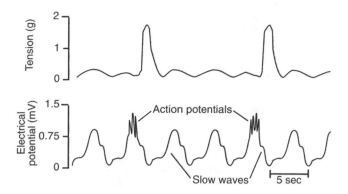

Figure 27–1 ■ Electrical slow waves are rhythmic changes in electrical potential that can be recorded with intracellular micro-electrodes or, as illustrated here, extracellular electrodes placed on the serosal surface of the small intestine (lower trace). Electrical slow waves trigger action potentials that occur at the crests of the slow waves. Action potentials trigger contractions of the intestinal muscle (top trace). A latent period separates the action potentials and onset of contraction.

but with the frequency gradient reversed: the lowest frequency is in the proximal colon and the highest is in the distal regions of the large bowel. The slow waves in the colon are not always uniform waveforms, as in the duodenum and jejunum, when recorded extracellularly at the serosal surface. The nervous system is not involved in the generation of electrical slow waves in the intestine. A current hypothesis is that slow waves are generated by a network of specialized cells called **interstitial cells of Cajal,** found between the longitudinal and circular muscle layers and in the sub-mucosal region of the intestine.

Not every slow wave triggers action potentials, and the frequency of the phasic contractions in an intestinal segment may not be the same as that of the slow waves. When not all slow waves are generating contractions, the intervals between contractions are multiples of the shortest slow wave interval (Fig. 27–1). We discuss later how nervous mechanisms determine whether a particular slow wave triggers action potentials and associated contractions.

Electrical slow waves occur synchronously around the circumference of the intestine. They behave as if they either travel very rapidly or are triggered instantaneously at all points around the intestinal segment. At the same time, they appear to travel at much slower velocity in the longitudinal direction along the intestine. The direction of travel is aboral in the small intestine, at a velocity of 5–15 cm/sec (depending on animal species) in the duodenum. Velocity of propagation decreases from the proximal to distal small intestine. Since slow waves trigger the action potentials that produce contractions, contractile waves tend to spread aborally with the same velocity as the slow waves.

■ Inhibitory Motor Innervation Plays an Important Role in Gastrointestinal Motility

Overall, the intestinal musculature behaves like a self-excitable electrical syncytium: action potentials starting anywhere in the muscle spread from muscle cell to muscle cell in three dimensions throughout the muscle. As mentioned, self-excitation results from a non-neural pacemaker system in the form of electrical slow waves. The circular muscle can be thought of as responsive to the slow waves.

One may ask why not every slow wave cycle triggers action potentials and contractions in the circular muscle, and why action potentials and contractions do not continue to spread in the syncytium throughout the entire length of intestine every time they occur. The answers lie in the function of inhibitory motor neurons in the enteric nervous system. The circular muscle can respond to slow waves only when the inhibitory neurons in the segment are turned off. Control of the on-off behavior of inhibitory motor neurons determines when the ever-running pacemaker initiates a contraction and the distance and direction of propagation once the contraction has started.

The inhibitory motor neurons for the circular muscle are continuously active. Muscle responses occur only when the inhibitory neurons are switched off by input from interneurons in the control circuits of the enteric nervous system (Fig. 27–2). In sphincters, the inhibitory neurons are normally off. They are turned on when the timing is appropriate for coordinating opening of the sphincter and physiologic events in adjacent regions.

In the circular muscle, the activity of inhibitory motor neurons determines how far within the syncytium action potentials spread, which in turn determines the length of the contracting segment. Segments in which inhibition is switched off can contract, whereas adjacent segments with ongoing inhibitory activity cannot (Fig. 27–3). The boundaries of the contracted segment reflect the transition zone from inactive to active inhibitory motor neurons. The directional sequence in which inhibitory motor neurons are switched off determines the direction of propagation of the contraction along the intestinal tube. Normally, they are turned off in sequence in the aboral direction, resulting in contractions that propagate in the anal direction.

The physiology of neuromuscular relations in the intestine predicts that uncontrolled muscle contraction will result from any condition in which the inhibitory motor neurons are destroyed. Without inhibitory control, the self-excitable syncytium of nonsphincteric regions contracts continuously and obstructs the passage of chyme through the intestinal lumen. This occurs because every slow wave cycle can trigger contractions, with no control over the distance or direc-

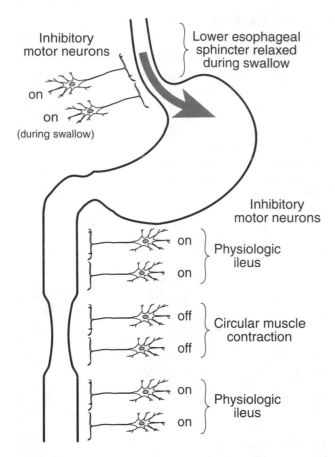

Figure 27–2 ■ Role of inhibitory motor neurons in control of the contractile activity of the smooth muscle along the gastrointestinal tract. Inhibitory motor neurons must be turned off for the circular muscle outside of sphincters to contract. Inhibitory motor neurons to the lower esophageal sphincter are normally off and must be turned on for the sphincter to relax.

tion of spread. Contractions spreading from opposite directions in the uncontrolled syncytium randomly collide, resulting in muscular fibrillation in the intestinal segment.

Basic Patterns of Gastrointestinal Motor Behavior

Gastrointestinal motility accounts for the propulsion, mixing, and reservoir functions necessary for the orderly processing of ingested food and elimination of waste products. **Propulsion** is the controlled movement of ingested foods, liquids, gastrointestinal secretions, and sloughed cells from the mucosa through the digestive tract. It moves the food from the stomach into the small intestine and along the small intestine with appropriate timing for efficient digestion and absorption. Propulsive forces move undigested material into the large intestine and eliminate waste through

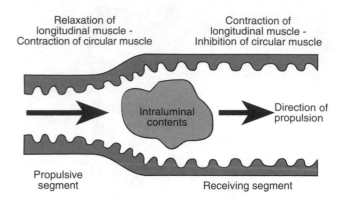

Figure 27–3 ■ Peristaltic propulsion is a stereotyped behavior pattern consisting of a propulsive segment and a receiving segment. Contraction of the longitudinal muscle and inhibition of the circular muscle occur in the receiving segment. In the propulsive segment, the longitudinal muscle is relaxed while the circular muscle contracts.

defecation. **Mixing** of ingested food decreases particle size, thereby increasing the surface area for action by digestive enzymes. Mixing also brings products of digestion into contact with the absorptive surfaces of the mucosa. **Reservoir** functions are performed by the stomach and colon. The body of the stomach stores ingested food and exerts steady mechanical forces, which are important determinants of gastric emptying. The colon holds material during the time required for absorption of excess water and stores the residual material until defecation is convenient.

■ The Enteric Nervous System Programs Several Patterns of Motility

Several patterns of motility are observed in the various specialized locations along the digestive tract. These patterns differ, depending on factors such as time after a meal, awake or sleeping state, and presence of disease. All patterns of gastrointestinal motility are based on three fundamental motor events programed by the enteric nervous system: peristalsis, mixing movements, and physiologic ileus.

■ Peristalsis Is a Stereotyped Propulsive Motor Response

Peristalsis is the organized propulsion of material over variable distances within the intestinal lumen. The muscle layers of the intestine behave in a stereotypical pattern during peristaltic propulsion (Fig. 27–3). This pattern is determined by the synaptic circuitry of the enteric nervous system. During peristalsis, the longitudinal muscle layer in the segment ahead of the advancing intraluminal contents contracts while the circular muscle layer simultaneously relaxes. The intestinal tube behaves like a cylinder with constant surface area. Shortening of the longitudinal axis of the cylinder is accompanied by widening of the cross-sectional diameter. Simultaneous shortening of the longitudinal axis and relaxation of the circular muscle results in expansion of the lumen. This prepares a **receiving segment** for the forward-moving intraluminal contents during peristalsis.

The second component of the peristaltic behavioral complex is contraction of the circular muscle in the segment behind the advancing intraluminal contents. The longitudinal muscle layer in this segment relaxes simultaneously with contraction of the circular muscle, resulting in conversion of this region to a **propulsive segment** that propels the luminal contents ahead, into the receiving segment. Intestinal segments ahead of the advancing front become receiving segments and then propulsive segments in succession as the peristaltic behavioral complex travels along the intestine.

■ Mixing Movements Are Characteristic of the Fed State

Mixing movements are characteristic of the **fed state** or **digestive state.** Mixing movements are also called **segmenting movements** (or segmentation), due to their appearance on x-ray films of the small intestine. In the mixing pattern, the behavior of the musculature is organized to propel luminal contents in both directions over short distances. This is in contrast to peristaltic propulsion, which moves the luminal contents in one direction over extended lengths of intestine. Mixing movements consist of circular muscle contraction in segments separated on either end by relaxed receiving segments (Fig. 27–4). This is undoubtedly a mechanism for the mixing and stirring of luminal contents in the receiving segments. The motor program for mixing is cyclic: receiving segments convert to contracting segments and contracting segments becoming receiving segments. Segmenting movements occur with the same intervals as electrical slow waves or at multiples of the shortest slow wave interval in the particular region of intestine.

■ Physiologic Ileus Reflects the Operation of a Neural Program

Physiological ileus—the absence of motility along the intestine—is a fundamental behavioral state of the intestine in which quiescence of motor function is neurally programmed. The state of physiologic ileus disappears after ablation of the enteric nervous system. When enteric neural functions are blocked (by anesthetics or when pathologic factors have destroyed the enteric nervous system), disorganized and nonpropulsive contractile behavior occurs continuously.

■ Sphincters Prevent the Reflux of Luminal Contents

Smooth muscle sphincters are found at the gastroesophageal junction, gastroduodenal junction, ileocolonic junction, and termination of the large intestine in the anus. They consist of rings of smooth muscle

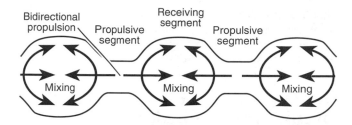
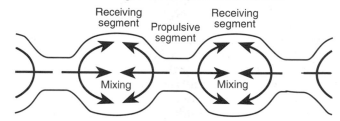

Figure 27–4 ■ The segmentation pattern of motility characteristic of the digestive state. Propulsive segments separated by receiving segments occur randomly at many sites along the small intestine. Mixing of the luminal contents occurs in the receiving segments. Receiving segments convert to propulsive segments, while propulsive segments become receiving segments. Segmental interconversion occurs continuously along the small intestine until digestion is complete. The intraluminal contents are moved slowly in the aboral direction by the segmentation motility pattern.

that remain in a continuous state of contraction. The effect of the tonic contractile state is to produce a high-pressure zone in the lumen, separating two specialized compartments along the digestive tract. With the exception of the internal anal sphincter, sphincters function to prevent the backward movement of intraluminal contents. The internal anal sphincter prevents uncontrolled movement of intraluminal contents through the anus.

The **lower esophageal sphincter** prevents reflux of gastric acid into the esophagus. Incompetence results in chronic exposure of the esophageal mucosa to acid, which can lead to heartburn and dysplastic changes that may become cancerous.

The **gastroduodenal sphincter,** or **pyloric sphincter,** prevents excessive reflux of duodenal contents into the stomach. Incompetence of this sphincter can result in the reflux of bile acids from the duodenum. Bile acids are damaging to the protective barrier in the gastric mucosa; prolonged exposure can lead to gastric ulcers.

The **ileocolonic sphincter** prevents reflux of colonic contents into the ileum. Incompetence can allow entry of bacteria into the ileum from the colon, which may result in bacterial overgrowth in the small intestine, which normally has low bacterial counts.

The ongoing contractile tone in the smooth muscle sphincters is generated by **myogenic mechanisms.** The contractile state is an inherent property of the muscle and independent of the nervous system. Transient relaxation of the sphincter to permit the forward passage of material is accomplished by activation of inhibitory motor neurons.

The Organization of Neural Circuits for Gastrointestinal Motility

Programing of peristaltic propulsion by the enteric nervous system is like the reflex behavior mediated by the central nervous system. Basic neural circuits with fixed connections automatically reproduce a stereotyped pattern of behavior each time the circuit is activated. Connections for the reflex remain, irrespective of destruction of adjacent regions of the spinal cord. The peristaltic reflex circuit is similar, but the basic circuit is repeated along and around the intestine.

Just as the monosynaptic reflex circuit of the spinal cord is the terminal circuit for production of almost all skeletal muscle movements (Chap. 5), the same basic peristaltic circuitry underlies all patterns of propulsive motility. Blocks of the same basic circuit are connected in series along the length of the intestine and repeated in parallel around the circumference. The basic peristaltic circuit consists of synaptic connections between sensory neurons, interneurons, and motor neurons; synaptic gates determine the distance of propagation of the peristaltic behavioral complex.

■ Driver Networks Synchronize Motor Events around the Circumference of an Intestinal Segment

Driver networks of interconnected neurons are believed to provide simultaneous synaptic drive to subpopulations of inhibitory or excitatory motor neurons to the muscles (Fig. 27–5). A driver network encircles each intestinal segment and is repeated in each consecutive segment all along the intestine. The driver networks are organized to activate motor neurons simultaneously around the circumference of a segment. Synaptic gates determine whether or not sequential driver networks are switched on for continued propagation of the peristaltic complex.

Multipolar interneurons with interconnections for **feedforward excitation** make up the driver networks. In feedforward excitation, the neurons in the network make recurrent excitatory synaptic connections with one another. This mechanism ensures simultaneous activation of the entire network around the circumference of the segment. Simultaneous activation is important for effective application of the forces necessary for peristaltic propulsion. Effective propulsion could not be achieved, for example, if contraction of the circular muscle occurred only part way around the circumference of the propulsive segment or if inhibition were in effect in only part of the circumference of the receiving segment.

Slow excitatory postsynaptic potentials (slow EPSPs; Chap. 26) are thought to provide the mechanism for coordinating discharge of the neurons in the driver networks. In Figure 27–5, the driver networks must be turned on by synaptic inputs that are appropriately timed for the correct spatial organization of muscle behavior along the intestine. Subgroups of Dogiel type I neurons are the most likely candidates for the interneurons that transmit timing signals to appropriate driver networks up or down the length of the intestine.

The morphology, direction of the long axonal projection, and localization of 5-hydroxytryptamine in a subpopulation of excitatory Dogiel type I myenteric neurons (Chap. 26) suggest that these neurons transmit descending timing signals. The slow EPSP-like action of applied 5-hydroxytryptamine is consistent with its being the neurotransmitter. Release of 5-hydroxytryptamine from these neurons evokes slow EPSPs in some of the neurons of the driver network, leading to rapid buildup of excitation in the entire network due to the feedforward connections. The morphology of neurons containing substance P and the slow EPSP-like actions of this peptide suggest that it is the excitatory transmitter involved in the feedforward connections of the driver networks.

Receiving segments in the peristaltic pattern convert to propulsive segments as the complex travels down the intestine. This requires that discharge of action potentials in the driver network of the receiving segment stop abruptly for conversion to a propulsive segment to occur. Excitation in the network is shut down by timed activation of inhibitory Dogiel type I neurons, which send their axons in the aboral direction in the myenteric plexus (Fig. 27–5). Release of inhibitory neurotransmitter from these neurons results in slow inhibitory postsynaptic potentials (slow IPSPs), which terminate ongoing slow synaptic excitation in the network.

■ Gating Mechanisms Determine the Distance of Propagation of the Peristaltic Behavioral Complex

Slow EPSPs "open the somal gates" of multipolar neurons in the driver networks (Chap. 26). With the gates open, inbound action potentials in dendrites are passed on by the cell body as outbound signals in the axons. These signals release slow excitatory transmitter at synapses on neighboring neurons, which relay the same signals to their neighbors. This form of gating results in multiplication of signals, producing rapid buildup of neuronal discharge in the driver network.

Conversion from a receiving segment to a propulsive segment requires a mechanism for closing the somal gates in that segment while the gates in the next segment along the propulsive path are opened. Slow IPSPs provide this mechanism and in this respect are the functional converse of the slow EPSPs. Some of the putative inhibitory neurotransmitters act presynapti-

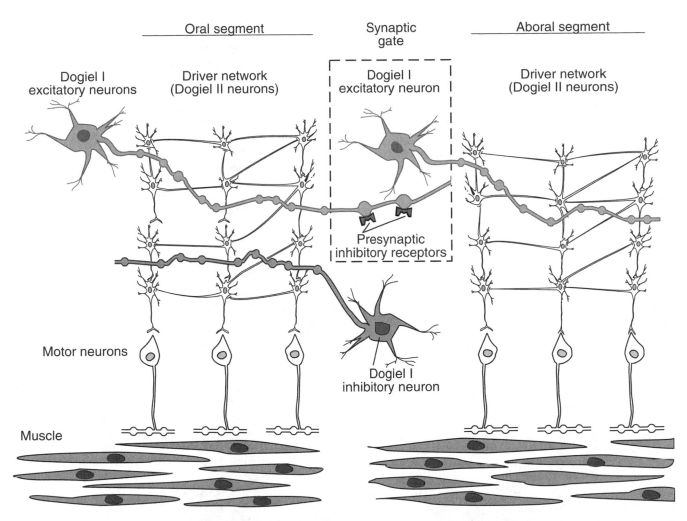

Figure 27–5 ■ Synaptic gates between two driver networks. Excitatory interneurons with Dogiel type I morphology turn the driver networks on, and inhibitory interneurons turn them off. Synaptic gates between driver networks are closed by activation of presynaptic inhibitory receptors, which stops peristaltic propulsion at that segment. (Modified from Wood JD: Neurophysiological theory of intestinal motility. *Jpn J Smooth Muscle Res* 23:143–186, 1987.)

cally to suppress synaptic transmission as well as inhibiting the excitability of the somal membranes. The peptide galanin, for example, blocks fast and slow EPSPs, which, in concert with galanin's inhibitory somal actions, effectively terminates recurrent excitation in a driver network.

Presynaptic mechanisms are involved in gating the transfer of signals between sequentially positioned driver networks. Synapses between the Dogiel type I neurons that carry excitatory signals to the driver networks of multipolar neurons function as gating points for control of the distance of propagation of peristaltic propulsion. Messengers that act presynaptically (Chap. 26) to block the release of transmitter from axons of Dogiel type I neurons close the gates for transfer of information.

The basic circuit for peristalsis is repeated serially along the intestine. Synaptic gates are thought to con-

nect the blocks of basic circuitry and to provide the mechanism for controlling the distance over which the peristaltic behavioral complex travels (Fig. 27–6). When the gates are opened, neural signals pass between successive blocks of the basic circuit resulting in propagation of the peristaltic event over extended distances. Long distance propulsion is prevented when all gates are closed.

■ The Basic Circuit for Peristalsis Contains Sensory Neurons, Interneurons, and Motor Neurons

The basic circuit responsible for peristaltic propulsion consists of synaptic connections between sensory neurons, interneurons, and motor neurons (Fig. 27-7). The primary requirement of the circuit is for the interneurons to be connected for generation of long-

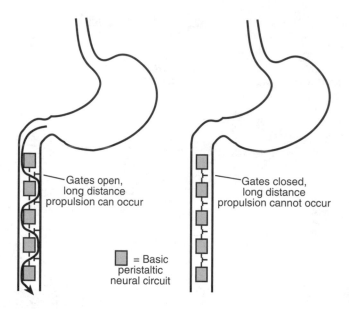

Figure 27–6 ■ Operation of synaptic gates between basic blocks of peristaltic circuitry. Opening of the gates between successive blocks of the basic circuit results in extended propagation of the propulsive event. Long-distance propulsion is prevented when all gates are closed.

lasting trains of action potentials in excitatory motor neurons to the longitudinal muscle and in inhibitory motor neurons to the circular muscle. This must occur ahead of the advancing luminal contents and be timed to occur synchronously around the circumference of the receiving segment. As the material in the lumen is propelled into the receiving segment, the same motor neurons must be abruptly switched off for the receiving segment to become a propulsive segment. Timed activation of excitatory motor neurons in the propulsive segment is unnecessary in the design of a minimal circuit, because the ever-present electrical slow waves are capable of driving circular muscle contraction after the inhibitory motor neurons have been turned off.

Experimental evidence suggests that an accessory circuit contains excitatory motor neurons to the circular muscle (Fig. 27-7). This circuit reinforces the strength of contraction in the propulsive segment as demanded by existing conditions in the intestine. For example, the accessory circuit may be brought into action to produce the powerful propulsive waves associated with luminal enterotoxins or food allergens.

The basic circuit may be switched on by input from sensory receptors in the region or by **gating neurons.** Sensory input may come from stretch receptors in the muscle or mechanoreceptors in the mucosa. Stretch receptors are responsible for triggering the peristaltic circuit in response to distension of the intestinal wall. Mucosal mechanoreceptors detect the movement of luminal contents as they brush by the mucosa and signal the progress of movement to the enteric processing circuits.

■ Ileus Reflects the Operation of a Neural Program

Quiescence of the intestinal circular muscle is believed to reflect the operation of a neural program in which all of the gates within and between basic peristaltic circuits are held shut. In this state, the inhibitory motor neurons are continuously active, and responsiveness of the circular muscle to the electrical slow waves is suppressed. This normal condition is in effect for various periods, depending on the digestive state.

The normal state of motor quiescence becomes pathologic when the gates for the particular motor patterns are rendered inoperative for abnormally long periods. In this state of **pathologic ileus** (paralytic ileus), the basic circuits are locked in an inoperable state while unremitting activity of the inhibitory motor neurons suppresses myogenic activity.

■ Intestinal Peristalsis Reverses During Emesis

The enteric neural circuits are programed to produce peristaltic propulsion in either direction along the intestine. If forward passage of the intraluminal bolus is impeded in the large intestine, reverse peristalsis propels the bolus over a variable distance away from the obstructed segment. **Retroperistalsis** then stops and forward peristalsis moves the bolus again in the direction of the obstruction. During the act of vomiting, retroperistalsis occurs in the small intestine. In this case, as well as in the obstructed intestine, the coordinated muscle behavior of peristalsis is the same except that it is organized by the nervous system to travel in the oral direction.

During emesis, powerful propulsive peristalsis starts in the jejunum and travels rapidly to the stomach. As a result, the small intestinal contents are propelled rapidly and continuously toward the stomach. As the propulsive complex advances toward the stomach, the gastroduodenal junction and the stomach wall relax and permit passage of the intestinal contents into the stomach. This occurs in association with contraction of the longitudinal muscle of the esophagus and dilation of the gastroesophageal junction. The overall result is the formation of a funnellike cavity that allows the free flow of gastric contents into the esophagus as intraabdominal pressure is increased by contraction of the diaphragm and abdominal muscles during retching.

■ Specialized Neural Circuits Relax or Contract Sphincters

Reflex responses of gastrointestinal sphincters reflect the presence of another specialized neural circuit in the enteric nervous system. This circuit coexists with the fundamental circuits for forward and reverse propulsion and operates in close coordination with the propulsive circuits.

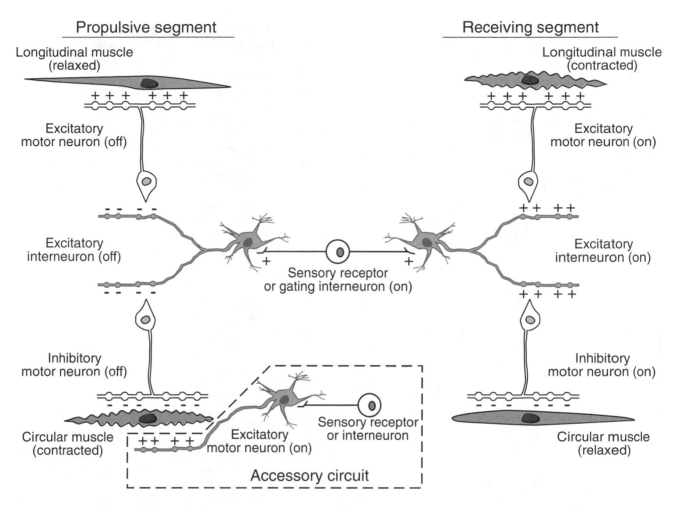

Figure 27–7 ■ The basic neural circuit responsible for peristaltic propulsion, showing the on-off states of the neurons required to produce the behavioral complex. The accessory circuit operates to satisfy special demands for increased strength of the propulsive event. (+) Release of excitatory neurotransmitter; (−) Release of inhibitory neurotransmitter.

Experimental distension of the rectosigmoid region of the large intestine with a balloon results in relaxation of the internal anal sphincter. The gastroduodenal junction contracts in response to distention of the duodenum, and the lower esophageal sphincter relaxes during propulsion of swallowed food in the esophagus.

The circuits that bring about reflex relaxation of these sphincters are specialized adaptations of the basic peristaltic circuit. Inhibitory motor neurons to the sphincteric muscles are homologous with those elsewhere in the system. They differ from other inhibitory motor neurons in the intestine in not being continuously active. Those to the sphincters remain silent until switched on by appropriately timed signals to relax the sphincter and permit passage of luminal contents. This is a function of synaptic connections that are specialized extensions of the basic circuit for unidirectional propulsive motility. Long distance projection

of information by axons of interneurons and mechanisms such as those in the intestine for rapid buildup of excitation in motor neurons around the circumference of the sphincter are involved in these circuits.

Motility in the Esophagus

The esophagus is divided into three functionally distinct regions: the upper esophageal sphincter, the esophageal body, and the lower esophageal sphincter (Fig. 27–8). Motor behavior of the esophagus involves striated muscle in the upper esophagus and smooth muscle in the lower esophagus.

When not involved in the act of swallowing, the muscles of the esophageal body are relaxed and the lower esophageal sphincter is tonically contracted. In contrast to the intestine, the relaxed state of the esophageal body is not produced by ongoing activity of inhibitory motor neurons. Excitability of the muscle

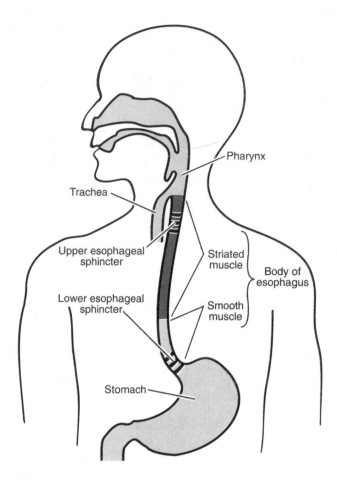

Figure 27–8 ■ Anatomy of the esophagus. Striated muscle is found in the upper esophagus and smooth muscle in the lower esophagus and lower esophageal sphincter.

is low and there are no electrical slow waves to trigger contractions. Activation of excitatory motor neurons, rather than myogenic mechanisms, accounts for the coordinated contractions of the esophagus during a swallow.

■ Peristalsis and Relaxation of the Lower Esophageal Sphincter Are the Main Motor Events in the Esophagus

The esophagus is a conduit for the transport of food from the pharynx to the stomach. Transport is accomplished by peristalsis produced by organized contractile behavior of the longitudinal and circular muscle coats.

Esophageal peristalsis may occur as **primary peristalsis** or **secondary peristalsis.** Primary peristalsis is initiated by the voluntary act of swallowing, irrespective of the presence of food in the mouth. Secondary peristalsis occurs when the primary peristaltic event fails to clear the swallowed material from the body of the esophagus. It is initiated by activation of mechanoreceptors and can be evoked experimentally by distending a balloon in the esophagus.

■ Manometric Catheters Are Used to Monitor Esophageal Motility

Esophageal motor disorders are diagnosed clinically with **manometric catheters** (multiple small catheters fused into a single assembly with pressure sensors positioned at various levels). They are placed into the esophagus via the nasal cavity. Manometric catheters record a distinctive pattern of motor behavior following a swallow (Fig. 27–9). At the onset of the swallow, the lower esophageal sphincter relaxes. This is recorded as a fall in pressure in the sphincter that lasts throughout the swallow and until the esophagus empties its contents into the stomach. Signals for relaxation of the lower esophageal sphincter are transmitted by the vagal nerves. The pressure-sensing ports along the catheter assembly show transient increases in pressure as the circular muscle contraction of the peristaltic pattern passes on its way to the stomach.

■ Dysphagia, Diffuse Spasm, and Achalasia Are Motor Disorders of the Esophagus

Failure of peristalsis in the esophageal body or failure of the lower esophageal sphincter to relax will result in **dysphagia**—difficulty in swallowing. Some humans show abnormally high pressure waves as peristalsis propagates past the recording ports on manometric catheters. This condition, called **nutcracker esophagus,** is sometimes associated with chest pain that may be experienced as anginalike pain.

In **diffuse spasm,** organized propagation of the peristaltic behavioral complex fails to occur after a swallow. Instead, the act of swallowing results in simultaneous contractions all along the smooth muscle esophagus. On manometric tracings, this is observed as a synchronous rise in intraluminal pressure at each of the recording sensors.

In **achalasia** of the lower esophageal sphincter, the sphincter fails to relax normally during a swallow. As a result, the ingested material does not enter the stomach and accumulates in the body of the esophagus. This leads to **megaesophagus,** in which distension and gross enlargement of the esophagus is evident. With advanced untreated achalasia, peristalsis does not occur in response to a swallow.

Achalasia appears to be a disorder of inhibitory motor neurons in the lower esophageal sphincter. The number of neurons in the lower esophageal sphincter is reduced and the level of the inhibitory neurotransmitter, vasoactive intestinal peptide, is diminished. It is a degenerative disease of the enteric nervous system that results in loss of the inhibitory mechanisms for relaxing the sphincter with appropriate timing for a successful swallow.

Motility in the Stomach

The main anatomic regions of the stomach are the **fundus, corpus (body), antrum,** and **pylorus** (Fig.

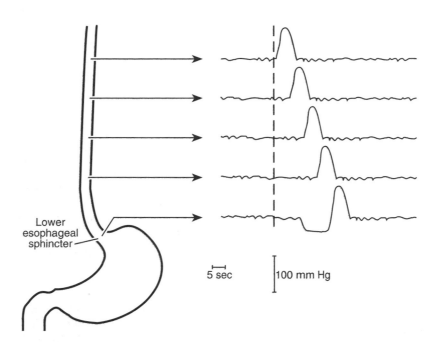

Figure 27–9 ■ Manometric recording of pressure events in the esophageal body and lower esophageal sphincter following a swallow. The propulsive segment of the peristaltic behavioral complex produces a positive pressure wave at each recording site in succession as the propulsive complex travels down the esophagus. Pressure falls in the lower esophageal sphincter shortly after the onset of the swallow, and the sphincter remains relaxed until the propulsive complex has transported the swallowed material into the stomach.

27–10). Functional regions of the stomach do not correspond to these anatomic regions. The stomach is divided into proximal and distal regions on the basis of distinct differences in functional motility between the two regions. The proximal region consists of the fundus and approximately one-third of the corpus; the distal region includes the caudal two-thirds of the corpus, the antrum, and the pylorus.

Differences in motility between the proximal and distal stomach reflect adaptations for different functions. The muscles of the proximal stomach are adapted for maintaining continuous contractile tone and usually do not contract phasically. Slow changes in the resting membrane potential and the absence of action potentials are the electrical correlates of the tonic contractile behavior of the muscles in the proximal stomach. In contrast, the muscles of the distal stomach contract phasically. The spread of phasic contractions along the distal stomach rapidly propels the gastric contents toward the gastroduodenal junction. Strong propulsive waves do not occur spontaneously in the proximal stomach.

■ Gastric Motility Is Specialized for Storage, Retention, Mixing, and Grinding of Food

The proximal region of the stomach is specialized for storage and retention of ingested food. This part

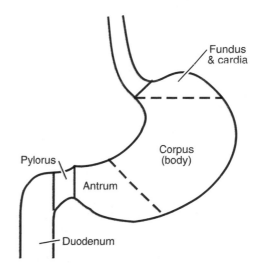

Figure 27–10 ■ Anatomy of the stomach.

of the stomach relaxes to accommodate increased intake of liquid or solid meals. As a result, intragastric pressure hardly rises during high rates of gastric filling. Filling a balloon in the proximal stomach to a volume of 200 ml elicits only a small increase in gastric pressure (Fig. 27–11). With an additional increase to 600 ml, little or no further increase in pressure occurs.

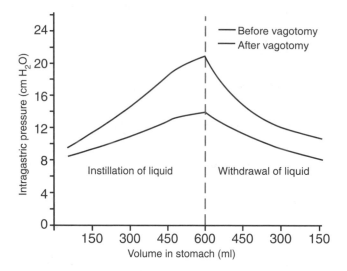

Figure 27–11 ■ The relationship of volume to intraluminal pressure in the stomach before and after vagotomy. In this human experiment, 600 ml of water was gradually instilled into the stomach and then withdrawn at about the same rate; the experiment was repeated after vagotomy for duodenal ulcer. After vagotomy, instillation of the water was accompanied by a sharper rise in the intragastric pressure.

This phenomenon, called **receptive relaxation,** is the mechanism by which the stomach adjusts to the intake of a large meal. When the proximal stomach is not filling, tonic contraction of the muscles exerts sustained compressive forces on the gastric contents. This results in steady pressure that delivers the contents to the distal stomach for grinding, mixing, and delivery to the small intestine.

Gastric receptive relaxation is controlled by command signals transmitted from the brainstem to the stomach by the vagal nerves. Relaxation of muscle contraction results from activation of inhibitory motor neurons in the enteric nervous system of the proximal stomach. Vasoactive intestinal peptide and nitric oxide are probable neurotransmitters released from the inhibitory motor neurons.

Receptive relaxation is impaired after surgical resection of the vagal nerves in selective vagotomy for treatment of ulcer disease. After vagotomy, filling a balloon in the stomach causes a sharp increase in intragastric pressure (Fig. 27-11). This can result in early activation of gastric stretch receptors during food ingestion and probably accounts for the early sensation of fullness and satiety experienced by patients following vagotomy.

The motility of the distal stomach is specialized for grinding and mixing functions. Grinding is the reduction of large masses of food by intragastric forces to progressively smaller particles, suitable for delivery to the small intestine. Mixing exposes the food to gastric secretions, including acid and digestive enzymes.

Grinding and mixing in concert with gastric secretion result in liquefaction of most food materials. Specialized propulsive contractile complexes accomplish the mixing and grinding functions.

Each propulsive contractile complex originates at the upper boundary of the distal stomach and travels in the caudal direction. This produces a ringlike indentation around the circumference of the stomach. As the ringlike contractions travel distally, they increase in strength and propagation velocity. Velocity in the corpus is about 0.5 cm/sec and progressively accelerates to 4 cm/sec as the contractile ring reaches the terminal antrum. As one contractile complex terminates in the antrum, another starts in the midstomach. Thus, contractile activity occurs in rhythmic cycles at a maximal frequency of 3/min in humans.

Grinding and mixing are produced by coordinated contractile behavior during each contractile cycle in the distal stomach. Each cycle consists of two separate ringlike contractions separated by intervals of 2–3 seconds as they propagate toward the pyloroduodenal junction (Fig. 27-12). The leading contraction, which is the weaker of the two, pushes the contents forward, resulting in delivery of liquefied chyme through the open pylorus into the upper duodenum. Arrival of the leading contraction at the pylorus closes the pyloric canal and halts the delivery of chyme into the duodenum. Two to three seconds later, the trailing contraction drives suspended particles of food against the closed pylorus and greatly elevates the pressure in the terminal antrum. The buildup of pressure propels the particles rapidly back toward the proximal stomach. As this occurs, shearing forces break up and reduce the size of the particles as they are jetted backward. The repetitive cycles of propulsion and retropulsion continue until digestible food particles are reduced to a size of about 0.1 mm, sufficient to be discharged into the duodenum.

■ Gastric Action Potentials Trigger Coordinated Contractions of the Antral Muscles

Like other gastrointestinal muscles, muscles of the distal stomach have functional syncytial properties and are active in the absence of chemical mediators. In contrast to intestinal muscle, the contractile complex of the antral muscles is associated with a single action potential.

Gastric action potentials are single electrical events consisting of a sharply rising upstroke depolarization, a plateau phase, and repolarization (Fig. 27-13). They originate spontaneously at the orad boundary of the distal stomach and propagate through the muscle coat to the gastroduodenal junction. At the site of origin, the action potentials are preceded by **pacemaker potentials,** which in the human stomach continuously initiate action potentials, at a frequency of 3/min. Electrical syncytial properties of the muscles account for

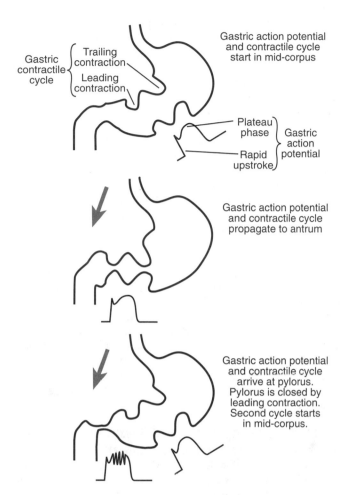

Figure 27–12 ■ The two-component contractile cycle of the distal stomach in relation to the gastric action potential. (Modified from Szurszewski JH: Electrical basis for gastrointestinal motility, in Johnson LR, Christensen J, Jackson M et al (eds): *Physiology of the Gastrointestinal Tract*. 2nd ed. New York, Raven, 1987, pp 383–422.)

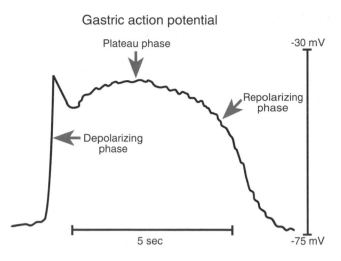

Figure 27–13 ■ The gastric action potential.

the propagation of action potentials from the site of origin to the gastroduodenal junction.

Gastric action potentials initiate the propulsive contractile complexes observed in the distal stomach and determine the duration and strength of the contractions. The leading contraction of each complex, which is relatively constant in strength, is associated with the upstroke of the action potential, whereas the trailing contraction of the complex occurs coincident with the plateau phase of the action potential (Fig. 27–12). The trailing contraction occurs only if the amplitude of the plateau potential is at level of depolarization beyond a specific threshold. Sometimes called the **mechanical threshold,** this threshold is about 30 mV more positive than the resting membrane potential of the muscle cell. With levels of depolarization greater than mechanical threshold, the strength of the trailing contraction increases in direct relation to the amplitude of the plateau potential.

The action potentials of the distal stomach are **myogenic,** because they are an inherent property of the muscle and occur in the absence of any neurotransmitters or other chemical messengers. Motor neurons, however, modulate their characteristics. Neurotransmitters affect primarily the amplitude of the plateau phase of the action potential and thereby control the strength of the contractile event triggered by the plateau phase. Neurotransmitters, such as acetylcholine from excitatory motor neurons, increase the amplitude of the plateau potential and of the contraction initiated by the plateau. The hormones gastrin and cholecystokinin have similar effects. Gastrin also increases the frequency of occurrence of gastric action potentials and the associated contractile complexes. Inhibitory neurotransmitters, such as epinephrine and vasoactive intestinal peptide, decrease the amplitude of the plateau and the strength of the associated contraction.

The magnitude of the effects of neurotransmitters and hormones increases with increasing concentration of the messenger substance. Through the actions of their neurotransmitters on the plateau phase of the action potential, the motor neurons determine whether the trailing contraction of the propulsive complex of the distal stomach occurs. When the amplitude of the plateau exceeds the threshold for contraction, the level beyond threshold and, therefore, the strength of contraction is determined by the amount of neurotransmitter released and present at receptors on the muscles.

The action potentials in the terminal antrum and pylorus differ somewhat in configuration from those in the more proximal regions. The principal difference is the occurrence of spikelike potentials in the plateau phase (Fig. 27–12), which trigger short-duration phasic contractions superimposed on the phasic contraction associated with the plateau.

The Nature of the Meal and Conditions in the Duodenum Determine the Rate of Gastric Emptying

In addition to the storage, mixing, and grinding of food, an important function of gastric motility is the orderly delivery of the gastric chyme to the duodenum at a rate that does not overload the digestive and absorptive functions of the small intestine. The rate of gastric emptying is adjusted to compensate for variations in the volume, composition, and physical state of the gastric contents.

The volume of liquid in the stomach is one of the important determinants of gastric emptying. The rate of emptying of isotonic, noncaloric liquids, such as water, is proportional to the initial volume or amount of distension of the proximal stomach. The larger the initial volume, the more rapid the emptying.

With a mixed meal in the stomach, liquids empty faster than solids. If an experimental meal consisting of solid particles of various sizes suspended in water is instilled in the stomach, emptying of the particles lags behind emptying of the liquid by 15–20 minutes (Fig. 27–14). Liquid emptying begins within 2–3 minutes of ingestion. With digestible particles, such as chunks of liver, the delayed emptying reflects the time required for the grinding action of the distal stomach to reduce the particle size. If the particles are plastic spheres of various sizes, the smallest spheres are emptied first; however, spheres up to 7 mm in diameter empty at a slow but steady rate when digestible food is in the stomach. Selective emptying of smaller particles first is referred to as the **sieving action** of the distal stomach. Inert spheres greater than 7 mm are not emptied while food is in the stomach: they empty at the start of the first migrating motor complex as the digestive tract enters the interdigestive state (see below).

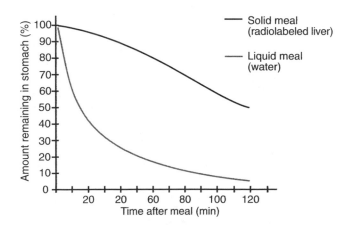

Figure 27–14 ▪ Liquid meals are emptied from the stomach at a faster rate than solid meals. Liquids in the stomach begin to empty immediately at a rapid rate, whereas solid particles empty slowly at first (a lag phase) and then more rapidly.

Osmolality, acidity, and caloric content of the gastric chyme are major determinants of the rate of gastric emptying. Hypotonic and hypertonic liquids empty more slowly than do isotonic liquids. The rate of gastric emptying decreases as the acidity of the gastric contents increases. Meals with a high caloric content empty from the stomach at a slower rate than meals with low caloric content. The mechanisms of control of gastric emptying keep the rate of delivery of calories to the small intestine within a narrow range, regardless of whether the calories are presented as carbohydrate, protein, fat, or a mixed meal. Of all of these, fat is emptied most slowly or, to state the converse, is the most potent inhibitor of gastric emptying. Part of the inhibition of gastric emptying by fats may involve release of the hormone cholecystokinin, which itself is a potent inhibitor of gastric emptying.

The intraluminal milieu of the small intestine is extremely different from that of the stomach (Chaps. 28, 29). Undiluted stomach contents have a composition that is poorly tolerated by the duodenum. Mechanisms of control of gastric emptying automatically adjust the delivery of gastric chyme to an optimal rate for the small intestine. This guards against overload of the small intestinal mechanisms for the neutralization of acid, dilution to isosmolality, and enzymatic digestion of the foodstuff.

Delayed Emptying and Rapid Emptying Are Disorders of Gastric Motility

Disorders of gastric motility can be divided into the broad categories of delayed and rapid emptying. The generalized symptoms of delayed and rapid emptying are given in Table 27–1.

Delayed gastric emptying is common in diabetes mellitus and may be related to pathology of the vagal nerves as part of a spectrum of autonomic neuropathy. Surgical vagotomy results in rapid emptying of liquids and delayed emptying of solids. As mentioned earlier, vagotomy impairs receptive relaxation and results in increased contractile tone in the proximal stomach. Increased pressure in the proximal stomach more forcefully presses liquids into the distal compartments. Paralysis with loss of propulsive motility in the antrum occurs after vagotomy. This results in **gastroparesis,**

TABLE 27–1 ▪ Symptoms of Abnormal Gastric Emptying

Delayed Emptying	Rapid Emptying
Nausea	Anxiety
Vomiting	Weakness
Bloating, fullness	Dizziness
Early satiety	Tachycardia
Epigastric pain	Sweating
Heartburn	Flushing
Anorexia	Decreased consciousness
Weight loss	Food avoidance

which can account for the delayed emptying of solids after vagotomy. When selective vagotomy is performed as treatment for peptic ulcer disease, the pylorus is enlarged **(pyloroplasty)** to compensate for postvagotomy gastroparesis.

Delayed gastric emptying with no demonstrable underlying condition is common. Up to 80% of patients with anorexia nervosa have delayed gastric emptying of solids. Another such condition is **idiopathic gastric stasis,** in which no evidence of an underlying condition can be found. **Prokinetic agents** (motility stimulating drugs) are used successfully in treatment of these patients. In children, **hypertrophic pyloric stenosis** impedes gastric emptying. This is a thickening of the muscles of the pyloric canal associated with loss of neurons in the enteric nervous system. Absence of inhibitory motor neurons and failure of the circular muscles to relax accounts for the obstructive stenosis.

Rapid gastric emptying often occurs in patients who have had vagotomy and antrectomy for treatment of peptic ulcer disease. These individuals have rapid emptying of solids and liquids. The pathologic effects (Table 27-1) constitute the **dumping syndrome,** which probably results from the "dumping" of large osmotic loads into the proximal small intestine.

Motility in the Small Intestine

Emptying of small radiopaque markers from the stomach and transit through the small intestine into the colon requires about 12 hours. Three fundamental patterns of motility occur in the small intestine, influencing the transit of material: the interdigestive pattern, the digestive pattern, and power propulsion.

■ The Migrating Motor Complex Characterizes the Interdigestive State

The **interdigestive pattern** of small intestinal motility begins 2–3 hours after a meal, after digestion and absorption of the nutrients. The contractile behavior is detected by placing pressure sensors in the lumen of the intestine or attaching electrodes to the intestinal surface. At a single recording site, the pattern consists of three consecutive phases: a silent period, **phase I,** which has no contractile activity; **phase II,** which consists of irregularly occurring contractions; and **phase III,** which consists of regularly occurring contractions. Phase I occurs after phase III, and the cycle is repeated.

With multiple sensors positioned along the intestine (Fig. 27-15), propagation of the phase II and phase III activity slowly down the intestine is evident. This behavioral pattern is thus called the **migrating motor complex** (MMC).

At a given time, the MMC occupies a limited length of intestine called the **activity front,** which has an upper and a lower boundary (Fig. 27-16). The activity front slowly advances (migrates) along the intestine at a rate that progressively slows as it approaches the ileum. Peristaltic propulsion of luminal contents in the aboral direction occurs in the activity front. Each peristaltic wave starts at the oral boundary and propagates to the aboral boundary. Successive peristaltic waves start a little further in the aboral direction and propagate a little beyond the boundary where the previous one stopped. Thus, the entire activity front slowly migrates down the intestine, sweeping the lumen clean as it goes.

The MMC is seen in most mammals, including humans, in conscious states and during sleep. It starts in the antrum of the stomach as an increase in the strength of the regularly occurring antral contractile complexes. From the antrum, the activity front progresses into the duodenum and down the small intestine. In humans and other large mammals, 80-120 minutes is required for the MMC to reach the ileum. As one activity front terminates in the ileum, another begins in the antrum. In humans, the time between cycles is longer during the day than at night. The activity front travels at about 3–6 cm/min in the duodenum and progressively slows to about 1–2 cm/min in the ileum. It is important not to confuse the speed of travel of the activity front of the MMC with that of the electrical slow waves, action potentials, and contractions within the activity front. Slow waves with associated action potentials and contractions travel about 10 times faster.

Phase III of the MMC occurs in the central region of the activity front as it slowly progresses along the intestine. Phase III lasts 8-15 minutes as the central region of the front passes a single recording site (Fig. 27-15, 27-16). Phase III is shortest in the duodenum and progressively increases as the complex migrates toward the ileum. Each slow wave, with action potentials and associated contraction, reflects a single propulsive behavioral complex occurring within the activity front. Phases I and II each occupy about half of the rest of the cycle as recorded at a single site (Fig. 27-16). Irregular electrical and contractile activity occur at the leading and trailing edges of the activity front. At the leading edge this reflects failure of some of the propulsive complexes to propagate as far as others. At the trailing edge, irregular activity indicates failure of some slow wave cycles to initiate propulsive complexes as the end of the front passes the recording site.

Cycling of the MMC continues until it is ended by the ingestion of food. A sufficient nutrient load terminates the MMC simultaneously at all levels of the intestine. Termination requires physical presence of a meal in the upper digestive tract; intravenous feeding does not end the fasting pattern. The speed with which the MMC is terminated at all levels of the intestine suggests a neural or hormonal mechanism. Both gastrin and cholecystokinin, which are released by a meal, when injected intravenously terminate the MMC in the stomach and upper small intestine, but not in the ileum.

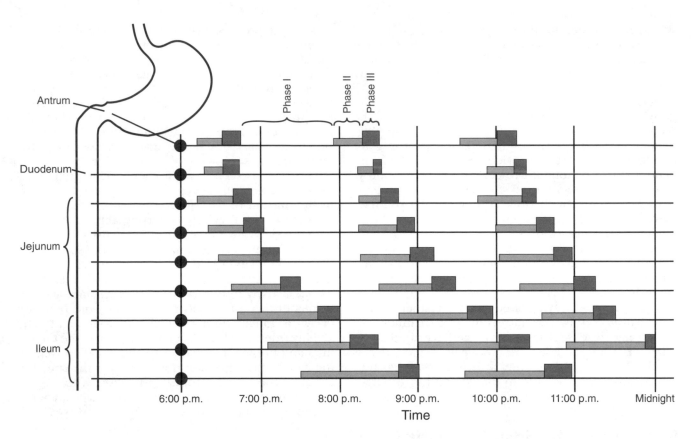

Figure 27–15 ■ The relative duration of each of the three phases of the migrating motor complex (MMC) in the dog. Sensors positioned along the intestine record the MMC as either phasic contractions or action potentials that trigger the contractions. Phase I is a period of no activity at the location of one sensor. Phase II is the occurrence of an occasional phasic contraction. Phase III follows phase II immediately and consists of continuous phasic contractions at the recording site. Phases II and III travel from the stomach to the ileum. The rate of travel of the MMC slows as it moves distally along the small intestine.

The MMC is organized by synaptic networks in the enteric nervous system. It continues in the small intestine after cutting off the extrinsic nervous input but stops when it reaches a region of the intestine where the enteric nervous system has been interrupted. Presumably, command signals to the enteric neural circuits are necessary for initiation of the interdigestive pattern, but, whether the commands are neural or hormonal or both is unknown. Although levels of the hormone **motilin** increase in the blood at the onset of the interdigestive state, it is unclear whether motilin triggers the MMC or is released as a consequence of it.

The adaptive significance of the MMC appears to be a mechanism for clearing undigestible debris from the intestinal lumen during the fasting state. Large undigestible particles are emptied from the stomach only during the interdigestive state. Bacterial overgrowth in the small intestine is associated with absence of the MMC. This suggests that the MMC may play a "housekeeper" role in preventing the overgrowth of microorganisms that might occur in the small intestine were the contents allowed to stagnate in the lumen.

■ Mixing Movements Characterize the Digestive State

Feeding interrupts the MMC and converts intestinal motor behavior to a **fed pattern** characteristic of the **digestive state.** The fed pattern is distinguished by peristaltic contractions that propagate for only very short distances. This activity occurs continuously along the length of the intestine. Since each short segmental contraction does not propagate far, it jets the chyme in both directions (Fig. 27–4). These contractions, spaced closely together as they are along the bowel, accomplish mixing of the luminal contents and over time, net aboral propulsion of the luminal chyme.

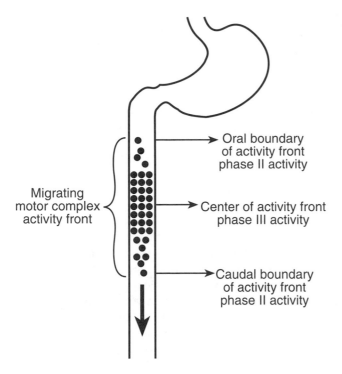

Figure 27–16 ■ The migrating motor complex (MMC) consists of an activity front as it travels slowly down the intestine. Peristaltic behavior within the activity front accounts for phases II and III of the MMC. Phase II reflects the leading edge, or caudal boundary, of the activity front as a few peristaltic behavioral complexes begin to arrive at the recording site on the bowel. As the center of the front passes the site, continuous peristaltic contractions give the appearance of phase III. As the trailing edge of the front passes the recording site, peristalsis is again occasional and briefly recorded as phase II-like activity.

■ Commands Transmitted from the Brain by the Vagal Nerves Sustain the Digestive Motility Pattern in the Small Intestine

Command signals transmitted from the brain to the intestine by the vagal nerves are important in the conversion from the fasting to the fed pattern. After the vagi are cut, a larger quantity of ingested food is necessary for interruption of the interdigestive motor pattern, and interruption of the MMCs is often incomplete. Evidence of vagal commands for the fed motor pattern has been obtained in dogs with cooling cuffs placed surgically around each vagal nerve. During the fed pattern, cooling and blockade of impulse transmission in the nerves results in interruption of the fed pattern of mixing movements. When the vagi are blocked during the fed pattern, MMCs reappear in the intestine but not in the stomach. With warming of the nerves and release of neural blockade, the fed pattern returns.

■ Power Propulsion Is a Response to Harmful Agents

Power propulsion involves strong, long-lasting contractions (**giant migrating contractions**) of the circular muscle that propagate for extended distances along the small intestine. The giant migrating contractions are considerably stronger than the phasic contractions during the MMC or fed pattern. Giant migrating contractions last 18–20 seconds and span several cycles of the electrical slow waves. They are a component of a highly efficient propulsive mechanism that rapidly strips the lumen clean as it travels at about 1 cm/sec over long lengths of intestine.

Intestinal power propulsion is a form of the stereotyped propulsive motor response of peristalsis described earlier. It differs from peristaltic propulsion during the MMC and fed state in that circular contractions in the propulsive segment are stronger and more open gates permit propagation over longer reaches of intestine. The circular contractions are not time-locked to the electrical slow waves and probably reflect strong activation of the muscle by excitatory motor neurons in accessory components of the basic peristaltic circuit (Fig. 27–7).

Power propulsion occurs in the retrograde direction during emesis and in the orthograde direction in response to noxious stimulation of the small intestine. Abdominal cramping sensations and sometimes diarrhea are associated with this motor behavior. Application of vinegar to the mucosa, the introduction of luminal parasites such as *Trichinella spiralis*, enterotoxins from pathogenic bacteria, and exposure to ionizing radiation all trigger the propulsive response. This suggests that power propulsion is an adaptation for rapid clearance of undesirable contents from the intestinal lumen. It may also accomplish mass movement of intraluminal material in normal states, especially in the large intestine.

■ Pseudo-Obstruction Is a Neuromuscular Disorder of the Small Intestine

Idiopathic intestinal pseudo-obstruction is characterized by symptoms of intestinal obstruction in the absence of a mechanical obstruction. The mechanisms for control of organized propulsive motility fail while the intestinal lumen is patent. This syndrome may result from abnormalities of the muscles or enteric nervous system. The general symptoms of colicky abdominal pain, nausea and vomiting, and abdominal distension simulate mechanical obstruction.

Idiopathic pseudo-obstruction is often associated with degenerative changes in the enteric nervous system. Failure of propulsive motility reflects the loss of the neural networks that program and control the organized motility patterns of the intestine. This can occur in varying lengths of intestine or for the entire

CHRONIC INTESTINAL PSEUDO-OBSTRUCTION

Chronic intestinal pseudo-obstruction (CIPO) presents with symptoms characteristic of mechanical obstruction but without blockage of the lumen due to tumors, strictures, or other factors. The symptoms, which include colicky abdominal pain, abdominal distension, abdominal tenderness, and nausea and vomiting, often cannot be distinguished from those of mechanical obstruction and require exploratory laparotomy to rule out the latter. A conclusive diagnosis of CIPO is made when surgical examination of the bowel reveals no signs of obstruction.

Some forms of CIPO are secondary to other disorders diagnosed earlier. These include diabetes mellitus, Chagas disease, and paraneoplastic syndrome, which reflect disordered physiology of the enteric nervous system. In other patients with CIPO no cause is apparent; this form of the disease is called chronic idiopathic intestinal pseudo-obstruction. Idiopathic CIPO may also result from enteric neuropathy and was insidious for many patients before recent basic research explained the pathophysiology.

In the past, patients with idiopathic CIPO would present with a clinical picture of obstruction serious enough to warrant surgery. No obstruction or lesion would be found during laparotomy, but the symptoms would persist. The patient would then undergo another operation in search of adhesions that might have developed from the first operation, and the ordeal would continue, with a scarred abdomen as evidence of a fruitless search for an obstruction.

Two forms of idiopathic CIPO are recognized. One is a degenerative disorder of the smooth muscle known as hollow visceral myopathy because the disease includes smooth muscle in other organs, such as the urinary bladder and ureters. Symptoms of the second form reflect a degenerative disorder of the enteric nervous system. Idiopathic CIPO runs in families, is generally manifest in young adulthood, and complicates life due to impaired food intake, often requiring total parenteral nutrition.

Biopsy and motility recording provide the definitive diagnosis for idiopathic CIPO. In the myopathic form, amplitude of contractions is reduced and biopsy samples show degenerated smooth muscle. In the neuropathic form, the contractions have normal or exaggerated amplitude but are uncoordinated and incapable of organized propulsion of the luminal content. Contractile activity in neuropathic bowel, unlike normal bowel, is continuous without any expression of physiologic ileus. Biopsy samples reveal reduced numbers or total loss of neurons in the enteric nervous system. Contractile activity in neuropathic CIPO reflects the normal autogenic properties of smooth muscle in the absence of the neural microcircuits that normally organize motor behavior. Inhibitory motor innervation of the musculature is lost in the neuropathic form, with the absence of inhibitory control manifest as uninterrupted contractile activity of the self-excitable muscle.

Symptoms of CIPO occur secondary to Chagas disease, small cell carcinoma in the lungs, and diabetes mellitus. Chagas disease is caused by the blood-borne parasite *Trypanosoma cruzi*. Some of the antigens at the surface of the parasite are similar to those found on enteric neurons. During the body's immune response to the infection, the antigens on the neurons look to the immune system like the parasite. An attack is launched on both parasites and neurons, with consequential destruction of the enteric neurons. A similar scenario occurs in paraneoplastic syndrome, where commonality of antigens results in immune attack simultaneously against the lung cancer and enteric neurons. In diabetics, the disruption of glucose and electrolyte balance is often accompanied by relentless degeneration of the peripheral nervous system, including the enteric nervous system. The mechanisms underlying the neuropathy are poorly understood, and the symptoms of associated CIPO are understood in terms of the loss of the normal neurophysiologic control of the intestinal muscles.

Many less severe cases of neuropathic, but not myopathic, forms of CIPO are treated successfully with prokinetic drugs, which enhance and strengthen propulsive motility. They include a class of compounds known as substituted benzamides and a class consisting of macrolide antibiotics, which includes erythromycin. Part of the mechanism of action of the drugs is to facilitate release of acetylcholine at nicotinic synapses, thereby "energizing" the enteric microcircuits responsible for propulsive motility.

length of small intestine. Contractile behavior of the circular muscle is hyperactive but disorganized in the denervated segments. This reflects the absence of inhibitory nervous control of the muscles, which are self-excitable when released from the braking action of enteric inhibitory motor neurons.

Paralytic ileus, another form of pseudo-obstruction, is characterized by prolonged motor inhibition. The electrical slow waves are normal but muscular action potentials and contractions are absent. Prolonged ileus can result from a mild and brief stimulus. For example, it occurs commonly after abdominal operations during which the intestine is handled but not incised.

Ileus results from suppression of the synaptic circuits that organize propulsive motility in the intestine. A probable mechanism is presynaptic inhibition and closure of synaptic gates, as discussed earlier. Coincident with suppression of the motor circuits is continuous discharge of the inhibitory motor neurons. This prevents the circular muscle from responding to the electrical slow waves, which are undisturbed in ileus.

Motility in the Large Intestine

In the large intestine, contractile activity occurs almost continuously. Whereas the contents of the small intestine move through sequentially with no mixing of individual meals, the large bowel contains a mixture of the remnants of several meals ingested over 3–4 days. The arrival of undigested residue from the ileum does not predict the time of its elimination in the stool.

The large intestine is subdivided into functionally distinct regions corresponding approximately to the ascending colon, transverse colon, descending colon, rectosigmoid region, and internal anal sphincter (Fig. 27–17). Transit of small radiopaque markers through the large intestine occurs in 36–48 hours, on average.

■ The Ascending Colon Is Specialized for Processing Chyme Delivered from the Terminal Ileum

Power propulsion in the terminal length of ileum may deliver relatively large volumes of chyme into the **ascending colon,** especially in the fed state. Neuromuscular mechanisms analogous to receptive relaxation in the stomach permit filling without large

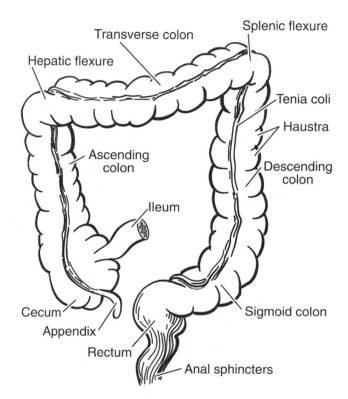

Figure 27–17 ■ Anatomy of the large intestine.

increases in intraluminal pressure. Chemoreceptors and mechanoreceptors in the cecum and ascending colon provide feedback information for control of delivery from the ileum, analogous to the feedback control of gastric emptying.

Dwell time of the material in the ascending colon is short. When radiolabeled chyme is instilled into the cecum in humans through a transgastrointestinal catheter, half of the instilled volume empties in an average time of 87 minutes. This is a long time compared to an equivalent length of small intestine but short compared with the transverse colon. It suggests that the ascending colon is not the primary site for storage, mixing, and removal of water from the feces.

The motor pattern of the ascending colon consists of orthograde or retrograde peristaltic propulsion. The significance of backward propulsion in this region is uncertain: it may be a mechanism for temporary retention of the chyme in the ascending colon. Forward propulsion in this region is probably controlled by feedback signals on the fullness of the transverse colon.

■ The Transverse Colon Is Specialized for the Storage and Dehydration of Feces

Radioscintigraphy shows that the labeled material is moved relatively quickly into the **transverse colon** (Fig. 27-18), where it is retained for about 24 hours. This suggests that the transverse colon is the primary location for removal of water and electrolytes and storage of solid feces in the large intestine.

A segmental pattern of motility accounts for the ultraslow forward movement of feces in the transverse colon. Ringlike contractions of the circular muscle divide the colon into chambers called **haustra** (Fig. 27-19). The motility pattern, called **haustration,** differs from segmental motility in the small intestine in that the contracting segment and the receiving segments on either side remain in their respective states for longer periods. In addition, there is uniform repetition of the haustra along the colon. The contracting segments in some places appear to be fixed and are marked by anatomic thickening of the circular muscle. The transverse colon is a sacculated structure; however, it is unclear if the contracting segment is always the circular muscle between "sacs," while the sacs per se are the "pockets" of the haustra.

Haustrations are dynamic in that they form and reform at different sites. The most common pattern in the fasting individual is for the contracting segment to propel the contents in both directions into receiving segments. This mixes and compresses the semiliquid feces in the haustral pockets and probably facilitates absorption of water without any net forward propulsion.

Net forward propulsion occurs when sequential migration of the haustra occurs along the length of bowel. The contents of one haustral pocket are propelled into the next region, where a second pocket is formed, and from there to the next segment, where the same events occur. This results in slow forward progression and is believed to be a mechanism for compacting the feces in storage.

Power propulsion is a different programed motor event in the transverse and descending colon. This motor behavior fits the general pattern of neurally coordinated peristaltic propulsion and results in the mass movement of feces over long distances. Mass movements may be triggered by increased delivery of ileal chyme into the ascending colon following a meal. The increased incidence of mass movements and generalized increase in segmental movements following a meal is called the **gastrocolic reflex.** Certain laxatives, such as castor oil, act at mucosal receptors to initiate the motor program for power propulsion in

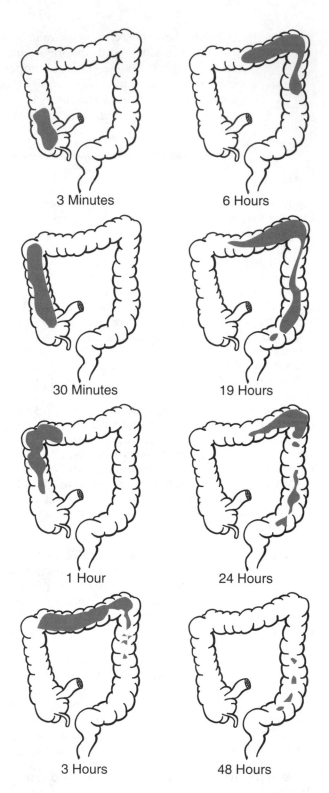

3 Minutes   6 Hours

30 Minutes   19 Hours

1 Hour   24 Hours

3 Hours   48 Hours

Figure 27–18 ■ Successive scintigrams reveal that the longest dwell time for intraluminal markers injected into the cecum is in the transverse colon. The image is faint after 48 hours, indicating that most of the marker has been excreted with the feces.

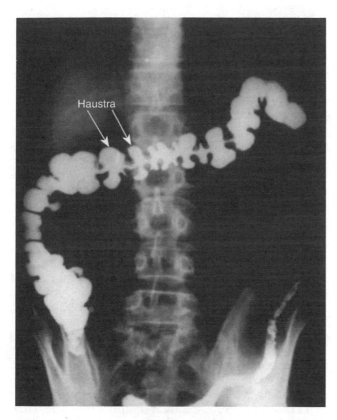

Figure 27–19 ■ Radiograph showing haustral contractions in the ascending and transverse colon of a human. Between the haustral "pouches" are segments of contracted circular muscle.

thc colon. The presence of threatening agents in the colonic lumen, such as parasites, enterotoxins, and food antigens, can also initiate power propulsion.

Mass movement of feces in the healthy bowel usually starts in the middle of the transverse colon and is preceded by relaxation of the circular muscle and downstream disappearance of haustral contractions. A large portion of the colon may be emptied as the contents are propelled at rates up to 5 cm/min as far as the rectosigmoid region. Haustration returns after passage of the power contractions.

■ **The Descending Colon Is a Conduit between the Transverse and Sigmoid Colon**

Radioscintigraphic studies in humans show that feces do not have long dwell times in the **descending colon.** Labeled feces begin to accumulate in the sigmoid colon and rectum about 24 hours after the label is instilled in the cecum. This is taken as evidence that the transverse colon is the main fecal storage reservoir and that the descending colon functions as a conduit

without long-term retention of the feces. This region has the neural program for power propulsion. Activation of the program can produce mass movements of feces into the sigmoid colon and rectum.

■ **The Physiology of the Rectosigmoid Region, Anal Canal, and Pelvic Floor Musculature Are Important in Maintenance of Fecal Continence**

The **sigmoid colon** and **rectum** are reservoirs with a capacity upward of 500 ml. Distensibility in this region is an adaptation for temporarily accommodating the mass movements of feces. The rectum begins at the level of the third sacral vertebra and follows the curvature of the sacrum and coccyx for its entire length (Fig. 27–20). Next is the anal canal, which is surrounded by the internal and external anal sphincters. The surrounding pelvic floor is formed by overlapping sheets of striated muscle fibers called **levator ani** muscles. This muscle group, which includes the **puborectalis muscle,** and the striated external anal sphincter comprise a functional unit that maintains continence. These are skeletal muscles that behave in many respects like the somatic muscles that determine posture elsewhere in the body.

The pelvic floor musculature can be viewed as an inverted funnel consisting of the levator ani and external sphincter muscles in a continuous sheet from the

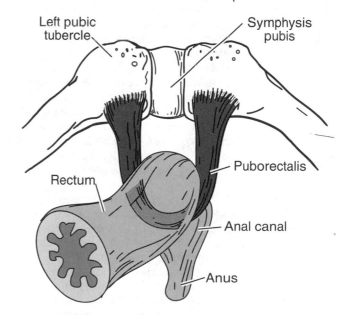

Pelvis viewed from top

Figure 27–20 ■ Structural relations of the anorectum and puborectalis muscle. Contraction of the puborectalis muscle helps form the anorectal angle, believed to be important in the maintenance of fecal continence.

bottom margins of the pelvis to the anal verge (the transition zone between mucosal epithelium and stratified squamous epithelium of the skin). After childbirth or defecation, the levator ani contract to restore the perineum to its normal position. Fibers of the puborectalis join behind the anorectum and pass around it on both sides to insert on the pubis. This forms a U-shaped sling that pulls the anorectal tube anteriorly, such that the long axis of the anal canal lies at nearly a right angle to that of the rectum (Fig. 27–20). Tonic pull of the puborectalis narrows the anorectal tube from side to side at the bend of the angle, resulting in a physiologic valve that is important in the mechanisms of continence.

The puborectalis sling and the upper margins of the internal and external sphincters form the anorectal ring, which marks the boundary of the anal canal and rectum. Surrounding the anal canal over a length of about 2 cm are the internal and external anal sphincters. The external sphincter is skeletal muscle attached to the coccyx posteriorly and the perineum anteriorly. When contracted, it compresses the anus into a slit, thereby closing the orifice. The internal anal sphincter is a modified extension of the circular muscle coat of the rectum. It is comprised of smooth muscle that contracts tonically to sustain closure of the anal canal.

Sensory innervation of the anorectum is important in the mechanisms of continence. Mechanoreceptors in the rectum detect distension and supply the enteric neural circuits with sensory information in much the same manner as occurs in the upper portions of the gastrointestinal tract. In contrast, the anal canal in the region of skin at the anal verge is innervated by somatosensory nerves that transmit sensory information to the central nervous system. This region has sensory receptors that detect touch, pain, and temperature with high sensitivity. Processing of information from these receptors allows the individual consciously to discriminate the presence of gas, liquid, and solids in the anal canal. In addition, stretch receptors in the muscles of the pelvic floor detect changes in the orientation of the anorectum as feces are propelled into the region. This accounts for the ability to experience the sensations of fecal presence in the region after surgical resection of the anorectum and ileoanal anastomosis.

Contraction of the **internal sphincter** and the puborectalis muscles blocks the passage of feces and maintains continence with small volumes in the rectum. When the rectum is distended, the **rectoanal reflex** or **rectosphincteric reflex** pathway is activated to relax the internal sphincter. Like other enteric reflexes, this involves a stretch receptor, enteric interneurons, and excitation of inhibitory motor neurons to the smooth muscle sphincter. Distension also results in the sensation of rectal fullness, mediated by central processing of information from mechanoreceptors in the pelvic floor musculature.

Relaxation of the internal sphincter allows contact of the rectal contents with the sensory receptors in the lining of the anal canal. This is believed to be how the individual discriminates between gas, liquid, and solid in the anorectum. It also provides an early warning of the possibility of a breakdown of the mechanisms of continence. When this occurs, continence is protected by voluntary contraction of the external sphincter and puborectalis muscle. The external sphincter closes the anal canal and the puborectalis sharpens the anorectal angle. An increase in the anorectal angle creates a "flap" valve: increases in intraabdominal pressure collapse the anterior rectal wall onto the upper end of the anal canal, thereby occluding the lumen.

Whereas the rectoanal reflex is mediated by the enteric nervous system, synaptic circuits for the neural reflexes of the external anal sphincter and other pelvic floor muscles reside in the sacral portion of the spinal cord. The mechanosensory receptors are thought to be muscle spindles and Golgi tendon organs like those found in skeletal muscles elsewhere. Sensory input from the anorectum and pelvic floor is transmitted over dorsal roots to the sacral cord, and motor outflow to these areas is in sacral root motor nerve fibers. The spinal circuits account for the reflex increases in contraction of the external sphincter and pelvic floor by behaviors that raise intraabdominal pressure, such as coughing, sneezing, and lifting weights.

■ Defecation Involves Neural Coordination of Muscles in the Large Intestine and Pelvic Floor

Distension of the rectum by the mass movement of feces or gas results in an urge to defecate or release flatus. Central nervous processing of mechanosensory information from the rectum is the underlying mechanism for this sensation. Local processing of the mechanosensory information in the enteric neural circuits activates the motor program for relaxation of the internal anal sphincter. At this stage of rectal distension, voluntary contraction of the external anal sphincter and the puborectalis muscles prevents leakage. The decision to defecate at this stage is determined by individual volition.

The act of defecation in normal individuals is a decision of the central nervous system. When the decision is finalized, commands from the brain to the sacral cord shut off the excitatory input to the external sphincter and levator ani muscles. Additional skeletal motor commands contract the abdominal muscles and diaphragm to increase intraabdominal pressure. Coordination of the skeletal muscle components of defecation results in straightening of the anorectal angle, descent of the pelvic floor, and opening of the anus.

Central commands initiate the large intestinal motor program for fecal transport in coordination with the programed behavior of the skeletal muscles. Transport of feces during defecation is achieved by powerful peristaltic propulsion in the sigmoid colon and rectum. The motor program for peristaltic propulsion resides in the synaptic circuitry of the enteric nervous system and can be initiated by both intrinsic commands and commands from the central nervous system. Pelvic nerves of the parasympathetic division of the autonomic nervous system transmit the commands from the spinal cord to the enteric nervous system. Parasympathetic ganglia located on the serosal surface of the colon relay and distribute the command signals to the enteric circuits along the length of bowel.

Programed behavior of the smooth muscle during defecation includes shortening of the longitudinal muscle coat in the sigmoid colon and rectum, followed by strong contraction of the circular muscle coat. This behavior corresponds to the basic stereotyped pattern of peristaltic propulsion. It represents **terminal intestinal peristalsis** in that the circular muscle of the distal colon and rectum become the final propulsive segment while the outside environment receives the forwardly propelled luminal contents.

A voluntary decision to resist the urge to defecate is eventually accompanied by relaxation of the circular muscle of the rectum. This is a form of receptive relaxation that accommodates the increased volume in the rectum. As wall tension relaxes, the stimulus for the rectal mechanoreceptors is removed and the urge to defecate subsides. Receptive relaxation of the rectum is accompanied by return of contractile tension in the internal anal sphincter, relaxation of tone in the external sphincter, and increased pull by the puborectalis muscle sling. When this occurs, the feces remain in the rectum until the next mass movement further increases the rectal volume and stimulation of mechanoreceptors again signals the neural mechanisms for defecation.

■ Hirschsprung's Disease and Incontinence Are Motor Disorders of the Large Intestine and Anorectum

Hirschsprung's disease is a developmental disorder that is present at birth but may not be diagnosed until later childhood. It is characterized by difficulty or failure of defecation. It is often called **congenital megacolon** because the proximal colon may become grossly enlarged with impacted feces or **congenital aganglionosis** because the ganglia of the enteric nervous system fail to develop in the terminal region of the large intestine. Enteric neurons may be absent in the rectosigmoid region only, in the descending colon, or in the entire colon. The aganglionic region appears constricted due to continuous contractile activity of the circular muscle, whereas the normally innervated intestine proximal to the aganglionic segment is distended with feces.

The constricted terminal segment in Hirschsprung's disease presents a functional obstruction to the forward passage of fecal material. Constriction and narrowing of the lumen of the segment reflects uncontrolled myogenic contractile activity of the circular muscle. In the absence of inhibitory motor neurons, a state of physiologic ileus is impossible and the muscle contracts continuously. The neural circuits for peristaltic propulsion are also missing in the segment, making it impossible to evoke the rectoanal distension reflex. Absence of the rectoanal reflex is one of the diagnostic hallmarks of the disease.

Incontinence is inappropriate leakage of feces and flatus to a degree that it disables the patient by disrupting routine daily activities. As discussed above, the mechanisms for maintenance of continence involve the coordinated interactions of a number of different components. Consequently, sensory malfunction, incompetence of the internal anal sphincter, or disorders of neuromuscular mechanisms of the external sphincter and pelvic floor muscles can be factors in the pathophysiology of incontinence.

Sensory malfunction renders the patient unaware of the filling of the rectum and stimulation of the anorectum, in which case he or she does not perceive the need for voluntary control over the muscular mechanisms of continence. This is tested clinically by distending an intrarectal balloon. The normal subject will perceive the distention with an instilled volume of 15 ml or less, whereas the sensory-deprived patient either will not report any sensation at all or will require much larger volumes before becoming aware of the distension.

Incompetence of the internal anal sphincter is usually related to a surgical or mechanical factor or perianal disease, such as prolapsing hemorrhoids. Incompetence of the neuromuscular mechanisms of the external sphincter and pelvic floor muscles may also result from surgical or mechanical trauma, such as can occur during childbirth. Physiologic deficiencies of the skeletal motor mechanisms can be a significant factor in the common occurrence of incontinence in the elderly. Whereas the resting tone of the internal anal sphincter does not seem to decrease with aging, the strength of contraction of the external anal sphincter does weaken with age. The striated muscles of the external anal sphincter and pelvic floor lose contractile strength with age. This occurs in parallel with deterioration of nervous function, reflected by decreased conduction velocity in fibers of the pelvic nerves. Clinical examination with intraanal manometry reveals decreased ability of the patient with disordered voluntary muscle function to increase intraanal pressure when asked to "squeeze" the intraanal catheter.

■ ■ ■ ■ ■ ■ ■ ■ ■ ■ ■ **REVIEW EXERCISES** ■ ■ ■ ■ ■ ■ ■ ■ ■ ■ ■

Identify Each with a Term

1. Low-resistance electrical junctions that account for fiber-to-fiber coupling in gastrointestinal muscles
2. Rhythmic cycles of depolarization and repolarization resembling sinusoidal waveforms in intestinal muscles
3. The leading segment of peristaltic propulsion in which the longitudinal muscle is contracted and the circular muscle inhibited
4. The normal absence of motility along the intestine
5. The predominant motor pattern of the small intestine in the interdigestive state

Define Each Term

6. Electrical syncytium
7. Lower esophageal sphincter
8. Neuronal driver network
9. Gastric receptive relaxation
10. Power propulsion

Choose the Correct Answer

11. Intestinal electrical slow waves:
    a. Function to trigger action potentials and associated contractions in the muscle
    b. Are generated by enteric neurons
    c. Have the highest frequency in the ileum
    d. Are found in the small but not the large intestine
12. Inhibitory enteric neurons:
    a. Are continuously active when they innervate sphincters
    b. Are normally inactive in the body of the intestine
    c. Are normally inactive when they innervate sphincters
    d. Use acetylcholine as an inhibitory neurotransmitter
13. During peristaltic propulsion, the:
    a. Direction of propulsion is always toward the anus
    b. Longitudinal muscle in the propulsive segment contracts
    c. Circular muscle in the propulsive segment contracts
    d. Circular muscle in the receiving segment contracts
14. Reflux of intraluminal contents at strategic points along the digestive tract is prevented by:
    a. Mixing movements
    b. Electrical slow waves
    c. Longitudinal muscle
    d. Sphincters
15. The basic neural circuit for peristalsis consists of:
    a. Sensory neurons
    b. Interneurons

c. Motor neurons
d. All of the above
16. During a swallow:
    a. The lower esophageal sphincter contracts
    b. Secondary peristalsis in the esophagus is initiated by the voluntary act of swallowing
    c. Pressure in the lower esophageal sphincter falls
    d. Smooth muscle is the first to contract as peristalsis starts in the uppermost portion of the esophageal body
17. In the stomach:
    a. The proximal region is adapted for phasic contractions
    b. The distal region is adapted for tonic contractions
    c. Receptive relaxation is controlled by command signals transmitted from the brainstem to the stomach by vagal nerves
    d. Receptive relaxation is produced by activation of excitatory motor neurons to the muscles in the proximal region
18. Gastric emptying is:
    a. Accelerated by the presence of acid in the duodenum
    b. Faster for liquid than for solid particles
    c. Slower for isotonic liquids than for hypo- or hypertonic liquids
    d. Increased as the caloric content of the nutrients in the duodenum increases
19. The migrating motor complex:
    a. Is triggered by ingestion of a meal
    b. Progressively increases in rate of travel as it approaches the ileum
    c. Has the most contractile activity during phase I
    d. Consists of peristaltic propulsion within an activity front

Briefly Answer

20. What are the general functions of the transverse colon?
21. What is the relation of inhibitory motor neurons to relaxation of sphincters?
22. What is the relation between electrical slow waves, inhibitory motor neurons, circular muscle action potentials, and contractions in the small intestine?
23. How do the specialized functions of the proximal and distal stomach differ?
24. How do excitatory motor neurons in the distal stomach influence the gastric action potential to control the strength of phasic contractions?
25. What is the adaptive significance of the migrating motor complex in the stomach and intestine?

■ ■ ■ ■ ■ ■ ■ ■ ■ ■ ■ **ANSWERS** ■ ■ ■ ■ ■ ■ ■ ■ ■ ■ ■

1. Gap junctions
2. Electrical slow waves
3. Receiving segment
4. Physiologic ileus

5. Migrating motor complex
6. A group of smooth muscle or other cell types in which electrical activity spreads directly from cell to cell to excite all the cells in the group

7. Ring of smooth muscle at the gastroesophageal junction, tonically contracted to produce an intraluminal high-pressure zone and barrier to reflux from stomach to esophagus

8. A group of synaptically interconnected neurons that function to provide simultaneous synaptic drive to sub-populations of inhibitory or excitatory motor neurons to the muscles

9. A phenomenon in which intragastric pressure hardly rises during high rates of gastric filling; it is a mechanism by which the stomach adjusts to the intake of a meal and is brought about by neurally mediated relaxation of the muscles in the proximal stomach

10. Strong, long-lasting contractions of the circular muscle that propagate for extended distances along the intestine; they function to eliminate rapidly potentially harmful agents from the intestinal lumen

11. a
12. c
13. c
14. d
15. d
16. c
17. c

18. b
19. d
20. The transverse colon stores and dehydrates feces.
21. Inhibitory motor neurons to sphincters are normally silent and are turned on transiently to relax the sphincter, with appropriate timing for passage of luminal material.
22. When inhibitory motor neurons are discharging, the circular muscle is inhibited and triggering of action potentials and associated contractions by the ongoing electrical slow waves is suppressed.
23. The proximal stomach is specialized for storage and retention of ingested food. Motility of the distal stomach is specialized for grinding and mixing functions.
24. Excitatory motor neurons in the distal stomach release acetylcholine, which increases the amplitude of the plateau phase of the gastric action potential. The strength of each contraction is determined by the amplitude of the plateau on the action potential. Contractile strength is increased by increased release of acetylcholine and consequent increase in the amplitude of the plateau.
25. The MMC provides a mechanism for clearing indigestible debris from the gastric and small intestinal lumen.

Suggested Reading

Furness JB, Costa M: *The Enteric Nervous System.* New York, Churchill Livingstone, 1987.

Johnson LR, Christensen J, Jacobson G et al (eds): *Physiology of the Gastrointestinal Tract.* 3rd ed. New York, Raven, 1994.

Kirsner JB: The Growth of Gastroenterologic Knowledge During the Twentieth Century. Philadelphia; Lea & Febiger, 1994.

Wood JD (ed): *Handbook of Physiology: The Gastrointestinal System—Motility and Circulation.* Bethesda, American Physiological Society, 1989.

Wood JD: Neurophysiological theory of intestinal motility. *Jpn J Smooth Muscle Res* 23:143-186, 1987.

CHAPTER

■ ■ ■ ■ ■ ■ ■ ■

28

Gastrointestinal Secretion

OBJECTIVES

After studying this chapter, the student should be able to:

1. Describe the salivon, acinus, and major salivary glands; describe the changes in electrolyte concentration with different rates of salivary secretion; and explain why secretion is hypotonic to plasma
2. List the functions of saliva; describe how parasympathetic stimulation directly and indirectly stimulates salivary secretion.
3. Define *parietal cells, chief cells, carbonic anhydrase,* and *alkaline tide*
4. Describe the functions and phases of gastric secretion and list the factors that inhibit gastric secretion
5. Define *acid tide* and *proenzyme;* describe the changes in electrolyte concentration with different rates of pancreatic secretion; list the major types of digestive enzymes produced by the pancreas
6. Describe the neural and hormonal control of pancreatic secretion; discuss the stimulation of pancreatic secretion by secretin, acetylcholine, and CCK; describe the cellular basis of potentiation
7. Define *primary* and *secondary bile acids, bile salts, bile acid-dependent bile flow, bile acid-independent bile flow;* describe the digestive function of bile and the mechanisms controlling its formation and secretion
8. Describe the enterohepatic circulation of bile salts and its physiologic significance

Secretions from the salivary glands, stomach, pancreas, and liver facilitate the digestion and absorption of nutrients by the gastrointestinal tract and protect the gastrointestinal mucosa from harmful effects of noxious agents. In this chapter we consider the relevant anatomy, mechanism and composition, and regulation of the rate of gastrointestinal secretion. Despite their differences, gastrointestinal secretions share a number of common features. A given secretion originates from individual groups of cells (e.g., the salivon in the salivary gland) before it pools with others. Secretions usually empty into smaller ducts, which in turn empty into larger ducts, and then into the lumen of the gastrointestinal tract. The ductal system not only serves as a conduit for gastrointestinal secretions but also is involved in modification of secretions. Finally, the enzyme carbonic anhydrase is involved in the formation of gastric, pancreatic, and intestinal secretions.

Salivary Secretion

Saliva is produced by a heterogeneous group of exocrine glands called the **salivary glands.** Saliva performs a number of functions: it facilitates chewing and swallowing by lubricating food, carries immunoglobulins that combat pathogens, and assists in carbohydrate digestion. The parotid, submandibular (submaxillary), and sublingual glands are the major salivary glands. They are drained by individual ducts into the mouth. The sublingual gland also has numerous small ducts that open into the floor of the mouth. The secretions of the major glands differ markedly: the parotid glands secrete a saliva rich in water and electrolytes, whereas the submandibular and sublingual glands secrete a saliva rich in mucin. There are also minor salivary glands located in the labial, palatine, buccal, lingual, and sublingual mucosae. The salivary glands are well endowed with a rich blood supply and receive innervations from both the parasympathetic and sympathetic nervous systems. Although hormones may modify the composition of saliva, their physiologic role is questionable,

and it is generally believed that salivary secretion is mainly under the control of the autonomic nervous system.

■ The Salivary Glands Consist of a Network of Acini and Ducts

A diagram of the human submandibular gland is shown in Figure 28-1. The basic unit (the **salivon**) consists of acinus, intercalated duct, striated duct, and excretory (collecting) duct. The **acinus** is a blind sac containing mainly pyramidal cells. Occasionally, there

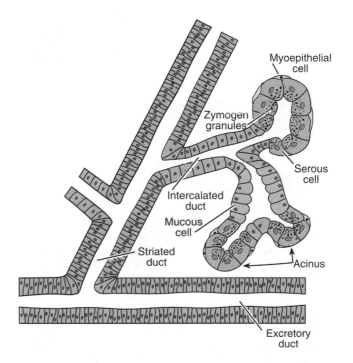

Figure 28-1 ■ An acinus and associated ductal system from the human submandibular gland. (Modified from Johnson LR, Christensen J, Jackson MJ et al (eds): *Physiology of the Gastrointestinal Tract.* New York, Raven, 1987.)

are stellate-shaped **myoepithelial cells** surrounding the large pyramidal cells. The cells of the acinus are not homogeneous; **serous cells** secrete digestive enzymes and **mucous cells** secrete mucin. The serous cells contain an abundance of rough endoplasmic reticulum, reflecting active protein synthesis, and numerous zymogen granules. Salivary amylase is an important digestive enzyme synthesized and secreted by the serous acinar cells.

An abundance of mucin droplets is stored in the mucous acinar cells. The mucin is composed of glycoproteins of various molecular weights. Both the serous and mucous acinar cells package proteins destined to be secreted into **zymogen granules** (Fig. 28–1). The zymogen granules are modified in the Golgi apparatus, and as they mature their appearance changes as a result of an increase in density. They are then released into the acinus lumen by exocytosis.

The **intercalated ducts** are lined with small cuboidal cells. The function of these ducts is unclear, but they may be involved in the secretion of proteins, since secretory granules are occasionally observed in their cytoplasm. The intercalated ducts are connected to the striated duct, which eventually empties into the excretory duct. The **striated duct** is lined with columnar cells. Its major function is to modify the ionic composition of the salivary secretions. The large **excretory ducts,** lined with columnar cells, also play a role in modifying the ionic composition of salivary secretions. Although the majority of proteins are synthesized and secreted by the acinar cells, the duct cells also synthesize a number of proteins, such as epidermal growth factor, ribonuclease, α-amylase, and proteases.

■ Saliva Contains Various Electrolytes and Proteins

Electrolytes in Saliva ■ The electrolyte composition of the **primary secretion** produced by the acinar cells probably resembles that of plasma (Fig. 28–2). Micropuncture samples have revealed that there is little modification of the electrolyte composition of the primary secretion in the intercalated duct. However, samples from the excretory (collecting) ducts are *hypotonic* relative to plasma, indicating modification of the primary secretion in the striated and excretory ducts. As shown in Figure 28–2, there is less Na^+, less Cl^-, more K^+, and more HCO_3^- in saliva than in plasma. This is because Na^+ is actively absorbed from the lumen by the ductal cells, whereas K^+ and HCO_3^- ions are actively secreted into the lumen. Chloride ions leave the lumen either in exchange for HCO_3^- ions or by passive diffusion along the electrochemical gradient created by Na^+ absorption.

The electrolyte composition of saliva depends on the rate of secretion (Fig. 28–2). As the secretion rate increases, the electrolyte composition of saliva approaches the ionic composition of plasma, but at low flow rates it differs markedly. At low secretion rates,

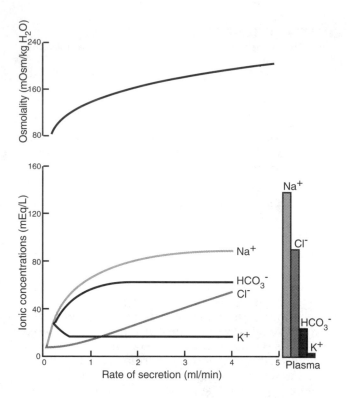

Figure 28–2 ■ The osmolality and electrolyte composition of saliva at different secretion rates. (Modified from Granger DN, Barrowman JA, Kvietys PR: *Clinical Gastrointestinal Physiology.* Philadelphia, WB Saunders, 1985.)

the ductal epithelium has more time to modify and thus reduce the osmolality of the primary secretion, so the saliva has a much lower osmolality than plasma. The opposite is true at high secretion rates.

Although absorption and secretion of ions may explain changes in the electrolyte composition of saliva, these processes do not explain why the osmolality of saliva is lower than that of the primary secretion of the acinar cells. Saliva is hypotonic to plasma because of a net absorption of ions by the ductal epithelium, a result of the action of Na^+/K^+ ATPase located at basolateral cell membrane of the cells. The Na^+/K^+ ATPase transports three Na^+ ions out of the cell in place of two K^+ ions taken up by the cell. The epithelial lining of the duct is not very permeable to water, so water does not follow the absorbed salt.

Proteins in Saliva ■ The two major proteins present in saliva are amylase and mucin. Amylase is produced predominantly by the parotid glands and mucin mainly by the sublingual and submandibular glands. Amylase is a hydrolytic enzyme involved in the digestion of starch. It is synthesized by the rough endoplasmic reticulum and then transferred to the Golgi apparatus, where it is packaged into zymogen granules. The zymogen granules are stored at the apical region of the acinar cells and released with appropriate stimuli. The action of amylase is described in Chapter 29.

Mucin is the most abundant protein in saliva; the term describes a family of glycoproteins, each associated with different amounts of different sugars. Mucin is responsible for most of saliva's viscosity. Also present in saliva are small amounts of **muramidase,** a lysozyme that can lyse the muramic acid of certain bacteria (e.g., *Staphylococcus*); **lingual lipase,** an enzyme important in digestion of milk fat; **lactoferrin,** a protein that binds iron; epidermal growth factor, which stimulates gastric mucosal growth; immunoglobulins (mainly IgA); and ABO blood group substances.

Saliva Performs Both Protective and Digestive Functions

Protective Functions ■ Saliva's pH is almost neutral (pH = 7), and it contains bicarbonate that can neutralize any acidic substance entering the oral cavity, including regurgitated gastric acid. Saliva plays an important role in the general hygiene of the oral cavity. The muramidase present in saliva combats bacteria by lysing the bacterial cell wall; the lactoferrin binds iron strongly, thus depriving these microorganisms of sources of iron vital to their growth. Saliva lubricates the mucosal surface, thus reducing the frictional damage caused by the rough surfaces of food. It helps small food particles stick together to form a bolus, which makes them easier to swallow. Moistening of the oral cavity with saliva facilitates speech. Saliva can dissolve sapid substances, thereby stimulating the different taste buds located on the tongue. Finally, saliva plays an important role in water intake; the sensation of dryness of the mouth due to low salivary secretion urges an individual to drink.

Digestive Functions ■ Salivary α-amylase (or **ptyalin**) catalyzes the hydrolysis of polysaccharides with 1,4-α-glycosidic linkages. Because usually some time passes before acids in the stomach can inactivate the amylase, a significant portion of the ingested carbohydrate can be digested before reaching the duodenum. **Salivary lingual lipase,** secreted by Von Ebner's glands located on the dorsal aspects of the tongue, is unusual in that its pH optimum is 4, which allows it to work in the acidic medium in the stomach. It prefers medium-chain triglycerides over long-chain triglycerides. Milk contains a large number of medium-chain triglycerides and thus is an excellent substrate for lingual lipase. Lingual lipase probably plays an important role in neonates, since pancreatic lipase is poorly developed at that age.

The Autonomic Nerves Are the Chief Modulators of Saliva Output and Content

Autonomic Control of Salivary Secretion ■ Salivary secretion is predominantly under the control of the autonomic nervous system. In the resting state, salivary secretion is low, amounting to about 30 ml/h. The submandibular glands contribute about two-thirds of resting salivary secretion, the parotid about one-fourth, and the sublingual the remainder. Stimulation increases the rate of salivary secretion, most notably in the parotid gland, up to 400 ml/h. The most potent stimuli for salivary secretion are sapid substances, such as citric acid. Other types of stimuli that induce salivary secretion include the smell of food and chewing. Secretion is inhibited by anxiety, fear, and dehydration.

Parasympathetic stimulation of the salivary glands results in increased activity of the acinar and ductal cells and increased salivary secretion. The parasympathetic nervous system plays an important role in controlling the secretion of saliva. The centers involved are located in the medulla oblongata. Preganglionic fibers from the inferior salivatory nucleus are contained in cranial nerve IX, synapse in the otic ganglion, and send postganglionic fibers to the parotid glands. Preganglionic fibers from the superior salivatory nucleus course with cranial nerve VII, synapse in the submandibular ganglion, and send postganglionic fibers to the submandibular and sublingual glands.

Blood flow to resting salivary glands is about 50 ml/min/100 g tissue and can increase as much as 10-fold when salivary secretion is stimulated. This increase in blood flow is under parasympathetic control. Parasympathetic stimulation induces the acinar cells to release the protease **kallikrein** (Fig. 28–3), which acts on a plasma globulin, **kininogen,** to release **lysylbradykinin,** which causes vasodilation of the blood vessels supplying the salivary glands. Atropine, an anticholinergic agent, is a potent inhibitor of salivary secretion. Agents that inhibit acetylcholinesterase (e.g., pilocarpine) enhance salivary secretion. Some parasympathetic stimulation also increases blood flow to the salivary glands directly, apparently via the release of the transmitter **vasoactive intestinal polypeptide** (VIP).

The salivary glands are also innervated by the sympathetic nervous system. Sympathetic fibers arise in the upper thoracic segments of the spinal cord and synapse in the **superior cervical ganglion.** Postganglionic fibers leave the superior cervical ganglion and innervate the acini, ducts, and blood vessels. Stimulation of the sympathetic nervous system tends to result in a short-lived and much smaller increase in salivary secretion than does parasympathetic stimulation. The increase in salivary secretion observed during sympathetic stimulation is mainly via β-adrenergic receptors; α-adrenergic receptors are more involved in stimulating contraction of myoepithelial cells.

Although both sympathetic and parasympathetic stimulation increase salivary secretion, the responses produced are different (Table 28–1).

Hormonal Control of Salivary Secretion ■ Mineralocorticoid administration reduces the sodium concentration of saliva, with a corresponding rise in potassium concentration. Mineralocorticoids act mainly on the striated and excretory ducts. Antidi-

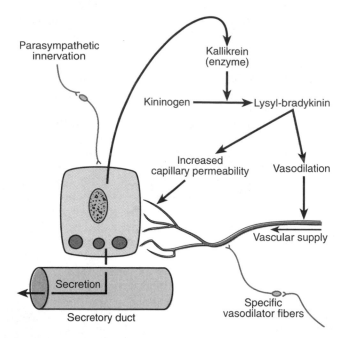

Figure 28–3 ■ The effect of parasympathetic innervation on blood flow to the salivary glands. (Modified from Sanford PA: *Digestive System Physiology.* Baltimore, University Park Press, 1982.)

TABLE 28–1 ■ The Effect of Parasympathetic and Sympathetic Stimulation on Salivary Secretion Responses

Responses	Parasympathetic	Sympathetic
Saliva output	Copious	Scant
Temporal response	Sustained	Transient
Composition	Protein-poor, high K^+ and HCO_3^-	Protein-rich, low K^+ and HCO_3^-
Response to denervation	Decreased secretion, atrophy	Decreased secretion

uretic hormone reduces the sodium concentration in saliva by increasing sodium reabsorption by the ducts. Some gastrointestinal hormones (e.g., VIP, and substance P) have been demonstrated experimentally to evoke salivary secretory responses.

Gastric Secretion

The major function of the stomach is storage, but it also absorbs water and lipid-soluble substances (e.g., alcohol and some drugs). An important function of the stomach is to prepare the chyme for digestion in the small intestine. **Chyme** is the semifluid gruel-like material produced by gastric digestion of food. This is achieved partly by the conversion of large solid particles into smaller particles via the combined peri-

staltic movements of the stomach and the pyloric sphincter. The propulsive, grinding, and retropulsive movements associated with antral peristalsis are discussed in Chapter 27. A combination of the squirting of antral content into the duodenum, the grinding action of the antrum, and retropulsion provides much of the mechanical action necessary for the emulsification of dietary fat, which plays an important role in fat digestion.

■ Numerous Types of Cells in the Stomach Contribute to the Gastric Secretions

The fundus of the stomach (Fig. 27–10) is relatively thin-walled and thus can be expanded with ingested food. The main body (corpus) of the empty stomach is composed of many longitudinal folds (**rugae gastricae**) and is called the **glandular gastric mucosa.** Its mucosal lining contains three main types of glands: **cardiac, pyloric,** and **oxyntic.** These glands contain surface mucous cells that secrete mucus and HCO_3^- ions, which protect the stomach from the acid in the stomach cavity. The **cardiac glands** are located in a small area adjacent to the esophagus and are lined by mucus-producing columnar cells. The **pyloric glands** are located in a larger area, adjacent to the duodenum. They contain cells similar to mucous neck cells but differ from cardiac and oxyntic glands in having many gastrin-producing cells, called **G cells.** The **oxyntic glands,** the most abundant glands in the stomach, are found in the fundus and the corpus.

The oxyntic glands contain **oxyntic cells** (parietal cells) and **chief cells,** numerous mucous neck cells, and some endocrine cells (Fig. 28–4). The surface mucous cells occupy the isthmus, but the majority are located in the neck region. The base of the oxyntic gland contains mainly chief cells, along with some parietal and endocrine cells. Mucous neck cells secrete mucus; parietal cells principally secrete hydrochloric acid (HCl) and **intrinsic factor;** and chief cells secrete **pepsinogen.** Intrinsic factor and pepsinogen are discussed in Chapter 29.

Parietal cells are the most distinctive cells in the stomach. Resting parietal cells have a rather unique morphology in that they have intracellular canaliculi as well as an abundance of mitochondria (Fig. 28–5). This network of intracellular canaliculi is composed of clefts and canals that are continuous with the outside milieu. There is also an extensive smooth endoplasmic reticulum, referred to as the tubulovesicular membranes. In active parietal cells (Fig. 28–5), the extensive tubulovesicular system is greatly diminished, with a concomitant increase in the intracellular canaliculi. The mechanism for these morphologic changes in the parietal cells is not well understood. Hydrochloric acid (HCl) is secreted across the parietal cell microvillar membrane and flows out of the intracellular canaliculi into the oxyntic gland lumen. The parietal cells also

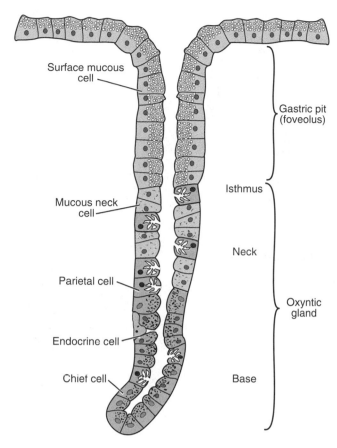

Surface mucous
cell

Gastric pit
(foveolus)

Isthmus

Mucous neck
cell

Neck

Parietal cell

Oxyntic
gland

Endocrine cell

Chief cell

Base

Figure 28–4 ■ A simplified diagram of the oxyntic gland in the corpus of a mammalian stomach. (Modified from Ito S: Functional gastric morphology, in Johnson LR, Christensen J, Jackson MJ et al (eds): *Physiology of the Gastrointestinal Tract.* New York, Raven, 1987.)

produce intrinsic factor, which is important for the absorption of vitamin B_{12}.

Surface mucous cells line the entire surface of the gastric mucosa and the openings of the cardiac, pyloric, and oxyntic glands (the gastric pits, or foveolae). These cells secrete mucus and HCO_3^- to protect the gastric surface from the acidic environment of the stomach. The distinguishing characteristic of a surface mucous cell is the presence of numerous mucous granules at its apex. The number of mucous granules in storage varies, depending on synthesis and secretion. The mucous neck cells of the oxyntic glands are similar in appearance to surface mucous cells.

Chief cells are distinguished morphologically principally by the presence of numerous zymogen granules in the apical region and an extensive endoplasmic reticulum. The zymogen granules contain pepsinogen, a precursor to the enzyme pepsin.

Also present in the stomach are numerous neuroendocrine cells, such as **G cells,** located predominantly in the antrum. These cells produce the hormone **gastrin,** which stimulates acid secretion by the stom-

ach. The **D cells,** also present in the antrum, produce **somatostatin,** another important gastrointestinal hormone.

■ Gastric Juice Contains Hydrochloric Acid, Electrolytes, and Proteins

The important constituents of human gastric juice are HCl, electrolytes, pepsinogen, and intrinsic factor. The pH is very low, about 2–5, which raises an interesting question: How does the gastric mucosa protect itself from this acidity? As mentioned earlier, the surface mucous cells secrete a fluid containing mucus and bicarbonate ions. The mucus forms a mucous layer covering the surface of the gastric mucosa. Bicarbonate trapped in the mucous layer neutralizes acid in the lumen, thus preventing damage to the mucosal cell surface.

Hydrochloric Acid Production ■ The HCl present in the gastric lumen is secreted by the parietal cells of the fundus and the corpus. The mechanism of HCl production is depicted in Figure 28-6. In the apical (luminal) cell membrane of the parietal cell is an **H^+-K^+ ATPase** that actively pumps H^+ out of the cell in exchange for K^+ entering the cell. The origin of the H^+ secreted is not fully understood but is believed to be largely derived from the dissociation of H_2O. To replenish the dissociated water molecule, the OH^- combines with H^+ derived from the dissociation of carbonic acid (Fig. 28-6). Carbonic acid is formed from CO_2 and H_2O in a reaction catalyzed by carbonic anhydrase. The CO_2 is provided by metabolic sources inside the cell and the blood.

For H^+-K^+ ATPase to work, an adequate supply of K^+ ions must exist outside the cell. Although the mechanism is still unclear, there is an increase in K^+ conductance (through K^+ channels) in the apical membrane of the parietal cells simultaneous with active acid secretion. This surge of K^+ conductance ensures plenty of K^+ in the lumen. The H^+-K^+ ATPase recycles K^+ ions back into the cell in exchange for H^+ ions. As shown in Figure 28-6, the basolateral cell membrane has an electroneutral Cl^--HCO_3^- exchanger that balances the entry of Cl^- ions into the cell with an equal amount of HCO_3^- ions entering the bloodstream. The Cl^- ions inside the cell then leak into the lumen through the Cl^- channels, down an electrochemical gradient. Consequently, HCl is secreted into the lumen. A large amount of HCl can be secreted by the parietal cells; this is balanced by an equal amount of HCO_3^- secreted into the bloodstream. The blood coming from the stomach during active acid secretion contains much HCO_3^-, a phenomenon called the **alkaline tide.** The osmotic gradient created by the HCl concentration in the gland lumen drives water passively into the lumen, thus maintaining the isoosmolality of the gastric secretion.

Electrolytes and Proteins in the Gastric Juice ■ Figure 28-7 depicts the changes in the electrolyte composition of gastric juice with different secretion

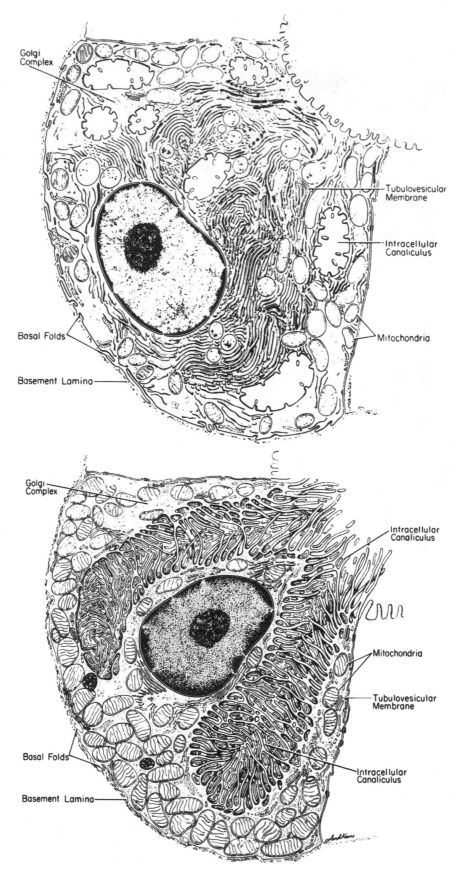

Figure 28–5 ■ (A) Morphology of a nonsecreting parietal cell. The cytoplasm is filled with tubulovesicular membranes, and the intracellular canaliculi have become internalized, distended, and devoid of microvilli. (B) Morphology of an actively secreting parietal cell. Compared to the resting parietal cell, the most striking difference is the abundance of long microvilli and the paucity of the tubulovesicular system, making the mitochondria appear more numerous. (From Ito S: Functional gastric morphology, in Johnson LR, Christensen J, Jackson MJ et al (eds): *Physiology of the Gastrointestinal Tract.* New York, Raven, 1987.)

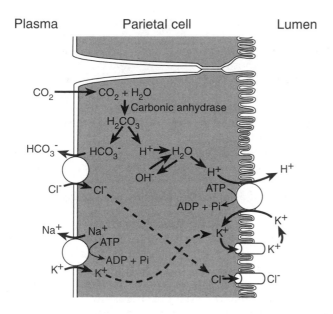

Plasma   Parietal cell   Lumen

Figure 28–6 ■ Mechanism of HCl secretion by parietal cells.

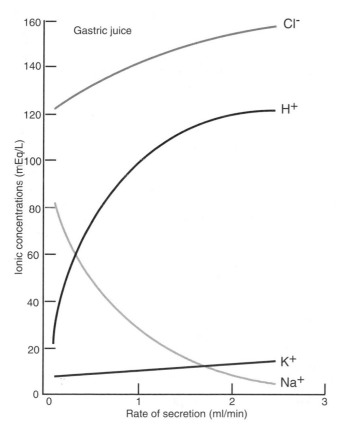

Figure 28–7 ■ The concentration of electrolytes in gastric juice of a normal young man as a function of rate of secretion. (Modified from Davenport HW: *Physiology of the Digestive Tract.* Chicago, Year Book, 1977.)

rates. At a low secretion rate, the gastric juice contains a large amount of NaCl and small amounts of K^+ and H^+. When the rate of secretion increases, the concentration of Na^+ ions decreases while that of H^+ increases markedly. Also coupled with this increase in gastric secretion is an increase in Cl^- concentration. To explain the changes in electrolyte composition of gastric juice with different secretion rates, it is important to remember that gastric juice is derived from the secretions of two major sources, the parietal cells and the nonparietal cells. Secretion from the nonparietal cells is probably quite constant. Thus, it is parietal secretion (HCl secretion) that contributes mainly to the changes in electrolyte composition with higher secretion rates.

Gastric juice contains a number of proteins, pepsinogens, pepsins, salivary amylase, lingual lipase, and intrinsic factor. Only pepsinogens and intrinsic factor are secreted by the stomach.

■ Gastric Secretions Perform Digestive, Protective, and Other Functions

Digestive Functions ■ The chief cells of the oxyntic glands release inactive pepsinogens. Pepsinogen is activated in the acidic gastric lumen to form the active enzyme **pepsin.** Further activation is mediated by the acidity in the gastric luminal content or by pepsin. Pepsin, an **endopeptidase,** cleaves protein molecules from the inside, resulting in the formation of smaller peptides. The optimal pH for pepsin activity is 1.8–3.5, so it is extremely active in the highly acidic medium of the gastric juices.

Protective Functions ■ The acidity of the gastric secretion poses a barrier to microbial invasion of the gastrointestinal tract.

Other Functions ■ Intrinsic factor, secreted by the parietal cells, binds to vitamin B_{12}, protecting it from gastric and intestinal digestion. In the ileum, the vitamin B_{12}-intrinsic factor complex is absorbed via receptor-mediated endocytosis. **Pernicious anemia** results from insufficient intrinsic factor production; vitamin B_{12} is important in hemoglobin synthesis. Past treatment for pernicious anemia involved eating raw liver. Investigators discovered that liver contains a large quantity of B_{12}, and eventually this led to identification of intrinsic factor. (Chapter 29 discusses intrinsic factor in more detail).

■ Gastric Secretion Is under Neural and Hormonal Control

Gastric acid secretion is mediated through neural and hormonal pathways. Vagal nerve stimulation is the neural effector, and histamine and gastrin are the hormonal effectors (Fig. 28–8). The stomach possesses special histamine receptors (**H_2 receptors**), whose stimulation results in increased acid secretion. Mast cells in the lamina propria of the stomach are believed to be the source of this histamine, but the mechanisms

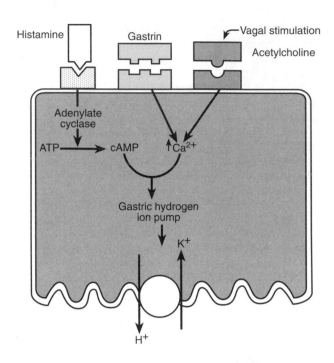

Figure 28–8 ■ The stimulation of acid secretion by histamine, gastrin, and acetylcholine, and potentiation of the process.

that stimulate them to release histamine are unknown. The importance of histamine as an effector of gastric acid secretion has been indirectly demonstrated by the effectiveness of cimetidine, an H$_2$ blocker, in reducing acid secretion. (For this reason, cimetidine is widely used for the treatment of gastric ulcer disease).

The effects of each of these three stimulants augment those of the others, a process known as **potentiation.** Potentiation is said to occur when the effect of two stimulants is greater than the effect of either stimulant alone. For example, the interaction of gastrin and acetylcholine molecules with their respective receptors results in an increase in intracellular Ca^{2+} concentration, and the interaction of histamine with its receptor results in an increase in cellular cAMP production. The increased intracellular Ca^{2+} and cAMP interact in numerous ways to stimulate the gastric hydrogen ion pump, which brings about the increase in acid

secretion (Fig. 28–8). Exactly how the increase in intracellular Ca^{2+} and cAMP greatly enhances the effect of the other in stimulating gastric acid secretion is not well understood.

Both mucin and pepsinogen secretion are stimulated by the parasympathetic nervous system. Secretin also stimulates pepsinogen secretion (see below).

Acid Secretion during a Meal ■ Stimulation of acid secretion as a result of ingestion of a meal can be divided into three phases (Table 28–2). The first phase, the **cephalic phase,** involves the central nervous system. Smelling or merely thinking of food and chewing and swallowing food send impulses via the vagus nerves to the parietal and G cells in the stomach. The nerve endings release acetylcholine, which directly stimulates acid secretion from parietal cells. The nerves also release **gastrin-releasing peptide** (GRP, or bombesin), which stimulates G cells to release gastrin, thus indirectly stimulating parietal cell acid secretion. The fact that the effect of GRP is atropine-resistant indicates that it works through a noncholinergic pathway. The cephalic phase probably accounts for about 40% of total acid secretion (Table 28–2).

The second phase, or **gastric phase,** is mainly a result of gastric distension and chemical agents such as digested proteins. Distension of the stomach stimulates mechanoreceptors, which stimulate the parietal cells directly through the short local (enteric) reflexes and by the long **vagovagal reflexes.** Vagovagal reflexes are those arising as a result of the afferent and efferent impulses carried by the vagus nerve. Digested proteins in the stomach are also potent stimulators of gastric acid secretion, an effect mediated through gastrin release. A number of other chemicals, such as alcohol and caffeine, stimulate gastric acid secretion through mechanisms that are not well understood. The gastric phase accounts for about 50% of the total gastric acid secretion.

During the third phase, the **intestinal phase,** protein digestion products in the duodenum stimulate gastric acid secretion through the action of the circulating amino acids on the parietal cells. Distension of the small intestine also stimulates gastric acid secretion, probably mediated by the release of the hormone **enterooxyntin** from the intestinal endo-

TABLE 28–2 ■ Mechanism of Stimulation of Acid Secretion after Ingestion of a Meal

Phase	Stimulus	Pathway	Stimulus to Parietal Cell
Cephalic	Thought of food, smell, chewing, swallowing	Vagus nerve to:	
		Parietal cells	Acetylcholine
		G cells	Gastrin
Gastric	Stomach distension	Local (enteric) reflexes and vagovagal and reflexes to:	
		Parietal cells	Acetylcholine
		G cells	Gastrin
	Digested peptides	G cells	Gastrin
Intestinal	Protein digestion products in duodenum	Amino acids in blood	Amino acids
	Distension	Intestinal endocrine cell	Enterooxyntin

ACID SECRETION AND DUODENAL ULCER

Ulcerative lesions of the gastroduodenal area are classified as **peptic ulcer disease.** Peptic ulcer disease is associated with a high rate of recurrence. The saying, "No acid, no ulcer" has withstood the test of time and is still accepted by most physicians and researchers as generally true. One possible cause of gastric and duodenal ulcers is reduced mucosal defense mechanisms. However, human and animal data have demonstrated that duodenal ulcers do not occur with reduced mucosal defense mechanisms alone, but also require the presence of sufficient amounts of acid. In one study, patients suffering from duodenal ulcer had a significantly increased mean number of parietal cells and appeared to have increased sensitivity to gastrin when compared with normal subjects. Stomach emptying is greatly increased in duodenal ulcer patients, but the reason is unknown. Another abnormality in duodenal ulcer patients is decreased inhibition of gastrin release by acid; however, it should be emphasized that a significant number of patients with duodenal ulcer do not have oversecretion of acid. Psychological and genetic factors may also play an interactive role in the development of duodenal ulcer.

An exciting recent development in the field of gastric and duodenal ulcer disease is the finding of a possible correlation between *Helicobacter pylori* infection and the incidence of gastric and duodenal ulcers. How *H. pylori* infection may play a role in the genesis of gastric and duodenal ulcers is unclear, but in a significant number of patients eradication of the bacteria reduces the rate of ulcer recurrence. *H. pylori* produces large quantities of urease, which hydrolyzes urea to produce ammonia. The ammonia neutralizes acid in the gastric lumen, thus protecting the bacteria from the injurious effects of hydrochloric acid.

Although the mechanism has not been elucidated, the presence of *H. pylori* in the stomach enhances the secretion of gastrin by the gastric mucosa during fasting and in the fed state. Whether increased gastrin release by the presence of *H. pylori* is responsible for the increased recurrence of gastric and duodenal ulcers in patients has yet to be proved. It has been demonstrated that H_2 receptor antagonists (cimetidine and ranitidine) have no effect on *H. pylori* infection. In contrast, omeprazole (an inhibitor of the H^+-K^+ ATPase) appears to be bacteriostatic. A combined therapy using omeprazole and amoxicillin appears to be quite effective in the eradication of *H. pylori* in 50–80% of patients with duodenal ulcer disease, resulting in a significant reduction of duodenal ulcer recurrence.

---

crine cells, which stimulates acid secretion. The intestinal phase accounts for only about 10% of total gastric acid secretion.

Inhibition of Gastric Acid Secretion ■ Inhibition of gastric acid secretion is physiologically important for two reasons. First, the secretion of acid is important only during the digestion of food. Second, excess acid can damage the gastric and the duodenal mucosal surfaces, thereby causing ulcerative conditions. The body has a rather elaborate system for regulating the amount of acid secreted by the stomach. Gastric luminal pH is a sensitive regulator of acid secretion. Proteins in the food offer excellent buffering in the lumen; consequently, the gastric luminal pH is usually above 3. However, if the buffering capacity of

protein is exceeded, then the pH of the gastric lumen will fall below 3. When this happens, the endocrine cells (D cells) in the antrum secrete somatostatin, which inhibits the release of gastrin and, thus, gastric acid secretion.

Another mechanism for inhibiting gastric acid secretion is acidification of the duodenal lumen. Acidification stimulates the release of **secretin,** which inhibits the release of gastrin, and a number of peptides released by endocrine cells collectively known as **enterogastrones.** Acid, fatty acids, or hyperosmolar solutions in the duodenum stimulate the release of enterogastrones, which inhibit gastric acid secretion. **Gastric inhibitory peptide** (GIP), an enterogastrone produced by the small intestinal GIP cells, inhibits parietal cell acid secretion. There are also a number of as-yet unidentified enterogastrones. The presence of such stimulants as acid, fatty acids, and hyperosmolar solution in the duodenum also elicits a local nervous response to inhibit acid secretion.

Pancreatic Secretion

One of the major functions of pancreatic secretion is to neutralize the acids in the chyme when it enters the duodenum from the stomach. This is important because pancreatic enzymes operate optimally near neutral pH. Another important function is the production of enzymes involved in the digestion of dietary carbohydrate, fat, and protein.

■ The Pancreas Consists of a Network of Acini and Ducts

The human pancreas is located in close apposition to the duodenum. It performs both endocrine and exocrine functions, but in this chapter we discuss only its exocrine function.

The exocrine pancreas is composed of numerous small, saclike dilatations called **acini,** composed of a single layer of pyramidal acinar cells (Fig. 28–9). These cells are actively involved in the production of enzymes; their cytoplasm is filled with an elaborate system of endoplasmic reticulum and Golgi apparatus. Zymogen granules are observed in the apical region of acinar cells. There are also a few **centroacinar cells** lining the lumen of the acinus. These cells, in contrast to the acinar cells, lack an elaborate endoplasmic reticulum and Golgi apparatus. Their major function seems to be modification of the electrolyte composition of the pancreatic secretion. Because the processes involved in the secretion or uptake of ions are active, the centroacinar cells have numerous mitochondria in their cytoplasm.

The acini empty their secretions into intercalated ducts, which join to form the intralobular and then the interlobular ducts. The interlobular ducts empty into two pancreatic ducts, a major duct, the duct of Wirsung, and a minor duct, the duct of Santorini. The duct of Santorini enters the duodenum more proximally than the duct of Wirsung, which enters the duodenum usually together with the common bile duct. There is a ring of smooth muscle, the **sphincter of Oddi,** surrounding the opening of these ducts in the duodenum. The sphincter of Oddi not only regulates the flow of bile and pancreatic juice into the duodenum, but also prevents the reflux of intestinal contents into the pancreatic ducts.

■ Pancreatic Secretions Are Rich in Bicarbonate Ions

The pancreas secretes about 1 L/day of bicarbonate-rich fluid. The osmolality of pancreatic fluid, unlike

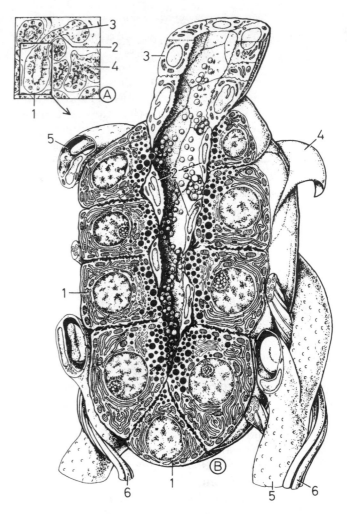

Figure 28–9 ■ The ultrastructure of a pancreatic acinus. (From Granger DN, Barrowman JA, Kvietys PR: *Clinical Gastrointestinal Physiology.* Philadelphia, WB Saunders, 1985.)

that of saliva, is equal to that of plasma at all secretion rates. The Na^+ and K^+ concentrations of pancreatic juice are the same as those in plasma, but unlike plasma, pancreatic juice is enriched with HCO_3^- and has a relatively low Cl^- concentration (Fig. 28–10). The HCO_3^- concentration increases with increases in secretion rate and reaches a maximal concentration of about 140 mEq/L, yielding a solution with a pH of 8.2. A reciprocal relationship exists between the Cl^- and HCO_3^- concentration in pancreatic juice. As the concentration of HCO_3^- increases with secretion rate, the Cl^- concentration falls accordingly, resulting in a combined total anion concentration that remains relatively constant (150 mEq/L), regardless of the pancreatic secretion rate.

Two separate mechanisms have been proposed to explain the secretion of a bicarbonate-rich juice by the pancreas and the HCO_3^- concentration changes.

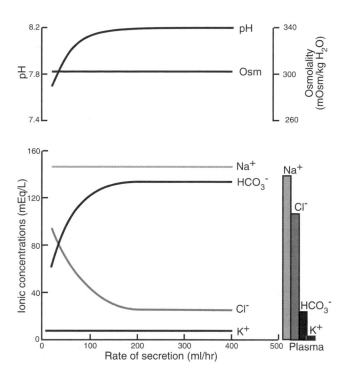

Figure 28–10 ■ The pH, osmolality, and electrolyte composition of pancreatic juice at different secretion rates. Plasma electrolyte composition is provided for comparison. (Adapted from Granger DN, Barrowman JA, Kvietys PR: *Clinical Gastrointestinal Physiology.* Philadelphia, WB Saunders, 1985.)

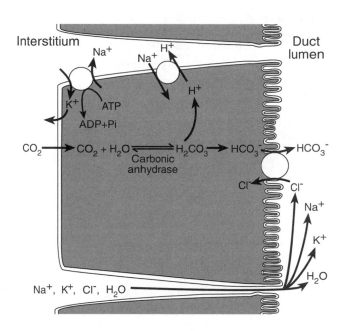

Figure 28–11 ■ Model for electrolyte secretion by pancreatic duct cells. (Modified from Exocrine pancreas: Pancreatitis, in *The Undergraduate Teaching Project in Gastroenterology and Liver Disease,* American Gastroenterological Association, 1984, unit 16.)

The first mechanism proposes that some cells, probably the acinar cells, secrete a plasmalike fluid containing predominantly Na^+ and Cl^-, while other cells, probably the centroacinar and duct cells, secrete a bicarbonate-rich solution when stimulated. Depending on the different rates of secretion from these three different types of cells, the pancreatic juice can be rich in either HCO_3^- or Cl^-. The second mechanism depicts the primary secretion as rich in HCO_3^-. As the HCO_3^- solution moves down the ductal system, HCO_3^- ions are exchanged for Cl^- ions. When the flow is fast, there is little time for this exchange, so the concentration of HCO_3^- is high. The opposite is true when the flow is slow.

The secretion of electrolytes by pancreatic ductal cells is depicted in Figure 28–11. There is a Na^+-H^+ exchanger located in the basolateral cell membrane. The energy required to drive the exchanger is provided by the Na^+/K^+ ATPase-generated sodium gradient. Carbon dioxide diffuses into the cell and combines with H_2O to form H_2CO_3, a reaction catalyzed by carbonic anhydrase, which dissociates to H^+ and HCO_3^-. The H^+ is extruded by the Na^+-H^+ exchanger and HCO_3^- is exchanged for luminal Cl^- via a Cl^--HCO_3^- exchanger. The Na^+/K^+ ATPase also removes cell Na^+ that enters through the Na^+-H^+ antiporter. Sodium follows the HCO_3^- by diffusing through a paracellular path (between the cells). Movement of water into the duct

lumen is passive, driven by the osmotic gradient. The net result of pancreatic HCO_3^- secretion is the release of H^+ into the plasma, thus pancreatic secretion is associated with an **acid tide** in the plasma.

■ Pancreatic Secretions Neutralize Luminal Acids and Digest Nutrients

One of the major functions of pancreatic secretion is to neutralize the acidic chyme presented to the duodenum. The enzymes present in intestinal lumen work best at a pH close to neutral, thus it is crucial to increase the pH of the chyme. As described above, pancreatic juice is highly basic because of its HCO_3^- content. Thus, the acidic chyme presented to the duodenum is rapidly neutralized by pancreatic juice.

The other major function of pancreatic secretion is the production of large amounts of pancreatic enzymes. Table 28–3 summarizes the various enzymes present in pancreatic juice. Some are secreted as proenzymes, which are activated in the duodenal lumen to form the active enzymes. Digestion of nutrients by these enzymes is discussed in Chapter 29.

The first major class is the proteolytic enzymes. These can be divided into **endopeptidases** and **exopeptidases.** Whereas endopeptidases split polypeptides interiorly into smaller peptides, exopeptidases split amino acids from the terminals of the peptide. Trypsin, chymotrypsin, and elastase are endopeptidases; carboxypeptidase and aminopeptidase are exopeptidases.

The major enzyme for the digestion of carbohydrates

TABLE 28–3 ■ Characteristics of Pancreatic Enzymes

Enzyme	Specific Hydrolytic Activity
Proteolytic	
Endopeptidases	
Trypsin(ogen)	Cleaves peptide linkages in which the carboxyl group is either arginine or lysine
Chymotrypsin(ogen)	Cleaves peptides at the carboxyl end of hydrophobic amino acids (e.g., tyrosine or phenylalanine)
(Pro)elastase	Cleaves peptide bonds at the carboxyl terminal of aliphatic amino acids
Exopeptidases	
(Pro)carboxypeptidase	Cleaves amino acids from the carboxyl end of the peptide
(Pro)aminopeptidase	Cleaves amino acids from the amino end of the peptide
Amylolytic	
α-Amylase	Cleaves α-1,4 glycosidic linkages of glucose polymers
Lipases	
Lipase	Cleaves the ester bond at the 1- and 3-position of triglycerides, producing free fatty acids and 2-monoglyceride
(Pro)phospholipase A_2	Cleaves the ester bond at the 2- position of phosphoglycerides
Carboxylester hydrolase (cholesterol esterase)	Cleaves cholesterol ester to free cholesterol
Nucleolytic	
Ribonuclease	Cleaves ribonucleic acids into mononucleotides
Deoxyribonuclease	Cleaves deoxyribonucleic acids into mononucleotides

The suffix -ogen or prefix pro- indicates that the enzyme is secreted in an inactive form.

is α-amylase, which hydrolyzes the α-1,4 glycosidic linkage of glucose polymers. Also present in pancreatic juice are three major lipolytic enzymes: **pancreatic lipase, phospholipase A_2** and **carboxylester hydrolase (cholesterol esterase)**. Pancreatic juice contains two nucleolytic enzymes, **ribonuclease** and **deoxyribonuclease**.

■ Pancreatic Secretion Is under Neural and Hormonal Control

Neural Control ■ Pancreatic secretion is stimulated by parasympathetic fibers in the vagus nerve that release acetylcholine. Stimulation of the vagus nerve results predominantly in an increase in enzyme secretion. Pancreatic fluid and HCO_3^- secretion are marginally stimulated or unchanged. Sympathetic nerve fibers mainly innervate the blood vessels supplying the pancreas, causing vasoconstriction. Stimulation of the sympathetic nerves either does not affect pancreatic secretion or inhibits it, probably as a result of the reduction in blood flow.

Hormonal Control ■ The secretion of electrolytes and enzymes by the pancreas is greatly influenced by circulating gastrointestinal hormones, particularly **secretin** and **cholecystokinin** (CCK). Secretin tends to stimulate pancreatic secretion rich in HCO_3^-; CCK stimulates a marked increase in enzyme secretion. Both hormones are produced by the small intestine, and the pancreas has receptors for them.

Structurally similar hormones have effects similar to those of secretin and CCK. For example, VIP, which is structurally similar to secretin, stimulates the secretion of HCO_3^- and H_2O. However, because VIP is much weaker than secretin, when it is given together with secretin it produces a weaker pancreatic response than secretin alone. Similarly, gastrin can stimulate pancreatic enzyme secretion because of its structural similar-

ity to CCK, but unlike CCK it is a weak agonist for pancreatic enzyme secretion.

■ Pancreatic Secretion Is Phasic

Seeing, smelling, tasting, chewing, swallowing, or thinking about food result in the stimulation of pancreatic secretion rich in protein. This is the **cephalic phase** of pancreatic secretion, in which stimulation is mediated by direct efferent impulses sent by vagal centers in the brain to the pancreas and by the indirect effect of parasympathetic stimulation of gastrin release.

The **gastric phase** of pancreatic secretion is initiated when food enters the stomach and distends it. Pancreatic secretion is then stimulated by vagovagal reflex. Gastrin may also be involved in this phase.

Last and most important is the **intestinal phase.** Entry of acidic chyme from the stomach into the small intestine stimulates the release of secretin by the **S cells** of the intestinal mucosa (Fig. 28–12). When the pH of the lumen in the duodenum decreases, the secretin concentration in plasma increases. This is followed by an increase in HCO_3^- output by the pancreas. The secretion of pancreatic enzymes is regulated by the amount of circulating CCK and by parasympathetic stimulation, through the vagovagal reflex. The release of CCK by the CCK cells of the intestinal mucosa is stimulated by exposure of the intestinal mucosa to long-chain fatty acids (lipid digestion products) and free amino acids. The regulation of pancreatic secretion by various hormonal and neural factors is summarized in Table 28–4.

Potentiation, as described for gastric secretion, also exists in the pancreas. Its effect in pancreatic secretion is a result of the different receptors used for acetylcholine, CCK, and secretin. Secretin binding triggers an increase in adenylate cyclase activity, which in turn

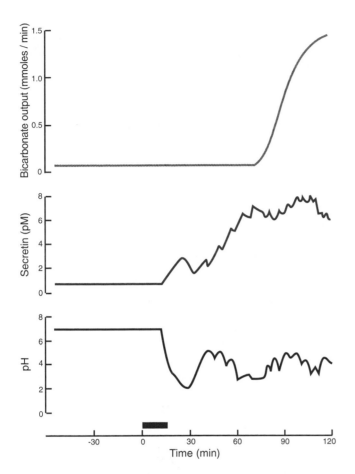

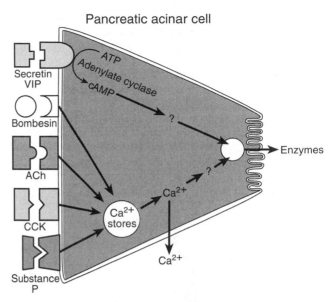

Figure 28–13 ■ Diagram showing the stimulation of pancreatic secretion by hormones and neurotransmitters.

Figure 28–12 ■ Intestinal phase of pancreatic secretion after duodenal acidification (shown by the bar in the bottom graph). Note marked increases in the plasma secretin level and bicarbonate secretion. (Modified from Granger DN, Barrowman JA, Kvietys PR: *Clinical Gastrointestinal Physiology.* Philadelphia, WB Saunders, 1985.)

stimulates the formation of cAMP (Fig. 28–13). Acetylcholine, CCK, and the neuropeptides bombesin and substance P bind to their respective receptors and trigger the release of Ca^{2+} from intracellular stores. The increase in intracellular Ca^{2+} and cAMP results in an increase in pancreatic enzyme secretion, by a mechanism that is not well understood.

Biliary Secretion

The human liver secretes 600–1200 ml/day of bile into the duodenum. Bile contains bile salts, bile pigments (e.g., bilirubin), cholesterol, phospholipids, and proteins and serves a number of important functions. For example, the **bile salts** play an important role in the intestinal absorption of lipid. Bile salts are conjugates of bile acids, with either taurine or glycine. Bile acids are derived from cholesterol and so constitute a path for its excretion. Biliary secretion is an important route for the excretion of bilirubin from the body.

■ The Biliary System Consists of Canaliculi, Bile Ducts, and the Gallbladder

The bile canaliculi are fine tubular canals running between the hepatocytes. Bile flows through the canaliculi to the bile ducts, which drain into the gallbladder. During the interdigestive period, the sphincter of

TABLE 28–4 ■ Factors Regulating Pancreatic Secretion after a Meal

Phase	Stimulus	Mediators	Response
Cephalic	Sight, smell, taste, chewing, and swallowing	Release of acetylcholine and gastrin by vagal stimulation	Increased secretion, with a greater effect on enzyme output
Gastric	Protein in food	Gastrin	Increased secretion, with a greater effect on enzyme output
	Gastric distension	Acetylcholine release by vagal stimulation	Increased secretion, with a greater effect on enzyme output
Intestinal	Acid in chyme	Secretin	Increased water and bicarbonate secretion
	Long-chain fatty acids	CCK and vagovagal reflex	Increased secretion, with a greater effect on enzyme output
	Amino acids and peptides	CCK and vagovagal reflex	Increased secretion, with a greater effect on enzyme output

Oddi, which controls the opening of the duct that carries biliary and pancreatic secretions, is normally contracted and the gallbladder is relaxed. Thus, most of the hepatic bile is stored in the gallbladder during this period. After ingestion of a meal, CCK is released into the blood, causing contraction of the gallbladder and resulting in the delivery of bile into the duodenum.

■ The Major Components of Bile Are Electrolytes, Bile Salts, and Lipids

The electrolyte composition of human bile collected from the common bile duct (Table 28-5) is similar to that of blood plasma, except the HCO_3^- concentration may be higher, resulting in an alkaline pH.

Bile acids are formed in the liver from cholesterol. During the conversion, hydroxyl groups and a carboxyl group are added to the steroid nucleus. Bile acids are classified as primary or secondary (Fig. 28-14). The primary bile acids are synthesized by the hepatocytes and include **cholic acid** and **chenodeoxycholic acid.** Bile acids are secreted in the form of bile salts,

the conjugates of taurine or glycine. The ratio of tauroconjugates to glycoconjugates is 3:1. When bile enters the gastrointestinal tract, bacteria present in the lumen act on the primary bile acids and convert them to secondary bile acids by dehydroxylation. Cholic acid is converted to **deoxycholic acid** and chenodeoxycholic acid to **lithocholic acid.** Lithocholic acid is poorly absorbed by the human small intestine.

Conjugated bile acids ionize more readily than the unconjugated bile acids and thus usually exist as salts of various cations (e.g., sodium glycocholate). Bile salts are much more polar than bile acids and thus have greater difficulty penetrating cell membranes. Consequently, bile salts are absorbed much more poorly by the small intestine than bile acids. This property of bile salts is important, because they play an integral role in the intestinal absorption of lipid. It is therefore important that they are absorbed by the small intestine after all the lipid has been absorbed.

The major lipids in bile are phospholipids and cholesterol. Of the phospholipids, the predominant species is phosphatidylcholine (lecithin). The phospholipid and cholesterol concentrations of hepatic bile are 0.3-11 mmol/L and 1.6-8.3 mmol/L, respectively. The concentration of these lipids in the gallbladder bile is even higher because of the absorption of water by the gallbladder. Cholesterol in bile is responsible for the formation of cholesterol gallstones.

TABLE 28–5 ■ Electrolyte Composition of Human Hepatic Bile

Constituent	Bile Concentration (mEq/L)	Plasma Concentration (mEq/L)
Na^+	140-170	150
K^+	4.0-6.0	4.5
Ca^{2+}	1.2-5.0	4.6
Mg^{2+}	1.5-3.0	1.6
Cl^-	95-125	100
HCO_3^-	15-60	25

■ Total Bile Secretion Consists of Three Components, One of Which Depends on Bile Acids

The total bile flow is composed of the ductular secretion and the canalicular bile flow (Fig. 28-15).

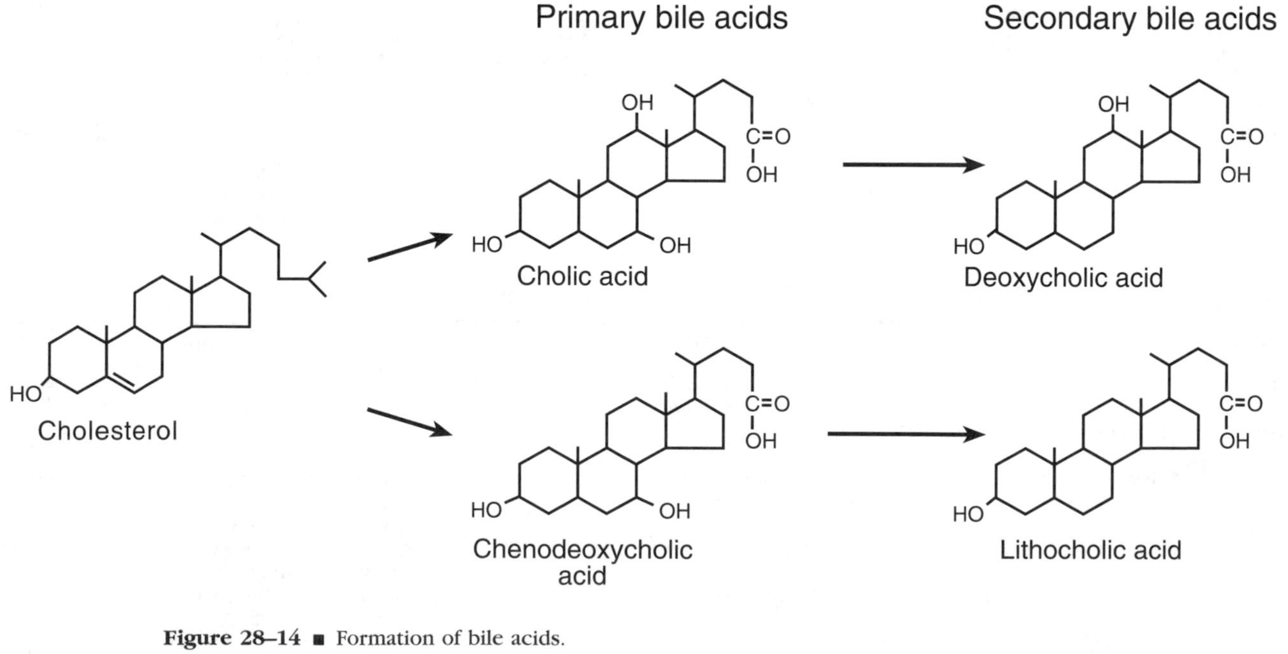

Figure 28–14 ■ Formation of bile acids.

The ductular secretion is from the cells lining the bile ducts; these cells actively secrete HCO_3^- ions into the lumen, resulting in the movement of water into the lumen of the duct. Another mechanism that may contribute to ductular secretion is the presence of an electroneutral NaCl pump that pumps NaCl into the lumen, accompanied by water movement.

The bile canalicular flow can be conceptually divided into two components: bile acid-dependent secretion and bile acid-independent secretion.

Bile Acid-Dependent Canalicular Flow ■ The uptake of free and conjugated bile acids is Na^+-dependent and mediated by a **bile acid-sodium symport** (Fig. 28-16). The energy required is provided by the transmembrane Na^+ gradient generated by the Na^+/K^+ ATPase. This mechanism is a secondary active transport, because the energy required for active uptake of bile acid or its conjugate is not directly provided by ATP but by an ionic gradient. The free bile acids are reconjugated with taurine or glycine before secretion as bile salts. The hepatocytes also make new bile acids from cholesterol. The bile salts are secreted by the hepatocytes by a carrier located at the canalicular membrane. This secretion is not Na^+-dependent but is instead driven by the electrical potential difference between the hepatocyte and the canaliculus lumen.

Other major components of bile, such as phospholipid and cholesterol, are secreted in concert with bile salts. Bilirubin is secreted by the hepatocytes by an active process. Although the secretion of cholesterol and phospholipid is not well understood, it is closely coupled to bile salt secretion. The osmotic pressure generated as a result of secretion of bile salts draws water into the canaliculus lumen through the paracellular pathway.

Bile Acid-Independent Canalicular Flow ■ As the name implies, this component of canalicular flow

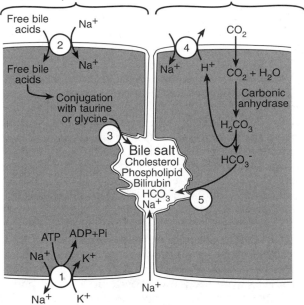

Figure 28–16 ■ Mechanism of bile acid secretion and bile flow. (1) Na^+/K^+ ATPase. (2) Bile acid-sodium symport. (3) Canalicular bile acid carrier. (4) Na^+/H^+ exchanger. HCO_3^- transport system.

is not dependent on the secretion of bile salts. The Na^+/K^+ ATPase plays an important role in bile acid-independent bile flow. This is clearly demonstrated by a marked reduction in bile flow when an inhibitor of this enzyme is applied. Another mechanism responsible for bile acid-independent flow is canalicular HCO_3^- secretion.

■ Bile Secretion Is Mainly Regulated by a Feedback Mechanism, but Hormonal and Neural Controls Also Participate

Feedback Control via Bile Salts in the Portal Blood ■ The major determinant of bile salt synthesis and secretion by the hepatocytes is the bile salt concentration in hepatic portal blood, which exerts a negative feedback effect over the synthesis of bile acids from cholesterol. The concentration of bile salts in portal blood also determines bile acid-dependent secretion. During interdigestive periods, the portal blood concentration of bile salts is usually extremely low. This results in increased bile acid synthesis but reduced bile acid-dependent flow. After a meal, there is increased delivery of bile salts in the portal blood, which not only inhibits bile acid synthesis but also stimulates bile acid-dependent secretion.

Hormonal Control ■ Cholecystokinin, a 33-amino acid gastrointestinal hormone, is secreted by the intestinal mucosa when fatty acids or amino acids are present in the lumen. It causes contraction of the gallbladder, which in turn causes an increase in the pressure in the bile ducts. As the bile duct pressure

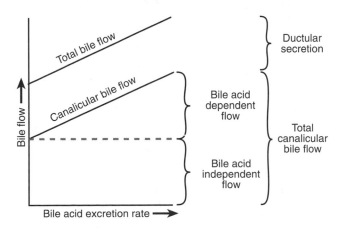

Figure 28–15 ■ Components of total bile flow: canalicular bile flow and ductular secretion. Total canalicular bile flow is composed of bile acid-dependent flow and bile acid-independent flow. (Modified from Scharschmidt BF: In Zakim D, Boyer T (eds): *Hepatology.* Philadelphia, WB Saunders, 1982.)

rises, the sphincter of Oddi relaxes and bile is delivered into the lumen.

When the mucosa of the small intestine is exposed to the acid in the chyme, it releases secretin into the blood. Secretin stimulates HCO_3^- secretion by the cells lining the bile ductules; as a result, bile contributes to the neutralization of acid in the duodenum.

Gastrin stimulates bile secretion directly by affecting the liver and indirectly by stimulating increased acid production that results in increased secretin release. Steroid hormones (e.g., estrogen and some androgens) are inhibitors of bile secretion, and reduced bile secretion is a side effect associated with the therapeutic use of these hormones. During pregnancy, the high circulating level of estrogen can reduce bile salt secretion. **Motilin,** a hormone released by the intestinal mucosa, causes contraction of the gallbladder and may play a role in stimulating bile secretion during the intestinal phase of digestion.

Neural Control ■ The biliary system is supplied by nerves from the parasympathetic and sympathetic nervous systems. Parasympathetic (vagal) stimulation results in contraction of the gallbladder and relaxation of the sphincter of Oddi as well as increased bile formation. Bilateral vagotomy results in reduced bile secretion following a meal, suggesting that the parasympathetic nervous system plays a role in mediating bile secretion. By contrast, stimulation of the sympathetic nervous system results in reduced bile secretion and relaxation of the gallbladder.

■ Gallbladder Bile Differs from Hepatic Bile

Gallbladder bile has a very different composition from hepatic bile, principally in that it is more highly concentrated. Water absorption is the major mechanism involved in concentrating hepatic bile by the gallbladder. Water absorption by the gallbladder epithelium is passive and is secondary to active Na^+ transport via a Na^+/K^+ ATPase in the basolateral surface of the epithelial cells lining the gallbladder. As a result of isotonic fluid absorption from the gallbladder bile, the concentration of the various unabsorbed components of hepatic bile increases dramatically, as much as 20-fold.

■ The Enterohepatic Circulation Recycles Bile Salts between the Intestine and the Liver

The **enterohepatic circulation** of bile salts is the recycling of bile salts between the small intestine and the liver. The total amount of bile acids in the body, primary or secondary, conjugated or free, at any time is defined as the **total bile acid pool.** In normal humans, the bile acid pool ranges from 2 to 4 g. The enterohepatic circulation of bile salts in this pool is physiologically extremely important. By cycling a number of times during a meal, a relatively small bile salt pool can provide the body with sufficient amounts of bile salts to promote lipid absorption. In a light eater, the bile salt pool may circulate 3-5 times a day; in a heavy eater, it may circulate 14-16 times a day. As described earlier, the intestine is normally extremely efficient in absorbing the bile salts by carriers located in the distal ileum. Inflammation of the ileum can lead to their malabsorption and result in the loss of large quantities of bile salts in the feces. Depending on the severity of illness, this may result in malabsorption of fat.

The bile salts in the intestinal lumen are absorbed via three pathways (Fig. 28–17). First, they are absorbed throughout the entire small intestine by passive diffusion; however, only a small fraction of the total amount of bile salts is absorbed in this manner. Second, and probably most important, bile salts are absorbed in the terminal ileum by an active carrier-mediated process, an extremely efficient process in which usually less than 5% of the bile salts escape into the colon. Third, bacteria in the terminal ileum and colon decon-

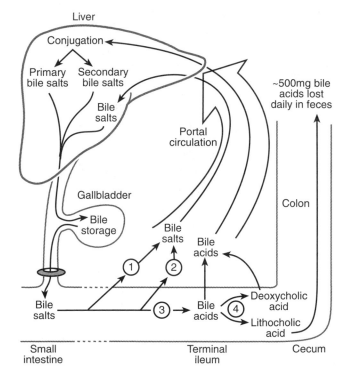

Figure 28–17 ■ Enterohepatic circulation of bile salts. Bile salts are recycled out of the small intestine in four ways: (1) passive diffusion along the small intestine (plays a relatively minor role); (2) carrier-mediated active absorption in the terminal ileum (the most important absorption route); (3) deconjugation to primary bile acids with subsequent passive absorption, also in the terminal ileum; (4) conversion of primary bile acids to secondary bile acids with subsequent absorption of deoxycholic acid.

GASTROINTESTINAL SECRETION

jugate the bile salts to form bile acids, which are much more lipophilic than bile salts and thus can be absorbed passively. These same bacteria are responsible for transforming the primary bile acids to secondary bile acids by dehydroxylation, forming deoxycholate and lithocholate.

Although bile salt and bile acid absorption is extremely efficient, some are nonetheless lost with every cycle of enterohepatic circulation. About 500 mg of bile acids is lost daily; the lost bile acids are replenished by synthesis of new bile acids from cholesterol. The loss of bile acid in feces is therefore an efficient way to excrete cholesterol.

Absorbed bile salts are transported in the portal blood bound to albumin or high-density lipoproteins (HDL). The uptake of bile salts by the hepatocytes is extremely efficient. In just one pass through the liver, more than 80% of the bile salts in the portal blood is removed. Once taken up by the hepatocytes, bile salts are secreted into bile. Bile acids are secreted by the hepatocytes only after they have been conjugated with glycine or taurine. The uptake of bile salts by the liver is a primary determinant of bile salt secretion by the liver.

■ The Liver Secretes Bile Pigments

The major pigment present in bile is the orange compound **bilirubin,** an end product of hemoglobin degradation in the monocyte-macrophage system in the spleen, bone marrow, and liver (Fig. 28–18). Hemoglobin is first converted to biliverdin, with the release of iron and globin. Biliverdin is then converted into bilirubin, which is transported in blood bound to albumin. The liver removes bilirubin from the circulation rapidly and conjugates it with glucuronic acid, which is secreted into the bile canaliculi through an active carrier-mediated process. In the small bowel, bilirubin glucuronide is poorly absorbed. However, in the colon, bacteria deconjugate it, and part of the bilirubin released is converted to the highly soluble, colorless compound called urobilinogen. Urobilinogen can be oxidized in the intestine to stercobilin or absorbed by the small intestine, and it is excreted either in urine or in bile. Stercobilin is responsible for the brown color of the stool.

■ Cholesterol Gallstones Form When Cholesterol Supersaturates the Bile

The bile salts and lecithin in the bile help solubilize cholesterol in bile. Normally cholesterol in bile is present as a solution, but when its concentration increases to the point that it cannot be solubilized, it starts to form crystals. Cholesterol crystallization results in the formation of cholesterol gallstones. Eventually calcium deposits form in the stones, increasing their optical opacity and making them easily detectable on x-ray images of the gallbladder.

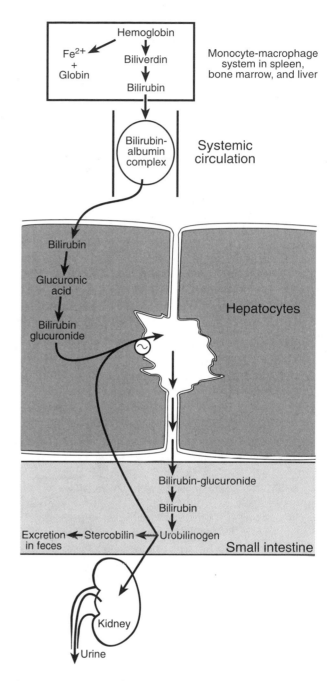

Figure 28–18 ■ Metabolism of bile pigment (bilirubin).

Intestinal Secretion

■ Intestinal Secretions Contain Electrolytes and Mucin and Serve Protective Functions

The small intestine secretes 2–3 L/day of isotonic alkaline fluid. This secretion is derived mainly from cells in the crypts of Lieberkühn, tubular glands located at the base of intestinal villi. Of the three major cell types in the crypts of Lieberkühn—argentaffin cells, Paneth cells, and undifferentiated cells—the undifferentiated cells are responsible for intestinal secretions.

Intestinal secretion probably helps maintain the fluidity of the chyme and may also play a role in diluting noxious agents and washing away infectious microorganisms. The HCO_3^- ions in intestinal secretions protect the intestinal mucosa by neutralizing any H^+ ions present in the lumen. This is important in the duodenum and also in the ileum, where bacteria degrade certain foods to produce acids (e.g., dietary fibers to short-chain fatty acids).

The fluid and electrolytes from intestinal secretions are usually absorbed by the intestine and colon, but if secretion surpasses absorption (e.g., in cholera) watery diarrhea may result. If uncontrolled, this can lead to the loss of large quantities of fluid and electrolytes, which can result in dehydration and electrolyte imbalances and, ultimately, death. A number of noxious agents, such as bacterial toxins (e.g., **cholera toxin**), can induce intestinal hypersecretion. Such toxins bind to the brush border membrane of crypt cells and increase intracellular adenylate cyclase activity. This causes a dramatic increase in intracellular AMP, which stimulates active Cl^- and HCO_3^- secretion into the lumen.

Also present in intestinal secretion are various mucins (mucoproteins) secreted by **goblet cells.** Mucins are glycoproteins that are high in carbohydrate and which form gels in solution. They are extremely diverse in structure and usually very large molecules.

Goblet cells secrete mucus into the intestinal secretions. The mucus lubricates the mucosal surface and protects it from mechanical damage by solid food particles. In the small intestine, it may also provide a physical barrier to the entry of microorganisms into the mucosa.

It is well documented that tactile stimuli or an increase in intraluminal pressure stimulates production of intestinal secretions. Other potent stimuli are certain noxious agents and the toxins produced by microorganisms. With the exception of toxin-induced secretions, our understanding of the normal control of intestinal secretions is meager. Vasoactive intestinal polypeptide is known to be a potent stimulator of intestinal secretions. This is demonstrated by a form of endocrine tumor of the pancreas that results in the secretion of large amounts of VIP into the circulation; in this condition intestinal secretion rates are high.

■ ■ ■ ■ ■ ■ ■ ■ ■ ■ ■ REVIEW EXERCISES ■ ■ ■ ■ ■ ■ ■ ■ ■ ■ ■

Identify Each with a Term
1. The recycling of bile salts between the small intestine and the liver
2. The enzyme responsible for active secretion of hydrogen ions by the parietal cells
3. When the effect of two stimulants is greater than the effect of either of the stimulants alone

Define Each Term
4. Bilirubin
5. Primary bile acids
6. Secondary bile acids
7. Alkaline tide
8. Endopeptidase
9. Bile acid-dependent bile flow

Choose the Correct Answer
10. Which of the following secretions is almost exclusively under neural control?
    a. Gastric secretion
    b. Intestinal secretion
    c. Pancreatic secretion
    d. Salivary secretion
11. Bile acid uptake by hepatocytes is dependent on:
    a. Calcium
    b. Iron
    c. Sodium
    d. Potassium
12. Gastric parietal cells secrete:
    a. Intrinsic factor
    b. Gastrin
    c. Somatostatin
    d. Cholecystokinin (CCK)
13. Which phase of stimulation of gastric acid secretion is responsible for the bulk of acid secretion?
    a. Cephalic
    b. Gastric
    c. Intestinal
14. The enzyme carbonic anhydrase catalyzes which of the following chemical reactions?
    a. The formation of carbon dioxide from carbon and oxygen
    b. The formation of carbonic acid from carbon dioxide and water
    c. The formation of bicarbonate ion from carbonic acid
    d. None of the above
15. The bile acid pool in the body is about:
    a. 10 g
    b. 25 g
    c. 3 g
    d. 1 g
16. Parasympathetic stimulation induces pancreatic acinar cells to release the protease:
    a. Bradykinin
    b. Kallikrein
    c. Kininogen
    d. Kinin

GASTROINTESTINAL SECRETION **549**

17. Which of the following proteins is absent in saliva?
   a. Lingual lipase
   b. Amylase
   c. Mucin
   d. Intrinsic factor
18. What is the pH of the gastric lumen just before a meal?
   a. 0.1–0.5
   b. 1–2
   c. 4–5
   d. 6–7
19. Salivary secretion is inhibited by:
   a. Atropine
   b. Pilocarpine
   c. Cimetidine
   d. Aspirin
20. The chief cells of the stomach secrete:
   a. Intrinsic factor
   b. Hydrochloric acid
   c. Pepsinogen
   d. Gastrin
21. The interaction of histamine with its receptor in the parietal cell results in:
   a. An increase in intracellular sodium concentration
   b. An increase in intracellular cAMP production
   c. An increase in intracellular cGMP production
   d. A decrease in intracellular calcium concentration
22. When the pH of the stomach lumen falls below 3, the antrum of the stomach releases a peptide that acts locally to inhibit gastrin release. The peptide is:
   a. Enterogastrone
   b. Gastric inhibitory peptide
   c. Secretin
   d. Somatostatin
23. Which of the following hormones tends to stimulate pancreatic secretion that is rich in bicarbonate?
   a. Somatostatin
   b. Secretin
   c. CCK
   d. Gastrin
24. Which of the following hormones causes the gallbladder to contract?
   a. Gastrin
   b. Secretin
   c. Somatostatin
   d. CCK

Briefly Answer

25. What are the protective functions of saliva?
26. Describe the mechanism that keeps salivary secretion hypotonic with respect to plasma.
27. Describe the gastric cause for pernicious anemia.
28. Describe the changes in electrolyte composition of pancreatic juice as the rate of secretion increases.
29. Describe the intestinal phase of the control of pancreatic secretion.

ANSWERS

1. Enterohepatic circulation of bile salts
2. H^+-K^+ ATPase
3. Potentiation
4. A bile pigment that is an end product of hemoglobin degradation
5. Bile acids synthesized by the hepatocytes (cholic and chenodeoxycholic acid)
6. Bile acids formed by dehydroxylation of primary bile acids by bacteria in the intestinal lumen
7. The surge in the alkalinity of venous blood from the stomach during active acid secretion due to diffusion of a large amount of bicarbonate into the blood
8. An enzyme that splits proteins and polypeptides interiorly into smaller peptides
9. The fraction of bile flow that depends on hepatic uptake and secretion of bile salts
10. d
11. c
12. a
13. b
14. b
15. c
16. b
17. d
18. b
19. a
20. c
21. b
22. d
23. b
24. d
25. Saliva protects the oral cavity by partially neutralizing ingested acidic substances and neutralizing acidic and corrosive regurgitated gastric acid and pepsin. It plays an important role in the general hygiene of the oral cavity by the antibacterial actions of muramidase and lactoferrin. Furthermore, saliva lubricates the mucosal surfaces, thus reducing frictional damage.
26. Saliva is more hypotonic than plasma because there is a net absorption of ions by the ductal epithelium, and the epithelium lining the duct is not very permeable to water.
27. Pernicious anemia can be caused by lack of production of intrinsic factor by the parietal cells, impairing absorption of vitamin B_{12}.
28. Pancreatic secretion is always isotonic. As the concentration of bicarbonate ions increases with secretion, the chloride concentration falls accordingly.

29. When the acidic chyme of the stomach enters the small intestine, the pH of the chyme stimulates the release of secretin by the S cells of the intestinal mucosa. This stimulates bicarbonate secretion by the pancreas. The secretion of pancreatic enzymes is regulated by the amount of circulating CCK and vagovagal reflexes. The release of CCK by the CCK cells of the intestinal mucosa and vagovagal reflexes are stimulated by the exposure of the intestinal mucosa to long-chain fatty acids and free amino acids.

Suggested Reading

Davenport HW: *Physiology of the Digestive Tract.* 5th ed. Chicago, Year Book, 1982.

Forte JG, Wolosin JM: HCl secretion by the gastric oxyntic cell, in Johnson LR (ed): *Physiology of the Gastrointestinal Tract.* 2nd ed. New York, Raven, 1987, pp 853-863.

Granger DN, Barrowman JA, Kvietys PR: *Clinical Gastrointestinal Physiology.* Philadelphia, WB Saunders, 1985.

Hobsley M: *Disorders of the Digestive Tract.* Baltimore, University Park Press, 1982.

Ito S: Functional gastric morphology, in Johnson LR (ed): *Physiology of the Gastrointestinal Tract.* 2nd ed. New York, Raven, 1987, pp 817-851.

Johnson LR: *Gastrointestinal Physiology.* 3rd ed. St. Louis, CV Mosby, 1985.

Sanford PA: *Digestive System Physiology.* Baltimore, University Park Press, 1982.

Schulz I: Electrolyte and fluid secretion in the exocrine pancreas, in Johnson LR (ed): *Physiology of the Gastrointestinal Tract.* 2nd ed. New York, Raven, 1987, pp 1147-1171.

Sernka T, Jacobson E: *Gastrointestinal Physiology.* 2nd ed. Baltimore, Williams & Wilkins, 1983.

Solomon TE: Control of exocrine pancreatic secretion, in Johnson LR (ed): *Physiology of the Gastrointestinal Tract.* 2nd ed. New York, Raven, 1987, pp 1173-1207.

Young JA, Cook DI, van Lennep EW, Roberts M: Secretion by the major salivary glands, in Johnson LR (ed): *Physiology of the Gastrointestinal Tract.* 2nd ed. New York, Raven, 1987, pp 773-815.

Digestion and Absorption

CHAPTER OUTLINE

I. THE DIGESTION AND ABSORPTION OF CARBOHYDRATES
   A. The diet contains both digestible and nondigestible carbohydrates
   B. Carbohydrates are digested in different parts of the gastrointestinal tract
   C. The enterocytes play an important role in carbohydrate absorption and metabolism
   D. Monosaccharides are transported in the portal blood
   E. The lack of some digestive enzymes impairs carbohydrate absorption
   F. Dietary fiber plays an important role in gastrointestinal motility

II. THE DIGESTION AND ABSORPTION OF LIPIDS
   A. The luminal lipid consists of both exogenous and endogenous lipids
   B. Different lipases carry out the hydrolysis of lipids
   C. Bile salt plays an important role in lipid absorption
   D. Enterocytes process absorbed lipid to form lipoproteins
   E. Enterocytes secrete chylomicrons and very-low-density lipoproteins
   F. Fatty acids can also travel in the blood bound to albumin
   G. Lack of pancreatic lipases or bile salts can impair lipid absorption

III. THE DIGESTION AND ABSORPTION OF PROTEINS
   A. Luminal protein derives from the diet, gastrointestinal secretions, and enterocytes
   B. Proteins are digested in the gastrointestinal tract, yielding amino acids and peptides
   C. Specific transporters in the small intestine take up the amino acids and peptides
   D. Amino acids, peptides, and proteins travel in the portal blood
   E. Defects in digestion and transport can impair protein absorption

IV. THE ABSORPTION OF VITAMINS
   A. The fat-soluble vitamins include A, D, E, and K
      1. Vitamin A
      2. Vitamin D
      3. Vitamin E
      4. Vitamin K

OBJECTIVES

After studying this chapter, the student should be able to:
1. List the monosaccharides present in lactose, sucrose, maltose, starch, and glycogen
2. Describe the luminal digestion, uptake, and transport of carbohydrates by the gastrointestinal tract
3. Describe the actions of acid lipase, colipase, pancreatic lipase, cholesterol esterase, and phospholipase A_2
4. Describe the critical micellar concentration of bile salts and the role of bile salts in the uptake of lipid digestion products (1-monoacylglycerol and fatty acids) by the small intestine; describe the chylomicrons and very-low-density lipoproteins made by the small intestine
5. Describe clinical conditions that result in fat malabsorption
6. Describe the protein requirement in humans; describe the digestion and absorption of proteins by the gastrointestinal tract
7. List the fat- and water-soluble vitamins and describe their absorption by the gastrointestinal tract
8. Describe the absorption of sodium, calcium, and iron by the gastrointestinal tract

CHAPTER OUTLINE (Continued)

The major function of the gastrointestinal tract is the digestion and absorption of nutrients. It also plays an important role in the absorption of fat- and water-soluble vitamins, electrolytes, water, and bile acids. The three major nutrients absorbed by the gastrointestinal tract are carbohydrate, fat, and protein. Although some drugs and medium-chain fatty acids are absorbed by the stomach, most digestion and absorption of nutrients occurs in the small intestine.

To ensure maximal absorption of nutrients, the gastrointestinal tract has a number of unique features. For instance, after a meal the small intestine undergoes rhythmic contractions called segmentations (Chap. 27), which ensure proper mixing of the small intestinal contents, exposure of the contents to digestive enzymes, and maximal exposure of digestion products to the small intestinal mucosa. The rhythmic segmentation has a gradient along the small intestine, with the highest frequency in the duodenum and the lowest in the ileum. This gradient ensures very slow but forward movement of intestinal contents toward the colon. Another unique feature of the small intestine is its architecture (Fig. 29–1). Spiral or circular concentric folds increase the surface area of the intestine about three times (compared to a cylinder). Fingerlike projections of the mucosal surface, called villi, further increase the surface area of the small intestine about 30 times. To amplify the absorptive surface further, each epithelial cell (enterocyte) is covered by numerous closely packed **microvilli**. This increases the total surface area to 600 times that of a cylinder. The various nutrients, vitamins, bile salts, and water are absorbed by the gastrointestinal tract by passive (down a concentration gradient), facilitated, or active transport. We will discuss the site and mechanism of absorption. The gastrointestinal tract has a large reserve for the digestion and absorption of various nutrients and vitamins. Malabsorption of nutrients is usually not detected unless a large portion of the small intestine has been removed due to disease.

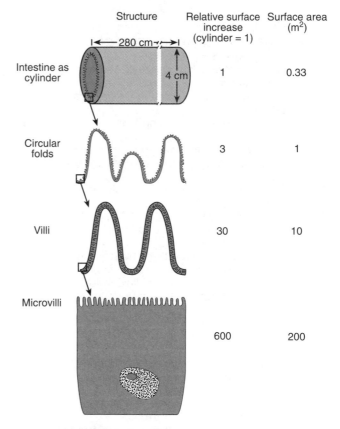

Structure	Relative surface increase (cylinder = 1)	Surface area (m^2)
Intestine as cylinder	1	0.33
Circular folds	3	1
Villi	30	10
Microvilli	600	200

Figure 29–1 ■ Amplification of the mucosal surface area by specialized features of the intestinal mucosa. (Adapted from Schmidt RF, Thews G: *Human Physiology.* Berlin, Springer-Verlag, 1993, p 602.)

Most nutrients and vitamins are absorbed by the duodenum and jejunum, but because bile salts are involved in the intestinal absorption of lipids it is important that they not be absorbed prematurely. For effective fat absorption, the small intestine has adapted to absorb the bile salts only in the ileum, through a receptor-mediated process. The intestinal epithelial cells along the villus that are involved in the absorption of nutrients are replaced every 2–3 days.

The Digestion and Absorption of Carbohydrates

■ The Diet Contains Both Digestible and Nondigestible Carbohydrates

Carbohydrates are compounds composed of carbon, hydrogen, and oxygen. Humans can digest most carbohydrates; those we cannot are the dietary fibers that form the roughage of the diet. Carbohydrate is present in food as monosaccharides, disaccharides, oligosaccharides, and polysaccharides. The monosaccharides are mainly hexoses (six-carbon sugars), and glucose is by far the most abundant of these. Glucose is obtained directly from the diet or from the digestion of disaccharides, oligosaccharides, or polysaccharides. The next most common monosaccharides are galactose, fructose, and sorbitol. Galactose is present in the diet only as milk lactose, a disaccharide composed of galactose and glucose. Fructose is present in abundance in fruit and honey and is usually present as disaccharides or polysaccharides. Sorbitol is derived from glucose and is almost as sweet as glucose, but sorbitol is absorbed much more slowly and thus maintains a high blood sugar level for a longer period when similar amounts are ingested. It has been used as a weight reduction aid to delay the onset of hunger sensation.

The major disaccharides in the diet are sucrose, lactose, and maltose. Sucrose, present in sugar cane and honey, is composed of glucose and fructose. Lactose, the main sugar in milk, is composed of galactose and glucose. Maltose is composed of two glucose units.

The digestible polysaccharides are starch, dextrins, and glycogen. Starch, by far the most abundant carbohydrate in the human diet, is made of amylose and amylopectin. Amylose is composed of a straight chain of glucose units, and amylopectin is composed of branched glucose units. Dextrins, formed from heating (e.g., toasting bread) or the action of the enzyme amylase, are intermediate products of starch digestion. Glycogen is a highly branched polysaccharide that stores carbohydrate in the body. The structure of glycogen is illustrated in Figure 29-2. Normally, about 300–400 g of glycogen is stored in the liver and muscle, with more stored in muscle than in the liver. Muscle glycogen is used exclusively by muscle, and liver glycogen is used to provide blood glucose during fasting.

Dietary fibers are polysaccharides that are usually poorly digested by the digestive enzymes in the small intestine. They have an extremely important physiologic function in that they provide the "bulk"

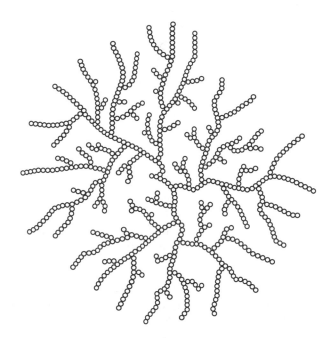

Figure 29–2 ■ The structure of glycogen.

that facilitates intestinal motility and function. Many vegetables and fruits are rich in fibers, and their frequent ingestion greatly decreases intestinal transit time.

■ Carbohydrates Are Digested in Different Parts of the Gastrointestinal Tract

The digestion of carbohydrate starts when food is mixed with saliva during chewing. The enzyme responsible is **salivary amylase.** Salivary amylase acts on the α-1,4-glycosidic linkage of amylose and amylopectin of polysaccharides to release the disaccharide maltose and oligosaccharides maltotriose and α-limit dextrins (Fig. 29-3). Because salivary amylase works best at neutral pH, its digestive action terminates rapidly after the food bolus mixes with acid in the stomach. However, if the food is thoroughly mixed with amylase during chewing, a substantial amount of complex carbohydrates are digested before this point. **Pancreatic amylase** continues the digestion of the remaining carbohydrates. However, the chyme must first be neutralized by pancreatic secretions, because pancreatic amylase works best at neutral pH. The products of pancreatic amylase digestion of polysaccharides are also maltose, maltotriose, and α-limit dextrins.

The digestion products of starch, together with other disaccharides (sucrose and lactose), are further digested by enzymes located at the brush border membrane. Table 29-1 lists the enzymes involved in the digestion of di- and oligosaccharides and the products of their action. The final products are glucose, fructose, galactose, and oligosaccharides.

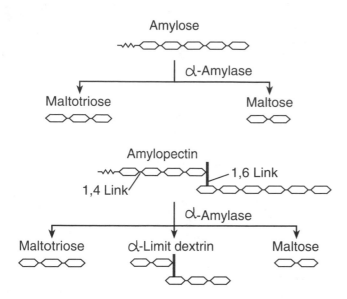

Figure 29–3 ■ The digestion products of starch after exposure to salivary or pancreatic α-amylase. Reducing sugar units are indicated by hexagons. (From Granger DN, Barrowman JA, Kvietys PR: *Clinical Gastrointestinal Physiology.* Philadelphia, WB Saunders, 1985.)

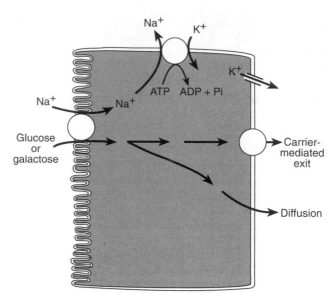

Figure 29–4 ■ The sodium-dependent carrier system for glucose and galactose.

TABLE 29–1 ■ Digestion of Dissacharides and Oligosaccharides by Brush Border Enzymes

Enzyme	Substrate	Site of Action	Products
Sucrase	Sucrose	α-1,2-glycosidic linkage	Glucose and fructose
Lactase	Lactose	β-1,4-glycosidic linkage	Glucose and galactose
Isomaltase	α-Limit dextrins	α-1,6-glycosidic linkage	Glucose, maltose, and oligosaccharides
Maltase	Maltose	α-1,4-glycosidic	Glucose

■ The Enterocytes Play an Important Role in Carbohydrate Absorption and Metabolism

Monosaccharides are absorbed by the intestinal epithelial cells (**enterocytes**) either actively or by facilitated transport. Glucose and galactose are absorbed via secondary active transport by a symporter (Chap. 2) that transports one molecule of monosaccharide with a Na[+] ion (Fig. 29–4). The movement of Na[+] into the cell down concentration and electrical gradients effects the uphill movement of glucose into the cell. The low intracellular Na[+] concentration is maintained by the basolateral membrane Na[+]/K[+] ATPase, which pumps three Na[+] out of the cell in exchange for two K[+] ions into the cell. The osmotic effects of sugars increases the Na[+]/K[+] ATPase activity and the K[+] conductance of the basolateral membrane. Sugars accumulate in the cell at a higher concentration than in plasma and leave the cell by Na[+]-independent facilitated transport or passive diffusion through the basolateral cell membrane. Glucose and galactose share a common transporter at the brush border membrane of entero-cytes and thus compete with each other during absorption.

Fructose is taken up by facilitated transport. Although facilitated transport is carrier-mediated, it is not an active process (Chap. 2). Fructose absorption is much slower than glucose and galactose absorption and is not Na[+]-dependent. Although in some animal species both galactose and fructose can be converted to glucose in the enterocytes, this is probably not important in humans.

■ Monosaccharides Are Transported in the Portal Blood

The sugars absorbed by the enterocytes are transported by the portal blood to the liver, where they are converted to glycogen or sent into the bloodstream. After a meal, the level of blood glucose rises rapidly, usually peaking at 30–60 minutes. The concentration of glucose can be as high as 150 mg/dl. Although the enterocytes can use glucose for fuel, glutamine is preferred. Both galactose and glucose can be used in the glycosylation of proteins in the Golgi apparatus of the enterocytes.

■ The Lack of Some Digestive Enzymes Impairs Carbohydrate Absorption

Impaired carbohydrate absorption caused by a lack of salivary or pancreatic amylase almost never occurs because these enzymes are usually present in great excess. Impaired absorption due to deficiency in membrane disaccharidases is rather common, however. Such deficiencies can be either genetic or acquired. Among congenital deficiencies, lactase deficiency is by far the most common, particularly among Asians, South Americans, and Africans. Affected individuals suffer from **lactose intolerance,** a condition in which ingestion of milk products results in severe osmotic (watery) diarrhea. The mechanism responsible is depicted in Figure 29–5. Undigested lactose in the bowel lumen increases the osmolality of the luminal contents. The osmolality is further increased by production of lactic acid from the action of intestinal bacteria on the lactose. The increase in luminal osmolality results in net water secretion into the lumen. The accumulation of fluid distends the small intestine and accelerates peristalsis, eventually resulting in watery diarrhea.

Congenital sucrase deficiency results in symptoms similar to those of lactase deficiency. Sucrase deficiency can be inherited or acquired through disorders of the small intestine, such as tropical sprue or Crohn's disease.

■ Dietary Fiber Plays an Important Role in Gastrointestinal Motility

Dietary fibers include indigestible carbohydrates and carbohydratelike components, mainly found in fruits and vegetables. The most common are cellulose, hemicellulose, pectins, and gums. Cellulose and hemi-

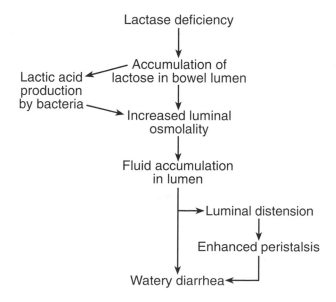

Figure 29–5 ■ Mechanism for osmotic diarrhea resulting from lactase deficiency.

cellulose, which are not soluble in water, are poorly digested by humans and provide the bulkiness of stool. Pectin, which absorbs water readily to form a gel, has been given clinically to slow gastric motility.

Because dietary fibers impart bulk to the bolus, their presence greatly shortens transit time. It has been proposed that dietary fibers reduce the incidence of colon cancer by shortening gastrointestinal transit time, which, in turn, reduces the formation of carcinogenic bile acids (e.g., lithocholic acid). Lithocholic acid is a secondary bile acid formed from the dehydroxylation of chenodeoxycholic acid by bacteria in the gut. Because dietary fibers also bind bile acids, which are formed from cholesterol, fiber consumption can result in lowering of blood cholesterol by promoting excretion.

The Digestion and Absorption of Lipids

Lipids are a concentrated form of energy. They provide 30–40% of the daily caloric intake in the West. Lipids are also essential for normal body functions, as they form part of cellular membranes, precursors of bile acids, hormones, prostaglandins, and leukotrienes. The human body is capable of synthesizing most of the lipids it requires, with the exception of the **essential fatty acids,** linoleic acid (C 18:2, 18-carbon long with two double bonds) and arachidonic acid (C 20:4). Both of these acids belong to the family of ω-6 fatty acids. Recently, researchers have provided convincing evidence that **eicosapentaenoic acid** (C 20:5) and **docosahexaenoic acid** (C 22:6) are also essential for the normal development of vision in newborns. Both of these acids belong to the family of ω-3 fatty acids, and they are abundant in seafood and algae.

■ The Luminal Lipid Consists of Both Exogenous and Endogenous Lipids

Lipids are comprised of a number of classes of compounds that are insoluble in water but soluble in organic solvents; by far the most abundant dietary lipids are triacylglycerols, or triglycerides. They consist of a glycerol backbone esterified in the three positions with fatty acids (Fig. 29–6A). More than 90% of the daily dietary lipid intake is in the form of triglycerides.

The other lipids in the human diet are cholesterol and phospholipids. Cholesterol is a sterol derived exclusively from animal fat. Humans also ingest a small amount of plants sterols, notably β-sitosterol and campesterol. The phospholipid molecule is similar to a triglyceride, with fatty acids occupying the first and second positions of the glycerol backbone (Fig. 29–6B). However, the third position of the phospholipid molecule is occupied by a phosphate group coupled to a nitrogenous base (e.g., choline or ethanolamine), by which each type of phospholipid molecule is named.

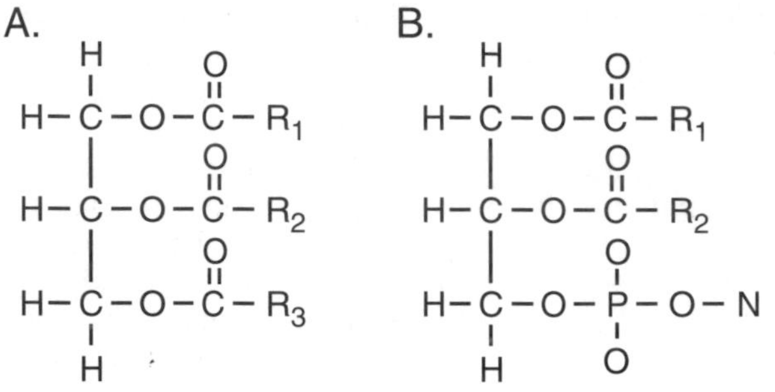

A.

B.

Figure 29–6 ■ (A) Triglyceride molecule. R₁, R₂, and R₃ are composed of different fatty acids. (B) Phospholipid molecule. The fatty acid occupying the first position (R₁) is usually a saturated fatty acid and that in the second position (R₂) is usually an unsaturated or polyunsaturated fatty acid. The third position after the phosphate group is occupied by a nitrogenous base (N), such as choline or ethanolamine.

The bile serves as an endogenous source of cholesterol and phospholipids. Bile contributes about 12 g/day of phospholipid to the intestinal lumen, most in the form of phosphatidylcholine (lecithin), whereas dietary sources contribute 2–3 g/day. Another important endogenous source of lipid is desquamated intestinal villus epithelial cells.

■ Different Lipases Carry Out the Hydrolysis of Lipids

The salivary secretion contains **lingual lipase,** secreted by von Ebner's glands located on the dorsal aspect of the tongue. Lingual lipase has an optimal pH of 4, so it works well even in the stomach. It is especially suited to digest the medium-chain triglycerides that are abundant in human milk and probably plays an important role in the digestion of milk fat during the neonatal period, before the maturation of pancreatic functions. In adults, it helps the partial digestion of triglycerides to form fatty acids and partial glycerides, which aids in the emulsification of fat in the stomach, facilitating its further digestion by pancreatic lipase. A gastric lipase that differs from lingual lipase has been found in humans; it is the predominant lipase in the lumen of the stomach.

Lipid digestion mainly occurs in the intestinal lumen. Humans secrete an overabundance of pancreatic lipase. Depending on the substrate being digested, pancreatic lipase has an optimal pH of 7–8, so it works well in the intestinal lumen after the acidic contents from the stomach have been neutralized by pancreatic bicarbonate secretion. **Pancreatic lipase** hydrolyzes the triglyceride molecules to monoglycerides and two fatty acids (Fig. 29–7). It works on the triglyceride molecule at the oil-water interface, thus the rate of

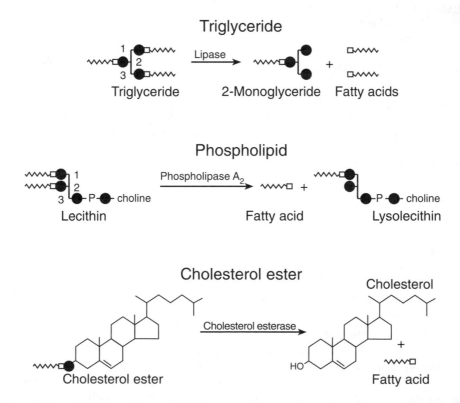

Figure 29–7 ■ Digestion of dietary lipids by pancreatic enzymes in the small intestine. Solid circles indicate oxygen atoms.

lipolysis depends on the surface area of the interface. The partial digestion of dietary triglyceride by lingual and gastric lipases helps increase the area of the oil-water interface. Pancreatic juice also contains the peptide **colipase,** which is necessary for the normal digestion of fat by pancreatic lipase. Colipase binds lipase at a molar ratio of 1:1, thereby allowing the lipase to bind to the oil-water interface where lipolysis takes place. Colipase also counteracts the inhibition of lipolysis by bile salt, which, despite its importance in intestinal fat absorption, prevents the attachment of pancreatic lipase to the oil-water interface.

Phospholipase A₂ is the major pancreatic enzyme for digesting phospholipids, forming lysophospholipids and fatty acids. For instance, phosphatidylcholine (lecithin) is hydrolyzed to form lysophosphatidylcholine (lysolecithin) and fatty acid (Fig. 29-7).

Dietary cholesterol is presented as both free and cholesterol ester. The hydrolysis of cholesterol ester is catalyzed by the pancreatic enzyme carboxyl ester hydrolase, also called **cholesterol esterase** (Fig. 29-7). The digestion of cholesterol ester is important because cholesterol can be absorbed only as the free sterol.

■ Bile Salt Plays an Important Role in Lipid Absorption

A layer of poorly stirred fluid, called the **unstirred water layer,** coats the surface of the intestinal villi (Fig. 29-8A). The unstirred water layer reduces the absorption of lipid digestion products because they are poorly soluble in water. They are rendered water-soluble by their **micellar solubilization** by bile salts in the small intestinal lumen. This greatly enhances the concentration of these products in the unstirred water layer (Fig. 29-8B). The lipid digestion products are then absorbed by the enterocytes, mainly by passive diffusion. Fatty acid and monoglyceride molecules are taken up individually. Similar mechanisms seem to operate for cholesterol and lysolecithin.

Bile salts are derived from cholesterol but are quite different from cholesterol in that they are water-soluble. Bile salts result from the conjugation of bile acids with taurine or glycine. They are essentially detergents—molecules that possess both hydrophilic and hydrophobic properties. Because bile salts are polar molecules, they penetrate cell membranes poorly. This is significant because it ensures their minimal absorption by the jejunum, where most fat absorption takes place. At or above a certain concentration of bile salts, the **critical micellar concentration,** they aggregate to form micelles; the concentration of luminal bile salts is usually well above the critical micellar concentration. When bile salts alone are present in the micelle, it is called a **simple micelle.** The simple micelles incorporate the lipid digestion products, monoglyceride and fatty acids, to form **mixed micelles.** This renders the lipid digestion products water-soluble by incorporation into mixed micelles.

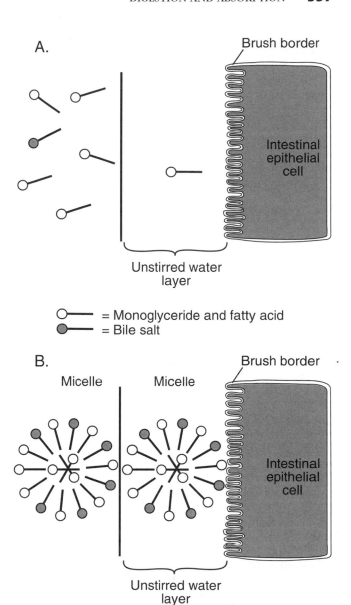

Figure 29–8 ■ Micellar solubilization of lipids enhances delivery of lipid to the brush border membrane. A = In the absence of bile salt; B = In the presence of bile salts.

The mixed micelles diffuse across the unstirred water layer and deliver lipid digestion products to the small intestinal surface for absorption.

■ Enterocytes Process Absorbed Lipid to Form Lipoproteins

After entering the enterocytes, the fatty acids and monoglycerides migrate to the smooth endoplasmic reticulum. A fatty acid-binding protein may be involved in the intracellular transport of fatty acids, but whether or not a protein carrier is involved in the intracellular transport of monoglycerides is unknown. In the smooth endoplasmic reticulum, monoglycerides and fatty acids are rapidly reconstituted to form triglycerides (Fig. 29-9). Fatty acids are first activated to form acyl-CoA, which is then used to esterify monoglyceride to form diglyceride, which is transformed into triglyceride. The lysolecithin absorbed by the enterocytes can be re-esterified in the smooth endoplasmic reticulum to form lecithin. Cholesterol can be transported out of the enterocytes as free cholesterol or as esterified cholesterol. The enzyme responsible for the esterification of cholesterol to form cholesterol ester is **acyl-CoA cholesterol acyltransferase** (ACAT).

■ Enterocytes Secrete Chylomicrons and Very-Low-Density Lipoproteins

The reassembled triglycerides, lecithin, cholesterol, and cholesterol ester are then packaged into lipoproteins and exported from the enterocytes. The intestine produces two major classes of lipoproteins, chylomicrons and very-low-density lipoproteins (VLDLs), triglyceride-rich lipoproteins with densities less than 1.006 g/ml. Chylomicrons are made exclusively by the small intestine. Their major function is to transport the large amount of dietary fat absorbed by the small intestine from the enterocytes to the lymph. Chylomicrons are large, spheric lipoproteins with diameters of 80-500 nm. They contain less protein and phospholipid than VLDLs and are therefore less dense than VLDLs. Very-low-density lipoproteins are made continuously by the small intestine, during both fasting and feeding, although the liver contributes significantly more VLDLs to the circulation.

Apoproteins apo A-I, apo A-IV, and apo B are among the major proteins associated with the production of chylomicron and VLDLs. Apo B is the only protein that seems to be necessary for the normal formation of intestinal chylomicrons and VLDLs. This protein is made in the small intestine; its molecular weight is 250,000 and it is extremely hydrophobic. Apo A-I is involved in a reaction catalyzed by the plasma enzyme **lecithin cholesterol acyltransferase** (LCAT). Plasma LCAT is responsible for the esterification of cholesterol in the plasma to form cholesterol ester, with the fatty acid derived from the 2 position of lecithin. After the chylomicrons and VLDLs enter the plasma, apo A-I is rapidly transferred from chylomicrons and VLDLs to high-density lipoproteins (HDLs). Apo A-I is the major protein present in plasma HDLs. Apo A-IV is made by the small intestine and the liver. It was recently shown that apo A-IV may be an important factor secreted by the small intestine that is responsible for the anorexia observed after fat feeding.

The newly synthesized lipoproteins in the smooth endoplasmic reticulum are transferred to the Golgi apparatus, where they are packaged in vesicles. The Golgi-derived vesicles carrying the chylomicrons and VLDLs undergo exocytosis and are released into the intercellular space (Fig. 29-10). From there they are transferred to the central lacteals (beginning of lymphatic vessels) by a process that is not well understood. Experimental evidence seems to indicate that the transfer probably occurs mostly by diffusion. Intestinal lipid absorption is associated with a marked increase in lymph flow, called the lymphagogic effect of fat feeding. This increase in lymph flow plays an important role in the transfer of lipoproteins from the intercellular spaces to the central lacteal.

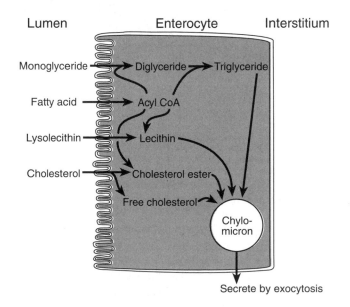

Figure 29-9 ■ Intracellular metabolism of absorbed lipid digestion products.

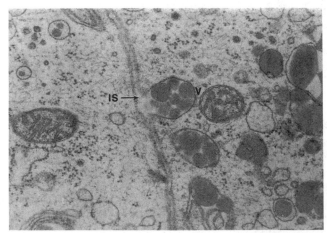

Figure 29-10 ■ The exocytosis of chylomicrons is evident in this electron micrograph. The nascent chylomicrons in the secretory vesicle are similar in size and morphology to those already present in the intercellular space (ICS). (From Sabesin SM, Frase S: Electron microscopic studies of the assembly, intracellular transport, and secretion of chylomicrons by rat intestine. *J Lipid Res* 18:496-511, 1977.)

Fatty Acids Can Also Travel in the Blood Bound to Albumin

While the majority of the long-chain fatty acids are transported from the small intestine as triglycerides packaged in chylomicrons and VLDLs, some are transported in the portal blood, bound to serum albumin. Most of the medium-chain (8–12 carbons) and all of the short-chain fatty acids are transported by the portal route.

Lack of Pancreatic Lipases or Bile Salts Can Impair Lipid Absorption

In a number of clinical conditions, lipid digestion and absorption are impaired, resulting in malabsorption of lipids and other nutrients and fatty stools (**steatorrhea**). Abnormal lipid absorption can result in numerous problems, because the body requires certain fatty acids (e.g., linoleic and arachidonic acid, precursors of prostaglandins) to function normally. These have been called essential fatty acids, because the human body cannot synthesize them and is therefore totally dependent on diet to supply them. Recent studies suggest that the human body may also require ω-3 fatty acids in the diet during development; these include linolenic, docosahexaenoic, and eicosapentaenoic acids. Linolenic acid is abundant in plants and docosahexaenoic and eicosapentaenoic acids are abundant in fish. Docosahexaenoic acid is an important fatty acid present in the retina and brain.

Pancreatic deficiency markedly reduces the ability of the exocrine pancreas to produce digestive enzymes. Because the pancreas normally produces an excess, enzyme production has to be reduced to about one-tenth of normal before symptoms of malabsorption develop. One characteristic of pancreatic deficiency is steatorrhea resulting from poor digestion of fat by the pancreatic lipase. Normally about 5 g/day of fat is excreted in human stool. With steatorrhea, as much as 50 g/day can be excreted.

Fat absorption subsequent to the action of pancreatic lipase requires solubilization by bile salt micelles. Acute or chronic liver disease can cause defective biliary secretion, resulting in bile salts concentrations lower than those necessary for micelle production, thereby inhibiting normal absorption of fat.

Abetalipoproteinemia, an autosomal recessive disorder, is characterized by a complete lack of apo B, which is required for the formation and secretion of chylomicrons and VLDLs. Apo B-containing lipoproteins in the circulation, including chylomicrons, VLDLs, and low-density lipoproteins (LDLs), are lacking. Plasma LDLs are also absent, because they are derived mainly from the metabolism of VLDLs. Since individuals with abetalipoproteinemia do not produce any chylomicrons or VLDLs in the small intestine, they are unable to transport absorbed fat, resulting in an accumulation of lipid droplets in the cytoplasm of enterocytes. They also suffer from a deficiency of fat-soluble vitamins (e.g., A, D, E, and K).

The Digestion and Absorption of Proteins

Proteins form the fundamental structure of cells and are the most abundant of all organic compounds in the body. Most proteins are found in muscle, with the remainder in blood, body fluids, and body secretions. Enzymes and many hormones are also proteins. Proteins are composed of amino acids and have molecular weights of a few thousand to a few hundred thousand. More than 20 common amino acids form the building blocks for proteins (Table 29-2). Of these, nine are considered essential. Although the **nonessential amino acids** are also required for normal protein synthesis, the body can synthesize them from other amino acids.

Complete proteins are those that can supply all of the essential amino acids in amounts sufficient to support normal growth and body maintenance. Examples are eggs, poultry, and fish. The proteins in vegetables and grains are called **incomplete proteins** because they do not provide all of the essential amino acids in amounts sufficient to sustain normal growth and body maintenance.

Luminal Protein Derives from the Diet, Gastrointestinal Secretions, and Enterocytes

The average American takes in 70–110 g/day of protein. The minimum protein requirement for adults is about 0.8 g/kg body weight. Thus, for a 70-kg person, the minimum daily protein requirement is 56 g. Pregnant or lactating women require 20–30 g above the recommended daily allowance to keep pace with the extra demand for protein. A lactating woman can lose as much as 12–15 g of protein per day as milk protein. Children need more protein for body growth, and the recommended daily allowance for infants is about

TABLE 29–2 ■ Common Amino Acids Found in Proteins

Essential	Nonessential
Histidine	Alanine
Isoleucine	Arginine
Leucine	Aspartic acid
Lysine	Citrulline
Methionine	Glutamic acid
Phenylalanine	Glycine
Threonine	Hydroxyglutamic acid
Tryptophan	Hydroxyproline
Valine	Norleucine
	Proline
	Serine
	Tyrosine

2 g/kg body weight. While most of the protein entering the gastrointestinal tract is dietary protein, there are also proteins derived from endogenous sources, such as pancreatic, biliary, and intestinal secretions, and the cells shed from the intestinal villi. Between 20 and 30 g/day of protein in pancreatic secretion and about 10 g/day in biliary secretion enters the intestinal lumen. The enterocytes of the intestinal villi are continuously shed into the intestinal lumen, and as much as 50 g/day of enterocyte proteins enter the intestinal lumen. Thus, an average of 150–180 g/day of total protein is presented to the small intestine, more than 90% of which is absorbed.

Proteins Are Digested in the Gastrointestinal Tract, Yielding Amino Acids and Peptides

The majority of the protein in the intestinal lumen is completely digested to either amino acids or di- or tripeptides before it is taken up by the enterocytes. The enzymes involved in the digestion of proteins are described in Chapter 28.

Protein digestion starts in the stomach with the action of **pepsin,** which is secreted as proenzyme and activated by acid in the stomach. Pepsin hydrolyzes protein to form smaller polypeptides. It is classified as an endopeptidase because it attacks specific peptide bonds inside the protein molecule. This phase of protein digestion is normally not very important except in individuals suffering from pancreatic exocrine deficiency.

The majority of the digestion of proteins and polypeptides takes place in the small intestine. Most of the proteases are secreted in the pancreatic juice as inactive proenzymes. When the chyme containing the polypeptides enters the duodenum, it stimulates the release of **enterokinase** by the duodenal mucosa. Enterokinase acts on the **trypsinogen** to release the active enzyme **trypsin,** which converts the other proenzymes to active enzymes.

The pancreatic proteases are classified as endopeptidases or exopeptidases (Table 29-3). Endopeptidases hydrolyze certain internal peptide bonds of proteins or polypeptides to release the smaller peptides. The three endopeptidases present in pancreatic juice are trypsin, chymotrypsin, and elastase. Trypsin splits off basic amino acids from the carboxyl terminal of a protein; chymotrypsin attacks peptide bonds with an aromatic carboxyl terminal; and elastase attacks peptide bonds with a neutral aliphatic carboxyl terminal. The exopeptidases in pancreatic juice are carboxypeptidase A and carboxypeptidase B. Like the endopeptidases, the exopeptidases are specific in their action. Carboxypeptidase A attacks polypeptides with a neutral aliphatic or aromatic carboxyl terminal; carboxypeptidase B attacks polypeptides with a basic carboxyl terminal. The final products of protein digestion are amino acid and small peptides.

TABLE 29-3 ■ Classification of Pancreatic Proteases

Enzyme	Primary Action
Endopeptidases	Hydrolyze internal peptide bonds of proteins and polypeptides
Trypsin	Attacks peptide bonds with basic carboxyl terminal
Chymotrypsin	Attacks peptide bonds with aromatic carboxyl terminal
Elastase	Attacks peptide bonds with neutral aliphatic carboxyl terminal
Exopeptidases	Hydrolyze external peptide bonds of proteins and polypeptides
Carboxypeptidase A	Attacks polypeptides with neutral aliphatic or aromatic carboxyl terminal
Carboxypeptidase B	Attacks polypeptides with basic carboxyl terminal

Specific Transporters in the Small Intestine Take Up the Amino Acids and Peptides

Amino acids are taken up by the enterocytes via secondary active transport. Four major amino acid carriers in the small intestine have been identified: dibasic, dicarboxylic, neutral amino acid, and imino acid/glycine transporters. The amino acid transporters favor the L form over the D form. As in the uptake of glucose, the uptake of amino acids is dependent on a Na^+ concentration gradient across the brush border membrane of the intestinal epithelial cells.

The absorption of peptides by enterocytes was once thought to be less efficient than amino acid absorption. However, subsequent studies in humans clearly demonstrated that dipeptide and tripeptide uptake is significantly more efficient than uptake of amino acids. The di- and tripeptides use transporters different from those used by amino acids. The peptide transporter prefers di- and tripeptides with either glycine or lysine residues. Furthermore, tetra- and higher peptides are only poorly transported by the peptide transporter. These peptides can be further broken down to di- and tripeptides by the peptidases (exopeptidases) located on the brush border of the enterocytes. Di- and tripeptides are given to individuals suffering from malabsorption, because they are absorbed more efficiently and are more palatable than the free amino acids. Another advantage of peptides over amino acids is the smaller osmotic stress created as a result of delivering them.

In adults, a negligible amount of protein is absorbed as undigested protein. In some individuals, however, intact or partially digested proteins are absorbed, and this can cause anaphylactic or hypersensitivity reactions. The pulmonary and cardiovascular systems are the major organs involved in anaphylactic reactions. For the first few weeks after birth, the newborn's small intestine absorbs considerable amounts of intact proteins. This is possible because of reduced proteolytic function in the stomach; reduced pancreatic

CELIAC SPRUE (GLUTEN-SENSITIVE ENTEROPATHY)

Celiac sprue, also called gluten-sensitive enteropathy, is a commonly encountered disease that involves a primary lesion of the intestinal mucosa. It is caused by the sensitivity of the small intestine to gluten. This disorder can result in the malabsorption of all nutrients, caused by shortening or total loss of intestinal villi, which reduces the mucosal enzymes for nutrient digestion and the mucosal surface for absorption. Celiac sprue occurs in about 1–6/10,000 individuals in the western world. The highest incidence is seen in western Ireland, where the prevalence is as high as 3/1000 individuals. The disease may occur at any age but is more common during the first few years and the third to fifth decades of life.

In patients with celiac sprue, the water-soluble gluten (present in cereal grains such as wheat, barley, rye, and oats) or its breakdown product interacts with the intestinal mucosa and causes the characteristic lesion. Precisely how the binding of gluten to the intestinal mucosa causes mucosal injury is unclear. It has been hypothesized that patients prone to celiac sprue may have a brush border peptidase deficiency and that the consequent incomplete digestion of gluten results in the production of a toxic substance, which aggravates the intestinal mucosa. This hypothesis is probably incorrect, however, because the intestinal brush border peptidases revert back to normal after healing of damaged intestinal mucosa. Another hypothesis is that immune mechanisms are involved. This is supported by the facts that the number and activity of plasma cells and lymphocytes increase during the active phase of celiac sprue and that usually antigluten antibodies are present. Furthermore, it has been demonstrated that the small intestine makes a lymphokinelike substance which inhibits the infiltration of leukocytes into the lamina propria of the intestinal mucosa when exposed to gluten. Unfortunately, it is not clear whether these immunologic manifestations are primary or secondary phenomena of the disease.

Withdrawal of diets containing gluten is a standard treatment for patients with celiac sprue. Occasionally, intestinal absorptive function and intestinal mucosal morphology of patients with celiac sprue are improved with glucocorticoid therapy. Presumably, glucocorticoid treatment is beneficial because of the immunosuppressive and antiinflammatory actions of these homones.

secretion of peptidases; and poor development of intracellular protein degradation by lysosomal proteases.

Absorption of immunoglobulins (predominantly IgG) plays an important role in the transmission of **passive immunity** from the mother's milk to the newborn in a number of animal species (e.g., ruminants and rodents). In humans the absorption of intact immunoglobulins does not appear to be an important mode of transmission of antibodies, for two reasons. First, passive immunity in humans is derived almost entirely from the intrauterine transport of maternal antibodies. Second, the human **colostrum**—the thin, yellowish, milky fluid secreted by the mammary gland a few days before or after parturition—contains mainly IgA, which is poorly absorbed by the small intestine.

The ability to absorb intact proteins is rapidly lost as the gut matures; this is called **closure.** Colostrum contains a factor that promotes the "closure" of the small intestine.

■ Amino Acids, Peptides, and Proteins Travel in the Portal Blood

After di- and tripeptides are taken up by the enterocytes, they are further broken down to amino acids by the peptidases in the cytoplasm. The amino acids are transported in the portal blood. The small amount of protein that is taken up by the adult intestine is largely degraded by lysosomal proteases, although some proteins escape degradation.

■ Defects in Digestion and Transport Can Impair Protein Absorption

Pancreatic deficiency has the potential to affect protein digestion but usually does not except in severe cases. Pancreatic deficiency seems to affect lipid digestion more than protein digestion. There are a number of extremely rare genetic disorders of the amino acids carriers. In **Hartnup's disease,** the membrane carrier for neutral amino acids (e.g., tryptophan) is defective. **Cystinuria** involves the carrier for basic amino acids (e.g., lysine and arginine) and the sulfur-containing amino acids (e.g., cystine). Cystinuria was once thought to involve only the kidneys, because of the excretion of amino acids such as cystine in urine, but the small intestine is involved as well. Because the peptide transport system remains unaffected, disorders of some amino acid transporters can be treated with supplemental dipeptides containing these amino acids. However, this alone is not effective if the kidney transporter is also involved, as in cystinuria.

The Absorption of Vitamins

A vitamin is an organic substance needed in small quantities for normal metabolic function and growth and maintenance of the body.

■ The Fat-Soluble Vitamins Include A, D, E, and K

The only feature shared by the fat-soluble vitamins is their lipid solubility; they are otherwise structurally very different. Most are absorbed passively (Table 29–4).

Vitamin A ■ The principal form of vitamin A is **retinol;** the aldehyde (retinal) and the acid (retinoic acid) are also active forms of vitamin A. Retinol can be derived directly from animal sources or through conversion from β-carotene (found abundantly in carrots) in the small intestine. One molecule of β-carotene yields two molecules of vitamin A, but the process is only about 50% efficient. Vitamin A is rendered water-soluble by micellar solubilization and is absorbed by the small intestine passively. It is converted in the small intestinal mucosa to an ester **(retinyl ester),** which is incorporated in the chylomicrons and is taken up by the liver. Vitamin A is stored in the liver and released to the circulation bound to **retinol binding protein** only when needed. Vitamin A is important in the production and regeneration of the visual purple of the retina and in the normal growth of the skin. Vitamin A-deficient individuals develop night blindness and skin lesions.

Vitamin D ■ Vitamin D is a group of fat-soluble compounds collectively known as the calciferols. Vitamin D_3 (also called cholecalciferol or activated dehydrocholesterol) in the human body is derived from two main sources: the skin, which contains a rich source of 7-dehydrocholesterol that is rapidly converted to cholecalciferol when exposed to ultraviolet light, and dietary vitamin D_3. Like vitamin A, vitamin D_3 is absorbed by the small intestine passively and is dependent on its micellar solubilization by bile salts. It is transported by the small intestine in free form, predominantly in association with chylomicrons. During the metabolism of chylomicrons, vitamin D_3 is transferred to a binding protein in plasma called the **vitamin D binding protein.** Unlike vitamin A, vitamin D is not stored in the liver but is distributed between the various organs, depending on their lipid content. In the liver, vitamin D_3 is converted to the biologically active metabolite **25-hydroxycholecalciferol,** which is five times as potent as vitamin D_3. 25-hydroxycholecalciferol is the predominant form of circulating vitamin D, but it is converted in the kidney to form 1,25-dihydroxycholecalciferol, which is 10 times more potent than vitamin D_3 and believed to be involved in enhancing calcium and phosphate absorption by the small intestine and mobilizing calcium and phosphate from bones. Vitamin D is essential for normal development and growth and the formation of bones and teeth. Vitamin D deficiency can result in rickets, a disorder of normal bone ossification mani-

TABLE 29–4 ■ Fat-Soluble Vitamins

Vitamin	RDA	Sources	Site and Mode of Absorption	Role
A	1000 RE	Liver, kidney, butter, whole milk, cheese, β-carotene (yields 2 molecules of retinol)	Small intestine; passive	Vision, bone development, epithelial development, reproduction
D	200 IU	Liver, butter, cream, vitamin D-fortified milk, conversion from 7-dehydrocholesterol by ultraviolet light	Small intestine; passive	Growth and development, formation of bones and teeth, stimulation of intestinal Ca^{2+} and phosphate absorption, mobilization of Ca^{2+} ions from bones
E	10 mg	Wheat germ, green plants, egg yolk, milk, butter, meat	Small intestine; passive	Antioxidant
K	70–100 μg	Green vegetables and intestinal gut flora	Phylloquinones from green vegetables are absorbed actively from the proximal small intestine; menaquinones from gut flora are absorbed passively	Blood clotting

RE, retinol equivalent; IU, international unit; 1 IU = 0.025 μg

fested by distorted bone movements during muscular action.

Vitamin E ■ The major dietary vitamin E is **α-tocopherol.** Vegetable oils are rich in vitamin E. It is absorbed by the small intestine by passive diffusion, dependent on micellar solubilization by bile salts. Vitamin E is transported in the free form by the small intestinal mucosa in association with chylomicrons. Unlike vitamins A and D, vitamin E is transported in the circulation associated with lipoproteins and erythrocytes. It is a potent antioxidant and therefore prevents lipid peroxidation. Tocopherol deficiency is associated with increased red cell susceptibility to lipid peroxidation, which may explain why the red cells are more fragile in vitamin E-deficient individuals than in normal individuals.

Vitamin K ■ Vitamin K can be derived from green vegetables in the diet or the gut flora. The vitamin K derived from green vegetables is in **phylloquinones,** and that derived from bacteria in the small intestine is in **menaquinones.** The phylloquinones are taken up by the small intestine by an energy-dependent process from the proximal small intestine. In contrast, the menaquinones are absorbed from the small intestine passively, dependent only on the micellar solubilization of these compounds by bile salts. Vitamin K is transported by the small intestine associated with chylomicrons. It is rapidly taken up by the liver and secreted together with the VLDLs. No carrier protein for vitamin K has been identified. It is required in the synthesis of various clotting factors by the liver. Vitamin K deficiency is associated with bleeding disorders.

■ The Water-Soluble Vitamins Include C, B$_1$, B$_2$, B$_6$, B$_{12}$, Niacin, Biotin, and Folic Acid

Most of the water-soluble vitamins are absorbed by the small intestine by both passive and active processes (Table 29–5).

Vitamin C ■ The major source of vitamin C **(ascorbic acid)** is green vegetables and fruits. It is easily destroyed by many food-processing procedures, especially being heated. It plays an important role in many oxidative processes by acting as a coenzyme or cofactor. It is absorbed mainly by active transport in the ileum. Vitamin C deficiency is associated with scurvy, a disorder manifested by weakness, fatigue, anemia, and bleeding of the gums.

Vitamin B$_1$ ■ Vitamin B$_1$ **(thiamine)** plays an important role in carbohydrate metabolism. Thiamine deficiency results in beriberi, characterized by anorexia and disorders of the nervous system and heart. Thiamine is absorbed by the jejunum by two different mechanisms. At low luminal concentration, it is absorbed by an active, carrier-mediated process. At high luminal concentration, it is absorbed by passive diffusion.

TABLE 29–5 ■ Water-Soluble Vitamins

Vitamin	RDA	Sources	Site and Mode of Absorption	Role
C	60 mg/day	Fruits, vegetables, organ (liver and kidney) meat	Active transport by the ileum	Coenzyme or cofactor in many oxidative processes
B$_1$ (thiamine)	1 mg/day	Yeast, liver, cereal grains	At low luminal concentrations, by active, carrier-mediated process; at high luminal concentrations, by passive diffusion	Carbohydrate metabolism
B$_2$ (riboflavin)	1.7 mg/day	Dairy products	Active transport in the proximal small intestine	Metabolism
Niacin	19 mg/day	Brewers' yeast, meat	At low luminal concentrations, by a Na$^+$-dependent, carrier-mediated, facilitated transport	A component of the coenzymes NAD(H) and NADP(H); metabolism of carbohydrates, fats, and proteins; synthesis of fatty acid and steroid
B$_6$ (pyridoxine)	2.2 mg/day	Brewer's yeast, wheat germ, meat, whole grain cereals, dairy products	By passive diffusion in small intestine	Amino acid and carbohydrate metabolism
Biotin	200 μg/day	Brewer's yeast, milk, liver, egg yolk	At low luminal concentrations, by active transport; at high luminal concentrations, by simple diffusion	Coenzyme for carboxylase, transcarboxylase, and decarboxylase enzymes; metabolism of lipids, glucose, and amino acids
Folic acid	0.5 mg/day	Liver, beans, dark green leafy vegetables	By Na$^+$-dependent facilitated transport	Nucleic acid biosynthesis, maturation of red blood cells, and promotion of growth
B$_{12}$	3 μg/day	Liver, kidney, dairy products, eggs, fish	Absorbed in terminal ileum By active transport involving binding to intrinsic factor	Normal cell division; bone marrow and intestinal mucosa most affected in deficiency state, characterized by pernicious anemia

Vitamin B₂ ■ Vitamin B₂ **(riboflavin)** is a component of the two groups of flavoproteins, flavin adenine dinucleotide (FAD) and flavin mononucleotide (FMN). Riboflavin plays an important role in metabolism. Riboflavin deficiency is associated with anorexia, impaired growth, impaired use of food, and nervous disorder. Riboflavin is absorbed by a specific, saturable, active transport system located in the proximal small intestine.

Niacin ■ Niacin plays an important role as a component of the coenzymes NAD(H) and NADP(H), which are involved in a wide variety of oxidation-reduction reactions involving hydrogen transfer. These reactions are involved in the release of energy from carbohydrates, fats, and proteins. NADPH also serves as a source of reducing equivalents essential for a number of biosynthetic reactions, such as the synthesis of fatty acids and steroids. Niacin deficiency is characterized by many clinical symptoms, including anorexia, indigestion, muscle weakness, and skin eruption. Severe deficiency leads to **pellagra,** characterized by dermatitis, dementia (a mental disorder characterized by a loss in memory, abstract thinking, and judgment and a change in personality), and diarrhea (the "3 Ds of pellagra"). At low concentrations, niacin is absorbed by the small intestine by a Na⁺-dependent, carrier-mediated facilitated transport system. At high concentrations, it is absorbed by passive diffusion. Niacin has been used to treat hypercholesterolemia in humans. It decreases plasma total cholesterol and LDL cholesterol, yet it increases plasma HDL cholesterol. It has been extremely useful in the treatment of hypercholesterolemia and the prevention of coronary disease.

Vitamin B₆ ■ Vitamin B₆ **(pyridoxine)** is involved in amino acid, protein, and carbohydrate metabolism. Vitamin B₆ is absorbed throughout the small intestine by simple diffusion. A deficiency of this vitamin is often associated with disorders of the central nervous system and anemia.

Biotin ■ Biotin acts as a coenzyme for the carboxylase, transcarboxylase, and decarboxylase enzymes, which play an important role in the metabolism of lipids, glucose, and amino acids. Biotin is so common in food that deficiency is rarely observed. At low luminal concentrations, biotin is absorbed by the small intestine by a Na⁺-dependent active transport system. At high concentrations, it is absorbed by simple diffusion.

Folic Acid ■ Folic acid is usually found in the diet as polyglutamyl conjugates (pteroylpolyglutamates). It is required for the formation of nucleic acids, maturation of red blood cells, and promotion of growth. Folic acid deficiency causes a fall in serum and red cell folic acid content and, in its most severe form, development of megaloblastic anemia, dermatologic lesions, and poor growth. An enzyme on the brush border degrades pteroylpolyglutamates to yield a monoglutamylfolate, which is taken up by the enterocyte by facilitated transport. Inside the enterocytes, the monoglutamyfo-

late is released directly into the bloodstream or converted to 5-methyltetrahydrofolate before exiting the cell. A folate-binding protein binds the free and methylated forms of folic acid.

Vitamin B₁₂ ■ The discovery of vitamin B₁₂ followed from the observation that patients with pernicious anemia who ate large quantities of raw liver recovered from the disease. Subsequent analysis of liver components isolated the cobalt-containing complex called **cobalamin,** which plays an important role in the production of red blood cells. Vitamin B₁₂ is present in food, bound to proteins. This linkage is broken and cobalamin is liberated during cooking or by coming in contact with acid or protease. A glycoprotein secreted by the parietal cells in the stomach, called the **intrinsic factor,** binds strongly with cobalamin to form a complex that is then absorbed in the terminal ileum through a receptor-mediated process (Fig. 29–11). The cobalamin is transported in the portal blood bound to the protein **transcobalamin.** Individuals who lack the intrinsic factor fail to absorb cobalamin and develop pernicious anemia.

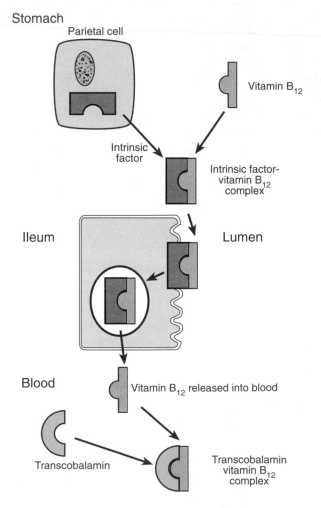

Figure 29–11 ■ Intestinal absorption of vitamin B₁₂.

The Absorption of Electrolytes and Minerals

■ Intestinal Absorption of Electrolytes and Minerals Involves Both Passive and Active Processes

Sodium ■ The gastrointestinal system is well equipped to handle the large amount of Na^+ ions entering the gastrointestinal lumen daily. On average, about 25-35 g of Na^+ ions enter the gastrointestinal tract daily, about 5-8 g of which are derived from the diet and the rest from salivary, gastric, bile, pancreatic, and small intestinal secretions. The gastrointestinal tract is extremely efficient in conserving this Na^+; only 0.5% of intestinal Na^+ is lost in the feces. The jejunum absorbs more than half of the total Na^+ and the ileum and colon absorb the remainder. The small intestine absorbs the bulk of the Na^+ presented to it, but the colon is most efficient in conserving Na^+.

Sodium is absorbed by a number of different mechanisms, operating at varying degrees in different parts of the gastrointestinal tract. When a meal that is hypotonic to plasma is ingested, considerable absorption of water from the lumen to the blood takes place, predominantly through tight junctions and intercellular spaces between the enterocytes. This results in the absorption of small solutes such as Na^+ and Cl^- ions. This mode of absorption, called **solvent drag**, is responsible for a significant amount of the Na^+ absorption by the duodenum and jejunum, but probably plays a minor role in Na^+ absorption by the ileum and colon, because more distal regions of the intestine are lined by a "tight" epithelium (Chap. 2).

In the jejunum, Na^+ ions are actively pumped out of the basolateral surface of the enterocytes by a Na^+/K^+ ATPase (Fig. 29-12). This creates a low intracellular Na^+ concentration and the luminal Na^+ ions enter enterocytes down the electrochemical gradient. This provides the energy for the extrusion of H^+ into the lumen (via a Na^+-H^+ exchanger). The H^+ ions then react with HCO_3^- ions in bile and pancreatic secretions in the intestinal lumen to form carbonic acid. Carbonic acid dissociates to form CO_2 and water. The CO_2 readily diffuses across the small intestine into the blood. Another mode of Na^+ uptake is via a carrier located in the brush border membrane of the enterocyte that transports Na^+ together with a monosaccharide (e.g., glucose) or an amino acid molecule (a symport type of transport).

In the ileum, the presence of a Na^+-K^+ ATPase at the basolateral membrane also creates a low intracellular Na^+ concentration, and luminal Na^+ ions enter enterocytes down the electrochemical gradient. Sodium absorption by the Na^+-coupled transport is not as great as in the jejunum because most of the monosaccharides and amino acids have already been absorbed by the small intestine (Fig. 29-13). Sodium chloride is transported via two exchangers located at the brush

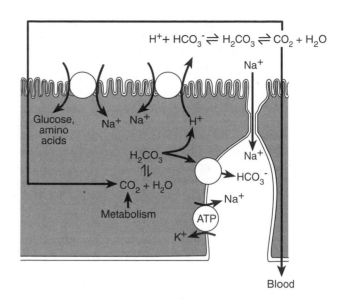

Figure 29–12 ■ Absorption of Na^+ by the jejunum.

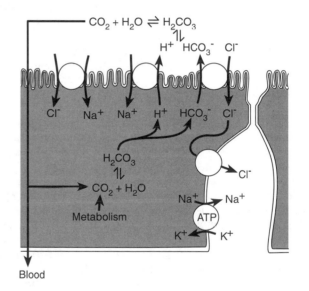

Figure 29–13 ■ Absorption of Na^+ by the ileum.

border membrane. One is a Cl^--HCO_3^- exchanger and the other is a Na^+-H^+ exchanger. The downhill movement of Na^+ into the cell provides the energy required for the uphill movement of the H^+ ions from the cell to the lumen. Similarly, the downhill movement of HCO_3^- ions out of the cell provides the energy for the uphill entry of Cl^- ions into the enterocytes. The Cl^- ions then leave the cell through facilitated transport. This mode of Na^+ uptake is called the Na^+-H^+, Cl^--HCO_3^- countertransport.

In the colon, the mechanisms for Na^+ absorption are mostly similar to those described for the ileum. There is no sugar- or amino acid-coupled Na^+ transport because most sugar and amino acids have already been absorbed. Sodium is also absorbed here via Na⁺-selective ion channels in the apical cell membrane (electrogenic Na⁺ absorption).

Potassium ■ The average daily intake of potassium is about 4 g; absorption takes place throughout the intestine by passive diffusion through the tight junctions and lateral spaces of the enterocytes. The driving force for K⁺ absorption is the difference between luminal and blood K⁺ concentration. The absorption of water results in an increase in luminal K⁺ concentration, resulting in K⁺ absorption by the intestine. In the colon, K⁺ can be absorbed or secreted, depending on the luminal K⁺ concentration. With diarrhea, considerable K⁺ can be lost. Prolonged diarrhea can be life-threatening because of the dramatic fall in extracellular K⁺ concentration, causing complications such as cardiac arrhythmias.

Chloride ■ Of the 2–3 g of Cl⁻ ions entering the gastrointestinal tract each day, only 0.1–0.2 g/day is excreted. Chloride absorption by the gastrointestinal tract involves both passive and active processes. In the jejunum, an electrochemical gradient generated by Na⁺/K⁺ ATPase provides the driving force for the absorption of Na⁺ ions by the enterocytes. This transepithelial Na⁺ movement generates a potential difference across the small intestinal mucosa, with the serosal side more positive than the lumen. Chloride ions follow this potential difference and enter the bloodstream via the tight junctions and lateral intercellular spaces. In the ileum and colon, Cl⁻ ions are taken up actively by the enterocyte via Cl⁻-HCO₃⁻ exchange, as discussed above. This absorption of Cl⁻ is inhibited by the presence of other halides.

Bicarbonate ■ Bicarbonate ions are absorbed in the jejunum together with Na⁺. In humans the absorption of HCO₃⁻ ions by the jejunum stimulates the absorption of Na⁺ and water (Fig. 29–12). Through the Na⁺-H⁺ exchanger, H⁺ ions are released into the intestinal lumen, where H⁺ and HCO₃⁻ react to form carbonic acid; this dissociates to form CO₂ and water. The CO₂ then diffuses into the enterocytes, where it reacts with water to form carbonic acid (catalyzed by carbonic anhydrase), which then dissociates to HCO₃⁻ and H⁺ ions. The HCO₃⁻ ions then diffuse into the blood.

In the ileum and colon, HCO₃⁻ ions are actively secreted into the lumen in exchange for Cl⁻ ions. This secretion of HCO₃⁻ ions is important in buffering the decrease in pH resulting from the short-chain fatty acids produced by bacteria in the distal ileum and colon.

Calcium ■ The amount of calcium entering the gastrointestinal tract is about 1 g/day, approximately half of which is derived from the diet. The majority of dietary calcium is derived from meat and dairy products. Of the calcium presented to the gastrointestinal tract, about 40% is absorbed. A number of factors affect Ca²⁺ absorption. For instance, the presence of

fatty acid can retard Ca²⁺ absorption by the formation of calcium soap. In contrast, bile salt molecules form complexes with Ca²⁺ ions, which facilitates their absorption. Calcium absorption takes place predominantly in the duodenum and jejunum, is mainly active, and involves two steps. Calcium ions are first taken into the enterocyte via the Ca²⁺-binding protein (CaBP) located in the brush border membrane. It is not certain whether this uptake is accomplished by active or facilitated transport. The level of CaBP in the brush border is regulated by 1,25-dihydroxy vitamin D₃ (1,25-dihydroxycholecalciferol). Once inside the cell, the Ca²⁺ ions are sequestered in the endoplasmic reticulum and Golgi membranes by binding to the CaBP in these subcellular organelles. The exit of Ca²⁺ requires energy and is mediated by the Ca²⁺ ATPase located in the basolateral membrane.

Calcium absorption by the small intestine is regulated by the circulating plasma Ca²⁺ concentration. Lowering of the concentration stimulates the release of parathyroid hormone, which stimulates, in the kidney, the conversion of vitamin D to its active metabolite, 1,25-dihydroxy vitamin D₃, which in turn stimulates the synthesis of CaBP and the Ca²⁺ ATPase by the enterocytes (Fig. 29–14). Since protein synthesis is

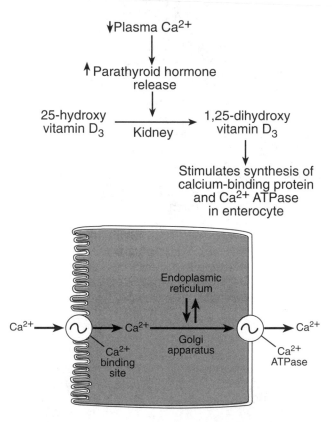

Figure 29–14 ■ Mechanism of absorption of Ca²⁺ by the enterocyte. Parathyroid hormone stimulates the conversion of vitamin D₃ in the kidney to its active metabolite 1,25-dihydroxycholecalciferol (1,25-diOH-vitamin D₃), which stimulates the synthesis of both Ca²⁺ binding protein and the Ca²⁺ ATPase.

involved in the stimulation of Ca^{2+} uptake by parathyroid hormone, there is usually a lapse of a few hours between the release of parathyroid hormone and the increase in Ca^{2+} absorption by the enterocytes.

Magnesium ■ Humans ingest about 0.4–0.5 g/day of magnesium. The absorption of magnesium seems to take place along the entire small intestine, and the mechanism involved seems to be passive.

Zinc ■ The average daily zinc intake by Americans is 10–15 mg, about half of which is absorbed, apparently primarily in the ileum. A carrier located in the brush border membrane actively transports zinc from the lumen into the cell, where it can be stored or transferred into the bloodstream. Zinc plays an important role in a number of metabolic activities. For example, a group of metalloenzymes (e.g., alkaline phosphatase, carbonic anhydrase, and lactic dehydrogenase) require zinc to function.

Iron ■ Iron plays an important role not only as a component of heme but as a participant in many enzymatic reactions. About 12–15 mg/day of iron enters the gastrointestinal tract, where it is absorbed mainly by the duodenum and upper jejunum. The mechanism of iron absorption is summarized in Figure 29–15. There are two forms of dietary iron: heme and nonheme. The heme iron is absorbed intact by the enterocytes. Nonheme iron absorption depends on both pH and concentration. Ferric (Fe^{3+}) salts are not soluble at pH 7; whereas ferrous (Fe^{2+}) salts are. Consequently, in the duodenum and upper jejunum, unless Fe^{3+} ion is chelated, it forms a precipitate. A number of compounds, such as tannic acid in tea and phytates in vegetables, form insoluble complexes with iron, which prevents them from being absorbed. Iron is absorbed by an active process via a carrier located in the brush border membrane.

Once inside the cell, heme iron is released by the action of xanthine oxidase and mixed with the intracellular free iron pool. Iron is either stored in the cytoplasm of the enterocytes bound to the storage protein **apoferritin** to form **ferritin** or is transported across the cell bound to transport proteins, which carry the iron across the cytoplasm and release it into the intercellular space. Iron is bound and transported in the blood by transferrin, a β-globulin synthesized by the liver.

Iron absorption is closely regulated by iron storage in the enterocytes and iron concentration in the plasma. Enterocytes are continuously shed into the lumen, and the ferritin contained within is also lost. Normally, iron in the enterocyte is derived from the lumen and the blood (Fig. 29–16). The amount of iron absorbed is regulated by the amount stored in the enterocytes. In iron deficiency, the circulating plasma iron concentration is low. This stimulates absorption of iron from the lumen and the transport of iron into the blood. Also in deficiency less iron is stored as ferritin in the enterocytes, so the loss of iron through this means is markedly reduced. In iron-loaded patients, there is less absorption of iron because of the large amount of mucosal iron storage, which markedly increases iron loss as a result of shedding of the enterocytes. Furthermore, because of the high level of circulating plasma iron, the transfer of iron from the enterocytes to the blood is markedly reduced. Through a combination of various mechanisms, body iron homeostasis is maintained.

The Absorption of Water

In human adults, the daily intake of water is about 2 liters. As shown in Table 29–6, secretions from the salivary glands, pancreas, liver, and gastrointestinal tract make up the majority of the fluid entering the gastrointestinal tract (about 7 liters). Despite this

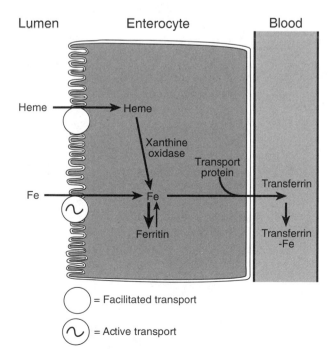

Figure 29–15 ■ Intestinal absorption of iron.

TABLE 29–6 ■ Water Intake, Absorption, and Excretion by the Gatrointestinal Tract

Water added to gastrointestinal tract	
Food and beverages	2000 ml
Salivary secretion	1000 ml
Biliary secretion	1000 ml
Gastric secretion	2000 ml
Pancreatic secretion	1000 ml
Intestinal mucosal secretion	2000 ml
Water absorbed or lost in feces	
Water absorbed	
Duodenum and jejunum	4000 ml
Ileum	3500 ml
Colon	1400 ml
Water loss in feces	100 ml

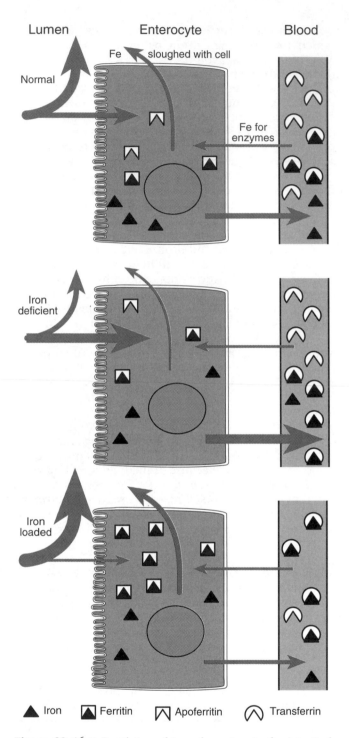

Lumen · Enterocyte · Blood

Fe · sloughed with cell

Normal

Fe for enzymes

Iron deficient

Iron loaded

▲ Iron · ◨ Ferritin · ⋀ Apoferritin · ⋒ Transferrin

Figure 29–16 ■ Regulation of iron absorption in the intestinal mucosa. In normal subjects the amount of iron that enters the enterocytes is regulated by the amount of iron in the cells and circulating in plasma. In iron-deficient subjects, little iron is incorporated into enterocytes and less is circulating in plasma, therefore absorption is increased and excretion is decreased. In iron-loaded subjects, the mucosal cells and transferrin are saturated. This limits absorption and increases excretion. (Adapted from Krause MV, Mahan LK (eds): *Food, Nutrition, and Diet Therapy.* Philadelphia, WB Saunders, 1984.)

rather large volume of fluid, only 100 ml of fluid is lost in the feces. Thus the gastrointestinal tract is extremely efficient in absorbing the water presented. Water absorption by the gastrointestinal tract is passive. The rate of absorption is dependent on both the region of the intestinal tract and the luminal osmolality of the chyme. The duodenum, jejunum, and ileum absorb the bulk of the water that enters the gastrointestinal tract daily. The colon normally absorbs about 1.4 liters and excretes about 100 ml. It is capable of absorbing considerably more water (about 4.5 liters), however, and watery diarrhea occurs only if this capacity is exceeded.

Since water absorption is determined by the difference in osmolality of the lumen and blood, water can move both ways in the intestinal tract (i.e., secretion and absorption). The osmolality of blood is about 300 mOsm. The ingestion of a hypertonic meal (e.g., 600 mOsm) leads to a net water movement from blood to lumen, but as the various nutrients and electrolytes are absorbed by the small intestine the luminal osmolality falls, resulting in the net water movement from lumen to blood. The water of a hypertonic meal is therefore absorbed mainly in the ileum and colon. In contrast, if a hypotonic meal is ingested (e.g., 200 mOsm), net water movement is immediately from the lumen to the blood, resulting in absorption of the majority of the water in the duodenum and jejunum.

■ ■ ■ ■ ■ ■ ■ ■ ■ ■ ■ ■ ■ ■ **REVIEW EXERCISES** ■ ■ ■ ■ ■ ■ ■ ■ ■ ■ ■ ■ ■

Identify Each with a Term

1. The concentration at or above which bile salt aggregates to form micelles

2. Fatty acids containing 8–12 carbons

3. The process whereby the ability to absorb intact proteins in the newborn is rapidly lost as the small intestine matures

4. An active metabolite of vitamin D that enhances intestinal absorption of calcium

Define Each Term

5. Dietary fibers

6. Amylose

7. Amylopectin

8. Triglyceride

9. Essential fatty acids

10. Biotin

Choose the Correct Answer

11. The major disaccharide present in milk is:
    a. Maltose
    b. Sucrose
    c. Lactose
    d. Amylose

12. Lactase hydrolyzes lactose to form:
    a. Glucose
    b. Glucose and galactose
    c. Glucose and fructose
    d. Galactose and fructose

13. Maltase hydrolyzes maltose to form:
    a. Glucose
    b. Glucose and galactose
    c. Glucose and fructose
    d. Galactose and fructose

14. Which of the following sugars is taken up by the small intestinal epithelial cell by facilitated diffusion?
    a. Glucose
    b. Galactose
    c. Fructose
    d. Xylose

15. Which membrane disaccharidase deficiency is most common?
    a. Maltase
    b. Lactase
    c. Isomaltase
    d. Sucrase

16. Lithocholic acid is derived from the dehydroxylation of:
    a. Cholic acid
    b. Chenodeoxycholic acid
    c. Deoxycholic acid
    d. Ursodeoxycholic acid

17. The life span of small intestinal epithelial cells is about:
    a. 12–24 hours
    b. 1 day
    c. 3 days
    d. 10 days

18. Pancreatic lipase hydrolyzes triglyceride to form mostly:
    a. Lysophosphatidylcholine and fatty acids
    b. Glycerol and fatty acids
    c. Diglyceride and fatty acid
    d. Monoglyceride and fatty acids

19. Dietary lipid absorbed by the small intestine is transported in the lymph mainly as:
    a. Very-low-density lipoproteins
    b. Free fatty acids bound to albumin
    c. Chylomicrons
    d. Low-density lipoproteins

20. Prior to absorption, protein must first be digested to form:
    a. Free amino acids
    b. Free amino acids, dipeptides, and tripeptides
    c. Pentapeptides
    d. Oligopeptides

21. Dietary protein is transported in portal blood mainly as:
    a. Free amino acids
    b. Di- and tripeptides
    c. Free amino acids and dipeptides
    d. Free amino acids and tripeptides

22. Which of the following is not fat-soluble?
    a. Vitamin A
    b. Vitamin D
    c. Vitamin K
    d. Vitamin B_1

23. Which of the following is important for the absorption of calcium by the small intestine?
    a. Vitamin E
    b. Vitamin D
    c. Vitamin A
    d. Vitamin K

24. Which of the following is transported in chylomicrons as an ester?
    a. Vitamin E
    b. Vitamin D
    c. Vitamin A
    d. Vitamin K

25. Potassium is absorbed in the jejunum by:
    a. Active transport
    b. Facilitated transport
    c. Passive absorption
    d. Active and passive transport

26. Why does it takes a few hours between the release of parathyroid hormone and an increase in calcium absorption by the small intestine?
    a. It takes time for parathyroid hormone to stimulate vitamin D absorption.
    b. It takes time for vitamin D to be converted to the 1,25-dihydroxycholecalciferol.
    c. It takes time for 1,25-dihydroxycholecalciferol to be taken up by the enterocytes.
    d. It takes time because the stimulation of intestinal calcium absorption by parathyroid hormone involves protein synthesis in the enterocytes.

Briefly Answer

27. Why are bile salts important for the uptake of lipid by the small intestinal epithelial cells?

28. Describe the digestion and absorption of proteins by the human small intestine.

29. Describe the absorption of dietary vitamin B_{12} by the gastrointestinal tract.

30. Describe the mechanism and regulation of iron absorption by the gastrointestinal tract.

■ ■ ■ ■ ■ ■ ■ ■ ■ ■ **ANSWERS** ■ ■ ■ ■ ■ ■ ■ ■ ■ ■ ■

1. Critical micellar concentration

2. Medium-chain fatty acids

3. Closure

4. 1,25-dihydroxycholecalciferol (1,25-dihydroxy vitamin D_3)

5. Polysaccharides that are usually poorly digested by the digestive enzymes in the small intestine; they serve an important physiologic function by providing the "bulk" that facilitates intestinal motility

6. A carbohydrate composed of a straight chain of glucose units

7. A carbohydrate composed of branched glucose units

8. The major (>90%) dietary lipid; it consists of a glycerol backbone esterified in all three positions with fatty acids, usually long-chain fatty acids

9. Fatty acids that are not synthesized by the body and that are necessary for essential body functions.

10. A ubiquitous member of the vitamin B complex required by the body; it is a coenzyme for the carboxylase, transcarboxylase, and decarboxylase enzymes

11. c

12. b

13. a

14. c

15. b

16. b

17. c

18. d

19. c

20. b

21. a

22. d

23. b

24. c

25. c

26. d

27. The unstirred water layer—a layer of poorly stirred fluid—coats the surface of the intestinal villi and prohibits the absorption lipid digestion products, which are poorly soluble in water. Before they can be absorbed, lipid digestion products must be rendered water-soluble. Bile salt plays an important role in the micellar solubilization of lipid, which greatly increases the concentration of lipid in the unstirred water layer.

28. Proteins are digested in the stomach and small intestine into free amino acids and di- and tripeptides. The amino acids are taken up by the enterocytes, together with Na^+ ions, via carrier-mediated processes. The di- and tripeptides are taken up by the enterocytes more efficiently. They also use Na^+-dependent transporters, but different from those used by amino acids.

29. The parietal cells in the stomach secrete the protein intrinsic factor, which binds with vitamin B_{12} to form a complex that is absorbed in the terminal ileum through a receptor-mediated process. The vitamin B_{12} is then transported in the portal blood bound to the protein transcobalamin. Individuals who lack intrinsic factor fail to absorb vitamin B_{12} and develop pernicious anemia.

30. There are two forms of dietary iron, heme and nonheme. The heme iron is absorbed intact by the enterocytes. In contrast, nonheme iron absorption is dependent on pH and concentration. Ferric ion is not absorbed but ferrous ion is. Once inside the cell, the heme iron is released by the action of xanthine oxidase. Iron is either stored in the cytoplasm of the enterocytes as ferritin or released as iron into the blood.

Suggested Reading

Alpers D: Digestion and absorption of carbohydrates and proteins, in Johnson LR (ed): *Physiology of the Gastrointestinal Tract.* 2nd ed. New York, Raven, 1987, pp 1469-1487.

Davenport HW: *Physiology of the Digestive Tract.* 5th ed. Chicago, Year Book, 1982.

Granger DN, Barrowman JA, Kvietys PR: *Clinical Gastrointestinal Physiology.* Philadelphia, WB Saunders, 1985.

Hobsley M: *Disorders of the Digestive Tract.* Baltimore, University Park Press, 1982.

Johnson LR: *Gastrointestinal Physiology.* 3rd ed. St. Louis, CV Mosby, 1985.

Rose RC: Intestinal absorption of water-soluble vitamins, in Johnson LR (ed): *Physiology of the Gastrointestinal Tract.* 2nd ed. New York, Raven, 1987, pp 1581-1596.

Sanford PA: *Disgestive System Physiology.* Baltimore, University Park Press, 1982.

Sernka T, Jacobson E: *Gastrointestinal Physiology.* 2nd ed. Baltimore, Williams & Wilkins, 1983.

Tso P, Weidman SW: Absorption and metabolism of lipids in humans, in Horisberger M, Bracco U (eds): *Lipids in Modern Nutrition.* New York, Raven, 1987, pp 1-15.

The Physiology of the Liver

OBJECTIVES

After studying this chapter, the student should be able to:
1. Discuss the arrangement of hepatocytes, endothelial cells, Kupffer cells, and fat-storing perisinusoidal Ito cells along the liver sinusoids
2. Describe the phase 1 and phase 2 reactions of drug metabolism by the liver
3. Discuss the use of glucose, fructose, and galactose by hepatocytes; describe the function of the glycolysis, gluconeogenesis, and pentose phosphate pathways
4. Describe the mechanism and regulation of the formation of glycogen from glucose in the liver
5. Describe the synthesis and metabolism of fatty acids by the liver
6. Describe the hepatic synthesis and function of the major plasma proteins
7. Discuss the role of liver in the synthesis and interconversion of amino acids
8. Describe the liver as a storage organ for fat-soluble vitamins and iron
9. Describe the endocrine functions of the liver

The liver is the largest organ in the body, constituting about 2.5% of adult body weight. It receives 25% of the cardiac output via the hepatic portal vein and hepatic artery. The hepatic portal vein carries the absorbed nutrients from the gastrointestinal tract to the liver, which is involved in the uptake, storage, and distribution of both nutrients and vitamins. The liver plays an important role in the maintenance of blood sugar level. It also regulates the circulating blood lipids by the amount of **very-low-density lipoproteins** (VLDLs) it secretes. Many of the circulating plasma proteins are synthesized by the liver. In addition, the liver takes up numerous toxic compounds and drugs from the portal circulation. It is well equipped to deal with the metabolism of drugs and toxic substances. The liver also serves as an excretory organ for bile pigments, cholesterol, and drugs. Finally, it performs important endocrine functions.

Anatomy of the Liver

■ The Arrangement of Hepatocytes along Liver Sinusoids Aids Rapid Exchange of Molecules

The hepatocytes are highly specialized cells (Fig. 30-1). The bile canaliculus is usually lined by two hepatocytes and is separated from the pericellular space by tight junctions, which are impermeable and thus prevent the mixing of contents between the bile canaliculus and the pericellular space. The bile formed at the bile canaliculus drains into a series of ductules and ducts and eventually joins the pancreatic duct and enters the duodenum via the sphincter of Oddi. The biliary tree and bile secretion are described in detail in Chapter 28.

The pericellular space, the space between two hepatocytes, is continuous with the perisinusoidal space. The perisinusoidal space, also known as the **space of Disse,** is separated from the sinusoidal space by a layer of **sinusoidal cells.** The hepatocytes possess numerous fingerlike projections that extend into the perisinusoidal space, thereby greatly increasing the surface area over which the hepatocytes come into contact with the fluid in the perisinusoidal space.

The majority of the sinusoidal cells are endothelial cells. Unlike those in other parts of the cardiovascular system, the endothelial cells of the liver lack a basement membrane. Furthermore, they have sievelike plates that permit the ready exchange of material between the perisinusoidal space and the sinusoid. Electron microscopy has demonstrated that even particles as big as chylomicrons (100–500 nm in diameter) can penetrate these porous plates. Although the barrier between the perisinusoidal space and the sinusoid seems to be quite permeable, it does have some sieving properties. For example, the protein concentration

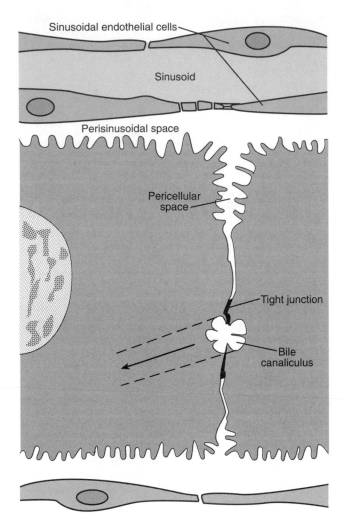

Figure 30–1 ■ The relationship of the hepatocytes, perisinusoidal space, and sinusoid. (Modified from Arias IM, Popper H, Schachter D, Shafritz DA (eds): *The Liver: Biology and Pathobiology.* New York, Raven, 1982.)

of hepatic lymph, assumed to be derived from the perisinusoidal space, is lower than that of plasma.

Kupffer cells also line the hepatic sinusoids. These are resident macrophages of the **fixed monocyte-macrophage system** that play an extremely important role in removing unwanted material (e.g., bacteria, virus particles, fibrin-fibrinogen complexes, damaged erythrocytes, and immune complexes) from the circulation. Endocytosis is the mechanism used to remove these particles.

Some sinusoidal cells contain distinct lipid droplets in the cytoplasm. These are the fat storage cells, or **Ito cells.** The lipid droplets contain vitamin A. These fat storage cells are closely associated with collagen bundles and appear to be similar to fibroblasts. They may be involved in the pathologic fibrosis of the liver.

■ The Liver Receives Venous Blood through the Portal Vein and Arterial Blood through the Hepatic Artery

Circulation to the liver was discussed in Chapter 17; here we only briefly describe some of its unique features. The hepatic portal vein provides 70–80% of the liver blood supply, the rest being provided by the hepatic artery. Hepatic portal blood is poorly oxygenated, whereas that from the hepatic artery is well-oxygenated. The portal vein branches repeatedly to form smaller venules that subsequently empty into the sinusoids. The hepatic artery branches to form arterioles and then capillaries, which also drain into the sinusoids. Consequently, liver sinusoids can be considered specialized capillaries. As mentioned earlier, the hepatic sinusoid is extremely porous and allows the rapid exchange of materials between the perisinusoidal space and the sinusoid. The sinusoids then empty into the central veins, which subsequently join to form the hepatic vein, which then joins the inferior vena cava.

Metabolism of Drugs and Xenobiotics

■ The Metabolism of Drugs and Xenobiotics Often Involves Conjugation

The hepatocytes in the liver play an extremely important role in the metabolism of drugs and **xenobiotics,** compounds that are foreign to the body, some of them toxic. Most drugs and xenobiotics are introduced into the body with food. The kidney ultimately disposes of these substances, but for effective elimination the drug or its metabolites must be made hydrophilic (polar, water-soluble). This is because reabsorption of a substance by the renal tubules is dependent on its hydrophobicity (nonpolar, lipid solubility); the more hydrophobic a substance, the more likely it will be reabsorbed. Many drugs and metabolites are not hydrophilic, and the liver plays an important role in converting these into hydrophilic compounds.

Two reactions (phase 1 and 2) catalyzed in different enzyme systems are involved in the conversion of xenobiotics and drugs into hydrophilic compounds. In phase 1 reactions, the parent compound is biotransformed into more polar compounds by the introduction of one or more polar groups. The common polar groups are hydroxyl (OH) and carboxyl (COOH). Most phase 1 reactions involve oxidation of the parent compound. The enzymes involved are mostly located in the smooth endoplasmic reticulum; some, are located in the cytoplasm. For example, alcohol dehydrogenase is located in the cytoplasm of the hepatocyte. In this reaction, alcohol is rapidly

taken up by the hepatocyte and rapidly converted to acetaldehyde, catalyzed by alcohol dehydrogenase. The acetaldehyde is then rapidly converted to form acetate, catalyzed by aldehyde dehydrogenase, located mainly in the mitochondria. The acetate is then metabolized by the hepatocytes.

The enzymes involved in phase 1 reactions of drug biotransformation are present as an enzyme complex composed of the NADPH-cytochrome P450 reductase (NADPH-cytochrome c reductase) and a series of hemoproteins called cytochrome P450 (Fig. 30-2). The drug combines with the oxidized cytochrome P450^{+3} to form the cytochrome P450^{+3}-drug complex. This complex is then reduced to cytochrome P450^{+2}-drug complex, catalyzed by the enzyme NADPH cytochrome P450 reductase. The reduced complex combines with molecular oxygen to form an oxygenated intermediate. One atom of the molecular oxygen then combines with two H$^+$ to form water. The other oxygen atom remains bound to the cytochrome P450^{+3}-drug complex and is transferred from the cytochrome P450^{+3} to the drug molecule. The drug product with an oxygen atom incorporated is released from the complex. The cytochrome P450^{+3} released can then be recycled for the oxidation of other drug molecules.

The products from phase 1 reaction then undergo conjugation with a number of compounds to render them more hydrophilic (phase 2 reactions). Glucuronic acid is the substance most commonly used for conjugation, and the enzymes involved are the glucuronyltransferases. Other molecules used in conjugation are glycine, taurine, and sulfates.

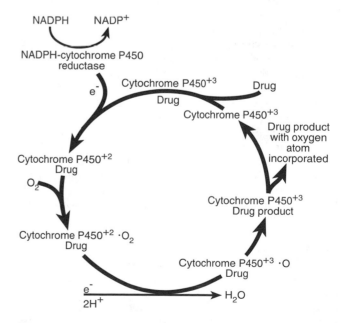

Figure 30–2 ■ Phase 1 reactions in the metabolism of drugs.

▪ Aging, Nutrition, and Genetics Influence Drug Metabolism

The enzyme systems in phase 1 and 2 reactions are age-dependent. These systems are poorly developed in human newborns, whose ability to metabolize any given drug is lower than that of adults. The elderly also have a lower capacity to metabolize drugs than young adults.

Nutritional factors can also affect the enzymes involved in phase 1 and 2 reactions. Insufficient protein in the diet to sustain normal growth results in production of fewer of the enzymes involved in drug metabolism.

Finally, it is well known that the drug-metabolizing enzymes can be induced by certain factors, such as polycyclic aromatic hydrocarbons. Smokers inhale polycyclic aromatic hydrocarbons, which stimulate drug-metabolizing enzymes, thereby increasing the metabolism of certain drugs, such as caffeine.

Genetic regulation of drug metabolism by the liver is less well understood. Briefly, the metabolism of drugs by the liver can be controlled by a single gene or a number of genes (polygenic control). Careful study of the metabolism of a certain drug by the population can provide important clues as to whether its metabolism is under single gene or polygenic control.

Energy Metabolism in the Liver

▪ The Intestine Supplies Nutrients to the Liver

The majority of water-soluble nutrients and water-soluble vitamins and minerals absorbed from the small intestine are transported via the portal blood to the liver. The nutrients transported in portal blood include amino acids and monosaccharides, and the fatty acids are predominantly short- and medium-chain forms. The short-chain fatty acids are largely derived from the fermentation of dietary fibers by bacteria in the colon. Some dietary fibers, such as pectin, are almost completely digested to form short-chain fatty acids (or volatile fatty acids), whereas cellulose is not well digested by the bacteria. Only a small amount of long-chain fatty acids bound to albumin is transported by the portal blood; the majority is transported in intestinal lymph as triglyceride-rich lipoproteins (chylomicrons).

▪ The Liver Is Important in Carbohydrate Metabolism

The liver is extremely important in maintaining an adequate supply of nutrients for cell metabolism. After ingestion of a meal, the blood glucose increases to a concentration of 120–150 mg/dl, usually in 1–2 hours. Glucose is taken up by the hepatocytes by a facilitated carrier-mediated process. Unlike in the small in-

testine, the uptake of glucose by the hepatocytes is not sodium-dependent.

Metabolism of Monosaccharides ▪ Monosaccharides are first phosphorylated by a reaction catalyzed by the enzyme hexokinase. In the liver, there is a specific enzyme for phosphorylation of glucose to form glucose-6-phosphate. Depending on the energy requirement, the glucose-6-phosphate is channeled to glycogen synthesis or used for energy production by the glycolytic pathway.

Fructose is taken up by the liver and phosphorylated by fructokinase to form fructose-1-phosphate. This is either isomerized to form glucose-6-phosphate or metabolized by the glycolytic pathway. Fructose-1-phosphate is used by the glycolytic pathway more efficiently than glucose-6-phosphate.

Galactose is an important sugar used not only to provide energy but also in the biosynthesis of glycoproteins and glycolipids. When galactose is taken up by the liver it is phosphorylated to form galactose-1-phosphate, which then reacts with uridine diphosphate-glucose (**UDP-glucose**) to form UDP-galactose and glucose-1-phosphate. The UDP-galactose can be used for glycoprotein and glycolipid biosynthesis or converted to UDP-glucose, which can then be recycled.

Gluconeogenesis ▪ Gluconeogenesis is the production of glucose from noncarbohydrate sources, such as fat, amino acids, and lactate. The process is energy-dependent, and the starting substrate is pyruvate. The energy required seems to be derived predominantly from the β-oxidation of fatty acids. Pyruvate can be derived from lactate and the metabolism of glucogenic amino acids, those that can contribute to the formation of glucose. The two major organs involved in the production of glucose from noncarbohydrate sources are the liver and the kidney. However, because of its size, the liver plays a far more important role than the kidney in the production of sugar from noncarbohydrate sources.

Gluconeogenesis is important in maintaining blood glucose and is prevalent during fasting. The red blood cells and renal medulla are totally dependent on blood glucose for energy, and glucose is the preferred substrate for the brain. Most amino acids can contribute to the carbon atoms of the glucose molecule, with alanine from the muscle being most important. The rate-limiting factor in gluconeogenesis is not the liver enzymes but the availability of substrates. Gluconeogenesis is stimulated by epinephrine and glucagon, but greatly suppressed by insulin. Thus, in type-1 diabetics, gluconeogenesis is greatly stimulated, thereby contributing to the hyperglycemia observed in these patients.

Glycogen Metabolism ▪ Glycogen is the main carbohydrate store in the liver. As much as 7–10% of the weight of a normal, healthy liver can be glycogen. Glycogen is derived from both carbohydrate and noncarbohydrate sources. After a meal, the liver is well supplied with glucose absorbed from the gastrointestinal tract. It is converted to **glucose-6-phosphate** and then UDP-glucose. The UDP-glucose can be used to add glucosyl units to the glycogen chain. Recent evidence in animals and humans seem to indicate that lactate produced peripherally is also an important source of the glucose-6-phosphate used for glycogen synthesis.

The addition of a glucosyl unit to the glycogen chain can be via the α-1,4- (to form a straight chain) or α-1,6-glycosidic linkage (to form a branched chain). Thus, the glycogen formed resembles a tree with numerous branches. The advantage of such a configuration is that the glycogen chain can be broken down at multiple sites, making release of glucose much more efficient than would be the case with a straight-chain polymer.

Glycogen is broken down by the enzyme **glycogen phosphorylase** to produce glucose-1-phosphate, which can be converted to glucose with glucose-6-phosphate as an intermediate. The glycogen phosphorylase acts only on the α-1,4-glycosidic linkage, resulting in a number of dextrans (polymers of glucose). To digest these dextrans further by the glycogen phosphorylase, the enzyme α-1,6-glucosidase is employed to break the α-1,6-glycosidic linkages.

Regulation of Carbohydrate Metabolism ■ The liver plays an important role in regulating blood glucose (Fig. 30-3). After a meal, blood glucose is high and it is taken up by the liver through a carrier-mediated process. It is converted to glucose-6-phosphate and then UDP-glucose, which can be used for glycogen synthesis **(glycogenesis).** It is generally believed that blood glucose is the major precursor of glycogen synthesis. However, recent evidence seems to indicate that the lactate in blood (from the peripheral metabo-

lism of glucose) is also a major precursor of glycogen synthesis. Amino acids (e.g., alanine) can supply pyruvate to synthesize glycogen.

During fasting, glycogen is broken down by **glycogenolysis** to form glucose-6-phosphate, which is then hydrolyzed to form glucose. The enzyme involved is **glucose-6-phosphatase** and is present in the liver. The glucose formed is released to the circulation. Glucose-6-phosphate is an important intermediate in carbohydrate metabolism because it can be channeled either to provide blood glucose or for glycogen formation.

Both glycogenolysis and glycogenesis are hormonally regulated. Insulin is secreted by the pancreas into the portal blood. Thus, the liver is the first organ to see changes in plasma insulin levels, to which it is extremely sensitive. For instance, a doubling of portal insulin concentration completely shuts down hepatic glucose production. The liver also removes about half of the insulin in portal blood in its first pass through. Insulin tends to lower blood glucose by stimulating glycogenesis and suppressing glycogenolysis and gluconeogenesis. Glucagon, in contrast, stimulates glycogenolysis. It also stimulates gluconeogenesis, thereby raising blood sugar levels. Epinephrine (also called adrenaline) stimulates glycogenolysis.

The liver regulates the blood glucose level within a narrow limit of 70–100 mg/dl. Although one might expect patients with liver disease to have difficulty regulating blood glucose, this is usually not the case, because of the relatively large reserve of hepatic function. However, those with chronic liver disease occasionally have reduced glycogen synthesis and reduced gluconeogenesis. Some patients with advanced liver disorder develop portal hypertension, which induces the formation of portosystemic shunting, resulting in elevated arterial blood levels of insulin and glucagon. Elevated levels of circulating insulin and/or glucagon can result in metabolic dysfunctions. For instance, in patients with **hyperglucagonemia** caused by portosystemic shunting, the fasting level of lactic acid and glycerol can be very high. The reduced gluconeogenesis usually seen in these patients can cause lactic acidosis.

■ The Liver Plays an Important Role in the Metabolism of Lipids

Oxidation of Fatty Acids ■ The liver takes up free fatty acids and lipoproteins—complexes of lipid and protein—from the plasma. Lipid is circulated in the plasma as lipoproteins because lipid and water are not miscible; the lipid droplets coalesce in an aqueous medium. The protein and phospholipid on the surface of the lipoprotein particles stabilize the hydrophobic triacylglycerol center of the particle.

The fatty acids derived from the plasma can be metabolized in the mitochondria of the hepatocytes by **β-oxidation** to provide energy. Fatty acids are broken down to form acetyl-CoA, which can be used in the

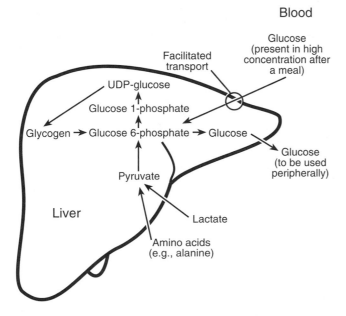

Figure 30-3 ■ Regulation of carbohydrate metabolism in the liver.

tricarboxylic acid cycle for ATP production; in the synthesis of fatty acids; and in the formation of ketone bodies.

Synthesis of Fatty Acids ■ Fatty acids are synthesized from acetyl-CoA. Thus, any substances that contribute acetyl-CoA, such as carbohydrate and protein sources, enhance fatty acid synthesis.

The liver is one of the main organs involved in fatty acid synthesis. Palmitic acid is synthesized in the hepatocellular cytosol; the other fatty acids synthesized in the body are derived by shortening, elongating, or desaturating the palmitic acid molecule.

Synthesis of Lipoproteins ■ One of the major functions of the liver in lipid metabolism is lipoprotein synthesis. The four major classes of circulating plasma lipoproteins are chylomicrons, very-low-density lipoproteins (VLDLs), low-density lipoproteins (LDLs) and high-density lipoproteins (HDLs). These lipoproteins, which differ in chemical composition, are usually isolated from plasma according to their flotation properties.

Chylomicrons are the lightest of the four lipoprotein classes, with a density of less than 0.95 g/ml (Table 30–1). They are made only by the small intestine and are produced in large quantities during fat ingestion. Their major function is to transport the large amount of absorbed fat to the bloodstream.

The VLDLs are denser and smaller than the chylomicrons (Table 30.1). Liver synthesizes about 10 times more of the circulating VLDLs than the small intestine. Like chylomicrons, VLDLs are triglyceride-rich and carry most of the triglyceride from the liver to the other organs. The triacylglycerol of VLDLs is broken down by **lipoprotein lipase** to yield fatty acids, which can be metabolized to provide energy. The human liver normally has a considerable capacity to produce VLDLs, but in acute or chronic liver disorders, this ability is markedly compromised. Liver VLDLs are associated with an important class of proteins, the apo B proteins. The two forms of circulating apo B are B_{48} and B_{100}. The human liver makes only apo B_{100}, which has a molecular weight of about 500,000. Apo B_{100} is important for the hepatic secretion of VLDL. This is underscored in **abetalipoproteinemia,** in which apo B synthesis, and therefore secretion of VLDLs, is blocked. Large lipid droplets can be seen in the cytoplasm of the hepatocytes of abetalipoproteinemic patients. Although considerable amounts of the circulating

plasma LDLs and HDLs are produced in the plasma, the liver also produces a small amount of these two cholesterol-rich lipoproteins. LDLs are heavier than VLDLs, and HDLs are denser than LDLs. The function of LDLs is to transport cholesterol ester from the liver to the other organs; HDLs are believed to remove cholesterol from the peripheral tissue and transport it to the liver.

The formation and secretion of lipoproteins by the liver is regulated by the supply of precursors and hormones such as estrogen and thyroid hormone. For instance, during fasting, the fatty acids in the VLDLs are derived mainly from the fatty acids mobilized from the adipose tissue. In contrast, during fat feeding, the fatty acid in the VLDLs produced by the liver is largely derived from chylomicrons.

As noted earlier, the fatty acids taken up by the liver can be used for β-oxidation and ketone body formation. The relative amounts of fatty acid channeled for these various purposes are largely dependent on the individual's nutritional and hormonal status. More fatty acid is channeled to ketogenesis or β-oxidation when the supply of carbohydrate is short (e.g., during fasting) or under conditions of high circulating glucagon or low circulating insulin (e.g., diabetes mellitus). In contrast, more of the fatty acid is used for synthesis of triacylglycerol for lipoprotein export when the supply of carbohydrate is abundant (e.g., during feeding) or under conditions of low circulating glucagon or high circulating insulin.

Catabolism of Lipoproteins ■ The importance of the liver in lipoprotein metabolism is exemplified in **familial hypercholesterolemia,** in which the liver fails to produce the LDL receptor. When LDL binds its receptor, it is internalized and catabolized in the hepatocyte. Consequently, the LDL receptor is crucial for the removal of LDL from the plasma. Individuals suffering from familial hypercholesterolemia usually have high plasma LDLs, which predisposes them to early coronary heart disease. Often the only effective treatment is a liver transplant.

The liver also plays an important role in the uptake of chylomicrons after their metabolism. After the chylomicrons produced by the small intestine enter the circulation, lipoprotein lipase in the endothelial cells of blood vessels acts on them to liberate fatty acids and glycerol from the triacylglycerols. As their metabolism progresses, the chylomicrons shrink, resulting in the

TABLE 30–1 ■ Characteristics of Human Plasma Lipoproteins

	Source	Density (g/ml)	Size (nm)	Protein	Lipid
Chylomicron	Intestine	< 0.95	80–500	1%	99%
VLDL	Intestine and liver	0.95–1.006	30–80	7–10%	90–93%
LDL	Chylomicron and VLDL	1.019–1.063	18–28	20–22%	78–80%
HDL	Chylomicron and VLDL	1.063–1.21	5–14	35–60%	40–65%

VLDL, very-low-density lipoprotein; LDL, low-density lipoprotein; HDL, high-density lipoprotein.

HUMAN GENE THERAPY

Current therapies for human genetic diseases are largely inadequate. For instance, familial hypercholesterolemia (FH) is inherited as a dominant trait. It is usually seen in heterozygotes, but there are a few rare cases (1 in 1 million) of homozygotes. Typically, FH homozygotes have extremely high plasma low-density lipoprotein (LDL) cholesterol concentrations. The clinical manifestations include deposition of cholesterol esters on the Achilles tendons and the extensor tendons of the hand and the presence of xanthomas (tumors composed of lipid-laden foam cells) over the thighs and buttocks. Usually FH homozygotes have advanced atherosclerosis at a young age, and they seldom live past their second decade. Modern molecular genetic techniques offer hope for these individuals. Liver-directed gene therapy is being actively investigated as a treatment for homozygous FH. One approach is the transplantation of autologous hepatocytes that have been genetically modified with recombinant retroviruses ex vivo. Hepatocytes are isolated from patients and exposed to recombinant retroviruses expressing the cDNA for the human LDL receptor. Success of the infection of the primary hepatocytes with the recombinant retroviruses is determined by the expression of the recombinant-derived proteins by cytochemical assays. Animal experiments thus far have shown that the procedure is feasible and that hepatocytes transduced with the LDL receptor gene express normal levels of LDL receptor proteins. This represents a significant step in the development of liver-directed gene therapy in humans.

Chowdhury and colleagues (*Science* 254:1802–1805, 1991) reported an important study using Watanabe heritable hyperlipidemic rabbits, an excellent model for human familial hypercholesterolemia. Autologous hepatocytes were isolated from affected rabbits and exposed to recombinant retroviruses expressing the cDNA for normal rabbit LDL receptor. These LDL receptor-transduced autologous hepatocytes were then transplanted into the rabbits. The procedure resulted in a 30–50% reduction in serum cholesterol level in the transplanted rabbits, which persisted for the entire duration of the experiment (122 days).

---

detachment of free cholesterol, phospholipid, and proteins, and the formation of HDL. Chylomicrons are converted to chylomicron remnants during metabolism, and chylomicron remnants are rapidly taken up by the liver via chylomicron remnant receptors.

Production of Ketone Bodies ■ Most organs, except the liver, can use ketone bodies as fuel. For example, although glucose is the preferred fuel for the brain, during prolonged fasting the brain shifts to using ketone bodies for energy. The two ketone bodies in the body are acetoacetate and β-hydroxybutyrate. Their formation by the liver is normal and physiologically important. For instance, during fasting there is rapid depletion of the glycogen stores in the liver, resulting in a shortage of substrates (e.g., oxaloacetate) for the citric acid cycle. There is also a rapid mobilization of fatty acids from adipose tissues to the liver. Under these circumstances, the acetyl-CoA formed from β-oxidation is channeled to ketone bodies.

The liver is efficient in producing ketone bodies; in humans it can produce half of its equivalent weight of ketone bodies per day. However, it lacks the ability to metabolize the ketone bodies formed, since it lacks the necessary enzyme (ketoacid-CoA transferase).

The level of ketone bodies circulating in the blood is usually low, but during prolonged starvation and in diabetes mellitus it is highly elevated, a condition known as **ketosis.** In diabetic patients, large amounts of β-hydroxybutyric acid can make the blood pH acidic, a state called **ketoacidosis.** Ketoacidosis is associated with a significant loss in urine of fluid and large amounts of sodium and other cations. Ketoacidotic patients may also suffer from emesis and dehydration and may require hospitalization.

Cholesterol Metabolism ■ The liver plays an important role in cholesterol homeostasis. Liver cholesterol is derived both from de novo synthesis and the lipoproteins taken up by the liver. Hepatic cholesterol can be used in the formation of bile acids, biliary cholesterol secretion, the synthesis of VLDLs, and the synthesis of liver membranes. Since the absorption of biliary cholesterol and bile acids by the gastrointestinal tract is incomplete, this is an important and efficient method of eliminating cholesterol from the body. The VLDLs secreted by the liver provide cholesterol to organs that need it for the synthesis of steroid hormones (e.g., adrenal glands, ovary, and testis).

Regulation of Lipid Metabolism ■ The liver plays a pivotal role in lipid metabolism in the body (Fig. 30–4). During fasting, fatty acids are mobilized from adipose tissue and are taken up by the liver. They are used by the hepatocytes to provide energy via β-oxidation, for the generation of ketone bodies, and to synthesize the triacylglycerol necessary for VLDL formation. After feeding, chylomicrons from the small intestine are metabolized peripherally, and the chylomicron remnants formed are rapidly taken up by the liver. The fatty acids derived from the triacylglycerols of the chylomicron remnants are used for the formation VLDLs or for energy production via β-oxidation.

The liver also plays an important role in cholesterol homeostasis. By secreting VLDLs secreted into the circulation, it provides cholesterol to various organs. Secretion of cholesterol and bile salts into bile provides a means for the body to excrete cholesterol.

Protein and Amino Acid Metabolism in the Liver

■ The Liver Produces Most of the Circulating Plasma Proteins

The liver synthesizes many of the circulating plasma proteins, albumin being the most important (Fig.

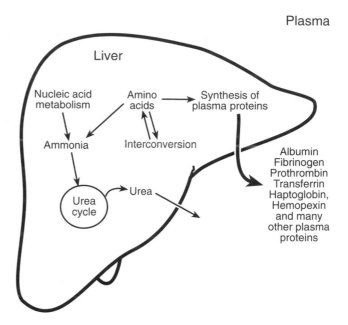

Figure 30–5 ■ Regulation of protein and amino acid metabolism in the liver.

30–5). It synthesizes about 3 g/day of albumin. Albumin plays an important role in maintaining plasma volume and tissue fluid balance, by maintaining the colloid osmotic pressure of plasma. This important function of plasma proteins is well illustrated by the fact that both liver disease and long-term starvation result in generalized edema and ascites. Plasma albumin plays a pivotal role in the transport of many substances in blood, such as free fatty acids and some drugs (penicillin and salicylates).

The other major plasma proteins synthesized by the liver are components of the complement system, components of the blood clotting cascade (fibrinogen, prothrombin), and proteins involved in iron transport (transferrin, haptoglobin, and hemopexin) (Chap. 11).

■ The Liver Produces Urea

Ammonia plays a pivotal role in nitrogen metabolism and is needed in the biosynthesis of nonessential amino acids and nucleic acids. It is derived from protein and nucleic acid catabolism. Ammonia metabolism is a major function of the liver, which has an ammonia concentration of about 0.7 mmol/L, 10 times higher than plasma ammonia levels. High circulating ammonia levels are highly neurotoxic, and a deficiency in hepatic function can lead to several distinct neurologic disorders, including coma in severe cases. Liver synthesizes most of the urea in the body. The enzymes involved in the urea cycle are regulated by protein intake. In humans, starvation stimulates these enzymes.

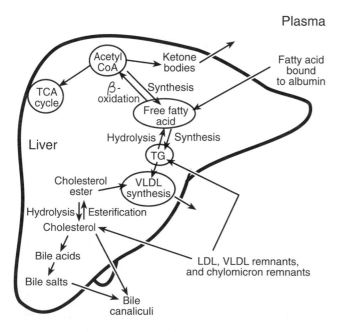

Figure 30–4 ■ Regulation of lipid metabolism in the liver. LDL, low-density lipoprotein; VLDL, very-low-density lipoprotein; TG, triglycerides; TCA, tricarboxylic acid.

■ The Liver Plays an Important Role in the Synthesis and Interconversion of Amino Acids

Humans can synthesize all but nine of the amino acids necessary for protein synthesis. These nine, the essential amino acids, must be supplied in the diet; they are histidine, methionine, threonine, tryptophan, isoleucine, leucine, valine, phenylalanine, and lysine. The liver is one of the major organs involved in the synthesis of nonessential amino acids from the essential amino acids. For instance, tyrosine can be synthesized from phenylalanine, and cysteine from methionine. Glutamic acid and glutamine play an important role in the biosynthesis of certain amino acids in the liver. Glutamic acid is derived from the amination of α-ketoglutarate by ammonia. This is an extremely important reaction, because ammonia is used directly in the formation of the α-amino group and constitutes a mechanism for shunting nitrogen from wasteful ureagenous (urea-forming) products. Glutamic acid can be used in the amination of other α-keto acids to form the corresponding amino acids. It can also be converted to glutamine by coupling with ammonia, a reaction catalyzed by glutamine synthetase. After urea, glutamine is the second most important metabolite of ammonia in the liver. It plays an extremely important role in the storage and transport of ammonia in the blood. Through the action of various transaminases, glutamine can be used to transaminate various keto acids to their corresponding amino acids. It also acts as an important oxidative substrate and in the small intestine is the preferred substrate for providing energy.

The Liver as a Storage Organ

Another important role of the liver is the storage of vitamins A, D, and K and iron, which it takes up from the bloodstream.

■ Fat-Soluble Vitamins Are Stored in the Liver

Vitamin A comprises a family of compounds related to retinol. It is important in vision, growth, maintenance of epithelia, and reproduction. The liver plays a pivotal role in the uptake, storage, and maintenance of circulating plasma vitamin A levels by mobilizing its vitamin A store (Fig. 30-6). Retinol (an alcohol) is transported in chylomicrons mainly as an ester of long-chain fatty acids (Chap. 29). When the chylomicrons enter the circulation, the triacylglycerol is rapidly acted on by lipoprotein lipase; the triglyceride contents of the particles are markedly reduced while the retinyl ester contents remain unchanged. Receptors in the liver mediate the rapid uptake of the chylomicron remnants, which are degraded, and the retinyl ester is stored.

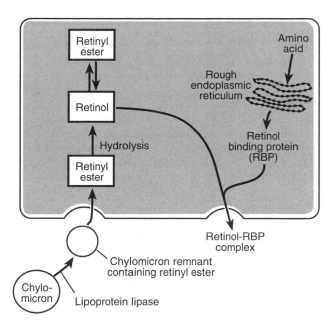

Figure 30-6 ■ Metabolism of vitamin A (retinol) by the hepatocyte.

When the vitamin A level in blood falls, the liver mobilizes the vitamin A store by hydrolyzing the retinyl ester (Fig. 30-6). The retinol formed is bound with **retinol-binding protein** (RBP), which is synthesized by the liver, before it is secreted into blood. The amount of RBP secreted into blood is dependent on vitamin A status; vitamin A deficiency markedly inhibits the release of RBP, whereas vitamin A loading stimulates its release.

Hypervitaminosis A develops when massive quantities of vitamin A are consumed. Since liver is the storage organ for vitamin A, hepatotoxicity is often associated with hypervitaminosis A. Continued ingestion of excessive amounts of vitamin A eventually leads to portal hypertension and cirrhosis.

Vitamin D is an important fat-soluble vitamin that plays a role in calcium and phosphate metabolism (Chap. 38). The principal form of dietary vitamin D is vitamin D_3, which can also be derived from 7-dehydrocholesterol when the skin is exposed to ultraviolet light. Unlike retinol, vitamin D_3 is transported in chylomicrons in nonesterified form. It has been proposed that vitamin D is located on the surface of the chylomicron particle and is released during metabolism of the particle. It is believed that a major portion of vitamin D is stored in skeletal muscle and adipose tissue. The liver plays a major role in vitamin D metabolism. It is responsible for the initial activation of vitamin D by converting vitamin D_3 to 25-hydroxycholecalciferol, and it synthesizes vitamin D-binding protein.

Vitamin K is a fat-soluble vitamin important in the hepatic synthesis of prothrombin. Prothrombin is synthesized as a precursor that is converted to the mature prothrombin, a reaction that requires the presence of vitamin K (Fig. 30–7). Vitamin K deficiency therefore leads to impaired blood clotting.

The metabolism of vitamin K in plasma is not well understood. The largest vitamin K store is in skeletal muscle, but the physiologic significance of this and other body stores is unknown. Although the liver has limited amounts of stored vitamin K, vitamin K deficiency is not uncommon in the Western world. The vitamin K requirement in the diet is extremely small and is adequately supplied by the average North American diet. Bacteria in the gastrointestinal tract also provide vitamin K. This appears to be an important source of vitamin K, because prolonged administration of wide-spectrum antibiotics sometimes results in hypoprothrombinemia. Since vitamin K absorption is dependent on normal fat absorption, any prolonged malabsorption of lipid can result in its deficiency. Since the liver vitamin K store is relatively limited, hypoprothrombinemia can develop within a few weeks. Parenteral administration of vitamin K usually provides a cure.

■ The Liver Is Important in the Storage and Homeostasis of Iron

The liver is important in iron metabolism. It is the major site of synthesis of a number of proteins involved in iron transport and metabolism. The protein transferrin plays an important role in the homeostasis of iron circulating in the blood and a pivotal role in the plasma transport of iron. The circulating plasma transferrin level is inversely proportional to the iron load of the body. The higher the concentration of ferritin in the hepatocyte, the lower the rate of transferrin synthesis. During iron deficiency, liver synthesis of transferrin is markedly stimulated, and this enhances the intestinal absorption of iron. **Haptoglobin,** a large

glycoprotein with a molecular weight of 100,000, binds free hemoglobin in the blood, and the hemoglobin-haptoglobin complex is rapidly removed by the liver, thus conserving iron in the body. **Hemopexin** is another protein synthesized by the liver that is involved in the transport of free heme in the blood. It forms a complex with free heme and the complex is removed rapidly by the liver.

The spleen is the major organ where red blood cells are removed. It removes the red blood cells that are slightly altered. Kupffer cells of the liver also have the capacity to remove damaged red blood cells, especially those that are moderately damaged (Fig. 30–8). The red cells taken up by the Kupffer cells are rapidly digested by the secondary lysosomes to release heme. Microsomal **heme oxygenase** releases iron from the heme, which then enters the free iron pool and is stored as ferritin or released into the bloodstream (bound to apotransferrin). Some of the ferritin iron may be converted to **hemosiderin granules.** It is unclear whether the iron from the hemosiderin granules is exchangeable with the free iron pool.

It was long believed that the Kupffer cells were the only cells involved in iron storage, but recent studies suggest that hepatocytes are the major sites of long-term iron storage. Transferrin binds to receptors on the surface of hepatocytes, and the entire transferrin-receptor complex is internalized and processed (Fig. 30–9). The apotransferrin (not containing iron) is recycled back to the plasma and the iron released enters a labile iron pool. The iron from transferrin is probably the major source of iron for the hepatocytes, but they also derive iron from haptoglobin-hemoglobin and hemopexin-heme complexes. When hemoglobin is released inside the hepatocytes, it is degraded in the

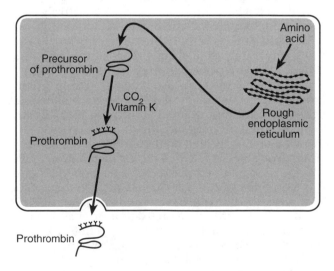

Figure 30–7 ■ Formation and secretion of prothrombin by the hepatocyte.

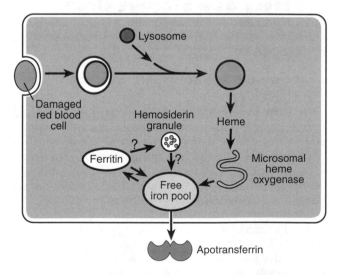

Figure 30–8 ■ The possible pathways following phagocytosis of damaged red blood cells by Kupffer cells. (Adapted from Young SP, Aisen P: The liver and iron, in Arias I, Jakoby WB, Popper H, et al (eds): *The Liver: Biology and Pathobiology.* New York, Raven, 1988.)

secondary lysosomes and heme is released. Heme is processed in the smooth endoplasmic reticulum and free iron released enters the labile iron pool. A significant portion of the free iron in the cytosol probably combines rapidly with apoferritin to form ferritin. Like the Kupffer cells, the hepatocytes may transfer some of the iron in ferritin to hemosiderin.

Iron is absolutely essential for survival, but iron overload can be extremely toxic, especially to the liver, where it can cause **hemochromatosis,** characterized by excessive amounts of hemosiderin in the hepatocytes. The hepatocytes in hemochromatosic patients are defective and fail to perform many normal functions.

Endocrine Functions of the Liver

■ The Liver Can Modify or Amplify Hormone Action

The liver appears to play a role in modifying or amplifying the action of hormones released by other organs. For example, vitamin D (cholecalciferol) is transported in blood and taken up rapidly by the liver, where it undergoes metabolic conversion to form

25-hydroxycholecalciferol. This metabolic conversion has two major consequences. First, since cholecalciferol is rapidly taken up by the liver and its metabolites are not, the plasma concentration of 25-hydroxycholecalciferol is about 2–30 times higher than that of cholecalciferol. Second, the 25-hydroxycholecalciferol is rapidly taken up by the kidney and converted to 1,25-dihydroxycholecalciferol, the most potent metabolite of cholecalciferol. It is 10 times more potent than vitamin D_3 (cholecalciferol) and plays an extremely important role in the regulation of calcium and phosphate absorption by the gastrointestinal tract. The liver facilitates this conversion of 25-hydrocholecalciferol by the kidney. In various types of liver disease the circulating level of 25-hydroxycholecalciferol is low, often resulting in bone disorders.

Thyroxine is a hormone secreted by the thyroid gland as tetraiodothyronine (T_4), which is converted peripherally to the biologically more potent T_3 through deiodination. The peripheral conversion of T_4 to T_3 is considered activation of a hormone, and the liver plays a major role in this function. The regulation of the hepatic T_4-T_3 conversion occurs at both the uptake step and the conversion step. Due to the relatively large reserve of the liver in converting T_4 to T_3, hypothyroidism is not common in patients with liver disease, but in advanced chronic liver disease signs of it (e.g., increased thyroid uptake of radioactive iodine and increased circulating levels of thyroid stimulating hormone) may be evident.

The liver modifies the function of growth hormone secreted by the pituitary gland. Secretion is stimulated by growth hormone releasing factor and inhibited by somatostatin. Some growth hormone actions are mediated by a class of small peptides called insulinlike growth factors (or somatomedins) made by the liver. They play an important role in cartilage function by promoting the uptake of sulfate and synthesis of collagen, which results in increased cartilage synthesis.

■ The Liver Removes Circulating Hormones

The liver plays an important role in the removal and degradation of many circulating hormones. Insulin is degraded in many organs, but the liver and kidneys are by far most important. There are insulin receptors on the surface of the liver, and there is strong evidence that binding of insulin to these receptors results in degradation of some insulin molecules. There is also degradation of insulin by proteases of the hepatocytes that do not involve the insulin receptor.

Glucagon is another hormone degraded mainly by the liver and kidneys. Growth hormone, with a circulating half-life of 20 minutes, is degraded by both the liver and kidneys. The liver may also play a role in metabolism of the gastrointestinal hormones (e.g., gastrin), but the kidneys and other organs probably play a much more significant role in metabolizing these hormones.

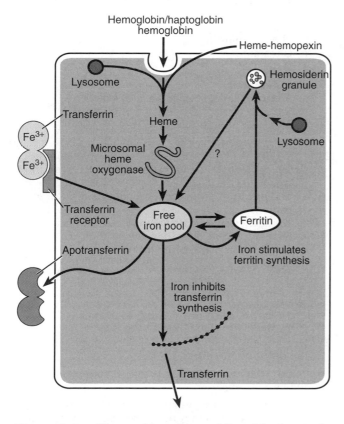

Figure 30–9 ■ The possible pathways followed by iron in the hepatocyte. (Adapted from Young SP, Aisen P: The liver and iron, in Arias I, Jakoby WB, Popper H, et al (eds): *The Liver: Biology and Pathobiology.* New York, Raven, 1988.)

■ ■ ■ ■ ■ ■ ■ ■ ■ ■ ■ **REVIEW EXERCISES** ■ ■ ■ ■ ■ ■ ■ ■ ■ ■ ■

Identify Each with a Term

1. The production of glucose from noncarbohydrate sources, such as fat, amino acids, and lactate

2. The pathway that produces the NADPH used for fatty acid synthesis

3. The formation of glycogen from glucose

4. The pathway that breaks down fatty acids to provide energy

5. The product of fatty acid biosynthesis from acetyl-CoA

Define Each Term

6. Kupffer cell

7. Gluconeogenesis

8. Ketoacidosis

9. Transferrin

10. Glutamic acid

Choose the Correct Answer

11. Which of the following is not a function of the liver?
    a. Albumin synthesis
    b. Urea production
    c. Chylomicron production
    d. Very-low-density lipoprotein production

12. The major enzyme involved in ethanol metabolism is:
    a. Cytochrome P450
    b. NADPH-cytochrome P450 reductase
    c. Alcohol oxygenase
    d. Alcohol dehydrogenase

13. The arterial blood glucose concentration in a normal human soon after a meal is:
    a. 30–50 mg/dl
    b. 50–70 mg/dl
    c. 120–150 mg/dl
    d. 180–220 mg/dl

14. The reason the liver, but not muscle, can supply glucose to the bloodstream is that:
    a. The liver has the enzyme glucose-6-phosphatase but the muscle does not
    b. The liver has glycogen but the muscle does not
    c. The liver has the enzyme glucose-1-phosphatase but the muscle does not
    d. The liver can convert UDP-glucose to glucose but the muscle cannot

15. The pathway in the liver which provides ribose-5-phosphate, which is required for nucleic acid synthesis is:
    a. Oxidative phosphorylation
    b. Kennedy pathway
    c. Embden-Meyerhof-Parnas pathway
    d. Pentose phosphate pathway

16. Carnitine plays an important role in β-oxidation of fatty acids in the liver because it:
    a. enhances the uptake of fatty acids by hepatocytes
    b. enhances the conversion of fatty acid to acyl-CoA
    c. is involved in the transport of acyl-CoA into mitochondria

d. decreases the production of acetyl-CoA from acyl-CoA in mitochondria

17. Fatty acid synthesis occurs in:
    a. Cytoplasm
    b. Mitochondria
    c. Nucleus
    d. Endosomes

18. The triglyceride-rich lipoproteins produced by the liver are:
    a. Chylomicrons
    b. Very-low-density lipoproteins
    c. Low-density lipoproteins
    d. High-density lipoproteins

19. During fasting, most of the fatty acid taken up by the liver is used for:
    a. The formation of ketone bodies
    b. The formation of very-low-density lipoproteins
    c. β-oxidation
    d. β-oxidation and the formation of ketone bodies

20. In blood, the major function of haptoglobin is to bind:
    a. Free hemoglobin
    b. Copper
    c. Zinc
    d. Magnesium

21. Other than urea, an important metabolite of ammonia in the liver is:
    a. Histidine
    b. Phenylalanine
    c. Glutamine
    d. Lysine

22. Hemosiderin is:
    a. An intracellular storage form of iron
    b. An intracellular storage form of copper
    c. A plasma protein that binds iron
    d. A plasma protein that binds copper

23. In patients with a portocaval shunt (connection between the portal vein and vena cava), the circulating glucagon level is extremely high because:
    a. The pancreas produces more glucagon in these patients
    b. The kidney is less efficient in removing the circulating glucagon in these patients
    c. The liver normally is the major site for the removal of glucagon
    d. None of the above

Briefly Answer

24. Discuss phase 2 reactions of drug metabolism.

25. Describe the removal of low-density lipoproteins from the blood by the liver.

26. Discuss the metabolism of ketone bodies produced by the liver.

27. Using vitamin D as an example, explain how the liver can amplify the potency of hormones released by other organs.

■ ■ ■ ■ ■ ■ ■ ■ ■ ■ ■ ■ ANSWERS ■ ■ ■ ■ ■ ■ ■ ■ ■ ■ ■ ■

1. Gluconeogenesis

2. Pentose phosphate pathway

3. Glycogenesis (glycogen synthesis)

4. β-oxidation (fatty acid oxidation)

5. Palmitic acid

6. Cells lining the hepatic sinusoids that are resident macrophages of the resident monocyte-macrophage system and play an important role in removing unwanted material from the circulation (e.g., bacteria, viruses, damaged erythrocytes, fibrin-fibrinogen complexes)

7. The synthesis of glucose from noncarbohydrate sources

8. The accumulation of large amounts of β-hydroxybutyric acid and acetoacetate in the blood, which renders the blood pH acidic

9. A protein synthesized mainly by the liver that plays a pivotal role in plasma iron transport

10. An amino acid derived from the amination of α-ketoglutarate by ammonia

11. c

12. d

13. c

14. a

15. d

16. c

17. a

18. b

19. d

20. a

21. c

22. a

23. c

24. The products of phase 1 reactions usually undergo conjugation with a number of compounds to render them more water-soluble. The substance most commonly used for conjugation is glucuronic acid. Other molecules used are glycine, taurine, and sulfates.

25. Low-density lipoproteins bind to the LDL receptors of the liver and are internalized and catabolized in the hepatocyte.

26. The liver can easily produce ketone bodies but lacks the ability to metabolize them because it lacks the necessary enzyme (ketoacid-CoA transferase). Ketone bodies are metabolized by the heart muscle, renal cortex, and brain.

27. Vitamin D (cholecalciferol) is rapidly taken up by the liver, where it is hydroxylated to form 25-hydroxycholecalciferol, which is secreted into plasma. 25-Hydroxycholecalciferol is not readily taken up by the liver, so the plasma concentration is many times higher than that of cholecalciferol. Furthermore, the kidneys hydroxylate 25-hydroxycholecalciferol to form 1,25-dihydroxycholecalciferol, the most potent metabolite of cholecalciferol.

Suggested Reading

Carey MC, Cahaland MJ: Enterohepatic circulation, in Arias IM, Jakoby WB, Popper H, et al (eds): *The Liver: Biology and Pathobiology.* 2nd ed. New York, Raven, 1988, pp 573–616

Glickman RM, Sabesin SM: Lipoprotein metabolism, in Arias IM, Jakoby WB, Popper H, et al (eds): *The Liver: Biology and Pathobiology.* 2nd ed. New York, Raven, 1988, pp 331–354.

Granger DN, Barrowman JA, Kvietys PR: *Clinical Gastrointestinal Physiology.* Philadelphia, WB Saunders, 1985.

Hobsley M: *Disorders of the Digestive Tract.* Baltimore, University Park Press, 1982.

Sanford PA: *Digestive System Physiology.* Baltimore, University Park Press, 1982.

Seifter S, England S: Energy metabolism, in Arias IM, Jakoby WB, Popper H, et al (eds): *The Liver: Biology and Pathobiology.* 2nd ed. New York, Raven, 1988, pp 279–316.

Ziegler DM: Detoxication: Conjugation and hydrolysis, in Arias IM, Jakoby WB, Popper H, et al (eds): *The Liver: Biology and Pathobiology.* 2nd ed. New York, Raven, 1988, pp 375–388.

PART EIGHT

Temperature Regulation and Exercise Physiology

The Regulation of Body Temperature*

---

*The views, opinions, and findings contained in this chapter are those of the author and should not be construed as official Department of the Army position, policy, or decision unless so designated by other official documentation. Approved for public release; distribution unlimited.

CHAPTER OUTLINE (Continued)

2. The effect of nonthermal inputs on thermoregulatory responses
  D. Physiologic and pathologic influences may change the thermoregulatory set point
  E. Peripheral factors modify the responses of skin blood vessels and sweat glands
    1. Skin temperature and cutaneous vascular and sweat gland responses
    2. Skin wettedness and the sweat gland response
VI. THERMOREGULATORY RESPONSES DURING EXERCISE
  A. Core temperature rises during exercise, triggering heat-loss responses
  B. Exercise in the heat can threaten cardiovascular homeostasis
    1. Impairment of cardiac filling during exercise in the heat
    2. Compensatory responses during exercise in the heat
VII. HEAT ACCLIMATIZATION
  A. Heat acclimatization includes adjustments in heart rate, temperatures, and sweat rate

VIII. RESPONSES TO COLD
  A. Blood vessels in the shell constrict to conserve heat
  B. Human cold acclimatization confers a modest advantage
    1. Metabolic changes in cold acclimatization
    2. Increased tissue insulation in cold acclimatization
    3. Cold-induced vasodilation and the Lewis hunting response
IX. CLINICAL ASPECTS OF THERMOREGULATION
  A. Fever enhances defense mechanisms
  B. Many factors, including physical fitness, age, drugs, and diseases, affect thermoregulatory responses and tolerance to heat and cold
  C. Heat stress causes or aggravates a number of disorders
    1. Heat syncope
    2. Water and salt depletion due to heat exposure
    3. Heat stroke
    4. Aggravation of disease states due to heat exposure
  D. Hypothermia occurs when the body's defenses against cold are disabled or overwhelmed

Importance of Tissue Temperature

Humans, like other mammals, are **homeotherms,** or warm-blooded animals, and regulate their internal body temperatures within a narrow range near 37°C (Fig. 31–1) in spite of wide variations in environmental temperature. Internal body temperatures of **poikilotherms,** or cold-blooded animals, by contrast, are governed by environmental temperature. The range of temperatures that living cells and tissues can tolerate without harm extends from just above freezing to nearly 45°C—far wider than the limits within which homeotherms regulate body temperature. What biologic advantage do homeotherms gain by maintaining a stable body temperature? As we shall see, tissue temperature is important for two reasons.

■ Temperature Extremes Injure Tissue Directly

High temperatures alter the configuration and overall structure of protein molecules, even though the sequence of amino acids is unchanged. Such alteration of protein structure is called **denaturation.** A familiar example of denaturation by heat is the coagulation of albumin in the white of a cooked egg. Since the biologic activity of a protein molecule depends on its configuration and charge distribution, denaturation inactivates a cell's proteins and injures or kills the cell. Injury occurs at tissue temperatures higher than about 45°C, which is also the point at which heating the

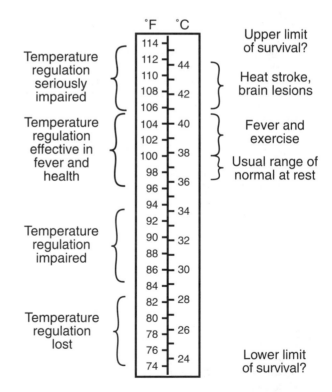

Figure 31–1 ■ Ranges of rectal temperature found in healthy persons, patients with fever, and persons with impairment or failure of thermoregulation. (Modified from Wenger CB, Hardy JD: Temperature regulation and exposure to heat and cold. In Lehmann JF (ed): *Therapeutic Heat and Cold.* 4th ed. Baltimore, Williams & Wilkins, 1990, pp 150–178.)

skin becomes painful. The severity of injury depends on the temperature to which the tissue is heated and how long the heating lasts.

As a water-based solution freezes, ice crystals consisting of pure water form, so that all dissolved substances in the solution are left in the unfrozen liquid. Thus as more ice forms, the remaining liquid becomes more and more concentrated. Freezing damages cells through two mechanisms. First, ice crystals themselves probably damage the cell mechanically. Second, the increase in solute concentration of the cytoplasm as ice forms denatures the proteins by removing their water of hydration, increasing the ionic strength of the cytoplasm, and other changes in the physicochemical environment in the cytoplasm.

■ Temperature Is a Fundamental Physicochemical Variable, Affecting Many Biologic Processes

Temperature changes profoundly alter biologic function, both through specific effects on such specialized functions as electrical properties and fluidity of cell membranes and through a general effect on most chemical reaction rates. In the physiologic temperature range, most reaction rates vary approximately as an exponential function of absolute temperature (T); increasing T by $10°K$ increases the reaction rate by a factor of 2–3. For any particular reaction, the ratio of the rates at two temperatures $10°K$ apart is called the Q_{10} for that reaction, and the effect of temperature on reaction rate is called the Q_{10} **effect.** The notion of Q_{10} may be generalized to apply to a group of reactions that have some measurable overall effect (such as O_2 consumption) in common and are thus thought of as comprising a physiologic process. The Q_{10} effect is clinically important in managing patients who have high fevers and are receiving fluid and nutrition intravenously. A commonly used rule is that a patient's fluid and calorie needs are increased 13% above normal for each $1°C$ of fever.

The profound effect of temperature on biochemical reaction rates is illustrated by the sluggishness of a reptile that comes out of its burrow in the morning chill and becomes active only after being warmed by the sun. Homeotherms avoid such a dependence of metabolic rate on environmental temperature by regulating their internal body temperatures within a narrow range. A drawback of homeothermy is that in most homeotherms, certain vital processes cannot function at low levels of body temperature that poikilotherms tolerate quite easily. Thus, for example, shipwreck victims immersed in cold water die of respiratory or circulatory failure (through disruption of the electrical activity of the brainstem or heart) at body temperatures of about $25°C$, even though such a temperature produces no direct tissue injury and fishes thrive in the same water.

Body Temperatures and Heat Transfer in the Body

■ The Body Is Divided into a Warm Internal Core and a Cooler Outer Shell

The temperature of the shell (Fig. 31–2) is strongly influenced by the environment, thus its temperature is not regulated within narrow limits as internal body temperature is, even though thermoregulatory responses strongly affect the temperature of the shell, especially the skin, its outermost layer. The thickness of the shell depends on the environment and the body's need to conserve heat. In a warm environment, the shell may be less than 1 cm thick, but in a subject conserving heat in a cold environment it may extend several centimeters below the skin. The internal body temperature that is regulated is the temperature of the vital organs inside the head and trunk, which, together with a variable amount of other tissue, comprise the warm internal core.

Heat is produced in all tissues of the body but is lost to the environment only from tissues in contact with the environment, predominantly from skin but to a lesser degree from the respiratory tract also. We

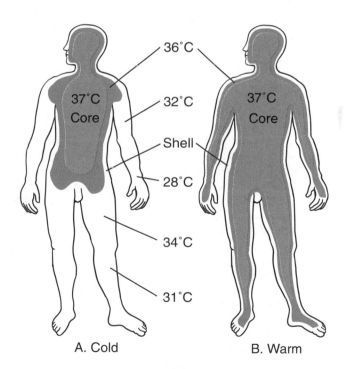

Figure 31–2 ■ Distribution of temperatures in the body and division of the body into core and shell during exposure to (A) cold and (B) warm environments. The temperatures of the surface and the thickness of the shell depend on environmental temperature, so that the shell is thicker in the cold and thinner in the heat. (Modified from Elizondo RS: Regulation of body temperature, in Rhoades RA, Pflanzer RG (eds): *Human Physiology.* Philadelphia, WB Saunders, 1989, pp 823–855.)

therefore need to consider heat transfer within the body, especially heat transfer (1) from major sites of heat production to the rest of the body and (2) from core to skin. Heat is transported within the body by two means: **conduction** through the tissues and **convection** by the blood, a process in which flowing blood carries heat from warmer tissues to cooler tissues.

Heat flow by conduction varies directly with the thermal conductivity of the tissues, the change in temperature over the distance the heat travels, and the area (perpendicular to the direction of heat flow) through which the heat flows. It varies inversely with the distance the heat must travel. As Table 31–1 shows, the tissues are rather poor heat conductors.

Heat flow by convection depends on the rate of blood flow and the temperature difference between the tissue and the blood supplying the tissue. Because the capillaries have thin walls and, taken together, a large total surface area, heat exchange between tissue and blood is most efficient in the capillary beds. Changes in skin blood flow in a cool environment change the thickness of the shell. When skin blood flow is reduced in the cold, the affected skin becomes cooler, and the underlying tissues—which in the cold may include most of the limbs and the more superficial muscles of the neck and trunk—become cooler as they lose heat by conduction to cool overlying skin and ultimately to the environment. In this way these underlying tissues, which in the heat were part of the body core, now become part of the shell. Thus in addition to the organs in the trunk and head, the core includes a greater or lesser amount of more superficial tissue—mostly skeletal muscle—depending on the body's thermal state.

Since the shell lies between the core and the environment, all heat leaving the body core—except heat lost through the respiratory tract—must pass through the shell before being given up to the environment. Thus the shell insulates the core from the environment. In a cool subject skin blood flow is low, so core-to-skin heat transfer is dominated by conduction; the

shell is also thicker and thus provides more insulation to the core, since heat flow by conduction varies inversely with the distance the heat must travel. Thus changes in skin blood flow, which directly affect core-to-skin heat transfer by convection, also indirectly affect core-to-skin heat transfer by conduction, by changing the thickness of the shell. In a cool subject, the subcutaneous fat layer contributes to the insulation value of the shell both because the fat layer increases the thickness of the shell and because fat has a conductivity about 0.4 times that of dermis or muscle (Table 31–1), and thus is a correspondingly better insulator. In a warm subject, on the other hand, the shell is relatively thin, and thus provides little insulation. Furthermore, a warm subject's skin blood flow is high, so heat flow from the core to the skin is dominated by convection. In these circumstances the subcutaneous fat layer—which affects conduction but not convection—has little effect on heat flow from core to skin.

■ Core Temperature Is Close to Central Blood Temperature

Core temperature varies slightly from one site to another depending on such local factors as metabolic rate, blood supply, and the temperatures of neighboring tissues. However, temperatures at different places in the core are all close to the temperature of the central blood and tend to change together. Thus the notion of a single uniform core temperature, though not strictly correct, is a useful approximation. The value of 98.6°F often given as the normal level of body temperature may give the misleading impression that body temperature is regulated so precisely that it is not allowed to deviate even a few tenths of a degree. In fact, 98.6°F is simply the Fahrenheit equivalent of 37°C and, as Figure 31–1 indicates, body temperature does vary somewhat. The effects of heavy exercise and fever are quite familiar; variation among individuals and such factors as time of day (Fig. 31–3), phase of the menstrual cycle, and acclimatization to heat can also cause differences of up to about 1°C in core temperature at rest.

To maintain core temperature within a narrow range, the thermoregulatory system needs continuous information about the level of core temperature. Temperature-sensitive neurons and nerve endings in the abdominal viscera, great veins, spinal cord, and, especially, the brain provide this information. We discuss how the thermoregulatory system processes and responds to this information later in the chapter.

Core temperature should be measured at a site whose temperature is not biased by environmental temperature. Sites used clinically include the rectum, the mouth, and, occasionally, the axilla. The rectum is well insulated from the environment, so its temperature is independent of environmental temperature, and it is a few tenths of 1°C warmer than arterial blood and other core sites. The tongue is richly supplied

TABLE 31–1 ■ Thermal Conductivities and Rates of Heat Flow through Slabs of Different Materials 1 m² in Area and 1 cm Thick, with a 1°C Temperature Difference between the Two Faces of the Slab

	Conductivity (kcal/(sec·m))	Rate of Heat Flow	
		kcal/h	W
Copper	0.09200	33,120	38,474
Epidermis	0.00005	18	21
Dermis	0.00009	32	38
Fat	0.00004	14	17
Muscle	0.00011	40	46
Oak (across grain)	0.00004	14	17
Glass fiber insulation	0.00001	3.6	4.2

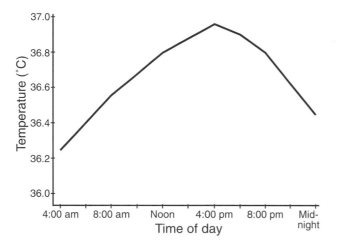

Figure 31–3 ■ Effect of time of day on internal body temperature of healthy resting subjects. (Drawn from data of Mackowiak PA, Wasserman SS, Levine MM: A critical appraisal of 98.6°F, the upper limit of normal body temperature, and other legacies of Carl Reinhold August Wunderlich. *JAMA* 268:1578–1580, 1992; and Stephenson LA, Wenger CB, O'Donovan BII, Nadel ER: Circadian rhythm in sweating and cutaneous blood flow. *Am J Physiol* 246:R321–R324, 1984.)

with blood, so oral temperature under the tongue is usually close to blood temperature (and 0.4–0.5°C below rectal temperature), but cooling the face, neck, or mouth can make oral temperature misleadingly low. If a patient holds his or her upper arm firmly against the chest to close the axilla, axillary temperature will eventually come reasonably close to core temperature. However, since this may take 30 minutes or more, axillary temperature is no longer widely used.

■ Skin Temperature Is Important in Heat Exchange and Thermoregulatory Control

Most heat is exchanged between the body and the environment at the skin surface. Skin temperature is much more variable than core temperature and is affected by thermoregulatory responses such as skin blood flow and sweat secretion, the temperatures of underlying tissues, and environmental factors such as air temperature, air movement, and thermal radiation. Skin temperature is one of the major factors determining heat exchange with the environment. For these reasons, it provides the thermoregulatory system with important information about the need to conserve or dissipate heat. Many bare nerve endings just under the skin are very sensitive to temperature. Depending on the relation of discharge rate to temperature they are classified as either warm or cold receptors (Chap. 4), with cold receptors being about 10 times as numerous as warm receptors. Furthermore, as the skin is heated, warm receptors respond with a transient burst of activity and cold receptors respond with a transient sup-

pression; the reverse happens as the skin is cooled. These transient responses at the beginning of heating or cooling give the central integrator almost immediate information about changes in skin temperature and may explain, for example, the intense, brief sensation of being chilled that occurs during a plunge into cold water.

Skin temperature usually is not uniform over the body surface, so a mean skin temperature ($\bar{T}_{sk}$) is frequently calculated from skin temperatures measured at several selected sites, usually weighting the temperature measured at each site according to the fraction of body surface area it represents. It would be prohibitively invasive and difficult to measure shell temperature directly. Instead, therefore, skin temperature is commonly used along with core temperature to calculate a mean body temperature and to estimate the quantity of heat stored in the body.

Balance between Heat Production and Heat Loss

All animals exchange energy with the environment. Some energy is exchanged as mechanical work, but most is exchanged as heat (Fig. 31-4). Heat is exchanged by conduction, convection, and radiation and as latent heat through evaporation or (rarely) condensation of water. If the sum of energy production and energy gain from the environment does not equal energy loss, the extra heat is "stored" in, or lost from, the body. This is summarized in the heat balance equation:

$$M = E + R + C + K + W + S \qquad (1)$$

where M is metabolic rate; E is rate of heat loss by evaporation; R and C are rates of heat loss by radiation and convection, respectively; K is the rate of heat loss by conduction; W is rate of energy loss as mechanical work; and S is rate of heat storage in the body, manifested as changes in tissue temperatures.

M is always positive, but the terms on the right side of equation 1 represent energy exchange with the environment and storage and may be either positive or negative. E, R, C, K, and W are positive if they represent energy losses from the body and negative if they represent energy gains. When $S = 0$, the body is in heat balance and body temperature neither rises nor falls. When the body is not in heat balance, its mean tissue temperature increases if S is positive and decreases if S is negative. This commonly occurs on a short-term basis and lasts only until the body responds to changes in temperature with thermoregulatory responses sufficient to restore balance; however, if the thermal stress is too great for the thermoregulatory system to restore balance, the body will continue to gain or lose heat until either the stress diminishes so that the thermoregulatory system can again restore the balance or the animal dies.

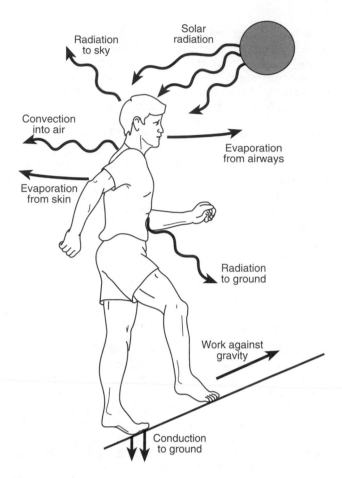

Figure 31–4 ■ Exchange of energy with the environment. This hiker gains heat from the sun by radiation and loses heat by conduction to the ground through the soles of his feet, convection into the air, radiation to the ground and sky, and evaporation of water from his skin and respiratory passages. In addition, some of the energy released by his metabolic processes is converted into mechanical work, rather than heat, since he is walking uphill.

The traditional units for measuring heat are a potential source of confusion, since the term *calorie* refers to two units differing by a thousandfold. The *calorie* used in chemistry and physics is the quantity of heat that will raise the temperature of 1 g of pure water by 1°C; it is also called the small calorie or gram calorie. The *Calorie* (capital C) used in physiology and nutrition is the quantity of heat that will raise the temperature of 1 kg of pure water by 1°C; it is also called the large calorie, kilogram calorie, or (the usual practice in thermal physiology) the **kilocalorie** (kcal). Since heat is a form of energy, it is now often measured in joules, the unit of work (1 kcal = 4186 J), and rate of heat production or heat flow in watts, the unit of power (1 W = 1 J/sec). This practice avoids confusion of calories and Calories. However, kilocalories are still used widely enough that it is necessary to be familiar with them, and there is a certain advantage to a unit based on water since the body itself is mostly water.

■ Heat Is a Byproduct of Energy-Requiring Metabolic Processes

Metabolic energy is used for active transport via membrane pumps, for energy-requiring chemical reactions such as formation of glycogen from glucose and proteins from amino acids, and for muscular work. Most of the metabolic energy used in these processes is converted into heat within the body. This may occur almost immediately, as with energy used for active transport or heat produced as a byproduct of muscular activity. Other energy is converted to heat only after a delay, as when the energy used in forming glycogen or protein is released as heat when the glycogen is converted back into glucose, or the protein back into amino acids.

Metabolic Rate and Sites of Heat Production at Rest ■ Among subjects of different body size, metabolic rate at rest varies approximately in proportion to body surface area. In a resting and fasting young man it is about 45 W/m² (81 W or 70 kcal/h for 1.8 m² body surface area), corresponding to an O_2 consumption of about 240 ml/min. About 70% of energy production at rest occurs in the body core—trunk viscera and brain—even though they comprise only about 36% of the body mass (Table 31-2). As a byproduct of their metabolic processes these organs produce most of the heat needed to maintain heat balance at comfortable environmental temperatures; only in the cold must such byproduct heat be supplemented by heat produced expressly for thermoregulation.

Factors besides body size that affect metabolism at rest include sex and age (Fig. 31-5), hormones, and digestion. The ratio of metabolic rate to surface area is highest in infancy and then declines with age, most rapidly in childhood and adolescence and more slowly thereafter. Children have high metabolic rates in relation to surface area because of the energy used to synthesize the fats, proteins, and other tissue components needed to sustain growth. Similarly, a woman's metabolic rate increases during pregnancy to supply the energy needed for the growth of the fetus. However, a nonpregnant woman's metabolic rate is 5–10% lower than that of a man of the same age and surface area, probably because a higher proportion of the

TABLE 31–2 ■ Relative Masses and Rates of Metabolic Heat Production of Various Body Compartments during Rest and Severe Exercise

	Body Mass (%)	Heat Production (%)	
		Rest	Exercise
Brain	2	16	1
Trunk viscera	34	56	8
Muscle and skin	56	18	90
Other	8	10	1

Modified from Wenger CB, Hardy JD: Temperature regulation and exposure to heat and cold, in Lehmann JF (ed): *Therapeutic Heat and Cold.* 4th ed. Baltimore, Williams & Wilkins, 1990, pp 150–178.

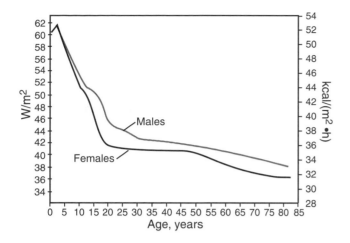

Figure 31–5 ■ Effects of age and sex on basal metabolic rate of normal subjects. Metabolic rate here is expressed as the ratio of energy consumption to body surface area. (Modified from Elizondo RS: Regulation of body temperature, in Rhoades RA, Pflanzer RG (eds): *Human Physiology.* Philadelphia, WB Saunders, 1989, pp 823–855.)

female body is composed of fat, a tissue with low metabolism.

Catecholamines and thyroxine are the hormones having the greatest effect on metabolic rate. Catecholamines cause glycogen to break down into glucose and stimulate many enzyme systems, thus increasing cellular metabolism. Hypermetabolism is a clinical feature of some cases of pheochromocytoma, a secreting tumor of the adrenal medulla. Thyroxine magnifies the metabolic response to catecholamines, increases protein synthesis, and stimulates oxidation by the mitochondria. The metabolic rate is typically 45% above normal in hyperthyroidism (but up to 100% above normal in severe cases) and 25% below normal in hypothyroidism (but 45% below normal with complete lack of thyroid hormone). Other hormones have relatively minor effects on metabolic rate.

A resting person's metabolic rate increases 10–20% after a meal. This effect of food, called the **specific dynamic action,** lasts several hours. The effect is greatest after eating protein and less after carbohydrate and fat; it appears to be associated with processing the products of digestion in the liver.

Measurement of Metabolic Rate ■ Since so many factors affect metabolism at rest, metabolic rate is often measured under a set of standard conditions to compare it with established norms. Metabolic rate measured under these conditions is called **basal metabolic rate** (BMR). The commonly accepted conditions for measuring BMR are that the person must have fasted for 12 hours; the measurement be made in the morning after a good night's sleep, beginning after the person has rested quietly for at least 30 minutes; and that the air temperature be comfortable, about 25°C (77°F). Basal metabolic rate is "basal" only during

wakefulness, since metabolic rate during sleep is somewhat less than BMR.

Heat exchange with the environment can be measured directly in a **human calorimeter,** an insulated chamber specially constructed so that heat can leave the chamber only in the air ventilating the chamber or, often, in water flowing through a heat exchanger in the chamber. By accurately measuring the flow of air and water and their temperatures as they enter and leave the chamber, one can accurately determine the subject's heat loss by conduction, convection, and radiation. By measuring also the moisture content of air entering and leaving the chamber, one can determine heat loss by evaporation. This technique is called **direct calorimetry,** and though conceptually simple, it is cumbersome and costly.

Metabolic rate is often estimated by **indirect calorimetry,** which is based on measuring a person's rate of O_2 consumption, since virtually all energy available to the body depends ultimately on reactions that consume O_2. Consumption of 1 liter of O_2 is associated with release of 21.1 kJ (5.05 kcal) if the fuel is carbohydrate, 19.8 kJ (4.74 kcal) if the fuel is fat, and 18.6 kJ (4.46 kcal) if the fuel is protein. An average value that is often used for metabolism of a mixed diet is 20.2 kJ (4.83 kcal) per liter of O_2. The ratio of CO_2 produced to O_2 consumed in the tissues is called the **respiratory quotient** (RQ). The RQ is 1.0 for oxidation of carbohydrate, 0.71 for oxidation of fat, and 0.80 for oxidation of protein. In a steady state where CO_2 is exhaled from the lungs at the same rate it is produced in the tissues, RQ is equal to the respiratory exchange ratio, R (Chap. 21). One can improve the accuracy of indirect calorimetry by also determining R and either estimating the amount of protein oxidized—which usually is small compared to fat and carbohydrate—or calculating it from urinary nitrogen excretion.

Skeletal Muscle Metabolism and External Work
■ Even during very mild exercise the muscles are the principal source of metabolic heat, and during intense exercise they may account for up to 90%. Moderately intense exercise by a healthy but sedentary young man may require a metabolic rate of 600 W (in contrast to about 80 W at rest), and intense activity by a trained athlete, 1400 W or more. Because of their high metabolic rate, exercising muscles may be almost 1°C warmer than the core. Blood perfusing these muscles is warmed and in turn warms the rest of the body; and consequently raises core temperature. Like steam and gasoline engines, muscles convert most of the energy in the fuels they consume into heat rather than mechanical work. During phosphorylation of ADP to form ATP, 58% of the energy released from the fuel is converted into heat, and only about 42% is captured in the ATP that is formed in the process. When a muscle contracts, some of the energy in the ATP that was hydrolyzed is converted into heat rather than mechanical work. The efficiency at this stage varies enormously; it is zero in isometric muscle contraction, in

which a muscle's length does not change while it develops tension, so that no work is done even though metabolic energy is required. Finally, some of the mechanical work produced is converted by friction into heat within the body. (This is, for example, the fate of all of the mechanical work done by the heart in pumping blood.) At best, no more than one-fourth of the metabolic energy released during exercise is converted into mechanical work outside the body, and the other three-fourths or more is converted into heat within the body.

■ Convection, Radiation, and Evaporation Are the Main Avenues of Heat Exchange with the Environment

Convection is transfer of heat due to movement of a fluid, either liquid or gas. In thermal physiology the fluid is usually air or water in the environment, or blood in the case of heat transfer inside the body. To illustrate, consider an object that is immersed in a cooler fluid. Heat passes from the object to the immediately adjacent fluid by conduction. If the fluid is stationary, conduction is the only means by which heat can pass through the fluid, and over time the rate of heat flow from the body to the fluid will diminish as the fluid nearest the object approaches the temperature of the object. In practice, however, fluids are rarely stationary. If the fluid is moving, heat will still be carried from the object into the fluid by conduction, but once the heat has entered the fluid it will be carried by the movement of the fluid itself—in other words, it will flow by convection. The same fluid movement that carries heat away from the surface of the object constantly brings fresh cool fluid to the surface, so the object gives up heat to the fluid much more rapidly than if the fluid were stationary. Although conduction plays a role in this process, convection so dominates the overall heat transfer that we refer to the heat transfer as if it were entirely convection. Therefore the conduction term (K) in the heat balance equation is restricted to heat flow between the body and other solid objects, and it usually represents only a small part of the total heat exchange with the environment.

Every surface emits energy as electromagnetic radiation, with a power output proportional to the area of the surface to the fourth power of its absolute temperature (i.e., measured from absolute zero) and to the **emissivity** (e) of the surface, a number between 0 and 1. (In this discussion the term *surface* is broadly defined, so that a flame and the sky, for example, are surfaces.) Such radiation, called thermal radiation, is largely in the infrared range at ordinary tissue and environmental temperatures. Most surfaces except polished metals have emissivities near 1 in this range of temperatures, and thus emit with a power output near the theoretical maximum. The emissivity of any surface is equal to the **absorptivity**—the fraction of

incident radiant energy that the surface absorbs. (For this reason an ideal emitter is called a **black body.**) If two bodies exchange heat by thermal radiation, radiation travels in both directions, but since each body emits radiation with an intensity that depends on its temperature, the net heat flow is from the warmer to the cooler body.

When 1 g of water is converted into vapor at 30°C, it absorbs 2425 J (0.58 kcal), the **latent heat of evaporation,** in the process. Evaporation of water is thus an efficient way of losing heat, and it is the body's only means of losing heat when the environment is hotter than the skin, as it usually is when the environment is warmer than 36°C. Evaporation must then dissipate both the metabolic heat and any heat gained from the environment by convection and radiation. Most water evaporated in the heat comes from sweat, but even in the cold the skin loses some water by evaporation of **insensible perspiration,** water that diffuses through the skin rather than being secreted. In equation 1 E is nearly always positive, representing heat loss from the body. However, E is negative in the rare circumstances (e.g., in a steam room) in which water vapor gives up heat to the body by condensing on the skin.

■ Heat Exchange Is Proportional to Surface Area and Obeys Biophysical Principles

Animals exchange heat with their environment through both skin and respiratory passages, but only the skin exchanges heat by radiation. In panting animals respiratory heat loss may be large and may be an important means of achieving heat balance. In humans, however, respiratory heat exchange is usually relatively small and (though hyperthermic subjects may hyperventilate) is not predominantly under thermoregulatory control. Therefore, we do not consider it further here.

Convective heat exchange between the skin and the environment is proportional to the difference between skin and ambient air temperatures, as expressed by the equation:

$$C = h_c \times A \times (\bar{T}_{sk} - T_a) \qquad (2)$$

where A is the body surface area, $\bar{T}_{sk}$ and T_a are mean skin and ambient temperatures, and h_c is the convective heat transfer coefficient.

The value h_c includes the effects of the factors other than temperature and surface area that affect convective heat exchange. For the whole body, air movement is the most important of these factors, and convective heat exchange (and thus h_c) varies approximately as the square root of the air speed (Fig. 31–6). Other factors that affect h_c include the direction of air movement and the curvature of the skin surface. As the radius of curvature decreases, h_c increases, so the

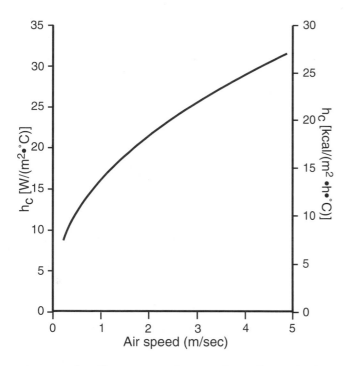

Figure 31–6 ■ The convective heat transfer coefficient, h_c, for a standing human as a function of air speed. The horizontal axis can be converted into English units by using the relation 5 m/sec = 16.4 ft/sec = 11.2 mph. The evaporative heat transfer coefficient, h_e, also increases with air speed, and $h_e = h_c \times 2.2°C/mm\,Hg$.

hands and fingers are effective in convective heat exchange out of proportion to their surface area.

Radiative heat exchange is proportional to the difference between the fourth powers of the absolute temperatures of the skin and of the radiant environment (T_r) and to the emissivity of the skin (e_{sk}): $R \propto e_{sk} \times (\bar{T}_{sk}^4 - T_r^4)$. However if T_r is close enough to $\bar{T}_{sk}$ that $\bar{T}_{sk} - T_r$ is much smaller than the absolute temperature of the skin, R is nearly proportional to $e_{sk} \times (\bar{T}_{sk} - T_r)$. Some parts of the body surface (e.g., inner surfaces of the thighs and arms) exchange heat by radiation with other parts of the body surface, so the body exchanges heat with the environment as if it had an area smaller than its actual surface area. This smaller area, called the **effective radiating surface area** (A_r), depends on the posture, being closest to the actual surface area in a spread-eagle posture and least in someone curled up. Radiative heat exchange can be represented by the equation

$$R = h_r \times e_{sk} \times A_r \times (\bar{T}_{sk} - T_r) \quad (3)$$

where h_r is the radiant heat transfer coefficient, 6.43 W/(m² · °C) at 28°C.

Evaporative heat loss from the skin to the environment is proportional to the difference between the water vapor pressure at the skin surface and the water

vapor pressure in the ambient air. These relations are summarized in the equation:

$$E = h_e \times A \times (P_{sk} - P_a) \quad (4)$$

where P_{sk} is the water vapor pressure at the skin surface, P_a is the ambient water vapor pressure, and h_e is the evaporative heat transfer coefficient.

Water vapor, like heat, is carried away by moving air, so geometric factors and air movement affect E and h_e in the same way they affect C and h_c. If the skin is completely wet, the water vapor pressure at the skin surface is the saturation water vapor pressure (Fig. 31–7) at skin temperature and evaporative heat loss is E_{max}, the maximum possible for the prevailing skin temperature and environmental conditions. This condition is described as:

$$E_{max} = h_e \times A \times (P_{sk,sat} - P_a) \quad (5)$$

where $P_{sk,sat}$ is the saturation water vapor pressure at skin temperature. When the skin is not completely wet, it is impractical to measure P_{sk}, the actual average water vapor pressure at the skin surface. Therefore a coefficient called skin **wettedness** (w) is defined as the ratio E/E_{max}, with $0 \le w \le 1$. Skin wettedness depends on the hydration of the epidermis and the fraction of the skin surface that is wet. We can now rewrite equation 4 as:

$$E = h_e \times A \times w \times (P_{sk,sat} - P_a) \quad (6)$$

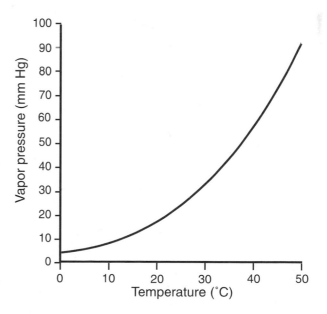

Figure 31–7 ■ The saturation vapor pressure of water as a function of temperature. For any given temperature, the water vapor pressure is at its saturation value when the air is "saturated" with water vapor (i.e., holds the maximum amount possible at that temperature).

Wettedness depends on the balance between secretion and evaporation of sweat. If secretion exceeds evaporation, sweat accumulates on the skin and spreads out to wet more of the space between neighboring sweat glands, so increasing wettedness and E; if evaporation exceeds secretion, the reverse occurs. If sweat rate exceeds E_{max}, then once wettedness becomes 1, the excess sweat drips from the body, since it cannot evaporate.

Note that P_a, on which evaporation from the skin directly depends, is proportional to the actual moisture content in the air. By contrast, the more familiar quantity **relative humidity** (rh) is the ratio between the actual moisture content in the air and the maximum moisture content possible at the temperature of the air. It is important to recognize that rh is only indirectly related to evaporation from the skin. For example in a cold environment, P_a will be low enough that sweat can easily evaporate from the skin even if rh = 100%.

■ Heat Storage Is a Change in the Heat Content of the Body

The rate of **heat storage** is the difference between heat production/gain and heat loss (equation 1); it can be determined experimentally from simultaneous measurements of metabolism by indirect calorimetry and heat gain or loss by direct calorimetry. Storage of heat in the tissues changes their temperature, and the amount of heat stored is the product of body mass, the body's mean specific heat, and a suitable mean body temperature (T_b). The body's mean specific heat depends on its composition, especially the proportion of fat, and is about 3.55 kJ/(kg·°C) [0.85 kcal/(kg·°C)]. Empirical relations of T_b to core temperature (T_c) and $\bar{T}_{sk}$, determined in calorimetric studies, depend on ambient temperature, with T_b varying from $0.65 \times T_c + 0.35 \times \bar{T}_{sk}$ in the cold to $0.9 \times T_c + 0.1 \times \bar{T}_{sk}$ in the heat. The shift from cold to heat in the relative weighting of T_c and $\bar{T}_{sk}$ reflects the accompanying change in the thickness of the shell (Fig. 31–2).

Heat Dissipation

Figure 31–8 shows rectal and mean skin temperatures, heat losses, and calculated shell conductances for nude resting men and women at the end of 2-hour exposures in a calorimeter to ambient temperatures from 23 to 36°C. Shell conductance represents the sum of heat transfer by two parallel modes—conduction through the tissues of the shell and convection by the blood—and is calculated by dividing heat loss through the skin (HF_{sk}) (i.e., total heat loss less heat loss through the respiratory tract) by the difference between core and mean skin temperatures:

$$C = HF_{sk}/(T_c - \bar{T}_{sk}) \qquad (7)$$

where C is shell conductance and T_c and $\bar{T}_{sk}$ are core and mean skin temperatures.

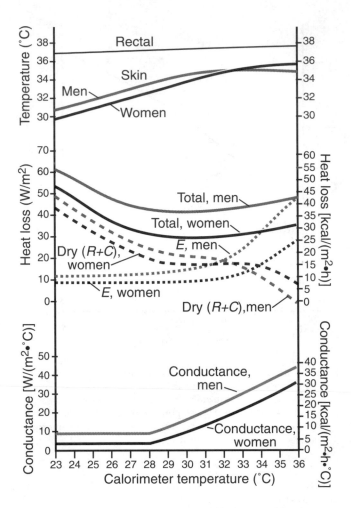

Figure 31–8 ■ Average values of rectal and mean skin temperatures, heat loss, and core-to-skin thermal conductance for nude resting men and women near steady state after 2 hours at different environmental temperatures in a calorimeter. (All energy exchange quantities in this figure have been divided by body surface area to remove the effect of individual body size.) Total heat loss is the sum of dry heat loss [by radiation (R) and convection (C)] and evaporative heat loss (E). Dry heat loss is proportional to the difference between skin temperature and calorimeter temperature, and it decreases with increasing calorimeter temperature. (Redrawn from data of Hardy JD, DuBois EF: Differences between men and women in their response to heat and cold. *Proc Natl Acad Sci USA* 26:389–398, 1940.)

From 23 to 28°C the subjects' conductance is minimal, because their skin is vasoconstricted and its blood flow is quite low. The minimal level of conductance attainable depends largely on the thickness of the subcutaneous fat layer, and the women's thicker layer allows them to attain a lower conductance than men. At about 28°C conductance begins to increase, and above 30°C conductance continues to increase and sweating begins.

For these nude subjects, the range 28–30°C is the zone of **thermoneutrality,** the range of comfortable environmental temperatures in which thermal balance is maintained without either shivering or sweating. In this zone, heat balance is maintained entirely by controlling conductance and $\bar{T}_{sk}$ and thus R and C. As equations 2-4 show, C, R, and E all depend on skin temperature, which, in turn, depends partly on skin blood flow. E depends also, through skin wettedness, on sweat secretion. Thus all of these modes of heat exchange are partly under physiological control.

■ Evaporation of Sweat Can Dissipate Large Amounts of Heat

In Figure 31-8 evaporative heat loss is nearly independent of ambient temperature below 30°C and is 9-10 W/m^2, corresponding to evaporation of about 13-15 $g/(m^2 \cdot h)$, of which about half is moisture lost in breathing and half is insensible perspiration. This evaporation occurs independent of thermoregulatory control. As the ambient temperature increases, the body depends more and more on evaporation of sweat to achieve heat balance.

There are two histologic types of sweat glands, **eccrine** and **apocrine.** In northern Europeans, apocrine glands are found mostly in the axilla and pigmented skin, such as the lips, but they are more widely distributed in other populations. Eccrine sweat is essentially a dilute electrolyte solution, but apocrine sweat also contains fatty material. Eccrine sweat glands, the dominant type in all human populations, are more important in human thermoregulation and number about 2,500,000. They are controlled through postganglionic sympathetic nerves, which release acetylcholine rather than norepinephrine. A healthy man unacclimatized to heat can secrete up to 1.5 L/h of sweat. Although the number of functional sweat glands is fixed before the age of 3, the secretory capacity of the individual glands can change, especially with endurance exercise training and heat acclimatization; a man well acclimatized to heat can secrete as much as 2.5 L/h. Such rates cannot be maintained, however; the maximum daily sweat rate is probably 12-15 liters.

The sodium concentration of eccrine sweat ranges from less than 5 to 60 mEq/L (versus 135-145 mEq/L in plasma), but even at 60 mEq/L sweat is the most dilute body fluid. In producing sweat that is hypotonic to plasma, the glands reabsorb sodium from the sweat duct by active transport. As sweat rate increases, the rate at which the glands reabsorb sodium increases more slowly, so that sodium concentration in the sweat increases. The sodium concentration of sweat is affected also by heat acclimatization and the action of mineralocorticoids.

■ Skin Circulation Is Important in Heat Transfer

Heat produced in the body must be delivered to the skin surface to be eliminated. When skin blood flow is minimal, core-to-skin thermal conductance (i.e., the conductance of the shell) is typically 5-9 W/°C per m^2 of body surface. For a lean resting subject with a surface area of 1.8 m^2, minimal whole body conductance of 16 W/°C [i.e., 8.9 $W/(°C \cdot m^2)$ × 1.8 m^2] and a metabolic heat production of 80 W, the temperature difference between core and skin must be 5°C (i.e., 80 W ÷ 16 W/°C) for the heat produced to be conducted to the surface. In a cool environment, T_{sk} may easily be low enough for this to occur. However, in an ambient temperature of 33°C, $\bar{T}_{sk}$ is typically about 35°C, and without an increase in conductance core temperature would have to rise to 40°C—a high though not yet dangerous level—for the heat to be conducted to the skin. If the rate of heat production were increased to 480 W by moderate exercise, the temperature difference between core and skin would have to rise to 30°C—and core temperature to well beyond lethal levels—to allow all the heat produced to be conducted to the skin. In these latter circumstances, the conductance of the shell must increase greatly for the body to reestablish thermal balance and continue to regulate its temperature. This is accomplished by increasing skin blood flow.

The Effectiveness of Skin Blood Flow in Heat Transfer ■ If we assume that blood on its way to the skin remains at core temperature until it reaches the skin, comes to skin temperature as it passes through the skin, and then stays at skin temperature until it returns to the core, we can compute the rate of heat flow (HF_b) due to convection by the blood as:

$$HF_b = SkBF \times (T_c - T_{sk}) \times 3.85 \text{ kJ}/(L \cdot °C) \quad (8)$$

where SkBF is the rate of skin blood flow, expressed in L/sec rather than the more usual L/min to simplify computing HF in W (i.e., J/sec); and 3.85 kJ/(L·°C) [0.92 kcal/(L·°C)] is the volume-specific heat of blood. Conductance due to convection by the blood (C_b) is calculated as:

$$C_b = HP_b/(T_c - T_{sk}) = SkBF \times 3.85 \text{ kJ}/(L \cdot °C) \quad (9)$$

Of course, heat continues to flow by conduction through the tissues of the shell, so total conductance is the sum of conductance due to convection by the blood plus that due to conduction through the tissues. Total heat flow is given by:

$$HF = (C_b + C_0) \times (T_c - T_{sk}) \quad (10)$$

in which C_0 is thermal conductance of the tissues when skin blood flow is minimal and thus is due predominantly to conduction through the tissues.

The assumptions made in deriving equation 8 are somewhat artificial and represent the conditions for maximum efficiency of heat transfer by the blood. In practice, blood exchanges heat also with the tissues through which it passes on its way to and from the skin. Heat exchange with these other tissues is greatest when skin blood flow is low, and in such cases heat flow to the skin may be much less that predicted by equation 8 (discussed further below). However, equation 8 is a reasonable approximation in a warm subject with moderate to high skin blood flow. It is not possible to measure whole body SkBF directly, but it is believed to reach several liters per minute during heavy exercise in the heat. The maximum obtainable is estimated to be nearly 8 L/min. If SkBF = 1.89 L/min (0.0315 L/sec), then according to equation 9 skin blood flow contributes about 121 W/°C to the conductance of the shell. If conduction through the tissues contributes 16 W/°C, total shell conductance is 137 W/°C, and if T_c = 38.5°C and T_{sk} = 35°C, then this will produce a core-to-skin heat transfer of 480 W, the heat production in our earlier example of moderate exercise. Thus, even a moderate rate of skin blood flow can have a dramatic effect on heat transfer.

When a person is not sweating, raising skin blood flow brings skin temperature nearer to blood temperature and lowering skin blood flow brings skin temperature nearer to ambient temperature. Under such conditions the body is able to control dry (convective and radiative) heat loss by varying skin blood flow and thus skin temperature. Once sweating begins, skin blood flow continues to increase as the person becomes warmer, but in these conditions the tendency of an increase in skin blood flow to warm the skin is approximately balanced by the tendency of an increase in sweating to cool the skin. Therefore after sweating has begun, further increases in skin blood flow usually cause little change in skin temperature or dry heat exchange and serve primarily to deliver to the skin the heat that is being removed by evaporation of sweat. Skin blood flow and sweating thus work in tandem to dissipate heat under such conditions.

Sympathetic Control of Skin Circulation ■ Blood flow in human skin is under dual vasomotor control. In most of the skin the vasodilation that occurs during heat exposure depends on sympathetic nervous signals that cause the blood vessels to dilate, and this vasodilation can be prevented or reversed by regional nerve block. Since it depends on the action of nervous signals, such vasodilation is sometimes referred to as active vasodilation. Active vasodilation occurs in almost all the skin, except in so-called acral regions—hands, feet, lips, ears, and nose. In the skin areas where active vasodilation occurs, vasoconstrictor activity is minimal at thermoneutral temperatures, and active vasodilation during heat exposure does not begin until close to the onset of sweating. Thus skin blood flow in these areas is not much affected by small temperature changes within the thermoneutral range. The neurotransmitter or other vasoactive substance responsible for active vasodilation in human skin has not been identified. However, since sweating and vasodilation operate in tandem in the heat, some investigators have proposed that the mechanism is somehow linked to the action of sweat glands.

Reflex vasoconstriction, occurring in response to cold and also as part of certain nonthermal reflexes such as baroreflexes, is mediated primarily through adrenergic sympathetic fibers distributed widely over most of the skin. Reducing the flow of impulses in these nerve fibers allows the blood vessels to dilate. In the acral regions and superficial veins (whose role in heat transfer is discussed below), vasoconstrictor fibers are the predominant vasomotor innervation, and the vasodilation that occurs during heat exposure is largely a result of the withdrawal of vasoconstrictor activity. Blood flow in these skin regions is sensitive to small temperature changes even in the thermoneutral range and may be responsible for "fine-tuning" heat loss to maintain heat balance in this range.

Thermoregulatory Control

In discussions of control systems (Chap. 1), the words *regulation* and *regulate* have meanings distinct from those of *control*. The variable that a control system acts to maintain within narrow limits (e.g., temperature) is called the *regulated* variable, and the quantities it controls to accomplish this (e.g., sweating rate, skin blood flow, metabolic rate, and thermoregulatory behavior) are called *controlled* variables.

Humans have two distinct subsystems to regulate body temperature: behavioral thermoregulation and physiologic thermoregulation. Physiologic thermoregulation is capable of fairly precise adjustments of heat balance but is effective only within a relatively narrow range of environmental temperatures. On the other hand, behavioral thermoregulation, through the use of shelter, space heating, and clothing, enables humans to live in the most extreme climates on earth, but it does not provide fine control of body heat balance.

■ Behavioral Thermoregulation Is Governed by Thermal Sensation and Comfort

Sensory information about body temperatures is an essential part of both behavioral and physiologic thermoregulation. The distinguishing feature of behavioral thermoregulation is the involvement of consciously directed effort to regulate body temperature. Thermal discomfort provides the necessary motivation for thermoregulatory behavior, and behavioral thermoregulation acts to reduce the discomfort and the physiologic burden imposed by a stressful thermal environment. For this reason, the zone of thermoneutrality is characterized by both thermal comfort and the absence of shivering and sweating.

Warmth and cold on the skin are felt as either comfortable or uncomfortable, depending on whether they decrease or increase the physiologic strain. Thus a shower temperature that feels pleasant after strenuous exercise may be uncomfortably chilly on a cold winter morning. The processing of thermal information in behavioral thermoregulation is not as well understood as is that in physiologic thermoregulation. However, perceptions of thermal sensation and comfort respond much more quickly than core temperature or physiologic thermoregulatory responses to changes in environmental temperature, and thus appear to anticipate changes in the body's thermal state. Such an anticipatory feature would be advantageous, since it would reduce the need for frequent small behavioral adjustments.

■ Physiologic Thermoregulation Operates through Graded Control of Heat-Production and Heat-Loss Responses

Familiar inanimate control systems, such as most refrigerators and heating and air conditioning systems, operate at only two levels—on and off. In a steam heating system, for example, when indoor temperature falls below the desired level, the thermostat turns on the burner under the boiler; when the temperature is restored to the desired level, the thermostat turns the burner off. Rather than operating at only two levels, most physiologic control systems produce a graded response according to the size of the disturbance in the regulated variable. In many instances, changes in the controlled variables are proportional to displacements of the regulated variable from some threshold value; such control systems are called **proportional control** systems.

The control of heat-dissipating responses is an example of a proportional control system. Figure 31–9 shows how reflex control of two heat-dissipating responses, sweating and skin blood flow, depends on body core temperature and mean skin temperature. Each response has a core-temperature threshold—a temperature at which the response starts to increase—and these thresholds depend on mean skin temperature. Thus at any given skin temperature the change in each response is proportional to the change in core temperature, and increasing the skin temperature lowers the threshold level of core temperature and increases the response at any given core temperature. In humans a change of 1°C in core temperature elicits about nine times as great a thermoregulatory response as a 1°C change in mean skin temperature. (Besides its effect on the reflex signals, skin temperature has a local effect that modifies the response of the blood vessels and sweat glands to the reflex signal, discussed later.)

Cold stress elicits increases in metabolic heat production through shivering and nonshivering thermogenesis. Shivering is a rhythmic oscillating tremor of skeletal muscles. The **primary motor center for shivering** lies in the dorsomedial part of the posterior hypothalamus and is normally inhibited by signals of warmth from the preoptic area of the hypothalamus. In the cold, these inhibitory signals are withdrawn, and the primary motor center for shivering sends impulses down the brainstem and lateral columns of the spinal

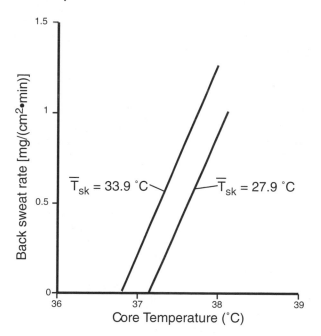

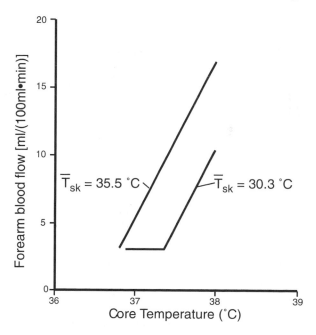

Figure 31–9 ■ The relations of back (scapular) sweat rate (left) and forearm blood flow (right) to core temperature and mean skin temperatures ($\overline{T}_{sk}$). In the experiments shown, core temperature was increased by exercise. (Left panel drawn from data of Sawka MN, Gonzalez RR, Drolet LL, Pandolf KB: Heat exchange during upper- and lower-body exercise. *J Appl Physiol* 57:1050–1054, 1984. Right panel modified from Wenger CB, Roberts MF, Stolwijk JAJ, Nadel ER: Forearm blood flow during body temperature transients produced by leg exercise. *J Appl Physiol* 38:58–63, 1975.)

cord to anterior motor neurons. These impulses are not rhythmic themselves but increase the tone of the muscles. The increased tone itself increases metabolic rate somewhat. Once tone exceeds a critical level, contraction of one group of muscle fibers stretches the muscle spindles in other groups of fibers in series with it, eliciting contractions from those groups of fibers via the stretch reflex, and so on; thus the rhythmic oscillations characteristic of frank shivering begin. Shivering tends to occur in bursts, and the "shivering pathway" is inhibited by signals from the cerebral cortex, so that voluntary muscular activity and attention suppress shivering. Since the limbs are part of the shell in the cold, there is a preferential recruitment of trunk and neck muscles for shivering (so-called **centralization of shivering**), to help retain the heat produced during shivering within the body core; it is a familiar experience that "chattering" of the teeth is one of the earliest signs of shivering. As with heat-dissipating responses, control of shivering depends on both core and skin temperature, but the details of its control in terms of these temperatures are not precisely understood.

In most laboratory mammals chronic cold exposure also causes **nonshivering thermogenesis,** an increase in metabolic rate that is not due to skeletal muscle activity. Nonshivering thermogenesis appears to be elicited through sympathetic stimulation and circulating catecholamines. It occurs in many tissues, especially the liver and **brown fat,** a tissue specialized for nonshivering thermogenesis whose color is imparted by high concentrations of iron-containing respiratory enzymes. Brown fat is found in human infants, and nonshivering thermogenesis is important for their thermoregulation. In human adults the existence of brown fat and nonshivering thermogenesis is controversial, though catecholamines do have a thermogenic effect.

■ The Central Nervous System Integrates Thermal Information from Core and Skin

Temperature receptors in the body core and skin transmit information about their temperatures through afferent nerves to the brainstem, especially the hypothalamus, where much of the integration of temperature information occurs. The sensitivity of the thermoregulatory responses to core temperature allows the thermoregulatory system to adjust heat production and heat loss to resist disturbances in core temperature. Their sensitivity to mean skin temperature allows the system to respond appropriately to mild heat or cold exposure with little change in body core temperature, so that changes in body heat content due to changes in environmental temperature take place almost entirely in the peripheral tissues (Fig. 31–2). For example, the skin temperature of someone who enters a hot environment rises and elicits sweating even if there is no change in core temperature.

On the other hand, an increase in heat production within the body, as occurs during exercise, elicits the appropriate heat-dissipating responses through a rise in core temperature.

Core temperature receptors that participate in the control of thermoregulatory responses are very unevenly distributed and are concentrated in the hypothalamus. In experimental mammals, temperature changes of only a few tenths of 1°C in the anterior preoptic area of the hypothalamus elicit changes in the thermoregulatory effector responses, and this area contains many neurons that increase their firing rate in response to either warming or cooling. Thermal receptors have been reported elsewhere in the core of laboratory animals, including the heart, pulmonary vessels, and spinal cord, but the thermoregulatory role of core thermal receptors outside the central nervous system is unknown.

Consider what happens when some disturbance—say, an increase in metabolic heat production due to exercise—upsets the thermal balance. Additional heat is stored in the body, and core temperature rises. The thermoregulatory controller receives information about these changes from the thermal receptors and responds by calling forth appropriate heat-dissipating responses. Core temperature continues to rise, and these responses continue to increase until they are sufficient to dissipate heat as fast as it is being produced, thus restoring heat balance and preventing further increases in body temperatures. In the language of control theory, the rise in core temperature that elicits heat-dissipating responses sufficient to reestablish thermal balance during exercise is an example of a **load error.** A load error is characteristic of any proportional control system that is resisting the effect of some imposed disturbance or "load." Although the disturbance in this example is exercise, parallel arguments apply if the disturbance is a decrease in metabolic rate or a change in the environment. However, if the disturbance is in the environment, most of the temperature change will be in the skin and shell rather than in the core, and if it produces a net loss of heat, the body will restore heat balance by decreasing heat loss and increasing heat production.

The Relation of Controlling Signal to Thermal Integration and Set Point ■ Both sweating and skin blood flow depend on core and skin temperatures in the same way, and changes in the threshold for sweating are accompanied by similar changes in the threshold for vasodilation. We may therefore think of the central integrator (Fig. 31–10) as generating one thermal command signal for the control of both sweating and skin blood flow. This signal is based on the information about core and skin temperatures that the integrator receives and on the thermoregulatory **set point.** We may think of the set point as the target level of core temperature, or the setting of the body's "thermostat." In the operation of the thermoregulatory system, it is a reference point that determines the

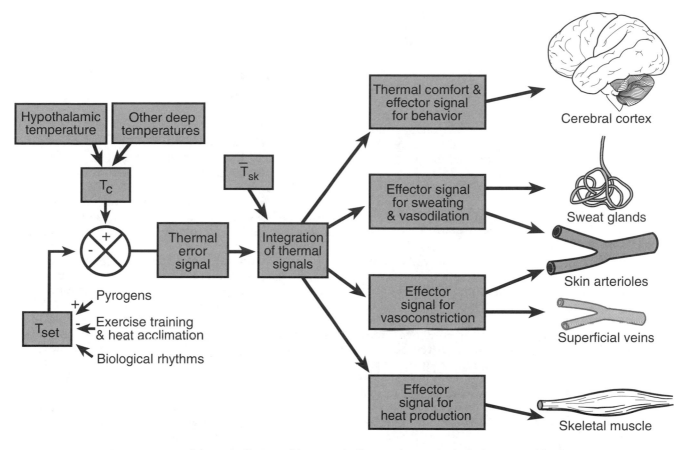

Figure 31–10 ■ Schematic diagram of the control of human thermoregulatory responses. The signs by the inputs to T_{set} indicate that pyrogens raise the set point and heat acclimation lowers it. Core temperature (T_C) is compared with the set point (T_{set}) to generate an error signal, which is integrated with thermal input from the skin to produce effector signals for the thermoregulatory responses. (Modified from Sawka MN, Wenger CB: Physiological responses to acute exercise-heat stress, in Pandolf KB, Sawka MN, Gonzalez RR (eds): *Human Performance Physiology and Environmental Medicine at Terrestrial Extremes.* Indianapolis, Benchmark, 1988, pp 97-151.)

thresholds of all of the thermoregulatory responses. Shivering and thermal comfort are affected by changes in the set point in the same way as sweating and skin blood flow. However, our understanding of the control of shivering is insufficient to say whether it is controlled by the same command signal as sweating and skin blood flow. (Thermal comfort, as we saw earlier, seems not to be controlled by the same command signal.)

The Effect of Nonthermal Inputs on Thermoregulatory Responses ■ Each of the thermoregulatory responses may be affected by inputs besides body temperatures and factors that affect the thermoregulatory set point. We have already noted that voluntary activity affects shivering and certain hormones affect metabolic heat production. In addition, nonthermal factors may produce a burst of sweating at the beginning of exercise, and emotional effects on sweating and skin blood flow are matters of common experience. Skin blood flow is the thermoregulatory re-

sponse most affected by nonthermal factors, because of its involvement in reflexes that function to maintain cardiac output, blood pressure, and tissue O_2 delivery during heat stress, postural changes, hemorrhage, and sometimes exercise, especially in the heat.

■ Physiologic and Pathologic Influences May Change the Thermoregulatory Set Point

Fever elevates core temperature at rest, heat acclimatization decreases it, and time of day and phase of the menstrual cycle change it in a cyclic fashion. Core temperature at rest varies in an approximately sinusoidal fashion with time of day. The minimum temperature occurs at night, several hours before awaking, and the maximum, which is 0.5-1°C higher, occurs in the late afternoon or evening (Fig. 31-3). This pattern coincides with patterns of activity and eating but does not depend on them, and it occurs even during bed rest in fasting subjects. This pattern is an example of a **circadian** rhythm, a rhythmic pattern in a physiologic function with a period of about 1 day. During the menstrual cycle core temperature is at its lowest point just before ovulation; over the next few days it rises 0.5-1°C to a plateau that persists through most of the luteal phase. Each of these factors—fever, heat acclimatization, the circadian rhythm, and the menstrual cycle—changes the core temperature at rest by changing the thermoregulatory set point, thus producing corresponding changes in the threshold for all of the thermoregulatory responses.

Peripheral Factors Modify the Responses of Skin Blood Vessels and Sweat Glands

The skin is the organ most directly affected by environmental temperature, and skin temperature affects heat loss responses not only through the reflex actions shown in Figure 31-9 but also through direct effects on the effectors themselves.

Skin Temperature and Cutaneous Vascular and Sweat Gland Responses ■ Local temperature changes act on skin blood vessels in at least two ways. First, local cooling potentiates (and heating weakens) the constriction of blood vessels in response to nervous signals and vasoconstrictor substances. (At very low temperatures, however, **cold-induced vasodilation** increases skin blood flow; see below.) Second, in skin regions where active vasodilation occurs, local heating causes vasodilation (and local cooling causes vasoconstriction) through a direct action on the vessels themselves, independent of nervous signals. The local vasodilator effect of skin temperature is especially strong above 35°C, and when the skin is warmer than the blood increased blood flow helps cool the skin and protect it from heat injury, unless this response is impaired by vascular disease. Local thermal effects on sweat glands parallel those on blood vessels, so local heating potentiates (and local cooling diminishes) the local sweat gland response to reflex stimulation or acetylcholine, and intense local heating elicits sweating directly, even in sympathectomized skin.

Skin Wettedness and the Sweat Gland Response ■ During prolonged (several hours) heat exposure with high sweat output, sweat rates gradually decline and the sweat glands' response to local application of cholinergic drugs is reduced. This reduction of sweat gland responsiveness is sometimes called sweat gland "fatigue." Wetting the skin makes the stratum corneum swell, mechanically obstructing the sweat duct and causing a reduction in sweat secretion, an effect called **hidromeiosis.** The glands' responsiveness can be at least partly restored if air movement increases or humidity is reduced to allow some of the sweat on the skin to evaporate. Sweat gland fatigue may involve processes besides hidromeiosis, since pro-

longed sweating also causes histologic changes, including depletion of glycogen, in the sweat glands.

Thermoregulatory Responses during Exercise

■ Core Temperature Rises during Exercise, Triggering Heat-Loss Responses

Exercise increases heat production, causing an increase in core temperature, which in turn elicits heat-loss responses. Core temperature continues to rise, until heat loss has increased enough to match heat production and core temperature and the heat-loss responses reach new steady-state levels. Since the heat-loss responses are proportional to the increase in core temperature, the increase in core temperature at steady state is proportional to the rate of heat production, and thus to the metabolic rate.

A change in ambient temperature causes a change in the level of sweating and skin blood flow necessary to maintain any given level of heat dissipation, but the change in ambient temperature also elicits, via direct and reflex effects of the accompanying skin temperature changes, a change of these responses in the right direction. For any given rate of heat production, there is a certain range of environmental conditions within which an ambient temperature change elicits the necessary changes in heat-dissipating responses almost entirely through the effects of skin temperature changes, with virtually no core temperature change. (The limits of this range of environmental conditions depend on the rate of heat production and such individual factors as skin surface area and state of heat acclimatization.) Within this range, the core temperature reached during exercise is nearly independent of ambient temperature; for this reason it was once believed that the increase in core temperature during exercise is caused by an increase in the thermoregulatory set point, just as during fever. As noted, however, the increase in core temperature with exercise is an example of a load error rather than an increase in set point. This difference between fever and exercise is shown in Figure 31-11. Note that although heat production may increase substantially (through shivering) when core temperature is rising early during fever, it need not stay high to maintain the fever; in fact, it returns nearly to prefebrile levels once the fever is established. During exercise, however, an increase in heat production not only causes the elevation in core temperature but is necessary to sustain it. Also, while core temperature is rising during fever, rate of heat loss is, if anything, lower than it was before the fever began, but during exercise the heat-dissipating responses and the rate of heat loss start to increase early and continue increasing as core temperature rises.

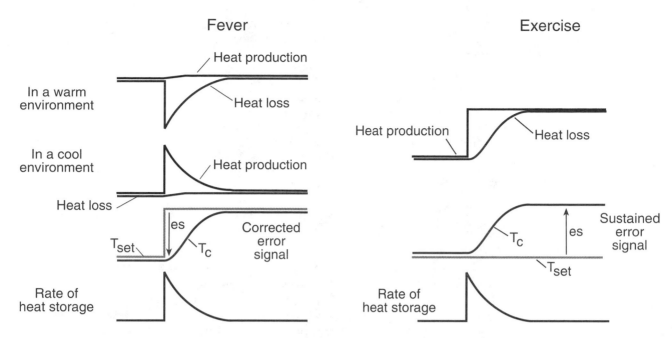

Figure 31–11 ■ Thermal events during the development of fever (left) and the increase in core temperature (T_c) during exercise (right). The error signal (es) is the difference between T_c and the set point (T_{set}). At the start of a fever, T_{set} has risen, so that T_{set} is higher than T_c and es is negative. At steady state, T_c has risen to equal the new level of T_{set} and es is corrected (i.e., it returns to zero.) At the start of exercise, $T_c = T_{set}$, so that es = 0. At steady state, T_{set} has not changed but T_c has increased and is greater than T_{set}, producing a sustained error signal, which is equal to the load error. (The error signal, or load error, is here represented with an arrow pointing downward for $T_c < T_{set}$ and with an arrow pointing upward for $T_c > T_{set}$.) (Modified from Stitt JT: Fever versus hyperthermia. *Fed Proc* 38:39-43, 1979.)

■ Exercise in the Heat Can Threaten Cardiovascular Homeostasis

The rise in core temperature during exercise increases the temperature difference between core and skin somewhat, but not nearly enough to match the increase in metabolic heat production. Therefore, as we saw earlier, skin blood flow must increase to carry all of the heat that is produced to the skin. In a warm environment, where the temperature difference between core and skin is relatively small, the necessary increase in skin blood flow may be several liters per minute.

Impairment of Cardiac Filling during Exercise in the Heat ■ The work of providing the skin blood flow required for thermoregulation in the heat may impose a heavy burden on a diseased heart, but in healthy subjects the major cardiovascular burden of heat stress results from impairment of venous return. As skin blood flow increases, the dilated vascular bed of the skin becomes engorged with large volumes of blood, thus reducing central blood volume and cardiac filling (Fig. 31–12). Stroke volume is decreased and a higher heart rate is required to maintain cardiac output. These effects are aggravated by a decrease in

plasma volume if the large amounts of salt and water lost in the sweat are not replaced. Since the main cation in sweat is sodium, disproportionately much of the body water lost in sweat is at the expense of extracellular fluid, including plasma, although this effect is mitigated if the sweat is dilute.

Compensatory Responses during Exercise in the Heat ■ Several reflex adjustments help maintain cardiac filling, cardiac output, and arterial pressure during exercise and heat stress. The cutaneous veins constrict during exercise; since most of the vascular volume is in the veins, constriction makes the cutaneous vascular bed less easily distensible and reduces peripheral pooling. Splanchnic and renal blood flow diminish in proportion to the intensity of the exercise or heat stress. The reduction of blood flow has two effects. First, it allows a corresponding diversion of cardiac output to skin and exercising muscle. Second, since the splanchnic vascular beds are very compliant, a decrease in their blood flow reduces the amount of blood pooled in them (Fig. 31–12), helping to compensate for decreases in central blood volume caused by reduced plasma volume and blood pooling in the skin. Because of the essential role of skin blood flow in thermoregulation during exercise and heat stress, the

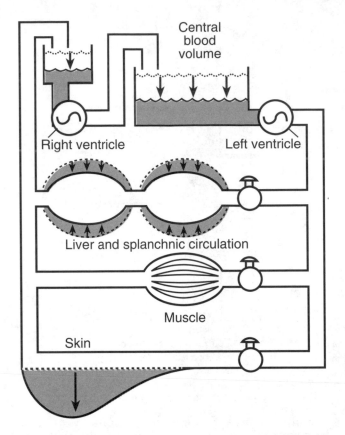

Figure 31–12 ■ Schematic diagram of the effects of skin vasodilation on peripheral pooling of blood and the thoracic reservoirs from which the ventricles are filled and the effects of compensatory vasomotor adjustments in the splanchnic circulation. The valves drawn at the right sides of liver/splanchnic, muscle, and skin vascular beds represent the resistance vessels that control blood flow through those beds. Arrows show the direction of the changes during heat stress. (Modified from Sawka MN, Wenger CB: Physiological responses to acute exercise-heat stress, in Pandolf KB, Sawka MN, Gonzalez RR (eds): *Human Performance Physiology and Environmental Medicine at Terrestrial Extremes.* Indianapolis, Benchmark, 1988, pp 97-151.)

body preferentially compromises splanchnic and renal flow for the sake of cardiovascular homeostasis. Above a certain level of cardiovascular strain, however, skin blood flow, too, will be compromised.

Heat Acclimatization

Prolonged or repeated exposure to stressful environmental conditions elicits significant physiologic changes, called **acclimatization,** that reduce the physiologic strain such conditions produce. (Such changes are usually called by the nearly synonymous term *acclimation* when produced in a controlled experimental setting.) Some degree of heat acclimatization is produced either by heat exposure alone or by regular strenuous exercise, which raises core temperature and

provokes heat-loss responses. Indeed, the first summer heat wave produces enough heat acclimatization that most people notice an improvement in their level of energy and general feeling of well-being after a few days. However, the acclimatization response is greater if heat exposure and exercise are combined, so as to cause a greater rise of internal temperature and more profuse sweating. Evidence of acclimatization appears in the first few days of combined exercise and heat exposure, and most of the improvement in heat tolerance occurs within 10 days. The effect of heat acclimatization on performance can be quite dramatic, so that acclimatized subjects can easily complete exercise in the heat that earlier was difficult or impossible.

■ Heat Acclimatization Includes Adjustments in Heart Rate, Temperatures, and Sweat Rate

Cardiovascular adaptations that reduce the heart rate required to sustain a given level of activity in the heat appear quickly and reach nearly their full development within 1 week. Changes in sweating develop more slowly. After acclimatization, sweating begins earlier and at a lower core temperature (i.e., the core temperature threshold for sweating is reduced). The sweat glands become more sensitive to cholinergic stimulation, and a given elevation in core temperature elicits a higher sweat rate; in addition, the glands become resistant to hidromeiosis and fatigue, so higher sweat rates can be sustained. These changes reduce the levels of core and skin temperatures reached during a given exercise-heat stress, increase the sweat rate, and enable one to exercise longer. The threshold for cutaneous vasodilation is reduced along with the threshold for sweating, so heat transfer from core to skin is maintained.

The lower heart rate and core temperature and the higher sweat rate (Fig. 31-13) are the three classical signs of heat acclimatization. Other physiologic changes also occur. During the first week, total body water and, especially, plasma volume increase. These changes likely contribute to the cardiovascular adaptations. Later, the fluid changes seem to diminish or disappear, though the cardiovascular adaptations persist. In an unacclimatized person, sweating occurs mostly on the chest and back, but during acclimatization, especially in humid heat, the fraction of sweat secreted on the limbs increases, to make better use of the skin surface for evaporation. An unacclimatized person who is sweating profusely can lose large amounts of sodium. With acclimatization, the sweat glands become able to conserve sodium by secreting sweat with a sodium concentration as low as 5 mEq/L. This effect is mediated through aldosterone, which is secreted in response to sodium depletion and to exercise and heat exposure. The sweat glands respond to aldosterone more slowly than the kidneys, requiring several days; unlike the kidneys, they do

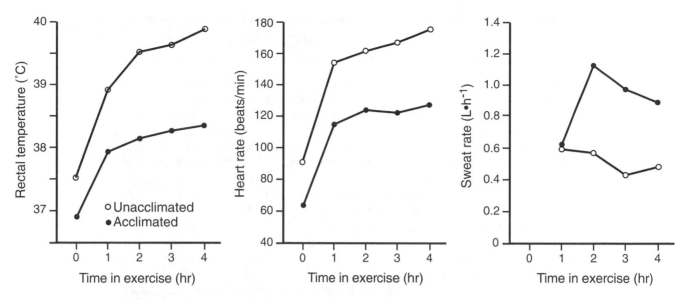

Figure 31–13 ■ Rectal temperatures, heart rates, and sweat rates during 4 hours' exercise (bench stepping, 35 W mechanical power) in humid heat (33.9°C dry bulb, 89% relative humidity, 35 mm Hg ambient vapor pressure) on the first and last days of a 2-week program of acclimation to humid heat. (Modified from Wenger CB: Human heat acclimatization, in Pandolf KB, Sawka MN, Gonzalez RR (eds): *Human Performance Physiology and Environmental Medicine at Terrestrial Extremes.* Indianapolis, Benchmark, 1988, pp 153–197.)

not escape the influence of aldosterone once sodium balance has been restored, but continue to conserve sodium as long as acclimatization persists.

Heat acclimatization is transient, disappearing in a few weeks if not maintained by repeated heat exposure. The components of heat acclimatization are lost in the order in which they were acquired, so that the cardiovascular changes decay more quickly than the reduction in exercise core temperature and sweating changes.

Responses to Cold

The body maintains core temperature in the cold by minimizing heat loss and, when this is insufficient, increasing heat production. Reducing core-to-skin thermal conductance is the chief physiologic means of heat conservation in humans. Furred or hairy animals also can increase the thickness of their coat, and thus its insulating properties, by making the hairs stand on end. This response, called **piloerection,** makes a negligible contribution to heat conservation in humans, but manifests itself as "goose flesh."

■ Blood Vessels in the Shell Constrict to Conserve Heat

Constriction of cutaneous arterioles reduces skin blood flow and core-to-skin thermal conductance. Constriction of the superficial limb veins further improves heat conservation by diverting venous blood

to the deep limb veins, which lie close to the major arteries of the limbs and do not constrict in the cold. (Many penetrating veins connect the superficial veins to the deep veins, so that venous blood from anywhere in the limb potentially can return to the heart via either superficial or deep veins.) In the deep veins, cool venous blood returning to the core can take up heat from the warm blood in the adjacent deep limb arteries. Thus some of the heat contained in the arterial blood as it enters the limbs takes a "short circuit" back to the core. When the arterial blood reaches the skin it is already cooler than the core and so loses less heat to the skin than it otherwise would. (When the superficial veins dilate in the heat, most venous blood returns via superficial veins so as to maximize core-to-skin heat flow.) The transfer of heat from arteries to veins by this "short circuit" is called **countercurrent** heat exchange, and it can cool the blood in the radial artery of a cool but comfortable subject to as low as 30°C by the time it reaches the wrist.

As we saw earlier, the shell's insulating properties increase in the cold, as its blood vessels constrict and its thickness increases. Furthermore, the shell includes a fair amount of skeletal muscle in the cold, and although muscle blood flow is believed not to be affected by thermoregulatory reflexes it is reduced by direct cooling. In a cool subject, the resulting reduction in muscle blood flow adds to the shell's insulating properties. As the blood vessels in the shell constrict, blood is shifted to the central blood reservoir in the thorax. This shift produces many of the same effects

HYPOTHERMIA

Hypothermia is classified according to the patient's core temperature as mild (32–35°C), moderate (28–32°C), or severe (below 28°C). Shivering is usually prominent in mild hypothermia, but diminishes in moderate hypothermia and is absent in severe hypothermia. The pathophysiology is characterized chiefly by the depressant effect of cold (via the Q_{10} effect) on multiple physiologic processes and differences in the degree of depression of each process. Apart from shivering, the most prominent features of mild and moderate hypothermia are due to depression of the central nervous system. These begin with mood changes (commonly apathy, withdrawal, and irritability) progressing, as hypothermia deepens, to confusion and lethargy and then ataxia and speech and gait disturbances, which may mimic a cerebrovascular accident ("stroke"). In severe hypothermia, voluntary movement, reflexes, and consciousness are lost and muscular rigidity appears. Cardiac output and respiration decrease as core temperature falls. Myocardial irritability increases in severe hypothermia, causing a substantial danger of ventricular fibrillation, with the risk increasing as cardiac temperature falls. The primary mechanism presumably is that cold depresses conduction velocity in Purkinje fibers more than in ventricular muscle, favoring the development of circus-movement propagation of action potentials. Myocardial hypoxia also contributes. In more profound hypothermia, cardiac sounds become inaudible and pulse and blood pressure are unobtainable, because of circulatory depression; the electrical activity of the heart and brain becomes unmeasurable; and extensive muscular rigidity may mimic **rigor mortis.** The patient may appear clinically dead, but patients have been revived from core temperatures as low as 17°C, so that "no one is dead until warm and dead." The usual causes of death during hypothermia are cessation of respiration and failure of cardiac pumping, because of either ventricular fibrillation or direct depression of cardiac contraction.

Depression of renal tubular metabolism by cold impairs reabsorption of sodium, causing a diuresis and leading to dehydration and hypovolemia. Acid-base disturbances in hypothermia are complex. Respiration and cardiac output typically are depressed more than metabolic rate, and a mixed respiratory and metabolic acidosis results, due to CO_2 retention and lactic acid accumulation and the cold-induced shift of the hemoglobin-O_2 dissociation curve to the left. Acidosis aggravates the susceptibility to ventricular fibrillation.

Treatment consists of preventing further cooling and restoring fluid, acid-base, and electrolyte balance. Patients in mild to moderate hypothermia may be warmed solely by providing abundant insulation to promote retention of metabolically produced heat; those who are more severely affected require active rewarming. The most serious complication associated with treating hypothermia is development of ventricular fibrillation. Vigorous handling of the patient may trigger this, but an increase in the patient's circulation (e.g., associated with warming or skeletal muscle activity) may itself increase the susceptibility to such an occurrence. This may happen as follows. Peripheral tissues of a hypothermic patient are, in general, even cooler than the core, including the heart, and acid products of anaerobic metabolism will have accumulated in underperfused tissues while the circulation was most depressed. As the circulation increases, a large increase in blood flow through cold, acidotic peripheral tissue may return enough cold, acidotic blood to the heart to cause a transient drop in the temperature and pH of the heart increasing its susceptibility to ventricular fibrillation.

The diagnosis of hypothermia is usually straightforward in a patient rescued from the cold but may be far less clear in a patient in whom hypothermia is the result of a serious impairment of physiologic and behavioral defenses against cold. A typical example is the elderly person, living alone, who is discovered at home, cool and obtunded or unconscious. The setting may not particularly suggest hypothermia, and when the patient comes to medical attention the diagnosis may easily be missed, since standard clinical thermometers are not graduated low enough (usually only to 34.4°C) to detect hypothermia, and in any case do not register temperatures below the level to which the mercury has been shaken. Because of the depressant effect of hypothermia on the brain, the patient's condition may be misdiagnosed as cerebrovascular accident or other primary neurologic disease. Recognition of this condition thus depends on the physician's considering it when examining a cool, obtunded patient and obtaining a true core temperature with a low-reading glass thermometer or other device.

as an increase in blood volume, including diuresis (so-called **cold diuresis**) as the kidneys respond to the increased central blood volume.

Once skin blood flow is near minimal, metabolic heat production increases. In human adults, nearly all of this increase occurs in skeletal muscles, as a result first of increased tone and later of frank shivering. Shivering may increase metabolism at rest by more than fourfold acutely, but only about half that amount can be sustained after several hours.

■ Human Cold Acclimatization Confers a Modest Advantage

The pattern of human cold acclimatization depends on the nature of the cold exposure. It is partly for this reason that the occurrence of cold acclimatization in humans was long controversial. Our knowledge of human cold acclimatization comes both from laboratory studies and from studies of populations whose occupation or life-style exposes them repeatedly to cold.

Metabolic Changes in Cold Acclimatization ■ At one time it was believed that humans must acclimatize to cold as laboratory mammals do—by increasing their metabolic rate. There are a few reports of increased basal metabolic rate, and sometimes also thyroid activity, in winter, and there is evidence for functioning brown fat in the neck and mediastinum of outdoor workers. More often, however, increased metabolic rate has not been observed in studies of human cold acclimatization. In fact, there are a number of reports of the opposite response, consisting of a lower core temperature threshold for shivering, with a greater fall in core temperature and a smaller metabolic response during cold exposure. Such a response would spare metabolic energy, and so might be advantageous in an environment that is not so cold that a blunted metabolic response would allow core temperature to fall to dangerous levels.

Increased Tissue Insulation in Cold Acclimatization ■ A lower core-to-skin conductance (i.e., increased insulation by the shell) has often been reported in studies of cold acclimatization in which a reduction in the metabolic response to cold occurred.

This increased insulation is not due to subcutaneous fat (in fact, it has been observed in very lean subjects), but apparently results from lower blood flow in the limbs or improved countercurrent heat exchange in the acclimatized subjects. In general, the cold stresses that elicit a lower core-to-skin conductance after acclimatization involve either cold water immersion or exposure to air temperatures that are chilly but not so cold as to risk freezing the vasoconstricted extremities.

Cold-Induced Vasodilation and the Lewis Hunting Response ■ As the skin is cooled below about 15°C, its blood flow begins to increase somewhat, a response called **cold-induced vasodilation** (CIVD). This response is elicited most easily in comfortably warm subjects and in skin rich in arteriovenous anastomoses (in the hands and feet). The mechanism has not been established but may involve a direct inhibitory effect of cold on contraction of vascular smooth muscle or on neuromuscular transmission. After repeated cold exposure, CIVD begins earlier during cold exposure, produces higher levels of blood flow, and takes on a rhythmic pattern of alternating vasodilation and vasoconstriction. This is called the **Lewis hunting response,** since the rhythmic pattern of blood flow suggests that it is "hunting" for its proper level. This response is often well developed in workers whose hands are exposed to cold, such as fishermen working with nets in cold water. Since the Lewis hunting response increases heat loss from the body somewhat, it is debatable whether it is truly an example of acclimatization to cold. It is, however, advantageous, since it keeps the extremities warmer, more comfortable, and functional and probably protects them from cold injury.

Clinical Aspects of Thermoregulation

Temperature is important clinically because of the presence of fever in many diseases, the effects of many factors on tolerance to heat or cold stress, and the effects of heat or cold stress in causing or aggravating certain disorders.

■ Fever Enhances Defense Mechanisms

Infection, inflammatory processes such as collagen vascular diseases, trauma, neoplasms, acute hemolysis, and diseases due to immune mechanisms release an assortment of proteins, protein fragments, and bacterial lipopolysaccharide toxins, collectively called **pyrogens** or **exogenous pyrogens.** Exogenous pyrogens stimulate monocytes and macrophages to release **endogenous pyrogen** (EP), a protein that causes the thermoreceptors in the hypothalamus (and perhaps elsewhere in the brain) to alter their firing rate and input into the thermoregulatory integrating centers, and thereby raises the thermoregulatory set point. This effect of EP is believed to be mediated by local release of prostaglandins. Aspirin and other drugs that inhibit the synthesis of prostaglandins also reduce fever.

Fever accompanies disease so frequently, and is such a reliable indicator of the presence of disease, that body temperature is probably the most commonly measured clinical index. Many of the body's defenses against infection and cancer are elicited by a group of proteins called **cytokines,** and EP is now identified with a member of this group, **interleukin 1** (IL-1). (However, other cytokines, particularly **tumor necrosis factor** and **interleukin-6,** are pyrogenic too.) Recent evidence indicates that elevated body temperature enhances the development of these defenses, contradicting the long-held belief that fever confers no benefit. (Although in this chapter *fever* means specifically an elevation in core temperature due to pyrogens, some authors use the term more generally to mean any significant elevation of core temperature.)

■ Many Factors, Including Physical Fitness, Age, Drugs, and Diseases, Affect Thermoregulatory Responses and Tolerance to Heat and Cold

Regular physical exercise and heat acclimatization increase heat tolerance and the sensitivity of the sweating response. Aging has the opposite effect; in healthy 65-year-old men the sensitivity of the sweating response is half that in 25-year-old men. Many drugs inhibit sweating, most obviously those used for their anticholinergic effects, such as atropine and scopolamine. In addition, some drugs used for other purposes, such as glutethimide (a sleep-inducing drug), tricyclic antidepressants, phenothiazines (tranquilizers and antipsychotic drugs), and antihistamines, also have some anticholinergic action. All of these and several others have been associated with heat stroke. Congestive heart failure and certain skin diseases (e.g., ichthyosis and anhidrotic ectodermal dysplasia) impair sweating, and in patients with these diseases heat exposure and, especially, exercise in the heat may raise body temperature to dangerous levels. Neurologic diseases that involve the thermoregulatory structures in the brainstem can impair thermoregulation. Although such disorders can produce **hypothermia** (abnormally low core temperature), **hyperthermia** (abnormally high core temperature) is more usual, and typically is characterized by loss of sweating and the circadian rhythm.

Certain drugs, such as barbiturates, alcohol, and phenothiazines, and certain diseases, such as hypothyroidism, hypopituitarism, congestive heart failure, and septicemia, may impair the defense against cold. (Thus septicemia, especially in debilitated patients, may be accompanied by hypothermia, instead of the usual febrile response to infection.) Furthermore, newborns and many healthy elderly persons are less able than older children and younger adults to maintain body temperature in the cold. This appears to be due to a reduced ability both to conserve body heat by reducing heat loss and to increase metabolic heat production in the cold.

■ Heat Stress Causes or Aggravates a Number of Disorders

The harmful effects of heat stress are exerted through cardiovascular strain, fluid and electrolyte loss, and, especially in heat stroke, tissue injury whose mechanism is uncertain. In a patient suspected of having hyperthermia secondary to heat stress, temperature should be measured in the rectum, since hyperventilation may render oral temperature spuriously low.

Heat Syncope ■ **Heat syncope** is circulatory failure due to pooling of blood in the peripheral veins, with a consequent decrease in venous return and diastolic filling of the heart, and thus a decrease in cardiac output, and a fall of arterial pressure. Symptoms range from lightheadedness and giddiness to loss of consciousness. Thermoregulatory responses are intact, so core temperature typically is not substantially elevated, and the skin is wet and cool. The large thermoregulatory increase in skin blood flow in the heat is probably the primary cause of the peripheral pooling. Heat syncope affects mostly those who are not acclimatized to heat.

Water and Salt Depletion Due to Heat Exposure ■ Water and salt can be lost rapidly in the sweat, and people exercising in the heat drink less water than they are losing if they drink only according to their thirst. Therefore, complete water replacement during heat stress is difficult even if ample water is available, and gradual progressive dehydration is likely to occur. Salt loss in the sweat is quite variable; some people lose enough salt to become salt-depleted even though water is replaced. Thus **heat exhaustion** may be associated either predominantly with salt depletion or predominantly with water depletion. Since body water is distributed so as to maintain osmotic balance between the intra- and extracellular spaces, the body water of salt-depleted patients is lost predominantly from the extracellular space, so that these patients are hypovolemic out of proportion to the degree of dehydration.

Like heat syncope, heat exhaustion is characterized by reduced diastolic filling of the heart, but hypovolemia plays a much greater role in its development and the baroreflex responses are usually sufficient to maintain consciousness in spite of the hypovolemia. The baroreflex responses may be manifested in nausea, vomiting, pallor, cool or even clammy skin, and tachycardia. The patient usually is sweating profusely. Heat exhaustion ranges from fairly mild disorders that respond well to rest in a cool environment and oral fluid replacement, to severe forms that require intravenous replacement of fluid and salt and may be accompanied by collapse, confusion, and hyperthermia. Unconsciousness is infrequent, but there may be vertigo, ataxia, headache, weakness, and low blood pressure. Dehydration impairs thermoregulation, and heat exhaustion may lead to heat stroke. Therefore, patients should be actively cooled if rectal temperature is 40.6°C (105°F) or higher.

Heat Stroke ■ The most severe and dangerous heat disorder is characterized by rapid development of hyperthermia and severe neurologic disturbances, with loss of consciousness and, frequently, convulsions. Hepatic and renal injury and disturbances of blood clotting are frequent accompaniments. The pathogenesis is not well understood. Factors besides hyperthermia probably contribute to its development, and several lines of evidence suggest that products of the bacterial flora in the gut—perhaps lipopolysaccharide endotoxins—enter the circulation and play an important role in the development of heat stroke.

Heat stroke occurs in two forms, classical and exertional. In the classical form, the primary factor is environmental heat stress that overwhelms an impaired thermoregulatory system; in exertional heat stroke the primary factor is high metabolic heat production. Patients with exertional heat stroke tend to be younger and more physically fit (typically, soldiers and athletes) than patients with the classical form. The traditional diagnostic criteria of heat stroke—coma, hot dry skin, and rectal temperature above 41.3°C (106°F)—are seen primarily with the classical form. Patients with exertional heat stroke tend to have somewhat lower rectal temperatures and may be sweating profusely. Heat stroke is a medical emergency, with a high mortality if not treated promptly and vigorously. Prompt lowering of core temperature is the cornerstone of treatment, and this is most effectively accomplished by immersion in cold water.

Malignant hyperpyrexia, or **malignant hyperthermia,** a rare process triggered by inhalational anesthetics or neuromuscular blocking agents, was once considered a form of heat stroke but is now known to be a distinct disorder that occurs in genetically susceptible individuals. Susceptibility may be associated with any of several myopathies or may occur as an autosomal dominant trait with no other clinical manifestations. In 90% of susceptible individuals, biopsied skeletal muscle tissue contracts on exposure to caffeine, halothane, or hexamethonium in concentrations having little effect on normal muscle. Reuptake of calcium ion by the sarcoplasmic reticulum is severely impaired, so that calcium concentration in the cytoplasm rises, activating myosin ATPase and leading to an uncontrolled hypermetabolic process that produces a rapid rise in core temperature. Treatment with dantrolene sodium, which appears to act by reducing release of calcium ion from the sarcoplasmic reticulum, has dramatically reduced the mortality rate of this disorder.

Aggravation of Disease States Due to Heat Exposure ■ Besides producing specific disorders, heat exposure aggravates a number of other diseases. Epidemiologic studies show that during unusually hot weather, mortality may be two to three times that normally expected for the months in which "heat waves" occur. Deaths ascribed to specific heat disorders account for only a small fraction of the excess mortality (i.e., the increase above the mortality expected for the month). Most of the excess mortality is accounted for by deaths from diabetes, various diseases of the cardiovascular system, and diseases of the blood-forming organs.

■ Hypothermia Occurs When the Body's Defenses against Cold Are Disabled or Overwhelmed

Hypothermia reduces metabolic rate via the Q_{10} effect and thus prolongs the time tissues can safely tolerate loss of blood flow. Since the brain is damaged by ischemia soon after circulatory arrest, controlled hypothermia is often used to protect the brain during surgical procedures in which its circulation is occluded or the heart is stopped. Much of our knowledge about the physiologic effects of hypothermia comes from observations on surgical patients.

During the initial phases of cooling, stimulation of shivering through thermoregulatory reflexes overwhelms the Q_{10} effect. Metabolic rate therefore increases, reaching a peak at a core temperature of 30-33°C. At lower core temperatures, however, metabolic rate is dominated by the Q_{10} effect, and thermoregulation is lost. A vicious circle develops, wherein a fall in core temperature depresses metabolism and allows core temperature to fall further, so that at 17°C, O_2 consumption is about 15% and cardiac output 10% of precooling values.

Hypothermia that is not induced for therapeutic purposes is called **accidental hypothermia.** It occurs in individuals whose defenses are impaired by drugs (especially ethanol, in the United States), disease, or other physical conditions and in healthy individuals who are immersed in cold water or become exhausted working or playing in the cold.

■ ■ ■ ■ ■ ■ ■ ■ ■ ■ ■ ■ **REVIEW EXERCISES** ■ ■ ■ ■ ■ ■ ■ ■ ■ ■ ■ ■

Identify Each with a Term

1. An abnormally low body core temperature

2. An abnormally high body core temperature

3. The process of heat transfer in a fluid (liquid or gas) that depends on the movement of the fluid that carries the heat

4. The heat required to evaporate a liquid (e.g., sweat on the skin)

5. The increase in hand blood flow that may occur when the hand is cooled to temperatures below about 15°C

6. A disorder characterized by dehydration and impairment of diastolic cardiac filling during heat exposure

7. The range of environmental conditions in which the body maintains heat balance without either shivering or sweating

8. A process by which heat is transferred from a warmer surface to a cooler surface by electromagnetic (typically, infrared) waves

9. The ratio of the rates at which a reaction or physiologic process occurs at two different temperatures 10°K apart

10. The histologic type of sweat gland that is most important in thermoregulation

Define Each Term

11. Endogenous pyrogen

12. Basal metabolic rate

13. Sweat gland fatigue

14. Shell

15. Homeotherm

16. Thermoregulatory set point

17. Circadian rhythm

18. Insensible perspiration

Choose the Correct Answer

19. Which statement best describes current knowledge about the postganglionic nerve fibers that control skin blood flow and sweating?
    a. Skin blood flow is controlled by adrenergic sympathetic nerves, and sweating is controlled by cholinergic (i.e., acetylcholine-secreting) parasympathetic nerves.
    b. Skin blood flow is controlled by adrenergic sympathetic nerves, and sweating is controlled by cholinergic sympathetic nerves.
    c. Skin blood flow is controlled by adrenergic sympathetic vasoconstrictor nerves and cholinergic sympathetic vasodilator nerves, and sweating is controlled by cholinergic sympathetic nerves.
    d. Skin blood flow is controlled by adrenergic sympathetic vasoconstrictor nerves and sympathetic vasodilator nerves that secrete an unidentified neurotransmitter(s), and sweating is controlled by cholinergic sympathetic nerves.
    e. Skin blood flow is controlled by adrenergic sympathetic vasoconstrictor nerves and cholinergic sympathetic vasodilator nerves, and sweating is controlled by cholinergic parasympathetic nerves.

20. The core temperature threshold for sweating is about 0.7°C higher at 4:00 p.m. than it is at 4:00 a.m. Which of the following best describes the threshold for cutaneous vasodilation, and the core temperature of a resting subject, at those two times?
    a. At 4:00 p.m., core temperature is about the same as at 4:00 a.m., but the threshold for cutaneous vasodilation is *lower* than at 4:00 a.m., because at night the body relies more on sweating and less on skin blood flow to dissipate heat than it does during the day.
    b. The threshold for cutaneous vasodilation is the same at 4:00 p.m. as at 4:00 a.m.; core temperature is the same also, unless the ambient temperature is high enough to make the subject sweat.
    c. At 4:00 p.m., both core temperature and the threshold for cutaneous vasodilation are also about 0.7°C higher than at 4:00 a.m., because all of these changes are due to a change in thermoregulatory set point, which affects all thermoregulatory responses the same.
    d. The threshold for cutaneous vasodilation is the same at 4:00 p.m. as at 4:00 a.m. Core temperature is higher at 4:00 p.m. than at 4:00 a.m. (though not as much as 0.7°C higher), depending on the balance between sweating and vasodilation.
    e. At 4:00 p.m., the threshold for cutaneous vasodilation is *lower* than at 4:00 a.m., because the skin is warmer than at 4:00 a.m. Core temperature is higher at 4:00 p.m. than at 4:00 a.m. by an amount depending on the balance between sweating and vasodilation.

21. Compared to an unacclimatized person, one who is acclimatized to cold has:
    a. A higher metabolic rate in the cold, to produce more heat
    b. A lower metabolic rate in the cold, to conserve metabolic fuel
    c. Lower peripheral blood flow in the cold, to retain heat
    d. Higher blood flow in the hands and feet in the cold, to preserve their function.
    e. a and d
    f. b and c
    g. Various combinations of the above, depending on circumstances

Calculate (see Table 31–3 for values needed)

22. Muscle blood flow is determined primarily by local metabolic needs. The blood flow needed to sustain aerobic metabolism depends on the O_2 content, which typically is no more than 200 ml/Liter of arterial blood (hemoglobin concentration in blood of 150 g/L × 1.34 ml O_2/g of saturated hemoglobin). (a) Assuming oxidation of a mixed diet, how much metabolic heat can be produced using the oxygen in 1 liter of blood? (b) The blood perfusing an organ comes into thermal equilibrium with the organ and so leaves the organ with the same temperature as the organ. If an organ is warmer than the incoming blood, the blood flow through the organ removes heat from the organ. Using the answer to question a and the specific heat of blood, calculate

TABLE 31–3 ■ Illustrative Values for Thermal Physiology

Measurement	Heat Units, S.I.*	Heat Units, Traditional
Energy equivalent of oxygen for a mixed diet	20.2 kJ/L	4.83 kcal/L
Heat of evaporation of water	2.43 kJ/g	0.58 kcal/g
Data for a "typical" healthy lean young man		
Mass	70 kg	
Body surface area	1.8 m²	
Mean specific heat of the body	3.55 kJ/(kg·°C)	0.85 kcal/(kg·°C)
Volume specific heat of blood	3.85 kJ/(L·°C)	0.92 kcal/(L·°C)
Maximum rate of O_2 consumption ($\dot{V}_{O_2}$ max)	3.5 L/min	
Metabolic rate at rest**	45 W/m²	52.3 kcal/(m²·h)
Core-to-skin conductance with minimal skin blood flow**	9 W/(m²·°C)	10.5 kcal/(m²·°C·h)

*Système Internationale (in which heat is expressed in units of work).
**Per m² of body surface area

the amount by which the temperature of the blood leaving the muscle (and thus the temperature of the muscle) exceeds the temperature of the inflowing blood under these circumstances. This example shows the greatest extent to which an organ's metabolism can heat the organ above the temperature of its blood supply. (Since most organs ordinarily extract only a fraction of the O_2 contained in their blood supply, they will be somewhat cooler than this answer indicates.)

23. A healthy 70-kg young man can readily exercise at a metabolic rate of 800 W (48 kJ/min). His exercise produces mechanical work at a rate of 140 W and his rate of dry heat loss ($R + C$) is 100 W. (a) Assuming oxidation of a mixed diet, how much O_2 per minute does he need to sustain this exercise? Compare your answer to the "typical" value of $\dot{V}_{O_2}$ max in Table 31–3. (b) At what rate must he dissipate heat by evaporation to achieve heat balance? Making the slightly artificial assumption that sweat provides all of the water he is evaporating, at what rate must he sweat to achieve heat balance? (c) If he cannot sweat, how long will it take his core temperature to rise by 4°C? (Use your answer to the first question in part b, and the mean specific heat of the body in Table 31–3.) This increase would bring his core temperature to about 41°C, a potentially dangerous level.

Briefly Answer

24. (a.) What are the two components of heat transfer through the shell? (Hint: see equation 10.) (b) Describe the effect of a subcutaneous layer of fat on the insulation value of the shell in the cold and in the heat. Explain any differences between cold and heat in this regard.

25. A subject dressed only in shorts sits on a stool in a climatic chamber with a temperature (both air and walls) of 32.5°C (90.5°F). The subject initially is in thermal balance. The chamber temperature is dropped to 10°C (50°F), and he continues sitting on the stool until he comes into heat balance at the new chamber temperature. Let us call phase 1 the period when the chamber

is at 32.5°C, phase 2 the period when the chamber is at 10°C and the subject has not yet reached thermal balance, and phase 3 the period when the chamber is at 10°C and the subject has reached thermal balance. In general terms (e.g., positive, negative, large, negligible, zero, increase, decrease): (a) Describe the subject's thermoregulatory responses, how they change, and the reflex mechanisms through which any changes are brought about, in phases 1, 2, and 3. (Assume that at thermal balance at 10°C, total rate of heat loss to the environment is higher than at 32.5°C.) (b) Describe the changes that occur in mean skin temperature, the thickness of the shell, and the average shell temperature, and relate these changes to the total heat content of the body. (c) In this example, $W = 0$, and K is negligible, so the heat balance equation can be simplified to $M = E + R + C + S$. Describe each term in the simplified heat balance equation, and how it changes, in phases 1, 2, and 3.

26. Describe the effects of acclimatization to heat on: (a) the levels of heart rate and rectal temperature reached during a given level of exercise in the heat; (b) the thermoregulatory set point and thresholds for sweating and cutaneous vasodilation; (c) sweating and sweat gland function, including any effect on sweat composition.

27. List four factors that change the thermoregulatory set point: give the direction(s) of the change in set point that each one produces.

28. A duck hunter falls into a chilly lake. Fortunately he is wearing a life vest and tries to remain fairly still so as not to stir the water and increase his rate of heat loss. When he is rescued, his core temperature has fallen to 29°C. What happens to his metabolic rate: (a) in the first few minutes after he falls in the water; (b) as his core temperature falls a few degrees (say, to 33°C or so); and (c) as his core temperature continues to fall to 29°C? What physiologic mechanism is responsible for the effect on metabolic rate in each period?

■ ■ ■ ■ ■ ■ ■ ■ ■ ■ ■ ■ **ANSWERS** ■ ■ ■ ■ ■ ■ ■ ■ ■ ■ ■ ■

1. Hypothermia

2. Hyperthermia

3. Convection

4. Latent heat of evaporation

5. Cold-induced vasodilation (Lewis hunting reaction is an acceptable answer, but implies a rhythmic pattern that was not mentioned as part of the question)

6. Heat exhaustion

7. Thermoneutrality or thermoneutral zone

8. Radiation

9. Q_{10}

10. Eccrine

11. A protein (or group of proteins) that acts on the thermoreceptors in the hypothalamus to raise the thermoregulatory set point and produce fever

12. The body's rate of energy use, measured at rest under standard conditions (i.e., in an awake, well-rested person after a 12-hour fast and at a comfortable room temperature). (Note that basal metabolic rate is not the body's minimum physiologic metabolic rate.)

13. A reduction of sweat gland responsiveness after prolonged heavy sweating (partly caused by, but not the same as, hidromeiosis, or mechanical obstruction of the sweat duct due to swelling of the stratum corneum)

14. The outer layer of the body, surrounding, and cooler than, the core, the thickness of which depends on the body's thermal state

15. An animal that maintains its core temperature within a relatively narrow range, in spite of changes in environmental temperature

16. The setting of the body's "thermostat"; more precisely, a reference point that determines the thresholds of all the thermoregulatory responses

17. A rhythmic pattern, with a period of about 24 hours, in a physiologic function

18. Moisture that diffuses through the skin and evaporates, as distinguished from water that evaporates after being secreted by sweat glands

19. d

20. c

21. g

22. (a) 20.2 kJ/L × 0.2 L = 4.04 kJ
    (b) 4.04 kJ/L of blood ÷ 3.85 kJ/ (L·°C) = 1.05°C

23. (a) 48 kJ/min ÷ 20.2 kJ/L = 2.38 L/min; 2.38 L/min ÷ 3.5 L/min = 68% $\dot{V}_{O_2}$ max
    (b) 800W − 140W − 100W = 560W; 0.56 kJ/sec ÷ 2.43 kJ/g = 0.23 g/sec = 13.8 g/min
    (c) 70 kg × 3.55 kJ/(kg·°C) × 4°C ÷ 0.56 kJ/sec = 1775 sec = 29.6 min

24. (a) Conduction through the shell and convection by blood flow to the skin
    (b) The layer of subcutaneous fat affects only the heat that is transferred through the shell by conduction. In the cold, the rate of heat transfer through the shell via convection by the blood is small, and most of the heat transfer through the shell occurs by conduction. Since the heat passing through the shell by conduction has to be conducted through the layer of subcutaneous fat, if that layer is fairly thick it contributes a great deal to the overall insulation value of the shell. In the heat, most of the heat transfer through the shell takes place via convection by the blood, which is unaffected by the subcutaneous fat layer. The subcutaneous fat layer affects only heat passing through the shell by conduction; since that is only a minor part of total heat transfer in the heat, the subcutaneous fat layer contributes little to the insulation value of the shell in the heat.

25. (a) In phase 1, the subject is sweating, skin blood flow is elevated, and metabolic rate is minimal. In phase 2, sweating drops and then stops altogether, skin blood flow drops, and eventually metabolic rate increases due to shivering. The reflex effect of the decrease in skin temperature is the chief mechanism producing these changes in sweating, skin blood flow, and shivering. (Whether or not core temperature changes in this example depends on the subject's individual characteristics, but if core temperature changes it, too, will affect these responses.) In phase 3, sweating has stopped, skin blood flow is minimal, and metabolic rate is above its resting value.
    (b) Due to the decrease in skin blood flow and the fall in the temperature of the environment, mean skin temperature falls and the boundary between core and shell moves inward, so that the thickness of the shell increases. Since all the tissue that makes up the shell in phase 3 is at a lower temperature than in phase 1, the total heat content of the body is reduced.
    (c) In phase 1, M is minimal and R and C are relatively small because of the warm environment. Since the subject is in thermal balance, $S = 0$ and E is elevated to compensate for the relatively low R and C. In phase 2, R and C increase as chamber temperature falls and E falls, eventually reaching a minimal level determined by respiratory water loss and insensible perspiration. S is negative, since the body is losing heat until M increases enough to match the increase in heat loss. Since total rate of heat loss to the environment eventually reaches a value higher than in phase 1, and since the subject does reach a new thermal balance, M starts to rise at some point during phase 2. In phase 3, M is elevated somewhat above its value in phase 1, R and C are elevated proportionately more so, E is minimal, and, since the subject has reached a new thermal balance, $S = 0$ once again.

26. (a) After heat acclimatization, the levels of heart rate and core temperature reached during any level of exercise in the heat are lower than before acclimatization.
    (b) The thermoregulatory set point and the thresholds for sweating and cutaneous vasodilation are all lower after acclimatization than before.
    (c) After acclimatization, the sweat glands are capable of secreting sweat at a higher rate than before acclimatization. They become more resistant to sweat gland fatigue, so they are able to maintain high sweat rates for a longer time, and they secrete sweat with a lower salt concentration.

27. Fever raises the thermoregulatory set point; heat acclimatization lowers it; and the circadian rhythm and the menstrual cycle alter it in a cyclic fashion, so that it is higher in the late afternoon than the early morning and higher in the luteal phase than just before ovulation.

28. (a) The metabolic rate rises due to onset of shivering soon after falling into the lake, as a result of the reflex effect of the decrease in skin temperature.

(b) As his core temperature falls, shivering and metabolic rate increase even more, since a reflex effect of the decrease in core temperature is added to the reflex effect of skin temperature.

(c) Lower temperatures tend to reduce all metabolic processes (including those of the central nervous system and skeletal muscle) via the Q_{10} effect. As his core temperature continues to fall, it reaches a point where the Q_{10} effect starts to "win out" over the reflex effect of core temperature on shivering. Once that point is reached, further decreases in core temperature cause a reduction in metabolic rate.

Suggested Reading

Dinarello CA: Biology of interleukin 1. *FASEB J* 2:108–115, 1988.

Gagge AP, Nishi Y: Heat exchange between human skin surface and thermal environment, in DHK Lee, Falk HL, Murphy SD (eds): *Handbook of Physiology: Reactions to Environmental Agents.* Bethesda, American Physiological Society, 1977, sect. 9, chap. 5, pp 69–92.

Gordon CJ, Heath JE: Integration and central processing in temperature regulation. *Annu Rev Physiol* 48:595–612, 1986.

Hubbard RW, Gaffin SL, Squire D: Heat-related illnesses, in Auerbach PS (ed): *Management of Wilderness and Environmental Emergencies,* 3rd ed. St. Louis, Mosby-Year Book, 1995, chap. 8, in press.

Knochel JP, Reed G: Disorders of heat regulation, in Maxwell HH, Kleeman CR, Narins RG (eds): *Clinical Disorders of Fluid and Electrolyte Metabolism.* 4th ed. New York, McGraw-Hill, 1989, chap. 47, pp. 1197–1232.

Mitchell D, Laburn HP: Pathophysiology of temperature regulation. *Physiologist* 28:507–517, 1985.

Pandolf KB, Sawka MN, Gonzalez RR (eds): *Human Performance Physiology and Environmental Medicine at Terrestrial Extremes.* Indianapolis, Benchmark, 1988.

Petersdorf RG: Hypothermia and hyperthermia, in Wilson JD, Braunwald E, Isselbacher KJ, et al (eds): *Harrison's Principles of Internal Medicine.* 12th ed. New York, McGraw-Hill, 1991, pp 2194–2200.

Root RK, Petersdorf RG: Fever and chills, in Wilson JD, Braunwald E, Isselbacher KJ, et al (eds): *Harrison's Principles of Internal Medicine.* 12th ed. New York, McGraw-Hill, 1991, pp 125–133.

Rowell LB: Cardiovascular aspects of human thermoregulation. *Circ Res* 52:367–379, 1983.

C H A P T E R

■ ■ ■ ■ ■ ■ ■ ■

32

Exercise Physiology

OBJECTIVES
After studying this chapter, the student should be able to:
1. List the aspects of "exercise" that must be defined before predicting physiologic responses
2. Define $\dot{V}o_2$ max, what physiologic factors limit it, and its usefulness in predicting work performance
3. Describe the shifts in regional blood flow that occur during dynamic and isometric exercise and the physiologic control of these changes
4. Define the effects of training on the heart and coronary circulation
5. Explain how the respiratory system responds to increased O_2 consumption and CO_2 production with exercise
6. Explain what causes muscle fatigue with exercise
7. Describe how chronic physical activity alters insulin sensitivity and glucose entry into cells, and the implications of these changes for persons with diabetes mellitus

CHAPTER OUTLINE (Continued)

VI. IMMUNE, PSYCHIATRIC, AND AGING RESPONSES

A. Acute exercise transiently elevates many circulating immune system markers, but the long-term effects of training on immune function are unclear

B. Exercise may help relieve depression in some patients, but its efficacy and neurochemical effects are uncertain

C. As persons age, the effects of exercise on functional capacity are much more profound than their effect on longevity

---

Exercise, or physical activity, is an ubiquitous physiologic state, so common in its many forms that true physiologic "rest" is indeed rarely achieved. Defined ultimately in terms of skeletal muscle contraction, exercise involves every organ system in coordinated response to increased muscular energy demands.

Exercise is as various as it is ubiquitous. A single episode of exercise, or "acute" exercise, may provoke responses quite different from the adaptations seen when activity is chronic (commonly called training). The forms of exercise vary as well. The amount of muscle mass at work (one arm? one finger? both legs?), the intensity of the effort, its duration, and the type of muscle contraction itself (isometric, rhythmic) all profoundly determine the organism's responses and adaptations.

These many aspects of exercise imply that its interaction with disease is multifaceted. There is no simple answer as to whether exercise promotes health. In fact, physical activity can be healthful, harmful, or irrelevant, depending on the patient, the disease, and the specific exercise in question.

Quantification of Exercise

■ The Most Common Method of Quantifying Aerobic Exercise Is with Reference to Maximum Oxygen Uptake ($\dot{V}o_2$ max)

Fundamental to any discussion of exercise is a description of its intensity. Since exercising muscle primarily uses energy in the long-term from oxidative metabolism, a century-old standard is to measure, by mouth, the oxygen consumption of an exercising subject. This measurement is limited to dynamic exercise and usually to the steady state, where exercise intensity and oxygen consumption are stable and no net energy is provided from nonoxidative sources. Two outgrowths of the original oxygen consumption measurements deserve mention. First, the apparent excess in oxygen consumption during the first minutes of recovery has been termed its **oxygen debt** (Fig. 32-1), a term often used as a rough synonym for "fatigue" or "dyspnea." In fact, the "excess" oxygen consumption of recovery results from a multitude of physiologic

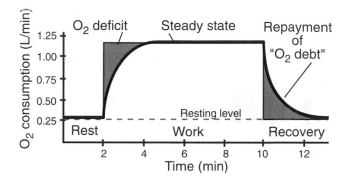

Figure 32–1 ■ Oxygen consumption during light steady-state exercise.

processes, and little usable information is obtained from its measurement. Second, and more useful, each person has a maximal oxygen consumption, a ceiling up to 20 times basal consumption, that cannot be exceeded, though it can be increased by appropriate training. This **maximal oxygen uptake ($\dot{V}o_2$ max)** is a useful but imperfect predictor of endurance athletic performance or, more generally, the ability to perform prolonged dynamic external work. $\dot{V}o_2$ max is decreased, all else equal, by age, bed rest, or increased body fat. Because maximal oxygen uptake represents a physiologic limit to oxygen transport and use, any defect in oxygen delivery to tissue or oxygen use by that tissue lowers $\dot{V}o_2$ max and hence endurance exercise capacity. For example, completion of Interstate 70 at 11,000 ft altitude in Colorado was delayed because the work capacities of laborers were reduced by the low atmospheric Po_2 and resultant low $\dot{V}o_2$ max.

$\dot{V}o_2$ max is also used to express relative work capacity. A world-champion cross-country skier obviously has a greater $\dot{V}o_2$ max than a novice, but when both work at an intensity requiring two-thirds of their respective oxygen uptake maxima (the world champion is moving much faster in doing this, due to higher capacity), both become exhausted at about the same time and for the same physiologic reasons (Fig. 32-2). In the discussions that follow, **relative** (to $\dot{V}o_2$ max) as well as **absolute** (expressed as L/min oxygen uptake) **work levels** are used in explaining physiologic re-

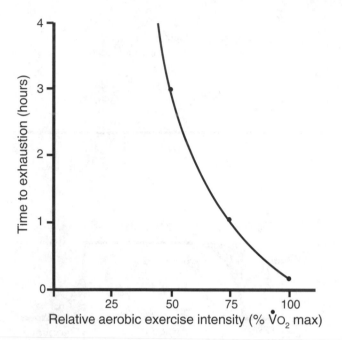

Figure 32–2 ■ Time to exhaustion during aerobic exercise is a predictable function of relative exercise intensity (% $\dot{V}O_2$ max).

sponses. The energy costs and relative demands of some familiar activities are listed in Table 32–1.

What causes oxygen uptake to reach a ceiling? Historically, many arguments claim primacy for either cardiac output (oxygen delivery) or muscle metabolic capacity (oxygen use) limitations. However, it may be that every link in the chain taking oxygen from the atmosphere to the mitochondrion reaches its capacity at about the same time. In practical terms, this means that any lung, heart, vascular, or musculoskeletal illness that reduces oxygen flow capacity will diminish a patient's functional capacity.

TABLE 32–1 ■ Absolute and Relative Costs of Daily Activities

Activity	Energy Cost (kcal/min)	%$\dot{V}O_2$max Sedentary 22-Year-Old	%$\dot{V}O_2$max Sedentary 70-Year-Old
Sleeping	1	6	8
Sitting	2	12	17
Standing	3	19	25
Dressing and undressing	3	19	25
Walking (3 mi/h)	4	25	33
Making a bed	5	31	42
Dancing	7	44	58
Gardening/shoveling	8	50	67
Climbing stairs	11	69	92
Crawl swimming (50 m/min)	16	100	
Running (8 mi/h)	16	100	

Modified from Astrand PO, Rodahl K: *Textbook of Work Physiology.* New York, McGraw-Hill, 1977, p 454.

For isometric exercise, work intensity is usually described in terms of the percentage used of the **maximal voluntary contractile force** (MVC). Analogous to work levels relative to $\dot{V}O_2$ max, the ability to endure isometric effort, and many physiologic responses to that effort, are predictable when the percentage of MVC among individuals is held constant.

Cardiovascular Responses

■ Blood Flow Is Preferentially Directed to Working Skeletal Muscle in Exercise

The increased energy expenditure with exercise demands more energy delivery. For prolonged work, this energy is supplied by the oxidation of foodstuffs, with the oxygen carried to working muscles by the cardiovascular system. Local control of blood flow is necessary, so that only working muscles receive increased blood and oxygen delivery. If the legs are active while the arms are not, leg muscle blood flow should increase while arm muscle blood flow remains unchanged or is reduced. Skeletal muscle normally receives only a small fraction of the cardiac output. In **dynamic exercise,** both total cardiac output and relative and absolute output directed to working skeletal muscle increase radically (Table 32–2).

Cardiovascular control in exercise, like the U.S. government, involves a federal system (the cardiovascular center in the brain, with its autonomic nervous output to the heart and resistance vessels) in tandem with local control. Because our ancestors for millennia have successfully used exercise both to escape being eaten and to catch food themselves, it is no surprise that cardiovascular control in exercise is complex and unique. It is as if a brain software package entitled "Exercise" were inserted each time work begins. It starts with activation of the motor cortex: the total neural activity is roughly proportional to the muscle mass and its work intensity. This neural activity communicates with the cardiovascular control center, reducing vagal tone on the heart (which raises heart

TABLE 32–2 ■ Blood Flow Distribution during Rest and Exercise in an Athlete

Area	Rest ml/min	Rest %	Heavy Exercise ml/min	Heavy Exercise %
Splanchnic	1,400	24.0	300	1.0
Renal	1,100	19.0	900	4.0
Brain	750	13.0	750	3.0
Coronary	250	4.0	1,000	4.0
Skeletal muscle	1,200	21.0	22,000	85.5
Skin	500	9.0	600	2.0
Others	600	10.0	100	0.5
Total cardiac output	5,800	100.0	25,650	100.0

Modified from Rhoades R, Pflanzer R: *Human Physiology.* 2nd ed. Fort Worth, Saunders College Publishing, 1992, p 969.

rate) and resetting the arterial baroreceptors to a higher level. In active muscle, reduced oxygen or associated local vasoactive factors dilate arterioles, preferentially enhancing flow. In mild work, these effects are sufficient to meet demand. In more intense work, sympathetic vasoconstriction is activated as a means of helping the system reach its blood pressure set point. The sympathetic drive further increases cardiac output via heart rate and contractility increases, and vasoconstriction everywhere except in regions where local controls predominate. Three favored sites are the brain, which requires constant flow independent of exercise; the heart, which is working hard and is vasodilated; and working skeletal muscle, where local vasodilator substances predominate over sympathetic drive. The sympathetic drive overwhelms opposing forces in nonexercising skeletal muscle, in the skin if thermoregulatory demands are absent, and in the viscera. Resting flows to these areas can fall as much as 75% if exercise is severe: vasoconstriction in these regions helps maintain blood pressure. Table 32–3 shows how, especially in dynamic exercise, a huge fall in systemic vascular resistance almost exactly matches the enormous rise in cardiac output.

Dynamic exercise, at its most severe level, forces the body to choose between maximal muscle vascular dilation and defense of blood pressure. Blood pressure is, in fact, maintained. With very hard work, sympathetic drive can begin to limit vasodilation in active muscle.

Isometric, or static, exercise yields a somewhat different response. Muscle blood flow increases relative to the resting condition, as does cardiac output, but the higher mean intramuscular pressure limits these flow increases much more than when exercise is rhythmic. Because the blood flow increase is blunted, inside a statically contracting muscle the fruits of hard work with too little oxygen appear quickly: a shift to anaerobic metabolism, production of lactic acid, and a rise in the ADP/ATP ratio. Static exercise is, consequently, quickly exhausting. Maintaining just 50% of the MVC is agonizing after about 1 minute and usually cannot be sustained after 2 minutes. A long-term sustainable level is only about 20% of maximum. These percentages are much less than the equivalent for dynamic work, which, as noted, is usually defined in terms of $\dot{V}O_2$ max. Rhythmic exercise requiring 70% $\dot{V}O_2$ max is sustainable in healthy individuals for about an hour and work at 50% $\dot{V}O_2$ max for several hours (Fig. 32–2).

The second consequence of the reliance of static exercise on anaerobic metabolism is a set of very different cardiovascular responses. Isometric exercise triggers muscle chemoreflex responses that raise blood pressure more, and cardiac output and heart rate less, than dynamic work (Fig. 32–3). Oddly, for dynamic exercise the elevation of blood pressure is most pronounced when a medium-sized mass is working (Fig. 32–3). This results from the combination of a relatively small dilated active muscle mass with relatively powerful central sympathetic vasoconstrictor drive. Arm exercise exemplifies a medium-sized muscle mass; shoveling snow is a notorious example of primarily arm and heavily isometric exercise. Shoveling snow can be a risky activity for persons in danger of stroke or heart attack, because it substantially raises systemic arterial pressure. The elevated pressure places compromised cerebral arteries at risk and presents an ischemic or failing heart with a greatly increased afterload.

■ Acute and Chronic Responses of the Heart and Blood Vessels Are Different

Vagal withdrawal and increases in sympathetic outflow elevate heart rate and contractility in rough proportion to exercise intensity. Equally crucial for development of high cardiac output in intense aerobic exercise are factors that enhance venous return: the "muscle pump," which compresses veins as muscles rhythmically contract, and the "respiratory pump," which increases oscillations in intrathoracic pressure

TABLE 32–3 ■ Cardiac Output, Mean Arterial Pressure, and Systemic Vascular Resistance Changes with Exercise

	Rest	Heavy Dynamic Exercise Large Muscle Mass
Cardiac output (L/min)	6	24
Mean arterial pressure (mm Hg)	93	105
Systemic vascular resistance (mm Hg/min/L)	15	4

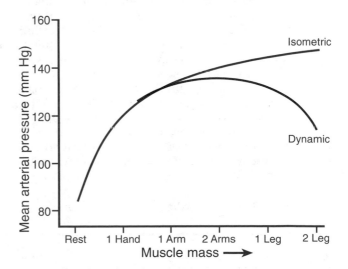

Figure 32–3 ■ Effect of active muscle mass on mean arterial pressure in exercise. Highest pressures during dynamic exercise occur when an intermediate-sized muscle mass is involved: pressure continues to rise in isometric exercise as more muscle is added.

(Chap. 18). Both of these factors increase with exercise intensity and help maintain stroke volume. Their importance is clear in patients with heart transplants who lack any extrinsic cardiac innervation. These persons are able to increase stroke volume as exercise intensity rises, due to increased venous return acting through the Frank-Starling mechanism (Chap. 14). In addition, circulating epinephrine and norepinephrine from the adrenal medulla and sympathetic nerve spillover eventually help increase heart rate and contractility.

The combined neural and mechanical influences on the healthy exercising heart increase both heart rate and stroke volume (Table 32–4). Maximal aerobic exercise yields a maximal heart rate: further vagal blockade (e.g., via pharmacologic means) cannot elevate heart rate further. Stroke volume, in contrast, reaches a plateau in moderate work and is unchanged as exercise reaches its maximal intensity. This plateau occurs in the face of ever-shortening filling time, testimony to the increasing effectiveness of the mechanisms that enhance venous return and those that promote cardiac contractility. Both sympathetic stimulation and tachycardia itself decrease the early diastolic values of left ventricular volume (due to increased ejection fraction) and pressure, leading to rapid mitral valve flow early in diastole. This helps maintain stroke volume as diastole shortens. Even in untrained persons, the ejection fraction (stroke volume as a percentage of end-diastolic volume) reaches 80% in severe exercise.

The increased blood pressure, heart rate, stroke volume, and cardiac contractility seen in exercise all increase myocardial oxygen demands. These demands are met by a linear increase in coronary flow during exercise that can reach a value five times the basal level. This increase in flow appears to be driven by metabolic factors (e.g., low Po_2, adenosine) acting on coronary resistance vessels in defiance of sympathetic vasoconstrictor tone. Coronary oxygen extraction, high at rest, increases further with exercise (up to 80% of delivered oxygen) in healthy persons. Also, there is no evidence of myocardial ischemia under any exercise condition, and there may be a coronary vasodilator reserve in even the most intense exercise.

Over longer periods of time, the heart adapts to exercise overload much as it does to high-demand pathologic states: by increasing left ventricular volume when exercise requires high flow and by left ventricular hypertrophy when exercise creates high systemic arterial pressure (high afterload). Consequently, the hearts of persons adapted to prolonged, rhythmic exercise that involves relatively low arterial pressure show large left ventricular volumes with normal wall thickness, while wall thickness is increased at normal volume in persons adapted to activities (e.g., weight lifting) involving isometric contraction and greatly elevated arterial pressure.

The larger left ventricular volume in persons chronically active in aerobic exercise leads directly to larger resting and exercise stroke volume; a simultaneous increase in vagal tone and decrease in β-adrenergic sensitivity enhances the resting and exercise bradycardia seen after training, so that in effect the heart operates further up the ascending limb of the Starling relationship. Resting bradycardia is a classic but overrated sign of endurance fitness: genetic factors explain a much larger proportion of variation among individuals in resting heart rate.

The effects of endurance training on coronary blood flow are partly mediated through changes in myocardial oxygen uptake. Since myocardial Vo_2 is roughly proportional to the rate-pressure product (heart rate × mean arterial pressure), and since heart rate falls after training at any absolute exercise intensity, coronary flow at a fixed submaximal workload is reduced in parallel. The peak coronary blood flow, however, is increased by training, as are cardiac muscle capillary density and peak capillary exchange capacity. The coronary vessels also undergo changes in responsiveness to adenosine, endothelium-mediated regulation, and control of intracellular free Ca^{2+}.

■ The Blood Lipid Profile Is Influenced by Exercise Training

Chronic, aerobic exercise is associated with increased circulating levels of high-density lipoproteins (HDLs) and unchanged total cholesterol, such that the HDL/total cholesterol ratio is increased. These changes in cholesterol fractions occur at any age if exercise is regular. Weight loss, so often a concomitant of increased chronic physical activity, undoubtedly contributes to these changes in cholesterol transport. Because exercise acutely and chronically enhances fat metabolism and cellular metabolic capacities for β-oxidation of free fatty acids, it is not surprising that over time regular activity increases both muscle and

TABLE 32–4 ■ Acute Cardiac Response to Graded Exercise in a 30-Year-Old Untrained Woman

Exercise Intensity	$\dot{V}o_2$ (L/min)	Heart Rate (beats/min)	Stroke Volume (ml/beat)	Cardiac Output (L/min)
Rest	0.25	72	70	5.0
Walking	1.0	110	90	10.0
Jogging	1.8	150	100	15.0
Running fast	2.5	190	100	19.0

adipose tissue lipoprotein lipase activity. Changes in lipoprotein lipase activity in concert with increased lecithin-cholesterol acyltransferase activity and apo A-I synthesis all enhance levels of circulating HDLs.

■ Exercise Has a Role in Prevention and Rehabilitation from Various Cardiovascular Diseases

The changes in HDL/total cholesterol ratio that take place with regular physical activity reduce the risk of atherogenesis and coronary artery disease in active as compared with sedentary persons. Lack of exercise is now established as a risk factor for coronary heart disease similar in magnitude to hypercholesterolemia, hypertension, and smoking. This reduced risk grows out of the changes in lipid profiles noted above, reduced insulin requirements and increased insulin sensitivity, and reduced cardiac β-adrenergic responsiveness and increased vagal tone. When coronary ischemia does occur, increased vagal tone may reduce the risk of fibrillation.

Regular exercise often, but not always, reduces resting blood pressure. Why some persons respond to chronic activity with a resting blood pressure decline and others do not remains unknown. Responders typically show diminished resting sympathetic tone, so that systemic resistance falls. In obesity-linked hypertension, a decline in insulin secretion as weight is lost and an increase in insulin sensitivity with exercise may explain the salutary effects of combining training with weight loss. Nonetheless, because some obese persons who exercise and lose weight show no blood pressure changes, exercise remains adjunctive therapy.

■ Pregnancy Shares Many Cardiovascular Characteristics with the Trained State

The physiologic demands and adaptations of pregnancy in some ways are similar to those of chronic exercise. Both increase blood volume, cardiac output, skin blood flow, and caloric expenditure. Exercise clearly has the potential to be deleterious to the fetus. Acutely, it increases body core temperature, causes splanchnic (hence uterine and umbilical) vasoconstriction, and shifts sharply the endocrinologic milieu; chronically, it increases caloric requirements. This last demand may be devastating if food shortages exist: the superimposed caloric demands of successful pregnancy and lactation are estimated at 80,000 kcal. Given adequate nutritional resources, however, there is to date little evidence of any other damaging effects of maternal exercise on any aspect of fetal development. The failure of exercise to harm well-nourished women may relate in part to the increased maternal/fetal mass and blood volume, which reduces specific heat loads, moderates vasoconstriction in the uterine and umbilical circulations, and diminishes the maternal exercise capacity.

At least in previously active women, even the most intense concurrent exercise regimen (unless associated with excessive weight loss) does not alter fertility, implantation, or embryogenesis. Continuation of exercise throughout pregnancy characteristically results in a normal term infant after relatively brief labor. These infants are normal in length and lean body mass but reduced in fat. Preliminary reports from the perinatal period suggest that the incidence of cord entanglement, abnormal fetal heart rate during labor, stained amniotic fluid, and low fetal responsiveness scores may all be reduced in women active throughout pregnancy.

Respiratory Responses

■ Ventilation in Exercise Matches Metabolic Demands, but the Exact Control Mechanisms Are Unknown

Increased breathing is perhaps the single most obvious physiologic response to acute exercise. Figure 32-4 shows that minute ventilation (the product of breathing frequency and tidal volume) rises linearly with work intensity through a range, then in supralinear fashion beyond that point. Ventilation of the lungs in exercise is linked to the twin goals of oxygen intake and carbon dioxide removal.

Exercise increases oxygen consumption and carbon dioxide production by working muscles, and the pulmonary response is precisely calibrated to maintain homeostasis of these gases in arterial blood. In mild or moderate work, arterial P_{O_2} (and hence oxygen content), P_{CO_2}, and pH all remain unchanged from resting levels (Table 32-5). The respiratory muscles accomplish this several-fold increase in ventilation pri-

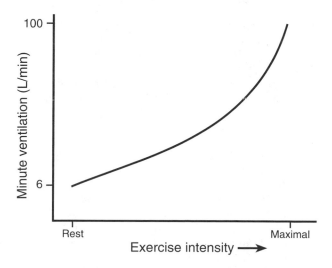

Figure 32–4 ■ Dependence of minute ventilation on the intensity of dynamic exercise. Ventilation initially rises linearly with intensity, becoming more pronounced as exercise nears maximal.

TABLE 32–5 ■ Acute Respiratory Response to Graded Aerobic Exercise in a 30-Year-Old Untrained Woman

Exercise Intensity	Ventilation (L/min)	$\dot{V}/\dot{Q}$	P_{AO_2} (mm Hg)	P_{aO_2} (mm Hg)	P_{aCO_2} (mm Hg)	pH
Rest	5	1	103	100	40	7.40
Walking	20	2	103	100	40	7.40
Jogging	45	3	106	100	36	7.40
Running fast	75	4	110	100	25	7.32

marily by increasing tidal volume, without provoking sensations of dyspnea. In fact, the increased tidal volume in mild work, attained while physiologic dead space remains constant, results in increased alveolar ventilation, since the dead space-to-tidal volume ratio falls.

More intense exercise presents the lungs with tougher problems. Near the halfway point from rest to maximal dynamic work, lactic acid formed in working muscles begins to appear in the circulation. This point, which depends on the type of work involved and the person's training status, is called the **anaerobic threshold** or **lactate threshold.** Lactate concentration gradually rises as work intensity increases, as more and more muscle fibers must rely on anaerobic metabolism. Almost fully dissociated, lactic acid drives down the arterial pH, a classic example of metabolic acidosis. Despite its already heavy burden of increased carbon dioxide removal, the healthy lung in exercise dutifully responds by further increasing ventilation, driving down the P_{CO_2} and helping restore pH toward normal. Technically speaking, increased ventilation in exercise that keeps P_{CO_2} constant is **hyperpnea;** further increases in breathing that drive down P_{CO_2} are **hyperventilation.** Certainly, this is not voluntary hyperventilation: the entire exercise respiratory response is involuntary and as work increases becomes less and less susceptible to voluntary control (hence the inability to eat or converse when work is hard). Through a range, the pH effects of lactic acid are fully respiratorily compensated, but eventually with the hardest work, now near exhaustion, ventilatory compensation becomes only partial and both pH and P_{CO_2} may fall well below normal (Table 32–5). Tidal volume continues to increase until pulmonary stretch receptors limit it, typically at or near half of vital capacity. Frequency increases at high tidal volume produce the remainder of the ventilatory volume increases.

The hyperventilation relative to carbon dioxide production that occurs with heavy exercise helps maintain arterial oxygenation. Keeping the arterial blood saturated with oxygen is more difficult with exercise than at rest, since the blood returned to the lung with exercise is more thoroughly depleted of oxygen. This depletion results from both active muscle removing more oxygen per unit blood and the preferential diver-

sion of blood flow from organs with low extraction (e.g., the kidneys) to working muscles. Because more of the blood entering the lung with exercise is completely deoxygenated, the blood which is shunted past ventilated areas can profoundly depress arterial oxygen content (Chap. 21). Besides having a diminished oxygen content, pulmonary arterial blood in exercise comes in a relative torrent, as cardiac output rises two, three, then fourfold over rest. In compensation, ventilation rises even faster than cardiac output, so the mean ventilation/perfusion ratio of the lung $(\dot{V}/\dot{Q})$ rises from near 1 at rest to greater than 4 with severe exercise (Table 32–5). Table 32–5 shows that healthy persons have no diminution of arterial P_{O_2} with acute exercise, although the alveolar (A) to arterial (a) P_{O_2} gradient does rise. This increase shows that despite the $\dot{V}/\dot{Q}$ increase, some areas of relative underventilation during the transit of blood through the pulmonary capillaries are not eliminated.

Despite intense scrutiny, researchers have been unable to explain the physiologic mechanisms that so effectively drive breathing as it maintains blood gas homeostasis in exercise. Where there are stimuli— such as in mixed venous blood returning to the lung, which is loaded with carbon dioxide and depleted of oxygen in proportion to exercise intensity—there are seemingly no receptors. Conversely, where there are receptors—in the carotid body, the lung parenchyma or airways, in the brainstem bathed by cerebrospinal fluid—no stimulus proportional to exercise demands exists. In fact, the central chemoreceptor is immersed in increasing alkalinity as exercise intensifies, a consequence of blood-brain barrier permeability to carbon dioxide but not hydrogen ions. Perhaps exercise ventilation proceeds from stimuli parallel to cardiovascular control: a central command in proportion to muscle activity that directly stimulates the respiratory center, coupled with feedback modulation from various peripheral receptor systems, including lung, ventilatory muscle, and chest wall mechanoreceptors and carotid body chemoreceptors.

■ The Respiratory System Is Largely Unchanged by Training

The effects of training on the pulmonary system are scant. Lung diffusing capacity, lung mechanics, and even lung volumes change little, if at all, with training. The widespread notion that training improves vital capacity is largely false; even very specific exercise designed to increase inspiratory muscle strength elevates vital capacity at most about 3%. The demands placed on ventilatory muscles increase their endurance, an adaptation that may reduce the sensation of dyspnea experienced in heavy exercise. Nonetheless, the primary respiratory changes with training are secondary to reduced lactate production that reduces ventilatory demands at once heavy absolute work levels.

In Lung Disease, Respiratory Limitations May Be Evidenced by Shortness of Breath or Decreased Oxygen Content of Arterial Blood

Any compromise in lung or chest wall function is much more readily apparent in exercise than at rest. One hallmark of lung disease is dyspnea during exertion, when this exertion normally was unproblematic. Restrictive lung diseases limit tidal volume, reducing the ventilatory reserve and hence exercise capacity. Obstructive lung diseases increase the work of breathing, exaggerating dyspnea and limiting capacity. The signs and symptoms of a respiratory limitation to exercise include exercise cessation with low maximal heart rate, desaturation of arterial blood, and severe shortness of breath. The prospects of exercise-based rehabilitation are modest, though muscle-based adaptations can reduce lactate production and consequently reduce ventilatory demands. Specific training of respiratory muscles to increase their strength and endurance is of slight benefit to patients with compromised lung function.

Exercise causes bronchoconstriction in nearly every asthmatic patient and is the sole apparent provocative agent in many persons. In health, catecholamine release from the adrenal medulla and sympathetic nerves normally dilates airways in exercise. This effect in asthmatic persons is overbalanced by constrictor effects, among them heat loss from airways (all else equal, cold, dry air is most potent as a bronchoconstrictor), increases in airway osmolality, and release of mediators. These effects of exercise on airways are due to hyperpnea per se; the exercise is incidental. Asthma patients are simply the most sensitive persons along a continuum; for example, breathing high volumes of cold, dry air provokes at least mild bronchospasm in all persons.

Muscle and Bone Responses

Muscle Fatigue Is Independent of Lactic Acid

The factors limiting exercise in patients with skeletal muscle disorders reveal much about the source of fatigue in exercise. Because severe exercise reduces intramuscular pH to values as low as 6.8 (arterial blood may fall to 7.2), an old idea is that hydrogen ion accumulation inhibits contractile activity. In fact, severe elevations in hydrogen ion concentrations may contribute to fatigue, especially in highly intense, short-term work. However, patients with defects in glycogenolysis or glycolysis (i.e., glycogen phosphorylase deficiency [McArdle's syndrome] or phosphofructokinase deficiency) easily fatigue during ischemic exercise yet reveal no lactate production and unchanged intramuscular pH. The best correlate of fatigue in these persons and in healthy subjects appears to be ADP accumulation in the face of normal or slightly reduced ATP, such that the ADP/ATP ratio is very high; however, such a correlation does not prove that a cause-and-effect relationship exists. The ADP/ATP ratio will rise whenever ATP usage exceeds ATP production capacity (via, primarily, anaerobic glycolysis, oxidative phosphorylation, and the creatine kinase reaction). Because the complete oxidation of glucose or glycogen to carbon dioxide and water is a major source of energy, it is not surprising that persons with defects in glycolysis or electron transport exhibit reduced ability to sustain submaximal exercise or to consume oxygen while attempting activity.

These metabolic defects are distinct from another group of disorders exemplified by the various muscular dystrophies. In these illnesses, loss of active muscle mass due to fat infiltration, cellular necrosis, or atrophy reduces exercise tolerance despite normal capacities (in healthy fibers) for ATP production. It is unclear whether fatigue in health ever occurs centrally (there is some evidence that pain from fatigued muscle feeds back to the brain to lower motivation and possibly directly to reduce central neural drive) or at the level of the motor neuron or the neuromuscular junction. In various disease states (e.g., amyotrophic lateral sclerosis) muscle atrophy leading to reduced strength and endurance stems from defects at these levels.

The Molecular Mechanisms That Underlie Muscle Atrophy or Hypertrophy Include Pretranslational and Posttranslational Events

Within skeletal muscle, enormous changes occur in response to training. Lessened activity (e.g., limb immobilization) results in atrophy; increased activity with low loads results in increased oxidative metabolic capacity without hypertrophy; increased activity with high loads produces muscle hypertrophy. The immediate changes in actin and myosin production with unloading or loading that lead to atrophy or hypertrophy are mediated at the posttranslational level; after a week or so, mRNA for these proteins is altered. Increased activity without overload increases capillary and mitochondrial density, myoglobin concentration, and

---

STRESS TESTING AND EXERCISE

As Americans become health-conscious, more are starting exercise programs. Unfortunately, many Americans, especially those over the age of 40, are at risk for coronary artery disease, and the advice to consult a physician before beginning a training program goes unheeded far too often.

To detect coronary artery disease in patients considering a new exercise program, physicians often perform an electrocardiogram (ECG), but at rest many disease sufferers have a normal-appearing ECG. To place a load on the heart and coronary circulation, a stress test is usually performed, in which an ECG is performed while the patient walks on a treadmill or rides a stationary bicycle. It is sometimes called an **exercise tolerance test.**

Normally, as heart rate increases during exercise, the line segments between the cardiac cycles become shorter, as do the distances between the start of the QRS complex and the T wave. In patients suffering from heart disease, other changes occur. Most common is a sloping of the line segment between the second lower peak of the QRS complex and the beginning of the T wave, known as ST segment depression. Note how the slope of the ST segment differs between the normal ECG during exercise and that of a diseased patient in Figure 32–5.

During the stress test, the ECG is monitored for such changes while blood pressure and heart rate are also monitored. At the start of the test, the exercise load is mild. The load is increased at regular intervals by increasing the speed and incline of the treadmill. The test is ended when the patient becomes exhausted, the heart rate reaches its safe maximum, significant pain occurs, or abnormal ECG changes are noted.

Generally, the stress test is a safe method for detecting subclinical coronary artery disease. Since the exercise load is gradually increased, the test can be stopped at the first sign of problems. With the gradual increase in exertion comes a greater demand by the heart for oxygen. If there are any problems with the blood supply, and therefore the supply of oxygen, to the heart, myocardial ischemia can be observed on the ECG as ST segment depression.

The admonition to seek medical advice before beginning an exercise program is important, especially for those at risk for coronary artery disease. A simple stress test can prevent a minor heart problem from becoming deadly serious during exercise. Without the test, novice athletes who are over 40, eat a high-fat diet, or who were previously sedentary may risk discovery of their previously undetected heart disease while on the track, in the pool, or on their bike.

---

virtually the entire enzymatic machinery for energy production from oxygen (Table 32–6). This training response is limited primarily by factors outside the muscle, since cross-innervation or chronic stimulation of muscles in animals can produce adaptations five times larger than those created by the most intense and prolonged exercise.

The biologic factors that link activity to changes in muscle of contractile proteins and enzymatic machinery are unknown: these factors may be linked to action potentials, intrinsic intracellular events, or, perhaps, neurotropic agents or growth factor induction. It is clear that the nature of the forces applied determines both the forms of contractile protein (slow or fast myosin isoforms, determining the twitch and fatigue characteristics of a fiber or motor unit) and the matching enzymatic machinery for energy production. For example, when chronic loading increases expression of slow as opposed to fast myosin, glycolytic enzymes also undergo down-regulation, while the enzymes of the Krebs cycle, the electron transport chain, and those involved in fatty acid oxidation are induced. This coordination of energy-producing and energy-utilizing systems in muscle ensures that even after atrophy the

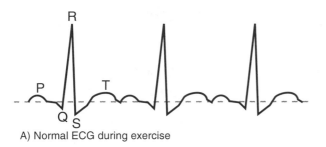

A) Normal ECG during exercise

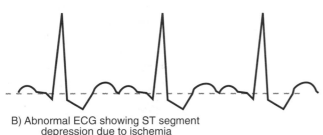

B) Abnormal ECG showing ST segment depression due to ischemia

Figure 32–5 ■ (A) Normal ECG during exercise. (B) Abnormal ECG, showing ST segment depression due to ischemia.

remaining contractile proteins are adequately supported metabolically (Table 32-6). In fact, the easy fatigability of atrophied muscle is due to the requirement that more motor units be recruited for identical external force; the fatigability per unit cross-sectional area is normal.

In practical terms, the local adaptations of skeletal muscle to activity have several applications. Because they lead to less reliance on carbohydrate as a fuel and allow more fat metabolism, they prolong endurance in heavy exercise and shift the carbon source away from species that can lead to lactic acid accumulation. This last adaptation, along with increased oxidative capacity, reduces circulating lactate and consequently the ventilatory demands of heavier work. Because metabolites accumulate less rapidly inside trained muscle, there is reduced chemosensory feedback to the central nervous system at any absolute work load. This, in

turn, reduces sympathetic outflow to the heart and blood vessels, enabling cardiac oxygen demands to be less at a fixed exercise level.

■ Muscle Hypertrophies in Response to Eccentric Contractions

Everyone knows it is easier to walk downhill than uphill, but the mechanisms underlying this commonplace phenomenon are complex. Muscle forces are identical in the two situations; the difference is that moving the body against gravity involves muscle shortening **(concentric contractions)**, while the reverse activity involves muscle tension development during lengthening **(eccentric contractions)**. All routine forms of physical activity, in fact, involve combinations of concentric, eccentric, and isometric contractions. Because less ATP is required for force development when external forces lengthen the muscle, the number of active motor units is reduced and energy demands are less for eccentric work. However, perhaps because the force exerted on the active motor units is greater in eccentric exercise, eccentric contractions cause a range of delayed responses after exercise that signify muscle damage. These include weakness (apparent the first day), soreness and edema (delayed 1–3 days in peak magnitude), and elevated plasma levels of intramuscular enzymes (delayed 2–6 days); histologic evidence of damage may persist 2 weeks. Damage is accompanied by an acute phase reaction: complement activation, increases in circulating cytokines, neutrophil mobilization, and trace metal redistribution are all characteristic. Eccentric contraction-induced muscle damage and its subsequent response may be the essential stimulus for muscle hypertrophy: while standard resistance exercise involves a mixture of contraction types, careful studies show that when one arm works purely concentrically and the other purely eccentrically at equivalent force, only the eccentric arm hypertrophies. Training adaptation to the eccentric components of exercise is efficient:

TABLE 32–6 ■ Effects of Training and Immobilization on the Human Biceps Brachii Muscle in a 22-Year-Old Woman

	Sedentary	After Endurance Training	After Strength Training	After 4 Months Immobilization
Total number of cells	300,000	300,000	300,000	300,000
Total cross-sectional area	10	10	13	6
Isometric strength (% control)	100	100	200	60
Fast-twitch fibers (% by numbers)	50	50	50	50
Fast-twitch fibers, average area ($\mu m^2 \times 10^2$)	67	67	87	40
Capillaries/fiber	0.8	1.3	0.8	0.6
Succinate dehydrogenase activity/unit area (% control)	100	150	77	100

Modified from Gollnich PD, Saltin B: Skeletal muscle physiology, in Teitz CC (ed): *Scientific Foundations of Sports Medicine.* Toronto, BC Decker, 1989, pp 185-242.

soreness after a second episode is minimal if it occurs within 6 weeks of a first episode.

■ Exercise Plays a Role in Ca²⁺ Homeostasis

Skeletal muscle contraction applies force to bone. Because the architecture of bone remodeling involves osteoblast and osteoclast activation in response to loading and unloading, physical activity is a major influence on bone mineral density and geometry. Repetitive physical activity can create excessive strain, leading to inefficiency in bone remodeling and the possibility of stress fracture. On the other hand, extreme inactivity leads to osteoclast dominance and bone loss.

Because the force applied to bone is ultimately related to the strength of the involved muscles, bone strength and density appear to be closely related to muscle strength. This suggests that programs to prevent or slow osteoporosis in aging persons should emphasize strength rather than endurance training. Even when activity is optimal, however, it is apparent that genetic contributions to bone mass are much greater than the superimposed exercise influence: perhaps 75% of the population variance is genetic and 25% due to different levels of activity. In addition, the predominant contribution of estrogen to homeostasis of bone in young women is apparent when amenorrhea occurs secondary to chronic heavy exercise.

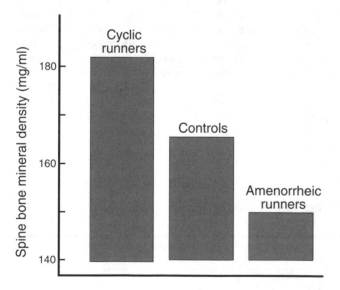

Figure 32–6 ■ Spine bone mineral density in 22-year-old women: nonathletes (controls) and cyclic and amenorrheic runners. Increased bone mineral density in cyclic runners indicates the role exercise can play, but reduced mineral density in amenorrheic runners shows the predominant role of estrogen. (Modified from Marcus R, Cann C, Madviy P, et al: Menstrual function and bone mass in elite women distance runners: Endocrine metabolic features. *Ann Intern Med* 102:158–163, 1985.)

These exceptionally active women exhibit low levels of circulating estrogens, low trabecular bone mass, and a high fracture risk (Fig. 32-6).

Gastrointestinal/Metabolic/ Endocrine Responses

■ Exercise Can Modify the Rate of Gastric Emptying and Intestinal Absorption

Many aspects of the effects of exercise on gastrointestinal function remain poorly understood. Specifically, it is unknown whether acute or chronic exercise alters the volume or composition of saliva. Little more is known about the role exercise may play in provoking gastroesophageal reflux. At least 20% of patients with reflux list exercise as a major precipitant; it is unknown whether exercise allows lower esophageal sphincter pressure to fall or if rises in intragastric pressure with hyperpnea are more severe in susceptible individuals.

Gastric emptying of liquids has been studied extensively in prolonged dynamic exercise, due to interest in fluid, electrolyte, and glucose replacement during long-term work. Dynamic exercise must be severe (>70% V_{O_2}) to slow gastric emptying of liquids, and, predictably, water is absorbed less quickly in the ingested solution as its glucose concentration rises. Little is known of the neural, hormonal, or intrinsic smooth muscular bases for this exercise effect. Gastric acid secretion is unchanged by acute exercise of any intensity; nothing is known of the effects of exercise on any other aspects of gastric or duodenal function relevant to development or healing of peptic ulcers.

Chronic physical activity leads to accelerated gastric emptying rates and rapid small bowel transit. These effects probably represent part of the adaptive response to chronically increased energy expenditure, leading as they do to more rapid processing of food and appropriately increased appetite. Animal models of hyperphagia routinely show specific adaptations in the small bowel (increased mucosal surface area, microvillar height, content of brush border enzymes and transporters) that lead to more rapid digestion and absorption; these same effects likely take place in humans rendered hyperphagic by regular physical activity.

Blood flow to the gut decreases in proportion to exercise intensity, a response that is part of cardiovascular regulation of regional flows. Water, electrolyte, and glucose absorption may be slowed in parallel, but malabsorption does not occur in healthy persons. It is unknown whether exercise alters symptomatology or disease progression in inflammatory bowel disease.

Although exercise is often recommended as treatment for postsurgical ileus (Chap. 27), uncomplicated constipation, or irritable bowel syndrome, little is known in these areas. Humans maintained on constant diets show no change in colonic transit time when exercise is regularly instituted; however, in "free living" persons chronic exercise increases food and fiber intake and accelerates colonic transit. In dogs, acute exercise provokes colonic giant migrating waves. A similar phenomenon may explain the increased urge to defecate anecdotally associated with activity.

■ Chronic Exercise Increases Appetite Slightly Less than It Increases Caloric Expenditure in Obese Persons

Anthropologic studies show that obesity is rare in primitive, relatively active populations but is common in more sedentary modernized cultures. The health risks of obesity are clear-cut: increased incidence of hypertension, heart disease, and diabetes. The metabolic cost of exercise averages about 100 kcal/mile walked. For exceptionally active persons exercise expenditure can exceed 3000 kcal/day added to the basal energy expenditure, which for a 55-kg woman averages about 1400 kcal/day. A glance at the bodies of the very active (e.g., competitive swimmers, cross-country skiers, farmers, or laborers without mechanized equipment) shows that appetite increases with activity to match the enormous caloric expenditure precisely. The biologic factors that allow this accurate match have never been defined. In the obese, modest increases in physical activity increase energy expendi-

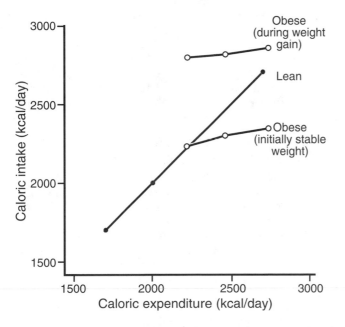

Figure 32–7 ■ Caloric intake as a function of exercise-induced increases in daily caloric expenditure. For lean individuals, intake matches expenditure over a wide range. For obese individuals during periods of weight gain or periods of stable weight, increases in expenditure are not matched by increases in caloric intake. (Modified from Pi-Sunyer FX: Exercise effects on caloric intake. *Hum Obesity* 499:94–103, 1987.)

ture more than food intake, so progressive weight loss can be instituted if exercise can be regularized (Fig. 32-7). This method of weight control is far superior to dieting (the $5 billion/year diet industry notwithstanding), since even moderate caloric restriction (>500 kcal/day) results in both a lowered basal metabolic rate and substantial loss of fat-free body mass. Exercise has subtle positive effects on the energy balance equation as well: a single episode may increase basal energy expenditure for several hours and may increase the thermic effect of feeding. The greatest practical problem remains compliance with even the most exact exercise "prescription"; patient dropout rates from even short-term programs typically exceed 50%.

■ Acute and Chronic Exercise Increases Insulin Sensitivity, Insulin Receptor Density, and Glucose Transport into Muscle

Though skeletal muscle is omnivorous, its work intensity and duration, training status, and inherent metabolic capacities determine its substrate choice. For very short-term exercise, stored phosphagens (ATP and creatine phosphate) are sufficient for crossbridge formation between actin and myosin; even maximal efforts lasting up to about 10–20 seconds require little or no glycolytic or aerobic energy production. When work to exhaustion is paced to be somewhat longer in duration, glycolysis is driven (particularly in high glycolytic fibers possessing fast myosin) by high intramuscular ADP concentrations, and this form of anaerobic metabolism, with its byproduct lactic acid, is the major energy source. The carbohydrate provided to glycolysis comes from stored, intramuscular glycogen or blood-borne glucose. Exhaustion from work in this intensity range (50–90% of $\dot{V}_{O_2}$ max) is associated with depletion of carbohydrate. Accordingly, factors that increase carbohydrate availability improve fatigue resistance. These include prior high dietary carbohydrate, training adaptations at the cellular level that increase the enzymatic potential for fatty acid oxidation (thereby sparing carbohydrate stores), and oral carbohydrate intake during exercise. Frank hypoglycemia rarely occurs in healthy persons during even the most prolonged or intense physical activity. When it does, it is usually in association with depletion of muscle and hepatic stores and a failure to supplement carbohydrate orally.

Exercise suppresses insulin secretion by increasing sympathetic tone at the pancreatic islets. Despite acutely falling levels of circulating insulin, both non-insulin-dependent and insulin-dependent muscle glucose uptake increase in exercise. For reasons not yet clear, exercise recruits glucose transporters from their intracellular storage sites to the plasma membrane of active skeletal muscle cells.

Because exercise increases insulin sensitivity, patients with type I diabetes (insulin-dependent) require less insulin when activity increases. This positive result can be treacherous, since exercise can accelerate hypoglycemia or even insulin coma in these persons. Chronic exercise, through its chronic reduction of insulin requirements (endogenous or exogenous), up-regulates insulin receptors throughout the body. This effect appears to be due less to training than simply to an oft-repeated acute stimulus: the effect is full-blown after 2–3 days of regular physical activity and can be lost as quickly. Consequently, healthy active persons show strikingly greater insulin sensitivity than do their sedentary counterparts (Fig. 32–8). In addition, up-regulation of insulin receptors and reduced insulin release of chronic exercise is ideal therapy in type II (non-insulin-dependent) diabetes, a disease characterized by high insulin secretion and low receptor sensitivity.

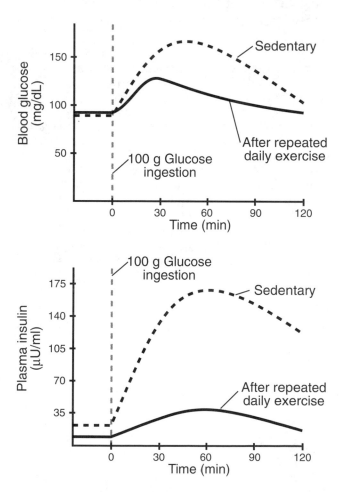

Figure 32–8 ■ Effect of repeated daily exercise on the blood glucose and insulin response to glucose ingestion. Both the responses are blunted by repeated exercise, demonstrating increased insulin sensitivity. These curves are for 60-year-old men. (Modified from Seals DR, Hagberg JM, Allen WK, et al: Glucose tolerance in young and older athlete and sedentary men. *J Appl Physiol* 56:1521–1525, 1984.)

Immune, Psychiatric, and Aging Responses

■ Acute Exercise Transiently Elevates Many Circulating Immune System Markers, but the Long-Term Effects of Training on Immune Function Are Unclear

In protein-calorie malnutrition, catabolism of body proteins for energy lowers immunoglobulin levels and decreases resistance to infection. Clearly, in this circumstance, exercise merely speeds the starvation process by increasing daily caloric expenditure and would be expected to diminish the immune response further. Nazi "labor" camps became death camps in part by combining severe food restrictions with incessant demands for physical work, the combination being a lethal acceleration of starvation with all its consequences.

If nutrition is adequate, it is less clear whether adopting an active versus a sedentary life-style alters immune responsivity. There is no doubt that in healthy people an acute episode of exercise briefly increases blood leukocyte concentration, and it may also transiently enhance neutrophil production of microbicidal-reactive oxygen species and natural killer cell activity. However, it remains unproved that regular exercise over time can protect against, for example, upper respiratory tract infections. In HIV-positive men, some with AIDS-related complex, standard aerobic and muscle strength training yields normal gains, without altering leukocyte, total lymphocyte, or lymphocyte subpopulation cell counts.

■ Exercise May Help Relieve Depression in Some Patients, but Its Efficacy and Neurochemical Effects Are Uncertain

In healthy people, prolonged exercise increases subsequent deep sleep, defined as stages 3 and 4 of non-rapid-eye-movement sleep (slow-wave sleep) (Chap. 7). This effect is apparently mediated entirely through the thermal effects of exercise, since equivalent passive heating produces the same result. It is unknown whether exercise can improve sleep in patients with insomnia.

Clinical depression is characterized by sleep and appetite dysfunction and profound changes in mood. It is uncertain whether acute or chronic exercise can help relieve depression. The two most prominent biologic theories of depression (dysregulation of central monoamine activity and dysfunction of the hypothalamic-pituitary-adrenal axis) have received almost no study with regard to the impact of exercise.

■ As Persons Age, the Effects of Exercise on Functional Capacity Are Much More Profound than Their Effect on Longevity

Regular aerobic exercise, as compared with sedentariness, increases **longevity** (average age at death) in rats and humans. In contrast, at least in rats, life span (the longest-lived individual animal) is not lengthened by exercise. There is no evidence for what was a serious line of thought in the 1920s, amusing today, that exercise increased the "rate of living" and shortened life span. The effects of exercise on longevity are modest; all-cause mortality is reduced, but only in amounts sufficient to increase longevity by 1-2 years.

More dramatic is the influence of exercise on strength and endurance at any comparable age. For example, while $\dot{V}o_2$ max declines, as a population average, steadily from age 20 onward, perhaps two-thirds of the American population decline is reversible by activity. While the $\dot{V}o_2$ ceiling gradually falls, the ability to train toward an age-appropriate ceiling is as intact at age 70 as it is at age 20 (Fig. 32-9). In fact, a highly active 70-year-old, otherwise healthy, will typically display an absolute exercise capacity greater than a sedentary 20-year-old (Fig. 32-9). Aging affects all the links in the chain of oxygen transport and use, so

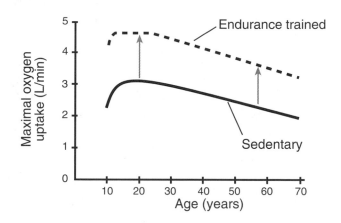

Figure 32–9 ■ Endurance-trained subjects display a higher maximal oxygen consumption than sedentary subjects, regardless of age. (Modified from Heath GW, Hagerg JM, Ehsani AA, et al: A physiological comparison of young and older endurance athletes. *J Appl Physiol* 51:634–640, 1981; and Astrand PO, Rodahl K: *Textbook of Work Physiology.* New York; McGraw-Hill, 1977, p 319.)

aging-induced declines in lung elasticity, lung diffusing capacity, cardiac output, and muscle metabolic potential take place in concert. Consequently, the physiologic mechanisms underlying fatigue are similar at all ages.

■ ■ ■ ■ ■ ■ ■ ■ ■ ■ ■ REVIEW EXERCISES ■ ■ ■ ■ ■ ■ ■ ■ ■ ■ ■

Identify Each with a Term

1. The highest amount of oxygen that can be used per unit time

2. A muscle contraction during which the muscle is lengthened by external force

3. That point in progressive exercise at which lactic acid begins to rise in blood

4. Muscle activity that involves rhythmic contraction and relaxation

Define Each Term

5. Training

6. Relative $\dot{V}o_2$

7. Isometric exercise

8. Hyperpnea

Choose the Correct Answer

9. Isometric exercise is usually quantified as:
    a. Relative $\dot{V}o_2$
    b. Mild, moderate, or severe
    c. Percentage usage of the MVC
    d. Anaerobic metabolism

10. Two persons, one highly trained and one not, each exercising at 75% $\dot{V}o_2$ max, become fatigued:
    a. For similar physiologic reasons
    b. Very slowly

    c. At quite different times
    d. While performing equally well

11. In dynamic exercise, systemic vascular resistance:
    a. Is constant
    b. Rises slightly
    c. Falls only if work is prolonged
    d. Falls dramatically

12. Vasoconstriction in the renal and splanchnic vascular beds in exercise:
    a. Rarely occurs
    b. Is an effect of training
    c. Helps maintain arterial blood pressure
    d. Parallels cerebral blood flow changes

13. The greatest risk in prolonged exercise for a healthy person is:
    a. Heat exhaustion and heat stroke
    b. Coronary ischemia
    c. Renal ischemia and anoxia
    d. Hypertension

14. Arterial blood pressure is maximized when exercise:
    a. Involves an intermediate-sized muscle mass
    b. Is prolonged
    c. Uses unfamiliar muscle groups
    d. Is dynamic as opposed to isometric

15. The baroreceptors, in exercise:
    a. Reset blood pressure to a lower level
    b. Are "turned off"

c. Are increased in sensitivity by training
d. Reset blood pressure to a higher level

16. Myocardial oxygen demands are increased in exercise:
   a. In proportion to the product of heart rate and total peripheral resistance
   b. Because left ventricular filling is compromised
   c. As heart rate and arterial pressure rise
   d. Due to decreased contractility

17. Exercise training:
   a. Elevates HDL and lowers total cholesterol
   b. Lowers total cholesterol; LDL and HDL remain in constant proportion
   c. Elevates HDL at constant total cholesterol
   d. Lowers LDL, allowing HDL to fall as well

18. Heavy, dynamic exercise in healthy persons causes:
   a. Elevated arterial Pa_{CO_2}
   b. Normal or reduced arterial Pa_{CO_2}
   c. Respiratory alkalosis
   d. Respiratory acidosis

19. The $\dot{V}/\dot{Q}$ ratio changes from rest to maximal exercise:
   a. From 1 to about 2
   b. From 1 to 4 or more
   c. From 1 to about 0.2
   d. Very little, since it is homeostatically regulated near 0.8–1.0

20. Intense, prolonged training can be expected:
   a. To double the vital capacity
   b. To have essentially no effect on lung elasticity
   c. To increase pulmonary diffusing capacity by about 200%
   d. To improve forced expiratory flow rates by about 25%

21. The best correlate of fatigue in an exercising muscle is:
   a. Muscle lactic acid concentration
   b. Muscle pH
   c. The intramuscular ADP/ATP ratio
   d. Depletion of intramuscular triglycerides

22. Eccentric muscle contractions:
   a. Prevent muscle soreness
   b. Produce muscle ischemia
   c. Rarely occur
   d. Are essential for hypertrophy

23. Atrophied muscle fatigues easily because:
   a. Less motor units are involved
   b. There is a relative excess of contractile protein
   c. Cells are small, so more cells are required to perform the same work

d. Oxidative energy-producing systems are up-regulated

24. Young women with low estrogen and amenorrhea:
   a. Have low trabecular bone mass
   b. Have weak muscles
   c. Are invariably physically inactive
   d. Are too young to be at risk for fractures

25. During exercise:
   a. Insulin secretion increases
   b. The pancreas is stimulated to release insulin and glucagon
   c. Muscle glucose uptake increases due to elevated insulin
   d. Insulin levels in blood decrease

26. In exercise lasting only a few seconds, energy requirements are met by:
   a. Using up O_2 stored in blood
   b. Lactic acid creation (via glycolysis)
   c. Stored ATP and creatine phosphate
   d. A moderate increase in ventilation and O_2 intake

27. Training reduces dependence on carbohydrate during prolonged work by:
   a. Increasing hepatic glycogen stores
   b. Reducing gastric emptying rates
   c. Reducing insulin sensitivity
   d. Increasing the enzymes of fatty acid oxidation

Calculate

28. What is the $\dot{V}_{O_2}$ max of an elderly man if walking ($\dot{V}_{O_2} = 0.9$ L/min) requires 75% of his aerobic capacity?

29. What is the coronary flow at rest and at maximal exercise if it represents 4% of resting output (6000 ml/min) and 3.5% of exercise output (24,000 ml/min)?

30. What is the energetic equivalent of walking of a successful pregnancy and lactation, assuming that walking 1 mile expends 100 kcal and the period of pregnancy-lactation is 26 months?

Briefly Answer

31. Why is it so critical to define the parameters of exercise?

32. List the physiologic systems involved in oxygen delivery in exercise.

33. List the risks and benefits to the fetus of exercise during pregnancy.

34. Does exercise help or harm the immune system?

■ ■ ■ ■ ■ ■ ■ ■ ■ ■ ■ ANSWERS ■ ■ ■ ■ ■ ■ ■ ■ ■ ■ ■

1. $\dot{V}_{O_2}$ max
2. Eccentric contraction
3. Anaerobic or lactate threshold
4. Dynamic exercise
5. Chronic exercise that provokes adaptive responses
6. Oxygen uptake expressed as a percentage of the maximal capacity
7. Muscle contraction without shortening or release of tension
8. Increased ventilation in proportion to increased CO_2 production
9. c
10. a
11. d

12. c

13. a

14. a

15. d

16. c

17. c

18. b

19. b

20. b

21. c

22. d

23. c

24. a

25. d

26. c

27. d

28. $0.75 \ (\dot{V}_{O_2} \ max) = 0.9 \ L/min$
    $\dot{V}_{O_2} \ max = 1.2 \ L/min$

29. $0.04 \times 6000 = 240 \ mL/min$ at rest
    $0.035 \times 24000 = 840 \ mL/min$ in exercise

30. 80,000 kcal/100 kcal/mi = 800 mi
    26 months = 26 months × 30 days = 780 days
    800 mi/780 days = approximately 1 mi/day

31. Physiologic responses and adaptations are entirely dependent on the mode, intensity, duration, and frequency of exercise and are extraordinarily wide-ranging.

32. Pulmonary ventilation, taking O_2 from atmosphere to alveolus; pulmonary diffusing capacity, moving O_2 into pulmonary capillary blood; cardiac output and muscle blood flow, delivering oxygenated blood to working skeletal muscle; muscle cellular O_2 uptake, using delivered oxygen and liberating energy

33. Especially if nutrition is inadequate, exercise during pregnancy increases energy expenditure and weight loss, increasing the risk of low birth weight. In adequately nourished women, the risks are low. The benefits of exercise in pregnancy are unclear at present.

34. Exercise clearly harms the immune system during protein caloric malnutrition, as its demand for energy further depletes body protein (e.g., immunoglobulin) stores. If nutrition is adequate, there is as yet no substantial evidence that exercise helps or harms immune function, even in HIV-seropositive patients.

Suggested Reading

Booth FW, Thomason DB: Molecular and cellular adaptation of muscle in response to exercise: Perspectives of various models. *Physiol Rev* 71:541-585, 1991.

Cheng C-P, Igarashi Y, Little WC: Mechanisms of augmented rate of left ventricular filling during exercise. *Circ Res* 70:9-19, 1992.

Holloszy JO (ed): *Exercise and Sport Sciences Reviews.* Baltimore, Williams & Wilkins, Vol.19(1991) and 20(1992).

Hudlicka O, Brown M, Egginton S: Angiogenesis in skeletal and cardiac muscle. *Physiol Rev* 72:369-418, 1992.

Lash JM, Bohlen HG: Perivascular and tissue P_{O_2} in contracting rat spinotrapezius muscle. *Am J Physiol* 252:H1192-H1202, 1987.

Laughlin MH, McAllister RM: Exercise training-induced coronary vascular adaptation. *J Appl Physiol* 73:2209-2225, 1992.

Lewis SF, Haller RG: Skeletal muscle disorders and associated factors that limit exercise performance. *Exercise Sport Sci Rev* 67-114, 1989.

Lewis SF, Snell PG, Taylor WF, Hamra M, et al: Role of muscle mass and mode of contraction in circulatory responses to exercise. *J Appl Physiol* 58:146-151, 1985.

Rowell LB, O'Leary DS: Reflex control of the circulation during exercise: Chemoreflexes and mechanoreflexes. *J Appl Physiol* 69:407-418, 1990.

Stone HL: Control of the coronary circulation during exercise. *Annu Rev Physiol* 45:213-227, 1983.

Teitz CC (ed): *Scientific Foundations of Sports Medicine.* Toronto, BC Decker, 1989.

Weibel ER, Taylor CR: Design of the mammalian respiratory system. *Resp Physiol* 44:1-164, 1981.

PART NINE

Endocrine Physiology

Endocrine Control Mechanisms

OBJECTIVES

After studying this chapter, the student should be able to:
1. Define *hormone* and describe the role hormones play in homeostasis
2. Explain why most if not all peptide and protein hormones are initially synthesized as preprohormones
3. List the six classes of steroid hormones and provide the common name for at least one hormone in each class
4. Describe the basic principles and limitations in RIA and ELISA
5. Explain the mechanisms by which hormones are transported in the bloodstream
6. List the different ways altered hormone/receptor interactions might contribute to the pathogenesis of endocrine abnormalities
7. Explain the second messenger model of hormone action and describe the four known systems
8. Describe the gene expression model of hormone action and the classes of hormones that use this mechanism to produce their biologic effects

Endocrinology is the branch of physiology concerned with the description and characterization of processes involved in the regulation and integration of cells and organ systems by a group of specialized chemical substances called hormones. Diagnosis and treatment of a large number of endocrine disorders is an important aspect of any general medical practice. Certain endocrine disease states, such as diabetes mellitus, thyroid disorders, and reproductive disorders, are fairly common in the general population, therefore it is likely that they will be encountered repeatedly in the practice of medicine. In addition, because hormones either directly or indirectly affect virtually every cell or tissue in the body, a number of other prominent diseases not primarily classified as endocrine diseases may have an important endocrine component. Atherosclerosis, certain forms of cancer, and even certain psychiatric disorders are examples of conditions in which an endocrine disturbance may contribute to the progression or severity of disease.

General Concepts of Endocrine Control

■ Hormones Are Blood-Borne Substances Involved in Regulating a Variety of Processes

The term *hormone* is derived from the Greek *hormaein*, which means to "excite" or to "stir up." The endocrine system forms an important communication system that serves to regulate, integrate, and coordinate a variety of different physiologic processes. The processes that hormones regulate fall into four areas: (1) the digestion, utilization, and storage of nutrients; (2) growth and development; (3) ion and water metabolism; and (4) reproductive function.

Working Definition of Hormones ■ It is difficult to describe hormones in absolute terms. As a working definition, however, it can be said that hormones serve as regulators and coordinators of various biologic functions in the animals in which they are produced. They are highly potent, specialized, organic molecules produced by endocrine cells in response to specific stimuli that exert their actions on specific target cells.

These target cells are equipped with receptors that bind hormones with high affinity and specificity; when bound, they initiate characteristic biologic responses by the target cells.

In the past, definitions or descriptions of hormones usually included a phrase indicating that these substances were secreted into the bloodstream and carried by the blood to a distant target tissue. Although many hormones travel by this mechanism, we now realize that there exists a whole spectrum of hormone or hormonelike substances that serve important roles in cell-to-cell communication that are not secreted directly into the bloodstream. Instead, these substances reach their target cells by diffusion through the interstitial fluid. Recall the discussion of autocrine and paracrine mechanisms discussed in Chapter 1 (Fig. 1–5).

Specificity of Hormone Receptors ■ In the endocrine system, a hormone molecule secreted into the blood is free to circulate and contact almost any cell in the body. However, only target cells, those cells that possess specific **receptors** for the hormone, will respond to that hormone. A hormone receptor is the molecular entity (usually a protein or glycoprotein) either outside or within a cell that recognizes and binds a particular hormone. When a hormone binds to its receptor, biologic effects characteristic of that hormone are initiated. Therefore, in the endocrine system the basis for specificity of cell-to-cell communication rests at the level of the receptor. Similar concepts apply to autocrine and paracrine mechanisms of communication.

A certain degree of specificity is ensured by the **restricted distribution** of some hormones. For example, several hormones produced by the hypothalamus regulate hormone secretion by the anterior pituitary. These hormones are carried via small blood vessels directly from the hypothalamus to the anterior pituitary, prior to entering the general systemic circulation. The anterior pituitary is therefore exposed to considerably higher concentrations of these hypothalamic hormones than is the rest of the body; as a result, the actions of these hormones focus on cells of the anterior pituitary. Another mechanism resulting in restricted distribution of active hormone involves the local transformation of a hormone within its target tissue from a less active to a more active form. An example is the formation of dihydrotestosterone from testosterone, occurring in certain androgen target tissues, such as the prostate gland. Dihydrotestosterone is a much more potent androgen than testosterone. Because the enzyme that catalyzes this conversion is found only in certain locations, its cell or tissue distribution partly localizes the actions of the androgens to these sites.

Thus, while receptor distribution is the primary factor in determining the target tissues for a specific hormone, other factors also may serve to focus the actions of a hormone on a particular tissue.

■ Feedback Regulation Is an Important Part of Endocrine Function

The endocrine system, like many other physiologic systems, is regulated by feedback mechanisms (Chap. 1). This is usually negative feedback, although a few positive feedback mechanisms have been found. Both types of feedback control occur because the endocrine cell, in addition to synthesizing and secreting its own hormone product, has the ability to sense the biologic consequences of secretion of that hormone. This enables the endocrine cell to adjust its rate of hormone secretion to produce the desired level of effect, thereby ensuring maintenance of homeostasis.

Hormone secretion may be regulated via very simple first-order feedback loops or more complex multilevel second- or third-order feedback loops. Since negative feedback is most prevalent in the endocrine system, only examples of this type are illustrated here.

Simple Feedback Loops ■ First-order feedback regulation is the simplest type and forms the basis for more complex modes of regulation. Figure 33–1A illustrates a simple first-order feedback loop. In this example, an endocrine cell secretes a hormone that produces a specific biologic effect in its target tissue. It also senses the magnitude of the biologic effect produced by the hormone. As the biologic response increases, the amount of hormone secreted by the endocrine cell is appropriately decreased.

Complex Feedback Loops ■ More commonly, feedback regulation in the endocrine system is multilevel, involving second- or third-order feedback loops. For example, multiple levels of feedback regulation may be involved in regulating hormone production by various endocrine glands under the control of the anterior pituitary (Fig. 33–1B). Regulation of target gland hormone secretion, such as adrenal steroids or thyroid hormones, begins with production of a releasing hormone by the hypothalamus. The releasing hormone stimulates production of a trophic hormone by the anterior pituitary, which in turn stimulates the production of the target gland hormone by the target gland. As indicated by the dashed lines in Figure 33–1B, the target gland hormone may have negative feedback effects to inhibit the secretion of both the trophic hormone from the anterior pituitary and the releasing

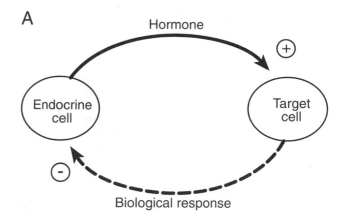

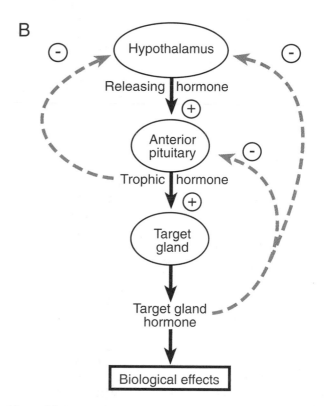

Figure 33–1 ■ Examples of simple and complex feedback loops. In the endocrine system, feedback relationships can be either (A) simple or (B) more complex, multilevel relationships, such as the feedback relationships of the hypothalamopituitary-target gland axis depicted here. Solid lines indicate stimulatory effects ⊕; dashed lines indicate inhibitory, negative feedback effects ⊖.

hormone from the hypothalamus. In addition, the trophic hormone may inhibit releasing hormone secretion from the hypothalamus, and there is evidence that in some cases the releasing hormone may inhibit its own secretion by the hypothalamus.

The more complex multilevel form of regulation (Fig. 33–1B) appears to provide certain advantages as compared to the simpler system (Fig. 33–1A). Theoretically, it permits a greater degree of fine-tuning of hormone secretion, and the multiplicity of regulatory steps minimizes changes in hormone secretion in the event that one component of the system is not functioning normally.

It is important to bear in mind the normal feedback relationships that control secretion of each individual hormone discussed in the following chapters. Clinical diagnoses are often made based on the evaluation of hormone-effector pairs relative to normal feedback relationships. For example, in the case of anterior pituitary hormones, measurement of both the trophic hormone and the target gland hormone concentration provides important information that may permit determination of whether a defect in hormone production exists at the level of the pituitary or at the level of the target gland. Furthermore, most dynamic tests of endocrine function performed clinically are based on our knowledge of these feedback relationships. Dynamic tests involve a prescribed perturbation of the feedback relationship(s); the range of response in a normal individual is well established, while a response outside the normal range is indicative of abnormal function at some level and greatly enhances information gained from static measurements of hormone concentrations.

■ Signal Amplification Is an Important Characteristic of the Endocrine System

Another important feature of the endocrine system is **signal amplification.** Blood concentrations of hormones are exceedingly low, generally, 10^{-9}–10^{-12} mol/L. Even at the higher concentration of 10^{-9} mol/L, only one hormone molecule would be present for roughly every 50 billion water molecules. Thus, for hormones to be effective regulators of biologic processes, amplification must be part of the overall mechanism of hormone action.

Amplification generally occurs as a result of activation of a series of enzymatic steps involved in hormone action. At each step, many times more signal molecules are generated than were present at the prior step, leading to a cascade of ever-increasing numbers of signal molecules. The self-multiplying nature of the hormone action pathways provides the molecular basis for amplification in the endocrine system.

■ Pleiotropic Hormone Effects and Redundancy of Regulation Are Also Characteristic of the Endocrine System

Most hormones have multiple actions in their target tissues and are therefore said to have **pleiotropic** effects. For example, insulin exhibits pleiotropic effects in skeletal muscle, where it stimulates glucose uptake, stimulates glycolysis, stimulates glycogenesis, inhibits glycogenolysis, stimulates amino acid uptake, stimulates protein synthesis, and inhibits protein degradation.

In addition, some hormones are known to have very different effects in several different target tissues. For example, testosterone, the male sex steroid, promotes normal sperm formation in the testes, stimulates growth of the sex accessory glands, such as the prostate and seminal vesicles, and promotes the development of several secondary sex characteristics, such as beard growth and deepening of the voice.

Multiplicity of regulation is also common in the endocrine system. The input of information from several sources allows a highly integrated response to a variety of stimuli, which is of ultimate benefit to the whole animal. For example, liver glycogen metabolism may be regulated or influenced by several different hormones, including insulin, glucagon, epinephrine, thyroid hormones, and adrenal glucocorticoids.

■ Hormones Are Often Secreted in Characteristic Patterns

Secretion of any one particular hormone is either stimulated or inhibited by a defined set of chemical substances in the blood or environmental factors. In addition to these specific **secretagogues,** many hormones are secreted in a defined, rhythmic pattern. These rhythms can take several forms. For example, they may be pulsatile episodic spikes in secretion lasting just a few minutes, or they may follow a daily, monthly, or seasonal change in overall pattern. Pulsatile secretion may occur in addition to other longer secretory patterns.

For these reasons, a single randomly drawn blood sample for determination of a certain hormone concentration may be of little or no diagnostic value. A dynamic test in which hormone secretion is specifically stimulated by a known agent often provides much more meaningful information.

The Nature of Hormones

■ Hormones Can Generally Be Classed as One of Three Chemical Types

Hormones can be categorized by a number of criteria. Grouping them by chemical structure is convenient; since in many cases hormones with similar structures also use similar mechanisms to produce their biologic effects, this also serves to group them according to their mechanism of action. In addition, hormones with similar chemical structures are usually produced by tissues with similar embryonic origins.

Amino Acid Derivatives ■ In terms of structure, the simplest hormones are derivatives of amino acids. This class of hormones is small in size and typically hydrophilic. These hormones are formed by conversion from a commonly occurring amino acid; epinephrine and thyroxine, for example, are derived from tyrosine. Each of these hormones is synthesized by a particular sequence of enzymes that are primarily localized in the endocrine gland involved in its production. The synthesis of amino acid-derived hormones can therefore be influenced in a relatively specific fashion by a variety of environmental or pharmacologic agents. The steps involved in the synthesis of these hormones are discussed in detail in later chapters.

Peptide and Protein Hormones ■ The size and complexity of hormones in the peptide and protein group are quite diverse. They may be as small as the tripeptide thyrotrophin releasing hormone (TRH) or as large as human chorionic gonadotrophin (hCG), which is composed of separate α and β subunits, has a molecular weight of approximately 34,000, and is a glycoprotein comprised of 16% carbohydrate by weight. To avoid any confusion in later sections, note that there is no universally accepted distinction between peptides (or polypeptides) and proteins. Generally, anything less than 50–100 amino acids is referred to as a peptide and those with 100 or more amino acids are categorized as a protein.

Within the peptide and protein class of hormones there exist a number of families of hormones, some of which are listed in Table 33–1. Hormones can be grouped into these families as a result of considerable homology with regard to amino acid sequence and structure. Presumably, the similarity of structure in these families results from the evolution of a single ancestral hormone into each of the separate and distinct hormones. In many cases there is also considerable homology among receptors for the hormones within a family.

Steroid Hormones ■ Steroids are lipid-soluble, hydrophobic molecules synthesized from cholesterol. They can be classified into six categories, based on their primary biologic activity. An example of each category is shown in Figure 33–2. **Glucocorticoids,** such as cortisol, are primarily produced in cells of the adrenal cortex and regulate processes involved in glucose, protein, and lipid homeostasis. Glucocorticoids generally produce effects that are catabolic in nature. Aldosterone, a primary example of a **mineralocorticoid,** is produced in cells of the outermost portion of the adrenal cortex. Aldosterone is primarily involved in regulating sodium and potassium balance by the kidneys and is the principal mineralocorticoid in the body. **Androgens,** such as testosterone, are primarily produced in the testes, but physiologically significant amounts can be synthesized by the adrenal

cortex as well. The primary female sex hormone is estradiol, a member of the **estrogen** family, produced by the ovaries and placenta. **Progestins,** such as progesterone, are involved in maintenance of pregnancy and are produced by the ovaries and placenta. The **calciferols,** such as 1,25-dihydroxycholecalciferol, are involved in regulation of calcium homeostasis. 1,25-Dihydroxycholecalciferol is the hormonally active form of vitamin D and is formed by a sequence of reactions occurring in skin, liver, and kidneys.

■ Peptide and Protein Hormones Are Synthesized in Advance of Need and Stored in Secretory Vesicles

Steroid hormones are synthesized and secreted on demand, but peptide and protein hormones are typically stored prior to secretion. Steroid hormone synthesis and secretion are discussed in Chapter 36; the discussion here is confined to synthesis and secretion of peptide and protein hormones.

Preprohormones and Prohormones ■ Like other proteins destined for secretion, peptide hormones are synthesized with a "pre-" or "signal" peptide at their amino terminal end that directs the growing peptide chain into the cisternae of the rough endoplasmic reticulum. Most, if not all, peptide hormones are synthesized as part of an even larger precursor, or **preprohormone.** To form the active hormone from the preprohormone, the prepeptide is cleaved on entry of the preprohormone into the rough endoplasmic reticulum to form the **prohormone.** As the prohormone is processed through the Golgi apparatus and packaged into secretory vesicles, the prohormone is proteolytically cleaved at one or more sites to yield active hormone. In many cases, preprohormones may contain the sequences for several different biologically active molecules. These active elements may in some cases be separated by inactive spacer segments of peptide. Examples of polyproteins that are the precursors for peptide hormones, which illustrate the multipotent nature of these precursors, are shown schematically in Figure 33–3. Note, for example, that proopiomelanocortin (POMC) actually contains the sequences for several biologically active signal molecules. Similarly, propressophysin serves as the precursor for the nine amino acid hormone ADH. The precursor for TRH contains five or more repeats of the TRH tripeptide in one single precursor molecule.

TABLE 33–1 ■ Examples of Peptide Hormone Families

Insulin Family	Glycoprotein Family	Growth Hormone Family	Secretin Family
Insulin	Luteinizing hormone (LH)	Growth hormone (GH)	Secretin
Insulinlike growth factor I	Follicle-stimulating hormone (FSH)	Prolactin (PRL)	Glucagon
Insulinlike growth factor II	Thyroid-stimulating hormone (TSH)	Chorionic somatomammotrophin	Gastrointestinal polypeptide
Relaxin	Chorionic gonadotrophin (hCG)	(hCS)	

Families listed are only examples; many other peptide hormone families exist.

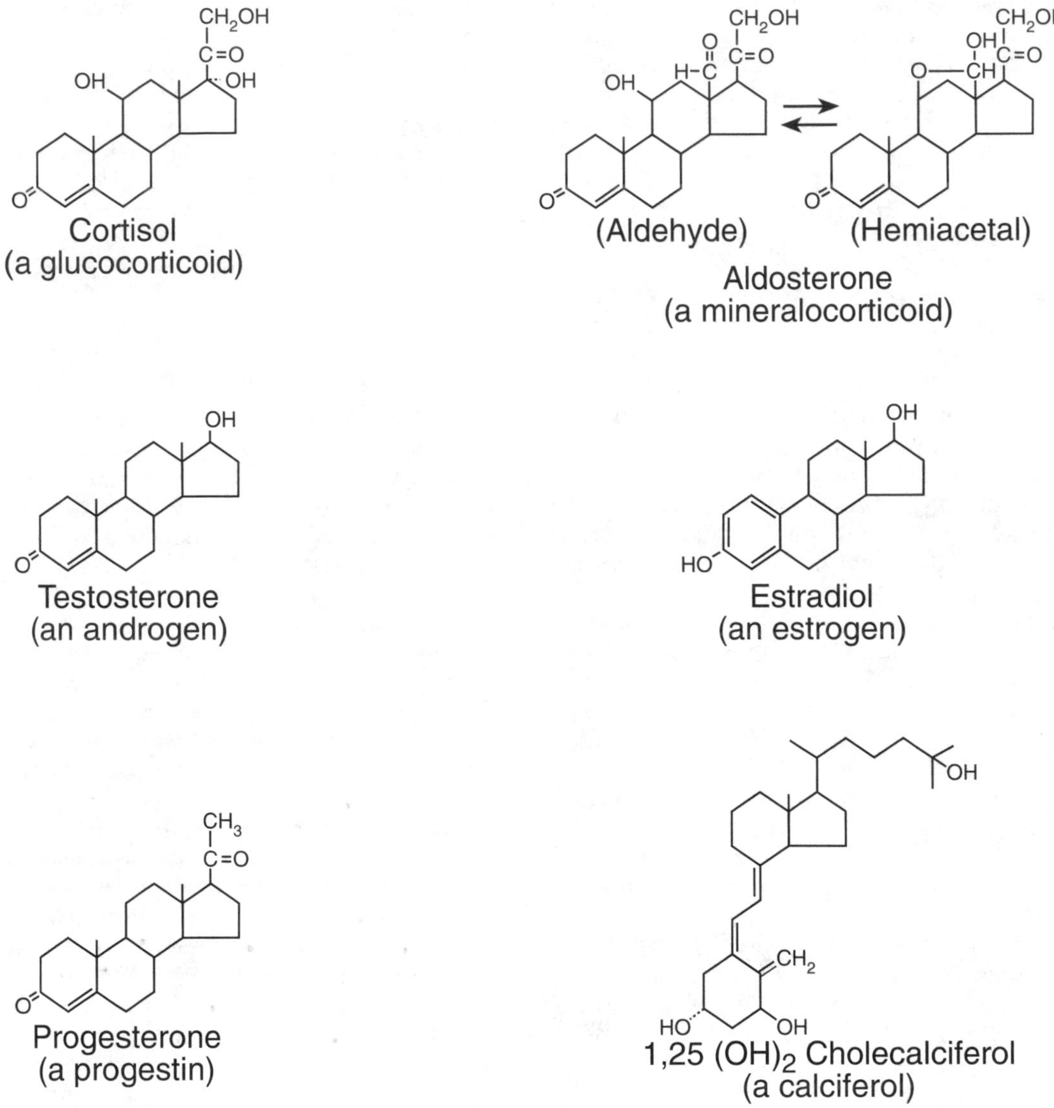

Figure 33–2 ■ Examples of the six types of naturally occurring steroids.

In general, two basic amino acid residues, either lys-arg or arg-arg, demarcate the point(s) at which the prohormone will be cleaved into its biologically active components. Presumably, these two basic amino acids serve as specific recognition sites for the trypsinlike endopeptidases thought to be responsible for cleavage of the prohormones. Although somewhat rare, there are documented cases of inherited diseases in which a point mutation involving an amino acid residue at the cleavage site results in an inability to convert the prohormone into active hormone, resulting in a state of hormone deficiency. Partially cleaved precursor having limited biologic activity may be found circulating in the blood in some of these cases.

In some disease states, large amounts of intact precursor molecules are found in the circulation. This situation may be the result of endocrine cell hyperac-

tivity or even uncontrolled production of hormone precursor by nonendocrine tumor cells. Although precursors usually have relatively low biologic activity, if they are secreted in sufficiently high amounts they may still produce biologic effects. In some cases, these effects may be the first recognized sign of neoplasia.

Tissue-specific differences in processing of prohormones are well known. Although the same prohormone gene may be expressed in different tissues, tissue-specific differences in the way the molecule is cleaved give rise to different final secretory products. For example, within alpha cells of the pancreas, proglucagon is cleaved at two positions to yield three peptides, illustrated in Figure 33–4 (left). Glucagon, an important hormone in the regulation of carbohydrate metabolism, is the best characterized of the three peptides. In contrast, in other cells of the gastrointes-

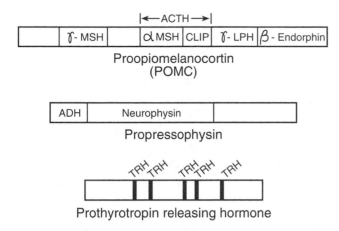

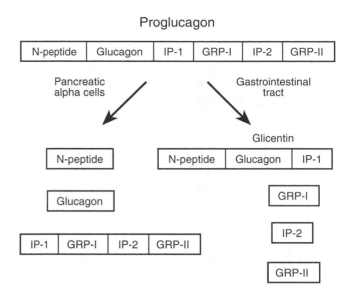

Proglucagon

Figure 33–3 ■ The structure of three prohormones. Relative sizes of individual peptides are only approximations.

Figure 33–4 ■ An example of differential processing of prohormones. In alpha cells of the pancreas (left) the major bioactive product formed from proglucagon is glucagon itself. It is not currently known whether the other peptides are processed to produce biologically active molecules. In intestinal cells (right), proglucagon is cleaved to produce the four peptides shown. Glicentin is the major glucagon-containing peptide in the intestine. IP-1, intervening peptide 1; IP-2, intervening peptide 2; GRP-I, glucagon-related peptide-I; GRP-II, glucagon-related peptide II.

tinal tract in which proglucagon is also produced, the molecule is cleaved at three different positions such that glicentin, glucagon-related peptide-I (GRP-I), and glucagon-related peptide-II (GRP-II) are produced (Fig. 33-4, right).

Intracellular Movement of Secretory Vesicles and Exocytosis ■ Upon insertion of the preprohormone into the cisternae of the endoplasmic reticulum, the pre- or signal peptide is rapidly cleaved from the amino terminal end of the molecule. The resulting prohormone is translocated to the Golgi apparatus, where it is processed and packaged for export. After processing in the Golgi apparatus, peptide hormones are stored in membrane-bound secretory vesicles. Secretion of the peptide hormone occurs by exocytosis: the secretory vesicle is translocated to the cell surface, its membrane fuses with the plasma membrane, and its contents are released into the extracellular fluid. Movement of the secretory vesicle and membrane fusion are triggered by an increase in cytosolic calcium stemming from an influx of calcium into the cytoplasm from internal organelles or the extracellular fluid. In some cells, an increase in cAMP and the subsequent activation of protein kinases is also involved in the stimulus secretion coupling process. Elements of the microtubule-microfilament system are thought to play a role in movement of secretory vesicles from their intracellular storage sites toward the cell membrane.

Cleavage of prohormone into active hormone molecules typically takes place during transit through the Golgi apparatus or, perhaps, soon after entry into secretory vesicles. Secretory vesicles therefore contain not only active hormone but also the excised biologically inactive fragments. When active hormone is released into the blood, a quantitatively equal amount of inactive fragment is also released. In some instances this forms the basis for an indirect assessment of hormone secretory activity. Other types of processing of peptide hormones that may occur during transit

through the Golgi apparatus include glycosylation and coupling of subunits.

■ Many Hormones Are Transported in the Bloodstream

According to the classical definition, hormones are carried by the bloodstream from their site of synthesis to their target tissues. However, the manner in which different hormones are carried in the blood varies.

Transport of Amine, Peptide, and Protein Hormones ■ Most amine, peptide, and protein hormones dissolve readily in the plasma, and thus no special mechanisms are required for their transport. Steroid and thyroid hormones are relatively insoluble in plasma. However, mechanisms are present to promote their solubility in the aqueous phase of the blood and ultimate delivery to a target cell site of action.

Transport of Steroid and Thyroid Hormones ■ In most cases, 90% or more of steroid or thyroid hormones in the blood are bound to plasma proteins. Some of the plasma proteins that bind hormones are specialized, in that they have a considerably higher affinity for one hormone than another, whereas others, such as albumin, bind a variety of hydrophobic hormones. The extent to which a hormone is protein-bound and the extent to which it binds to specific versus nonspecific transport proteins varies from one hormone to another. The principal binding proteins involved in specific and nonspecific transport of steroid and thyroid hormones are listed in Table 33–2. These proteins are synthesized and secreted by the liver, and their production is influenced by changes in various nutritional and endocrine factors.

Typically, for hormones that bind to carrier proteins, only 1–10% of the total hormone present in the plasma exists free in solution. However, only this free hormone is biologically active. Bound hormone cannot directly interact with its receptor and thus is part of a temporarily inactive pool. However, free hormone and carrier-bound hormone are in a dynamic equilibrium with each other (Fig. 33-5). The size of the free hormone pool, and thus the amount available to receptors, is influenced not only by changes in the rate of secretion of the hormone but also by the amount of carrier protein available for hormone binding and the rate of degradation or removal of the hormone from the plasma.

In addition to increasing the total amount of hormone that can be carried in plasma, transport proteins also provide a relatively large reservoir of hormone that buffers rapid changes in free hormone concentrations. As unbound hormone leaves the circulation and enters cells, additional hormone dissociates from transport proteins and replaces free hormone that is lost from the free pool. Similarly, following a rapid increase in hormone secretion or therapeutic administration of a large dose of hormone, the majority of newly appearing hormone is bound to transport proteins, since under most conditions these are present in considerable excess.

Protein binding greatly slows the rate of clearance of hormones from plasma. It not only slows entry of hormones into cells, thus slowing the rate of hormone degradation, but also prevents loss by filtration in the kidneys.

From a diagnostic standpoint it is important to recognize that most hormone assays are reported in terms of total concentration (i.e., the sum of free and bound hormone), not just free hormone concentration. The amount of transport protein and thus total plasma hormone content is known to change under certain physiologic or pathologic conditions, while the free

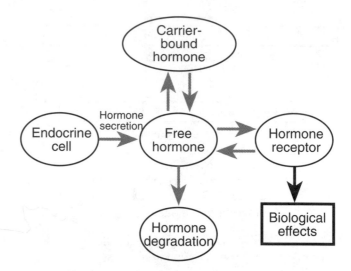

Figure 33–5 ■ The relationship between hormone secretion, carrier protein binding, and hormone degradation in determining the amount of free hormone available for receptor binding and production of biologic effects.

hormone concentration may remain relatively normal. For example, increased concentrations of binding proteins are seen during pregnancy and decreased concentrations are seen with certain forms of liver or kidney disease. Assays of total hormone content might therefore be misleading, since free hormone concentrations may be in the normal range. In such cases, it is helpful to determine the extent of protein binding, so free hormone concentrations can be estimated.

The relative distribution of hormone that is free, bound to a specific transport protein, and bound to albumin varies depending on its solubility, its relative affinity for the two classes of transport proteins, and the relative abundance of the transport proteins. For example, the affinity of cortisol for corticosteroid-binding globulin (CBG) is more than 1000 times greater than its affinity for albumin, but albumin is present in much higher concentrations than is CBG. Thus about 70% of plasma cortisol is bound to CBG, 20% is bound to albumin, and the remaining 10% is free in solution. Aldosterone also binds to CBG, but with a much lower affinity, such that only 17% is bound to CBG, 47% associates with albumin, and 36% is free in solution.

As this example indicates, more than one hormone may be capable of binding to a specific transport protein. When several such hormones are present simultaneously, they compete for a limited number of transport proteins and thus a limited number of binding sites. For example, cortisol and aldosterone compete for CBG binding sites. Increases in plasma cortisol result in displacement of aldosterone from CBG, thus raising the unbound (active) concentration of aldosterone in the plasma. Similarly, prednisone, a widely used synthetic corticosteroid, can displace about 35%

TABLE 33–2 ■ Circulating Transport Proteins

	Principal Hormone Transported
Specific transport proteins	
Corticosteroid binding globulin (CBG, transcortin)	Cortisol, aldosterone
Thyroxine binding globulin (TBG)	Thyroxine, triiodothyronine
Testosterone/estrogen binding globulin (TeBG)	Testosterone, estrogen
Nonspecific transport proteins	
Albumin	Most steroids, thyroxine, triiodothyronine
Transthyretin (prealbumin)	Thyroxine, some steroids

of the cortisol normally bound to CBG. As a result, with prednisone treatment the free cortisol concentration is higher than might be predicted from measured concentrations of total cortisol and CBG.

■ Peripheral Transformation, Degradation, and Excretion of Hormones in Part Determines Their Activity

Peripheral Transformation of Hormones ■ As a general rule, hormones are produced by their gland or tissue of origin in an active form. However, in a few notable exceptions peripheral transformation of a hormone plays a very important part in its action. It is well documented that if these specific transformations are impaired, for example as a result of a congenital enzyme deficiency or drug-induced inhibition of enzyme activity, endocrine abnormalities can result. Well-known examples are the conversion of testosterone to dihydrotestosterone (Chap. 39) and conversion of thyroxine to triiodothyronine (Chap. 35). Other examples are the formation of the octapeptide angiotensin II from its precursor, angiotensinogen (Chap. 36) and the formation of 1,25-dihydroxycholecalciferol from cholecalciferol (Chap. 38).

Mechanisms of Hormone Degradation ■ As in any regulatory control system, it is necessary that the hormonal signal dissipate or disappear once appropriate information has been transferred and the need for further stimulus has ceased. As described earlier, steady-state plasma concentrations of hormone are determined not only by the rate of secretion but also by the rate of degradation. Thus, any factor that markedly alters degradation of a hormone can potentially alter its circulating concentration. Commonly, however, secretory mechanisms can compensate for altered degradation such that plasma hormone concentrations remain within the normal range. Processes of hormone degradation show little if any regulation; alterations in rates of synthesis or secretion of hormones in most cases provide the primary mechanism for altering circulating hormone concentrations.

For most hormones, the liver is quantitatively the most important site of degradation; for a few others the kidneys play a significant role as well. Diseases of the liver and kidneys may therefore indirectly influence endocrine status as a result of altering the rates at which hormones are removed from the circulation. Various drugs also alter normal rates of hormone degradation, thus the possibility of indirect drug-induced endocrine abnormalities also exists. In addition to the liver and kidneys, target tissues may take up and degrade quantitatively smaller amounts of hormone. In the case of peptide and protein hormones, this occurs via receptor-mediated endocytosis.

The nature of specific structural modification(s) involved in hormone inactivation and degradation differs for each hormone class. As a general rule, however, specific enzyme-catalyzed reactions are involved. Inactivation and degradation may involve complete metabolism of the hormone to entirely different products or may be limited to a simpler process involving one or two steps, such as a covalent modification to inactivate the hormone. Urine is the primary route of excretion of hormone degradation products, but small amounts of intact hormone may also appear in the urine. In some cases, measurement of the urinary content of a hormone or hormone metabolite provides a useful, indirect, noninvasive means of assessing endocrine function.

Degradation of peptide and protein hormones has been studied only in a limited number of cases. However, it appears that peptide and protein hormones are inactivated in a variety of tissues by proteolytic attack. The first step(s) appears to involve attack by specific peptidases that results in formation of several distinct hormone fragments. These fragments are then metabolized by a variety of nonspecific peptidases to yield the constituent amino acids, which can be reused by the cell.

Metabolism and degradation of steroid hormones has been studied in much more detail. The primary organ involved is the liver, although some metabolism also takes place in the kidneys. Complete metabolism of steroids generally involves a combination of one or more of five general classes of reactions: reduction, hydroxylation, side chain cleavage, oxidation, and esterification. Reduction reactions are the principal reactions involved in the conversion of biologically active steroids to forms that possess little or no activity. Esterification (or conjugation) reactions are also particularly important. Groups added in esterification reactions are primarily glucuronides and sulfates. Addition of such charged moieties enhances the water solubility of the metabolites, facilitating their excretion. Steroid metabolites are eliminated from the body primarily via the urine, although smaller amounts also enter the bile and leave the body in the feces.

At times, quantitative information concerning the rate of hormone metabolism is clinically useful. One index of the rate at which a hormone is removed from the blood is the **metabolic clearance rate** (MCR). The metabolic clearance of a hormone is analogous to that of renal clearance to remove a substance from the plasma (Chap. 23). Metabolic clearance rate is the volume of plasma cleared of the hormone in question per unit time. It is calculated from the equation MCR = Hormone removed per unit time (mg/min) ÷ Plasma concentration (mg/ml) and is expressed in ml/min.

One approach to measuring MCR involves injecting a small amount of radioactive hormone into the subject and then collecting a series of timed blood samples to determine the amount of radioactive hormone remaining. Based on the rate of disappearance of hormone from the blood, its half-life and MCR can be calculated. The MCR and half-life are inversely related;

the shorter the half-life, the greater the MCR. The half-lives of different hormones vary considerably, from 5 minutes or less for some to several hours for others. The circulating concentration of hormones with short half-lives can vary dramatically over a short period of time. This is typical of hormones that regulate processes on an acute minute-to-minute basis, such as a number of those involved in regulating blood glucose. Hormones for which rapid changes in concentration are not required, such as those with seasonal variations in metabolism and those that regulate the menstrual cycle, typically have longer half-lives.

■ Measurement of Hormone Concentrations Is an Important Tool in Endocrinology

Often, the concentration of hormone present in a biologic fluid is measured to make a clinical diagnosis in suspected endocrine disease or to study basic endocrine physiology. Substantial advancements have been made in measuring hormone concentrations.

Bioassays ■ Even before hormones were chemically characterized, they were quantitated in terms of biologic responses they produced. Thus, early assays for measuring hormones were bioassays that depended on a hormone's ability to produce a characteristic biologic response. As a result, hormones came to be quantitated in terms of units, defined as an amount sufficient to produce a response of specified magnitude under a defined set of conditions. A unit of hormone is thus arbitrarily determined. Although bioassays are rarely used today for diagnostic purposes, many hormones are still standardized in terms of **biologic activity units.** For example, commercial insulin is still sold and dispensed based on the number of units in a particular preparation, rather than by weight or the number of moles of insulin.

Bioassays in general suffer from a number of shortcomings, including a relative lack of specificity and a lack of sensitivity. In many cases they are slow and cumbersome to perform, and quite often they are expensive, since biologic variability often requires the inclusion of many animals in the assay. Despite these limitations, however, all of the newer hormone assay techniques use something other than biologic activity as the end point, thus bioassay offers a distinct advantage in this regard.

Radioimmunoassay ■ Development of the radioimmunoassay (RIA) in the late 1950s and early 1960s was a major step forward in clinical and research endocrinology. Without this assay, much of our current knowledge of endocrinology would not be possible. An RIA or closely related assay is now available for virtually every known hormone. In addition, RIAs have been developed to measure circulating concentrations of a variety of other biologically relevant proteins, drugs, and vitamins.

The RIA is a prototype for a larger group of assays termed **competitive binding assays.** These are modi-

fications and adaptations of the original RIA that rely to a large degree on the principle of competitive binding on which the RIA is based. It is beyond the scope of this text to describe in detail the competitive binding assays currently used to measure hormone concentrations, but the principles are the same as those for the RIA.

The two key components of an RIA are a specific antibody (Ab) that has been raised against the hormone in question and radioactively labeled hormone (H*). If the hormone being measured is a peptide or protein, the molecule is commonly labeled with a radioactive iodine atom (^{125}I or ^{131}I), which can be readily attached to tyrosine residues of the peptide chain. For substances lacking the tyrosine residues, such as steroids, labeling may be accomplished by incorporating radioactive carbon (^{14}C) or hydrogen (^{3}H). In either case, use of radioactive hormone permits detection and quantitation of very small amounts of the substance.

The RIA is performed in vitro using a series of test tubes. Fixed amounts of Ab and of H* are added to all tubes (Fig. 33–6A). A volume of the sample (plasma, urine, cerebrospinal fluid, etc.) to be measured is added to one series of tubes. To another series of identical tubes is added varying known concentrations of unlabeled hormone (the standards). The basis of the assay, as indicated in Figure 33–6B, is that unlabeled hormone in the unknown sample competes with H* for binding to a limited amount of antibody present in the tube. The principle of the RIA is that labeled and unlabeled hormone compete for a limited number of antibody binding sites. The amount of each that is bound to antibody is a proportion of that present in

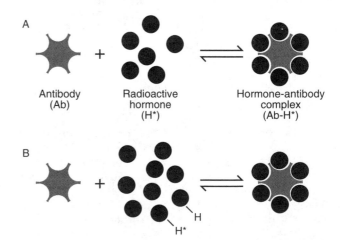

Figure 33–6 ■ The principles of radioimmunoassay. (A) Specific antibodies (Ab) bind with radioactive hormone (H*) to form hormone-antibody complexes (Ab-H*). (B) When unlabeled hormone (H) is also introduced into the system, less radioactive hormone binds to the antibody. (Modified from Hedge GA, Colby HD, Goodman RL: *Clinical Endocrine Physiology.* Philadelphia, WB Saunders, 1987.)

solution. In a sample containing a high concentration of hormone, less radioactive hormone will be able to bind to the antibody, and vice versa. In each case, the amount of radioactivity present as antibody-bound H* is determined. The response produced by the standards is used to generate a **standard curve,** such as that shown in Figure 33–7. Responses produced by the unknown samples are then compared to the standard curve to determine the amount of hormone present in the unknowns (dashed lines in Fig. 33–7).

One major limitation of RIAs is that they measure immunoreactivity, rather than biologic activity. Presence of an immunologically related but different hormone or heterogeneous forms of the same hormone can complicate interpretation of the results. For example, POMC, the precursor form of ACTH, is often present in high concentrations in the plasma of patients with bronchogenic carcinoma. Antibodies for ACTH may cross-react with POMC. Results of an RIA for ACTH in which such an antibody is used may suggest high concentrations of ACTH, when actually POMC is being detected. Because POMC has less than 5% of the biologic potency of ACTH, there may be little clinical evidence of markedly elevated ACTH. If appropriate measures are taken, however, such possible pitfalls can be overcome in most cases and reliable results from the RIA can be obtained.

One important modification of the RIA is the **radioreceptor assay,** which uses specific hormone receptors rather than antibodies as the hormone binding reagent. In theory, this method measures biologically active hormone, since receptor binding rather than

antibody recognition is assessed. However, the need to purify hormone receptors and the somewhat more complex nature of this assay limits its usefulness for routine clinical measurements. It is more likely to be used in a research setting.

ELISA ▪ The enzyme-linked immunosorbent assay (ELISA) is a solid-phase, enzyme-based assay whose use and application has increased considerably over the past decade. A typical ELISA is a colorimetric or fluorometric assay, and thus the ELISA, unlike the RIA, does not produce radioactive waste, an advantage due to environmental concerns and the rapidly increasing cost of radioactive waste disposal. In addition, because it is a solid-phase assay, the ELISA can be automated to a large degree, which reduces costs. Figure 33–8 shows a relatively simple version of an ELISA. More complex assays using similar principles have been developed to overcome a variety of technical problems, but the basic principle remains the same. In recent

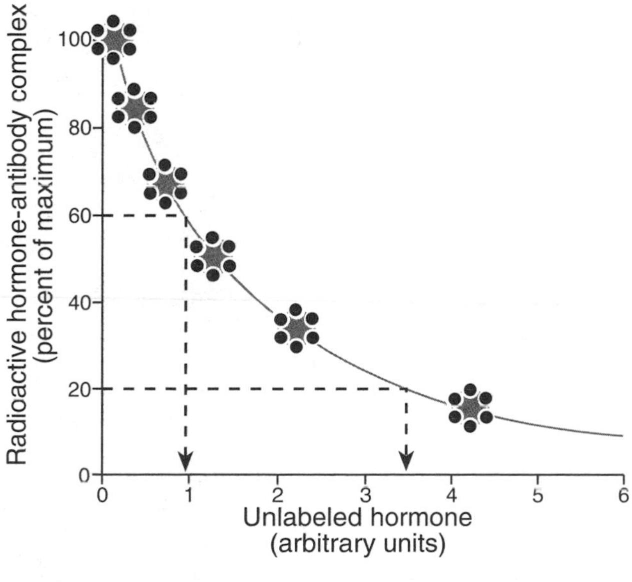

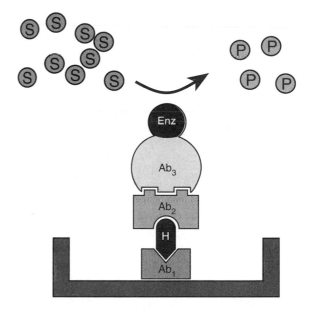

Figure 33–8 ▪ Components of an ELISA. A typical ELISA is performed in a 3 × 5-inch plastic plate containing 96 small wells. Each well is precoated with an antibody (Ab₁) that is specific for the hormone (H) being measured. Unknown samples or standards are introduced into the wells, followed by a second hormone-specific antibody (Ab₂). A third antibody (Ab₃), which recognizes Ab₂, is then added. Ab₃ is coupled to an enzyme that will convert an appropriate substrate (S) into a colored or fluorometric product (P). The amount of product formed can be determined using optical methods. After addition of each antibody or sample to the wells, the plates are incubated for an appropriate period of time to allow antibodies and hormones to bind. Any unbound material is washed out of the well before the addition of the next reagent. The amount of colored product formed is directly proportional to the amount of hormone present in the standard or unknown sample. Concentrations are determined using a standard curve. For purposes of illustrating the components of an ELISA, only one AB₁ molecule is shown in the bottom of the well, when in fact there is an excess of Ab₁ relative to the amount of hormone (H) to be measured.

Figure 33–7 ▪ A typical RIA standard curve. As indicated by the dashed lines, the hormone content in unknown samples can be deduced from the standard curve. (Modified from Hedge GA, Colby HD, Goodman RL: *Clinical Endocrine Physiology.* Philadelphia, WB Saunders, 1987.)

years the RIA has been the primary assay used clinically, but use of ELISA has expanded considerably and will likely be the predominant assay in the future due to the advantages listed above.

Mechanisms of Hormone Action

As indicated earlier, hormones are one mechanism by which cells communicate with one another. Fidelity of communication in the endocrine system depends on each hormone's ability to interact with a specific receptor in its target tissues. This interaction results in activation (or inhibition) of a series of specific events in cells that results in precise biologic responses characteristic of that hormone.

■ Binding with a Receptor Is the First Step in Hormone Action

As indicated above, binding of a hormone to its receptor and subsequent activation of the receptor is the first step in hormone action and also the point at which specificity is determined within the endocrine system. Abnormalities concerning interaction of hormones with their receptors has been shown to be involved in pathogenesis of a number of endocrine disease states, therefore considerable attention has been paid to this aspect of hormone action.

Kinetics of Hormone-Receptor Binding ■ The probability that a hormone-receptor interaction will occur is related to both the abundance of cellular receptors and the receptor's affinity for the hormone relative to the ambient hormone concentration. The more receptors available to interact with a given amount of hormone, the greater the likelihood of a response. Similarly, with a higher affinity of receptor for hormone, greater is the likelihood that an interaction will occur. The circulating hormone concentration is, of course, a function of the rate of hormone secretion relative to hormone degradation.

Association of a hormone with its receptor generally behaves as if it were a simple, reversible chemical reaction that can be described by the following kinetic equation:

$$[H] + [R] \rightleftharpoons [HR] \qquad (1)$$

where [H] is the free hormone concentration, [R] is the unoccupied receptor concentration, and [HR] is the hormone-receptor complex (also referred to as bound hormone or occupied receptor).

Assuming a simple chemical equilibrium, it follows that:

$$K_a = \frac{[HR]}{[H] \times [R]} \qquad (2)$$

If R_0 is defined as the total receptor number (i.e., [R] + [HR]), then after substituting and rearranging, we obtain the following relationship:

$$\frac{[HR]}{[H]} = -K_a[HR] + K_aR_0 \qquad (3)$$

Literally translated, this equation states:

$$\frac{\text{Bound hormone}}{\text{Free hormone}} = -K_a \times \text{Bound hormone} + K_a \times \text{Total receptor number} \qquad (4)$$

Notice that equations 3 and 4 have the general form of an equation for a straight line: $y = mx + b$.

To obtain information regarding a particular hormone-receptor system, a fixed number of cells (and therefore a fixed number of receptors) is incubated in vitro in a series of test tubes with increasing amounts of hormone. At each higher hormone concentration, the amount of receptor-bound hormone is increased until all receptors are occupied by hormone. Receptor number and affinity can be obtained by using the relationships given in equation 4 above and plotting the results as the ratio of receptor-bound hormone to free hormone ([HR]/[H]) as a function of the amount of bound hormone ([HR]). This type of analysis is known as a **Scatchard plot** (Fig. 33-9). In theory, a Scatchard plot of simple, reversible equilibrium binding is a straight line (Fig. 33-9A), with the slope of the line being equal to the negative of the association constant ($-K_a$) and the x-intercept being equal to the total receptor number (R_0). Other equally valid mathematical and graphic methods can be used to analyze hormone-receptor interactions, but the Scatchard plot is probably the most widely used.

In practice, Scatchard plots are not always straight lines but instead can be curvilinear (Fig. 33-9B). Insulin is a classic example of a hormone that gives curved Scatchard plots. One interpretation of this result is that cells contain two separate and distinct classes of receptors, each with a different binding affinity. Typically, one receptor population has a higher affinity but is fewer in number compared to the second population. Thus, as indicated in Figure 33-9B, $K_{a1} > K_{a2}$, but $R_{02} > R_{01}$. Computer analysis is often required to fit curvilinear Scatchard plots accurately to a two-site model.

Another explanation for curvilinear Scatchard plots is that occupied receptors influence the affinity of adjacent, unoccupied receptors by **negative cooperativity.** According to this theory, as one hormone molecule binds to its receptor it causes a decrease in the affinity of nearby unoccupied receptors, making it more difficult for additional hormone molecules to bind. The greater the amount of hormone bound, the greater the decrease in affinity of unoccupied

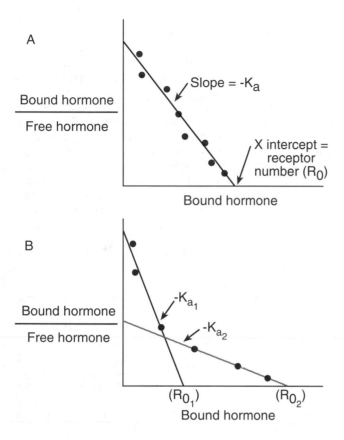

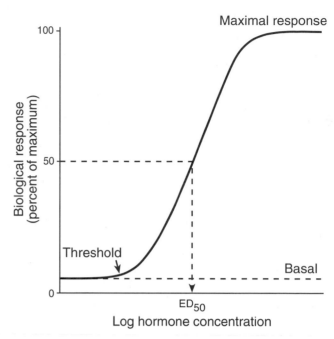

Figure 33–10 ■ A normal dose-response curve of hormone biologic activity.

Figure 33–9 ■ Scatchard plots of hormone-receptor binding data. (A) A straight line plot typical of hormone binding to a single class of receptors. (B) A curvilinear Scatchard plot typical of some hormones. As described in the text, several models have been proposed to account for nonlinearity of Scatchard plots.

receptors. Thus, as shown in Figure 33-9B, as bound hormone increases, the affinity (slope) steadily decreases. Whether curvilinear Scatchard plots in fact result from two-site receptor systems or from negative cooperativity between receptors is unknown.

Dose-Response Curves and Responsiveness versus Sensitivity ■ Hormone effects are generally not all-or-none phenomena; that is, they generally do not switch from totally off to totally on, and then back again. Instead, target cells exhibit graded responses proportional to the concentration of free hormone present. The dose-response relationship for a hormone generally exhibits a sigmoid shape when plotted as the biologic response on the y-axis versus the log of the hormone concentration on the x-axis (Fig. 33-10). Regardless of the biologic pathway or process being considered, cells typically exhibit an intrinsic **basal** level of activity in the absence of added hormone, even well after any previous exposure to hormone. As the hormone concentration surrounding the cells increases, a minimal **threshold** concentration must be present before any measurable increase in the cellular

response can be produced. At higher hormone concentrations, a **maximal response** by the target cell is produced and no greater response can be elicited by increasing the hormone concentration. The concentration of hormone required to produce a half-maximal response (ED_{50}; i.e., effective dose for 50% of maximal response) is a useful index of the **sensitivity** of the target cell for that particular hormone (Fig. 33-10).

For some peptide hormones, the maximal response may occur when only a small percentage (5-10%) of the total receptor population is occupied by hormone. The remaining 90-95% of the receptors have been termed "spare receptors," since on initial inspection they do not appear necessary to produce a maximal response. This term is unfortunate, since the receptors are not "spare" in the sense of being unused. While at any one point in time only 5-10% of the receptors may be occupied, hormone-receptor interactions are an equilibrium process and thus hormones continually dissociate and reassociate with their receptors. Thus, from one point in time to the next, different subsets of the total population of receptors may be occupied, but presumably all receptors participate equally in producing the biologic response.

Physiologic or pathophysiologic alterations in target tissue responses to hormones can take one of two general forms, as indicated by changes in their dose-response curve (Fig. 33-11). Although changes in dose-response curves are not routinely assessed in the clinical setting, they can serve to distinguish between a **receptor** or **postreceptor abnormality** in hormone

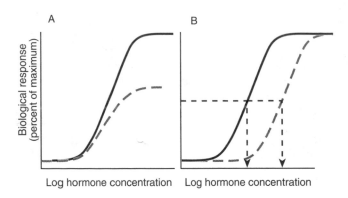

Figure 33–11 ■ Dose-response curves showing the effects of (A) decreased target tissue responsiveness and (B) decreased target tissue sensitivity.

action and therefore provide useful information regarding the underlying cause of a particular disease state. A change in responsiveness (Fig. 33–11A) is indicated by an increase or decrease in the maximal response of the target tissue and may be the result of one or more factors. Altered responsiveness can be caused by a change in the number of functional target cells in a tissue, by a change in the number of receptors per cell for the hormone in question, or, if receptor function itself is not rate-limiting for hormone action, by a change in the specific rate-limiting postreceptor step in the hormone action pathway.

A change in sensitivity (Fig. 33–11B) is reflected as a right or left shift in the dose-response curve and thus a change in the ED_{50}; a right shift indicates decreased sensitivity and a left shift indicates increased sensitivity for that hormone. Changes in sensitivity reflect an alteration in receptor affinity or, if submaximal concentrations of hormone are present, an increase in receptor number will enhance the sensitivity of the cell to that hormone. Dose-response curves may also reflect combinations of changes in responsiveness and sensitivity in which there is both a right/left shift of the curve (a sensitivity change) and a change in maximal biologic response to a lower or higher level (a change in responsiveness).

Cells can regulate their receptor number and/or function in several ways. Exposing cells to an excess of hormone for a sustained period typically results in a decreased number of receptors for that hormone per cell. This phenomenon is referred to as **down-regulation.** In the case of peptide hormones, which have receptors on cell surfaces, a redistribution of receptors from cell surface to intracellular sites usually occurs as part of the process of down-regulation. Thus, there may be fewer total receptors per cell and a smaller percentage may be available for hormone binding on the cell surface. Although somewhat less prevalent than down-regulation, **up-regulation** may occur when certain conditions or treatments cause an increase in receptor number as compared to that nor-

mally present. Changes in rates of synthesis of receptors may also contribute to long-term down- or up-regulation.

In addition to changing receptor number, many target cells can regulate receptor function. Chronic exposure of cells to a hormone may cause the cells to become less responsive to subsequent exposure to the hormone by a process termed **desensitization.** If the exposure of cells to a hormone has a desensitizing effect on further action by that same hormone, the effect is termed **homologous desensitization.** If exposure of cells to one hormone has a desensitizing effect with regard to the action of a different hormone, the effect is termed **heterologous desensitization.**

■ The Second Messenger Model Describes How Hydrophilic Hormones Act

Due to their relatively low solubility in the lipid environment of cell membranes, peptide hormones and catecholamines cannot readily penetrate cell membranes. As a result, they must interact with receptors on the exterior surface of the cell. Coupling of the receptor to the intracellular enzymatic machinery of the cell results from generation of an intracellular signal or message. Intracellular signals are considered **second messengers,** while the hormone is considered the first messenger (Chap. 1).

The Adenylate Cyclase/Cyclic AMP Second Messenger System ■ As a result of binding to specific cell surface receptors, many peptide hormones and catecholamines produce an almost immediate increase in the intracellular concentration of **3′, 5′ cyclic adenosine monophosphate** (cAMP). cAMP activates an enzyme, **cAMP-dependent protein kinase** (protein kinase A), which in turn catalyzes the phosphorylation of various cellular proteins. This alters the activity or function of the proteins and ultimately leads to a desired cellular response. The intracellular signal provided by cAMP is terminated by its hydrolysis to 5′AMP by a group of enzymes known as **phosphodiesterases,** which have also been shown to be regulated by hormones in some instances. In addition to those hormones that act via the mechanism described above, some hormones act to decrease cAMP formation and therefore have intracellular effects opposite those described.

cAMP is perhaps the most widely distributed of the second messengers and has been shown to mediate various cellular responses to both hormonal and non-hormonal stimuli, not only in higher organisms but also in various primitive life forms, including slime molds and yeast. cAMP is formed from cellular ATP by the enzyme **adenylate cyclase,** a membrane-bound enzyme located on the inner aspect of the membrane and exposed to the cytoplasmic compartment of the cell.

Adenylate cyclase is functionally coupled to various

cell surface receptors by a family of guanosine nucleotide-binding regulatory proteins known as G proteins (Fig. 33–12). These proteins are named for their requirement for GTP binding and hydrolysis and have been shown to serve a broad role in linking various membrane receptors to membrane-bound effector systems. G proteins are heterotrimers, and consist of a 39–52 kd GTP-binding α subunit, a 35 kd β subunit, and a 9 kd γ subunit. The β and γ subunits of most G proteins are identical; the specificity of a particular G protein predominantly resides in its α subunit. Adenylate cyclase is coupled to cell surface receptors by two types of G protein. Hormones that bind to receptors coupled to adenylate cyclase through the G_s (s = stimulatory) protein promote increased adenylate cyclase activity and an increase in intracellular cAMP. Hormones binding to receptors coupled to the G_i (i = inhibitory) protein promote the inhibition of adenylate cyclase and a decrease in cAMP. Thus, bidirectional regulation of adenylate cyclase is achieved by coupling different classes of cell surface receptors to the enzyme by either G_s or G_i. G proteins serve as transducers in a variety of receptor-effector

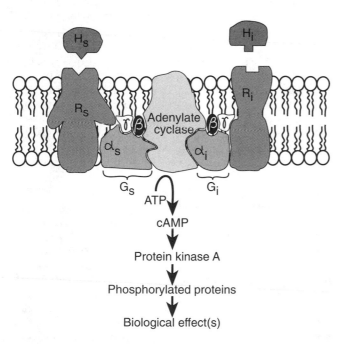

Figure 33–12 ■ The functional relationship of G proteins that regulate adenylate cyclase. Stimulatory (G_s) and inhibitory (G_i) G proteins couple hormone binding to the receptor with activation or inhibition of adenylate cyclase, respectively. Each G protein is a trimer consisting of α, β, and γ subunits. The β and γ subunits in G_s and G_i appear to be similar; the α subunits are distinct in each and provide the specificity for activation or inhibition of adenylate cyclase. Hormones (H_s) that stimulate adenylate cyclase interact with receptors (R_s) that are coupled to adenylate cyclase through stimulatory G proteins (G_s). Conversely, hormones (H_i) that inhibit adenylate cyclase interact with receptors (R_i) that are coupled to adenylate cyclase through inhibitory G proteins (G_i).

systems and thus are an important part of cellular communication mechanisms.

The protein kinases with which cAMP interacts consist of a catalytic and regulatory subunit (Fig. 33–13). The kinase activity of these enzymes resides in their catalytic subunits. When cAMP concentrations in the cell are low, the regulatory subunits bind to and inactivate the catalytic subunits. When cAMP is formed in response to hormonal stimulation, it binds to the regulatory subunits, which then dissociate from the catalytic subunits. This relieves the inhibition on catalytic subunits and allows them to catalyze the phosphorylation of appropriate substrates, which produces appropriate biologic responses to the hormone (Fig. 33–13).

A similar, parallel second messenger to cAMP is cGMP, formed much like cAMP by the enzyme guanylate cyclase. The role of cGMP as a second messenger is not as great as that of cAMP, but it does serve an important role in the action of several hormones.

The Phosphatidylinositol Second Messenger System ■ When certain hormones or neurotransmitters bind to their cell surface receptors, they activate one or more of a family of specific **phospholipase** enzymes present in the membrane. These enzymes catalyze the hydrolysis of specific membrane phospholipids, the breakdown of which results in the generation of at least two different second messengers in the cell. The signaling mechanism begins with activation of **phosphatidylinositol-specific phospholipase C**, an enzyme that cleaves a minor membrane phospholipid, **phosphatidylinositol 4,5 bisphosphate** (PIP_2), which is present primarily in the inner leaflet of the cell membrane lipid bilayer. Hydrolysis of PIP_2 results in formation of **1,2-diacylglycerol** (DAG) and **inositol trisphosphate** (IP_3) (Fig. 33–14). Both DAG and IP_3 serve as second messengers in the cell. Analogous to the situation seen in the adenylate cyclase-cAMP system, cell surface receptors are coupled to phospholipase C by a G protein, in this case designated G_p.

In its second messenger role, DAG activates the

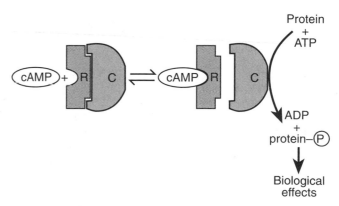

Figure 33–13 ■ The mechanism of cAMP activation of protein kinase A. R, regulatory subunit; C, catalytic subunit.

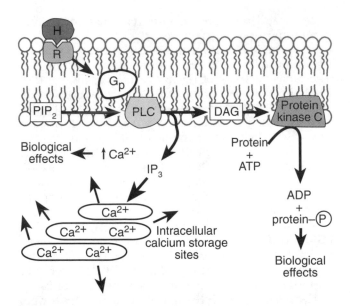

Figure 33–14 ■ The phosphatidylinositol second messenger system: the generation and second messenger roles of inositol trisphosphate and diacylglycerol. H, hormone; R, receptor; G_p, a G protein that couples receptors to activation of phospholipase C; PLC, phospholipase C; PIP_2, phosphatidylinositol 4,5, bisphosphate; DAG, diacylglycerol; IP_3, inositol trisphosphate.

membrane-bound calcium- and lipid-sensitive enzyme **protein kinase C** (Fig. 33–14). When activated, this enzyme catalyzes the phosphorylation of specific proteins in the cell to produce appropriate biologic effects. Several tumor-promoting **phorbol esters** have been shown to activate protein kinase C and can therefore bypass the receptor and activate this part of the hormone action pathway.

In the endoplasmic reticulum and perhaps also in mitochondria, IP_3 promotes the release of calcium ions into the cytoplasm. The concentration of free calcium ions in the cytoplasm of most cells is in the range of 0.1 μmol/L. With appropriate stimulation, the concentration may abruptly increase 1000 times or more. Almost every cell contains the protein **calmodulin,** which serves as an intracellular "receptor" for calcium. Calcium-calmodulin complexes bind to and activate a variety of cellular proteins, resulting in altered cellular function.

Mechanisms exist to reverse the effects of DAG and IP_3 by rapidly removing them from the cytoplasm. The IP_3 is dephosphorylated to inositol, which can be reused for phosphoinositide synthesis. The DAG is converted to phosphatidic acid by the addition of a phosphate group to carbon number 3. Like inositol, phosphatidic acid can be used for resynthesis of membrane inositol phospholipids. On removal of the IP_3 signal, calcium is very quickly pumped back into its storage sites, restoring cytoplasmic calcium concentrations to their low prestimulus levels. Lowering cytoplasmic calcium concentrations shifts the equilibrium

in favor of the release of calcium from calmodulin. Calmodulin then dissociates from the various proteins that were activated, and the cell returns to its basal state.

In addition to IP_3, other, perhaps more potent, phosphoinositols, such as IP_4 or IP_5, may also be produced in response to stimulation. These are formed by hydrolysis of appropriate phosphatidylinositol phosphate precursors found in the cell membrane. The exact role of these other phosphoinositols has yet to be fully defined. There is also some evidence to suggest that the hydrolysis of other phospholipids, such as phosphatidylcholine, may play an analogous role in hormone signaling processes.

Ion Channels and Second Messenger Systems ■ In some cells, binding of hormones or neurotransmitters (Chap. 3) directly opens gated ion channels in the cell membrane. The influx of ions serves as an intracellular signal to convey the hormonal message to the cell interior.

In many instances, activation of hormone receptors leads to opening of calcium channels in the cell membrane. Increasing cytoplasmic calcium may result in direct activation of calcium-sensitive enzymes or, as described above, formation of calcium-calmodulin complexes with all their attendant effects.

Coupling of receptors and ion channels may involve a G protein, as is the case for muscarinic cholinergic receptors. In other instances the receptor may directly interact with an ion channel or both the ion channel and receptor function may reside in the same protein, as is the case for the nicotinic cholinergic receptor.

The Tyrosine Kinase System ■ The mechanism of action of insulin, epidermal growth factor (EGF), and other related growth factors has not been fully elucidated. However, in recent years it has become apparent that the receptors for these agents not only possess the capacity to bind ligand but also exhibit tyrosine-specific protein kinase activity. This activity is intrinsic to the receptor molecule itself. These receptors consist of a hormone-binding portion that is exposed to the extracellular fluid; the tyrosine kinase activity is located in that part of the receptor exposed to the interior of the cell (Fig. 33–15). As is the case for the EGF receptor, the receptor may be a single polypeptide with an extracellular ligand-binding domain, a transmembrane domain, and an intracellular tyrosine kinase domain. Alternatively, it may consist of subunits, with its hormone binding and tyrosine kinase functions located in different subunits, as is the case for the insulin receptor. In either case, it is convenient and appropriate to think of the hormone as an allosteric modifier of the tyrosine kinase enzyme activity of the receptor, much as allosteric modifiers are described in biochemistry.

The general scheme for this hormone action pathway begins with hormone binding to the extracellular portion of the receptor and subsequent activation of the tyrosine kinase portion of the molecule. The acti-

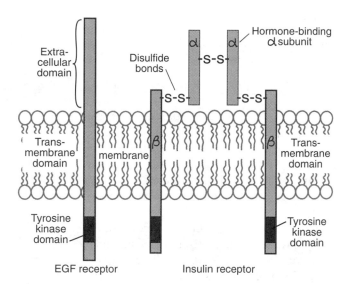

Figure 33–15 ■ The structures of the epidermal growth factor (EGF) receptor and the insulin receptor. The EGF receptor is a single-chain, transmembrane protein consisting of an extracellular region containing the hormone-binding domain, a transmembrane domain, and an intracellular region that contains the tyrosine kinase domain. The insulin receptor is a heterotetramer, consisting of two α and two β subunits held together by disulfide bonds. The α subunits are entirely extracellular and involved in insulin binding. The β subunits are transmembrane proteins and contain the tyrosine kinase activity of the receptor in the intracellular portion of the subunit.

vated kinase then catalyzes phosphorylation of a limited number of tyrosine residues of a few cellular proteins. In most instances, the identity and function of these phosphotyrosine-containing proteins has yet to be determined. However, it is clear from the study of these receptors as well as their oncogene-associated analogs that tyrosine phosphorylation is intimately associated with the regulation of cell growth. Products of several oncogenes have been shown to be modified hormone receptors that generally lack the capacity for regulation which results from hormone binding as normally observed with intact receptors. For example, the erb-B oncogene product has been shown to be a truncated version of the EGF receptor lacking the extracellular EGF binding domain. Absence of this domain of the molecule results in continuous activation of the tyrosine kinase portion of the receptor, as if the EGF signal were continuously present.

For hormones whose receptor displays tyrosine kinase activity, it is tempting to conclude the actions of the hormone result from activation of the kinase activity and subsequent formation of phosphotyrosine-containing cellular proteins. However, at present, one cannot rule out the possibility that these hormones may also participate in generation of other as yet unidentified second messengers or that receptors may directly interact with and activate other membrane components, such as glucose or amino acid transport

proteins, and produce insulinlike effects without involvement of tyrosine kinase activity of receptors. Another point to consider is that actions of hormones are not necessarily limited to just one of the hormone action pathways described above.

■ The Gene Expression Model Describes How Hydrophobic Hormones Act

Unlike peptide and protein hormones, which cannot readily penetrate cell membranes, steroid and thyroid hormones readily enter cells and thus do not require cell surface receptors and second messengers to carry hormone signals to the cell interior. Instead, these hydrophobic hormones interact with intracellular receptors and produce their effects by regulating expression of specific genes in the cell nucleus.

Steroid and Thyroid Hormone Receptors ■ Using molecular biology techniques, researchers have determined putative structures of receptors for steroid and thyroid hormones in recent years. It has been determined that receptors for many steroid and thyroid hormones share considerable homology, suggesting that perhaps each of these hormone-receptor systems evolved from a single master system. The prototypical receptor for this class of hormones has three major domains (Fig. 33–16): a carboxy terminal hormone binding region; a DNA binding region located approximately 50 amino acids upstream from the steroid binding region toward the amino terminal end of the protein; and an amino terminal peptide of variable length whose function is unknown.

The central DNA binding region of steroid receptors is rich in basic amino acids, which presumably facilitates binding of this region to negatively charged DNA. This region is also highly conserved, as approximately 60% of the amino acid sequence is identical. Basic amino acids in this region are arranged in clusters of 9–15 residues flanked by two pairs of cysteines. It has been postulated that these cysteines bind to a metal ion, such as zinc, and force the intervening peptide

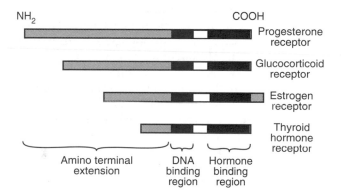

Figure 33–16 ■ The structures of receptors for representative steroid and thyroid hormones. Assignment of DNA and hormone binding domains in each receptor is based on amino acid sequence data deduced from gene and mRNA sequence information.

CELLULAR COMMUNICATION AND ONCOGENES

Elucidation of the mechanisms of cellular communication and signal transduction has been an area of considerable research interest over the past several years. Attention has been directed toward this area with the anticipation that by understanding normal mechanisms of regulation we might also understand the basic underlying pathophysiology in conditions of abnormal cellular regulation, such as cancer. This has proven true with the finding that altered versions of proteins involved in signal transduction are present in certain disease states and certain forms of cancer. These altered proteins are known as **oncoproteins,** and the genes that encode them are termed **oncogenes** (from the Greek *onkos,* meaning mass or tumor.)

An oncogene can arise when a virus picks up either a fragment or a corrupted copy of a normal gene from its host cell. The altered gene is then inserted into subsequent host cells, where an altered gene product is produced. Typically, an oncoprotein is an altered version of a normal cell protein involved in cellular communication and regulation, but in which the normal regulatory elements of that protein have been lost. The result is that normal signal transduction pathways capable of activation and deactivation in a cell may be left in a state in which they do not respond appropriately to outside signals coming into the cell.

For example, epidermal growth factor (EGF) is an important signal peptide involved in regulating growth in certain cells. It initiates its biologic effects by interacting with a cell surface receptor that possesses an intrinsic tyrosine kinase activity. The external portion of the receptor contains the EGF-binding domain, and the tyrosine kinase domain is contained in the intracellular portion. The **erb-B** oncogene (the name is derived from the virus in which it was first discovered) codes for a protein similar to the normal EGF receptor, with the very significant difference that it lacks the extracellular domain that normally binds EGF. In a normal receptor, the extracellular EGF-binding domain inhibits the intracellular tyrosine kinase domain; binding of EGF to the receptor relieves the inhibition and activates the kinase. In the protein encoded by erb-B, however, the extracellular portion of the receptor is absent, thus the intracellular kinase domain remains in a constantly activated state, and EGF-stimulated pathways of signal transduction also remain constantly activated.

There are other situations in which production of a variant form of a protein normally involved in signal transduction may lead to altered cellular regulation. For example, the **ras** oncogene encodes a protein homologous to normal G proteins, except that a mutation results in a continuously activated G protein and therefore continuous activation of those processes that the G protein normally regulates. The **src** family of oncogenes encodes proteins with tyrosine kinase activity that lack certain regulatory components, and thus remain in a continuously activated state. Further examples include the **erb-A** oncoprotein and the **raf/mil** oncoproteins. The oncogene erb-A encodes a protein that is an unregulated version of the thyroid hormone receptor and thus continuously activates pathways of thyroid hormone action, even in the absence of thyroid hormones. The raf/mil oncoproteins are a group of cytoplasmic serine/threonine kinases that are deficient in the components normally involved in regulation, and thus represent a group of unregulated protein kinases.

The study of oncogene biology is a good example of how the study of abnormal physiology can help us understand normal physiology and basic changes that lead to the disease state. As a result, we now have a more detailed understanding of the function of several components of signal transduction mechanisms and are closer to understanding the basic biology of a number of disease states.

sequence into loops or DNA-binding "fingers." These fingers are thought to bind to DNA in a highly specific manner, such that transcription of certain genes is regulated by each class of receptor.

The hormone-binding carboxy terminal portion of the receptors exhibits a modest (50–60%) degree of homology between receptors for closely related steroids, such as the receptors for glucocorticoids and mineralocorticoids. Lesser homology (approximately 20%) exists between the receptors for more dissimilar steroids, such as glucocorticoids and estrogens. It is thought that in the absence of hormone, the steroid binding region exerts an inhibitory influence over its DNA binding domain and prevents interaction of the receptor with DNA. Binding of steroid to the receptor removes this inhibition and permits binding of receptor to DNA.

Regulation of Gene Expression ■ The model of steroid hormone action shown in Figure 33–17 is generally applicable to all steroid and thyroid hormones. These hormones are typically delivered to their target cells bound to specific carrier proteins. The hormones freely diffuse through both plasma and nuclear membranes. Receptors for these hormones are located largely in the cell nucleus and possibly to some extent in the cytoplasm as well. Interaction of hormone and receptor leads to **activation** (or **transformation**) of receptors into forms with an increased affinity for binding to specific **response elements** or acceptor sites on the chromosomes. The molecular basis of activation in vivo is unknown but appears to involve a decrease in apparent molecular weight or in the aggregation state of receptors, as determined by density gradient centrifugation. Binding of hormone-receptor complexes to chromatin results in alterations in RNA polymerase activity that lead to either increased or decreased transcription of specific portions of the genome. As a result, mRNA is produced that leads to production of new cellular proteins or changes in rates of synthesis of preexisting proteins. These newly synthesized proteins/enzymes have effects on cellular metabolism and function that comprise the responses attributable to that particular steroid or thyroid hor-

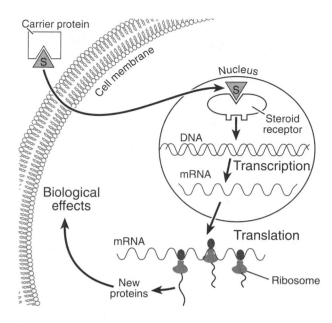

Figure 33–17 ■ The general mechanism of action of steroid hormones (S). Thyroid hormones are thought to act by a similar, if not identical, mechanism.

mone. Thus, the end result of stimulation by steroid hormones is a change in the readout or transcription of the genome. While most effects involve increased production of specific proteins, repression of production of certain proteins by steroid hormones can also occur.

The molecular mechanism of steroid hormone receptor activation/transformation, how the hormone-receptor complex activates transcription, and how the hormone-receptor complex recognizes specific response elements of the genome are not well understood but are under active investigation. Steroid hormone receptors are also known to undergo phosphorylation/dephosphorylation reactions. The effect of this covalent modification is also an area of active research.

■ ■ ■ ■ ■ ■ ■ ■ ■ ■ ■ ■ **REVIEW EXERCISES** ■ ■ ■ ■ ■ ■ ■ ■ ■ ■ ■ ■

Identify Each with a Term

1. A situation in which one particular hormone has many different effects in its target tissue
2. The hormone concentration at which a half-maximal biologic effect is produced
3. The group of proteins that couple certain receptors with activation or inactivation of adenylate cyclase

Define Each Term

4. Preprohormone
5. Progestins

6. ELISA
7. Metabolic clearance rate
8. Radioimmunoassay (RIA)
9. Homologous desensitization
10. 1,2-Diacylglycerol (DAG)

Choose the Correct Answer

11. A shift to the right in the biologic activity dose-response curve for a hormone with no accompanying change in the maximal response indicates:
    a. Decreased responsiveness *and* decreased sensitivity
    b. Increased responsiveness only

c. Decreased sensitivity only
d. Increased sensitivity *and* decreased responsiveness

12. IP_3 is thought to produce its biologic effects by:
    a. Directly activating calmodulin
    b. Directly stimulating protein kinase C
    c. Promoting release of calcium from intracellular stores into the cytoplasm
    d. Activating the α subunit of G proteins

13. Within the endocrine system, specificity of communication is determined by:
    a. The chemical nature of the hormone
    b. The distance between the endocrine cell and its target cell(s)
    c. The presence of specific receptors on target cells
    d. Anatomic connections between the endocrine and target cells

14. The principal mineralocorticoid in the body is:
    a. Aldosterone
    b. Testosterone
    c. Progesterone
    d. Prostaglandin E

15. An index of the affinity of a hormone for binding to its receptor can be obtained by examining:
    a. The y-intercept of a Scatchard plot
    b. The slope of a Scatchard plot
    c. The maximum point on a biologic dose-response curve
    d. The x-intercept of a Scatchard plot

16. G proteins are or may be involved in the coupling of receptors and their respective effector systems with regard to activation of:
    a. Ion channels
    b. Adenylate cyclase
    c. Phospholipase C
    d. All of the above

Briefly Answer

17. Compare and contrast mechanisms for peptide and steroid hormone action.
18. Compare and contrast bioassays, radioimmunoassays, and ELISA.
19. Describe the phosphatidylinositol second messenger system.

ANSWERS

1. Pleiotropic effects
2. ED_{50}
3. G proteins
4. The initial gene product formed in the production of several peptide and protein hormones; they are specifically cleaved to yield active hormone molecules
5. The general class of steroid hormones involved in processes related to the maintenance of pregnancy
6. Enzyme-linked immunosorbent assay (ELISA), a test used to determine hormone concentrations in a variety of biologic fluids
7. A theoretical volume of plasma from which a particular hormone is totally removed per unit of time (usually expressed as ml/min); it provides a relative index of how fast a hormone is degraded or removed from the blood
8. A test used to measure hormone concentrations in a variety of biologic fluids
9. The condition in which application of hormone to a tissue results in a reduced biologic response to a second or repeated exposure to the same hormone
10. A second messenger, formed as a result of PIP_2 hydrolysis by phospholipase C; it activates protein kinase C
11. c
12. c
13. c
14. a
15. b
16. d

17. Peptide hormones cannot pass through the cell membrane and interact with receptors on the cell surface to generate one or more of several second messengers, which act inside the cell to produce the biologic effects characteristic of that particular hormone. Steroid hormones are lipid-soluble and can pass through the cell membrane to interact with specific receptors in the cytoplasm or nucleus. This interaction results in changes in transcriptional rates for specific genes in the target cell.

18. Bioassays measure hormone concentrations ultimately as a result of interaction of hormones with their receptors in target cells, but they often suffer from lack of sensitivity and specificity and are not used in clinical practice. Radioimmunoassays and ELISA are superior in that they typically exhibit a high degree of sensitivity and specificity, are generally rapid, can be used to measure hormones in various biologic fluids, and are less expensive to perform. Of note, however, both RIA and ELISA measure immunoreactivity rather than biologic activity, thus in some instances a strong positive signal may result from the presence of high levels of biologically inactive hormone, rather than the authentic hormone itself.

19. When certain hormones bind to their receptors, they activate the phosphatidylinositol second messenger system. Hormone binding activates phospholipase C, which cleaves PIP_2 into IP_3 and DAG. The IP_3 acts by increasing calcium influx into the cytoplasm from intracellular storage sites, which, in turn, promotes a variety of cellular processes. An increase in membrane DAG activates protein kinase C, which also serves to regulate a variety of cell functions.

Suggested Reading

Goodman HM: *Basic Medical Endocrinology.* 2nd ed. New York, Raven, 1994.

Griffin JE, Ojeda SR: *Textbook of Endocrine Physiology.* 2nd ed. New York, Oxford University Press, 1992.

Hedge GA, Colby HD, Goodman RL: *Clinical Endocrine Physiology.* Philadelphia, WB Saunders, 1987.

Martin CR: *Endocrine Physiology.* New York, Oxford University Press, 1985.

Norman AW, Litwack G: *Hormones.* Orlando, Academic, 1987.

Rasmussen H (ed): *Cell Communication in Health and Disease.* New York, WH Freeman, 1991.

Wilson JD, Foster DW: *Williams Textbook of Endocrinology.* 8th ed. Philadelphia, WB Saunders, 1992.

C H A P T E R

■ ■ ■ ■ ■ ■ ■

34

The Hypothalamus and Pituitary Gland

CHAPTER OUTLINE

I. THE HYPOTHALAMOPITUITARY AXIS
   A. The human pituitary is composed of two morphologically and functionally distinct glands
   B. Anterior pituitary hormones are synthesized and secreted in response to hypothalamic releasing hormones carried in the hypothalamohypophyseal portal circulation
   C. Posterior pituitary hormones are synthesized by hypothalamic neurosecretory cells whose axons terminate in the posterior lobe

II. HORMONES OF THE ANTERIOR PITUITARY
   A. The anterior pituitary secretes six protein hormones
   B. ACTH regulates adrenal cortical function
      1. Structure and formation of ACTH
      2. CRH and ACTH secretion and synthesis
      3. Glucocorticoids and ACTH synthesis and secretion
      4. Stress and ACTH secretion
      5. ADH and ACTH secretion
      6. The sleep-wake cycle and ACTH secretion
   C. TSH regulates the functions of the follicles of the thyroid gland
      1. Structure and formation of TSH
      2. TRH and TSH secretion and synthesis
      3. Thyroid hormones and TSH secretion and synthesis
      4. Other factors affecting TSH secretion
   D. Gonadotrophins regulate reproduction
   E. Prolactin regulates milk synthesis
   F. Growth hormone regulates body growth during childhood
      1. Structure and synthesis of human GH
      2. Actions of GHRH and somatostatin on GH secretion
      3. GH and insulinlike growth factor I
      4. Feedback effects of GH on its own secretion
      5. Pulsatile secretion of GH
      6. Actions of GH

III. HORMONES OF THE POSTERIOR PITUITARY
   A. ADH and oxytocin are closely related peptides produced in the hypothalamus
   B. ADH increases the reabsorption of water by the kidneys
   C. Oxytocin stimulates the contraction of smooth muscle in the mammary gland and uterus

OBJECTIVES

After studying this chapter, the student should be able to:

1. List the structures that compromise the hypothalamopituitary axis
2. Describe the hypophyseal portal circulation and its physiologic function; explain the physiologic functions of the hypothalamic releasing hormones; list the hypothalamic releasing hormones
3. Explain the role of the CNS in regulating anterior pituitary function
4. List the six anterior pituitary hormones of known physiologic importance; define *trophic hormone*
5. Describe the hypothalamopituitary-adrenal axis; explain the roles of CRH, ACTH, and glucocorticoids in this control loop; describe how this mechanism is influenced by physical and emotional stress, ADH, and the sleep-wake cycle
6. Describe the hypothalamopituitary-thyroid axis; explain the roles of TRH, TSH, and thyroid hormones in regulating this control loop; describe how this mechanism is affected by exposure to a cold environment
7. Describe the mechanisms that regulate GH secretion; explain the roles of GHRH, somatostatin, IGF-I, and GH in regulating GH secretion; describe the effects of aging, onset of deep sleep, stress, exercise, and hypoglycemia on GH secretion
8. Describe the growth promoting, lipolytic, diabetogenic, and insulinlike actions of GH
9. Describe the mechanism involved in the synthesis and secretion of ADH and oxytocin; list the physiologic signals that trigger secretion of these hormones; list the main physiologic actions of ADH and oxytocin

The pituitary gland is the most complex of the endocrine organs. It secretes an array of peptide hormones that have important actions on almost every aspect of body function. Some pituitary hormones influence key cellular processes involved in preserving the volume and composition of body fluids. Others bring about changes in body function, which enable the individual to grow, reproduce, or respond successfully to trauma. The pituitary hormones produce these physiologic effects either by acting directly on their target cells or by stimulating other endocrine glands to secrete hormones, which in turn bring about changes in body function.

Stimuli that affect the secretion of pituitary hormones may originate within or outside the body. These stimuli are perceived and processed by the brain, which signals the pituitary gland to increase or decrease the rate of secretion of a particular hormone. Thus, the brain links the pituitary gland to events occurring within or outside the body, which call for changes in pituitary hormone secretion. This important functional connection between the brain and the pituitary, in which the hypothalamus plays a central role, is called the **hypothalamopituitary axis.**

The Hypothalamopituitary Axis

■ The Human Pituitary Is Composed of Two Morphologically and Functionally Distinct Glands

The pituitary, or **hypophysis,** is located at the base of the brain and is connected to the hypothalamus by a stalk. It sits in a depression in the sphenoid bone of the skull called the **sella turcica.** The two morphologically and functionally distinct glands comprising the human pituitary are the **adenohypophysis** and the **neurohypophysis** (Fig. 34–1). The adenohypophysis consists of the **pars tuberalis,** which forms the outer covering of the pituitary stalk, and the **pars distalis** or **anterior lobe.** The neurohypophysis is composed of the **median eminence** of the hypothalamus, the **infundibular stem,** which forms the inner part of the stalk, and the **infundibular process** or **posterior lobe.** The anterior and posterior lobes of the adult human pituitary are separated by a thin, diffuse region of cells that is the vestige of the anatomically distinct **pars intermedia** or **intermediate lobe** of the pituitary glands of lower vertebrates.

The adenohypophysis and neurohypophysis have different embryologic origins. The adenohypophysis is formed from an evagination of the oral ectoderm called **Rathke's pouch.** The neurohypophysis forms as an extension of the developing hypothalamus, which fuses with Rathke's pouch as development proceeds. The posterior lobe is therefore composed of neural tissue and is a functional part of the hypothalamus.

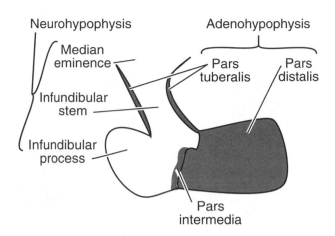

Figure 34–1 ■ Midsagittal section of a human pituitary gland (Modified from Goodman HM: *Basic Medical Endocrinology.* New York, Raven, 1988.)

■ Anterior Pituitary Hormones Are Synthesized and Secreted in Response to Hypothalamic Releasing Hormones Carried in the Hypothalamohypophyseal Portal Circulation

The anterior lobe contains clusters of histologically distinct types of cells closely associated with blood sinusoids that drain into the venous circulation. These cells produce the anterior pituitary hormones and secrete them into the blood sinusoids. The six well-known anterior pituitary hormones are produced by separate kinds of cells. **Adrenocorticotrophic hormone** (ACTH) is secreted by **corticotrophs, thyroid stimulating hormone** (TSH) by **thyrotrophs, growth hormone** (GH) by **somatotrophs, prolactin** (PRL) by **lactotrophs,** and **follicle stimulating hormone** (FSH) and **luteinizing hormone** (LH) by the **gonadotrophs.**

The cells that produce the anterior pituitary hormones are not innervated by secretomotor fibers and hence are not under direct neural control. Rather, their secretory activity is regulated by **releasing hormones** or **hypophysiotropic hormones** produced in the hypothalamus. These releasing hormones are synthesized by neurosecretory neurones (neuroendocrine cells) located in the hypothalamus. Granules containing the releasing hormones are stored in the axon terminals of these cells on capillary networks in the median eminence of the hypothalamus and lower infundibular stem. These capillary networks give rise to the principal blood supply to the anterior lobe of the pituitary.

The blood supply to the anterior pituitary is shown in Figure 34–2. Arterial blood is brought to the hypothalamic-pituitary region by the **superior** and **inferior hypophyseal arteries.** The superior hypo-

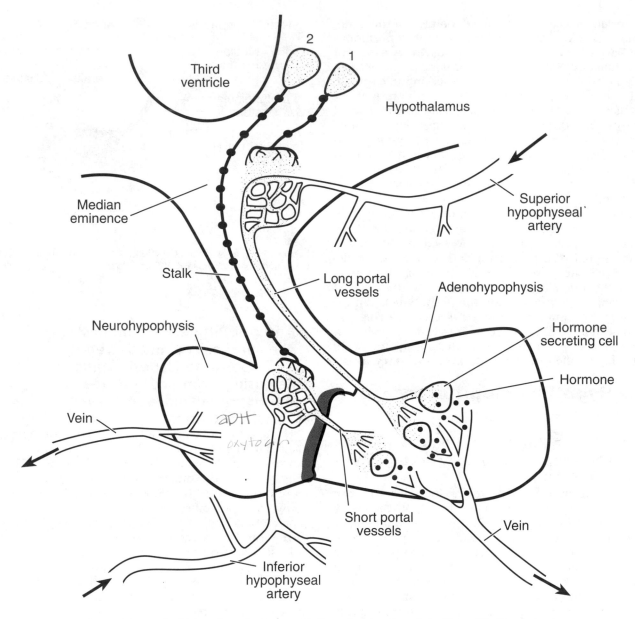

Figure 34–2 ■ Sagittal section of a human pituitary, showing the relationship of its blood supply to the hypothalamic neurosecretory cells that produce releasing hormones. Neurosecretory neurons (1 and 2) are shown secreting releasing factors into capillary networks giving rise to the long and short hypophyseal portal vessels, respectively. The releasing hormones are shown reaching the hormone-secreting cells of the anterior lobe via the portal vessels. (Modified from Gay VL: The hypothalamus: Physiology and clinical use of releasing factors. *Fertil Steril* 23:51–63, 1972.)

physeal arteries give rise to a rich capillary network in the median eminence of the hypothalamus (Fig. 34-2). These capillaries then converge into long veins that run down the pituitary stalk and empty into the blood sinusoids in the anterior lobe. These veins are considered to be portal veins, because they deliver blood to the anterior pituitary rather than joining the venous circulation that carries blood back to the heart. Therefore, they are called the **long hypophyseal portal vessels.** The inferior hypophyseal arteries pro-

vide arterial blood to the posterior lobe. They also penetrate into the lower infundibular stem, where they form another important capillary network. The capillaries of this network converge into **short hypophyseal portal vessels,** which also deliver blood into the sinusoids of the anterior pituitary. This special blood supply to the anterior lobe of the pituitary gland is known as the **hypothalamohypophyseal portal circulation** or simply the **hypophyseal portal circulation.**

When a neurosecretory neuron is stimulated to secrete, the releasing hormone is discharged into the hypophyseal portal circulation (Fig. 34–2). The releasing hormones travel only a short distance before they come in contact with their target cells in the anterior lobe. Only the amount of releasing hormone needed to control anterior pituitary hormone secretion is delivered to the hypophyseal portal circulation by the neurosecretory neurons. Consequently, the releasing hormones are almost undetectable in systemic blood.

A releasing hormone either stimulates or inhibits the secretion of a particular anterior pituitary hormone. A stimulatory releasing hormone also stimulates the synthesis of the anterior pituitary hormone. Although separate stimulatory and inhibitory releasing hormones controlling the secretion of each anterior pituitary hormone have been proposed, only a few of these substances have been isolated (Table 34–1). For example, individual releasing hormones that stimulate the secretion of ACTH, TSH, GH, and the gonadotropins have been found. They are called, respectively, **corticotrophin-releasing hormone** (CRH), **thyrotrophin-releasing hormone** (TRH), **growth hormone-releasing hormone** (GHRH), and **gonadotrophin-releasing hormone** (GnRH), the stimulatory releasing hormone for FSH and LH. The inhibitory releasing hormone for GH secretion is **somatostatin** or **somatotrophin release-inhibiting factor** (SRIF). All of these substances are peptides. Once discovered, they were produced synthetically and are now undergoing intense study for use in the diagnosis and treatment of diseases of the endocrine system. For example, synthetic GnRH is now used for the treatment of infertility in women. In addition to these peptide releasing hormones, the catecholamine **dopamine** is thought to be the inhibitory releasing hormone for PRL.

Releasing hormones are secreted in response to neural inputs from other areas of the central nervous system (CNS). These neural inputs are generated by external events that affect the body or by changes occurring within the body itself. For example, sensory nerve excitation, emotional or physical trauma, biologic rhythms, changes in sleep pattern or in the sleep-wake cycle, and changes in the circulating levels of certain hormones or metabolites all affect the secretion of particular anterior pituitary hormones. Signals generated in the CNS by such events are transmitted to the neurosecretory neurons in the hypothalamus. Depending on the nature of the event and the signal generated, the secretion of a particular releasing hormone may be either stimulated or inhibited. This, in turn, affects the rate of secretion of the appropriate anterior pituitary hormone. The neural pathways involved in transmitting these signals to the neurosecretory neurons in the hypothalamus are not well defined.

■ Posterior Pituitary Hormones Are Synthesized by Hypothalamic Neurosecretory Cells Whose Axons Terminate in the Posterior Lobe

The infundibular stem of the pituitary gland contains bundles of nonmyelinated nerve fibers, which terminate on the capillary bed in the posterior lobe (Fig. 34–3). These fibers are the axons of neurons that originate in the **supraoptic** and **paraventricular nuclei** in the hypothalamus. The cell bodies of these neurons are quite large compared to those of other hypothalamic neurons, hence they are called **magnocellular neurons.** The hormones ADH and oxytocin are synthesized as parts of larger precursor proteins in the cell bodies of these cells. The precursor proteins are then packaged into granules and enzymatically processed to produce ADH and oxytocin. The granules are transported down the axons by axoplasmic flow. These granules accumulate at the

TABLE 34–1 ■ Hypothalamic Releasing Hormones

Hormone	Chemistry	Actions on Anterior Pituitary
Corticotrophin-releasing hormone (CRH)	Single chain of 41 amino acids	Stimulates secretion of ACTH by corticotrophs; stimulates expression of gene for POMC in corticotrophs
Thyrotrophin-releasing hormone (TRH)	Peptide consisting of 3 amino acids	Stimulates secretion of TSH by thyrotrophs; stimulates expression of genes for α and β subunits of TSH in thyrotroph; stimulates synthesis of PRL by lactotrophs
Gonadotrophin-releasing hormone (GnRH), luteinizing hormone-releasing hormone (LHRH)	Single chain of 10 amino acids	Stimulates secretion of FSH and LH by gonadotrophs
Growth hormone-releasing hormone (GHRH)	Two forms in human: single chain of 44 amino acids, single chain of 40 amino acids	Stimulates secretion of GH by somatotrophs; stimulates expression of gene for GH in somatotroph
Somatostatin, somatotrophin release-inhibiting hormone (SRIF)	Single chain of 14 amino acids	Inhibits secretion of GH by somatotrophs
Dopamine	Catecholamine	Inhibits biosynthesis and secretion of PRL by lactotrophs

ACTH, adrenocorticotrophic hormone; FSH, follicle stimulating hormone; GH, growth hormone; LH, luteinizing hormone; POMC, proopiomelanocortin; PRL, prolactin; TSH, thyroid stimulating hormone.

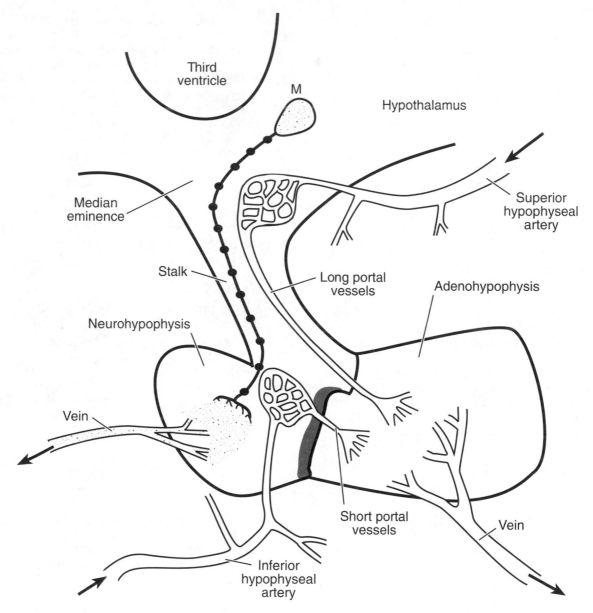

Figure 34–3 ■ Sagittal section of a human pituitary, showing the relationship of its blood supply to the magnocellular neurons of the supraoptic and paraventricular nuclei of the hypothalamus. The neuron labeled M represents a magnocellular neuron releasing ADH or oxytocin at its axon terminals into the capillaries giving rise to the venous drainage of the posterior lobe. (Modified from Gay VL: The hypothalamus: Physiology and clinical use of releasing factors. *Fertil Steril* 23:51–63, 1972.)

axon terminals in the posterior lobe. Stimuli for the secretion of the posterior lobe hormones may be generated by events occurring within or outside the body. These stimuli are processed by the CNS and the signal for the secretion of ADH or oxytocin is then transmitted to the neurosecretory neurons in the hypothalamus. Secretory granules containing the hormone are then released into the nearby capillary circulation, from which they are carried into the systemic circulation.

Hormones of the Anterior Pituitary

■ The Anterior Pituitary Secretes Six Protein Hormones

The six anterior pituitary hormones are all small proteins, ranging in molecular size from 4.5 to 29 kd. Their chemical and physiologic features are given in Table 34-2.

TABLE 34–2 ■ Hormones of the Anterior Pituitary

Hormone	Chemistry	Physiologic Actions
Corticotrophin (ACTH)	Single chain of 39 amino acids; 4.5 kd	Stimulates production of glucocorticoids and androgens by the adrenal cortex; maintains size of zona fasciculata and zona reticularis of cortex
Thyrotrophin, thyroid stimulating hormone (TSH)	Glycoprotein having two subunits, α and β; 28 kd	Stimulates production of thyroid hormones, T_4 and T_3, by thyroid follicular cells; maintains size of follicular cells
Follicle stimulating hormone (FSH)	Glycoprotein having two subunits, α and β; 28-29 kd	Stimulates development of ovarian follicles; regulates spermatogenesis in the testis
Luteinizing hormone (LH)	Glycoprotein having two subunits, α and β; 28-29 kd	Causes ovulation and formation of corpus luteum in the ovary; stimulates production of estrogen and progesterone by the ovary; stimulates testosterone production by the testis
Prolactin (PRL)	Single chain of 198 amino acids; 23 kd	Essential for milk production by lactating mammary gland
Growth hormone (somatotrophin; GH)	Single chain of 191 amino acids; 22 kd	Stimulates postnatal body growth; stimulates triglyceride lipolysis; inhibits actions of insulin on carbohydrate and lipid metabolism

Four of the anterior pituitary hormones have effects on the morphology and secretory activity of other endocrine glands. Because of this they are called **tropic** (Greek for "to turn to") or **trophic** (Greek for "to nourish") hormones. For example, ACTH maintains the size of certain cells in the adrenal cortex and stimulates these cells to synthesize and secrete the **glucocorticoid** hormones, **cortisol** and **corticosterone.** Similarly, TSH maintains the size of the cells of the thyroid follicles and stimulates these cells to produce and secrete the thyroid hormones **thyroxine** and **triiodothyronine.** The two other trophic hormones, FSH and LH, are called gonadotropins, because both act on the ovary and testis. Follicle stimulating hormone stimulates the development of follicles in the ovary and regulates the process of **spermatogenesis** in the testis. Luteinizing hormone causes **ovulation** and **luteinization** of the ovulated **Graafian follicle** in the ovary of the human female and stimulates the production of the female sex hormones **estrogen** and **progesterone** by the ovary. In the male, LH stimulates the **Leydig cells** of the testis to produce and secrete the male sex hormone, **testosterone.**

The two remaining anterior pituitary hormones, GH and PRL, are not usually thought of as trophic hormones because their main target organs are not endocrine glands, at least in humans. As is discussed later, however, these two hormones have certain effects that can be regarded as "trophic". The main physiologic action of GH is its stimulatory effect on the growth of the body during childhood. In humans, PRL is essential for the synthesis of milk by the mammary gland during **lactation.**

The following discussion focuses on ACTH, TSH, and GH. Regulation of the secretion of the gonadotropins and PRL and descriptions of their actions are given in Chapters 39–41.

■ ACTH Regulates Adrenal Cortical Function

The adrenal cortex produces the glucocorticoid hormones, cortisol and corticosterone, in the cells of its two inner zones, the **zona fasciculata** and the **zona reticularis.** These cells also synthesize **androgens,** or male sex hormones, the main one being **dehydroepiandrosterone.**

Glucocorticoids act on many processes, mainly by altering gene transcription and thereby changing the protein composition of their target cells. Glucocorticoids permit metabolic adaptations during fasting, which prevent the development of **hypoglycemia,** or low blood glucose level. They also play an essential role in the body's response to physical and emotional traumas. Other actions of glucocorticoids include their inhibitory effect on inflammation, their ability to suppress the immune system, and their regulation of vascular responsiveness to norepinephrine.

Adrenocorticotrophic hormone is the physiologic regulator of the synthesis and secretion of glucocorticoids by the zona fasciculata and zona reticularis. It stimulates the synthesis of these steroid hormones and promotes the expression of the genes for various enzymes involved in steroidogenesis. It also maintains the size and functional integrity of the cells of the zona fasciculata and zona reticularis.

The other physiologically important hormone made by the adrenal cortex is **aldosterone,** produced by the cells of the outer zone of the cortex, the **zona glomerulosa.** It acts to stimulate sodium reabsorption by the kidney. Adrenocorticotrophic hormone is not an important regulator of aldosterone synthesis and secretion.

The actions of ACTH on glucocorticoid synthesis and secretion and the details of the physiologic effects of glucocorticoids are described in Chapter 36.

Structure and Formation of ACTH ■ Adrenocorticotrophic hormone, the smallest of the six anterior pituitary hormones, consists of a single chain of 39 amino acids and has a molecular size of 4.5 kd.

The corticotrophs do not synthesize ACTH directly. Instead, they produce a 30 kd prohormone called **proopiomelanocortin** (POMC). The 39 amino acid residues comprising ACTH are part of the amino acid sequence of POMC, which also contains the amino acid sequence of another substance, **β-lipotropin** (Fig. 34–4). β-Lipotropin has effects on lipid metabolism, but its physiologic function in humans has not been established. Adrenocorticotrophic hormone and β-lipotropin are produced in the corticotroph when POMC is broken into discrete fragments by proteolytic enzymes. Proopiomelanocortin also contains the amino acid sequences of certain other hormones, such as β-endorphin, but only ACTH and β-lipotropin are produced from POMC in the human corticotroph. Proteolytic processing of POMC occurs after it is packaged into secretory granules. When the corticotroph receives a signal to secrete, ACTH and β-lipotropin are released into the bloodstream in a 1:1 molar ratio.

Proopiomelanocortin is also synthesized by cells of the intermediate lobe of the pituitary gland. These cells also cleave POMC into fragments. However, in these cells the ACTH sequence of POMC is fragmented further to release a small peptide, α-melanocyte stimulating hormone (α-MSH), from its N-terminal end. Very little ACTH is therefore produced by these cells. α-Melanocyte stimulating hormone acts in lower vertebrates to produce temporary changes in skin color by causing the dispersion of melanin granules in pigment cells. As noted earlier, the adult human has only a vestigial intermediate lobe of the pituitary gland. Thus, it does not produce and secrete significant amounts of α-MSH or other hormones derived from POMC. α-Melanocyte stimulating hormone is relevant to human physiology, however. Because ACTH contains the α-MSH amino acid sequence at its N-terminal end, it has melanocyte stimulating activity when present in the bloodstream at high concentrations. Thus, humans who have high levels of ACTH in the blood due to **Addison's disease** or an ACTH-secreting tumor are often hyperpigmented.

CRH and ACTH Secretion and Synthesis ■ Corticotrophin-releasing hormone is the main physiologic regulator of ACTH secretion and synthesis. The CRH produced by humans consists of 41 amino acid residues in a single peptide chain.

Corticotrophin-releasing hormone is synthesized in the paraventricular nuclei of the hypothalamus by a group of neurons with small cell bodies, called **parvicellular neurons.** The axons of the parvicellular neurons terminate on the capillary networks, giving rise to the hypophyseal portal vessels. Secretory granules containing CRH are stored in the axon terminals of these cells. Upon receiving the appropriate stimulus, these cells secrete CRH into the capillary network, which then enters the hypophyseal portal circulation and is delivered to the anterior pituitary gland.

Corticotrophin-releasing hormone binds to receptors on the plasma membranes of the corticotrophs. These receptors are coupled to adenylate cyclase by stimulatory G proteins. Interaction of CRH with its receptor stimulates adenylate cyclase, causing the concentration of **cAMP** to rise in the corticotroph (Fig. 34–5). The rise in cAMP concentration activates protein kinase A, which then phosphorylates cell proteins. By unknown mechanisms, protein phosphorylation stimulates the corticotroph to secrete ACTH and β-lipotropin.

The increase in cAMP produced in the corticotroph by CRH also stimulates the expression of the gene for POMC, thereby increasing the level of POMC mRNA in these cells (Fig. 34–5). Thus, CRH not only stimulates ACTH secretion but also maintains the capacity of the corticotroph to synthesize the precursor for ACTH.

Glucocorticoids and ACTH Synthesis and Secretion ■ The rise in glucocorticoid concentration in the blood resulting from the action of ACTH on the adrenal cortex inhibits the secretion of ACTH (Fig. 34–6). Thus, glucocorticoids have a negative feedback effect on ACTH secretion, which, in turn, reduces the rate of secretion of glucocorticoids by the adrenal cortex. If the glucocorticoid level begins to fall in the blood for some reason, this negative feedback effect is reduced, thereby stimulating ACTH secretion and hence restoring the blood glucocorticoid level. This interactive relationship (Fig. 34–6) is called the **hypothalamopituitary-adrenal axis.** Because of this control loop, the level of glucocorticoids in the blood remains relatively stable in the resting state, although there is some diurnal variation in glucocorticoid secretion. However, as is discussed later, this control mechanism can be greatly influenced by physical and emotional stresses.

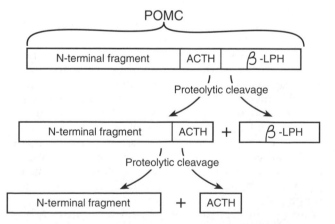

Figure 34–4 ■ The proteolytic processing of proopiomelanocortin (POMC) by the human corticotroph. β-LPH, β-lipotropin; ACTH, adrenocorticotrophic hormone. (Modified from Eipper BA, Mains RE: Structure and biosynthesis of pro-adrenocorticotropin/endorphin and related peptides. *Endocrin Rev* 1:1–27, 1980.)

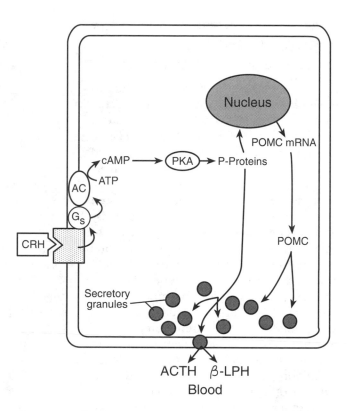

Figure 34–5 ■ The main actions of corticotrophin-releasing hormone (CRH) on the corticotroph. CRH binds to membrane receptors, which are coupled to adenylate cyclase (AC) by stimulatory G proteins (G_s). Adenylate cyclase is stimulated and cAMP rises in the cell. cAMP activates protein kinase A (PKA), which then phosphorylates proteins (P-Proteins) involved in stimulating ACTH secretion and the expression of the POMC gene. β-LPH, β-lipotropin.

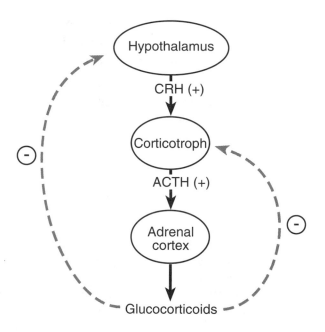

Figure 34–6 ■ The hypothalamopituitary-adrenal axis. The negative feedback actions of glucocorticoids are shown in red. CRH, corticotrophin-releasing hormone; ACTH, adrenocorticotrophic hormone.

The negative feedback effect of glucocorticoids on ACTH secretion results from actions on both the hypothalamus and the corticotroph (Fig. 34-6). When the concentration of glucocorticoids rises in the blood, CRH secretion is inhibited. As a result, the stimulatory effect of CRH on the corticotroph is reduced and the rate of ACTH secretion falls. It is not clear whether the parvicellular neurons that secrete CRH respond to glucocorticoids directly, or whether some other component of the CNS monitors glucocorticoid level and relays the information to these cells. It is known that glucocorticoid receptors are present in the paraventricular nuclei. Glucocorticoids also inhibit ACTH secretion by acting directly on the corticotroph to inhibit the action of CRH. As a result, CRH becomes less effective in stimulating ACTH secretion.

If the concentration of glucocorticoids in the blood remains high for a long period of time, expression of the gene for POMC is inhibited. As a result, the amount of POMC mRNA falls in the corticotroph and gradually the production of ACTH and the other POMC peptides declines as well. Since CRH stimulates POMC gene expression and glucocorticoids inhibit CRH secretion,

it has been proposed that the inhibitory effect of glucocorticoids on POMC gene expression is due mainly to their ability to suppress CRH secretion. However, it has been shown that POMC gene expression is inhibited when isolated anterior pituitary cells are cultured in the presence of glucocorticoids. Therefore, glucocorticoids may suppress POMC gene expression both by inhibiting CRH secretion and by having a direct effect on the corticotroph itself.

The negative feedback actions of glucocorticoids are essential for the normal operation of the hypothalamopituitary-adrenal axis. This is vividly illustrated by the disturbances that occur when blood glucocorticoid levels are changed drastically by disease or glucocorticoid administration. For example, if an individual's adrenal glands have been surgically removed or damaged by disease (e.g., Addison's disease), the resultant lack of glucocorticoids stimulates the corticotrophs to secrete large amounts of ACTH. As noted earlier, this may result in hyperpigmentation due to the melanocyte stimulating activity of ACTH. Individuals with glucocorticoid deficiency caused by inherited genetic defects affecting certain enzymes involved in steroid hormone synthesis by the adrenal cortex have high blood ACTH levels because of the lack of the negative feedback effects of glucocorticoids on ACTH secretion. The high concentration of ACTH in the blood causes hypertrophy of the adrenal glands. Because of this, these genetic diseases have been collectively called **congenital adrenal hyperplasia** (Chap. 36). In contrast, the adrenal cortex atrophies in individuals treated chronically with large doses of glucocorti-

BIOSYNTHETIC HUMAN GROWTH HORMONE

The fact that growth hormone (GH) is species-specific caused a serious problem in the past for the treatment of GH deficiency in children. The only human GH available for treating such children was a very limited amount made from human pituitaries obtained at autopsy. In 1963, the National Institutes of Health and the College of American Pathologists established the National Pituitary Agency, which then collected human pituitary glands from around the world for the preparation of GH. These pituitaries were processed in several university laboratories, and the small amount of purified hormone obtained was distributed to physicians for the treatment of children with growth defects. A small amount of human GH was also prepared and sold by pharmaceutical companies. However, there was never enough human GH available to treat all of the children who might benefit from it. Also, it was impossible to determine whether the hormone might have other beneficial uses in humans, due to its protein anabolic and fat-mobilizing actions. The solution to this problem came when the gene for human GH was cloned in 1979 and then expressed in bacteria. It became possible to produce large amounts of a biosynthetic version of human GH, which proved to have all the activities of the natural substance. During the 1980s, careful clinical trials established that biosynthetic human GH was safe to use in GH-deficient children to promote growth. The hormone was approved for clinical use and is now produced and sold worldwide. Aside from its value in promoting growth in children with short stature, the question remains whether GH might have other therapeutic uses in humans. For example, studies are in progress to learn whether biosynthetic GH can be used as an effective treatment for weight reduction in obesity, since it has lipolytic activity. Also, there is some evidence that the biosynthetic hormone may have some beneficial effects on the aging process in humans. However, because of its diabetogenic action, GH can cause insulin resistance when present in greater than normal amounts, thus posing a risk for diabetes in susceptible individuals. This may ultimately limit the use of biosynthetic human GH to treating children with growth deficiencies.

coids, because the high level of glucocorticoids in the blood inhibits ACTH secretion, resulting in the loss of its trophic influence on the adrenal cortex.

Stress and ACTH Secretion ■ The hypothalamo-pituitary-adrenal axis is greatly influenced by stress. When an individual experiences physical or emotional trauma, ACTH secretion is increased. As a result, the level of glucocorticoids rises rapidly in the blood. Stress stimulates the hypothalamopituitary-adrenal axis regardless of the glucocorticoid concentration prevailing in the bloodstream. This occurs because neural activity generated in higher levels of the CNS in response to the stress stimulates the parvicellular neurons in the paraventricular nuclei to secrete CRH at a greater rate. Thus stress can override the normal operation of the hypothalamopituitary-adrenal axis. If the stress persists, the glucocorticoid level remains high in the bloodstream, because the glucocorticoid negative feedback mechanism functions at a higher set point.

ADH and ACTH Secretion ■ Glucocorticoid deficiency and certain stresses also increase the concentra-

tion of antidiuretic hormone (ADH) in hypophyseal portal blood. The physiologic significance is that ADH, like CRH, can stimulate the corticotroph to secrete ACTH. Acting along with CRH, ADH amplifies the stimulatory effect of CRH on ACTH secretion.

Antidiuretic hormone interacts with receptors in the plasma membranes of the corticotrophs. These receptors are coupled to the enzyme **phospholipase C (PLC)** by G proteins. The interaction of ADH with its receptor activates PLC, which, in turn, hydrolyzes phosphatidylinositol 4,5 bisphosphate (PIP_2) present in the plasma membrane. This generates the intracellular messengers **inositol trisphosphate** (IP_3) and **diacylglycerol** (DAG). They are thought to mediate the stimulatory effect of ADH on ACTH secretion.

As noted earlier, ADH and oxytocin are produced by magnocellular neurons of the supraoptic and paraventricular nuclei of the hypothalamus. These neurons terminate in the posterior lobe, where they secrete ADH and oxytocin into capillaries that feed into the systemic circulation. However, parvicellular neurons in the paraventricular nuclei also produce ADH, which they secrete into hypophyseal portal blood. It appears that much of the ADH secreted by the parvicellular neurons is made in the same cells that produce CRH. It is assumed that the ADH in hypophyseal portal blood comes from these cells and a small number of ADH-producing magnocellular neurons whose axons pass through the median eminence of the hypothalamus on their way to the posterior lobe.

The Sleep-Wake Cycle and ACTH Secretion ■ Under normal circumstances, the hypothalamopituitary-adrenal axis of the human functions in a pulsatile manner, resulting in a number of bursts of secretory activity over a 24-hour period. This appears to be due to rhythmic activity in the CNS, which causes bursts of CRH secretion and, in turn, bursts of ACTH and glucocorticoid secretion (Fig. 34-7). A diurnal oscillation in secretory activity of the axis is thought to be due to a diurnal change in the sensitivity of CRH-producing neurons to the negative feedback action of glucocorticoids, thereby altering their rate of CRH secretion in a diurnal fashion. As a result, there is a diurnal oscillation in the rate of ACTH and glucocorticoid secretion. This **circadian rhythm** is reflected in the daily pattern of glucocorticoid secretion. In individuals who are awake during the day and sleep at night, the blood glucocorticoid level begins to rise during the early morning hours, reaches a peak some-

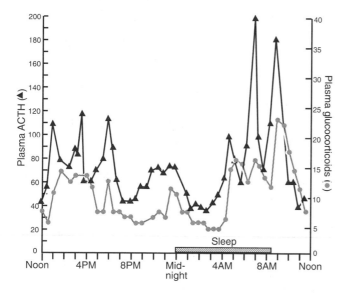

Figure 34–7 ■ Pulsatile changes in the concentrations of adenocorticotrophic hormone (ACTH) and glucocorticoids in the blood of a young woman over a 24-hour period. Note that the amplitude of the pulses in ACTH and glucocorticoids is lower during the evening hours and then increases greatly during the early morning hours. This is due to the diurnal oscillation of the hypothalamo-pituitary-adrenal axis. ACTH concentration is plotted as picograms per milliliter of plasma. Glucocorticoid concentration is plotted as micrograms per 100 ml of plasma. (Modified from Krieger DT: Rhythms in CRF, ACTH and corticosteroids, in Krieger DT (ed): *Endocrine Rhythms.* New York, Raven, 1979, pp 123-142.)

time before noon, and then falls gradually to a low level around midnight (Fig. 34-7). This pattern is reversed in individuals who sleep during the day and are awake at night. This inherent biologic rhythm is superimposed on the normal operation of the hypothalamo-pituitary-adrenal axis.

■ TSH Regulates the Functions of the Follicles of the Thyroid Gland

The thyroid gland is composed of aggregates of many spheric **follicles,** which are formed from a single layer of cells. These follicular cells produce and secrete thyroxine (T_4) and triiodothyronine (T_3), thyroid hormones that are iodinated derivatives of the amino acid tyrosine. The thyroid hormones act on many cells by changing the expression of certain genes, thus changing the capacity of their target cells to produce particular proteins. These changes are thought to bring about the important actions of the thyroid hormones on the differentiation of the CNS, on body growth, and on the pathways of energy and intermediary metabolism.

Thyroid stimulating hormone is the physiologic regulator of T_4 and T_3 synthesis and secretion by the thyroid gland. It also promotes nucleic acid and protein synthesis in the cells of the thyroid follicles, thereby maintaining their size and functional integrity. The actions of TSH on thyroid hormone synthesis and secretion and the physiologic effects of the thyroid hormones are described in Chapter 35.

Structure and Formation of TSH ■ Thyroid stimulating hormone is a glycoprotein consisting of two structurally different subunits, called α and β. The **α subunit** of human TSH is a single peptide chain of 92 amino acid residues with two carbohydrate chains linked to its structure. The **β subunit** is a single peptide chain of 112 amino acid residues, to which a single carbohydrate chain is linked. The α and β subunits are held together by noncovalent bonds. The two subunits combined give the TSH molecule a molecular weight of about 28,000. Neither subunit has significant TSH activity by itself. The two subunits must be combined in a 1:1 ratio to form an active hormone. The gonadotropins FSH and LH are also composed of two noncovalently combined subunits. The α subunits of TSH, FSH, and LH are structurally very similar. The β subunits are different, however; it is the β subunit that gives each hormone its particular set of physiologic activities.

The thyrotroph synthesizes the peptide chains of the α and β subunits of TSH from separate mRNA molecules, which are transcribed from two different genes. The peptide chains of the α and β subunits begin to combine and undergo glycosylation in the rough endoplasmic reticulum. The combination and glycosylation processes are completed as the TSH molecules pass through the Golgi apparatus and are packaged into secretory granules. Normally, the thyrotroph makes more α subunits than β subunits. As a result, secretory granules contain excess α subunits. When the thyrotroph is stimulated to secrete TSH, it releases both TSH and free α subunits into the bloodstream. In contrast, very little free TSH β subunit is in the blood.

TRH and TSH Secretion and Synthesis ■ Thyrotrophin-releasing hormone is the main physiologic stimulator of TSH secretion and synthesis by the thyrotroph. It is a small peptide consisting of three amino acid residues produced by neurons in various areas of the hypothalamus. These neurons terminate on the capillary networks giving rise to the hypophyseal portal vessels. Normally, these neurons secrete TRH into the hypophyseal portal circulation at a constant or tonic rate. It is assumed that the concentration of TRH in the blood perfusing the thyrotrophs does not change greatly, hence the thyrotrophs are continuously exposed to TRH.

Thyrotrophin-releasing hormone binds to receptors on the plasma membranes of the thyrotrophs. These receptors are coupled to PLC by G proteins (Fig. 34–8).

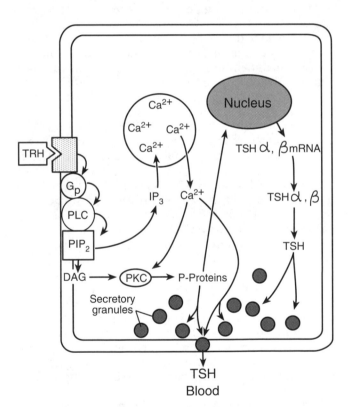

Figure 34–8 ■ The main actions of thyrotrophin-releasing hormone (TRH) on the thyrotroph. TRH binds to membrane receptors, which are coupled to phospholipase C (PLC) by G proteins (G_p). PLC hydrolyzes phosphatidylinositol 4,5 bisphosphate (PIP_2) in the plasma membrane, generating inositol trisphosphate (IP_3) and diacylglycerol (DAG). IP_3 mobilizes intracellularly bound Ca^{2+}. The rise in Ca^{2+} stimulates TSH secretion. Ca^{2+} and DAG activate protein kinase C (PKC), which phosphorylates proteins (P-Proteins) involved in stimulating TSH secretion and the expression of the genes for the α and β subunits of TSH.

The interaction of TRH with its receptor is thought to activate PLC, causing the hydrolysis of PIP_2 in the membrane. This releases the intracellular messengers IP_3 and DAG. Inositol trisphosphate causes the concentration of Ca^{2+} in the cytosol to rise, which stimulates the secretion of TSH into the blood. The rise in cytosolic Ca^{2+} and the increase in DAG activate protein kinase C (PKC) in the thyrotroph. The PKC phosphorylates proteins that are somehow also involved in stimulating TSH secretion (Fig. 34–8).

Thyrotrophin-releasing hormone also stimulates the expression of the genes for the α and β subunits of TSH (Fig. 34–8). As a result, the amounts of mRNA for the α and β subunits are maintained in the thyrotroph, thereby maintaining the production of TSH.

Thyroid Hormones and TSH Secretion and Synthesis ■ Thyrotrophin-releasing hormone has a tonic stimulatory effect on TSH secretion. The main factor that regulates the secretion of TSH is the concentration of thyroid hormones circulating in the bloodstream.

The thyroid hormones exert a direct negative feedback effect on the thyrotroph, which modulates its sensitivity to TRH. For example, when the concentration of thyroid hormones is high in the blood, the sensitivity of the thyrotroph to TRH is reduced. As a result, the rate of TSH secretion falls. In turn, its stimulatory effect on the follicular cells of the thyroid is reduced, resulting in a decrease in T_4 and T_3 secretion. On the other hand, when the circulating levels of T_4 and T_3 are low, their negative feedback effect on the thyrotroph is reduced. As a result, the thyrotroph becomes more sensitive to TRH and secretes TSH at a greater rate. This, in turn, increases the rate of thyroid hormone secretion. This control system comprises the **hypothalamopituitary-thyroid axis** (Fig. 34–9).

The thyroid hormones also have a direct negative feedback effect on TSH synthesis. They not only blunt the stimulatory action of TRH on TSH synthesis but also inhibit TSH production directly.

The negative feedback effects of the thyroid hormones on the thyrotroph are produced in the following way. Most of the T_4 and T_3 circulating in the blood is bound to plasma proteins (Chap. 35). Only a small fraction (less than 1%) is free. The free T_4 and T_3 molecules are taken up by the thyrotroph and affect its functions. Thyroxine is converted to T_3 by the enzymatic removal of one iodine atom from its structure (Chap. 35). The newly formed T_3 molecules and those taken up directly from the blood enter the nucleus, where they bind to thyroid hormone receptors in the chromatin. The interaction of T_3 with its receptors changes the expression of certain genes in the thyrotroph, which decreases the ability of the thyrotroph to produce and secrete TSH. For example, T_3 inhibits the expression of the genes for the α and β subunits of TSH, thereby directly decreasing the synthesis of TSH. Also, T_3 changes the expression of other unidentified genes that code for proteins which decrease the

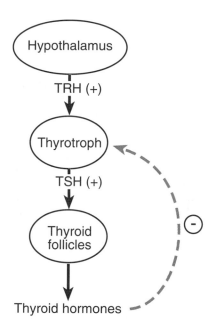

Figure 34–9 ■ The hypothalamopituitary-thyroid axis. The negative feedback effect of thyroid hormones on the thyrotroph is shown in red. TRH, thyrotrophin-releasing hormone; TSH, thyroid stimulating hormone.

sensitivity of the thyrotroph to TRH. The loss in sensitivity is thought to be partly due to a reduction in the number of TRH receptors in thyrotroph plasma membranes.

Whether the secretion of TRH by hypothalamic neurons increases or decreases in response to changes in circulating thyroid hormones is unclear. There is some evidence that thyroid hormones can inhibit TRH gene expression in the hypothalamus of the rat.

Other Factors Affecting TSH Secretion ■ Exposure of some animals to a cold environment stimulates TSH secretion. This makes sense from a physiologic perspective, since the thyroid hormones are important in regulating body heat production (Chap. 35). Brief exposure of experimental animals to a cold environment stimulates the secretion of TSH, presumably due to enhanced TRH secretion. The newborn human behaves much the same way, in that it responds to brief cold exposure with an increase in TSH secretion. This response to cold does not occur in the adult human. Prolonged exposure of the adult human to cold brings about some increase in thyroid hormones in the blood after several days. This response is believed to be a compensatory reaction of the hypothalamopituitary-thyroid axis to the increased destruction of circulating thyroid hormones that results from the elevated rate of metabolism in the cold-exposed individual.

The hypothalamopituitary-thyroid axis, like the hypothalamopituitary-adrenal axis, follows a diurnal circadian rhythm in humans. Peak TSH secretion occurs in the early morning and a low point is reached in the evening.

In humans, TSH secretion does not appear to be changed in any consistent way by emotional or physical stresses.

■ Gonadotrophins Regulate Reproduction

The testis and ovary each has two essential functions in human reproduction. The first is to produce the sperm and ovum, respectively. The second is to produce an array of steroid and peptide hormones, which influence virtually every aspect of the reproductive process. The gonadotrophic hormones FSH and LH regulate both of these functions. The production and secretion of the gonadotrophins by the anterior pituitary is, in turn, regulated by the hypothalamic releasing hormone GnRH and the hormones produced by the testis and ovary in response to gonadotrophic stimulation.

The regulation of human reproduction by this **hypothalamopituitary-gonad axis** is discussed in Chapters 39 and 40. Here we describe the chemistry and formation of the gonadotropins.

Like TSH, human FSH and LH are composed of two structurally different glycoprotein subunits called α and β, which are held together by noncovalent bonds. The α subunits of FSH and LH are thought to be identical to the α subunit of TSH. However, the β subunits differ from that of TSH and from each other.

The β subunit of human FSH consists of a peptide chain of 110 amino acid residues, to which two chains of carbohydrate are attached. The β subunit of human LH is a peptide of 121 amino acid residues. It is also glycosylated with two carbohydrate chains. The combined α and β subunits of FSH and LH give these hormones a molecular size of about 28–29 kd.

As with TSH, the individual subunits of the gonadotrophins have no hormonal activity. They must be combined with each other in a 1:1 ratio to have activity. Again, it is the β subunit that gives the gonadotrophin molecule either FSH or LH activity.

Follicle stimulating hormone and LH are produced by the same gonadotrophs in the anterior pituitary. There are separate genes for the α subunit and for the FSH and LH β subunits in the gonadotroph. Hence, the peptide chains of these subunits are translated from separate mRNA molecules. Glycosylation of these chains begins as they are synthesized and before they are released from the ribosome. The folding of the subunit peptides into their final three-dimensional structure, the combination of an α with a β subunit, and the completion of glycosylation all occur as these molecules pass through the Golgi apparatus and are packaged into secretory granules. As with the thyrotroph, the gonadotroph produces an excess of α subunits over FSH and LH β subunits. Therefore, the rate of β subunit production is considered to be the rate-limiting step in gonadotrophin synthesis.

The synthesis of FSH and LH is regulated by the hormones of the hypothalamopituitary-gonad axis. For example, gonadotrophin production is stimulated by GnRH. It is also affected by the steroid and peptide hormones produced by the gonads in response to stimulation by the gonadotrophins. Such hormonally regulated changes in gonadotrophin production are caused mainly by changes in the expression of the genes for the gonadotrophin subunits. More information about the regulation of gonadotrophin synthesis and secretion is found in Chapters 39 and 40.

■ Prolactin Regulates Milk Synthesis

Lactation is the final phase of the process of human reproduction. During pregnancy, the **alveolar cells** of the mammary glands develop the capacity to synthesize milk in response to stimulation by a variety of steroid and peptide hormones. Milk synthesis by these cells begins shortly after childbirth. To continue to synthesize milk, these cells must be stimulated periodically by PRL. This is thought to be the main physiologic function of PRL in the human female. What role, if any, PRL has in the human male is unclear. It is known to have some supportive effect on the action of androgenic hormones on the male reproductive tract, but whether this is an important physiologic function of PRL is not established.

Human PRL is a globular protein consisting of a single peptide chain of 198 amino acid residues with three intrachain disulfide bridges. Its molecular size is about 23 kd. Human PRL has considerable structural similarity to human GH and to a PRL-like hormone produced by the human placenta called **placental lactogen** (hPL). It is thought that these hormones are structurally related because their genes evolved from a common ancestral gene during the course of vertebrate evolution. Because of its structural similarity to human PRL, human GH has substantial PRL-like or **lactogenic activity.** However, PRL and hPL have little GH-like activity. Human placental lactogen is discussed further in Chapter 41.

Prolactin is synthesized and secreted by the lactotrophs in the anterior pituitary. It is synthesized in the rough endoplasmic reticulum as a somewhat larger peptide. Its N-terminal signal peptide sequence is then removed and the 198 amino acid protein passes through the Golgi apparatus and is packaged into secretory granules.

The synthesis of PRL is stimulated by estrogenic hormones and certain other hormones, such as TRH, which stimulate the expression of the PRL gene. On the other hand, the catecholamine, dopamine, inhibits the synthesis of PRL. Dopamine produced by hypothalamic neurons plays a major role in the regulation of PRL synthesis and secretion by the hypothalamopituitary axis.

The regulation of the synthesis and secretion of PRL and its physiologic actions are discussed in Chapter 41.

■ Growth Hormone Regulates Body Growth during Childhood

As its name implies, growth hormone (somatotropin) promotes growth of the human body. It does not affect fetal growth, nor is it an important growth factor during the first few months after birth. Thereafter, it is essential for the normal rate of body growth during childhood and adolescence. For example, a child with untreated GH deficiency fails to grow to normal height or stature and becomes a dwarf.

Growth hormone is secreted by the anterior pituitary throughout life and remains physiologically important even after growth has stopped. In addition to its growth-promoting action, GH has effects on many aspects of carbohydrate, lipid, and protein metabolism. For example, GH is thought to be one of the physiologic factors that counteract and thus modulate some of the actions of insulin on the liver and peripheral tissues (see below).

Structure and Synthesis of Human GH ■ Human GH is a globular protein consisting of a single chain of 191 amino acid residues with two intrachain disulfide bridges. It has a molecular size of about 22 kd. As noted earlier, human GH has considerable structural similarity to human PRL and placental lactogen.

The anterior pituitaries of all vertebrates produce GH, but over the course of vertebrate evolution the GH gene has undergone considerable mutation. As a result, the amino acid sequence of the GH molecule changed greatly. The significance of this is that only the growth hormones of the human and other primates are active growth-promoting agents in humans. It is believed that the GH molecules of lower animals differ enough in structure from human GH that they cannot be recognized by human GH receptors. Thus, GH has been called **species-specific.** This is not true for other hormones of the endocrine system.

Growth hormone is produced in the somatotrophs of the anterior pituitary. It is synthesized in the rough endoplasmic reticulum as a somewhat larger prohormone consisting of an N-terminal signal peptide and the 191 amino acid residue hormone. The signal peptide is then cleaved from the prohormone, and the hormone traverses the Golgi apparatus and is packaged in secretory granules.

The hypothalamic GHRH stimulates the production of GH by stimulating the expression of the GH gene in somatotrophs. Expression of the GH gene is also stimulated by thyroid hormones. As a result, the normal rate of production of GH is dependent on these hormones. For example, a thyroid hormone-deficient individual is also GH-deficient. This important action of thyroid hormones is discussed further in Chapter 35.

Actions of GHRH and Somatostatin on GH Secretion ■ The secretion of GH is regulated by two opposing hypothalamic releasing hormones. GHRH stimulates GH secretion, and SRIF inhibits GH secretion by inhibiting the action of GHRH. The rate of GH secretion is determined by the net effect of these counteracting hormones on the somatotroph (see below). When GHRH predominates, GH secretion is stimulated. When somatostatin predominates, GH secretion is inhibited.

Human GHRH is a peptide composed of a single chain of 44 amino acid residues. A slightly smaller version of GHRH consisting of 40 amino acid residues is also present in humans. Growth hormone-releasing hormone is synthesized in the cell bodies of neurosecretory neurons in the **arcuate** and **ventromedial nuclei** of the hypothalamus. The axons of these cells project to the capillary networks giving rise to the portal vessels. When these neurons receive a stimulus for GHRH secretion, they discharge GHRH from their axon terminals into the hypophyseal portal circulation.

Growth hormone-releasing hormone binds to receptors in the plasma membranes of the somatotrophs. These receptors are coupled to adenylate cyclase by the stimulatory G protein, G_s. The interaction of GHRH with its receptors stimulates adenylate cyclase, causing the concentration of cAMP to rise in the somatotroph. The rise in cAMP activates protein kinase A, which, in turn, phosphorylates proteins that stimulate GH secretion and GH gene expression (Fig. 34–10). The stimulation of GH secretion by GHRH is also thought to be Ca^{2+}-dependent. In addition, some evidence suggests that GHRH may stimulate PLC, causing the hydrolysis of membrane PIP_2 in the somatotroph. How important this phospholipid pathway is for the stimulation of GH secretion by GHRH is not established.

Somatostatin is a small peptide consisting of 14 amino acid residues. Although made by neurosecretory neurons in various parts of the hypothalamus, somatostatin neurons are especially abundant in the **anterior periventricular region** (i.e., close to the third ventricle). The axons of these cells terminate on the capillary networks giving rise to the hypophyseal portal circulation, where they release somatostatin into the blood.

Somatostatin binds to receptors in the plasma membranes of the somatotrophs. These receptors, like those for GHRH, are also coupled to adenylate cyclase, but they are coupled by an inhibitory G protein (Fig. 34–10). Interaction of somatostatin with this receptor inhibits the stimulatory action of GHRH on adenylate cyclase. As a result, GHRH cannot stimulate GH secretion. Thus, somatostatin has a negative modulating influence on the action of GHRH.

GH and Insulinlike Growth Factor I ■ Although GH is not considered a trophic hormone, it does stimulate the production of another hormone called **insulinlike growth factor I** (IGF-I), a potent **mitogenic agent** that mediates the growth-promoting action of GH (see below). Because of its role in promoting growth, IGF-I has also been called a **somatomedin** or somatotrophin-mediating hormone (somatomedin C, to be specific).

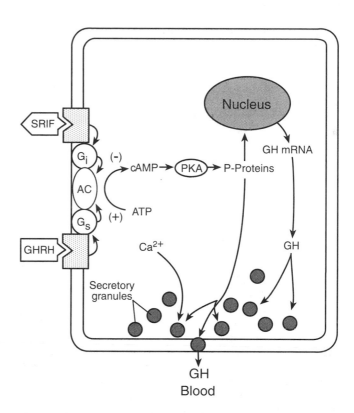

Figure 34–10 ■ The actions of growth hormone-releasing hormone (GHRH) and somatostatin on the somatotroph. GHRH binds to membrane receptors, which are coupled to adenylate cyclase (AC) by stimulating G proteins (G_s). Adenylate cyclase is stimulated, and cAMP rises in the cell. cAMP activates protein kinase A (PKA), which then phosphorylates proteins (P-Proteins) involved in stimulating growth hormone (GH) secretion and the expression of the gene for GH. Ca^{2+} is also involved somehow in the action of GHRH on GH secretion. The possible involvement of the phosphatidylinositol pathway in GHRH action is not shown. Somatostatin (SRIF) binds to membrane receptors, which are coupled to adenylate cyclase by inhibitory G proteins (G_i). This inhibits the ability of GHRH to stimulate adenylate cyclase, thereby blocking its action on GH secretion.

Insulinlike growth factor I is a protein consisting of a single chain of 70 amino acids. It has a molecular size of 7.5 kd. It has some structural similarity to proinsulin. As a result, IGF-I can produce some effects like those of insulin. It is produced by many cells of the body and released into the bloodstream. However, the liver is the main source of the IGF-I in the blood. Most IGF-I in the blood is bound to specific IGF-binding proteins; only a small amount circulates in the free form. The bound form of circulating IGF-I has little insulinlike activity, so it does not play a physiologic role in the regulation of blood glucose level.

Growth hormone stimulates the expression of the gene for IGF-I in various tissues and organs, such as the liver (see below). As a result, the production and release of IGF-I by these cells is stimulated by GH. Thus, individuals secreting excessive amounts of GH have a greater than normal amount of IGF-I in the blood. Those with GH deficiency have lower than normal levels, but there is still some IGF-I present, since the production of IGF-I by cells is regulated by a variety of hormones and other factors in addition to GH.

Insulinlike growth factor I has a negative feedback effect on the secretion of GH (Fig. 34–11). It acts directly on the somatotroph to inhibit the stimulatory action of GHRH on GH secretion. It also inhibits GHRH secretion and stimulates the secretion of somatostatin by neurosecretory neurons. The net effect of these actions is the inhibition of GH secretion. Thus, by stimulating IGF-I production, GH inhibits its own secretion. This is analogous to the way ACTH and TSH regulate their own secretion through the respective negative feedback effects of the glucocorticoid and thyroid hormones. This interactive relationship involving GHRH, somatostatin, GH and IGF-I comprises the **hypothalamopituitary-GH axis.**

Feedback Effects of GH on Its Own Secretion ■ An increase in the concentration of GH in the blood has feedback effects on its own secretion, with the net result being a decrease in GH secretion. These effects of GH are due to the inhibition of GHRH secretion and stimulation of somatostatin secretion by hypothalamic neurosecretory neurons (Fig. 34–11). Growth hormone circulating in the blood can enter the interstitial spaces of the median eminence of the hypothala-

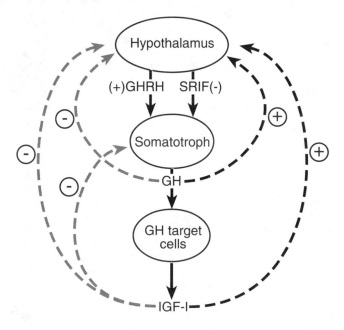

Figure 34–11 ■ The hypothalamopituitary-GH axis. Growth hormone-releasing hormone (GHRH) stimulates and somatostatin inhibits growth hormone (GH) secretion by acting directly on the somatotroph. The feedback loops (−) shown on the left in red inhibit GHRH secretion and action on the somatotroph, causing a decrease in GH secretion. The feedback loops (+) shown on the right stimulate somatostatin secretion, causing a decrease in GH secretion. IGF-I, insulinlike growth factor I.

mus, because there is no blood-brain barrier in this area. Some evidence suggests that GH exerts these feedback effects directly on the neurons that produce GHRH and somatostatin.

Pulsatile Secretion of GH ■ In humans, GH is secreted in periodic bursts, which produce large but short-lived peaks in GH concentration in the blood. Between these episodes of high GH secretion, the somatotrophs release very little GH into the blood. As a result, GH concentration falls to a very low level between the bursts. It is believed that these periodic bursts of GH secretion are caused by an increase in the rate of GHRH secretion and a fall in the rate of somatostatin secretion. The intervals between bursts, when GH secretion is suppressed, are thought to be caused by increased somatostatin secretion. These changes in GHRH and somatostatin secretion result from neural activity generated in higher levels of the CNS, which affects the secretory activity of the GHRH and somatostatin neurons in the hypothalamus.

Normally, a growing child has bursts of GH secretion during both the awake and sleep periods of a day (Fig. 34–12). The bursts of GH secretion during sleep occur 1–2 hours after onset of deep sleep (stages III–IV of slow-wave sleep). Arousal from deep sleep inhibits GH secretion. As the child ages, episodes of GH secretion during the awake period become less frequent. By adulthood, bursts of GH secretion normally occur only in response to deep sleep (Fig. 34–12). As the adult human ages, the amplitude of the GH peaks in the blood declines.

A variety of factors in addition to this episodic pattern affect the rate of GH secretion in humans. These factors are thought to work by changing the secretory activity of GHRH and somatostatin neurons in the hypothalamus. For example, emotional or physical stress causes a great increase in the rate of GH secretion. Vigorous exercise also stimulates GH secretion. The effects of exercise on GH secretion are more pronounced in women than in men, but the reason is unknown.

Changes in the circulating levels of certain metabolites also affect the rate of GH secretion. A decrease in blood glucose concentration stimulates GH secretion, whereas **hyperglycemia** inhibits it. Growth hormone secretion is also stimulated by an increase in the blood concentration of certain amino acids, such as arginine and leucine.

Actions of GH ■ The cells of many tissues and organs of the body have receptors for GH in their plasma membranes. Interaction of GH with these receptors produces its growth-promoting and other metabolic effects, but the mechanisms that produce these effects are not established. For example, the intracellular messengers that mediate the actions of GH on its target cells have not been identified.

Although the exact mechanism by which GH stimulates body growth is unknown, there is good evidence that its growth-promoting action is exerted on **pro-**

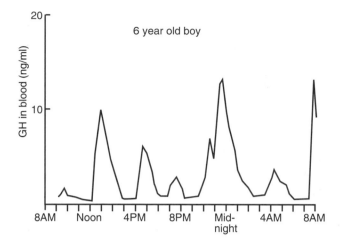

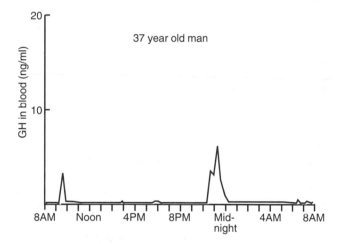

Figure 34–12 ■ Pulsatile growth hormone (GH) secretion in a child and adult. (Modified from (top) Albertson-Wikland K et al: Spontaneous secretion of growth hormone and serum levels of insulinlike growth factor I and somatomedin binding protein in children of different growth rates; and (bottom) Thorner MO, et al: Some physiological and therapeutic considerations of GHRH in the regulation of growth, in Isaksson O, Binder C, Hall K, Hökfelt B (eds): *Growth Hormone: Basic and Clinical Aspects.* Amsterdam, Elsevier 1987, pp 163–175 and 153–162.)

genitor or stem cells, such as the prechondrocytes in the growth plates of bone and the satellite cells of skeletal muscle. Growth hormone stimulates such precursor cells to differentiate into cells with the capacity to undergo cell division. An important action of GH on the differentiation of precursor cells is stimulation of the expression of the gene for IGF-I. As a result, IGF-I is produced and released by the cells.

Insulinlike growth factor I is a potent mitogenic agent. It is thought that the IGF-I produced by these differentiating cells has either an autocrine mitogenic action on the cell that produced it or a paracrine action on neighboring cells. These cells undergo division, causing the affected tissues to grow mainly because of cell replication.

Growth hormone deficiency in childhood causes a decrease in the rate of body growth and, if untreated, results in so-called **pituitary dwarfism.** Pituitary dwarfs may be deficient in GH only or may have multiple anterior pituitary hormone deficiencies. Growth hormone deficiency can be caused by a defect in the mechanisms that control GH secretion or in the ability of the somatotrophs to produce GH. In some individuals, the target cells for GH fail to respond normally to the hormone. Pituitary dwarfs have short stature but normal body proportions.

Excessive secretion of GH during childhood, caused by a defect in the mechanisms regulating GH secretion or a GH-secreting tumor, results in **gigantism.** Affected individuals may reach a height of 7–8 feet. When excessive GH secretion occurs in an adult, further linear growth does not occur because the growth plates of the long bones have calcified. Instead, it causes the bones of the face, hands, and feet to become thicker and certain organs, such as the liver, to undergo hypertrophy. This condition, known as **acromegaly,** can also be caused by the chronic administration of excessive amounts of GH to adults.

Although the main physiologic action of GH is on body growth, it also has important effects on certain aspects of fat and carbohydrate metabolism. Its main action on fat metabolism is to stimulate the mobilization of triglycerides from the fat depots of the body by the process of **triglyceride lipolysis.** This involves the hydrolysis of triglycerides to fatty acids and glycerol by the enzyme **hormone-sensitive lipase.** The fatty acids and glycerol are then released from the adipocytes and enter the bloodsteam. How GH stimulates lipolysis is not understood, but most evidence suggests that it causes adipocytes to be more responsive to other lipolytic stimuli, such as fasting and catecholamines.

Growth hormone is also thought to function as one of the counterregulatory hormones that limit the actions of insulin on muscle, fat, and the liver. For example, GH inhibits glucose use by muscle and fat and increases glucose production by the liver. These effects are opposite those of insulin. Also, GH makes muscle and fat cells resistant to the action of insulin itself. Thus, GH normally has a tonic inhibitory effect on the actions of insulin, much like the glucocorticoid hormones (Chap. 36).

These insulin-opposing actions of GH can produce serious metabolic disturbances in individuals who secrete excessive amounts of GH (e.g., acromegalics) or are given large amounts of GH for an extended period of time. They may have increased insulin resistance and an elevated insulin level in the blood. They may also have hyperglycemia caused by the underutilization and overproduction of glucose. These disturbances are much like those in individuals with **non-insulin-dependent (type II) diabetes mellitus.** For this reason, this metabolic response to excess GH is called its **diabetogenic action.**

In GH-deficient individuals, GH has a transitory **insulinlike action.** For example, intravenous injection of GH in a GH-deficient person produces hypoglycemia. The hypoglycemia is caused by the ability of GH to stimulate the uptake and use of glucose by muscle and fat and to inhibit glucose production by the liver. After about 1 hour, the blood glucose level returns to normal. If this person is given a second injection of GH, hypoglycemia does not occur. This is because the person has become insensitive or **refractory** to the insulinlike action of GH and remains so for some hours. Normal individuals do not respond to the insulinlike action of GH, presumably because they are always refractory from being exposed to their own endogenous GH.

In GH-deficient experimental animals, GH has stimulatory effects on amino acid uptake into cells and on the synthesis of a variety of proteins. These changes occur within minutes. It is thought that these effects are also caused by the insulinlike action of GH.

The above actions of GH are summarized in Table 34-3.

Hormones of the Posterior Pituitary

■ ADH and Oxytocin Are Closely Related Peptides Produced in the Hypothalamus

Antidiuretic hormone and oxytocin are produced by magnocellular neurons in the supraoptic and paraventricular nuclei of the hypothalamus. Individual neurons make either ADH or oxytocin, but not both. As described earlier, the axons of these neurons form the infundibular stem and terminate on the capillary network in the posterior lobe, where they discharge ADH and oxytocin into the systemic circulation.

Antidiuretic hormone and oxytocin are closely related small peptides, each consisting of nine amino acid residues. Two forms of ADH, one containing argi-

nine and the other containing lysine, are made by mammals. **Arginine ADH** is the main form of the hormone made by humans. Although ADH and oxytocin differ by only two amino acid residues, these structural differences are sufficient to give these two molecules very different hormonal activities. They are similar enough, however, for ADH to have slight oxytocic activity and for oxytocin to have slight antidiuretic activity.

The genes for ADH and oxytocin are located near one another on the same chromosome. They code for much larger precursor peptides that contain the amino acid sequences for ADH or oxytocin and for a 93-amino acid peptide called **neurophysin.** The neurophysin coded by the ADH gene has a somewhat different structure than that coded by the oxytocin gene. The precursor peptides for ADH and oxytocin are synthesized in the cell bodies of the magnocellular neurons and transported in secretory granules to the axon terminals in the posterior lobe, as described earlier. During the passage of the granules from the Golgi apparatus to the axon terminals, the precursor peptides are cleaved by proteolytic enzymes to produce ADH or oxytocin and their associated neurophysins.

When the magnocellular neurons receive neural input signaling for ADH or oxytocin secretion, action potentials are generated in these cells. These action potentials travel down the axons and trigger the release of ADH or oxytocin and neurophysin from the axon terminal. These substances diffuse into nearby capillaries and then enter the systemic circulation.

■ ADH Increases the Reabsorption of Water by the Kidneys

Two physiologic signals—a rise in the osmolality of the blood and a decrease in blood volume—generate the CNS stimulus for ADH secretion. The main physiologic action of ADH is to increase water reabsorption by the collecting ducts of the kidneys. This results in decreased water excretion and formation of an osmotically concentrated urine (Chap. 23). This action of ADH works to counteract the conditions that stimulate its secretion. For example, reducing water loss in the urine limits a further rise in the osmolality of the blood and conserves blood volume.

■ Oxytocin Stimulates the Contraction of Smooth Muscle in the Mammary Gland and Uterus

Two physiologic signals stimulate the secretion of oxytocin by hypothalamic magnocellular neurons. Suckling of the mammary gland by the nursing infant stimulates sensory nerves in the nipple. The afferent nerve impulses generated enter the CNS and eventually stimulate the oxytocin-secreting magnocellular neurons. These neurons then fire in synchrony and release a bolus of oxytocin into the bloodstream. The oxytocin stimulates the contraction of **myoepithelial cells,**

TABLE 34–3 ■ Actions of Growth Hormone

Growth-promoting
    Stimulates IGF-I gene expression by target cells; IGF-I produced by these cells has an autocrine or paracrine stimulatory effect on cell division, resulting in growth
Lipolytic
    Stimulates mobilization of triglycerides from fat deposits
Diabetogenic
    Inhibits glucose use by muscle and fat and increases glucose production by the liver
    Inhibits the action of insulin on glucose and lipid metabolism by muscle and fat
Insulinlike
    Transitory stimulatory effect on the uptake and use of glucose by muscle and fat in GH-deficient individuals
    Transitory inhibitory effect on glucose production by the liver of GH-deficient individuals

which surround the milk-laden alveoli in the lactating mammary gland, aiding in milk ejection.

Oxytocin secretion is also stimulated by neural input from the female reproductive tract during childbirth. Cervical dilation before the beginning of labor stimulates stretch receptors in the cervix. The afferent impulses generated pass through the CNS to the oxy-

tocin-secreting neurons. Oxytocin release stimulates contraction of smooth muscle cells in the uterus during labor, thus aiding in the delivery of the fetus and placenta. The actions of oxytocin on the mammary gland and the female reproductive tract are discussed further in Chapter 41.

■ ■ ■ ■ ■ ■ ■ ■ ■ ■ ■ REVIEW EXERCISES ■ ■ ■ ■ ■ ■ ■ ■ ■ ■ ■

Identify Each with a Term

1. The blood supply that links the hypothalamus to the anterior lobe of the pituitary

2. Hypothalamic substances that stimulate or inhibit the secretion of anterior pituitary hormones

3. The preprohormone molecule from which ACTH is produced by proteolytic cleavage

4. Short body stature caused by GH deficiency during childhood

5. The other product formed when the precursor peptides for ADH and oxytocin are proteolytically cleaved

Define Each Term

6. Magnocellular neurons

7. Hypothalamopituitary-thyroid axis

8. Lactogenic activity

9. Acromegaly

Choose the Correct Answer

10. Which of the following decreases the rate of ACTH secretion?
    a. An increase in ADH in hypophyseal portal blood
    b. Stress
    c. Destruction of the parvicellular neurons in the paraventricular nuclei of the hypothalamus
    d. A decrease in glucocorticoids in the blood

11. GH secretion in the human is stimulated by:
    a. Insulinlike growth factor I (IGF-I)
    b. Hyperglycemia
    c. GH
    d. Stress

12. Which of the following is mediated by a rise in cAMP?
    a. Inhibition of GH secretion by somatostatin
    b. Stimulation of GH gene expression by GHRH
    c. Stimulation of TSH secretion by TRH
    d. Inhibition of TSH α and β subunit gene expression by TRH

13. Which of the following is not an effect of human GH?
    a. Stimulation of triglyceride lipolysis in fat cells
    b. Production of insulin resistance
    c. Stimulation of milk synthesis by the lactating mammary gland
    d. Stimulation of fetal growth

14. Which of the following pairs of peptide hormones are structurally related to each other?
    a. FSH and TSH
    b. GH and IGF-I
    c. ADH and CRH
    d. Oxytocin and PRL

15. Which of the following is not a part of the proopiomelanocortin (POMC) molecule?
    a. α-MSH
    b. ACTH
    c. β-Endorphin
    d. CRH

16. Glucocorticoids:
    a. Stimulate the expression of the gene for CRH
    b. Have a direct negative feedback effect on the corticotrophs to inhibit ACTH secretion
    c. Stimulate the expression of the gene for POMC in corticotrophs
    d. Stimulate the secretion of CRH

17. Which of the following releasing hormones binds to a receptor that is coupled to phospholipase C by a G protein?
    a. CRH
    b. TRH
    c. Somatostatin
    d. GHRH

18. Thyroid hormones:
    a. Increase the sensitivity of the thyrotroph to TRH
    b. Stimulate the expression of the genes for the α and β subunits of TSH in the thyrotroph
    c. Increase the secretion of TSH by the thyrotroph
    d. Stimulate the expression of the gene for GH in the somatotroph

19. Which of the following hormones aids in the ejection of milk from the lactating mammary gland?
    a. PRL
    b. ADH
    c. Oxytocin
    d. Human GH

Briefly Answer

20. What is the physiologic function of the hypophyseal portal circulation?

21. Why are humans who have an ACTH-secreting tumor often hyperpigmented?

22. What is the role of ADH in the response to stress?

23. What is the physiologic function of PRL in the female?

24. What metabolic effects of GH account for its diabetogenic action?

■ ■ ■ ■ ■ ■ ■ ■ ■ ■ ■ ■ **ANSWERS** ■ ■ ■ ■ ■ ■ ■ ■ ■ ■ ■

1. Hypophyseal portal circulation
2. Releasing or hypophysiotropic hormones
3. Proopiomelanocortin (POMC)
4. Pituitary dwarfism
5. Neurophysin
6. The neurons with large cell bodies in the supraoptic and paraventricular nuclei of the hypothalamus that make and secrete ADH and oxytocin
7. The control system involving the action of TRH on the thyrotroph; the action of TSH on the thyroid follicular cell; and the negative feedback effect of circulating thyroid hormones on the thyrotroph, which regulates the secretion of T_4 and T_3 by the thyroid gland.
8. The ability of a hormone such as PRL or human GH to stimulate milk synthesis by the mammary gland
9. The condition produced when excessive amounts of GH are present in the adult human, which is characterized by thickening of the bones of the face, hands, and feet and hypertrophy of organs such as the liver
10. c
11. d
12. b
13. d
14. a
15. d
16. b
17. b
18. d
19. c
20. The axons of the neurosecretory cells in the hypothalamus that produce releasing hormones terminate on the capillary networks that give rise to the hypophyseal portal blood vessels. The releasing hormones are discharged into these capillaries and then conveyed by the portal vessels directly to their target cells in the anterior pituitary gland.
21. Adrenocorticotrophic hormone contains the amino acid sequence of α-MSH at its N-terminal end. As a result, ACTH has melanocyte-stimulating activity when present in the bloodstream at high concentrations, as in an individual with an ACTH-secreting tumor.
22. Antidiuretic hormone is produced by the parvicellular neurons in the paraventricular nuclei that also produce CRH; like CRH, ADH is secreted by these cells in response to stress. It interacts with receptors on the corticotrophs and stimulates them to secrete ACTH, thereby amplifying the effect of CRH on ACTH secretion.
23. The alveolar cells of the mammary gland must be stimulated periodically by PRL to continue to synthesize milk once lactation has started after childbirth.
24. When present in excess, GH inhibits the utilization of glucose by muscle and fat and increases glucose production by the liver. These actions can result in hyperglycemia, which causes the level of insulin to rise in the blood, as in some forms of non-insulin-dependent diabetes mellitus.

Suggested Reading

Antoni FA: Hypothalamic control of adrenocorticotropin secretion: Advances since the discovery of 41-residue corticotropin-releasing factor. *Endocrin Rev* 7:351–378, 1986.

Baumann G: Growth hormone heterogeneity: Genes, isohormones, variants, and binding proteins. *Endocrin Rev* 12:424–449, 1991.

Frohman LA, Jannson J-O: Growth hormone-releasing hormone. *Endocrin Rev* 7:223–253, 1986.

Gershengorn MC: Mechanism of thyrotropin releasing hormone stimulation of pituitary hormone secretion. *Annu Rev Physiol* 48:515–526, 1986.

Gharib SD, Wierman ME, Shupnik MA, et al: Molecular biology of the pituitary gonadotropins. *Endocrin Rev* 11:177–199, 1990.

Goodman HM: *Basic Medical Endocrinology.* 2nd ed. New York, Raven, 1994.

Isaksson OGP, Lindahl A, Nilsson A, et al: Mechanism of the stimulatory effect of growth hormone on longitudinal bone growth. *Endocrin Rev* 8:426–438, 1987.

Lundblad JR, Roberts JL: Regulation of proopiomelanocortin gene expression in pituitary. *Endocrin Rev* 9:135–158, 1988.

Page RB: The anatomy of the hypothalamo-hypophyseal complex, in Knobil E, Neill JD (eds): *The Physiology of Reproduction.* New York: Raven, 1988, pp 1161–1233.

Patel YC, Srikant CB: Somatostatin mediation of adenohypophyseal secretion. *Annu Rev Physiol* 48:551–567, 1986.

Reichlin S: Neuroendocrinology, in Wilson JD, Foster DW (eds): *Williams Textbook of Endocrinology.* 8th ed. Philadelphia, WB Saunders, 1992, pp 135–219.

Sara V, Hall K: Insulin-like growth factors and their binding proteins. *Physiol Rev* 70:591–614, 1990.

Shupnik MA, Ridgway EC, Chin WW: Molecular biology of thyrotropin. *Endocrin Rev* 10:459–475, 1989.

C H A P T E R

■ ■ ■ ■ ■ ■ ■

35

The Thyroid Gland

CHAPTER OUTLINE

I. FUNCTIONAL ANATOMY OF THE THYROID GLAND
- A. Thyroxine and triiodothyronine are synthesized and secreted by the thyroid follicle
- B. Parafollicular cells are the sites of calcitonin synthesis

II. SYNTHESIS, SECRETION, AND METABOLISM OF THE THYROID HORMONES
- A. Follicular cells synthesize iodinated thyroglobulin
  1. Synthesis and secretion of the precursor for thyroglobulin
  2. Iodide uptake
  3. Formation of iodothyronine residues
- B. Thyroid hormones are formed from hydrolysis of thyroglobulin
  1. Uptake of thyroglobulin by follicular cells
  2. Secretion of free T_4 and T_3
  3. Binding of T_4 and T_3 to serum proteins
- C. Thyroid hormones are metabolized by peripheral tissues
  1. Conversion of T_4 to T_3
  2. Deiodinations that inactivate T_4 and T_3
  3. Regulation of 5' deiodination
  4. Conjugation with glucuronic acid
  5. Metabolism of the iodothyronine side chain
- D. TSH regulates thyroid hormone synthesis and secretion
  1. TSH receptors and cAMP
  2. TSH and thyroid hormone formation and secretion
  3. TSH and thyroid size
- E. Dietary iodide is essential for synthesis of the thyroid hormones

III. MECHANISM OF THYROID HORMONE ACTION

IV. ROLE OF THYROID HORMONES IN DEVELOPMENT, GROWTH, AND METABOLISM
- A. Thyroid hormones are essential for development of the central nervous system
- B. Thyroid hormones are essential for normal body growth
  1. Thyroid hormones and the gene for GH
  2. Other effects of thyroid hormones on growth
- C. Thyroid hormones regulate the basal energy economy of the body
  1. Basal oxygen consumption and body heat production
  2. The calorigenic action of thyroid hormones
  3. Tissues affected by the calorigenic action of thyroid hormones

OBJECTIVES

After studying this chapter, the student should be able to:
1. Describe the functional anatomy of the thyroid gland
2. List the thyroid hormones of physiologic importance; describe the processes involved in the synthesis and secretion of these hormones by thyroid follicular cells; explain the importance of dietary iodide for the synthesis of the thyroid hormones
3. Describe the processes by which thyroid hormones are metabolized by peripheral tissues; explain the physiologic significance of the enzymatic conversion of thyroxine to triiodothyronine by target cells
4. Describe the mechanism by which TSH regulates the synthesis and secretion of thyroid hormones
5. Explain the current concept of how thyroid hormones act at the cellular level
6. Describe the actions of the thyroid hormones on development of the CNS, body growth, and the basal energy economy of the body; explain the effects of thyroid hormone deficiency and excess on these processes

The development of the human body from embryogenesis to adulthood is an orderly, programed process. The timing of events during development is remarkably constant from one individual to the next, with developmental milestones reached at about the same time in all of us. For example, the early development of motor skills, body growth, the start of puberty, and final sexual and physical maturation occur within rather narrow time frames during our life span.

At the level of the individual cell, the timing or rate of metabolic processes is also tightly regulated. For example, energy metabolism occurs at the rate needed to make the amount of ATP required for activities such as maintaining osmotic integrity, excitability, secretion, and countless biosynthetic processes. Thus, the cell not only meets its basic metabolic housekeeping needs but also remains poised to do its own special work in the body, such as conducting nerve impulses, contracting, absorbing, or secreting. This is because during its life span, the cell continues to make the enzymatic and structural proteins that ensure that an appropriate rate of metabolism will be maintained.

The thyroid hormones, thyroxine and triiodothyronine, play key roles in the regulation of body development and governing the rate at which metabolic housekeeping occurs in individual cells. These hormones are not essential for life. Without them, however, life loses its orderly nature. Body development fails to proceed on time and to come to completion. Cellular housekeeping moves at a slower pace, eventually influencing the ability of individual cells to carry out their physiologic functions. We now know that the thyroid hormones exert their regulatory functions by altering gene expression, thereby affecting the developmental program and the amount of cellular constituents needed for the normal rate of metabolism.

Functional Anatomy of the Thyroid Gland

The human thyroid gland consists of two lobes attached to either side of the trachea by connective tissue. The two lobes are connected by a band of thyroid tissue or isthmus, which lies just below the cricoid cartilage. A normal adult human thyroid gland weighs about 20 g.

Each lobe of the thyroid receives its arterial blood supply from a superior and an inferior thyroid artery, which arise from the external carotid and subclavian artery, respectively. Blood leaves the lobes of the thyroid by a series of thyroid veins that drain into the external jugular and innominate veins. This circulation provides a rich blood supply to the thyroid gland, giving it a higher rate of blood flow per gram than even that of the kidneys.

The thyroid gland receives adrenergic innervation from the cervical ganglia and cholinergic innervation from the vagus nerve. It is not clear whether this innervation to the thyroid has more than just a vasomotor function.

■ Thyroxine and Triiodothyronine Are Synthesized and Secreted by the Thyroid Follicle

The lobes of the thyroid gland consist of aggregates of many spheric **follicles** that consist of a wall composed of a single layer of epithelial cells (Fig. 35-1). The apical membranes of the follicular cells, which face the lumen, are covered with microvilli. Pseudopods formed from the apical membrane extend into the lumen. The lateral membranes of the follicular cells are connected by tight junctions, which provides a seal for the contents of the lumen. The basal membranes of the follicular cells are close to the very rich capillary network that penetrates the stroma between the follicles (Fig. 35-1).

The lumen of the follicle contains a thick, gel-like substance called **colloid** (Fig. 35-1). The colloid is a solution of primarily one large protein, **thyroglobulin,** which is a storage form of the thyroid hormones. The high viscosity of the colloid is due to the very high concentration (10-25%) of thyroglobulin.

The thyroid follicle produces and secretes the two thyroid hormones, **thyroxine** and **triiodothyronine.** The structures of these hormones are shown in Figure 35-2. Thyroxine and triiodothyronine are iodinated derivatives of the amino acid tyrosine. They are formed by the coupling of the phenyl rings of two iodinated tyrosine molecules to form an ether linkage. The resulting structure is called an **iodothyronine.** The mechanism of this process is discussed in detail later.

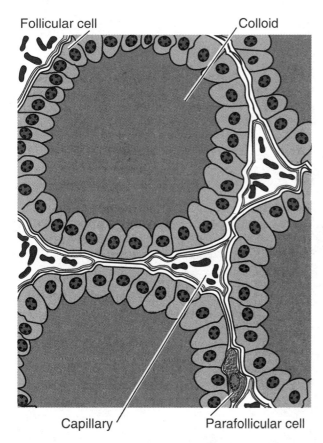

Follicular cell   Colloid

Capillary   Parafollicular cell

Figure 35–1 ■ Histologic cross-section through a portion of a normal thyroid gland. (Modified from diFiore MSH: *Atlas of Normal Histology.* 6th ed. Philadelphia, Lea & Febiger, 1989.)

Thyroxine contains four iodine atoms on the 3, 5, 3′, and 5′ positions of the thyronine ring structure, whereas triiodothyronine has only three iodine atoms, at ring positions 3, 5, and 3′ (Fig. 35–2). As a consequence, thyroxine is usually abbreviated as T_4 and triiodothyronine as T_3. Because T_4 and T_3 contain the element iodine, their synthesis by the thyroid follicle is dependent on an adequate supply of iodine in the diet.

■ Parafollicular Cells Are the Sites of Calcitonin Synthesis

In addition to the epithelial cells that secrete T_4 and T_3, the wall of the thyroid follicle contains the **parafollicular cell,** or **C cell** (Fig. 35–1). It is a large cell present in only small numbers in human thyroid follicles. The parafollicular cell is usually embedded in the wall of the follicle, inside the basal lamina surrounding the follicles. However, its plasma membrane does not form part of the wall of the lumen. The parafollicular cell produces and secretes the hormone calcitonin. Calcitonin and its effects on calcium metabolism are discussed in Chapter 38.

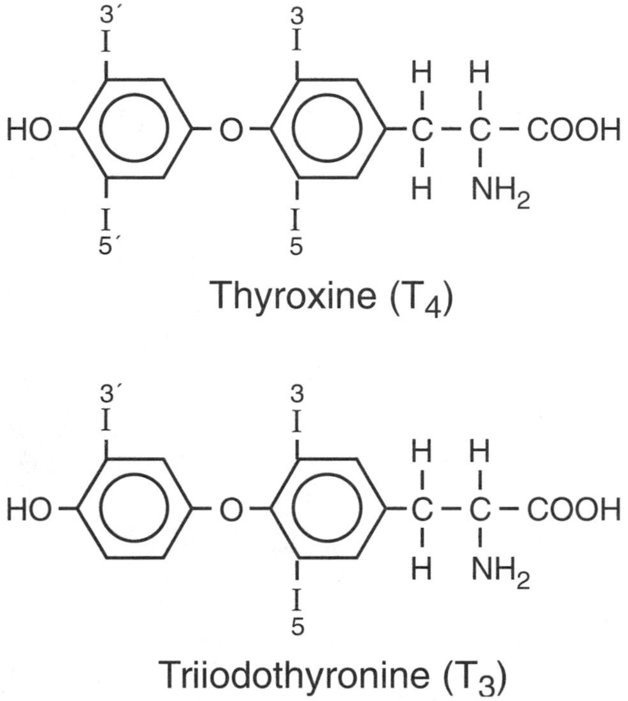

Thyroxine (T$_4$)

Triiodothyronine (T$_3$)

Figure 35–2 ■ Structures of the thyroid hormones. The numbering of the iodine atoms on the iodothyronine ring structure is shown in red.

Synthesis, Secretion, and Metabolism of the Thyroid Hormones

Thyroxine and T_3 are not directly synthesized by the thyroid follicle in their final form (Fig. 35–2); rather, they are formed by the chemical modification of tyrosine residues in the peptide structure of thyroglobulin, as it is secreted by the follicular cells into the lumen of the follicle. Consequently, the T_4 and T_3 formed by this chemical modification are actually a part of the amino acid sequence of thyroglobulin.

The high concentration of thyroglobulin in the colloid provides a large reservoir of stored thyroid hormones for later processing and secretion by the follicle. The synthesis of T_4 and T_3 is completed when thyroglobulin is retrieved by the follicular cells by pinocytosis of the colloid. The thyroglobulin so taken up is then hydrolyzed by lysosomal enzymes to its constituent amino acids. This releases T_4 and T_3 molecules from their peptide linkage in the thyroglobulin structure, and they are then secreted into the blood.

■ Follicular Cells Synthesize Iodinated Thyroglobulin

The steps involved in the synthesis of iodinated thyroglobulin are shown in Figure 35–3.

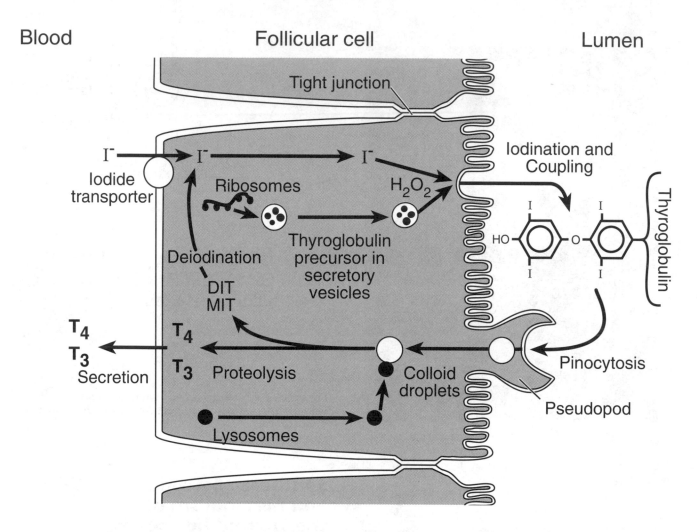

Figure 35–3 ■ Thyroid hormone synthesis and secretion. (Modified from van den Hove-Vanden-broucke M-F: Secretion of thyroid hormones, in de Visscher M (ed): *The Thyroid Gland.* New York, Raven, 1980, pp 61-79.)

Synthesis and Secretion of the Precursor for Thyroglobulin ■ The synthesis of the protein precursor for thyroglobulin is the first step in the formation of T_4 and T_3. This substance is a 600 kd glycoprotein composed of two similar 300 kd subunits held together by disulfide bridges. They are synthesized by ribosomes on the rough endoplasmic reticulum and then undergo dimerization and glycosylation in the smooth endoplasmic reticulum. The completed glycoprotein is then packaged into vesicles by the Golgi apparatus. These vesicles migrate to the apical membrane of the follicular cell and fuse with it. The thyroglobulin precursor protein is then extruded onto the apical surface of the cell, where its iodination is thought to take place.

Iodide Uptake ■ The iodide used for iodination of the thyroglobulin precursor protein comes from the blood perfusing the thyroid gland. The basal plasma membranes of the follicular cells, which are quite close to the capillaries that supply the follicle, contain **iodide transporters.** These transporters move iodide

across the basal membrane and into the cytosol of the follicular cell. The iodide transporter is an active transport mechanism that requires ATP, is saturable, and can also transport certain other anions, such as bromide, thiocyanate, and perchlorate. It enables the follicular cell to concentrate iodide many times over the concentration of iodide present in the blood. Thus, the follicular cells are very efficient extractors of the small amount of iodide circulating in the human body. Once inside the follicular cells, the iodide ions diffuse rapidly to the apical membrane, where they are used for iodination of the precursor of thyroglobulin.

Formation of Iodothyronine Residues ■ The next step in the formation of thyroglobulin is the addition of one or two iodine atoms to certain tyrosine residues in the precursor protein. The precursor of thyroglobulin contains 120 tyrosine residues, but only a small fraction of these become iodinated. A typical thyroglobulin molecule contains only 20–30 atoms of iodine.

The iodination of thyroglobulin is catalyzed by a **thyroid peroxidase** enzyme, which is bound to the apical membranes of follicular cells. The exact chemical mechanism by which the iodination takes place is uncertain. It is thought that thyroid peroxidase binds an iodide ion and a tyrosine residue in the thyroglobulin precursor, bringing them in close proximity. The enzyme then oxidizes the iodide ion and the tyrosine residue to short-lived free radicals, using hydrogen peroxide that has been generated by the follicular cells. The free radicals then undergo addition. The product formed is a **monoiodotyrosine** (MIT) residue, which remains in peptide linkage in the thyroglobulin structure. A second iodine atom may be added to an MIT residue by this same enzymatic process, thereby forming a **diiodotyrosine** (DIT) residue.

Iodinated tyrosine residues that are close to each other in the thyroglobulin precursor molecule then undergo a **coupling reaction,** which forms the iodothyronine structure. Again, the chemical mechanism involved is not established. The thyroid peroxidase enzyme is thought to oxidize neighboring iodinated tyrosine residues to short-lived free radicals, using hydrogen peroxide. These free radicals then undergo addition as shown in Figure 35-4. The addition reaction produces an iodothyronine residue and a **dehydroalanine residue,** both of which remain in peptide linkage in the thyroglobulin structure. For example, when two neighboring DIT residues couple by this mechanism, T_4 is formed (Fig. 35-4). After being iodinated, the thyroglobulin molecule is stored as part of the colloid in the lumen of the follicle.

Only about 20-25% of the DIT and MIT residues in the thyroglobulin molecule become coupled to form iodothyronines. For example, a typical thyroglobulin molecule contains five to six uncoupled residues of DIT and two to three residues of T_4. However, T_3 is formed in only about one of every three thyroglobulin molecules. As a result, the thyroid secretes substantially more T_4 than T_3. Small amounts of other products are formed by the coupling reaction, such as 3, 3', 5' triiodothyronine or **reverse T_3** (rT_3) and **3, 3' diiodothyronine.** These minor products are also secreted by the thyroid gland, but they are not hormonally active.

■ Thyroid Hormones Are Formed from Hydrolysis of Thyroglobulin

Uptake of Thyroglobulin by Follicular Cells ■ When the thyroid gland is stimulated to secrete thyroid hormones, vigorous pinocytosis occurs at the apical membranes of follicular cells. Pseudopods form from the apical membrane and reach into the lumen of the follicle, engulfing bits of the colloid (Fig. 35-3). The endocytotic vesicles or **colloid droplets** formed by this pinocytotic activity migrate toward the basal region of the follicular cell. Lysosomes, which are mainly located in the basal region of resting follicular cells, migrate toward the apical region of the stimulated

Figure 35–4 ■ Theoretical model for the coupling reaction between two diiodotyrosine (DIT) residues in iodinated thyroglobulin. This model is based on free radical formation catalyzed by thyroid peroxidase. (Adapted from Nunez J: Iodination and thyroid hormone synthesis, in de Visscher M (ed): *The Thyroid Gland.* New York, Raven, 1980, pp 39–59.)

cells. The lysosomes fuse with the colloid droplets and hydrolyze the thyroglobulin in them to its constituent amino acids. As a result, T_4 and T_3 and the other iodinated amino acids are released into the cytosol.

Secretion of Free T_4 and T_3 ■ Thyroxine and T_3 formed from thyroglobulin traverse the basal membrane of the follicular cell and enter the nearby capillary circulation. The mechanism involved in the transport of T_4 and T_3 across the basal membrane has not been defined. The DIT and MIT generated by the hydrolysis of thyroglobulin are deiodinated in the follicular cell. The free iodide formed is then used again by the follicular cell for iodination of thyroglobulin (Fig. 35-3).

Binding of T_4 and T_3 to Serum Proteins ■ Most

of the T_4 and T_3 molecules that enter the bloodstream become bound to serum proteins. About 70% bind noncovalently to **thyroxine-binding globulin** (TBG), a 54 kd glycoprotein that is synthesized and secreted by the liver. A molecule of TBG has a single binding site for a thyroid hormone molecule. Nearly all of the remaining T_4 and T_3 in the blood is bound either to albumin or another serum protein, **thyroxine-binding prealbumin** (TBPA). Less than 1% of the T_4 and T_3 in blood is in the free form, and it is in equilibrium with the large protein-bound fraction. It is this small amount of free thyroid hormone that interacts with target cells.

The protein-bound form of T_4 and T_3 represents a large reservoir of preformed hormone that can replenish the small amount of circulating free hormone as it is cleared from the bloodstream. This reservoir provides the body with a buffer against drastic changes in circulating thyroid hormone levels due to sudden changes in the rate of secretion of T_4 and T_3. Also, the protein-bound T_4 and T_3 molecules are protected from metabolic inactivation and excretion in the urine. As a consequence of these factors, the thyroid hormones have long half-lives in the bloodstream. The half-life of T_4 is about 7 days and that of T_3 about 1 day.

■ Thyroid Hormones Are Metabolized by Peripheral Tissues

Conversion of T_4 to T_3 ■ As noted earlier, T_4 is the major secretory product of the thyroid gland and is the predominant thyroid hormone in the blood. However, about 40% of the T_4 secreted by the thyroid gland is converted to T_3 by enzymatic removal of the iodine atom at position 5' of the thyronine ring structure (Fig. 35-5). This reaction is catalyzed by a **5'-deiodinase.** It occurs in many tissues of the body but has been best studied in the liver, kidney, and anterior pituitary. The T_3 formed by this deiodination and that secreted by the thyroid then reacts with thyroid hormone receptors in target cells. Thus, T_3 can be considered the physiologically active form of the thyroid hormones. It should be noted, however, that T_4 itself can interact to a limited extent with thyroid hormone receptors and produce physiologic responses. The fact that T_3 is the primary thyroid hormone that binds to thyroid hormone receptors probably explains its greater potency on a molar basis when it is administered to humans or animals.

Deiodinations That Inactivate T_4 and T_3 ■ Whereas the 5' deiodination of T_4 to produce T_3 can be viewed as a metabolic activation process, both T_4 and T_3 undergo enzymatic deiodinations, particularly in the liver and kidneys, which inactivate them. For example, about 40% of the T_4 secreted by the human thyroid gland is deiodinated at the 5 position on the thyronine ring structure. This produces reverse T_3 (Fig. 35-5). Since reverse T_3 has little or no thyroid hormone activity, this deiodination reaction is a major pathway for the metabolic inactivation or disposal of T_4. Triiodo-

thyronine and reverse T_3 also undergo deiodination to yield 3, 3' diiodothyronine (Fig. 35-5). This inactivate metabolite may be further deiodinated before being excreted.

Regulation of 5' Deiodination ■ The 5' deiodination reaction is a regulated process influenced by certain physiologic and pathologic factors. The result is a change in the relative amounts of T_3 and reverse T_3 produced from T_4. For example, the human fetus produces less T_3 from T_4 than does the child or adult, because the 5' deiodination reaction is less active in the fetus. Also, 5' deiodination is inhibited during fasting, particularly in response to carbohydrate restriction, but it can be restored to normal when the individual is fed again. Trauma and most acute and chronic illnesses also suppress the 5' deiodination reaction. Under all of these circumstances, the amount of T_3 produced from T_4 is reduced and its concentration falls in the bloodstream. On the other hand, the amount of reverse T_3 rises in the circulation, mainly because its conversion to 3,3' diiodothyronine by 5' deiodination is reduced. Thus, a rise in reverse T_3 in the blood may signal that the 5' deiodination reaction is suppressed. Note that during fasting or in the disease states mentioned above the secretion of T_4 is usually not increased, despite the decrease of T_3 in the circulation. This indicates that, under these circumstances, the decrease in T_3 in the blood does not stimulate the hypothalamopituitary-thyroid axis.

Conjugation with Glucuronic Acid ■ Thyroxine and, to a lesser extent, T_3 are also metabolized by conjugation with glucuronic acid in the liver. The conjugated hormones are then secreted into the bile. Bacteria in the intestine can hydrolyze the conjugated hormones, thereby releasing free T_4 and T_3. In humans, the free T_4 and T_3 are then mainly excreted in the feces. In some animals a significant fraction of the regenerated T_4 and T_3 can be absorbed from the intestine back into the blood. This so-called enterohepatic circulation of thyroid hormones does not appear to be an important salvage pathway for T_4 and T_3 in humans.

Metabolism of the Iodothyronine Side Chain ■ Many tissues also metabolize thyroid hormones by modifying the three-carbon side chain of the iodothyronine structure. These modifications include decarboxylation and deamination. The derivatives formed, such as **tetraiodoacetic acid** (tetrac) from T_4 (Fig. 35-5), may also undergo deiodinations before being excreted.

■ TSH Regulates Thyroid Hormone Synthesis and Secretion

When the concentrations of free T_4 and T_3 fall in the blood, the anterior pituitary gland is stimulated to secrete **thyroid stimulating hormone** (TSH), thereby raising the concentration of TSH in the blood. This causes an increase in interactions between TSH and its receptors on thyroid follicular cells.

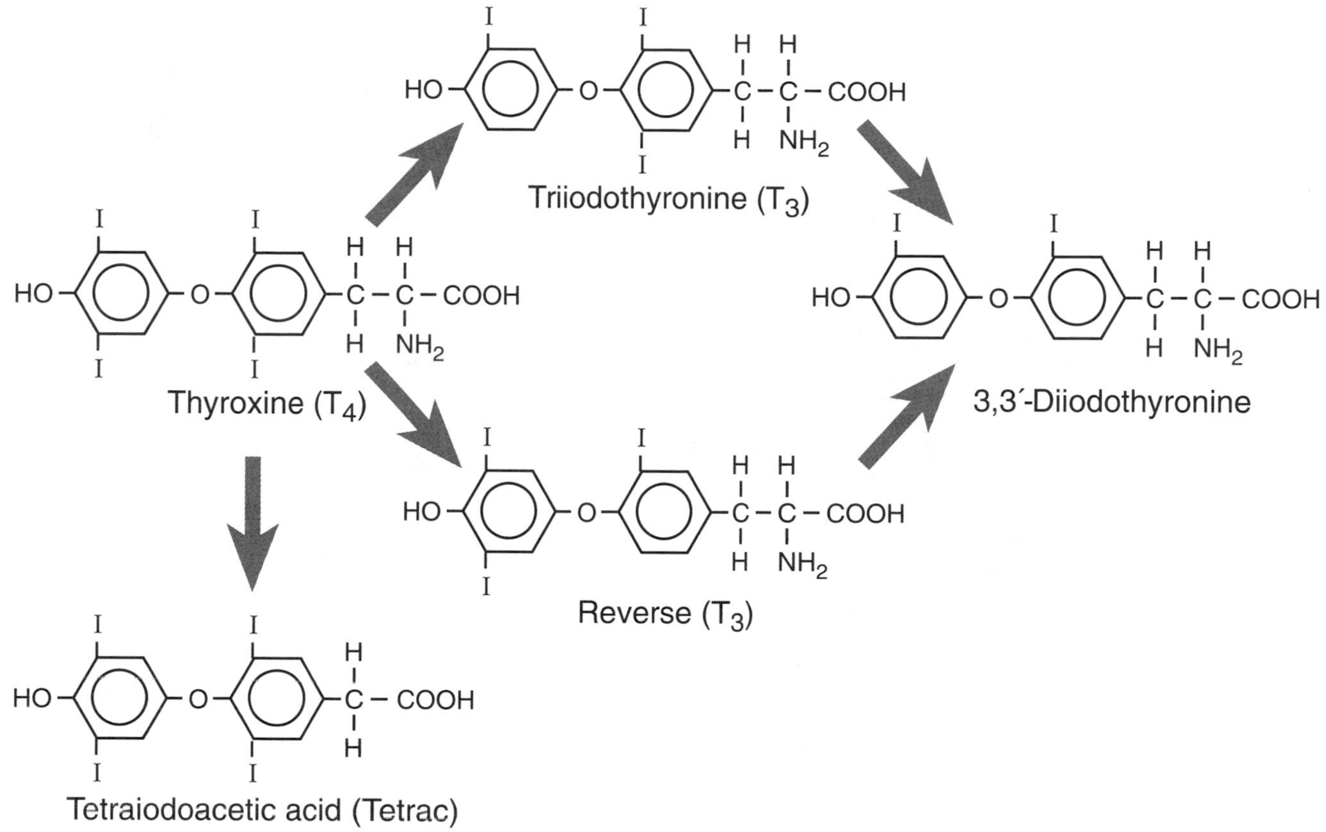

Figure 35–5 ■ Metabolism of thyroxine. Thyroxine is deiodinated by a 5′ deiodinase to form T₃, the physiologically active thyroid hormone. Some T₄ is also enzymatically deiodinated at the 5 position to form the inactive metabolite, reverse T₃. T₃ and reverse T₃ undergo additional deiodinations (e.g., to 3, 3′ diiodothyronine) before being excreted. A minor amount of T₄ is also decarboxylated and deaminated to form the metabolite, tetraiodoacetic acid (tetrac). Tetrac may then be deiodinated before being excreted.

TSH Receptors and cAMP ■ The receptor for TSH is a transmembrane glycoprotein thought to be located on the basal plasma membrane of the follicular cell. There are about 1000 of these receptors on each follicular cell. These receptors are coupled to the enzyme **adenylate cyclase** by stimulatory **G proteins** (G$_s$ proteins). Occupation of the extracellular domain of the TSH receptor by TSH stimulates adenylate cyclase, causing a rapid increase in **cAMP** concentration in the follicular cell. The rise in cAMP level activates **protein kinase A** in the follicular cell, which, in turn, phosphorylates key proteins that produce the actions of TSH on this cell.

The human TSH receptor is also coupled to the enzyme **phospholipase C** (PLC) by G proteins. Thus, TSH can also stimulate PLC in the follicular cell. This causes the hydrolysis of **phosphatidylinositol bisphosphate** in the plasma membrane, producing the intracellular messengers **inositol trisphosphate** and **diacylglycerol.** Inositol trisphosphate increases the concentration of Ca^{2+} in the cytosol of the follicular cell. This rise in Ca^{2+} and diacylglycerol activate **pro-

tein kinase C,** which leads to the phosphorylation of proteins that produce some of the effects of TSH on the follicular cell. The physiologic importance of this action of TSH on phospholipid metabolism is still unclear, however, because PLC is activated in human follicular cells only by very high concentrations of TSH.

TSH and Thyroid Hormone Formation and Secretion ■ Thyroid stimulating hormone stimulates most of the processes involved in thyroid hormone synthesis and secretion by the follicular cell. The rise in cAMP produced by TSH is believed to cause many of these effects. Thyroid stimulating hormone stimulates the uptake of iodide by follicular cells, usually after a short interval during which iodide transport is actually depressed. Also, TSH stimulates the iodination of tyrosine residues in the thyroglobulin precursor and the coupling of iodinated tyrosines to form iodothyronines. Moreover, it stimulates the pinocytosis of colloid by the apical membranes, resulting in a great increase in endocytosis of thyroglobulin and its hydrolysis. The overall result of these effects of TSH is an increase in

T_4 and T_3 release into the bloodstream. In addition to its effects on thyroid hormone synthesis and secretion, TSH has rapid stimulatory effects on the energy metabolism of the thyroid follicular cell.

TSH and Thyroid Size ■ Over the long term, TSH promotes protein synthesis in thyroid follicular cells, thereby maintaining their size and structural integrity. Evidence of this **trophic effect** of TSH is seen in the hypophysectomized individual, in whom the thyroid atrophies due, in large part, to a reduction in the height of the follicular cells. On the other hand, chronic exposure of an individual to excessive amounts of TSH causes the thyroid gland to increase in size. This enlargement is due to an increase in follicular cell height and number. Such an enlarged thyroid gland is called a **goiter.** These trophic and proliferative effects of TSH on the thyroid are thought to be mediated by cAMP.

■ Dietary Iodide Is Essential for Synthesis of the Thyroid Hormones

Since iodine atoms are constituent parts of the T_4 and T_3 molecules, a continual supply of iodide is required for the synthesis of these hormones. If an individual's diet is severely deficient in iodide, as it is in some parts of the world, T_4 and T_3 synthesis is limited by the amount of iodide available to the thyroid gland. As a result, the concentrations of T_4 and T_3 in the blood fall. This causes a chronic stimulation of TSH secretion, which in turn produces a goiter. Enlargement of the thyroid gland increases its capacity to accumulate iodide from the blood and to synthesize T_4 and T_3. However, the degree to which the enlarged gland can produce thyroid hormones to compensate for their deficiency in the blood depends on the severity of the deficiency of iodide in the diet. To prevent iodide deficiency and the consequent formation of goiter in the human population, iodide is added to the table salt (iodized salt) sold in most developed countries.

Mechanism of Thyroid Hormone Action

Most cells of the body are targets for the action of thyroid hormones. Their sensitivity or responsiveness to the thyroid hormones correlates to some degree with the number of receptors for these hormones in a given type of cell. The cells of the central nervous system (CNS) appear to be an exception to this rule. As will be discussed later, the thyroid hormones play an important role in CNS development during fetal and neonatal life. Hence, developing nerve cells in the brain are important targets for thyroid hormones. In the adult, however, brain cells show little responsiveness to the metabolic regulatory action of thyroid hormones, although they have numerous receptors for these hormones. The reason for this discrepancy is unclear.

Thyroid hormone receptors are located in the nuclei of target cells and are bound to DNA in the chromatin. They are protein molecules of about 50 kd that have some structural similarities to the receptors for steroid hormones and vitamin D. Although thyroid hormone receptors can interact with both T_3 and T_4, they have a much greater affinity for T_3.

The free forms of T_3 and T_4 circulating in the blood are taken up by target cells by an as yet undefined mechanism. Once inside the cell, T_4 may be deiodinated to T_3, which enters the nucleus of the cell and binds to its receptor in the chromatin. This T_3-receptor interaction alters the expression of certain genes in the target cell. As a result, the production of mRNA for certain proteins is either increased or decreased, thereby changing the cell's capacity to make these proteins. Thus, T_3 can influence differentiation by regulating the kinds of proteins produced by its target cells. It can influence growth and metabolism by changing the amounts of structural and enzymatic proteins present in the cells. The mechanisms by which T_3 alters gene expression are currently under investigation.

The signs of T_3 action are slow to appear: when it is given to an animal or human, several hours elapse before its physiologic effects can be detected. This delayed action undoubtedly reflects the time required for changes in gene expression and consequent changes in the synthesis of key proteins to occur. When T_4 is administered, its course of action is usually slower than that of T_3. This is due in part to the additional time required for the body to convert T_4 to T_3.

Role of Thyroid Hormones in Development, Growth, and Metabolism

■ Thyroid Hormones Are Essential for Development of the Central Nervous System

The human brain undergoes its most active phase of growth during the last 6 months of fetal life and the first 6 months of postnatal life. During the second trimester of pregnancy, the multiplication of neuroblasts in the fetal brain reaches a peak and then declines. As pregnancy progresses and the rate of neuroblast division drops, neuroblasts differentiate into neurons and begin the process of synapse formation that extends into postnatal life.

Thyroxine and T_3 first appear in the blood of the human fetus during the second trimester of pregnancy, when neuroblast replication reaches its peak. The levels of these hormones continue to rise during the remaining months of fetal life. Also, thyroid hormone receptors increase about 10-fold in the fetal brain at about the time the concentrations of T_4 and T_3 begin to rise in the blood. These events are critical for normal

AUTOIMMUNE THYROID DISEASE

Certain diseases affecting the function of the thyroid gland occur when an individual's immune system fails to recognize particular thyroid proteins as "self" and reacts to them as it would to foreign proteins. This usually triggers both humoral and cellular immune responses. As a result, antibodies to these proteins are generated and thyroid follicular cells may begin to function abnormally or be killed. One such autoimmune disease that occurs relatively often in humans is chronic lymphocytic thyroiditis, or Hashimoto's disease. The thyroid gland becomes enlarged and infiltrated with lymphocytes. The follicular epithelium is damaged and may become replaced with fibrous tissue. Many individuals with Hashimoto's disease eventually become thyroid-hormone-deficient, and some may have symptoms of thyroid hormone excess during early stages of the disease. The thyroid protein thought for many years to be the main autoantigen involved in Hashimoto's thyroiditis is a membrane substance called thyroid microsomal antigen, whose chemical identity remained unclear. Recently, the gene for human thyroid peroxidase was cloned and found to code for a 933-amino acid transmembrane glycoprotein (see McLachlan SM, Rapoport B in Suggested Reading). It was expressed in Chinese hamster ovary cells, and the recombinant protein was found to have the immunologic properties of the thyroid microsomal antigen. Thus, the thyroid peroxidase enzyme that catalyzes the iodination and coupling reactions involved in thyroid hormone synthesis is most likely the main autoantigen involved in Hashimoto's thyroiditis. Recombinant human thyroid peroxidase is now used as the antigen in immunoassays designed to detect the presence of autoantibodies in humans suspected of having Hashimoto's disease.

brain development, since thyroid hormones are essential for the timing of the decline in nerve cell division and initiation of differentiation and maturation of these cells.

If thyroid hormones are deficient during this prenatal and postnatal period of differentiation and maturation of the brain, mental retardation occurs. This is thought to be due to inadequate development of the neuronal circuitry of the CNS. Thyroid hormone therapy must be given to a thyroid hormone-deficient child during the first few months of postnatal life to prevent mental retardation. Starting thyroid hormone therapy after behavioral deficits have occurred cannot reverse the mental retardation (i.e. thyroid hormone must be present when differentiation normally occurs). Thyroid hormone deficiency in infancy causes mental retardation and growth impairment (see below). Fortunately, this occurs rarely today, because thyroid hormone deficiency is usually detected in newborn infants and hormone therapy is given at the proper time.

The exact mechanism by which thyroid hormones influence differentiation of the CNS is unknown. It has been found in animal experiments that thyroid hormones inhibit nerve cell replication in the brain and stimulate growth of nerve cell bodies, the branching of dendrites, and the rate of myelinization of nerve fibers. These effects of thyroid hormones

are presumably due to their ability to regulate the expression of genes involved in nerve cell replication and differentiation. However, the details, particularly in the human, are unclear.

■ Thyroid Hormones Are Essential for Normal Body Growth

The thyroid hormones are important factors regulating the growth of the entire body. For example, the thyroid hormone-deficient individual who has not been given thyroid hormone therapy during childhood will have short stature.

Thyroid Hormones and the Gene for GH ■ A major way thyroid hormones promote normal body growth is by stimulating the expression of the gene for growth hormone (GH) in the somatotrophs of the anterior pituitary gland. In a thyroid hormone-deficient individual, GH synthesis by the somatotrophs is greatly reduced and consequently GH secretion is impaired. Thus, a thyroid hormone-deficient individual will also be GH deficient. If this occurs in a child, it will cause growth retardation, which is largely due to the lack of the growth-promoting action of GH (Chap. 34).

Other Effects of Thyroid Hormones on Growth ■ The thyroid hormones have additional effects on growth. In tissues such as skeletal muscle, heart, and liver, the thyroid hormones have direct effects on the synthesis of a variety of structural and enzymatic proteins. For example, they stimulate the synthesis of structural proteins of mitochondria, as well as the formation of many enzymes involved in intermediary metabolism and oxidative phosphorylation. Presumably, the thyroid hormones stimulate the expression of the genes for these proteins.

Thyroid hormones also promote the calcification, and hence the closure, of the cartilagenous growth plates of the bones of the skeleton. This action limits further linear body growth. How the thyroid hormones promote calcification of the growth plates of bones is not understood.

■ Thyroid Hormones Regulate the Basal Energy Economy of the Body

When the body is at rest, about half of the ATP produced by its cells is used to drive energy-requiring membrane transport processes. The remainder is used in involuntary muscular activity, such as respiratory movements, peristalsis, and contraction of the heart, and in many metabolic reactions requiring ATP, such as protein synthesis. The energy required to do this work is eventually evolved as body heat.

Basal Oxygen Consumption and Body Heat Production ■ The major site of ATP production is the mitochondria, where the oxidative phosphorylation of ADP to ATP takes place. The rate of oxidative phosphorylation is dependent on the supply of ADP for electron transport. The ADP supply is, in turn, a function of the amount of ATP used to do work. For example, when more work is done per unit time, more ATP is used and more ADP is generated, thereby increasing the rate of oxidative phosphorylation.

The rate at which oxidative phosphorylation occurs is reflected in the amount of oxygen consumed by the body, since oxygen is the final electron acceptor at the end of the electron transport chain.

Activities that occur when the body is not at rest, such as voluntary movements, use additional ATP for the work involved. Thus the amounts of oxygen consumed and body heat produced depend on total body activity.

The Calorigenic Action of Thyroid Hormones ■ The thyroid hormones regulate the basal rate at which oxidative phosphorylation takes place in cells. As a result, they set the basal rate of body heat production and of oxygen consumed by the body. This is called the **calorigenic action** of the thyroid hormones.

Thyroid hormone levels in the blood must be within normal limits for basal metabolism to proceed at the rate needed for a balanced energy economy of the body. For example, if thyroid hormones are present in excess, oxidative phosphorylation is accelerated, and body heat production and oxygen consumption are abnormally high. The converse occurs when the blood concentrations of T_4 and T_3 are lower than normal. The fact that the thyroid hormones affect the amount of oxygen consumed by the body has been used clinically to assess the status of thyroid function. Oxygen consumption is measured under resting conditions and compared to the rate expected of a similar individual with normal thyroid function. This is the **basal metabolic rate (BMR)** test.

Tissues Affected by the Calorigenic Action of Thyroid Hormones ■ Not all tissues are sensitive to the calorigenic action of thyroid hormones. Tissues and organs that give this response include skeletal muscle, heart, liver, and kidney, which are also tissues in which thyroid hormone receptors are abundant. The adult brain, skin, lymphoid organs, and gonads show little calorigenic response to the thyroid hormones. With the exception of the adult brain, these contain few thyroid hormone receptors, which may explain their poor response to this effect of the thyroid hormones.

Mechanisms of the Calorigenic Action of Thyroid Hormones ■ The calorigenic action of the thyroid hormones is poorly understood at the molecular level. The calorigenic effect takes many hours to appear after administration of thyroid hormones to a human or animal (Fig. 35–6), probably because of the

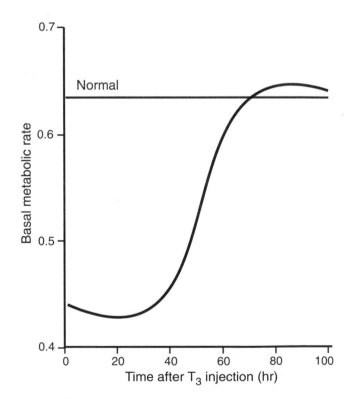

Figure 35–6 ■ The effect of a single injection of T_3 on the basal metabolic rate (BMR) of thyroidectomized rats, over time. The average BMR for normal rats is indicated by the horizontal line. Note that it is higher than the initial (0 time) BMR of the thyroid hormone-deficient thyroidectomized rats. BMR is given on the ordinate of the graph as milliliters of oxygen consumed per hour per gram of body weight. (Modified from Tata JR, Ernster L, Lindberg O, et al: The action of thyroid hormones at the cell level. *Biochem J* 86:408–428, 1963.)

time required for changes in the expression of genes involved. Triiodothyronine is known to stimulate the synthesis of cytochromes and cytochrome oxidase in certain cells. This suggests that T_3 may regulate the number of respiratory units in these cells, thereby affecting their capacity to carry out oxidative phosphorylation. Also, the thyroid hormones stimulate the synthesis of Na^+/K^+ ATPase in a variety of cells. On the basis of this and related findings, it has been suggested that the thyroid hormones regulate the rate of ion transport in cell membranes, a major user of ATP in cells. Regulating this process would control the rate of ADP supply to the oxidative phosphorylation process and hence the rate of oxidative phosphorylation. Further, the energy provided by ATP for the work involved in transporting ions is dissipated as body heat. By regulating the rate of use of ATP, the thyroid hormones would influence body heat production. Whether this can actually explain the calorigenic action of thyroid hormones remains to be proved.

■ Thyroid Hormones Stimulate Intermediary Metabolism

In addition to their ability to regulate the rate of basal energy metabolism, thyroid hormones influence the rate at which most of the pathways of intermediary metabolism operate in their target cells. When thyroid hormones are deficient, pathways of carbohydrate, lipid, and protein metabolism are slowed and their responsiveness to other regulatory factors, such as other hormones, is decreased. On the other hand, these same metabolic pathways run at an abnormally high rate when thyroid hormones are present in excess. Thus, thyroid hormones can be viewed as amplifiers of cellular metabolic activity. How they exert this overall amplifying effect on intermediary metabolism is not fully understood; however, their action on the expression of genes for enzymes involved in the pathways of intermediary metabolism is undoubtedly involved.

■ Thyroid Hormones Regulate Their Own Secretion

An important action of the thyroid hormones is their ability to regulate their own secretion. As discussed in Chapter 34, T_3 exerts an inhibitory effect on TSH secretion by the thyrotrophs in the anterior pituitary gland by decreasing the sensitivity of the thyrotrophs to thyrotrophin-releasing hormone (TRH). Consequently, when the circulating concentration of free thyroid hormones is high, the thyrotrophs are relatively insensitive to TRH, and the rate of TSH secretion therefore decreases. The consequent fall of TSH in the blood reduces the rate of thyroid hormone secretion. On the other hand, when the free thyroid hormone level falls in the blood, the negative feedback effect of T_3 on the thyrotroph is reduced, and the rate of TSH secretion increases. The rise in TSH in the blood stimulates the thyroid gland to secrete thyroid hormones at a greater rate. This action of T_3 on the thyro-

trophs is thought to be due to changes in gene expression in these cells.

The physiologic actions of the thyroid hormones described above are summarized in Table 35–1.

Deficiency and Excess of Thyroid Hormones

■ Thyroid Hormone Deficiency Causes Nervous, Growth, and Metabolic Disorders

Thyroid hormone deficiency in humans is caused by a variety of conditions. For example, iodide deficiency may result in a reduction in thyroid hormone production, as noted earlier, and autoimmune disease involving circulating antibodies against the thyroid gland (e.g., **Hashimoto disease**) may impair thyroid hormone biosynthesis. Other causes of thyroid hormone deficiency include inheritable diseases that affect certain steps in the biosynthesis of thyroid hormones and hypothalamic or pituitary diseases that interfere with TRH or TSH secretion. Obviously, surgical removal of the thyroid gland also causes thyroid hormone deficiency. **Hypothyroidism** is the disease state that results from thyroid hormone deficiency.

As described earlier, a deficiency of thyroid hormones at birth that is not treated during the first few months of postnatal life causes irreversible mental retardation. Thyroid hormone deficiency later in life can also influence the function of the nervous system. For example, certain reflexes are slowed due to a reduction in nerve conduction velocity, and mentation may be impaired and gradual mental deterioration may occur. These changes can usually be reversed with thyroid hormone therapy.

A child with thyroid hormone deficiency grows at a slower than normal rate due to the loss of the direct growth-promoting effects of thyroid hormones and to

TABLE 35–1 ■ Physiologic Actions of Thyroid Hormones

Development of central nervous system
  Inhibit nerve cell replication
  Stimulate growth of nerve cell bodies
  Stimulate branching of dendrites
  Stimulate rate of myelinization
Body growth
  Stimulate expression of gene for growth hormone in somatotrophs
  Stimulate synthesis of many structural and enzymatic proteins
  Promote calcification of growth plates of bones
Basal energy economy of body
  Regulate basal rates of oxidative phosphorylation, body heat production, and oxygen consumption (calorigenic action)
Intermediary metabolism
  Stimulate synthetic and degradative pathways of carbohydrate, lipid, and protein metabolism
Thyroid stimulating hormone (TSH) secretion
  Inhibit TSH secretion by decreasing sensitivity of thyrotrophs to thyrotrophin-releasing hormone

GH deficiency. If the child is not given thyroid hormone therapy, it will not reach normal adult body height.

The thyroid hormone-deficient individual is generally hypometabolic. Basal metabolic rate is reduced, and, as a consequence, body heat production is reduced. Vasoconstriction occurs in the skin as a compensatory maneuver to conserve body heat. Heart rate and cardiac output are reduced. Food intake is reduced, and the synthetic and degradative processes of intermediary metabolism slow down. In severe hypothyroidism, a substance consisting of hyaluronic acid and chondroitin sulfate complexed with protein is deposited in the extracellular spaces of the skin, causing water to accumulate osmotically. This gives a puffy appearance to the face, hands, and feet, called **myxedema.** All of the above disorders can be normalized with thyroid hormone therapy.

■ Thyroid Hormone Excess Produces Nervous and Other Disorders

The most common cause of excessive thyroid hormone production in the human is **Graves' disease,** an autoimmune disease thought to be caused by antibodies formed against certain components of the plasma membranes of thyroid follicular cells. These antibodies are believed to interact with their respective antigens in the follicular cell plasma membrane, a result of which is the activation of adenylate cyclase. The consequent rise in cAMP in the follicular cells produces effects similar to those caused by the action of TSH. For example, the thyroid gland enlarges to form a **diffuse toxic goiter,** which synthesizes and secretes thyroid hormones at an accelerated rate, causing thyroid hormones to be chronically elevated in the blood. Less common conditions that cause chronic elevations in circulating thyroid hormones include adenomas of the thyroid gland that secrete thyroid hormones and excessive TSH secretion caused by malfunctions of the hypothalamopituitary-thyroid axis. The disease state that develops in response to excessive thyroid hormone secretion, called **hyperthyroidism** or **thyrotoxicosis,** is characterized by many changes in the functioning of the body that are the opposite of those caused by thyroid hormone deficiency.

Hyperthyroid individuals are nervous and irritable. They also experience physical weakness and fatigue. Basal metabolic rate is increased and, as a result, body heat production is increased. Vasodilation in the skin and sweating occur as compensatory mechanisms to dissipate excessive body heat. Heart rate and cardiac output are increased. All metabolic processes speed up, with degradative processes predominating. There is destruction of body tissue, a gradual depletion of vital cofactors and enzymes, and weight loss, despite increased food intake. All of these changes can be reversed by therapy directed toward reducing the rate of thyroid hormone secretion, either with drugs or by surgical removal of thyroid tissue.

■ ■ ■ ■ ■ ■ ■ ■ ■ ■ REVIEW EXERCISES ■ ■ ■ ■ ■ ■ ■ ■ ■ ■ ■

Identify Each with a Term

1. The storage form of the thyroid hormones in the colloid
2. The enzyme that catalyzes the conversion of thyroxine to triiodothyronine in target cells
3. A thyroid gland that is enlarged from chronic stimulation by TSH
4. The effect of the thyroid hormones on basal energy metabolism
5. The puffy appearance of the face, hands, and feet of individuals with severe hypothyroidism

Define Each Term

6. Parafollicular cell
7. Iodide transporter
8. Coupling reaction
9. Thyroxine-binding globulin

Choose the Correct Answer

10. Triiodothyronine (T_3):
    a. Is produced in greater amounts by the thyroid gland than T_4
    b. Binds to receptors in the cytosol of target cells
    c. Can be produced from T_4 by pituitary thyrotrophs
    d. Has a half-life of a few minutes in the bloodstream

11. Which of the following is *not* an effect of TSH on thyroid follicular cells?
    a. Stimulation of endocytosis of thyroglobulin stored in the colloid
    b. Release of a large pool of T_4 and T_3 stored in secretory granules at the basal cell membrane
    c. Stimulation of the uptake of iodide from the blood
    d. Increase in cell height

12. Chronic iodide deficiency in the diet would increase all of the following *except:*
    a. Thyroid follicle cell number
    b. Concentration of TSH in the blood
    c. Thyroid follicle cell height
    d. Concentration of T_4 in the blood

13. Which of the following would be decreased in an individual producing an excess of thyroid hormones?
    a. Body weight
    b. Basal metabolic rate
    c. Heart rate
    d. Food intake

Briefly Answer

14. A child is born with a rare disorder in which the thyroid gland does not respond to TSH. What would be the predicted effects on mental ability, body growth rate,

and thyroid gland size when the child reaches 6 years of age?

15. If the 6-year-old child is treated with thyroid hormones, how would mental ability, body growth rate, and thyroid gland size be affected?

16. At what age should thyroid hormone therapy be started, and why?

17. What substance is formed when T_4 is deiodinated at ring position 5, and what effect does this have on hormone activity?

■ ■ ■ ■ ■ ■ ■ ■ ■ ■ ■ ANSWERS ■ ■ ■ ■ ■ ■ ■ ■ ■ ■ ■ ■

1. Thyroglobulin

2. 5′ deiodinase

3. Goiter

4. Calorigenic action

5. Myxedema

6. A large cell embedded in the wall of the thyroid follicle that produces and secretes calcitonin

7. The active transport mechanism in the basal plasma membrane of the follicular cell that enables the cell to concentrate iodide from the blood

8. The reaction catalyzed by thyroid peroxidase that couples two neighboring iodotyrosine residues in the thyroglobulin precursor to form an iodothyronine residue and a dehydroalanine residue

9. The serum glycoprotein to which most of the T_4 and T_3 in the blood is bound

10. c

11. b

12. d

13. a

14. Mental ability would be impaired, body growth rate would be slowed, and thyroid gland size would be smaller than normal.

15. Mental ability and thyroid gland size would not be affected by thyroid hormone treatment, but body growth rate would be restored toward normal.

16. Thyroid hormone therapy should be started during the first few months of postnatal life to prevent irreversible mental retardation.

17. Reverse T_3 is formed, and it lacks thyroid hormone activity.

Suggested Reading

Dussault JH, Ruel J: Thyroid hormones and brain development. *Annu Rev Physiol* 49:321-334, 1987.

Goodman HM: *Basic Medical Endocrinology.* 2nd ed. New York, Raven, 1994.

Larsen PR, Ingbar SH: The thyroid gland, in Wilson JD, Foster DW (eds): *Williams Textbook of Endocrinology.* 8th ed. Philadelphia, WB Saunders, 1992, pp 357-487.

McLachlan SM, Rapoport B: The molecular biology of thyroid peroxidase: Cloning, expression and role as autoantigen in autoimmune thyroid disease. *Endocrin Rev* 13:192-206, 1992.

Nunez J, Pommier J: Formation of thyroid hormones. *Vitamin Horm* 39:175-230, 1986.

Oppenheimer JH, Schwartz HL, Mariash CN, et al: Advances in our understanding of thyroid hormone action at the cellular level. *Endocrin Rev* 8:288-308, 1987.

Samuels HH, Forman BM, Horowitz ZD, Ye Z-S: Regulation of gene expression by thyroid hormone. *Annu Rev Physiol* 51:623-639, 1989.

Vassart G, Dumont JE: The thyrotropin receptor and the regulation of thyrocyte function and growth. *Endocrin Rev* 13:596-611, 1992.

C H A P T E R

■ ■ ■ ■ ■ ■ ■

36

The Adrenal Gland

OBJECTIVES

After studying this chapter, the student should be able to:
1. Describe the functional anatomy of the adrenal gland
2. List the hormones of physiologic importance that are secreted by the adrenal cortex; list the sources of cholesterol used for the synthesis of these hormones; describe the pathways for the synthesis of glucocorticoids, androgens, and aldosterone from cholesterol; explain how genetic defects in these steroidogenic pathways affect hormone secretion by the adrenal cortex
3. Explain how adrenal steroid hormones are metabolized and inactivated by the liver
4. Describe the mechanism by which ACTH regulates the synthesis of glucocorticoids and androgens by the adrenal cortex
5. Describe the mechanism by which angiotensin II regulates the synthesis of aldosterone by the adrenal cortex
6. Explain the current concept of how glucocorticoids act at the cellular level
7. Describe the role of glucocorticoids in the response of the body to fasting, injury, and stress
8. List the hormones of physiologic importance that are secreted by the adrenal medulla
9. Explain the current concept of how the catecholamines act at the cellular level
10. List the stimuli that cause secretion of catecholamines by the adrenal medulla; explain the role of catecholamines in the fight-or-flight response and in metabolic response to hypoglycemia

To remain alive, the organs and tissues of the human body must have a finely regulated extracellular environment. This environment must contain the correct concentrations of ions to maintain body fluid volume and to enable excitable cells to function. Also, it must have an adequate supply of the metabolic substrates used by cells to generate ATP. Normally, salts, water, and some organic substances are continually lost from the body as a result of perspiration, respiration, and excretion. Also, substrates are used by metabolic processes in the cells. Ordinarily, these critical constituents of the extracellular environment of the body are replenished from the food and liquids taken during meals. Yet a human can survive for days, and in some cases weeks, on little else but water. This is because the human body has the remarkable capacity to adjust the functions of its organs and tissues to preserve body fluid volume and composition.

The adrenal glands play a key role in making these adjustments. This is readily apparent from the fact that an adrenalectomized animal, unlike its normal counterpart, cannot survive prolonged fasting. Its blood glucose supply diminishes, ATP generation by the cells becomes inadequate to support life, and the animal eventually dies. Even when fed a normal diet, an adrenalectomized animal typically loses body sodium and water over time and eventually dies of circulatory collapse. Death of the adrenalectomized animal is due to the lack of certain steroid hormones that are produced and secreted by the cortex of the adrenal gland.

The glucocorticoid hormones, **cortisol** and **corticosterone,** play essential roles in adjusting the metabolism of carbohydrates, lipids, and proteins in the liver, muscle, and adipose tissue during fasting, which assures an adequate supply of glucose and fatty acids for energy metabolism despite the lack of food intake. Another steroid hormone produced by the adrenal cortex, the mineralocorticoid **aldosterone,** stimulates the kidneys to conserve sodium and hence body fluid volume.

The glucocorticoids also enable the body to cope with physical and emotional traumas or stresses. The great physiologic importance of this action of the glucocorticoids is emphasized by the fact that adrenalectomized animals lose their ability to cope with physical or emotional stresses. Even when given an appropriate diet to prevent blood glucose and body sodium depletion, an adrenalectomized animal may die when exposed to traumas that are not fatal to normal animals.

Hormones produced by the other endocrine component of the adrenal gland, the medulla, are also involved in compensatory reactions of the body to trauma or life-threatening situations. These hormones are the catecholamines, **epinephrine** and **norepinephrine,** which have widespread effects on the cardiovascular system, muscular system, and carbohydrate and lipid metabolism in liver, muscle, and adipose tissue.

Functional Anatomy of the Adrenal Gland

■ Adrenal Cortex and Adrenal Medulla Have Different Embryologic Origins

The adrenal glands of the human are paired, pyramid-shaped organs located on the upper surfaces of each kidney. The adrenal gland is actually a composite of two separate endocrine organs, one inside the other, each secreting separate hormones and each regulated by different mechanisms. The outer portion, or **cortex,** of the adrenal gland completely surrounds the inner portion, or **medulla,** and makes up most of the bulk of the gland. During embryologic development, the cortex forms from mesoderm. The medulla, on the other hand, arises from neural ectoderm.

■ The Adrenal Cortex Consists of Three Distinct Zones

In the adult human, the adrenal cortex consists of three histologically distinct zones or layers (Fig. 36–1). The outer zone, which lies immediately under the capsule of the gland, is called the **zona glomerulosa** and consists of small clumps of cells that produce the mineralocorticoid aldosterone (Fig. 36–2). The middle and thickest layer of the cortex is the **zona fasciculata,** which consists of cords of cells oriented radial to the center of the gland. The inner layer of cortex is comprised of interlaced strands of cells called the **zona reticularis.** The zona fasciculata and zona reticularis both produce the physiologically important glucocorticoids, cortisol and corticosterone (Fig. 36–2).

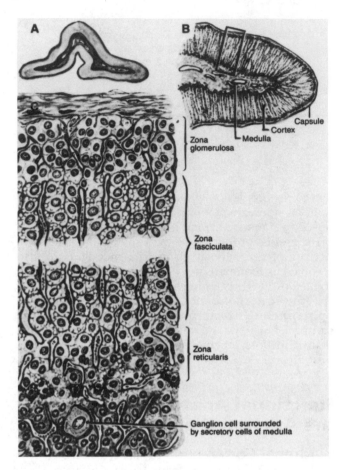

Figure 36–1 ■ Diagrammatic representation of the human adrenal gland. (A) Gross structure. (B) Low-power view. (C) Area marked in B, at a higher magnification. Note the three zones of the adrenal cortex. (From Cormack DH: *Ham's Histology.* 9th ed. Philadelphia, JB Lippincott, 1987.)

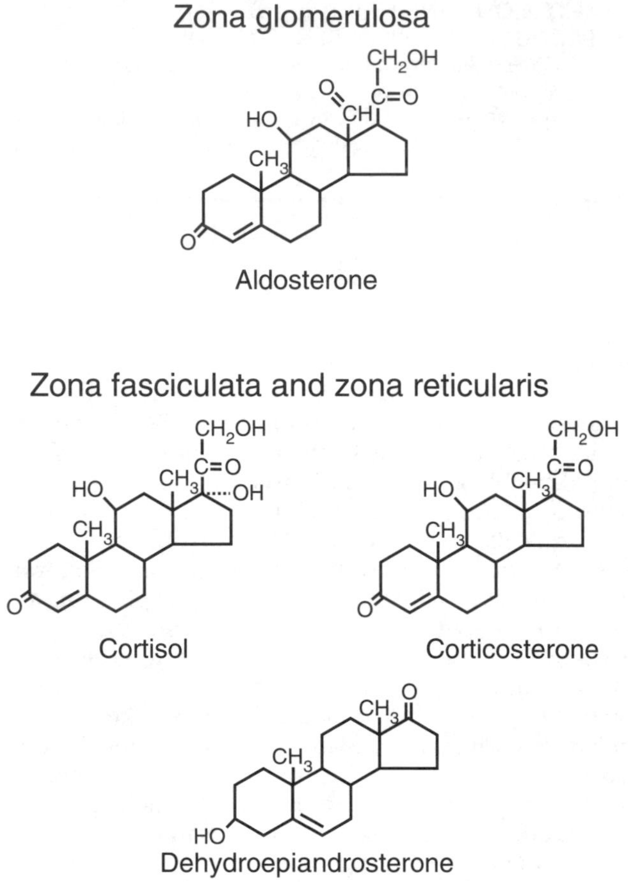

Zona glomerulosa

Aldosterone

Zona fasciculata and zona reticularis

Cortisol   Corticosterone

Dehydroepiandrosterone

Figure 36–2 ■ Physiologically important hormones secreted by the adrenal cortex.

These layers of the cortex also produce the androgen **dehydroepiandrosterone** (Fig. 36–2), which is related chemically to the male sex hormone **testosterone** (Fig. 33–2).

Like all endocrine organs, the adrenal cortex is well vascularized. Many small arteries branch from the aorta and renal arteries and enter the cortex. These give rise to capillaries that course radially through the cortex and terminate in venous sinuses in the zona reticularis and adrenal medulla (Fig. 36–1). Thus the hormones produced by the cells of the cortex have ready access to the circulation.

The cells of the adrenal cortex contain abundant lipid droplets in the cells, especially those in the zona fasciculata. This stored lipid is functionally significant because cholesterol esters present in the droplets are an important source of the cholesterol used as precursor for the synthesis of the steroid hormones.

■ The Adrenal Medulla Is a Modified Sympathetic Ganglion

The adrenal medulla, formed from neural ectoderm, can be considered a modified sympathetic ganglion. The medulla consists of clumps and strands of **chromaffin cells** interspersed with venous sinuses (Fig. 36–1). The chromaffin cells can be viewed as modified postganglionic neurons that receive sympathetic preganglionic cholinergic innervation from the splanchnic nerves. The chromaffin cells produce catecholamine hormones, principally epinephrine and norepinephrine (Fig. 3–19). Epinephrine and norepinephrine are stored in granules in the chromaffin cells and discharged into the venous sinuses of the adrenal medulla when the adrenal branches of the splanchnic nerves are stimulated.

Hormones of the Adrenal Cortex

Only small amounts of the glucocorticoids, aldosterone, and adrenal androgens are found in the cells of the adrenal cortex at a given time, because these cells produce and secrete these hormones on demand, rather than storing them. Table 36-1 shows the amounts of these hormones typically secreted during a 24-hour period by a normal adult human under resting (unstimulated) conditions. Since the molecular weights of these substances are not greatly different, comparison of the amounts secreted (mg/day) gives a fair indication of the relative number of molecules of each hormone produced daily. It is clear that humans secrete about 10 times more cortisol than corticosterone during an average day. Given that and the fact that corticosterone has only one-fifth the glucocorticoid activity of cortisol (Table 36-2), cortisol is considered the physiologically important glucocorticoid in humans. In comparison to the glucocorticoids, a much smaller amount of aldosterone is secreted each day (Table 36-1).

Because of similarities in their structures, the glucocorticoids and aldosterone have overlapping actions. For example, cortisol and corticosterone have some mineralocorticoid activity and, conversely, aldosterone has some glucocorticoid activity (Table 36-2). However, given the amounts of these hormones secreted under normal circumstances and their relative activities, glucocorticoids are not physiologically important mineralocorticoids, nor does aldosterone function physiologically as a glucocorticoid.

As we discuss in detail later, the amounts of glucocorticoids and aldosterone secreted by an individual can vary greatly from those given in Table 36-1. The amount secreted depends on the individual's physiologic state. For example, in an individual subjected to severe physical or emotional trauma, the rate of cortisol secretion may be 10 times greater than the resting rate given in Table 36-1. Certain diseases of the adrenal cortex that involve steroid hormone biosynthesis can markedly increase or decrease the amount of hormones produced.

The adrenal cortex also produces and secretes substantial amounts of androgenic steroids. Dehydroepiandrosterone (DHEA) in both the free and sulfated form (DHEAS) is the main androgen secreted by the cortex of both men and women (Table 36-1). Lesser

TABLE 36–1 ■ Normal Adult Daily Production of Hormones by the Adrenal Cortex

Hormone	Amount Produced (mg/day)
Cortisol	20.0
Corticosterone	2.0
Aldosterone	0.1
Dehydroepiandrosterone	30.0

Modified from Liddle GW, Melmon KL: The adrenals, in Williams RH (ed): *Textbook of Endocrinology.* 5th ed. Philadelphia, WB Saunders, 1974, pp 233-322.

TABLE 36–2 ■ Comparison of Shared Activities of Adrenal Cortical Hormones

Hormone	Glucocorticoid Activity[a]	Mineralcorticoid Activity[b]
Cortisol	100	0.25
Corticosterone	20	0.5
Aldosterone	10	100

[a]Percent activity, with cortisol being 100%.
[b]Percent activity, with aldosterone being 100%.
Modified from Liddle GW, Melmon KL: The adrenals, in Williams RH (ed): *Textbook of Endocrinology.* 5th ed. Philadelphia, WB Saunders, 1974, pp 233-322.

amounts of other androgens are also produced. The adrenal cortex is the main source of the androgens in the human female's blood. In the human male, however, androgens produced by the testis and adrenal cortex contribute to the male sex hormones circulating in the blood. Aside from playing a role in developmental events before the start of puberty in both girls and boys, adrenal androgens normally have little physiologic effect. This is because their male sex hormone activity is quite weak. Exceptions occur in individuals who produce inappropriately large amounts of certain adrenal androgens due to diseases affecting the pathways of steroid biosynthesis in the adrenal cortex.

Adrenal Steroid Hormones Are Synthesized from Cholesterol

Cholesterol is the starting material for the synthesis of steroid hormones.

Structure of Cholesterol ■ Cholesterol consists of four interconnected rings of carbon atoms and a side chain of eight carbon atoms extending from one ring (Fig. 36-3). In all, there are 27 carbon atoms in cholesterol, numbered as shown in the figure. To indicate the three-dimensional structure of steroids, substituents on the ring structure that would project upward toward the viewer are typically drawn with a solid line and designated β. An example is the 3β-hydroxyl group on carbon 3 of cholesterol in Figure 36-3. Substituents that would project downward from the ring structure, away from the viewer, are drawn with a dashed line and designated α. The carbon 18 and 19 methyl groups are often simply drawn as solid lines (Fig. 33-2).

Sources of Cholesterol ■ The immediate source of the cholesterol used in the biosynthesis of steroid hormones is the abundant lipid droplets in adrenal cortical cells. The cholesterol present in these lipid droplets is mainly in the form of **cholesterol esters,** single molecules of cholesterol esterified at the 3β-hydroxy position to single fatty acid molecules (Fig. 36-3). The free cholesterol used in steroid biosynthesis

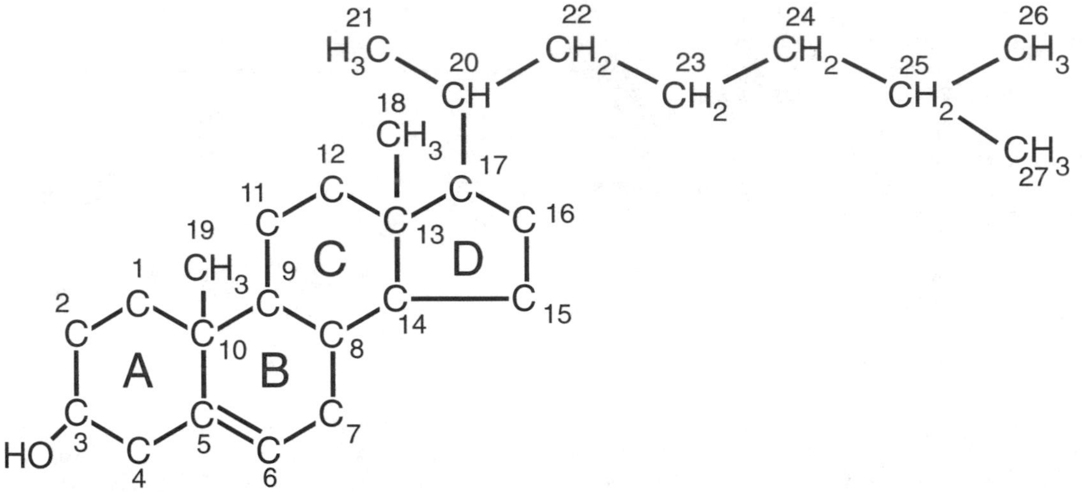

Figure 36-3 ■ Chemical structure of cholesterol. Note how the four rings are lettered and how the carbons are numbered. The hydrogen atoms on the carbons composing the rings are omitted from the figure.

is generated from these cholesterol esters by the action of **cholesterol esterase (cholesterol ester hydrolase;** CEH), which hydrolyzes the ester bond. (Recent evidence suggests that this cholesterol esterase in the adrenal cortex is identical to hormone-sensitive lipase, which is responsible for triglyceride lipolysis in adipose tissue.) The free cholesterol generated by that cleavage enters mitochondria located in close proximity to the lipid droplet. The process of remodeling the cholesterol molecule into steroid hormones is then initiated.

The cholesterol that has been removed from the lipid droplets for steroid hormone biosynthesis is replenished in two ways (Fig. 36-4). Most of the cholesterol converted to steroid hormones by the human adrenal gland comes from cholesterol esters contained in **low-density lipoprotein** (LDL) particles circulating in the blood. The LDL particles consist of a core of cholesterol esters surrounded by a coat of cholesterol and phospholipids. A 400 kd protein molecule called **apoprotein B-100** is also present on the surface of the LDL particle; it is recognized by LDL receptors on the plasma membranes of adrenal cortical cells (Fig. 36-4). The apoprotein binds to the LDL receptor, and the LDL particle is then taken up by the cell by endocytosis. The LDL particles are then degraded by the cell. The cholesterol esters in the core of the particle are hydrolyzed to free cholesterol and fatty acid by the action of cholesterol esterase. Any cholesterol not immediately used by the cell is converted again to cholesterol esters by the action of the enzyme **acyl-CoA:cholesterol acyltransferase** (ACAT). The esters are then stored in the lipid droplets of the cell to be used later. When steroid biosynthesis is proceeding at a high rate, cholesterol delivered to the adrenal cell by this process may be diverted directly to the mito-

chondria for steroid production rather than re-esterified and stored. When steroidogenesis occurs at a high rate over a long period of time, the delivery of cholesterol to the adrenal by the uptake of LDL particles becomes critical for steroid formation to continue unimpeded.

In humans, cholesterol that has been synthesized de novo from acetate by the adrenal is a significant but minor source of cholesterol for steroid hormone formation. The rate-limiting step in this process is catalyzed by the enzyme **3-hydroxy-3-methylglutaryl CoA reductase** (HMG CoA reductase). The newly synthesized cholesterol is then incorporated into cellular structures, such as membranes, or converted to cholesterol esters through the action of ACAT and stored in lipid droplets (Fig. 36-4).

Pathways for the Synthesis of Steroid Hormones ■ The conversion of cholesterol into steroid hormones begins with the formation of free cholesterol from the cholesterol esters stored in intracellular lipid droplets (Fig. 36-4). The free cholesterol molecules then enter mitochondria, which are located in close proximity to the lipid droplets, by a mechanism that is not well understood. There is evidence that free cholesterol must first associate with a small protein called **sterol carrier protein 2,** which facilitates its entry into the mitochondrion in some manner. Also, cAMP mediates the process, but exactly what it does is unknown.

Once inside the mitochondria, single cholesterol molecules bind to the enzyme **cytochrome P450scc, or cholesterol side chain cleavage (SCC) enzyme,** which is embedded in the inner mitochondrial membrane. After a cholesterol molecule binds to a cytochrome P450scc molecule, the first and rate-limiting reaction in steroidogenesis occurs. This reaction re-

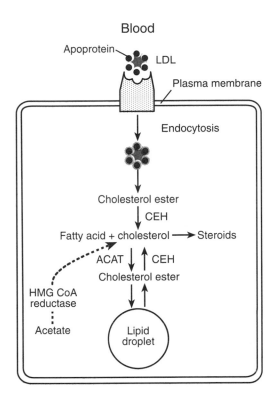

Figure 36–4 ■ Sources of cholesterol for steroid biosynthesis by the adrenal cortex. Most comes from low-density lipoprotein (LDL) particles in the blood, which bind to receptors in the plasma membrane and are taken up by endocytosis. The cholesterol in the LDL particle is either used directly for steroidogenesis or stored in lipid droplets for later use. Some cholesterol is also synthesized directly from acetate. CEH, cholesterol ester hydrolase; ACAT, acyl-CoA:cholesterol acyltransferase.

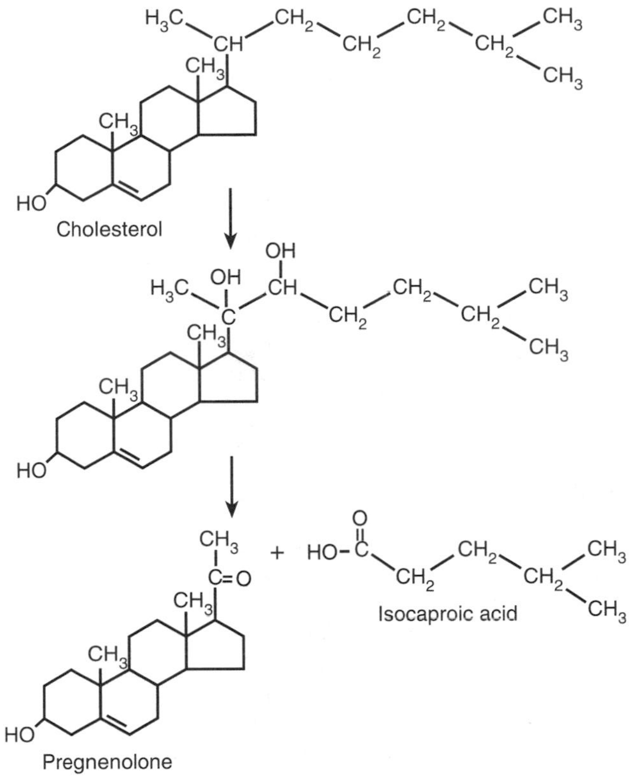

Figure 36–5 ■ Formation of pregnenolone from cholesterol by the action of cytochrome P450scc.

models the cholesterol molecule into a 21-carbon steroid intermediate called **pregnenolone.** The reaction occurs in three steps (Fig. 36–5). The first two steps consist of the hydroxylation of carbons 20 and 22 of cholesterol by cytochrome P450scc. Then the enzyme cleaves the side chain of cholesterol between carbons 20 and 22, yielding pregnenolone and **isocaproic acid.**

Cytochrome P450scc belongs to a large family of cytochrome P450 oxidases, all of which have a molecular weight of about 50,000 and a single heme group. Cytochrome P450scc hydroxylates cholesterol at carbons 20 and 22 by reducing molecular oxygen with electrons donated from NADPH:

$$RH + O_2 + H^+ + NADPH \xrightarrow{P450scc} ROH + H_2O + NADP^+$$

The overall reaction is somewhat more complicated than this and involves the participation of two other mitochondrial proteins, as shown in Figure 36–6. Electrons are first transferred from NADPH to the flavoprotein enzyme **adrenodoxin reductase,** which is

present on the inner surface of the inner mitochondrial membrane. The electrons are then transferred to a small (~12 kd), freely diffusible protein present in the mitochondrial matrix called **adrenodoxin.** It contains two atoms of iron and two atoms of noncystine sulfur coordinated within its structure. Adrenodoxin transfers the electrons to cytochrome P450scc. The electrons are used for the reduction of molecular oxygen. The oxidized adrenodoxin then shuttles back to accept new electrons from reduced adrenodoxin reductase. In all, three pairs of electrons are transferred from NADPH to cytochrome P450scc during the side chain cleavage reaction (i.e., three molecules of NADPH are required to produce one molecule of pregnenolone).

Once formed, pregnenolone molecules dissociate from cytochrome P450scc, leave the mitochondria, and enter smooth endoplasmic reticulum near the mitochondria. The mechanism involved in the movement of pregnenolone into endoplasmic reticulum is not understood. At this point, the further remodeling of pregnenolone into steroid hormones can vary, depending on whether the process occurs in the zona fasciculata and reticularis or the zona glomerulosa. We first consider what occurs in the zona fasciculata and zona reticularis. These biosynthetic events are summarized in Figure 36–7.

In cells of the zona fasciculata and zona reticularis,

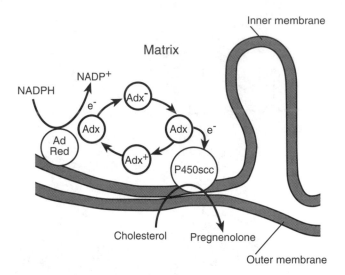

Matrix

Inner membrane

Outer membrane

Cholesterol   Pregnenolone

Figure 36–6 ■ The mitochondrial electron transport chain involved in the hydroxylation and side chain cleavage reactions that form pregnenolone from cholesterol. Adrenodoxin reductase accepts electrons (e^-) from NADPH. These are passed to adrenodoxin, which is free to move in the matrix of the mitochondrion. Adrenodoxin donates these electrons to cytochrome P450scc. It carries out the hydroxylations and side chain cleavage. Oxidized adrenodoxin returns to adrenodoxin reductase to gain another pair of electrons. Three such cycles (three pairs of electrons) are involved in producing one molecule of pregnenolone. Ad red, adrenodoxin reductase; Adx, adrenodoxin; P450scc, cytochrome P450scc. (Modified from Miller WL: Molecular biology of steroid hormone synthesis. *Endocrin Rev* 9:295–318, 1988.)

most of the pregnenolone is converted to cortisol and the main adrenal androgen DHEA. The pregnenolone molecules bind to the enzyme **cytochrome P450c17 (17α-hydroxylase cytochrome P450),** which is embedded in the membranes of the endoplasmic reticulum. Cytochrome P450c17 hydroxylates pregnenolone at carbon 17. Again, the electrons involved in the hydroxylation reaction are donated by NADPH, but in this case they are transferred to cytochrome P450c17 by the flavoprotein **NADPH cytochrome P450 reductase,** which is different from adrenodoxin reductase. Adrenodoxin is not involved. The product formed by this reaction is **17α-hydroxypregnenolone** (Fig. 36–7).

Cytochrome P450c17 has an additional enzymatic action, which becomes important at this step in the steroidogenic process. Once it has hydroxylated carbon 17, it has the ability to lyse or cleave the carbon 20-21 side chain from the steroid structure. Some molecules of 17α-hydroxypregnenolone undergo this reaction as well and are converted to the 19-carbon steroid DHEA (Fig. 36–7). This action of cytochrome P450c17 is essential for the formation of androgens (19 carbon steroids) and estrogens (18 carbon steroids), which lack the carbon 20-21 side chain. Hence this **lyase** activity of cytochrome P450c17 is very important in the gonads, where androgens and estrogens are pri-

marily made. Cytochrome P450c17 does not exert significant lyase activity in children before the age of 7–8 years. As a result, young boys and girls do not secrete significant amounts of adrenal androgens. The appearance of significant adrenal androgen secretion in children of both sexes is termed **adrenarche.** It is not related to the onset of puberty, since it normally occurs before the activation of the hypothalamopituitary-gonad axis, which initiates puberty. The adrenal androgens produced as a result of adrenarche are a stimulus for the growth of pubic and axillary hair.

Those molecules of 17α-hydroxypregnenolone that dissociate as such from cytochrome P450c17 bind next to another enzyme in the endoplasmic reticulum, **3β-hydroxysteroid dehydrogenase.** This enzyme acts on 17α-hydroxypregnenolone to isomerize the double bond in ring B to ring A and to dehydrogenate the 3β-hydroxy group, thus forming a 3-keto group. The product formed is **17α-hydroxyprogesterone** (Fig. 36–7). This intermediate then binds to the enzyme **cytochrome P450c21 (21-hydroxylase cytochrome P450),** which hydroxylates it at carbon 21. The mechanism of this hydroxylation is similar to that performed by cytochrome P450c17. The product formed is **11-deoxycortisol,** which is the immediate precursor for cortisol.

To be converted to cortisol, 11-deoxycortisol molecules must be transferred back into the mitochondria to be acted on by **cytochrome P450c11 (11β-hydroxylase cytochrome P450),** embedded in the inner mitochondrial membrane. This enzyme hydroxylates 11-deoxycortisol on carbon 11, thereby converting it into cortisol. The 11β-hydroxyl group is the molecular feature that confers glucocorticoid activity on the steroid. Cortisol is then secreted into the bloodstream.

As is evident in Figure 36–7, some of the pregnenolone molecules generated in cells of the zona fasciculata and zona reticularis first bind to 3β-hydroxysteroid dehydrogenase when they enter the endoplasmic reticulum. As a result, they are converted to **progesterone.** Some of these progesterone molecules are hydroxylated by cytochrome P450c21 to form the mineralocorticoid **11-deoxycorticosterone** (DOC) (Fig. 36–7). The 11-deoxycorticosterone formed may be either secreted or transferred back into the mitochondrion. There it is acted on by cytochrome P450c11 to form corticosterone, which is then secreted into the circulation.

Progesterone may also undergo 17α-hydroxylation in the zona fasciculata and zona reticularis. It is then converted either to cortisol or to the adrenal androgen **androstenedione.**

Cytochrome P450c17 is not present in cells of the zona glomerulosa, therefore pregnenolone does not undergo 17α-hydroxylation in these cells and cortisol and adrenal androgens are not formed by these cells. Instead, the enzymatic pathway leading to the formation of aldosterone shown in Figure 36–8 is followed.

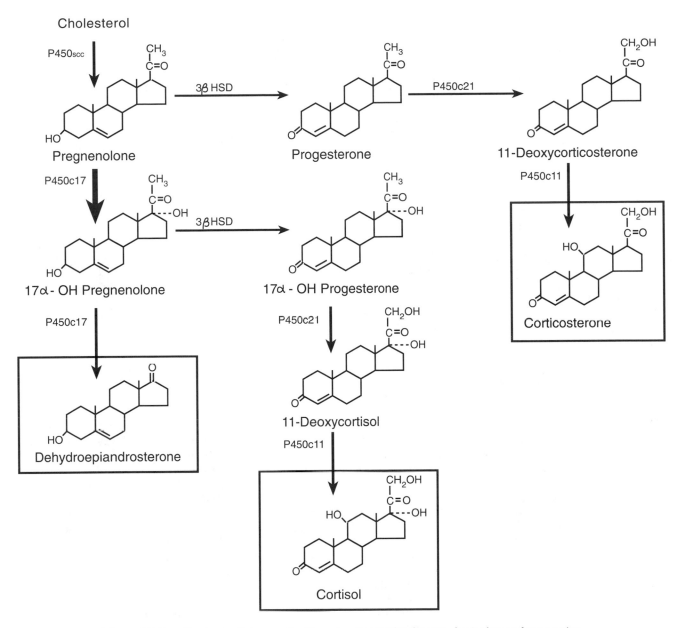

Figure 36–7 ■ Synthesis of glucocorticoids and androgens by the zona fasciculata and zona reticularis of the adrenal cortex.

Pregnenolone is converted to 11-deoxycorticosterone by enzymes in the endoplasmic reticulum. This biosynthetic intermediate then moves into the mitochondrion, where it is converted to aldosterone. This conversion involves three steps: the hydroxylation of carbon 11 to form corticosterone, the hydroxylation of carbon 18 to form 18-hydroxycorticosterone, and the oxidation of the 18-hydroxymethyl group to form aldosterone. In humans, these three reactions are catalyzed by a single enzyme, **aldosterone synthase,** an isozyme of cytochrome P450c11 expressed only in glomerulosa cells. The cytochrome P450c11 enzyme, which is expressed in the zona fasciculata and zona reticularis, although closely related to aldosterone synthase, cannot catalyze all three reactions involved in the conversion of 11-deoxycorticosterone to aldosterone. Hence, aldosterone is not synthesized in the zona fasciculata and zona reticularis of the adrenal cortex.

Genetic Defects in Adrenal Steroidogenesis ■ Inherited genetic defects can cause relative or absolute deficiencies in the enzymes involved in the steroid hormone biosynthetic pathways. The immediate consequences of these defects are changes in the types and amounts of steroid hormones secreted by the adrenal cortex. The end result is disease.

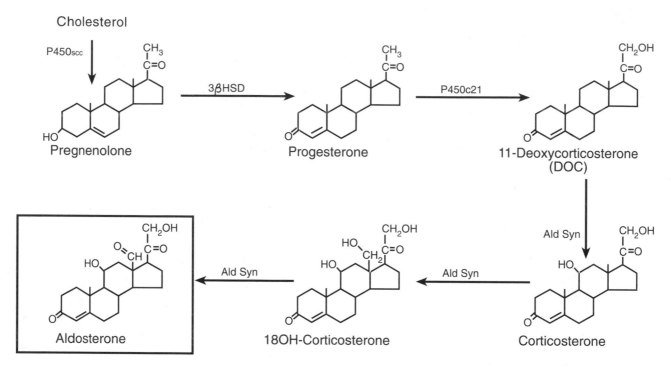

Figure 36–8 ■ Synthesis of aldosterone by the zona glomerulosa of the adrenal cortex. Ald Syn, aldosterone synthase.

Most of the genetic defects affecting the steroidogenic enzymes impair the formation of cortisol. As discussed in Chapter 34, a drop in cortisol concentration in the blood is the stimulus for the secretion of adrenocorticotrophic hormone (ACTH) by the anterior pituitary. The consequent rise in ACTH in the blood exerts a trophic or growth-promoting effect on the adrenal cortex, resulting in adrenal hypertrophy. Because of this mechanism, individuals with genetic defects affecting adrenal steroidogenesis usually have hypertrophied adrenal glands. These diseases are collectively called **congenital adrenal hyperplasia.**

In humans, inherited genetic defects occur that affect cytochrome P450scc, cytochrome P450c17, 3β-hydroxysteroid dehydrogenase, cytochrome P450c21, cytochrome P450c11, and aldosterone synthase. The most common defect involves mutations in the gene for cytochrome P450c21 and occurs in 1 of every 7000 people. The gene for cytochrome P450c21 may be deleted entirely or mutant genes may code for forms of cytochrome P450c21 with impaired enzyme activity. The consequent reduction in the amount of active cytochrome P450c21 in the adrenal cortex interferes with the formation of cortisol, corticosterone, and aldosterone, all of which are hydroxylated at carbon 21. Because of the reduction of cortisol (and corticosterone) secretion in these individuals, ACTH secretion is stimulated. This, in turn, causes hypertrophy of the adrenal glands and stimulates the glands to produce steroids. Since 21-hydroxylation is impaired by the cytochrome P450c21 deficiency, the ACTH

stimulus causes pregnenolone to be converted to adrenal androgens in inappropriately high amounts. Thus, women afflicted with cytochrome P450c21 deficiency exhibit **virilization** due to the masculinizing effects of excessive adrenal androgen secretion. In severe forms of this disease, the deficiency in aldosterone production can lead to body sodium depletion, dehydration, vascular collapse, and death, if appropriate hormone therapy is not given.

Addison's Disease ■ Glucocorticoid and aldosterone deficiency also occur as a result of pathologic destruction of the adrenal glands by microorganisms or autoimmune disease. This disorder is called **Addison's disease.** If much adrenal cortical tissue is lost, the resulting decrease in aldosterone production can lead to vascular collapse and death, unless hormone therapy is given.

Transport of Steroid Hormones in Blood ■ As noted earlier, steroid hormones are not stored to any extent by cells of the adrenal cortex but are continually synthesized and secreted. The rate of secretion may change dramatically, however, depending on stimuli received by the adrenal cortical cells. The process by which steroid hormones are secreted is not well studied. It has been assumed that the accumulation of the final products of the steroidogenic pathways creates a concentration gradient for steroid hormone between cells and blood. This gradient is thought to be the driving force for diffusion of the lipid-soluble steroids through cellular membranes and into the circulation.

GENETIC DEFECTS IN THE ISOZYMES OF CYTOCHROME P450c11

Mutations in the genes coding for the enzymes of the steroidogenic pathway can produce changes in the amounts and types of steroid hormones secreted by the adrenal cortex. The disturbances in hormone balance produced by these mutations are responsible for a number of endocrine disorders. In humans, the most frequent mutations occur in the gene for cytochrome P450c21, resulting in the common form of congenital adrenal hyperplasia. Because of a relative or absolute lack of P450c21 activity, such individuals produce little or no cortisol, corticosterone, or aldosterone and instead make excessive amounts of adrenal androgens. A much less common form of congenital adrenal hyperplasia is caused by mutation of the gene for cytochrome P450c11 in which the adrenal cortex is incapable of producing normal amounts of cortisol and corticosterone and instead produces excessive amounts of adrenal androgens and various biosynthetic intermediates, such as 11-deoxycorticosterone. 11-Deoxycorticosterone has substantial mineralocorticoid activity and consequently produces hypertension and potassium loss in many individuals with P450c11 deficiency. Mutations in cytochrome P450c11 do not produce aldosterone deficiency, because this enzyme is expressed only in the zona fasciculata and zona reticularis of the adrenal cortex, which lack the complete set of enzymatic activities necessary to convert 11-deoxycorticosterone to aldosterone. A closely related isozyme of cytochrome P450c11, aldosterone synthase, is expressed in the zona glomerulosa and is responsible for catalyzing the conversion of 11-deoxycorticosterone to aldosterone. In humans, cytochrome P450c11 and aldosterone synthase are encoded by two genes that reside in the same region of chromosome 8. A recent discovery involving these two genes provides an explanation at the molecular level for an endocrine disorder called **glucocorticoid-suppressible hyperaldosteronism.** Individuals with this disease produce excessive amounts of aldosterone and are hypertensive. If they are given glucocorticoid treatment, which suppresses ACTH secretion, their blood level of aldosterone falls and blood pressure returns toward normal. It was suspected that they inappropriately produce aldosterone in the zona fasciculata and zona reticularis of the adrenal cortex and that this aldosterone production is somehow regulated by ACTH. It is now known (see White and Pascoe in Suggested Reading) that the genes for cytochrome P450c11 and aldosterone synthase can undergo an intergenic recombination to produce a hybrid gene that is expressed in the zona fasciculata and zona reticularis. The hybrid gene consists of the 5' end of the cytochrome P450c11 gene, which includes the regulatory sequences that respond to ACTH stimulation, and the 3' end of the aldosterone synthase gene, which contains the sequence for its enzymatic activity. Thus, the zona fasciculata and zona reticularis cells of individuals bearing this hybrid gene express the mutant enzyme in response to stimulation by ACTH and can therefore produce and secrete substantial amounts of aldosterone.

A large fraction of the steroid molecules that enter the bloodstream become bound noncovalently to certain plasma proteins. One of these is **corticosteroid binding globulin** (CBG), a glycoprotein produced by the liver. It binds both glucocorticoids and aldosterone but has a greater affinity for the glucocorticoids. **Serum albumin** also binds adrenal steroids to some extent. The binding of a steroid hormone to a circulating protein molecule prevents it from being taken up by cells or being excreted in the urine. Circulating steroid hormone molecules not bound to plasma proteins are free to interact with receptors on cells and hence are cleared from the blood. As this occurs, bound hormone dissociates from its binding protein and replenishes the circulating pool of free hormone. Because of this process, adrenal steroid hormones have long half-lives in the body, ranging from many minutes to hours.

Metabolism of Adrenal Steroids in the Liver ■ Adrenal steroid hormones are eliminated from the body primarily by excretion in the urine after they have been structurally modified to destroy their hormone activity and increase their water solubility. Although many cells are capable of carrying out these modifications, this occurs primarily in the liver.

The most common structural modifications made in the adrenal steroids involve reduction of the double bond in ring A and conjugation of the resultant hydroxyl group formed on carbon 3 with glucuronic acid. Figure 36–9 shows how cortisol is modified in this manner to produce a major excretable metabolite, **tetrahydrocortisol glucuronide.** Also, 21-carbon steroids, which have a 17α-hydroxyl group and a 20-keto group, such as cortisol, may undergo lysis of the carbon 20-21 side chain as well. The resultant metabolite, with a keto group on carbon 17, appears

as one of the **17-ketosteroids** in the urine. Adrenal androgens are also 17-ketosteroids. They are usually conjugated with sulfuric acid or glucuronic acid before being excreted and normally comprise the bulk of the 17-ketosteroids in the urine. Before the development of specific methods to measure androgens and 17α-hydroxycorticosteroids in body fluids, the amount of 17-ketosteroids in urine was used clinically as a crude indicator of the production of these substances by the adrenal gland.

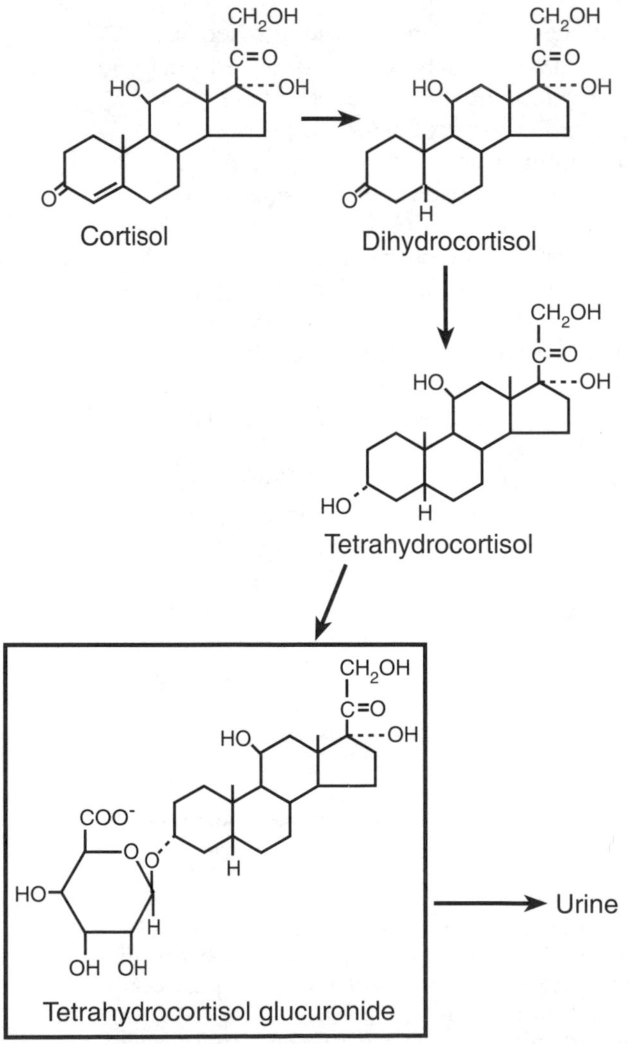

■ ACTH Regulates Synthesis of Adrenal Steroids

Adrenocorticotrophic hormone is the physiologic regulator of the synthesis and secretion of glucocorticoids and androgens by the zona fasciculata and zona reticularis. It has a very rapid stimulatory effect on steroidogenesis in these cells, which can result in a great rise in blood glucocorticoids within seconds. It also exerts several long-term trophic effects on these cells, all directed toward maintaining the cellular machinery necessary to carry out steroidogenesis at a high, sustained rate. These actions of ACTH are summarized in Figure 36–10.

Role of cAMP ■ When the level of ACTH in the blood rises, increased numbers of ACTH molecules interact with receptors on the plasma membranes of adrenal cortical cells. These ACTH receptors are coupled to the enzyme adenylate cyclase by stimulatory guanine nucleotide-binding proteins (G_s proteins). The binding of ACTH to its receptor stimulates adenylate cyclase. In turn, the production of cAMP from ATP greatly increases, and the concentration of cAMP rises in the cells. cAMP activates protein kinase A, which then phosphorylates proteins that regulate steroidogenesis.

The rapid rise in cAMP produced by ACTH stimulates the mechanism that transfers cholesterol into the inner mitochondrial membrane. This provides abundant cholesterol for cytochrome P450scc, which carries out the rate-limiting step in steroidogenesis. As a result, the rate of steroid hormone formation and secretion rises greatly.

Expression of Genes for Steroidogenic Enzymes ■ Adrenocorticotrophic hormone maintains the capacity of the cells of the zona fasciculata and zona reticularis to produce steroid hormones by stimulating the expression of the genes for many of the enzymes involved in steroidogenesis. For example, the genes for cytochrome P450scc, cytochrome P450c17, cytochrome P450c21, cytochrome P450c11, adrenodoxin reductase, adrenodoxin, and NADPH cytochrome P450 reductase are all transcribed at a greater rate several hours after adrenal cortical cells have been stimulated by ACTH. Since normal individuals are continually exposed to episodes of ACTH secretion (Fig. 34–7), the mRNA for these enzymes is well maintained in the cells. Again, this long-term or maintenance effect of ACTH is due to its ability to increase cAMP in the cells (Fig. 36–10).

Figure 36–9 ■ Metabolism of cortisol to tetrahydrocortisol glucuronide in the liver. The reduced and conjugated steroid is inactive. Being more water-soluble than cortisol, it is easily excreted in the urine.

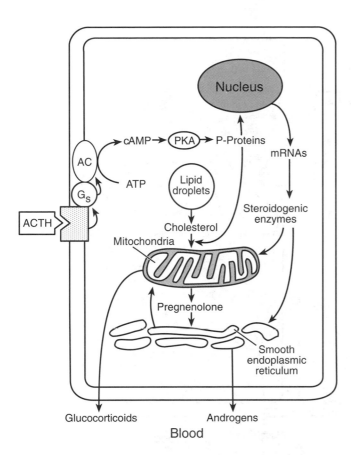

Figure 36–10 ■ The main actions of ACTH on steroidogenesis. ACTH binds to plasma membrane receptors, which are coupled to adenylate cyclase (AC) by stimulatory G proteins (G_s). cAMP rises in the cells and activates protein kinase A (PKA), which then phosphorylates certain proteins (P-Proteins). These proteins presumably initiate steroidogenesis and stimulate the expression of genes for steroidogenic enzymes.

The importance of ACTH in gene transcription becomes evident in hypophysectomized animals or humans with ACTH deficiency. An example of the latter is a human treated chronically with large doses of cortisol or related steroids, which causes prolonged suppression of ACTH secretion by the anterior pituitary. The chronic lack of ACTH decreases the expression of the genes for steroidogenic enzymes, causing a deficiency in these enzymes in the adrenals. As a result, administration of ACTH to such an individual does not cause a marked increase in glucocorticoid secretion. Chronic exposure to ACTH is required to restore mRNA levels for the steroidogenic enzymes, and hence the enzymes themselves, to obtain normal steroidogenic responses to ACTH. Thus, a patient receiving long-term treatment with glucocorticoid may suffer serious glucocorticoid deficiency if hormone therapy is halted abruptly. Withdrawing glucocorticoid therapy gradually allows time for endogenous ACTH to restore steroidogenic enzyme levels to normal.

Effects on Cholesterol Metabolism ■ Adreno-corticotrophic hormone has several long-term effects on cholesterol metabolism that support steroidogenesis in the zona fasciculata and zona reticularis. It increases the abundance of LDL receptors and the activity of the enzyme HMG-CoA reductase in these cells. These actions increase the availability of cholesterol for steroidogenesis. It is not clear whether ACTH exerts these effects directly. The abundance of LDL receptors in the plasma membrane and the activity of HMG-CoA reductase in most cells is inversely related to the amount of cholesterol in the cells. By stimulating steroidogenesis, ACTH reduces the amount of cholesterol in adrenal cells. Thus, the increased abundance of LDL receptors and high HMG-CoA reductase activity in ACTH-stimulated cells may merely result from the normal compensatory mechanisms that function to maintain cell cholesterol levels.

Also, ACTH stimulates the activity of cholesterol esterase in adrenal cells. This promotes the hydrolysis of the cholesterol esters stored in the lipid droplets of these cells, making free cholesterol available for steroidogenesis. As noted earlier, the cholesterol esterase in the adrenal cortex appears to be identical to hormone-sensitive lipase, which is activated when it is phosphorylated by a cAMP-dependent protein kinase. Thus, the rise in cAMP concentration produced by ACTH might account for its effect on the enzyme.

Trophic Action on Adrenal Cortical Cell Size ■ Adrenocorticotrophic hormone maintains the size of the two inner zones of the adrenal cortex, presumably by stimulating the synthesis of structural elements of the cells. On the other hand, it does not affect the size of the cells of the zona glomerulosa. This trophic effect of ACTH is clearly evident in states of ACTH deficiency or excess. In hypophysectomized or ACTH-deficient individuals, the cells of the two inner zones atrophy. Chronic stimulation of these cells with ACTH causes them to hypertrophy. The mechanisms involved in this trophic action of ACTH are undefined.

ACTH and Aldosterone Production ■ The cells of the zona glomerulosa have ACTH receptors, which are coupled to adenylate cyclase. cAMP increases in these cells in response to ACTH, resulting in some increase in aldosterone secretion by these cells, but ACTH is not the important physiologic regulator of aldosterone secretion.

■ Aldosterone Secretion Is Regulated by Angiotensin II

Angiotensin II is the primary physiologic regulator of aldosterone synthesis and secretion by the zona glomerulosa. Other factors, such as an increase in serum potassium or ACTH, can also stimulate aldosterone secretion, but normally they play only a secondary role.

Angiotensin II Formation ■ Angiotensin II is a short peptide consisting of eight amino acid residues. It is formed in the bloodstream by the proteolysis of the α_2-globulin **angiotensinogen,** which is secreted by the liver. The formation of angiotensin II occurs in two stages (Fig. 36–11). Angiotensinogen is first cleaved at its N-terminal end by the circulating protease **renin,** releasing the inactive decapeptide **angiotensin I.** Renin is produced and secreted by cells of the granular juxtaglomerular apparatus in the kidneys (Chap. 23). A dipeptide is then removed from the C-terminal end of angiotensin I, producing angiotensin II. This cleavage is performed by the protease **converting enzyme,** present on the enothelial cells lining the vasculature. This step usually occurs as angiotensin I molecules traverse the pulmonary circulation. The rate-limiting factor for the formation of angiotensin II is the renin concentration of the blood.

Action of Angiotensin II on Aldosterone Secretion ■ Angiotensin II stimulates aldosterone synthesis by promoting the rate-limiting step in steroidogenesis (i.e., the movement of cholesterol into the inner mitochondrial membrane and its conversion to pregnenolone.) The primary mechanism believed to be involved is shown in Figure 36–12.

The stimulation of aldosterone synthesis is initiated when angiotensin II binds to its receptors on the plasma membranes of zona glomerulosa cells. The signal generated by the interaction of angiotensin II with its receptors is transmitted to phospholipase C (PLC) by a G protein, and the enzyme becomes activated. The PLC then hydrolyzes phosphatidylinositol 4,5 bisphosphate (PIP$_2$) in the plasma membrane, producing the intracellular second messengers inositol trisphosphate (IP$_3$) and diacylglycerol (DAG). The IP$_3$ mobilizes calcium, which is bound to intracellular structures, thereby increasing the calcium concentration in the cytosol. This increase in intracellular calcium and DAG activates protein kinase C. The rise in intracellular calcium also activates **calmodulin-dependent protein kinase.** These enzymes phosphorylate proteins, which then become involved in initiating steroidogenesis.

Signals for Increased Angiotensin II Formation ■ Although angiotensin II is the final mediator

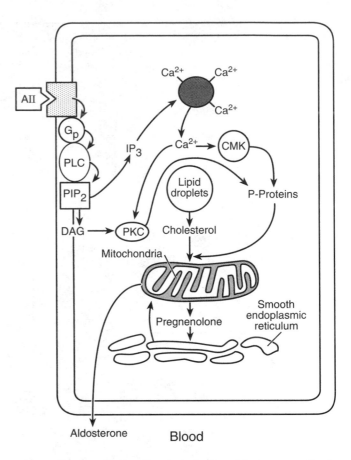

Figure 36–12 ■ Action of angiotensin II on aldosterone synthesis. Angiotensin II (AII) binds to receptors on plasma membranes of zona glomerulosa cells. This activates phospholipase C (PLC), which is coupled to the angiotensin II receptor by G proteins (G$_p$). PLC hydrolyzes phosphatidylinositol 4,5 bisphosphate (PIP$_2$) in the plasma membrane, producing inositol trisphosphate (IP$_3$) and diacylglycerol (DAG). IP$_3$ mobilizes intracellularly bound Ca^{2+}. The rise in Ca^{2+} and DAG activates protein kinase C (PKC) and calmodulin-dependent protein kinase (CMK). These enzymes phosphorylate proteins (P-Proteins) involved in initiating aldosterone synthesis.

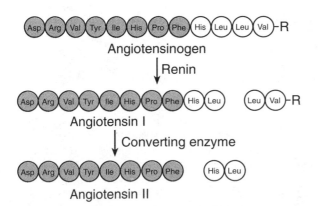

Figure 36–11 ■ The formation of angiotensin II from angiotensinogen.

in the physiologic regulation of aldosterone secretion, its formation from angiotensinogen is dependent on the secretion of renin by the kidneys. Hence, the rate of renin secretion ultimately determines the rate of aldosterone secretion. As noted earlier, renin is secreted by the granular cells in the walls of the afferent arterioles of renal glomeruli. These cells are stimulated to secrete renin by three signals that indicate a possible loss of body fluid: a fall in blood pressure in the afferent arterioles of the glomeruli, a drop in sodium in renal tubular fluid at the macula densa, and an increase in renal sympathetic nerve activity (Chap. 23, 24). The increase in renin secretion results in an increase in angiotensin II formation in the blood. This, in turn, stimulates aldosterone secretion by the zona glomerulosa. This series of events tends to conserve body fluid volume, since aldosterone stimulates sodium reabsorption by the kidneys.

Extracellular Potassium Concentration and Aldosterone Secretion ■ Aldosterone secretion is also stimulated by an increase in the potassium concentration in extracellular fluid. This is due to a direct effect of potassium on the glomerulosa cells. The glomerulosa cells are very sensitive to this effect of extracellular potassium and increase their rate of aldosterone secretion in response to very small increases in blood and interstitial fluid potassium concentration. This signal for aldosterone secretion is appropriate from a physiologic point of view, since aldosterone promotes renal excretion of potassium (Chap. 24).

A rise in extracellular potassium depolarizes glomerulosa cell membranes. This, in turn, activates voltage-dependent calcium channels in the membranes. The consequent rise in cytosolic calcium is thought to stimulate aldosterone synthesis by the mechanisms described above for the action of angiotensin II.

Aldosterone and Sodium Reabsorption by Kidney Tubules ■ The physiologic action of aldosterone is to stimulate sodium reabsorption by the distal tubule and collecting duct of the nephron and to promote the excretion of potassium and hydrogen ions. The mechanism of action of aldosterone on the kidney and its role in water and electrolyte balance are discussed in Chapter 24.

■ Glucocorticoids Play a Role in the Reactions to Fasting, Injury, and Stress

Glucocorticoids widely influence physiologic processes. In fact, most cells have receptors for glucocortoids and hence are potential targets for their actions. As a consequence, glucocorticoids have been used extensively as therapeutic agents, and much is known about their pharmacologic effects.

Actions on Gene Expression ■ Unlike many other hormones, glucocorticoids influence physiologic processes slowly, sometimes taking hours to produce their effects.

Glucocorticoids, which are free in the blood, diffuse through the plasma membranes of target cells; once inside, they bind tightly but noncovalently to receptor proteins. The interaction between the glucocorticoid molecule and its receptor molecule produces an activated glucocorticoid-receptor complex. These complexes bind to specific regions of genomic DNA called **steroid response elements,** which are near glucocorticoid-sensitive target genes. The binding triggers events that either stimulate or inhibit the transcription of the target gene. As a result of the change in gene transcription, amounts of mRNA for certain proteins are either increased or decreased. This, in turn, affects the abundance of these proteins in the cell, which produces the physiologic effects of the glucocorticoids. The apparent slowness of glucocorticoid action is due to the time required by the mechanism to change the protein composition of a target cell.

Glucocorticoids and the Metabolic Response to Fasting ■ Most humans have a daily feeding-fasting cycle. There are two or three periods during the day when metabolic substrate supply is abundant, but for much of the time in a given day the individual is fasting. During these fasting periods, metabolic adaptations prevent hypoglycemia. These metabolic adaptations are not fully expressed in the course of daily life, since the individual feeds before the metabolic changes fully develop. Full expression of these changes is seen only after many days to weeks of fasting.

At the onset of a prolonged fast, there is a gradual decline in the concentration of glucose in the blood. Within 1–2 days, the blood glucose level stabilizes at a concentration of 60–70 mg/dL, where it remains even if the fast is prolonged for many days (Fig. 36–13). Maintenance of the blood glucose concentration at this level assures the brain of an adequate supply of glucose for ATP production, until the blood level of ketone bodies (acetoacetate, β-hydroxybutyrate) rises sufficiently to provide an alternative energy source. The brain is capable of using ketone bodies for a portion of its energy metabolism (but it must still metabolize a certain amount of glucose). The level of ketones in the blood during the first few days of a fast is too low to meet even a small portion of the brain's energy needs. The blood glucose level is stabilized through production of glucose by the body and restriction of its use by tissues other than the brain.

A limited supply of glucose is available from glycogen stored in the liver. During the first days of a fast, liver glycogen is mobilized by the action of **glycogen phosphorylase,** and the glucose formed is delivered to the bloodstream. In humans, the liver glycogen stores are never depleted, however, and some glycogen (10–30% of that present initially) remains in the liver for the duration of the fast. The more important source of blood glucose during the first days of a fast

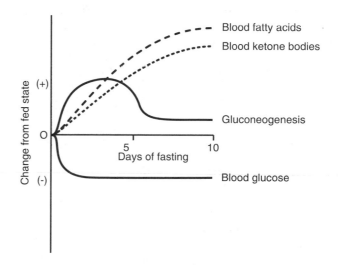

Figure 36–13 ■ Changes in the concentrations of blood glucose, fatty acids, and ketone bodies and the rate of gluconeogenesis during the course of a prolonged fast. Only the *direction* of change—increase (+) or decrease (−)—over time is indicated.

is gluconeogenesis in the liver and, to some extent, in the kidneys.

Gluconeogenesis begins several hours after the start of a fast. Amino acids derived from tissue protein are the main substrates. Fasting results in net protein breakdown in the peripheral tissues of the body and accelerated release of amino acids into the bloodstream. The precise events that trigger breakdown of tissue protein are unknown. Protein breakdown and protein accretion in adult humans are regulated by at least two opposing hormones, insulin and glucocorticoids:

$$\text{Protein} \underset{\text{Glucocorticoids}(-)}{\overset{\text{Insulin}(-)}{\rightleftharpoons}} \text{Amino acids}$$

During fasting, insulin secretion is suppressed and the inhibitory effect of insulin on protein breakdown is lost. Also, glucocorticoids are known to inhibit the synthesis of many proteins in peripheral tissues. This effect of glucocorticoids persists during fasting and tends to inhibit the reuse of amino acids derived from tissue proteins for new protein synthesis. Releasing the brake (insulin) on protein breakdown while maintaining the brake (glucocorticoids) on protein accretion favors net breakdown of proteins to their constituent amino acids.

During fasting, amino acids released into the blood by the peripheral tissues are extracted from the blood at an accelerated rate by the liver and kidney. The amino acids then undergo metabolic transformations in these tissues, leading to the synthesis of glucose. In the case of the amino acid alanine, for example, the first step is transamination of the amino acid to form pyruvate, which enters the mitochondria and is carboxylated to oxaloacetate by the action of pyruvate carboxylase. The enzymes that convert the oxaloacetate to glucose are in the cytosol. Since oxaloacetate cannot pass through the mitochondrial membrane, it is converted to malate and aspartate, which pass through the mitochondrial membrane and into the cytosol. There they undergo reconversion to oxaloacetate. This oxaloacetate then serves as substrate for phosphoenolpyruvate carboxykinase, resulting in the production of phosphoenolpyruvate. The phosphoenolpyruvate undergoes metabolic transformations (i.e., reversal of glycolysis), which results in its synthesis into glucose. This glucose is delivered to the bloodstream.

The glucocorticoids are essential for the acceleration of gluconeogenesis during fasting. They play a **permissive** role in this process by maintaining the expression of the genes, and therefore the intracellular concentrations, of many of the enzymes needed to carry out gluconeogenesis in the liver and kidneys. For example, glucocorticoids maintain the amounts of transaminases, pyruvate carboxylase, phosphoenolpyruvate carboxykinase, fructose-1,6-diphosphatase, fructose-6-phosphatase, and glucose-6-phosphatase needed to carry out gluconeogenesis at an accelerated rate. In an untreated, glucocorticoid-deficient individual, the amounts of these enzymes in the liver are greatly reduced. As a consequence, the individual cannot respond to fasting with accelerated gluconeogenesis and will die from hypoglycemia. In essence, the glucocorticoids maintain the liver and kidney in a state that makes them capable of carrying out accelerated gluconeogenesis should the need arise.

Gluconeogenesis from body protein is accelerated only for the first few days of a fast (Fig. 36–13). Thereafter, the rate of gluconeogenesis from protein declines gradually to a minimum level for the remainder of the fast. The burst of gluconeogenesis from protein early in the fast provides the glucose required by the brain when the level of ketone bodies in the blood is too low to provide enough substrate for the energy metabolism of the brain. The later reduction in gluconeogenesis from protein appears to be designed to spare the structural elements of cells from destruction and thus to prolong survival.

The other important metabolic adaptations that occur during fasting involve the mobilization and use of stored fat. Within the first few hours of the start of a fast, the concentration of free fatty acids (complexed to albumin) rises in the blood (Fig. 36–13). This is due to the acceleration of lipolysis in the fat depots, as a result of the activation of hormone-sensitive lipase (HSL):

$$\text{Triglyceride} \xrightarrow{\text{HSL}} \text{Fatty acids} + \text{Glycerol}$$

Hormone-sensitive lipase is activated when it is phosphorylated by a cAMP-dependent protein kinase. As the level of insulin falls in the blood during fasting, the inhibitory effect of insulin on cAMP accumulation in the fat cell diminishes. There is a rise in the cellular level of cAMP, and HSL is activated. The glucocorticoids are essential for maintaining fat cells in an enzymatic state that permits lipolysis to occur during a fast. This is evident from the fact that accelerated lipolysis does not occur when a glucocorticoid-deficient individual is fasted.

The abundant fatty acids produced by lipolysis are taken up by many tissues. These fatty acids enter the mitochondria, undergo β-oxidation to acetyl-CoA, and then become the substrate for ATP synthesis. The enhanced use of fatty acids for energy metabolism by the peripheral tissues spares the blood glucose supply. There is also significant gluconeogenesis from the glycerol released from triglyceride by lipolysis. In prolonged fasting, when the rate of glucose production from body protein has declined, a significant fraction of blood glucose is derived from triglyceride glycerol.

Within a few hours of the start of a fast, the ability of liver mitochondria to take up fatty acids is increased. This is thought to be due to the rise in the glucagon-to-insulin ratio in the blood, which stimulates the activity of the carnitine acyltransferase system in the membranes of liver mitochrondria. This system transfers fatty acids across the mitochondrial membranes into the mitochondrial matrix, where β-oxidation occurs. This increase in fatty acid uptake by liver mitochondria, and the subsequent β-oxidation of the acetyl-CoA moieties, results in the production of substantial amounts of acetyl-CoA. Some of this is funneled into the enzymatic pathway leading to the production of the ketone bodies. As a result of these events in the liver, there is a gradual rise in ketone bodies in the blood as a fast continues over many days (Fig. 36–13). This provides the ketone bodies used by the CNS for energy metabolism during the later stages of fasting.

The increased use of fatty acids for energy metabolism by peripheral tissues also leads to the inhibition of phosphorylation and hence use of glucose in these cells. The mechanisms thought to be involved are shown in Figures 36–14 and 36–15. The increased oxidation of fatty acids by the mitochondria of these cells leads to inhibition of pyruvate dehydrogenase, which reduces the decarboxylation of pyruvate to acetyl-CoA and slows the flow of glucose through the glycolytic pathway (Fig. 36–14). Also, because of the high production of acetyl-CoA from the oxidation of fatty acids, substantial amounts of citrate are produced in the mitochondrion through the condensation of acetyl-CoA with oxaloacetate (Fig. 36–15). Citrate diffuses out of the mitochondria into the cytosol, where it accumulates. The high cytosolic citrate level inhibits the phosphofructokinase reaction, believed to be the rate-limiting step in glycolysis. Inhibition of this enzyme causes fructose-6-phosphate and glucose-6-phosphate to accumulate in the cells. This accumulation impairs the cell's ability to phosphorylate additional glucose. As a result, the uptake and use of glucose from the blood is greatly reduced.

In summary, the strategy behind the metabolic adaptation to fasting is to provide the body with glucose produced primarily from protein until the ketone bodies become abundant enough in the blood to be a principal source of energy for the brain. From that point on, the body uses mainly fat for energy metabolism, and it can survive until the fat depots are exhausted.

In a normal individual, the glucocorticoids do not trigger these metabolic adaptations to fasting but only provide the metabolic machinery necessary for the adaptations to occur. When present in excessive amounts, however, glucocorticoids alone can cause these changes. **Cushing's disease** is an example of such a pathologic **hypercortisolic state.** Cushing's disease is usually caused by a corticotroph adenoma, which secretes excessive ACTH and hence stimulates

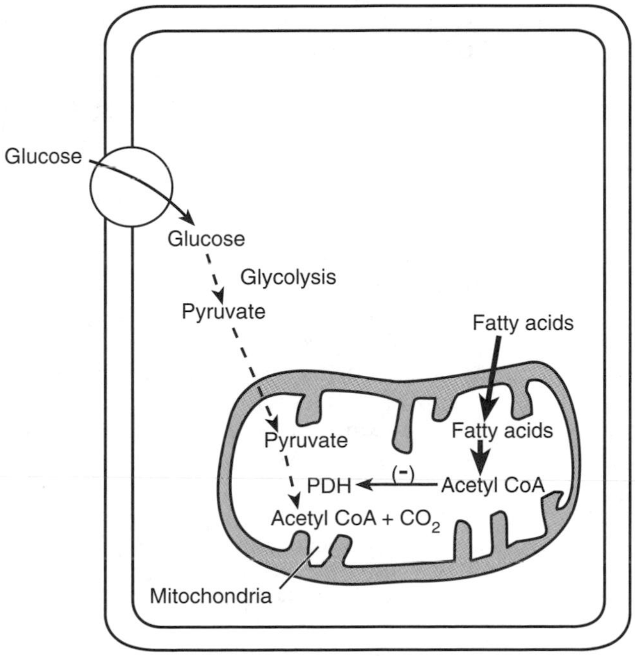

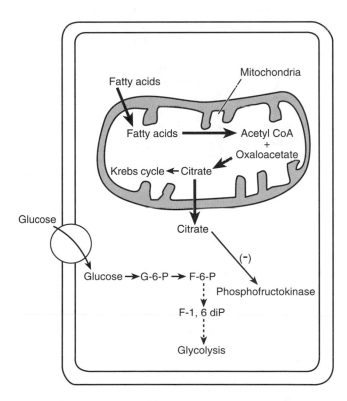

Figure 36–14 ■ Proposed mechanism for the slowing of glucose flow through the glycolytic pathway in peripheral tissues caused by the inhibition of the pyruvate dehydrogenase reaction (PDH) due to increased fatty acid oxidation.

Figure 36–15 ■ Proposed mechanism for the slowing of glycolysis in peripheral tissues caused by the inhibition of the phosphofructokinase reaction due to increased fatty acid oxidation. G-6-P, glucose-6-phosphate; F-6-P, fructose-6-phosphate; F-1,6-diP, fructose-1, 6-diphosphate.

the adrenal cortex to produce large amounts of cortisol. Prolonged exposure of the body to large amounts of glucocorticoids causes breakdown of peripheral tissue protein, increased glucose production by the liver, and mobilization of lipid from the fat depots. (In Cushing's disease, there is also an abnormal deposition of fat on the abdomen, between the shoulders, and in the face.) The increased mobilization of lipid provides abundant fatty acids for metabolism by peripheral tissues. The increased oxidation of these fatty acids by peripheral tissues reduces their ability to phosphorylate and hence use glucose (Fig. 36–15). The underuse of glucose coupled with increased glucose production by the liver results in hyperglycemia, which, in turn, stimulates the pancreas to secrete insulin. In this instance, however, the rise in insulin is not very effective in reducing the blood glucose concentration, because glucose uptake and use are decreased in the peripheral tissues. There is also evidence that excessive glucocorticoids decrease the affinity of insulin receptors for insulin. The net result is that the individual becomes insensitive or resistant to the action of insulin and little glucose is removed from the blood, despite the high level of circulating insulin. The persisting hyperglycemia continually stimulates the pancreas to secrete insulin. The result is a form of "diabetes" similar in many respects to type II diabetes mellitus (Chap. 37).

The opposite situation occurs in the glucocorticoid-deficient individual. Very little lipid mobilization and use occur, so there is little restriction on the rate of glucose use by peripheral tissues. The glucocorticoid-deficient individual is very sensitive to insulin in the sense that a given concentration of blood insulin is more effective in clearing the blood of glucose than it is in a normal person. Consequently, the administration of even small doses of insulin to such individuals may produce hypoglycemia.

Antiinflammatory Action of Glucocorticoids ∎ Tissue injury triggers a complex mechanism called **inflammation** that precedes the actual repair of damaged tissue. A host of chemical mediators are released into the damaged area by neighboring cells, adjacent vasculature, and phagocytic cells that migrate to the damaged site. Mediators known to be released under these circumstances include **prostaglandins, leukotrienes, kinins, histamine, serotonin,** and **lymphokines.** These substances exert a multitude of actions at the site of injury and either directly or indirectly promote the local vasodilation, increased capillary permeability, and edema formation that characterize the inflammatory response (Chap. 11).

Since glucocorticoids inhibit the inflammatory response to injury, they have been used extensively as therapeutic antiinflammatory agents. The mechanisms by which glucocorticoids inhibit the inflammatory response are not well defined. Best understood is their regulation of the production of prostaglandins and leukotrienes. These substances play a major role in

mediating the inflammatory reaction. They are synthesized from the unsaturated fatty acid arachidonic acid, which is released from plasma membrane phospholipids by the hydrolytic action of **phospholipase A$_2$.** Glucocorticoids stimulate the synthesis of the protein **macrocortin** in their target cells. Macrocortin inhibits the activity of phospholipase A$_2$, thereby reducing the amount of arachidonic acid available for conversion to prostaglandins and leukotrienes.

Effects on the Immune System ∎ The glucocorticoids have little influence on the human immune system under physiologic conditions. When administered in very large doses over a prolonged period, however, they can suppress antibody formation and interfere with cell-mediated immunity. As a consequence, glucocorticoid therapy has been used to suppress the rejection of surgically transplanted organs and tissues.

Immature T cells in the thymus and immature B and T cells in the lymph nodes can be killed by exposure to high concentrations of glucocorticoids, decreasing the number of circulating lymphocytes. Destruction of immature T and B cells by glucocorticoids also causes some reduction in the size of the thymus and lymph nodes.

Maintenance of the Vascular Response to Norepinephrine ∎ Glucocorticoids are required for normal responses of vascular smooth muscle to the vasoconstrictor action of norepinephrine. Norepinephrine is much less active on vascular smooth muscle in the absence of glucocorticoids. This is another example of the permissive action of glucocorticoids.

Glucocorticoids and Stress ∎ Perhaps the most interesting but least understood of all the actions of glucocorticoids is their ability to protect the body against stress. All that is really known is that the body cannot cope successfully with even mild stresses in the absence of glucocorticoids. One must presume that the processes which enable the body to defend itself against physical or emotional trauma must require glucocorticoids to occur. This, again, emphasizes the permissive role they play in physiologic processes.

Stresses stimulate the secretion of ACTH, which increases the secretion of glucocorticoids by the adrenal cortex (Chap. 34). In humans, this increase in glucocorticoid secretion during stress appears to be important for the appropriate defense mechanisms to be put into place. For example, it is well known that glucocorticoid-deficient individuals receiving replacement therapy require larger doses of glucocorticoid to maintain their well-being during periods of stress.

Regulation of Glucocorticoid Secretion ∎ A very important physiologic action of glucocorticoids is their ability to regulate their own secretion. This is accomplished through negative feedback effects exerted by glucocorticoids on the secretion of corticotrophin-releasing hormone (CRH) and ACTH and on proopiomelanocortin (POMC) gene expression (Chap. 34).

Products of the Adrenal Medulla

■ Catecholamines Are Synthesized and Stored in Chromaffin Cells of the Adrenal Medulla

The catecholamines, epinephrine and norepinephrine, are the two hormones synthesized by the chromaffin cells of the adrenal medulla. The human adrenal medulla produces and secretes about four times more epinephrine than norepinephrine. Postganglionic sympathetic neurons also produce and release norepinephrine from their nerve terminals, but they do not produce epinephrine.

Epinephrine and norepinephrine are formed in the chromaffin cells from the amino acid tyrosine. The synthetic pathway for formation of catecholamines is illustrated in Figure 3–19.

■ Trauma, Exercise, and Hypoglycemia Stimulate the Medulla to Release Catecholamines

Epinephrine and some norepinephrine are released from chromaffin cells by the fusion of secretory granules with the plasma membrane. The contents of the granules are extruded into the interstitial fluid. The catecholamines diffuse into capillaries and are transported in the bloodstream.

Neural stimulation of the cholinergic preganglionic fibers that innervate the chromaffin cells triggers the secretion of catecholamines. Stimuli such as injury, anger, anxiety, pain, cold, strenuous exercise, and hypoglycemia generate impulses in these fibers, causing rapid discharge of the catecholamines into the bloodstream.

■ Catecholamines Have Rapid, Widespread Effects

Most cells of the body have receptors for catecholamines and thus are their target cells. There are four structurally related forms of catecholamine receptors, all of which are transmembrane proteins: α_1, α_2, β_1, and β_2. All can bind epinephrine or norepinephrine, to varying extents (Chap. 3).

The Fight-or-Flight Response ■ Epinephrine and norepinephrine produce widespread effects on the cardiovascular system, muscular system, and carbohydrate and lipid metabolism in liver, muscle, and adipose tissue. In response to a sudden rise in catecholamines in the blood, the heart rate accelerates, coronary blood vessels dilate, and blood flow to muscles is increased due to vasodilation (but vasoconstriction occurs in the skin). Smooth muscle in the airways of the lungs, gastrointestinal tract, and urinary bladder relaxes. Muscles in the hair follicles contract, causing piloerection. Blood glucose level also rises. This overall reaction to the sudden release of catecholamines is known as the "fight-or-flight response" (Chap. 6).

Catecholamines and the Metabolic Response to Hypoglycemia ■ Catecholamines secreted by the adrenal medulla and norepinephrine released from sympathetic, postganglionic nerve terminals are key agents used to defend against hypoglycemia. Catecholamine release usually starts when the blood glucose concentration falls to the low end of the physiologic range (60–70 mg/dl). A further decline in blood glucose concentration into the hypoglycemic range produces marked catecholamine release. Hypoglycemia can be produced by a variety of situations, such as overdosage with insulin, catecholamine antagonists, or drugs that block fatty acid oxidation. It is always a dangerous condition, since the CNS will die of ATP deprivation in extended hypoglycemia. The length of time profound hypoglycemia can be tolerated depends on its severity and the individual's sensitivity.

When blood glucose concentration drops toward the hypoglycemic range, CNS receptors monitoring blood glucose concentration are activated. This stimulates the neural pathway leading to the fibers innervating the chromaffin cells. As a result, the adrenal medulla discharges catecholamines and sympathetic nerve terminals release norepinephrine.

Catecholamines act on the liver to stimulate glucose production. They activate glycogen phosphorylase, resulting in the hydrolysis of stored glycogen, and stimulate gluconeogenesis from lactate and amino acids. Both of these actions of catecholamines on the liver are exerted through the α_1 catecholamine receptors, which are coupled to phospholipase C and the polyphosphoinositide pathway.

Catecholamines also activate glycogen phosphorylase in skeletal muscle and adipose cells by interacting with β receptors, thereby activating adenylate cyclase and increasing cAMP in the cells. The elevated cAMP activates glycogen phosphorylase. The glucose-6-phosphate generated in these cells is metabolized, however, since the cells lack glucose-6-phosphatase. In muscle, glucose-6-phosphate is converted by glycolysis to lactate, much of which is released into the blood. The lactate taken up by the liver is converted to glucose via gluconeogenesis and returned to the blood.

In adipose cells, the rise in cAMP produced by catecholamines activates hormone-sensitive lipase, causing hydrolysis of triglycerides and release of fatty acids and glycerol into the bloodstream. These fatty acids provide an alternative substrate for energy metabolism in peripheral tissues. In being used, they also block the phosphorylation and metabolism of glucose, by mechanisms described earlier.

Thus, during profound hypoglycemia, the rapid rise in blood catecholamine levels triggers some of the same metabolic adjustments that occur more slowly during fasting. During fasting these adjustments are triggered mainly in response to the gradual rise in the ratio of glucagon to insulin in the blood. The ratio also rises during profound hypoglycemia, reinforcing the actions of the catecholamines on glycogenolysis, gluconeogenesis, and lipolysis. Moreover, the catecholamines released during hypoglycemia are thought to be partly responsible for the rise in the glucagon-to-insulin ratio by directly influencing the secretion of these hormones by the pancreas. Catecholamines stimulate the secretion of glucagon by the α cells and inhibit the secretion of insulin by β cells (Chap. 37). These catecholamine-mediated responses to hypoglycemia are summarized in Table 36–3.

TABLE 36–3 ■ Catecholamine-Mediated Responses to Hypoglycemia

Liver
　Stimulation of glycogenolysis
　Stimulation of gluconeogenesis
Skeletal muscle
　Stimulation of glycogenolysis
Adipose tissue
　Stimulation of glycogenolysis
　Stimulation of triglyceride lipolysis
Pancreatic islets
　Inhibition of insulin secretion by β cells
　Stimulation of glucagon secretion by α cells

■ ■ ■ ■ ■ ■ ■ ■ ■ ■ ■ **REVIEW EXERCISES** ■ ■ ■ ■ ■ ■ ■ ■ ■ ■ ■

Identify Each with a Term

1. The outer layer of cells of the adrenal cortex that lie immediately under the capsule of the adrenal gland
2. The main androgen secreted by the adrenal cortex of both men and women
3. The enzyme that catalyzes the hydrolysis of cholesterol esters in adrenal cortical cells
4. Diseases caused by genetic defects that affect adrenal steroidogenesis
5. The circulating glycoprotein produced by the liver that binds glucocorticoids with high affinity
6. The circulating protease that acts on angiotensinogen to produce angiotensin I

Define Each Term

7. Chromaffin cell
8. Cytochrome P450scc
9. Adrenarche
10. Addison's disease
11. Steroid response element
12. β-Hydroxybutyrate

Choose the Correct Answer

13. Which of the following sources of cholesterol is most important for sustaining adrenal steroidogenesis when it occurs at a high rate over a long period of time?
　a. De novo synthesis of cholesterol from acetate
　b. Cholesterol esters in LDLs
　c. Plasma membrane cholesterol
　d. Cholesterol esters in lipid droplets within adrenal cortical cells
14. Which of the following would increase in an individual given large doses of cortisol for a prolonged period?
　a. Sensitivity of peripheral tissues to insulin
　b. Blood glucose concentration
　c. Blood DHEA concentration
　d. Prostaglandin synthesis
15. In a person with a cytochrome P450c11 deficiency, which of the following hormones would be increased in the blood?
　a. Corticosterone

　b. Aldosterone
　c. 11-Deoxycorticosterone
　d. Cortisol
16. The rate of aldosterone secretion would decrease in response to:
　a. A decrease in renin secretion by the kidney
　b. A rise in serum potassium
　c. A fall in blood pressure in the kidney
　d. A decrease in tubule fluid sodium concentration at the macula densa
17. Which of the following is *not* mediated by a rise in cAMP?
　a. The stimulation of triglyceride lipolysis by epinephrine
　b. The stimulation of the expression of the gene for cytochrome P450scc by ACTH
　c. The stimulation of glycogen phosphorylase in skeletal muscle by epinephrine
　d. The stimulation of the conversion of cholesterol to pregnenolone in adrenal glomerulosa cells by angiotensin II
18. A cortisol-deficient individual:
　a. Develops atrophied lymph nodes
　b. Has reduced sensitivity to insulin
　c. Has an increased ability to mobilize stored triglyceride
　d. Has reduced resistance to stress

Briefly Answer

19. Why are adrenal androgens not produced by the zona glomerulosa?
20. What are the structural modifications made in adrenal steroid hormones by the liver that destroy their hormone activity and increase their water solubility?
21. Why may a patient receiving long-term glucocorticoid therapy become glucocorticoid-deficient if hormone therapy is halted suddenly?
22. What is meant by "permissive action" in reference to the role of glucocorticoids in accelerating gluconeogenesis during fasting?
23. How do catecholamines act on liver and skeletal muscle to stabilize blood glucose concentration during sudden hypoglycemia?

■ ■ ■ ■ ■ ■ ■ ■ ■ ■ ■ **ANSWERS** ■ ■ ■ ■ ■ ■ ■ ■ ■ ■ ■

1. Zona glomerulosa

2. Dehydroepiandrosterone (DHEA)

3. Cholesterol esterase (cholesterol ester hydrolase)

4. Congenital adrenal hyperplasia

5. Corticosteroid-binding globulin

6. Renin

7. The cells in the adrenal medulla that synthesize and secrete epinephrine and norepinephrine

8. The mitochondrial enzyme in adrenal cortical cells that converts cholesterol to pregnenolone

9. The age at which significant amounts of adrenal androgens begin to be secreted by the adrenal cortex

10. A condition caused by pathologic destruction of the adrenal glands that results in glucocorticoid and aldosterone deficiency

11. A region of genomic DNA that binds the activated steroid hormone-receptor complex

12. One of the ketone bodies produced by the liver and used by the CNS as a major substrate for energy metabolism during the later stages of fasting

13. b

14. b

15. c

16. a

17. d

18. d

19. Androgens are 19-carbon steroid molecules. The zona glomerulosa lacks the enzyme cytochrome P450c17, which cleaves the carbon 20-21 side chain from the steroid structure to yield 19-carbon steroids.

20. The double bond in ring A of the steroid structure is reduced and the resultant hydroxyl group on carbon 3 is conjugated with glucuronic acid. Also, 21-carbon steroids that have a 17α-hydroxyl group and a 20-keto group, like cortisol, undergo lysis of the carbon 20-21 side chain to form 17-ketosteroids.

21. Glucocorticoid therapy exerts a negative feedback effect on ACTH secretion, resulting in ACTH deficiency. The chronic lack of ACTH decreases the expression of genes for the steroidogenic enzymes, thereby impairing the ability of the adrenal glands to synthesize glucocorticoids.

22. Glucocorticoids maintain the expression of the genes, and therefore the intracellular concentrations, of many of the enzymes needed to carry out gluconeogenesis in the liver and kidney. Therefore, glucocorticoids maintain the liver and kidney in a state that makes them capable of accelerated gluconeogenesis when fasting occurs.

23. Catecholamines stimulate glycogenolysis and gluconeogenesis in the liver, causing glucose to be formed and released into the blood. Catecholamines also cause glycogenolysis in skeletal muscles, resulting in the release of lactate into the blood. This lactate can be taken up by the liver and converted by gluconeogenesis into glucose, which is then released into the blood.

Suggested Reading

Burnstein KL, Cidlowski JA: Regulation of gene expression by glucocorticoids. *Annu Rev Physiol* 51:683–699, 1989.

Goodman HM: *Basic Medical Endocrinology.* 2nd ed. New York, Raven, 1994.

Miller WL: Molecular biology of steroid hormone synthesis. *Endocrin Rev* 9:295–318, 1988.

O'Malley B: The steroid receptor superfamily: More excitement predicted for the future. *Mol Endocrinol* 4:363–369, 1990.

Quinn SJ, Williams GH: Regulation of aldosterone secretion. *Annu Rev Physiol* 50:409–426, 1988.

Simpson ER, Waterman MR: Regulation of the synthesis of steroidogenic enzymes in adrenal cortical cells by ACTH. *Annu Rev Physiol* 50:427–440, 1988.

White PC, Pascoe L: Disorders of steroid 11β-hydroxylase isozymes. *Trends Endocrinol Metab* 3:229–234, 1992.

CHAPTER

■ ■ ■ ■ ■ ■ ■
37

The Endocrine Pancreas

Development of mechanisms to allow storage of large amounts of metabolic fuel was an important step in the evolution of higher organisms. As might be expected, the complex processes involved in digestion, storage, and use of fuels require a high degree of regulation and coordination. The pancreas, which plays a vital role in these processes, consists of two functionally different groups of cells. As described in earlier chapters, cells of the **exocrine** pancreas produce and secrete digestive enzymes and fluids into the upper part of the small intestine. The **endocrine** pancreas, an anatomically small portion of the pancreas (1–2% of the total mass), produces hormones involved in regulating fuel storage and use. For convenience, functions of the exocrine and endocrine portions of the pancreas are usually discussed separately. While this chapter focuses primarily on hormones of the endocrine pancreas, the reader should bear in mind that the overall function of the pancreas is to coordinate and direct a wide variety of processes related to the digestion, uptake, and use of metabolic fuels.

Synthesis and Secretion of Islet Hormones

The endocrine pancreas consists of numerous groups of cells located throughout the pancreatic mass.

■ The Islets of Langerhans Are the Functional Units of the Endocrine Pancreas

The islets of Langerhans are discrete clusters of cells that contain from a few hundred to several thousand hormone-secreting endocrine cells. They are found throughout the pancreas but are most abundant in the tail region of the gland. The human pancreas contains, on average, about 1 million islets, which vary in size from about 50 to 300 μm in diameter. Each islet is surrounded by a connective tissue sheath that anatomically separates the islet from surrounding acinar tissue.

Islets are composed of four hormone-producing cell types: insulin-secreting beta cells, glucagon-secreting alpha cells, somatostatin-secreting delta cells, and pan-creatic polypeptide-secreting F cells. Immunofluorescent staining techniques have shown the four cell types to be arrayed in each islet in a manner suggestive of a highly organized cellular community in which paracrine influences may play an important role in determining hormone secretion rates (Fig. 37–1). Further evidence that cell-to-cell communication within the islet may play a role in regulating hormone secretion comes from the finding that islet cells have both gap junctions and tight junctions. Gap junctions link

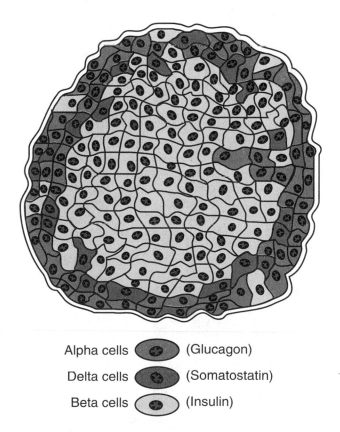

Alpha cells   (Glucagon)
Delta cells   (Somatostatin)
Beta cells   (Insulin)

Figure 37–1 ■ Major cell types in a typical islet of Langerhans. Note the distinct anatomic arrangement of cell types. (Modified from Orci L, Unger RH: Functional subdivision of islets of Langerhans and possible role of D cells. *Lancet* 2:1243–1244, 1975.)

different cell types in the islets cells and potentially provide a means for transfer of ions, nucleotides, or bioelectric current between cells. The presence of tight junctions between outer membrane leaflets of contiguous cells could result in formation of microdomains in the interstitial space, which may also be important for paracrine communication. It should be emphasized, however, that although the existence of gap junctions and tight junctions in pancreatic islets is well documented, their exact function has not been fully defined.

Arrangement of the vascular supply to islets is also consistent with paracrine involvement in regulating islet secretion. Afferent blood vessels penetrate nearly to the center of the islet before branching out and returning to the surface of the islet. The innermost cells of the islet therefore receive arterial blood, while those cells nearer the surface receive blood containing secretions from inner cells. Since there is a definite architectural arrangement of cells in the islet (Fig. 37-1), one cell type could affect the secretion of others. In general, the effluent from smaller islets passes through neighboring pancreatic acinar tissue before entering into the portal venous system. By contrast, the effluent from larger islets passes directly into the venous system without first perfusing adjacent acinar tissue. Therefore, islet hormones arrive in high concentrations in some areas of the exocrine pancreas before reaching peripheral tissues. However, the exact physiologic significance of these arrangements is unknown.

Neural inputs also influence islet cell hormone secretion. Islet cells receive sympathetic and parasympathetic innervation. Responses to neural input occur as a result of activation of various adrenergic and cholinergic receptors (described below). Other neuropeptides coreleased with the classical neurotransmitters may also be involved in regulating hormone release.

Beta Cells ■ In the early 1900s, M. A. Lane established a histochemical method by which two kinds of islet cells could be distinguished. He found that alcohol-based fixatives dissolved the secretory granules in most of the islet cells but preserved them in a small minority of cells. Water-based fixatives had the opposite effect. He named cells containing alcohol-insoluble granules A, or alpha, cells, and those containing alcohol soluble granules B, or beta, cells. Many years later other investigators used immunofluorescence techniques to demonstrate that beta cells produce insulin and alpha cells produce glucagon.

Insulin-secreting beta cells are the most numerous cell type of the islet, comprising 70–90% of the endocrine cells. Beta cells typically exhibit a more central location in the islets (Fig. 37-1). They are generally 10–15 μm in diameter and contain secretory granules that measure 0.25 μm.

Alpha Cells ■ Alpha cells comprise the bulk of the remaining cells of the islets. They are generally located near the periphery, where they form a cortex of cells surrounding more centrally located beta cells (Fig. 37-1). Blood vessels pass through the outer zone of the islet before extensive branching occurs. Inward extensions of the cortex may be present along the axes of blood vessels toward the center of the islet, giving the appearance that the islet is subdivided into small lobules.

Secretory granules in alpha cells have a distinct appearance when viewed with the electron microscope. The centers of the granules are more electron-dense than are the outer regions, giving the granules a halo-like appearance. Immunologic studies suggest that the central core of granules contains mature glucagon, while the outer halo region contains glucagon precursors (see below) that are apparently in the process of conversion into mature glucagon.

Delta Cells ■ Delta cells are the site of production of somatostatin in the pancreas. These cells are typically located in the periphery of the islet, often between beta cells and the surrounding mantle of alpha cells. Somatostatin produced by pancreatic delta cells is identical to that previously described in a neurotransmitter role (Chap. 3) and as a hypothalamic hormone that inhibits growth hormone secretion by the anterior pituitary (Chap. 34).

F Cells ■ F cells are the least abundant of the hormone-secreting cells of islets, representing only about 1% of the total islet cell population. Distribution of F cells is generally similar to that of D cells. F cells secrete pancreatic polypeptide, the role of which is not well understood.

■ Increased Blood Glucose Stimulates Secretion of Insulin

Although a variety of factors, including other pancreatic hormones, are known to influence insulin secretion, the primary physiologic regulator of insulin secretion is the blood glucose concentration.

Synthesis of Proinsulin ■ The structural gene for insulin is located on the short arm of chromosome 11 in humans. Like other hormones and secretory proteins, insulin is first synthesized by ribosomes of the rough endoplasmic reticulum as a larger precursor peptide that is then converted to the mature hormone prior to secretion (Chap. 33).

The insulin gene product is a 110-amino acid peptide, preproinsulin. Proinsulin consists of 86 amino acids (Fig. 37-2); residues 1–30 constitute what will form the B chain, residues 31–65 form the connecting peptide, and residues 66–86 constitute the A chain. (Note that "connecting peptide" should not be confused and used interchangeably with "C peptide"; see below.)

In the process of conversion of proinsulin to insulin, two pairs of basic amino acid residues are clipped out of the proinsulin molecule, resulting in the formation of insulin and C peptide (Fig. 37-2), which are

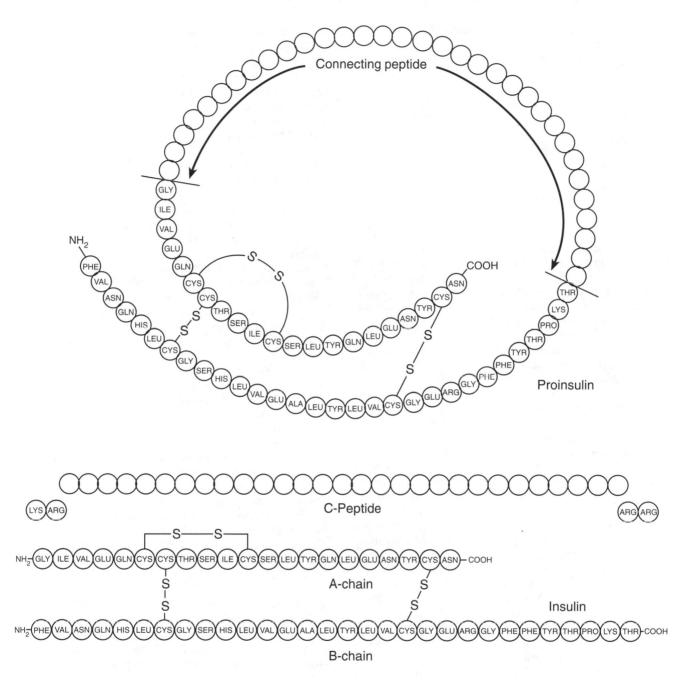

Figure 37–2 ■ The structure of proinsulin (top), C peptide (center), and insulin (bottom). Note that with removal of two pairs of basic amino acids, proinsulin is converted into insulin and C peptide.

ultimately secreted from the beta cell in equimolar amounts.

It is of clinical significance that insulin and C peptide are cosecreted in equal amounts. Measurement of circulating C peptide levels may sometimes provide important information regarding beta cell secretory capacity that could not be obtained by measuring circulating insulin levels alone.

Secretion of Insulin ■ Table 37–1 lists the physiologically relevant regulators of insulin secretion. As indicated previously, an elevated blood glucose level is the most important regulator of insulin secretion. In humans, the threshold value for glucose-stimulated insulin secretion is approximately 100 mg of glucose per 100 ml of plasma (100 mg/dl).

Based on studies using isolated animal pancreas

TABLE 37–1 ■ Factors Affecting Insulin Secretion from the Pancreas

Stimulatory agents or conditions
 Hyperglycemia
 Amino acids
 Fatty acids, especially long-chain
 Gastrointestinal hormones, especially gastrin and secretin
 Acetylcholine
 Sulfonylureas
Inhibitory agents or conditions
 Somatostatin
 Norepinephrine
 Epinephrine

preparations maintained in vitro, it has been determined that insulin is secreted in a biphasic manner in response to a marked increase in blood glucose. There is an initial burst of insulin secretion that may last 5–15 minutes, a result of secretion of preformed insulin secretory granules. This is followed by more gradual and sustained insulin secretion that results largely from biosynthesis of new insulin molecules.

In addition to glucose, several other factors serve as important regulators of insulin secretion (Table 37-1). These include dietary constituents, such as amino acids and fatty acids, as well as hormones and drugs. Among the amino acids, arginine is the most potent secretagogue for insulin. Among the fatty acids, long-chain fatty acids (16–18 carbons) generally are considered the most potent stimulators of insulin secretion. Several hormones secreted by the gastrointestinal tract, including gastrin and secretin, promote insulin secretion. An oral dose of glucose produces a greater increment in insulin secretion than an equivalent intravenous dose, because oral glucose promotes secretion of gastrointestinal hormones that augment insulin secretion by the pancreas. Direct infusion of acetylcholine into the pancreatic circulation stimulates insulin secretion, reflecting the role of parasympathetic innervation in regulating insulin secretion. Sulfonylureas, a class of drugs used orally in the treatment of non-insulin-dependent diabetes (see below), promote insulin's action in peripheral tissues but also directly stimulate insulin secretion.

In addition to factors that stimulate insulin secretion, there are several potent inhibitors (Table 37-1). Exogenously administered somatostatin is a strong inhibitor. It is presumed that pancreatic somatostatin plays a role in regulating insulin secretion, but the importance of this effect has not been fully established. Epinephrine and norepinephrine, the primary catecholamines, are potent inhibitors of insulin secretion. These responses would appear appropriate, since during periods of stress and high catecholamine secretion the desired response is mobilization of glucose and other nutrient stores. Insulin generally promotes the opposite response, and by inhibiting insulin secretion the catecholamines produce their full effect without the opposite actions of insulin.

■ Decreased Blood Glucose Stimulates Secretion of Glucagon

Like insulin, glucagon is first synthesized as part of a larger precursor protein. Glucagon secretion is regulated by many of the factors that regulate insulin secretion, but in most cases these factors have the opposite effect on glucagon secretion.

Synthesis of Proglucagon ■ Glucagon is a simple 29-amino acid peptide. The initial gene product for glucagon, preproglucagon, is a much larger peptide. Like other peptide hormones, the "pre-" piece is removed in the endoplasmic reticulum, and the prohormone is converted into a mature hormone as it is packaged and processed in secretory granules (Chap. 33).

Preproglucagon is synthesized in cells other than pancreatic alpha cells, including cells of the gastrointestinal tract and certain parts of the brain. In these cells, however, the precursor peptide is cleaved in a different manner and peptides other than glucagon are produced. Thus, depending on the exact nature of the proteolytic cleavage process, different bioactive peptides can be generated from the same prohormone precursor.

Secretion of Glucagon ■ The principal factors that influence glucagon secretion are listed in Table 37-2. With a few exceptions, this table is nearly a mirror image of Table 37-1, which lists the factors that regulate insulin secretion. The primary regulator of glucagon secretion is blood glucose; specifically, a decrease in blood glucose below about 100 mg/dl promotes glucagon secretion. As with insulin, amino acids, especially arginine, are potent stimulators of glucagon secretion. Somatostatin inhibits glucagon secretion, as it does insulin secretion. This provides additional evidence that delta cells in the pancreas play an important role in the regulation of both insulin and glucagon secretion, although the exact role has yet to be fully determined.

■ Increased Blood Glucose and Glucagon Stimulates Secretion of Somatostatin

The exact role of somatostatin in regulating islet hormone secretion has not been fully established. Somatostatin clearly inhibits both glucagon and insulin secretion from the alpha and beta cells of the pancreas,

TABLE 37–2 ■ Factors Regulating Glucagon Secretion

Stimulatory agents or conditions
 Hypoglycemia
 Amino acids
 Acetylcholine
 Norepinephrine
 Epinephrine
Inhibitory agents or conditions
 Fatty acids
 Somatostatin
 Insulin

respectively, when it is given exogenously. The anatomic and vascular arrangement of the delta cells relative to the alpha and beta cells further suggest that somatostatin may play a role in regulating both insulin and glucagon secretion. Although many of the data are circumstantial, it is generally accepted that somatostatin plays a paracrine role in regulating insulin and glucagon secretion from the pancreas.

Synthesis of Preprosomatostatin ■ Somatostatin is first synthesized as a larger peptide precursor, preprosomatostatin. The hypothalamus also produces this protein, but the regulation of somatostatin secretion from the hypothalamus is independent of that from the pancreatic delta cells. Upon insertion of preprosomatostatin into the rough endoplasmic reticulum, it is initially cleaved and converted to prosomatostatin. The prohormone is converted into active hormone during packaging and processing in the Golgi.

Secretion of Somatostatin ■ Factors that stimulate pancreatic somatostatin secretion include hyperglycemia, glucagon, and amino acids. Glucose and glucagon are generally considered the most important regulators of somatostatin secretion.

■ Paracrine Interactions May Also Regulate Secretion of Islet Cell Hormones

As described above, each of the three major cell types of the islets of Langerhans is in some manner responsive to hormones produced by one or both other cell types. Vascular and anatomic arrangements suggest the possibility of paracrine interactions in regulating pancreatic hormone secretion. It should be emphasized that while potential interactions are well documented in experimental situations, the exact physiologic role they play in short-term or long-term regulation of islet cell secretion in humans has not been established.

Metabolic Effects of Insulin and Glucagon

The overall role of the pancreas is to coordinate digestion and use of nutrients. The endocrine pancreas secretes hormones that direct storage and use of fuels during times of nutrient abundance (fed state) and nutrient deficiency (fasting). Insulin is secreted in the fed state and is thus called the "hormone of nutrient abundance." In contrast, glucagon is secreted in response to an overall deficit in nutrient supply. These two hormones play an important role in directing the flow of metabolic fuels.

■ Insulin Affects Metabolism of Carbohydrates, Lipids, and Proteins in Liver, Muscle, and Adipose Tissue

The primary targets for insulin are liver, skeletal muscle, and adipose tissue. Insulin has multiple individual actions in each of these tissues, the net result of which is fuel storage.

Mechanism of Insulin Action ■ Although insulin was one of the first peptide hormones to be identified, isolated, and characterized, its exact mechanism of action remains elusive. The insulin receptor is a heterotetramer consisting of a pair of α-β subunit complexes held together by disulfide bonds (Fig. 37-3). The α subunit is an extracellular protein containing the insulin-binding component of the receptor. The β subunit is a transmembrane protein that couples the extracellular event of insulin binding to its intracellular actions.

Activation of the β subunit of the insulin receptor results in autophosphorylation involving phosphorylation of a few selected tyrosine residues in the intracellular portion of the receptor. It is thought that this event further activates the tyrosine kinase portion of the β subunit, leading to tyrosine phosphorylation of specific intracellular substrates. Presumably, this initiates a cascade of events leading to the pleiotropic actions of insulin in its target cells. While tyrosine phosphorylation events appear to be the early steps in insulin action, serine/threonine phosphorylation or dephosphorylation is involved in many of the final steps of insulin action.

Insulin and Glucose Transport ■ Perhaps one of the most important actions of insulin is to promote uptake of glucose from blood into cells. Uptake of glucose into many cell types is by facilitated diffusion. A specific cell membrane carrier is involved but no

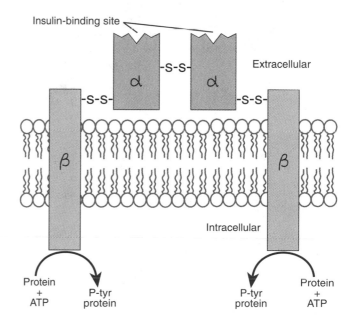

Figure 37–3 ■ Structure of the insulin receptor, a heterotetramer consisting of two extracellular insulin-binding α subunits linked by disulfide bonds to two transmembrane β subunits. The β subunits contain an intrinsic tyrosine kinase activity that is activated upon insulin binding to the α subunit.

energy is required, and the process cannot move glucose against a concentration gradient. The carriers simply shuttle glucose across the membrane faster than would occur by diffusion alone. Considerable recent work has revealed not just one but a family of perhaps six different glucose transporters, commonly called GLUT 1 to GLUT 6 (GLUT being an acronym for glucose transporter). These transporters are expressed in different tissues and in some cases at different times during fetal development.

GLUT 4, the insulin-stimulated glucose transporter, is the primary form of the transporter present in skeletal muscle and adipose tissue. It is present in plasma membrane and inactive intracellular pools in smooth endoplasmic reticulum (Fig. 37–4). In certain target cells, the effect of insulin is to promote translocation of GLUT 4 transporters from an inactive intracellular pool into plasma membranes. As a result, more transporters are available in the plasma membrane and glucose uptake by target cells is increased.

Insulin and the Synthesis of Glycogen ▪ Besides promoting glucose uptake into target cells, insulin promotes its storage. Glucose is stored in two primary forms: glycogen and (by metabolic conversion) triglycerides. Glycogen is a short-term storage form that plays an important role in maintaining normal blood glucose level. The primary glycogen storage sites are the liver and skeletal muscle; other tissues, such as adipose, also store glycogen but in quantitatively small amounts. Insulin promotes glycogen storage primarily through two enzymes (Fig. 37–5). It activates **glycogen synthase** by promoting its dephosphorylation and concomitantly inactivates **glycogen phosphorylase,** also by promoting its dephosphorylation. The net result is that glycogen synthesis is promoted and glycogen breakdown is inhibited.

Insulin and Glycolysis ▪ Insulin also enhances glycolysis. In addition to increasing glucose uptake and therefore providing a mass action stimulus for glycolysis, insulin activates the enzymes glucokinase, phosphofructokinase, pyruvate kinase, and pyruvate dehydrogenase of the glycolytic pathway.

Lipogenic and Antilipolytic Effects of Insulin ▪ In adipose tissue and liver, insulin promotes lipogenesis and inhibits lipolysis (Fig. 37–6). Insulin has similar actions in muscle, but since muscle is not a major site of lipid storage the discussion here focuses on actions in adipose and liver. By promoting the flow of intermediates through glycolysis, insulin promotes formation of α-glycerol phosphate and fatty acids necessary for

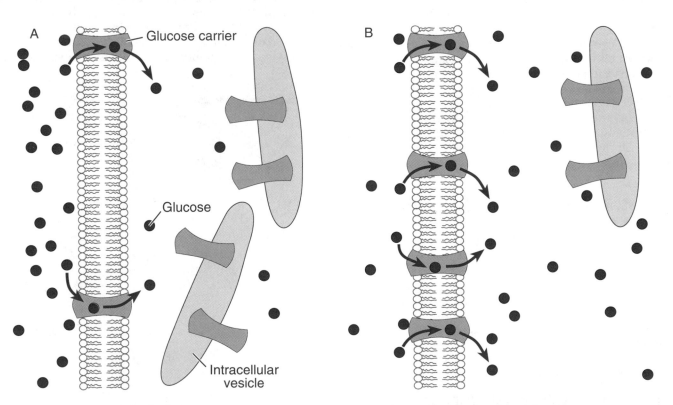

Figure 37–4 ▪ Mechanism of insulin-stimulated glucose uptake. (A) The situation that might exist in the fasted state, in which most glucose carriers are located in inactive intracellular sites. (B) The situation in the fed state, when insulin levels are high and promote the transfer of glucose carriers from intracellular sites to the plasma membrane. Due to the increased number of carriers in the plasma membrane, insulin promotes glucose uptake into its target cells.

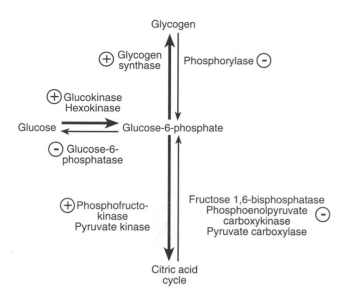

Figure 37–5 ▪ Insulin stimulation of glycogen synthesis and glucose metabolism. Insulin promotes glucose uptake into target tissues, stimulates glycogen synthesis, and inhibits glycogenolysis. In addition, it promotes glycolysis in its target tissues. + = stimulated by insulin; − = inhibited by insulin

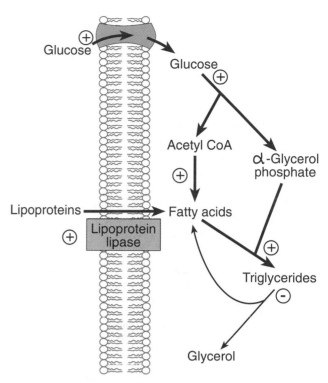

Figure 37–6 ▪ Lipogenic and antilipolytic effects of insulin in adipose tissue. In adipose tissue, insulin promotes glucose uptake; in adipose and liver it promotes metabolism of glucose to fatty acids and α-glycerol phosphate, which are esterified to form triglycerides. Insulin has antilipolytic effects and inhibits triglyceride breakdown. It also increases the activity of lipoprotein lipase, making more fatty acids available inside the cell. For simplicity, lipoprotein lipase is depicted as being present on the surface of the adipose cell; however, this enzyme actually resides on the surface of endothelial cells in the vasculature of adipose tissue. The effects of insulin on the pathways shown may be the result of changes in the rate of enzyme synthesis, covalent modifications, or other mechanisms. + = processes stimulated by insulin; − = processes inhibited by insulin.

triglyceride formation. In addition, it stimulates fatty acid synthase, leading directly to increased fatty acid synthesis. Insulin inhibits the breakdown of triglycerides by inhibiting hormone-sensitive lipase (Fig. 37–6), which is activated by a variety of counterregulatory hormones, such as epinephrine and adrenal glucocorticoids. By inhibiting this enzyme, insulin promotes accumulation of triglycerides in adipose tissue.

In addition to promoting de novo fatty acid synthesis in adipose tissue, insulin increases the activity of lipoprotein lipase, which is responsible for uptake of fatty acids from the blood into adipose tissue (Fig. 37–6). As a result, lipoproteins synthesized in liver are taken up by adipose tissue and fatty acids are ultimately stored as triglycerides.

Effects of Insulin on Protein Synthesis and Protein Degradation ▪ Insulin promotes protein accumulation in its primary target tissues—liver, adipose tissue, and muscle—in three specific ways (Fig. 37–7). First, it stimulates amino acid uptake, primarily by stimulating activity of amino acid transport system A. Second, it increases the activity of several factors involved in protein synthesis. For example, it increases the activity of protein synthesis initiation factors, thus promoting the start of translation and increasing the efficiency of protein synthesis. Insulin also promotes phosphorylation of ribosomal protein L6. Although the role of this phosphorylation is not entirely clear, it appears to result in increased protein synthesis. Insulin also increases the amount of protein synthetic machinery in cells by promoting ribosome synthesis. Third, insulin inhibits protein degradation by reducing lysosome activity and possibly other mechanisms as well.

▪ Glucagon Primarily Affects Liver Metabolism of Carbohydrates, Lipids, and Proteins

The primary physiologic actions of glucagon are exerted in liver. Numerous effects of glucagon have been documented in other tissues, primarily adipose, when the hormone has been added at unphysiologically high concentrations in experimental situations. While these effects may play a role in certain abnormal situations, the normal daily effects of glucagon occur primarily in liver.

Mechanism of Glucagon Action ▪ Glucagon initiates its biologic effects by interacting with one or more types of cell surface receptor. Glucagon receptors are coupled to G proteins and promote increased intracellular cAMP via activation of adenylate cyclase or elevated cytosolic calcium as a result of phospholipid breakdown to form IP_3.

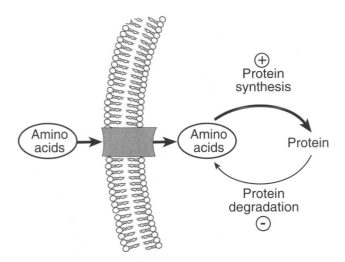

Figure 37–7 ■ Effects of insulin on protein synthesis and protein degradation. In liver, skeletal muscle, and adipose tissue, insulin promotes the accumulation of protein by stimulating protein synthesis and inhibiting protein degradation. By stimulating amino acid transport in these tissues, insulin also provides additional substrate for protein synthesis. + = processes stimulated by insulin; − = processes inhibited by insulin.

Glucagon and Glycogenolysis ■ Glucagon is an important regulator of hepatic glycogen metabolism. It produces a net effect of glycogen breakdown by increasing intracellular cAMP levels, initiating a cascade of phosphorylation events that ultimately results in phosphorylation of phosphorylase b and its activation by conversion into phosphorylase a. Similarly, glucagon promotes the net breakdown of glycogen by promoting the inactivation of glycogen synthase (Fig. 37–8).

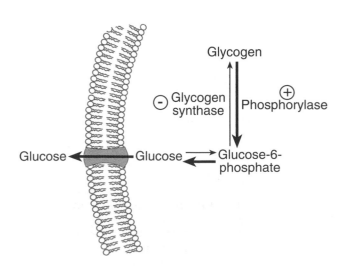

Figure 37–8 ■ Glucagon promotes glycogenolysis and glucose production by liver.

Glucagon and Gluconeogenesis ■ In addition to promoting hepatic glucose production by stimulating glycogenolysis, glucagon stimulates hepatic gluconeogenesis (Fig. 37-9). It does this principally by increasing transcription of mRNA coding for the enzyme phosphoenolpyruvate carboxykinase (PEPCK), a key rate-limiting enzyme in gluconeogenesis. Glucagon also stimulates amino acid transport into liver and degradation of hepatic proteins, thereby aiding in provision of substrates for gluconeogenesis.

Glucagon and Ureagenesis ■ The glucagon-enhanced conversion of amino acids into glucose leads to increased formation of ammonia. Glucagon assists in the disposal of ammonia by increasing the activity of the urea cycle enzymes in liver (Fig. 37-9).

Glucagon and Lipolysis ■ Glucagon promotes lipolysis in liver (Fig. 37-10), although the quantity of lipids stored in liver is small compared to that in adipose tissue.

Glucagon and Ketogenesis ■ Glucagon promotes ketogenesis, the production of ketones (Fig. 37-10), by lowering the levels of malonyl CoA, thus relieving an inhibition of palmitoyl transferase and allowing fatty acids to enter the mitochondria for oxidation to ketones. Ketones are an important source of fuel for tissues such as muscle and heart during times of starvation, thus sparing blood glucose for

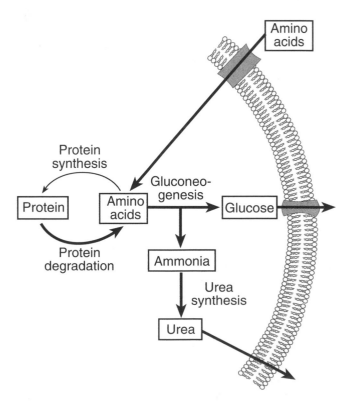

Figure 37–9 ■ Glucagon promotes gluconeogenesis and ureagenesis in liver. The processes stimulated by glucagon are indicated by heavy arrows and those inhibited by glucagon by light arrows.

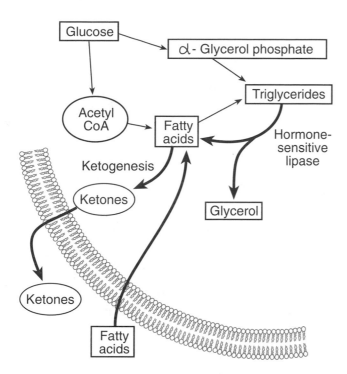

Figure 37–10 ■ Glucagon action on lipolysis and ketogenesis in liver. Steps accelerated by glucagon are indicated by heavy arrows and those inhibited by glucagon by light arrows.

other tissues that are obligate glucose users, such as the central nervous system. During prolonged starvation, the brain adapts its metabolism to use ketones as a fuel source, thereby lessening the overall need for hepatic glucose production (Chap. 36).

■ The Insulin/Glucagon Ratio Determines Metabolic Status

In most instances, insulin and glucagon produce effects that oppose one another. It should not therefore be surprising that the net physiologic response is determined by the relative levels of both hormones in the blood.

The Insulin/Glucagon Ratio in the Fed, Postabsorptive, and Fasted States ■ The **insulin/glucagon ratio** (I/G ratio) may vary 100-fold or more, since the amount of each hormone can vary considerably in different nutritional states. In the fed state, the molar I/G ratio is approximately 30. After an overnight fast it may fall to about 2, and with prolonged fasting to as low as 0.5.

Inappropriate Insulin/Glucagon Ratio in Diabetes ■ A good example of the profound influence of the I/G ratio on metabolic status is in insulin-deficient diabetes. Insulin levels are low, so pathways that insulin stimulates operate at a reduced level. However, insulin is also necessary for alpha cells to sense blood glucose appropriately, so in the absence of insulin the secretion of glucagon is inappropriately elevated. The result is an imbalance in the I/G ratio and accentuation

of glucagon effects well above what would be seen in normal states of low insulin, such as in fasting.

Diabetes Mellitus

Examination of the conditions that result from over- or underproduction of the hormone involved in diabetes mellitus serves to illustrate basic biologic effects of the hormone. Diabetes mellitus is the most common endocrine disorder. Some 10 million people may suffer from the disease in the United States; the exact number is unknown since many people have a borderline, subclinical disease form. Diabetes is the third leading health-related cause of death in the United States and the second leading cause of blindness. Many deaths attributed to cardiovascular disease are in fact the result of complications of diabetes.

Diagnosis of diabetes mellitus is not difficult. Symptoms usually include frequent urination, increased thirst, increased food consumption, and weight loss. The standard criterion for diagnosis of diabetes is elevation of the plasma glucose level after an overnight fast on at least two separate occasions. A glucose value above 140 mg/dl is often used as the diagnostic value.

Diabetes mellitus is a heterogeneous disorder. The causes, symptoms, and general medical outcomes are variable. Generally, the disease takes two forms, **insulin-dependent diabetes mellitus** (IDDM) or **non-insulin-dependent diabetes mellitus** (NIDDM). The terms **type I diabetes** and **type II diabetes** are often used interchangeably for IDDM and NIDDM, respectively. Although type I and type II diabetes are the major subtypes within IDDM and NIDDM, the terms usually imply specific etiologies of the disease state.

Earlier attempts to classify diabetes focused on presumed age of onset, distinguishing juvenile onset diabetes (primarily IDDM) from maturity-onset diabetes (primarily NIDDM). It is now clear that either form of the disease may occur at any age, so current classifications eliminate this distinction.

■ Most Forms of Insulin-Dependent Diabetes Mellitus Involve an Autoimmune Disorder

Pathogenesis of IDDM ■ Insulin-dependent diabetes mellitus is typified by an inability of beta cells to produce physiologically appropriate amounts of insulin. In some instances this may result from a mutation in the preproinsulin gene. However, the most common form of IDDM results from an autoimmune disorder in which pancreatic beta cells are destroyed by the immune system. This condition constitutes what is known as type I diabetes. The initial pathologic event is **insulitis,** involving a lymphocytic attack on beta cells. Antibodies to beta cell surface antigens have also been found in the circulation of many type I diabetics.

The Role of Genetics and Environment in Type I Diabetes ■ Studies of identical twins have provided

THE DIABETIC FOOT

Despite efforts to control their disease and maintain a normal glycemic state, most diabetics eventually suffer from one or more secondary complications of the disease. These complications may be somewhat subtle in onset and slow in progression, but they nonetheless are responsible for the majority of morbidity and mortality. While the specific mechanisms involved remain areas of debate and research activity, it is clear that most secondary complications are either vascular or neural.

Vascular complications may involve atherosclerotic-like lesions in the large blood vessels or impaired function in the microcirculation. Damage to the basement membrane of capillaries in the eye (**diabetic retinopathy**) or kidney (**diabetic nephropathy**) is commonly seen. Although there is no satisfactory direct treatment for diabetic vascular disease, its progression is often monitored closely as an indirect indicator of the overall diabetic state.

Diabetic neuropathy typically involves symmetric sensory loss in the distal lower extremities, or autonomic neuropathy, leading to impotence, gastrointestinal dysfunction, or anhidrosis (lack of sweating) in the lower extremeties. The underlying cause of diabetic neuropathy is not clear. One theory is that excess metabolism of glucose via the polyol pathway is the key initiating event. In this pathway, glucose is reduced to sorbitol by the enzyme **aldose reductase.** Accumulation of sorbitol in tissues may lead to osmotic swelling and subsequent tissue damage. Agents that inhibit aldose reductase are partially effective in reducing or slowing diabetic neuropathy and are often prescribed to patients.

The diabetic foot is an example of several complicating factors of diabetes exacerbating one another. From 50% to 70% of all nontraumatic amputations in the United States each year are due to diabetes. Breakdown of the foot in diabetics is commonly due to a combination of neuropathy, vascular impairment, and infection. In a typical scenario, small lesions or ulcers of the foot result from dryness of the skin due to a combination of neural and vascular complications. Impairments in sensory nerve function may result in these small lesions going unnoticed by the patient until a severe infection or gangrene has become well established.

Loss of the affected foot or limb often can be avoided with patient and physician education. The focus on management of diabetic patients often is simply to maintain a normal blood glucose level, avoiding primary complications such as diabetic ketoacidosis or hyperosmolar coma and the initial secondary complications, such as diabetic retinopathy. However, there is increasing awareness that the feet of the diabetic patient should be assessed at each patient visit. Results of one study show that the likelihood of amputation is reduced by half if diabetic patients simply remove their shoes for foot inspection during every outpatient clinic visit. Thus, while the underlying physiologic mechanisms of the problem may be complex, the problem can be avoided or delayed relatively simply.

important information regarding the genetic basis of Type I diabetes. If one twin develops type I diabetes, the odds that the second will develop the disease are much higher than for any random individual in the population, even when the twins are raised apart under different socioeconomic conditions. In addition, individuals with certain cell surface HLA antigens bear a higher risk for the disease than others.

Environmental factors are involved as well, since the development of diabetes in one twin predicts only a 50% or less chance that the second will develop the disease. The environmental factor(s) has not been identified. Much evidence implicates viruses. Thus it appears that a combination of genetics and environment are strong contributing factors to the development of type I diabetes.

Treatment of IDDM ■ Because the primary defect in IDDM is the inability of beta cells to secrete adequate amounts of insulin, these patients must be treated with injections of insulin. In an attempt to match insulin concentrations in the blood with the metabolic requirements of the individual, various formulations of insulin with different duration of action have been developed. Patients inject an appropriate amount of these different insulin forms to match their dietary and life-style requirements. Long-term control of IDDM depends on maintaining a balance between three factors: insulin, diet, and exercise. To promote strict control of blood glucose, patients are advised to control their diet and level of physical activity as well as insulin dosage. Exercise per se increases glucose uptake into muscle, much like insulin. The diabetic patient must take this into account and make appropriate adjustments in diet or insulin whenever general exercise levels change dramatically.

■ Non-Insulin-Dependent Diabetes Mellitus Primarily Originates in the Target Tissue

Non-insulin-dependent diabetes mellitus (NIDDM) results primarily from an impairment in the ability of insulin target tissues to respond to the hormone. There are multiple forms of the disease, each with a different etiology. In some cases it is a permanent, lifelong disorder and in others is the result of secretion of counterregulatory hormones in a normal (e.g., pregnancy) or pathophysiologic (e.g., Cushing's syndrome) state. In addition to an abnormal target cell response, insulin secretory response by the pancreas in response to glucose is often blunted in magnitude and delayed in time as compared to a normal individual, although the exact contribution of these changes to the development of the disease is not well established. The most common form of NIDDM is type II diabetes.

Insulin Resistance in Type II Diabetes ■ In most cases of type II diabetes, normal or higher than normal amounts of insulin are present in the circulation. Thus, there is no impairment in the secretory capacity of pancreatic beta cells but only in the ability of target cells to respond to insulin. In some instances it has been demonstrated that the fundamental defect is in the insulin receptor. In most cases, though, receptor function appears normal, so the impairment in insulin action is ascribed to a postreceptor defect. Since the exact mechanism of insulin action has not been determined, it is difficult to explore the causes of insulin resistance in much greater depth than this simple receptor/postreceptor scheme.

Genetics, Environment, and Type II Diabetes ■ As with type I diabetes, key information on the influence of genetics and environmental factors in type II diabetes comes from studies of identical twins. These studies indicate that there is a very strong genetic component to development of type II diabetes and that environmental factors, including diet, play a considerably lesser role. If one identical twin develops type II diabetes, chances are nearly 100% that the second will as well, even if they are raised apart under entirely different environmental and socioeconomic conditions.

Many type II diabetics are overweight, and often the severity of their disease can be lessened simply by weight loss. However, no strict cause-and-effect relationship between these two conditions has been established. Clearly, not all type II diabetics are obese, and not all obese individuals develop diabetes.

Treatment Options for Type II Diabetes ■ In milder forms of type II diabetes, dietary restriction leading to weight loss may be the only treatment necessary. Commonly, however, dietary restriction is supplemented by treatment with one of several orally active agents, most often of the sulfonylurea class. These drugs appear to act in two ways. First, they promote insulin action in target cells, thereby lessening insulin resistance in tissues. Second, they correct or reverse a somewhat sluggish response of pancreatic beta cells often seen in type II diabetes, normalizing insulin secretory responses to glucose. The exact mechanisms of these effects are unknown. In some cases, type II diabetics may also be treated with insulin, although in the majority of cases oral agents and dietary manipulation is effective.

■ Complications of Diabetes Mellitus Present Major Health Problems

If left untreated or if glycemic control is poor, diabetes leads to acute complications that may prove fatal. However, even with reasonably good control of blood glucose, over a period of years most diabetics suffer from secondary complications of the disease that result in tissue damage, primarily involving the cardiovascular and nervous systems.

Acute Complications of Diabetes ■ The nature of acute complications that develop in type I and type II diabetics differs. Poorly controlled type I diabetics often exhibit hyperglycemia, glucosuria, dehydration, and ketoacidosis (**diabetic ketoacidosis**). As blood glucose becomes elevated above the renal plasma threshold, glucose appears in the urine. Due to osmotic effects, water follows glucose, leading to polyuria, excessive loss of fluid from the body, and dehydration. With fluid loss, circulating blood volume is reduced, compromising cardiovascular function, which may lead to circulatory failure.

Excessive ketone formation leads to acidosis and electrolyte imbalance in type I diabetics. If uncontrolled, ketones may be elevated in the blood to such an extent that acetone (one of the ketones) can be smelled on the breath. Production of the primary ketones, β-hydroxybutyrate and acetoacetate, results in generation of excess hydrogen ions, leading to acidosis. Ketones may accumulate in the blood to such a degree that they exceed renal transport capacities and appear in the urine. Due to osmotic effects, water is also lost in the urine. In addition, the pK of ketones is such that even with the most acidic urine a normal kidney can produce, roughly half of the excreted ketones are in the salt (or base) form. To ensure electrical neutrality, these must be accompanied by a cation, usually either sodium or potassium. Therefore, loss of ketones in the urine also results in a loss of important electrolytes. Thus, excessive ketone production in type I diabetes results in acidosis, loss of cations, and loss of fluids. Emergency room procedures are directed toward immediate correction of these acute problems and usually involve the administration of base, fluids, and insulin.

The rather complex sequelae of events that can result from uncontrolled type I diabetes are shown in Figure 37–11. If left unchecked, many of these complications can have an additive effect to further the severity of the disease state.

Type II diabetics are generally not ketotic and thus do not develop acidosis or the electrolyte imbalances characteristic of type I diabetes. Hyperglycemia leads to fluid loss and dehydration. Severe cases may result in hyperosmolar coma due to excessive fluid loss. The initial objective of treatment in these individuals is administration of fluids to restore fluid volumes to normal and eliminate the hyperosmolar state.

■ Chronic Secondary Complications of Diabetes

With good control of their disease, most diabetics can avoid the acute complications described above. However, it is rare that a diabetic will not suffer from some of the chronic secondary complications of the disease. In most instances, this will ultimately lead to reduced life expectancy.

Most lesions occur in the circulatory system, although the nervous system is also often affected. Large vessels often show changes similar to those in atherosclerosis, with deposition of large fatty plaques in arteries. However, most of the circulatory complications in diabetes occur in microvessels. The common finding in affected vessels is a thickening of the basement membrane. This leads to impaired delivery of nutrients and hormones to the tissues and removal of waste products, resulting in irreparable tissue damage.

Some of the more disabling consequences of diabetic circulatory impairment are deterioration of blood flow to the retina of the eye, causing retinopathy

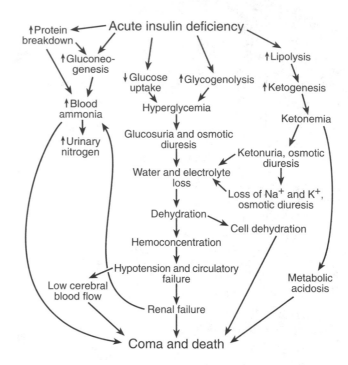

Figure 37–11 ■ Events resulting from acute insulin deficiency in IDDM. If left untreated, insulin deficiency may lead to a number of complications, which may have additive or confounding effects that may ultimately result in death. Adapted from: Tepperman, J. and Tepperman, H.M., *Metabolic and Endocrine Physiology*, Fifth Edition, Year Book Medical Publishers, Chicago, p. 284, 1987.

and blindness, deterioration of blood flow to the extremities, causing in some cases the need for amputation of a foot or leg, and deterioration of glomerular filtration in the kidneys, leading to renal failure.

Peripheral neuropathy is also a common complication of long-standing diabetes. This usually involves sensory nerves and those of the autonomic nervous system. Many diabetics suffer from diminished sensation in the extremities, especially in the feet and legs, which compounds the problem of diminished blood flow to these areas. Often, diabetics are unaware of severe ulcerations of their feet caused by reduced blood flow because of impaired sensory nerve function. Men may develop impotence, and both sexes may have impaired bladder and bowel function.

■ ■ ■ ■ ■ ■ ■ ■ ■ ■ REVIEW EXERCISES ■ ■ ■ ■ ■ ■ ■ ■ ■ ■ ■

Identify Each with a Term

1. Clusters of cells that produce hormones and are the basic functional unit of the endocrine pancreas

2. Cell type responsible for production and secretion of glucagon

3. The initial product of the insulin gene

4. Primary target tissue(s) of insulin in the body
5. Form of diabetes mellitus typically caused by an auto-immune disorder

Define Each Term

6. Pancreatic delta cells
7. C peptide
8. Somatostatin
9. GLUT 4
10. Type II diabetes

Choose the Correct Answer

11. Which of the following stimulate the secretion of *both* insulin and glucagon from the pancreas?
    a. Epinephrine
    b. Amino acids
    c. Acetylcholine
    d. b and c
12. Insulin:
    a. Inhibits amino acid uptake into skeletal muscle
    b. Stimulates glucose uptake into all tissues in the body
    c. Inhibits protein degradation in skeletal muscle
    d. Stimulates hormone-sensitive lipase in adipose tissue
13. Glucagon:
    a. Inhibits insulin secretion by pancreatic β cells.

b. Exerts its most important physiologic effects on adipose tissue
c. Promotes gluconeogenesis and urea synthesis in liver
d. Indirectly stimulates ketogenesis in liver by inhibiting pancreatic somatostatin secretion

14. Type I diabetes is characterized by:
    a. Insulin resistance
    b. Treatment with exogenous insulin
    c. Sulfonylurea treatment
    d. Virtual absence of secondary complications
15. Type II diabetes:
    a. Bears a strong genetic component to the development of the disease
    b. Is typified by low or negligible circulating insulin
    c. Occurs only in obese individuals
    d. Is treated in the same manner as type I diabetes

Briefly Answer

16. Summarize the effects of insulin on adipose tissue metabolism.
17. Summarize the effects of glucagon on hepatic protein metabolism.
18. Describe the pathogenesis and general characteristics of the primary forms of type I and type II diabetes.

ANSWERS

1. Islets of Langerhans
2. Alpha cells
3. Preproinsulin
4. Liver, muscle, and adipose tissue
5. Type I diabetes
6. Cells of the islets of Langerhans responsible for the production of somatostatin
7. A part of the proinsulin molecule that is released when proinsulin is converted into insulin
8. A 14-amino acid peptide produced in delta cells of the pancreatic islets, hypothalamus, and other tissues
9. An insulin-responsive glucose carrier found primarily in adipose tissue and skeletal muscle
10. An insulin-resistant form of diabetes often treated with a combination of weight reduction and sulfonylureas
11. d
12. c
13. c
14. b
15. a
16. Insulin promotes glucose uptake into adipose and its conversion into α-glycerol phosphate and fatty acids. It promotes fatty acid production by increasing the activity of fatty acid synthase. In addition, it increases the activity of lipoprotein lipase, resulting in increased delivery of fatty acids to the adipose cell. Esterification of fatty acids to α-glycerol phosphate to form triglycerides is increased, while breakdown of triglycerides by hormone-sensitive lipase is inhibited by insulin. The net result is increased lipogenesis and fat storage, accompanied by reduced fat breakdown.

17. The net effect of glucagon is to promote conversion of hepatic (and extrahepatic) protein into glucose. Glucagon accomplishes this by stimulating gluconeogenesis and the conversion of amino acids into glucose. It provides substrates for gluconeogenesis by increasing the transport of amino acids into liver cells, inhibiting hepatic protein synthesis, and promoting protein degradation. Conversion of amino acids into glucose results in increased ammonia nitrogen; glucagon promotes urea production and thus increases nitrogen excretion.

18. Type I diabetes mellitus is caused by an autoimmune disorder. Cells of the immune system attack pancreatic beta cells, eventually rendering the individual incapable of secreting insulin. Since insulin is lacking, these individuals are treated with exogenous insulin via injections. Type II diabetes, the most prevalent form, is typified by normal or higher than normal circulating insulin. Target tissues of insulin are generally unresponsive to the hormone. Treatment by weight loss and oral hypoglycemic agents such as sulfonylureas is most common.

Suggested Reading

Hedge GA, Colby HD, Goodman RL: *Clinical Endocrine Physiology.* Philadelphia, WB Saunders, 1987.

National Diabetes Data Group: Classification and diagnosis of diabetes mellitus and other categories of glucose intolerance. *Diabetes* 28:1039–1057, 1979.

Diabetes in America. National Diabetes Data Group; U.S. Department of Health and Human Services, National Institutes of Health Publ. no. 85–1468, Bethesda, 1985.

Olefsky JM: The insulin receptor: A multifunctional protein. *Diabetes* 39:1009–1016, 1990.

Rifkin H, Porte D Jr: *Ellenberg and Rifkin's Diabetes Mellitus: Theory and Practice.* 4th ed. New York, Elsevier, 1990.

Wilson JD, Foster DW (eds): *Williams Textbook of Endocrinology.* 8th ed. Philadelphia, WB Saunders, 1992.

capacity of these cells to transport calcium actively (Chap. 29).

Phosphate Absorption ■ The small intestine is also a primary site for phosphate absorption. Uptake occurs by both active transport and passive diffusion, but active transport is the primary mechanism. As indicated in Figure 38-2, phosphate is efficiently absorbed from the small intestine: typically 80% or more of ingested phosphate is absorbed. However, phosphate absorption from the small intestine is regulated very little. To a minor extent, active transport of phosphate is coupled to calcium transport. Therefore, when active transport of calcium is low, as with vitamin D deficiency, phosphate absorption is also low.

■ The Kidneys Play an Important Role in Regulating Plasma Concentrations and Excretion of Calcium and Phosphate

Renal Handling of Calcium ■ As discussed in Chapter 24, filterable calcium comprises about 60% of the total calcium in the plasma and consists of calcium ion and calcium bound to a filterable anion such as bicarbonate or citrate. The remaining 40% of the total calcium circulates bound to proteins and thus is not filterable. Ordinarily, only about 1% of filtered calcium is eventually excreted in the urine, with the remaining 99% reabsorbed and returned to the plasma. Reabsorption occurs in both the proximal and distal tubules and in the loop of Henle. Active transport mechanisms are involved in uptake from the proximal and distal tubules. Approximately 60% of filtered calcium is reabsorbed in the proximal tubule, 30% from the loop of Henle, and 9% from the distal tubule; the remaining 1% is excreted in the urine. Calcium excretion is regulated by the kidney primarily at the distal tubule. Parathyroid hormone stimulates calcium reabsorption in the distal tubule, thus promoting calcium retention and lowering urinary calcium. Parathyroid hormone is an important regulator of plasma calcium concentration.

Renal Handling of Phosphate ■ The majority of ingested phosphate is absorbed from the gastrointestinal tract, and the primary route of excretion of this phosphate is via the urine. Kidneys therefore play a key role in regulating body phosphate homeostasis. Ordinarily about 85% of filtered phosphate is reabsorbed and 15% is excreted in the urine. Phosphate reabsorption occurs via active transport processes at two sites along the nephron: the proximal tubule and the distal tubule. The majority (70%) of phosphate reabsorption occurs in the proximal tubule and 15% is reabsorbed in the distal tubule. Parathyroid hormone inhibits phosphate reabsorption in the proximal tubule and thus has a major regulatory effect on phosphate homeostasis. As a result of its actions, parathyroid hormone increases urinary phosphate excretion **(phosphaturia)** with an accompanying decrease in plasma phosphate concentration.

■ Each Day, Substantial Amounts of Calcium and Phosphate Enter and Leave Bone

The Composition of Bone ■ Mature bone can be simply described as inorganic mineral deposited on an organic framework. The mineral portion of bone is composed largely of calcium phosphate in the form of **hydroxyapatite crystals,** which have the general chemical formula $Ca_{10}(PO_4)_6(OH)_2$. The mineral portion of bone typically comprises about 25% of its volume, but because of its high density the mineral fraction is responsible for roughly half the weight of bone. Bone also contains considerable amounts of carbonate, magnesium, and sodium, in addition to calcium and phosphate (Table 38-2).

The organic matrix of bone on which the bone mineral is deposited is called **osteoid. Type I collagen** is the primary constituent of osteoid, comprising 95% or more. Collagen in bone is similar to that in skin and tendons, but bone collagen exhibits some biochemical differences that impart increased mechanical strength. The remaining noncollagen portion (5%) of organic matter is referred to as **ground substance.** Ground substance consists of a mixture of various **proteoglycans,** high-molecular-weight compounds consisting of different types of polysaccharides linked to a polypeptide backbone. Typically they are 95% or more carbohydrate.

Electron microscopic study of bone reveals needle-like hydroxyapatite crystals lying alongside collagen fibers. This orderly association of hydroxyapatite crystals with the collagen fibers is responsible for the strength and hardness characteristic of bone. Loss of either bone mineral or organic matrix greatly affects the mechanical properties of bone. Complete demineralization of bone leaves a flexible collagen framework, and complete removal of organic matrix leaves a bone with its original shape, but extremely brittle.

Cell Types Involved in Bone Formation and Bone Resorption ■ The three principal cell types involved in bone formation and bone resorption are **osteoblasts, osteocytes,** and **osteoclasts** (Fig. 38-3).

Osteoblasts are located on the bone surface and are responsible for osteoid synthesis. Like many cells actively synthesizing protein for export, osteoblasts have an abundant rough endoplasmic reticulum and Golgi complex. Cells actively engaged in osteoid synthesis are cuboidal, while those less active are more flattened. Numerous cytoplasmic processes connect adjacent osteoblasts on the bone surface and connect osteoblasts with osteocytes deeper in the bone. Osteoid produced by osteoblasts is secreted into the space adjacent to the bone. Eventually, new osteoid becomes mineralized, and in the process osteoblasts are surrounded by mineralized bone.

As osteoblasts are progressively engulfed by mineralized bone, they lose much of their bone-forming ability

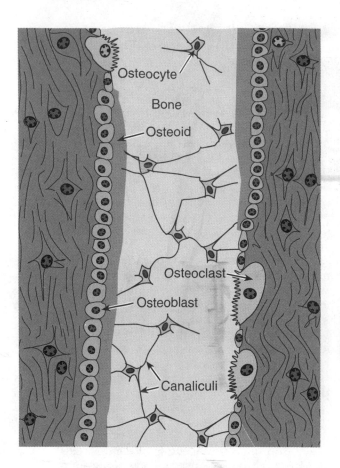

Figure 38–3 ■ Location and relationship of the three primary cell types involved in bone metabolism.

and become quiescent. At this point they are called osteocytes. Many of the cytoplasmic connections in the osteoblast stage are maintained into the osteocyte stage. These connections become visible channels, or **canaliculi,** that provide direct contact for osteocytes deep in bone with other osteocytes and with the bone surface. It is generally believed that these canaliculi provide a mechanism for transfer of nutrients, hormones, and waste products between the bone surface and its interior.

Osteoclasts are cells responsible for bone resorption. They are large, multinucleated cells located on bone surfaces. Osteoclasts promote bone resorption by secreting acid and proteolytic enzymes into the space adjacent to the bone surface. Surfaces of osteoclasts facing bone are ruffled to increase their surface area and promote bone resorption. Bone resorption is a two-step event. First, osteoclasts create a local acidic environment that increases solubility of surface bone mineral. Second, proteolytic enzymes secreted by osteoclasts degrade organic matrix of the bone.

Bone Formation and Bone Remodeling ■ Early in fetal development, the skeleton consists of little more than a cartilaginous model of what will later form the bony skeleton. The process of replacing this cartilaginous model with mature, mineralized bone begins in the center of the cartilage and progresses toward the two ends of what will later form the bone. As mineralization progresses, the bone increases in thickness and in length. The **epiphyseal plate** is a region of growing bone of particular interest, since it is here that elongation and growth of bones occurs after birth. Histologically, the epiphyseal plate shows considerable differences between its leading and trailing edges. The leading edge consists primarily of **chondrocytes,** which are actively engaged in synthesis of cartilage of the epiphyseal plate. These cells gradually become engulfed in their own cartilage, and these trapped chondrocytes are replaced by new cells on the surface of the cartilage to allow the process to continue. The cartilage gradually becomes calcified and the embedded chondrocytes die. The calcified cartilage then begins to erode, and osteoblasts migrate into the area. Osteoblasts secrete osteoid, which eventually becomes mineralized, and new mature bone is formed. In the epiphyseal plate, therefore, the continuing processes of cartilage synthesis, calcification, erosion, and osteoblast invasion results in a zone of active bone formation that moves away from the middle or center of the bone toward the end.

Chondrocytes of epiphyseal plates are regulated by hormones. Insulinlike growth factor I (IGF-I), primarily produced by the liver in response to growth hormone, serves as a primary stimulator of chondrocyte activity and ultimately of bone growth. Insulin and thyroid hormones provide an additional stimulus for chondrocyte activity.

Beginning a few years after puberty, the epiphyseal plates in long bones (e.g., as in the thighs and arms) gradually become less responsive to hormonal stimuli and eventually are totally unresponsive. This phenomenon is referred to as **closure of the epiphyses.** In most individuals, epiphyseal closure is complete by about age 20; adult height is reached at this point, since further linear growth is impossible. Not all bones undergo closure. For example, those in the fingers, feet, skull, and jaw remain responsive, which accounts for the skeletal changes seen in acromegaly, the condition of growth hormone overproduction (Chap. 34).

The flux of calcium and phosphate into and out of bone each day (Figs. 38–1, 38–2) reflects a turnover of bone mineral and changes in bone structure generally referred to as **remodeling.** Bone remodeling occurs along most of the outer surface of the bone, making it either thinner or thicker, as required. In long bones, remodeling can also occur along the inner surface of the bone shaft, next to the marrow cavity. Remodeling is a beneficial adaptive process that allows bone to be reshaped to meet changing mechanical demands placed on the skeleton. It also allows the body to store or mobilize calcium rapidly.

CYTOKINES, ESTROGENS, AND OSTEOPOROSIS

It is well established that a decline in circulating levels of 17β-estradiol is a major contributing factor in development of osteoporosis in postmenopausal women. Until just recently, specific mechanisms by which estradiol might influence bone metabolism was largely unknown. Recent studies suggest that estradiol influences the production and/or modulates the activity of several cytokines involved in regulating bone remodeling.

Normal bone remodeling involves a regulated balance between the processes of bone formation and bone resorption. Parathyroid hormone (PTH) is the primary systemic influence on bone resorption. However, at the local cellular level several cytokines influence bone resorption, and results of recent studies suggest a link between estrogen levels and the influence of these cytokines on bone resorption.

Osteoclast-mediated bone resorption involves two processes: activation of mature functional osteoclasts and recruitment and differentiation of osteoclast precursors. In addition to PTH, the cytokines interleukin-1 (IL-1) and tumor necrosis factor (TNF) are involved in activation of mature osteoclasts to cause bone resorption. For maturation of osteoclast precursors, the cytokines macrophage-colony stimulating factor (M-CSF) and interleukin-6 (IL-6) appear to be involved. Estradiol plays a role in bone remodelling by suppressing formation of these cytokines. As a result of its ability to interact with bone cells and their precursors to regulate local paracrine signaling mechanisms, estradiol produces antiosteoporotic effects in bone.

When estradiol is present, as in a premenopausal state, it acts as a governor to reduce cytokine production and limit osteoclast activity. When estradiol levels are reduced, the governor is lost, secretion of these cytokines increases, and osteoclast formation and activity increase, resulting in increased bone resorption.

Current research efforts attempt to define more clearly the specific source(s) and roles of the cytokines involved. It is anticipated that elucidation or these factors might allow development of diagnostic tools, such as the assessment of cytokine levels, to monitor osteoporosis. In addition, such knowledge should facilitate development of drugs that might interfere with cytokine action and potentially be of value in the treatment of osteoporosis.

Regulation of Plasma Calcium and Phosphate Concentrations

Regulatory mechanisms for calcium include rapid nonhormonal mechanisms with limited capacity and somewhat slower hormonally regulated mechanisms with much greater capacity.

■ Nonhormonal Mechanisms Can Rapidly Buffer Small Changes in Plasma Concentrations of Free Calcium

Protein-Bound Calcium ■ The association of calcium with proteins is a simple, reversible, chemical equilibrium process. Protein-bound calcium therefore has the capacity to serve as a buffer of free plasma calcium concentrations. This effect is rapid and does not require complex signaling pathways, but its capacity is quite limited and it cannot serve a long-term role in calcium homeostasis.

Readily Exchangeable Pool of Calcium in Bone ■ Recall that approximately 99% of total body calcium is present in bone, and there is roughly 1–2 kg of calcium in a normal adult (Table 38–1, Fig. 38–1). The majority of the calcium in bone exists as mature hardened bone mineral that is not readily exchangeable but can be moved into the plasma via hormonal mechanisms (described below). However, approximately 1% (or 10 g) of the calcium in bone is in a simple chemical equilibrium with plasma calcium. This readily exchangeable source of calcium is primarily located on the surface of newly formed bone. Any change in free calcium in the plasma or extracellular fluid, therefore, results in a shift of calcium either into or out of bone mineral until a new equilibrium is reached. Although this mechanism, like that described above, provides for a rapid defense against changes in free calcium concentrations, it, too, is limited in its capacity and thus can only provide for very short-term fine-tuning of calcium homeostasis.

■ Hormonal Mechanisms Provide High-Capacity, Long-Term Regulation of Plasma Calcium and Phosphate

The hormonal mechanisms described here have a large capacity and the ability to make long-term adjustments in calcium and phosphate fluxes, but they do not respond instantaneously. It may take several minutes or hours for the response to occur and adjustments to be made. However, these are the principal mechanisms by which plasma calcium and phosphate concentrations are regulated.

Chemistry of Parathyroid Hormone, Calcitonin, and 1,25-Dihydroxycholecalciferol and Regulation of Their Production ■ One of the primary regulators of plasma calcium concentrations is **parathyroid hormone** (PTH, or parathormone). Parathyroid hormone is an 84-amino acid peptide produced by the parathyroid glands. Synthetic peptides containing the first 34 amino terminal residues appear to be as active as the native hormone.

There are two pairs of parathyroid glands, located on the dorsal surface of the left and right lobes of the thyroid gland. Because of this close proximity, damage to the parathyroid glands themselves or to their blood supply may occur during surgical removal of the thyroid gland.

The primary physiologic stimulus for parathyroid hormone secretion is a *decrease* in plasma calcium. Figure 38–4 shows the relationship between parathyroid hormone secretion and total plasma calcium. It is actually a decrease in the **ionized calcium** concentration that triggers an increase in PTH secretion. Note in this figure that as the calcium concentration decreases from a normal value of 10 mg/100 ml to approximately 8.5–9.0 mg/100 ml, there is a significant increase in PTH secretion. The net effect of PTH is to increase the flow of calcium into plasma, and thus return concentrations toward normal (see below).

Calcitonin (CT, or thyrocalcitonin) is a 32-amino acid peptide produced by **parafollicular cells** of the thyroid gland (Fig. 35–1). Unlike PTH, for which only the initial amino terminal segment is required for biologic activity, the full peptide is required for calcitonin activity. Salmon calcitonin differs from human calcitonin in 9 of the 32 amino acid residues but in humans has a potency of about 100-fold higher. The higher potency may be due to higher affinity for receptors and slower degradation by peripheral tissues. Calcitonin used clinically is often a synthetic peptide matching the sequence of salmon calcitonin.

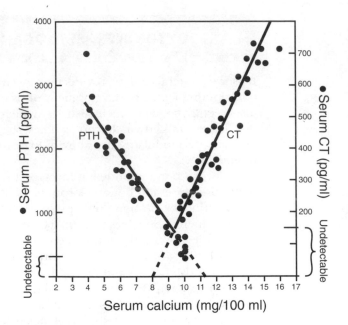

Figure 38–4 ■ The effect of changes in plasma calcium on parathyroid hormone (PTH) and calcitonin (CT) secretion. (Modified from Johnson LR: *Essential Medical Physiology.* New York, Raven, 1992.)

In contrast to PTH, calcitonin secretion is stimulated by an increase in plasma calcium (Fig. 38–4). Hormones of the gastrointestinal tract, especially gastrin, also promote calcitonin secretion. Since the net effect of calcitonin is to promote calcium deposition in bone, stimulation of calcitonin secretion by gastrointestinal hormones provides an additional mechanism facilitating calcium uptake into bone after ingestion of a meal.

The third key hormone involved in regulating plasma calcium is **vitamin D_3 (cholecalciferol).** More precisely, a metabolite of vitamin D_3 serves as a hormone in calcium homeostasis. The D vitamins, a group of lipid-soluble compounds derived from cholesterol, have long been known to be effective in the prevention of rickets. Research over the past 25 years indicates that vitamin D exerts it effects through a hormonal mechanism.

Figure 38–5 shows the structure of vitamin D_3 and the related compound **vitamin D_2 (ergocalciferol).** Ergocalciferol is the form principally found in plants and yeasts and is the primary form used to supplement human foods because of its relative availability and lesser cost. Although it is less potent on a mole-per-mole basis, vitamin D_2 undergoes the same metabolic conversion steps and ultimately produces the same biologic effects as does vitamin D_3. The physiologic actions of vitamin D_3, therefore, also apply to vitamin D_2.

Vitamin D_3 can be provided by the diet or formed in the skin by the action of ultraviolet light on a precursor, **7-dehydrocholesterol,** derived from cholesterol

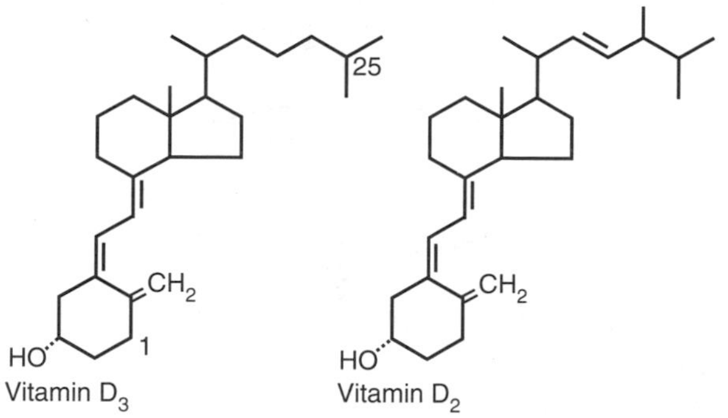

Figure 38–5 ■ Structures of vitamin D_3 and vitamin D_2. Note that they differ only by a double bond between carbons 22 and 23 and a methyl group at position 24.

(Fig. 38–6). In many countries where food is not systematically supplemented with vitamin D, this pathway provides the major source of vitamin D. Because of the number of variables involved, it is difficult to specify a minimum exposure time, but typically exposure to moderately bright sunlight for 30–120 min/day can provide enough vitamin D to supply the body's needs without any dietary supplementation.

Vitamins D_3 and D_2 are by themselves relatively inactive. However, they undergo a series of transformations in the liver and kidneys that converts them into powerful calcium-regulatory hormones (Fig. 38–6). The first step occurs in liver and involves addition of a hydroxyl group to carbon 25, to form 25-hydroxy-cholecalciferol (25-OH-cholecalciferol). This reaction is largely unregulated, although certain drugs and liver diseases may affect this step (see below). 25-Hydroxycholecalciferol is released into the blood, and in the kidney it undergoes a second hydroxylation reaction on carbon

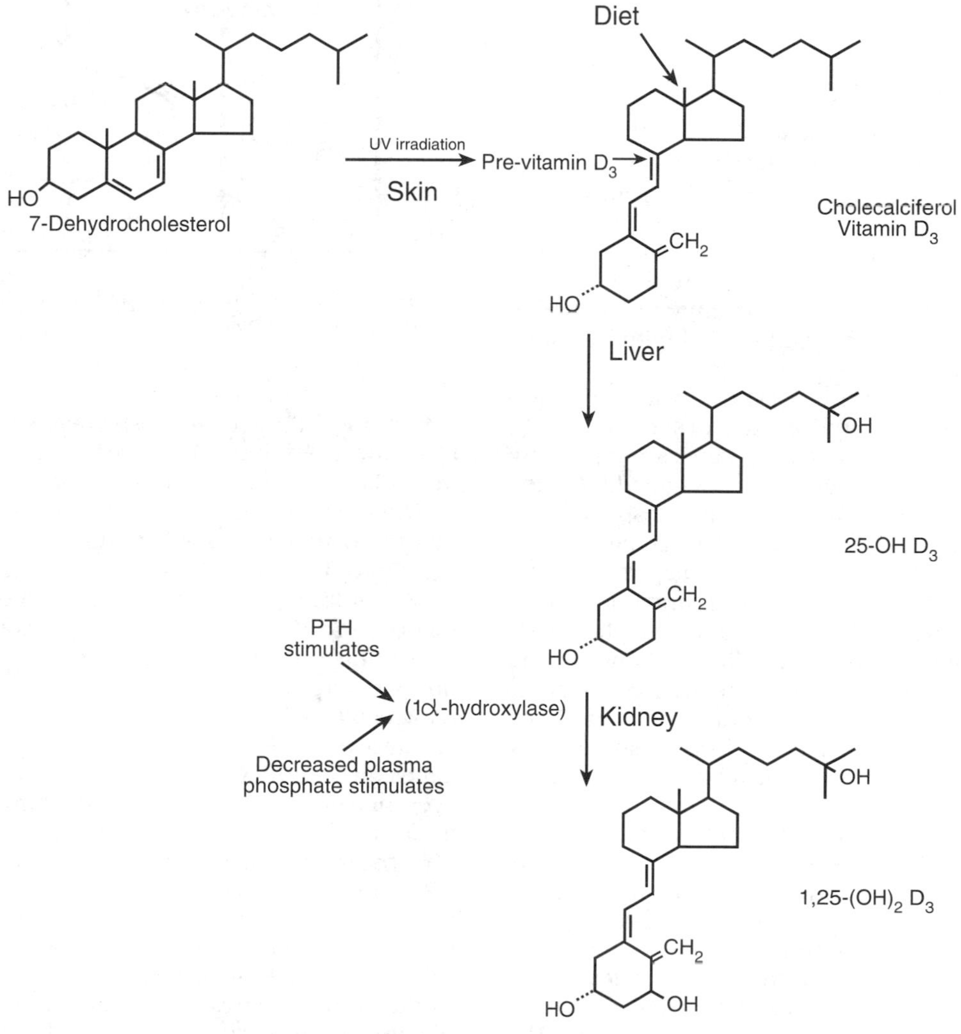

Figure 38–6 ■ Pathway for conversion of vitamin D_3 into 1,25-dihydroxycholecalciferol.

1. The product is **1,25-dihydroxycholecalciferol,** the principal hormonally active form of the vitamin. The biologic activity of 1,25-dihydroxycholecalciferol is approximately 100–500 times greater than that of 25-hydroxycholecalciferol. The reaction in the kidney is catalyzed by the enzyme 1α-hydroxylase, located in proximal tubule cells.

The final step in 1,25-dihydroxycholecalciferol formation is highly regulated. Activity of 1α-hydroxylase is regulated primarily by PTH, which stimulates its activity. Thus, if plasma calcium levels fall, PTH secretion increases and PTH in turn promotes formation of 1,25-dihydroxycholecalciferol. In addition, enzyme activity increases in response to a decrease in plasma phosphate. This does not appear to involve any intermediate hormonal signals but apparently involves direct activation of either the enzyme or cells in which the enzyme is located. Therefore, both a decrease in plasma calcium, which triggers PTH secretion, and a decrease in circulating phosphate result in activation of 1α-hydroxylase and an increase in 1,25-dihydroxycholecalciferol.

■ Actions of Parathyroid Hormone, Calcitonin, and 1,25-Dihydroxycholecalciferol

Most hormones generally improve the quality of life and improve the chances for survival when an animal is placed in a physiologically challenging situation. However, PTH is essential for life. Complete absence of PTH causes death from hypocalcemic tetany within just a few days, but this can be avoided with hormone replacement therapy.

The net effects of PTH on plasma calcium and phosphate and its sites of action are described in Figure 38–7. Parathyroid hormone causes an increase in plasma calcium concentration while decreasing plasma phosphate. This decrease in phosphate concentration is important with regard to calcium homeostasis. At normal plasma concentrations, calcium and phosphate are at or near chemical saturation levels. If PTH were to increase *both* calcium and phosphate levels, they would simply crystalize in bone or soft tissues as calcium phosphate, and the necessary increase in plasma calcium concentration would not occur. Thus, the effect of PTH to lower plasma phosphate is an important aspect of its role in regulating plasma calcium.

Parathyroid hormone has several important actions in the kidneys (Fig. 38–7). It stimulates calcium reabsorption in the thick ascending limb and distal tubule, thereby decreasing calcium loss in the urine and increasing plasma concentrations. It also inhibits phosphate reabsorption in the proximal tubule, leading to increased urinary phosphate excretion and a decrease in plasma phosphate. Another important effect of PTH is to increase the activity of kidney 1α-hydroxylase, which is involved in forming active vitamin D.

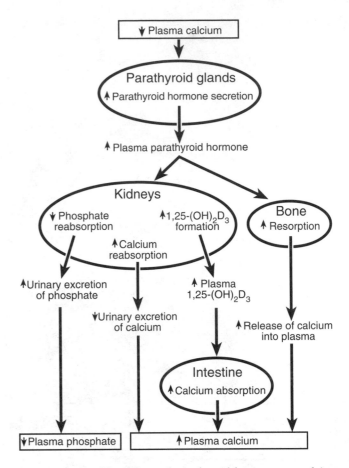

Figure 38–7 ■ The effects of parathyroid hormone on calcium and phosphate metabolism.

In bone, PTH activates osteoclasts to increase bone resorption and thus the delivery of calcium from bone into plasma (Fig. 38–7). In addition to stimulating existing osteoclasts, PTH stimulates maturation of immature osteoclasts into mature, active osteoclasts. Parathyroid hormone also inhibits collagen synthesis by osteoblasts, resulting in decreased bone matrix formation and decreased flow of calcium from plasma into bone mineral. The actions of PTH to promote bone resorption are augmented by 1,25-dihydroxycholecalciferol.

Parathyroid hormone does not appear to have any major direct effects on the gastrointestinal tract. However, because it increases active vitamin D formation, it ultimately increases absorption of both calcium and phosphate from the gastrointestinal tract (Fig. 38–7).

Calcitonin is very important in a number of lower vertebrates, but despite its many demonstrated biologic effects in humans it appears to play only a minor role in calcium homeostasis. This conclusion generally stems from two lines of evidence. First, loss of calcitonin following surgical removal of the thyroid gland (and therefore removal of calcitonin-secreting parafollicular cells) does not lead to overt clinical abnormalities of calcium homeostasis. Second, calcitonin hypersecretion, such as from thyroid tumors involving parafollicular cells, does not cause any overt problems. On a daily basis, calcitonin probably only fine-tunes the calcium regulatory system.

The overall action of calcitonin is to decrease *both* calcium and phosphate concentrations in plasma (Fig. 38–8). The primary target of calcitonin is bone, although some lesser effects also occur on kidney. In the kidneys, calcitonin decreases tubular reabsorption of calcium and phosphate. This leads to an increase in urinary excretion of both calcium and phosphate and ultimately to decreased levels of both ions in the plasma. In bone, calcitonin opposes the action of PTH on osteoclasts by inhibiting their activity. This leads to decreased bone resorption and an overall net transfer of calcium from plasma into bone. Calcitonin has little or no direct effect on the gastrointestinal tract.

The net effect of 1,25-dihydroxycholecalciferol is to increase both calcium and phosphate concentrations in plasma (Fig. 38–9). The activated form of vitamin D primarily influences the gastrointestinal tract, although it has actions in bone and kidney as well.

In the kidneys, 1,25-dihydroxycholecalciferol increases tubular reabsorption of calcium and phosphate, promoting retention of both ions in the body. However, this is a weak and probably only minor effect of the hormone. In bone, the hormone promotes actions of PTH on osteoclasts, thereby increasing bone resorption (Fig. 38–9).

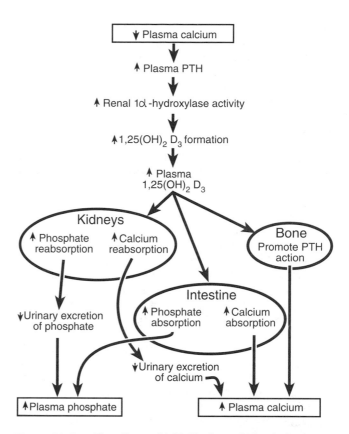

Figure 38–9 ■ The effects of 1,25-dihydroxycholecalciferol (1,25(OH)₂D₃) on calcium and phosphate metabolism.

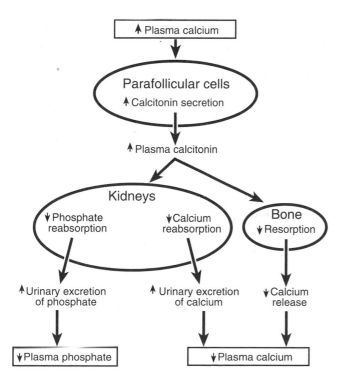

Figure 38–8 ■ The effects of calcitonin on calcium and phosphate metabolism.

In the gastrointestinal tract, 1,25-dihydroxycholecalciferol stimulates calcium and phosphate absorption by the small intestine, increasing plasma concentrations of both. This effect is mediated by increased production of calcium transport proteins resulting from gene transcription events and usually requires several hours to appear.

Abnormalities of Bone Mineral Metabolism

There are a number of **metabolic bone diseases,** all typified by ongoing disruption of the normal processes of either bone formation or bone resorption. The conditions most frequently encountered clinically are osteoporosis, osteomalacia, and Paget's disease.

■ Osteoporosis Is a Reduction in Bone Mass

Osteoporosis is a major health problem, particularly since the elderly are more prone to this disorder and the average age of the population is increasing. Osteoporosis involves a reduction in total bone mass with an equal loss of both bone mineral and organic matrix. A number of factors are known to contribute directly to osteoporosis. Long-term dietary calcium deficiency can lead to osteoporosis, since bone mineral is mobilized to maintain plasma calcium levels. Vitamin C deficiency also can result in net loss of bone, since vitamin C is required for normal collagen synthesis to occur. A defect in matrix production and the inability to produce new bone eventually result in net loss of bone. For reasons that are not entirely understood, a reduction in the mechanical stress placed on bone can lead to bone loss. Immobilization or disuse of a limb, such as with a cast or paralysis, can result in localized osteoporosis of the affected limb. Space flight can produce a type of disuse osteoporosis owing to the conditions of weightlessness.

Most commonly, osteoporosis is associated with advancing age in both men and women and cannot be assigned to any specific definable cause. For several reasons, women are more prone to develop the disease than men. Figure 38-10 shows the average bone mineral content (as grams of calcium) for men and women versus age. Until roughly the time of puberty, males and females have similar bone mineral content. However, at this time males begin to acquire bone mineral at a greater rate, such that peak bone mass may be approximately 20% greater than that of women. Maximum bone mass is attained between 30 and 40 years of age and then tends to decrease in both sexes. Initially this occurs at a roughly equivalent rate, but women begin to experience a more rapid rate of bone mineral loss at the time of menopause (roughly 50 years of age). This appears to result from the decline in estrogen secretion that occurs at menopause. Low-dose estrogen supplementation of postmenopausal women is usually effective in combating osteoporosis without causing undue side effects. This condition of increased bone loss in women after menopause is called **postmenopausal osteoporosis.**

■ Osteomalacia and Rickets Result from Inadequate Bone Mineralization

Osteomalacia and rickets are characterized by inadequate mineralization of new bone matrix, such that the ratio of bone mineral to matrix is reduced. As a result, bones may have reduced strength and are subject to distortion in response to mechanical loads. Osteomalacia is the name given to the disease when it occurs in adults and rickets when it occurs in children. In children, the condition often produces bowing of long bones in the legs. In adults, it is often associated with severe bone pain.

The primary cause of osteomalacia and rickets is a deficiency in vitamin D activity. Vitamin D may be deficient in the diet, it may not be converted into its hormonally active form, or target tissues may not adequately respond to the active hormone (Table 38-3). Dietary deficiency is generally not a problem in the United States, where vitamin D is added to many foods, but is a major health problem in other parts of the world. Because the liver and kidneys are involved in conversion of vitamin D_3 into its hormonally active

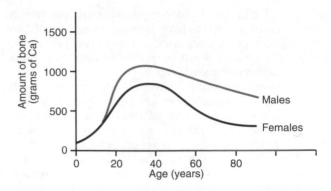

Figure 38–10 ■ Changes in bone calcium content as a function of age in males and females. These changes can be roughly extrapolated into changes in bone mass and bone strength.

form, primary disease of either of these organs may result in vitamin D deficiency. Impairments in action of vitamin D are somewhat rare but can be produced by certain drugs. In particular, some anticonvulsants used in the treatment of epilepsy may produce osteomalacia or rickets after prolonged treatment.

■ Paget's Disease Leads to Disordered Bone Formation

Paget's disease affects about 3% of those older than 40 years. It is typified by disordered bone formation and resorption (remodeling) and may occur at a single local site or at multiple sites in the body. Radiographs of affected bone often exhibit increased density, but the abnormal structure makes the bone weaker than normal. Often those with Paget's disease experience considerable pain, and in severe cases there may be crippling deformities leading to serious neurologic complications. The cause of the disease is not well understood. Both a genetic basis and an environmental basis (probably viral) appear to be important factors. Several therapies are available for treatment of the disease, including treatment with salmon calcitonin, but these typically offer only temporary relief from pain and complications.

TABLE 38–3 ■ Causes of Osteomalacia and Rickets

Inadequate availability of vitamin D
Dietary deficiency or lack of exposure to sunlight
Fat-soluble vitamin malabsorption syndrome
Defects in metabolic activation of vitamin D
25-Hydroxylation (liver)
Liver disease
Certain anticonvulsants, such as phenobarbital
1-Hydroxylation (kidney)
Renal failure
Hypoparathyroidism
Impaired action of 1,25-OH vitamin D_3 on target tissues
Certain anticonvulsants
Uremia

■ ■ ■ ■ ■ ■ ■ ■ ■ ■ ■ REVIEW EXERCISES ■ ■ ■ ■ ■ ■ ■ ■ ■ ■

Identify Each with a Term

1. A characteristic spasm of the muscles of the arm resulting from hypocalcemia in which there is flexion of the wrist and thumb with extension of the fingers

2. The primary form in which phosphorus circulates in the plasma

3. The nephron segments where parathyroid hormone acts to regulate urinary calcium excretion

4. The organic matrix of bone

5. Cells responsible for osteoid synthesis

6. Cells of the thyroid gland responsible for calcitonin synthesis

7. Metabolic bone disease in which there is equal loss of organic matrix and mineral from bone

Define Each Term

8. Chvostek's sign

9. Hydroxyapatite

10. Ground substance

11. Canaliculi

12. Osteoclasts

13. Epiphyseal plate

14. 1,25-Dihydroxycholecalciferol

15. Osteomalacia

Choose the Correct Answer

16. What percentage of the total plasma calcium is normally present as the free Ca^{2+} ion?
    a. 1%
    b. 50%
    c. 60%
    d. 100%

17. The major route by which ingested calcium leaves the body is via the:
    a. Urine
    b. Sweat
    c. Feces
    d. Bile

18. The major route by which ingested phosphate leaves the body is via the:
    a. Urine
    b. Sweat
    c. Feces
    d. Bile

19. Which of the following is *not* either directly or indirectly involved in formation of the active metabolite of vitamin D_3?
    a. Bone
    b. Skin
    c. Kidney
    d. Liver

20. The kidney enzyme 1α-hydroxylase catalyzes the conversion of:
    a. Cholecalciferol to dehydrocholesterol
    b. Vitamin D_3 to vitamin D_2
    c. 25-Hydroxycholecalciferol to 1,25-dihydroxycholecalciferol
    d. Calcium to hydroxyapatite

Briefly Answer

21. What is the mechanism of hypocalcemic tetany?

22. Explain how kidney disease can result in osteomalacia and rickets.

23. What actions of parathyroid hormone lead to increased plasma calcium and decreased plasma phosphate?

■ ■ ■ ■ ■ ■ ■ ■ ■ ■ ■ ANSWERS ■ ■ ■ ■ ■ ■ ■ ■ ■ ■

1. Trousseau's sign

2. Orthophosphate (as HPO_4^{2-} and $H_2PO_4^-$)

3. Thick ascending limb and distal tubule

4. Osteoid

5. Osteoblasts

6. Parafollicular cells

7. Osteoporosis

8. A spasm of the facial muscles elicited by tapping the facial nerve at the point where it crosses the angle of the jaw

9. The crystalline form of calcium phosphate in bone

10. An amorphous mixture of proteoglycans that constitutes about 5% of the organic matrix of bone

11. Channels in bone that are largely the remnants of earlier cellular processes connecting adjacent osteoblasts

12. Large, multinucleated cells responsible for bone resorption

13. Regions of the long bones that are responsive to hormones involved in promoting growth; the responsive cells are chondrocytes that increase collagen synthesis, which ultimately results in increased bone formation

14. The hormonally active form of vitamin D; it acts primarily on the small intestine to increase calcium uptake, but also acts on bone to promote PTH action

15. A form of metabolic bone disease resulting from inadequate mineralization of bone; the juvenile form is called rickets.

16. b

17. c

18. a

19. a

20. c

21. Extracellular calcium influences sodium permeability in nerve cells. In hypocalcemic conditions, sodium in-

flux into nerve may be increased to the extent that spontaneous action potentials are generated. If a motor nerve is involved, this results in muscle twitches and, in more severe cases, tetanization of the muscle.

22. The final step in formation of hormonally active vitamin D (1,25-dihydroxycholecalciferol) occurs in the kidney, catalyzed by 1α-hydroxylase. The enzyme activity is regulated or influenced by PTH and by the plasma phosphate concentration. Certain forms of kidney disease can result in an inability to form 1,25-dihydroxycholecalciferol and thus lead to osteomalacia or rickets.

23. Parathyroid hormone primarily exerts its effects in three target tissues: bone, kidney, and the gastrointestinal tract. In bone, PTH promotes resorption by increasing the activity of osteoclasts. In kidney, it promotes calcium reabsorption and phosphate excretion and increases the activity of 1α-hydroxylase, leading to increased vitamin D activation. The effects of PTH that increase calcium and phosphate uptake in the gastrointestinal tract are indirect, resulting from its role in promoting active vitamin D formation. The net result of these individual effects is to increase plasma calcium and decrease plasma phosphate.

Suggested Reading

Aurbach GD, Marx SJ, Spiegel AM: Metabolic bone disease, in Wilson JD, Foster DW (eds): *Williams Textbook of Endocrinology.* 8th ed. Philadelphia, WB Saunders, 1992, pp 1497–1517.

Aurbach GD, Marx SJ, Spiegel AM: Parathyroid hormone, calcitonin, and the calciferols, in Wilson JD, Foster DW (eds): *Williams Textbook of Endocrinology.* 8th ed. Philadelphia, WB Saunders, 1992, pp 1397–1496.

Goodman HM: *Basic Medical Endocrinology.* 2nd ed. New York, Raven, 1994.

Martin CR: *Endocrine Physiology.* New York, Oxford University Press, 1985.

Norman AW, Litwack G: *Hormones.* Orlando, Academic, 1987.

Reichel H, Koeffler HP, Norman AW: Role of the vitamin D endocrine system in health and disease. *N Engl J Med* 320:980–991, 1989.

PART TEN

Repro-ductive Physiology

The Male Reproductive System

CHAPTER OUTLINE

OBJECTIVES

After studying this chapter, the student should be able to:
1. Identify the endocrine glands involved in male reproduction
2. Describe the gross anatomy and cellular structures of the testes, duct system, and accessory glands
3. Outline the process of spermatogenesis, including its time course, spatial distribution, and role of supporting cells and hormones; identify the morphologic and functional features of a mature spermatozoon
4. Describe the biosynthetic pathways leading to testosterone synthesis; discuss the enzymes involved in steroidogenesis, their intracellular compartmentalization, and regulation by gonadotrophins
5. Describe the negative feedback interactions between the hypothalamic, pituitary, and testicular hormones
6. List the effects of androgens on target organs, secondary sex characteristics, libido, and sexual behavior
7. Identify the causes of major reproductive dysfunctions in men

The production of viable progeny depends on the proper functioning of highly specialized organs, the gonads. In the male, the testes fulfill a dual function. One is **spermatogenesis,** the process by which the sperm are produced. The second is **steroidogenesis,** the synthesis of sex steroids. Both processes are subject to regulation by gonadotrophins from the anterior pituitary. Steroidogenesis in both sexes proceeds along the same biochemical pathway, but with a different emphasis. Testosterone is the primary sex hormone in the male, whereas estradiol and progesterone are the predominant sex steroids in the female. Testosterone is essential for spermatogenesis and also has profound effects on practically every system in the body, being responsible for sexual dimorphism in many physiologic functions and affecting male sexual behavior.

Overview of the Male Reproductive System

A schematic overview of the male reproductive system is presented in Figure 39–1. The system is divided into: brain centers, which control endocrine functions and sexual behavior; endocrine glands, which secrete hormones; gonadal structures, which produce sperm and hormones; a duct system, which stores and transports sperm; and accessory glands, which support sperm viability.

The endocrine glands include the hypothalamus, anterior pituitary, and testes. The hypothalamus processes information obtained from the external and internal environment and relays it to the anterior pituitary by means of the **gonadotrophin-releasing hormone** (GnRH). The GnRH acts on the gonadotrophs of the anterior pituitary and stimulates the secretion of **luteinizing hormone** (LH) and **follicle stimulating hormone** (FSH). The targets for LH and FSH are the Leydig and Sertoli cells, respectively. The **Leydig cells** reside in the interstitium of the testis and produce the sex steroids. The **Sertoli cells** are located in the seminiferous tubules and support spermatogenesis.

Testosterone belongs to a class of hormones, the **androgens,** that promote "maleness." It carries out multiple functions, including feedback on the hypothalamus and anterior pituitary, support of spermatogenesis, regulation of sexual behavior, and support of secondary sex characteristics. The testes also produce a glycoprotein hormone, **inhibin,** which regulates FSH release.

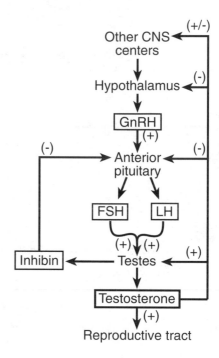

Figure 39–1 ■ Overall regulation of reproduction in the male. The main reproductive hormones are shown in boxes. Positive and negative regulation are depicted by + and − signs, respectively.

The duct system includes the epididymis, vas deferens, urethra, and penis. The sperm acquire motility and fertility in the epididymis and are stored in the epididymis and vas deferens. They are transported via the urethra to the penis, to be expelled by ejaculation. The accessory glands include the prostate, seminal vesicles, and bulbourethral glands. These contribute many constituents to the seminal fluid that are necessary for maintaining sperm viability.

Hypothalamic Regulation of the Testes

■ Neurons That Produce GnRH Are Distributed in Several Hypothalamic Sites

Information arising from a variety of external cues and from the internal environment is fed into the brain.

Through a complex network of neurotransmitters and neuropeptides, this information is processed, amplified, integrated, and transduced to the hypothalamus. The hypothalamus translates these signals into the production and pulsatile secretion of a humoral messenger, the decapeptide GnRH, that influences the anterior pituitary. Neurons that produce GnRH develop initially in the olfactory placode and then migrate during embryonic development toward the hypothalamus. The GnRH neurons in adults are scattered throughout the preoptic area, septum, and diagonal band of Broca. In humans, their highest concentration is in the medial basal hypothalamus, in the region of the infundibulum and arcuate nucleus. Patients with **Kallmann syndrome** are anosmic (unable to smell) and hypogonadal due to GnRH deficiency resulting from a failure of GnRH neurons to migrate from the olfactory bulbs.

Most projections of the GnRH neurons terminate in the median eminence. The GnRH released from these terminals is picked up by the capillaries of the portal vessels and transported to the anterior pituitary. Some GnRH neurons also send projections to extrahypothalamic areas, including the mesencephalic central gray matter, midbrain, hippocampus, amygdala, and olfactory tubercle. Several of these structures mediate olfactory and visual cues that play important roles in reproduction. The limbic system, which contains receptors for GnRH, participates in coordination of sexual behavior.

Gonadotrophin-releasing hormone is initially synthesized as a large precursor molecule (Fig. 39–2). The preproGnRH consists of a 23-amino acid signal peptide, a 10-amino acid GnRH, and a 56-amino acid GnRH-associated peptide (GAP). The GnRH is flanked by pairs of basic amino acids that designate enzymatic cleavage sites. The α amino and γ carboxyl groups of glutamine at the amino terminal of GnRH form an internal peptide bond, producing pyroglutamic acid (pGlu). The glycine residue on the carboxy terminal is amidated. As a consequence, both ends of the GnRH molecule are "blocked" (i.e., protected from degradation by exopeptidases). The mRNA sequence encoding GnRH is highly conserved among various species, underlying its importance in fulfilling an essential physiologic function.

■ GnRH Secretion Is Pulsatile and Is Regulated by Gonadal Hormones and Neurotransmitters

The hormones of the hypothalamopituitary-gonadal axis are secreted in a pulsatile manner, and this appears to be necessary for their proper functioning. In primates, GnRH is released into portal blood at a frequency of one pulse per hour to one pulse per 3–4 hours, each pulse lasting for only few minutes. The discrete bursts of GnRH secretion are followed by pulses of LH release (Fig. 39–3). Follicle stimulating hormone is also released in a pulsatile manner, but the lack of complete synchrony with LH suggests that its release is regulated by additional factors. Although the exact identity of the cells responsible for generating GnRH pulsatility is unknown, the presence of a **pulse generator** has been postulated. The putative pulse generator resides in the medial basal hypothalamus and is responsible for the synchronized and rhythmic firing of a population of neurons. The activity of the pulse generator is modified by several factors. For example, castration in either males or females causes a large increase in frequency and amplitude of LH pulses. This suggests that the pulse generator is tonically inhibited by gonadal steroids. However, GnRH neurons lack receptors for gonadal steroids, suggesting that steroidal effects are mediated by other agents. The most likely candidates are norepinephrine, dopamine, and opioid peptides.

The mode of action of GnRH is unique among the hypothalamic releasing hormones. Constant exposure

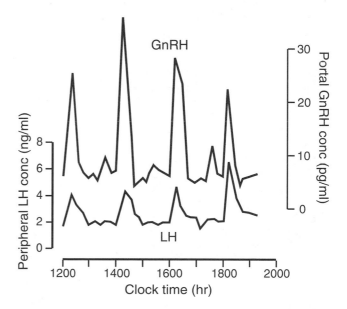

Figure 39–3 ■ Coordinate pulsatile release of gonadotrophin-releasing hormone (GnRH) in portal blood and luteinizing hormone (LH) in peripheral blood in sheep. (Modified from Clarke IJ, Cummins JT: The temporal relationship between gonadotropin releasing hormone (GnRH) and luteinizing hormone (LH) secretion in ovariectomized ewes. *Endocrinology* 111:1737–1739, 1982.)

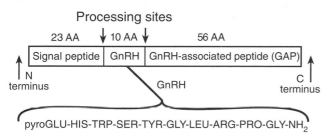

Figure 39–2 ■ The precursor molecule that contains gonadotrophin-releasing hormone (preproGnRH) and the amino acid sequence of GnRH.

of the gonadotrophs to GnRH results in desensitization of GnRH receptors, leading to a decline in LH and FSH release. Thus, the pulsatile pattern of GnRH release serves an important physiologic function. Administration of GnRH at an improper frequency (too slow or too fast) results in a decline in hormone release or a substantial change in the ratio of LH/FSH release. The realization that gonadotrophin secretion necessitates pulsatile delivery of GnRH has helped in designing an effective therapy for some reproductive disorders. Individuals who are GnRH-deficient remain sexually juvenile, since they do not undergo normal puberty. Treatment with a long-acting GnRH analog that provides constant stimulation of the pituitary is ineffective. When they are fitted with a pump that delivers GnRH intermittently, pubertal maturation is induced and normal reproductive function is often restored.

■ LH and FSH Are Glycoproteins Produced by Anterior Pituitary Gonadotrophs

Luteinizing hormone and FSH are produced and secreted by the same cell, although some gonadotrophs are monohormonal. The gonadotrophs are morphologically heterogeneous, reflecting either biochemical-functional specialization or different stages in the cell cycle. They are scattered throughout the anterior pituitary and constitute 5–10% of the total cell population. Like other hormone-producing cells, they have prominent rough endoplasmic reticulum, well-developed Golgi complexes, and many secretory granules. The secretory granules are small relative to other pituitary cells, ranging in diameter from 250 to 350 nm. Neither the number nor the morphology of the gonadotrophs is constant, but change with the physiologic conditions. Castration, for example, induces a two- to threefold increase in the number of gonadotrophs, presumably due to increased cell division. It also results in the appearance of "castration cells," hypertrophied gonadotrophs characterized by dilation of the endoplasmic reticulum and the presence of large vacuoles.

Although named for their function in females, LH and FSH in the male are identical in structure to their counterparts in the female. As discussed in Chapter 34, each hormone is composed of α and β subunits. The α subunits in LH and FSH are identical and the β subunits differ. The β subunit confers the biologic activity of the hormone. A single gene codes for the common α subunit, but separate genes code for the different β subunits.

■ GnRH-Stimulated Gonadotrophin Release Requires Intracellular Mediators

Gonadotrophin-releasing hormone binds to a single class of high-affinity receptors on the plasma membrane of the gonadotrophs. The affinity of the receptor for GnRH remains relatively constant, whereas the number of receptors fluctuates and is subject to hormonal regulation. Prolonged exposure of gonadotrophs to GnRH results in desensitization. This involves a reduction in the number of surface receptors, uncoupling of the receptor from intracellular mediators, and postreceptor inactivation mechanisms. Conversely, intermittent exposure to GnRH results in increased sensitivity. The self-priming effect of GnRH induces an increase in the number of receptors, thus enhancing pituitary responsiveness to subsequent exposures to GnRH.

The binding of GnRH to the receptor elicits a biphasic rise in intracellular calcium. The first phase is sharp and transient, whereas the second is prolonged and of lower magnitude. The first phase involves the phosphatidylinositol pathway. Activation of phospholipase C via a guanine-nucleotide binding protein (G protein) causes the cleavage of phosphatidylinositol 4,5-bisphosphate into inositol trisphosphate (IP_3) and diacylglycerol (DAG). The IP_3 stimulates a rapid release of calcium from intracellular stores, primarily in the endoplasmic reticulum, which contains specific receptors for IP_3. Diacylglycerol, on the other hand, increases the activity of calcium-dependent protein kinase C (Fig. 39–4).

Protein kinase C is a cytosolic enzyme. On stimulation, it shifts to the plasma membrane and binds DAG, calcium, and phospholipids. The activated enzyme phosphorylates several classes of proteins by transferring a phosphate group from ATP to hydroxyl groups of serine or threonine of target proteins. These include cytoskeletal proteins and proteolytic enzymes important for exocytosis, membrane channels that facilitate ion transport, and DNA-binding proteins (transcription factors), which regulate gene expression. Thus, activation of protein kinase C by GnRH results in a cascade of events lasting various periods of time that culminate in increased hormone release, followed, in time, by enhanced hormone synthesis.

Gonadotrophin-releasing hormone binding also induces an influx of extracellular calcium. This results in the second, prolonged phase of cytosolic calcium elevation. Calcium enters the cells via voltage-gated or ligand-gated channels. Ligand-gated channels predominate in the gonadotrophs. These channels are activated by direct coupling to ligand-occupied receptors or indirectly through phosphorylation by protein kinases.

The GnRH-stimulated LH release occurs by exocytosis, which is always preceded by a large rise in intracellular calcium and increased activity of calcium-calmodulin-dependent protein kinase and protein kinase C. The calcium level is returned to its normal intracellular value by three mechanisms: sequestration into intracellular organelles, such as mitochondria, secretory granules, and endoplasmic reticulum; neutralization by binding to cytosolic proteins serving as internal buffers; and transport out of the cell by an

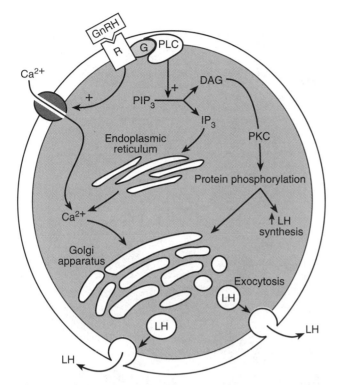

Figure 39–4 ■ Proposed intracellular mechanism by which gonad-otrophin-releasing hormone (GnRH) stimulates luteinizing hormone (LH) release from pituitary gonadotrophs. Binding of GnRH to membrane receptors increases intracellular Ca^{2+} levels and phosphorylation of proteins, which together lead to increased LH release. The rise in intracellular Ca^{2+} occurs via two mechanisms: stimulation of release of Ca^{2+} from intracellular storage sites and increased Ca^{2+} influx from the extracellular fluid. R, receptor; G, G protein; PLC, phospholipase C; PKC, protein kinase C; PIP_2, phosphatidylinositol bisphosphate; IP_3, inositol trisphosphate; DAG, diacylglycerol.

energy-dependent calcium-ATPase or a Na^+-Ca^{2+} countertransporter.

■ Gonadal Hormones Exert Negative Feedback on Gonadotrophin Release

Gonadotrophins have one peripheral target tissue, the gonads. Gonadal products, in turn, provide feedback information to the gonadotrophs. Testosterone inhibits LH release by decreasing the amount of GnRH reaching the anterior pituitary and, to a lesser extent, by reducing pituitary sensitivity to the peptide. Testosterone primarily slows the frequency of GnRH pulses and only slightly reduces the pulse amplitude. Part of its action is carried out after conversion to estradiol by the enzyme P450arom (aromatase cytochrome P450). Acute testosterone treatment does not alter pituitary responsiveness to GnRH, but prolonged exposure significantly reduces the secretory response to GnRH.

Removal of the testes results in increased circulating levels of LH and FSH. Replacement therapy with physiologic doses of testosterone restores LH to precastra-

tion levels but does not completely correct FSH levels. This observation led to a search for a gonadal factor that specifically inhibits FSH release. The polypeptide hormone **inhibin** was eventually isolated from both seminal fluid and ovarian tissue. Inhibin has a molecular weight of 32,000 and is composed of two dissimilar subunits (α and β) held together by disulfide bonds. Inhibin acts directly on the anterior pituitary and inhibits the secretion of FSH but not LH.

The Male Reproductive Organs

■ The Testis Is the Site of Sperm Formation

The testes are paired ovoid organs that lie suspended in the scrotal sac (Fig. 39–5). Shortly before birth,

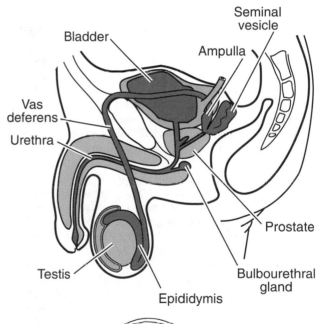

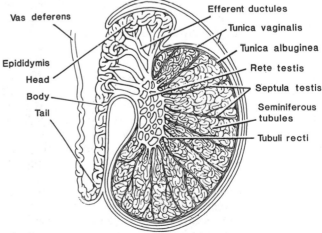

Figure 39–5 ■ The male reproductive organs (top) and cutaway drawing of the testis, epididymis, and vas deferens (bottom).

they descend from the abdominal cavity through the inguinal canal into the scrotum. The location of the testes in the scrotum is important for sperm production, which is optimal at 2–3°C below core body temperature. Two systems help maintain the testes at a cooler temperature. One is the **pampiniform plexus** of blood vessels, which serves as a countercurrent heat exchanger between warm arterial blood reaching the testes and cooler venous blood leaving the testes. The second is the **cremasteric muscle,** which responds to changes in temperature by moving the testes closer or farther away from the body. Prolonged exposure of the testes to elevated temperature, fever, or dysfunction of thermoregulation can lead to temporary or permanent sterility due to failure of spermatogenesis, whereas steroidogenesis is unaltered.

The testes are encapsulated by a thick fibrous layer, the tunica albuginea, made of connective tissue and smooth muscle. Each testis contains hundreds of tightly packed **seminiferous tubules,** ranging in diameter from 150 to 300 μm and in length from 30 to 70 cm. The tubules are arranged in lobules, separated by extensions of the tunica albuginea, and open on both ends into the rete testis. Examination of a cross-section of the testis reveals distinct morphologic compartmentalization. Sperm production is carried out in the seminiferous tubules, whereas testosterone is produced by the **interstitial cells of Leydig,** which are scattered in the loose connective tissue between the tubules.

Each seminiferous tubule is surrounded by a basement membrane and neighboring myoid, or smooth muscle cells. The perimeter of the tubule is lined by large, irregularly shaped **Sertoli cells,** which span the width of the tubule from the basement membrane to the lumen (Fig. 39-6). Sertoli cells are attached to one another by tight junctions (Fig. 39-7). The junctional complexes limit the transport of fluid and macromolecules from the interstitial space into the tubular lumen, forming the **blood-testis barrier.** Each tubule is effectively divided into a basal compartment, whose constituents are exposed to circulating agents, and an adluminal compartment, which is isolated from blood-borne elements.

Located between the nonproliferating Sertoli cells are germ cells at various stages of division and differentiation. The **spermatogonia,** or diploid progenitors of spermatozoa, are located in the basal compartment (Fig. 39-7). As spermatogenesis proceeds, the dividing cells move across the junctional complexes into the adluminal compartment, where they mature into **spermatozoa,** or **gametes.** The adluminal compartment constitutes an immunologically privileged site. Hence, spermatozoa, which develop only after puberty, are not recognized as "self" by the immune system, which completes its organization during early neonatal life. Consequently, males can be immunized against their own sperm. Antibodies against sperm can

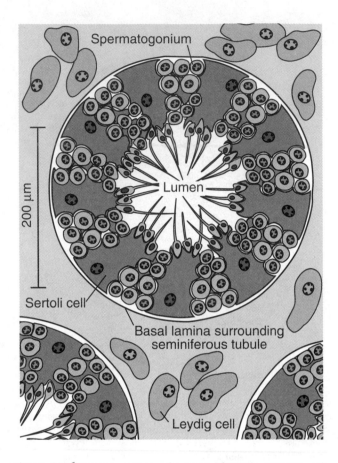

Figure 39–6 ■ Cross-section of a testis, showing anatomic relationship between Leydig cells, basement membrane, seminiferous tubules, Sertoli cells, and sperm. (Modified from Alberts B, Bray D, Lewis M, et al: *Molecular Biology of the Cell.* 2nd ed. New York, Garland, 1989.)

cause infertility and are often present after vasectomy or testicular injury and in some autoimmune diseases.

■ Sertoli Cells Have Multiple Activities

Sertoli cells fulfill numerous functions that change in intensity throughout life. The Sertoli cells undergo active mitosis during fetal life and are affected by agents that regulate the organogenesis of the testes and by Müllerian inhibiting factor, which suppresses the development of female internal genitalia (Chap. 41). During early postnatal life, Sertoli cells lose the ability to multiply and increase their capacity for binding FSH. Receptors for FSH, present only on Sertoli cells, are oligomeric glycoproteins linked to adenylate cyclase via a G protein. Follicle stimulating hormone exerts multiple effects on the Sertoli cell, all of which are mediated by cAMP and protein kinase A (Fig. 39-8). Among these are stimulation of the production of **androgen binding protein** (ABP), increased secretion of inhibin, induction of P450arom activity, production

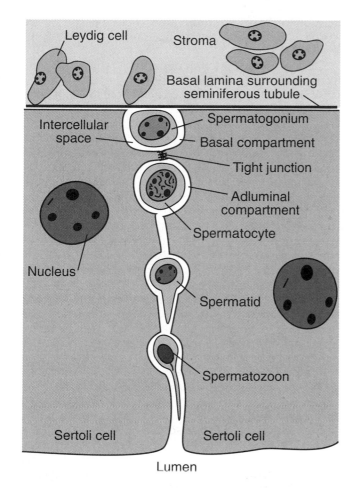

Figure 39–7 ■ Sertoli cells are connected by tight junctions, which divide the intercellular space into a basal compartment and an adluminal compartment. The spermatogonia are located in the basal compartment and maturing sperm in the adluminal. Spermatocytes bud from the spermatogonia and cross the tight junctions into the adluminal compartment, where they mature into spermatozoa. (Modified from Fawcett DW: Ultrastructure and function of the Sertoli cell, in Hamilton DW, Greep RO (eds). *Handbook of Physiology.* Sect. 7, vol. 5. Washington, DC, American Physiological Society, 1975.)

of plasminogen activator, and increased cell proliferation.

Androgen binding protein is a 90 kd protein, made of a heavy and a light chain, which has a high binding affinity for dihydrotestosterone and testosterone. It is similar in function, with some homology in structure, to another binding protein, sex steroid binding globulin (SSBG), which is of liver origin. Androgen binding protein is found at high concentrations in the human testes and epididymis. It serves as a carrier of testosterone in the Sertoli cell, as a storage protein for androgens in the seminiferous tubules, and as a carrier of testosterone from the testes to the epididymis.

Another specific product of the Sertoli cell is in-

hibin, which suppresses FSH release from the pituitary gonadotrophs. The pituitary gonadotrophs and testicular Sertoli cells form a classical negative feedback loop in which FSH stimulates inhibin secretion and inhibin suppresses FSH release. Inhibin also functions as a paracrine agent in the testes.

The intimate juxtaposition between the germ and Sertoli cells suggests that the latter play a role in supporting and nurturing the gametes. Sertoli cells identify and phagocytize damaged germ cells and assist in **spermiation,** or the final detachment of mature spermatozoa from the Sertoli cell into the lumen. The latter function is carried out by **plasminogen activator,** which converts plasminogen to plasmin, a proteolytic enzyme. Sertoli cells also synthesize large amounts of transferrin, an iron-transport protein important for sperm development.

■ Leydig Cells Produce Testosterone

Similar to other steroid-producing cells, **Leydig cells** are of mesenchymal origin. These large polyhedral cells are often found in clusters near blood vessels in the interstitium between seminiferous tubules. They have numerous mitochondria, a prominent smooth endoplasmic reticulum, and conspicuous lipid droplets. They have only small Golgi complexes and very few secretory granules. Leydig cells do not require the processing-packaging function of the Golgi or the storage capability of granules because steroids are lipid-soluble and are not stored in the producing cells.

Leydig cells undergo significant changes in quantity and activity throughout life. In the human fetus, the period from week 8 to week 18 is marked by active steroidogenesis, which is obligatory for differentiation of the male genital ducts. Leydig cells at this time are prominent and very active, reaching their maximal steroidogenic activity at about 14 weeks, when they constitute more than 50% of the testicular volume. Because the fetal hypothalamopituitary axis is still underdeveloped, steroidogenesis is controlled by **human chorionic gonadotrophin** (hCG) from the placenta, rather than by LH from the fetal pituitary (Chap. 41). After this period, Leydig cells slowly regress. At about 2–3 months of life, male infants have a significant rise in testosterone production (infantile testosterone surge), the regulation and function of which are unknown. Leydig cells remain quiescent throughout childhood but increase in number and activity at the onset of puberty.

The Leydig cell is a target for LH (Fig. 39–8). The major effect of LH is to stimulate steroidogenesis via a cAMP-dependent mechanism. The main product of the Leydig cell is testosterone, but two other androgens of less biologic activity, dehydroepiandrosterone (DHEA) and androstenedione, are also produced. Luteinizing hormone also exerts a trophic effect on the Leydig cell, and its presence is required for maintaining overall cell function.

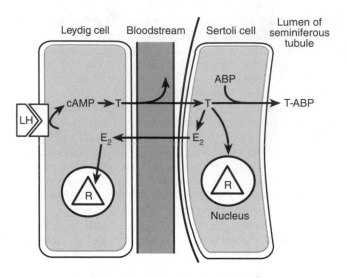

Figure 39–8 ▪ Regulation, hormonal products, and interactions between Leydig and Sertoli cells. ABP, androgen-binding protein; E_2, estradiol; T, testosterone; R, receptor.

There are bidirectional interactions between Sertoli and Leydig cells (Fig. 39–8). The Sertoli cell is incapable of producing testosterone, but it contains testosterone receptors as well as an FSH-dependent P450arom. The Leydig cell does not produce much estradiol but contains receptors for it. Testosterone diffuses from the Leydig to the Sertoli cell, binds to ABP, and can reach very high local concentrations. Testosterone is obligatory for spermatogenesis and the proper functioning of the Sertoli cell. In the Sertoli cell, testosterone also serves as a precursor for estradiol production. The role of estradiol in the functioning of the Leydig cell is unclear. Other compounds, such as opioids, a GnRH-like peptide, vasopressin, oxytocin, and several growth factors and neurotransmitters, previously thought to be produced only by the brain, have been detected in the testes and are proposed to function as paracrine agents.

▪ The Duct System Functions in Sperm Maturation, Storage, and Transport

After formation in the seminiferous tubules, spermatozoa are transported to the rete testes and from there through the efferent ductules to the **epididymis** (Fig. 39-5). This is accomplished by ciliary movement, flow of fluid, and muscle contraction. The epididymis is a single, tightly coiled duct ranging from 4 to 5 meters long. It is composed of a head (caput), body (corpus), and tail (cauda). A significant portion of sperm maturation is carried out in the caput, whereas sperm is stored in the cauda. Consequently, heightened sexual activity and very frequent ejaculation can result in the appearance of immature and immotile sperm in the ejaculate. The ampulla of the **vas deferens** serves as an accessory storage site for sperm. Disconnection of the vas in the scrotal area, or vasectomy, is an effective method of male contraception. Because sperm are stored in the ampulla, men remain fertile for 4–5 weeks after vasectomy.

Successful reproduction depends on the male's ability to achieve and maintain penile erection. During sexual arousal, afferent stimuli derived from the genitalia, together with nerve signals originating in the limbic system, trigger motor impulses in the spinal cord. These are carried by the nervi erigentes to the penis, causing vasodilation of the arterioles as well as venoconstriction. As a result, blood is trapped in the surrounding erectile tissue, leading to engorgement of the penis and erection. Sperm is expelled by a reflex process, divided into two sequential phases. **Emission** causes the sperm to move from the cauda epididymis to the urethra. Efferent stimuli originating in the lumbar area of the spinal cord and mediated by adrenergic sympathetic nerves induce contraction of smooth muscles surrounding the epididymis, vas deferens, and ejaculatory duct. This propels the sperm into the urethra and closes the internal urethral sphincter, thus preventing retrograde ejaculation into the bladder. **Ejaculation** involves expulsion of the sperm from the penile urethra. This phase is also controlled by a reflex mechanism and involves rhythmic contractions of the striated muscles surrounding the urethra. Unejaculated spermatozoa are eventually phagocytosed and reabsorbed.

The secretions of the accessory glands promote sperm survival and fertility. The semen contains only 10% sperm by volume, with the remainder consisting of the combined secretions of the accessory glands. The **seminal vesicles** contribute about 60% of the seminal fluid. Their secretion contains fructose (the principal substrate for sperm glycolysis), ascorbic acid, and prostaglandins. In fact, prostaglandins were first discovered in seminal fluid and were mistakenly considered the product of the prostate. The **prostate** produces polyamines, citric acid, cholesterol, fibrinolysin, and acid phosphatase.

Spermatogenesis

▪ Spermatogenesis Is an Ongoing Process from Puberty to Senescence

Spermatogenesis is divided into three distinct phases: **mitosis**, or cell replication; **meiosis**, or reductional division; and **spermiogenesis**, or cell differentiation (Fig. 39-9). Although spermatogenesis begins only at puberty, germ cells are not dormant during fetal life. Primordial germ cells are identified in the dorsal yolk sac by 24-25 days of gestation. The cells migrate by ameboid movement to the location of the developing testes and become embedded in its medulla. Between 3 and 6 months, they undergo rapid proliferation, followed by a massive loss of cells. The mechanisms responsible for the dramatic rise and fall in the number of germ cells are not understood.

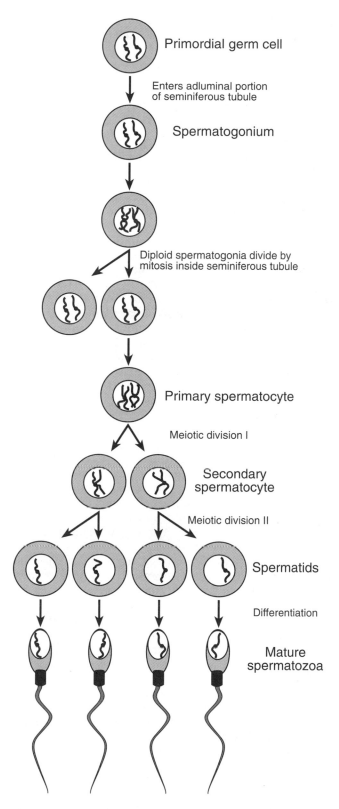

Figure 39–9 ■ The process of spermatogenesis, showing successive cell divisions and remodeling leading to the formation of haploid spermatozoa. (Modified from Alberts B, Bray D, Lewis J, et al: *Molecular Biology of the Cell.* 2nd ed. New York, Garland, 1989.)

[figure labels: Primordial germ cell; Enters adluminal portion of seminiferous tubule; Spermatogonium; Diploid spermatogonia divide by mitosis inside seminiferous tubule; Primary spermatocyte; Meiotic division I; Secondary spermatocyte; Meiotic division II; Spermatids; Differentiation; Mature spermatozoa]

Throughout childhood, the seminiferous tubules are quiescent. Spermatogenesis is initiated shortly before puberty, under the influence of the rising levels of gonadotrophins, and continues throughout life, with only a slight decline during old age.

Like other cells undergoing active division (e.g., bone marrow, embryonic cells), sperm cells are sensitive to mitogens. Chemical carcinogens, certain drugs, toxins, irradiation, or extremes of temperature can reduce the number of replicating germ cells or cause chromosomal abnormalities in individual cells. While defective somatic cells are normally detected and destroyed by the immune system, the blood-testis barrier isolates advanced germ cells from immune surveillance. If a defective sperm fertilizes an egg, the defect can have profound and long-lasting effects, since it will be transmitted to all the cells of the fetus.

The duration of the spermatogenic cycle is 65–70 days in humans. Hormones and cytotoxic agents can alter the number of spermatozoa but do not affect the duration of the cycle. Spermatogenesis occurs along the length of each seminiferous tubule in successive cycles. New cycles are initiated at regular time intervals (every 2–3 weeks) before the previous ones are completed. Consequently, cells at different stages of development are spaced along each tubule in a "spermatogenic wave." Such a succession ensures continuous and uninterrupted production of fresh spermatozoa. Approximately 200×10^6 spermatozoa are produced daily in the human testes. This is about the same number of sperm present in a normal ejaculate.

■ Spermatogonia Undergo Mitotic and Meiotic Divisions and Become Spermatids

The cell cycle consists of a long period of interphase and a very short period of cell division. Cell division progresses in two sequential steps: mitosis (nuclear division) and cytokinesis (cytoplasmic division). Human somatic and primordial germ cells are diploid, each consisting of 46 chromosomes. Every chromosome exists in two copies, called homologs, derived from the male or the female parent. In preparation for mitosis, each chromosome duplicates and the sister chromatids remain attached by a centromere. Mitosis then begins and the cell undergoes several sequential phases: prophase, metaphase, anaphase, and telophase.

During prophase, the chromosomes condense, a mitotic spindle is formed, and the nuclear envelope disappears. At metaphase, the chromosomes migrate toward the center of the cell, line up halfway between the spindle poles, and are held in position by kinetochore microtubules. At anaphase, the sister chromatids separate and are pulled to opposite poles by shortening of the kinetochores. During telophase, the chromosomes decondense and the nuclear envelope reforms. Cytokinesis then follows, and the cytoplasm and its contents are divided equally by cleavage. The end result is the formation of two identical diploid cells (Fig. 39-9).

Meiosis is unique to the male and female gametes. It differs from mitosis by three criteria. One is the halving of the number of chromosomes, so that each daughter cell is **haploid,** with only 23 chromosomes. This ensures reconstitution of the diploid status on unification with the gamete of the opposite sex. The second involves an exchange of genetic material between homologous chromosomes during **crossing over.** The third is a random segregation of paternal and maternal chromosomes between daughter cells. Consequently, the number of genetically distinct gametes that can be produced by an individual is almost unlimited. Each germ cell entering meiosis undergoes two consecutive cell divisions but only one replication of the chromosomes. The first cell division is dominated by a very prolonged prophase; the second resembles regular mitosis except that it is not preceded by DNA replication (Fig. 39-9).

Each **spermatogonium** entering the meiotic process first undergoes four mitotic divisions, forming 16 daughter diploid clones connected by cytoplasmic bridges. The clones migrate away from the walls of the seminiferous tubules; from then on, their microenvironment is provided by the Sertoli cells. Upon entering the first meiotic division, the cells are called **primary spermatocytes.** The prophase of the first meiotic division takes about 24 days. At this point, each chromosome consists of two closely apposed sister chromatids, the result of the preceding DNA replication. The homologous chromosomes recognize each other, become juxtaposed, and form structures called tetrads (or bivalents). The tetrads pair along their entire length and crossing over occurs between maternal and paternal chromatids. At this stage, the spermatocytes are especially susceptible to chemical or radiation damage, which can cause extensive cell degeneration.

Similar to mitosis, the remainder of the first meiotic division progresses through metaphase, anaphase, and telophase. At the end, the chromosomal complement has been reduced from tetraploidy to diploidy. Each homolog, composed of sister chromatids joined by a centromere, has segregated to a new cell, the **secondary spermatocyte.** Segregation of a maternal or paternal chromosome to each daughter cell is random. Note that secondary spermatocytes are already haploid but still contain double the amount of DNA.

The second meiotic division begins after a transient interphase with no chromosomal replication. This is a rapid process in which the chromosomes divide one more time. The separated chromatids are then distributed into new cells, the **spermatids.** Of every four spermatids emanating from a primary spermatocyte, two contain X chromosomes and two have Y chromosomes. Clinical disorders resulting from a faulty segregation of sex chromosomes, or nondisjunction, are discussed in later chapters. Because of the

four mitotic divisions and meiosis, each spermatogonium committed to meiosis should have yielded 64 spermatids, if all cells survive. All spermatids derived from a single spermatogonium are connected by cytoplasmic bridges.

■ Formation of the Mature Spermatozoon Requires Extensive Cell Remodeling

Spermatids are small, round, and nondistinct cells. During the second half of the spermatogenic cycle they undergo considerable restructuring to form mature spermatozoa. Notable changes include alterations in the nucleus, formation of a tail, and a massive loss of cytoplasm. The nucleus becomes eccentric and decreases in size, and the chromatin becomes condensed. The **acrosome,** a lysosomelike structure unique to spermatozoa, buds from the Golgi complex, flattens, and covers most of the nucleus. The centrioles, located near the Golgi complex, migrate to the caudal pole and form a long axial filament made of nine peripheral doublet microtubules surrounding a central pair (9 + 2 arrangement). This becomes the **axoneme,** or major portion of the tail. Throughout this reshaping process, the cytoplasmic content is redistributed and discarded. During spermiation, most of the remaining cytoplasm is shed in the form of residual bodies.

The reasons for this lengthy and metabolically costly process become apparent when the unique functions of this cell are considered. Unlike other cells, the spermatozoon serves no apparent purpose in the organism itself. Its only function is to reach, recognize, and fertilize an egg. Hence, it must fulfill several prerequisites: it should possess an energy supply and means of locomotion; it should be able to withstand a foreign and even hostile environment; it should be able to recognize and penetrate an egg; and it must carry all the genetic information necessary to create a new individual. The mature spermatozoon exhibits a remarkable degree of structural and functional specialization well adapted to carry out these functions. The cell is small, compact, and streamlined, with a diameter of about 1-2 μm and length that can exceed 50 μm in humans. It is packed with specialized organelles and long axial fibers but contains only few of the normal cytoplasmic constituents, such as ribosomes, endoplasmic reticulum, or Golgi apparatus. It has a very prominent nucleus, a flexible tail, numerous mitochondria, and an assortment of proteolytic enzymes.

As shown in Figure 39-10, the spermatozoon consists of three main parts: the head, the middle piece, and the tail. The two major components in the head are the condensed chromatin and the acrosome. The haploid chromatin is transcriptionally inactive throughout the life of the sperm until fertilization, when the nucleus decondenses and becomes a pronucleus. The acrosome contains proteolytic enzymes, such as hyaluronidase, acrosin, neuraminidase, phos-

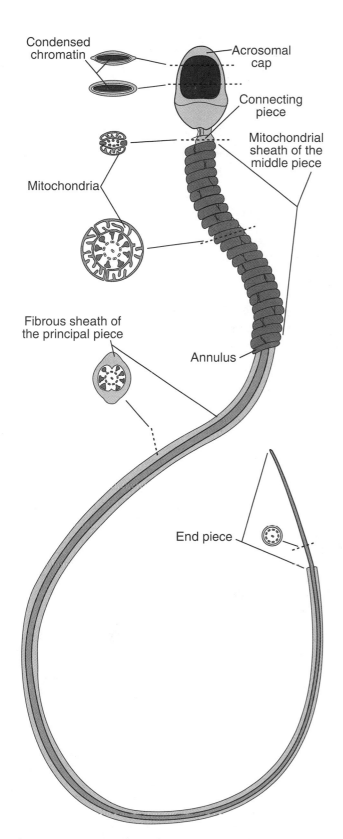

Figure 39–10 ■ Structure of a mature spermatozoon. (Modified from Fawcett DW: Ultrastructure and function of the Sertoli cell, in Hamilton DW, Greep RO (eds): *Handbook of Physiology.* Section 7, vol. 5. Washington, DC, American Physiological Society, 1975.)

pholipase A, and esterases. They are inactive until the **acrosome reaction** occurs upon contact of the sperm head with the egg (Chap. 41). Their proteolytic actions enables sperm to penetrate through the egg membranes. The middle piece contains spiral sheaths of mitochondria that supply energy for sperm metabolism and locomotion. The tail is composed of a 9 + 2 arrangement of microtubules, which is typical of cilia and flagella, and is surrounded by a fibrous sheath that provides some rigidity. The tail propels the sperm by a twisting motion involving interactions between tubulin fibers and dynein side arms and requiring ATP and magnesium.

■ Testosterone Is Essential for Sperm Production and Maturation

Spermatogenesis requires the presence of FSH and high intratesticular levels of testosterone. The FSH and testosterone act primarily through the Sertoli cell, which contains receptors for each. Their exact action at each point of cell division or differentiation is unknown, largely because there is no suitable in vitro model for spermatogenesis. The rate and number of mitotic divisions of the spermatogonia are probably hormone-independent and may be programed by the cell genome. On the other hand, FSH appears to stabilize the cohorts of dividing cells and prevent their degradation. Upon entering meiosis, spermatogenesis depends on the availability of FSH and testosterone. Follicle stimulating hormone is required for the initiation of spermatogenesis before puberty or after a long cessation. When adequate sperm production has been achieved, LH alone (by way of stimulating testosterone production) or testosterone alone will suffice.

Sperm recovered from the rete testes are immobile and incapable of fertilizing an egg. They must pass through the epididymis to become fertile and motile, via extensive morphologic and physiologic modifications. These include changes in the composition of the cell membrane, a switch in substrate use, changes in ion transport, and increased adenylate cyclase and protein kinase activities. While in the epididymis, sperm are exposed to lower pH, increased K^+/Na^+ ratio, altered substrate concentrations, and specific epididymal products, such as enzymes, glycoproteins, and glycolipids. The ability of the epididymis to provide such an environment is androgen-dependent and is lost on withdrawal of androgens.

During their passage through the epididymis, sperm acquire the ability for forward motion, that depends on the length of time they have resided in the epididymal duct. In the epididymis, sperm also develop the ability to bind to the **zona pellucida** of the egg (Chap. 40). This binding depends on the presence of species-specific surface recognition molecules. Normally, sperm are incapable of binding to the zona of an egg from a different species. This fact has helped in designing a functional test for human fertility. Human sperm is exposed to hamster eggs whose zona pellucidae have been digested by proteolytic enzymes. The percentage of sperm penetrating the eggs and initiating the first cell division is scored. This zona-less penetration test, while imperfect, provides a diagnostic tool for evaluating sperm that depends on functional, rather than morphologic, criteria.

Although testosterone plays a key role in spermatogenesis, infertile men with a low sperm count do not benefit from testosterone treatment. Unless given at supraphysiologic doses, exogenous testosterone cannot achieve the required local high concentration. In fact, exogenous testosterone inhibits endogenous LH release, leading to a suppression of testosterone production by the Leydig cells and a further decrease in testicular testosterone. Another adverse effect is the induction of feminization because of peripheral conversion of testosterone to estradiol.

Testicular Steroidogenesis

■ Testosterone Production Requires Two Intracellular Compartments and Several Enzymes

Steroid hormones are produced from cholesterol by the adrenal cortex, ovaries, testes, and placenta. Each organ uses the same general biosynthetic pathway, but the relative amount of the final products depends on the particular subset of enzymes expressed in that tissue. The main product of the testes is testosterone, but other androgens as well as a small amount of estrogens are also produced.

Pregnenolone is a key intermediate for all classes of steroid hormones (Fig. 39–11). After formation in the mitochondria, it is transported to the smooth endoplasmic reticulum, where the remainder of the biosynthetic reactions take place. Pregnenolone can be converted to testosterone via the delta 5 or delta 4 pathway. These refer to the stage at which the double bond in ring B (position 5,6) is switched to ring A (position 3,4). The delta 5 intermediates include 17 α-hydroxypregnenolone, DHEA, and androstenediol, while the delta 4 intermediates are progesterone, 17 α-hydroxyprogesterone, and androstenedione. The **delta 5** pathway is the major route for testosterone production in the human testis, whereas the delta 4 pathway predominates in the ovary.

The conversion of C-21 steroids to androgens (e.g., pregnenolone to DHEA) proceeds in two steps: 17 α-hydroxylation and C-17,20 cleavage. This is accomplished by a single enzyme, cytochrome P450c17, which requires molecular oxygen and NADPH. The DHEA is converted to androstenedione by another two-step enzymatic reaction: dehydrogenation in position 3 (3β hydroxysteroid dehydrogenase) and shifting of the double bond from ring B to A (δ4,5-ketosteroid isomerase). It is unclear whether these involve a single

enzyme with two activities or two separate enzymes. The final reaction yielding testosterone is carried out by 17-ketosteroid reductase, which substitutes the keto group in position 17 with a hydroxyl group. Unlike all the preceding enzymatic reactions, this is a reversible step.

Although estrogens are only minor products of testicular steroidogenesis, they are present as a normal circulating constituent in men. Androgens are converted to estrogens by the action of the enzyme complex P450arom. The products of aromatization of testosterone and androstenedione are estradiol and estrone, respectively (Fig. 39–11). In men, there is some aromatase activity in both Leydig and Sertoli cells, but aromatization mainly takes place in peripheral tissues. Aromatization is a complex reaction involving removal of the methyl group in position 19 and rearrangement of ring A into an unsaturated ring.

■ LH Binds to Receptors Coupled to G Proteins

The action of LH on the Leydig cell is mediated through specific membrane receptors of high affinity and low capacity. A single Leydig cell possesses about 15,000 LH receptors. Occupancy of less than 5% of these is sufficient to elicit maximal steroidogenic response. This is an example of "spare receptors" (Chap. 33). Excess receptors may increase target cell sensitivity to low circulating levels of hormones by raising the probability that enough receptors will be occupied. Alternatively, the total pool of receptors is functionally heterogeneous, and occupancy of different subgroups is required to elicit specific biologic responses. After exposure to a high LH concentration, the number of LH receptors declines and testosterone output, in response to subsequent challenges with LH, is decreased. This so-called desensitization involves two postulated mechanisms: loss of surface receptors due to internalization and receptor modification (e.g., phosphorylation) resulting in uncoupling from the adenylate cyclase system.

The LH receptor is a single 93 kd glycoprotein that shares sequence homology with other G protein-coupled receptors, such as rhodopsin and the β-adrenergic receptor. Its structure reveals three functional domains: an unusually large and glycosylated extracellular hormone-binding domain, a transmembrane spanning domain that contains seven noncontiguous segments, and an intracellular domain. Coupling of the receptor to a G_s protein occurs on a loop of one of the transmembrane segments. Under basal conditions G_s is inactive. Occupancy of the receptor by LH causes the replacement of GDP by GTP and activates G_s. The α subunit dissociates from the rest of the complex and binds to adenylate cyclase (Fig. 39–12). Adenylate cyclase then catalyzes the formation of cAMP from ATP; it continues to do so until the α subunit is inactivated by its intrinsic GTPase.

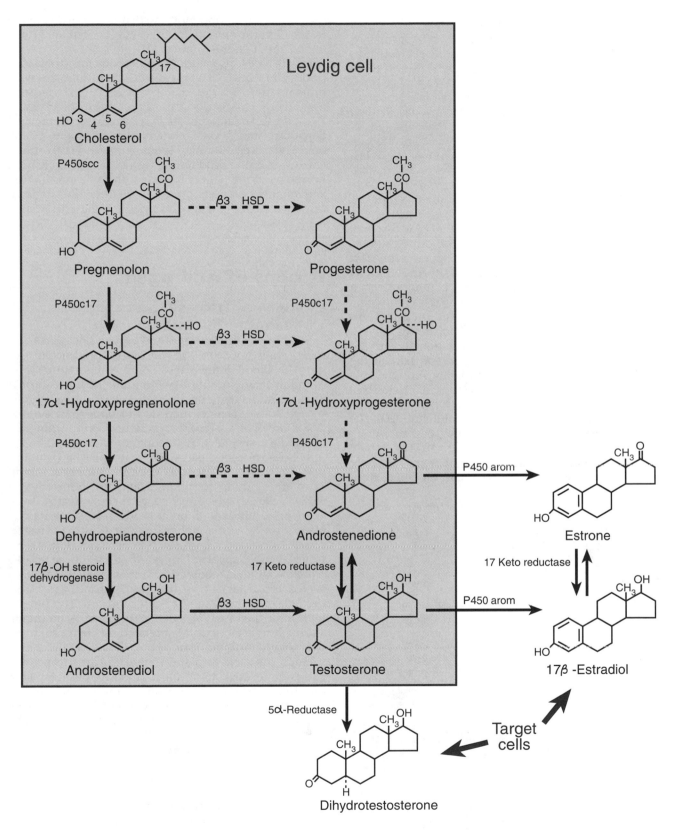

Figure 39–11 ■ Steroidogenesis in Leydig cells and further modifications of androgens in target cells. Solid arrows represent the delta 5 pathway. Dashed arrows represent the delta 4 pathway, which predominates in the ovary.

■ Cyclic AMP and Protein Kinase A Mediate the Action of LH

Cyclic AMP, produced in response to LH, activates protein kinase A (Fig. 39–12). Although by itself quite stable, cAMP is cleaved and inactivated by **phosphodiesterase.** Low doses of LH can stimulate testosterone production without detectable changes in total cell cAMP concentration. However, the amount of cAMP bound to the regulatory subunit of protein kinase A increases in response to such low doses of LH. This emphasizes the importance of compartmentalization or microenvironment for both enzymes and substrates in mediating hormonal action. It is unclear whether other intracellular mediators, such as the phosphatidylinositol system or calcium, play essential roles in regulating steroidogenesis. The proteins phosphorylated by protein kinase A are specific for each cell type. Some of these, such as cAMP response element binding protein (CREB), which functions as a DNA-binding protein, have recently been identified.

Steroid biosynthesis can be regulated at each of several levels, including availability of substrates and cofactors, induction of enzymes, alteration in enzyme activity, and transport of intermediates. Luteinizing hormone stimulates steroidogenesis by acting at two principal sites. One is phosphorylation of cholesterol esterase, which releases cholesterol from its intracellular stores and transports it to the mitochondria. The other is activation of the rate-limiting enzyme, mitochondrial cytochrome P450scc (Fig. 39–12), which yields pregnenolone from cholesterol. Some evidence indicates that the second side chain cleavage, from

C-21 to C-19 compounds, mediated by the enzyme P450c17 in the smooth endoplasmic reticulum, is also under direct stimulation by LH.

Leydig cells also contain receptors for **prolactin** (PRL) and GnRH. Hyperprolactinemia in men with pituitary tumors is associated with decreased testosterone levels. This is due to a direct effect of elevated PRL on the Leydig cells or is secondary to decreased gonadotrophin levels. Under nonpathologic conditions, however, PRL may synergize with LH to stimulate testosterone production. Gonadotrophin-releasing hormone alone stimulates testosterone production and augments the action of LH. Neither the origin of testicular GnRH nor its exact physiologic importance is known.

Actions of Androgens

■ Peripheral Tissues Process and Metabolize Testosterone

Unlike neurotransmitters and protein hormones, testosterone is not stored in the Leydig cell but diffuses into the blood immediately after being synthesized. An adult man produces 6–7 mg/day of testosterone. This amount slowly declines after age 50 to 4 mg/day by age 70–80. Hence, men do not undergo a sudden cessation of sex steroid production upon aging, as women do during their postmenopausal period.

Testosterone circulates bound to plasma proteins, with only 2–3% present as the free hormone. About 30–40% is bound to albumin and the remainder to SSBG, a 94 kd glycoprotein produced by the liver. Sex steroid binding globulin binds both estradiol and testosterone, with a higher binding affinity for testosterone. Because its production is increased by estrogens and decreased by androgens, plasma SSBG concentration is higher in women than in men. It serves as a reservoir for free testosterone, since only the unbound hormone can enter the cell. The physiologic importance of SSBG in healthy adults is uncertain, although its production is reduced by cirrhosis of the liver and hypothyroidism.

Once released into the circulation, testosterone's fate is quite variable. In most target tissues, testosterone functions as a prohormone and is converted to the biologically active derivatives dihydrotestosterone (DHT), estradiol, and androstenedione (Fig. 39–13). Skin, hair follicles, and most of the male reproductive tract contain an active **5α reductase.** The enzyme irreversibly catalyzes the reduction of the double bond in ring A and generates DHT (Fig. 39–11). Dihydrotestosterone has a strong binding affinity for the androgen receptor and is two to three times more potent than testosterone. Congenital deficiency of 5α reductase in males results in ambiguous genitalia, since DHT is critical for normal development of the external genitalia during embryonic life (Chap. 41).

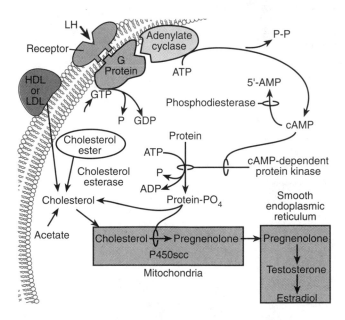

Figure 39–12 ■ Proposed intracellular mechanism by which luteinizing hormone (LH) stimulates testosterone synthesis. (Modified from Patton HD, Fuchs AF, Hille B, et al: *Textbook of Physiology.* 21st ed. Philadelphia, WB Saunders, 1989.)

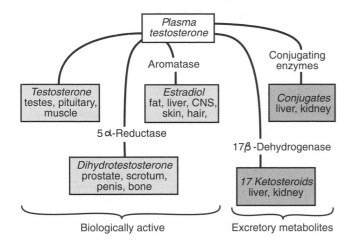

Figure 39–13 ■ Conversion of testosterone to different products in extratesticular sites.

Aromatization of testosterone to estradiol occurs in fat, liver, skin, and brain tissue. Circulating levels of total estrogens (estradiol plus estrone) in men approach those of women during the early follicular phase, yet men are protected from feminization as long as production of and tissue responsiveness to androgens are normal. Treatment of hypogonadal patients with high doses of testosterone, intake of anabolic steroids by athletes, abnormal reduction in testosterone secretion, estrogen-producing testicular tumors and tissue insensitivity to androgens can lead to **gynecomastia** (breast enlargement). All of these conditions are characterized by a decrease in the testosterone-to-estradiol ratio.

Androgens are metabolized in the liver to biologically inactive derivatives suitable for excretion by the kidney. The major products of testosterone metabolism are the 17-ketosteroids androsterone and etiocholanolone. These, as well as native testosterone, are conjugated in position 3 to form sulfates and glucuronides, which are water-soluble and excreted into the urine (Fig. 39–13). Other urinary 17-ketosteroids are the breakdown products of adrenal steroids.

■ Androgens Have Widespread Effects on Reproductive and Nonreproductive Tissues

An **androgen** is a substance that stimulates growth of the male reproductive tract or development of secondary sexual characteristics. Androgens have profound effects on almost every body tissue: alterations in primary sex structures (i.e., testes and genital tract); stimulation of secondary sex structures (i.e., accessory glands); development of secondary sex characteristics responsible for masculine phenotypic expression; and effects on psychology and sexual behavior. At each level, androgens can participate in cell differentiation, maturation, secretion, or maintenance of structural

and functional integrity. The relative potency ranking of androgens is DHT > testosterone >> androstenedione > DHEA. The action of sex steroids on somatic tissues such as liver and muscle has been termed "anabolic," although they are mediated by the same molecular mechanisms as those resulting in virilization.

Between weeks 8 and 18 of fetal life, androgens mediate the differentiation of the male genitalia. The organogenesis of the wolffian ducts into the epididymis, vas deferens, and seminal vesicles is directly influenced by testosterone, which reaches these target tissues by diffusion rather than by a systemic route. The differentiation of the urogenital sinus and the genital tubercle into the penis, scrotum, and prostate gland depends on testosterone being converted to DHT. Toward the end of fetal life, the descent of the testes into the scrotum is promoted by testosterone (Chap. 41).

The onset of puberty is marked by enhanced androgenic activity. Androgens promote growth of the penis and scrotum, stimulate growth and secretory activity of the epididymis and accessory glands, and increase pigmentation of the genitalia. Enlargement of the testes occurs under the influence of the gonadotrophins. Spermatogenesis, which is initiated during puberty, is absolutely dependent on adequate amounts of testosterone. Throughout adulthood, androgens are responsible for maintaining the structural and functional integrity of all reproductive tissues. Castration of adult men results in regression of the reproductive tract and involution of the accessory glands.

■ Androgens Are Responsible for Secondary Sex Characteristics and Masculine Phenotype

Androgens affect changes in hair distribution, skin texture, pitch of voice, bone growth, and muscle development. Hair is classified by its sensitivity to androgens into: nonsexual (eyebrows and extremities); **ambisexual** (axilla), which is responsive to low levels of androgens; and **sexual** (face, chest, upper pubic triangle), which is responsive only to high androgen levels. Hair follicles metabolize testosterone to DHT or androstenedione. Androgens stimulate growth of facial, chest, and axillary hair but, together with genetic factors, promote temporal hair recession. Normal axillary and pubic hair growth in women is also under androgenic control, whereas excess androgen production in women causes **hirsutism** (excessive growth of sexual hair).

The growth and secretory activity of the sebaceous glands on the face, upper back, and chest are stimulated by androgens and inhibited by estrogens. The active androgen is DHT. Increased sensitivity of target cells to androgenic action, especially during puberty, is the cause of **acne vulgaris** in both males and females. Skin derived from the urogenital ridge (e.g., prepuce, scrotum, clitoris, and labia majora) remains sensitive to androgens throughout life and contains an active 5α reductase. Growth of the larynx and thickening of the vocal cords are also androgen-dependent. Eunuchs maintain the high-pitched voice typical of prepubertal boys.

The growth spurt of adolescent males is influenced by a complex interplay between androgens, growth hormone (GH), and genetic factors. This includes growth of the vertebrae, long bones, and shoulders. The mechanism by which androgens (likely DHT) alter bone metabolism is unclear, but they also accelerate closure of the epiphyses in the long bones, eventually limiting further growth. Because of the latter, precocious puberty is associated with a final short adult stature, whereas delayed puberty or eunuchoidism usually results in tall stature. Androgens have multiple effects on skeletal and cardiac muscle. Since 5α reductase activity in muscle is very low, the androgenic action is due to testosterone itself. Testosterone stimulates muscle hypertrophy, thus increasing muscle mass, but has minimal or no effect on muscle hyperplasia. Testosterone, in synergy with GH, causes a net increase in protein synthesis. Whether androgens affect energy metabolism is controversial, raising a question as to the benefit of anabolic steroids for improving athletic performance.

Other nonreproductive organs are affected, directly or indirectly, by androgens. These include the liver, kidney, adipose tissue, and hematopoietic and immune systems. The kidneys are larger in males, and some renal enzymes (e.g., β glucuronidase and ornithine decarboxylase) are induced by androgens. High-density lipoprotein levels are lower and triglyceride concentrations higher in men as compared to premenopausal women. This may explain the higher prevalence of atherosclerosis in men. Androgens increase red blood cell mass (and hence hemoglobin levels) by stimulating erythropoietin production and by increasing stem cell proliferation in the bone marrow.

■ The Brain Is Also a Target Site for Androgen Action

One of the most interesting, but least understood, effects of androgens is on the differentiation, maturation, and operation of the central nervous system. Many sites in the brain contain androgen receptors, with the highest density in the hypothalamus, preoptic area, septum, and amygdala. Most of these areas contain cytochrome P450arom, and many of the androgenic actions in the brain are attributed to estradiol. The pituitary also has abundant androgen receptors but no cytochrome P450arom. The enzyme 5α reductase is widely distributed in the brain, but its activity is generally higher during the prenatal period than in adults. Sexual dimorphism in the size, number, and arborization of neurons in the preoptic area, amygdala, and superior cervical ganglia is well characterized in birds and rodents and has been recently recognized in humans.

Male sexuality is also controlled by androgens. In most species, courtship behavior and sexual activity decline on withdrawal of androgens and can be restored to normal when androgens are administered. In male hierarchal animal societies, the position of the dominant male is strongly dependent on testosterone and is greatly compromised by castration. The situation in humans, however, is far more complex. Unlike most species, which mate only to produce young, sexual activity and procreation in humans are not tightly linked. Superimposed on the basic reproductive mechanisms dictated by hormones are numerous psychologic and societal factors. In normal men, no correlation is found between circulating testosterone levels and sexual drive, frequency of intercourse, or sexual fantasies. Similarly, there is a lack of relationship between the levels of testosterone and impotence or homosexuality. Castration of adult men results in a slow decline, but not complete elimination, of sexual interest and activity.

It is evident that androgens regulate a wide spectrum of actions, from cell division to organogenesis and from metabolic activity to phenotypic expression. This raises an intriguing question as to the mechanism by which a small, relatively simple molecule can elicit such diverse actions. The common denominator is the receptor. Upon activation by a bound hormone, steroid receptors associate with the chromatin and induce or suppress a select set of genes (Chap. 33). Ultimately, the hormonal effects depend on the type of tissue expressing these genes and the commitment of the target cells to division, differentiation, secretion, or metabolic activity.

Reproductive Dysfunctions

■ Hypogonadism Can Result from Defects at Several Levels

Several questions should be examined when evaluating male reproductive function: Is there a defect in spermatogenesis alone, steroidogenesis alone, or both? Is it a primary defect in the testes or secondary to a hypothalamopituitary dysfunction? Did the onset of gonadal failure occur before or after puberty? There are several considerations in answering these questions. First, normal spermatogenesis almost never occurs with defective steroidogenesis, but normal steroidogenesis can be present with defective spermatogenesis. Second, primary testicular failure removes feedback inhibition from the hypothalamopituitary axis, resulting in elevated plasma gonadotrophins. In contrast, hypothalamic and/or pituitary failure is almost always accompanied by decreased levels of gonadotrophins. Third, gonadal failure before puberty results in absence of secondary sexual characteristics, creating a distinctive clinical presentation called **eunuchoidism.** On the other hand, postpubertal tes-

IMPOTENCE: DIAGNOSIS AND THERAPY

Impotence is generally defined as an inability to achieve or maintain erection or ejaculation. Difficulty to achieve erection on a regular basis occurs in approximately 10–12% of men and infrequently in about 35%. The etiology of impotence can be psychogenic or traced to specific pathologies resulting from vascular defects, diabetes mellitus, or neuronal dysfunction. Psychogenic impotence is the most common cause and is often associated with alternating periods of potency and impotence. Recent advances have enabled more precise diagnosis and management of impotence for many patients.

Patients who present for evaluation of impotence should undergo a careful, systematic evaluation. The patient's history—medical, social, and sexual—is examined in detail. The onset, duration, and degree of erectile dysfunction and whether it can be traced to a psychosexual event (partner-specific), trauma, or onset of a specific disease are clarified. Risk factors for impotence include age, coronary artery disease, hypertension, peripheral vascular disease, peripheral neuropathy, smoking, alcohol consumption, and medications. Physical examination includes evaluation of secondary sexual characteristics, motor and sensory functions, and size and symmetry of external genitalia. Potential endocrine problems can be uncovered by the determination of serum prolactin and testosterone. The degree of erectile integrity is evaluated by monitoring nocturnal erections. This can be performed by the patient at home, using a snap gauge consisting of a ring with three plastic bands that break successively under increasing force, or with more sophisticated devices during a hospital stay. Finally, penile blood flow can be assessed using angiography under anesthesia or ultrasonography using a blood pressure cuff placed at the penile base following injection of a vasoactive agent.

Once a diagnosis is made, several strategies are available for therapy. If the origin appears to be psychogenic, counseling is the method of choice. However, some improvement of libido (sexual desire) and erectile function in patients with psychogenic impotence has been achieved with yohimbine, a centrally acting α_2-adrenergic blocking agent. Corrective surgery can be employed in selected cases of local lesions and vascular defects. Endocrine causes, such as hyperprolactinemia or hypoandrogenism, can be treated, respectively, by reducing blood prolactin levels with dopamine agonists or by careful administration of testosterone. Self-injection therapy is also available, consisting of vasodilators such as papaverine, prostaglandin E_1, vasoactive intestinal peptide, and phentolamine, alone or in combination. Finally, vacuum constriction devices have proved especially effective in patients capable of partial rigidity.

ticular failure is associated with retention of most masculine features but reduced or absent ability to produce functional sperm.

To establish the cause(s) of reproductive dysfunction, physical examination and medical history, semen analysis, hormone determinations, hormone stimulation tests, and genetic analysis are performed. Physical examination should establish whether eunuchoidal features (i.e., infantile appearance of external genitalia and poor or absent development of secondary sex characteristics) are present. In men with adult-onset reproductive dysfunction, physical examination can uncover problems such as cryptorchidism (nondescendent testes), testicular injury, varicocele (abnormality of the spermatic vasculature), testicular tumors, prostatic inflammation, or gynecomastia. Medical and family history help determine delayed puberty, anosmia (inability to smell, often associated with GnRH dysfunction), previous fertility, changes in sexual performance, ejaculatory disturbances, or impotence (inability to achieve erection).

Semen analysis is performed to evaluate fertility. It is best done with a specimen collected after 3–5 days of sexual abstinence. Initial examination includes determination of viscosity, liquefaction, and semen volume. The sperm are then counted and the percentage of sperm showing forward motility scored. The spermatozoa are evaluated morphologically, with attention to abnormal head configuration and defective tails. Chemical analysis can provide information on the secretory activity of the accessory glands, which is considered abnormal if semen volume is too low or sperm motility is impaired. Fructose and prostaglandin levels are determined to assess the function of the seminal vesicles and levels of zinc, magnesium, and acid phosphatase levels to evaluate the prostate. Terms used in evaluating fertility include *aspermia* (no semen), *hypo-* and *hyperspermia* (too small or too large semen volume), *azoospermia* (no spermatozoa), and *oligozoospermia* (less than 15 million spermatozoa per milliliter of semen).

Serum testosterone, estradiol, LH, and FSH analysis is performed by radioimmunoassay. Free and total testosterone levels should be measured; because of the pulsatile nature of LH release, several consecutive blood samples are needed. Dynamic hormone stimulation tests are most valuable for establishing the site of abnormality. Failure to increase LH release upon treatment with **clomiphene** likely indicates a hypothalamic abnormality. Clomiphene blocks the inhibitory effects of estrogen and testosterone on endogenous GnRH release. An absence of or blunted testosterone rise after hCG treatment suggests primary testicular defect. Genetic analysis is employed when congenital defects are suspected. The presence of Y chromosome can be revealed by karyotyping of cultured peripheral lymphocytes or direct detection of specific Y antigens on cell surfaces.

■ Reproductive Disorders Are Associated either with Hypogonadotrophic or Hypergonadotrophic States

Endocrine factors are responsible for approximately 50% of hypogonadal or infertility cases. The remainder are of unknown etiology or due to injury, deformities, and environmental factors. Endocrine-related hypogonadism can be classified as hypothalamic-pituitary defects (hypogonadotrophic), primary gonadal defects (hypergonadotrophic), and defective androgen action. Each of these is further subdivided into several categories, but only a few examples are discussed here.

Hypogonadotrophic hypogonadism can be congenital, idiopathic, or acquired. The most common congenital form is Kallmann syndrome, which results from decreased or absent GnRH secretion. It is often associated with anosmia or hyposmia and is transmitted as an autosomal dominant trait. Patients do not undergo pubertal development and have eunuchoidal features. Plasma LH, FSH, and testosterone are low, and the testes are immature with no spermatozoa. There is no response to clomiphene, but intermittent treatment with GnRH can produce sexual maturation and full spermatogenesis. A puzzling disorder is the

isolated LH deficiency, or **fertile eunuch syndrome.** Plasma FSH is near normal, with low LH and testosterone levels. In spite of the eunuchoidal features, viable sperm are often found in the ejaculate. Such patients can be treated with hCG or testosterone to promote virilization and testicular maturation.

Another category of hypogonadotrophic hypogonadism, **panhypopituitarism,** or **pituitary failure,** can occur before or after puberty and is usually accompanied by a deficiency of other pituitary hormones. **Hyperprolactinemia,** whether due to hypothalamic disturbance or pituitary adenoma, often results in decreased GnRH production, hypogonadotrophic state, impotence, and decreased libido. It can be treated with dopaminergic agonists (e.g., bromocryptine), which suppress PRL release (Chap. 40). Excess androgens can also result in suppression of the hypothalamic-pituitary axis, resulting in lower LH levels and impaired testicular function. This often results from **congenital adrenal hyperplasia** and increased adrenal androgen production due to cytochrome P450c21 (21 hydroxylase) deficiency.

Hypergonadotrophic hypogonadism usually results from impaired testosterone production, which can be congenital or acquired. The most common disorder is **Klinefelter syndrome,** a congenital abnormality with an incidence of approximately 0.2% of live male births. It results from sex chromosomal nondisjunction, characterized by the presence of XXY chromosomes in the testes and often in most other tissues. In the classic form, clinical manifestations include small, firm testes; azoospermia; gynecomastia; and inadequate masculinization. Patients have some eunuchoidal features, with low testosterone levels and elevated gonadotrophins.

Several other disorders are associated with an hypergonadotrophic state. One is germinal cell aplasia, or **Sertoli cell only syndrome.** This is caused by abnormality in the seminiferous tubules and is associated with azoospermia, normal LH and testosterone levels, but elevated FSH levels. Another disorder is **testicular agenesis,** in which patients have no recognizable testicular tissue but normal male internal genitalia, suggesting that functional testes were present at the time of sexual differentiation. Postpubertal hypergonadotrophic states can be the consequence of testicular damage due to infections, mumps, irradiation, and toxic agents.

■ Male Pseudohermaphroditism Often Results from Resistance to Androgens

A pseudohermaphrodite is an individual with gonads of one sex and genitalia of the other. One of the most interesting causes of male reproductive abnormalities is an end organ insensitivity to androgens. The best characterized syndrome is **testicular feminization,** an X-linked recessive disorder caused by a defect in the testosterone receptor. In the classical form, patients are male pseudohermaphrodites with a female phenotype and an XY male genotype. They have abdominal testes that secrete testosterone but no other internal genitalia of either sex (Chap. 41). They commonly have female external genitalia, but with a short vagina ending in a blind pouch. Breast development is typical of a female (due to peripheral aromatization of testosterone), but axillary and pubic hair, which are androgen-dependent, are scarce or absent. Testosterone levels are normal or elevated, estradiol levels are above the normal male range, and circulating gonadotrophins levels are high. The testes usually have to be removed because of increased risk of cancer. After orchiectomy, patients are treated with estradiol to maintain a normal female phenotype.

Another example of male pseudohermaphroditism is **5α reductase deficiency,** an autosomal recessive disorder that results from an inability to convert testosterone to DHT. Differentiation of the internal genitalia, which is testosterone-dependent, proceeds normally along the male line. The external genitalia, which depend on DHT, develop along the female line. Such individuals are often identified as females at birth with bilateral inguinal masses, but undergo marked masculinization during puberty in response to rising levels of testosterone (Chap. 41).

REVIEW EXERCISES

Identify Each with a Term

1. The temporal pattern of GnRH release
2. Common term for LH and FSH
3. The hormonal relationship between the secretion of testosterone and LH
4. The process by which germ cells become haploid
5. The masculine features that are affected by androgens

Define Each Term

6. α and β subunits of LH
7. Cryptorchidism
8. Semen
9. Aromatization
10. Androgens
11. Hypogonadism
12. Male pseudohermaphrodite

Choose the Correct Answer

13. During spermatogenesis, the spermatogonia undergo active mitosis:
    a. Under the influence of FSH
    b. Under the influence of testosterone
    c. Under the influence of LH
    d. Without specific hormonal stimulation
14. The process from spermatogonia to mature spermatozoa:
    a. Takes 32 days
    b. Takes 70 days

c. Is prolonged by elevated temperature
d. Is shortened by testosterone administration

15. The acrosome:
    a. Is located in the middle section of the spermatozoon
    b. Is part of the tail section of the spermatozoon
    c. Constitutes the central portion of the centriol
    d. Represents a modified lysosome
16. Cholesterol has:
    a. 25 carbon atoms
    b. 27 carbon atoms
    c. A double bond in ring A 3-4.
    d. a hydroxyl group in position 17
17. The first enzymatic reaction, which is the rate-limiting step, in the production of testosterone:
    a. Occurs in the mitochondria
    b. Occurs in the ribosomes
    c. Involves aromatization
    d. Generates progesterone as the immediate derivative
18. Testosterone in the general circulation:
    a. Is bound to high-density lipoproteins
    b. Is bound to androgen-binding protein
    c. Is converted to dihydrotestosterone in the prostate
    d. Is converted to 17-hydroxyprogesterone in the liver

Briefly Answer

19. What are the major intracellular mediators for GnRH?
20. What are the mechanisms for maintaining the testes at a lower temperature?
21. What are the major products of Sertoli cells?
22. At which stage of spermatogenesis does chromosomal crossover occur?

ANSWERS

1. Pulsatile
2. Gonadotrophins
3. Negative feedback

4. Meiosis
5. Secondary sex characteristics
6. Two peptide chains that comprise the LH molecule; the

α subunit is identical in LH, FSH, and TSH, whereas the β subunit is unique to LH and confers its biologic activity

7. A condition in which one or two of the testes did not descend into the scrotum during late fetal life

8. The combined secretory products of the testes (sperm) and accessory glands, including primarily the seminal vesicles and prostate

9. The process by which androgens are converted to estrogens by the action of the enzyme complex aromatase; this can occur both in the testes and in peripheral tissues

10. Substances that stimulate the growth of the male reproductive tract or affect male secondary sex characteristics

11. Dysfunction of male reproduction that may affect spermatogenesis, steroidogenesis, or both and either results from a primary testicular defect or is secondary to hypothalamopituitary dysfunction

12. An individual with testes but having external female-appearing genitalia

13. d

14. b

15. d

16. b

17. a

18. c

19. The main intracellular mediators of GnRH are the two products of phosphatidylinositol cleavage, IP_3 and DAG, which increase intracellular calcium concentrations and activate calcium-dependent protein kinase C, respectively.

20. Scrotal temperature is maintained at a 2–3 °C below core body temperature by the countercurrent heat exchanger in the pampiniform plexus and the temperature-sensitive cremasteric muscle.

21. The major products of Sertoli cells are androgen-binding protein (ABP), which serves as an intracellular carrier for testosterone, and inhibin, which suppresses FSH release from the pituitary gland.

22. Chromosomal crossover occurs during prophase of the first meiotic division.

Suggested Reading

Burger H, De Krester D. (eds): *The Testes*. New York, Raven, 1981.

Conn PM, McArdle CA, Andrews WM, Huckle WR: The molecular basis of gonadotropin-releasing hormone (GnRH) action in the pituitary gonadotroph. *Biol Reprod* 36:17–35, 1987.

Crowley WF, Filicori M, Spratt DI, Santoro NF: The physiology of gonadotropin-releasing hormone (GnRH) in men and women. *Rec Prog Hormone Res* 41:473–526, 1985.

Goodman MH: *Basic Medical Endocrinology*. 2nd ed. New York, Raven, 1994.

Griffin JE: The physiology of the testes and male reproductive tract and disorders of testicular function, in Carr BR, Blackwell RE. (eds): *Textbook of Reproductive Medicine*. London, Prentice-Hall, 1993, pp 221–245.

Hershman JM: *Endocrine Pathophysiology*. 3rd ed. Philadelphia, Lea & Febiger, 1988.

Lipshultz LI, Howards SS. (eds): *Infertility in the Male*. New York, Churchill Livingstone, 1983.

Martin JB, Reichlin S: *Clinical Neuroendocrinology*. 2nd ed. Philadelphia, FA Davis, 1987.

Nooradian AD, Morley JE, Korenman SG: Biological actions of androgens. *Endocrin Rev* 8:1–28, 1987.

Veldhuis J: The hypothalamic-pituitary-testicular axis, in Yen SSC, Jaffe RB. (eds): *Reproductive Endocrinology*. 3rd ed. Philadelphia, WB Saunders, 1991, pp 409–459.

The Female Reproductive System

OBJECTIVES
After studying this chapter, the student should be able to:
1. Explain the differences between tonic and surge modes of gonadotrophin release
2. Discuss the regulation of release and functions of prolactin and oxytocin
3. Describe the gross anatomy of the female reproductive organs
4. Discuss the process of oogenesis and follicular development from the embryonic stage to adulthood
5. Identify the cell types of a mature follicle and corpus luteum and describe their functions; discuss the role of theca, granulosa, and corpus luteum cells in steroidogenesis
6. Describe the process of ovulation, corpus luteum formation, and atresia
7. Outline the temporal relationship between the hormones of the hypothalamus, pituitary, and ovary during the different phases of the reproductive cycle; discuss the hormonal mechanisms leading to the generation of the midcycle LH surge
8. Discuss the process leading to, and the cause of, menstrual bleeding
9. Identify and discuss primary and secondary amenorrhea
10. List the effects of ovarian steroids on reproductive and nonreproductive tissues

The reproductive capability of the female, unlike that of the male (Chap. 39), is intermittent, and the production of mature germ cells, or **ova,** is not continuous. In contrast to the millions of germ cells produced daily by the male, only one or a few ova are produced by the female at any given time. The cyclic release of ova is driven by cyclic alterations in hormone secretion. These, in turn, affect profound cyclic changes in the structure and function of the reproductive organs. Ovarian steroids exert both negative and positive feedback effects on the hypothalamus and on pituitary gonadotrophs, generating the cyclic pattern of gonadotrophin release characteristic of the female reproductive system. Given the dependency on precise synchronization of multiple components, reproduction in the female is readily affected by stress, environmental, psychological, and social factors. The human ovulatory cycle, or **menstrual cycle,** is characterized by **menses,** resulting from withdrawal of hormonal support and shedding of the superficial layer of the uterine lining at the end of each cycle. Menstrual cycles begin during **puberty,** are interrupted during **pregnancy** and **lactation,** and cease at the time of **menopause.**

Overview of the Female Reproductive System

An overview of the female reproductive system is shown in Figure 40-1. The basic system consists of the brain, pituitary, and ovaries. The hypothalamus produces gonadotrophin releasing hormone (GnRH), which controls the release of luteinizing hormone (LH) and follicle stimulating hormone (FSH). Like the male gonads, the ovaries have a dual function: production of germ cells (oogenesis) and steroidogenesis. Each germ cell is enclosed within a larger structure, the follicle. After extrusion of the ovum during ovulation, the follicle is transformed into a new endocrine structure, the corpus luteum. Follicle stimulating hormone is primarily involved in stimulating the growth of ovarian follicles, while LH controls ovulation and regulates steroidogenesis. The female has two important sex steroids. **Estradiol** is the main product of the follicle, and **progesterone** is produced by the corpus luteum. The two steroids differ in structure and function, and their actions can be additive, synergistic, or antagonistic. The ovary also produces the polypeptide hormone **inhibin,** which regulates FSH release.

Pregnancy and lactation necessitate the participation of additional reproductive organs, such as the placenta and mammary glands. The placenta assumes the function of both the pituitary and the ovaries by producing protein and steroid hormones. These help in adapting the maternal organism for pregnancy, support fetal development, and participate in parturition. The mammary glands provide the young with nutritional support and immunologic protection. Lactation depends on two main hormones: **prolactin** from the

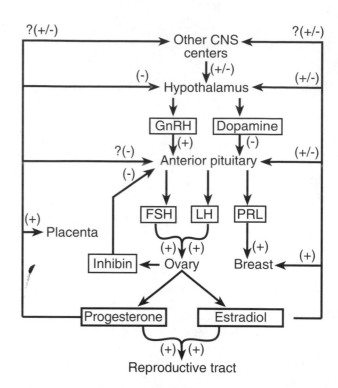

Figure 40–1 ■ Overall regulation of reproduction in the female. The main reproductive hormones are shown in boxes. Positive and negative regulation are depicted by + and − signs, respectively. Note that dopamine inhibits prolactin release but has no effect on LH or FSH. GnRH, gonadotrophin-releasing hormone; FSH, follicle stimulating hormone; LH, luteinizing hormone; PRL, prolactin.

anterior pituitary, which regulates milk production, and **oxytocin** from the posterior pituitary, which induces milk ejection.

The Hypothalamopituitary Axis

■ The Pattern of Reproductive Hormone Release Changes in a Cyclic Manner

The secretory pattern of LH and FSH in the female is more complex than that in the male (Chap. 39). It is influenced by hypothalamic neurotransmitters and neuropeptides, is affected by the pattern of pulsatile release of GnRH, and is driven by the changing circulating levels of ovarian steroids. Gonadotrophin secretion in the female has two distinct modes of release: a **tonic** mode and a **surge** mode. The tonic mode is similar to that operating in the male and involves low but pulsatile release of LH. The surge mode is unique to the female and involves periodic massive releases of LH and, to a lesser extent, FSH.

The human **menstrual cycle** is approximately 28 days long and is divided into three phases: follicular, ovulatory, and luteal. The **follicular phase,** during the first half of the cycle, is dominated by growing follicles that secrete estradiol. The second half of the cycle, the **luteal phase,** is dominated by the corpus luteum, which secretes progesterone. During both phases, gonadotrophin release is tonic. The **ovulatory phase** is a very short period (24–36 hours) in the middle of the cycle when the midcycle surge of LH and FSH occurs, followed by ovulation. In this section, we discuss the regulation of tonic LH release. The mechanism leading to the surges of LH and FSH is discussed in the section on the menstrual cycle.

The pulsatile release of LH varies with the stage of the menstrual cycle. Table 40-1 lists several mean parameters of LH pulsatility in normal women during the tonic phases of the cycle. The early follicular phase is characterized by high frequency and low amplitude. In the late follicular phase, the frequency increases slightly (i.e., pulse interval decreases) without a significant change in amplitude. During the early luteal phase, the pulse frequency slows but the amplitude increases. The late luteal phase is characterized by infrequent LH pulses of low amplitude.

Such a pattern of LH release is seen only in adult women with functional ovaries. A schematic representation of LH release throughout the life span is shown in Figure 40-2. During childhood, LH is released at low and steady rates. Pulsatile release begins with the onset of puberty and for several years is expressed

TABLE 40–1 ■ Luteinizing Hormone Pulsatility in Women

Phase	Pulse Interval (min)	Pulse Amplitude (mIU/ml)
Follicular		
Early	94	6.5
Late	71	7.2
Luteal		
Early	103	14.9
Late	217	7.6

Modified from Filicori M et al: Characterization of the physiological pattern of episodic gonadotropin secretion throughout the human menstrual cycle. *J Clin Endocrinol Metab* 62:1136-1144, 1986.

only during sleep. Upon establishment of functional menstrual cycles, LH pulsatility prevails throughout the 24-hour period, changing in a monthly cyclic manner. In postmenopausal women whose ovaries are nonfunctional, mean circulating LH levels are high and pulses occur at a high frequency. Similarly, removal of the ovaries during the reproductive years causes large increases in both frequency and amplitude of LH. Collectively, such observations indicate that ovarian steroids exert negative feedback effects on the pulsatile release of LH.

The hypothalamus serves as one target for negative feedback by estradiol. The GnRH pulse generator is located in the medial basal hypothalamus in the vicinity of the arcuate nucleus (Chap. 39). Implantation of crystalline estradiol into this region suppresses LH pulses in ovariectomized monkeys. However, estradiol does not affect the GnRH neurons directly; its actions are mediated by neurotransmitters. Infusion of norepinephrine into discrete hypothalamic sites quickens LH pulses, whereas α-adrenergic receptor blockers produce the opposite effect. Dopamine and serotonin appear to inhibit GnRH release. Estradiol also has a direct negative feedback action on the anterior pituitary. Evidence for this is derived from experiments in monkeys with lesions in their arcuate nuclei. In such monkeys, pulsatile LH release was restored on application of exogenous pulses of GnRH, but infusion of estrogen reduced pituitary responsiveness to each exogenous GnRH pulse.

Progesterone also plays a role in regulating LH pulsatility. The luteal phase is characterized by LH pulses of slower frequency and higher magnitude (Table 40-1). Evidence that this is due to progesterone comes from the marked slowing of the LH pulses when women are treated with progesterone during the follicular phase. The actions of progesterone are exerted on the brain and are mediated by opioid peptides. Treatment with naloxone, an opioid receptor antagonist, during the luteal phase increases LH pulse frequency. Endogenous opioid activity is highest during the luteal phase. The opioids do not act directly on the anterior pituitary but

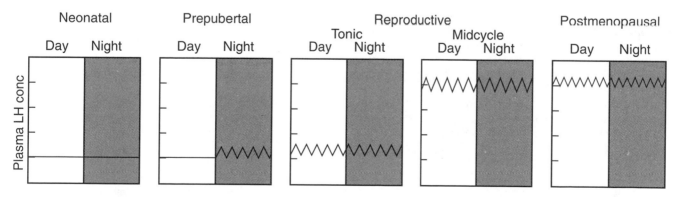

Figure 40–2 ■ Luteinizing hormone (LH) pulsatility and relative levels of LH release throughout life in the female. (Modified from Yen SSC et al, in Ferin M et al (eds). *Biorhythms and Human Reproduction.* New York, Wiley, 1974.)

interact with the GnRH neurons or catecholaminergic neurons that affect GnRH.

■ Prolactin Takes Part in Several Reproductive Processes

Prolactin is involved in a broad spectrum of functions, including lactation, reproduction, and growth. In all mammals, prolactin is essential for the preparation and maintenance of lactation. It also plays a permissive role in the regulation of gonadal function in both sexes.

Human prolactin is comprised of a single polypeptide chain of 199 amino acids with three disulfide bonds. It shares structural homology and overlapping functions with growth hormone (GH) and a nonpituitary hormone, placental lactogen. The three hormones originated from a common ancestral gene. Prolactin is secreted by lactotrophs, which constitute 20–30% of the anterior pituitary cell population. The number and size of the lactotrophs increases markedly during pregnancy under the influence of estrogens. The somatomammotrophs can secrete both prolactin and GH and probably function as progenitor cells that can be directed toward production of either hormone under the proper stimulation.

Unlike other endocrine cells of the pituitary, lactotrophs have an intrinsic high rate of production and secretion and require constant inhibition. Pituitary trophic hormones (e.g., LH, FSH, thyroid stimulating hormone [TSH], and adrenocorticotrophic hormone [ACTH]) generally have a single target organ, and their secretory rate represents a balance between stimulation by a hypothalamic releasing hormone and feedback inhibition by a target organ hormone. In contrast, prolactin has multiple peripheral target organs, including the mammary glands, liver, adrenals, and gonads, but no identifiable target organ hormone(s) for transmitting information to the lactotrophs. Instead, the brain provides the major regulation over prolactin release.

The hypothalamus exerts predominant inhibitory control over prolactin release. Disruption of connections between the hypothalamus and pituitary or hypothalamic lesions results in increased prolactin release and reduced secretion of other trophic hormones. The major physiologic inhibitor of prolactin release is dopamine (Fig. 40-1), which binds to high-affinity receptors located on the plasma membrane of lactotrophs. These receptors are a D_2 receptor subtype that is negatively linked to adenylate cyclase via an inhibitory G protein (G_i). Dopamine inhibits prolactin release primarily by reducing intracellular calcium concentration. This is accomplished, in part, by changing the membrane potential of lactotrophs and reducing calcium influx. Dopamine also affects the redistribution of intracellular calcium either by reducing inositol trisphosphate levels or by interfering with the binding of calcium to calmodulin.

Several substances stimulate prolactin secretion. Some, such as opioids and serotonin, affect prolactin indirectly by inhibiting the dopaminergic system. Others, such as thyrotrophin-releasing hormone, vasoactive intestinal peptide, and oxytocin, directly stimulate the lactotrophs. Much evidence suggests the presence of a distinct prolactin-releasing factor, but its identity is yet unknown. Prolactin also autoregulates its own release. Upon reaching the arcuate nuclei by retrograde blood flow from the pituitary gland, elevated prolactin increases dopamine synthesis and release. This relationship comprises a short-loop negative feedback mechanism and compensates for the lack of a long-loop negative feedback from the periphery, which the other anterior pituitary hormones have.

Basal circulating prolactin levels in males and females are similar (Fig. 40-3). Prolactin increases substantially during pregnancy in response to estrogen and rises during lactation in response to the suckling stimulus. Many drugs that interfere with dopamine metabolism, such as neuroleptics, also affect prolactin secretion. The prevalence of abnormal secretion of prolactin is clinically important. More than half of the pituitary adenomas exhibit abnormal secretion of prolactin. A majority of women with highly elevated plasma prolactin levels are amenorrheic and anovulatory. Some patients with hyperprolactinemia (both men and women) have **galactorrhea,** or inappropriate milk secretion. The release of prolactin in response to suckling and its role during lactation are discussed in Chapter 41.

■ Oxytocin Forms in

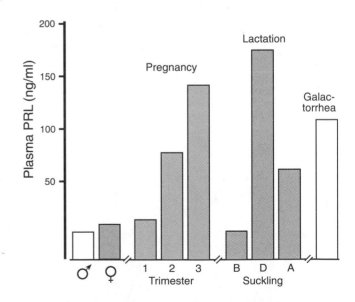

Figure 40-3 ■ Plasma prolactin (PRL) levels during different physiologic conditions. B, before; D, during; A, after. (Modified from Jacobs LS, Mariz IK, Daughaday WH: A mixed heterologous radioimmunoassay for human prolactin. *J Clin Endocrinol Metab* 34:484-490, 1972.)

the Hypothalamus but Is Secreted by the Posterior Pituitary

Another hypothalamopituitary hormone that participates in reproductive processes is oxytocin, a small peptide composed of 8 amino acids that is produced by the magnocellular neurons of the hypothalamus. In females, oxytocin increases the contraction of the uterine muscle during parturition and causes ejection of milk from the mammary glands during lactation. The functions of oxytocin in males are unclear.

The oxytocinergic neuron represents a hybrid between a classical neuron and an endocrine cell. Like any other neuron, it can be depolarized or hyperpolarized by input from the many synapses that impinge on it. Upon reaching a critical threshold, the cell body generates a self-propagating action potential that travels down the axon. As the action potentials reach the terminals in the neural lobe of the pituitary, they cause a transient influx of sodium ions, followed by an opening of calcium channels. Calcium is driven into the terminal by the large concentration gradient between the extracellular and intracellular compartments.

Following the increase in intracellular calcium, secretory granules containing oxytocin translocate toward the terminal membrane, fuse with it, and discharge their content to the outside. Unlike a classical neuron, but resembling an endocrine cell, the secretory product of the oxytocin neuron is not confined to a synaptic cleft but is released into the extracellular space, from where it diffuses into capillaries and the general circulation. The function of oxytocin in the process of milk let-down is discussed in Chapter 41.

The Female Reproductive Organs

■ The Ovaries and Duct System Make Up the Female Reproductive Tract

The female reproductive tract has two main components. One is the ovaries, which produce the mature egg and secrete the female sex hormones. The other is the duct system, which facilitates the transport and union of the sperm and egg and maintains the developing conceptus until delivery. The morphology and function of these structures change in a cyclic manner under the influence of the reproductive hormones.

Unlike the testes, which descend into the scrotum, the **ovaries** remain in the abdominal cavity and do not require cooler temperature for normal function. The ovaries are located in the pelvic cavity on both sides of the uterus and are anchored by ligaments (Fig. 40-4). Each ovary in adult women weighs 8-12 g and consists of an outer cortex and an inner medulla, without a sharp demarcation. The cortex is surrounded by a fibrous tissue, the tunica albuginea, covered by a single layer of surface epithelium. The cortex contains **ova** enclosed in **follicles** of various sizes, **corpora lutea,** corpora albicantia, and stroma cells. The medulla contains connective and interstitial tissues. Blood vessels, lymphatics, and nerves enter the medulla via the hilus.

The **oviducts** (Fallopian tubes) provide an environment for further maturation of the sperm and egg, for fertilization, and for development of the fertilized egg (zygote). They are 10-15 cm long and composed of infundibulum, ampulla, and isthmus. The infundibu-

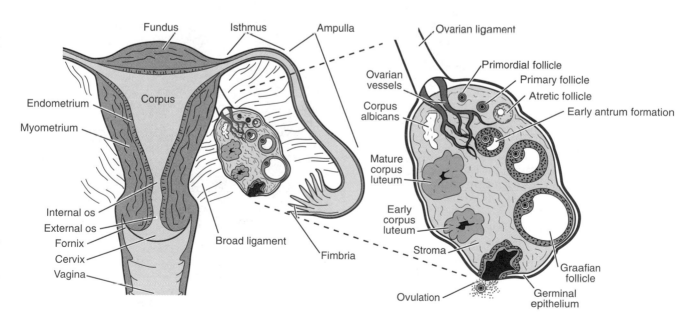

Figure 40–4 ■ The female reproductive organs. (Modified from Patton BM: *Human Embryology.* New York, McGraw-Hill, 1976.)

lum is trumpet-shaped and opens to the peritoneal cavity. Its thin walls are covered with densely ciliated projections, the fimbria, which facilitate ovum uptake during ovulation. The ampulla is the site of fertilization. It has a thin musculature and well-developed mucosal surface. The isthmus is located at the uterotubal junction and has a narrow lumen surrounded by smooth muscle. It has sphincterlike properties and presents an anatomic barrier to the passage of germ cells. The oviducts transport the germ cells in two directions: sperm ascend toward the ampulla and the zygote descends toward the uterus. This requires coordination between smooth muscle contraction, ciliary movement, and fluid secretion, all of which are under hormonal and neuronal control.

The **uterus** is a hollow organ located in the center of the pelvic cavity. It provides the proper environment for the growing fetus and generates the mechanical force necessary for delivery. It grows from the size of a fist in a nonpregnant woman to a very large structure that accommodates a 3–4 kg fetus. In cross-section, the uterus is composed of two types of tissue. The outer part is the **myometrium,** composed of multiple layers of smooth muscle. The inner part is the **endometrium,** which contains a deep stromal layer next to the myometrium and a superficial epithelial layer. The stroma is permeated by **spiral arteries** and contains much connective tissue. The epithelial layer is interrupted by **uterine glands,** which also penetrate the stromal layer and are lined by columnar secretory cells.

The **cervix** is a narrow muscular canal connecting the vagina and the uterus. It is lined with columnar epithelium that secretes mucus. The cervix presents a mechanical barrier to the entry of sperm and provides protection against infection. The **vagina** is about 10 cm long, is well innervated, and has a rich blood supply. It is lined by several layers of epithelium that changes in histology during the menstrual cycle. The vagina contains glands that produce a viscous secretion that acts as a lubricant during coitus.

■ The Follicle and Corpus Luteum Share a Common Origin

Close examination of an adult ovary reveals a mixture of structures of different histologic appearance scattered throughout the ovary. These include follicles of varying sizes, **atretic** (degenerating) follicles, and corpora lutea.

The follicle is essential for reproduction. It contains an oocyte and several layers of supporting cells that protect and nourish the oocyte and provide the proper hormonal environment. **Primordial follicles** are the smallest and most numerous and constitute an inactive resting pool from which all follicles develop (Fig. 40–5). They are located in the outer cortex beneath the tunica and are composed of a **primary oocyte** (less than 20 μm in diameter) surrounded by a single layer of undifferentiated flattened **granulosa cells.** A

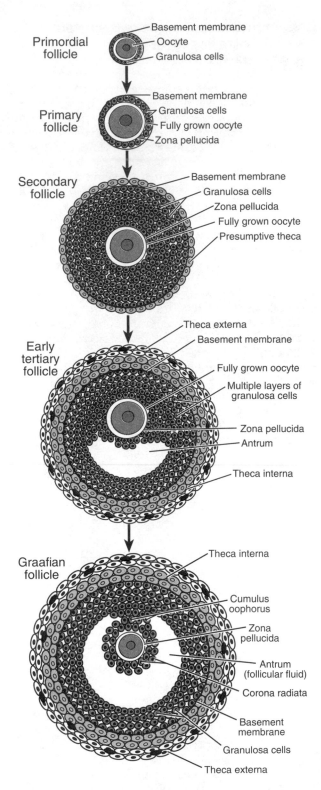

Figure 40–5 ■ Structure of the developing follicle, from primordial through graafian. (Modified from Erickson GF, in Sciarra JJ and Speroff L (eds). *Reproductive Endocrinology, Infertility, and Genetics.* New York, Harper & Row, 1981.)

basement membrane separates the primordial follicle from the adjacent ovarian stroma.

The **primary follicle** represents the second stage in follicular development. Its oocyte is much larger (80–100 μm in diameter), but it is still surrounded by a single layer of cuboidal granulosa cells. The **zona pellucida** forms a concentric layer that separates the oocyte from the granulosa cells. It is composed of mucopolysaccharides secreted by the granulosa cells and glycoproteins secreted by the oocyte.

The **secondary (preantral) follicle** has migrated deeper into the cortex. It contains multiple layers of granulosa cells formed by cell proliferation. They send cytoplasmic projections that communicate with the oocyte through gap junctions located at contact points between their plasma membranes. Outside the basement membrane, stromal cells differentiate and form concentric layers around the follicle, the **theca** layer. The theca layer is richly vascularized and contains lymphatic and nerve terminals. They do not penetrate the basement membrane. The oocyte and enclosing granulosa cells obtain nutrients only by diffusion through the basement membrane.

The **tertiary (antral) follicle** contains even more granulosa cells, which have started to secrete a fluid. The **follicular fluid** accumulates in spaces in the granulosa layer. These cells eventually coalesce, forming a single, fluid-filled cavity, the **antrum.** The granulosa cells develop numerous gap junctions and become electrically and metabolically coupled. The theca cells have further differentiated into **theca interna** and **theca externa.**

The **preovulatory (graafian) follicle** constitutes the final stage in follicular development. It has reached its largest size (2–2.5 cm in diameter), and the follicular fluid fills the entire antrum. The oocyte has been displaced to an eccentric position and is attached to the follicular wall by a stalklike structure. The innermost cells surrounding the oocyte are called **corona radiata,** and the next layer, which also forms the stalk, is called the **cumulus oophorus.**

Atretic follicles constitute the largest fraction of the total population of follicles. They are formed throughout follicular development and include both small and large follicles. In small follicles, atresia begins with lysis of the oocyte. In medium-sized follicles, the granulosa layer becomes pyknotic and disorganized. In large follicles, the thecal layer is hyperplastic and highly vascularized, and there is an increased proliferation of connective tissue.

The **corpus luteum** develops from the evacuated graafian follicle. After extrusion of the oocyte with its surrounding granulosa cells and follicular fluid, the wall of the follicle collapses and becomes convoluted. Blood vessels and connective tissue invade the granulosa layer and the antral cavity fills with blood. This is later replaced by the granulosa layer, which has undergone significant proliferation and differentiation by the process called **luteinization.** The corpus luteum has one of the richest blood supplies in the body.

After degeneration, the luteinized cells are replaced by a fibrous tissue, creating a nonfunctional structure, the **corpus albicans.**

■ Oogonia Begin Developing during Fetal Life, but Development Arrests in Midgestation

Primordial germ cells originate in the endoderm of the yolk sac of the embryo. During the first 3–5 weeks of fetal life they migrate by ameboid movement to the genital ridges. About 2000 of these, now called **oogonia,** invade the cortex of the developing ovary and start proliferating by active mitosis. By 8 weeks, their number reaches 600,000. Between 8–13 weeks, some oogonia cease mitosis and begin the first meiotic division. Others continue to proliferate, and by 5–6 months of fetal life the ovary contains 6–7 million oogonia. By this time, all the germ cells, now classified as **primary oocytes,** have entered meiosis (Table 40–2). From then on, the population of germ cells declines markedly, to about 1–2 million at birth. This decline is due to atresia. The survival of an oocyte in the fetal ovary depends on its contact and encapsulation by granulosa cells. Oocytes failing to form primordial follicles undergo atresia and degenerate. The number of oocytes continues to decline with increasing age. At puberty, the ovary contains about 400,000 oocytes. Only a few oocytes remain in the ovary at menopause.

The process of meiosis extends over many years and is interrupted twice, first during fetal life and then at the time of ovulation (Table 40–2). After the germ cells begin meiosis at midfetal life, they proceed to prophase of the first meiotic division. This involves alignment of the homologous chromosomes, pairing, formation of chiasmata, crossing over, and exchange of genetic material. Primary oocytes remain arrested in this phase of meiosis until ovulation, 13–50 years later. The mechanisms regulating entry of oogonia into meiosis and their prolonged arrest are not well understood. The presence of two X chromosomes, one of which is later inactivated (Barr body), appears necessary for oogonia to enter meiosis and survive. The presence of both a meiotic stimulating substance and a meiotic inhibiting factor has been postulated.

Throughout meiotic arrest, primary oocytes are not quiescent. They exhibit transcriptional activity and synthesize several classes of stable RNA. From the time of formation to ovulation, primary oocytes grow from less than 20 μm to about 100 μm in diameter. They also synthesize several products unique to oocytes, such as cortical granules and components of the zona pellucida. The granulosa cells that surround the oocytes are metabolically coupled to the oocyte and are essential for its growth.

The resumption of meiosis (oocyte maturation) is triggered by the preovulatory LH surge (see below). Luteinizing hormone is believed to act by neutralizing the effect of a **meiosis inhibiting factor.** The resumption of meiosis is characterized by disappearance

TABLE 40–2 ■ Different Stages in the Development of Ova and Follicles

Stage	Process	Ovum	Follicle
Fetal life	Migration	Primordial germ cells	Primordial follicle
	Mitosis	Oogonia	↓
	First meiotic division begins	Primary oocyte	Primary follicle
Birth	Arrest in prophase		↓
	Growth of oocyte and follicle		
Puberty	Follicular maturation		Secondary follicle
			↓
Cycle			Antral follicle
			↓
Ovulation	Resumption of meiosis	Secondary oocyte	Graafian follicle
	Emission of first polar body		
	Arrest in metaphase		↓
			Corpus luteum
Fertilization	Second meiotic division complete	Ootid (zygote)	
	Emission of second polar body		
Implantation	Mitotic divisions		↓
	Blastocyst formation	Embryo	

of the nuclear membrane, condensation of the chromosomes, and nuclear dissolution (germinal vesicle breakdown). A meiotic spindle is then formed and migrates to the cell periphery. An unequal division of the cell cytoplasm yields a large **secondary oocyte** and a small, nonfunctional cell, the **first polar body.**

The secondary oocyte is formed several hours before ovulation. It rapidly begins the second meiotic division and proceeds through a short prophase to become arrested in metaphase of the second meiotic division. At this stage, the secondary oocyte is expelled from the graafian follicle. The second arrest period is relatively short. In response to penetration by a spermatozoon during fertilization, meiosis resumes and is rapidly completed. A second unequal cell division soon follows, producing a small **second polar body** and a large fertilized egg, the zygote. The first polar body resulting from the first cell division either degenerates or divides, yielding two small nonfunctional cells.

■ Female and Male Germ Cells Differ in Ontogeny, Quantity, and Morphology

Oocytes share many characteristics with spermatozoa, including a common embryonic origin, reliance on supporting cells (Sertoli or granulosa cells), partial regulation by gonadotrophins, and dependence on steroid hormones. Moreover, both represent the only cells in the body that undergo meiosis and become haploid. There are, however, significant differences between the two germ cells in terms of ontogeny, duration of meiosis, quantity, and morphology.

The ontogeny of the female germ cells underscores several fundamental differences from the male. Meiosis in the male begins at puberty and takes 65–70 days to complete. Men produce fresh spermatozoa throughout life. Women, on the other hand, ovulate eggs that have remained arrested in the first meiotic division for many years. During this time, the nonreplicating DNA is exposed to a variety of environmental insults. Consequently, older women, but not older men, have an increased probability of producing defective germ

cells, resulting in an increased incidence of babies with birth defects such as Down syndrome. The male produces 100–200 million spermatozoa every day, and its stem cell population (spermatogonia) continues to divide with only a small decline from puberty to old age. The female, on the other hand, produces only several hundred mature ova throughout the reproductive years, and this process ceases at menopause.

Other differences between male and female germ cells involve size, morphology, and association with supporting cells. Spermatozoa are very small, with a diameter of 1–2 μm. Their highly specialized shape is adapted for locomotion, egg recognition, and penetration. Each spermatogonium undergoing meiosis produces four functional spermatozoa of similar structure. With a diameter of 100 μm, the ovum is the largest cell in the body; it is round and appears nondistinct and nonspecialized. Each oogonium yields one functional ovum and two to three polar bodies. The ovum maintains its large size in part by consuming most of the cytoplasm of the polar bodies.

Spermatozoa and ova depend on supporting cells, Sertoli and granulosa cells, respectively, for nutritional support. However, during spermiation, the spermatozoa detach from the Sertoli cells and function independently, whereas the ovum is shed surrounded by the corona radiata. Given the large size of the ovum and its surrounding granulosa cells, the egg constitutes a large and distinct target for recognition and penetration by the sperm cells.

Folliculogenesis, Steroidogenesis, and Ovulation

■ Folliculogenesis Is Controlled by Hormonal and Nonhormonal Factors

Folliculogenesis is the process by which follicles develop and mature. At any given time, follicles are found under four conditions: resting, growing, degenerating, or ready to ovulate. Some of these processes occur without hormonal intervention, while others are controlled by an intricate relationship between the gonadotrophins, steroids, and local ovarian factors.

Progression from primordial to primary follicles occurs at a relatively constant rate throughout fetal, juvenile, prepubertal, and adult life. Once primary follicles leave the resting reservoir, they are committed for further development. Most become atretic at some point of their development, and typically only one will ovulate in the human female. The conversion from primordial to primary follicles is believed to be independent of gonadotrophins. The exact signal that recruits a follicle from a resting to a growing pool is unknown; it could be programed by the cell genome or influenced by local ovarian factors.

Development beyond the primary follicle is gonadotrophin-dependent. It begins at puberty and continues in a cyclic manner throughout the reproductive years. Maturation of primary follicles to the preantral stage takes about 3 weeks. The critical hormone responsible for progression from preantral to antral stage is FSH. However, pure preparations of FSH are less effective than those containing some LH, indicating that both hormones, at a certain ratio, are required. The granulosa cells of secondary follicles acquire receptors for FSH and start producing small amounts of estrogen. Mitosis of the granulosa cells is stimulated by FSH and estradiol. As the number of granulosa cells increases, production of estrogens, binding capacity for FSH, size of the follicle, and volume of the follicular fluid all increase markedly.

In addition to blood-borne hormones, antral follicles are exposed to a unique microenvironment. The follicular fluid contains different concentrations of pituitary hormones, steroids, peptides, and growth factors. Some are present in the follicular fluid at a concentration 100–1000 times higher than in the circulation. Table 40–3 lists some parameters of human follicles at successive stages of development. There is a 5-fold increase in follicular size and a 25-fold rise in the number of granulosa cells. As the follicle matures, the intrafollicular concentration of FSH does not change much, that of LH increases, and that of prolactin declines. Of the steroids, the concentrations of estradiol and progesterone increase 20-fold while androgen levels remain unchanged.

The follicular fluid contains other substances, including inhibin, activin, GnRH-like peptide, growth factors, opioid peptides, oxytocin, and plasminogen activator. Inhibin and activin inhibit and stimulate, respectively, the release of FSH from the anterior pituitary. Inhibin consists of two dissimilar α and β subunits linked by disulfide bonds. It exists in two forms, A and B, each having an identical α subunit and a distinct β subunit. **Activin** is composed of two β subunits. Inhibin is secreted by the granulosa cells into the circulation, where it acts as a distant hormone, and into the follicular fluid, where it acts as a paracrine factor. The functions of other peptides and growth factors in the follicular fluid are unclear. Plasminogen activator participates in the process of follicular rupture during ovulation.

■ Both Granulosa and Theca Cells Take Part in Steroidogenesis

Ovarian steroidogenesis proceeds along the same biosynthetic pathway as in the testes. The main product of the follicle is estradiol, while progesterone is produced by the corpus luteum. Small amounts of androgens are produced by both structures and by ovarian stromal cells. Ovarian steroidogenesis depends on the availability of cholesterol, which is produced locally from acetate or taken up from the circulation via low-density lipoprotein (LDL) receptors. Conversion of cholesterol to pregnenolone by side chain cleavage is a rate-limiting step regulated by the gonadotrophins. From pregnenolone to androgens (C-19 steroids), ovarian steroidogenesis proceeds primarily along the delta 4 pathway (Fig. 39–11). Unlike the testes, the ovary has an active cytochrome P450arom; and the main products of the follicle are estrogens, rather than androgens.

The steroidogenic output of the follicle requires cooperation between granulosa and theca cells and coordination between FSH and LH. An understanding of the two-cell, two-gonadotrophin concept (Fig. 40–6) requires recognition of several characteristics that distinguish theca from granulosa cells. These include distribution and abundance of hormone receptors, expression of steroidogenic enzymes, and effective microenvironment.

The actions of FSH are restricted to granulosa cells, since all other ovarian cell types lack FSH receptors.

TABLE 40–3 ■ Different Parameters of Follicles during the First Half of the Cycle in Women

Cycle (day)	Diameter (mm)	Volume (ml)	Granulosa Cells (10^6)	FSH	LH	PRL	A	E$_2$	P$_4$
1	4	0.05	2	2.5	—	60	800	100	—
4	7	0.15	5	2.5	—	40	800	500	100
7	12	0.50	15	3.6	2.8	20	800	1000	300
12	20	0.50	50	3.6	6.0	5	800	2000	2000

A, androstenedione; E$_2$, estradiol; P$_4$, progesterone; FSH, follicle stimulating hormone; LH, luteinizing hormone; PRL, prolactin.
Modified from Erickson GF: An analysis of follicle development and ovum maturation. *Semin Reprod Endocrinol* 4:233–254, 1986.

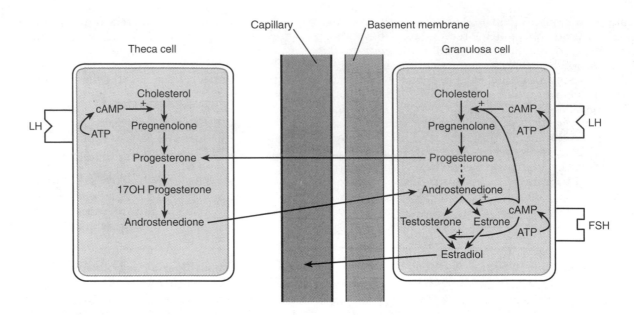

Figure 40–6 ■ The two-hormone, two-cell theory of ovarian steroidogenesis. The dashed line represents the inability of granulosa cells to convert progesterone to androstenedione. LH, luteinizing hormone; FSH, follicle stimulating hormone.

Luteinizing hormone actions, on the other hand, are exerted on theca and granulosa cells, stromal (interstitial) cells, and the corpus luteum. The expression of LH receptors is time-dependent. Theca cells acquire LH receptors at a relatively early stage, whereas LH receptors on the granulosa cells are induced by a combined action of FSH and estradiol only in the maturing follicle.

The biosynthetic enzymes are differentially expressed in the two cells. Cytochrome P450arom is expressed mainly in granulosa cells, and both its activation and induction are regulated by FSH in synergism with estradiol. Granulosa cells are deficient in cytochrome P450c17 and cannot proceed beyond C-21 steroids to generate C-19 androgenic compounds. Consequently, estrogen production by granulosa cells depends on an adequate supply of exogenous aromatizable androgens, provided by the theca cells. Under LH regulation, theca cells produce androgenic substrates, primarily androstenedione, which reaches the granulosa cells by diffusion. Androstenedione is then converted to testosterone by 17-keto reductase, followed by aromatization to estradiol.

Theca and granulosa cells are exposed to a different microenvironment. Vascularization is restricted to the theca layer, since blood vessels cannot penetrate the basement membrane. Theca cells, therefore, have better access to circulating cholesterol, which enters the cells via LDL receptors. Granulosa cells, on the other hand, primarily produce cholesterol from acetate, a less efficient process than uptake. In addition, they are bathed in follicular fluid and exposed to paracrine control by locally produced peptides and growth factors.

Follicle stimulating hormone acts on the granulosa cells by a cAMP-dependent mechanism and exhibits a broad range of activities, including increased mitosis and cell proliferation, stimulation of progesterone synthesis, induction of aromatase, and increased inhibin synthesis. As the follicle matures, the number of receptors for both gonadotrophins increases. Follicle stimulating hormone stimulates the formation of its own receptors and induces the appearance of LH receptors. The combined activity of the two gonadotrophins greatly amplifies estrogen production.

Androgens are produced by theca and stromal cells. They serve as precursors for estrogen synthesis and also have a distinct local action. At low concentrations, androgens enhance aromatase activity, thus promoting estrogen production. At high concentrations, androgens are converted by 5α reductase to more potent androgens, such as dihydrotestosterone (DHT). When follicles are overwhelmed by androgens, the intrafollicular androgenic environment antagonizes estrogen-induced cell proliferation and leads to atresia.

■ From a Pool of Growing Follicles, One Is Selected for Ovulation

The number of ovulating eggs is species-specific and is influenced by genetic, nutritional, and environmental factors. In humans, normally only one follicle will ovulate, but **superovulation** can be induced by timed administration of gonadotrophins. The mechanism by which one follicle is selected from a cohort of growing follicles is poorly understood. It occurs during the first few days of the cycle, immediately after the onset of menstruation. Once selected, the follicle begins to grow and differentiate at an exponential rate and becomes "dominant."

In parallel with the growth of the dominant follicle, the rest of the preantral follicles undergo atresia. Two main factors contribute to atresia in the nonselected follicles. One is the suppression of plasma FSH in response to increased estradiol secretion by the dominant follicle. The decline in FSH support decreases aromatase activity and estradiol production and interrupts granulosa cell proliferation. The dominant follicle is protected from a fall in circulating FSH levels because it has accumulated FSH in the follicular fluid and increased the density of FSH receptors on its granulosa cells. The high concentration of estradiol in the follicular fluid of the dominant follicle also amplifies FSH actions in terms of mitogenesis and increased aromatase activity. The second factor is the accumulation of androgenic compounds, especially those that are nonaromatizable, such as DHT, in the nonselected follicles. This changes the intrafollicular ratio of estrogen to androgen and antagonizes the actions of FSH and estradiol.

As the dominant follicle grows, vascularization of the theca layer increases. On day 9–10 of the cycle, the vascularity of the dominant follicle is twice that of the other antral follicles. This permits a more efficient delivery of cholesterol to the theca cells and better exposure to circulating gonadotrophins. At this time, the main source of circulating estradiol is the dominant follicle. Since estradiol is the primary regulator of LH and FSH secretion, the dominant follicle ultimately determines its own fate.

The LH surge, generated by factors discussed below, induces multiple changes in the dominant follicle, which occur within a relatively short time. These include vascular alterations, resumption of meiosis, granulosa cell differentiation and transformation, activation of proteolytic enzymes, increased production of prostaglandins, histamine, and other local factors, and a shift in steroidogenesis. Within 30–36 hours after the onset of the LH surge, this coordinated series of biochemical and morphologic events culminates in follicular rupture and ovulation. The midcycle FSH surge is not essential for ovulation, as an injection of either LH or human chorionic gonadotrophin (hCG) before the endogenous gonadotrophin surge can induce normal ovulation. However, only follicles that have been adequately primed with FSH will ovulate.

One of the earliest responses of the ovary to a rise in LH is increased blood flow, resulting from an LH-mediated release of vasodilatory substances such as histamine, bradykinin, and prostaglandins. The highly vascularized dominant follicle becomes hyperemic and edematous. There is also an increased production of follicular fluid, disaggregation of granulosa cells, and detachment of the oocyte-cumulus complex from the follicular wall. The basement membrane separating the theca from granulosa cells begins to disintegrate and the granulosa cells begin to undergo luteinization.

The resumption of meiosis is triggered by LH. Up to this point, the primary oocyte has been protected from premature resumption of division by a local factor, **oocyte maturation inhibitor,** produced by the granulosa cells. Luteinizing hormone-induced cAMP activity overcomes this inhibition and the oocyte resumes meiosis, undergoes the first cell division, and extrudes the first polar body.

As ovulation approaches, the follicle enlarges and protrudes from the surface of the ovary. In response to the surge, plasminogen activator is produced by theca and granulosa cells of the dominant follicle and converts plasminogen to plasmin. **Plasmin** is a proteolytic enzyme that acts directly on the follicular wall and stimulates the production of collagenase enzymes, which digest the connective tissue matrix. The thinning and increased distensibility of the wall facilitates the rupture of the follicle. The extrusion of the oocyte-cumulus cell mass is aided by smooth muscle contraction. The levels of prostaglandins of the E and F series increase markedly after the LH surge in response to increased cAMP levels. Prostaglandin E_2 causes vasodilation and smooth muscle relaxation, whereas prostaglandin $F_2\alpha$ induces venomotor smooth muscle contraction.

The LH surge also causes a large but short-lived increase in steroid biosynthesis. Luteinizing hormone stimulates the conversion of pregnenolone to progesterone and enhances the production of progesterone as well as androgens and estrogens. Progesterone acts locally to promote ovulation by increasing the distensibility of the follicular wall and enhancing the activity of the proteolytic enzymes. As LH levels reach their peak, plasma estradiol levels plunge because of down-regulation by LH of its own receptors.

■ The Corpus Luteum Has a Finite Life Span

The corpus luteum is a transient endocrine structure formed from the evacuated graafian follicle. It serves as the main source of circulating steroids during the postovulatory phase of the cycle and is essential for maintenance of pregnancy during the first trimester. The process of luteinization begins before ovulation. The granulosa cells, after acquiring a high concentration of LH receptors, respond to the LH surge by undergoing morphologic and biochemical transformation. This involves cell enlargement (hypertrophy) and development of smooth endoplasmic reticulum and lipid inclusions, typical of steroid-secreting cells. Unlike the nonvascular granulosa cells, luteal cells have a rich blood supply. Invasion by capillaries starts immediately after the LH surge and is facilitated by the dissolution of the basement membrane. Peak vascularization is reached 7–8 days after ovulation.

The corpus luteum of many species is composed of transformed granulosa cells and secretes progesterone only. In women, differentiated theca and stroma cells are incorporated into the corpus luteum, and all three classes of steroids—androgens, estrogens, and progestins—are synthesized. Although some progesterone is secreted before ovulation, peak progesterone production is reached 6–8 days after the LH surge. The life span of the corpus luteum is limited. Unless pregnancy occurs, it degenerates within 9-11 days after formation. The function of the corpus luteum is maintained by LH.

Regression of the corpus luteum is not due to removal of LH, since it occurs even when LH levels are maintained artificially; rather, it is induced by luteolytic agents. Since removal of the uterus (hysterectomy) in primates does not prolong the cycle, uterine factors do not appear to serve as luteolysins. Several candidates, such as estrogen, oxytocin, and GnRH, have been proposed, but their role as luteolysins is controversial. The corpus luteum is rescued from degeneration once implantation of the developing embryo has occurred by the action of hCG produced by the trophoblast (Chap. 41).

The Menstrual Cycle

■ Menstruation Is Unique to Primates

The most conspicuous feature of female reproduction is interrupted fertility. Under normal conditions, ovulation occurs at timed intervals. Once pregnancy ensues, ovulation ceases. After delivery, lactation imposes a period of interrupted or reduced fertility.

Most female mammals exhibit a definite period of sexual receptivity near the time of ovulation, termed **estrus** ('heat'). At any other time during the estrous cycle, the female refuses advances by the male. Estrous behavior is triggered by estradiol, supported by progesterone, and mediated by GnRH. The female becomes receptive to the male, displays a species-specific stereotypic behavior, and often emits olfactory cues by means of pheromones to signal readiness to mate. Thus, sexual receptivity ensures that the female is mated at the time of ovulation. In contrast, a woman is potentially receptive to a man at any time during the cycle and does not display behavioral, physical, or olfactory signs that indicate ovulation. In the absence of correlation between sexual receptivity and ovulation, the most obvious event that indicates cyclicity is menstruation. Menstruation signifies a failure to conceive and results from regression of the corpus luteum and withdrawal of steroidal support of the superficial endometrial layer.

Menstrual cycles begins at adolescence. The first few cycles are usually irregular and anovulatory, due to delayed maturation of the positive feedback by estradiol. The average cycle length in adult women is 28 days, with upper and lower limits of 35 and 23 days. The interval from ovulation to the onset of menstruation (the luteal phase) is relatively constant, averaging 14 ± 2 days in most women, and is dictated by the fixed life span of the corpus luteum. In contrast, the interval from the onset of menses to ovulation (the follicular phase) is more variable and accounts for differences in cycle lengths among ovulating women. The cycles become irregular as menopause approaches and cease thereafter. During the repro-

ductive years, menstrual cyclicity is interrupted by conception and lactation and is subjected to modulation by physiologic, psychological, and social factors. Selected causes of **amenorrhea** (cessation of menstruation) are discussed below.

■ The Hormonal Cycle Requires Synchrony among the Ovary, Brain, and Pituitary

Thus far we have discussed separately the mechanisms that regulate the synthesis and release of the reproductive hormones. Now we put them together in terms of sequence and interaction. For this purpose, we use a hypothetical cycle of 28 days (Fig. 40-7), divided as follows: menses (days 0-5), follicular phase (days 0-13), LH surge and ovulation (days 13-14), and luteal phase (days 14-28).

During the menses, estrogen and progesterone levels are very low due to corpus luteum regression and low steroid synthesis by immature follicles. In response to removal of negative feedback, plasma FSH levels are elevated while mean LH levels are low. Follicle stimulating hormone acts on a cohort of follicles recruited 20-25 days earlier from resting primordial follicles. The size of the follicles on days 3-5 averages 4-6 mm in diameter, and they are propelled by FSH to the preantral stage. The granulosa cells begin to proliferate, aromatase activity increases, and plasma estradiol levels rise slightly between days 3 and 7. The designated dominant follicle is selected between days 5 and 7, starts growing, and increases in size and steroidogenic activity. Between days 8 and 10, plasma estradiol levels rise sharply, reaching peak levels above 200 pg/ml on day 12, the day before the LH surge.

During the early follicular phase, LH pulsatility is of low amplitude and high frequency (Table 40-1). As estradiol levels rise, the pulse frequency further increases, without a change in amplitude. This results in a small, gradual rise in mean plasma LH level, which supports follicular steroidogenesis. At mid- to late follicular phase, estradiol suppresses FSH release. The decline in FSH, together with an accumulation of nonaromatizable androgens, induces atresia in the nonselected follicles. The dominant follicle is saved by virtue of its high density of FSH receptors, accumulation of FSH and estradiol in its follicular fluid (Table 40-3), and acquisition of LH receptors by the granulosa cells.

The midcycle surge of LH is rather short, lasting 24-36 hours (Fig. 40-7). For the LH surge to occur, estrogen must be maintained at a critical concentration (about 200 pg/ml) for a sufficient duration (36-48 hours). Any deviation (i.e., prevention of the estradiol rise or a rise that is too small or too short) eliminates the surge. Paradoxically, then, although it exerts negative feedback on LH release most of the time, positive feedback by estradiol is required to generate the midcycle surge.

The timing of the surge in primates is determined by the ovary. Estrogen exerts its effects directly on the

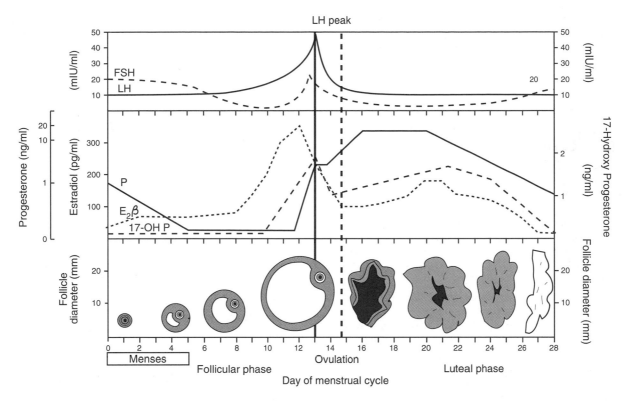

Figure 40–7 ■ Hormonal and ovarian events during the menstrual cycle. FSH, follicle stimulating hormone; LH, luteinizing hormone; P, progesterone; $E_2\beta$, estradiol; 17-OH P, 17-hydroxy progesterone. (Modified from Erickson GF, Yen SSC: New data on follicle cells in polycystic ovaries: A proposed mechanism for the genesis of cystic follicles. *Semin Reprod Endocrinol* 2:231–243, 1984.)

anterior pituitary, with GnRH playing a permissive, albeit mandatory, role. This concept is derived from experiments in monkeys whose medial basal hypothalamus, including the GnRH neurons, was lesioned, resulting in a marked decrease in plasma LH levels. Administration of exogenous GnRH at a fixed frequency restored LH release. When estradiol was given at an optimal concentration for an appropriate time, an LH surge was generated in spite of maintaining steady and unchanging pulses of GnRH.

The mechanism that transforms estradiol from a negative to a positive regulator of LH release is enigmatic. Part of the positive action involves an increase in the number of GnRH receptors on the gonadotrophs, thereby increasing pituitary responsiveness to GnRH. Another is the conversion of a storage pool of LH (perhaps within a subpopulation of gonadotrophs) to a readily releasable pool. Estrogen may also increase GnRH release, serving as a fine-tuning or fail-safe mechanism. There is a small but distinct rise in progesterone before the LH surge. This rise is important for augmenting the LH surge and, together with estradiol, promoting a concomitant surge in FSH. There are indications that the midcycle FSH surge is important for inducing enough LH receptors on granulosa cells, stimulating plasminogen activator, and activating a cohort of follicles destined to develop in the next cycle. A midcycle

rise of 17-hydroxy progesterone is also observed (Fig. 40–7), but its function is unclear.

The LH rise inhibits the cytochrome P450c17 enzyme, decreasing androstenedione production by the dominant follicle. Estradiol levels decline, 17-hydroxy progesterone increases, and progesterone levels plateau. The prolonged exposure to high LH levels downregulates the LH receptors, accounting for the immediate postovulatory suppression of estradiol. As the corpus luteum matures, it increases progesterone production and reinitiates that of estradiol. Both reach high plasma concentrations on days 20–23, about 1 week after ovulation.

During the luteal phase, circulating FSH levels are suppressed by the elevated steroids. The LH pulse frequency is reduced during the early luteal phase, but the amplitude is higher than that during the follicular phase (Table 40–1). Luteinizing hormone is important at this time for maintaining the function of the corpus luteum and sustaining steroid production. In late luteal phase, both LH pulse frequency and amplitude are reduced by a progesterone-dependent, opioid-mediated suppression of the GnRH pulse generator. In addition to acting on the brain, progesterone acts in the ovary to suppress new follicular growth. Consequently, ovulation in the next cycle usually occurs in the ovary contralateral to the previous corpus luteum, resulting in alternate ovulations in consecutive cycles.

Following the demise of the corpus luteum on days 24–26, estradiol and progesterone levels plunge. This results in withdrawal of support from the uterine endometrium, culminating within 2–3 days in menstruation. The reduction in ovarian steroids acts centrally to remove feedback inhibition. The FSH level begins to rise, and a new cycle is initiated.

■ Sequential Release of Hormones Causes Cyclic Changes in the Reproductive Tract

In response to the changing levels of ovarian steroids, the female reproductive tract undergoes cyclic alterations. The most notable changes occur in the histology of the endometrium, the composition of cervical mucus, and the cytology of the vagina (Fig. 40-8). At the time of ovulation, there is also a small but detectable rise in basal body temperature, caused by progesterone. All of the above parameters are clinically useful for diagnosing menstrual dysfunction and infertility.

The endometrial cycle is divided into four phases: proliferative, secretory, ischemic and menstrual. The **proliferative stage** coincides with the mid- to late follicular phase. Under the influence of the rising estradiol, the stroma and epithelial layers of the uterine endometrium undergo hyperplasia and hypertrophy and increase in size and thickness. The endometrial glands elongate and are lined with columnar epithelium. The endometrium becomes vascularized and more spiral arteries develop. Estradiol also induces the formation of progesterone receptors and increases myometrial excitability and contractility.

The **secretory phase** begins on the day of ovulation and coincides with the early to midluteal phase. Under the combined action of progesterone and estrogen, the endometrial glands become coiled, store glycogen, and secrete large amounts of carbohydrate-rich mucus. The stroma increases in vascularity and becomes edematous, and the spiral arteries become tortuous (Fig. 40-8). Peak secretory activity, edema formation, and overall thickness of the endometrium are reached on days 6–8 after ovulation in preparation for implantation of the zygote. Progesterone antagonizes the effect of estrogen on the myometrium and reduces spontaneous myometrial contractions.

The **ischemic phase** is initiated by the declining levels of steroids due to corpus luteum regression. Necrotic changes occur in the mucosa layer and pseudodecidual reactions become apparent in the stroma, near the spiral arteries. The arteries constrict, reducing the blood supply to the superficial endometrium. Leukocytes and macrophages invade the stroma and begin to phagocytize the ischemic tissue. The leukocytes persist in large numbers throughout menstruation, providing resistance against infection to the denuded endometrial surface.

Desquamation and sloughing of the entire functional layer of the endometrium occurs during the **men-**

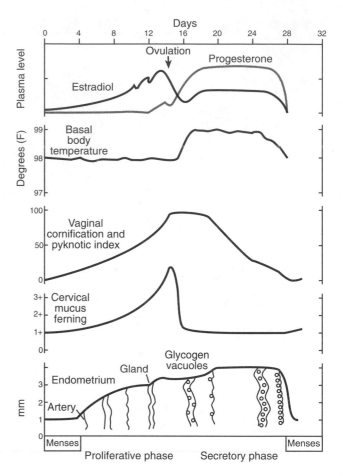

Figure 40–8 ■ Cyclic changes in the uterus, cervix, vagina, and body temperature during the menstrual cycle. (Modified from Odell WD: The reproductive system in women, in Degroot LJ et al (eds): *Endocrinology.* vol. 3. New York, Grune & Stratton, 1979.)

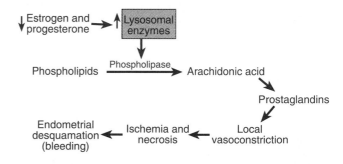

Figure 40–9 ■ Proposed mechanism leading to menstruation.

strual phase. The mechanism leading to necrosis is only partly understood (Fig. 40-9). The reduction in steroids destabilizes lysosomal membranes in endometrial cells, resulting in liberation of proteolytic enzymes and increased production of prostaglandins from stored phospholipids. The prostaglandins induce vasospasm of the spiral arteries, and the proteolytic enzymes digest the tissue. Eventually, the blood vessels rupture and blood is released together with cellular debris. The menstrual flow lasts 4–5 days and averages 30–50 ml. It does not clot because of the presence of fibrinolysin.

The **cervical mucus** undergoes cyclic changes in composition and volume. During the follicular phase, estrogen increases the quantity, alkalinity, viscosity, and elasticity of the mucus. At the time of ovulation, mucosal elasticity, or spinnbarkeit, is greatest; this can be observed on stretching a small drop of mucus. A characteristic ferning pattern is seen when the mucus is placed on a glass slide. These changes in the properties of the mucus promote the survival and transport of sperm. With the rising progesterone either after ovulation or during pregnancy, the quantity and elasticity of the mucus decline; it becomes thicker and does not form a fern pattern. Under these conditions, the mucus provides better protection against infections. Under the influence of estrogen, the vaginal epithelium proliferates. Basophilic cells predominate early in the follicular phase and then become cornified. During the postovulatory period, progesterone induces the formation of thick mucus, and the epithelium becomes infiltrated with leukocytes.

■ Estrogen and Progesterone Have Synergistic and Antagonistic Actions

The principal female sex steroids are estrogens and progestins. Three estrogens are present in significant quantities—**estradiol, estrone,** and **estriol.** Estradiol is the most abundant and is 12 and 80 times more potent than estrone and estriol, respectively. Much of estrone is derived from peripheral conversion of androgens. During pregnancy, large quantities of estriol are produced by the fetoplacental unit (Chap. 41). The most important of the progestins is progesterone. It is secreted in significant amounts only during the luteal phase. During pregnancy, the corpus luteum secretes progesterone throughout the first trimester, and the placenta continues progesterone production until parturition. Small amounts of 17-hydroxy progesterone are secreted along with progesterone. Circulating androgens in the female, primarily androstenedione and dihydroepiandrosterone (DHEA), originate from the adrenal cortex and ovarian theca and stroma cells. Peripheral conversion from prehormones provides an additional source of testosterone and 4-dihydrotestosterone (DHT).

Estrogens and progesterone circulate bound to plasma albumins and specific estrogen- and progesterone-binding globulins. Progesterone shows some pulsatility, which roughly corresponds to that of LH. The action of steroids on their target cells is mediated via nuclear receptors, as was described for testosterone. Steroids are also capable of fast, nongenomic actions, but the mechanism of these is poorly understood.

The main function of estrogens is to promote cellular differentiation and growth of the primary sex structures and other tissues related to reproduction. Unlike testosterone, whose function is critical for sexual dif-

ferentiation of the male fetus, estrogens have little effect on fetuses or on prepubertal girls. During adolescence, estrogens induce growth and maturation of the oviducts, uterus, vagina, and external genitalia. They are responsible for breast enlargement at puberty by inducing the development of stromal tissue, growth of the ductule system, and fat deposition. Similar to androgens in boys, estrogens cause the spurt of growth in girls during puberty by increasing osteoblastic activity, but they also induce closure of the epiphyses of the long bones.

In addition to their effects on the reproductive tract already discussed, estrogens influence the mucosal lining of the oviducts, causing proliferation of the glandular tissue and increasing the number of ciliated epithelial cells. The activity of the cilia is enhanced by estrogens, helping to propel the ovum toward the uterus. Estrogens exert positive and negative effects on LH and FSH release and also stimulate prolactin synthesis. They cause hypertrophy and hyperplasia of the pituitary during pregnancy, primarily by affecting proliferation of the lactotrophs.

Estrogens are considered the "feminizing hormones." Some of the "feminine" characteristics, such as narrow shoulders, broad hips, characteristic fat and hair distribution, and high pitched voice, are also seen in castrate males. Hence, these occur simply because of the absence of testicular androgens. Estrogens cause some salt and water retention and have an effect on the sebaceous glands and skin texture. They lower blood cholesterol levels, thus providing some protection against myocardial infarction and other vascular diseases. In most mammals, estrogens are responsible for sexual receptivity, but there is no evidence for that in human females.

Progesterone is considered the "hormone of pregnancy," since its primary action is geared toward preparation for and maintenance of pregnancy. The principal targets of progesterone are the uterus, breasts, and brain. The progestational effects on the uterine endometrium, cervix, and vagina have already been described. Progesterone antagonizes the actions of estrogens on the myometrium by decreasing myometrial excitability, spontaneous electrical activity, and sensitivity to oxytocin. It also reduces the number of estrogen receptors in the endometrium. Progesterone promotes the development of lobules and alveoli in the breast and induces differentiation of the duct tissue to become secretory. Progesterone acts on hypothalamic thermoregulatory centers and accounts for the rise in basal body temperature at the time of ovulation.

Androgens in the female promote hair growth in the pubic triangle, axillae, and distal extremities. Abnormal androgen production in females, whether from adrenal (e.g., congenital adrenal hyperplasia) or ovarian (e.g., polycystic ovary syndrome) origin, often results in hirsutism and virilism. There are some indications that androstenedione, whose plasma levels normally rise in midcycle, stimulates libido in females.

HIRSUTISM

Hirsutism is broadly defined as excessive growth of hair and is used clinically in reference to women with excessive growth of terminal hair in a male pattern. Hirsutism is one of the most common endocrine disorders, affecting as many as 10% of women in the United States. Almost invariably, it results from abnormal androgens (hyperandrogenism). It can represent a serious pathologic condition or a nonmedical cosmetic problem that often causes significant emotional agony.

The causes of hirsutism can be divided into four categories. One is exogenous androgens, often resulting from intake of androgens or anabolic steroids to increase performance in female athletes. Second is androgens of adrenal origin, resulting from tumors, congenital adrenal hyperplasia, or Cushing syndrome. Third is excess androgen production by the ovary, resulting from ovarian tumors or polycystic ovary syndrome. Fourth is idiopathic, when causes are unknown but increased 5α reductase activity or increased androgen receptor sensitivity is suspected.

Clinical evaluation of hirsutism begins with qualitative and quantitative approaches. Male-pattern terminal hair is coarse, curly, and pigmented and is graded in several areas, including the upper lip, chin, chest, back, abdomen, arms, thighs, and extremities. The axilla and pubis are excluded because the growth of terminal hair here is androgen-dependent in all women. In hirsute women, additional signs of hyperandrogenism should be sought such as temporal hair recession, acne, menstrual irregularities, increased size of clitoris, masculine voice, increased muscle mass, and psychological changes. The next objective is to confirm hyperandrogenism and attempt to identify the source of excess androgens. Hyperandrogenism is confirmed by measuring total and unbound serum testosterone levels and dehydroepiandrosterone sulfate. This is followed by a differential diagnosis. For example, a challenge with dexamethasone, a synthetic corticosteroid, and subsequent determination of circulating androgens can help differentiate between androgens of ovarian and adrenal origin.

Therapies for hirsutism include cosmetic measures aimed at reducing or masking excessive hair growth and systemic intervention directed at reducing androgen production or suppressing their actions. Among the cosmetic measures, bleaching, plucking, waxing, shaving, chemical depilatories, and electrolysis are used with varying degrees of success. Systemic therapies include low-dose oral contraceptives to suppress ovarian androgen production, dexamethasone to suppress adrenal androgen production, and androgen receptor blockers, such as spironolactone and flutamide. Although none of these therapies is perfect, careful administration can result in significant benefit to the patient.

Female Infertility

■ Menopause Is an Age-Related Physiologic Cessation of Ovulation

A complete, age-related cessation of ovulation is unique to the human female. Females of other species show a decline in but not a complete termination of reproductive functions with age. Menopause, or permanent cessation of menses, occurs between the ages of 48 and 52. This has not changed considerably throughout history, although menarche occurs at younger ages today than 100 years ago and life expectancy has increased.

The stock of oocytes is nonrenewable and is depleted throughout life by atresia and ovulation. From about age 45 on, the frequency of ovulation begins to decline, resulting in irregular menses and reduced fertility. This is accompanied by a gradual decrease in estradiol production and consequently an elevation in plasma gonadotrophin levels. The increase in FSH precedes that of LH and is of greater magnitude. Eventually, all follicles are exhausted and ovarian estradiol production becomes almost undetectable. The elevated LH stimulates ovarian stroma cells to continue producing androstenedione. Estrone, derived almost entirely from peripheral conversion of adrenal and ovarian androstenedione, becomes the dominant estrogen (Table 40-4). Since the ratio of estrogens to androgens decreases, some women have hirsutism due to androgen excess.

There are many consequences of the marked decline in estradiol, including hot flashes, changes in the reproductive tract and secondary sex characteristics, a decline in bone mass, increased risk of cardiovascular disease, and psychological alterations. Hot flashes occur in approximately 70% of women at menopause and can last less than a year to several years. They are due to declining estrogen levels, rather than complete estrogen deficiency. Hot flashes are associated with episodic increases in upper body and skin temperature, peripheral vasodilation, and sweating. They occur concurrently with LH pulses but are not caused by the gonadotrophins, since they are evident in hypophysectomized women. Hot flashes reflect temporary disturbances in the hypothalamic thermoregulatory centers, which are somehow linked to the GnRH pulse generator.

Several other organs undergo changes following menopause. Secretion of vaginal mucus decreases and the vaginal epithelium atrophies, often leading to pain and irritation. Similar changes in the urinary tract may give rise to urinary disturbances. The skin epidermis becomes thinner and less elastic and the breasts atrophy. A major cause of morbidity is increased incidence of osteoporosis. The loss of bone density results from a combination of increased bone resorption and decreased bone formation. This reduces skeletal strength and increases susceptibility to fractures, especially in the hips and vertebrae. The decline in estrogens also causes increased cholesterol levels and a higher incidence of coronary artery disease. There may also be emotional changes, including insomnia, depression, and anxiety.

Many of the physiologic and some psychological symptoms of postmenopausal years can be relieved by estrogen therapy. Treatment with estrogen, however, can cause adverse effects, such as vaginal bleeding, nausea, and headache. Estrogen therapy is contraindicated in cases of existing reproductive tract carcinomas or hypertension and other cardiovascular disease. The prevailing opinion is that treatment of postmenopausal women with a combination of estrogen and progesterone for limited periods does not increase the risk of developing breast or endometrial carcinomas.

■ Genetic, Hormonal, Physical, and Psychological Factors Can Cause Female Infertility

In light of the complexity of hormones and organ systems involved in reproduction, it is not surprising that infertility can be caused by multiple factors affecting regulation at different levels. Environmental factors, abnormal function of the central nervous system, hypothalamic disease, pituitary disorders, and ovarian abnormalities can interfere with ovulation. Once ovulation has occurred, reproductive tract and hormonal disorders can prevent fertilization, impede transport and implantation of the embryo, or interfere with pregnancy.

Disorders of the menstrual cycle can be divided into two broad categories: **amenorrhea** (absence of menses) and **oligomenorrhea** (infrequent menstruation). **Primary amenorrhea** is a condition in which menstruation has never occurred. One example is **Turner's syndrome,** or gonadal dysgenesis, a congenital abnormality caused by a nondisjunction of one of the X chromosomes, resulting in a 45 X0 chromosomal karyotype. Since the two X chromosomes are necessary for normal oocyte and ovarian development during fetal life, such an individual has only rudimentary gonads and fails to undergo normal puberty. Because of steroidal deficiency, secondary sex characteristics remain prepubertal and plasma gonadotrophins are elevated. Other abnormalities include short stature, webbed neck, coarctation of the aorta, and renal disorders.

Another congenital form of primary amenorrhea is hypogonadotrophism with anosmia, similar to Kallmann syndrome in males (Chap. 39). Patients do not undergo normal puberty and have low and nonpulsatile gonadotrophin levels, normal stature, female karyotype, and anosmia. The disorder is believed to involve failure of olfactory lobe development and GnRH deficiency. Primary amenorrhea can also be caused by congenital malformation of reproductive tract structures originating from the müllerian duct, including absence or obstruction of the vagina, cervix, or uterus.

Secondary amenorrhea is defined as cessation of menstruation for longer than 6 months. Pregnancy, lactation, and menopause are among the most common physiologic causes of secondary amenorrhea.

TABLE 40–4 ■ Steroid Levels (pg/ml) in Premenopausal and Postmenopausal Women

Stage	Estrone	Estradiol	Testosterone	Androstenedione
Premenopausal				
Early follicular phase	52	64	290	1690
Late follicular phase	27	20	130	997
Postmenopausal				
With ovaries	34	13	230	830
Following ovariectomy	38	14	100	720

Clinically, these disorders can be grouped into premature ovarian failure, polycystic ovary syndrome, hyperprolactinemia, and hypopituitarism.

Premature ovarian failure is defined as occurrence of amenorrhea, low estrogen, and high gonadotrophin levels before the age of 40 years. The symptoms are similar to those of menopause. The etiology is quite variable, including chromosomal abnormalities, lesions due to irradiation, chemotherapy or viral infections, and autoimmune conditions. The latter are often associated with autoantibodies to other glands, such as thyroid and adrenals.

Polycystic ovary (Stein-Leventhal syndrome) is a heterogeneous group of disorders manifested as amenorrhea or anovulatory bleeding, elevated LH/FSH ratio, high androgen levels, hirsutism, and obesity. It has been proposed that the syndrome is initiated by excessive adrenal androgen production during puberty or following stress. Androgens are converted peripherally to estrogens and stimulate LH release. Excess LH, in turn, increases ovarian stromal androgen production and impairs follicular maturation. The hyperstimulated ovaries are enlarged and contain many small atretic follicles and hyperplastic and luteinized theca cells. The elevated plasma androgen levels cause hirsutism, increased activity of sebaceous glands, and other signs of virilization, such as clitoral hypertrophy.

Hyperprolactinemia is a common cause of secondary amenorrhea, accounting for 20–30% of all cases. Galactorrhea, a persistent milklike discharge from the nipple in nonlactating individuals, is a frequent symptom. The etiology of hyperprolactinemia is variable. Pituitary prolactinomas account for about 50% of cases. Other causes are hypothalamic disorders, trauma to the pituitary stalk, and psychotropic medications, all of which are associated with a reduction in dopamine release. Hypothyroidism, chronic renal failure, and hepatic cirrhosis are additional causes of hyperprolactinemia. The mechanism by which elevated prolactin levels suppress ovulation is not entirely clear. It has been postulated that prolactin may inhibit GnRH release, affect LH release in response to GnRH stimulation, or act directly at the level of the ovary (Fig. 40-10).

Oligomenorrhea is often caused by exercise and nutritional, psychological, and social factors. **Anorexia nervosa,** a severe behavioral disorder afflicting more females than males, is characterized by extreme malnutrition and endocrine changes secondary to psychological and nutritional disturbances. About 30% of patients develop amenorrhea that does not alleviate with weight gain. Strenuous exercise, especially by competitive athletes and dancers, frequently causes menstrual irregularities. Two main factors, a low level of body fat and the effect of stress itself, appear responsible. There are no consistent changes in plasma gonadotrophins or ovarian steroids. Other forms of stress, such as relocation, college examinations, general illness, and job-related stresses, have been known to induce some forms of oligomenorrhea.

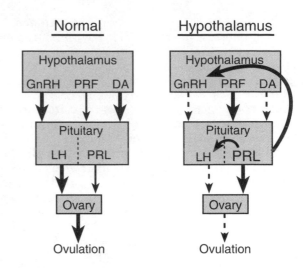

Figure 40–10 ■ The effect of hyperprolactinemia on ovulation and fertility. Under normal conditions, prolactin (PRL) release is low and the GnRH-LH-ovarian axis operates normally. During hyperprolactinemia, elevated PRL levels suppress GnRH and LH release and also directly inhibit ovulation. Heavy arrows indicate predominant effect; dashed arrows indicate suppressed effect. GnRH, gonadotrophin-releasing hormone; PRF, prolactin releasing factor; DA, dopamine; LH, luteinizing hormone.

■ Diagnosis and Treatment of Female Infertility Require a Systematic Approach

The diagnosis and treatment of amenorrhea present a challenging problem. For a successful outcome, it is important first to categorize amenorrhea as primary or secondary and to exclude conditions such as menopause, pregnancy, and lactation. The next step is to determine whether the disorder originates in one of the following compartments: hypothalamus and central nervous system, anterior pituitary, ovary, and reproductive tract. After a positive diagnosis is made, hormonal therapy, corrective surgery, or in vitro fertilization may be performed to achieve feminine appearance for congenital disorders or pregnancy for patients who desire it.

Several treatments can alleviate infertility problems. For example, some success has been achieved in hypothalamic disease with pulsatile administration of GnRH. When hypogonadotrophism is the cause of infertility, sequential administration of FSH and hCG is a common treatment for inducing ovulation, although the risk of ovarian hyperstimulation and multiple ovulations is quite high. Hyperprolactinemia can be treated surgically by removing the pituitary adenoma or pharmacologically with bromocriptine, a dopaminergic agonist. Treatment with clomiphene, an antiestrogen, can induce ovulation in women with endogenous estrogens in the normal range by reducing negative feedback and augmenting endogenous FSH and LH release. When the cause of infertility is reproductive tract lesions, corrective surgery or in vitro fertilization is the treatment of choice.

■ ■ ■ ■ ■ ■ ■ ■ ■ ■ ■ REVIEW EXERCISES ■ ■ ■ ■ ■ ■ ■ ■ ■ ■

Identify Each with a Term

1. The rise of LH that precedes ovulation
2. Abnormal secretion of prolactin associated with infertility
3. Cells that have FSH receptors
4. The process by which more than 99% of the follicles are eliminated
5. The stage at which primary oocytes are arrested during cell division
6. The process by which androgens are converted to estrogens

Define Each Term

7. Myometrium
8. Graafian follicle
9. Granulosa cells
10. Luteolysis
11. Menstruation
12. Proliferative stage of the endometrium
13. Primary amenorrhea
14. Oligomenorrhea

Choose the Correct Answer

15. Stimulation of steroid synthesis by the mature graafian follicle involves:
    a. Binding of FSH to receptors on the theca cells
    b. Binding of FSH to receptors on the stroma
    c. Activation of adenylate cyclase
    d. Increased calcium influx via calcium-activated channels
16. Granulosa cells cannot produce estradiol from cholesterol because they do not have an active:

a. cytochrome P450c17
b. Aromatase
c. 5α reductase
d. Sulfatase

17. The anovulation and amenorrhea typical of postmenopausal women are associated with:
    a. Inability of the ovaries to respond to prolactin
    b. An increase in plasma FSH levels
    c. Excessive corpora lutea
    d. Increased number of cornified cells in the vagina
18. Treatment with a long-lasting progesterone preparation in an ovariectomized woman will result in:
    a. Decreased prolactin secretion
    b. Increased FSH release
    c. Induction of ovulation
    d. Increase in basal body temperature by 0.5–1.0°C
19. The theca interna cells of the graafian follicle are characterized by:
    a. Their capacity for converting progesterone to androstenedione
    b. Having both LH and FSH receptors
    c. Exhibiting active aromatization of testosterone to estradiol
    d. Their ability to produce inhibin

Briefly Answer

20. Define Turner's syndrome.
21. What are the phases of the ovarian cycle?
22. List two physiologic functions of prolactin.
23. When is the second polar body produced?
24. What are the physiologic functions of androgens in females?

■ ■ ■ ■ ■ ■ ■ ■ ■ ■ ■ ANSWERS ■ ■ ■ ■ ■ ■ ■ ■ ■ ■ ■

1. Preovulatory surge
2. Hyperprolactinemia
3. Granulosa cells of the follicle
4. Atresia
5. Prophase of the first meiotic division
6. Aromatization
7. The outer layer of the uterus, which primarily consists of smooth muscle fibers
8. The dominant preovulatory follicle with a large antrum and a displaced oocyte in an eccentric position
9. The inner layer (or layers) of cells in the follicle closest to the oocyte; granulosa cells also form the corona radiata and cumulus oophorus
10. The process by which the corpus luteum undergoes degeneration and becomes the corpus albicans
11. Sloughing of the functional layer of the endometrium

resulting from corpus luteum regression and withdrawal of steroids
12. Increased size and thickness of the stromal and epithelial layers of the endometrium under the influence of rising estradiol during the mid- to late follicular phase
13. Pathologic condition of infertility characterized by a total absence of menstruation
14. Infrequent menstruation
15. c
16. a
17. b
18. d
19. a
20. Turner's syndrome is a congenital abnormality resulting in ovarian dysgenesis and primary amenorrhea due to a lack of one X chromosome (45, X0 karyotype).

21. The phases of the ovarian cycle are: follicular, under the influence of the growing follicles during the first half of the cycle; luteal, under the influence of the corpus luteum during the second half of the cycle; and ovulatory, a short period during midcycle.
22. The physiologic functions of prolactin are support of lactation and modulation of fertility.
23. The second polar body is produced during the process of fertilization, as a result of sperm penetration and resumption of the second meiotic division.
24. Androgens in females support the growth of pubic and axillary hair and possibly are involved in libido.

Suggested Reading

Austin CR, Short RV: *Hormonal Control of Reproduction.* 2nd ed. London, Cambridge University Press, 1984.

Beulieu EE, Kelly PA: *Hormones.* New York, Hermann, 1990.

Bolander FF: *Molecular Endocrinology.* New York, Academic, 1989.

Carr BR, Blackwell RE: *Textbook of Reproductive Medicine.* Norwalk, CT, Appleton & Lange, 1993.

Goodman MH: *Basic Medical Endocrinology.* 2nd ed. New York, Raven, 1994.

Hershman JM: *Endrocrine Pathophysiology.* 3rd ed. Philadelphia, Lea & Febiger, 1988.

Martin JB, Reichlin S: *Clinical Neuroendocrinology.* 2nd ed. Philadelphia, FA Davis, 1987.

Yen SSC, Jaffe RB: *Reproductive Endocrinology.* 3rd ed. Philadelphia, WB Saunders, 1991.

CHAPTER

■ ■ ■ ■ ■ ■ ■

41

Conception, Pregnancy, and Fetal Development

CHAPTER OUTLINE

I. FERTILIZATION AND IMPLANTATION
   A. Both egg and sperm are transported to the oviduct
   B. A multitude of cellular events occur at the time of fertilization
   C. Implantation involves interactions between the endometrium and embryo
   D. Contraceptives interfere with ovulation, fertilization, or implantation

II. PREGNANCY
   A. The placenta is a unique organ interposed between the mother and fetus
   B. The recognition and maintenance of pregnancy depend on hormones
   C. Steroid production during pregnancy involves the fetoplacental unit
   D. Maternal physiology changes substantially throughout gestation

III. FETAL DEVELOPMENT AND PARTURITION
   A. The fetal endocrine system is largely autonomous
   B. The sex chromosomes dictate the development of the fetal gonads
   C. Differentiation of the genital ducts is determined by hormones
   D. A complex interplay between maternal and fetal factors induces parturition

IV. THE POSTPARTUM AND PREPUBERTAL PERIODS
   A. Mammogenesis and lactogenesis are regulated by multiple hormones
   B. The suckling stimulus maintains lactation and inhibits ovulation
   C. The onset of puberty depends on maturation of the hypothalamic GnRH pulse generator
   D. Disorders of sexual development can manifest before or after birth

OBJECTIVES

After studying this chapter, the student should be able to:
1. Describe the transport of germ cells to the site of fertilization; list the events that occur from the time of sperm penetration to the first division of the zygote
2. Discuss the events and time course of embryonic development, implantation, and placenta formation
3. Describe the different categories of contraceptives and evaluate their mechanism of action, reliability, and reversibility
4. Identify the roles of the corpus luteum and trophoblast in the maintenance of pregnancy and the function of the fetoplacental unit in hormone production
5. Outline the hormonal events associated with pregnancy
6. Describe the adjustments in the maternal and fetal endocrine systems in adapting to intrauterine growth and development
7. Discuss the process of sexual differentiation; identify at least two disorders leading to ambiguous sexuality
8. List the events leading to parturition
9. Discuss the role of hormones in breast development, milk production, and lactation
10. Outline the hormonal, physiologic, and physical changes from the neonatal period to puberty

Pregnancy begins with the union of a sperm and an egg. The life span of germ cells is rather short, requiring their rapid transport to the oviduct, where fertilization occurs. Immediately after fertilization, the zygote begins to divide. Because it contains only a limited energy supply, the embryo has to reach the uterus within a short time. Implantation occurs only in a receptive uterus that has been primed by gonadal steroids. Upon implantation, the embryo generates a hormonal signal, human chorionic gonadotrophin (hCG), informing the woman that she is pregnant. This signal extends the life of the corpus luteum and prevents the onset of the next ovulatory cycle and menstruation. The placenta, an organ that exists only during pregnancy, serves as the lungs, kidneys, and food supply of the fetus. It also produces protein and steroid hormones, thus duplicating the functions of the pituitary and gonads. Some of the fetal endocrine glands function before birth and participate in unique functions such as sexual differentiation.

Parturition represents the final stage of gestation. Its onset is triggered by signals emanating from both the fetus and the mother and involves biochemical and mechanical changes in the myometrium and cervix. The reproductive process does not terminate with delivery. To provide nutrition to the newborn, the mammary glands must be fully developed and ready to secrete milk. Milk is produced and secreted in response to the suckling stimulus. The act of suckling also postpones new ovulatory cycles. The woman is thus provided with a natural contraceptive until she regains metabolic balance, which has been taxed by pregnancy and lactation. Sexual maturity is attained during puberty, at 12–14 years of age. It requires changes in the sensitivity, activity, and function of several organs along the hypothalamic-pituitary-gonadal axis.

Fertilization and Implantation

■ Both Egg and Sperm Are Transported to the Oviduct

Aging of germ cells in subprimates is not of concern, since mating and ovulation occur within a short time interval due to behavioral and hormonal modifications. Primates, on the other hand, mate throughout the ovulatory cycle. To increase the probability that the sperm and egg will meet at an optimal time, the female reproductive tract facilitates sperm transport during the follicular phase of the menstrual cycle (estrogen dominance), whereas during the luteal phase (progesterone dominance) sperm survival and access to the oviduct is decreased.

The volume of ejaculate in fertile men is 2–6 ml, and it contains 50–250 million spermatozoa per milliliter. Seminal plasma coagulates after ejaculation but liquefies within 20–30 minutes due to proteolytic enzymes secreted by the prostate. The coagulum forms a temporary reservoir of sperm and reduces expulsion from the vagina. During intercourse, some spermatozoa are immediately propelled into the cervical canal. Those remaining in the vagina do not survive long because of the acidic environment (pH 5.7), although some protection is provided by the alkalinity of the seminal plasma. The cervical canal constitutes a more favorable environment, enabling sperm survival for several hours. Under estrogen dominance, mucin molecules in the cervical mucus become oriented in parallel and facilitate sperm migration. Sperm stored in the cervical crypts constitute a pool for slow release into the uterus.

Sperm survival in the uterine lumen is short due to phagocytosis by leukocytes. The uterotubal junction also presents an anatomic barrier limiting the passage of sperm into the oviducts. Abnormal or dead spermatozoa may be prevented from entry to the oviduct. Of the hundreds of millions of sperm deposited in the vagina, only a few hundred, usually spaced in time, will reach the ampulla. Major losses of sperm occur in the vagina, uterus, and uterotubal junction. Sperm that survives can reach the ampulla within 10–20 minutes of coitus. The innate motility of sperm alone cannot account for this rapid transit; transport is assisted by muscular contractions of the vagina, cervix, and uterus and ciliary movement, peristaltic activity, and fluid flow in the oviducts.

There is no evidence for chemotactic interactions between the egg and sperm. Rather, the sperm arrive in the vicinity of the egg at random and some exit through the open fimbria into the abdominal cavity. Although sperm remain motile for up to 4 days, their fertilizing capacity is limited to 1–2 days in the female reproductive tract. Sperm of most mammals can be cryopreserved for years, provided that agents such as glycerol are used to prevent crystal formation during freezing.

Freshly ejaculated sperm cannot immediately penetrate an egg. During maturation in the epididymis, the sperm acquire surface glycoproteins that act as stabilizing factors but also prevent sperm-egg interactions. To bind to the zona pellucida, the sperm must undergo **capacitation,** a reversible process that involves an increase in sperm motility, removal of surface proteins, and loss of lipids from the membrane. This results in redistribution of membrane constituents, increased membrane fluidity, and a rise in calcium permeability. Capacitation takes place along the female genital tract and lasts 1 to several hours. It is not species-specific: sperm of one species can be capacitated in the genital tract of another. Sperm can be capacitated in a chemically defined medium, a fact that has enabled in vitro fertilization.

The ovary and oviduct are not connected, requiring an active "pickup" of the ovulating egg from the surface of the ovary by the fimbria. This is done by ciliary movement and muscle contractions, both activated by estrogen at the time of ovulation. Because of the open arrangement, eggs can get lost in the abdominal cavity; if they are fertilized **ectopic pregnancy** may result. Egg transport from the fimbria to the ampulla is accomplished by coordinated ciliary activity and depends on the presence of the granulosa cells surrounding the egg. The fertilizable life of the human ovum is about 24 hours.

■ A Multitude of Cellular Events Occur at the Time of Fertilization

The interaction of the sperm and egg reconstitutes the full chromosomal complement and initiates the development of a new individual. Fertilization involves recognition of the egg by the sperm; regulation of sperm entry into the egg; prevention of polyspermy; completion of the second meiotic division; unification of the male and female pronuclei; and initiation of the first cell division (Fig. 41–1).

To ensure that fusion does not occur between germ and nongerm cells, specific recognition mechanisms have evolved. The zona pellucida contains glycoproteins that serve as sperm receptors. These are species-specific and account, in part, for the failure of interspecies fertilization. Contact between the sperm and the egg triggers the **acrosome reaction,** a process that enables sperm penetration. It begins with multiple fusions between the plasma membrane and acrosomal membrane of the sperm. Vesicles are then formed, releasing their content of proteolytic enzymes to dissolve the granulosa cell matrix and zona pellucida. After creating a narrow slit, the sperm penetrates the zona pellucida, aided by the propulsive force of the tail. Upon reaching the perivitelline space, the sperm head becomes anchored to the surface of the egg. Microvilli protruding from the oolemma (membrane of the egg) extend and clasp the sperm, and the oocyte membrane bulges and surrounds the sperm. Eventually the whole head, and later the tail, are incorporated into the ooplasm.

Shortly after fusion, several events occur in rapid succession. An immediate event is the prevention of polyspermy. In well-studied species (e.g., sea urchin), this proceeds in two stages: a fast, incomplete block involving membrane depolarization, followed by a late and permanent block. Although transient changes in membrane potential during sperm penetration have been observed in mammalian eggs, it is doubtful that these constitute a fast block to polyspermy. It has been argued that unlike the sea urchin egg, which is surrounded by thousands of sperm, in mammals only a few spermatozoa simultaneously reach the egg during internal fertilization, rendering a fast blockade less critical. The incidence of polyspermy increases during in vitro fertilization since the egg is exposed to many spermatozoa.

The complete block to polyspermy is provided by the cortical reaction, which induces the **zona reaction.** Cortical granules are lysosomelike organelles located underneath the oolemma which contain pro-

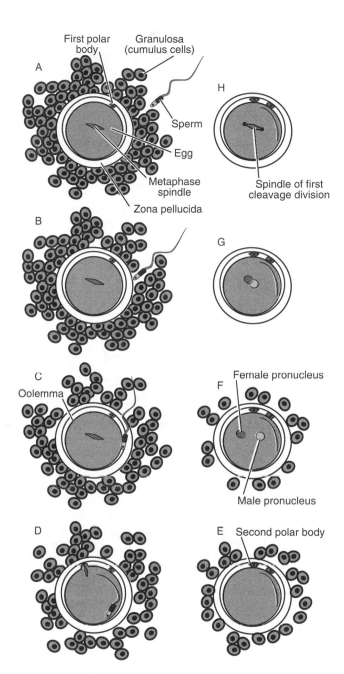

Figure 41–1 ■ The process of fertilization. (A) Sperm is approaching an egg. (B) Contact between sperm and zona pellucida. (C) Entry of sperm and contact with oolemma. (D) Resumption of second meiotic division. (E) Completion of meiosis. (F) Formation of male and female pronuclei. (G) Migration of pronuclei to center of cell. (H) The zygote is ready for the first mitotic division.

teolytic enzymes. The cortical reaction involves fusion of the cortical granules with the oolemma. Fusion starts at the point of sperm attachment and propagates over the entire egg surface. The content of the granules is released into the perivitelline space and diffuses into the zona pellucida, inducing the zona reaction. The sperm receptors are inactivated by proteolysis

IN VITRO FERTILIZATION

Good candidates for in vitro fertilization are women with tubal disease, unexplained infertility, or endometriosis (implantation of endometrial cells in ectopic locations) and those whose partners are infertile (e.g., low sperm count). Ovulation is induced with GnRH analogs, clomiphene, or FSH. Follicular growth is monitored by serum estradiol measurement and ultrasound imaging. When the leading follicle is 16–17 mm in diameter and/or estradiol level is greater than 300 pg/ml, hCG is injected to induce final follicular maturation. About 34–36 hours later, the largest oocytes are retrieved by laparoscopy or a transvaginal approach. Oocyte maturity is judged by the morphology of the cumulus-corona cells and the presence of germinal vesicle and first polar body. Oocytes are then placed in culture media supplemented with maternal serum or fetal cord serum.

Sperm are prepared by washing, centrifuging, and collecting the most motile sperm. About 100,000 spermatozoa are added for each oocyte. After 24 hours, the eggs are examined for the presence of two pronuclei. Embryos are grown to the four- to eight-cell stage, about 60–70 hours after retrieval. As many as three to five embryos are often deposited in the uterine lumen, to increase the chance for pregnancy. Others are cryopreserved. To ensure a receptive endometrium, daily progesterone administrations begin on the day of retrieval. A success rate of 15–25% has been reported by many groups, which compares rather favorably with the 30–40% efficacy of natural human pregnancy.

and the zona becomes hardened. Consequently, once the first spermatozoon triggers the zona reaction, other sperm cannot bind or penetrate the zona.

The sperm-induced influx of calcium into the egg, together with massive release of calcium from intracellular stores, activates the dormant egg. Completion of the second meiotic division occurs first. The chromosomes separate and half of the chromatin is extruded with the second polar body. The remaining haploid nucleus is transformed into a female pronucleus. Soon after being incorporated into the ooplasm, the nuclear envelope of the sperm disintegrates and the male pronucleus is formed and increases four to five times in size. The two pronuclei migrate to the center of the cell, assisted by contractions of microtubules and microfilaments. DNA replication begins in both pronuclei. Pores are formed in their nuclear membranes, and the pronuclei fuse. The **zygote** (fertilized egg) then enters the first mitotic (cleavage) division; the two-cell stage is attained within 24–36 hours.

■ Implantation Involves Interactions between the Endometrium and Embryo

Oogenesis is characterized by cellular growth without cell division, generating a large mature oocyte. In contrast, rapid cell division after fertilization occurs without growth. Thus, the cells of the early embryo become progressively smaller, reaching the dimension of somatic cells after several cell divisions. The embryo continues to cleave as it moves from the ampulla toward the uterus (Fig. 41-2). Until implantation, the embryo is enclosed in the zona pellucida. Retention of an intact zona is necessary for embryo transport, protection against mechanical damage or adhesion to the oviduct wall, and prevention of immunologic rejection by the woman.

After the first cell division, the embryo consists of two cells of unequal size. The larger one divides first, followed a few hours later by the smaller one. After additional cell divisions, the embryo becomes a solid ball of cells, the **morula.** At the 20–30-cell stage, a fluid-filled cavity (blastocoele) appears and then enlarges until the embryo becomes a hollow sphere, the **blastocyst.** The cells of the blastocyst have undergone significant differentiation. A single peripheral layer of flattened cells, linked by tight junctions, make up the **trophoblast.** These will participate in implantation, form the placenta and embryonic membranes, and produce hCG. A cluster of smaller eccentric cells comprise the **embryoblast.** These will give rise to the organs of the new individual.

The blastocyst reaches the uterus 3–4 days after ovulation. It remains suspended in the uterine cavity for 2–3 days, being nourished by constituents of the uterine fluid. **Implantation** (attachment to the uter-

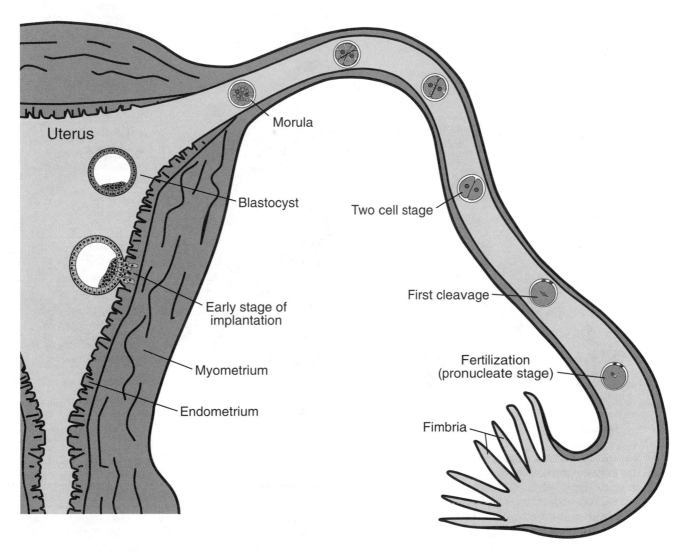

Figure 41–2 ■ Transport of the developing embryo from the oviduct, the site of fertilization, to the uterus, the site of implantation.

ine wall) begins on days 7–8 after ovulation and requires proper priming of the uterus by estrogen and progesterone. The first step of implantation involves removal of the zona pellucida. The zona is ruptured by expansion of the blastocyst and lysed by enzymes in the uterine fluid. The denuded trophoblast becomes negatively charged and adheres to the endometrium via surface glycoproteins. Microvilli from the trophoblastic cells then interdigitate and form junctional complexes with the epithelial cells.

In response to the attaching blastocyst, the endometrium undergoes the **decidual reaction.** This involves dilation of blood vessels, increased capillary permeability, edema formation, and increased proliferation of endometrial, glandular, and epithelial cells. Priming by progesterone is a prerequisite for the decidual reaction. The exact embryonic signals that trigger this

reaction are unclear; carbon dioxide production and release of histamine, steroids, prostaglandins, and pregnancy-associated proteins have been proposed.

Invasion of the endometrium is facilitated by the release of plasminogen activator by the trophoblast. By 8–12 days after ovulation, the human conceptus has penetrated the uterine epithelium and is embedded in the stroma (Fig. 41–3). The trophoblastic cells have differentiated into large polyhedral **cytotrophoblasts** and peripheral **syncytiotrophoblasts** lacking distinct cell boundaries. Maternal blood vessels dilate and vacuoles appear and fuse, forming blood-filled lacunae. Between weeks 2 and 3, villi are formed, and a functional communication is established between the maternal blood and the developing embryonic vascular system (Chap. 17). At this time, the embryoblast has differentiated into three layers: **ectoderm,** destined

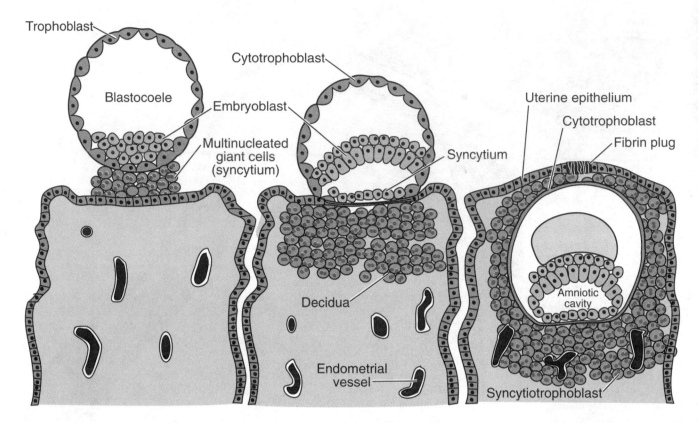

Figure 41–3 ■ The process of embryo implantation and the decidual reaction.

to form the epidermis, its appendages (nails and hair), and the entire nervous system; **endoderm,** which will give rise to the epithelial lining of the digestive tract and associated structures; and **mesoderm,** which will form the bulk of the body, including connective tissue, muscle, bone, blood, and lymph.

■ Contraceptives Interfere with Ovulation, Fertilization, or Implantation

Control of fertility can be achieved by interfering with the association between the sperm and oocyte, preventing gametogenesis or implantation, or aborting an established pregnancy. Contraceptive methods may also be divided into reversible and irreversible. Most current methods regulate fertility in women, with only a few contraceptives available for men (Table 41-1).

Methods based on preventing contact between the germ cells include coitus interruptus (withdrawal before ejaculation), rhythm, and barriers. The failure rate of the first two methods is substantial (20-30%). The high failure of coitus interruptus is due to a lack of control by the male partner or an escape of small amounts of sperm before ejaculation. The rhythm method requires abstinence during the fertile period, extending 3-4 days before and after the midcycle. The major problem is that the time of ovulation in individual women cannot be predicted with certainty,

TABLE 41–1 ■ Contraceptive Use in the United States and Efficacy Rates

Method	Estimated Use (%)	% Accidental Pregnancy in Year 1
Pill	32	3
Female sterilization	19	0.4
Condom	17	12
Male sterilization	14	0.15
Diaphragm	4-6	2-23
Spermicides	5	20
Rhythm	4	20
Intrauterine device	3	6

From *Developing New Contraceptives: Obstacles and Opportunities.* Washington, D.C., National Academy Press, 1990.

even when they monitor their daily basal body temperature and inspect the cervical mucus.

Barrier methods include condoms, diaphragms, and cervical caps. When combined with spermicidal agents, they approach the high success rate of oral contraceptives. Condoms are the most widely used reversible contraceptives for men. Since they also provide protection against transmission of venereal diseases and AIDS, their use has increased in recent years. Diaphragms and cervical caps seal off the cervix and have to be fitted by a physician. Spermicides are available as foams, jellies, creams, sponges, or suppositories. When used without mechanical devices, their efficacy is not high. Postcoital douching, with or without spermicides, is not an effective contraceptive since some sperm enter the uterus very rapidly.

Vasectomy and tubal ligation are considered irreversible methods of fertility control. Vasectomy is performed under local anesthesia; the vasa deferentia are cut after making an incision in the scrotum. Vasectomy prevents sperm from passing into the ejaculate; after being trapped within the epididymis, they are digested and resorbed. Testicular hormone output, volume of ejaculate, and sexual activity are not affected. There is an increased incidence of sperm antibodies after vasectomy, but its consequences are unknown. Vasectomy can be reversed by microsurgery, with a success rate of 50–60%. Tubal ligation is performed by laparoscopy under general or spinal anesthesia. The oviducts are closed by ligation or cautery. The success rate of restorative surgery is limited.

Steroidal contraceptives became available with the development of orally administered synthetic steroids. The combination-type pill contains different ratios of estrogen and progesterone, with a recent trend to lower the estrogen content. The pill is taken daily for 3 of every 4 weeks. Omission of pills for more than 2 days reduces their efficacy. Oral steroids prevent ovulation by inhibiting the midcycle gonadotrophin surge; their effectiveness is also increased by affecting the reproductive tract. Estrogen stabilizes the endometrium to prevent bleeding and induces progesterone receptors. Progesterone alters cervical mucus and oviductal peristalsis, thereby impeding gamete transport.

Noncontraceptive benefits of the pill include reduction in excessive menstrual bleeding, alleviation of premenstrual syndrome, and some protection against pelvic inflammatory disease. Adverse effects include nausea, headache, breast tenderness, water retention, and weight gain, some of which disappear after prolonged use. The pill is contraindicated in cardiovascular and liver disease, in heavy smokers older than 35 years, and in women with known or suspected breast or estrogen-dependent carcinomas. There is no evidence for reduction in fertility on discontinuation of the pill.

Several contraceptives act by interfering with zygote transport or implantation. Among these are long-acting progesterone preparations, very high doses of estrogen, and progesterone receptor antagonists (RU 486). The intrauterine device (IUD) also prevents implantation by provoking sterile inflammation of the endometrium and prostaglandin production. The contraceptive efficacy of IUDs, especially those impregnated with progestins, copper, or zinc, is high. The drawbacks include a high rate of expulsion, uterine cramps, excessive bleeding, perforation of the uterus, and increased incidence of ectopic pregnancy. Established pregnancy can be interrupted by surgical (dilatation and curettage) or pharmacologic (prostaglandin administration) means. Newer contraceptive methods include gonadotrophin-releasing hormone (GnRH) analogs, hCG antibodies, inhibin preparations, and inhibitors of spermatogenesis (gossypol).

Pregnancy

■ The Placenta Is a Unique Organ Interposed between the Mother and Fetus

In the human placenta, the fetal and maternal components are interdigitated. The functional units of the placenta, the **chorionic villi,** form on days 11–12 and extend tissue projections into the maternal lacunae that form from endometrial blood vessels during implantation. By week 4 the villi are spread over the entire surface of the **chorionic sac.** As the placenta matures, it becomes discoid. During the third month the chorionic villi are confined to the area of the **decidua basalis.** The decidua basalis and chorionic plate together form the placenta proper (Fig. 41–4).

The decidua capsularis around the conceptus and decidua parietalis on the uterine wall fuse and occlude the uterine cavity. The yolk sac becomes vestigial and the amniotic sac expands, pushing the chorion against the uterine wall. From the fourth month onward, the fetus is enclosed within the amnion and chorion and is connected to the placenta by the umbilical cord. Fetal blood flows through two umbilical arteries to capillaries in the villi, is brought into juxtaposition with maternal blood in the sinuses, and returns to the fetus through a single umbilical vein. The two circulations do not mix (Chap. 17).

A major function of the placenta is to deliver nutrients and remove waste products from the fetus. Oxygen diffuses toward the fetal blood down a gradient of 60–70 mm Hg. The oxygen-transporting capacity of fetal blood is enhanced by fetal hemoglobin, which has a high affinity for oxygen. The Pco_2 of fetal arterial blood is 2–3 mm Hg higher than that of maternal blood, allowing diffusion of carbon dioxide toward the maternal compartment. Other compounds, such as glucose, amino acids, free fatty acids, electrolytes, vitamins, and some hormones, are transported by diffusion, facilitated diffusion, or pinocytosis. Waste products, such as urea and creatinine, diffuse away from the fetus down their concentration gradients. Large proteins, including most polypeptide hormones, do not readily cross the placenta, whereas the lipid-soluble steroids do. The **blood-placental barrier** allows the transfer of some immunoglobulins, viruses, and drugs from the mother to the fetus (Fig. 41–5).

■ The Recognition and Maintenance of Pregnancy Depend on Hormones

The placenta is a remarkable endocrine organ, incorporating many of the biosynthetic capabilities and functions of the hypothalamopituitary-gonadal axis. It

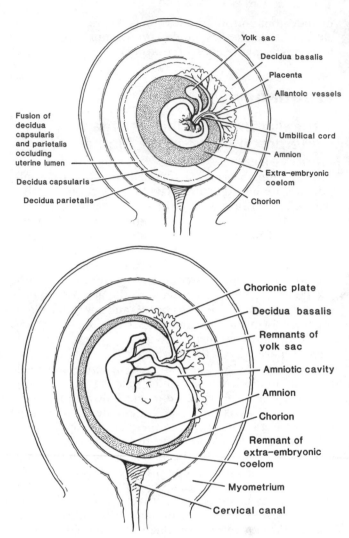

Figure 41–4 ■ Two stages in the development of the placenta, showing the origin of the membranes around the fetus.

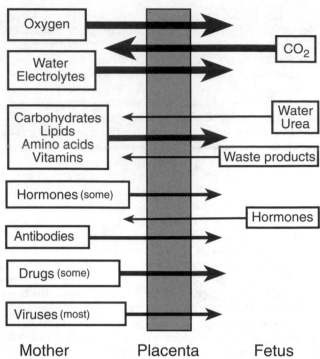

Figure 41–5 ■ The role of the placenta in the bidirectional interactions between the fetal and maternal compartments. Size of arrows indicates amount of exchange between the compartments.

produces GnRH (a "hypothalamic" peptide) and progesterone and estrogens (ovarian steroids). These hormones are essential for the continuance of pregnancy. The placenta also produces protein hormones unique to pregnancy, such as human placental lactogen (hPL). Several peptides, including corticotrophin-releasing hormone (CRH) and insulinlike growth factors, are also synthesized by the placenta and function as paracrine agents.

During the menstrual cycle, the corpus luteum regresses within 11–12 days of formation and stops producing estrogen and progesterone. After losing its steroidal support, the superficial endometrial layer is expelled, resulting in menstruation. This is incompatible with pregnancy. To prevent it, the implanting embryo signals its presence and extends the life of the corpus luteum. This process has been termed "maternal recognition of pregnancy." The human trophoblast starts making hCG almost as soon as it is formed, 8–9

days after ovulation. The hCG stimulates progesterone production by the corpus luteum, preventing the next menses. Appearance of hCG in the woman's urine is used as a test for pregnancy.

Human chorionic gonadotrophin is a glycoprotein made of two dissimilar subunits; it belongs to the same hormone family as luteinizing hormone (LH), follicle stimulating hormone (FSH), and thyroid stimulating hormone (TSH). The α subunit is made of the same 92 amino acids as the other glycoprotein hormones. The β subunit is made of 145 amino acids, with six N- and O-linked oligosaccharide units. It resembles the LH β subunit but has a 24-amino acid extension at the C-terminus. Because of heavy glycosylation, the half-life of hCG in the circulation is longer than that of LH. Like LH, the major function of hCG is stimulation of steroidogenesis. Both bind to the same, or similar, membrane receptors and increase the formation of pregnenolone from cholesterol by a cAMP-dependent mechanism.

Blood hCG reaches peak levels at about 10 weeks of gestation. It is reduced by about 80% by 20 weeks and remains at that level until term (Fig. 41–6). During the first trimester, locally produced GnRH appears to regulate hCG production by a paracrine mechanism. The suppression of hCG release during the second half of pregnancy is attributed to negative feedback by placental progesterone or steroids produced by the fetus. Progesterone secretion by the corpus luteum is

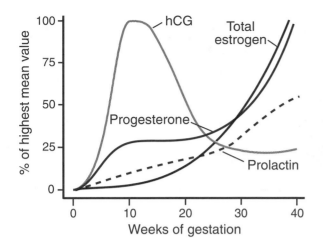

Figure 41–6 ■ Profiles of human chorionic gonadotrophin (hCG) progesterone, total estrogens, and prolactin in the maternal blood throughout gestation.

maximal 4–5 weeks after conception and then declines, although hCG levels are still rising. Corpus luteum refractoriness to hCG results from receptor desensitization and the rising levels of placental estrogens. From week 7 to 10 of gestation, steroid production by the corpus luteum is gradually replaced by steroid production by the placenta. Removal of the corpus luteum after week 10 does not result in abortion.

Human chorionic gonadotrophin exerts other important functions. Sexual differentiation of the male fetus depends on testosterone production by the fetal testes. Peak production of testosterone occurs 11–17 weeks after conception. This time predates the functional maturity of the fetal hypothalamopituitary axis and coincides with peak hCG production. Human chorionic gonadotrophin appears to regulate Leydig cell proliferation as well as testosterone biosynthesis. There are some indications for an association between elevated hCG and maternal "morning sickness."

Human placental lactogen, also called human chorionic somatomammotropin (hCS), is synthesized by the syncytiotrophoblast and secreted into the maternal circulation, where its levels gradually rise until term. It is composed of a single chain of 191 amino acids with two disulfide bridges; its structure and function resemble those of prolactin and growth hormone. Human placental lactogen promotes cell specialization in the mammary gland but is less potent than prolactin in stimulating milk production. Its main function is to alter fuel availability by antagonizing maternal glucose consumption and enhancing fat mobilization. The amniotic fluid also contains large amounts of prolactin, which is indistinguishable from pituitary prolactin. This prolactin is produced by the decidua, but its function and regulation are unclear.

The fetus constitutes a foreign body in the maternal body, since half of its chromosomes come from the father. Yet, even though the maternal immune system functions properly against invasion by foreign antigens, the fetus is protected from rejection. During transport to the uterus, the embryo is enclosed by the zona pellucida, which prevents direct contact between maternal immune cells and the conceptus. Once implanted, several mechanisms operate to protect the fetus. These include reduced production of histocompatibility antigens by the trophoblast and coating of its surface antigens by sialomucin and pregnancy-specific proteins, thus reducing immunologic recognition. In addition, placental hormones affect the maternal immune system. Progesterone acts locally to suppress T cell activation, affects macrophage function, and reduces leukocyte numbers; hCG depresses maternal immunoreactivity.

■ Steroid Production during Pregnancy Involves the Fetoplacental Unit

Sex steroids fulfill many essential functions throughout gestation. Estrogens increase the size of the uterus, induce receptors for progesterone and oxytocin, stimulate maternal hepatic protein secretion, and increase the mass of breast and adipose tissue. Progesterone is essential for maintaining placental implantation, inhibits myometrial contraction, and suppresses maternal immunologic response to fetal antigens. It also serves as a precursor for steroid production by the fetal adrenal and plays a role in the onset of parturition.

From 7 to 9 weeks of gestation, steroidogenesis is carried out by the placenta, an active, though incomplete, steroid-producing organ. It cannot make significant amounts of cholesterol from acetate and obtains cholesterol from the maternal blood via low-density lipoproteins. The placenta also lacks the P450c17 enzyme for converting pregnenolone to androgens or estrogens. Maternal 17-hydroxy progesterone can be measured during the first trimester and serves to assess corpus luteum function, since the placenta cannot make this compound. The levels of progesterone in the maternal circulation rise during the first 6–8 weeks, reaching a plateau during the transition period from corpus luteal to placental production. Progesterone levels progressively increase toward term, attaining 10-fold higher concentrations than the peak production by the corpus luteum (Fig. 41–6).

The production of estrogens (estradiol, estrone, and estriol) during gestation requires cooperation between the maternal, placental, and fetal compartments, termed **the fetoplacental unit** (Fig. 41–7). To produce estrogens, the placenta uses androgenic substrates derived from both the fetus and the mother. The primary androgenic precursor is **dehydroepiandrosterone sulfate** (DHEAS), which is produced by the fetal zone of the fetal adrenal gland. The fetal adrenal gland is extremely active in the production of steroid hormones, but because it lacks 3β-hydroxysteroid dehydrogenase, δ4,5 isomerase activity, it cannot

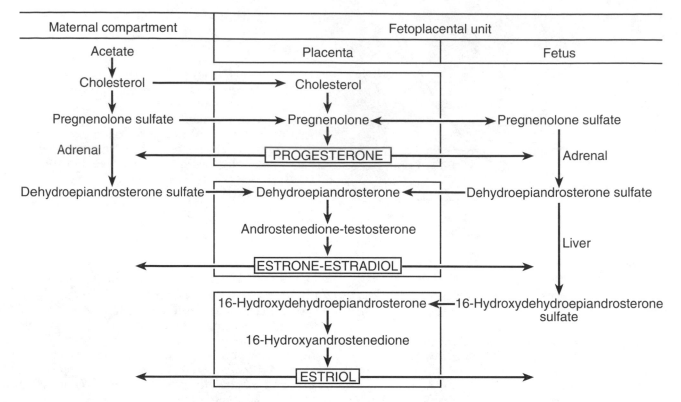

Figure 41–7 ■ The fetoplacental unit and steroidogenesis. Note that estriol is the product of reactions occurring in the fetal adrenal, fetal liver, and placenta. (Modified from Goodman HM: Basic Medical Endocrinology. New York, Raven, 1988.)

make progesterone and uses progesterone from the placenta. Conjugation of androgenic precursors to sulfates ensures greater water solubility, aids in their transport, and reduces their biologic activity while in the fetal circulation. The DHEAS diffuses into the placenta and is cleaved by a sulfatase to yield a nonconjugated androgenic precursor. The placenta has an active aromatase that converts androgenic precursors to estradiol and estrone.

The major estrogen produced during human pregnancy is **estriol,** which has relatively weak estrogenic activity. Estriol is produced by a unique biosynthetic pathway involving 16-hydroxylation of DHEAS by the fetal liver and, to a lesser extent, the fetal adrenal gland. This is followed by deconjugation by placental sulfatase and conversion to estriol by placental aromatase. Since 16-OH-DHEAS is not made by the maternal ovary, the levels of estriol in the amniotic fluid or maternal urine are used as an index of fetal well-being.

■ Maternal Physiology Changes Substantially throughout Gestation

The pregnant woman provides nutrients for the growing fetus, is the sole source of its oxygen, and removes fetal waste products. These functions necessitate significant adjustments in the woman's pulmonary, cardiovascular, renal, metabolic, and endocrine sys-

tems. Among the most notable changes during pregnancy are hyperventilation, reduced arterial blood P_{CO_2} and osmolality, increased blood volume and cardiac output, increased renal blood flow and glomerular filtration rate, and substantial weight gain. These are brought about by the rising levels of estrogens, progesterone, hPL, and other placental hormones and by mechanical factors such as the expanding size of the uterus and the development of uterine and placental circulations.

The maternal endocrine system undergoes significant adaptations. The hypothalamopituitary-gonadal axis is suppressed by the high levels of sex steroids. Consequently, circulating gonadotrophins are low and ovulation does not occur. In contrast, the rising levels of estrogens stimulate prolactin release. Prolactin levels begin to rise during the first trimester, increasing gradually to reach a level 10 times higher near term. Pituitary lactotrophs undergo hyperplasia and hypertrophy and mostly account for the 50% enlargement of the pituitary gland of the pregnant woman. Thyroid and parathyroid glands also enlarge. The release of parathyroid hormone increases mostly during the third trimester. It enhances calcium mobilization from maternal bone stores in response to the growing demands for calcium by the fetus. The rate of adrenal secretion of mineralocorticoids and glucocorticoids increases, and plasma free cortisol is higher because of its displacement from transcortin, its binding globulin, by progesterone.

Maternal metabolism responds in several ways to the increasing nutritional demands of the fetus. The major net weight gain of the mother occurs during the first half of gestation, mostly due to fat deposition. This is attributed to progesterone, which increases appetite and diverts glucose into fat synthesis. The extra fat stores are used as an energy source later in pregnancy, when the metabolic requirements of the fetus are at their peak, and also during periods of starvation. Several maternal and placental hormones act together to provide a constant supply of metabolic fuels to the fetus. Toward the second half of gestation, the mother develops a resistance to insulin. This is brought about by a combined action of hormones antagonistic to insulin, such as growth hormone, pro- lactin, hPL, glucagon, and cortisol, and an impairment of insulin action at the target cells. As a result, maternal glucose use declines and gluconeogenesis increases, thereby maximizing the glucose gradient across the placenta. The development of insulin resistance can unmask latent diabetes in some women. Gestational diabetes affects as many as 4–6% of pregnant women and is a common cause of high-risk pregnancy.

Fetal Development and Parturition

■ The Fetal Endocrine System Is Largely Autonomous

The protective intrauterine environment postpones initiation of some physiologic functions that are essen- tial for life after birth. For example, the fetal lungs and kidneys do not act as organs of gas exchange and excretion, respectively, since their functions are car- ried out by the placenta. Constant isothermal sur- roundings alleviate the need to expend calories to maintain body temperature. The gastrointestinal tract does not carry out digestive activities, and fetal bones and muscles do not support weight or locomotion. Being exposed to low levels of external stimuli and environmental insults, the fetal nervous and immune systems develop slowly. Homeostasis in the fetus is regulated by hormones, and the fetal endocrine system plays a vital role in its growth and development.

Given that most protein and polypeptide hormones are excluded from the fetus by the blood-placental barrier, the maternal endocrine system has little direct influence on the fetus. Instead, the fetus is almost self-sufficient in its hormonal requirements. Notable exceptions are some of the steroid hormones, which are produced by the fetoplacental unit, cross easily between the different compartments, and carry out integrated functions in both the fetus and the woman. By and large, fetal hormones fulfill the same functions as in the adult, but they also subserve unique proc-

esses, such as sexual differentiation and initiation of labor.

The fetal hypothalamopituitary system is well dev- eloped by midgestation. The hypophyseal portal vas- culature is established by 20–21 weeks of gestation; well-differentiated hormone-producing cells in the an- terior pituitary are also apparent at this time. Whether the fetal pituitary is tightly regulated by hypothalamic hormones or possesses some autonomic activity is unclear. Experiments with long-term catheterization of monkey fetuses indicate that by the last trimester, both LH and testosterone increase in response to GnRH administration.

The fetal adrenal glands are unique in both structure and function. At 4 months of gestation, they are larger than the kidneys, due to the development of a fetal zone that constitutes 75–80% of the whole gland. The outer definitive zone will form the adult adrenal cor- tex, whereas the deeper fetal zone involutes after birth. The fetal zone produces large amounts of DHEAS and provides androgenic precursors for estrogen syn- thesis by the placenta (Fig. 41–7). The definitive zone produces cortisol, which has multiple functions dur- ing fetal life, including promotion of pancreas and lung maturation, induction of liver enzymes, promotion of intestinal tract cytodifferentiation, and initiation of labor.

The rate of fetal growth increases markedly during the last trimester. Surprisingly, growth hormone of maternal, placental, or fetal origin has little effect on fetal growth, as judged by normal weight of hypopitu- itary dwarfs or anencephalic fetuses. Fetal insulin is the most important hormone in regulating fetal growth. Glucose is the main metabolic fuel for the fetus. Fetal insulin, produced by the pancreas by week 12 of gesta- tion, regulates tissue glucose use, controls liver glyco- gen storage, and facilitates fat deposition. It does not control the supply of glucose, however; this is deter- mined by maternal gluconeogenesis and placental glu- cose transport. The release of insulin in the fetus is relatively constant, increasing only slightly in response to a rapid rise in blood glucose levels. When blood glucose levels are chronically elevated, as in diabetic women, the fetal pancreas becomes enlarged and cir- culating insulin levels increase. Consequently, fetal growth is accelerated, and infants of uncontrolled dia- betic women are overweight (Fig. 41–8).

■ The Sex Chromosomes Dictate the Development of the Fetal Gonads

The process of sexual differentiation is complex and prolonged. It begins at the time of fertilization by a random unification of an egg with an X- or Y-bearing spermatozoon and continues during early embryonic life with the development of male or female gonads. Up to this point, the process is under genetic control. From then on, sexual differentiation is dictated by gonadal hormones that act at critical times during organogenesis. Testicular hormones impose masculi-

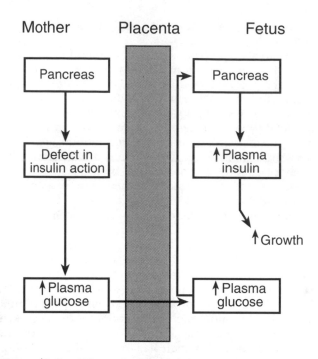

Mother   Placenta   Fetus

Figure 41–8 ■ Effects of maternal diabetes on fetal growth.

nization on an inherent female state, which does not require hormonal intervention. The process is incomplete at birth; secondary sexual characteristics and attainment of functional reproductive capacity are achieved only at puberty.

Human somatic cells have 44 autosomes and 2 sex chromosomes. The female is homogametic (having two X chromosomes) and produces similar X-bearing ova. The male is heterogametic (having one X and one Y chromosome) and generates two populations of spermatozoa. The X chromosome is large, containing 80–90 genes responsible for many vital functions. The Y chromosome is much smaller, carrying only few genes responsible for testicular development and normal spermatogenesis. Mutation of genes on an X chromosome results in transmission of X-linked traits, such as hemophilia and colorblindness, to male offspring, which, unlike females, cannot compensate with an unaffected allele.

Theoretically, by having two X chromosomes the female has an advantage over the male, which has only one. This does not occur, however, because the second X chromosome is inactivated at the morula stage. Each cell randomly inactivates either the paternally or maternally derived X chromosome, and this continues throughout its progeny. The inactivated X chromosome is recognized cytologically as the sex chromatin, or Barr body. In males with more than one X chromosome or in females with more than two, extra X chromosomes are inactivated and only one remains functional. This does not apply to the germ cells. The single active X chromosome of the spermatogonium

becomes inactivated during meiosis, and a functional X chromosome is not necessary for the formation of fertile sperm. The oogonium, on the other hand, reactivates its second X chromosome, and both are functional in oocytes and important for normal oocyte development.

Testicular differentiation requires a Y chromosome and ensues even in the presence of two or more X chromosomes. Gonadal sex determination is regulated by a testis-determining gene designated SRY (or Tdy) which is located on the short arm of the Y chromosome. SRY encodes a DNA-binding protein which binds to the target DNA in a sequence-specific manner. The presence or absence of SRY in the genome determines whether male or female gonadal differentiation takes place. Thus, in normal XX (female) fetuses, which lack a Y chromosome, ovaries, rather than testes, develop.

Whether possessing an XX or an XY karyotype, every embryo goes initially through an ambisexual stage and has the potential to acquire either masculine or feminine characteristics. A 4–6-week-old human embryo possesses indifferent gonads; a pair of wolffian ducts capable of forming internal male genital ducts; a pair of müllerian ducts serving as the anlage of internal female genital ducts; bipotential structures of the external genitalia; and undifferentiated pituitary, hypothalamus, and higher brain centers.

The indifferent gonad consists of a genital ridge, derived from coelemic epithelium and underlying mesenchyme, and primordial germ cells, which migrate from the yolk sac to the genital ridges. Depending on genetic programing, the inner medullary tissue will become the testicular components and the outer cortical tissue will develop into an ovary. The primordial germ cells will become oogonia or spermatogonia. In an XY fetus, the testes differentiate first. Between 6 and 8 weeks of gestation, the cortex regresses, the medulla enlarges, and the seminiferous tubules become distinguishable. Sertoli cells line the basement membrane of the tubules and Leydig cells undergo rapid proliferation. Development of the ovary begins only in weeks 9–10 of gestation. Primordial follicles, composed of oocytes surrounded by a single layer of granulosa cells, are discernible in the cortex between 11 and 12 weeks and reach maximal development by week 20–25.

■ Differentiation of the Genital Ducts Is Determined by Hormones

During the indifferent stage, the internal genitalia consist of mesonephric (wolffian) and paramesonephric (müllerian) ducts. In the normal male fetus, the wolffian ducts give rise to the epididymis, vas deferens, seminal vesicles, and ejaculatory ducts while the müllerian ducts become vestigial. In the normal female fetus, the müllerian ducts fuse in the midline and develop into the oviducts, uterus, cervix, and upper portion of the vagina, while the wolffian ducts regress (Fig. 41-9).

The fetal testes differentiate between 6 and 8 weeks of gestation. The Leydig cells, either autonomously or under regulation by hCG, start producing testosterone. The Sertoli cells produce two nonsteroidal compounds. One is **müllerian inhibiting factor** (MIF), a large glycoprotein with a sequence homology to inhibin and transforming growth factor β, which inhibits cell division of the müllerian ducts. The second is androgen binding protein (ABP), which binds testosterone. Peak production of these hormones occurs between weeks 9 and 12, coinciding with the time of differentiation of the internal genitalia along the male line. The ovary, which differentiates later, does not produce hormones and has a passive role.

The primordial external genitalia include the genital tubercle, genital swellings, urethral folds, and urogenital sinus. Differentiation of the external genitalia also occurs between 8 and 12 weeks of gestation and is determined by the presence or absence of male sex hormones. Differentiation along the male line requires active 5 α-reductase, the enzyme that converts testosterone to DHT. Without DHT, regardless of the genetic,

gonadal, or hormonal sex, the external genitalia develop along the female pattern. The structures that develop from the primordial structures are illustrated in Figure 41-10, and a summary of sexual differentiation during fetal life is shown in Figure 41-11. Androgen-dependent differentiation occurs only during fetal life and is thereafter irreversible. However, exposure of females to high androgens either before or after birth can cause clitoral hypertrophy. Testicular descent into the scrotum, which occurs during the third trimester, is also controlled by androgens.

■ A Complex Interplay between Maternal and Fetal Factors Induces Parturition

The duration of pregnancy in women averages 270 ± 14 days from the time of fertilization. The onset of

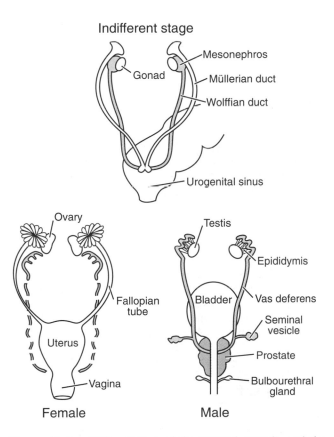

Figure 41-9 ■ Differentiation of the internal genitalia and the primordial ducts. (Modified from Georege FW, Wilson JD: Embryology of the urinary tract, in Walsh PC, Retik AB, Stamey TA, et al (eds): *Campbell's Urology*. 6th ed. Philadelphia, WB Saunders, 1992, p 1496.)

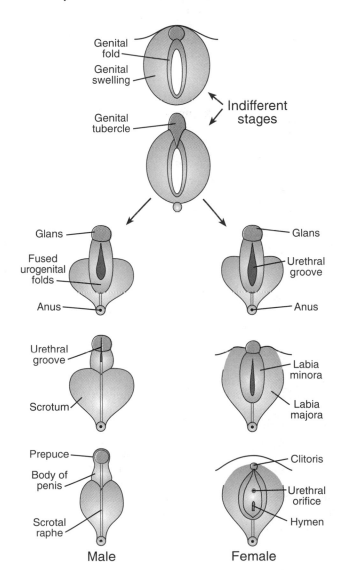

Figure 41-10 ■ Differentiation of the external genitalia from bipotential primordial structures.

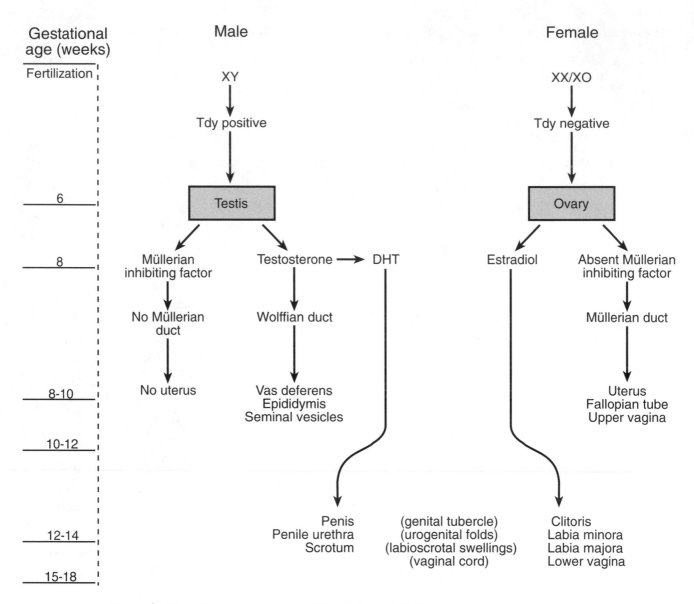

Figure 41–11 ■ The process of sexual differentiation and its time course.

birth, or **parturition,** is regulated by interactions of fetal and maternal factors. Uncoordinated uterine contractions start about 1 month before the end of gestation. The termination of pregnancy is initiated by strong rhythmic contractions that may last several hours and eventually generate enough force to evacuate the conceptus. The contraction of the uterine muscle is regulated by hormones and mechanical factors. The hormones include progesterone, estrogen, prostaglandins, oxytocin, and relaxin. The mechanical factors include distension of the uterine muscle and stretching or irritation of the cervix.

Progesterone hyperpolarizes myometrial cells, lowers their excitability, and suppresses uterine contractions. It also prevents the release of phospholipase A_2,

the rate-limiting enzyme in prostaglandin synthesis. Estrogen, in general, has the opposite effects. Maintenance of uterine quiescence throughout gestation, preventing premature delivery, has been termed the "progesterone block." In many species, a sharp decline in the circulating levels of progesterone and a concomitant rise in estrogen precede birth. In humans, progesterone does not fall significantly before delivery. However, its effective concentration may be altered by a rise in placental progesterone-binding protein or by a decline in the number of progesterone receptors.

Prostaglandins $F_{2\alpha}$ and E_2 are potent stimulators of uterine contraction. They increase intracellular calcium concentration of myometrial cells and activate the actin-myosin contractile apparatus. Shortly before the onset of parturition, the concentration of prostaglandins in amniotic fluid rises abruptly. Prostaglandins are produced by myometrium, decidua, or chorion. Aspirin and indomethacin, inhibitors of prostaglandin synthesis, delay or prolong parturition. Oxytocin is a potent stimulator of uterine contractions, and its release from both maternal and fetal pituitaries increases during labor. Relaxin assists parturition by softening the cervix, thus permitting the eventual passage of the fetus, and by increasing oxytocin receptors. However, the relative role of relaxin in parturition in humans is unclear, since its levels do not rise toward the end of gestation.

The fetus itself plays a role in initiating labor. This is best documented in sheep, in which the concentration of ACTH and cortisol in the fetal plasma rise during the last 2–3 days of gestation. Ablation of the fetal lamb pituitary or removal of the adrenals prolongs gestation, while administration of ACTH or cortisol leads to premature delivery. Cortisol accelerates the conversion of progesterone to estradiol, thus changing the progesterone to estrogen ratio, and increases the production of prostaglandins. The role of cortisol, however, has not been established in humans. Given that anencephalic fetuses, which lack a pituitary and have atrophied adrenal glands, have an unpredictable length of gestation, the endocrine status in the fetus may play a role in fine-tuning the onset of parturition in humans.

The Postpartum and Prepubertal Periods

■ Mammogenesis and Lactogenesis Are Regulated by Multiple Hormones

Lactation (the secretion of milk) represents the final phase of the reproductive process. The mammary gland is a complex endocrine target organ, and many hormones participate in its differentiation and growth and in the production and delivery of milk. **Mammogenesis** refers to the differentiation, growth, and maturation of the mammary glands. **Lactogenesis** designates milk production and **galactopoiesis** the maintenance of an established lactation. **Milk ejection** is the process by which stored milk is released.

Mammogenesis is divided into three periods: embryonic, pubertal, and gestational. The mammary glands begin to differentiate during the weeks 7–8 of fetal life. The mammary buds are derived from surface epithelium, which invades the underlying mesenchyme. During the fifth month, the buds elongate, arborize, and send out sprouts, which become the **lactiferous ducts.** The ducts unite, grow, and extend to the site of the future nipple. The primary buds give rise to secondary buds, which are separated into lobules by connective tissue. These become surrounded by **myo-**

epithelial cells derived from epithelial progenitors. The nipple and areola are formed during the eighth month of gestation. The development of the mammary glands in utero is independent of hormones but is influenced by paracrine interactions between the mesenchyme and epithelium.

The mammary glands of male and female infants are identical. Although underdeveloped, they have the capacity to respond to hormones. This is revealed by the secretion of small amounts of milk (witch's milk) in many newborns, resulting from responsiveness of the fetal mammary tissue to lactogenic hormones and the withdrawal of placental steroids at birth. Sexual dimorphism in breast development begins at the onset of puberty. Estrogen exerts a major influence on breast growth at puberty, but the development of estrogen receptors requires prolactin. The first response to estrogen is an increase in size and pigmentation of the areola and accelerated deposition of adipose and connective tissue. Estrogen stimulates growth and branching of the ducts, whereas progesterone acts primarily on the alveolar components. The action of both hormones, however, requires synergism with prolactin, growth hormone, insulin, cortisol, and thyroxine.

The mammary glands undergo significant changes during pregnancy. The ducts become elaborate during the first trimester, and new lobules and alveoli are formed in the second trimester. The terminal alveolar cells differentiate into secretory cells, replacing most of the connective tissue. Development of secretory capability requires estrogen, progesterone, prolactin, and placental lactogen. Their action is supported by insulin, cortisol, and several growth factors. Lactogenesis begins during the fifth month of gestation, but only **colostrum** (initial milk) is initially produced. Full lactation during pregnancy is prevented by elevated progesterone levels, which antagonize the action of prolactin. Thus, the ovarian steroids synergize with prolactin in stimulating mammary growth but antagonize its actions in promoting milk secretion.

Lactogenesis is fully expressed only after parturition, on withdrawal of placental steroids. Women produce up to 600 ml/day of milk, increasing to 800–1100 ml/day by the sixth postpartum month. Milk is isosmotic with plasma, and its main constituents include proteins, such as casein and lactalbumin, lipids, and lactose. The composition of milk changes with the stage of lactation. Colostrum, produced in small quantities during the first postpartum days, is higher in protein, sodium, and chloride content and lower in lactose and potassium than normal milk. Colostrum also contains secretory immunoglobulin A, macrophages, and lymphocytes, which provide passive immunity to the infant by acting on its gastrointestinal tract. During the first 2–3 weeks, the protein content of milk decreases, whereas that of lipids, lactose, and water-soluble vitamins increases.

The **alveolar cells** are specialized for secretion,

both morphologically and biochemically (Fig. 41–12). They form a single layer of cells, which are joined by junctional complexes. The bases of the cells abut on the contractile myoepithelial cells and their luminal apex is enriched with microvilli. They have a well-developed endoplasmic reticulum and Golgi apparatus and numerous mitochondria and lipid droplets. The alveolar cells contain plasma membrane receptors for prolactin, which can be internalized after binding to the hormone. Prolactin, in synergism with insulin and glucocorticoids, is critical for lactogenesis. It promotes mammary cell division and differentiation and increases the synthesis of milk constituents. Prolactin stimulates the synthesis of casein by increasing its transcription rate and stabilizing its mRNA. Prolactin also stimulates enzymes that regulate production of lactose.

■ The Suckling Stimulus Maintains Lactation and Inhibits Ovulation

The suckling stimulus is central to the maintenance of lactation in that it coordinates the release of prolactin and oxytocin and delays the onset of ovulation. Lactation involves two components—milk synthesis

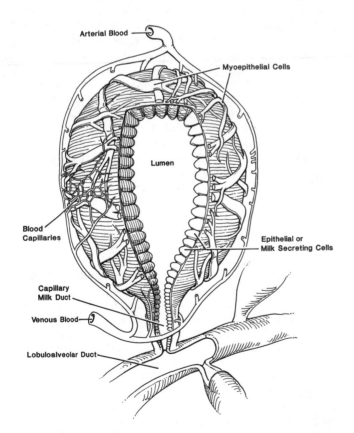

Figure 41–12 ■ Structure of a lactating mammary gland. (Modified from Goodman HM: *Basic Medical Endocrinology.* New York, Raven, 1988, p 317.)

Arterial Blood

Myoepithelial Cells

Lumen

Blood Capillaries

Epithelial or Milk Secreting Cells

Capillary Milk Duct

Venous Blood

Lobuloalveolar Duct

and milk removal—which are regulated independently. Milk synthesis and secretion is a continuous process, whereas milk removal is intermittent. Milk secretion involves the synthesis of milk constituents by the alveolar cells, their intracellular transport, and the subsequent release of formed milk into the alveolar lumen. Prolactin is the major regulator of milk secretion in women and most other mammals. Oxytocin is responsible for milk removal by activating milk ejection, or "let-down."

Stimulation of sensory nerves in the breast by the infant initiates the **suckling reflex.** Unlike ordinary reflexes with only neural components, the afferent arc of the suckling reflex is neural and the efferent arc is hormonal. The suckling stimulus increases the release of prolactin, oxytocin, and ACTH and inhibits the secretion of gonadotrophins (Fig. 41–13). The neuronal component is composed of sensory receptors in the nipple that initiate nerve impulses in response to breast stimulation. These impulses reach the hypothalamus via ascending fibers in the spinal cord and then via the mesencephalon. Fibers terminating in the supraoptic and paraventricular nuclei trigger the release of oxytocin from the posterior pituitary into the general circulation (Chap. 34). On reaching the mammary glands, oxytocin induces contraction of myoepithelial cells, thereby increasing intramammary pressure and forcing the milk into the main collecting ducts. The milk ejection reflex can be conditioned; milk ejection can occur because of anticipation or in response to a baby's cry.

Basal prolactin levels, which are elevated by the end of gestation, decline by 50% within the first postpartum week and decrease to near pregestation levels by 6 months. Suckling elicits a rapid and significant rise in plasma prolactin. The amount released is determined by the intensity and duration of nipple stimulation. The exact mechanism by which suckling triggers prolactin release is unclear, but suppression of dopamine, the major inhibitor of prolactin release, and stimulation of prolactin-releasing factor(s) have been considered. Lactation can be terminated by dopaminergic agonists that reduce prolactin or by discontinuation of suckling. The swollen alveoli depress milk production by exerting local pressure, resulting in vascular stasis, and alveolar regression.

Lactation is associated with suppression of cyclicity and anovulation. The contraceptive effect of lactation is strong in some species (e.g., rodents) but moderate in humans. In non-breast-feeding women, cyclicity is resumed within 1 month after delivery, whereas fully lactating women have a period of several months of lactational amenorrhea, with the first few menstrual cycles being anovulatory. Cessation of cyclicity results from a combined effect of the act of suckling and elevated prolactin levels. Prolactin suppresses ovulation by acting at several sites, including inhibition of pulsatile GnRH release, suppression of pituitary re-

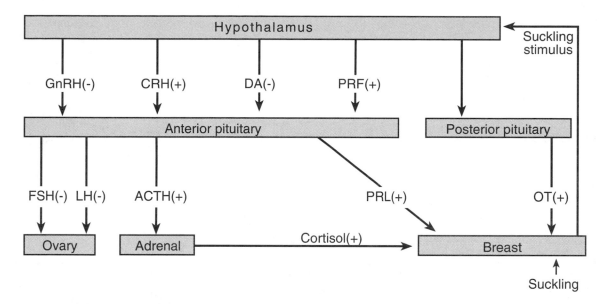

Figure 41–13 ■ Effect of suckling on hypothalamic, pituitary, ovarian, and adrenal hormones. GnRH, gonadotrophin-releasing hormone; CRH, corticotrophin-releasing hormone; DA, dopamine; PRF, prolactin-releasing factor; FSH, follicle stimulating hormone; LH, luteinizing hormone; ACTH, adrenocorticotrophic hormone; PRL, prolactin; OT, oxytocin.

sponsiveness to GnRH, and inhibition of ovarian activity (Fig. 40–10).

■ The Onset of Puberty Depends on Maturation of the Hypothalamic GnRH Pulse Generator

The onset of puberty depends on a sequence of maturational processes that begin during fetal life. The hypothalamopituitary-gonadal axis undergoes a prolonged and multiphasic activation-inactivation process. By midgestation, LH and FSH levels in fetal blood are elevated, reaching near adult values. Experimental evidence suggests that the hypothalamic GnRH pulse generator is operative at this time and the gonadotrophins are released in a pulsatile manner. The levels of FSH are lower in males than in females, likely due to suppression by the fetal testosterone at midgestation. As the levels of placental steroids increase, they exert negative feedback on GnRH release, lowering LH and FSH to very low levels toward the end of gestation.

After birth, the newborn is deprived of maternal and placental steroids. The reduction in steroidal negative feedback stimulates gonadotrophin secretion, the levels of which exhibit pulsatility with wide fluctuations during the first few months. These stimulate the gonads, resulting in transient increases in serum testosterone in male infants and estradiol in females. Follicle stimulating hormone levels in females are usually higher than those in males. At approximately 3 months of age, the levels of both gonadotrophins and gonadal

steroids are at the low normal adult range. Circulating gonadotrophins then decline to very low levels by 6–7 months in males and 1–2 years in females and remain suppressed until the onset of puberty.

Throughout childhood, the gonads are quiescent and plasma steroid levels are low. Yet gonadotrophin release is also suppressed. The prepubertal restraint of gonadotrophin secretion is explained by two mechanisms, both of which affect the hypothalamic GnRH pulse generator. One is a sex steroid-dependent mechanism that renders the pulse generator extremely sensitive to negative feedback by steroids. The other is an intrinsic central nervous system (CNS) inhibition of the GnRH pulse generator. Together, they suppress the amplitude, and probably the frequency, of GnRH pulses, resulting in diminished secretion of LH, FSH, and gonadal steroids. Throughout this period of quiescence, the pituitary and the gonads can respond to exogenous GnRH and gonadotrophins, but at a relatively low sensitivity (Fig. 41–14).

The hypothalamopituitary unit becomes reactivated during the late prepubertal period. This involves a decrease in hypothalamic sensitivity to sex steroids and a reduction in effectiveness of the intrinsic CNS inhibition over the GnRH pulse generator. The mechanisms underlying these changes are unclear but might involve endogenous opioids. As a result of disinhibition, the frequency and amplitude of GnRH pulses increase. Initially, pulsatility is most prominent at night, entrained by deep sleep; later it becomes established throughout the 24-hour period. The GnRH acts

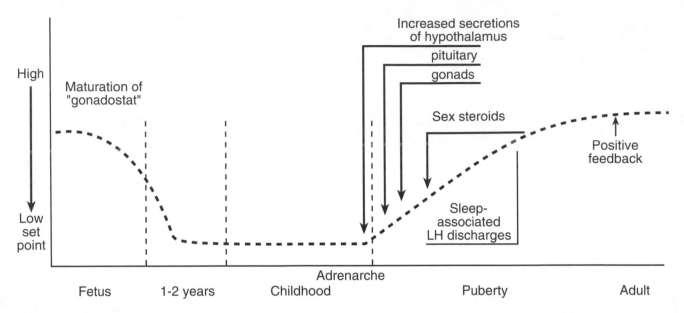

Figure 41–14 ■ Changes in the sensitivity and activity of the hypothalamopituitary-gonadal axis throughout life. LH, luteinizing hormone.

on the gonadotroph cells of the anterior pituitary as a self-primer. It increases the number of GnRH receptors (up-regulation) and augments the synthesis, storage, and secretion of the gonadotrophins. The increased responsiveness of FSH to GnRH in females occurs earlier than that of LH, accounting for a higher FSH/LH ratio at the onset of puberty than that during late puberty and throughout adulthood. A reversal of the ratio is seen again after menopause.

The increased pulsatile GnRH release initiates a cascade of events. The sensitivity of gonadotrophs to GnRH is increased, the secretion of LH and FSH is augmented, the gonads become more responsive to the gonadotrophins, and the secretion of gonadal hormones is stimulated. The rising circulating levels of gonadal steroids induce a progressive development of the secondary sex characteristics and establish an adult pattern of negative feedback on the hypothalamo-pituitary unit. Activation of the positive feedback mechanism in females and the capacity to exhibit an estrogen-induced LH surge is a late event, expressed in mid- to late puberty.

The onset of puberty in humans begins at age 10–11. It is a slow process, lasting 3–5 years, that involves development of secondary sex characteristics, growth spurt, and acquisition of fertility. The timing of puberty is determined by genetic, nutritional, climatic, and geographic factors. Over the last 150 years, the age of puberty has declined by 2–3 months per decade; this appears to correlate with improvements in nutrition and general health.

The first physical signs of puberty in girls are breast budding **(thelarche)** and appearance of pubic hair. Axillary hair growth and peak height spurt occur within 1-2 years. **Menarche,** or establishment of men-strual cyclicity, is a late event, achieved at a median age of 12.8 years in American girls. The first few cycles are usually anovulatory. The first sign of puberty in boys is enlargement of the testes, followed by the appearance of pubic hair and enlargement of the phallus. The peak growth spurt and appearance of axillary hair in boys usually occurs 2 years later than in girls. Growth of facial hair, deepening of the voice, and broadening of the shoulders are late events in male pubertal maturation (Fig. 41-15).

Puberty is also regulated by hormones other than gonadal steroids. The adrenal androgens DHEA and DHEAS are primarily responsible for development of pubic and axillary hair. Adrenal maturation, or **adrenarche,** precedes gonadal maturation, or **gonadarche,** by 2 years. The pubertal growth spurt requires a concerted action of sex steroids and growth hormone. The principal mediator of growth hormone is insulinlike growth factor-I (IGF-I). Plasma concentration of IGF-I increases markedly during puberty, with peak levels observed earlier in girls than in boys. Insulinlike growth factor-I is essential for accelerated growth. The gonadal steroids appear to act primarily by augmenting pituitary growth hormone release which, in turn, stimulates the production of IGF-I in liver and other tissues.

■ Disorders of Sexual Development Can Manifest Before or After Birth

Normal sexual development depends on a complex, orderly sequence of events that begins during early fetal life and is completed only at puberty. Any deviation can result in infertility, sexual dysfunction, or various degrees of intersexuality **(hermaphroditism).** A true hermaphrodite possesses both ovarian and testic-

at birth. The extra X chromosome, however, interferes with the development of the seminiferous tubules, which show extensive hyalinization and fibrosis, whereas the Leydig cells are hyperplastic. Such males have small testes, are azoospermic, and often exhibit some eunuchoidal features. Because of having only one X chromosome, an individual with 45,XO karyotype will have no gonadal development during fetal life and is presented at birth as a phenotypic female. Given the absence of ovarian follicles, such patients have very low levels of estrogens, primary amenorrhea, and do not undergo normal pubertal development.

Female pseudohermaphrodites are 46,XX females with normal ovaries and internal genitalia but with a different degree of virilization of the external genitalia, resulting from exposure to excessive androgens in utero. The most common cause is **congenital adrenal hyperplasia,** an inherited abnormality in adrenal steroid biosynthesis, with most cases of virilization due to cytochromes P450c21 and P450c11 deficiency. In such cases, cortisol production is low, causing increased production of ACTH by activating

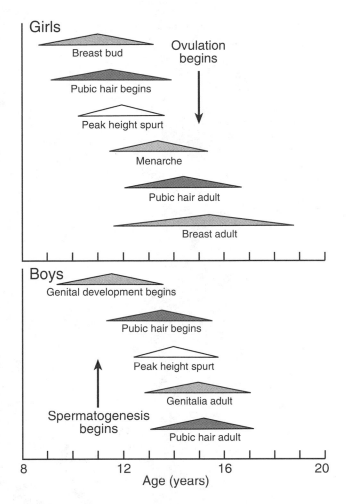

Figure 41–15 ■ Peripubertal maturation of secondary sex characteristics in boys and girls.

ular tissues, either separate or combined as ovotestes. A **pseudohermaphrodite** has one type of gonads but a different degree of sexuality of the opposite sex. Sex is normally assigned according to the type of gonads. Disorders of sexual differentiation can be classified as gonadal dysgenesis, female pseudohermaphroditism, male pseudohermaphroditism, or true hermaphroditism. Selected cases and their manifestations are briefly discussed here, as are disorders of pubertal development (see also Chaps. 39, 40).

Gonadal dysgenesis refers to incomplete differentiation of the gonads and is usually associated with sex chromosome abnormalities. These result from errors in the first or second meiotic division and occur by chromosomal nondisjunction, translocation, rearrangement, or deletion. The two most common disorders are **Klinfelter syndrome** (47,XXY) and **Turner's syndrome** (45,XO). Because of a Y chromosome, an individual with 47,XXY karyotype has normal testicular function in utero in terms of testosterone and MIF production and no ambiguity of the genitalia

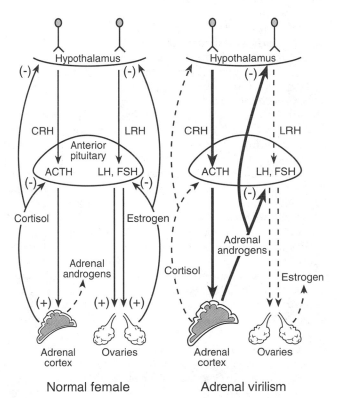

Figure 41–16 ■ Hormonal interactions along the ovarian and adrenal axes during the normal female development as compared to adrenal virilism. Dashed lines indicate decreased production of the hormone. Heavier lines indicate increased hormone production. CRH, corticotrophin-releasing hormone; LRH, gonadotrophin releasing hormone; ACTH, adrenocorticotrophic hormone; LH, luteinizing hormone; FSH, follicle stimulating hormone. (+) and (−) indicate positive and negative effects, respectively.

the hypothalamopituitary axis (Fig. 41–16). The elevated ACTH levels induce adrenal hyperplasia and abnormal production of androgens and corticosteroid precursors. These infants are born with ambiguous external genitalia (i.e., clitoromegaly, labioscrotal fusion, or phallic urethra). The degree of virilization depends on the time of onset of excess fetal androgen production. When aldosterone levels are also affected, there is a life-threatening salt-wasting disease. Untreated patients with congenital adrenal virilism develop progressive masculinization, amenorrhea, and infertility.

Male pseudohermaphrodites are 46,XY individuals with differentiated testes but underdeveloped and/or absent wolffian-derived structures and inadequate virilization of the external genitalia. These result from defects in testosterone biosynthesis, metabolism, or action. The **5α-reductase deficiency** is an autosomal recessive disorder caused by the inability to convert testosterone to DHT. Such infants have ambiguous or female external genitalia and normal male internal genitalia (Fig. 41–17). They are often raised as females but undergo a complete or partial testosterone-dependent puberty, including enlargement of the penis, testicular descent, and development of male psychosexual behavior. Azoospermia is common.

The **testicular feminization syndrome** is an X-linked recessive disorder caused by end-organ insensitivity to androgens, usually because of absent or defective androgen receptors. These 46,XY males have abdominal testes that secrete normal testosterone levels. Because of androgen insensitivity, the wolffian ducts regress and the external genitalia develop along the female line. The presence of MIF in utero causes regression of the müllerian ducts. Thus, they have neither male nor female internal genitalia and phenotypic female external genitalia, with the vagina ending in a blind pouch. They are reared as females and undergo feminization during puberty because of peripheral conversion of testosterone to estradiol.

Disorders of puberty are classified into **precocious puberty,** defined as sexual maturation before the age of 8 years, and **delayed puberty,** when menses does not start by age 17 or testicular development is delayed beyond age 20. **True precocious puberty** results from premature activation of the hypothalamopituitary-gonadal axis, leading to development of secondary sexual characteristics as well as gametogenesis. The most frequent causes are CNS lesions or infections, hypothalamic disease, and hypothyroidism. **Pseudoprecocious puberty** is defined as early development of secondary sexual characteristics without gameto-

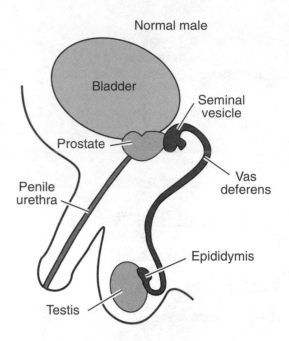

Normal male

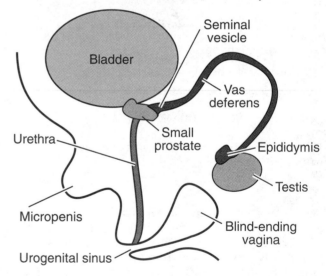

Male with 5α-reductase deficiency

Figure 41–17 ■ Effects of 5α -reductase deficiency on differentiation of the internal and external genitalia.

genesis. It can result from abnormal exposure of immature boys to androgens and of immature girls to estrogens. Augmented steroid production can be of gonadal or adrenal origin.

■ ■ ■ ■ ■ ■ ■ ■ ■ ■ **REVIEW EXERCISES** ■ ■ ■ ■ ■ ■ ■ ■ ■ ■

Identify Each with a Term

1. The process that enables sperm to bind to the egg

2. The embryonic cells that invade the endometrium

3. A contraceptive method that depends on abstinence during the ovulatory phase

4. A glycoprotein that affects sexual differentiation of the internal genitalia

Define Each Term

5. Acrosome reaction
6. Zona pellucida
7. Myoepithelial cells
8. Fetoplacental unit
9. Barr body
10. Wolffian ducts
11. Mammogenesis

Choose the Correct Answer

12. The mammary gland of the newborn:
    a. Consists primarily of connective tissue
    b. Has differentiated lobuloalveolar structures
    c. Can secrete milk in response to hormones
    d. Can produce only colostrum

13. The suckling stimulus:
    a. Has afferent hormonal and efferent neuronal components
    b. Increases placental lactogen secretion
    c. Increases the release of dopamine from the arcuate nucleus
    d. Triggers the release of oxytocin by stimulating the supraoptic nuclei

14. Implantation occurs:
    a. On day 4 after fertilization
    b. Only after priming of the uterine endometrium by estrogen
    c. When the embryo is at the morula stage
    d. After the endometrium undergoes a decidual reaction

15. Upon contact between the sperm head and the zona pellucida, penetration is allowed because of:
    a. The acrosome reaction
    b. The zona reaction
    c. Sperm motility
    d. Pronuclei formation

16. The next ovulatory cycle after implantation is prevented because of:
    a. High levels of prolactin
    b. Production of hCG by the trophoblast
    c. Production of prostaglandins by the corpus luteum
    d. Depletion of oocytes in the ovary

Briefly Answer

17. How is polyspermy prevented?
18. List the embryonic stages, from the time of fertilization to the time of implantation.
19. What is the mechanism by which IUD acts as a contraceptive?
20. List two main functions of hCG.
21. Which steroid can be used as an index for fetal well-being? Why?
22. What are the causes of lactational amenorrhea?

ANSWERS

1. Capacitation
2. Trophoblast cells
3. Rhythm
4. Müllerian inhibiting factor (MIF)
5. Release of proteolytic enzymes that enable the sperm to digest the cumulus cells and penetrate the zona pellucida
6. A thin layer of proteins and polysaccharides that surrounds the oocyte and provides protection during transport in the female genital tract
7. Contractile cells that surround the mammary alveoli and contract in response to oxytocin, causing milk let-down
8. Combined contributions of the maternal, placental, and fetal compartments to the production of estrogens during pregnancy
9. The inactivated X chromosome that can be detected by cytology
10. The indifferent fetal mesonephric ducts that develop into the epididymis, vas deferens, seminal vesicles, and ejaculatory ducts on exposure to testosterone
11. The process by which the mammary glands differentiate, grow, and mature

12. c
13. d
14. d
15. a
16. b
17. Polyspermy is prevented primarily by the zona reaction, which causes destruction of sperm receptors and hardening of the zona pellucida.
18. The embryonic stages are zygote, morula, and blastocyst.
19. The IUD induces sterile inflammation in the uterus and speeds the passage of blastocysts, preventing implantation.
20. Two main functions of hCG are prevention of corpus luteum regression and promotion of testosterone production by the fetal testes.
21. Estriol can be used as an index of fetal well-being, because its synthesis requires contribution by the fetal liver as well as the adrenals.
22. Lactational amenorrhea is caused by the act of suckling itself and the high levels of prolactin, which has an antigonadotrophic effect.

Suggested Reading

Austin CR, Short RV: *Embryonic and Fetal Development.* 2nd ed. London, Cambridge University Press, 1982.

Austin CR, Short RV: *Germ Cells and Fertilization.* 2nd ed. London, Cambridge University Press, 1982.

Knobil E, Neill JD: *The Physiology of Reproduction.* New York, Raven, 1988, chaps. 4-7, 48-59.

Mulchahey JJ, Diblasio AM, Martin MC, et al: Hormone production and peptide regulation of the human fetal pituitary gland. *Endocrin Rev* 8:406–425, 1987.

Speroff L, Glass RH, Kase NG: *Clinical Gynecologic Endocrinology and Infertility.* 5th ed. Baltimore, Williams & Wilkins, 1994.

Yen SSC, Jaffe RB: *Reproductive Endocrinology.* 3rd ed. Philadelphia, WB Saunders, 1991, chaps. 14, 15, 23, 25, 26, 27.

Index

Creep phenomenon, of smooth muscle, 188
Cremasteric muscle, 742
CRH. *See* Corticotrophin-releasing hormone (CRH)
Critical micellar concentration, 557
Crossbridge cycle, 156-158, 156*f*
    chemical processes, 158, 158*f*
    metabolite build-up in, 179
    of muscle contraction, 156-158, 156*f*
    smooth muscle, 186, 187
Crossbridges, and sliding filament theory, 155
Crossing over, during meiosis, 746
CT. *See* Calcitonin (CT)
Cumulus oophorus, in female reproductive system anatomy, 763
Cupula, inner ear, 83, 83*f*
Curare, and postsynaptic blockade of neuromuscular junction, 168
Cushing's disease, 701
Cutaneous sensation
    localization of, 69
    nociceptors, 70
    tactile receptors, 69, 69*f*
    thermal receptors, 69-70
Cyanosis, 213
    enterogenous, 213
Cyclic adenosine monophosphate (cAMP). *See* cAMP (cyclic adenosine monophosphate)
Cylindric lens, for astigmatism correction, 72
Cystic fibrosis, 24
Cystic fibrosis transmembrane conductance regulator (CFTR), 24
Cystinuria, 562
Cytochrome(s)
    P450c11, 692
        genetic defects in isozymes of, 695
    P450c17, 692
    P450c21, 692, 695
    P450scc, 690
Cytokines, 219
    fever and, 608
    in hematopoiesis, 212
    and osteoporosis, 727
    production, 219
Cytoplasmic calcium pool, cardiac muscle contractility and, 204, 204*f*
Cytoskeleton, neuronal, 55
Cytosol
    composition, 15
    gap junction and, 8, 9*f*
Cytotoxic T cells, formation, 220*f*
Cytotrophoblasts, 781

DA. *See* Dopamine (DA)
DAG. *See* Diacylgycerol (DAG)
Dalton's law, 353, 387
Dark adaptation, 76
Day/night cycle. *See* Circadian rhythm
Daytime vision, 74
D cells, 535
Dead space, 351-352, 352*f*. *See also* Alveolar dead space; Anatomic dead space; Physiologic dead space
    determination, 353
Dead space-tidal volume ratio (VD/VT), 349*t*, 351
Dead space ventilation, 352-353, 352*t*
Decidua basalis, 783
Decidual reaction, of endometrium, 781

Decidua spongiosa, 318
Deep pain, 70
Defecation, 526-527. *See also* Feces
Defense reaction, cardiovascular effects of, 328
Defensins, 218
Dehydration, intracellular, thirst and, 452
Dehydration reaction, 469
Dehydroalanine residue, 677
7-dehydrocholesterol, 728
Dehydroepiandrosterone, 659. *See also* Adrenocortical hormones
    chemical structure, 688, 688*f*
Dehydroepiandrosterone sulfate (DHEAS), 785
5′ deiodinase, 678
Deiod2znination, thyroid hormones, 678
Deiters cells, 79-80, 80*f*
Delayed emptying, 518-519, 518*t*
Delayed pain, 70
Delayed puberty, 796
Delayed-type hypersensitivity, 218, 223
Delta cells, of islets of Langerhans, 707, 707*f*, 708
Delta waves, EEG recording, 134, 134*f*
Denaturation, 588
Dendrites, in neuron anatomy, 34, 35*f*
    postsynaptic potentials integration at, 42-45, 44*f*, 45*f*
Denervation, acute renal, 455
Denervation hypersensitivity, 330
Dense bands, of smooth muscle tissue, 181
Dense bodies, of smooth muscle tissue, 181
Deoxycholic acid, 544
11-deoxycorticosterone (DOC), 692
11-deoxycortisol, 692
Deoxyhemoglobin (Hb), 391
Deoxyribonuclease, 542
Depolarization, 245
    atrial, 251
    ear hair bundle deflection, 81, 81*f*
    endplate, at neuromuscular junction, 168
    late, 245
    of myocardial cells, 243
    threshold, 245
    ventricular, 251
Depolarization phase
    action potential, 37, 37*f*, 38
    postsynaptic membrane potential, 42-43, 43*f*
Depolarizing blockers, of neuromuscular junction, 168
Depression
    exercise effects, 626
    limbic system involvement, 142
Depressions, in cell membrane, 16, 17*f*
    coated pits, 17, 17*f*
Depth perception, 73
Dermal circulation, 314-315
    regulation of body temperature, 315
Dermatitis, dementia, 564
Descending colon, large intestine motility and, 525
Descending motor tracts
    brainstem, 101-102, 101*f*
        outdated terminology, 104
    corticospinal. *See* Corticospinal tract, descending motor tract
Desensitization, 646
Desmosomes, cardiac muscle cells, 195
Detector, in negative feedback control system, 5-6, 6*f*

Detrusor muscle, 460
Deuterium oxide, in body water measurement, 448
DHEAS (dehydroepiandrosterone sulfate), 785
Diabetes insipidus
    central, 450, 451
    nephrogenic, 451
Diabetes mellitus
    complications
        acute, 717-718, 718*f*
        chronic secondary, 716, 718
    glucose levels in, 426
    insulin-dependent (type I)
        genetic and environmental factors, 715-716
        insulin/glucagon ratio in, 715
        pathogenesis, 715
        treatment, 717
    non-insulin-dependent (type II), 669, 715-718
        genetic and environmental factors, 717
        insulin resistance, 717
        origination, 717
        treatment, 717
    in pregnancy, fetal growth and, 787, 788*f*
Diabetic foot, 716, 718
Diabetic ketoacidosis (DKA), 717
Diabetic nephropathy, 716, 718
Diabetic neuropathy, 716, 718
Diabetic retinopathy, 716, 718
Diabetogenic action, of GH, 669
Diacylglycerol (DAG), 647, 648*f*, 679
    ADH and, 662
    production of, 42
    as second messenger, 11
Dialysis, and renal transplantation, 431
Diaphragm, in respiration, 344-345, 402, 402*f*
    movement of, 344, 344*f*
Diastole, 238
    atrial, 261
    cardiac blood flow during, 306-307
    ventricular, 261-262
Diastolic pressure (P$_d$), 238-239
Dichromatic vision, 74-75
Diencephalon, hypothalamus and, 129, 129*f*
Diet. *See* Food ingestion
Dietary fiber, 553
    role in gastrointestinal motility, 555
Diffuse spasm, 514
Diffusing capacity (DL), of lung, 390-391, 390*f*
    and hematocrit, 391
    interpretation, factors affecting, 391
    and pulmonary capillary blood volume, 391
Diffusion, 239
    capillary, 239
        rate of, 295-296
    gas, 387, 388, 388*f*
        across alveolar-capillary membrane, 388-389, 388*f*
    solute movement via
        carrier proteins facilitating, 19-20, 19*f*
        diffusive membrane transport, 18-19, 18*f*
        ion channels facilitating, 20-21, 20*f*, 21*f*
        simple, 17-18, 18*f*

Macula densa, 421
    sodium chloride delivery to, 427, 427f
Maculae, inner ear, 84–85, 84f
Macula lutea, 72
Magnesium (Mg)
    in body fluids, 447t
    intestinal absorption, 567
Magnocellular neurons, of hypothalamus, 129, 130f, 657
Malabsorption, of lipids, 559
Male reproductive system, 737–755
    anatomy, 764
    androgens, actions of, 750–752, 751f
    dysfunctions
        hypogonadism, 752–754
        hypogonadotrophic/hypergonadotrophic states and, 754
        impotence, diagnosis and therapy, 753
        infertility, 752–754
        pseudohermaphroditism, 754–755
    regulation, 738, 738f
    spermatogenesis, 748–750, 749f, 750f
    testes
        duct system, 744
        hypothalamic regulation, 738–741, 739f, 741f
        Leydig cells, 743–744, 743f
        Sertoli cells, 742–743, 743f
        sperm formation, 741–742, 741f–743f
        testosterone production, 743–744, 743f
Males
    puberty onset, 793–794, 794f, 795f
    reproductive system. See Male reproductive system
    sexual development disorders, 794–796, 795f, 796f
Malignant hyperpyrexia, 609
Malignant hyperthermia, 609
Malleus bone, middle ear, 78
Mammary gland
    lactating, structure of, 792f
    oxytocin and, 670–671
Mammillothalamic tract, 130, 130f
Mammogenesis, hormone regulation, 791
Manic-depressive disorder, limbic system involvement, 142
Manometric catheters, monitoring esophageal motility, 514, 515f
MAO (monoamine oxidase), 48–50, 49f
MAP. See Mean arterial pressure (MAP)
MAPs (microtubule-associated proteins), 55
Marrow stroma, 219
Masking, auditory signal, 77
Mass balance principle, renal medulla, 441–442, 442f
Mast cells, 223
Mathematical ability, cortical area controlling, 137
Maximal oxygen uptake (Vo₂ max), 615
Maximal voluntary contractile force (MVC), 616
MCH (mean cell hemoglobin), 213
MCHC (mean cell hemoglobin concentration), 213
MCR (metabolic clearance rate), 641
MCV (mean cell volume), 213, 215
Mean arterial pressure (MAP), 236–238, 279, 281
    baroreceptor reflex and, 325, 325f, 326, 326f

defined, 238, 238f
    factors affecting, 279–280
Mean cell hemoglobin (MCH), 213
Mean cell hemoglobin concentration (MCHC), 213
Mean cell volume (MCV), 213, 215
Mechanical stimuli, 62
Mechanical threshold, 517
Mechanoreceptor(s), 62
    hypothetical, 63, 65f
    involuntary organs and, 116
Medulla
    breathing-related cells in, 402–403, 403f
        dorsal respiratory group, 402–403, 403f
        ventral respiratory group, 402–403, 403f
    cardiovascular reflex integration in, 324
    central pattern generator for breathing, 402–406
    chemosensitive neurons, response to local H⁺, 407
    renal. See Renal medulla
Medulla oblongata, vagal nerves and, 499, 501
Medullary cardiovascular center, 324, 325
Megacolon, congenital, 527
Megaesophagus, 514
Megaloblasts, 215
Meiosis, 744
Meiosis inhibiting factor, 763, 763t
Meiotonic muscle contraction, 173
Meissner's corpuscles, as tactile receptors, 69, 69f
Meissner's plexus, of enteric nervous system, 490–491
Melanin, in pigment epithelium, 74, 74f
Membrane length (space) constant, 44
Membrane potential
    during an action potential, 37f, 38
    cardiac, 243–246
        effect of ionic permeability on, 244, 244f
        resting, 246
        threshold, 245
    electrochemical, 29
    electrotonic potential and, 43–44, 44f
    equilibrium, 29
    Ohm's law, 36
    of postsynaptic neuron, 42
    resting, 30. See also Resting membrane potential
    steady state concept, 28–29, 29f
Membrane system, of skeletal muscle, 154f, 155
Membrane vesicles, in endothelial cells, 291f, 292, 294
Memory
    cerebral cortex and limbic system involvement, 142–143
    immunologic, 220, 222
Memory cells, 222
    formation, 220f
Menaquinones, 563
Menopause, 771–773, 773t
Menstrual cycle, hypothalamic regulation, 131
Menstrual phase, of endometrial cycle, 770
Menstruation. See Female reproductive system, menstrual cycle
Mercury, column, pressure expressed as height of, 235, 236f

Merkel's discs, as tactile receptors, 69, 69f
Meromyosin, of thick myofilaments, 154, 154f
Mesangial cells, glomerular and extraglomerular, 421
Mesencephalic locomotor region, 101
Mesocortical system
    of dopaminergic neurons, 140
    neuroleptic drugs and, 140, 142
Mesoderm, 782
Mesolimbic system
    of dopaminergic neurons, 140
    neuroleptic drugs and, 140, 142
Messenger molecules, 9. See also Hormones
Messengers, hormones as, 9–10, 9f
    hydrophilic hormones, 646–649, 647f–649f
    second messengers, 10, 12f. See also Second messengers
Metabolic acidosis
    chemical buffering, 480
    compensatory responses, 478t
        renal, 480–481
        respiratory, 480
    defined, 477, 480
    ventilatory response to, 410
Metabolic alkalosis
    defined, 477
    vomiting and, 479
Metabolic cascade, 225
Metabolic clearance rate (MCR), 641
Metabolic processes
    in cold acclimatization, 606
    exercise effects, in obese persons, 625, 625f
    in fasting, glucocorticoids and, 699–702, 699f, 701f
    hepatic. See under Liver
    producing body heat, 592–594, 592t, 593f
    thyroid hormones and, 683, 684
Metabolic reactions, cell function and, 5
Metabolic substrates, mobilization of, fight-or-flight response, 125
Metabolism, as source of hydrogen ions, 467
Metallic receptors, 86
Metarhodopsin II, 75
Methemoglobin (metHb), 213
N-methyl-D-aspartate (NMDA) receptors, 51
Mg. See Magnesium (Mg)
Micellar solubilization, 557, 557f
Micelles, formation, 557
Microcirculation, 290–292. See also Microvasculature (microvessels)
    defined, 290
    functions of, 290
Microcytes, 215, 216f
Microfilaments, of neuronal cytoskeleton, 55
Microtubule-associated proteins (MAPs), 55
Microtubules, of neuronal cytoskeleton, 55
Microvascular pressures, regulation of, 299–300
Microvascular resistance, 299–300
    regulation of, 300–303
Microvasculature (microvessels)
    arterial, 290–291
    branching structure of, 294
    and hypertension, 298
    number of, and diffusion distance between cells and blood, 294–295
    surface area of, 294

Pyridoxine. *See* Vitamin B₆
Pyrogens, defined, 608

Q₁₀ effect, 589
Quantal number, synaptic transmission, 166

R. *See* Resistance (R); Respiratory exchange ratio (R)
Race, blood pressure and, 279
Radiant heat and light, 62
Radiation
  electromagnetic. *See* Light
  external heat exchange via, 594
  principles, 595
Radioimmunoassay (RIA), of hormones, 642-643, 642f, 643f
Radioiodinated serum albumin, measuring plasma volume, 448
raf/mil oncoproteins, 650
Rage reaction, in cats, 324-325, 328
Ranvier, nodes of, action potential propagation, 38-39, 40f
Rapid adaptation, of sensory receptors, 66, 67f
Rapid emptying, 518-519, 518t
Rapid-eye-movement (REM) sleep, 135, 135f
  wakening during, 135-136
Rarefaction, sound waves, 77
ras oncogene, 650
RBF. *See* Renal blood flow (RBF)
RBP (retinol-binding protein), 579
Reabsorption
  of filtered HCO₃⁻, 473-474, 474f
  renal tubule. *See* Tubular reabsorption
Reactive oxygen species (ROS), lung injury caused by, 377-378
Real image, converging light rays producing, 71, 71f
Receiving segment, in peristalsis, 509
Receptive relaxation, ingestion and, 516
Receptor adaptation, 326, 326f
Receptor-mediated endocytosis, 17, 17f
Receptor (generator) potential, in hypothetical mechanoreceptor, 63, 65f
Receptors
  adrenergic, 50
  cell communication and, 9, 9f
  cholinergic, 47
  GABA, 51
  glutamate
    role in nerve cell death, 52
    subtypes, 51
  for hormones, 634
  neurotransmitters binding to, 34, 42
  in postsynaptic membrane, 41f, 42
  sensory. *See* Sensory receptor(s)
  serotonin, 50
  types, 42t
  up-regulation, 330
Reciprocal innervation/inhibition, muscle functions, 92, 99
Reciprocating tachycardias, 252
Rectoanal reflex, 526
Rectosigmoid region, fecal continence and, 525-526
Rectosphincteric reflex, 526
Rectum, 525
  temperature ranges in, 588f
    during exercise, 605f
    under nude conditions, 596f
Red blood cells. *See* Erythrocytes

Red muscle fibers, skeletal muscle, 178-179, 178f
  and white muscle fiber proportions, 179
Red nucleus, descending motor tract, 101f, 102
Red-sensitive pigment, 74
5α-Reductase deficiency, 755, 796, 796f
Referred pain, 69, 116
Reflex(es)
  acoustic, 79
  activity mediated by spinal cord, 98-101, 99f, 100f
  areflexia, 101
  autonomic, 122-123
  in autonomic and somatic nervous systems, 116, 116f
  from chest wall, 406-407
  coordination by central nervous system, 123
  enteroenteric, 502
  flexor withdrawal, 100, 100f
  generation and response, 98-99
  inhibitory, enteric nervous system and, 502, 502f
  inverse myotatic, 99-100, 100f
  local axon, 122, 122f
  loss of, 101
  from lungs, 406-407
  micturition. *See* Micturition reflex
  myotatic (stretch). *See* Myotatic reflex
  postural adjustment and, 103
  stretch, bladder distention initiating, 460
  vagovagal, 501, 538
  vestibuloocular, 85
Reflex functions, spinal motor systems serving, 98, 98f
Reflex responses, of enteric nervous system, 489
Refraction, errors of, visual, 72
Refractive index, light, 70
Refractory period
  after action potential initiation, 40, 40f
  of cardiac muscle cells, 246
Regeneration, of nerve fibers, 56
Regenerative process, action potential, 38
Regulated pathway, exocytosis and, 17
Regulated variable, in negative feedback control system, 5-6, 6f
Regulation
  actin-linked. *See* Actin-linked regulation
  of cell volume, 27-28, 28f
  hypothalamus role in. *See* Hypothalamus, regulatory functions
  myosin-linked. *See* Myosin-linked regulation
  physiologic, 8-12
  of renal blood flow. *See* Autoregulation, renal blood flow
Regulatory light chains, smooth muscle, 186
Regulatory volume mechanisms
  decrease, 27-28, 28f
  increase, 28, 28f
Reissner's membrane, 79, 80f
Relative humidity, 596
Relative refractory periods, after action potential initiation, 40, 40f
Relaxation
  in cardiac muscle contraction cycle
    cytoplasmic calcium concentration during, 204, 204f
    isometric, 197, 198f-199f

in fast-twitch muscle fibers, 179
  smooth muscle, 189
    biochemical control, 184-186, 185f
Releasing hormones, of hypothalamus. *See* Hypothalamic releasing hormones
Remodeling, of bone, 724, 726
REM sleep. *See* Rapid-eye-movement (REM) sleep
Renal acid excretion, 471-472
Renal artery, 419, 421, 421f
Renal blood flow (RBF), 426-428
  autoregulation, 427-428, 427f
  capacity, 426
  hormones changing, 428
  renal plasma flow estimating, 424-425
  renal sympathetic nerves changing, 428
  vascular components, 419, 421, 421f
Renal clearance, 422
Renal compensation
  in metabolic acidosis, 480-481
  in metabolic alkalosis, 481
  in respiratory acidosis, 478-479
  in respiratory alkalosis, 480
Renal cortex, 418-419, 419f
  blood flow rate in, 426-427, 426f
Renal lymphatics, 421
Renal medulla, 418-419, 419f
  blood flow rate in, 426-427, 426f
  mass balance principle applied to, 441-442, 442f
  osmolality gradient in, 439, 439f
Renal transplantation, 431
Renal tubule. *See also* Tubular entries
  pH regulation in, 471-476
  reabsorption. *See* Tubular reabsorption
  in renal anatomy, 419, 420f
  secretion. *See* Tubular secretion
  transport. *See* Tubular transport; Tubular transport maximum (Tm)
Renin, 698
  influencing tubular sodium reabsorption, 453-454, 453t, 454f
  production of, 421-422
Renin-angiotensin-aldosterone system
  cardiovascular effects, 330-331, 330f
  influencing tubular sodium reabsorption, 453-454, 453t, 454f
Repolarization phase, 245
  action potential, 37, 37f, 38
  postsynaptic membrane potential, 42-43, 43f
  ventricular, 252
Reproduction, gonadotrophins regulating, 665-666
Reproductive physiology, 738-796
  conception, pregnancy, and fetal development, 778-796. *See also* Pregnancy
  female. *See* Female reproductive system
  male. *See* Male reproductive system
Reservoir functions, of stomach and colon, 508
Residual volume (RV), 348, 348f, 349f, 349t, 350f
Residual volume-total lung capacity ratio (RV/TLC), 349t
Resistance (R)
  airway (Raw), 363-364, 363f
  arterial, effect of vascular disease on, 232
  to flow, 233
  nerve cell membrane, 36
  in parallel, 283, 284f

Sleep—*Continued*
  responses to respiratory stimuli in, 411–412
  rapid eye movement (REM)
    breathing pattern in, 411
    responses to respiratory stimuli in, 411–412
  upper airway patency in, 412
Sleep apnea
  central, 410
  obstructive, 410, 412
Sleep apnea syndromes, 410
Sleepiness, daytime, with sleep apnea, 410, 412
Sleep/wake cycle. *See* Circadian rhythm
Sliding filament theory, 155–156, 155*f*, 156*f*
Slit pore, in glomerular filtration membrane, 428
Slow adaptation, of sensory receptors, 66, 67*f*
Slow axoplasmic transport, 55
Slow drift, eye movement, 73
Slow inward current, calcium ion flux, 196
Slow-twitch muscle fibers, 95–96, 178–179, 178*f*
Slow wave potential change, smooth muscle, 183
Slow waves, in intestinal muscle, 506–507, 507*f*
Slow-wave sleep, 135, 135*f*
Small intestine
  blood flow to, 306*t*
  calcium and phosphate absorption, 724–725
  circulation in, 310–312
  enterohepatic circulation of bile salts and, 546–547, 546*f*
  functions of, 310
  glucose absorption in, 25, 26*f*
  intestinal secretion, composition and function, 547–548
  layers of, 310
  motility. *See* Small intestine motility
  oxygen consumption, 306*t*
  pseudo-obstruction, 521–523
  smooth muscle arrangement in, 180, 180*f*
  transporters in amino acid and peptide absorption, 560–561
Small intestine motility, 519–523
  migrating motor complex, 519–520, 520*f*, 521*f*
  mixing movements, 520
  power propulsion, 521
  vagal nerves and, 521
Small-molecule transmitters, 46
Smell sensation. *See* Olfactory sensation
Smooth muscle, 151, 165–192
  anatomy
    structural specializations, 180, 180*f*
    tissue organization, 181–182, 181*f*, 182*f*
  contractile and force transmission in, 181–182, 182*f*
  electric coupling in, 182
  hypertrophic changes, 189–190
  mechanical activity in, 186–189
    contraction, 187–189, 187*f*, 188*f*
    relaxation, 189
  oxytocin and, 670–671
  regulation and control, 182–186
    activation, 183
    biochemical, 184–186, 185*f*

calcium role in contraction, 183–184, 184*f*
  innervation, 182–183
tasks performed by, 180
vascular
  contraction, 290
  relaxation, 322–323
  viscoelasticity of, 188–189
Snoring, 412
So₂ (oxyhemoglobin saturation), 391–393
Sodium (Na)
  in body fluids, 447*t*
  in cardiac membrane potential, 243–244
  intake, 452
    and excretion, balance between. *See* Sodium balance
  intestinal absorption, 565–566, 565*f*
  in proximal tubular fluid, 433–434, 433*f*
  reabsorption
    excretion of hydrogen ions and, 476
    in proximal tubular fluid, 434–435, 434*f*
    renal, aldosterone and, 699
  renal excretion of, 452
    factors influencing, 453–455
    and intake, balance between. *See* Sodium balance
  secretion in cortical collecting duct, 438, 438*f*
Sodium appetite, 452
Sodium balance, 335–336, 336*f*, 452–455
  factors influencing, 453–455
    hormones, 453–455. *See also* Hormone(s), renal sodium excretion influenced by
    physical, 453
    renal hemodynamics, 453
    renal sympathetic nerve activity, 455
    renin-angiotensin-aldosterone system, 453–454, 453*t*, 454*f*
  hyponatremia and, 451
  regulatory mechanisms, 452, 453*f*
Sodium channel(s)
  action potential and
    cardiac, 245–246, 245*f*
    gating mechanisms, 38, 39*f*
    propagation and speed, 38, 40*f*
  cardiac membrane, 243–244, 244*t*
  in pacemaker potential, 246
Sodium chloride (NaCl), delivery to macula densa, 427, 427*f*
Sodium equilibrium potential, 244, 244*f*
Sodium pump (Na⁺/K⁺-ATPase), 21, 22*f*
  resting membrane potential and, 28, 29*f*, 30
  in secondary active transport, 23, 25
Solutes, in plasma, 211, 211*t*
Solute transport
  across cell membrane
    mechanisms
      active movement, 21–23, 25
      endocytosis, 16–17, 17*f*
      exocytosis, 17, 17*f*
      osmotic pressure differences, 26–27
      passive movement, 17–21
      volume regulation, 27–28, 28*f*
    resting membrane potential and, 28–30
  in proximal convoluted tubule, 433–434, 433*f*
    sodium reabsorption and, 434–435, 434*f*

Solutions
  osmolality and tonicity of, 27
  osmotic pressure of, 27
Solvent drug, 565
Soma
  in neuron anatomy, 34, 35*f*
    postsynaptic potentials integration at, 42–45, 44*f*, 45*f*
  protein synthesis in, 54–55
Somatic nervous system
  versus autonomic nervous system, 115, 116
  reflexes in, 116, 116*f*
Somatic pain, 70
Somatic structures, sympathetic division innervation, 118
Somatomedin, 667
Somatosensory cortex, 106
Somatostatin, 535
  chemistry and action on anterior pituitary, 657, 657*t*
  GH and, 666–667, 667*f*
  glucagon and insulin secretion and, 710–711
  preprosomatostatin synthesis, 711
  secretion, 711
Somatotopic maps, 104, 105*f*
Somatotrophs, 655
  GH, GHRH, and somatostatin acting on, 667*f*
Somatotrophin release-inhibiting hormone (SRIF), chemistry and action on anterior pituitary, 657, 657*t*
Sound, nature of, 76–77
Sound pressure, 77, 77*t*
Sound transmission
  through external eye, 77–78
  through middle ear, 78–79, 78*f*
Sound wave(s)
  and pitch discrimination, 82–83
  basilar membrane deformation, 82, 82*f*
  characteristics, 77
  defined, 76–77
  in environment, 62
  morphology, 77
Sour sensation
  receptors for, 86
  substances producing, 86
Space constant, membrane, 44
Space of Disse, 572
Spasm
  diffuse, 514
  hemifacial, 169
Spasmodic dysphonia, 169
Spastic neurogenic bladder, 461
Spatial summation, postsynaptic potentials, 43, 45*f*
Special senses. *See* Sensory receptor(s)
Specific dynamic action, 593
Spectral sensitivity, of photopigments, 74
Speech. *See* Language and speech
Sperm
  maturation, storage, and transport, 744
  transport to oviduct, 778–779
Spermatids, formation, 745–746
Spermatocytes, formation, 746
Spermatogenesis, 738, 744–745, 745*f*
  cell division, 745–746
  cell remodeling, 746–747
  FSH and testosterone in, 747–748
  site, 741–742, 741*f*–743*f*

and CNS development, 680–681
growth effects, 681–682
intermediary metabolism, 683
mechanism of, 680
self-regulation, 683
TSH secretion and synthesis, 664–665, 665*f*
cell communication and, 11–12
deficiencies, disorders of, 683–684
excesses, disorders of, 684
liver action on, 581
receptors, 649, 649*f*, 651
structure, 674, 675*f*
synthesis and secretion
dietary iodide and, 680
hydrolysis of thyroglobulin, 677–678
iodinated thyroglobulin, 675–677, 676*f*, 677*f*
metabolic process, 678
TSH regulating, 678–680
thyroid follicle and. *See* Thyroid follicle
transport, 639–641, 640*f*, 640*t*
Thyroiditis, chronic lymphocytic, 681
Thyroid peroxidase, 677, 681
Thyroid stimulating hormone (TSH), 655
action, 663
chemistry and action, 659*t*
secretion and synthesis
thyroid hormones and, 664–665, 665*f*, 678–680
TRH and, 664, 664*f*
structure and formation, 663–664
Thyrotoxicosis, 684
Thyrotrophin-releasing hormone (TRH)
chemistry and action on anterior pituitary, 657, 657*t*
TSH secretion and synthesis and, 664, 664*f*
Thyrotrophs, 655
Thyroxine (T₄). *See* Thyroid hormones
Thyroxine-binding globulin (TBG), 678
Thyroxine-binding prealbumin (TBPA), 678
Tidal volume (Vᴛ), 348, 348*f*, 349*t*
distribution of, 351–352, 352*f*
Tight junctions
in capillaries, 291*f*, 292
epithelial cells, 25, 26*f*
Time cues, biological rhythm research, 131–132
Tissue
androgenic effects, 751–752
and calorigenic action of thyroid hormones, 682
conditions required for optimal functioning, 4
Tissue factor, 226, 226*t*
Tissue growth factors, in cell communication, 10
Tissue hydrostatic pressure, 297, 297*f*
Tissue injury, glucocorticoids and, 702
Tissue insulation, in cold acclimatization, 606
Tissue plasminogen activator (TPA), 227
Tissue temperature, 588–589, 588*f*
extremes injuring, 588–589
as physicochemical variable, 589
Tissue thromboplastin, 226
Titratable acid, 471
excretion of, 474, 475*f*
in urine, 471–472
T lymphocytes, 218
activation, 223

cytotoxic, 223
formation, 220*f*
functions of, 223
helper, 218, 223
formation, 220*f*
subsets of, 218
suppressor, 218, 223
formation, 220*f*
Tm. *See* Tubular transport maximum (Tm)
Tn-C subunit, of thin myofilaments, 153*f*, 154
Tn-I subunit, of thin myofilaments, 153*f*, 154
Tn-T subunit, of thin myofilaments, 153*f*, 154
α-Tocopherol, 563
Tones. *See* Pitch
Tongue
taste buds on, 85, 85*f*
taste receptor localization on, 85–86
Tonic contractions, 506
smooth muscle, 183, 187, 187*f*
Tonicity, of solution, 27
Tonic receptors, 66
Tonotopic organization, 82–83
Total bile acid pool, 546
Total bile flow, 544–545, 545*f*
Total body water, 447
measurement of, 448–449
Total lung capacity (TLC), 348, 348*f*, 349*f*, 349*t*, 350*f*
Toxic cardiomyopathy, 203
TPA (tissue plasminogen activator), 227
Trachea, 342, 343*f*
Tracheal air, 354
partial pressure of respiratory gases in, 388, 388*t*
Tracking movements, eye, 73
Transairway pressure (Pta), 346, 346*f*
during forced expiration, 365, 365*f*
Transcellular transport, across epithelial cells, 25, 26*f*
Transcobalamin, 564
Transducer
sensory receptors as, 63. *See also* Sensory receptor(s)
visual system, biochemical steps, 75, 75*f*
Transducin, in visual transduction process, 75, 75*f*
Transection, of spinal cord, 101
Transferrin, 216
transport of, 17
Transit time, capillary, 389, 390
Transmission line, for sensory signals, 68, 68*f*
Transmitters. *See* Neurotransmitters
Transmural pressure (Ptm), 236, 346, 346*f*
Transneuronal transport, of trophic substances, 56
Transplantation
bone marrow. *See* Bone marrow transplantation
heart, 266
neural, in Parkinson's disease, 108
renal, 431
Transport
active systems. *See* Active transport systems
diffusive membrane, 18–19, 18*f*
renal tubule. *See* Kidney, tubular transport; Tubular transport; Tubular transport maximum (Tm)

of solutes. *See* Solute transport, across cell membrane
of trophic substances, 56
water movement. *See* Water movement, across cell membrane
Transporters, in small intestine, 560–561
Transport proteins, 639–641, 640*f*, 640*t*
Transpulmonary pressure (Pʟ), 346, 346*f*, 347, 356
gravitational effects on, 358*f*, 359
Transverse colon, large intestine motility and, 524–525, 524*f*, 525*f*
Transverse tubular system, 194
Transverse tubules. *See* T tubules
Trauma. *See* Injury
Traveling wave, sound transmission, 82, 82*f*
Tremor
extraocular muscles, 73
intention (action), 111
in Parkinson's disease, 108
TRH. *See* Thyrotrophin-releasing hormone (TRH)
Trigeminal (V) cranial nerve, olfactory sensation and, 87
Triglyceride lipolysis, 669
Triglycerides, 555, 556*f*. *See also* Lipids
Trigone, in urinary bladder anatomy, 459
Triiodothyronine (T₃). *See* Thyroid hormones
Trophic substances, transneuronal transport, 56
Trophoblast, 780
Tropic factors, secretion of, 56
Tropomyosin, of thin myofilaments, 153–154, 153*f*
Troponin, of thin myofilaments, 153–154, 153*f*
subunits, 153*f*, 154
Troponin-tropomyosin complex
calcium and, 158–159, 158*f*
cardiac muscle contraction, 196
Trousseau's sign, 722
True precocious puberty, 796
Trypsin, 560
Trypsinogen, 560
TSH. *See* Thyroid stimulating hormone (TSH)
T tubules, in skeletal muscle membrane system, 155
excitation-contraction coupling and, 159
T tubule system, in cardiac muscle, 194
Tuberoinfundibular system, of dopaminergic neurons, 140
Tubular reabsorption
body fluid compartments regulation, 449–459
calcium, 456–457, 457*f*
magnesium, 457–458, 458*f*
phosphate, 458–459, 458*f*, 459*f*
potassium, 455–456, 455*f*
sodium. *See* Sodium balance
water. *See* Water balance
calculation, 425
defined, 422, 422*f*
excretion of hydrogen ions and, 476
glucose titration assessing, 425–426, 425*f*
Tubular secretion
calculation, 425
defined, 422, 422*f*
Tubular transport
in distal nephron, 437–438
in loop of Henle, 436–437